John M. Gottman

Volume 1

Howell • Fulton • Ruch • Patton

Textbook of PHYSIOLOGY

Excitable Cells and Neurophysiology

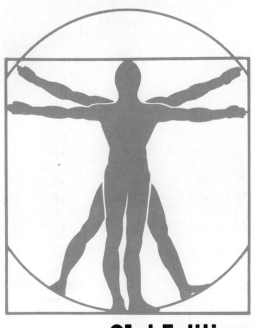

21st Edition

Edited by:

Harry D. Patton, M.D., Ph.D., LL.D.

Albert F. Fuchs, Ph.D.

Bertil Hille, Ph.D.

Allen M. Scher, Ph.D.

Robert Steiner, Ph.D.

University of Washington
School of Medicine
Seattle, Washington

1989
W.B. SAUNDERS COMPANY
Harcourt Brace Jovanovich, Inc.

Philadelphia • London • Toronto • Montreal • Sydney • Tokyo

W. B. SAUNDERS COMPANY
Harcourt Brace Jovanovich, Inc.

The Curtis Center
Independence Square West
Philadelphia, PA 19106

Library of Congress Cataloging-in-Publication Data

Textbook of physiology.

Rev. ed. of: Physiology and biophysics / Howell-Fulton.
20th ed. 1973–1982.

Includes bibliographies and indexes.
1. Human physiology.   I. Patton, Harry D., 1918–
II. Howell, William Henry, 1860–1945.   Physiology and
biophysics.   [DNLM: 1. Physiology.   QT 104 T3553]

QP31.2.T47   1988   612   87–9539

ISBN 0–7216–1990–8 (set)

ISBN 0–7216–2523–1 (v. 1)

ISBN 0–7216–2524–X (v. 2)

Editor: John Dyson
Designer: Karen O'Keefe
Production Manager: Carolyn Naylor
Manuscript Editors: Lorraine Zawodny and Kate Mason
Cover Designer: Michelle Maloney
Illustration Coordinator: Lisa Lambert
Indexer: Ann Cassar

Textbook of Physiology:
Excitable Cells and   Volume 1   ISBN 0–7216–2523–1
Neurophysiology—Volume 1   2 Volume Set ISBN 0–7216–1990–8

Last digit is the print number: 9   8   7   6   5   4   3   2   1

Contributors

WOLFHARD ALMERS, Ph.D.
Professor, Department of Physiology and
Biophysics
University of Washington School of Medicine
Seattle, Washington
Excitation-Contraction Coupling in Skeletal Muscle

MARJORIE E. ANDERSON, Ph.D.
Professor, Departments of Physiology and
Biophysics and of Rehabilitation Medicine
University of Washington School of Medicine
Seattle, Washington
*The Neural Control of Movement; Spinal and
Supraspinal Control of Movement and Posture; The
Cerebellum; The Basal Ganglia*

ALBERT J. BERGER, Ph.D.
Professor, Department of Physiology and
Biophysics
University of Washington School of Medicine
Seattle, Washington
Control of Respiration

MARC D. BINDER, Ph.D.
Professor, Department of Physiology and
Biophysics
University of Washington School of Medicine
Seattle, Washington
*The Neural Control of Movement; Properties of Motor
Units; Peripheral Motor Control: Spinal Reflex
Actions of Muscle, Joint, and Cutaneous Receptors;
Functional Organization of the Motoneuron Pool;
Spinal and Supraspinal Control of Movement and
Posture*

MARK BOTHWELL, Ph.D.
Associate Professor, Department of Physiology
and Biophysics
University of Washington School of Medicine
Seattle, Washington
Growth Factors

GEORGE L. BRENGELMANN, Ph.D.
Professor, Department of Physiology and
Biophysics

University of Washington School of Medicine
Seattle, Washington
Body Temperature Regulation

ARTHUR C. BROWN, Ph.D.
Professor, Department of Physiology and
Pharmacology
Oregon Health Sciences University
Portland, Oregon
*Introduction to Sensory Mechanisms; Somatic
Sensation: Peripheral Aspects; Pain and Itch*

JOHN D. BRUNZELL, M.D.
Professor of Medicine,
University of Washington School of Medicine
Division of Metabolism, Endocrinology, and
Nutrition
Seattle, Washington
Energy Balance, Storage, and Transport

ALISON M.J. BUCHAN, Ph.D.
Associate Professor, Department of Physiology
University of British Columbia
Vancouver, B.C.
Gastrointestinal Motility; Digestion and Absorption

JOHN BUTLER, M.D.
Professor of Medicine,
University of Washington School of Medicine
Medical Staff, University of Washington
Hospital, Harborview Medical Center, Seattle
Veterans Administration Medical Center, and
Pacific Medical Center
Seattle, Washington
The Circulation of the Lung

JUDY L. CAMERON, Ph.D.
Assistant Professor of Psychiatry, Physiology,
and Behavioral Neuroscience
University of Pittsburgh School of Medicine
Pittsburgh, Pennsylvania
Endocrine Control of Reproduction

DONALD K. CLIFTON, Ph.D.
Associate Professor, Department of Obstetrics
and Gynecology
University of Washington School of Medicine
Seattle, Washington
The Anterior Pituitary

DANIEL L. COOK, M.D., Ph.D.
Research Associate Professor
Division of Metabolism, Department of Medicine;
Department of Physiology and Biophysics
University of Washington School of Medicine
Veterans Administration Medical Center
Seattle, Washington
*The Cellular Biology of the Endocrine System: An
Overview*

LAWRENCE E. CORNETT, Ph.D.
Associate Professor, Physiology and Biophysics
University of Arkansas for Medical Sciences
Little Rock, Arkansas
The Adrenal Medulla

WAYNE E. CRILL, M.D.
Professor and Chairman, Department of
Physiology and Biophysics
University of Washington School of Medicine
Seattle, Washington
*Synaptic Transmission; Transformation of Synaptic
Input Into Spike Trains in Central Mammalian
Neurons; The Milieu of the Central Nervous System*

SUSANNA L. CUNNINGHAM, Ph.D.
Associate Professor and Associate Dean for
Community Relations
University of Washington
Seattle, Washington
The Physiology of Body Fluids

RALPH G. DACEY, Jr., M.D.
Professor and Chief, Division of Neurological
Surgery
University of North Carolina at Chapel Hill
North Carolina Memorial Hospital
Chapel Hill, North Carolina
Cerebral Circulation

PETER B. DETWILER, Ph.D.
Professor, Department of Physiology and
Biophysics
University of Washington School of Medicine
Seattle, Washington
Sensory Transduction; Synaptic Transmission

ROBERT A. DOBIE, M.D.
Professor, Department of Otolaryngology
University of Washington School of Medicine
Director, Otolaryngology Clinic, University
Hospital
Attending Physician, University Hospital,
Veterans Administration Medical Center,
Harborview Medical Center, Children's Hospital
and Medical Center, and Providence Medical
Center
Seattle, Washington
*The Auditory System: Acoustics, Psychoacoustics, and
the Periphery; The Auditory System: Central Auditory
Pathways*

DANIEL M. DORSA, Ph.D.
Research Associate Professor, Departments of
Medicine and Pharmacology
University of Washington School of Medicine
Associate Director, Geriatric Research, Education,
and Clinical Center, Veterans Administration
Medical Center
Seattle, Washington
*Neurohypophyseal Hormones; Neuropeptides as
Neurotransmitters*

ERIC O. FEIGL, M.D.
Professor, Department of Physiology and
Biophysics
University of Washington School of Medicine
Seattle, Washington
*The Heart: Introduction and Physical Principles;
Cardiac Mechanics; The Arterial System; Coronary
Circulation*

EBERHARD E. FETZ, Ph.D.
Professor, Department of Physiology and
Biophysics
University of Washington School of Medicine
Seattle, Washington
*The Neural Control of Movement; Motor Functions of
Cerebral Cortex*

JO ANN E. FRANCK, Ph.D.
Research Assistant Professor, Department of
Neurosurgery
University of Washington School of Medicine
Seattle, Washington
The Limbic System

ALBERT F. FUCHS, Ph.D.
Professor, Department of Physiology and
Biophysics
University of Washington School of Medicine
Seattle, Washington

Somatic Sensation: Central Processing; The Visual System: Optics, Psychophysics, and the Retina; The Visual System: Neural Processing Beyond the Retina; Gustation and Olfaction; The Neural Control of Movement; The Vestibular System; Association Cortex; The Limbic System; Physiological Bases of Learning, Memory, and Adaptation

WAYNE R. GILES, PH.D.
Professor, Department of Medical Physiology
University of Calgary
Calgary, Alberta, Canada
Intracellular Electrical Activity in the Heart

CHARLES J. GOODNER, M.D.
Professor of Medicine
Head, Division of Endocrinology, Harborview
Medical Center
University of Washington School of Medicine
Seattle, Washington
Intermediary Metabolism and Regulation of Fuel Metabolism During Adaptation to Changing Environmental and Physiological Conditions

ALBERT M. GORDON, PH.D.
Professor, Department of Physiology and
Biophysics
University of Washington School of Medicine
Seattle, Washington
Molecular Basis of Contraction; Contraction of Skeletal Muscle; Contraction in Smooth Muscle and Nonmuscle Cells

WILLIAM L. GREEN, M.D.
Professor of Medicine
State University of New York Health Science
Center at Brooklyn
Associate Chief of Staff for Research and
Development, Veterans Administration Medical
Center
Brooklyn, New York
The Thyroid Gland

JACOB HILDEBRANDT, PH.D.
Professor, Department of Physiology and
Biophysics
University of Washington School of Medicine
Seattle, Washington
Structural and Mechanical Aspects of Respiration

BERTIL HILLE, PH.D.
Professor, Department of Physiology and
Biophysics

University of Washington School of Medicine
Seattle, Washington
Introduction to Physiology of Excitable Cells; Transport Across Cell Membranes: Carrier Mechanisms; Membrane Excitability: Action Potential Propagation in Axons; Voltage-Gated Channels and Electrical Excitability; Neuromuscular Transmission

MICHAEL P. HLASTALA, PH.D.
Professor, Departments of Physiology and
Biophysics and of Medicine; Adjunct Professor of
Bioengineering
University of Washington School of Medicine
Seattle, Washington
Gas Transport and Exchange

THOMAS F. HORNBEIN, M.D.
Professor and Chairman, Department of
Anesthesiology; Professor, Department of
Physiology and Biophysics
University of Washington School of Medicine
Professor, University Hospital, Harborview
Medical Center, Veterans Administration
Hospital, and Childrens Hospital and Medical
Center
Seattle, Washington
Control of Respiration

GUY A. HOWARD, PH.D.
Research Associate Professor, Departments of
Medicine and Oral Biology
University of Washington School of Medicine
Seattle, Washington
Associate Research Career Scientist, Veterans
Administration Medical Center
Tacoma, Washington
Calcium Metabolism and Physiology of Bone

LEE L. HUNTSMAN, PH.D.
Professor and Director, Center for
Bioengineering, University of Washington School
of Medicine
Seattle, Washington
Cardiac Mechanics

JOHN M. JOHNSON, PH.D.
Associate Professor, Department of Physiology
University of Texas Health Science Center
San Antonio, Texas
Circulation to Skeletal Muscle; Circulation to the Skin

ROBERT H. KNOPP, M.D.
Professor of Medicine; Adjunct Professor of
Obstetrics and Gynecology

University of Washington School of Medicine
Attending Physician, Harborview Medical Center
and University Hospital
Seattle, Washington
Pregnancy and Parturition

DONNA J. KOERKER, Ph.D.
Professor of Medicine, Harborview Medical
Center and the University of Washington;
Professor, Department of Physiology and
Biophysics
University of Washington School of Medicine
Seattle, Washington
*Intermediary Metabolism and Regulation of Fuel
Metabolism During Adaptation to Changing
Environmental and Physiological Conditions*

M. SCOTT MAGEE, M.D.
Acting Instructor in Medicine
University of Washington School of Medicine
Attending Physician, Harborview Medical Center
Seattle, Washington
Pregnancy and Parturition

MARC R. MAYBERG, M.D.
Assistant Professor, Department of Neurological
Surgery
Harborview Medical Center
University of Washington School of Medicine
Seattle, Washington
Cerebral Circulation

CHARLES H. MULLER, Ph.D.
Research Assistant Professor, Departments of
Obstetrics and Gynecology and of Biological
Structure
University of Washington School of Medicine
Seattle, Washington
Germ Cell Development and Fertilization

HARRY D. PATTON, M.D., Ph.D., LL.D.
Professor and Chairman Emeritus, Department
of Physiology and Biophysics
University of Washington School of Medicine
Seattle, Washington
The Autonomic Nervous System

JAMES O. PHILLIPS, Ph.D.
Departments of Psychology and of Physiology
and Biophysics
University of Washington School of Medicine
Seattle, Washington
*Somatic Sensation: Central Processing; Gustation and
Olfaction; Association Cortex; The Limbic System*

EUGENE M. RENKIN, Ph.D.
Professor and Chairman, Department of Human
Physiology
University of California at Davis
Davis, California
Microcirculation and Exchange

ALAN DAVID ROGOL, M.D., Ph.D.
Professor, Departments of Pediatrics and
Pharmacology
University of Virginia School of Medicine
Attending Physician, University of Virginia
Hospital
Charlottesville, Virginia
*Sexual Differentiation and Pubertal Maturation of the
Reproductive System in Humans*

EDWIN W RUBEL, Ph.D.
Professor, Departments of Otolaryngology,
Physiology and Biophysics, and Neurological
Surgery; Adjunct Professor, Department of
Psychology
University of Washington School of Medicine
Seattle, Washington
*Auditory System: Acoustics, Psychoacoustics, and the
Periphery; The Auditory System: Central Auditory
Pathways*

ALLEN M. SCHER, Ph.D.
Professor, Department of Physiology and
Biophysics
University of Washington School of Medicine
Seattle, Washington
*The Heart: Introduction and Physical Principles; The
Electrocardiogram; Events of the Cardiac Cycle:
Measurements of Pressure, Flow, and Volume; The
Veins and Venous Return; Cardiovascular Control*

ROBERT S. SCHWARTZ, M.D.
Associate Professor, Departments of Internal
Medicine and of Geriatrics
University of Washington School of Medicine
Chief, Geriatric Medicine
Veterans Administration Medical Center
Seattle, Washington
Energy Balance, Storage, and Transport

PHILIP A. SCHWARTZKROIN, Ph.D.
Professor, Departments of Neurological Surgery
and of Physiology and Biophysics
University of Washington School of Medicine
Seattle, Washington
*The Limbic System; Physiological Bases of Learning,
Memory, and Adaptation*

PETER C. SCHWINDT, Ph.D.
Professor, Department of Physiology and
Biophysics
University of Washington School of Medicine
Seattle, Washington
*Transformation of Synaptic Input into Spike Trains in
Central Mammalian Neurons*

JOHN B. SIMPSON, Ph.D.
Professor, Department of Psychology;
University of Washington School of Medicine
Seattle, Washington
*The Circumventricular Organs and Brain Barrier
Systems*

M. SUSAN SMITH, Ph.D.
Professor, Department of Physiology
University of Pittsburgh School of Medicine
Pittsburgh, Pennsylvania
Lactation

ROBERT A. STEINER, Ph.D.
Professor, Departments of Obstetrics and
Gynecology, and of Physiology and Biophysics
University of Washington School of Medicine
Seattle, Washington
Endocrine Control of Reproduction

ROBERT B. STEPHENSON, Ph.D.
Associate Professor, Department of Physiology
Michigan State University
East Lansing, Michigan
The Splanchnic Circulation; The Renal Circulation

CHARLES E. STIRLING, Ph.D.
Professor, Department of Physiology and
Biophysics

University of Washington School of Medicine
Seattle, Washington
Epithelial Transport; The Kidney

GERALD J. TABORSKY, Jr., Ph.D.
Research Associate Professor of Medicine
University of Washington School of Medicine
and Veterans Administration Medical Center
Seattle, Washington
The Endocrine Pancreas: Control of Secretion

CHARLES W. WILKINSON, Ph.D.
Research Assistant Professor, Department of
Psychiatry and Behavioral Sciences
University of Washington School of Medicine
Seattle, Washington
Research Physiologist, Geriatric Research,
Education, and Clinical Center
American Lake Veterans Administration Medical
Center
Tacoma, Washington
*Endocrine Rhythms and the Pineal Gland; The
Adrenal Cortex*

H. RICHARD WINN, M.D.
Professor and Chairman, Department of
Neurological Surgery
University of Washington School of Medicine
Chief of Service, Harborview Medical Center and
University Hospital
Seattle, Washington
Cerebral Circulation

J. WALTER WOODBURY, Ph.D.
Professor, Department of Physiology
University of Utah School of Medicine
Salt Lake City, Utah
Body Acid-Base State and Its Regulation

Preface

The 21st Edition of Howell's *Textbook of Physiology* has been nearly completely rewritten. We provide a text that is introductory but scholarly, one that is clear yet sophisticated enough for the advanced student and teacher. This new edition continues to emphasize the unity of physiology from the molecular to the systems level. It has been divided into two volumes to permit the first volume to be used in courses of neurophysiology. We are deeply indebted to our colleagues, staff, and students for much help in the preparation of this book. We thank the authors for their insightful chapters, and we are particularly grateful to Helen Halsey and Marj Domenowski for much fine artwork and to Patrick Roberts and Susan Usher for excellent photographs.

THE EDITORS

Contents

VOLUME 1
Excitable Cells and Neurophysiology

VOLUME 2
Circulation, Respiration, Body Fluids, Metabolism, and Endocrinology

Section I

BERTIL HILLE
Section Editor

Membranes and Ions

Chapter 1

Bertil Hille

Introduction to Physiology of Excitable Cells

INTRODUCTION

Physiology. Animal physiology is the study of body function. The ancient Greeks used the word *physiologia* to denote the study of all natural phenomena (*physis*, nature + *logos*, discourse), but as the accumulating knowledge grew ever larger, the all-embracing "natural philosophy" became increasingly subdivided into disciplines: physics, chemistry, medicine, anatomy, physiology, pharmacology, zoology, botany, and so forth. These divisions of natural science remain complementary and overlapping. Each has unique ways of thinking, but none can be pursued in isolation.

The characteristic approaches of physiology evolved from the need to analyze function in terms of the laws of physics and chemistry applied to body parts. The body can be regarded as a machine—with vessels to be described in terms of pressure and flow, muscles to be described in terms of force and velocity, and electrical signals to be described in terms of voltage and current. This approach often leads to reductionism, the analysis of effects and causes at progressively simpler and lower levels, and eventually to the molecular level.

At the same time, if physiology is to explain locomotion, emotion, or the control of circulation, functions of the whole animal, it must also use an integrative approach. The pieces of the puzzle must be assembled to reveal the whole. One major theme here is regulation of body processes through internal feedback systems and the simultaneous involvement of many organs in the overall response. Inhibitory and excitatory hormonal and nervous messages work together to keep major

1

processes in balance and responsive to the needs of the organism. The physiologist wants to know not only which interactions exist but also how important each is to the overall result. Thus physiology has a strong quantitative orientation.

Themes of Cell Physiology. The first 12 chapters of this volume use the reductionist approach. They describe nerve and muscle action at the cellular and subcellular levels. Thus the nerve impulse is reduced to a problem of ion flows in molecular pores of cell membranes; muscular contraction is reduced to a problem of macromolecular arms moving on filamentous proteins; and neurotransmitter release is reduced to a problem of membrane fusion. These areas of cell physiology introduce fundamental biological mechanisms whose origins predate the first appearance of *animals* on Earth. Five themes that weave through Chapters 1 to 10 are

1. The cytoplasm of each cell is completely surrounded by a lipid-protein *membrane* that is a barrier to the movement of many molecules and across which various transport devices are constantly maintaining concentration gradients of ions and small molecules. The membrane is also the detector of most extracellular signals.
2. Electrical signals are generated across the cell membrane by the opening and closing of ion-selective pores, the *ionic channels*, in the membrane. Ionic channels make the cell excitable, responsive to stimuli. Different channels are permeable to monovalent cations, to monovalent anions, or even to Ca^{2+} ions.
3. Many cellular functions are regulated by the intracellular free Ca^{2+} ion concentration. *Calcium* ions may be called *internal second messengers*, since they serve as intracellular links between electrical or chemical excitation and cell responses. Other important internal second messenger molecules are cyclic adenosine monophosphate and inositol trisphosphate.
4. Secretory products—neurotransmitters, hormones, digestive enzymes, and so forth—are prepackaged in membrane-bounded *vesicles* in the cytoplasm of secretory cells. They are released by the process of *exocytosis*, a kind of inverse pinocytosis in which the membrane of the vesicle fuses with the cytoplasmic membrane, suddenly placing the vesicle contents in the extracellular medium. Exocytosis is one of the many processes that may be regulated by the internal free Ca^{2+} ion concentration.
5. Cell motility and shortening are accomplished by *filamentous proteins*, notably actins and myosins, that *pull* and *slide* by each other at the expense of metabolic energy. The interaction of the contractile proteins is regulated by the intracellular free Ca^{2+} ion concentration and by other internal second messengers.

Of these five themes, the first is basic to all autonomous living forms. From the bacteria to the highest animals, all cells have a cytoplasmic membrane that maintains gradients of small molecules. The cell membrane must have evolved over 3.5 billion years ago, when the earliest prokaryotic cells appeared. Animal cell membranes and their strategies for creating concentration gradients are described in this chapter and the next.

The remaining four themes concern processes of which most have not been found in any bacterial cell. Apparently, they evolved about 1.4 billion years ago with the emergence of the first eukaryotes—cells with a true nucleus that were ancestral to the modern protozoa, slime molds, fungi, algae, higher plants, and animals. Ionic channels for signaling, Ca^{2+} as a second messenger, actin and myosin, and exocytosis are innovations of the transition from prokaryotic to eukaryotic life. Here we emphasize their roles in animal nervous and muscular systems, but one should remember that they play roles in all cells of the body and even in cells of protozoa, plants, and the other divisions of eukaryotes.

The rest of this chapter contains background information from biology, chemistry, and physics that is essential to any detailed discussion of physiological mechanisms at the cellular level. The topics to be considered are the cell membrane, second messengers, types of transport across membranes, ions, rules of ionic electricity, energy calculations for transport processes, and principles of diffusion. We start with membranes.

CELL MEMBRANES

Properties. The contents of living cells are prevented from floating away into the surrounding medium by a cytoplasmic membrane, often called the *plasma membrane*. Because it forms the boundary between the living cell and "inanimate" extracellular space, the plasma membrane is the primary transducer of most intercellular signaling and is the cell component most studied by physiologists. This membrane, far too thin to be seen with a light microscope, was first recognized phenomenologically by the *permeability barrier* that it presents to many classes of molecules, particularly charged molecules and those with several polar groups. For example, if charged fluorescent dyes such as lucifer yellow or fluorescein are injected through micropipettes into the cytoplasm, they diffuse readily throughout the cell and to its surface, but there they stop, giving the boundary a sharp, fluorescent contour. However, if the surface is torn by a quick scratch from a microneedle, the dye and the cytoplasmic contents begin to pour out. Studies with tracer-labeled molecules confirm that the permeability of the intact membrane is low for many hydrophilic ("water-loving") substances such as salts, carbohydrates, and amino acids. On the other hand, hydrophobic ("water-fearing") molecules that prefer to dissolve in oils or lipids can cross the membrane. The impermeability to many hydrophilic molecules and the permeability to hydrophobic ones led Ernst Overton to propose correctly as early as 1899 that cells are surrounded by a surface layer made of "fatty oils or cholesterol."

The plasma membrane can be visualized with the electron microscope. After fixation with glutaraldehyde or osmic acid and staining with uranyl or lead salts, it appears in cross section as two parallel, electron-dense lines about 2.5 nm (25 Å) wide with a 2.5 to 4.0 nm electron-translucent band between. Within the cell there are membranes as well, delimiting the intracellular organelles—the nucleus, mitochondria, Golgi apparatus, endoplasmic reticulum, secretory vesicles, and so forth. Since under the electron microscope the intracellular and cytoplasmic membranes are nearly indistinguishable, it is generally supposed that they have the same architecture. Indeed, they have qualitatively similar general chemical composition.

Membranes can be isolated for chemical analysis from gently homogenized tissues. The technique of differential centrifugation on sucrose or other density gradients can separate cell homogenates into fractions enriched in the various cell components. Because different membranes bear different enzymatic activities, these activities can be used as biochemical *markers* to determine which fractions contain the membranes of interest. For example, the plasma membrane carries the $(Na^+ + K^+)$-dependent ATPase, the enzymatic activity of the Na^+-K^+ pump that is described in Chapter 2. Membranes purified this way contain mostly lipid and protein with some sugar residues. As Table 1–1 shows, the ratio of protein to lipid and even the types of lipids vary. Thus while membranes have a compositional similarity dictated by the minimum requirements to form a membrane, they are otherwise specialized. Membranes with many surface receptors, transport sites, or enzymatic

Table 1–1 Protein and Lipid Composition of Membranes of Mammalian Cells

Component	Liver Plasma Membrane	Brain Myelin	Sarcoplasmic Reticulum
	% by Weight		
Protein	60	31	60
Total lipid	40	69	40
Cholesterol	16	27	<2
Phospholipid	39	45	93
Other	45	28	5

functions, such as the liver plasma membrane, the inner mitochondrial membrane, or the sarcoplasmic reticulum of muscle, have higher protein content—the proteins confer all functions except the barrier function to membranes. On the other hand, myelin, a compacted multilayer of membranes whose primary role seems to be to provide an electrically insulating protection around myelinated axons, has a low protein content and a high lipid content—the lipids are good insulators.

Lipid Bilayer. The major lipids of membranes are phospholipids, cholesterol, and glycolipids. The plasma membrane has more cholesterol than intracellular membranes. All membrane lipids are *amphipathic*, meaning that one part of the molecule, the hydrophilic "head group," prefers to remain in water, while another, the hydrophobic "tail," prefers to remain in a nonpolar environment. For example, consider the phospholipid phosphatidyl choline (Fig. 1–1A). The highly polar head group comprises a glycerol phosphocholine moiety and the ester-linked carboxyl groups of two fatty acids. The tail comprises the hydrocarbon chains of the two fatty acids. In schematic drawings, the head group is often drawn as a ball with two wiggly chains hanging from it (Fig. 1–1B). The head groups of the common phospholipids either have a net negative charge (phosphatidyl serine, phosphatidyl inositol) or are zwitterionic, having positive and negative charge together (phosphatidyl choline, phosphatidyl ethanolamine, sphingomyelin).

Because of their amphipathic character, phospholipids in water aggregate spontaneously to form molecular bilayers with head groups on either side interacting freely with water while the tails interact with themselves in a hydrophobic sheet (Fig. 1–1C). Like soap bubbles, such sheets are self-sealing and tend to form closed structures. Lipid bilayers are easily formed in the laboratory, and their properties are studied by the tracer flux and electrical methods of cell physiology. They have many of the characteristics of biological mem-

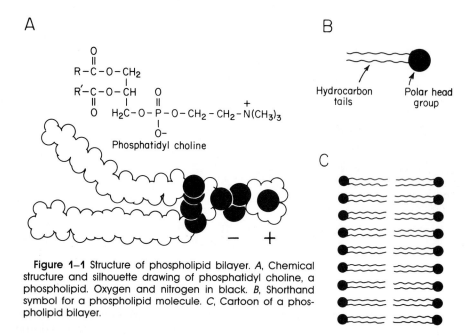

Figure 1–1 Structure of phospholipid bilayer. *A*, Chemical structure and silhouette drawing of phosphatidyl choline, a phospholipid. Oxygen and nitrogen in black. *B*, Shorthand symbol for a phospholipid molecule. *C*, Cartoon of a phospholipid bilayer.

branes. They are 3.5 to 5.0 nm thick, impermeable to hydrophilic molecules, and permeable to hydrophobic ones.

The hydrocarbon core of lipid bilayers is nearly liquid. Heterogeneity of the lipid chains prevents the kind of regular packing that might lead to "freezing" of the bilayer. Because of thermal agitation, each fatty acid chain bends and sways rapidly, and whole lipid molecules spin about their long axes as well as diffusing laterally, changing places with their neighbors. This constant motion gives the bilayer flexibility and permits lipid-soluble substances to move readily across it. However, one motion called flip-flop (transfer between inner and outer membrane leaflets) is rare. Presumably because of their polar head group, phospholipids do not flip from one side of the bilayer to the other. Therefore it is possible for stable lipid bilayers to be asymmetric, i.e., for the lipid compositions of the two lipid leaflets (half bilayers) to differ.

The lipid bilayer is a major structural feature of all cell membranes. It gives the membrane fluidity to conform to the changing shapes of cells, and it fills in the gaps between the protein molecules of the membrane. The bilayer areas underlie the general impermeability to polar substances and the permeability to nonpolar substances. The bilayer bears a fixed negative charge, owing to acidic phospholipids, and is believed to be asymmetric. In the red blood cell, for example, the major part of the negatively charged lipid faces the cyto-plasm. The glycolipids, bearing sugar residues, face only the outside.

Proteins. If lipids enable membranes to partition space into functional compartments, then proteins enable membranes to act on molecules within compartments and to transmit signals between compartments. They serve transport, signaling, catalytic, and structural roles (Table 1–2). Animal physiology emphasizes the transport and signaling roles, so they appear repeatedly in this volume.

Proteins that are an integral part of the membrane are immersed in the lipid bilayer in definite and fixed orientations (Fig. 1–2A). Their orientation can be assayed by functional tests. Thus receptor sites for acetylcholine at the neuromuscular junction always face away from the cytoplasm to sense the neurotransmitter in the extracellular medium, while the active site of the adenylate cyclase enzyme always faces the cytoplasm to convert intracellular adenosine triphosphate (ATP) to intracellular cyclic adenosine

Table 1–2 Roles of Membrane Proteins

Function	Example
Gated pore	Na channel
Active transport	Na^+-K^+ pump
Exchange	Cl^-/Cl^- exchanger
Enzyme	Adenylate cyclase
Agonist receptor	β-adrenergic receptor
Attachment site	Glycophorin

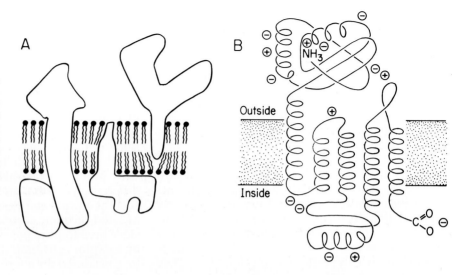

Figure 1–2 Protein molecules in the lipid bilayer of biological membranes. *A,* Cartoon showing how mass of protein molecules may hang out into intracellular or extracellular space. *B,* Exploded view, showing the backbone of a protein chain threaded several times across the bilayer. The part in the bilayer is wound in α-helices.

monophosphate (cAMP). Probably every membrane protein has only one orientation, so at the molecular level, membranes are truly asymmetric. The disposition of the component polypeptide chains of membrane proteins can be determined biochemically. Enzymes or reactive reagents can be applied to one side of a membrane to find which domains of the protein are accessible to attack from that side. Monoclonal antibodies can be made to specific amino-acid sequences to ask, with electron microscopic immunocytochemistry, on which side those sequences appear. In this way the polypeptide chains of some transmembrane proteins have been shown to loop back and forth across the bilayer multiple times (Fig. 1–2B). For example, the large, multiple-subunit acetylcholine receptor protein of the neuromuscular junction is believed to have 25 different α-helical segments traversing the membrane.

Most integral membrane proteins are synthesized on membrane-bound polyribosomes, with the nascent peptide chain being fed directly into the membrane as synthesis proceeds. During this intracellular synthesis—occurring on the rough endoplasmic reticulum—the protein may fold and achieve its correct orientation at once. If it has multiple subunits, the separately synthesized subunits need to find each other by diffusion within the plane of the membrane. Eventually the protein—still in membrane—makes its way through the Golgi apparatus, is packaged as small membrane vesicles, and is delivered as a patch of membrane to be inserted at its final destination. Membrane proteins that extend into the central lumen of the endoplasmic reticulum during syn-

thesis can become glycosylated (sugar residues are added) on that side. Additional glycosylation occurs in the Golgi apparatus. In the Na channel, a gated pore of excitable membranes, as much as 30% of the dry weight of the macromolecule is carbohydrate (oligosaccharide residues). The orientation of membrane proteins is preserved by hydrophobic and polar forces. Runs of hydrophobic amino acids in the primary sequence give the membrane-crossing segments strong hydrophobic interactions with the fatty acid chains of the lipid bilayer. Polar and charged amino acids lock the intracellular and extracellular domains of the peptide in the aqueous phase. Covalently linked sugar residues add an additional polar component to extracellular domains.

According to Singer and Nicholson's[14] "fluid-mosaic model" for cell membranes, the proteins are viewed as icebergs floating in a fluid sea of lipid. Their model emphasizes that membrane proteins are mobile, always diffusing about in the plane of the membrane. Recent experiments show that some proteins are freely mobile and others are tied down.[1, 12] Some receptor proteins mediating signal transduction require mobility to accomplish their physiological function. Consider the action of the neurotransmitter-hormone epinephrine (adrenalin),* which strengthens the force of contraction of the heart by a mechanism using cyclic AMP as an internal second messenger. As

---

*In the United States only, epinephrine and norepinephrine are the correct technical names for the hormones known as adrenalin and noradrenalin to the rest of the world. Our vocabulary was precipitated by a proprietary use of the name adrenalin in this country.

we see in the next section, the response begins with a sequence of membrane events. Extracellular norepinephrine binds to a receptor protein in the membrane, which in turn interacts with a second membrane protein, which in turn interacts with a third membrane protein, adenylate cyclase, to start the synthesis of cAMP. These proteins are not permanently associated with one another, but rather find each other by diffusion in the plane of the membrane.

Other receptor proteins are clearly not mobile. Consider the acetylcholine receptor protein at a vertebrate neuromuscular junction, which mediates the excitatory action of the neurotransmitter acetylcholine. Tens of millions of copies of this membrane protein are tightly clustered as a nearly crystalline two-dimensional array in the muscle membrane just opposite a nerve terminal. They lie within 0.05 to 0.5 μm of sites of release of acetylcholine from the nerve terminal, a position that permits them to detect released transmitter in a fraction of a millisecond. Far from the nerve terminal the muscle surface may have less than 0.1% of the acetylcholine sensitivity that is found under the terminal. To remain so sharply localized, these receptor proteins must be tied to the extracellular matrix and to the intracellular cytoskeleton or to both by stable bonds. We shall see that in fully differentiated cells there are many examples of topographically organized membrane specializations that would require a restriction of the mobility of their constituent proteins.

Cellular Signals and Second Messengers. The cell surrounds itself by a barrier yet needs to remain aware of chemical messages in the environment and to translate them into meaningful physiological actions. Being at the interface, the plasma membrane contains the receptors for all hormones and neurotransmitters except for the few hormones, such as the steroid hormones, that are membrane permeant. Receptors for acetylcholine, epinephrine, insulin, growth factors, and so forth are membrane glycoproteins that initiate cellular signals in response to binding of their extracellular ligand. On different cells and even on the same cell, there may be several kinds of receptor for the same extracellular chemical transmitter, each producing its own intracellular signal. The different receptors are distinguished by names such as α_1, α_2, and β-adrenergic receptors or M_1, M_2, and nicotinic acetylcholine receptors. We consider here four classes of cellular signals produced by receptor activation (Fig. 1–3). Each of them can be considered an internal second messenger.

Some receptors for the neurotransmitters acetylcholine, γ-aminobutyric acid, glycine, and glutamate typify the fastest cellular signaling system, changes of electrical potential across membranes. The neurotransmitter receptor macromolecule in this case includes a gated ionic channel that opens when ligand is bound, letting a few thousand ions cross the membrane. The opening can occur within 10 to 100 μs of the time when the neurotransmitter binds, so an electrical signal resulting from ion movement starts with little delay. Membrane potential changes are signals that can spread to neighboring regions of the surface membrane where they are detected by other ionic channels whose gates are controlled by potential changes rather than by ligand binding. Membrane potential changes are not directly detectable to molecules that are not in the membrane.

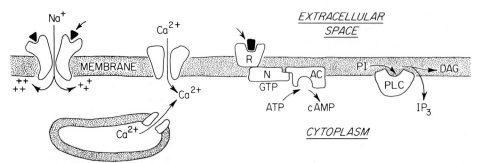

Figure 1–3 Four common second messenger systems. From left to right: (i) Membrane potential changes are produced when ionic channels are opened or closed. The channel shown is gated open in response to the binding of two molecules of neurotransmitter. The resulting inflow of positively charged ions makes the cell interior more positive. (ii) A rise of internal free Ca^{2+} ions occurs when voltage-gated Ca channels of the plasma membrane are opened or when Ca^+ is released from internal stores—shown here hypothetically as release through another ionic channel. (iii) Internal cAMP is synthesized from ATP when adenylate cyclase (AC) is activated by a nucleotide-binding protein (N) that in turn is activated by a receptor (R) with bound agonist. (iv) IP_3 and diacylglycerol (DAG) are synthesized from phosphatidylinositol-bisphosphate (PI) by the enzyme phospholipase C (PLC).

The first chemical second messenger to attract the attention of cell physiologists was the Ca^{2+} ion. Sydney Ringer[13] showed in 1883 that the isolated frog heart continues beating only if the bathing saline solution contains some calcium salts. The fertilization of eggs and transmission at chemical synapses each requires Ca^{2+} ions in the extracellular solution. Their role is actually within the cell as regulators of a broad spectrum of Ca-dependent processes. At the moment of excitation, Ca^{2+} ions enter the cell through voltage-dependent Ca channels or are released from intracellular stores into the cytoplasm. There they interact with specific Ca-binding proteins, calmodulin, troponin, and others to turn on or off different cellular functions. The contraction of skeletal muscle is initiated by the release of Ca^{2+} from intracellular stores (the sarcoplasmic reticulum), and the release of neurotransmitter vesicles is initiated by entry of Ca^{2+} from the external medium. Several kinds of membrane pumps remove Ca^{2+} ions from the cytoplasm to terminate the excitation. While Ca^{2+} ions do diffuse in cytoplasm, the binding and pumping mechanisms are so rapid that the site of action of the ions is usually not far from where they enter the cytoplasm.

We have mentioned already that the internal second messenger, cyclic AMP (cAMP), is produced in response to epinephrine. Three mobile membrane proteins are required in the first stages of this process: the β-adrenergic receptor, a nucleotide-binding protein called G_S (or N_S), and adenylate cyclase. The cyclase is normally inactive because it requires a short-lived, activated form of the G_S protein to operate. When norepinephrine binds to its external receptor, the receptor becomes capable of activating G_S proteins that it encounters by diffusion in the membrane. A bound receptor molecule may eventually activate many G_S proteins, giving catalytic amplification to the signal. The activated G_S proteins in turn encounter adenylate cyclase molecules and stimulate the synthesis of cAMP from ATP. Cyclic AMP is a small molecule that diffuses readily throughout the cytoplasm. Its only well-documented role is to activate an enzyme, cAMP-dependent protein kinase, that phosphorylates intracellular proteins and thereby alters their function. Thus between epinephrine and its physiological effects we can identify many intermediates: β-adrenergic receptor, G_S, adenylate cyclase, cAMP, cAMP-dependent protein kinases, and phosphorylation of proteins. The number of proteins whose functions are changed by phosphorylation is large.[4] One gets the impression that it may include the majority of the proteins accessible to cytoplasmic protein kinases.

A variety of hormones increases the activity of adenylate cyclase, most presumably through receptors interacting with a stimulatory protein like G_S. Antagonistic hormones *decrease* the activity of adenylate cyclase. Their receptors seem to interact with another nucleotide-binding protein, G_i, that is inhibitory to the cyclase. For generality we should mention that another cyclic nucleotide, cyclic guanosine monophosphate (cGMP), has second messenger roles. The action of cyclic nucleotides is terminated when they are cleaved by intracellular phosphodiesterase enzymes, enzymes that themselves are regulated by stimuli and second messengers.

Another set of second messengers, broadly recognized only in the 1980s, are the cleavage products of a membrane inositol-phospholipid.[3, 10, 15] Several receptors, including the V_1 vasopressin and the α_1 adrenergic receptors of liver, activate the membrane-associated enzyme phospholipase C, which cleaves phosphatidylinositol 4,5-bisphosphate, a phospholipid of the inner leaflet of the plasma membrane. Both products, inositol 1,4,5-trisphosphate (IP_3) and diacylglycerol, act as second messengers. IP_3 diffuses through the cytoplasm and initiates release of Ca^{2+} ions from the lumen of the endoplasmic reticulum by mechanisms yet to be elucidated. It is termed a "calcium-mobilizing" second messenger. Simultaneously, diacylglycerol diffuses in the plane of the membrane and activates a Ca^{2+}- and phospholipid-dependent protein kinase, protein kinase C, that in turn modulates cellular activities by phosphorylating certain intracellular proteins. Potent intracellular enzymes terminate the action of IP_3 and diacylglycerol and recycle the products to resynthesize phosphatidylinositol bisphosphate. Receptors activating the phosphodiesterase, phospholipase C, to make IP_3 and diacylglycerol are therefore said to stimulate inositol phospholipid *turnover*.

In summary, at least four classes of intracellular signals are used in translating extracellular messages into cellular action: changes of membrane potential, internal free Ca^{2+}, cyclic nucleotides, and the products of phosphatidylinositol turnover. The translation occurs at the plasma membrane. These second-messenger systems do not act completely independently. Thus internal free Ca^{2+} is affected by membrane potential acting on channels and by IP_3 acting on intracellular stores; ionic channels, including Ca channels, and several ion pumps can be phosphorylated by cAMP-dependent protein kinase; and so forth. These interactions of intracellular regulatory responses produce a complexity reminiscent of that encountered by systems physiologists when dealing with the integrative responses of organ systems interacting through extracellular messages. The presence of second messengers is detected by other regulatory proteins that inhibit or stimulate enzymes, pumps, and ionic channels by direct binding and unbinding or by phosphorylating and dephosphorylating the target. Second messengers are said to *modulate* cellular activities.

Transport Across Membranes. Like any manufacturing plant, cells, organs, and whole organisms are characterized by continual utilization of energy, consumption of raw materials, and output of finished products and wastes. These processes require flow of materials across cell membranes, a subject called *transport*. In common English usage, the word "transport" always implies movement by some conveyance; however, in technical usage a far broader range of net movements is meant, including the flow of air from a region of high pressure to one of low pressure or the diffusion of sugar molecules down their concentration gradient in water. Transport studies, the measurement of molecular influx and efflux from cells, have been a favorite tool of cell physiologists and have had a profound influence on our view of what a membrane is.

Based on the early impression that *no* hydrophilic molecules cross the membrane, the initial proposal of cell physiologists was that membranes could be considered as homogeneous hydrophobic phases, slabs of oily material, separating two aqueous compartments. However, this hypothesis had to be modified as fluxes of ions, sugars, amino acids, and other metabolites were described. Physiologists then postulated specialized transport processes, which from their apparent functional properties were given names such as "pore" and "carrier." These would explain movements of hydrophilic molecules. Such concepts predated the knowledge that (i) cell membranes contain proteins, (ii) proteins are macromolecules with a defined sequence of amino acids, and (iii) proteins are the molecular machines underlying virtually all substrate-specific catalysis and transport in living cells. Today we expect membrane transport devices to be large protein macromolecules analogous to enzymes, but so far, very few of them have been purified and studied to the point at which structure and function can be correlated. Therefore, membrane transport remains a largely phenomenological field poised on the threshold of molecular understanding. Figure 1–4 lists the primary transport mechanisms that interest us here.

The simplest kind of membrane transport is *solubility-diffusion*. Here the permeating molecule

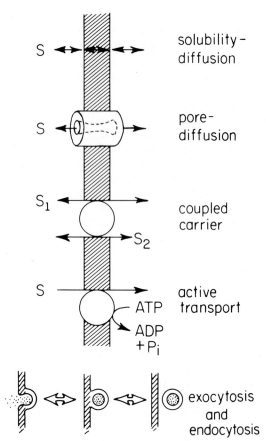

Figure 1–4 Five mechanisms for transport across cell membranes. Lipid-soluble molecules cross by dissolving and diffusing in the membrane. Pore diffusion is used primarily for signaling with small ions. Coupled carriers and active transport move essential metabolites and establish ionic gradients. Exocytosis effects multimolecular secretion of prepackaged products.

dissolves in the lipid bilayer and diffuses to the other side without the assistance of any transport macromolecules. Net movement will always be down a concentration gradient, and permeation rate will be roughly proportional to lipid solubility—permeation is favored by solubility in the membrane. This mechanism suffices to explain the observed permeability of lipid-soluble hormones and drugs, such as steroid hormones, prostaglandins, aspirin, or local anesthetics as well as of small neutral molecules such as O_2, CO_2, and, in some cases, H_2O.

A second, conceptually simple permeation mechanism is *pore-diffusion*. Here the permeating molecule passes through a water-filled molecular tunnel in the membrane. The only clearly known examples in cell membranes are ion-selective pores, the *ionic channels* that play a central role in electrical excitability. They are discussed in length in Chapters 3 and 4. The five ionic channels that have been isolated biochemically turn out to be large, membrane-spanning glycoproteins. Some cells may also use pores to control water movements. Like free diffusion, diffusion in pores permits molecules to move *down* their concentration gradients. (As is shown later, for ions a more correct statement is that they move down their electrochemical potential gradient.) Pore diffusion should not be possible for molecules whose size exceeds that of the available pores.

Another class of mechanisms is called *carrier mediated*. Several of these are described in Chapter 2. The operational intent of this term can be described, but as yet no carrier has been described in sufficient molecular detail to explain how the transport is accomplished. Consider sugar fluxes into a red blood cell. D-Glucose is an essential metabolic substrate that the red blood cell receives continuously from the plasma. A couple of millimoles are consumed by a liter of packed cells each hour, implying a permeation rate far higher than could occur by solubility diffusion alone. The following observations suggest that the sugar molecule must be bound to a stereospecific binding site in order to be transported into the cell: (i) D-Glucose, 2-deoxy-D-glucose, and D-mannose can enter, but L-glucose and D-mannitol cannot. (ii) The entry of D-glucose does not speed up linearly with increasing bath glucose concentration. Instead, the flux saturates once the outside glucose concentration reaches a few millimolar. (iii) The influx of glucose is competitively depressed by adding other permeant sugars (2-deoxyglucose or D-mannose) to the bathing medium. The three characteristics of carrier-mediated fluxes, specificity, saturation, and competition, are formally identical to the kinetics of enzyme reactions, kinetics that are always explained in terms of stoichiometric binding of substrates to specific regions of the enzyme active site.

The simplest carrier-mediated processes move one type of molecule down a concentration gradient. They can be called *facilitated diffusion*. Others move two or more types of molecules simultaneously and require that all of them be present before any transport occurs. An example of such *coupled* carriers would be the Na$^+$-dependent choline transporter of cholinergic nerve terminals, a device that moves Na$^+$ ions and choline together from extracellular fluid into the cytoplasm of the presynaptic neuron, where the choline is needed for the synthesis of the neurotransmitter, acetylcholine. It is a *cotransport* device inasmuch as its two substrates are carried in the same direction. Other coupled carriers are *exchange* or *countertransport*

devices that move their obligate substrates in opposite directions. Some carriers couple the flux of substrates to the breakdown of high energy bonds in ATP. Some devices can perform *active transport*, the movement of substrates against their concentration gradient, or more precisely, against electrical and concentration gradients. A major active transport device, the Na$^+$-K$^+$ pump, is described in Chapter 2.

Coupled carrier mechanisms typically move their substrates in a fixed stoichiometric ratio. Thus the Na$^+$-K$^+$ pump moves 3 Na$^+$ ions out of the cell and 2 K$^+$ ions into the cell for each ATP high-energy bond hydrolyzed. The stoichiometry presumably reflects the number of binding sites on the pump macromolecule that must be loaded up before a pump cycle occurs.

A final pair of transport processes, *exocytosis* and *endocytosis*, move large numbers of molecules in a single event. Exocytosis is the standard mechanism of secretion of neurotransmitters, hormones, or enzymes. The secretory products are prepackaged in the cell inside membrane-bounded secretory vesicles. When the appropriate stimulus is received, the membrane of such an intracellular vesicle fuses with the cytoplasmic membrane, and suddenly the entire content of the vesicle is delivered to the extracellular space as in Figure 1–4. Exocytosis is used to secrete soluble molecules, and it also adds new membrane to the cytoplasmic membrane. This is how membrane proteins destined to sit on the cell surface reach their destination after synthesis within the cell. Endocytosis is like exocytosis in reverse. It is used to withdraw patches of the cytoplasmic membrane back into the cell, where it may be reused for packaging secretory products or where the membrane proteins may be degraded.

Having introduced the broad classes of transport mechanisms, we turn now to review physicochemical principles needed in quantitative descriptions of membrane transport and excitability. The topics to be considered in the rest of the chapter are chemical and electrical energy calculations, the origin of the resting potential of excitable cells, and the principles of diffusion. The material is somewhat mathematical in nature. Some courses may want to skip the next section, Energy Calculations.

ENERGY CALCULATIONS

The purpose of this section is to review energy calculations relevant to transport problems. Suppose we wanted to know how steep a concentration gradient the Na⁺-K⁺ pump might be able to establish. We would have to calculate the amount of energy that could be made available when an ATP high-energy bond is broken and how much work it would take to move Na⁺ and K⁺ ions across a membrane. The answers to these thermodynamic questions are readily obtained from the basic relations between Gibbs free energy changes, concentration changes, and work at equilibrium. In order to develop a deeper understanding of these ideas, we now review the desired relations, starting with the work of expanding ideal gases.

Work, Gibbs Free Energy, and Chemical Potential. An ideal gas is defined as one made up of point masses that do not interact with each other—the only energy is kinetic energy of thermal motion. If the gas is contained in a volume V, it exerts a pressure, P, on the walls of the container, where pressure, volume, and temperature are related by the ideal gas laws of Boyle, Gay-Lussac, and Avogadro: $PV = nRT$. Here, n is the number of moles of gas particles, R is called the gas constant (Table 1–3), and T is the absolute (Kelvin) temperature.

Suppose an ideal gas is placed in a cylinder where it can expand, doing work against a loaded piston as in Figure 1–5. The cylinder is immersed in a large water bath to keep the temperature of the gas constant. How much work can the gas do in an expansion of volume from P_1 to P_2? By definition, the mechanical work, dw, done in a small step, dx, of the piston is the product of the force against which work is done times the distance moved. Writing the force in terms of the gas pressure and integrating from P_1 to P_2 gives the answer (see Appendix to Chap. 1 for the derivation).

$$w_T = nRT \ln \left(\frac{P_1}{P_2} \right) \qquad (1)$$

where ln stands for the natural logarithm ($\log_e$). The subscript T on w_T is a notational convention of thermodynamics signifying to us that the temperature is being held constant. The work of expansion depends on the *ratio* of the initial and final pressure. As expected, work can be done *by* the gas when the final pressure is lower than the initial pressure. We have actually calculated the maximum possible work by assuming that the expanding gas was pushing against the largest possible load. This means that the expansion process was kept as close as possible to equilibrium and occurred extremely slowly—a condition called *reversible* expansion in thermodynamics. If the gas were allowed to expand against no load at all, irreversible expansion, it would have done no work, i.e., w = 0.

The first and second laws of thermodynamics are the theoretical foundations for energy calculations in biology. The first law says that energy U is conserved in any process. The second law says that the entropy, S, of the universe is always increasing until equilibrium is reached, and then entropy doesn't change further. These ideas are developed further in the appendix following this chapter.

Despite their theoretical importance, energy and entropy are not the most useful thermodynamic functions for analyzing practical questions such as how much work a chemical reaction can do under biological conditions. J. Willard Gibbs discovered a new thermodynamic function, now called the *Gibbs free energy*, G, that has the desired properties. It is simply an algebraic combination of well-known thermodynamic functions:

$$G \equiv U + PV - TS$$

If a process takes place in a system, the Gibbs free energy of the system would ordinarily change.

Table 1–3 Useful Physical Constants

Avogadro's number	$N = 6.022 \times 10^{23}$ mol⁻¹
Elementary charge	$e = 1.602 \times 10^{-19}$ C
Faraday's constant	$F = 9.649 \times 10^4$ C
Absolute temperature	$T(K) = 273.15 + T$ (°Celsius)
Gas constant (in mechanical units)	$R = 8.314$ J K⁻¹ mol⁻¹
Gas constant (in electrical units)	$R = 8.314$ V C K⁻¹ mol⁻¹
RT at 20°C	$RT = 2.437$ kJ mol⁻¹
2.303 RT at 20°C	$RT = 5.612$ kJ mol⁻¹
RT/F at 20°C	$RT/F = 25.26$ mV
2.303 RT/F at 20°C	2.303 $RT/F = 58.16$ mV
1 Joule = 1 kg m² s⁻²	= 1 V C = 1 W s = 0.239 cal

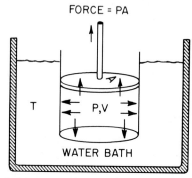

FORCE = PA

Figure 1–5 An expanding ideal gas at temperature T exerts a force PA against a piston of area A.

The sign and magnitude of the change ΔG have immediate predictive significance in systems held at a *constant temperature* and exposed to a *constant ambient pressure*—the conditions of most biological reactions. The two important rules, derived in the Appendix, are (i) If ΔG_{PT} of a process is negative, the process can occur spontaneously; if zero, the process is at equilibrium; and if positive, the process will not occur spontaneously. (ii) The maximum work that a given process can do on the external world is $-\Delta G_{PT}$.

$$w_{PT} \leq -\Delta G_{PT} \tag{2}$$

Hence if ΔG_{PT} of a process is negative, the process can supply useful external work, and if ΔG_{PT} is positive, work needs to be done from another source to get the process to go. A corollary of these rules is that a spontaneous process, with a negative ΔG_{PT} (e.g., ATP breakdown), can be *coupled* to a "forbidden" process, with a positive ΔG_{PT} (e.g., muscle contraction) to make it occur. With a perfectly efficient coupling, the overall coupled process can occur whenever the sum of the separate ΔG_{PT}s is negative.

The Gibbs free energy, like entropy and energy, is an extensive property, meaning that G for a system is the sum of the Gs for each of the components.

$$G_{system} = \sum G_{parts}$$

Further, because G for each component is proportional to the amount of that component present, we can speak of the Gibbs free energy per mole of substance, a concept so useful that it is given a name, *chemical potential*, symbolized μ. The chemical potential, like pressure or temperature, is an intensive property that can be found listed in tables for standardized states of common substances. Thus the chemical potential of *pure* liquid water at 25°C is 237 kJ/mol. The Gibbs free energy of n_i moles of substance i is

$$G_i = n_i \mu_i$$

When the concentration of a component is decreased, its chemical potential decreases too. The concentration dependence is easily derived from the formula we have already given for the maximum work an ideal gas can do while expanding at constant pressure (Equation 1). Setting $-\Delta G_T$ equal to this work, gives

$$-\Delta G_T = w_T = nRT \ln(P_1/P_2)$$

Therefore the chemical potential change for an ideal gas expanding at constant temperature is

$$\Delta \mu = RT \ln (P_2/P_1)$$

Similarly for an ideal solution of an ideal gas, the dependence on concentration is

$$\Delta \mu = RT \ln (c_2/c_1) \tag{3}$$

In careful thermodynamic work, this ideal equation can be shown to be slightly inaccurate for real solutions in which the dissolved molecules may interact with each other, and one would prefer to speak of "activities" instead of concentrations, but we will ignore the slight deviation here. When one component of a system has come to equilibrium, the chemical potential, and hence the activity, of that component is the same in all parts of the system.

We can now make simple energy calculations relevant to transport. Suppose that substance X, a neutral molecule, exists at 5 mM concentration inside a cell and only 0.5 mM outside. There is an outwardly directed gradient of concentration and chemical potential for X. The chemical potential difference $\Delta \mu_X$ calculated from Equation 3 is RTln (5/0.5) or 5.6 kJ/mol at 20°C. If X is diffusing out of the cell, down its concentration gradient, then the efflux could make up to 5.6 kJ/mol of free energy available to be coupled to some other transport. Conversely, if X is being transported into the cell, against its concentration gradient, at least 5.6 kJ/mol of free energy would have to come from another source to do this work.

Chemical Reactions.[9] A natural next step is to ask whether hydrolysis of ATP could provide enough free energy to pump substance X. How do you calculate the free energy change for a chemical reaction? Making use of the additive nature of Gibbs free energies, we can write

$$\Delta G_{reaction} = G_{products} - G_{reactants} \tag{4}$$

The ΔGs of common reactions, including the breakdown of ATP, have been determined under standardized conditions and listed in thermodynamic tables. For the reaction

$$ATP \xrightarrow{\quad H_2O \quad} ADP + P_i$$

the tables say that $\Delta G° = -30$ kJ/mol at 20°C and pH 7, where $\Delta G°$, the *standard Gibbs free energy change*, has a superscript to indicate that it is calculated for standard conditions of 1 M concentrations (activities) of the reactants and products. The G under cellular conditions is a little different. Recalling how chemical potentials vary with concentration (Equation 3), we must add adjustments to account for the dilution of the products relative to the 1 M standard conditions and, according to Equation 4, subtract adjustments for dilution of the reactants.

$$\Delta G = \Delta G° + \{RT \ln[ADP] + RT \ln[P_i]\} - \{RT \ln[ATP]\}$$

Gathering the logarithms makes the expression more compact.

$$\Delta G = \Delta G° + RT \ln \frac{[ADP]\,[P_i]}{[ATP]} \qquad (5)$$

If actual cellular concentrations are 4 mM ATP, 1 μM ADP, and 2 mM P_i, the correction term would be $RT\ln(5 \times 10^{-7})$, or -35 kJ/mol ($-30\ -35 = -65$). The ΔG of ATP breakdown in the cell becomes -65 kJ/mol, enough to pump more than 10 moles of substance X per mole of ATP hydrolysed.

Because of its usefulness, we also give the more general form of the free energy change for a chemical reaction. Consider the reaction

$$aA + bB \rightarrow cC + dD$$

in which a moles of A react with b moles of B, and so forth. Writing the effect of dilutions and gathering terms as before gives the general equation:

$$\Delta G = \Delta G° + RT \ln \frac{[C]^c[D]^d}{[A]^a[B]^b}$$

Recapitulation of Thermodynamics. This ends our foray into chemical thermodynamics. Since the path to the desired free-energy functions was a long one, it is worthwhile to summarize the most important steps: An expanding ideal gas could be made to do mechanical work, the maximum work being done if the expansion was performed reversibly—against a constantly adjusted maximal load. Problems involving work, chemical reactions, and the directions of process can conveniently be solved using a new thermodynamic function, the Gibbs free energy, providing one is willing to keep temperature and pressure constant. The two most important relations we have derived are the use of Gibbs free energy to deter-

mine the maximum work that can be done at equilibrium

$$w_{PT} = -\Delta G_{PT} \qquad (2)$$

and the concentration dependence of chemical potential.

$$\Delta\mu = RT \ln(c_2/c_1) \qquad (3)$$

We turn now to ions, charges, and potentials in cells.

IONS AND MEMBRANE POTENTIAL

Ions and Their Gradients. Salts like NaCl, KCl, and $CaCl_2$ dissociate fully in water to form free cations Na^+, K^+, and Ca^{2+} and anions Cl^-. The dissociation of salts is easily recognized by the ability of electrolyte solutions to conduct electricity and by their higher osmotic pressure in solution as compared with that of an equimolar nonelectrolyte solution. Salts dissociate because the dipoles of water molecules interact with free ions in water about as strongly as cations interact with anions in the dry crystal. The interaction with water molecules, called *hydration*, is strong but rapidly changing. In fact water molecules in contact with an ion are replaced about 10^9 times per second by new water molecules. Ions do not enter a hydrophobic region such as the core of a lipid bilayer because all of the stabilizing water dipoles would have to be left behind at a prohibitive energy cost of several hundred kilojoules per mole of ion.

The common inorganic ions serve major physiological roles. The number of osmotically active particles and hence the fluid volume of individual cells or even of the whole body are dominated by the quantity of K^+, Na^+, and Cl^- ions. Electrical signals are generated by movements of Na^+, K^+, Cl^-, and Ca^{2+} ions down their concentration gradients, and the free energy of movements of Na^+, Cl^-, and H^+ ions also may be used by coupled carriers to help drive accumulation of desired organic metabolites. The pH of cytoplasm and body fluids is controlled by H^+, HPO_4^{2-}, and HCO_3^- ions. Intracellular Ca^{2+} ions have a regulatory role as an internal second messenger. Enzymes whose substrates contain high-energy phosphate bonds require Mg^{2+} as a cofactor. These ideas are all expanded in later chapters.

The ionic composition of the cytoplasm is quite different from that of the extracellular fluid. Already with the bacterial level of evolution, the

Table 1–4 Free Ionic Concentrations and Equilibrium Potentials for Mammalian Skeletal Muscle

Ion	Extracellular Concentration (mM)	Intracellular Concentration (mM)	$\frac{[\text{Ion}]_o}{[\text{Ion}]_i}$	Equilibrium Potential[a] (mV)
Na^+	145	12	12	+67
K^+	4	155	0.026	−98
Ca^{2+}	1.5	$<10^{-7}$ M	>15,000	>+128
Cl^-	123	4.2[b]	30[b]	−90[b]

[a]Calculated from Equation 1–7 at 37°C.
[b]Calculated assuming a −90 mV resting potential for the muscle membrane and assuming that Cl^- ions are at equilibrium at rest.

cytoplasm became a region of elevated K^+ ion concentration and depleted Ca^{2+} concentration. To this the animals added an extracellular fluid that, like sea water, had high Na^+ and Cl^- concentrations. Table 1–4 lists the plasma and cytoplasmic ionic concentrations for a mammalian skeletal muscle cell. The major gradients of Na^+, K^+, Cl^-, and Ca^{2+} ions, set up by active transport mechanisms, influence so much of cell physiology that the student will need to learn their directions and the approximate concentrations.

Definition of Electrical Potential. This section is a first review of electrical ideas needed to discuss membrane potentials of living cells. All materials contain a nearly equal number of negative and positive charges. A mole of hydrogen atoms contains 6×10^{23} electrons and 6×10^{23} protons. The charge on a mole of protons, 96,500 coulombs, is called Faraday's number, F.

Opposite charges are difficult to separate because of their mutual attraction. The attractive forces keep all bodies nearly electroneutral. Nevertheless, it is possible to remove a minute amount of charge from one body and transfer it to another. Such charge separation would automatically create an electrical potential difference between the bodies as well as an electric field oriented so as to attract the missing charges back home. This is familiar in the electrical *capacitor* formed by two parallel conducting plates. Charges may be moved from one plate and "stored" on the other. The first charges are easy to move, but as a potential difference builds up, the charges become harder and harder to move against the electric field that draws them back. Potential difference E, measured in volts, is *defined* by the work it takes to move a test charge between two points. The electrical work to move a Faraday of charge through a potential difference E is FE. The electrical work to move n_S moles of ion S is

$$w_{\text{electrical}} = n_S z_S FE \qquad (6)$$

where z_S is the valence of the ion, -1 for Cl^{-1}, $+2$ for Ca^{2+}, and so forth.

Membrane Potential. Virtually all eukaryotic cells are electrically polarized—the cytoplasm is negative with respect to the extracellular medium. Measurement of this potential difference requires some kind of intracellular probe, most commonly a glass capillary micropipette, and a sensitive voltage-recording amplifier. The pipette is prepared by pulling a heated capillary tube to a needle-sharp point (< 0.5 μm) that still has an opening at the tip (Fig. 1–6) and filling the capillary with a conducting solution, such as 3 M KCl. An Ag-AgCl electrode, connected to the amplifier, is placed into the wide end of the pipette. This whole arrangement is commonly called a glass *microelectrode*. To complete the recording circuit, the ground lead of the amplifier is attached via another Ag-AgCl electrode to the bath solution. The objective is to monitor electrical potential changes at the tip of the capillary as it is inserted into a cell. The glass microelectrode is first dipped into the bath, and the amplifier is zeroed. Then the microelectrode is carefully advanced to a cell until suddenly it pops in. Abruptly, the amplifier reports a negative potential at the tip of the pipette. If the microelectrode is advanced through the cytoplasm, the recorded potential remains the same until the tip pops out of the far side of the cell and the recorded potential returns to zero. The step of potential occurs across the cytoplasmic membrane and is called the *membrane potential*, symbolized E_M.

Electrical potentials must be measured as potential differences between two points. Membrane potentials of isolated cells are always defined as inside potential minus outside, so when we say

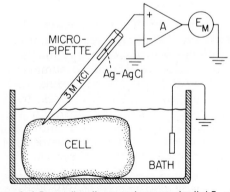

Figure 1–6 Recording the membrane potential E_M of a cell with an intracellular micropipette connected to an amplifier (conventionally symbolized by a triangle). The amplifier measures the potential difference between the cell and the bath.

the membrane potential is negative we mean that the inside is negative with respect to the outside. The membrane potential of quiescent animal cells, the *resting potential*, is typically in the range −40 to −95 mV, but during activity it changes. As we shall see, the resting potential arises because of a tendency for K^+ ions to diffuse outwards through K channels in the plasma membrane.

Nernst Equation. The existence of a negative resting potential requires charge separation across the membrane. There must be an excess of negative charges inside and of positive charges outside, implying that, for example, some cations have been removed from the cytoplasm and placed in the extracellular space. How can a charge separation come about? We have already shown that it would require electrical work to achieve. Specifically from Equation 6, 8.7 kJ/mol of electrical work would have to be done to move monovalent cations out against a 90-mV potential difference. Where did the free energy come from?

The answer to these questions comes from the theoretical work of Walther Nernst in 1889. Consider the experiment in Figure 1–7. In the left panel is a vessel with two fluid compartments separated by a porous membrane. A voltmeter connected through two pipette electrodes records the potential difference between the compartments. At time zero the compartments are filled with electrolyte solutions, a concentrated solution of KA (A for arbitrary anion) on the left side and a dilute solution on the right. At this time both solutions are electroneutral, and the membrane potential is zero.

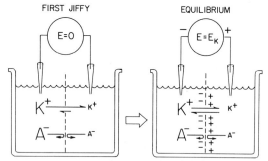

Figure 1–7 Origin of the membrane potential in a purely K^+-permeable membrane. The porous membrane separates unequal concentrations of the dissociated salt K^+A^- (A^- means any impermeant anion). At the moment that the salt solutions are added, the membrane potential E_M, recorded by the electrodes, is zero. Immediately, K^+ ions begin to diffuse to the right down their concentration gradient. Since the anion A^- cannot cross the membrane, a net charge separation builds up. At equilibrium, the E_M becomes equal to the Nernst potential, E_K, and fluxes of K^+ become equal in the two directions.

Now suppose that the membrane pores are permeable to K^+ ions but not to A^- ions. Because of their concentration gradient across the membrane, more K^+ ions diffuse from the left to the right than in the opposite direction. As soon as they do, however, there will be a charge separation since the A^- ions are unable to follow. As positive charge accumulates on the right side, an electric field is built up, opposing further movement of K^+ ions. This whole process might take only a fraction of a millisecond before reaching the electrochemical *equilibrium* condition shown on the right panel of the figure. Diffusion of K^+ ions down their concentration gradient has created a minor charge imbalance sufficient to set up a stable equilibrium membrane potential. This is a quite general explanation for the resting membrane potential of animal cells: The steep outwardly directed concentration gradient for K^+ ions and the presence of K channels in the plasma membrane mean that the cell interior becomes negative as a *small* number of free K^+ ions leave the cell without an accompanying anion.

The potential, E_K, recorded in our idealized experiment is called the *potassium equilibrium potential* or the *Nernst potential* for K^+ ions. A formula for the equilibrium potential of any ion is easily derived from our thermodynamic equations. Permeant* ions move across the membrane in response to two forces, a "chemical" or concentration force and an electrical force. The equilibrium condition occurs when the forces exactly balance. Equation 2 is useful here. Applied to this problem, it says that at equilibrium the chemical potential change from diffusion of permeant ion S down a concentration gradient should match the electrical work done.

$$w_{electrical} = -n_S \Delta \mu_S$$

Substituting Equation 3 for chemical potential changes and Equation 6 for electrical work gives

$$z_S F(E_2 - E_1) = -RT \ln \frac{[S]_2}{[S]_1} = RT \ln \frac{[S]_1}{[S]_2}$$

Rearrangement gives the formula for the equilibrium potential difference, the *Nernst equation*.

$$E_2 - E_1 = \frac{RT}{z_S F} \ln \frac{[S]_1}{[S]_2}$$

---

*The words *permeable* and *permeant* should be distinguished. The membrane is *permeable* (able to be permeated) and the ion is *permeant* (able to permeate).

We can rewrite the Nernst equation, taking into account the biological convention that membrane potentials are defined as inside minus outside,

$$E_S = \frac{RT}{z_S F} \ln \frac{[S]_o}{[S]_i} \tag{7}$$

where $[S]_i$ and $[S]_o$ refer to inside and outside concentrations (more correctly, "activities"). Converting to $\log_{10}$ and calculating RT/F for 20°C (Table 1–3) gives a practical form.

$$E_S = \frac{58.2 \text{ mV}}{z_S} \log_{10} \frac{[S]_o}{[S]_i} \tag{8}$$

From inspection we see that a 10-fold concentration gradient of a monovalent ion could produce a 58.2 mV membrane potential. The sign of the equilibrium potential changes if the sign of the permeant ion is changed and if the direction of the gradient is changed. For the 39-fold outwardly directed gradient of K^+ ions in a mammalian muscle (Table 1–4), the predicted value of E_K at 20°C is -92 mV. For the 12-fold *inwardly* directed gradient of Na^+ ions, the predicted E_{Na} is $+67$ mV. Because the membrane is normally far more permeable to K^+ ions than to Na^+ ions, the actual resting potential recorded from a muscle cell is closer to E_K than to E_{Na}.

The equilibrium potential concept is pivotal in thinking about the ionic origin of membrane potentials and needs to be considered carefully. Three useful statements about the equilibrium potential for K^+ ions are the following: (i) If a membrane is exclusively permeable to K^+ ions, then small movements of K^+ ions will occur in such a way as to change E_M until it reaches E_K. (ii) If a membrane is held at E_K, there will be no *net* flux of K^+ ions unless some kind of active transport device is operating. (iii) There is no net flux of K^+ ions at E_K because the sum of the electrical and chemical concentration forces on them is zero.

Thermodynamic problems involving both concentration differences and electrical potentials are often more conveniently discussed in terms of changes of *electrochemical* potential $\bar{\mu}$ rather than chemical potential μ. Electrochemical potential changes are defined by

$$\Delta\mu \equiv -w_{electrical}/n + \Delta\mu = zF(E_2 - E_1) + RT \ln(c_2/c_1) \tag{9}$$

Expanding on Equation 2 gives for the maximum work at equilibrium,

$$\Delta\bar{\mu} = -w_{nonelectrical}/n$$

In these terms, the Nernst equation could be derived directly by noting that for an ion at equilibrium across a cell membrane, its electrochemical potential μ would be equal on the two sides and $\Delta\bar{\mu}$ for that ion would be zero.

Goldman-Hodgkin-Katz Voltage Equation.[7, 8] In 1902, Julius Bernstein[2] proposed that resting potentials arise across a surface membrane that is selectively permeable to K^+ ions. His membrane hypothesis could not be tested properly until almost 40 years later, when reliable methods for intracellular recording had been developed.

One quantitative prediction of the hypothesis is that the resting potential should change with the bathing K^+ ion concentration $[K]_o$, following E_K exactly. *As $[K]_o$ is raised, E_M should depolarize* (become less negative), finally reaching zero when the internal and external K^+ concentrations are made equal, i.e., when there is no gradient for K^+ ions. The circles in Figure 1–8 show measured membrane potentials in such an experiment done on the squid giant axon—a favorite cell to study because of its exceptional size. The points are plotted on semilogarithmic axes so that the predictions of the Nernst equation become a straight line—labeled E_K. At high $[K]_o$, theory and experi-

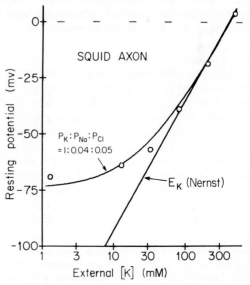

Figure 1–8 Dependence of resting membrane potential on the external K^+ ion concentration in a squid giant axon. The sum of $[Na^+]_o$ and $[K^+]_o$ were kept constant as $[K^+]_o$ was varied. The line labeled E_K shows the expected Nernst potential for K^+ ions, and the curved line is a solution of the Goldman-Hodgkin-Katz voltage equation (Equation 10), assuming that $P_K:P_{Na}:P_{Cl}$ of the axon membrane is 1.0:0.04:0.05. (Data after Curtis and Cole. *J. Cell Comp. Physiol.* 19:135–144, 1942.)

ment agree well, but at physiological values of [K]$_o$, the membrane potential is less negative than predicted.

The hypothesis that the axon membrane is K-selective is correct but incomplete. Potassium ions are by far the most permeant, but Na$^+$ and Cl$^-$ ions are slightly permeant, too, and when [K]$_o$ falls to very low values, the contribution of the other ions becomes apparent. Goldman, Hodgkin, and Katz (GHK) derived an equation to describe the concentration-dependence of the membrane potential when several ions are permeant.[7, 8] It takes into account the relative permeabilities P to each ion.

$$E_M = \frac{RT}{F} \ln \frac{P_K[K]_o + P_{Na}[Na]_o + P_{Cl}[Cl]_i}{P_K[K]_i + P_{Na}[Na]_i + P_{Cl}[Cl]_o} \quad (10)$$

The GHK voltage equation looks similar to the Nernst equation and says that ions with high concentration-permeability products make the most important contributions to E_M. If the permeabilities to all but one ion are set to zero, the expression reduces to the Nernst equation for the permeant ion exactly.

The curved line in Figure 1–8 is a solution of the GHK equation, assuming that the membrane permeability ratios $P_K:P_{Na}:P_{Cl}$ are 1:0.04:0.05 in the squid giant axon. This assumption describes the results much better than the Nernst equation alone. Experiments with radioactive isotope fluxes confirm that K$^+$ ions permeate most easily, but other ions (especially Na$^+$, Cl$^-$, Ca^{2+}, Mg^{2+}, HPO$_4^{2-}$) cross the resting axon membrane as well. The GHK equation with fixed permeability coefficients does not give a perfect description of the observations partly because the number of open K channels increases as the membrane depolarizes. As we see in Chapter 3, K channels and others have voltage-dependent gates. Opening and closing of these gates are fundamental aspects of excitation of nerve and muscle. The *resting* potential of excitable cells arises because K-selective ionic channels, K channels, are open in the resting cell membrane, and the potential changes in *excited* nerve and muscle cells arise when gates of Na channels are opened by a stimulus.

Muscle cells also become depolarized when external [K] is elevated. However, the resting permeability ratios are clearly different from those of axon membranes, for isotope fluxes and electrical measurements on muscle agree that P_{Cl} can be even higher than P_K. In a typical mammalian twitch muscle, the permeability ratios $P_K:P_{Na}:P_{Cl}$ are 1:0.01:8. Evidently, the resting *muscle* mem-

brane has many open Cl channels in addition to open K channels.

DIFFUSION AND MEMBRANE PERMEABILITY

The thermal energy of molecules of matter keeps them in a constant state of agitation. At room temperature, water molecules and small solutes have average velocities on the order of hundreds of meters per second. In aqueous solutions, they suffer collisions after moving less than an atomic diameter, and zoom off in another direction. This microscopic buffeting promotes mixing and breaks down local concentration gradients, producing the macroscopic process of *diffusion*.

Diffusion Equation. The physiologist Adolph Fick—motivated by his interest in biological membranes—was the first to give a mathematical description of diffusion fluxes. Consider two stirred compartments separated by a permeable "membrane" barrier such as a thick piece of filter paper (Fig. 1–9). There is a solute concentration difference Δc across the membrane of thickness Δx, and therefore there is a concentration *gradient* $\Delta c/\Delta x$ between the compartments. In analogy with the flow of heat between hot and cold bodies, Fick said correctly that the flux M_S of solute S is proportional to the concentration gradient.

$$M_S = -D_S A \frac{\Delta c_S}{\Delta x} \quad (11)$$

Here *flux* is defined as the net number of moles per second of solute crossing a hypothetical plane parallel to the membrane; A is the area of the plane; and D_S is the diffusion coefficient for S, a measure of the mobility of S molecules in the membrane. The minus sign signifies that the net

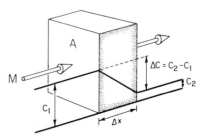

Figure 1–9 Definition of diffusion flux M across an area A and down a gradient $\Delta c/\Delta x$. The stippled slab of material is represented as a diffusion barrier between two well-stirred compartments.

movement is down rather than up the concentration gradient.

Two aspects of Fick's first law (Equation 11) can be noted from inspection. It is a *linear* law. No matter how high the concentrations become, the flux remains proportional to the gradient, i.e., free diffusion does not saturate. Second, the law implies *independence* of fluxes. The fluxes of solute S are independent of fluxes of other solutes Y and Z—there is no coupling.

Microscopic Viewpoint. Albert Einstein introduced molecular thinking about the nature of diffusion (1905–1908).[5] He derived a relation between diffusion coefficients and molecular sizes, and he explained the connection between diffusion and thermal agitation. Recognizing that diffusion is thermal motion impeded by frictional forces, Einstein proposed that the macroscopic rule for friction on spherical particles moving in a viscous medium might be applied at the molecular level. Stokes's law says that a macroscopic sphere of radius a moving in a medium of viscosity η experiences friction of $6\pi\eta a$. From this Einstein predicted that the diffusion coefficient for a spherical particle is

$$D_S = \frac{RT}{6\pi\eta aN} \tag{12}$$

Amazingly, the Stokes-Einstein relationship holds within a factor of 2 for molecules as small as H_2O. It says that the rate of diffusion varies inversely with the molecular radius: small molecules diffuse fastest. Since radius varies only with the cube root of molecular weight (for molecules of similar density), the rate of diffusion depends only weakly on molecular weight as is borne out

Table 1–5 Aqueous Diffusion Coefficients at 20°C

Molecule	Diffusion Coefficient (10^{-5} cm^{2}s)
H_2O	2.2
O_2	2.0
Cl^-	2.0
K^+	2.0
Na^+	1.3
Glycine	1.0
Glucose	0.6
Lactose	0.4
Hemoglobin	0.07

by the list of diffusion coefficients in Table 1–5. A useful rule of thumb is to remember that aqueous diffusion coefficients of small molecules are in the range of 10^{-5} cm^2/s.

Einstein's other contribution was to view diffusion (and Brownian motion) as a *random walk*. Individual molecules moving independently have no knowledge of a concentration gradient. They move to the left or to the right at random with *equal* probability, as if flipping a coin before making each step. This is a consequence of the molecular agitation already discussed. If all the particles are started at the same point, they gradually spread out in both directions with time, forming a bell-shaped (Gaussian) distribution of concentration in space, centered on the origin (Fig. 1–10). Einstein gave a beautifully simple expression for the mean squared distance $\overline{r^2}$ of each molecule from the origin in a one-dimensional case.

$$\overline{r^2} = 2Dt \tag{13}$$

For two- or three-dimensional diffusion, the right-hand side would be $4Dt$ or $6Dt$.

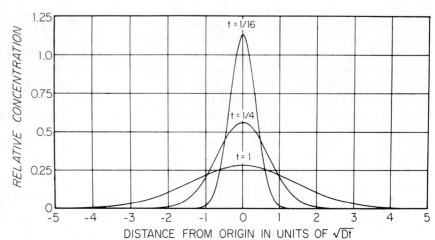

Figure 1–10 Concentration profiles of diffusion in one dimension. At $t = 0$ a unit amount of material is deposited at the origin and begins to diffuse to the left and right. Distances and times in normalized units. If the three curves represent time of $t/16$, $t/4$ and t, then the distance axis is marked off in units of $\sqrt{Dt}$. Thus if $t = 1$ ms, the distance axis is in units of 1 μm for a typical molecule with $D = 10^{-5}$ cm/s. (After Crank. *The Mathematics of Diffusion.* Oxford, Clarendon Press, 1956.)

Equation 13 is handy for obtaining quick answers to physiological questions concerning how far and how fast a diffusion mechanism can move molecules. A typical problem might be this: How close do acetylcholine receptor molecules need to be to the site of release of the neurotransmitter to respond within 100 μs? The answer, using the equation for two-dimensional free diffusion, is 0.6 μm. Note that Equation 13 says that the diffusion time goes up as the square of the distance to travel, rather than linearly, a consequence of the inefficiency of random forward and backward steps. Therefore, diffusion over long distances is extremely slow. For example, free diffusion of a small molecule over the 1 m distance between a motoneuron cell body in the spinal cord and the terminals of the same cell in the toes would take 16 years. For this reason, the axons of nerve cells have an internal conveyor system ("axoplasmic transport") that uses metabolic energy and can deliver proteins to distant terminals in 2.5 days.

Solubility-Diffusion Theory.[6, 11] Fick's law in its simplest form is suitable for aqueous diffusion in bulk solution or through water-filled "barriers" like the piece of filter paper. However, when the diffusion process requires permeating molecules to partition into a different phase, such as a lipid bilayer, and the barrier thickness is not known ahead of time, Fick's law is hard to apply. Biologists therefore use a simpler empirical expression for diffusion of neutral molecules across membranes.

$$M_S = -AP_S \, \Delta c_S \qquad (14)$$

Equation 14 defines P_S, the membrane *permeability coefficient* for a neutral molecule. When there is no membrane potential, the same definition of P_S applies to ions as well, but in the presence of an electrical driving force, a more complicated expression, the Goldman-Hodgkin-Katz[7, 8] flux equation is needed.

Standard methods to determine membrane permeability use measurements of influx or efflux of labeled molecules from cells. For small cells and highly permeant molecules, this may require rapid mixing of cells with the test solution and sampling within a fraction of a second to determine the initial rate of entrance or exit of label. For ions, the permeability is often determined from electrical measurements of membrane currents as is described in Chapter 3. Typical permeabilities of plasma membranes are listed in Table 1–6. The permeability of cell membranes for water turns out to be 10^3 to 10^5 times lower than the value

Table 1–6 Plasma Membrane Permeability Coefficients for Small Molecules

Cell	Molecule	Permeability (cm/s)
Lipid bilayer	Salicylic acid	1
Axon	Local anesthetic	~1
Various	H_2O	0.3 to 50 × 10^{-4}
Muscle	Cl^-	4 × 10^{-6}
Muscle (resting)	K^+	2 × 10^{-6}
Red blood cell	K^+	2 × 10^{-10}

predicted for a slab of water of equal thickness. In general, the permeability depends on the test molecule, the type of membrane, and the physiological state of the cell, which can alter the function of transport macromolecules in the membrane.

As Overton originally noted, membrane permeability increases with lipid solubility. It is possible to use this idea to predict permeability coefficients in lipid bilayers. Consider the steps of membrane permeation in Figure 1–11, and let the bilayer/water *partition coefficient* for S be β_S^*—a tiny number for hydrophilic substances and a larger number for lipid-soluble ones. Molecules on side one at concentration c_1 partition into the membrane, reaching an equilibrium concentration of $\beta_S^* c_1$ just within the bilayer-water boundary, and molecules on side two partition at a concentration

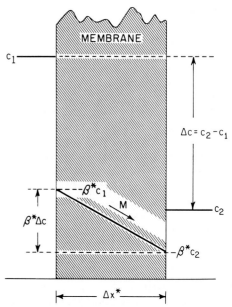

Figure 1–11 Solubility-diffusion theory, showing normal diffusion within the membrane and partitioning of molecule between bulk phase and membrane at the boundaries. The partition coefficient β^* is assumed to be 0.3., i.e., the molecule is slightly less soluble in the membrane than in water.

Table 1–7 Test of the Solubility-Diffusion Theory for Nonelectrolytes Crossing Lipid Bilayers

Molecule	Measured β_{hc} (10^{-5})	Measured D (10^{-5} cm^2/s)	Predicted P = D β_{hc}/5 nm (10^{-4} cm/s)	Measured P (10^{-4} cm/s)
Codeine	4250	0.63	5360	1400
Butyric acid	784	1.0	1570	640
H_2O	4.2	2.44	10.2	22
Acetamide	2.1	1.32	5.5	1.7
Formamide	0.79	1.7	2.7	1.03
Urea	0.35	1.38	0.97	0.04
Glycerol	0.20	1.09	0.44	0.054

Data from Orbach, E. and Finkelstein, A. *J. Gen. Physiol.* 75:427–436, 1980.
Membrane: Egg lecithin-*n*-decane planar bilayers at 25°C.
   P: Diffusional permeability coefficient defined by Equation 14 or predicted by Equation 15.
   β_{hc}: Water-hexadecane partition coefficient.
   D: Diffusion coefficient of test molecule in water. The diffusion coefficients in hydrocarbon should have been used but have not been measured. Because decane and water have nearly the same viscosity, the aqueous *D* should be nearly correct.

of $\beta_s^* c_2$. Now, if the membrane is like a liquid with the permeant molecule dissolved in it, we can write down the diffusion flux within the membrane

$$M_s = -AD_s^* \frac{\beta_s^* c_2 - \beta_s^* c_1}{\Delta x^*} = -AD_s^* \beta_s^* \frac{\Delta c_s}{\Delta x^*}$$

where D_s^* is the diffusion coefficient within the bilayer, and Δx^*, the thickness of the bilayer. Comparison of this expression with Equation 14 shows that we have derived an experimentally verifiable theory for the permeability coefficient.

$$P_s = \frac{D_s^* \beta_s^*}{\Delta x^*} \qquad (15)$$

A test of the theory is given in Table 1–7. Permeability coefficients for several molecules passing through a lipid bilayer are predicted assuming (i) the partition coefficient into membrane is the same as that into liquid hexadecane, a hydrocarbon, (ii) the diffusion coefficient in the bilayer is the same as in water, since the viscosity of hexadecane is the same as that of water, and (iii) the thickness of the bilayer is 5 nm. The predictions are compared with permeability coefficients measured from fluxes in phosphatidylcholine lipid bilayers. Considering the crude nature of these assumptions, the degree of agreement of experiment and theory over a four order-of-magnitude range of permeabilty coefficients is remarkable. This solubility-diffusion theory, although naively simple, gives a useful understanding of how lipid-soluble substances cross the membranes of living cells and a neat description of the relationship between lipid solubility and permeation.

SUMMARY

Living cells are surrounded by a plasma membrane containing protein molecules, with a bilayer of lipids filling in all the gaps. The lipid acts as a fluid sealant, giving a flexible but impermeable barrier to polar molecules. Hydrophobic substances can dissolve in the lipid and diffuse to the other side almost as in a liquid hydrocarbon phase. The proteins confer functions on the membrane. They are carriers, ionic channels, receptors, and enzymes. Carriers permit essential substrates and metabolites to enter and leave the cell, and some carriers establish concentration gradients across the membrane. Ionic channels gate open and closed, changing the membrane potential of cells. Receptor proteins detect extracellular chemical messages and convert them to cellular signals, internal second messengers such as changes of membrane potential, internal free Ca^{2+}, cyclic nucleotides, and the metabolites of inositol phospholipids.

In biology, the Gibbs free energy is usually the most useful measure of work. It allows us to relate energy changes in chemical reactions, mechanical work, dilution and concentration problems, and movements of electric charges. The free energy change of a chemical reaction or of a dilution process varies with the logarithm of the concentration of the reactants and products. At equilibrium, the chemical potential of a substance is the same everywhere and the ΔG for any process is zero.

Electrical potentials arise because cells have a plasma membrane with selective permeability to ions that have a concentration gradient between the cytoplasm and the extracellular medium. In animal cells, the K^+ ion dominates the membrane potential. It is concentrated inside the cell and therefore tends to diffuse out through K channels, establishing a negative resting potential. Were there no channels other than K-selective ones, the resting potential would be equal to the equilibrium potential for K^+ ions given by the Nernst equation.

Diffusion of molecules arises from thermal agitation pushing molecules against the frictional resistance of the suspending fluid. Since diffusion is a kind of random walk, the net distance of travel of one molecule grows only as the square root of time, rather than linearly. Nevertheless the net flux of molecules down a concentration gradient obeys a simple rule, Fick's law.

This concludes our review of scientific principles needed to undertake our approach to cellular physiology.

Appendix: Derivation of Thermodynamic Relations

This Appendix gives the derivation of thermodynamic relations used in the section ENERGY CALCULATIONS.

Maximum Work of Expanding Ideal Gases. We want to calculate the work that an ideal gas can do while expanding from pressure P_1 to P_2 at a constant temperature, as in Figure 1–5. By definition, the mechanical work dw done in a small step dx of the piston is the product of the force against which work is done times the distance moved. The maximum force the piston could actually move against is the product of the gas pressure times the area A of the piston.

$$dw = (force) \cdot dx = (PA) \cdot dx$$

The product of $A \cdot dx$ equals the volume change of the gas, and according to the ideal gas law, P may be substituted by nRT/V, giving

$$dw = PA \cdot dx = PdV = nRTdV/V$$

The total work done by the expanding gas is the sum (integral) of the work done in each small step.

$$w_T = \int_1^2 dw = nRT \int_1^2 \frac{dV}{V} = nRT \int_1^2 d\ell nV$$

Integrating and substituting the ideal gas law once again gives the answer.

$$w_T = nRT \ln \left(\frac{V_2}{V_1} \right) = nRT \ln \left(\frac{P_1}{P_2} \right) \quad (1)$$

Now we raise what may seem like a paradox. According to statistical mechanics, the energy U

of n moles of ideal gas is 3nRT/2, since the particles have three degrees of freedom and no energy other than kinetic energy. Hence the energy of the gas does not change during an expansion at constant temperature. How could a gas do work without losing energy? The answer requires the First Law of Thermodynamics.

First and Second Laws of Thermodynamics. The first law, the law of conservation of energy, says that the energy change dU of a system is equal to the heat dq put into the system minus the work done by the system (Fig. 1–12).

$$dU = dq - dw \quad (16)$$

The first law explains the apparent paradox. We know that dU for the expanding gas was zero; therefore $dq = dw$. Evidently all the energy needed to do the work came as heat from the constant-temperature bath. Taking $dU = 0$ and

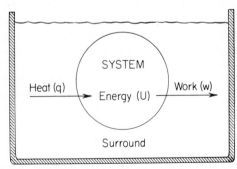

Figure 1–12 The first law of thermodynamics says that the energy of a closed system increases in proportion to the heat received from the surround and decreases in proportion to the work done on the surround.

inserting Equation 1, the heat transferred from the bath during the reversible expansion would have been

$$q_{rev} = w_T = nRT \ln(P_1/P_2) \qquad (17)$$

Had the expansion been done irreversibly against no load, no heat would have been drawn from the bath. In general, a process done irreversibly draws less heat from the bath and produces less work than the same process done reversibly.

The first law tells us how energy would flow if a process were to occur, but it does not tell us if a process will occur. We know that all systems are proceeding towards equilibrium, and we need to know which direction it is. We need a criterion that says whether gases would tend to expand or contract, whether heat flows from hot to cold bodies or from cold to hot, and ultimately, whether a biological reaction will occur spontaneously. This is the second law.

The second law concerns changes of *entropy* (symbolized S), whose definition is straightforward but whose meaning requires sustained exposure to appreciate. Statistical mechanics shows that entropy is related to disorder. Entropy changes dS are defined by

$$dS \equiv \frac{dq_{rev}}{T} \qquad (18)$$

in which dq_{rev} is the heat that would be taken up from the surroundings if the process occurring were done reversibly, i.e., very close to equilibrium.

The second law states that the *entropy of the universe increases* until equilibrium is reached, i.e.,

$$dS_{universe} \geq 0$$

At equilibrium there is no further entropy change. The law that entropy increases also applies to isolated systems, subsets of the universe that exchange neither energy nor matter with the rest of the universe. We now apply the second law in this form to gas expansion and heat flow between bodies to show that it agrees with observations: (i) Substituting Equation 17 into the definition of entropy changes gives the entropy change of an expanding ideal gas

$$\Delta S = \frac{q_{rev}}{T} = nR \ln(P_1/P_2) = nR \ln(V_2/V_1) \qquad (19)$$

If entropy of an isolated system always increases, then Equation 19 predicts that an isolated gas will

expand spontaneously in agreement with observations. (ii) For two bodies at temperature T_A and T_B, the entropy change from heat flow from A to B is

$$dS = \frac{dq}{T_B} - \frac{dq}{T_A} = dq \left(\frac{1}{T_B} - \frac{1}{T_A} \right)$$

If entropy of an isolated system always increases, then A, from which the heat is flowing, is warmer than B, also in agreement with observations. Hence the second law gives appropriate answers for the direction that processes will take.

Gibbs Free Energy. The second law is still not a practical expression for use in biology because it requires measurements of entropy changes of the entire universe or of completely isolated systems. We need thermodynamic criteria that require only properties that can be measured in our system, be it a cell in the body or a test tube in a water bath. This is the *Gibbs free energy*

$$G \equiv U + PV - TS \qquad (20)$$

in which each of the quantities is measured *in the system under study*.

The Gibbs free energy is a practical predictor of work and the direction of processes. To reduce it to the appropriate form will now require several lines of algebra and use of both the first and second laws. To consider small changes of G, we take derivatives of Equation 20 and then substitute in the first law.

$$\begin{aligned} dG &= dU + PdV + VdP - TdS - SdT \\ &= (dq - dw) + PdV + VdP - TdS - SdT \end{aligned}$$

Then we restrict ourselves to laboratory conditions where the temperature and ambient pressure do not change:

$$dG_{PT} = dq - dw + PdV + 0 - TdS - 0$$

Next we apportion the external work dw into two components, that which results from volume changes of the system and all other forms of work, including work against electrical, gravitational, elastic, and surface forces. The volume work is equal to PdV, and the remainder will be called dw_{nonPV}.

$$dG_{PT} = dq - (dw_{nonPV} + PdV) + PdV - TdS$$

Canceling PdV terms and substituting Equation 18 for dS gives

$$dG_{PT} = dq - TdS - dw_{nonPV}$$
$$= dq - dq_{rev} - dw_{nonPV}$$

Finally, noting from the second law that in an irreversible process the actual heat drawn from the temperature bath is less than q_{rev}, gives

$$dG_{PT} < -dw_{nonPV}$$

except at equilbrium (reversible process), where q = q_{rev}.

$$dG_{PT} = -dw_{nonPV}$$

This is a practical relationship between the maximum work that, for example, a chemical reaction can do and completely measurable thermodynamic properties of the reactants and products. When no work is done, processes at constant P and T will always proceed in the direction that decreases G of the system, and when we finally reach equilibrium, dG becomes zero. These properties make the Gibbs free energy the basic thermodynamic tool of biologists.

Chemical Potential Change of Ideal Gas Expansion. Consider the Gibbs free energy change for the isothermal, reversible expansion described previously, remembering that for ideal gases neither U nor the *product* PV should change. Starting with Equation 20, the result is

$$\Delta G_T = \Delta U_T + \Delta(PV)_T - \Delta(TS)_T$$
$$= 0 + 0 - T\Delta S$$
$$= -T \cdot nR \ln(P_1/P_2) = nRT \ln(P_2/P_1)$$

Therefore the chemical potential change for an ideal gas is

$$\Delta\mu_T = RT \ln(P_2/P_1)$$

ANNOTATED BIBLIOGRAPHY

1. Alberts, B.; Bray, D.; Lewis, J.; Raff, M.; Roberts, K.; Watson, J.D. *Molecular Biology of The Cell*. New York, Garland Publishing, 1983.
   A large modern cell biology text with good sections on membranes and second messengers.

2. Darnell, J.; Lodish, H.; Baltimore, D. *Molecular Cell Biology*. New York, W.H. Freeman, 1986.
   Another large modern cell biology text with good sections on membranes and second messengers.

3. Andreoli, T.E.; Hoffman, J.F.; Fanestil, D.D.; Schultz, S.G., eds. *Physiology of Membrane Disorders*. New York, Plenum Medical Book Co., 1986.
   A collection of instructive essays on membrane proteins, transport theory, pumps, carriers, and channels.

REFERENCES

1. Almers, W.; Stirling, C.E. The distribution of transport proteins over animal cell membranes. *J. Membr. Biol.* 77:169–186, 1984.
2. Bernstein, J. Untersuchungen zur Thermodynamik der bioelektrischen Ströme. Erster Theil. *Pflügers Arch.* 92:521–562, 1902.
3. Berridge, M.J.; Irvine, R.F. Inositol trisphosphate, a novel second messenger in cellular signal transduction. *Nature* 312:315–321, 1984.
4. Cohen, P. The role of protein phosphorylation in neural and hormonal control of cellular activity. *Nature* 296:613–620, 1982.
5. Einstein, A. The elementary theory of Brownian motion. *Z. Electrochem.* 14:235–239, 1908. Republished translation. In Einstein. A. *Investigations on the Theory of the Brownian Movement*, 68–85. New York, Dover Publications, 1956.
6. Finkelstein, A. Water and nonelectrolyte permeability of lipid bilayer membranes. *J. Gen. Physiol.* 68:127–135, 1976.
7. Goldman, D.E. Potential, impedance, and rectification in membranes. *J. Gen. Physiol.* 27:37–60, 1943.
8. Hodgkin, A.L.; Katz, B. The effect of sodium ions on the electrical activity of the giant axon of the squid. *J. Physiol. (Lond.)* 108:37–77, 1949.
9. Kushmerick, M.J. Energetics of muscle contraction. *Handbk. Physiol.*, Sec. 10, 189–236, 1983.
10. Nishizuka, Y. The role of protein kinase C in cell surface signal transduction and tumour promotion. *Nature* 308:693–698, 1984.
11. Orbach, E.; Finkelstein, A. The nonelectrolyte permeability of planar lipid bilayer membranes. *J. Gen. Physiol.* 75:427–436, 1980.
12. Poo, M. Mobility and localization of proteins in excitable membranes. *Annu. Rev. Neurosci.* 8:369–406, 1985.
13. Ringer, S. A further contribution regarding the influence of the different constituents of the blood on the contraction of the heart. *J. Physiol. (Lond.)* 4:29–42, 1883.
14. Singer, S.J.; Nicholson, G.L. The fluid mosaic model of the structure of cell membranes. *Science* 175:720–731, 1972.
15. Williamson, J.R.; Cooper, R.H.; Joseph, S.K.; Thomas, A. P. Inositol trisphosphate and diacylglycerol as intracellular second messengers in liver. *Am. J. Physiol.* 248, Cell Physiol. 17:C203–C216, 1985.

Transport Across Cell Membranes: Carrier Mechanisms

INTRODUCTION

One of the major jobs of cell membranes is to prevent or to facilitate the movement of essential small molecules from one closed compartment to another. Hence membrane physiology is intimately concerned with measuring and understanding the *transport properties* of biological membranes, in which transport refers to any net movement of a substance, whether by simple diffusion or by more specific active or passive mechanisms.

This chapter is concerned with carrier mechanisms. In the classic view, carriers were regarded as mobile and highly selective binding sites that could shuttle small molecules across the membrane much as a ferry boat shuttles people across a river. The carriers would make small molecules "soluble" in the lipid membrane for the journey. Carriers were recognized by four criteria: (i) Carriers could distinguish among very similar molecules, reflecting the *specificity* of the binding site. (ii) Carrier fluxes had a maximum velocity, reflecting a *saturation* of the shuttling mechanism. (iii) Likewise, addition of one transported species might reduce the transport of another, reflecting a *competition* for binding sites. (iv) Finally, in many instances, there was an obligate *coupling* of the fluxes of two transported species, as if the carrier mechanism waited until it was loaded with a specific stoichiometric combination of substrates before starting the trip.

The four classic criteria still are the most useful description of carrier kinetics. Nevertheless, two caveats must be stated: First, specificity, saturation, competition, and coupling are each theoretically possible and indeed have been demonstrated for transport mechanisms that are unambiguously pores.[28] Therefore, some additional criteria need to be used. Second, the classic concept of the carrier as a lipid-soluble ferryboat making journeys across the membrane by a kind of random walk has to be abandoned. Those carrier molecules that have now been identified biochemically are found to be large protein macromolecules whose peptide chains extend back and forth across the membrane several times. Such molecules already face both solutions and are much too large for the whole molecule to rotate or move in and out of the membrane during each transport event. Therefore, the motions underlying carrier kinetics are now considered to be smaller conformational changes within a protein that spans the membrane, and a contemporary mechanistic definition of a carrier is "a device that exposes a binding site alternately to the internal and to the external medium." There is as yet, however, no carrier molecule whose mechanism is understood in structural detail, so what we know is still based primarily on the physiological method of designing flux experiments to distinguish among various possible but hypothetical mechanisms.

Most of this chapter is devoted to the sodium pump, the best studied of the carrier systems in the cell surface membrane.

ACTIVE TRANSPORT OF SODIUM AND POTASSIUM IONS

The previous chapter shows that the principal cation of the extracellular fluid, the Na^+ ion, is much less concentrated in the intracellular fluid than expected for a system at equilibrium. This concentration gradient for Na^+ ions is maintained throughout the lifetime of cells, even though cell membranes are at least somewhat sodium permeable. Apparently, some mechanism removes Na^+ ions from the intracellular fluid as fast as they diffuse in through the cell membrane. This transport mechanism is often called the *sodium pump*, since it pumps Na^+ ions out of the cell. A more complete name, *Na^+-K^+ pump*, expresses the fact that the same mechanism brings K^+ ions into the cell.

The surface membranes of all animal cells have a sodium pump that seems to be substantially the same. For the cell physiologist, the sodium pump is a macromolecular machine with the job of maintaining steep gradients of Na^+ and K^+ ions across the plasma membrane. As we shall see, it is a multisubunit, membrane enzyme that transports its substrates at the expense of a stoichiometric breakdown of ATP. Moving ions against their electrochemical gradients makes this *active transport*. Cleaving high-energy bonds directly in the process makes it *primary* active transport. This section describes properties of the pump and progress toward understanding the mechanism of active transport. Many of the physiological experi-

ments are done with squid giant axons or mammalian red blood cells because of their many advantages in the measurement of ionic fluxes.

Ionic gradients maintained by continuous operation of the pump are essential to cellular function. By controlling the distribution of the most important osmotically active particles (ions), the Na^+-K^+ pump, together with the passive ionic "leak," regulates cell volume. In many cells, the free energy stored in the sodium gradient is used to drive an uphill transport of amino acids, sugars and other vital ions and substrates into the cell, a process called *secondary* active transport since the coupled carrier is not directly fueled by ATP. In addition, many cytoplasmic enzymes are stimulated by K^+ ions and inhibited by Na^+ ions. Whether this stimulation plays any regulatory role in cell function is largely unknown. The pump has a special significance in electrically excitable cells like nerve and muscle, because as explained in Chapter 3, these cells use the gradients of Na^+ and K^+ ions to produce electric currents for signaling. Nerve and muscle also use the energy stored in the sodium gradient to drive Ca^{2+} ions out of the cell. Finally, pumping of Na^+ and K^+ ions underlies the major salt- or water-secreting and resorbing properties of a wide variety of epithelia, including those of the gut lining, kidney tubules, avian salt glands, amphibian skin, teleost gills, and elasmobranch rectal glands. In some specialized tissues, such as the brain, the kidney, and red blood cells, 30 to 90 per cent of all energy metabolism goes toward providing ATP for the active transport of Na^+ and K^+ ions.

Physiological Properties of the Sodium Pump. The two most important features of the sodium pump are the need for metabolic energy and the linkage between the active transport of Na^+ ions outward and of K^+ ions inward. Both features are revealed by the classic work of Hodgkin and Keynes[29] on giant axons of the cuttlefish, *Sepia* (a close relative of squids). Figure 2–1 shows the time course of sodium efflux measured in one experiment. An axon was soaked in a solution containing radioactive $^{24}Na^+$ ions, so that the axoplasm (cytoplasm) became loaded with radioactive $^{24}Na^+$. Then the external radioactive solution was washed away, and the sodium efflux was measured by counting the rate of appearance of $^{24}Na^+$ in the solution bathing the outside of the axon. In normal seawater, containing about 10 mM K^+, the efflux was 90 to 120 counts/minute per minute. Most of this Na^+ efflux represents the operation of the sodium pump. In a K^+-free solution, the Na^+ efflux dropped at once to one third of the

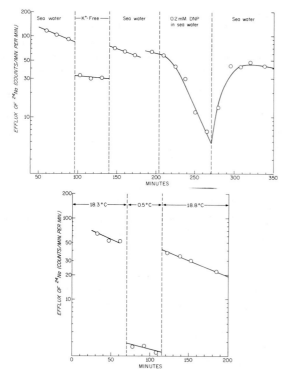

Figure 2–1 Na⁺ efflux from a resting giant axon of *Sepia* (cuttlefish) following a period of stimulation in a sea water containing radioactive $^{24}Na^+$. Ordinates are the efflux of $^{24}Na^+$, measured as the number of counts per minute appearing in the external medium over 1-minute intervals. Total Na⁺ efflux (radioactive plus normal) is proportional to radioactive efflux except for gradual dilution of intracellular $^{24}Na^+$ by unlabeled Na⁺ (the gradual fall shown under sea water).

Above, Effects of various bathing media. Second column from left shows that reducing $[K^+]_o$ from 10 mM to zero immediately reduces Na⁺ efflux to about one third that in artificial sea water; this effect is immediately reversed when the axon is returned to artificial sea water (middle column). Next, adding the metabolic inhibitor 2,4-DNP reduces Na⁺ efflux to values near zero within 1 to 2 hours; this effect is slowly reversible (right column) when the inhibitor is washed away.

Below, Reduction of temperature from 18° to 0.5° C immediately reduces Na⁺ efflux to near zero; raising the temperature immediately restores efflux. (From Hodgkin, A.L.; Keynes, R.D. *J. Physiol. [Lond.]* 128:28–60, 1955.)

control value. When external K⁺ was restored, the efflux immediately returned to normal. Thus external K⁺ ions are essential for a large component of the efflux of Na⁺ ions across the axon membrane.

Addition of 0.2 mM 2,4-dinitrophenol (DNP) reduced Na⁺ efflux to very low levels, although only after some delay. This poison stops the replenishment of cellular energy stores by uncoupling oxidative phosphorylation in mitochondria, thereby stopping the resynthesis of adenosine

triphosphate (ATP) from adenosine diphosphate (ADP) and inorganic phosphate (P_i). Independent experiments show that DNP stops ATP production quickly and that the apparent delay in stopping the sodium pump arises because a large store of high-energy compounds in the axoplasm must be used up by the pump and by other energy-requiring processes before the ATP runs out. In any case, all methods of blocking the synthesis of high-energy compounds ultimately bring the pump to a halt. The pump can be started again by injecting high-energy compounds, including ATP, into the axoplasm or as Figure 2–1 shows, by removing the uncoupling agent.[10] The lower part of Figure 2–1 shows that the pumping slows considerably as the axon is cooled. The threefold slowing for an 18°C temperature drop means that the active transport of Na⁺ ions has a fairly high activation energy, like many enzymatic reactions and unlike simple diffusion.

Coupling of Na⁺ and K⁺ Movements.[17, 20, 22, 40] External K⁺ is needed for active transport of sodium, because the pump moves K⁺ ions into a cell at the same time as it moves Na⁺ ions out. In the absence of external K⁺, little Na⁺ is pumped out, and in the absence of internal Na⁺, no K⁺ is pumped in. The two fluxes are said to be tightly *coupled*, hence the name Na⁺-K⁺ pump. Coupling is well demonstrated in different experiments by T.I. Shaw, I.M. Glynn, R.L. Post, and others with red blood cells. The experiment in Figure 2–2 demonstrates coupling in human red cells. Before the experiment, the cells were stored for nine days in a K⁺-free medium at 2°C. This treatment loaded the cells with Na⁺ and depleted them of K⁺. The normal ionic gradients had dissipated. At the beginning of the experiment, the cells were placed in a fresh medium containing sources of metabolic energy (glucose and adenosine) at 37°C, and at various intervals, the Na⁺ and K⁺ content of a sample of cells was determined. As Figure 2–2 shows, the cation content did not change while the cells were in a K⁺-free medium, even in the presence of nutrients, i.e., there was no pumping; but as soon as a small amount of KCl was added outside, raising the external K⁺ concentration to 21 mM, there was a brisk and simultaneous loss of Na⁺ and gain of K⁺. The net efflux of Na⁺ and net influx of K⁺ are unambiguously active transport fluxes, because they proceed despite the build-up of substantial concentration gradients across the red cell membrane. The requirement for external K⁺ ions seen here is similar to the result with squid giant axon shown earlier in Figure 2–1.

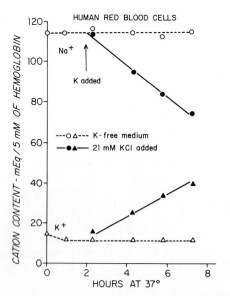

Figure 2–2 The requirement for external K$^+$ ion in active cation transport. Transport was studied by measuring the K$^+$ content and Na$^+$ content of samples of human red blood cells incubated for various times. Before the experiments, the cells were loaded with Na$^+$ and depleted of K$^+$ by long storage in a K$^+$-free medium at 2°C. Then the cells were placed at 37°C in a medium containing nutrients but no K$^+$. No active transport of Na$^+$ was seen in these conditions. At 2 hours, enough KCl was added to half the cells to bring [K$^+$]$_o$ to 21 mM. Active transport started at once, as seen by the rapid loss of Na$^+$ and gain of K$^+$. The same amount of NaCl was added to the control cells. (After Post, R.L.; Jolly, P.C. Biochim. Biophys. Acta [Amst.] 25:118–128, 1957.)

The coupling between Na$^+$ and K$^+$ active fluxes is so constant that it can usually be described by the simple stoichiometric equation.:

$$3\ Na_{in}^+ + 2\ K_{out}^+ \xrightarrow{pump} 3\ Na_{out}^+ + 2\ K_{in}^+$$

The equation says that 3 Na$^+$ ions are pumped out for every 2 K$^+$ ions pumped in. This *stoichiometry* is found in the experiment of Figure 2–2, in which the slopes of the two lines are in the ratio of 3:2. The equation applies over a broad range of internal and external cation concentrations, with red blood cells containing either low levels or high levels of high-energy compounds. However, there are some nonphysiological circumstances in which the coupling ratio differs. For example, in the complete absence of external K$^+$ ions and external Na$^+$ ions, the red blood cell exhibits a slow, energy-dependent, uncoupled Na$^+$ efflux. Pump fluxes in the squid giant axon also deviate from the 3:2 coupling ratio in nonphysiological conditions. Reliable measurements show that when breakdown products (ADP) of

high-energy compounds (ATP) are allowed to accumulate, the active Na$^+$ efflux may become independent of the K$^+$ ion, giving coupling ratios of 1:0 in the giant axon. Apparent coupling ratios of 1:1, 3:2, and 1:0 have been reported in various other tissues; but for technical reasons, accurate flux measurements are harder to make in cells other than red blood cells, and the coupling ratios are often not accurately determined. In the rest of this chapter, a coupling ratio of 3:2 is assumed for all animal cells under physiological conditions, although there may be a few cases in which this is not correct.

Saturation Kinetics.[22, 40, 47, 52] The rate of pumping depends on the concentrations of the ions to be pumped. Consider how pumping rate varies with [K$^+$]$_o$ in human red blood cells (Fig. 2–3). The rate first rises as [K$^+$]$_o$ is raised, but it then levels off at a maximum. Other experiments show that active transport has a similar dependence on [Na$^+$]$_i$. Kinetics of this type, showing a maximum velocity, are called *saturation kinetics*.

Saturation kinetics usually reflect binding to a limited number of sites—in this instance, a binding of Na$^+$ and K$^+$ ions to a limited number of pump sites. By analogy with the kinetics of enzymatic reactions, a preliminary approach to diagramming the steps of sodium pumping might be

$$E + Na_i^+ \underset{\text{fast binding}}{\rightleftharpoons} E \cdot Na^+$$

$$\xrightarrow{\text{slow translocation}} E + Na_o^+$$

in which E stands for the pump, a membrane enzyme. The suggested reaction has two steps, the first, a reversible binding to the pump; the second, a slower translocation (transport) of the ion to the outside. A similar equation could be written for binding and translocation of the K$^+$ ions. The rate equation, derived from the Law of Mass Action for the above reaction scheme, is called the *Michaelis-Menten equation* after the biochemists who first used it to explain the kinetics of enzyme action. Its derivation and limitations are given in all textbooks of biochemistry. If the Michaelis-Menten mechanism described the transport of Na$^+$ ions, the concentration dependence of the flux of sodium, M_{Na}, would be given by

$$M_{Na} = \frac{[Na^+]_i}{[Na^+]_i + K_{Na}} \cdot M_{Na,max}$$

in which $M_{Na,max}$ is the maximum flux (saturating velocity) and K_{Na} is the concentration required for

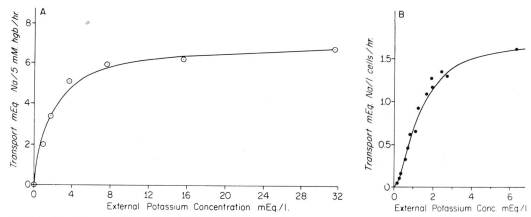

Figure 2–3 The dependence of Na⁺ pumping on the external K⁺ concentration in human red blood cells. The rate of active transport of Na⁺ out of a liter of cells is plotted as a function of $[K^+]_o$. The observations show saturation kinetics, since the rate of transport levels off at a maximum rate for high values of $[K^+]_o$. A, Cells loaded with Na⁺ by storage in the cold in a K-free medium. The curve is a solution of the Michaelis-Menten equation, with $M_{Na,max}$ = 7.1 mmol per hour, and K_K = 2.2 mM. $M_{Na,Max}$ is higher than in normal cells, because $[Na^+]_i$ is unusually high. (From Post, P.L.; Jolly, P.C. *Biochem. Biophys. Acta (Amst.)*25:118–128, 1957.) B, Normal cells containing about 9.3 mM Na⁺. The experiment focuses on the activation curve with low values of $[K^+]_o$. The smooth curve is a solution of an equation for transport requiring two bound K⁺ ions to activate and assuming $M_{Na,max}$ = 1.8 mmol per hour. (After Sachs, J.R.; Welt, L.G. *J. Clin. Invest.* 46:65–76, 1967.)

half saturation. According to the equation, for very low $[Na^+]_i$ the rate of active transport is linearly related to internal Na⁺ concentration and is given approximately by $[Na^+]_i \cdot M_{Na,max}/K_{Na}$. When $[Na^+]_i$ is equal to K_{Na}, the rate is one half of $M_{Na,max}$, and as $[Na^+]_i$ is increased further, the rate saturates at $M_{Na,max}$. A similar equation involving $[K^+]_o$ and K_K could be written for the activation of active transport by external K⁺.

The Michaelis-Menten equation has been used to summarize the observed kinetics in many studies of the transport of physiological substances. Except at very low pumping rates, such equations describe the observed rates of sodium pumping as a function of $[Na^+]_i$ and $[K^+]_o$, and thus provide evidence for the assumed rapid equilibrium binding of ions prior to a slow translocation. Typical values of the half-saturation constants K_{Na} and K_K are 20 mM for internal Na^+_i and 2 mM for external K^+_o. As is true of many carrier mechanisms, typical physiological concentrations $[Na^+]_i$ and $[K^+]_o$ are near or slightly below that required for half-maximal activation, so increases or decreases of $[Na^+]_i$ and $[K^+]_o$ produced by other cellular functions immediately speed or slow the rate of pumping. Careful work shows that whereas the apparent value of K_K is 2 mM in the presence of normal concentrations of external Na⁺, the value of K_K can drop to 0.4 mM when external Na⁺ is replaced by choline ions; i.e., K⁺ ions bind more readily to the external binding site in the absence of external Na⁺ ions. The observations are fitted by the standard equations for competitive inhibition in enzyme kinetics, providing evidence that external Na⁺ ions compete with external K⁺ ions for binding to the pump molecule.

Although the Michaelis-Menten equation fits at high pumping rates, the observed kinetics of activation actually show an upward curvature (sigmoid shape) for increasing ion concentration at very low $[K^+]_o$ or $[Na^+]_i$ (Fig. 2–3B), rather than the linear relation predicted. The deviation reflects the requirement for more than one bound K⁺ or Na⁺ ion to activate the transport mechanism. If there are *n* potassium binding sites on the pump, all of which must be filled before the translocation step occurs, the corresponding rate equation would give the sigmoid concentration dependence needed to fit the observations. Taking *n* as 2 for K⁺ and 3 for Na⁺ would correspond in a neat way to the stoichiometry of pumping. Indeed, rate equations for multiple site kinetics have been used in many recent studies (see the smooth curve in Fig. 2–3B) in preference to the Michaelis-Menten equation.

The single-site and multisite kinetic equations are useful but still empirical descriptions of the observations derived from a great simplification of the actual steps involved in pumping. As shown later, numerous sequential steps are involved in the overall process of active transport. Any more exact description of the pumping rates would necessarily be more complex than the equations discussed here. At the very least, the equations would need to take into account that both K^+ and Na^+ ions are needed for continuous transport, and one would need to know in what sequence and with what interactions the loading and unloading of several sites occurs.

Cyclic Models for the Pump.[22] In 1954 in a ground-breaking study, T.I. Shaw described the saturation kinetics and coupling of Na^+ and K^+ movements in red blood cells of the horse. He also presented the classic carrier model shown in Figure 2–4A (the stoichiometry has been brought up to date in the diagram). In his model, X is a form of the carrier substance that can bind K^+ ions, and Y is a form that can bind Na^+ ions. The complexes K_2X and Na_3Y can "move across" the membrane, carrying their preferred ion. He felt that the carrier makes ions "soluble" in the membrane. For this scheme to give active transport, X

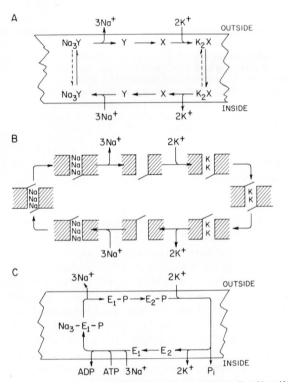

A

B

C

Figure 2–4 Diagrams of three cyclic models for the Na^+-K^+ pump. Each diagram represents a process capable of carrying three Na^+ ions out and two K^+ ions in across the membrane at the expense of some energy. The points where energy might be used in A and B are not shown. A, Shaw's carrier hypothesis with interconvertible lipid soluble carriers, X and Y. (After Glynn, J. Physiol. [Lond.] 134:278–310, 1956.) B, A gate-pore hypothesis with inside and outside gates controlling access to Na^+- and K^+-selective sites in the pore. C, The reaction sequence deduced from chemical experiments showing interconvertible enzyme forms E_1 and E_2 and their phosphorylation and dephosphorylation. (Partly after Sen, A.K.; Tobin, T.; Post, P.R. J. Biol. Chem. 244:6596–6604, 1969.)

was assumed to be converted quickly to Y at the inside edge of the membrane, and Y was assumed to be converted quickly to X at the outside edge. In this way, the external concentration of free potassium carrier is kept higher than that of free sodium carrier, so that external K^+ ion will be picked up in preference to external Na^+ ion, and conversely for the inside. At least one step requires the injection of metabolic energy.

Shaw's model was of great importance because it summarized neatly the observed saturation kinetics and it explained coupling. Some features of the model are still used today. These are that the pump molecule has ion-binding sites and alternates between two forms, now usually called E_1 and E_2; that E_1 and E_2 differ in their affinities for Na^+ and K^+ ions; and that energy is injected to make the cycle run. Other features of the Shaw model have been abandoned. Specifically, the idea that the carrier forms a lipid-soluble complex with ions that then plunges into the bilayer to ferry the ion across is no longer tenable. Rather, as described later, the pump is a large transmembrane glycoprotein with a small domain that is exposed permanently to the extracellular medium and major parts (including the ATP binding domain) that remain permanently in the intracellular medium. The protein does not move completely across the lipid bilayer while holding an ion; instead, the protein has to be regarded as relatively fixed—although readily capable of E_1 to E_2 conformational changes—and the ion moves through the protein.

A more acceptable yet kinetically similar model is the cyclic gate system given in Figure 2–4B. There the ions move by diffusion into a pocket of the pump protein. The pocket has binding sites that become alternately sodium- and potassium-selective. The pump is like an airlock with inner and outer doors or gates, only one of which can be open at a time. Active transport is achieved by the opening and closing of the gates in a sequence, coordinated with the alternations of binding selectively. Again, metabolic energy must be used to change the selectivity and to swing the gates. Kinetically, this model has the same sequence of steps as the mobile carrier model, although the meaning attached to each step is quite different. Conceptually, the most important change is that the pump here does not *carry* ions across the membrane. Rather, the pump makes ion-binding sites alternately accessible to the inner medium and to the outer medium.

The alternating-gate model and Shaw's carrier model envision a *cyclic alteration* of ionic selectivity with the same site first preferring Na⁺ ions, then K⁺ ions. Another class of models suggests a mechanism in which three Na⁺ ions and two K⁺ ions are transported *simultaneously*. This type of pump would wait for all five ions to be bound before moving any of them. Distinction among such models and additional hybrids between them has been the subject of a large and complex, but still not conclusive, series of kinetic investigations. The balance of evidence tilts towards cyclic alteration.

Inhibition by Cardiac Glycosides.[19, 23, 26] Sodium pumping can be stopped by drugs called *cardiac glycosides*, or *cardiotonic steroids*. The drugs get their name because many of them are glycoside derivatives of steroids that are widely used clinically to increase the force of contraction of the failing heart. These beneficial but potentially highly toxic natural products are extracted from plants of the genera *Digitalis* (foxglove), *Strophanthus*, and *Urginea* (formerly *Scilla*), and have been part of the herbal pharmacopoeia for centuries. Examples of cardiac glycoside preparations are ouabain, strophanthin, digitalis, digitoxin, and scillaren A. The mode of action of all of these compounds is identical, and ouabain has been the most popular for experimental work. Figure 2–5 shows that with as little as 300 nM strophanthin, the loss of Na⁺ and gain of K⁺ due to active transport of Na⁺ and K⁺ stops rapidly. When applied at these low concentrations, glycosides are thought to block the pump selectively with little direct effect on other physiological mechanisms.

Their specific blocking action makes cardiac glycosides a valuable tool for physiologists. One can distinguish ionic fluxes mediated by the Na pump from other fluxes, such as diffusion, by subtracting fluxes in the presence of glycoside from those in the absence of drug. The coupling of active transport to energy metabolism can be determined by metabolic studies with and without glycosides. The significance of the normal concentration gradients of Na⁺ and K⁺ to other cellular processes can be determined by blocking the pump long enough to permit the gradients to run down.

Still another use of cardiac glycosides has been in estimating the number of pump molecules on the membrane.[2, 32, 51] The binding of radioactive glycoside to red blood cell membranes closely parallels the inhibition of active transport. When the inhibition is complete, 250 to 500 glycoside molecules are bound to each red cell, so that with 100 to 140 μm^2 of membrane per cell, only two to four inhibitor molecules are bound per square micrometer of membrane. The experiments show that pump sites are extremely sparsely distributed

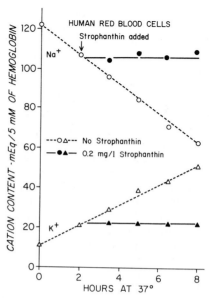

Figure 2–5 Rapid block of Na⁺ and K⁺ active transport by cardiac glycoside in human red blood cells. Cells have been prepared as in Figure 2–2. Active transport is started by placing the cells at 37°C in a medium containing glucose and some K⁺. Strophanthin added at 2 hours quickly blocks active transport, as shown by the level curves of Na⁺ and K⁺ content after treatment with glycoside. (After Post, R.L. *In* (Shanes, A.M., ed. *Biophysics of Physiological and Pharmacological Actions.* Englewood Cliffs, N.J.: Prentice-Hall, 1964.)

on the red cell surface. Cell membrane fragments washed clean of cytoplasm have as many glycoside binding sites as intact cells, showing that the pump is indeed an intimate part of the membrane and that the binding sites all face the extracellular space.

Most cells have to pump Na⁺ ions faster than red blood cells, because their membranes have a far larger passive inflow ("leak") of Na⁺ than in red cells. For example, a kidney cell may normally pump Na⁺ 1000 times faster than a red cell. Experiments with radioactive glycosides show that the same cells also bind 1000 times more glycoside than red cells. Binding site densities as high as 6,500 per μm^2 of membrane may be estimated for Na⁺-resorbing cells of the thick ascending limb of the loop of Henle.[51] A similar parallelism between capacity to pump and glycoside binding is found in the vagus nerve, in brain slices, in the heart, and in the liver. Evidently, individual pump sites can operate at a similar maximum rate in all tissues, and *the difference in pumping activity among tissues is primarily due to a difference in the number of functioning pump sites in the membrane.* In cells for which both the maximum pump flux, $M_{Na,max}$, and the number of glycoside binding sites are known,

$M_{Na,max}$ corresponds very roughly to 300 Na^+ ions transported per second at each site (37°C).

The mechanism of the block of sodium transport by cardiac glycosides has been studied extensively. Structure-activity studies show that a variety of natural and synthetic molecules have inhibitory action. Several parts of the drug molecule contribute to the potency and rates of action. The block is always reversible. Some glycoside derivatives dissociate rapidly from their receptor, but many glycosides take hours to dissociate. Ouabain can take three hours, but the kinetics depend on the tissue, the animal species, and the temperature. Glycosides and external K^+ ions seem to *compete* for the same form of the transport system, because a higher glycoside concentration is needed to block when the K^+ concentration is high and the block develops more slowly with high $[K^+]_o$. Conversely, the K^+ concentration needed for half-maximal activation of the pump is raised in the presence of glycoside. As appropriate for an external binding site, glycoside injected inside giant axons does not block. In the kinetic diagram of pumping (Fig. 2–4B), the E_2 form of the pump with K binding sites accessible to the external medium could be a target for apparent competitive binding of K^+ and glycosides. Despite the apparent competition for the same *form* of the pump (E_2), there is no evidence that K^+ ions and cardiac glycosides actually bind at the same *site* on the pump. More information is given in the section on the pump as an ATPase.

The therapeutic effect of cardiac glycosides is believed to arise indirectly from the influence of the Na^+ ion gradient on Ca^{2+} transport in the heart (see later section on Na^+-Ca^{2+} exchange).[26] Because intracellular free calcium is the ultimate regulator of contraction in muscle (Chap. 8), a reduction of the Na gradient that leads to a secondary accumulation of intracellular Ca^{2+} could augment contraction of the heart. The binding competition between K^+ ions and cardiac glycosides explains why controlled elevation of plasma potassium is a useful clinical treatment for an accidental overdose of cardiac glycoside.

The vanadate ion, the +5 oxidation state of the transition metal vanadium, is a potent inhibitor of the sodium pump in nanomolar concentrations. In contrast with cardiac glycosides, vanadate acts from the cytoplasmic side of the membrane only. This inhibitor is discussed further in the section on chemical properties of the sodium pump.

ATP as the Energy Source.[4, 10, 50] Early experiments on squid giant axon (Fig. 2–1) showed that active transport of sodium stops some time after the addition of metabolic inhibitors. Pumping can be started again by removal of the metabolic inhibition or by direct injection of various high-energy phosphate compounds into the axoplasm of the axon. The experiments show that the pump uses a chemical fuel as its energy source; but, because injected compounds are quickly converted to many other compounds in an intact cell, it is more difficult to decide which compound is the crucial one.

Adenosine triphosphate (ATP) is now known to be the fuel. One proof comes from experiments with red cell "ghosts" and perfused or dialyzed squid giant axons. Ghosts are prepared by placing red cells in dilute solutions with very low osmotic pressure. The red cells swell as water is drawn into them and stretch their membrane (without gross tearing) until virtually all the hemoglobin, soluble enzymes and small molecules leak out—a process called reversible osmotic lysis. Metabolites to be tested as fuels can be allowed to leak into the cell at the same time. The membranes now called ghosts remain intact and can be induced to reseal with the new contents and return to the original biconcave disc shape of the red cell. Resealed red-cell ghosts continue to pump sodium and potassium, provided there is some K^+ outside and Na^+, Mg^{2+}, and ATP inside. Analogous experiments can be done with the squid giant axon. In these experiments, the ends of the axon are cut, and much of the axoplasm is removed by coring, or squeezing it out. Test solutions are then perfused continuously through the inside of the axon from one end to the other, or alternatively, solutions are perfused through tiny dialysis tubing threaded through the length of the axon. Again, if the internal solution contains Na^+, Mg^{2+}, and ATP, active transport resumes. Experiments of this kind on cell membranes without cytoplasm are definitive proof of the membrane location of the pump.

Does ATP have enough free energy to run the pump? As in so many other processes driven by this universal energy source, the overall energy-yielding reaction is the cleavage of the terminal phosphate bond to give adenosine diphosphate (ADP) and "inorganic" orthophosphate (P_i). For a 100% efficient coupling, this reaction should be able to do 65 kJ/mol of work at the concentrations of ATP, ADP and P_i typical in living cells (Chap. 1). Careful measurements of the rate of consumption of ATP reveal a stoichiometry of about 1 phosphate bond broken per 3 Na^+ ions transported in nerves, muscles, and red cells. Thermodynamics tells us that the work needed to pump 1 mole of Na^+ out of a cell with a membrane potential of E_M is by definition the electrochemical difference ($\Delta\bar{\mu}_{Na}$) for Na^+ ions, taken with a minus sign (see Equation 9, Chap. 1).

$$G = -\Delta\bar{\mu}_{Na} = -FE_M + 2.303\ RT\ \log_{10}\frac{[Na^+]_o}{[Na^+]_i}$$

in which 2.303 RT is 5.6 kJ/mol at 20°C. The first term is the electrical work and the second term, the chemical work of pumping out 1 mole of Na^+ ions. The free energy needed to pump in 1 mole of K^+ ions is given by a similar equation, but with the signs reversed and with potassium rather than sodium concentrations. Altogether, the energy needed to pump 3 moles of Na^+ out and 2 moles of K^+ in is $2\ \Delta\bar{\mu}_K - 3\ \Delta\bar{\mu}_{Na}$, which for the mammalian muscle fiber in Table 1–4 totals 44 kJ, or two thirds of the free energy provided by hydrolyzing a mole of phosphate bond in ATP. The calculation shows that the pump can be efficient from the point of view of energetics and, conversely, that it would not be energetically possible to achieve very much higher gradients with a device having the 3:2:1 stoichiometry of the sodium pump.

Chemical Properties of the Sodium Pump

Na+ + K+-ATPase in Fragmented Cell Membranes.[24, 41, 42, 55] Although the physiological usefulness of sodium pumping requires an intact, almost impermeable cell membrane to maintain the ionic gradients, much recent progress in understanding the pump has come from studying fragmented cell membranes and solubilized membrane proteins. Here, the pump must be assayed as an enzyme catalyzing the cleavage of chemical bonds in the test tube rather than as a transport mechanism.

The sodium pump has the enzymatic activity of an ATPase, a name applied to enzymes whose mechanism includes the hydrolysis of ATP. As first shown by J.C. Skou,[54] isolated and fragmented cell membranes exhibit an ATPase activity requiring Mg^{2+}, Na^+, and K^+ ions and blocked by low concentrations of cardiac glycosides. This is the $(Na^+ + K^+)$-activated ATPase or $Na^+ + K^+$-ATPase for short. Activation of the ATPase activity by ions in the medium shows saturation kinetics and reaches half maximum at approximately the same concentrations of Na^+ and K^+ that are needed for half-maximal activation of pumping in intact cells. The ATPase activity is blocked by the same concentrations of cardiac glycosides that block pumping and also by vanadate. The amount of $(Na^+ + K^+)$-activated ATPase activity recovered from different tissues is in direct proportion to the pumping capacity and to the ouabain-binding capacity of the intact tissue. All of these properties show that this ATPase is derived from the sodium pump.

Like almost all ATPases, transport ATPases have a multistep reaction. In one step, the terminal phosphate of ATP is transferred onto an amino acid residue in the active site of the pump, and ADP is released. This phosphorylation requires Na^+ ions and Mg-ATP. In another step, the phosphate is cleaved off the transport ATPase and released. This dephosphorylation is activated by K^+ ions. The reactions may be written

$$ATP + E \underset{}{\overset{Na^+,\ Mg^{2+}}{\rightleftharpoons}} E\text{-}P + ADP \quad \text{(phosphorylation)}$$

$$E\text{-}P \xrightarrow{K^+,\ Mg^{2+}} E + P_i \quad \text{(dephosphorylation)}$$

in which E symbolizes the $Na^+ + K^+$-ATPase and E-P is the phosphoenzyme (phosphorylated form). The evidence for these reactions is that phosphate in the form of ^{32}P label becomes covalently bound to membrane fragments that have been incubated with $[^{32}P]$-ATP, Mg^{2+}, and Na^+ and that the label is subsequently released as orthophosphate. This dephosphorylation is greatly accelerated by addition of K^+ ions. Labeling with ^{32}P serves as a second convenient measure of the number of pump sites on a membrane. The number of phosphorylation sites found this way equals the number of cardiac glycoside–binding sites found in other experiments.

The reactions of the transport ATPase were dissected by penetrating biochemical studies done primarily in the laboratories of Post and Albers.[1, 41, 42] They demonstrated a sequence of substrate binding steps and changes in the pump protein molecule by measuring the rates and amounts of binding of ATP, phosphate, and ouabain under a wide variety of conditions and after treatment with different inhibitors. Others have looked at molecular changes by following changes in ion binding, in susceptibility to enzymatic attack, or in the fluorescence of covalently attached fluorescent probes. According to these studies, the pump molecule alternates between the two major conformational states, E_1 and E_2, which correspond to the kinase and the phosphatase reactivities of the active center. E_1 and E_2 in turn can interact with a variety of ligands. The cyclic reaction scheme in Figure 2–4C shows a few of the postulated chemical states of the pump. A Na^+-activated phosphorylation of the pump in the E_1 conformation gives the phosphoenzyme E_1-P(Na$_3$) in a form with "occluded" bound Na^+ ions. In the occluded state, the Na^+ ions seem to be buried within the protein and do not exchange with intracellular or extracellular Na^+ ions. This high-energy intermediate spontaneously undergoes a conformational change to the more stable phosphoenzyme form, E_2-P, which releases the Na^+ ions, a transformation that can be blocked by treatment with the sulfhydryl reagent, N-ethylmaleimide or by oligomycin. Cardiac glycosides bind most readily to the form E_2-P. Dephosphorylation of E_2-P is accelerated by K^+ ions (and by Cs^+, Rb^+, Tl^+ and NH_4^+ ions). The K^+ ions pass through a transient occluded state in the process. In both E_1-P and E_2-P, the phosphate is bound as an acyl phosphate to the β-carboxyl group of a single aspartyl residue. The inhibitor, vanadate, binds to the phosphate-binding site and thus interrupts the reaction cycle.

The resulting diagram of chemical states (Fig. 2–4C) is similar to the cyclic gate hypothesis, with phosphorylation standing for closing the inner gate and E_1 and E_2 standing for states with Na- and K-specific binding sites, respectively.

Purified (Na^+ + K^+)-Activated ATPase.[12, 24, 35, 56] Only in the late 1960s did biochemists develop general methods to solubilize and purify integral proteins from biological membranes using detergents. The sodium pump was among the first transport molecules to be isolated in nearly pure form. A high, specific activity of (Na^+ + K^+)-activated ATPase has been purified from mammalian renal medulla and brain, electric eel electric organ, and shark rectal gland. Actually, as with several other membrane proteins, the ATPase protein purified with detergent is inactive but regains activity if appropriate phospholipids are added to the assay medium. Presumably, enzymatic activity requires a "membrane-like" environment. The molecular weight of the smallest active protein particles has been variously estimated to be from 280,000 to 380,000.

Further dissociation of the active protein particles into their peptide subunits yields equal molar amounts of two major peptides, a large one, α, of 95,000 to 130,000 daltons and a smaller glycopeptide, β, of 36,000 to 55,000 daltons. The large peptide contains the phosphorylatable aspartyl residue and the ouabain-binding site. The messenger RNA for these peptides have been cloned so that the amino acid sequence of both chains are now known.[53] From the sequence and from studies with antibodies, covalently bound inhibitors, and chemical modifiers, it is concluded that the polypeptide chain of the α subunit may thread back and forth across the membrane six to eight times

with the NH_2-terminal end lying in the cytoplasm. About 30% of the mass of this chain may be in the membrane; 65% sticks out into the cytoplasm, and only a tiny bit protrudes into the extracellular space. Two isoforms, termed α and $\alpha(+)$, of the α subunit have been discovered in various mammalian tissues.[57] They differ slightly in molecular weight and considerably in ouabain-binding affinity and in their distribution among different cell types. The (+) form with high ouabain–binding affinity is not present in rat kidney, while both forms are present in brain. This difference lends some tissue specificity to the blocking actions of glycosides. The smaller β subunit is heavily glycosylated (30% by weight), and portions of it certainly face the extracellular medium. There may also be a still smaller γ subunit.

Because of the uncertainty of the molecular weights of the active particles, the peptide composition of the pump is not definitely known yet. A popular hypothesis is that the native Na^+-K^+ pump consists of the tetramer $\alpha_2\beta_2$. Sites have been counted biochemically by measuring the binding of four different substances: ouabain, ATP, vanadate, and phosphate. The number of sites is definitely the same for each ligand and is equivalent to one per α subunit. Contemporary modeling starts from the idea that the phosphorylation step of the pump cycle induces a folding, or turning, of domains of the large cytoplasmic extension of the pump. This change may exert torsion on some of the transmembrane helices, turning ion-binding sites facing the cytoplasm to face into a channel-like protein column that provides a path for ions across the lipid bilayer.[9]

Before one accepts that the extracted peptides are the *entire* sodium pump, one would like to know if they suffice to reconstitute full physiological pumping in a membrane system. This has now been demonstrated with several preparations of Na^+ + K^+-ATPase. For example, Goldin[25] used the enzyme purified from the renal medulla and originating primarily from the thick ascending limb of the loops of Henle. Enzyme with detergent was mixed vigorously with tiny bilayer vesicles (40 to 60 nm diameter) made from the common membrane phospholipid, phosphatidyl choline. After centrifugation to remove aggregates and extensive dialysis to remove detergent, the enzyme was found to be incorporated into the vesicle membranes, and ATPase activity and fluxes of $^{22}Na^+$ and $^{42}K^+$ across the vesicles could be observed. Some of the enzyme molecules may have been oriented one way and some, the other, but the assay using external ATP tested only for *inside-*

out enzymes (ATP site outward). Mg-ATP added to the *external* medium was broken down to ADP and P_i as Na^+ ions were moved in and K^+ moved out of the vesicles, creating a 1.5- to 2-fold concentration gradient for both ions within 5 minutes (Fig. 2–6). The ATP breakdown and ionic movements did not occur if the vesicles were formed with ouabain inside. The stoichiometry of Na^+ moved:K^+ moved:ATP hydrolysed was approximately 3:2:1. All of these properties confirm that active transport of Na^+ and K^+ ions is fully accounted for by a peptide macromolecule made up of some combination of α and β peptide chains (and possibly γ), combined with a phospholipid membrane, and fueled by Mg-ATP.

Further Functional Properties of the Sodium Pump

Electrogenic Pumping.[19, 44, 58] The resting membrane potential in nerve and muscle cells is primarily a diffusion potential arising because the membrane is selectively permeable to certain ions (especially K^+ ions) that are unequally distributed across the membrane (see Chaps. 1 and 3). However, after periods of intense stimulation, nerves and muscles often exhibit a more negative internal potential than is expected from diffusion potentials. This membrane hyperpolarization is caused by operation of the sodium pump.

Figure 2–7 shows the effects of electrical stimulation on the mean membrane potential recorded extracellularly from the frog sciatic nerve. At first, the nerve is at rest. Then the many large myelinated fibers (A fibers) in the bundle are stimulated to fire by shocks applied at 50 per second for 25 minutes (with a brief interruption once every min-

ute to measure the membrane potential). Between stimulations and for more than an hour afterward, the mean membrane potential is more negative than normal. In the next chapter, we show that stimulation of nerve causes the axons to become loaded with Na^+ ions. Measurements in parallel with the experiment of Figure 2–7 show that the hyperpolarization lasts just as long as it takes to pump out the extra Na^+ ions let in by the stimulation. This observation is similar to others on the unmyelinated fibers in sympathetic nerves, on molluscan and vertebrate ganglion cells, on the crayfish stretch receptor, and on vertebrate skeletal and cardiac muscle. In several nerves and muscles, the hyperpolarization is immediately abolished by cardiac glycosides and reduced by washing away most of the external K^+ ions or by cooling the tissue. The hyperpolarization does not occur after stimulation in a sodium-free, lithium Ringer's solution. Lithium substitutes for sodium in the action potential mechanism, but Li^+ ions are pumped out of cells more slowly than Na^+ ions. Hence the hyperpolarization arises from active transport, and the pump is called *electrogenic* (or in older literature, *rheogenic*).

The pump is electrogenic because it produces an electric current. By moving three Na^+ ions out and only two K^+ ions in, the active transport mechanism transfers one net positive charge outward per cycle. Removal of internal positive charge leaves the inside more negative, as shown in Figure 2–7. The most quantitative test of this explanation for the hyperpolarization during active transport comes from Thomas's experiments with a giant ganglion cell in the small central

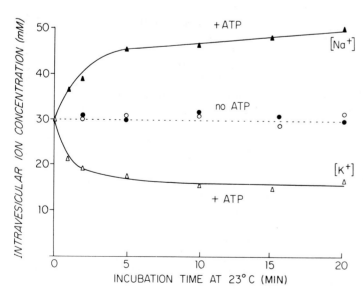

Figure 2–6 Reconstitution of the Na^+-K^+ pump from purified molecules. The Na^++K^+ ATPase was purified from canine renal medullary membranes and dialyzed into phospholipid vesicles. Addition of ATP outside the vesicles caused Na^+ ions to be brought into, and K^+ ions out of, the vesicles in an active transport catalyzed by Na^++K^+ ATPase molecules oriented in an "inside–out" direction (when compared with those of a living cell). (After Goldin, S.M. *J. Biol. Chem.* 252:5630-5642, 1977.)

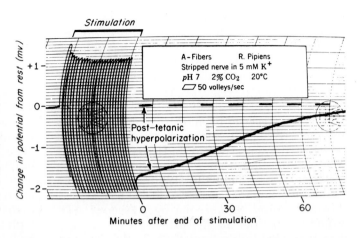

Figure 2–7 Long after hyperpolarization following stimulation of large myelinated nerve fibers in the sciatic nerve of a frog. The nerve is stimulated for 25 minutes, but the stimulus is turned off for 5 seconds each minute so that post-tetanic hyperpolarization can be observed. The curve shows the mean value of the membrane potential recorded with a device too slow to follow fast action potentials. The initial resting membrane potential is defined as 0 mV. At the end of the stimulus period, electrogenic Na^+ pumping hyperpolarizes the axons for more than an hour. The maximum recorded hyperpolarization is only 2 mV, but the extracellular recording method used involves much attenuation, so this hyperpolarization may actually be as high as 14 mV across the axon membranes. (After Connelly, C.M. *Rev Mod. Phys.* 31:475–484, 1959.)

nervous system.[58] Thomas injected Na^+ ions into the cell through glass micropipettes and found ouabain-sensitive hyperpolarizations of up to 20 mV that lasted 10 to 15 minutes following the injection. Then, using a voltage clamp (see Chap. 3), he measured the extra outward electric current generated by the pump across the membrane during this 15-minute period. This current corresponded to one positive charge transferred out for every 3.6 Na^+ ions originally injected, not far from the expected 1:3 ratio. Finally, using an internal sodium-selective glass microelectrode to measure the internal concentration (activity) of Na^+ ions, Thomas showed that $[Na^+]_i$ was abruptly elevated by the Na^+ injection and gradually returned to its original resting value (by the operation of the pump) over the 15-minute period.

The overall importance of the electrogenic feature of the pump in the nervous system remains to be worked out. Any process capable of changing the membrane potential can alter the excitability and firing rates of cells. Thus, the pump may significantly modulate excitability during and after periods of intense activity, especially in small cells and cell processes with large surface-to-volume ratios. In addition, drugs and hormones that accelerate or slow the pump can in this way have a direct electrogenic effect distinct from the long-term effect of altering steady-state ionic gradients. For example, with rat soleus muscle the addition of 6 μM epinephrine (adrenalin) to the bathing medium increases ^{22}Na efflux and ^{42}K influx and increases (hyperpolarizes) the resting membrane potential by 7 mV.[15]

Exchange Diffusion.[24, 61] The discussion so far has glossed over a major source of difficulty in measuring pump fluxes, namely, sorting out from the total flux of, for example, labeled Na^+ ions the component due to active transport. One compo-

nent of flux obeys the laws of passive diffusion and is insensitive to ouabain and to metabolic inhibitors. This might be called a *leak* and probably flows through various ionic channels. Another component discussed in this chapter obeys saturation kinetics, needs ATP, is blocked by ouabain, and usually shows a 3:2 stoichiometric Na^+-K^+ coupling. Finally, some components neither obey the laws of diffusion nor break down ATP. Some of these components show saturation kinetics and might be blocked by ouabain although they do not represent net active transport. These fractions of the total fluxes are carried in part by the pump operating in modes different from the full cyclic mode of active transport and in part by other Na^+-coupled transport mechanisms.

One of these modes of operation is seen in frog muscle and also becomes prominent in red blood cells or giant axons bathed in a K^+-free medium. The pump begins to exchange internal Na^+ ions for external Na^+ ions without hydrolyzing ATP in corresponding amount. Ussing[61] gave this kind of exchange the name *exchange diffusion* to distinguish it from ordinary diffusion and active transport. Surprisingly, ATP and ADP must be present although they are not used up. Ouabain and oligomycin block the exchange in red cells and giant axons. This Na^+-Na^+ exchange probably represents the reversible operation of half of the pump cycle. In Figure 2–4C, internal Na^+ and ATP probably load onto the E_1 form of the pump. The reaction proceeds until Na^+ is released to the outside. Then, in the absence of external K^+, other Na^+ ions bind to E_1-P and the whole process reverses, giving stoichiometric exchange of Na^+ and returning the ATP to the cytoplasm. A different mode of operation of the pump is a ouabain-sensitive efflux of K^+ ions that requires internal phosphate. In Figure 2–4C, this efflux may correspond to a reversed operation of the step going from E_2-P to E_2. These partial reactions of the pump are of interest because they can be exploited in laboratory investigations of the steps of pumping, but Na^+-Na^+ exchange and extra K^+ effluxes probably have little relevance to the normal life of a cell.

Complete Reversal of Active Transport.[21, 24] Two observations already discussed suggest that a reversal of the entire cyclic transport mechanism, synthesizing ATP while bringing Na^+ in and K^+ out, is possible. First, the step that carries Na^+ out and the step that carries K^+ in can be reversed separately. And second, the hydrolysis of ATP ordinarily does not supply vastly more free energy than is needed to transport the ions, so a modest change of conditions could tip the thermodynamic balance to reverse the reaction. The following conditions would favor reversal from an energetic viewpoint: high $[Na^+]_o$, zero $[K^+]_o$, high $[K^+]_i$, zero $[Na^+]_i$, high $[P_i]_i$, high $[ADP]_i$, and zero $[ATP]_i$. Indeed, red cell ghosts prepared with approximately these conditions by resealing after hemolysis show a ouabain-sensitive synthesis of ATP. Thus all the reactions in the diagrams of Figure 2–4 are reversible in principle and in practice. Of course, reversal of the pump is observed only with extreme laboratory conditions and is not thought to occur physiologically. In normal cells, the tendency of the reactions to run in the direction of the arrow is guaranteed by the high $[ATP]_i$ and low $[ADP]_i$ and $[P_i]_i$.

Mechanism of the Sodium Pump. Can we say how the sodium pump moves ions? Unfortunately, the answer is still No. The same may be said of all other physiological carrier systems. For the sodium pump, detailed functional and kinetic information has been obtained, and the molecule is available in the test tube and cloned and sequenced, but more must be learned about the molecular structure before radically different mechanisms can be distinguished and a realistic model can be developed. The hydrolysis of one high-energy bond in ATP is coupled to the transport of 3 Na^+ ions and 2 K^+ ions. Whether the Na^+ ions are moved first and the K^+ ions afterward or all ions are moved at once is not known. Transport must entail some motion within the macromolecular complex, and indeed the pump does alternate between conformational states E_1 and E_2. The changes seem to be driven both by phosphorylation and by binding of transported ions. For thermodynamic and mechanical reasons, complete rotations of entire peptide subunits within the membrane cannot occur hundreds of

times a second. But conformational changes as small as rotations of side chains or as large as shifts of the monomer units with respect to each other or rotation of α-helical columns can. Stationary sites with movable "gates" or movable sites exposed alternately to the internal and external media are possible. However, kinetic experiments will not suffice to make the distinction. Three-dimensional structure must be determined.

The combination of a phosphorylation and a dephosphorylation controlling the activity and specificity of an enzymatic process is a familiar pattern in metabolism. Many examples are known, of which the best one studied concerns the control of phosphorylase, the enzyme that initiates the breakdown of glycogen as a fuel for energy metabolism. Phosphorylase is a large, multisubunit enzyme existing in two forms, one with low activity and one with high. The low-activity form is converted to the more active form by a protein kinase that transfers phosphate groups to the enzyme. A phosphatase reverses the activation. Activation is accompanied by conformational changes and reaggregation of the component subunits. The sodium pump is also an oligomer of protein subunits. It phosphorylates and dephosphorylates itself in each transport cycle. Phosphorylation and dephosphorylation of a subunit induce conformational changes required for translocation and for a change of ionic specificity. The phosphatase and kinase reactions would be "allosterically" activated by the Na^+ and K^+ ions binding to their transport sites. These words do not explain pumping, but they show that many of the questions regarding the mechanism of active transport are common questions now actively studied in enzyme systems.

OTHER CARRIERS

During the 1970s, many biological disciplines converged in studying carrier mechanisms: physiology, biochemistry, bioenergetics, microbiology, and biophysics. Numerous carriers were found. The most revolutionary discoveries were those of P. Mitchell concerning the "chemiosmotic" mechanism of oxidative phosphorylation in mitochondria and microorganisms and of photosynthetic phosphorylation in chloroplasts and photosynthetic bacteria.[27, 39] Briefly, the electron transport chain of both types of phosphorylating systems was shown to be a proton-pumping mechanism (carrier) that builds up steep electrochemical gradients for H^+ ions across a membrane. Like the water behind a hydroelectric dam, the proton gradient stores the free energy from oxidative metabolism or absorbed light until it can be converted into another form. An amount $\Delta\bar{\mu}_H$ of free energy could be made available to cellular processes from each mole of stored protons. Two classes of mechanisms tap this energy source. (i) It is used to synthesize high-energy bonds. In a process resembling active transport run in reverse, a membrane enzyme called the F_1 ATPase transfers two protons "downhill" across the membrane and phosphorylates an ATP to synthesize ATP. (ii) It is used to power secondary active transport. A host of coupled carriers catalyze the cotransport or exchange of essential substrates or ions with protons. In prokaryotes (bacteria), the electron-transport chain that generates proton electrochemical gradients and the ATPase and coupled carriers that make use of the gradients are all in the cytoplasmic surface membrane. In eukaryotes, the inner mitochondrial and chloroplast membranes have the major proton-based energy coupling, so the plasma membrane is freed for other purposes.

The discoveries in bioenergetics led to a new and far wider appreciation of the energy storage and coupling possibilities of membrane transport systems. It also led to a new terminology. What classical biologists had called carriers were dubbed *porters* or *transporters*, cotransport was termed *symport*, and exchange or countertransport was designated *antiport*. Thus the Na^+-K^+ pump is an Na^+-K^+ antiporter. Although physiologists tend to use the older terminology, the interdisciplinary nature of transport studies requires familiarity with both.

Studies with bioenergetics have been aided by the finding of ion-transporting antibiotics. Many microorganisms synthesize small antibiotic molecules that act by making membranes permeable to certain ions and thus discharge (collapse) essential ionic gradients. Some of these substances, including gramicidin A, alamethicin, and nystatin, form ionic channels (pores) in membranes and artificial lipid bilayer systems. Others, including valinomycin, nonactin, and nigericin, are small lipid-soluble carriers that act as mobile ion-binding sites, shuttling back and forth across lipid bilayers. Uncouplers of oxidative phosphorylation such as 2,4-dinitrophenol and FCCP (carbonylcyanide *p*-trifluoro-methoxyphenylhydrazone) act as proton carriers that collapse the stored proton gradient of mitochondria. All of these carrier antibiotics have practical uses in physiology, and they are simple enough substances so that detailed mechanistic analysis of the transport is possible; however, they cannot be considered prototypes of typical biologic carriers for amino acids, sugars, or even ions, because they are much too small to span the membrane and must act by the diffusing-ferryboat principle.

The remainder of this chapter is devoted to a brief overview of some well-characterized carrier systems of the gut, kidneys, excitable tissues, and red blood cells. We will find that the major carrier mechanisms can be grouped into a small number of classes, quite possibly reflecting their evolutionary origins. The emphasis in the following sections is definitely on mammalian cell membranes.

Other Transport ATPases.[49] Now we recognize that the Na^+-K^+ pump is one of a *family* of transport ATPases. In animal cells, there are at least five classes of ATPases (Fig. 2–8) that perform active transport of small cations out of the cytoplasm. Because hydrolysis of the terminal phosphate of ATP can yield 65 kJ/mol of free energy, these pumps all establish steep electrochemical gradients of the ions they pump. So far there are no *unambiguous* examples of similar ATPases catalyzing active transport of anions, of organic molecules or of water. Four of the transport devices in Figure 2–8 (excluding the proton pump in secretory vesicles) have many functional similari-

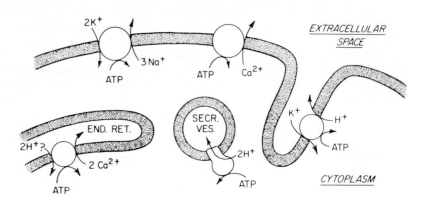

Figure 2–8 Five classes of ATP-dependent active transport of small cations. Three of them act across the plasma membrane and two, across intracellular membranes. All break down cytoplasmic ATP to ADP and P_i.

ties and are regarded as a family of related proteins. They use a mechanism with E_1-E_2 conformational changes and a phosphoenzyme intermediate. In three cases, it is known that the high-energy phosphate is transferred to the β-carboxyl group of an aspartyl residue exposed on the cytoplasmic side. Pumping is inhibited by vanadate and sulfhydryl reagents. Only the Na^+ + K^+-ATPase is blocked by cardiac glycosides, and for most of the others we still lack highly selective blocking agents.

Ca²⁺-Activated ATPases.[13, 30, 38] The cytoplasmic Ca^{2+} ion concentration of animal cells is kept low by the parallel operation of several primary and secondary (Na^+-coupled) active transport systems that excrete Ca^{2+} to the extracellular medium or sequester it within intracellular, membrane-bound compartments. Despite an extracellular Ca^{2+} concentration of several millimolar, these transport devices maintain internal free Ca^{2+} concentrations on the order of 10 to 300 nM in resting cells. Cytoplasm also has many Ca^{2+}-binding proteins that buffer changes in free Ca^{2+} concentration.

The best studied Ca^{2+}-pump was discovered in experiments to understand how a contracted muscle relaxes at the end of a twitch. The experimental design was to seek an intracellular factor that would dissociate crude actomyosin preparations in a test tube. By cell fractionation experiments, the "relaxing factor" was found to be in the microsomal fraction, membrane vesicles derived from the sacroplasmic reticulum, a prominent internal membrane system of muscle. As we know now, muscle contraction is regulated by internal Ca^{2+} ions, and the interaction of contractile proteins requires Ca^{2+}. The sarcoplasmic reticulum contains a concentrated store of Ca^{2+} that it liberates into the cytoplasm at the appropriate stimulus, initiating contraction (Chap. 7). So long as ATP is present, the sarcoplasmic reticulum membrane then quickly pumps the liberated Ca^{2+} ions back home again, permitting relaxation. In the absence of ATP, Ca^{2+} is not pumped back and muscle relaxation does not occur, a condition called rigor.

Purified vesicles of sarcoplasmic reticulum avidly accumulate Ca^{2+} ions from the suspending medium at the expense of ATP. Up to 90% of the vesicle membrane protein is a Ca-dependent ATPase. It has been solubilized, purified, reconstituted, cloned, and sequenced. The pump may actually perform an exchange of two Ca^{2+} ions, possibly for two H^+ ions per ATP high-energy bond broken. A maximum of 10 to 20 Ca^{2+} ions can be transported per second (25°C) per 110,000-dalton subunit, and half-maximal transport rates are achieved when the cytoplasmic free Ca^{2+} concentration is only 0.5 μM.

The functional protein might be a dimer or a trimer of the basic subunit. The amino-acid sequence of the subunit is consistent with a polypeptide chain that loops back and forth across the bilayer perhaps eight times.[9, 37] Many stretches of the amino-acid sequence show strong homology with that of the Na^+ + K^+-ATPase α chain, and for both molecules, the bulk of the molecular mass extends into the cytoplasm, where binding of ATP occurs. Kinetic and biochemical studies suggest that the transport mechanism involves a cyclic sequence of events much like those for the Na^+-K^+ pump (Fig. 2–4C), with Ca^{2+} ions taking the place of Na^+ and, perhaps, H^+ taking the place of K^+. Many nonmuscle cells—even liver—release and reaccumulate Ca^{2+} ions from their endoplasmic reticulum in response to stimuli. These cells presumably also have a Ca^{2+}-ATPase like that in muscle. The enzyme may be present inside all cells.

Another calcium pump fueled directly by ATP was first demonstrated in red blood cell ghosts resealed with an elevated $[Ca^{2+}]_i$. The Ca load is extruded against a concentration gradient at a maximum initial rate $M_{Ca,max}$ = 10 mmol per liter of cells per hour if the ghosts contain Mg-ATP. ATP is hydrolyzed simultaneously at a rate corresponding to approximately one phosphate bond per Ca^{2+} ion extruded. Without ATP, there is no detectable Ca movement during an hour. Because the stoichiometry is only one Ca^{2+} per ATP rather than 2:1, as in the endoplasmic reticulum, more free energy is available per ion moved, and the pump should be able to establish steeper electrochemical gradients. This makes sense, because the plasma membrane pump has to transport the ions *against* the large electric field of the resting potential while the endoplasmic reticulum pump probably pumps against no electric field—K^+ ions are believed to be highly permeant in this membrane, in which there are no pumps for K^+ ions.

As might be expected, fragmented red cell membranes have a Ca-activated ATPase activity corresponding to the Ca pump. The membranes can be phosphorylated from ATP in the presence of Ca^{2+} ions. Like the Na^+, K^+-ATPase, this Ca^{2+}-ATPase seems to have a kinase form, E_1, and a phosphatase form, E_2. The affinity for Ca^{2+} ions and hence the pump rate can be enhanced by addition of the cytoplasmic Ca^{2+}-dependent regulatory protein, calmodulin. Indeed, calmodulin binding is so strong that a chromatography column containing immobilized calmodulin is the preferred tool for purifying the plasma membrane Ca^{2+}-ATPase. Such an affinity column successfully extracts Ca^{2+}-ATPase from the plasma membranes of a variety of cells, so this pump may be ubiquitous. The pump probably suffices to handle the small Ca^{2+} loads of quiescent and "tight" cells such as red blood cells. However, in other tissues that respond to stimuli by significant elevations of $[Ca^{2+}]_i$, higher-capacity transport systems are needed at least to clear the peak load. These include a Ca^{2+},Na^+-exchange carrier to be described later and the endoplasmic reticulum Ca^{2+}-ATPase already described. In addition, because of their strongly negative internal potential, mitochondria can accumulate Ca^{2+}, which then precipitates as calcium phosphate in the intramitochondrial matrix. However, many physiologists think that mitochondrial calcium uptake is significant only in pathological states.

Proton Pumps.[14, 18] The most important proton gradients in eukaryotic cells are those across the inner mitochondrial and chloroplast membranes. Nevertheless, there are several proton transport mechanisms on nonmitochondrial membranes that play prominent roles in animal physiology. We consider here two proton-ATPases that serve to acidify noncytoplasmic compartments, and later we discuss a class of exchange-carrier mechanisms that help regulate intracellular pH.

Certainly, the largest pH gradient of the vertebrate body is that made by the stomach. Parietal cells of the gastric mucosa have an intracellular pH of 7.2 to 7.4, as in other kinds of cells, yet they are able to secrete acid with a pH as low as 0.8 into the gastric juices. Neglecting any electrical component, the chemical work of transporting protons against this tremendous gradient is 2.303 $RT \times (7.4 - 0.8)$ or 37 kJ/mol. This calculation uses Equation 3 of Chapter 1, noting that pH values are already logarithms (negative logarithms to the base 10). Pumping is performed by a $H^+ + K^+$-ATPase that seems to perform an electroneutral exchange of one proton for one K^+ ion per high-energy bond cleaved. The functioning device may be a trimer or tetramer of 100-kdalton polypeptide subunits, each of which has a phosphorylatable aspartyl residue at the nucleotide binding site. Again, the kinetics are analogous to those of the Na^+-K^+ pump, with protons taking the role of Na^+ ions.

Secretion of acid into the stomach is partly controlled by changes in topology of the parietal cell membranes, leading to withdrawal of membranes with functioning pump molecules from the surface. Antibodies to the $H^+ + K^+$-ATPase, used in conjunction with immunoelectron microscopy, reveal pump molecules on canalicular invaginations of the cell surface during active secretion. When the stimulus for secretion stops, the infoldings seem to pinch off from the surface, turning into a set of intracellular vesicles with no path to export protons. Upon restimulation, the internal membranes regain their surface connection and resume net transport of acid.

A more modest pH gradient is formed across secretory vesicles that lie in the cytoplasm of secretory cells. The best studied are the "chromaffin granules," large epinephrine-containing secretory vesicles of adrenal medullary chromaffin cells. These are the source of the epinephrine that is secreted into the circulation in response to fright and stress. Chromaffin granules have an acid interior, a pH near 5.5, and also an electrical potential 80 mV more positive than that of the cytoplasm. Hence there is an electrochemical potential difference for protons across the granule membrane of almost 20 kJ/mol. The pH gradient and the positive intragranule electrical potential are set up by an electrogenic proton pump that moves protons from the cytoplasm into the granule with an apparent stoichiometry of two protons per ATP high-energy bond. The 2:1 stoichiometry improves the efficiency of using a 65 kJ/mol fuel to generate a modest electrochemical gradient.

This proton ATPase differs from the other transport ATPases we have described. It does not operate through a phosphoenzyme intermediate with a cycle of E_1-E_2 conformational changes. Rather than being a close relative of the Na^+-K^+ pump, it may be more allied with the mitochondrial F_1 ATPase mentioned previously. The proton pump of chromaffin granule membranes has similar subunit structure, antibody recognition sites, inhibitors, stoichiometry, and mechanism to that of mitochondrial F_1 ATPase. They are nonetheless different molecules.

Why should secretory vesicles have a proton electrochemical gradient across them? One answer for chromaffin granules is that the proton gradient provides energy for secondary active transport of amines into the granule. Thus dopamine, a precursor of epinephrine, is brought into the granule by a proton-dopamine exchange transporter. Two protons leave the granule per dopamine entering. The accumulation can be demonstrated with membrane "ghosts" prepared from granules, provided that an outwardly directed proton gradient is set up. In other secretory tissues, much less information is available on the packaging of small molecules into secretory vesicles, but as a working hypothesis, it has been proposed that many hormones and neurotransmitters might be packaged by such a combination of proton-ATPase and proton-transmitter exchanger in the vesicle membrane.

Other intracellular compartments as well, including endosomes (vesicles just brought in from the surface membrane), lysosomes, and the Golgi apparatus, seem to be much more acidic than the cytoplasm. Their membranes have ATP-dependent proton pumps.

Na-Dependent Cotransport Devices.[48, 60] In animal plasma membranes, active transport of organic metabolic substrates is not accomplished by ATP-ase fueled pumps. From his studies of nutrient absorption from the gut, R.K. Crane[16] proposed instead that the gradient of Na^+ ions across the plasma membrane of intestinal epithelial cells serves as the power source for active transport of nutrients—the *Na-gradient hypothesis.* Today, a large number of Na-dependent cotransport systems (symport) have been identified, and the Na-gradient hypothesis has been extensively confirmed. See also the discussion in Chapter 55.

Amino Acids. Amino acids can be accumulated inside most cells at concentrations far above that in the medium by a Na-dependent transport process. The entry of amino acids shows saturation kinetics with respect to the amino acid concentration. Even with neutral amino acids such as glycine or alanine, the entry is electrogenic, making the cell interior more positive, as if a cation moves into the cell together with the amino acid. Indeed, the addition of amino acids to the medium does stimulate an extra influx of Na^+ ions into the cell. If the external Na^+ concentration is lowered, the rate of entry of amino acids falls, and the ability to accumulate amino acids against a gradient is lost. If cells are loaded with Na^+ by a prolonged treatment with ouabain and then the external Na^+ concentration is lowered (so that $[Na^+]_i > [Na^+]_o$), amino acids can be *extruded* from the cells against a concentration gradient. Ouabain itself does not have a direct effect on amino acid transport but acts only indirectly through the resulting rundown of the Na^+ ion gradient.

These observations are explained by the Na-gradient hypothesis: The amino acid carrier is a cotransport mechanism for amino acids and Na^+ ions (Fig. 2–9). Because of this coupling and the Law of Mass Action, the free energy in the electrochemical gradient for Na^+ ions powers the accumulation of amino acids against a gradient. For every mole of Na^+ ions moved, a free energy

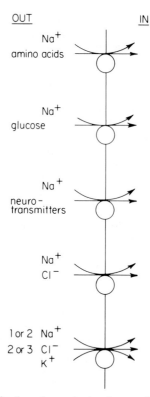

Figure 2–9 Sodium-dependent cotransport carriers operating across the cytoplasmic membrane. Each uses the free energy of the Na^+ gradient to transport other molecules into the cell against a concentration gradient. For most of them, the stoichiometry is not well established.

of $\Delta\widetilde{\mu}_{Na}$ is available to move amino acids. The free energy for pumping amino acids can be traced indirectly to the breakdown of ATP, since ATP fuels the sodium pump. However, the amino acid carrier itself is not thought to break high-energy bonds while transporting Na^+ ions and amino acids. This uphill transport is secondary active transport.

Sodium-coupled, secondary active transport of amino acids is found in most cells of the body and has been studied extensively in the intestine and kidney. Competition experiments suggest that there may be at least five different sodium-coupled amino acid carriers that distinguish among anionic, cationic, and zwitter ionic amino acids. Some cells, e.g., mammalian red blood cells, lack the sodium-coupled mechanism but do have a facilitated diffusion system for amino acids (which cannot transport against an electrochemical gradient).

Sugars. Hexose sugars such as D-glucose are also transported against concentration gradients by a Na-dependent carrier mechanism in cells of the intestinal epithelium and kidney tubules. For example, absorption of sugar from the intestinal lumen requires Na^+ ions in the lumen, and absorption of Na^+ ions from the lumen is speeded by transportable sugars in the lumen. Sugars will be *extruded* if the normal Na gradient is reversed. The movement of a mole of sugar is accompanied by the parallel movement of 1 to 3 moles of Na^+ ion and is electrogenic. Again, a coupled (sugar + Na^+)-cotransport mechanism has been postulated. This secondary active transport of sugars is competitively inhibited by a plant glycoside, *phlorizin*, at 10^{-6} M concentrations. (See also Chap. 55.)

Sodium-dependent sugar transport may be restricted to gut, kidneys, and a few other organs where it is used to drive net transport across epithelia (see later section on epithelial transport). Most body cells, including muscle and red blood cells, receive their glucose via a Na-independent, facilitated diffusion mechanism in their surface membranes—the glucose transporter.[62] This equilibrating carrier readily carries glucose in and out of cells and cannot establish a concentration gradient. It is inhibited by the phlorizin aglycone, *phloretin*. The glucose transporter purified from red blood cell ghosts is a glycoprotein of about 55,000 daltons and has been used to reconstitute glucose-carrying activity in phospholipid vesicles. The amino acid sequence determined from cloned DNA copies of the RNA message corresponds to the peptide of 46,000 daltons. The functioning protein is probably made up of several identical, membrane-spanning polypeptide subunits. Antibodies to red-cell glucose transporter cross-react with membranes of skeletal muscle, adipocytes, fibroblasts, and placenta, showing a basic similarity among the transporters of different tissues. Glucose transport is enhanced manyfold by insulin in skeletal muscle and adipocytes by a mechanism believed to involve recruitment of more transporters to the surface membrane by fusion of vesicles.

Neurotransmitters. Nerve terminals release neurotransmitters to communicate with other cells at synapses. Once released, the neurotransmitter may act on membrane receptors of the target cell and frequently is taken back up by a Na^+-dependent, high-affinity uptake system in the nerve terminal. The reuptake terminates transmitter action and replenishes the store in the presynaptic terminal. Sodium-dependent uptake systems have been demonstrated for choline, γ-aminobutyric acid, glutamate, glycine, serotonin, norepinephrine, and dopamine.

Chloride. Some cells of the intestinal mucosa and the thick ascending loop of Henle concentrate

Cl^- ions by a Na-dependent secondary active transport. The electroneutral cotransport is blocked by the potent and widely used clinical diuretic agent furosemide. The stoichiometry, which has been said to be 1:1, is hard to determine in complex tissues like kidney and intestine. The same or a closely related furosemide-sensitive cotransport device has been studied with more resolution in squid giant axons.[46] There, K^+ ions are cotransported in addition at a stoichiometry of 2 Na^+:K^+: 3 Cl^-. Yet other stoichiometries have been reported in other systems. Whether there actually is a variety of $Na^+ + K^+ + Cl^-$–cotransporters or whether experimental difficulties account for the apparent lack of full accord remains to be determined. Further mechanisms for chloride transport are described later in this chapter.

Na-Coupled Exchange Devices. The inwardly directed gradient of Na^+ ions not only provides free energy for cotransport of Cl^- and of many classes of substrates into cells, but it also provides free energy for countertransport of ions, notably Ca^{2+} and H^+, out of the cell (Fig. 2–10). While there could also be Na-dependent countertransport (antiport) of organic substrates in the plasma membrane, no clear examples in animals are yet known.

Na^+-Ca^{2+} Exchange.[5, 6, 13, 43] In most cells of the body the intracellular free Ca^{2+} ion concentration is kept very low, usually below 10^{-7} M, despite an extracellular Ca^{2+} concentration of 1 to 4 mM and a negative membrane potential that favors Ca^{2+} entry. Two transport mechanisms account for a steep Ca gradient: the primary active transport systems for Ca^{2+} ions, already described, and a secondary active transport system, described here.

In axons, synaptosomes, heart muscle, and smooth muscle, the steady-state intracellular cal-

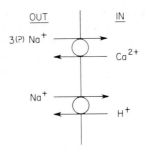

Figure 2–10 Sodium-dependent exchangers operating across the cytoplasmic membrane. They use the free energy stored in the Na^+ gradient to transport Ca^{2+} or H^+ ions out of the cell, against their electrochemical gradients. The stoichiometry of Na^+-Ca^+ exchange is not well established.

cium is inversely related to $[Na^+]_o$. For example, squid axons injected with aequorin, the luminescent, Ca-indicator protein, glow weakly (low $[Ca^{2+}]_i$) in normal seawater, but when all the external NaCl is replaced by LiCl, the glow may increase fivefold (indicating a rise of $[Ca^{2+}]_i$) within 1 minute (Fig. 2–11). The increase of $[Ca^{2+}]_i$ reverses within a minute when normal seawater is returned. Within seconds after the external Na^+ ion is restored, the calcium influx, measured in other experiments with ^{45}Ca as a tracer, slows down and the efflux speeds up. Internal ATP does increase the affinity of the carrier for internal Ca^{2+} ions, but if a squid giant axon is injected with apyrase, an enzyme that quickly breaks down all the axoplasmic ATP, a low $[Ca^{2+}]_i$ is still maintained. Thus ATP high-energy bonds are not essential. Ouabain has no direct effect on the Ca transport.

These and many other experiments show that the surface membrane has a coupled Na^+-Ca^{2+} exchange (countertransport) carrier mechanism that accomplishes secondary active transport of Ca^2 ions out of the cell at the expense of free energy stored in the electrochemical gradient for Na^+ ions. The stoichiometry of the coupling is estimated to be 3 Na^+ ions entering per Ca^{2+} ion leaving. Such high coupling ratios are required to account for the heavy free-energy cost of moving a Ca^{2+} ion against the extreme Ca electrochemical gradient found in living cells.*

Another divalent ion, Mg^{2+}, is at a lower free concentration inside cells (about 1 mM) than would be expected for equilibrium. At least in axon membranes, there is an ATP-independent, Na^+-Mg^{2+} exchange carrier that maintains this low $[Mg^{2+}]_i$.[11] A Na^+:Mg^{2+} stoichiometry of 2:1 would be sufficient to explain the Mg gradient.

Ca-extruding mechanisms must exist in all cell surface membranes. However, without labels to count sites reliably, we cannot say whether ATP-dependent mechanisms or Na^+-Ca^{2+} exchange mechanisms are the most common, nor can we generalize about which class of tissues tends to have one or the other. The two mechanisms are known to coexist in some cells. The exchange

mechanism is prominent in excitable cells, which have on the average more negative resting potentials than other types of cells as well as heavier, pulsatile Ca^{2+} loads. Perhaps most cells with small resting potentials will be found to use an ATP-dependent mechanism, since without a large negative resting potential, the exchange mechanism could not establish so large a Ca electrochemical gradient.

Na^+-H^+ Exchange.[3, 7, 31, 45] Physiologists have long been aware that the pH of blood is well regulated against dietary intake and metabolic production of acid. More recently it has been realized that intracellular pH is far from equilibrium with the plasma pH and that it too is strongly regulated. Transport devices moving protons and others moving bicarbonate ions contribute to pH regulation both at the level of single cells and in transport organs such as the kidney.

The cytoplasm of vertebrate cells has nearly the same pH as the interstitial fluid, between 7.1 and 7.4. If protons were in electrochemical equilibrium across the plasma membrane, pH_i should instead be approximately one unit lower than pH_o because of the negative membrane potential. Resting membrane potentials of -40 to -90 mV would correspond to equilibrium pH_i values 0.7 to 1.6 units lower than pH_o. Hence all cells require active transport of acid equivalents from cytoplasm to plasma.

Regulation of pH_i can be demonstrated in the laboratory with cells given an acid load. For example, using intracellular pH-sensitive glass microelectrodes to record pH_i, Aickin and Thomas[3] followed the recovery of normal pH_i after mouse skeletal fibers were acidified. The half time for recovery from 0.5 pH units of acidification was only 3 to 5 min. The rate of recovery was not changed by removing external K^+ ions, by removing 90% of the external Ca^{2+} ions, or by adding 10^{-4} M ouabain. However, recovery was markedly slowed by lowering the external Na^+ concentration or by adding 10^{-4} M of the diuretic agent amiloride to the bath. Simultaneous measurements of intracellular $[Na^+]$ with an intracellular Na-sensitive glass microelectrode showed that $[Na^+]_i$ rose during the period of extra acid removal. This rise was also blocked by amiloride. Removal of acid is therefore a Na-dependent process, not requiring external Ca^{2+} or K^+, independent of the Na^+-K^+ pump, and accompanied by inward transport of Na^+ ions. These are the signs of Na^+-H^+ exchange (antiport).

A Na^+-H^+ exchanger probably operates in the plasma membrane of all animal cells. Exchange is tightly coupled, reversible, usually sensitive to

---

*For a 3:1 stoichiometry, the "equilibrium" concentration ratios would be

$$\frac{[Ca^{2+}]_o}{[Ca^{2+}]_i} = \left(\frac{[Na^+]_o}{[Na^+]_i}\right)^3 \exp(-E_M F/RT)$$

Thus in a cell with a 12-fold concentration gradient for Na^+ ions and a membrane potential, E_M, of -70 mV, the internal free $[Ca^{2+}]$ could be lower than the external value by a factor of 3×10^4.

Figure 2–11 Na⁺-Ca⁺ exchange in the squid giant axon. The axoplasm has been injected with the protein aequorin, whose glow (luminescence) reports the free $[Ca^{2+}]$ inside the cell. The light recorded as current in a photomultiplier tube changes as the ions in the *external* medium (artificial sea water) are changed, reflecting net inward or outward fluxes of Ca^{2+} across the membrane. Internal free $[Ca^{2+}]$ drops when the external $[Ca^+]$ is lowered from 112 mM to 11 mM; it rises again when 50%, 75%, and finally 100% of the external Na⁺ ions are replaced by Li⁺; and it falls quickly when the normal external [Na⁺] is restored. The experiment shows that the normal Na gradient is necessary to keep the internal free $[Ca^{2+}]$ at a low physiological level. (From Baker, P.F.; Hodgkin, K.L.; Ridgway, E.B. *J. Physiol. [Lond.]* 218:709–755, 1971.)

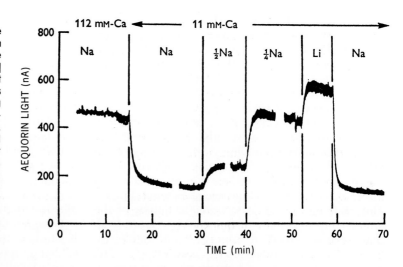

amiloride, and seems to serve at least three functions: (i) It keeps the cytoplasmic proton concentration well below the equilibrium level at the expense of the Na gradient—secondary active transport. (ii) It moves acid equivalents into the urine for excretion. (iii) It is one link of an intracellular signaling pathway based on variations of pH_i: Because many cytoplasmic enzymes are steeply pH sensitive, their activities can be regulated by pH_i, as by a second messenger system. Some cells enter a relatively quiescent state, shutting down their activities to a basal level by letting pH_i drop to 6.6–7.0. Then upon a suitable physiological stimulus, the Na⁺-H⁺ exchanger is activated, acid is excreted, and the depressed activities resume. This type of control was first clearly demonstrated in the sea urchin egg.[31] Before fertilization, pH_i is low and the egg is quiescent. Upon fertilization, there is first a rapid rise of internal free Ca^{2+} attributed to Ca^{2+} mobilization by IP_3, followed within 2 minutes by a burst of Na⁺-H⁺ exchange that raises the cytoplasmic pH and helps bring the machinery for protein synthesis and cell division into operation. The exchanger is believed to become activated by a protein–kinase C-catalyzed phosphorylation. Similar pH_i changes occur in the activation of sea urchin sperm and might help control cell division in a variety of cells.

Anion Exchangers. Several cellular functions require transport of anions across the cell membrane. For net secretion or reabsorption of NaCl across epithelial membranes, the furosemide-sensitive Na⁺ + (K⁺) + Cl⁻ cotransporters already discussed are well suited. However, for secretion of bicarbonate-containing fluids and for bicarbon-

ate-linked pH regulation, another family of carriers, the anion exchangers, are required (Fig. 2–12).

Cl⁻-HCO₃⁻ Exchange.[7, 63] One dramatic example of anion movements occurs in red blood cells twice during each transit around the circulatory system. It is the "chloride shift." In normal respiration, CO_2 generated by metabolism in the body is carried back to the lungs by the blood, mostly as bicarbonate ions, HCO_3^-. Once it enters the blood stream, the CO_2, a small neutral molecule, diffuses readily across the red cell membrane to the inside, where the enzyme carbonic anhydrase catalyzes the hydration reaction.

$$CO_2 + H_2O \xrightarrow{\text{carbonic anhydrase}} H_2CO_3 \leftrightarrow HCO_3^- + H^+$$

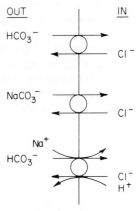

Figure 2–12 Anion exchangers operating across the cytoplasmic membrane. Each is electrically silent, moving no net charge in a full cycle.

The resulting intracellular HCO_3^- ions are then returned to the blood plasma in exchange for plasma Cl^- ions. This exchange, the chloride shift, greatly increases the effective CO_2-carrying capacity of the circulation. At the lungs, the whole process is reversed to liberate free CO_2 into the exhaled gases.

Anion fluxes underlying the chloride shift show many of the classic characteristics of a reversible carrier mechanism. At 37°C, a 50% turnover of all of the intracellular Cl^- ion takes only about 100 ms in the red cell, but as the temperature is lowered, the turnover time becomes much longer. The exchange rate is a saturable function of the anion concentrations, and somewhat more than half maximal activation occurs at the normal physiological concentrations of Cl^- and HCO_3^-. The opposite movements of transported anions are in a strict one-to-one stoichiometric ratio.* While an exchange of Cl^- and HCO_3^- may be one of the more important physiological functions of the underlying transport mechanism, the same device also can be shown in the laboratory to catalyze Cl^--Cl^- self-exchanges (which could have no physiological significance) and to transport I^-, Br^-, SO_4^{2-} and PO_4^{2-}. These other ions compete with Cl^- and HCO_3^- for transport. The exchange fluxes can be blocked quickly by the disulfonic stilbene reagents, SITS and DIDS, that react with protein amino groups.

In biochemical analyses of the red-cell membrane proteins after reaction with radioactively labeled stilbene inhibitors, most of the label is found in a protein called band 3 after its position in electrophoretic gels. Band 3, the anion exchanger molecule, accounts for as much as 25% of the membrane protein of the erythrocyte. It has been purified and sequenced by recombinant DNA methods.[34] Anion exchange can be reconstituted in lipid vesicles from the purified glycoprotein. The exchanger is a dimer or tetramer of 103,000-dalton subunits whose polypeptide chain threads back and forth up to 12 times across the lipid bilayer. The protein has a major intracellular domain that serves as a binding site for cytoskeletal elements, anchoring the cytoskeleton to the plasma membrane.

Anion exchange mechanisms sensitive to disulfonic stilbene inhibitors have been found in many cells. They play a role in intracellular pH regulation, removing HCO_3^- from the cell in response to an intracellular alkaline load and bringing HCO_3^- into the cell in response to an acid load. In effect, removing one HCO_3^- ion from the cell followed by the hydration of another CO_2 molecule in the cell is equivalent to production of an intracellular proton. We previously learned of pH_i regulation by the Na^+-H^+ exchanger. Apparently Na^+-H^+ and Cl^--HCO_3^- exchangers act in parallel to set pH_i.[3, 45] Some workers feel that Na^+-H^+ exchange is most effective against an intracellular acid load, and Cl^--HCO_3^- exchange against an alkaline load.

Complex Cl^- Exchangers. When pH_i regulation is studied in some cells, SITS-inhibitable anion exchange mechanisms are found that require cations as well (Fig. 2–12). Snail neurons[59] seem to exchange HCO_3^- + Na^+ for Cl^- + H^+ and squid axons[8] may exchange $NaCO_3^-$ for Cl^-. Both mechanisms as written are electroneutral, as is the simple anion exchange.

EPITHELIAL TRANSPORT

Thus far we have emphasized the role of the sodium pump and other transport devices to regulate the solute composition of the cytoplasm of individual cells. However, they also serve to drive net solute fluxes *across* entire sheets of cells in the many transporting epithelial tissues of the body. (See Chap. 55.) Thus, these carrier mechanisms are essential in the uptake of foodstuffs from the intestine, in the formation of urine, and in the secretion of saliva, digestive juices, cerebrospinal fluid, and vitreous humor. As was originally suggested by Koefoed-Johnson and Ussing,[33] for salt absorption by the frog skin, polarized transport across epithelial sheets requires that cell membranes facing the mucosal side of the epithelium have different pumps and permeability properties from those of the membrane facing the serosal side. Since the typical transporting epithelium is only one cell layer thick, polarized transport requires polarized cells. Transported solutes must enter the cell at one surface and leave at the other.

Consider the polarization of transport sites in a generalized absorptive epithelial cell, shown in Figure 2–13. The drawing shows several sodium-transport devices in the cell surface membrane, with their preferential orientations toward the mucosal or the serosal surface of the epithelium. The sodium pump sites are in the membrane of the serosal side of the cell (basal, or tissue, side). They

---

*Because the movements of charge are exactly balanced with a one-to-one stoichiometry, they carry no electric current and make no contribution to the electrical conductance of the membrane. Hence the exchange is sometimes called "electrically silent anion exchange."

expel Na⁺ ions into the serosal medium and bring K⁺ ions into the cell. As a consequence, an inwardly directed Na concentration gradient, and perhaps also an electrical gradient, develop at the mucosal side of the cell (luminal, apical, or external side). Depending on the epithelium, the mucosal membrane allows Na⁺ ions to flow in by several mechanisms. In frog skin, the Na⁺ ions enter from the external environment (pond water) via Na-selective pores (P_{Na}) that are specifically blocked by the diuretic agent amiloride if it is applied at 1 μM concentration from the external side.* The external pores together with the serosal sodium pump and a high serosal K⁺ ion permeability (not shown) are the three elements needed for net transport of Na⁺ ions across the frog skin from pond water to the tissue fluid.[33] On the other hand, some regions of the kidney and intestine are specialized for net reabsorption of sugars and amino acids from the lumen to the circulation. Here, the luminal membrane is well supplied with Na-dependent sugar and amino acid carriers. Again, serosal sodium pumps make a Na gradient. Luminal carriers use the gradient to bring sugars and amino acids into the cell (together with Na⁺ ions), and at the serosal surface, equilibrating facilitated diffusion carriers (not shown) let the accumulated sugar and amino acids escape passively from the cell. The sugar efflux on this side of the cell can be blocked with phloretin. Finally, in the thick ascending loop of Henle and in parts of the intestine, net NaCl uptake resembling active chloride transport is accomplished by a mucosal (Na + Cl)-cotransport and a serosal sodium pump.

While all the devices shown in Figure 2–13 may not typically be found in one cell, the sidedness indicated is well established by a variety of experiments on absorptive epithelia. The transport mechanisms have been localized by the sidedness of the actions of inhibitors and by autoradiography with labeled inhibitors.[2, 51] In addition, the luminal (brush border) surfaces of intestinal and kidney tubular cells have been isolated by cell fractionation in quite pure form as closed vesicles that are some of the best material for studying Na-dependent amino acid and sugar transport.

This brief excursion into epithelial transport physiology by no means exhausts the list of agents that are moved and coupled devices that have

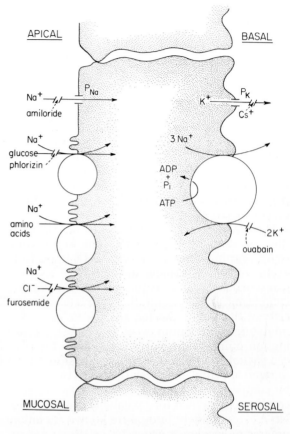

Figure 2–13 Polarized, Na-coupled transport mechanisms in the apical (mucosal) and basal (serosal) membranes of a generalized absorptive epithelial cell. On the apical membrane, from top to bottom, are amiloride-sensitive pores, phlorhizin-sensitive Na-glucose cotransport, Na–amino acid cotransport (several different mechanisms at the molecular level), and furosemide-sensitive Na-Cl cotransport. On the basal surface (and also on the lateral surface) there is the ouabain-sensitive Na⁺-K⁺ pump. These and other transport mechanisms drive the net flux of solutes from left to right across the epithelial cell.

been postulated. Rather, it is meant simply to show how, by appropriate differentiation of mucosal and serosal cell membranes, net transepithelial transport of small molecules can be accomplished. The diagram of Figure 2–13 is appropriate only for net absorptive transport (from mucosa to serosa). In secretory epithelia, the sodium pump is still found on the serosal surface, but the disposition of other transport devices is different.

SUMMARY

Physiologists have found that the fluxes of many small molecules across the cell membrane obey

---

*Note that the clinical diuretic amiloride and its congeners have two uses in the laboratory. They block the "amiloride-sensitive Na channel" of epithelia at 1 μM and the Na⁺-H⁺ exchanger at one hundredfold higher concentrations.

carrier kinetics characterized by stereoselectivity, saturation, and competition. These fluxes are an essential part of the metabolic economy of individual cells and also serve in epithelia to transfer substances between body compartments. The transport involves binding of the small molecules to specific sites on large intrinsic membrane proteins. Although some of these proteins have been isolated and used to reconstitute transport, the molecular details of how the bound small molecule moves from one aqueous compartment to the other are still unknown. Considering the many kinds of molecules that can be transported, the plasma membrane must contain scores of different carrier macromolecules.

The Na^+-K^+ pump of animal cell membranes is the best studied carrier mechanism. The hydrolysis of one high-energy bond in ATP leads to the export of 3 Na^+ ions and the import of 2 K^+ ions. Continual activity of the pump maintains the high cytoplasmic K^+ concentration characteristic of all cells of the body. Cardiac glycosides are specific inhibitors of the pump cycle. Other ATP-consuming ion pumps are specialized to lower the pH of cell vacuoles, to secrete acid into the stomach and to remove cytoplasmic free Ca^{2+} ions by export or by compartmentalization. These pumps establish conditions essential for excitability and digestion. No ATP-dependent pumps are known for anions, organic molecules or water.

In several tissues, the transmembrane gradient of Na^+ ions is used to provide power for Na-coupled cotransport of amino acids, sugars, chloride ions, or neurotransmitters. The gradient may also power Na-coupled exchange of H^+ or Ca^{2+}. Anions exchange across the membrane as well, sometimes coupled to the movement of cations. All of these cotransport and exchange devices can catalyze secondary active transport. In epithelia, the vectorial organization of carriers in the apical and basolateral membranes can produce impressive secretion or absorption of small molecules.

BIBLIOGRAPHY

1. Martonosi, A.N., ed. *The Enzymes of Biological Membranes*, 2nd ed., vol. 3. New York, Plenum Press, 1985.
   A collection of reviews of contemporary research on pumps and transporters.
2. Andreoli, T.E.; Hoffman, J.F.; Fanestil, D.D.; Schultz, S.G., eds. *Physiology of Membrane Disorders*. New York, Plenum Medical Book Co., 1986.
   A collection of instructive essays on membrane proteins, transport theory, pumps, carriers and channels.
3. Brandl, C.J.; Green, N.M.; Korczak, B.; MacLennan, D.H.

Two Ca^{2+} ATPase genes: homologies and mechanistic implications of deduced amino acid sequences. *Cell* 44:597–607, 1986.

A nice example of gene cloning and attempts to deduce mechanisms from amino acid sequences.

REFERENCES

1. Albers, R.W. The (sodium plus potassium)-transport ATPase. In Martinosi, A.N., ed. *The Enzymes of Biological Membranes*, vol. 3. New York, Plenum Press, 1976.
2. Almers, W.; Stirling, C.E.: The distribution of transport proteins over animal cell membranes. *J. Membr. Biol.* 77:169–186, 1984.
3. Aickin, C.C.; Thomas, R.C. An investigation of the ionic mechanism of intracelliular pH regulation in mouse soleus muscle fibres. *J. Physiol. (Lond.)* 273:295–316, 1977.
4. Baker, P.F.; Foster, R.F.; Gilbert, D. S.; Shaw, T.I. Sodium transport by perfused giant axons of *Loligo. J. Physiol. (Lond.)* 219:487–506, 1971.
5. Baker, P.F.; Hodgkin, A.L.; Ridgway, E.B. Depolarization and calcium entry in squid giant axons. *J. Physiol. (Lond.)* 218:709–755, 1971.
6. Blaustein, M.P.; Nelson, M.T. Sodium-calcium exchange: its role in the regulation of cell calcium. In Carafoli, E., ed. *Membrane Transport of Calcium*, 217–236. New York, Academic Press, 1982.
7. Boron, W.F. Transport of H^+ and of ionic weak acids and bases. *J. Membr. Biol.* 72:1–16, 1983.
8. Boron, W.F. Intracellular pH-regulating mechanism of the squid axon. Relation between the external Na^+ and HCO_3^- dependences. *J. Gen. Physiol.* 85:325–345, 1985.
9. Brandl, C.J.; Green, N.M.; Korczak, B.; MacLennan, D.H. Two Ca^{2+} ATPase genes: homologies and mechanistic implications of deduced amino acid sequences. *Cell* 44:597–607, 1986.
10. Caldwell, P.C.; Hodgkin, A.L.; Keynes, R.D.; Shaw, T.I. The effects of injecting "energy-rich" phosphate compounds on the active transport of ions in the giant axons of *Loligo. J. Physiol. (Lond.)* 152:561–590, 1960.
11. Caldwell-Violich, M.; Requena, J. Magnesium content and net fluxes in squid giant axons. *J. Gen. Physiol.* 74, 739–752, 1979.
12. Cantley, L.C. Structure and mechanism of the (Na,K)-ATPase. *Curr. Top. Bioenerg.* 11:201–237, 1981.
13. Carafoli, E. Biochemistry of plasma-membrane calcium-transporting systems. In Martonosi, A.N., ed. *The Enzymes of Biological Membranes*, 2nd ed., vol. 3, 235–248. New York, Plenum Press, 1985.
14. Carty, S. E.; Johnson, R.G.; Scarpa, A. H^+-translocating ATPase and other membrane enzymes involved in the accumulation and storage of biological amines in chromaffin granules. In Martonosi, A. N., ed. *The Enzymes of Biological Membranes*, 2nd ed., vol. 3. 449–495. New York, Plenum Press, 1985.
15. Clausen, T.; Flatman, J.A. The effect of catecholamines on Na-K transport and membrane potential in rat soleus muscle. *J. Physiol. (Lond.)* 270:383–414, 1977.
16. Crane, R.K. Na-dependent transport in the intestine and other animal tissues. *Fed. Proc.* 24:1000–1005, 1965.
17. De Weer, P. Effects of intracellular adenosine-5'-diphosphate and orthophosphate on the sensitivity of sodium efflux from squid axon to external sodium and potassium. *J. Gen. Physiol.* 56:583–620, 1970.
18. Faller, L.D.; Smolka, A.; Sachs, G. The gastric H,K-ATPase.

In Martonosi, A. N., ed. *The Enzymes of Biological Membranes,* 2nd ed., vol. 3, 431–448. New York, Plenum Press, 1985.

19. Gadsby, D.C. The Na/K pump of cardiac cells. *Annu. Rev. Biophys. Bioeng.* 13:373–398, 1984.
20. Garrahan, P.J., and Glynn, I.M. The stoichiometry of the sodium pump. *J. Physiol. (Lond.)* 192:217–235, 1967.
21. Garrahan, P.J., and Glynn, I.M. The incorporation of inorganic phosphate into adenosine triphosphate by reversal of the sodium pump. *J. Physiol. (Lond.)* 192:237–256, 1967.
22. Glynn, I.M. Sodium and potassium movements in human red cells. *J. Physiol. (Lond.)* 134:278—310, 1956.
23. Glynn, I.M. The action of cardiac glycosides on sodium and potassium movements in human red cells. *J. Physiol. (Lond.)* 136:148–173, 1957.
24. Glynn, I.M. The Na^+,K^+-transporting adenosine triphosphatase. In Martonosi, A.N., ed. *The Enzymes of Biological Membranes*, 2nd ed., vol. 3, 35–114. New York, Plenum Press, 1985.
25. Goldin, S.M. Active transport of sodium and potassium ions by the sodium and potassium ion–activated adenosine triphosphatase from renal medulla. Reconstruction of the purified enzyme into a well-defined in vitro transport system. *J. Biol. Chem.* 252:5630–5642, 1977.
26. Hansen, O. Interaction of cardiac glycosides with $(Na^+ + K^+)$-activated ATPase. A biochemical link to digitalis-induced inotropy. *Pharmacol. Rev.* 36:143–163, 1984.
27. Harold, F.M. Ion currents and physiological functions in microorganisms. *Ann. Rev. Microbiol.* 31:181–203, 1977.
28. Hille, B. *Ionic Channels of Excitable Membranes.* Sunderland, Mass., Sinauer Associates, 1984.
29. Hodgkin, A.L.; Keynes, R.D. Active transport of cations in giant axons from *Sepia* and *Loligo. J. Physiol. (Lond.)* 128:28–60, 1955.
30. Inesi, G.; de Meis, L. Kinetic regulation of catalytic and transport activities in sarcoplasmic reticulum ATPase. In Martonosi, A.N., ed. *The Enzymes of Biological Membranes*, 2nd ed, vol. 3, 157–191. New York, Plenum Press, 1985.
31. Johnson, J.D.; Epel, D.; Paul, M. Intracellular pH and activation of sea urchin eggs after fertilisation. *Nature* 262:661–664, 1976.
32. Joiner, C.H.; Lauf, P.K. The correlation between ouabain binding and potassium pump inhibition in human and sheep erythrocytes. *J. Physiol. (Lond).* 283:155–175, 1978.
33. Koefoed-Johnsen, V.; Ussing, H.H. The nature of the frog skin potential. *Acta Physiol. Scand.* 42:298–308, 1958.
34. Kopito, R.R.; Lodish, H.F. Primary structure and transmembrane orientation of the murine anion exchange protein. *Nature* 316:234–238, 1985.
35. Kyte, J.: Molecular considerations relevant to the mechanism of active transport. *Nature* 292:201–204, 1981.
36. Lee, C.O. 200 years of digitalis: the emerging central role of the sodium ion in the control of cardiac force. *Am. J. Physiol.* 249 or *Cell Physiol.* 18:C367–C378, 1985.
37. MacLennan, D.H.; Brandl, C.J.; Korczak, B.; Green, N.M. Amino-acid sequence of a Ca^{2+} $+Mg^{2+}$-dependent ATPase from rabbit muscle sarcoplasmic reticulum deduced from its complementary DNA sequence. *Nature* 316:696700, 1985.
38. Michalak, M. The sarcoplasmic reticulum membrane. In Martonosi, A.N., ed. *The Enzymes of Biological Membranes*, 2nd ed., vol. 3, 115–155. New York, Plenum Press, 1985.
39. Mitchell, P. Keilin's respiratory chain concept and its chemiosmotic consequences. *Science* 206:1148–1158, 1979.
40. Post, R.L.; Jolly, P.C. The linkage of sodium, potassium, and ammonium active transport across the human erythrocyte membrane. *Biochim. Biophys. Acta (Amst.)* 25:118–128, 1957.
41. Post, R.L.; Kume, S.; Tobin, T.; Orcutt, B.; Sen, A.K. Flexibility of an active center in sodium-plus-potassium adenosine triphosphatase. *J. Gen. Physiol.* 54(Suppl.):306s–326s, 1969.
42. Post, R.L.; Merritt, C.R.; Kinsolving, C.R.; Albright, C.D. Membrane adenosine triphosphatase as a participant in the active transport of sodium and potassium in the human erythrocyte. *J. Biol. Chem.* 235:1796–1802, 1960.
43. Requena, J.; Mullins, L.J. Calcium movement in nerve fibers. *Q. Rev. Biophys.* 12:3, 371–460, 1979.
44. Ritchie, J.M. Electrogenic ion pumping in nervous tissue. *Curr. Top. Bioenerg.* 4:327–356, 1971.
45. Roos, A.; Boron, W.F. Intracellular pH. *Physiol. Rev.* 61:296–434, 1981.
46. Russell, J.M. Cation-coupled chloride influx in squid axon. Role of potassium and stoichiometry of the transport process. *J. Gen. Physiol.* 81:909–925, 1983.
47. Sachs, J.R.; Welt, L.G. The concentration dependence of active potassium transport in the human red blood cell. *J. Clin. Invest.* 46:65–76, 1967.
48. Schultz, S.G. Sodium-coupled solute transport by small intestine: A status report. *Am. J. Physiol.* 233, *Gastrointest. Physiol.* 2:E249-E254, 1977.
49. Schuurmans Stekhoven, F.; Bonting, S.L. Transport adenosine triphosphatases: properties and functions *Physiol. Rev.* 61:1–76, 1981.
50. Sen, A.K.; Post, R.L. Stoichiometry and localization of adenosine triphosphate–dependent sodium and potassium transport in the erythrocyte. *J. Biol. Chem.* 239:345–352, 1964.
51. Shaver, J.L.; Stirling, C.E. Ouabain binding to renal tubules of the rabbit. *J. Cell Biol.* 76:278–292, 1978.
52. Shaw, T.I. Potassium movements in washed erythrocytes. *J. Physiol. (Lond).* 129:464–475, 1955.
53. Shull, G.E.; Schwartz, A.; Lingrel, J.B. Amino-acid sequence of the catalytic subunit of the $(Na^+ + K^+)$ATPase deduced from a complementary DNA. *Nature* 316:691–695, 1985.
54. Skou, J.C. The influence of some cations on an adenosine triphosphatase from peripheral nerves. *Biochim. Biophys. Acta (Amst.)* 23:394–401, 1957.
55. Skou, J.C. The $(Na^+ + K^+)$-activated enzyme system and its relationship to transport of sodium and potassium. *Q. Rev. Biophys.* 7(3):401–434, 1975.
56. Stahl, W.L. The Na^+,K^+-ATPase of nervous tissue. *Neurochem. Int.* 8:449–476, 1986.
57. Sweadner, K.J. Two molecular forms of $(Na^+ + K^+)$-ATPase in brain. *J. Biol. Chem.* 254:6060–6067, 1979.
58. Thomas, R.C. Membrane current and intracellular sodium changes in a snail neurone during extrusion of injected sodium. *J. Physiol. (Lond.)* 201:495–514, 1969.
59. Thomas, R.C. The role of bicarbonate, chloride and sodium ions in the regulation of intracellular pH in snail neurones. *J. Physiol. (Lond.)* 273:317–338, 1977.
60. Ullrich, K.J. Sugar, amino acid, and Na^+ cotransport in the proximal tubule. *Ann. Rev. Physiol.* 41:181–195, 1979.
61. Ussing, H.H. Transport of ions across cellular membranes. *Physiol. Rev.* 29:127–155, 1949.
62. Wheeler, T.J., Hinkle, P.C. The glucose transporter of mammalian cells. *Annu. Rev. Physiol.* 47:503–517, 1985.
63. Wieth, J.O.; Andersen, O.S.; Brahm, J.; Bjerrum, P.J.; Borders, C.L., Jr. Chloride-bicarbonate exchange in red blood cells: physiology of transport and chemical modification of binding sites. *Philos. Trans. R. Soc. Lond. (Biol.)* 299:383–399, 1982.

Excitable Cells

Membrane Excitability: Action Potential Propagation in Axons

INTRODUCTION

Living cells have the property of *excitability*—they respond to stimuli in their environment. Swimming protozoa back away from objects they bump into, pancreatic beta cells release the hormone insulin when blood glucose concentration is high, a sensory nerve fiber generates impulses when the skin is touched. The examples cited are relatively rapid responses, all mediated by the opening and closing of specific ionic channels—ion-selective pores—in the cell surface membrane. In multicellular animals, a network of specialized excitable cells, the nervous system, has evolved to coordinate and regulate activities of the organism. The coordination affects not only external actions, such as locomotion and ingestion, but, equally importantly, the regulation of the internal environment, such as control of the salt, acid, and nutrient content of the blood. These goals are accomplished by a continual interplay of chemical and electrical messages generated in receptors, processed and transmitted by neurons, and executed by effectors.

A simple example illustrates the operation of a few elements of the nervous system. If the hand is accidentally placed on a hot stove, it will be withdrawn within a fraction of a second by a powerful *reflex* requiring no voluntary decisions. The anatomical pathway for such a reflex, shown in Figure 3–1, involves the spinal cord but requires no higher centers of the nervous system. Receptors in the skin are stimulated by the heat and set up a propagating barrage of impulses in *afferent nerve fibers* going from the arm to the spinal cord. This message is communicated at nerve-nerve synapses within the cord to a group of *interneurons*, and once again a barrage of impulses is initiated and propagates to the nerve terminals of the interneurons. Then, the message is communicated

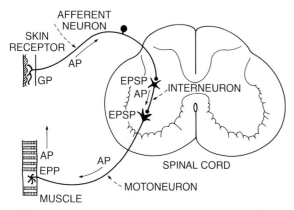

Figure 3–1 Schematic diagram of the path of excitation in a simple spinal reflex. The message originates at a receptor terminal in the skin and is relayed along the afferent neuron to an interneuron in the spinal cord, thence to a motoneuron, and finally to a muscle fiber. The electrical form of the message changes as the impulse proceeds, originating as a generator potential (GP), propagating as an action potential (AP), and reappearing on the postsynaptic side of each junction as an excitatory postsynaptic potential (EPSP) or as an end-plate potential (EPP).

across synapses to *motoneurons* in the spinal cord, setting up propagating impulses that leave the cord in *efferent nerve fibers* going to the arm. Finally, the message is communicated across neuromuscular junctions to the appropriate muscles, which propagate impulses over their lengths and begin to contract. Reflex arcs like this one, monitoring the environment and feeding back to effector organs to give adaptive or regulatory responses, are a mainstay of our understanding of the integrative function of the nervous system. They are encountered in the study of the physiology of each of the organ systems of the body.

The example of the reflex arc raises two major classes of questions. The first includes How is the nervous system "wired up" so that the stimuli produce the appropriate response? How are the messages directed to the correct muscles or effector organs? How does the system determine which movement will remove the hand from danger? These questions of neural circuitry are discussed starting in Chapter 13. The second class of major questions is at a lower level of inquiry, often called *cell biophysics*. These questions have to do with the nature of the messages and responses. How are stimuli detected and transduced to make a signal in the nervous system? What is the impulse that travels down a nerve? How does the synapse work? How do muscles contract? These topics are considered in this and the following nine chapters. Specifically, this chapter is concerned with the nature of electrical signals in axons, the prototype of an electrically excitable membrane. The following chapter generalizes these ideas to many other excitable cells and seeks to relate excitability to the laws of physical chemistry and to the molecular structure of the cell membrane components involved.

POTENTIALS AND THE IONIC HYPOTHESIS

Slow Potentials vs. Action Potentials. The electrical messages of excitable cells are changes of the electrical potential difference across the cell surface membrane. In favorable cases, the membrane potential changes can be measured with intracellular micropipette electrodes (see Chap. 1). At rest, the membrane potential, defined as inside minus outside potential, would lie between -70 mV and -95 mV in each of the four types of excitable cells in the reflex arc we have discussed (Fig. 3–1). If, while we were recording from any of these cells, the hand were placed on a hot surface, we would see rapid variations of the recorded membrane potential corresponding to the message traversing that point of the reflex arc. Indeed, until the muscle contracts, changes in membrane potential would be one of the few detectable signs of activity in the reflex arc. By convention, a cell is said to *depolarize* if the membrane potential becomes more positive, to *repolarize* when it returns toward the original negative value, and to *hyperpolarize* if the membrane potential goes more negative than the resting value. In these terms, the cells of the reflex arc we have described undergo numerous transient depolarizations during activity.

As a practical matter, many signals in excitable cells are brief changes of membrane potential, lasting less than 1.0 ms (10^{-3} s), and must be detected and displayed by devices capable of following small, fast signals. With modern electronics, the signal on the recording electrode may be amplified by a preamplifier made from quite inexpensive integrated circuits, and the final recording can be done in several ways. The most common way is with the cathode ray oscilloscope, by which an electron beam draws a graph of membrane potential against time on a phosphor-coated screen. The electron beam sweeps horizontally under control of the time-base generator of the oscilloscope and is simultaneously deflected in the vertical dimension by the signal to be displayed. The experimenter can observe this display and take photographs of it. Another increasingly practical form of recording involves a digital computer with an "analogue-to-digital converter" that turns the membrane potential signal into a stream of numbers recorded in the computer memory. The numbers represent the successive values of potential at fixed time points and can be used to reconstruct a graph of the time course of the membrane potential at some later time. In some experiments, rapid electrical signals must be recorded from structures too small or inaccessible even for a microelectrode. L.B. Cohen and his colleagues have developed optical methods to detect such signals, using light beams and photodetectors instead of electrodes.[11] Most of the methods involve vital staining of the tissue with dyes. Membrane potential changes are seen as changes of light absorption, of fluorescence, or of the polarization of light. Again, preamplifiers and oscilloscopes or digital computers permit practical rapid measurements.

In analyzing the electrical signals of the nervous system, neurophysiologists distinguish between *action potentials* and *slow potentials*. Action potentials are used to send messages rapidly over long distances, such as along the axons of a nerve trunk

or along the length of muscle fibers in a muscle. They travel as brief, depolarizing, pulselike changes of membrane potential, moving at a constant speed and with no loss of amplitude. Oscilloscope photographs of the time course of action potentials in nerve cells of several animals are shown in Figure 3–2. Because of their shape, action potentials are often called *spikes* or *impulses*, and the three terms are used interchangeably.* The exact shape depends on the cell and the temperature, but in each example in Figure 3–2 the membrane potential overshoots 0 mV and returns toward rest in 0.2 to 2 ms. Since action potentials have only one size and shape in any axon, all of the information carried by a nerve trunk is coded in the timing of the pulses and by the pattern of synaptic connections made by each fiber. The *only* stimulus for an action potential is a membrane depolarization larger than a certain threshold level.

In contrast with action potentials, slow potentials do not propagate at constant amplitude. They are responses of highly localized transducing mechanisms, graded in amplitude and duration in accordance with the amplitude and duration of the stimulus being transduced. Slow potentials have no threshold. They arise at chemical synapses in response to the chemical transmitter and at sensory and other receptors in response to their appropriate stimuli. They sum spatially and temporally with other slow potentials in the same cell. Some slow potentials are depolarizing, and others are hyperpolarizing. Whenever an action potential is initiated in normal body functions, it is triggered by a depolarizing slow potential of sufficient amplitude.

To review these concepts, consider again the reflex arc in Figure 3–1. The figure legend indicates four successive alternations of slow potentials and action potentials as the message travels around the arc. Later parts of this chapter discuss the mechanism of action potentials, and we return again to slow potentials at sensory receptors in Chapter 5, at the neuromuscular junction in Chapter 6, and at central synapses in Chapter 11.

The Ionic Hypothesis. The existence of electrical signals in nerve and muscle has been known since the late eighteenth century from the famous experiments of Galvani on "animal electricity." During the nineteenth century, Helmholtz determined the conduction velocity of the action potential in

---

*In precise usage, *impulse* has a broader meaning than *action potential*, referring to all events, chemical and physical, that accompany the electrical changes of the action potential.

ACTION POTENTIALS

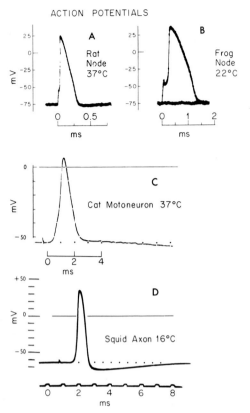

Figure 3–2 Resting and action potentials recorded intracellularly from different nerve cells. In each case, the intracellular resting potential is negative, and the action potential is a brief spike overshooting 0 mV. *A* and *B*, Records from nodes of Ranvier of single, dissected myelinated nerve fibers in response to a brief stimulus applied in the same node, i.e., not a propagating response. (After Nonner, W.; Horackova, M.; Stämpfli, R. 1967, unpublished data; Dodge, F. A., Jr. A Study of Ionic Permeability Changes Underlying Excitation in Myelinated Nerve Fibers of the Frog, Ph.D. thesis, The Rockefeller University, 1963.) *C*, Record from a cat lumbar spinal motoneuron excited antidromically by stimulation of a motor axon. (Courtesy of W. E. Crill.) *D*, Propagating response in a squid giant axon stimulated 2 cm away from the recording site. (After Baker, P.F.; Hodgkin, A.L.; Shaw, T.I. *J. Physiol. [Lond.]* 164:330–354, 1962.)

nerve, and Bernstein succeeded in measuring its time course, using only extracellular electrodes. Simultaneously, Arrhenius, Nernst, and Planck were formulating the theory of dissociation of salts into ions in solution and the theory of diffusion potentials for thick membranes. The completion of this pivotal physicochemical background by the 1890s set the stage for Julius Bernstein[4, 5] to propose the "membrane theory," or "ionic hypothesis," in 1902. His quite general assumptions were (i) potentials are generated across a membrane surrounding the cell; (ii) the membrane is selectively permeable to certain ions; (iii) the potentials arise because of diffusion of ions down their concentration gradients; and (iv) potential changes are caused by membrane permeability changes. Many ingenious quantitative experiments have now proved the correctness of the ionic hypothesis, and it is today the starting point for any discussion of electrical excitability.

Bernstein also made specific proposals for the origin of the resting potential and action potential.[4, 5] Knowing that cells have a higher concentration of K[+] ions inside than out and that K[+] salts applied externally depolarize cells, he suggested that the resting membrane is selectively permeable to K[+] ions, as discussed in Chapter 1. Bernstein guessed further that during the propagated action potential, the membrane becomes *more* permeable to the ions that had been held back. Modern experiments have refined this idea in showing that during action potentials (and also most slow potentials) the membrane becomes more permeable to *certain cations*.

We know now that excitation opens ionic channels, macromolecular pores in the plasma membrane. Several of the channels have been purified chemically from the membranes of mammalian nerve and muscle and the electric organs of electric fish. They are large glycoproteins with structures as complex as, for example, regulatory enzymes. The amino acid sequences and even clones of the genes for some of them have been obtained. Channels open and close in response to appropriate stimuli, a process called *gating*. For the action potential of axon membranes, the most important channels are Na channels and one of the types of K channels. They are named after the physiological ions they let through most easily. Cells that make action potentials also have Ca channels. These three types of channels respond to electrical stimuli, i.e., their gates open and close in response to changes of the membrane potential, so membranes containing these channels are often called *electrically excitable membranes*. The channels are called voltage gated. In other types of membranes, there are channels open by other stimuli, e.g., mechanical stimuli, chemical stimuli. An open ionic channel permits certain ions to rush across the membrane down their electrochemical gradients and is thus a molecular source of electric current in the excitable membrane. Some channels are highly selective for one physiological ion species and others seem to require only that the charge of the ion have the right sign (anion or cation) and that the size be not too big.

The next section, Passive Electrical Properties of

Membranes, is devoted to the electrical consequences of electric currents flowing across cell membranes. It is a lengthy but necessary diversion before we can consider the origins of action potentials in the subsequent section, Experiments on Action Potentials.

PASSIVE ELECTRICAL PROPERTIES OF MEMBRANES

Resistance and Capacitance. As we have seen in Chapter 1, cell membranes are exceedingly thin yet good insulators when compared with the cytoplasm or with the interstitial fluid. Hence they exhibit a relatively high electrical resistance and high electrical capacitance, properties that profoundly influence the spatial spread and time course of any electrical response within a cell. Before we continue with the questions of excitability, we review the definitions of electric current, resistance, and capacitance and consider how they are measured and what they do to the responses of the cell. This material is basically physics rather than biology, but without some discussion of electricity, it would be difficult to explain the electrical excitability of cells.

Electric Current and Ohm's Law. Current is defined as the net movement of charge, and its direction of flow is, by a convention of Benjamin Franklin, in the direction of movement of positive charges. Current always circulates in a closed loop, carried by whatever charged particles happen to be mobile in different parts of the circuit, and adding up so that the same net charge crosses each cross section of the circuit. Suppose that an electrolyte solution is placed between two metal plate electrodes attached to a battery. The battery supplies current to the positive electrode by drawing electrons from the wire; current in the electrolyte flows by the simultaneous movement of the positive ions (cations) toward the negative electrode (cathode) and of the negative ions (anions) toward the positive electrode (anode); and the battery draws current from the negative electrode by supplying electrons to the wire.

Ohm's law states that the current, I, flowing through a conductor is proportional to the voltage applied, E, and the electrical conductance, g, of the material (Fig. 3–3A). Conductance is a measure of the ease, and resistance, R, the reciprocal of conductance, a measure of the difficulty of passing steady current.

$$I = E/R = gE$$

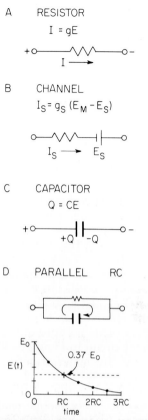

Figure 3–3 Electrical equivalent circuit elements used to describe membranes and ionic channels. *A,* A resistor conducts current according to Ohm's law. *B,* A channel passes ionic current with a reversal potential that depends on the permeant ion. *C,* A capacitor stores the charge. *D,* A parallel RC circuit discharges along an exponential time course as the charge equalizes between the plates.

where R is the resistance and g the conductance. In this equation, I is in amperes, E is in volts, R is in ohms, and g is in reciprocal ohms, called *siemens*. The conductance of a salt solution increases when more salts are added, and the conductance of a membrane increases when more salts are added to the bathing solutions and when more ionic channels open in the membrane. Two other useful rules are (i) the conductance of a system of two conductors in *parallel* is equal to the sum of the separate conductances, and (ii) the resistance of a system of two conductors in *series* is equal to the sum of the separate resistances. These rules follow from Ohm's law, because with two conductors in parallel, each experiences the full applied voltage and carries the same current as when it was not in parallel, and the net current therefore is the sum of the separate currents; whereas with two in series, only part of the potential appears across each conductor, and the single current that flows successively across them is therefore smaller.

In ionic channels, there are *two* driving forces on ions: the electrical force of an electric field and the "chemical" or statistical force of a concentration gradient. Hence when there is a concentration gradient, we need to modify Ohm's law to take its effects into account. With Ohm's law, the current goes to zero when E is zero. In a K^+ ion selective pore, we know from Chapter 1 that current (flux of K^+ ions) goes to zero when E_M equals E_K, the potassium equilibrium potential or Nernst potential. Therefore Ohm's law can be rewritten for a K-selective channel:

$$I_K = g_K(E_M - E_K)$$

where I_K and g_K are called the potassium current and the potassium conductance. Instead of viewing the channel as just a conductor, we consider it to be a conductor with an internal electromotive force—a battery. This is represented by the electrical equivalent circuit in Figure 3–3B.

Capacitors. If the material between the plate electrodes is a nonconductor such as oil or fat, no steady current flows in the circuit from the battery—there are no freely mobile charged components in oil or fat. However, at the moment when the battery is first connected, the circuit delivers a transient flow of current to "charge" up the plates to the potential of the battery. This *capacity current* flows as the atoms between the plates become electrically polarized—their electron clouds are displaced slightly in the direction of the positive electrode; their atomic nuclei, toward the negative electrode—and molecules containing polar bonds, such as C-O bonds, become slightly oriented in the field. Each small charge movement corresponds to an electric current flowing while the material between the plates develops its polarization. In the external circuit, charges are also moved up the wire to the plates, so that the plates bear opposite net charges stabilized by their mutual attraction across the polarized insulating material.

Because the plates have the capacity to store charges, they are said to form a capacitor (Fig. 3–3C). Capacity current is defined as the rate of accumulation of charge on the conductors of a capacitor, and the *capacitance*, C, is defined as the charge stored per volt. Hence the charge, Q, is given by Q = CE. Here, the capacitance is measured in farads and charge in coulombs, and the capacity current is given by

$$I_c = dQ/dt = C \, dE/dt$$

The capacitance between two parallel conductors can be increased by increasing their surface area, by decreasing the distance between them (so the internal field is higher), and by increasing the polarizability or dielectric constant of the medium between them.

If the battery connected to the plates of a capacitor is removed, the opposite charges remain until the polarized capacitor is allowed to discharge by connecting the plates together with a conductor (Fig. 3–3D). The charges then pass through the conductor until the plates become neutral again. A useful way to measure a capacitance is to measure how long it takes to be discharged through a known resistance in a parallel RC circuit. Simple circuit theory shows that the charge decays away exponentially in this case and that after a time equal to the product of resistance and capacitance, RC, there remains only 37% of the original charge (0.367 is the reciprocal of the number e, the basis of the natural logarithms). After a time 2 × RC, there remains 0.37 × 0.37, or 14% of the original charge. This characteristic time constant, $\tau = RC$, for discharge of a capacitor through a parallel resistor is used routinely in electrophysiological studies of cell membranes, as is described in the next section. Mathematically, the decay of voltage on a discharging capacitor is expressed

$$E = E_0 \exp(-t/RC) = E_0 \exp(-t/\tau)$$

where t is time and E_0 is the voltage at t = 0. In this formula, time, resistance and capacitance are expressed in units of seconds, ohms, and farads.

Measurement of Membrane Properties in Round Cells. The resting membrane resistance, R_M, and the membrane capacitance, C_M, can be measured relatively easily in those nearly spherical cells that are large enough and accessible enough to have electrodes inserted in them, e.g., egg cells, neuron cell bodies disconnected from their axons and dendrites, isolated cells in tissue culture, and some protozoa. The procedure is illustrated in Figure 3–4A. A protozoan, *Paramecium*, is impaled with two micropipette electrodes, one for measuring intracellular potential with respect to the bath potential and a second for injecting current pulses into the cell. The current applied from the intracellular current electrode spreads throughout the cell, passes uniformly out across all parts of the cell membrane, and then returns to the stimulator circuit via the grounded bath electrode. Membrane potential traces, E_M, and membrane current traces, I_M, are shown in Figure 3–6B as photographed from the face of an oscilloscope in an actual experiment. The three groups of records

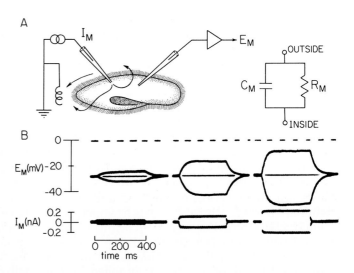

Figure 3–4 An experiment demonstrating that the cell membrane acts as a resistor and capacitor in parallel, as shown in the inset in *A*. *A*, A *Paramecium* is impaled with two micropipette electrodes. *B*, Current pulses, I_M, passed through one intracellular electrode to the bath ground, change the membrane potential, E_M, measured as the difference between the intracellular potential and the bath. The voltage trace rises or falls approximately exponentially each time the current is stepped on or off. (After Kung, C.; Eckert, R. *Proc. Natl. Acad. Sci. USA.* 69:93–97, 1972.)

are double exposures of a trial, with a 400-ms pulse of outward current and a trial with a similar pulse of inward current applied across the membrane. Applied currents change the membrane potential. For example, in the rightmost frame of Figure 3–4B, an applied current of −0.23 nA (1 nanoampere = 10^{-9} A) causes a hyperpolarization of the membrane by 23 mV from its resting potential of −28 mV.

Membrane Resistance. The relatively steady polarization of the membrane during each current pulse in Figure 3–4B means that the *Paramecium* cell membrane acts as a resistor in the stimulating circuit. Its effective resistance may be calculated from Ohm's law in the form $R_M = V_M/I_M$.* For the example with V_M = 23 mV and I_M = 0.23 nA, the calculated resistance is 23 × 10^{-3}/0.23 × 10^{-9} = 100 MΩ (1 megohm = 10^6 ohm). The experiment actually measures the resistance in the entire current path from the intracellular current electrode to the extracellular current electrode; however, as the effective resistance of the current path in the cytoplasm and in the bath are far smaller than $10^8 \Omega$, it is correct to attribute the measured value to the membrane alone.

Paramecia are normally electrically excitable and in response to depolarizations give action potentials as large as those in Figure 3–4, but the cell used in the experiment shown here was a genetic mutant that had lost its excitability through a loss of Ca channels. Nevertheless, even this cell shows a small asymmetry in the responses to equal and opposite applied currents, indicating a higher resistance to inward current than to outward current flow across the membrane. The common property of having different resistance to inward and outward currents is called *rectification*, in analogy to rectification in electrical devices such as diodes. In axons, rectification occurs because a membrane depolarization gradually opens more K^+ channels and a hyperpolarization gradually closes them, i.e., gating processes in these channels respond to the electric field in the membrane. Therefore, to make reliable membrane resistance measurements, one must avoid large perturbations that change the resistance; the current applied, I, and the voltage step developed, V, must be kept as small as possible.

Membrane Capacitance. The membrane potential traces in Figure 3–4B show that the polarization of the membrane takes many milliseconds to develop and to decay when the current is turned on and off. The slowed rise and fall of potential indicates the presence of electrical capacitance between the cytoplasm and the external medium. This is the membrane capacitance, a simple geometric consequence of separating the internal and external conducting media by a thin insulating membrane. If the membrane is represented as an electrical equivalent circuit, the capacitance should be drawn in parallel with the membrane resistance, as in the inset of Figure 3–4A. According to the physical chemistry of ionic solutions, the charges on the membrane capacitor lie in the electrolyte solutions as a cloud hovering within 1.0 nm of the membrane surface. In this very limited region only, there is a slight imbalance of the charge contributed by cations and anions in the solution. Membrane capacitance slows the

---

*In this chapter, the symbol E_M is used for membrane potential differences measured on an absolute scale, and the symbol V_M is used for deflections of the membrane potential from the resting potential (a relative scale).

electrical responses of the membrane, since charges have to be moved to and from the surface layer in order to change the membrane potential. If I_C is the current that is supplied by a stimulator or by the cell to discharge the membrane capacitance, C_M, the rate of change of membrane potential is

$$dV_M/dt = I_C/C_M$$

In the experiment with the *Paramecium*, all of the applied current acts initially to change the charge on the membrane capacitor, but as the membrane voltage changes from its normal resting value, the electrochemical gradient acting to drive ions through channels in the membrane increases as well. Eventually, the depolarization reaches a value at which all of the current supplied by the intracellular electrode is carried out of the cell by ions, and none goes to further charge the membrane, i.e., the voltage stops changing when $V_M = I_M R_M$.

When small rectangular pulses of current are applied to biological membranes, the membrane potential charges or discharges exponentially to a new steady level. As with electronic RC circuits, the time required to achieve 63% of the final potential change is the product $R_M C_M$. For a biological membrane, this time is simply referred to as the *membrane time constant*, τ_M. Measurements on the traces of Figure 3–4B show that the polarization of the *Paramecium* membrane rises and decays roughly exponentially with a time constant of 50 ms when the polarizing current is changed. Using 10^8 Ω as the membrane resistance, we can now calculate the membrane capacitance: $C_M = \tau_M/R_M = 50 \times 10^{-3}$ s/10^8 Ω $= 500$ pF (1 picofarad $= 10^{-12}$ farad).

Specific Membrane Properties. In discussions of membranes, one often compares the membranes of one cell with those of another. This is done by comparing the properties of membrane patches of equal area, usually 1 cm², giving the *specific* membrane capacitance in units of farads per square centimeter and the *specific* membrane resistance in units of ohms · centimeters squared. Note that the units of specific resistance are not ohms per square centimeter, since a 2-cm² patch of membrane has one half rather than twice the resistance of a 1-cm² patch; conductance, however, is proportional to membrane area, and specific conductance has the units siemens per square centimeter (S/cm²).

Specific Membrane Capacitance.[1] Specific membrane properties have been measured in a wide variety of excitable and nonexcitable cells by the simple RC analysis, by more sophisticated A.C. impedance techniques, and by other methods appropriate to elongated cells. The classic measurements in this area were made by K.S. Cole and collaborators.[8] The experiments reveal a remarkable constancy of the specific membrane capacitance. For cell surface membranes, C_M is always near 1 μF/cm², a value that would be obtained if the insulating region of the membrane had an average thickness of 2.6 nm and an average dielectric constant of 3.0.* For comparison, pure phospholipid bilayer membranes (a model system) have specific capacitances near 0.8 μF/cm². The similarity of this number to the biological value is another line of evidence that all cell membranes are fundamentally similar in construction and contain a large proportion of lipid bilayer. We return later to the constant membrane capacitance as a factor limiting the rate of rise of the action potential and setting a lower limit on the number of ions that must flow across the membrane during an electrical response.

In some cells, the apparent specific capacitance may be up to 15 times higher than the standard value, but this discrepancy is only apparent and indicates an underestimation of the amount of membrane contributing to the cell surface. For example, in the skeletal muscle cells, typical values might be 6 μF/cm², referred to the visible surface. Electron microscopy of muscle reveals, however, a complex intracellular network of tiny tubules that are membranous invaginations of the surface, and correcting for the extra area contributed by these internal membranes brings the specific membrane capacitance down to 1 μF/cm² again (see Chap. 7). Similarly, the specific capacitance of the *Paramecium* in Figure 3–4 seems to be 5 μF/cm² until one takes into account the considerable extra membrane area of the motile cilia and the sculpturing of the cell surface found with the electron microscope. Again, correcting for the extra area lowers C_M to near 1 μF/cm². The opposite kind of discrepancy is observed with myelinated nerve fibers, in which the specific capacitance of a large fiber is less than 0.01 μF/cm². Again, the reason lies in the anatomy. As described later, a myelinated nerve fiber 17 μm in diameter (Fig. 3–5) is in fact a 12-μm cylindrical axon surrounded by a 2.5-μm–thick sheath of insulating myelin formed

---

*The membrane thickness Δx can be calculated from the formula for a parallel plate capacitor, $C_M = \epsilon \epsilon_o/\Delta x$, where ϵ is the membrane dielectric constant and ϵ_o, the permittivity of free space, is 8.85×10^{-12} CV^{-1}m^{-1}.

Figure 3–5 Schematic drawing of three different types of nerve fibers, showing axons *(A)* enveloped in Schwann cells (SC). The vertebrate C fiber usually has several very small axons tended by one Schwann cell. The vertebrate A or B fiber has a modest axon heavily wrapped in myelin sheath formed by a spiral wrapping of Schwann-cell membranes. At periodic intervals, at nodes of Ranvier (N), the myelin sheath comes to an end, leaving a short stretch of axon uncovered. The invertebrate giant fiber contains a large axon covered with a thin sheet of Schwann cells closely interlocking in a brickworklike arrangement. Although the figure shows only one layer of cells forming the sheet, often there are several. See also Figure 3–15 for an electron micrograph of vertebrate nerve.

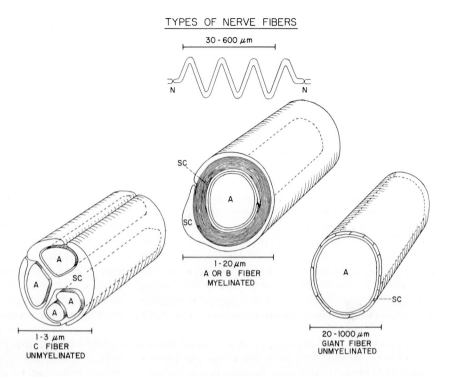

TYPES OF NERVE FIBERS

30 - 600 μm

N   N

SC

SC

A

1 - 20 μm
A OR B FIBER
MYELINATED

A

SC

20 - 1000 μm
GIANT FIBER
UNMYELINATED

A

A

A

A

SC

1 - 3 μm
C FIBER
UNMYELINATED

by a tight spiral wrapping of Schwann cell membrane around the axon. Any electric current flowing into the axon through the sheath has to cross about 300 Schwann cell membranes in series in addition to the axon membrane, making the electrical resistance of this barrier 300 times larger and the electrical capacitance 300 times smaller than that for a single membrane.

Specific Membrane Resistance. In contrast with the constancy of C_M, the electrical time constant, τ_M, for membrane charging ranges over at least four orders of magnitide in different cells. Recalling the definition of τ_M as the product $R_M C_M$ shows that the specific membrane resistance must then also vary among cells. A membrane time constant of 100 μs in electric organ cells of electric eels and in nodes of Ranvier of myelinated nerve corresponds to $R_M = 100 \times 10^{-6}/10^{-6} = 100 \ \Omega cm^2$. Similarly, a τ_M of 1 ms in squid giant axon corresponds to $R_M = 1 \ k\Omega cm^2$; a τ_M of 30 ms in frog fast skeletal muscle corresponds to $R_M = 30 \ k\Omega cm^2$, and a τ_M of 350 ms in frog slow skeletal muscle fiber to $R_M = 350 \ k\Omega cm^2$. These differences in R_M correspond crudely to differences in the number of open ionic channels per unit area of membrane. The values given are all for resting membranes. During excitation many additional ionic channels are opened, and the membrane resistance falls far below the resting value. From

the functional viewpoint, those cells that must generate large currents or must fire at high frequency seem to have the highest number of channels and shortest membrane time constants. The surface density of ionic channels ranges from 0.5 to 15,000 channels μm^{-2}.

This concludes the discussion of passive electrical properties of cell membranes, and we now return to the more biological question of how excitable cells make action potentials.

EXPERIMENTS ON ACTION POTENTIALS

Preparations. Action potentials are the unitary electrical messages carried in single nerve fibers of the nervous system. They have been studied best in large axons of the peripheral nervous system. Figure 3–5 shows the gross structure of the three common types of nerve fibers, in order of increasing size. With a few exceptions, peripheral axons are long cylindrical cell processes of a neuron surrounded by a sheath of one or more nonneuronal cells—glial cells, called in this instance *Schwann cells*. Glia adhere too intimately to the axon to be removed in any practical experiments.

Early in this century, the favorite tissue for studying action potentials was the whole sciatic nerve of frogs and toads. This preparation contains roughly a thousand myelinated nerve fibers and a thousand unmyelinated fibers in a nerve trunk 1 mm in diameter and 5 to 10 cm long. The nerve is easy to dissect and continues to function outside the animal for at least a day when bathed in Ringer's solution (physiological saline). Despite these advantages, progress was slowed by the difficulty of using a multi-fiber preparation without any method for intracellular recording. Then in 1936, J.Z. Young reported that the mantle muscles of squid are innervated by giant axons up to 1 mm in diameter and 30 cm long. The advantages of this spaghetti-sized axon were evident to K.S. Cole and H.J. Curtis, who began membrane biophysical investigation in Woods Hole, Massachusetts, followed soon by A.L. Hodgkin, A.F. Huxley, and B. Katz, working in Plymouth, England, initiating a renaissance of discovery culminating in the Hodgkin-Huxley model of the action potential in 1952. The squid giant axon is large enough so that 100-μm–diameter glass pipettes or metal-wire electrode assemblies may be threaded axially for several centimeters into the axoplasm from a cut end. The axoplasm may be collected for chemical analysis by gently squeezing it out of a cut end with a little roller. An axon with the axoplasm rolled out may be reinflated and can then conduct action potentials with appropriate test solutions perfused through the inside. These possibilities make the squid giant axon a versatile preparation for biophysical work.

Fortunately, a wide variety of other practical vertebrate and invertebrate preparations are also now available. Late in the 1920s, techniques were being developed to isolate and record from single vertebrate myelinated axons. By 1946, G.N. Ling and R.W. Gerard developed the glass micropipette electrode, which made it possible to record intracellularly from many cells. In the period from 1975 to 1980 E. Neher and B. Sakmann at Göttingen perfected another type of glass microelectrode, the *patch* pipette. Instead of being sharpened to pierce the cell membrane, the tip is fire polished to press against the cell surface, covering and sealing to a small patch of the plasma membrane. It permits recordings to be made from the small number of ionic channels that lie under the pipette (see Chapter 4). In the past 20 years, the popular preparations for membrane biophysics have included the giant axons of squid, giant ganglion cells in gastropod and nudibranch molluscs, vertebrate myelinated nerve fibers, barnacle giant muscle fibers, and vertebrate skeletal and cardiac Purkinje muscle fibers. Recently, many cell types dissociated from intact tissue by enzymatic treatments and established cell lines in cell culture have become accessible through the patch electrode technique. All this work reveals an underlying uniformity in how excitable membranes work, with an overlay of specialization allowing each cell type to perform its special functions. Specifically, in all axons studied, the propagation of normal action potentials depends on the sequential opening and closing of Na and K channels, and the properties of these channels depend very little on the animal species studied. These channels generate an orderly sequence of inward, then outward, ionic current to shape the spike. In nonaxonal excitable membranes, e.g., muscle or even nerve cell body and dendrites, many other channels make important contributions to inward or outward currents.

Propagating Wave. An action potential propagates along the axon as a true wave of membrane potential change. Its shape as a function of time, as displayed on the oscilloscope from a recording electrode, is the same as its shape as a function of distance, as might be "seen" in a hypothetical "snapshot" of the instantaneous spatial distribution of potential. Figure 3–6A shows the membrane potential of a giant squid axon recorded with two intracellular microelectrodes 16 mm apart. An electric shock applied to a pair of external stimulating wires excites a propagating action potential that depolarizes the membrane at recording site 1 within 0.3 ms and reaches site 2 with no loss of amplitude or change of shape 0.7 ms later. Had a third recording electrode been placed another 16 mm along the axon, it would have recorded the action potential arriving with an additional 0.7-ms delay.

The conduction speed of the wave is the distance traveled divided by the time, or in this case, 16 mm/0.7 ms = 23 mm/ms = 23 m/s. The physical length of the wave is about 16 mm at any instant, since the membrane just repolarizes at site 1 as it begins to depolarize at site 2. Hence the action potential in this 500-μm–diameter giant fiber at 20°C could be represented in a cartoon as a triangle 16 mm wide moving uniformly at 23 m/s away from the site of stimulation. The wavelength in any propagating wave is equal to the conduction velocity multiplied by the wave duration. For example, the largest mammalian myelinated nerve fibers (diameter 18 μm) have conduction velocities up to 120 m/s at 37°C with an action potential duration of 0.3 to 0.4 ms, and therefore their action potential wavelength is 36 to 48 mm.

Stimulation by Depolarization

Threshold. There is a sharp threshold for artificial stimulation of the action potential with electric currents. Hyperpolarizing stimuli do not normally

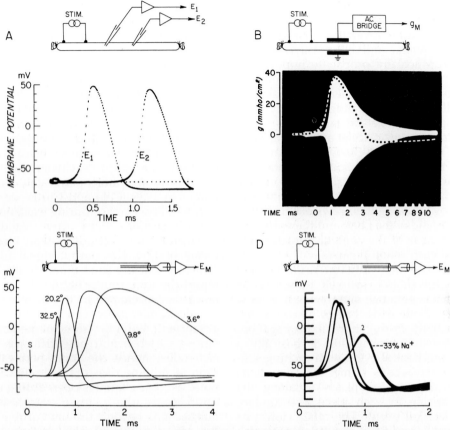

Figure 3–6 Four fundamental properties of the propagating action potential demonstrated with squid giant axons. *A,* The action potential recorded from two points maintains constant shape and size as it propagates. T = 20°C. (After del Castillo, J.; Moore, J.W. *J. Physiol. [Lond.]* 148:665–670, 1959.) *B,* The electrical conductance of the membrane (a measure of ionic permeability), displayed as a white band from an impedance bridge, increases transiently as the action potential (dotted line) propagates under the electrodes. (After Cole, K. S.; Curtis, H. J. *J. Gen. Physiol.* 22:649–670, 1939.) *C,* The conduction velocity and time course of the action potential slow as the axon is cooled. (After Hodgkin, A. L.; Katz, B. *J. Physiol. [Lond.]* 109:240–249, 1949.) *D,* The amplitude, rate of rise, and conduction velocity decrease reversibly when the external [Na⁺] is reduced by diluting the sea water with isotonic dextrose. (After Hodgkin, A.L.; Katz, B. *J. Physiol. [Lond.]* 108:37–77, 1949.)

excite; subthreshold depolarizing stimuli also produce no propagated response; and all suprathreshold stimuli elicit the same stereotyped wave traveling with the same velocity. The action potential is therefore said to be an *all-or-nothing* response. An adequate stimulus must depolarize the excitable membrane by at least 8 to 50 mV, depending on the cell, long enough to turn on the regenerative permeability changes of the action potential mechanism. Empirically, it is found that a brief stimulus current (e.g., one lasting only 20 µs) must be more intense than a longer stimulus (e.g., one lasting 4 ms) to reach threshold. This observation reflects, in part, the need to deposit a certain quantity of charge "on the membrane capacitance" to depolarize a cell by a given amount. However, the full relation between strength and duration of just-threshold stimuli is a complex

function of both the passive properties of the membrane and the kinetics of gating in the ionic channels.

Refractoriness. For a brief time following an action potential, even a very large stimulus will not excite a second action potential; this is called the *absolutely refractory period.* The cell is simply inexcitable. Then there follows a short *relatively refractory period* when the stimulus threshold is elevated above normal. The refractoriness remains until the membrane ionic channels return to their resting state. The absolutely refractory period sets an upper limit on the maximum rate of repetitive firing of an axon. For example, large mammalian myelinated nerve fibers have an action potential duration of approximately 0.4 ms and most cannot conduct impulses at a higher frequency than 1000 Hz, i.e., an interspike interval of 1 ms.[29] The refractoriness of the membrane behind advancing action potentials also plays a major role in preventing impulses from reversing their direction spontaneously. Axons and electrically excitable muscle cells can actually propagate impulses in either direction, so that an electrical stimulus applied to the middle of a long fiber will initiate two waves of excitation, one traveling each way. This property is useful for skeletal muscles in which synaptic excitation takes place at a neuromuscular junction near the center of a muscle fiber and the action potential spreads the excitation to both ends (Fig. 3–1). However, in normal use, axons are meant to conduct only in one direction.* They gain their polarity by receiving their physiological stimulus only at one end of the fiber, and they preserve the polarity because of the zone of refractoriness trailing behind each action potential.

Local Circuit Currents. Although artificial stimulation by electric shocks was well known in the 1930s, there remained much disagreement on how an excited region of axon brings an unexcited region into activity during normal propagation. One view was that propagation is basically a chemical or a mechanical event with exciting substances, chemical reactions, or mechanical motions advancing through the axoplasm or along the membrane, causing in each place a secondary measurable electrical event. The other view was that the electrical event itself, the action potential, acts as an electrical stimulus spread by "cable properties" of the axon† to bring neighboring patches of membrane into activity. Hodgkin[14] demonstrated the correctness of this electrical theory by showing that the small electric currents normally flowing between an excited region of axon and the neighboring unexcited region depolarize the unexcited membrane to its firing threshold exactly as if these currents had been supplied by an electric stimulator. Such currents are called *local circuit currents.* They flow inside the axon from the depolarized membrane to the resting membrane, out through that membrane, and back in a circuit to the active membrane through the extracellular space, obeying the rules for current flow in resistors and capacitors and giving a passive spread of depolarization ahead of the action potential.

Ionic Basis

Permeability Increases. A constant membrane permeability to K^+ ions could not explain how a patch of membrane gives rise to an all-or-nothing action potential and how this excitation can travel as a constant amplitude electrical wave over a distance of several meters. Instead, as Bernstein originally proposed, excitation involves specific changes in the ionic permeability of the membrane. In 1939, Cole and Curtis[9] demonstrated a permeability increase during excitation by placing a short section of an active squid giant axon between plate electrodes of an electrical impedance bridge. The output signal of the bridge, drawn as a white band on an oscilloscope, would report the time course of any changes of conductance or capacitance in the membranes between the electrodes. As is seen in Figure 3–6B, there is a transient, 40-fold increase of membrane conductance as each action potential passes under the electrodes; hence the ionic permeability of the axon membrane rises dramatically during the impulse as ionic channels open during excitation. In squid axons, the increase even lasts for several milliseconds beyond the time of repolarization of the action potential. The next question was to decide which ions are involved.

An essential clue to the identity of the ions came from measurements of the size of the action potential. Recall that opening of an ion-selective channel would tend to bring E_M closer to the equilibrium potential for that channel (Chap. 1). In 1939, Hodgkin and Huxley[18] reported that the peak of the action potential overshoots zero by up to 40 mV, meaning that the responsible ion(s) would have to have equilibrium potential(s) more positive than +40 mV. Table 1–4 in Chapter 1 reveals that Na^+ and Ca^{2+} ions are the major candidates. Either one would tend to enter the cell, driven by a large chemical and electrical gradient, and could in theory account for the rising phase and observed overshoot of the action potential. Hodgkin and Katz[24] established the overwhelming importance of Na^+ ions in axons by showing that reducing the external Na^+ concentration to 33% of normal decreases the amplitude and slows the conduction and the rate of rise of the propagated action potential (Fig. 3–6D). The low-Na solution is prepared by substituting osmotically equivalent amounts of the salt choline chloride, or of the nonelectrolyte sucrose, for some

---

*The normal direction of travel is termed the *orthodromic direction*, and artificially excited impulses traveling the other way are said to travel in the *antidromic direction*.

†An analogy is often drawn between the "passive" spread of small potential changes in axons and in long electrical cables. In either system, a potential change applied at one point spreads with attenuation and delay to distant points. In typical large axons, a passively spread signal may be attenuated to 37% (1/e) of its original value at a distance of only a few millimeters down the length of the axon. Cable properties of biological cells are discussed again later in this chapter.

of the NaCl. This experiment works on every type of axon tested, and in most cases, conduction stops when $[Na^+]_o$ is reduced to about 20% of normal; conduction returns when $[Na^+]_o$ is increased again. On the basis of such experiments, Hodgkin[15] suggested that depolarization causes the Na^+ permeability to increase, allowing Na^+ ions to enter and depolarize the fiber further, in a regenerative cycle that has been dubbed the *Hodgkin cycle*:

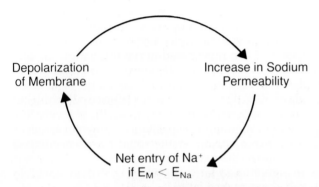

Depolarization of Membrane → Increase in Sodium Permeability → Net entry of Na^+ if $E_M < E_{Na}$ → Depolarization of Membrane

Today, we say that depolarization opens Na channels in the membrane, providing a pathway for Na^+ ions to flow into the cell. The net inward current carried by Na^+ ions depolarizes the membrane toward E_{Na} at a rate directly proportional to the current I_{Na} and inversely proportional to the membrane capacitance to be discharged. The Hodgkin cycle is a property of each patch of a squid axon membrane, indicating that Na channels are distributed all over the surface membrane.

The cycle works because activation of Na channels is voltage-dependent. The cycle fails to start if $[Na^+]_o$, and hence E_{Na}, are too low or if Na channels have been blocked by one of many pharmacological agents. The most important blocking agents are *local anesthetics* such as lidocaine and procaine and two natural toxins—the pufferfish toxin, *tetrodotoxin*, and the paralytic shellfish poison, *saxitoxin* (frequently abbreviated TTX and STX, respectively). Local anesthetics are important for their medical uses in blocking impulse conduction, and TTX and STX are valuable experimental tools for selectively blocking Na channels in electrophysiological investigations and as diagnostic reagents for counting and labeling Na channels in biochemical work (see Chapter 4). The two toxins are also a public health hazard since they can lead to respiratory paralysis if pufferfish or tainted shellfish are eaten. The paralysis is a consequence of blocked action potentials.

What brings an excited cell back to the resting potential? Two possibilities suggest themselves. One could imagine that Na channels close again spontaneously at the peak of the action potential so the membrane would repolarize as K^+ ions leave the cell through the same K channels that were open in the original resting state. Alternatively, the action potential might open up additional K channels, permitting an outward rush of K^+ ions to begin to repolarize the cell despite a simultaneous influx of Na^+ ions in Na channels. The Na channels might then close because of the repolarization. In fact, both of these mechanisms are used: At the peak of the action potential, Na channels begin to close spontaneously, a process called *inactivation*, and additional K channels begin to open. Like the activation of Na channels, the activation of K channels is favored by depolarization, but K-channel activation is a slower process. In the squid giant axon, so many K channels open during the spike that the membrane actually hyperpolarizes to near E_K for a period following the action potential (Figs. 3–2, 3–6), in what is called the *hyperpolarizing afterpotential*, or *afterhyperpolarization*.*

The subsequent decay of this raised permeability can be seen in the narrowing of the trace representing membrane conductance in the Cole-Curtis experiment (Fig. 3–6B). In some other excitable cells, such as mammalian myelinated nerve, the increase of potassium permeability is not large, and there may be no afterpotential or there may even be a *depolarizing afterpotential*.

As may be expected, the falling phase of the spike can be slowed and the spike prolonged by two classes of pharmacological agents. One blocks K channels and includes Cs^+, Ba^{2+}, tetraethylammonium ions, and 4-aminopyridine. The other slows the inactivation of Na channels and includes the venoms of scorpions and sea anemones. The painful and lethal effects of these venoms are the consequences of artificially prolonged action potentials that lead to undesired firing rhythms in nerve and muscle.

Ionic Fluxes.[7, 28] Further direct evidence for the role of ions comes from measurements of extra movements of ions during periods of activity. The extra fluxes are so small that they would be difficult to detect by chemical analysis of axoplasm,

---

*Another now confusing term, *positive afterpotential*, was once used for the hyperpolarizing afterpotential. It was coined in the earlier era of extracellular recording, when hyperpolarization was detected as an external positivity.

but they have been measured with ^{24}Na, ^{42}K, and ^{45}Ca as tracer ions. Calcium fluxes have also been studied by injection of indicator substances whose absorption spectrum, fluorescence, or luminescence reports the intracellular free calcium concentration. In the late 1940s, Keynes[28] developed techniques for using radioactive tracer ions with squid axons and found a net extra sodium entry of 3.5 pmol/cm² of axon membrane per impulse and a roughly corresponding loss of potassium. The following simple calculation from the rules of electricity shows that these small observed fluxes are adequate to account for the size of the action potential. Since the membrane has capacitance C_M, a change of membrane potential of magnitude V_M requires a net movement of $C_M V_M/F$ moles of charge per square centimeter. Taking, for example, a 100-mV signal and using the universal value of specific capacitance, $C_M = 1$ μF/cm², gives $10^{-6} \cdot 10^{-1}/10^5 = 10^{-12}$ mole of charge per square centimeter. The calculation shows that every time a membrane generates a 100-mV signal such as an action potential, 1 pmol/cm² of net positive charge must flow into the axon during the upstroke and an equal net amount must leave during the downstroke. Thus the observed influx of 3.5 pmol/cm² of Na^+ and efflux of 3.5 pmol/cm² K^+ ions could do the job. In the squid giant axon, the net extra entry of Ca^{2+} per impulse is only 1% of the Na^+ entry, confirming the general rule that the entry of Ca^{2+} ions plays an insignificant electrical role during the rising phase of action potentials in *axons*. However, even a small entry of Ca^{2+} ions can act as a *chemical* signal—an internal second messenger—to turn on or off specific intracellular physiological activities in a variety of cells. In certain muscles and nerve cell bodies, the inward current carried by Ca^{2+} ions is much larger and is electrically important—see Chapter 4.

Ionic fluxes accompanying each action potential occur at the expense of the ionic gradients set up by the Na^+-K^+ pump (Chap. 2). If an isolated nerve is stimulated repetitively for several minutes, one can readily detect an elevation of its metabolic rate for many minutes thereafter, during the period of elevated pumping of Na^+ and K^+ ions needed to restore the gradients. However, so long as the gradients are not dissipated too much, electrical activity can persist even if the Na^+-K^+ pump is blocked by an inhibitor. As we have seen, ionic fluxes accompanying action potentials are near the theoretical minimum, with a small inefficiency arising because Na channels are not all closed before extra K channels begin to open and work against them.

The impact of these fluxes on the ionic gradients is in direct proportion to the surface-to-volume ratio of the fiber. Hence the relative effect of an impulse is least for large axons. A 500-μm squid giant axon loses only one millionth part of the K^+ ions stored in its axoplasm per impulse and can still fire 10^5 action potentials in experiments in which the Na^+-K^+ pump is blocked. On the other hand, a 0.1-μm unmyelinated nerve fiber, 5000 times smaller, would be expected to lose 0.5 to 1% of the available K^+ in one impulse and, without a functioning Na^+-K^+ pump, might become inexcitble after only 50 impulses. By considering the surface-to-volume ratio alone, one would guess that a 15-μm myelinated nerve fiber might be able to fire no more than 6500 impulses with the pump blocked, but the experimental value is actually above 200,000.[2] This greater efficiency is due to the capacity-reducing myelin sheath. The very low effective membrane capacitance of myelin achieves a dramatic saving in the ionic fluxes needed to depolarize the axon and adapts myelinated axons to signaling at far less energy cost than unmyelinated axons (see later).

Perfusion and Gating. The electrical events of the action potential are powered by ionic gradients built up by the Na^+-K^+ pump. How about the permeability changes—what energy source drives them? Work must be done to open and close ionic channels in an orderly sequence. One hypothesis that has now been disproved was that metabolic energy is directly involved in gating. This idea would have been consistent with the fairly strong temperature dependence of gating as revealed by a lengthening of the action potential and slowing of the conduction velocity as axons are cooled (Fig. 3–6C). A 10°C drop in temperature increases the duration 2- to 5-fold and slows the velocity 1.8- to 3-fold in a variety of axons. However, the hypothesis was proven wrong by experiments with axons perfused internally with inorganic salt solutions. Most of the axoplasm may be removed mechanically from squid giant axons and the remaining traces digested away from perfusion of proteolytic enzymes through the inside. When simple salt solutions such as isotonic K_2SO_4 or even the metabolic poison KF are then perfused through the axon, the system is capable of propagating more than 10^5 action potentials.[3] About the only requirements for excitability are to have a K^+ salt inside and a Na^+ salt and a little Ca^{2+} outside.

Such experiments prove that axons *with ionic gradients maintained artificially* have no immediate requirement for ATP or other small molecules, or even for enzymes or other macromolecules, in the axoplasm to fire many normal-looking action potentials. Hence all the machinery for gating is in the membrane, and the gating process makes no direct use of metabolism. Of course, in the long run, metabolism is essential for *maintaining* the integrity and function of all parts of the cell. The gates of axonal Na and K channels are opened and closed by direct action of the membrane electric field on the channel molecule. The macromolecule has charges and dipolar components that are forced to move within the membrane by changes of the electric field, and these motions make the channels sensitive to potential changes and provide the energy input for voltage-dependent gating. Gating requires conformational changes of the protein.

Summary of Events of the Action Potential. We have now discussed qualitatively the major events of the propagated action potential. The essential ingredients are (i) the appropriate combination of ionic channels and ionic gradients for regenerative excitability, (ii) an initial depolarizing stimulus, and (iii) spreading of the wave of depolarization forward to unexcited regions of the cell membrane. Figure 3–7 summarizes the membrane events described so far: *A*, A few K channels (circles) are open in the resting membrane. *B*, The resting membrane begins to be depolarized by local circuit current from a neighboring excited region. The current flows longitudinally inside the axon and returns through the external medium. *C*, Some Na channels (triangles) are opened by the depolarization, and the Hodgkin cycle is initiated as the inward sodium current further depolarizes the membrane by discharging the membrane capacitance. *D*, The membrane becomes strongly depolarized, and the many open Na channels supply extra current in local circuits to depolarize unexcited membrane ahead and to retard the repolari-

zation of membrane behind. *E*, K channels begin to be opened by the sustained depolarization. They pass an outward K⁺ current, which eventually exceeds the inward Na⁺ current. At the same time, the sustained depolarization begins to close Na channels by inactivation. *F*, Outward K⁺ current repolarizes the membrane by recharging the membrane capacitance. *G*, The extra open K channels supply local circuit current that opposes the depolarizing effects of Na channels in the excited region. *H*, K channels close and Na channels recover from inactivation to their resting state.

VOLTAGE CLAMP

Rationale and Methods. Our quantitative understanding of electrical excitability in terms of ionic permeability changes of axons or muscle membranes is derived almost exclusively from experiments in which the voltage-clamp method is employed. As the name implies, *to voltage clamp* is to control the membrane potential electrically. Instead of passing fixed steps of *current* and observing the membrane potential response, as in a typical action potential measurement, the experimenter forces the membrane to follow step changes of *potential* and measures the membrane current that flows in response.

The voltage clamp is the biophysicist's assay to determine when ionic channels open and close. As we have seen, electric current flows across membranes both as ionic current, I_i, and as capacity current, I_C. However, with *step* changes of potential, capacity current flows only at the transition from one potential to another, because CdV_M/dt is zero at all other times. Therefore, except at the step, the recorded current can be identified as I_i. Two important advantages of volt-

Figure 3–7 Schematic diagram of the membrane events during a propagating action potential. The excitation begins on the left and finishes on the right. Open K channels are shown as circles and open Na channels as triangles. The arrows through the channels indicate electric (ionic) current, which flows in closed circuits, crossing the membrane a second time either through other open channels or as capacity current that changes the net charge on the membrane.

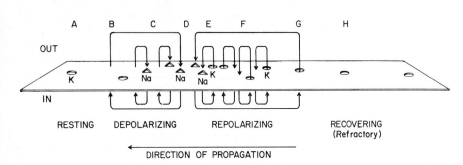

age clamp over conventional voltage recording are (i) the recorded current is carried by ions and gives a direct measure of the ionic permeability changes of the membrane, and (ii) the gating of Na, K, and Ca channels depends on membrane potential, so that manipulation of this variable permits the kinetics of gating to be controlled and analyzed. The first voltage clamp was tried for squid giant axons in 1947 by K.S. Cole and refined and used by A.L. Hodgkin, A.F. Huxley, and B. Katz in the period from 1948 to 1950 to give a complete quantitative description of Na and K permeability changes. The Hodgkin-Huxley[22] model of excitation has been the cornerstone of our understanding of action potentials. Hodgkin and Huxley received the Nobel Prize for these studies.

By now there exist many voltage-clamp techniques designed for different cell types. Each requires a reliable measure of membrane current, as well as rapid potential control by electronic feedback. In addition, all parts of the membrane to be studied must be at the same potential so that the applied current passes uniformly across the membrane rather than longitudinally in any local circuits. Automatic control of the membrane potential is achieved by monitoring the membrane potential with a recording electrode and a follower circuit, electronically amplifying any deviation from the desired potential (called the *command voltage*), and applying the amplified "error signal" via a current-delivering electrode to the membrane in such a way as to decrease the error (negative feedback). The current applied is measured as the potential drop in a resistor placed somewhere in the current-delivering circuit.

Five common voltage-clamp methods are illustrated schematically in Figure 3–8. The original technique is the axial-wire voltage clamp for giant axons. A platinum-wire electrode several centimeters long (labeled I') is inserted down the axis of the axon to eliminate longitudinal potential gradients by short-circuiting the length of the cell interior. The other methods reduce longitudinal potential gradients by a study of small patches of

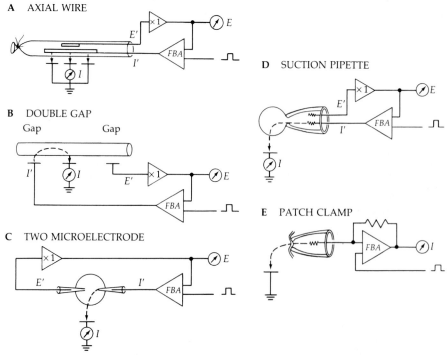

Figure 3–8 Five common voltage clamp methods. Intracellular potential is detected by the voltage electrode, E'; transmembrane current is applied intracellularly by the current electrode, I'; and the bath is grounded through a current-measuring system, I. The unity gain follower (x1) drives the voltage-indicating circuit, E, and the feedback amplifier (FBA) supplies the membrane current required to keep the membrane potential equal to the command voltage (shown as a square pulse). Today, the most popular methods are the suction pipette, applied to whole cells, and the patch clamp, applied to small areas of membrane.

membrane. The double-gap method isolates a short cylinder of membrane to be studied between two gaps of air, oil, sucrose, grease, or other insulator. The inactive ends of the fiber act as leads for recording voltage and passing current. This technique has been useful with numerous muscles, myelinated nerves, and some larger axons. The two-microelectrodes and the suction pipette methods take advantage of the inherent spatial uniformity of small round cells. They have been most useful for nerve cell bodies with their processes tied off or removed, for cultured or enzymatically dissociated cells, and for egg cells. The last method is the patch clamp. Here, a very small area (0.2 to 100 μm^2) of membrane is isolated by sucking it against a small, fire-polished pipette. This method can be sensitive enough to record currents from a single ionic channel.[12]

Separation of Currents.[16, 19, 20, 23] When Hodgkin, Huxley, and Katz began voltage-clamp experiments in the late 1940s, they believed correctly that the axon membrane underwent an orderly sequence of permeability changes, first to Na^+ ions and then to K^+ ions, during the action potential. Therefore, they hoped to measure with the voltage clamp an ionic current, I_i, that could be dissected into components I_{Na} and I_K. Figure 3–9A shows records of total ionic current flowing when

the membrane potential of a squid giant axon is abruptly stepped away from the resting value of -65 mV to the values indicated. According to the usual biophysical conventions, outward membrane current is considered positive (up); inward current, negative (down). Three qualitative conclusions can be drawn at once. First, the membrane is asymmetrical: A large hyperpolarizing voltage step to -130 mV elicits only the tiny inward ionic current (curve 1) expected from the low resting membrane permeability, whereas all depolarizations elicit larger ionic currents (curves 2 to 6), reflecting an elevated membrane permeability. Second, the membrane permeability changes with time after the start of the depolarization, since the currents change with time. Third, more than one type of ionic channel is involved, since during several of the depolarizing steps, the current has first an inward and then an outward phase (curves 2 to 4).

Ionic Substitution. The biphasic time course of ionic current in curves 2, 3, and 4 is expected from the experiments on action potentials. The inward phase should represent an early influx of Na^+ ions and the outward phase, a later efflux of K^+ ions. Hodgkin and Huxley[19] confirmed this hypothesis by studying the effect of replacing Na^+ ions in the bathing medium by the impermeant cation cho-

Figure 3–9 The voltage-clamp experiments of Hodgkin and Huxley, demonstrating the time course of membrane permeability changes to Na^+ and K^+ ions in squid giant axons. The membrane potential, E_M, was forced to follow step changes by an axial-wire voltage clamp (Fig. 3–8A), and the ionic current, I_i, was recorded (brief capacity currents have been omitted). A, Ionic current in response to a hyperpolarizing step to -130 mV and to different depolarizing steps. As traces 2 through 6 show, depolarizations elicit the rapid permeability changes (seen as time-varying ionic currents) that in an unclamped axon would generate an action potential. B, top, Separation of the Na^+ and K^+ components of the total membrane current during a step from -65 mV to -9 mV. I_i is the total current with the axon bathed in sea water. I'_i is the current seen after 90% of the external NaCl has been replaced by impermeant choline chloride. This delayed outward current is carried by K^+ ions leaving the fiber. I_{Na} is the difference $I_i - I'_i$ that represents the transient inward movements of Na^+ ions in standard sea water. Bottom, Calculation of the time courses of g_{Na} and g_K from the separated currents I_{Na} and I_K. (After Hodgkin, A.L.; Huxley, A.F. *J. Physiol. [Lond.]* 116:467–506 and 117:500–544, 1952.)

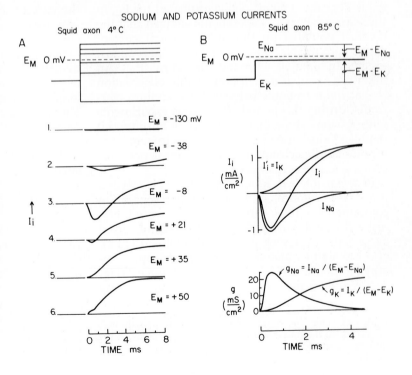

line. Figure 3–9B shows the result. Trace I_i, recorded in the control medium (seawater), shows the normal biphasic ionic current. Trace I_i', recorded after 90% of the NaCl was replaced by choline chloride, has no inward phase. Provided that choline does not itself have pharmacological effects on the membrane, the algebraic difference, $I_i - I_i'$, should represent the time course of sodium current in the Na-containing solution. This difference, labeled I_{Na}, is also drawn in Figure 3–9B. Similar ion-substitution experiments have been done with internally perfused cells. There, replacing all the internal K^+ ion with impermeant Cs^+ ion eliminates the delayed outward current I_i', confirming that I_i' is carried by K^+ ions.

Reversal Potentials.[19] A more fundamental method for identifying which ion carries a current is based on use of the Nernst equation. For example, if the membrane potential is clamped to the Na equilibrium potential, the electrochemical driving force on Na^+ ions goes to zero and I_{Na} should vanish.* At potentials below E_{Na}, I_{Na} should be inward, and at potentials above E_{Na}, it should be outward. Thus in experiments one can change the membrane potential systematically to find the *reversal potential* for a response—the potential at which the response changes from inward to outward. In Figure 3–9A, the early transient current is inward at +21 mV, absent at +35 mV, and outward at +50 mV; therefore its reversal potential is near the +40 mV value estimated for E_{Na} in these axons. Further experiments showed that the reversal potential for the early transient current varies with the Na^+ content of the medium almost exactly in parallel with the theoretical E_{Na}. An often-used verbal description of this observation is that the membrane behaves "as a sodium electrode" during this period. Analogous experiments with K^+ ions show that later, during an imposed depolarization, the membrane comes to behave as a potassium electrode, i.e., the reversal potential of the late current is near E_K.

Pharmacological Block. In the 1960s, several practical pharmacological methods for separating components of ionic current were discovered. Figure 3–10 shows the effects of two widely used agents, the pufferfish poison, tetrodotoxin (TTX), and the simple quaternary ammonium ion, tetraethylammonium (TEA). The control traces (Fig. 3–10A and C) show normal ionic currents flowing

in a node of Ranvier of a frog myelinated nerve fiber in response to eleven different depolarizing steps. This is an experiment like that in Figure 3–9A, but all the records are superimposed. As in the squid giant axon, the current has an inward sodium component and an outward potassium component for most depolarizing steps. However, after an exposure to 100 nM TTX, I_{Na} vanishes, while I_K is absolutely untouched (Fig. 3–10B). Conversely, after an exposure to 8 mM TEA, I_{Na} is untouched and I_K vanishes (Fig. 3–10D). The family of currents recorded in TEA shows very clearly the reversal of I_{Na} at E_{Na} and the outward I_{Na} at the largest depolarizations. The discovery of drugs able to block I_{Na} or I_K quite selectively had profound theoretical and practical consequences. The drugs are active in far too low a concentration to have any effect on the activity of free Na^+ or K^+ ions, and instead one has to accept the view that (i) the drugs act on relatively sparse special transport sites, the ionic channels, and (ii) the channels for different ionic components of current are separate, independent entities in the membrane.

Conductance Changes.[16, 20, 21, 22] Now that the current components I_{Na} and I_K have been separated and their reversal at E_{Na} and E_K has been determined, it is desirable to define a quantitative electrical measure of the number of Na and K channels open. Hodgkin and Huxley found that the current in open channels obeys Ohm's law if one uses the quantity $E_M - E_{reversal}$ as the electrochemical driving force. Therefore, if each open molecular pore contributes a fixed elementary conductance, γ_{Na} or γ_K, a good measure of the number of open Na and K channels would be the total ionic conductances, g_{Na} and g_K, of the membrane. These conductances can be calculated from ionic currents by the definitions $g_{Na} = I_{Na}/(E_M - E_{Na})$ and $g_K = I_K/(E_M - E_K)$. The time courses of g_{Na} and g_K for a depolarization to −9 mV are calculated in the bottom graph of Figure 3–9B. They show that Na channels open rapidly, but with a small lag (sigmoid time course), and then close again during a depolarization. In the Hodgkin-Huxley theory, the Na channels are said to "activate" and then "inactivate." At the same time, K channels open 10 times more slowly and with more lag. They activate but do not inactivate during brief imposed depolarizations.

When the calculations of g_{Na} and g_K are repeated for other potentials with currents such as those in Figure 3–9A and 3–10, the time courses are found to be steeply voltage-dependent. Any small depolarization opens some channels. The larger the

---

*Any movement of Na^+ ions that is coupled to the hydrolysis of ATP or to the movement of some other molecules with a different transmembrane electrochemical gradient does not need to go to zero at E_{Na}.

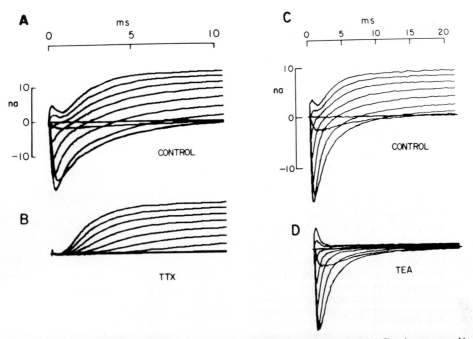

Figure 3–10 Pharmacological demonstration that Na and K channels are independent. The traces are Na^+ and K^+ ionic currents recorded from a node of Ranvier under voltage clamp (myelinated axon from a frog) at 11° to 13°C. The families of currents represent superimposed responses to 10 voltage steps spaced at 15-mV intervals from −60 to +75 mV, each starting at t = 0. *A* and *C*, Control measurements with normal I_{Na} and I_K time courses. *B*, A few minutes after the addition of 300 nM tetrodotoxin to the bathing solution. I_{Na} is totally blocked and I_K is untouched. *D*, Another fiber a few minutes after the addition of 6 mM tetraethylammonium chloride to the bathing solution. I_K is blocked and I_{Na} is normal. (*A* and *B*, Modified from Hille, B. *Nature* 210:1220–1222, 1966, by permission. Copyright © 1967, Macmillan Magazines Limited; *C* and *D*, Modified from *The Journal of General Physiology*, 1967, vol. 210, pp. 1287–1302 by copyright permission of The Rockefeller University Press.)

depolarization, the larger the fraction that activates, the faster the channels open, and, for Na channels, the faster they become inactivated. The gating steps leading to *activation* and *inactivation* evidently include very steeply voltage-dependent processes that are intensified by depolarization. The voltage dependence of these processes are drawn in Figure 3–11.

Recent experiments using the patch clamp method of recording currents from single ionic channels show that individual Na and K channels open and shut abruptly (Fig. 3–12A). Depolarization increases the probability that a steplike opening will occur at the microscopic scale, passing a current on the order of one picoampere. Only because the usual macroscopic records contain the summed contribution of thousands to millions of channels do activation and inactivation appear smooth. One can show that even the single channel has a smoothly increasing and decreasing *probability* of opening by averaging currents recorded in repeated trials on the same channel (Fig. 3–12B).

Reconstruction of the Action Potential.[13, 16, 22] Qualitatively, it was clear that the voltage-clamp

experiments confirmed the hypotheses developed to explain the electrical excitability of axons. As a definitive quantitative test of their theories, Hodgkin and Huxley sought to predict the full-time course of an action potential from voltage-clamp data alone. First, they developed kinetic equations to *approximate* the observed voltage and time dependence of g_{Na} and g_K. Because of its mathematical convenience, they chose a kinetic model equivalent to having four independent gates operating in series to control the permeability of each channel. The Na channel is assigned three rapidly moving "m-gates," whose opening upon depolarization accounts for activation, and one slowly moving "h-gate," whose closing upon depolarization accounts for inactivation. The fraction of Na channels open is proportional to m^3h, where m is the time-varying probability that any m-gate is open, and h is the time-varying probability that an h-gate is open. Similarly, K channels are assigned four slowly moving "n-gates," and the fraction of K channels open is n^4. With these definitions, rate constants for opening and closing of the hypothetical m-, h-, and n-gates can be determined at each voltage from the measured

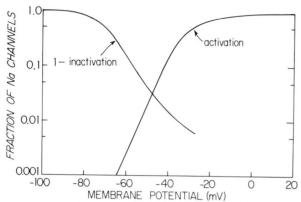

Figure 3–11 Voltage dependence of opening and inactivation of the Na channels in an axon membrane. The peak fraction of channels is plotted semilogarithmically against membrane potential and labeled "activation." The fraction of channels still available—not becoming inactivated—after a long pulse (>25 ms) is labeled "1 – inactivation." These curves describe voltage-dependent conformational transitions of the Na channel protein. Note that any small depolarization opens more Na channels and eventually also increases the number that are inactivated. Inactivated channels cannot be opened by further depolarization unless the inactivation is first removed by restoring the membrane potential to near rest or to hyperpolarized values.

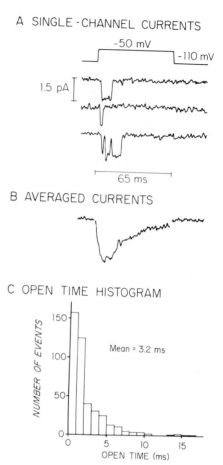

Figure 3–12 Ionic currents from single Na channels recorded by the patch-clamp method (Fig. 3–8) from embryonic rat muscle cells. *A*, Depolarizing voltage steps elicit pulselike inward current steps, each step arising from the opening and shutting of a single Na channel pore. Three successive depolarizations elicit different patterns of opening. *B*, When currents from 144 trials are averaged, the current time course looks smoother and resembles the activation-inactivation sequence seen when recording from large numbers of channels. *C*, Measurements of the duration of each channel opening in a large number of records show a nearly exponential distribution and a mean open time of 3.2 ms. The open times are long because the temperature is only 10°C. (Reproduced from Patlak, J.; Horn, R. *The Journal of General Physiology*, 1982, vol. 79, pp. 333–351 by copyright permission of The Rockefeller University Press.)

time courses of g_{Na} and g_K. For example, the *steady-state* voltage dependences of h and m^3 would be chosen to describe the curves labeled inactivation and activation in Figure 3–11, and in any step of potential, h and m would start at values appropriate to the starting potential and then relax towards the steady-state value for the new potential. Then, if the maximum conductance with all Na channels open is called $\bar{g}_{Na}$ and that with all K channels open, $\bar{g}_K$, the ionic currents in the model can be calculated from

$$I_{Na} = m^3 h \bar{g}_{na}(E_M - E_{Na})$$

and

$$I_K = n^4 \bar{g}_K (E_M - E_K)$$

Finally, to include a small, fixed background permeability through channels other than Na and K channels, a third current, called the *leak current*, is defined as follows:

$$I_L = g_L(E_M - E_L)$$

When the Hodgkin-Huxley model for I_{Na}, I_K, and I_L is combined with the cable equations for an axon having membrane capacitance and internal resistance, many of the phenomena of excitability are well reproduced. Figures 3–13 and 3–14 show some calculated action potentials. In Figure 3–13, a depolarizing "stimulus" is applied at x = 0, and the calculations give an action potential propagating at uniform velocity away from the stimulus site. Figure 3–14 shows a calculated propagating action potential together with the time courses of the underlying Na and K permeability increases. Na channels open very quickly after local-circuit currents from a preceding active patch depolarize

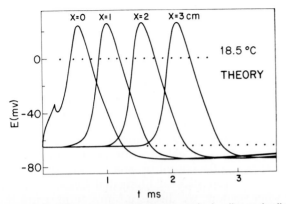

Figure 3–13 Time course of the propagated action potential calculated from the Hodgkin-Huxley model. In the calculation, a 0.2-ms current stimulus is applied at x = 0 starting at t = 0. The curves show the resulting membrane potentials at the origin and at three different distances from the stimulus. (Modified from Cooley, J.W.; Dodge, F.A., Jr. *The Biophysical Journal*, 1966, vol. 6, pp. 583–599 by copyright permission of The Rockefeller University Press.)

the membrane to near −50 mV. The membrane potential then swings rapidly toward E_{Na} as Na^+ ions rush in through open channels. But soon, as Na channels begin to inactivate more and more

and K channels activate, the efflux of K^+ ions repolarizes the membrane (and in the squid axon even hyperpolarizes the membrane toward E_K). The Hodgkin-Huxley equations also reproduce properties such as threshold and refractoriness well. Threshold is not the potential at which Na channels *begin* to open, since some (very few) Na channels are open even at rest (see Fig. 3–11), and even the smallest depolarization will open some more. Rather, threshold is the potential at which *enough* Na channels open to produce an inward current more than sufficient to counteract the repolarizing tendency of nearby open K and other channels (Cl) and of the cable connections to other unexcited regions of membrane. Refractoriness is accounted for by nearly complete inactivation of Na channels and by the excess of open K channels in the wake of an action potential. If axons are kept depolarized, for example by an elevation of $[K]_o$, Na channels inactivate (Fig. 3–11), and the axons become inexcitable. The membrane of a mammalian axon has to repolarize for at least 0.5 ms at 37°C before Na channels are brought back into the functioning pool.

The early voltage-clamp experiments with the axial-wire method were done on squid giant ax-

Figure 3–14 Opening and closing of Na and K channels during the propagated action potential, calculated from the Hodgkin-Huxley model. Because the action potential is a nondecrementing wave, the diagram shows equivalently the time course of events at one point on the axon or the spatial distribution of events at one time as the excitation propagates from right to left. The absolute calibration in terms of channels per μm² is only approximate. The lower diagram shows the local circuit current flowing during the action potential in an axon with greatly exaggerated diameter. (After Hodgkin, A.L.; Huxley, A.F. *J. Physiol. [Lond.]* 117:500–544, 1952.)

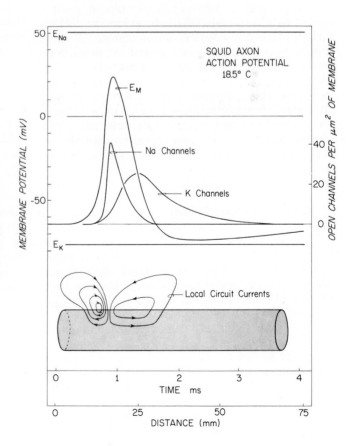

ons, but as other voltage clamp methods (Fig. 3–8) were perfected, other excitable cells could be studied. By now, numerous axons of four major animal phyla—Mollusca, Annelida, Arthropoda, and Vertebrata—have been studied.[13] They all use squid-like Na and K channels, and their propagating action potentials are successfully simulated by kinetic models differing only in detail from that for the squid giant axon. Voltage-clamp experiments in amphibian myelinated axons are shown in Figure 3–10. Myelinated axons of warm-blooded vertebrates do show a qualitative difference in having a high density of Na channels and K-permeable "leak channels" and relatively fewer voltage-sensitive K channels at their nodes of Ranvier. The rapid fall of the action potential in these fibers is due to rapid inactivation of Na channels and to strong repolarizing effects of the fixed leakage channels.

The work of Hodgkin, Huxley, and Katz was a landmark in membrane biophysics. The success of the quantitative predictions proved that local circuits, cable properties, and voltage-dependent permeability changes to Na^+ and K^+ ions are the basis of excitation in axons. It should be emphasized that the Hodgkin-Huxley equations are a kinetic *approximation* to the currents that were measured, and as with any kinetic model, even a close fit did not show that, for example, a four-gate mechanism actually exists. Indeed, further kinetic measurements done 20 and 30 years later have revealed several gating properties of Na channels that the equations do not adequately describe. They apparently do well at describing the net currents from the *population* of Na and K channels in the membrane but do not correspond closely to what is happening microscopically at the single-channel level. This is an area of continuing active investigation today.

EFFECTS OF FIBER GEOMETRY

Thus far we have emphasized experiments on the largest available axon, the squid giant axon. These 1000-μm fibers are ideal for biophysical analysis of membrane mechanisms, because they can be dissected free from other axons, probed with several intracellular electrodes, and perfused internally and externally with a wide range of test solutions. The giant axon even has Na and K channels similar to those in the membranes of vertebrate axons. Nevertheless, both in its size and in the anatomy of its Schwann cell sheath, the giant axon is quite different from axons of

vertebrate nervous systems (Fig. 3–5). Cross sections of vertebrate nerves reveal a spectrum of fiber sizes, unmyelinated axons with diameters ranging from 0.1 to 1.3 μm and myelinated fibers with outer diameters 1 to 20 μm (Fig. 3–15). Myelination and size affect the conduction of action potentials. The following sections show how these physical factors account for differences in conduction velocity, ease of stimulation, and other electrical properties of different axons. We start with the passive electrical properties, treating an axon as a collection of resistors and capacitors.

Cable Properties of Elongated Cells.[25, 30] In an earlier section, we discussed how to measure the passive membrane properties of a Paramecium using two intracellular microelectrodes, one to apply small currents across the membrane and the other to record the resulting potential changes (Fig. 3–4). In compact, nearly spherical cells, applied currents flow uniformly across all parts of the membrane, and the membrane potential changes are the same everywhere in the cell. The cell is said to be *isopotential* inside. With elongated cells like axons or muscle fibers, this is no longer true. Applied currents flow more intensely across the membrane nearest the current electrode and may never reach the most distant parts of the cell. Potential changes are largest near the current electrode and decay away with distance. The high resistance of the long, narrow column of cytoplasm between the current electrode and the distant parts of the membrane limits the intracellular spread of current. Over a century ago, L. Hermann noted the physical and mathematical analogies between the passive spread of potentials in elongated cells and the spread of telegraph signals in undersea cables, and we still use the name *cable properties* in this connection. Potentials spreading this way, without a boost from regenerative permeability changes, are said to spread *electrotonically* or *passively*.

Measurement. Two intracellular microelectrodes can be used to measure the effects of current flow on the membrane potentials of large axons or muscle fibers (Fig. 3–16A). An abrupt step of current (I_M) is passed out through one electrode, flowing into the axoplasm and out through the membrane by the paths of lowest resistance. The other electrode records changes in membrane potential (V_M) at several distances from the current electrode. Results obtained with frog sartorius muscle fibers in Figure 3–16B are typical of those for any long, thin cell. In response to a step of current near the current electrode ($x = 0$), V_M rises rapidly at first and then gradually levels off to a

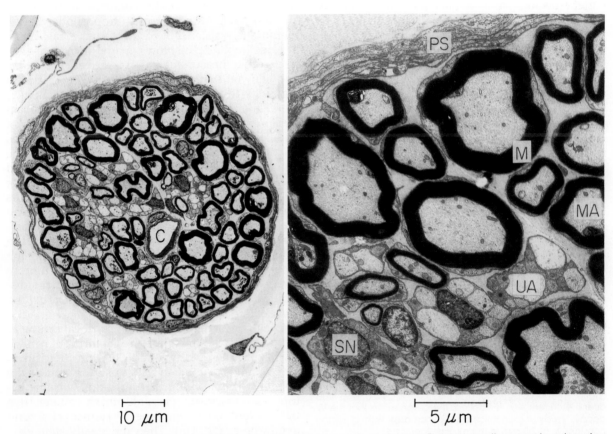

┣━━━━━━┫
10 μm

┣━━━━━━┫
5 μm

Figure 3–15 Electron micrograph of a small branch of the tibial nerve of an adult rat. The cross section reveals various sizes of myelinated axon (MA), surrounded by thick, electron-dense myelin (M), and smaller unmyelinated axons (UA) gathered in groups of 1 to 20 axons invaginated into thin Schwann cells. A Schwann cell nucleus is labeled (SN). The entire nerve bundle with 68 myelinated fibers and over 100 unmyelinated axons is surrounded by a perineurial sheath (PS) and is served by a single capillary (C). The picture on the right is from the same field but is taken with 3 times greater enlargement. (Courtesy of Dr. Margaret R. Byers, University of Washington.)

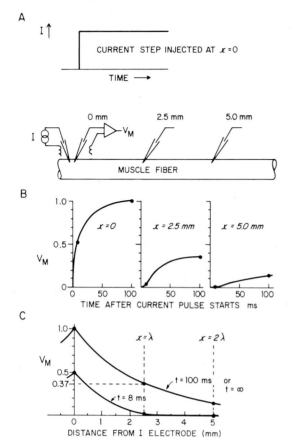

Figure 3–16 Measurement of the cable properties of a single muscle fiber. *A*, A step of current I is injected via an intracellular micropipette at the left, x = 0 mm. A second intracellular pipette is used to record membrane voltage deflections, V_M, at three distances from the point of current injection. *B*, Time course of V_M at the three recording points. Note that the perturbation of the resting membrane potential is smaller and rises more slowly as the distance from the current-delivering electrode increases. *C*, Points on the records in *B* are replotted as a function of distance to show the spatial spread of potential at 8 ms, 100 ms, and a long time after the current step is applied. The graph shows that the V_M decays away exponentially as x increases with a space constant, λ, of 2.5 mm in this muscle fiber.

fixed value. As the recording electrode is moved 2.5 or 5 mm from the current electrode in either direction, the electrotonic potential rises progressively more slowly and reaches a smaller final value. The observations are replotted in Figure 3–16C to show how voltage changes vary with distance from the current-applying electrode. After an 8-ms application of current, the voltage changes are confined to the region of the current electrode, and even in the steady-state after a long application, they spread only a few millimeters in each direction.

Theory. Qualitatively, the time course of V_M at x = 0 in a cylindrical cell (Fig. 3–16B) is similar to that in a compact round cell (Fig. 3–4), and the general explanation is the same. At first, the applied current serves as capacity current to change the membrane potential, and later, V_M rises to a steady level at which all the membrane current is ionic. However, because in a cable much current flows longitudinally down the axis of the fiber to more distant areas of membrane, the time course of V_M is not exactly an exponential response, as it would be for a parallel RC circuit, and the steady V_M at x = 0 is not linearly proportional to the membrane resistance.

Hodgkin and Rushton[25] made such two-electrode cable measurements on large crayfish axons and showed that the classical equivalent circuit in Figure 3–17 describes the results quantitatively. Here, the cytoplasm is represented as a chain of resistors, and each section of membrane as a parallel RC circuit. The equivalent circuit expresses the distributed nature of cytoplasm and membrane in a long cylinder. When current is first applied at the origin, the capacitors nearby charge up, producing a local rise of membrane potential. The potential change there forces a fraction of the current to exit as ionic current through the membrane and another fraction to flow down the cytoplasmic resistance, beginning to charge the next capacitor, and so forth. After some time, a steady-state spatial distribution of potential is established, and all the applied current is exiting somewhere as ionic current.

Cable measurements are a practical route to determining the membrane resistance of cylindrical fibers. The theory is derived from the equivalent circuit of Figure 3–17. We state some of the results without giving the derivation. Let the cytoplasmic resistance per unit length of fiber be r_i (Ω/cm) and the membrane resistance of a unit length be r_M (Ω · cm). Then the steady-state electrotonic potential distribution (Fig. 3–16C) falls off exponentially with distance from the origin $V_M(x) = V_M(0)\exp(-x/\lambda)$, where the *space constant* λ is $\sqrt{r_M/r_i}$. The space constant is the distance at which V_M falls to 37% (= 1/e) of its original value and is equal to the fiber length at which resistance of the membrane covering that fiber segment is equal to the longitudinal cytoplasmic resistance of the segment. In large muscle fibers, λ is of the order of a few millimeters; in the example in Figure 3–16, λ is 2.5 mm. The distance of spread increases if the membrane resistance is increased or the cytoplasmic resistance is decreased.

Another result of the cable theory concerns the

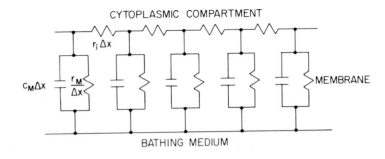

Figure 3–17 Simplified electrical equivalent circuit of the cable properties of an axon. Each element of membrane is represented as a parallel RC circuit connected inside via the axoplasm, represented as a resistor, and outside by the extracellular space, represented as a perfect conductor.

"input resistance" of a fiber. This is the effective electrical resistance encountered by current flowing from the intracellular microelectrode through cytoplasm and membrane to ground. The theory shows that input resistance of a infinite cable is $0.5 \sqrt{r_M r_i}$. Therefore, unlike the small round cell, a cable does not have an input resistance equal to or even directly proportional to the membrane resistance. Both cytoplasm and membrane contribute.

Significance. The cable properties of elongated cells are of major importance both to the proper functioning of the nervous system and in the experimental analysis of excitable systems. A first clear physiological conclusion is that if most cells have space constants for passive spread of current of less than 3 mm, then passive spread would be quite useless by itself for sending electrical messages like action potentials over hundreds of millimeters in the nervous system. On the other hand, passive spread *is* the mechanism of local spread of slow potentials, such as receptor potentials and synaptic potentials. For example, in a pacinian corpuscle, mechanical stimuli initiate current flow at the nerve terminal, and the resulting receptor potential spreads down the terminal a short distance to the regions of axonal membrane capable of initiating propagating action potentials. Similarly, in motoneurons the synaptic inputs produce currents out on dendritic processes of the cell. These currents spread mainly passively through the dendritic tree, with some active boost from voltage-gated channels, into the cell soma (cell body) and out again to the "spike initiation region" on the attached axon. The analysis of cable properties of dendritic trees and of receptors or primary sense cells is still an active research area. The propagation of action potentials also involves a continuous cable spread of the propagating signal through local circuits to depolarize unexcited membrane ahead, but then, as we have seen, a quite different chain of events takes over

to boost and reshape the advancing wave. Finally, cable spread also occurs from one cell to another in cases in which special intercellular pores form low-resistance junctions ("gap junctions"), coupling the cytoplasm of one cell to the next. In some tissues nearly all cells are electrically coupled, and the whole tissue acts as a continuous cable (see Chaps. 10 and 11).

Cable properties set constraints on experiments. Frequently, it is easier to impale a large cell body with a microelectrode than to impale its finer processes, yet if the electrical event to be studied does not actually arise in this convenient recording spot, the time course and amplitude will not be correctly measured. Thus, cable properties require one to record near the event to be studied. They also make it more difficult to determine the membrane capacity and membrane resistance, since the cell geometry has to be accurately known before calculation can begin. Fortunately, the shape of many axons and muscle fibers approximates an infinite cylinder, the prototype cable, for which measurements like those in Figure 3–16 suffice for values to be determined for C_M, R_M, and the resistance of the cytoplasm. The appropriate equations and graphs for cylindrical and other geometries are found in the original literature.[25, 27, 30]

Determinants of Conduction Velocity. Several factors determine the speed of impulse conduction in excitable fibers.

Physical Factors. (i) An increase in the number of available Na channels would increase conduction velocity. The larger the sodium current, the steeper the rate of rise of the action potential. In turn, a faster rising action potential means a steeper spatial voltage gradient along the fiber and hence higher local circuit currents, faster excitation of adjacent regions, and a higher conduction speed. Conversely, a low dose of a drug (TTX or local anesthetic) that blocks some Na channels decreases the maximum inward current, slows the rate of rise, and slows conduction. (ii) A reduction in the amount of depolarization to reach threshold would increase conduction speed, other things being equal, because the local current would not have to flow for as long to excite an inactive region. This can happen if the external divalent ion concentration is decreased (Chap 4). (iii) The membrane capacity per unit area determines the amount of charge stored on the membrane per unit area for a given voltage and hence the length of time a current must flow to depolarize to threshold. Smaller capacity, as in a myelinated fiber, means higher speed, and larger capacity, as in a muscle fiber, means lower speed. (iv) Similarly, the magnitude of local-circuit flow is limited by the resistance of the axoplasm. Other things being equal, the larger the cross-sectional area and the higher the concentration of highly mobile intracellular ions, the greater the current flow for a given voltage. Large axons conduct faster. (v) Temperature has large effects on the rate of increase of sodium conductance; Na channels open and close faster at higher temperature, and conduction speed is higher too (see Fig. 3–6C).

Myelinated Nerve and Saltatory Conduction.[16, 26, 32] Invertebrates have achieved high conduction speed by developing giant axons. However, the sheer physical size of these axons prevents an animal from having large numbers of them. Myelination allows rapid conduction with smaller fiber diameters. The myelinated nerve fiber is one of the functional developments that make possible large body size and sophisticated postural and locomotory mechanisms among vertebrates.

Roughly 99.9% of the axon surface area of a myelinated fiber is covered by the *myelin sheath* formed by Schwann cells. One Schwann cell covers a 2-mm length of axon in the largest-diameter fibers (20 μm), wrapping itself many times around the fiber in a tight spiral during development. The Schwann cell membranes become pressed closely together in a multilayered roll, with all Schwann cell cytoplasm squeezed out of them and an axon passing through the core. At periodic intervals along the axon, there is a gap in the sheath where one Schwann-cell wrapping ends and a new one starts (Fig. 3–5). Here a short stretch (0.5–2 μm) of the axon membrane is uncovered and in free communication with the interstitial fluid. This interruption in the myelin sheath is the *node of Ranvier*, and the sheathed portion is the *internode*.

Since the myelin sheath is composed of many layers of closely packed membranes, it is a good insulator. Each wrapping of the Schwann cell adds two membranes and 18 nm of thickness to the sheath. In the largest fibers, the myelin may be 3 μm thick, containing over 300 membranes in series. The electrical resistance to radial current flow through 1 cm^2 of such a sheath is 300 times the resistance of one membrane. Similarly, the electrical capacity of 1 cm^2 of such a sheath is 300 times smaller than that of one layer of membrane, and thus the amount of charge stored on the whole 2-mm internodal region is little more than on the 1-μm nodal region.

In mammalian myelinated nerve, resting and action potentials are generated by densely packed Na channels and K-permeable "leak" channels in the nodal membrane. The membrane of the axon has remarkable regional specialization. Evidently, there are few Na channels in the internodal axon membrane, but there are voltage-gated K channels.[33] What the K channels do there is not clear, since the insulating myelin covering them would keep them from making much contribution to normal excitation. The sodium currents generated at the node charge up the internodal membrane capacity, too. If one node is active and an adjacent node is inactive, there is a local-circuit flow between the nodes, depolarizing the inactive node. The length of time taken for an active region to depolarize an adjacent inactive one to threshold is determined by the amount of charge that must be removed and the resistance of the axoplasm between the regions. Since the effective membrane capacitance is greatly reduced by the myelin sheath, the conduction speed of a myelinated nerve fiber is many times higher than that of an unmyelinated fiber of the same diameter and membrane properties. Impulse propagation in myelinated nerve fibers can be summarized as a successive excitation of node after node via local circuits passing through well-insulated internodes. Propagation by this means is called *saltatory conduction* (from the Latin *saltare*, to dance). The discontinuous nature of saltatory conduction should not be emphasized too strongly, however, because in a typical mammalian myelinated nerve fiber, about 30 nodes of Ranvier are participating simultaneously in some phase of the action potential. Every 15 μs, a new node ahead becomes excited and another is left behind by the advancing wave of excitation.

The debilitating effects of human demyelinating diseases underscore the importance of myelin for proper conduction. Even incomplete demyelination of one internode suffices to block conduction, as seen in a computer simulation in Figure 3–18. Here, the equations for excitable nodes and the cable equation for internodes have been combined to simulate conduction in a myelinated nerve fiber.

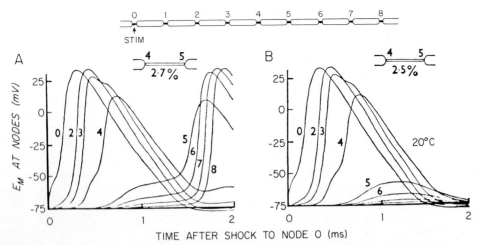

Figure 3–18 Computer simulation of action potentials propagating in a partially demyelinated nerve fiber at 20°C. The calculation uses a Hodgkin-Huxley–like model suitable for the excitable nodal membranes and cable equations appropriate for myelinated internodes. The time course of the predicted membrane potential changes is plotted for nine successive nodes of Ranvier. *A,* Internodal myelin between nodes 4 and 5 is assumed to be only 2.7% of normal thickness. *B,* The same internode, assumed to be 2.5% of normal thickness. (From Koles, Z.J.; Rasminsky; M. *J. Physiol. [Lond.]* 227:351–364, 1972.)

Calculated responses in nine successive nodes are labeled 0 to 8. Between nodes 4 and 5, the myelin is assumed to be damaged, with only 2.7% remaining in part A and 2.5% in part B. In A, the action potential proceeds almost normally between nodes 0, 1, 2, and 3. Charging up of node 4 is however slowed, since demyelination has raised the effective capacity near node 4. Then charging of node 5 is severely delayed because of the compromised cable properties of the internode between 4 and 5. The membrane potential at node 5 barely reaches threshold before the action potential at 4 dies out. Once 5 has fired, almost normal conduction resumes. In B, just a slight further worsening of the cable between 4 and 5 prevents the membrane potential at node 5 from ever reaching threshold. The impulse is blocked.

Such blocking lesions can be caused by multiple sclerosis in the central nervous system and by Guillain-Barré syndrome in the peripheral nervous system. Fortunately, a Guillain-Barré patient may recover fully after some months when Schwann cells remyelinate the damaged internodes. Multiple sclerosis, however, is a downward spiral, since the oligodendrocytes, as central myelin-forming glia are called, seem unable to perform repair. Nevertheless, between crises of severe debilitation, the patient has periods of improvement. Conduction seems to be partially restored at this time by the insertion of new Na channels into the newly uncovered internodal membranes, permitting impulses to travel slowly, as in an unmyelinated axon, through the lesion.[6, 33]

Elevations of body temperature, as in fever or exercise, exacerbate the symptoms of demyelination. The problem is explained by the temperature dependence of the action potential duration (Fig. 3–6C). A 2°C rise of temperature speeds gating and shortens the action potential by 20%. When conduction is already just marginal (Fig. 3–17A), a shortening of the action potential will shorten the time that local-circuit currents would be available to attempt to bring a "node 5" to its firing threshold. The extra exacerbation is immediately relieved in patients by cooling. In principle, marginal conduction can also be restored by prolonging action potentials with K-channel blockers or agents that slow inactivation of Na channels. It could also be restored by treatments that reduce the firing threshold such as lowered $[Ca^{2+}]_o$. However, unlike cooling, these are not now practical medical treatments.

Scaling: Velocity vs. Diameter. How does the velocity of conduction vary with fiber diameter? For myelinated fibers, the relationship is nearly linear (Fig. 3–19). A simple rule is that every micrometer of outer diameter adds 6 m/s to the conduction velocity at 37°C. At the smallest fiber sizes, the velocity predicted this way tends to be a little high.[29] Vertebrate unmyelinated axons are bundled together in a manner that prevents convenient dissection (Figs. 3–5 and 3–15), and to measure their diameter requires electron microscopy. The larger ones (~1.2 μm diameter) have conduction velocities of 1.8 to 2.2 m/s. The velocity is believed to vary with a power between 0.5 and

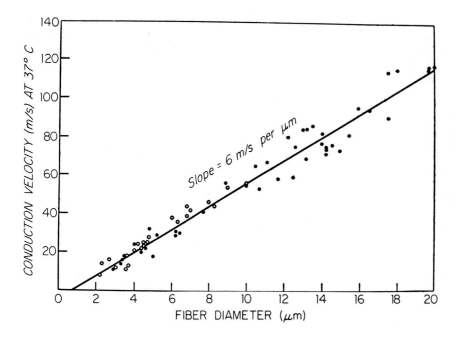

Figure 3–19 Linear relation between conduction velocity and outer diameter of mammalian myelinated nerve fibers at 37°C. Slope of line is approximately 6 m/s per micrometer of diameter. (After Hursh, J.B. after Gasser, H.S. *Ohio J. Sci.* 41:145–159, 1941.)

1.0 of the axon diameter. Powers between 0.57 and 0.8 have been found for various invertebrate unmyelinated axons.

Because myelinated and unmyelinated fibers have different velocity-diameter relations, we cannot express the velocity gain due to myelination as a single scaling factor. A squid giant axon has a conduction speed of about 20 m/s at 20°C for a diameter of 500 μm. On the basis of the square-root rule, a squid axon 20 μm in diameter would have a speed of about 4 m/s. In a frog, a myelinated fiber this size has a speed of 40 m/s, 10 times higher than that of the unmyelinated fiber. Perhaps a better comparison is that a 10-μm frog fiber has the same conduction speed as a 500-μm squid axon and that nearly 2500 10-μm fibers can be packed into the same volume as the giant axon. It is this characteristic—fast conduction in small axons—that makes possible the fast, precise control of muscle contraction necessary in warm-blooded vertebrates for maintaining an erect posture. A typical mammalian muscle nerve is about 1 mm in diameter and contains about 1000 large fibers 10 to 20 μm in diameter. If a similar nerve were composed of unmyelinated nerve fibers having the same conduction speeds, it would be about 38 mm in diameter.

The nearly linear velocity-diameter relation of myelinated fibers is accounted for by their anatomical scaling properties. Rushton[31] recognized the required idealized properties, all of which obtain only approximately in real fibers: When the diameter of the idealized fiber is doubled, the myelin thickness and internodal length should double, while the nodal gap (gap between adjacent Schwann cells) should be invariant. In addition, the nature of the axoplasmic solution and the nature and number per unit area of channels at the node have to remain invariant. Given these conditions, the area, conductance, and capacitance of the nodal membrane would double, and because of the simultaneous increase in myelin thickness, its total capacitance and conductance would only double; the longitudinal conductance of the cylinder of axoplasm within one internode would also double; and the cable space constant would double. The action potential duration and the time to jump from node to node would not change, because the scaling properties keep all time constants invariant and provide exactly twice the nodal membrane area to pass the current needed to charge twice the capacitance in the following internode and node through a local circuit path with twice the conductance. Hence, if two idealized myelinated fibers of different diameter were stimulated simultaneously, exactly the same action potential would be recorded from each of them with intracellular electrodes placed the *same number of nodes* away from the site of stimulation. Therefore, velocity varies linearly with diameter. A similar equivalency applies to passive spread in myelinated nerve. A steady, small signal applied at node 0 would be attenuated to 50% at node 1, to 25% at node 2, and so forth, independent of the size of the fiber.

The geometry of myelinated nerve fibers has been perfected by evolution to give the highest conduction velocity. The ratio of myelin thickness to fiber diameter and the ratio of internodal length to diameter both have optimal values.[26, 31] They would be kept optimal by the idealized scaling properties. Action potentials of the smallest myelinated fibers are actually somewhat longer than those of large fibers.[29] Therefore, real fibers do not scale entirely ideally. For example, the nodal gap might be wider and Na channels less dense at the node in the smallest fibers.

Analogous scaling rules can be derived for unmyelinated axons. A clear difference is that the capacitance per unit length increases with diameter in unmyelinated fibers. Therefore, velocity increases as a power less than 1.0 of the diameter. Hodgkin[16] showed that velocity would increase as the square root of diameter if as the axon diameter increased, the specific membrane and axoplasmic properties remained invariant. The same assumptions lead to a space constant that depends on the square root of diameter. Hence, in such idealized fibers passive and active distances scale with the 0.5 power of diameter. Since the actual dependence is somewhat steeper than this square-root relation, smaller unmyelinated axons probably have fewer Na channels per unit area than large ones.

Extracellular Stimulation and Recording. In biophysical work, cells are usually stimulated with an intracellular electrode, and the cell can be excited when that electrode deposits a positive charge inside, passing an outward current across the membrane to the external bath electrode. On the other hand, in most neurophysiological studies and in a neurological test, it may be necessary to stimulate cells with a pair of extracellular electrodes, e.g., two metal wires touching a nerve trunk. Again, current flows from the positive to the negative electrode (from anode to cathode), most of it passing directly through the external medium and only a small fraction taking the higher resistance pathway into the cell near the anode and out again near the cathode. This transcellular current hyperpolarizes the membrane near the anode and depolarizes it near the cathode. Hence, with external stimulation, the action potential arises at the cathode and the stimulus currents required to reach threshold are much larger than with intracellular stimulation. The efficiency of extracellular stimulation is improved if the extracellular resistance between the electrodes is raised by lifting the electrodes and nerve into air or mineral oil to force a larger fraction of the stimulus current to pass into the cells across the excitable surface membrane.

With the idealized scaling assumptions discussed in the previous section, the membrane depolarization required to bring an axon to its firing threshold would not depend on the fiber diameter, yet the *extracellular* stimulus would. Again, we can think of equivalent points along the fiber. The ability of a given shock to stimulate a myelinated fiber depends on the extracellular field it induces expressed in units of volts per node rather than volts per centimeter. Therefore smaller fibers, with more closely spaced nodes, require a larger applied voltage per centimeter to depolarize their membranes to threshold. In electrically stimulating a bundle of fibers in a nerve, it is straightforward to stimulate only the largest fibers with a modest shock amplitude. By raising the shock strength progressively, one can recruit fibers of smaller and smaller diameter. It is much harder to stimulate the small fibers selectively. To repeat, this convenient difference in "threshold" pertains only to the artificial extracellular shocks used by neurophysiologists, and does not indicate any inherent difference in excitability of the membrane of these cells.

In most neurophysiological research, action potentials are monitored with extracellular recording electrodes. The signals recorded in this way are not the membrane potential changes themselves but the voltage drops set up in the extracellular medium by the local-circuit currents of propagation. When the extracellular space is large in comparison with the cross section of the active tissue, the minute local circuit currents of one axon can make only a tiny signal of some tens of microvolts in the low resistance of the extracellular space (consider Ohm's law). Therefore, when possible, neurophysiologists often lift a nerve into oil or air at the recording electrodes to restrict the extracellular space. This gives a higher external resistance and hence a larger recorded extracellular spike.

The extracellularly recorded action potential frequently does not look like the intracellular one. If it is recorded with a point electrode in a large conducting volume such as the brain, the action potential of an axon looks triphasic—a brief positivity, a larger negativity, and another brief positivity. If it is recorded with two side-by-side electrodes on a nerve in oil or air, it looks diphasic. As the action potential sweeps by the electrodes, first one goes negative then the other. The diphasic action potential turns into a larger monophasic one if conduction of the impulse is interrupted between the recording electrodes, for example, by crushing the nerve with a pair of forceps. This mode of recording gives the largest and clearest picture of the action potential and is the preferred method when interruption of conduction does not interfere with the experiment.

In this discussion, we have ignored a basic fact of extracellular recording, namely that it usually involves a multifiber preparation. If a whole nerve*

---

*A nerve is the anatomical name for a natural bundle of nerve fibers, as in the sciatic nerve. In careful usage, one should differentiate the three terms nerve, nerve fiber, and axon.

is stimulated by a large shock, action potentials start simultaneously in each axon, propagating slowly with small local-circuit currents in small axons and rapidly with larger currents in large axons. The action potentials would arrive at a distant point with a dispersion, the fast ones first and the slow ones later.

A monophasic recording made in such an experiment on a cutaneous nerve (Fig. 3–20A) yields a *compound* action potential—compound because it is compounded of the responses of many axons. As J. Erlanger and H.S. Gasser first showed 50 years ago, the compound action potential contains several "elevations" that may be understood in terms of the diameter spectrum of the nerve (Fig. 3–20B). The elevation, labeled α*, comes from larger myelinated fibers with diameters from 6 to 14 μm in the spectrum, and δ comes from those with diameters from 1 to 5 μm. The α elevation clearly is conducted more rapidly and may be elicited in isolation by turning down the shock intensity until it is below the threshold for small fibers and above the threshold for large. Another elevation (not shown), from the tiny unmyelinated C fibers, can be elicited by large shocks. At the 34 mm conduction distance used here, the C elevation would appear about 30 ms after the shock, well beyond the end of our picture.

CONCLUSION

The propagation of action potentials requires that each excited region of a nerve fiber bring the next region into action in a cycle that progresses at high speed along the axon. Excitation involves opening of voltage-gated pores, first to Na^+ ions and then to K^+ ions. When a small depolarization opens enough Na channels, Na^+ ions enter the cell and depolarize the membrane further, starting the regenerative growth of the spike. As the depolarization eventually inactivates Na channels and opens more K channels, the membrane repolarizes, and the spike is terminated.

The gating properties of Na and K channels are best studied by the voltage-clamp method. With such experiments, Hodgkin and Huxley explained the firing threshold, shape, conduction, and refractoriness of action potentials. When these ideas

*Erlanger, Gasser, and their contemporaries classified axons in different nerves according to arbitrary size classes called Aα, Aδ, B, C, or I, II, III, and so forth.

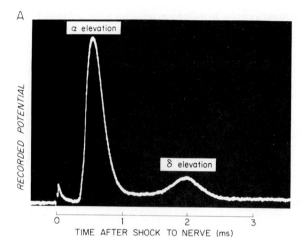

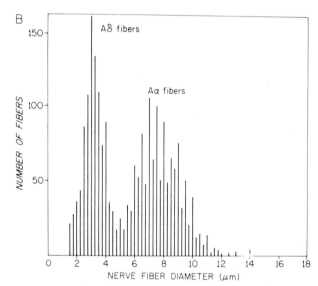

Figure 3–20 Action potential conduction in the myelinated fibers of cat saphenous nerve. A, Compound action potential recorded extracellularly 34 mm from site of stimulation. T = 37°C. (Courtesy of Dr. H.S. Gasser.) B, Distribution of fiber diameters determined from a stained histological specimen. (After Gasser, H.S.; Grundfest, H. *Am. J. Physiol.* 127:393–414, 1939.)

are combined with the cable properties for axons of different diameters, we can explain how conduction velocity depends on fiber size. Myelination is a significant evolutionary adaptation to speed up conduction in the larger axons of vertebrates.

This classic description of action potential propagation opened the door for detailed investigation of ionic channels, a subject that is continued in the following chapter.

ANNOTATED BIBLIOGRAPHY

1. Hodgkin, A.L.; Huxley, A.F. A quantitative description of membrane current and its application to conduction and excitation in nerve. *J. Physiol. (Lond.)* 117:500–544, 1952.
   The final classic synthesis of the HH model.
2. Hodgkin, A.L. *The Conduction of the Nervous Impulse.* Springfield, Ill., Charles C Thomas, 1964
   A short, readable account from the master.
3. Hille, B. Ionic basis of resting and action potentials. *Handbk. Physiol.*, Sec. 1, 1:99–136, 1977.
   A more complete description of the HH model and other biophysical aspects of action potentials than in the present chapter.

REFERENCES

1. Almers, W. Gating currents and charge movements in excitable membranes. *Rev. Physiol. Biochem. Pharmacol.* 82:96–190, 1979.
2. Asano, T.; Hurlbut, W.P. Effects of potassium, sodium, and azide on the ionic movements that accompany activity in frog nerves. *J. Gen. Physiol.* 41:1187–1203, 1958.
3. Baker, P.F.; Hodgkin, A.L.; Shaw, T.I. Replacement of the axoplasm of giant nerve fibres with artificial solutions. *J. Physiol. (Lond.)* 164:330–354, 1962.
4. Bernstein, J. Untersuchungen zur Thermodynamik der bioelektrischen Strom. Erster Theil. *Arch. ges. Physiol.* 92:521–562, 1902.
5. Bernstein, J. *Elektrobiologie.* Braunschweig, Vieweg, 1912.
6. Bostock, H.; Sears, T.A. The internodal axon membrane: electrical excitability and continuous conduction in segmental demyelination. *J. Physiol. (Lond.)* 280:273–301, 1978.
7. Cohen, L.B.; De Weer, P. Structural and metabolic processes directly related to action potential propagation. *Handbk. Physiol.* Sec. 1, 1:137–159, 1977.
8. Cole, K.S. Membranes, ions, and impulses: a chapter of classical biophysics. Berkeley, University of California Press, 1968.
9. Cole, K.S., and Curtis, H.J. Electric impedance of the squid giant axon during activity. *J. Gen. Physiol.* 22:649–670, 1939.
10. Goldman, D.E. Potential, impedance, and rectification in membranes. *J. Gen. Physiol.* 27:37–60, 1943.
11. Grinvald, A. Real-time optical mapping of neuronal activity from single growth cones to the intact mammalian brain. *Annu. Rev. Neurosci.* 8:263–305, 1985.
12. Hamill, O.P.; Marty, A.; Neher, E.; Sackman, B.; Sigworth, F.J. Improved patch-clamp techniques for high-resolution current recording from cells and cell-free membrane patches. *Pflugers Arch.* 391:85–100, 1981.
13. Hille, B. Ionic basis of resting and action potentials. *Handbk. Physiol.* Sec. 1, 1:99–136, 1977.
14. Hodgkin, A.L. Evidence for electrical transmission in nerve. Part I. *J. Physiol. (Lond.)* 90:183–210, 1937.
15. Hodgkin, A.L. The ionic basis of electrical activity in nerve and muscle. *Biol. Rev.* 26:339–409, 1951.
16. Hodgkin, A.L. A note on conduction velocity. *J. Physiol. (Lond.)* 125:221–224, 1954.
17. Hodgkin, A.L. *The Conduction of the Nervous Impulse.* Springfield, Ill., Charles C Thomas, 1964.
18. Hodgkin, A.L.; Huxley, A.F. Action potentials recorded from inside a nerve fibre. *Nature* 144:710–711, 1939.
19. Hodgkin, A.L.; Huxley, A.F. Currents carried by sodium and potassium ions through the membrane of the giant axon of *Loligo. J. Physiol. (Lond.)* 116:449–472, 1952.
20. Hodgkin, A.L.; Huxley, A.F. The components of membrane conductance in the giant axon of *Loligo. J. Physiol. (Lond.)* 116:473–496, 1952.
21. Hodgkin, A.L.; Huxley, A.F. The dual effect of membrane potential on sodium conductance in the giant axon of *Loligo. J. Physiol. (Lond.)* 116:497–506, 1952.
22. Hodgkin, A.L.; Huxley, A.F. A quantitative description of membrane current and its application to conduction and excitation in nerve. *J. Physiol. (Lond.)* 117:500–544, 1952.
23. Hodgkin, A.L.; Huxley, A.F.; Katz, B. Measurement of current-voltage relations in the membrane of the giant axon of *Loligo. J. Physiol. (Lond.)* 116:424–448, 1952.
24. Hodgkin, A.L.; Katz, B. The effect of sodium ions on the electrical activity of the giant axon of the squid. *J. Physiol. (Lond.)* 108:37–77, 1949.
25. Hodgkin, A.L.; Rushton, W.A.H. The electrical constants of a crustacean nerve fibre. *Proc. R. Soc. Lond. [Biol]* 133:444–479, 1946.
26. Huxley, A.F.; Stämpfli, R. Evidence for saltatory conduction in peripheral myelinated nerve fibers. *J. Physiol. (Lond.)* 108:315–339, 1949.
27. Jack, J.J.B.; Noble, D.; Tsien, R.W. *Electric Current Flow in Excitable Cells.* Oxford, Clarendon Press, 1983.
28. Keynes, R.D. The ionic movements during nervous activity. *J. Physiol. (Lond.)* 114:119–150, 1951.
29. Paintal, A.S. Conduction properties of normal peripheral mammalian axons. In Waxman, S.G., ed. *Physiology and Pathology of Axons.* New York, Raven Press, 1978.
30. Rall, W. Core conductor theory and cable properties of neurons. *Handbk. Physiol.* Sec. 1, 1:39–97, 1977.
31. Rushton, W.A.H. A theory of the effects of fiber size in medullated nerve. *J. Physiol. (Lond.)* 115:101–122, 1951.
32. Tasaki, I. Conduction of the nerve impulse. *Handbk. Physiol.* Sec. 1, 1:75–121, 1959.
33. Waxman, S.G.; Ritchie, J.M. Organization of ion channels in the myelinated nerve fiber. *Science* 228:1502–1507, 1984.

Voltage-Gated Channels and Electrical Excitability

INTRODUCTION

Since the mid 1960s, there has been a renaissance of interest in excitable membranes, a search for molecular understanding of ionic permeability. The concept of ionic channels as discrete, ion-selective, molecular pores has become universally accepted. The pores have been counted, their single-channel conductance has been measured, new types have been discovered and, in a few instances, a molecular purification has been accomplished. The genetics, developmental biology, adaptation, comparative physiology, pathology, pharmacology, and chemical modifications of ionic channels have been studied. These studies have led to the viewpoint illustrated in Figure 4–1. Ionic

channels such as the Na channels are thought to be integral membrane proteins lying in the fluid, lipid bilayer of the membrane and forming an aqueous pore lined by polar groups and charged particles. At the *selectivity filter*, the pore narrows to dimensions comparable to those of permeant ions so that it can make the interactions needed for selecting among permeant ions. Gated channels are macromolecules with different conformations—open and closed—but not with a continuum of partially open states. In each type of channel, the conformational changes of gating are driven by specific forces. These may be electrical forces acting on a charged or dipolar *voltage sensor* of the channel, chemical forces of the binding of external neurotransmitter molecules and internal second messengers, or forces exerted by sensory inputs.

Voltage-gated ionic channels occur nearly ubiquitously in animal cells. Their importance in the nervous system has led to a focus on that role, but their significance is far broader. Channels as we know them probably evolved when the eukaryotic style of cellular organization first appeared.[24] With the introduction at this time of calmodulin and other calcium-regulated proteins, ionic channels developed as a system for controlling the flow of Ca^{2+} ions into the cytoplasm. They were later exploited in the evolution of nervous systems, but also are well known in protozoa, algae, plants, and many nonnervous cells of the animal body. To this day, membrane potential changes must be translated into changes of cytoplasmic free-calcium concentration before they produce an interesting output. Thus, while electricity is a ubiquitous and easily measured aspect of nervous signaling, it is not biologically meaningful except as a regulator of calcium fluxes.

Several types of voltage-dependent K and Ca channels are common to protozoa, algae, and

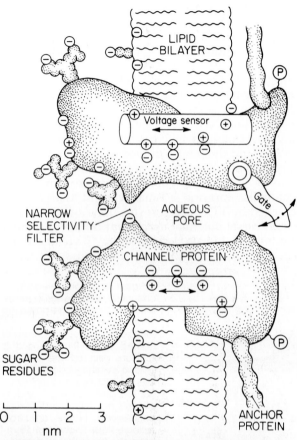

Figure 4–1 Working hypothesis for a channel. The channel is drawn as a transmembrane macromolecule with a hole through the center. The external surface of the molecule is glycosylated. The functional regions, selectivity filter, gate, and voltage sensor are deduced from voltage-clamp experiments but are only beginning to be charted by structural studies. We have yet to learn how they actually look. (From Hille, B. *Ionic Channels of Excitable Membranes.* Sinauer Associates, Inc., 1984.)

animals. These are channels that we presume may be essential for all eukaryotic cells. On the other hand, Na channels are not found until a nervous system evolves in multicellular animals. They appear abruptly in phylogeny with the Cnidaria (coelenterates: jellyfish, corals, and so forth), simultaneously with the first axon. An overriding similarity of voltage-dependent Na, K, and Ca channels suggests that they originated by evolutionary radiation from an early voltage-dependent cation channel. Chloride-permeable channels are also widely present in eukaryotic cells. Their function and origins are not well known.

After having learned of the elegant 2-channel mechanism for conducting action potentials in axons, one might wonder why Nature has use for more than two kinds of channels. The answer is

that the axons have an exceptionally simple task in comparison with nerve cell bodies, nerve terminals, or muscle. They need only propagate impulses rapidly without dropping any in the process. The other excitable membranes need to make decisions, weighing inputs, generating rhythms, or regulating nonelectrical events. These types of membranes typically use five or more channels to do their jobs.

This chapter concerns functional and molecular properties of voltage-gated ionic channels. We consider what is known mechanistically about flux, selectivity and gating at the molecular level and ask how a variety of excitable cells use channels. The area has been reviewed in book form.[24]

Na CHANNELS

Voltage-gated Na channels of excitable tissues are recognized by their (i) preference for Na^+ ions over K^+ ions, (ii) kinetics of rapid activation and inactivation, and (iii) sensitivity to a battery of neurotoxins, particularly tetrodotoxin. In different tissues and even in the same cell membrane, one can find voltage-gated Na channels that differ in detail in their gating kinetics and pharmacological profile. However, we know neither the molecular nature nor the adaptive advantages of the differences, so physiologists still tend to think loosely about *the* Na channel as if there were only one kind. We know more about Na channels than about any other voltage-gated channel.

Pore and Selectivity. What is the evidence that ionic channels are actually pores? To see a pore of atomic dimensions will probably require x-ray crystallography, and such work with channel macromolecules is only beginning. Therefore, the arguments for pores are still based on what channels can do. They include the high flux developed in a single channel, a permeability to ions up to a fixed size, an impermeability to ions above a fixed size, and blocking by drugs that act as if they are entering and clogging a narrow tunnel in the membrane. Consider first the argument based on single-channel fluxes.

Today the single-channel flux has been measured for many channels by the patch-clamp method.[21, 24] For example, 1.4 pA single-channel currents i_{Na} in Na channels are seen in Figure 3–11. When the channel is conducting, Na^+ ions are moving at a rate $i_{Na}N/F = i_{Na}/e = 1.4 \times 10^{-12}/1.6 \times 10^{-19} = 8.8$ million ions per second (see list of physical constants, Table 1–3). With a larger electrical driving force and higher $[Na]_o$, the rate can easily be brought up to 20 × 10⁶ ions/s. In some potassium channels, fluxes i_K as high as 25 pA are seen, corresponding to a throughput of 150 × 10⁶ ions/s. Evidently, ions can pass through a channel in less than 10 ns (10^{-8} s). The argument for a molecular pore is based on logical exclusion, namely that no known mechanism other than an aqueous pore could account for such prodigious fluxes. For comparison, the two fastest known enzymes, catalase and carbonic anhydrase, have turnover numbers of 5 × 10⁶ H_2O_2 molecules/s and 1.4 × 10⁶ CO_2 molecules/s, respectively, and the Na^+-K^+ pump can move at most 300 Na^+ ions/s. On the other hand, known model pores of molecular dimensions, such as the gramicidin A pore, have turnover numbers of the right magnitude. Flux in channels also has a relatively low

temperature coefficient, a Q_{10} of 1.5, like that for aqueous diffusion and unlike the typical Q_{10} of 3 for enzymes and pumps.

The minimum pore size of a channel can be gauged by the size of the ions that will pass through it. Thus the ionic selectivity of Na channels may be studied by replacing all of the external Na^+ ions by some other test ion. In such experiments, Li^+, NH_4^+, H^+, guanidinium, and a few organic cations will cross the Na channel (Fig. 4-2A), and the minimum pore size is a slit slightly larger than 0.3 × 0.5 nm (Fig. 4-2B).[23, 24] Even K^+ ions (r = 0.133 nm) are slightly permeant in Na channels, carrying a small current that can be blocked by TTX. When several ions are permeant, the reversal potential for current can be calculated from the Goldman-Hodgkin-Katz[19, 27] voltage equa-

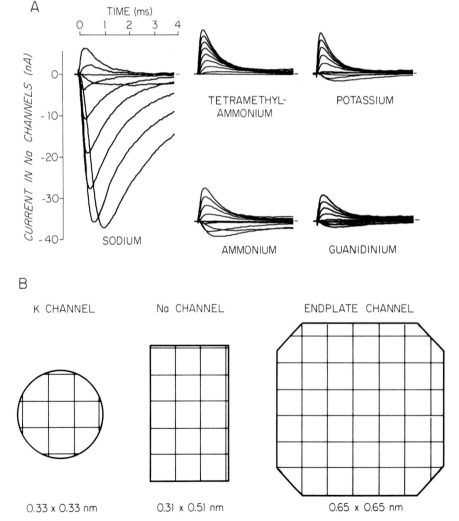

Figure 4-2 Ionic selectivity of channels. *A,* Permeability measurements on voltage-gated Na channels. Voltage-clamp currents of frog node of Ranvier bathed in Na Ringer's and in solutions with all the Na^+ replaced by other cations. K channels are blocked by 6 mM external TEA, and leak and capacity currents are subtracted. The superimposed traces are from voltage steps spaced at 15 mV intervals. The reversal potential for current in Na channels falls in the sequence Na^+ > NH_4 > guanidinium > K^+ > TMA. Only TMA gives no inward current. The outward currents are carried by K^+ ions moving out through Na channels. T = 5°C. *B,* Proposed dimensions of ionic selectivity filters in voltage-gated K and Na channels and in the acetylcholine-gated channel of the neuromuscular junction. Outline of minimum pore size that will pass the known permeant ions in frog nerve and muscle. Grid marks in 0.1-nm (1-Å) steps. Sizes were evaluated from space-filling models of the permeant and impermeant ions. (*A,* Modified from Hille, B. *The Journal of General Physiology,* 1971, vol. 58, pp. 599–619; 1972, vol. 59, pp. 637–658, by copyright permission of the Rockefeller University Press. *B,* Reproduced from Dwyer, T.M.; Adams, D.J.; Hille, B. *The Journal of General Physiology,* 1980, vol. 75, pp. 469–492 by copyright permission of The Rockefeller University Press.)

tion (Equation 10, Chap. 1), which for Na^+ and K^+ is

$$E_r = \frac{RT}{F} \ln \frac{P_{Na}[Na]_o + P_K[K]_o}{P_{Na}[Na]_i + P_K[K]_i}$$

In the Na channel, the permeabiity ratio P_K/P_{Na} is about 1:13, and the reversal potential in physiological conditions therefore is a few millivolts more negative than E_{Na}, the Nernst potential for Na^+. Larger ions, including methylammonium and all other methylated amines, are not permeant in Na channels. They would be excluded by a 0.3×0.5 nm minimum pore size. Calcium ions, whose crystal diameter is close to that of Na^+ ions, also pass through Na channels, but the permeability relative to Na is very low, $P_{Ca}/P_{Na} = 0.1$ to 0.01.

Ionic selectivity cannot be explained on the basis of geometry and steric interactions alone. When an ion enters the narrow part of the pore, some water molecules are shed from the ion and some dipolar or charged groups (probably carbonyl and carboxyl groups) from the channel touch the ion instead. In effect, the pore wall substitutes for some waters of hydration. The less the energy needed to do this, the more often it will happen. Thus the energetics of the ionic interaction with water and channel govern the selectivity of the channel. According to the ionic selectivity theory of Eisenman,[15, 24] the Na channel ought to have a negatively charged site of small radius (an ionized acid group) in the selectivity filter to permit Na^+ ions to pass more easily than K^+ ions. This would lower the energy of a partly naked Na^+ ion (itself of small crystal radius) more than it would lower the energy of a partly naked K^+ ion (of larger crystal radius). The energies of such interactions are, however, so sensitive to the details of the

geometry that any theoretical calculations have to wait for a high-resolution structure of the channel. The need for an ionized acidic group is borne out by the finding that permeability to Na becomes blocked at low pH; it titrates away with an apparent pK_a of 5 to 5.4 as the extracellular pH is lowered.

Pharmacology of Na Channels.[11, 13] The Na channel plays such a central role in the activities of animals that it has been selected numerous times during evolution as the target for neurotoxins of animals and plant origin. For scientists, the two most useful toxins have been tetrodotoxin (TTX), a water-soluble molecule of 319 daltons found in the tissues of some pufferfish and salamanders, and saxitoxin (STX), a water-soluble molecule of 299 daltons synthesized by dinoflagellates of the genus *Gonyaulax*, and accumulated in the tissues of filter-feeding shellfish (mussels and clams) during "red tides." Both toxins have paralytic, lethal effects and act by blocking Na channels reversibly, with a typical dissociation constant of 1 to 3 nM (Fig. 3–10). They compete for the same binding site on the outer face of the membrane (see label "TTX" in Fig. 4–3).

TTX and STX molecules have been made radioactive and are often used as chemical markers for Na channels.[11, 24, 36] By these criteria there are 35 to 550 Na channels per μm^2 of surface membrane in adult mammalian, molluscan, and arthropod axons and in amphibian and mammalian skeletal muscle. While Na channels usually are half blocked by a few nanomolar TTX, some types require as much as several micromolar to be blocked. Such "low-affinity" binding is common for Na channels of heart and of embryonic or denervated mammalian skeletal muscle. Whether the loss of affinity is due to expression of a

Figure 4–3 Schematic portrayal of the functional elements of a Na channel *(A)* and of the associated drug-receptor sites *(B)*. The drawings are fanciful, since the three-dimensional structure of the Na channel is only partially known. The drugs shown are tetrodotoxin (TTX), scorpion toxin (ScTx), batrachotoxin (BTX), local anesthetic (LA), and calcium ion (Ca^{2+}). (Modified from Hille, B. *The Journal of General Physiology*, 1978, vol. 22, pp. 283–306 by copyright permission of The Rockefeller University Press.)

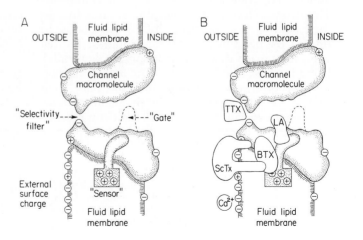

different Na channel gene or to only a minor change in phosphorylation, glycosylation, or other processing of the channel is not known. Animals making TTX (pufferfish, salamanders) have completely TTX-resistant Na channels. Numerous shellfish that must survive annual encounters with red tide (STX) have also evolved resistance. In many cases, resistance to TTX goes hand in hand with resistance to STX, but not always. These animals have not stopped using Na channels. Rather, their channels have lost the pattern of surface groups needed to bind the toxins.

Several classes of neurotoxins modify the voltage-dependent gating of Na channels, making them stay open too long. As a consequence, repetitive or exceptionally long action potentials are developed in nerves and muscles, and the result can be stinging, burning pain, cardiac arrhythmia, or "excitatory paralysis." One such class of toxins are the polypeptide venoms of scorpion stings and sea anemone nematocysts. They act on two sites that face the external medium (Fig. 4–3). The most common type prolongs the channel open time by vastly slowing the Na inactivation process. The peptide toxins iodinated with ^{125}I have been useful for counting Na channels and labeling their protein subunits. Another class of toxins includes a range of highly lipid-soluble, polycyclic molecules. This class includes the Colombian arrowfrog poison, batrachotoxin, and several plant toxins such as veratridine, aconitine, grayanotoxins, pyrethroid insecticides, and DDT. They probably bind to Na channels from the lipid phase, causing activation gating to require less depolarization than usual and nearly eliminating inactivation gating so that Na currents last much longer than usual.

Local anesthetic agents, lidocaine and procaine, are used clinically to block conducted action potentials in sensory and motor nerves. As might be anticipated, they act by blocking Na channels.* The drug molecules are ionizable amines that in their neutral form are sufficiently hydrophobic to diffuse readily through cell membranes. In this way, they reach their site of action on axons after being injected into tissue a few millimeters away. The receptor seems to be within the pore itself (Fig. 4–3), and experiments with membrane-impermeant derivatives show that the drug reaches its receptors from the axoplasmic, not from the extracellular, side of the channel.

*Local anesthetics should not be confused with general anesthetics (ether, halothane, etc.) that act on the brain to induce unconsciousness. The mode of action of general anesthetics and the seat of consciousness are not known.

The drug-receptor reaction has interesting kinetic complexities that explain a second therapeutic use of local anesthetics. When applied at subanesthetic concentrations, the drug blocks only a fraction of the Na channels. However, surprisingly, the number of channels blocked can be transiently increased by passage of one impulse or by application of a depolarizing voltage-clamp pulse, *use-dependent* block. The extra block accumulates to even higher levels if repetitive depolarizations are given at frequencies above 1 Hz, and it dies away if the axon is rested for several seconds. Use-dependent block makes local anesthetics and related molecules effective in the treatment of some cardiac arrhythmias: A potentially fatal condition exists if the ventricular half of the cardiac pump contracts more than once in each cardiac cycle, a sign that extra action potentials are being initiated in the ventricle. With very low doses of local anesthetics, an extra use-dependent block of Na channels persists for a fraction of a second after each normal action potential in the ventricular muscle and prevents an extra action potential from arising. The use-dependent block decays away, however, before the next normal beat and thus does not disturb normal function.

The mechanism of use-dependent block has been studied extensively by voltage-clamp methods. In brief, binding of the drug depends on the conformational state of the Na channel protein. The access and equilibrium for drug binding to the receptor are found to be strongly influenced by whether Na channels are in their resting, open, or inactivated forms and whether the local anesthetic molecule is in its neutral or positively charged form. Open channels give the best access of drug to its receptor within the channel. When the gates close, drug molecules may be trapped inside. A similar *modulation* of drug-receptor interactions by the gating machinery has been discovered for many types of drugs acting on many types of channels and shows that opening and closing of channels involve conformational changes throughout the channel protein, affecting sites facing the external medium and the internal medium as well as the conductance of the pore.

Gating in Na Channels.[10] The gating mechanisms of ionic channels are not understood in specific molecular terms. Historically, Hodgkin and Huxley's[26] kinetic study of macroscopic voltage-clamp currents provided the major ideas we have today: steeply voltage-dependent activation followed by inactivation. They observed that activation has a slight delay, so it must have several underlying steps before the channel opens, and they noted that voltage dependence requires that charged or dipolar domains of the channel be moved in the membrane during gating. The required movement of the voltage sensors would produce a tiny electric current, *gating current*, carried, not by permeation of ions, but by rearrangement of parts of the channel macromolecule within the membrane (protein conformational changes).

Since 1952, several new approaches have contributed to the study of gating. Already mentioned is the far-reaching conclusion from single-channel studies (Fig. 3–11) that the open-close transition is all-or-nothing in almost every kind of channel studied. Also, we have described how local anesthetic molecules in the axoplasm have access to a site within the pore when gates are open and may be trapped there when gates are closed. These results establish that the physical gate is at the axoplasmic end of the pore (Fig. 4–1). In addition, some toxins selectively stop inactivation gating. This shows that inactivation may be prevented without interfering with activation. Several internally perfused amino acid–modifying reagents and proteolytic enzymes (pronase) selectively remove inactivation as well. Together with the amino acid sequence of the Na channel, the reactivity spectrum of inactivation-modifying reagents will help to define key accessible amino acids that play a role in gating.

More new conclusions come from gating-current and patch-clamp experiments. In the early 1970s, the gating current predicted by Hodgkin and Huxley was finally detected by C.M. Armstrong, F. Bezanilla, R.D. Keynes and E. Rojas.[7, 10] It was recorded under voltage clamp from axons bathed in media containing only impermeant ions and channel-blocking drugs, so that scarcely any ionic current remained. After the total membrane current was corrected for the small residual ionic leak and symmetrical capacity currents, there remained a small charge movement coming primarily from the fast conformational changes of activation. Analysis of the "on" and "off" gating current at the beginning and end of variable-length depolarizations revealed that activation and inactivation gating are kinetically *inter*dependent, a conclusion contrasting with the *in*dependent "m³" and "h" processes of the Hodgkin-Huxley model. Patch-clamp analysis of inactivation steps in single Na channels has led to a similar conclusion.[6] Today, we think of gating as resulting from a sequence of conformational changes in a flexible channel macromolecule, and we seek to identify the conformational states in kinetic diagrams like those used to describe other types of chemical reactions. We

might represent the states and transitions of Na channel gating approximately as follows:

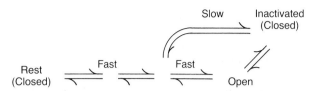

The steps proceeding to the right occur with depolarization and steps to the left occur with re- or hyperpolarization. Multiple arrows between "Rest" and "Open" signify that several rearrangements occur before opening, as implicit in Hodgkin and Huxley's original description. All such schemes are still tentative and hypothetical. The actual number of steps and their kinetic properties are still under vigorous investigation at the single-channel level.

The Na Channel Protein.[5, 11] The Na channel has been chemically purified from electric eel electric organs, mammalial skeletal muscle, and mammalian brain nerve endings (synaptosomes). The laboratories of M.A. Raftery, W.A. Catterall, and R.L. Barchi solubilized Na channels from the membranes with detergent, purified them with column chromatography, and reconstituted into phospholipid membranes TTX-sensitive Na fluxes that are activated by neurotoxins such as batrachotoxin. All preparations contain an unusually large glycopeptide of more than 200,000 daltons, and mammalian preparations contain two smaller peptides as well. The electric organ preparation contains only the large peptide that suffices to reconstitute TTX-sensitive channel activity.

A messenger RNA for the large peptide has been cloned and sequenced.[33] Remarkably, it consists of four major repetitive domains, each with at least four membrane-crossing, hydrophobic, helical segments as well as regions of high-charge density. It is tempting to imagine that stepwise conformational changes of one domain at a time could correspond to the high-order, m³h-like transitions of gating, and that regions with long sequences of acidic or basic amino acids form the voltage sensors. Further study of the structure ought to inspire major advances in understanding the physical nature of gating.

Ca CHANNELS

Ionic channels may have originated in the transition from prokaryotes to eukaryotes hand-in-hand with the emerging enzyme-regulatory sys-

Table 4–1 Processes Controlled by [Ca²⁺]ᵢ

Process	Regulatory Protein
Muscle contraction	Troponin C
	Regulatory light chain
	Light-chain kinase through calmodulin
Secretion by exocytosis of vesicles	Unknown
Protein phosphorylation	Protein kinase C
	Phosphorylase kinase through calmodulin
Increase of K(Ca) channel	Unknown
Inactivation of Ca current	Unknown
Increase of Ca²⁺ pumping	Calmodulin

tems based on intracellular free Ca^{2+} ions. Hence Ca channels evolved early and probably appear in all eukaryotes. When Ca channels in the plasma membrane open, Ca^{2+} ions move into the cell, down their steep electrochemical gradient, with two major consequences, (i) the intracellular free $[Ca^{2+}]$ might be transiently raised from a low basal level of 10–100 nM up to a level (500–2000 nM) sufficient to modulate Ca^{2+}-sensitive processes in the cell (Table 4–1), and (ii) the electric current carried by Ca^{2+} ions may be sufficient to depolarize the cell. In principle, a perfectly selective Ca channel would generate an electromotive force, E_{Ca}, more positive than $+100$ mV in a vertebrate cell bathed in a physiological medium containing 2 to 4 mM Ca^{2+} (Table 1–4).

Gating and Action Potentials.[20, 31, 34, 40] Like voltage-gated Na channels and the K channels of axons, voltage-gated Ca channels open with depolarization. The gating steps of Ca channels can be summarized as follows:

$$\text{Rest (closed)} \underset{\text{Fast}}{\overset{\text{Moderate}}{\rightleftharpoons}} \text{Open} \underset{}{\overset{\text{Very slow}}{\rightleftharpoons}} \text{Inactivated (closed)}$$

Again, steps to the right are favored by depolarization and steps to the left by re- or hyperpolarization. Typically, the activation and inactivation of Ca channels are slower than the activation and inactivation of Na channels.

There are several classes of Ca channels, differing in their gating kinetics.[31, 34] In one common class, the cell membrane has to be depolarized strongly (to 0 mV) to open many channels, and inactivation hardly occurs even during depolarizing pulses lasting hundreds of milliseconds. This type of Ca channel seems well suited to a messenger function, permitting Ca^{2+} ions to enter the cytoplasm as a second messenger when other

channels have generated a large depolarization. In another class of Ca channels, depolarizations to only -60 mV might suffice to initiate activation, and the channels will then inactivate if the depolarization is maintained for between ten and a few hundred milliseconds. This type of channel is suited to initiation of electrical activity and generating spikes.

Those Ca channels that do inactivate can be further differentiated. In one type, inactivation is a direct consequence of depolarization, as it is in Na channels. Inactivation is rapid. Stronger depolarization makes faster and more complete inactivation. In a second type, inactivation is typically slow, and it becomes weaker when large depolarizing pulses that approach E_{Ca} are given. In this type of Ca channel, inactivation is also prevented or slowed when a chelator for Ca^{2+} ions, EGTA, is injected into the cytoplasm. Eckert and Tillotson[14] noted a unifying theme in these two conditions: With pulses to E_{Ca}, no Ca^{2+} enters the cytoplasm, and with internal EGTA, Ca^{2+} may enter the cytoplasm for ordinary depolarizations, but because of the chelating action, the rise of $[Ca^{2+}]_i$ is minimized. They therefore reasoned that inactivation in the second type of Ca channels must be caused by increases of $[Ca^{2+}]_i$. In effect, there is negative feedback so that this Ca-entry mechanism shuts itself off, when enough Ca^{2+} has entered. Again, it should be recalled that not all Ca channels do this. The theme that there are several classes of Ca channels will recur.

Ca channels were first discovered in crustacean muscle, a tissue that can be induced to make action potentials without any Na channels (Fig. 4–4). Instead, Ca channels shape the rising phase of the action potential, making a regenerative cycle fully analogous to the Hodgkin cycle (Chap. 3). A fairly good generalization is that Na spikes are found in all axons and in other cells that give brief or quickly rising action potentials, whereas Ca spikes, or at least a calcium component, are the rule in cells giving long or slow responses (Table 4–2 and Fig. 4–4). Na channels and Ca channels coexist in many excitable membranes. For example, in the ventricle of the heart, the action potential must propagate rapidly yet last from 200 to 500 ms so that the muscular contraction is long enough to eject the blood. Rapidly inactivating Na channels could not depolarize the membrane this long. Therefore, this tissue uses a high density of Na channels to get rapid conduction and then very slowly inactivating Ca channels to take over the job of providing sustained inward current within a few milliseconds of the upstroke of the action potential.

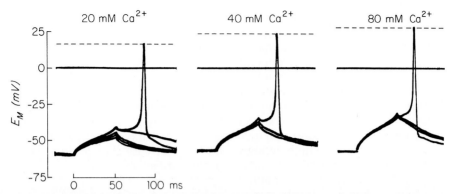

Figure 4–4 Calcium action potentials recorded intracellularly from a barnacle muscle fiber injected with EGTA solution to chelate myoplasmic free calcium. Extracellular calcium was 20, 40, or 80 mM, and Mg concentration was 0. The upper trace in each record shows the zero potential level. Depolarizing current pulses of an intensity just below or above the threshold were applied, and potential changes for these pulses were superposed in each record. The dashed lines marking the peak of the overshoot show how the height of the Ca spike increases as $[Ca^{2+}]_o$ is raised. (Modified from Hagiwara, B.; Takahashi, K. *The Journal of General Physiology,* 1967, vol. 50, pp. 583–601, by copyright permission of The Rockefeller University Press.)

Selectivity and Block.[8, 20, 40] Cells making Ca spikes stop firing if external calcium is removed, but excitability can be restored if Sr^{2+} or Ba^{2+} are added to the medium instead. The restoration is particularly striking with barium because it enhances spikes in three ways: Not only is it often a better current carrier than calcium in Ca channels, but it blocks K channels and, upon entering the cell, does not induce Ca-dependent inactivation of Ca channels. The permeant ions Ca^{2+}, Ba^{2+}, and Sr^{2+} compete for passage through the Ca channel, evidently binding in transit to a sequence of saturable binding sites in the pore. The open pore may have one or two divalent ions occupying it at most times, pausing on their way through. Magnesium is not permeant and may even block the channel weakly. Surprisingly when *all* divalents are removed from the medium, Ca channels become permeable to monovalent cations, an effect attributed to extraction of the Ca^{2+} ions normally resident within the open pore.

Ca spikes and Ca channels are insensitive to tetrodotoxin and local anesthetics, but they can be blocked by many transition metals, including Ni^{2+}, Mn^{2+}, Cd^{2+}, and Co^{2+}, and in many cases also by several clinically useful "Ca antagonist" drugs. Because their action is competitive with external Ca^{2+}, transition metal cations are believed to enter the pore, displacing Ca^{2+} ions from their binding sites and binding there instead. Among the organic Ca antagonists, one class has actions reminiscent of blocking of Na channels by local anesthetics. This includes verapamil and D-600, lipid-soluble compounds with ionizable amine groups. Their action is use dependent—blocking is enhanced by repetitive depolarization—a property that suits them for clinical use against cardiac arrhythmias originating in Ca-spiking atrial pacemaker cells. Experiments with membrane-impermeant D-600 derivatives show that the drug molecules reach their receptor on the Ca channel only from the cytoplasmic side and that opening of channel gates promotes binding. Hence as in Na channels, the voltage-dependent gates of Ca channels may open and close the cytoplasmic end of the pore. Another class of organic Ca^{2+} antagonists are the dihydropyridines, nifedipine, nitrendipine, and relatives, lipid-soluble drugs without ionizable amine groups. They block Ca^{2+} currents without use dependence and probably act at sites different from that for verapamil and D-600, judging by lack of competition in binding.

Quite surprisingly, some dihydropyridine analogues (Bay K 8644 and CGP 28 392) have just the opposite effect.[22] They augment Ca spikes by

Table 4–2 Distribution of Na and Ca Spikes

Propagation By Na Spikes
Axons of all animal phyla
Vertebrate twitch muscle
Cardiac atrial and ventricular muscle[1]
Cardiac Purkinje fibers[1]
Propagation By Ca Spikes
Nearly all invertebrate muscle
Cardiac sinoatrial and atrioventricular nodes
Vertebrate smooth muscle
Dendrites of central neurons[2]

[1]Na current produces the rapid rise and then Ca current maintains the depolarization during the long action potential.

[2]Some regions of the dendritic tree may also have local concentrations of Na channels. In either case the density of channels is supposed to be low so if spikes exist they are weak.

prolonging the open time of Ca channels. Such molecules have been called Ca agonists. They are useful tools to enhance Ca^{2+} entry and secondarily to potentiate intracellular processes that depend on Ca^{2+} entry. Since the single-channel conductance is unchanged by agonists, we suppose that the dihydropyridine binding site is not in the pore itself but rather at another position on the channel that interacts with conformational changes of gating. The dihydropyridines bind with high enough affinity to be used as labels for counting Ca channels in membranes and for following the channel macromolecule through chemical purifications. The different functional classes of Ca channels have different sensitivities to organic Ca antagonist and agonist drugs.[34] Calcium channels that have rapid, voltage-dependent inactivation tend to be less sensitive to organic blockers.

MANY K CHANNELS

Diversity in Gating.[1] By far the largest diversity of voltage-gated channel types occurs among the K-selective ionic channels. The K^+ ions they pass have no chemical messenger role, but their effect on electrical signals has been exploited repeatedly. Different K-permeable channels have been deployed to set the resting potential, to decrease cell excitability, to slow the rhythm of repetitive firing, to repolarize from large depolarizations and to hyperpolarize and terminate periods of Ca^{2+} influx. New K-permeable channels are being found each year, so any scheme of classification is tentative. Nevertheless, we will group K channels into classes (Table 4–3), according to their gating characteristics. The names given have historical significance and reflect the lack of any uniform guidelines for coining names of channels.

K channels with gating characteristics like those in squid giant axons are called *delayed rectifier* K channels—"rectifier" because they account for classical observations that the membrane conducts better during depolarizations than during hyperpolarizations, and "delayed" because they open with an appreciable delay after a step depolarization (n^4 kinetics of Hodgkin and Huxley). Their gating must have multiple steps, presumably reflecting stepwise changes in some repeated feature of their molecular structure.

$$\underset{\text{(closed)}}{\text{Rest}} \quad \overset{\text{Slow}}{\underset{\text{Slow}}{\rightleftharpoons \rightleftharpoons \rightleftharpoons \rightleftharpoons}} \quad \text{Open}$$

Although delayed rectifier K channels are found in most excitable cells, their speed of gating varies so much that they are probably not always channels of identical structure. Their most common role is to terminate brief action potentials. Most axons have delayed rectifiers and no other voltage-gated K channels.

Another class, the Ca^{2+}-*dependent* K channels, also activate with a delay during a depolarization, but their activation is profoundly enhanced by a rise in the internal free Ca^{2+} concentration. At normal resting $[Ca^{2+}]_i$, the K(Ca) channels are silent at all physiological potentials, while at 10 μM Ca^{2+}, their activation curve is similar to that of delayed rectifier K channels (Fig. 4–5). Several types of K(Ca) channels can coexist in the same cell, differing in single-channel conductance by as much as tenfold and in their sensitivity to block by various drugs. These channels open to terminate long Ca spikes or to terminate barrages of short Ca spikes once $[Ca^{2+}]_i$ has built up to a level sufficient to activate them. They seem to be ubiquitous in excitable cells, except that neurons do not have them in their *axonal* membrane.

The A current K channel[12] opens with depolarization, but is subject to strong inactivation even near the resting potential. The time course of the A current therefore is rapid activation followed by inactivation. This channel is present in membranes that turn depolarizing slow potentials into trains of action potentials, encoding the strength of sensory or synaptic inputs into the nervous system's frequency code—low frequency for weak inputs and high frequency for strong ones (Fig. 4–6). Rates of firing are governed by the membrane-potential trajectory between spikes. After a spike, the membrane might hyperpolarize and then be gradually driven positive by the stimulus current until firing threshold is reached. When the depolarization starts, A-current channels activate, slow-

Table 4–3 Voltage-Gated K Channels

Type	Abbreviation	Characteristics
Delayed rectifier	I_K	Squid axon type, opening with a delay after depolarization
Calcium-dependent	$I_{K(Ca)}$	Internal Ca^{2+} and depolarization synergistically promote opening
A current	I_A	Opens transiently with depolarization and remains inactivated at $E_M > -50$ mV
M current	I_M	Noninactivating, slow, delayed rectifier that is shut down by muscarinic agonists
Inward rectifier	I_{ir}	Conducts inward K current well but shuts off quickly with depolarization

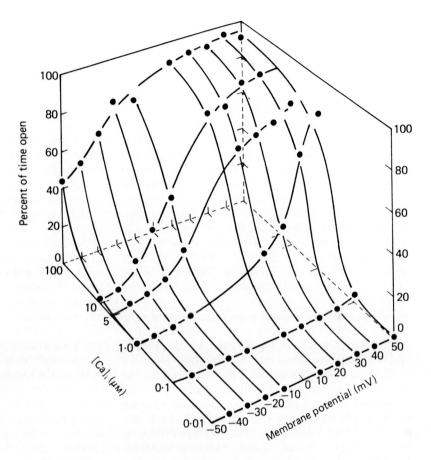

Figure 4–5 Voltage and Ca dependence of K(Ca) channel. Percentage of time open versus membrane potential and free [Ca²⁺]ᵢ for single K(Ca) channels. Calculated from long records with membrane patches excised from rat myotubes so that known [Ca²⁺] could be readily applied to the intracellular face. T = 21°C. (From Barrett, J.N.; Magleby, K.L.; Pallotta, B.S. *J. Physiol. [Lond.]* 331:211–230, 1982.)

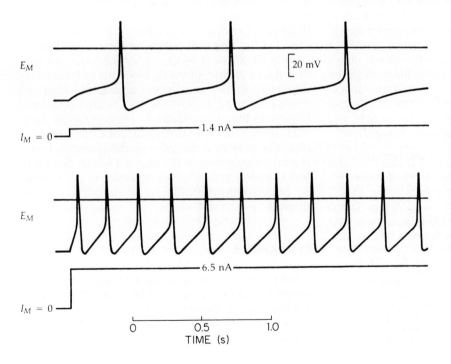

Figure 4–6 Repetitive firing of a neuron isolated from a marine mollusc. Action potentials recorded with an intracellular microelectrode from a nudibranch *(Anisodoris)* ganglion cell whose axon has been tied off. A second intracellular microelectrode passes a step of current *(I)* across the soma membrane, initiating a train of action potentials whose frequency depends on the strength of the current stimulus. T = 5°C. (From Connor, J.A.; Stevens, C.F. *J. Physiol. [Lond.]* 213:1–19, 1971.)

ing the depolarization, then inactivate, letting the depolarization continue. By transiently retarding weak excitatory effects, A currents help encoding membranes to achieve very low rates of firing (Chap. 12).

The excitability of some cells is hormonally modulated. Sympathetic ganglion cells, for example, receive excitatory inputs from preganglionic cells and encode the slow potentials into a spike-frequency code. The output frequency for a given excitatory input can be modulated by other receptors on the cell surface, a muscarinic acetylcholine receptor and a peptide hormone (GnRH-like peptide) receptor.[2] Activation of either receptor produces a lasting increase of the membrane resistance and hence an increase in excitability. The channel that is shut is the K-selective, delayed rectifying *M* channel, in which "M" stands for *muscarinic*. M channels are also found in central neurons.

Cardiac muscle, skeletal muscle, and some eggs express K channels that conduct at rest and shut down when the cell is depolarized by 10–20 mV. The conductance is maximal at potentials negative to E_K where K^+ ions would flow into the cell. In short, their potassium currents flow inward easily but outward only with difficulty, especially during strong depolarizations. This type of channel, the *inward*, or "anomalous," *rectifier* stabilizes the resting potential yet is shut off once a spike is initiated, saving energy. In cardiac pacemaker cells, the number of functioning inward rectifier channels is increased by acetylcholine action on muscarinic receptors, thereby slowing the rate of pacemaking. This is how stimulation of the vagus nerve, a cholinergic nerve that innervates the heart, slows the rate of beating. (Note that this muscarinically modulated channel becomes *active* when muscarinic receptors are occupied, while the M channel of neurons becomes *silent*.) In skeletal muscle and even in cardiac ventricular muscle, inward rectifiers are not under hormonal control.

Conductance, Selectivity, and Block.[1, 24, 29] The K channels are the narrowest and most ion-selective channels. The majority are permeable to K^+, NH_4^+, Tl^+ and Rb^+ (diameter = 0.296 nm*) ions but not to Cs^+ (diameter = 0.338 nm), results that require a minimum pore diameter of 0.33 nm (Fig. 4–2B). Inward rectifiers are usually blocked by Rb^+ and hence may have a slightly smaller minimum diameter. Despite their strong ionic selectivity, the

single-channel conductances of K channels are not small, typically 5 to 50 pS, and one of the types of K(Ca) channels has nearly the highest conductances known for any channel, 200 pS.

In a classical biophysical study comparing influxes and effluxes of ^{42}K in delayed rectifier K channels, Hodgkin and Keynes[28] discovered phenomena commonly called the "long-pore effect" or "single-file diffusion."[25] If ions move independently by electrodiffusion through a channel, one can readily predict the ratio of simultaneous forward and backward tracer fluxes in the channel.[41] Thus in a channel held at 0 mV and bathed by 100 mM K on one side and 10 mM on the other, the tracer flux from the 100 mM side should be ten times that from the 10 mM side. However, when the measurements were made in molluscan giant axons, the corresponding ratio was more like 500:1. Hodgkin and Keynes recognized that a pore containing several ions in a row, moving in single file, would behave this remarkable way. A stream of ions would be moving in the downhill direction, and ions attempting to cross the other way would be pushed back where they came from before completing the trip. The correct flux ratio is predicted if the channel has at least three occupied K^+ binding sites within it. Delayed rectifier and inward rectifier K channels show the long-pore effect, and as we have already mentioned, so do Ca channels.

While K channels are easy to block, it is not always possible to block one type selectively. Common blocking agents include tetraethylammonium (TEA) and relatives, Cs^+, Ba^{2+}, and 4-aminopyridine (4AP). The first three, all cations, produce a voltage-dependent block. When TEA, Cs^+ or Ba^{2+} are placed inside the cell, blocking is favored by depolarizing the membrane and unblocking, by hyperpolarizing it. Unblocking is also promoted by raising the extracellular K^+ concentration. From such observations, Armstrong[9] concluded that the blockers pass into the pore of the channel, blocking at a narrowing of the tunnel, where they may feel the electric field of the membrane and where K^+ ions coming from the opposite side can knock them out.

Cl CHANNELS

For more than 40 years, physiologists have been aware that skeletal muscle fibers are permeable to anions, indeed often more permeable to Cl^- than to K^+ ions at rest. The biophysical analysis of anion-permeable channels has lagged behind that

---

*Since ions trade many of their waters of hydration for oxygen dipoles of the channel wall, the relevant diameters are the crystal diameters.

of Na, K, or Ca channels, probably because anion channels rarely produce major permeability changes. (Clear exceptions are the anion channels opened by γ-aminobutyric acid and glycine at inhibitory synapses, Chap. 11.)

Various anion channels have been observed in the membranes of skeletal muscle, cardiac muscle, and electrical organs as well as in epithelial cells. Some seem strictly chloride-selective, and others are permeable to five or more different anions. Some have steep voltage-dependent gating, and others are relatively voltage independent. At the single-channel level, some show clear open-close transitions to a single open conductance, and others seem to have multiple open states. One gets the impression of a heterogeneous collection of anion channels. New work needs to be done to define the physiological roles of these ubiquitous yet little-known channels.

OTHER CHANNEL PROPERTIES

Hormonal Modulation of Channels.[30] In some membranes, the excitability is relatively fixed and in others, it can be tuned to varying requirements. Ultimately, when we understand the molecular basis of learning and memory, we will surely see numerous examples of modification of channel excitability in response to experience. Axon membranes, which only repeat impulses provided to them, have an excitability that probably needs no rapid adjustments. The axoplasm can be replaced by flowing salt solution, and the axon membrane still performs well, so cytoplasmic regulatory systems are not essential for conduction. On the other hand, the electrical properties of heart, smooth muscle, and the soma membrane of neurons adjust on a minute-by-minute basis to the demands and experience of the organism. Signals from various membrane receptors alter the properties particularly of certain K and Ca channels, changing the responses of these cells. The signals involve cytoplasmic second messengers, cofactors, coupling proteins, and enzymes and often would not occur in cells perfused internally with salts alone.

The meaning of the term *modulation* is broad, evolving, and not well defined operationally. It tends to be used when the pathway for changing channel function is not simply a direct action of the physiological stimulus on the channel molecule—when the receptor for the stimulus is a remote molecule that acts through intermediaries on the channel. This definition rules out ordinary

voltage-dependent gating, where the voltage sensor is part of the channel, or activation of nicotinic acetylcholine receptor by ACh, where the ACh binding sites are part of the channel molecule. Modulation follows the stimulus with a delay as the message makes its way from remote receptors to the channel, and modulation often outlasts the duration of the stimulus since it depends on possible long-lasting intermediate messages and modifications of the channel. It may last from seconds to hours. These ideas can be illustrated by some examples.

Epinephrine increases the force of cardiac contraction in part by increasing the Ca^{2+} influx during each beat.[35] Voltage-clamp and patch-clamp experiments after addition of epinephrine show a larger peak I_{Ca} and a larger number of voltage-gated Ca channels open during a depolarizing pulse. The conductance of individual cardiac Ca channels is not changed, but the probability of opening is increased, and perhaps the number of functioning channels is increased. As might be expected for β-adrenergic actions (see pp. 5–7), the effect of epinephrine can be mimicked by introducing cyclic AMP or its analogues into the cell or by injecting the catalytic subunit* of cAMP-dependent protein kinase. Hence the modulatory pathway involves intracellular cAMP and protein phosphorylation. This modulation probably occurs by direct phosphorylation of the Ca channel protein.

Many modulatory effects are exerted on K channels. We have already mentioned modulatory effects of muscarinic acetylcholine receptor activation on inward rectifiers of cardiac pacemaker cells and on M channels in sympathetic and central neurons. The former effect is exerted through an intermediary GTP-binding protein. The coupling between muscarinic receptors and M channels is not yet determined. Another striking example of K channel modulation has been described by the laboratory of E.R. Kandel in a sensory pathway of the *Aplysia* nervous system.[38] When one sensory pathway is stimulated, serotonin and other neurotransmitters are released onto the nerve terminals of another sensory cell, causing a K channel in the second cell to be shut. The input resistance of the nerve terminals rises, and the cell becomes more responsive to excitatory inputs from its sensory pathway, a phenomenon called behavioral

---

*This is the protein-phosphorylating subunit of protein kinase, which when artificially separated from its regulatory subunit is fully active even without the usual cAMP second messenger.

sensitization. Serotonin activates adenylate cyclase in these cells, and the channel that is eventually closed, called the S channel, can be silenced by intracellular injection of cAMP. Again, modulation occurs via a cAMP-dependent protein kinase that might phosphorylate the S channel. The shutting of the S channel, which sensitizes the cell for seconds, can be viewed as a molecular correlate of a type of short-term memory in *Aplysia*.

A majority of the substances now called neurotransmitters, including epinephrine, dopamine, serotonin, and peptide transmitters, can act through modulatory receptors to alter the functions of excitable cells.

Expression and Turnover. The life history of a channel protein follows the same steps as many other membrane proteins. RNA is transcribed from the appropriate genes and processed to mature messages in the nucleus. The messages are translated in the cytoplasm on the rough endoplasmic reticulum, and the nascent proteins are threaded into correct membrane orientations during their synthesis. In the endoplasmic reticulum and the Golgi apparatus, glycosylation and other modifications are made. Finally, from 15 minutes to several hours later, finished channels are ready in the membrane of small vesicles for delivery to their destination. In neurons, protein synthesis takes place only in the cell body, and the membrane proteins probably are moved in vesicles by rapid axoplasmic transport down to remote axonal membranes. At the end of their lifetime, the membrane proteins are internalized by endocytosis (pinching in of vesicles) and degraded. Each of these steps is subject to control, so that on the time scale of days, even cells that have no rapid modulation of channel function can adjust the placement and mix of channels in response to forces that we still know little about.

Dramatic changes of the mix of channels can occur in development, after trauma, and when cells are placed in tissue culture. In the most detailed in vivo observation of the maturation of a neuron, the Rohon-Beard neuron of frog embryo is found to proceed through stages in which it is inexcitable at 21 hours after the egg is fertilized, makes a long Ca spike at 27 hours, a mixed Na-Ca spike at 35 hours, and a brief Na spike at 3 days.[39] In the maturation of *Drosophila* flight muscle during the pupal transformation of a maggot into a fly, there are no functioning voltage-gated K channels in the muscle membrane 55 hours after puparium formation; transient A-current channels appear at 72 hours; delayed rectifier K channels are added at 90 hours; and Ca channels are evident a few hours later as the fly emerges from the pupal case.[37] While we presume that such sequential insertion of various channels is important for the orderly maturation of the adult response, we do not understand what factors regulate it. The ionic channels of adult mammalian muscle change within a few days if the natural pattern of excitatory input is halted by section of the motor nerve or other interruption of normal signaling. Tetrodotoxin-insensitive Na channels appear, the population of Cl channels is reduced, and vast numbers of new acetylcholine receptors are inserted all over the muscle surface. In addition, muscle contractile proteins and ion pumps are changed. This effect of "denervation" is described further in Chapter 6 as a reversion to a more "embryonic" pattern of gene expression.

Divalent Ions and Firing Threshold.[18, 24] The external calcium concentration has long been known to affect the electrical excitability of nerve and muscle. Cells that make Na spikes become more excitble if $[Ca^{2+}]_o$ is reduced twofold and less excitable if $[Ca^{2+}]_o$ is doubled. Such changes are manifest in hypoparathyroidism, which leads to reduction in serum calcium levels and a tendency for muscles to begin twitching spontaneously, and in hyperparathyroidism, which leads to elevated serum calcium and a tendency toward muscle weakness. Voltage-clamp experiments reveal that the external divalent ion concentration shifts the voltage dependence of gating in Na, K, and Ca channels along the voltage axis (Fig. 4–7). Hence with lowered $[Ca^{2+}]_o$, the probability that a Na channel will be opened at any voltage is increased, and a smaller depolarization is needed to bring the excitable membrane to its firing threshold.

Shifts of gating with divalent ions can be explained by the electrostatic interactions of cations in the external solution with fixed negative charges that are exposed on the external face of the Na, K, and Ca channel macromolecules. We suppose that the resting channel exposes 25 to 100 negatively charged groups (ionized carboxylic acid groups) to the external solution. These tend to attract multivalent cations in the bathing medium with two consequences, both of which reduce excitability. First, the voltage sensor of the channel is sensitive to the electric field across the channel, a field that comes from the membrane potential difference and also from any local charges near the channel molecule. Divalent cations attracted to the outside of the channel would produce a local electrical field that biases the voltage sensor in the same way that a true membrane hyperpolarization would. The second closely related effect

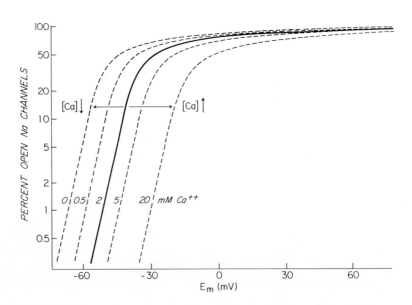

Figure 4–7 Effect of external calcium concentration on the voltage dependence of gating. Gating is measured as the percentage of Na channels opened by a step voltage clamp depolarization to the given membrane potential. The solid line is for a normal calcium concentration of about 2 mM. Elevating $[Ca^{2+}]_o$ shifts the curve to the right. Decreasing $[Ca^{2+}]_o$ shifts the curve to the left. These results come from a frog skeletal muscle fiber, but similar shifts are obtained for Na channels of axons and neuron cell bodies. (Modified from Campbell, D.T.; Hille, B. *The Journal of General Physiology*, 1976, vol. 67, pp. 309–323 by copyright permission of The Rockefeller University Press.)

is that the conformational changes of gating may involve withdrawal of negatively charged channel groups from the external surface, therefore stabilization of these groups by external counter ions would make them harder to withdraw from the surface.

Various Electrically Excitable Membranes. This chapter has explored the details of individual ionic channels. We end with brief examples of how a few cells use their channels for electrical excitability. The principles, advanced in Chapter 3, remain simple: To have a negative resting potential, you need to have open K channels, possibly supplemented by Cl channels at rest. For a regenerative excitatory response, you first need to open an adequate number of voltage-gated channels carrying inward current (Na^+ or Ca^{2+} ions). The response can be terminated by inactivating the Na or Ca channels, and it can be further shortened by activating K channels. We now suppose that every cell of the body has some voltage-gated channels. Probably, most cells do not make action potentials but many must have voltage-gated channels to regulate Ca-dependent processes, and perhaps all have K channels that give them negative resting potentials.

Neurons. A neuron is an entire nerve cell and comprises dendrites, cell body (soma), axon, and nerve terminal. In the vertebrate nervous system, the cell body of a neuron has the task of summing the influence of thousands of excitatory and inhibitory inputs and deciding what pattern of action potentials to send down its axon to other synapses—a decision-making role. Chapter 12 describes in detail some of the complex electrical properties of mammalian central neurons. There are many kinds of neurons, and each part of a neuron has its own repertoire of ionic channels. Neurons illustrate well the general problem of how a cell can synthesize and sort in the cell body membrane protein components that then are delivered in correct proportion to remote, specialized membrane regions. Consider the regional differentiation of a molluscan neuron. The axon, as we know from Hodgkin and Huxley, uses high densities of steeply voltage-gated Na and delayed rectifier K channels to make propagated action potentials. They operate against a resting background of relatively constant and low chloride permeability. On the other hand, cell bodies, from which axons arise, have many more types of channels, although the density per unit area is lower than in axons.[1] Prominent channels identified in voltage clamp of the cell bodies of molluscan neurons are voltage-gated Na, Ca, delayed rectifier K, A-current, and K(Ca) channels. In addition, several other very slow and small inward and outward voltage- and calcium-dependent currents and numerous neurotransmitter-gated and modulated channels are known in these cells. In mammalian central neurons, the density of Na channels is highest at the axon hillock where spikes are assumed to arise, and there may only be a few on most parts of the dendritic tree.

Given such an extensive collection of channels, it is no surprise that neurons develop complex firing patterns. For example, *bursting pacemaker* activity can be seen in certain molluscan neurose-

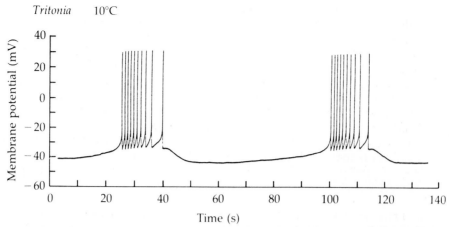

Figure 4–8 Bursting pacemaker activity in a molluscan neuron. The membrane potential trajectory of a *Tritonia* neuron shows bursts of spontaneous action potentials alternating with quiet intervals. No current input or nervous input is required for this rhythmic firing. T = 10°C. (From Smith, S.J. *The Mechanism of Bursting Pacemaker Activity in Neurons of the Mollusc* Tritonia diomeda. Ph.D. thesis, University of Washington. Ann Arbor, Mich, University Microfilms.)

cretory neurons even after they have been deprived of all synaptic inputs (Fig. 4–8). The cell alternates between periods of silence and periods of rhythmic, spontaneous firing. Presumably, every burst would normally cause a phasic output of neurosecretory product at the axon terminals of this cell. The cyclic alternation of activity is believed to be controlled by entering Ca^{2+} ions acting as an intracellular second messenger to modify the gating of several types of K and Ca channels. Cyclic changes of intracellular free Ca^{2+} can be demonstrated by injecting a single cell with a dye whose absorption or fluorescence spectrum is a sensitive indicator of ionized Ca^{2+}. The cell is then studied under a light microscope simultaneously

recording the membrane potential and the absorbance or fluorescence of the dye, which represents the Ca signal (Fig. 4–9). During each action potential of a burst, voltage-gated Ca channels open, and the average intracellular Ca^{2+} climbs up a few nanomolar (Fig. 4–9A). When $[Ca^{2+}]_i$ reaches a high level, the burst is shut off, possibly because some types of Ca channels become inactivated and K(Ca) channels become more active. Over the next 10 to 15 s, Ca^{2+} is pumped out of the cytoplasm, and when the free concentration becomes low again, the electrical activity resumes.

Muscle. The task of muscle cells is to contract, which they do under control of membrane potential changes of their surface membrane. Vertebrate

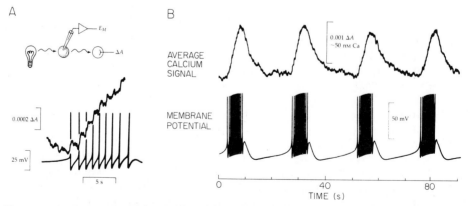

Figure 4–9 Changes of intracellular $[Ca^{2+}]$ during bursting. Simultaneous recordings of membrane potential and light absorbance changes from an arsenazo III–injected *Aplysia* neuron. *A,* During each spontaneous burst of action potentials, $[Ca^{2+}]_i$ (reported by the color of the dye) rises. In the quiet intervals, it falls. *B,* At higher resolution in another cell, the Ca^{2+} buildup is seen to occur during individual action potentials. The spikes of this cell clearly involve opening of voltage-gated Ca channels. T = 16°C. (After Gorman, A.L.F.; Thomas, M.V. *J. Physiol. [Lond.]* 275:357–376, 1978.)

muscle is conventionally divided into three categories, skeletal, cardiac, and smooth. The excitability of skeletal muscle seems the simplest.[3, 4] Two channels, an inward rectifying K channel and a Cl channel, establish the resting potential. The stimulus to contract is mediated by nicotinic acetylcholine receptors, which depolarize the muscle membrane locally at the neuromuscular junction. In twitch muscle, the depolarization triggers a regenerative Na spiking mechanism that propagates an action potential in both directions over the entire muscle fiber. Underlying the spike are TTX-sensitive Na channels and TEA-sensitive delayed rectifier K channels, as in axons. However, the delayed rectifier channels of skeletal muscle are clearly distinguishable from those in axons since they require 20 times higher TEA concentration to be blocked, and they inactivate fully during 1 s voltage-clamp depolarizations. Again the Na channels are in highest density around the neuromuscular junction, where the action potential has to be initiated. In addition, a slow, dihydropyridine-sensitive Ca channel and several slow K channels are known, one of them possibly a K(Ca) channel. The roles of the slow channels are not known. More information on skeletal muscle channels appears in Chapter 7.

Cardiac muscle uses many more channels. The heart is highly regionally differentiated. Pacemaking, conducting, and contracting are optimized in different cells, and the single action potential that underlies one heartbeat changes shapes many times as it sweeps through electrically coupled cells over the whole heart. A recent review lists 14 different current pathways in cardiac membranes.[32] The list includes all the voltage-gated channels we have discussed except M channels. In broad outline, cardiac action potentials are long (100 to 700 ms) and require Ca channels (several types) in addition to Na channels. To improve the efficiency of a system that spends a significant percentage of its time depolarized, the membrane conductance is kept low during the long plateaus by having few K channels (many types but few copies of each) open. Also, the resting K permeability comes from inward rectifiers that conduct no outward current during the plateau. Many cardiac channels are subject to modulation by the opposing influences of sympathetic and parasympathetic neural inputs. Further details of cardiac excitability are given in Chapter 37.

Smooth muscles make diverse shapes of action potentials that are calcium-dependent and they also generate slow waves of depolarization. They have many types of channels and may also use electrogenic pumps and carrier transport. What little is known about their electrical properties is described in Chapter 10.

Other Excitable Cells. All eukaryotic cells have ionic channels. In many animal cells—nerve, muscle, endocrine gland, exocrine gland, white blood cells, eggs, and more—obvious membrane potential changes are associated with normal physiological activities. The recognition of electrical excitability outside the nervous and muscular systems is new, and therefore much remains to be learned. Current signals that have been recorded in glands, blood cells, and eggs with patch-clamp and whole-cell voltage-clamp methods resemble those from neurons and are sensitive to conventional channel blockers. Thus the pancreatic B cell, which secretes insulin when the blood glucose is high, shows bursting pacemaking activity like that in the neuron of Figure 4–8 and has a similar repertoire of channels (Chap. 77). The chromaffin cell of the adrenal medulla secretes epinephrine and enkephalins into the circulation in moments of stress. Its activities are controlled by action potentials in the sympathetic splanchnic nerve almost in the same way as a skeletal muscle is controlled by action potentials in its motor nerve. The splanchnic nerve releases acetylcholine, which binds to nicotinic acetylcholine receptors on chromaffin cells, initiating a depolarization that fires a Na spike. The Na spike also opens voltage-gated Ca channels that let external Ca^{2+} ions enter the cell to trigger secretion.[16, 17]

SUMMARY

Ionic channels are believed to be pores because their elementary permeability is too high to account for by enzyme- and carrierlike mechanisms. They select ions partly on the basis of being small enough to pass through the channel and partly on the difference between the energies of partial dehydration and of interaction with the pore walls. Ions that stick too tightly or that proceed until they reach a narrowing of the pore can block the flow of permeant ions.

Voltage-dependent gating involves multiple sequential conformational changes in the channel macromolecule, finally leading to a rearrangement that opens the pore. The available evidence suggests that the gate(s) is(are) at the cytoplasmic end of the pore and that the narrowest part of the open pore is near the extracellular end. Toxins and drugs can alter the voltage-dependence and time course of gating.

There are many kinds of voltage-gated channels suited to the timing, integrating, coding, and damping functions that give excitable cells their biological utility. Particularly diverse are the K channels. Some open transiently and others for a long time, some open rapidly and others slowly, most open with depolarization, a few with hyperpolarization, and many are subject to modulation by internal Ca^{2+} ions or other receptor-controlled signals.

Channels are present in all eukaryotic cells and will undoubtedly be recognized as major control mechanisms of cell biology when more of their roles have been elucidated.

ANNOTATED BIBLIOGRAPHY

1. Hille, B. *Ionic Channels in Excitable Membranes.* Sunderland, Mass., Sinauer Associates, 1984.
   A didactic textbook reviewing the major ideas of the field.
2. Nowycky, M.C.; Fox, A.P.; Tsien, R.W. Three types of neuronal calcium channel with different calcium agonist sensitivity. *Nature* 316:440–443, 1985.
   A clear research paper illustrating how to use voltage clamp, patch clamp, and pharmacology to demonstrate subtypes of Ca channels in cultured neurons.
3. Kameyama, M.; Hescheler, J.; Hofmann, F.; Trautwein, W. Modulation of Ca current during the phosphorylation cycle in the guinea pig heart. *Pflugers Arch.* 407:123–128, 1986.
   One of a series of research papers studying the role of phosphorylation in adrenergic modulation of Ca channels.

REFERENCES

1. Adams, D.J.; Smith, S.J.; Thompson, S.H. Ionic currents in molluscan soma. *Annu. Rev. Neurosci.* 3:141–167, 1980.
2. Adams, P.R.; Brown, D.A.; Constanti, A. M-currents and other potassium currents in bullfrog sympathetic neurones. *J. Physiol. (Lond.)* 330:537–572, 1982.
3. Adrian, R.H.; Chandler, W.K.; Hodgkin, A.L. Voltage clamp experiments in striated muscle fibers. *J. Physiol. (Lond.)* 208:607–644, 1970.
4. Adrian, R.H.; Chandler, W.K.; Hodgkin, A.L. Slow changes in potassium permeability in skeletal muscle. *J. Physiol. (Lond.)* 208:645–668, 1970.
5. Agnew, W.S. Voltage-regulated sodium channel molecules. *Annu. Rev. Physiol.* 46:517–530, 1984.
6. Aldrich, R.W.; Corey, D.P.; Stevens, C.F. A reinterpretation of mammalian sodium channel gating based on single channel recording. *Nature* 306:436–441, 1983.
7. Almers, W. Gating currents and charge movements in excitable membranes. *Rev. Physiol. Biochem. Pharmacol.* 82:96–190, 1978.
8. Almers, W.; McCleskey, E.W. The nonselective conductance due to calcium channels in frog muscle: calcium selectivity in a single-file pore. *J. Physiol. (Lond.)* 353:585–608, 1984.
9. Armstrong, C.M. Interaction of tetraethylammonium ion derivatives with the potassium channels of giant axons. *J. Gen. Physiol.* 58:413–437, 1971.
10. Armstrong, C.M. Sodium channels and gating currents. *Physiol. Rev.* 61:644–683, 1981.
11. Catterall, W.A. Neurotoxins that act on voltage-sensitive sodium channels in excitable membranes. *Annu. Rev. Pharmacol. Toxicol.* 20:15–43, 1980.
12. Connor, J.A.; Stevens, C.F. Voltage clamp studies of a transient outward membrane current in gastropod neural somata. *J. Physiol. (Lond.)* 213:21–30, 1971.
13. Courtney, K.R.; Strichartz, G.R. Structural elements which determine local anesthetic activity. In Strichartz, G.R., ed. *Handbook of Experimental Pharmacology: Local Anesthetics.* New York, Springer-Verlag, 1985.
14. Eckert, R.; Tillotson, D.L. Calcium-mediated inactivation of the calcium conductance in caesium-loaded giant neurons of *Aplysia californica. J. Physiol. (Lond.)* 314:265–280, 1981.
15. Eisenman, G. Cation selective glass electrodes and their mode of operation. *Biophys. J.* 2 (Suppl. 2):259–323, 1962.
16. Fenwick, E.M.; Marty A.; Neher, E. A patch-clamp study of bovine chromaffin cells and of their sensitivity to acetylcholine. *J. Physiol. (Lond.)* 331:577–597, 1982.
17. Fenwick, E.M.; Marty, A.; Neher, E. Sodium and calcium channels in bovine chromaffin cells. *J. Physiol. (Lond.)* 331:599–635, 1982.
18. Frankenhaeuser, B.; Hodgkin, A.L. The action of calcium on the electrical properties of squid axons. *J. Physiol. (Lond.)* 137:218–244, 1957.
19. Goldman, D.E. Potential, impedance, and rectification in membranes. *J. Gen. Physiol.* 27:37–60, 1943.
20. Hagiwara, S.; Byerly, L. Calcium channel. *Annu. Rev. Neurosci.* 4:69–125, 1981.
21. Hamill, O.P.; Marty, A.; Neher, E.; Sakmann, B.; Sigworth, F.J. Improved patch-clamp techniques for high-resolution current recording from cells and cell-free membrane patches. *Pflugers Arch.* 391:85–100, 1981.
22. Hess, P; Lansman, J.B.; Tsien, R.W. Different modes of Ca channel gating behaviour favoured by dihydropyridine Ca agonists and antagonists. *Nature* 311:538–544, 1984.
23. Hille, B. The permeability of the sodium channel to organic cations in myelinated nerve. *J. Gen. Physiol.* 58:599–619, 1971.
24. Hille, B. *Ionic Channels in Excitable Membranes.* Sinauer Associates, Sunderland, Mass, 1984.
25. Hille, B.; Schwarz, W. Potassium channels as multi-ion single-file pores. *J. Gen. Physiol.* 72:409–442, 1978.
26. Hodgkin, A.L.; Huxley, A.F. A quantitative description of membrane current and its application to conduction and excitation in nerve. *J. Physiol. (Lond.)* 117:500–544, 1952.
27. Hodgkin, A.L.; Katz, B. The effect of sodium ions on the electrical activity of the giant axon of the squid. *J. Physiol. (Lond.)* 108:37–77, 1949.
28. Hodgkin, A.L.; Keynes, R.D. The potassium permeability of a giant nerve fibre. *J. Physiol. (Lond.)* 128:61–88, 1955.
29. Latorre, R.; Miller, C. Conduction and selectivity in potassium channels. *J. Memb. Biol.* 71:11–30, 1983.
30. Levitan, I.B. Phosphorylation of ion channels. *J. Memb. Biol.* 87:177–190, 1985.
31. Matteson, D.R.; Armstrong, C.M. Properties of two kinds of calcium channels in clonal pituitary cells. *J. Gen. Physiol.* 87:161–182, 1986.
32. Noble, D. The surprising heart: a review of recent progress in cardiac electrophysiology. *J. Physiol. (Lond.)* 353:1–50, 1984.
33. Noda, M.; Shimizu, S.; Tanabe, T.; Takai, T.; Kayano, T.; Ikeda, T.; Takahashi, H.; Nakayama, H.; Kanaoka, Y.; Minamino, N.; Kangawa, K.; Matsuo, H.; Raftery, M.A.;

Hirose, T.; Inayama, S.; Hayashida, H.; Miyata, T.; Numa, S. Primary structure of *Electrophorus electricus* sodium channel deduced from cDNA sequence. *Nature* 312:121–127, 1984.

34. Nowycky, M.C.; Fox, A.P.; Tsien, R.W. Three types of neuronal calcium channel with different calcium agonist sensitivity. *Nature* 316:440–443, 1985.

35. Reuter, H. Calcium channel modulation by neurotransmitters, enzymes and drugs. *Nature* 301:569–574, 1983.

36. Ritchie, J.M.; Rogart, R.B. The binding of saxitoxin and tetrodotoxin to excitable tissue. *Rev. Physiol. Biochem. Pharmacol.* 79:1–50, 1977.

37. Salkoff, L.; Wyman, R. Genetic modification of potassium channels in *Drosophila* shaker mutants. *Nature* 293:228–230, 1981.

38. Shuster, M.J.; Camardo, J.S.; Siegelbaum, S.A.; Kandel, E.R. Cyclic AMP-dependent protein kinase closes the serotonin-sensitive K^+ channels of *Aplysia* sensory neurones in cell-free membrane patches. *Nature* 313:392–395, 1985.

39. Spitzer, N.C. Ion channels in development. *Annu. Rev. Neurosci.* 2:363–397, 1979.

40. Tsien, R.W. Calcium channels in excitable cell membranes. *Annu. Rev. Physiol.* 45:341–358, 1983.

41. Ussing, H.H. The distinction by means of tracers between active transport and diffusion. The transfer of iodide across the isolated frog skin. *Acta Physiol. Scand.* 19:43–56, 1949.

Sensory Transduction

INTRODUCTION: GENERAL PROPERTIES

The Parthenon, the Iliad, and countless other examples of ancient Grecian art provide vivid evidence that the early Greeks were among the world leaders in the development of aesthetics. They made remarkable artistic progress in areas that exploited a wide range of sensory inputs: sculpture, music, theater, and bacchanalia. An aesthetics, or artistry of the mind, was also apparent in the teachings of Socrates and Plato. Ironically, the fathers of Greek philosophy, and their followers for centuries to come, had little regard for information obtained from the senses. They pointed out that it was often misleading or illusionary and argued that pure thought was possible only in the absence of sensory input. This view was challenged in the late eighteenth century when David Hume, the Scottish originator of empiricism, postulated the hypothetical situation of a child born devoid of all senses—no sight, no hearing, no touch, no taste, no smell—and by whatever means necessary kept alive for 18 years. He argued that without mechanisms to receive sensation from the outside world, the mind of such an 18-year-old would contain no thoughts, no information—nothing—and for all practical purposes would be dead.

We raise this issue for two reasons. First, it points out the importance of sensory input as the origin of all we know and justifies our inquiry into how receptors work. Second, it is related to one of the major themes of this chapter, an unwritten law of biology well understood by Hume: The life of every living thing—whole animals, plants, single living cells, bacteria, a virus looking for a host—depends on the ability to sense the surrounding environment. The first organisms to appear in the primordial soup were those that were able to locate and utilize nutrients to fuel their own metabolism. Cells that evolved from this ancestral stock and now form the building blocks of more complex living systems have not lost their need to detect life-essential morsels and messages. Although all living cells have the ability to translate physical events into biologically relevant signals, some cells exist solely to provide the nervous system with information about the detection and discrimination of different stimuli. These cells are commonly called sensory receptors, and the purpose of this chapter is to describe how they work and the basic principles governing their behavior.

The job of sensory receptors is conceptually straightforward: they receive physical stimuli and transduce (convert) them into biological signals. They are input-output devices. The input can be mechanical, chemical, electrical, thermal, or electromagnetic (light). The output can also vary. In all vertebrate sensory receptors studied to date, the biological signal produced is a change in membrane potential. This is a logical step toward the goal of providing the central nervous system with sensory information. The change in membrane potential may spread passively as an electrotonic slow potential or propagate as an action potential to the synaptic terminal of the receptor, where it influences the release of transmitter and relays the signal to the next cell in the afferent pathway. In lower animals and unicellular organisms, in which sending information to a central nervous system is not a relevant factor, the biological signal (the output) may be highly varied and is not easily categorized. We will consider an example later.

Transduction. Transduction of a physical stimulus into a biological signal is the keystone of sensory receptor function. The underlying mechanism may be thought of as a clockwork made of molecular gear wheels, pulleys, springs, and strings. Although many pieces of the molecular machinery have not been identified, in many cases the major events in transduction are known and are generally similar to cellular coupling processes you have already learned about. Neuromuscular transmission is a good example. We will use it to introduce the principal elements involved in the design of all sensory receptors.

Skeletal muscle may be considered a highly specialized chemoreceptor; it is a sophisticated acetylcholine detector. It wasn't until the early 1970s that methods to assay acetylcholine using gas chromatography–mass spectrometry were developed that matched or exceeded the sensitivity of bioassay techniques based on acetylcholine-induced contraction of muscle. Muscle's exquisite sensitivity to acetylcholine depends on a specialized transmembrane protein, the acetylcholine receptor, that changes conformation when acetylcholine binds to sites on the outside surface of the molecule (Chap. 6). The conformational change involves a molecular rearrangement that converts the protein into a nonselective cation pore through which flows a brief net influx of positive charge. In a similar manner, sensory receptors depend on specialized proteins to detect the presence of specific elements in the environment. They are analogous to the acetylcholine receptor or the insulin receptor or any other type of a large variety of membrane receptors. They may be thought of as the receptor cell's receptor, and to avoid confusion in terminology, we will refer to them as *sensors* or *receptor proteins*.

Specific sensors exist that can selectively detect chemical stimuli, mechanical events, electromagnetic radiation (photons), temperature, and electric fields. The first four are the basis of the classic senses: smell, taste, touch, hearing, sight, and temperature. Electric field sensors, although less familiar, are widely distributed in the animal king-dom. Voltage-sensitive ion channels such as the sodium channels in excitable membrane may be thought of as highly specialized sensors that monitor the transmembrane electric field. Sensors that have been adapted to detect extracellular electric fields are well developed in certain fish, in which they are known as electroreceptors, and play a role in navigation and the location of prey and predators.

Interaction of the physical stimulus with the sensor triggers a structural rearrangement of the protein that culminates in a change in the ionic permeability of the cell. The membrane permeability change releases energy stored in the form of ion concentration gradients to produce an amplified signal. The linkage between the stimulus-induced conformational change of the sensor and the change in the ionic permeability of the surface membrane is called the *transduction process*. It is the least understood aspect of sensory receptor function and varies from one receptor modality to another. It may simply be a conformational change that permits the sensor to double as an ion channel, as in the case of the acetylcholine receptor, or it may be a complex sequence of events that uses intracellular second messengers to modulate many independent permeation sites. The final effect of transduction may be either an increase or a decrease in ion permeability, depending on the type of sensory receptor. Typically, the permeation pathway controlled is, like the acetylcholine receptor channel, a relatively nonselective cation-permeable pore through which flows a net inward current with an equilibrium potential in the vicinity of 0 mV. The transduction-controlled channel is often referred to by the physical stimulus that operates it even if the coupling is indirect: e.g., light-sensitive channel in photoreceptors, mechanosensitive channel in mechanoreceptors, chemosensitive channel in chemoreceptors, and so forth.

The amplitude of the potential change produced by opening or closing a stimulus-sensitive channel depends on the size of the single-channel current and the passive electrical properties of the cell. Although there is little direct information about the size of the single-channel event, in most sensory receptors it is expected to be small. This makes it necessary to sum the voltage displacements produced by many channels in order to generate an appreciable change in membrane potential. Since the peak amplitude of the potential change declines with time and distance as it spreads from its site of generation, it is important for effective summation that channels be physically close together. For this reason, stimulus-sensitive channels are concentrated in a small region of the cell, rather than distributed diffusely over the entire cell surface. The part of the cell that contains the stimulus-sensitive channels is called the *transduction site*. It is a distinct region of the cell and is analogous to the postsynaptic or end-plate region of skeletal muscle.

Efficient summation of single channel events also requires that they operate as much as possible in unison; the more synchronously individual channels behave, the more their separate contributions are able to sum. In many sensory receptors synchronization is aided by *accessory structures* that act as an interface between the external environment and the sensory receptor. The analogous structure at the neuromuscular junction is the presynaptic nerve terminal that delivers a pulse of acetylcholine to the acetylcholine-sensitive postsynaptic region of the muscle. The accessory structures associated with sensory receptors are highly varied. They can be part of the receptor, e.g., the oil droplet of certain cone photoreceptors, or completely separate elements that are either simple, e.g., the mucous covering of a taste receptor, or very elaborate, e.g., the mechanical linkage that couples sound pressure to stimulation of the hair cell receptors in the auditory system. Accessory structures also perform a wide range of functions. These most commonly include a dual role of protecting the delicate transduction site of the receptor against injury from inappropriate stimuli while at the same time focusing, filtering, or modifying the delivery of appropriate stimuli. By having some control over the nature, form and intensity of the physical stimulus that reaches the transduction site, the accessory structures are able to influence many aspects of receptor performance. They are often a major source of diversity in receptor sensitivity. There are examples, principally in the area of mechanoreception, in which the design of the accessory structure is changed to permit the same type of receptor to detect a different aspect of a physical stimulus.

Receptor Potential. The tiny voltage changes produced by individual stimulus-sensitive channels add together to produce a larger change in membrane potential that is analogous to the endplate potential and is called the *receptor potential*. This a local potential change that is graded in amplitude according to the strength of the physical stimulus. The relationship between stimulus intensity and response is an important aspect of sensory receptor function and is qualitatively similar in many types of receptors. The *stimulus-response curve* for a typical sensory receptor is plotted in Figure 5–1. The receptor's response increases from the smallest detectable response to the largest response, which cannot be exceeded no matter how strong the stimulus. A physical stimulus that evokes a maximum response is called a saturating stimulus. Since the amplitude of the receptor potential cannot be used to discriminate between two saturating stimuli, the useful operating range of the receptor, often called its dynamic range, extends from the smallest detectable response to saturation.

In many kinds of sensory receptors, the growth and ultimate saturation of the response with increasing stimulus intensity may be approximated mathematically by a rectangular hyperbola, referred to as a Michaelis equation, having the form

$$V = V_{max} \frac{S}{S + K} \quad (1)$$

in which V is the measured response, V_{max} is the maximum response, S is the stimulus strength, and K is a constant. This empirical equation describes the behavior of systems with a limited number of at least one of the elements necessary for generating the response. Saturation occurs when the supply of this element has been exhausted. The limiting element could be the receptor protein, the stimulus-sensitive ion channel, or one of the reactants in the transduction chain that couples the receptor protein to the operation of the channel. Consider the case in which the receptor protein is the origin of saturation. The total number of receptor proteins exists as either free receptors (R) or as receptors occupied by the stimulus they are designed to detect (RS). This can be written as

$$R_{total} = R + RS \quad (2)$$

The combination of the stimulus with the receptor protein is a reversible reaction that obeys the law of mass action

$$R + S \rightleftarrows RS$$

with a "dissociation constant" given by

$$K = \frac{R \cdot S}{RS} \quad (3)$$

Combining equations 2 and 3 gives

$$RS = R_{total} \frac{S}{S + K} \quad (4)$$

Figure 5–1 Stimulus-response curve of a typical sensory receptor. The main graph plots response—expressed as a fraction of the maximum response (V_{max})—against a linear measure of stimulus intensity. Measured responses (points) extend from the smallest detectable response to the maximum or saturating response. The difference in intensity between the stimulus that evokes the smallest detectable response and the one that evokes a saturating response is the dynamic range of the receptor. It defines the range of intensities over which the receptor can provide useful information about strength of the stimulus. The smooth curve fitted to the data is a Michaelis equation (Equation 1). In order to plot linearly the full range of stimulus intensities on graph paper of this size, it is necessary to change the scale of the abscissa (indicated by breaks in the record). The insert shows that this is not necessary when the same data are plotted against the logarithm of stimulus intensity. The use of the log scale compresses the abscissa and transforms the rectangular hyperbola in the main graph to an S-shaped curve. The lower panel shows the rising phase of the stimulus reponse curve on an expanded scale. The slope of the curve, i.e, sensitivity, decreases with increasing stimulus intensity. Weak stimuli that evoke responses in the vicinity of the smallest detectable response do so by utilizing the maximum sensitivity of the receptor. The sensitivity with which the receptor generates larger responses decreases progressively with increasing stimulus strength.

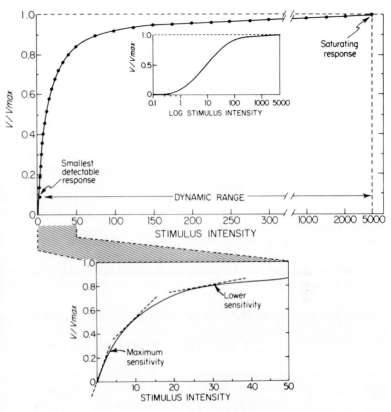

If the measured response (V) is directly proportional to the number of occupied receptor proteins (RS) and the maximum response occurs when all the receptors are occupied, then equation 4 becomes

$$V = V_{max} \frac{S}{S + K}$$

This illustrates that production of a response in a sensory receptor can be like other ligand-receptor–mediated processes in biology. It also raises some questions about the meaning of threshold in a sensory receptor. Each combination of a stimulus with a vacant receptor protein initiates in an all-or-none manner some discrete change that may be considered a response, regardless of whether or not it can be detected by an observer. Since a response defined in this way can be produced by any stimulus intensity greater than zero, it can be argued that such a response has no meaningful threshold. Threshold refers to the point at which a stimulus is just strong enough to produce a particular response, i.e., one that is *detectable*, such as an action potential, or a sensory perception.

The existence of a threshold indicates that the stimulus intensity can be greater than zero but still too weak to evoke the desired level of response. The initial step involving the combination of the stimulus with a receptor protein has no threshold, whereas subsequent steps have various thresholds, depending on which aspect of receptor function is considered.

The stimulus-response curve also contains information about the sensitivity of the receptor, which may be defined as the change in the response produced by a small change in the strength of the stimulus. Sensitivity is the slope of the stimulus response curve. The curve drawn in Figure 5–1 has a slope that changes with the stimulus intensity, as can be shown graphically by drawing tangential line segments to different points along the curve, or analytically by taking the derivative of Equation 1

$$\frac{dV}{dS} = \frac{K\,V_{max}}{(K + S)^2} \qquad (5)$$

Either method shows that sensitivity (slope) is greatest for the weakest stimuli and declines with increasing stimulus strength. To verify this, calculate dV/ds from Equation 5 when S = 0.1 K, S = K, and S = 10 K. Such adjustment of sensitivity according to stimulus strength is often called automatic gain control. It allows the receptor to provide useful information over a wide range of stimulus intensities. If, instead, the receptor operated at maximum sensitivity for all stimulus intensities, the response would go from threshold to saturation with a very small increase in stimulus intensity, and the dynamic range of the receptor would be greatly reduced.

Adaptation. Sensory receptors also show time-dependent changes in sensitivity. The response to a long-lasting stimulus decays over time from an early peak value. Such a decline in the amplitude of the response to a constant stimulus is called adaptation. Adaptational changes occur in many other biological systems. An analogous phenomenon at the neuromuscular junction is called de-

sensitization; the postsynaptic response to a maintained level of acetylcholine declines with time. Inactivation of voltage-sensitive sodium channels is another sort of adaptational change; sodium channels close in the face of constant depolarization. Adaptation is a property of all sensory receptors, but receptors vary according to their rates of adaptation. This fact has been used to classify receptors into three categories depending upon whether they show rapid adaptation, slow adaptation, or a mixture—fast followed by slow adaptation (Fig. 5–2).

The mechanism of adaptation in any sensory receptor is not fully understood. One reason for this is that every step from the arrival of a stimulus at the accessory structure to the production of the measured response is a likely source of adaptation; every step can change its performance over time. To understand adaptation fully requires identifying every event in the entire process and knowing

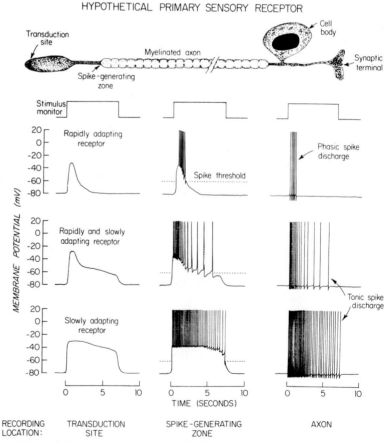

Figure 5–2 Effect of adaption on spike discharge in a primary sensory receptor. The generalized morphology of a primary receptor is shown in the upper drawing. A long-duration physical stimulus applied to the transduction site produces a depolarizing receptor potential that spreads passively to the spike-generating zone where action potentials are generated when the depolarization exceeds spike threshold. Once threshold is exceeded, the frequency of action potential production increases linearly with the amplitude of the depolarizing receptor potential. Action potentials are conducted by the axon to the synaptic terminal. Responses recorded at three different locations—(left to right) transduction site, spike-generating zone, and axon—from receptors showing three different rates of adaptation—(top to bottom) rapid, rapid/slow, and slow—are illustrated in the columns below the drawing. *Top row:* In the rapidly adapting receptor, the receptor potential is a transient event that briefly exceeds spike threshold and generates a short-lasting (phasic) spike discharge. *Middle row:* In the receptor showing mixed rates of adaptation, the receptor potential reaches an initial peak that decays rapidly to a plateau, which subsequently adapts slowly toward baseline. The receptor potenial exceeds spike threshold for the full duration of the stimulus, generating a tonic spike discharge with an initial high-frequency burst of action potentials corresponding to the large initial peak depolarization, followed by a steady lower frequency discharge during the plateau. *Bottom row:* In a slowly adapting receptor, the receptor potential declines gradually. It exceeds spike threshold over its entire duration and generates a tonic spike discharge without a prominent high-frequency burst at the onset of the stimulus.

its time-dependent behavior. This kind of information does not yet exist for any sensory receptor, so most discussions about adaptation are phenomenological rather than mechanistic. For this reason, adaptational changes, which are certainly very important to the daily behavior of a receptor, are not extensively discussed in this chapter.

Thus far, we have considered the sequence of events leading to the production of the receptor potential and some of its basic properties. The job of the sensory receptor, however, does not end with the appearance of the receptor potential. Its last task is to relay the information to the central nervous system by regulating the release of neural transmitter at the synapse with the next cell (commonly termed the second order cell). The synaptic terminal of the receptor cell is located at the opposite end of the cell from the transduction site. (The different functional specializations of opposite poles of the receptor are reminiscent of those in epithelia, and in fact, many sensory receptors are modified epithelial cells.) The coupling of the receptor potential to transmitter release is accomplished in either of two ways. (i) In some cells, the receptor potential modulates transmitter release directly, release being increased by a depolarizing receptor potential and decreased by a hyperpolarizing receptor potential. Since the receptor potential is a local potential change that decreases with distance, it is necessary that the transducing region where it arises be close to the part of the cell that releases transmitter.* Consequently, sensory receptors of this type are relatively small cells, on the order of 10 to 100 μm in length. Notable examples include photoreceptors, hair cells of the ear, and taste receptors. (ii) In other cases, the transducing region and synaptic terminal of the sensory receptor are separated by long distances, in some instances up to a meter or more. Examples in this category include muscle stretch receptors and joint receptors. Problems associated with the spatial attenuation of a local potential change are avoided in these cells by using the receptor potential to trigger one or more action potentials in a fashion analogous to the way the end-plate potential triggers a muscle action potential at the neuromuscular junction. In this way, information contained in the receptor poten-

tial is encoded in a train of action potentials that are conducted to the synaptic terminal, where they provide instructions for the release of specific amounts of transmitter. Receptors that generate action potentials are called *primary sensory receptors*, and their receptor potential is commonly referred to as a *generator potential*; it generates action potentials. Receptors that are incapable of action potential production are termed *secondary or tertiary sensory receptors*, depending on whether action potentials are first elicited in the second or third cell in the afferent pathway. Hair cells and taste cells are secondary receptors. Vertebrate photoreceptors, retinal rods and cones, are the only known examples of tertiary sensory receptors (i.e., the third cell in the visual pathway—the ganglion cell—is the first cell to generate action potentials).

Most sensory receptors are primary receptors that produce a depolarizing receptor potential, generating, when it exceeds threshold, action potentials at a frequency proportional to the amplitude of the depolarization. As a result of the proportionality between receptor potential amplitude and action potential frequency, long-lasting stimuli give rise to a spike train that changes frequency according to the adaptational changes in the amplitude of the receptor potential. The terminology used to describe the expected spike train produced by primary receptors that adapt at different rates is shown in Figure 5–2.

The schematic drawing in Figure 5–3 presents a blueprint of a generalized sensory receptor and summarizes the major events in the transduction of a physical stimulus into a biological signal. The details of transduction in different types of sensory receptors will be dealt with by considering specific examples. The order of discussion proceeds from best to least understood sensory modality. This will provide an overall impression of the field of sensory transduction and will illustrate that there are still many unsolved mysteries about receptor function.

PHOTORECEPTORS

Photosynthesis and respiration are the most fundamental energy transformations in biology. The former is driven by energy in the form of light and leads to the production of carbohydrate from carbon dioxide and water. The latter is the reverse reaction and liberates energy that is channeled through other reactions to perform cell work. All living systems run on energy that originates in light and is stored in chemical bonds. In view of

---

*In this discussion of *close*, the appropriate unit of measure is the cell's electrical space constant (the distance it takes a passive potential to decay to 1/e, or 37% of its maximum amplitude; Chap. 3). In order to preserve small signals it would seem appropriate that the transducing and synaptic regions of the cell be separated by less than a space constant.

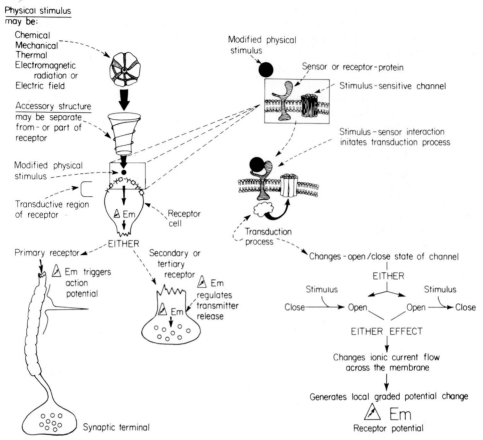

Figure 5–3 A summary of receptor function. The schematic diagram illustrates the principle events involved in the conversion of a physical stimulus into a biological signal. Accessory structures relay a modified form of the original physical stimulus to the transductive region of the receptor. A magnified view of the modified stimulus, the specialized sensor protein in the receptor membrane, and the stimulus-sensitive channel is shown on the right. The transduction process initiated by the combination of the stimulus and sensor protein results in either the opening or the closing—depending on the type of sensory receptor— of the stimulus-sensitive ion channel, causing a change in ionic current flow and production of a receptor potential. In secondary and tertiary sensory receptors, the receptor potential directly regulates the release of neurotransmitter from the synaptic terminal. The receptor potential in primary receptors does this indirectly by influencing the generation of action potentials that are conducted along an axon to the synaptic terminal.

the crucial role of light in biology and because light is an ideal sensory stimulus[12]—it travels in straight lines at a high speed and can be made to form an image that contains information about present and impending events—it is not surprising that all forms of life have some type of light detector. As a result, the topic of photoreception is an enormous one. Our discussion is restricted to the photoreceptors found in vertebrates, which include two distinct classes: rods and cones. Rods are more numerous and larger than cones, and for these reasons have been studied more extensively. Since we know more about them, this section deals almost exclusively with their physiology.

Structure. Vertebrate photoreceptors are of course found in the eye, but rather surprisingly, they are located in the back of the eye, in the deepest layer of the retina. Light focused by the lens (an accessory structure) on the retina must first pass through its inner layers before reaching the photoreceptors. This backward design probably has to do with the photoreceptors' dependence on a close association with the pigment epithelium for nourishment and maintenance.

Morphologically, photoreceptors may be divided into three parts: synaptic terminal, inner segment, and outer segment (Fig. 5–4). Light entering the pupil first passes the output end or the synaptic terminal of the photoreceptor. There are several interesting things about the synaptic connections of the photoreceptor, one of which has to do with the structure of the synapse. Dendritic processes of retinal second-order cells insert into an invagination in the base of the receptor terminal that contains a specialized intracellular structure called a synaptic ribbon. The ribbon consists of a pentalaminar bar with a row of synaptic vesicles on either side that evokes the image of a conveyor belt delivering vesicles to their release site; the true function of the synaptic ribbon is not known. It is found in many kinds of sensory receptors, hair cells and electroreceptors, for example, and is thought to occur where there is a need for prolonged periods of continuous transmitter release.

The synaptic terminal is connected by a process to the inner segment, which houses the metabolic machinery of the cell. It contains the nucleus, endoplasmic reticulum, Golgi, and a dense collection of mitochondria called the ellipsoid, which is located near the boundary between the inner and outer segments. The inner and outer segments are joined by a short, modified cilium called the ciliary neck. It is about 0.1 μm in diameter and contains a peripheral ring of nine pairs of microtubules emerging from a basal body in the inner segment. The presence of this structure frequently provokes questions as to why a cilium would develop into a photoreceptor. This question might be easier to consider if it were rephrased: Why would a photoreceptor develop a cilium? Here the answer is obvious: So the receptor can orient toward the light. Photoreceptors do in fact move and are capable of phototaxis.[7] Psychophysical studies of humans demonstrate that photoreceptors in different parts of the globe of the eye are set at different angles to optimize the collection of axial light rays from the pupil.[65] In other words, photoreceptors "look" at the pupil. The photoreceptors in a human subject wearing an opaque contact lens with an artificial, displaced pupil will adjust their angle to "look" at the position of the new pupil.[17] The mechanism of reorientation is not known, and it is quite possible it has nothing to

Figure 5–4 Photoreceptor morphology. Schematic drawing of vertebrate rod-and-cone type of photoreceptors surrounded by pigment epithelial cells. (Adapted from Young, R.W. *Sci. Am.* 223:81–91, October 1970.) The outer segment, inner segment, and synaptic terminal make up the three morphologically distinct regions of the receptor. Electron micrographs on the left present details of the ribbon synapse (from Lasansky, S. *Invest. Ophthal.* 11:265–275, 1972) in the synaptic terminal, the connecting cilium, and the outer segment disks, including an inset showing the separation between disks and surface membrane. (From Hogan, M.J.; Alvarado, J.A.; Weddell, J.E. *Histology of the Human Eye.* Philadelphia, W.B. Saunders Co., 1971.)

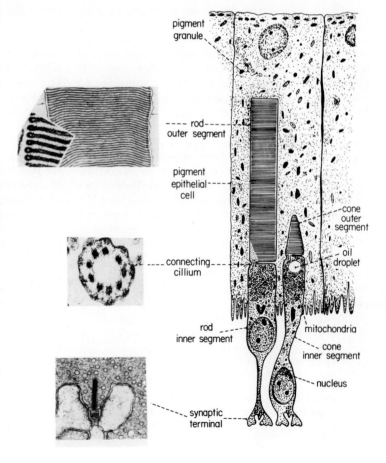

do with the ciliary neck.* Phototaxis was mentioned only to illustrate one of the amazing things photoreceptors do. There are, in fact, no accepted answers to questions about the functional relationship between cilia and photoreception. It is an intriguing problem that is made more so by the presence of ciliary structures in other kinds of sensory receptors, olfactory and mechanoreceptors for example. This may be a natural consequence of their common epithelial cell origin, or it may be an indication that the transduction processes performed by the different receptors have some common components or requirements.

The outer segment is the transduction site. It is the input or the light-sensitive region of the photoreceptor and, ironically, the last part of the cell to see the light. The principal structural difference between rods and cones concerns the morphology of their outer segments. Cone outer segments are cone-shaped, covered by a highly infolded surface membrane, and devoid of intracellular organelles. Rods, on the other hand, are cylindrical, covered by a smooth surface membrane, and filled with saclike intracellular organelles called disks. There are about 1000 disks per outer segment of rod. They are flat pancake-like structures that are not connected to the surface membrane, which as we shall see, imposes an important restraint on the transduction process. Cone infoldings and rod disks most likely represent different solutions to the common desire to improve light collection by increasing membrane surface area.

Rod and cone outer segments nestle next to the pigment epithelium, which extends processes that fill most of the extracellular space between the distal portion of neighboring receptors. A continual exchange of materials between the photoreceptors and the pigment epithelium is essential for the long-term survival of the receptors. One important exchange has to do with the removal of old outer-segment membrane. Photoreceptors, like many other kinds of sensory receptors, engage in continual membrane renewal. New disks are formed at the base of the outer segment at a rate of 1 every 30 minutes[71] in mammals by pinching off invaginations of the surface membrane. With the formation of each new disk, the stack of old disks is displaced toward the distal tip. The trip from base to tip takes about three weeks. Old disks are discarded on a daily basis by invagination

of the surface membrane and pinching off the distal tips, which are then phagocytosed by the pigment epithelium and ultimately degraded. The shedding process occurs in both rods and cones and is synchronized with diurnal light-dark cycles; in brief, we eat the tips of our rods for breakfast and the tips of our cones for dinner. If disk shedding is impaired, as in the hereditary retinal disease retinitis pigmentosa, photoreceptors degenerate and blindness results.

This completes a description of the main structural features of the photoreceptor and some of the dynamic changes it undergoes on a daily basis. The photoreceptor may be viewed as a magnificent machine that can locate a source of illumination, steadily discharge neurotransmitters, and constantly make new parts while discarding old ones.

Receptor Potential. How does the photoreceptor work? How does it know when the lights are on? A friend's response to this question was, ''The electric meter is running.'' This is an excellent answer, because it provokes a vivid image of how the photoreceptor could be designed. The only problem is it is backwards; photoreceptors know when the lights are on because the electric meter is off. Let us explain.

In darkness, an ionic current circulates steadily through the photoreceptor, flowing into the outer segment and out of the inner segment (Fig. 5–5A). The current entering the outer segment is predominantly carried by Na^+ ions moving down their electrochemical gradient through light-sensitive channels in the surface membrane. *The light-sensitive channels are open in darkness.* The steady influx of sodium into the outer segment keeps the receptor relatively depolarized at a ''resting potential'' of approximately -40 mV in darkness. The current flowing out of the inner segment is carried by K^+ ions moving down their electrochemical gradient through inner segment K channels. If an electric meter were placed on the receptor within the ciliary connection between the inner and outer segments, it would be running in darkness. The meter would measure a steady current that flows in darkness and is called the dark current. The photoreceptor pays its electric bills with ATP which is used to drive Na^+-K^+ pumps in the ellipsoid region of the inner segment that maintain the ion gradients necessary to sustain the dark current.

Approximately 50 ms after a bright light is turned on, the light-sensitive channels in the outer segment close, sodium influx ceases and the dark current shuts off (the electric meter stops running; Fig. 5–5B). Potassium continues to leave the inner

---

*The connecting cilium of the photoreceptor lacks the central pair of microtubules characteristic of motile cilia. In this regard, it resembles the primary cilium commonly observed in developing cells.

Figure 5–5 Ionic current flow in vertebrate photoreceptors. *A*, In darkness cations (predominately sodium) flow into the outer segment through open light-sensitive channels in the surface membrane while K^+ ions flow out of the inner segment through open light-insensitive K channels. Na^+-K^+ pumps in the inner segment maintain the concentration gradients necessary for the steady flow of current. The influx of Na keeps the rod partially depolarized in the dark (see plot) at approximately −40 mV. *B*, Light initiates the transduction process, which culminates in closure of light-sensitive channels. Potassium continues to leave the inner segment, causing the cell to hyperpolarize. The record plotted on the right illustrates the change in membrane potential evoked by a brief flash of light (indicated by upper line segment). As light-sensitive channels reopen, the intracellular potential decreases (depolarizes) and returns to the original dark value.

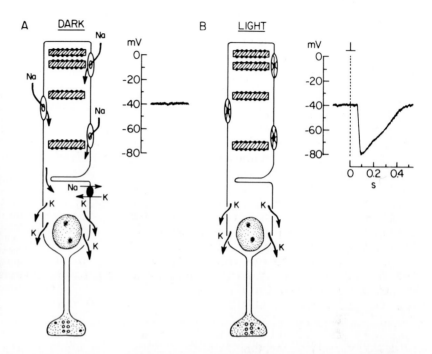

segment; its electrochemical gradient still exists, and the K^+ channels it moves through are not closed by light. The efflux of potassium causes the receptor to hyperpolarize (Fig. 5–5B). Transmitter release from the synaptic terminal of the photoreceptor is thought to be controlled by membrane potential, as it is at other chemical synapses. Release occurs continuously in the dark when the receptor is depolarized and decreases in light when the membrane potential of the receptor is more hyperpolarized. In principle, light could cause a depolarizing or hyperpolarizing synaptic potential in a postsynaptic cell, depending upon whether the transmitter released steadily in darkness has a hyperpolarizing or depolarizing influence on the membrane potential of the postsynaptic cell. Shutting off the release of a hyperpolarizing transmitter, for example, will cause the postsynaptic cell to depolarize. Indeed, both types of synaptic response exist (Chap. 19). Synaptic potentials evoked in second-order cells by the photoreceptor's light response eventually generate signals that are relayed to the central nervous system, where they provoke the perception that the lights are on, even though the electric meter is off.

The suppression of the dark current and the amplitude of the hyperpolarizing receptor potential are graded according to the number of absorbed photons (Fig. 5–6). The stimulus-response curve for a flash of light extends from 1 to 1000 photons (in rods) and is well described by a Michaelis equation.

The foregoing provides a broad view of photoreception and pertains to both rods and cones.[43] The question to be considered next is How does light close the light-sensitive channel? Or more specifically, What is the transduction process that couples light to a decrease in the ionic conductance of the surface membrane?

Rhodopsin. It has been known since the 1800s that the eye contains a light-sensitive compound. Investigators at that time discovered that the retinal lining of a living eye removed from a "volunteer" and cut in half in darkness had a deep robust purple color that turned sickly gray when exposed to light. They called the light-sensitive material visual purple and described it as being "bleached" by light. In subsequent years, this material was localized to the receptor layer, and methods were developed to study its chemistry. We now know that visual purple is the rod photopigment rhodopsin.[52] It consists of two parts: a 41,000 dalton integral membrane protein called opsin and a small ($\approx$ 500 dalton) vitamin A–based chromophore that acts as a light-operated switch. The chromophore is a specific geometric isomer of vitamin A aldehyde called 11-cis retinal. Vitamin A consists of a 6-carbon ring and a 9-carbon side chain made up of alternating double and single bonds. The side chain of the 11-cis isomer is bent at the eleventh carbon. Energy transferred to rho-

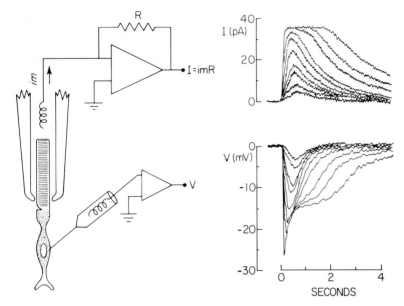

Figure 5–6 Simultaneous current and voltage recordings from a single rod. The recording arrangement is on the left. The outer segment is drawn into a suction pipette, and current flow across the membrane is measured by the amplifier while an intracellular microelectrode records membrane potential from the inner segment. Brief light flashes of increasing intensity delivered at time 0 suppress a standing inward dark current, producing an apparent outward photocurrent (uppermost set of traces) and a hyperpolarizing receptor potential (lower traces). Current and voltage are plotted relative to dark levels of −35 pA and −35 mV. The appearance of a peak-and-plateau phase in the voltage but not the current response to bright flashes is due to the presence of voltage-sensitive channels in the inner segment but not in the outer segment. (From Baylor, D. A., and Nunn, B. *J. Physiol.* 371:115–145, 1986.)

dopsin by absorption of a photon causes photoisomerization of 11-cis to all-trans retinal, which has a straight side chain. The bent and straight forms of the side chain of retinal thus represent the two positions of the light-activated switch. Removing the bend, i.e., turning the switch, is the only action of light in the entire visual process. It occurs in approximately a picosecond.* The structural change of the retinal destabilizes the opsin protein, causing a series of conformational changes that are driven by thermal energy and take place in the dark. The sequence of dark reactions can be followed by measuring changes in the color of the molecule (changes in absorption spectra) during the bleaching process. The details of the photochemistry of rhodopsin are not fully understood, but a list of the principal reactions is shown in Figure 5–7. After the formation of the last intermediate, metarhodopsin III, all-trans retinal dissociates from the membrane-bound opsin, and through an unknown mechanism, is converted back to 11-cis retinal, which in turn combines with a free opsin molecule to regenerate rhodopsin.

Rhodopsin is found only in the outer segment, where it accounts for approximately 98% of the protein. The molecule contains alternating hydrophilic and hydrophobic regions and loops back and forth across the membrane seven times. The transmembrane portions of the molecule are helical, with the chromophore located roughly in the

middle of the seventh helix. Rhodopsin has the overall shape of a prolate spheroid (a football) approximately 3 nm in diameter at its widest dimension and 9 nm long. It is mobile in the membrane. Rhodopsin spins about its long axis like a top and moves laterally within the plane of the membrane with as high a diffusion coefficient as any known membrane protein. There are about 10^9 rhodopsin molecules per rod outer segment, which corresponds to a density of $30,000/\mu m^2$ of disk membrane, or roughly 1 rhodopsin molecule for every 60 lipid molecules. As a result of their close packing and relatively free lateral diffusion, rhodopsin molecules bump into each other about a million times per second. As discussed later, its ability to move freely and collide with other membrane proteins is considered to play an important role in transduction.

Intracellular Second Messenger. It was first thought that the photochemistry of rhodopsin was the whole transduction process. There are several points, however, that make it clear that transduction involves additional elements. First, the various photochemical intermediates appear either too soon or too late to be directly responsible for closure of the light-sensitive channel. The reactions preceding the appearance of metarhodopsin II are extremely fast, with time constants in the microsecond or millisecond range, whereas subsequent reactions are slow, with time constants of hundreds of seconds. Second, 95% of all the rhodopsin in a retinal rod is found in disk membranes, which are separate from the plasma membrane,

---

*A picosecond is to 1 second as 1 second is to 31,709 years, or as 20 cents is to the 1986 national deficit.

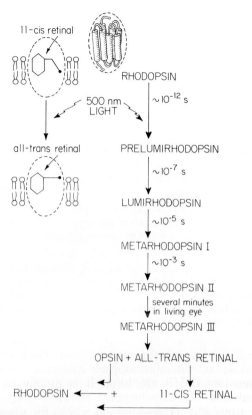

Figure 5–7 Rhodopsin cycle. Rhodopsin consists of a membrane protein (opsin) and a chromophore (11-cis retinal). The upper sketch shows the molecule as a prolate spheroid (a football) with cylinders representing the seven transmembrane segments. The absorption of a photon isomerizes 11-cis retinal to all-trans retinal and in so doing removes a bend from the 9-carbon side chain of retinal. This simple structural change in the molecule initiates a sequence of dark reactions that culminates in the dissociation of all-trans retinal from opsin. The cycle is completed by the reisomerization of all-trans to 11-cis retinal and subsequent regeneration of rhodopsin. The times associated with reactions indicate the approximate time for appearance of the next intermediate.

where the light-sensitive channels are located. This means that a diffusible cytoplasmic transmitter (an intracellular second messenger) is needed to inform the surface membrane that a rhodopsin molecule in a disk membrane has absorbed a photon. At the present time, metarhodopsin II is thought to be the photochemical intermediate that leads to the production of a second messenger.

Further evidence for an intracellular transmitter can be traced to psychophysical measurements in the 1940s[30] showing that the human visual system is able to detect single photons.* This suggests

*It is ironic that we can see a subatomic particle—a photon—whereas we are unable to see atoms.

that the absorption of a single photon—the isomerization of a single rhodopsin molecule—evokes a highly amplified response. It is difficult to account for the amplification required for the single photon sensitivity and still to explain the large dynamic range of the receptor without involving a second messenger. This can be illustrated by considering what would happen if each rhodopsin molecule were a large open channel that closed when the molecule was photoisomerized. Experiments by Baylor, Lamb, and Yau[3] on isolated rods from lower vertebrates indicate that a single photon absorption reduces the 40 pA standing dark current by 1 pA. If each isomerized rhodopsin molecule resulted in closure of a single channel that reduced the dark current by 1 pA the receptor response would be expected to reach its maximum response (saturation) with the absorption of 40 photons. The fact that the rod can give graded responses over an intensity range that extends from 1 to 1000 absorbed photons indicates that absorption of a single photon influences more than a single light-sensitive channel. In order for this to happen, a single isomerized rhodopsin molecule must generate a signal, mediated by many molecules of second messenger.

There have been two contrasting proposals for what light does to an internal transmitter. According to one model, light *releases* a transmitter substance (calcium) that blocks the light-sensitive channel.[24] In the other, model light *destroys* a transmitter substance (cyclic guanosine monophosphate, cGMP) that normally keeps the channels open.[34] The calcium-release hypothesis was formulated by Hagins and Yoshikami in the early 1970s.[25] It was based on the observation that the size of the dark current depends on external calcium concentration; increasing external calcium decreases the dark current. Since disks were known to contain millimolar amounts of calcium, the light-sensitive channel was postulated to be blocked by intracellular calcium released from the disks by light. At that time, there was much to favor this proposal. It was the natural choice. An extensive list of other possible second messengers did not exist, and the established role Ca^{2+} played in triggering secretion and skeletal muscle contraction provided a precedent for its use as an intracellular signal.

Shortly after the announcement of the calcium hypothesis, new interest in the biochemistry of photoreception, going beyond the classical photochemistry of rhodopsin, developed. This originated in the discovery of a unique series of light-sensitive enzyme reactions in photoreceptors. An enzyme that was extracted from rods was inactive and insensitive to light except when purified rhodopsin was added to the extract. In the presence of rhodopsin, enzymatic activity could be turned on by light. The enzyme was eventually identified as cGMP-specific phosphodiesterase—it cleaves cGMP. An alternative to the calcium hypothesis based on the postulate that cGMP keeps light-sensitive channels opened in darkness was proposed. Light-activation of phosphodiesterase reduces the cGMP level and causes channels to close.

In the following years, evidence for and against both the calcium and cGMP models was reported, and the apparent lead in the race for intracellular transmitter changed many times. Three arguments now favor the cGMP hypothesis:

(i) Experimental evidence for light-evoked intracellular calcium release, which is crucial to the calcium hypothesis, has not been clearly obtained. Many research groups have tried, using a variety of techniques, and the experimental findings have varied enormously. In different laboratories, the stochiometry of light-evoked calcium release (the number of calcium ions released into the cytoplasm per isomerized rhodopsin molecule) have ranged from 10,000 to 0.01. In the early 1980s, the first clear, although indirect, evidence for calcium release was published,[20, 70] and the calcium hypothesis was reinstated. These results were based on the measurement of a light-evoked increase in the concentration of calcium in the external solution bathing the outside surface of the rod outer segment. The interpretation of the observation depended on the assumption that changes in extracellular calcium arose from changes in intracellular calcium. Others argued that the extracellular calcium change was the consequence rather than the cause of closure of the light-sensitive channel. Further investigation showed that in darkness Ca leaks into the rod through open light-sensitive channels that have a small but finite permeability to divalent ions. Sodium-calcium exchange sites in the outer segment membrane use energy derived from the Na electrochemical gradient to move Ca out of the cell.[69] The resulting calcium efflux is light-insensitive and maintains the intracellular Ca concentration at approximately 0.2 µM in darkness. Light, by closing light-sensitive channels, reduces the influx of Ca without effecting its efflux via Na-Ca exchange. As a result, the concentration of Ca in the extracellular space surrounding the rod increases during illumination while the intracellular Ca concentration decreases. The discovery that light was associated with a decrease in intracellular Ca[8] was incompatible with the Ca hypothesis, which was based on the postulate that light triggered the release of Ca from intracellular stores.

(ii) Meanwhile information about an elaborate biochemical mechanism for the activation of phosphodiesterase by light was rapidly falling into place. It provided adequate amplification, fit the kinetic requirements of transduction, and drew indirect support from its similarity to the hormonal activation of adenylate cyclase (Chap. 1). The coupling of light to enzyme activity (Fig. 5–8) takes place in three steps.[66] The *first step* is the production of an activated form of rhodopsin (Rh*), which corresponds to one of the photochemical intermediates that appear after absorption of a photon (Fig. 5–7). The activated intermediate is thought to be metarhodopsin II; it appears at about the right time and stays around long enough to finish the job. In the *second step*, Rh* diffuses in the plane of the membrane and collides with a second protein, which is a GTP-binding protein called G protein, or transducin. Transducin is a peripheral membrane protein located on the cytoplasmic surface of the disk membrane. Its transient interaction with an activated rhodopsin catalyzes the exchange of GDP for GTP, causing activation of the transducin molecule. After catalyzing GDP-GTP exchange Rh* is free to interact with another molecule of transducin. A single Rh* molecule serially activates approximately 500 transducin molecules before it is phosphorylated by rhodopsin kinase and inactivated. (Phosphorylation apparently does not change the spectroscopic properties of metarhodopsin II.) The repeated activation of transducin represents the first stage of amplification. In the *third step*, cGMP phosphodiesterase is turned on by interaction with the activated form of transducin. The turnover number of phosphodiesterase is approximately 1000 per second; activation of the enzyme represents a second stage of amplification. Thus isomerization of a single rhodopsin molecule can lead to the hydrolysis of approximately 500,000 molecules of cGMP in a second. The breakdown of cGMP stops when transducin is inactivated by hydrolysis of its GTP to GDP. This is followed in time by recovery of the cGMP level through resynthesis from GTP in a reaction catalyzed by guanyl cyclase.

(iii) The light-evoked changes in cGMP were well documented before the mechanism coupling cGMP to the gating of the light-sensitive channel was established. It was originally thought, largely by analogy with other systems using cyclic nucleotide second messengers, that cGMP might activate a protein kinase that opened the channel by phos-

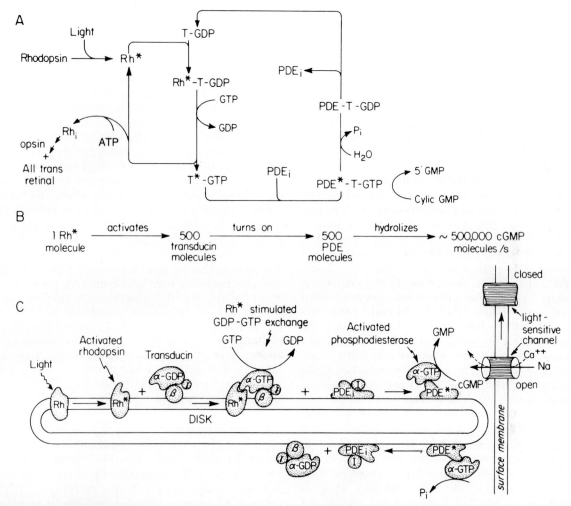

Figure 5–8 Light-regulated enzyme cascade. *A,* Following the absorption of light, rhodopsin undergoes a series of dark reactions leading to the production of an activated intermediate (metarhodopsin II). Activated rhodopsin (Rh*) readily diffuses in the plane of the membrane and through random collisions with transducin molecules (T) catalyzes a GDP-GTP exchange resulting in activation of its alpha subunit. Before it is inactivated by phosphorylation, a single Rh* molecule will activate approximately 500 molecules of transducin. The activated alpha subunit of transducin dissociates from beta and gamma subunits of the molecule and activates phosphodiesterase (PDE*) by removing an inhibitory constraint. Phosphodiesterase hydrolizes about 1000 cGMP molecules/s. The amounts of T and PDE in a rod relative to rhodopsin, which is about 10^9/rod, is 1/10Rh and 1/100Rh, respectively. *B,* Through two stages of amplification, the isomerization of one rhodopsin molecule leads to the hydrolysis of 500,000 cGMP molecules/s. *C,* Schematic drawing showing elements of the nucleotide cascade as integral (rhodopsin) and peripheral (T and PDE) proteins in the disk membrane. Activation of PDE by T_α-GTP results in cGMP hydrolysis and closure of light-sensitive channels. In the open state, the channels are permeable to Na and, to a much smaller extent, Ca. Na^+-K^+ pumps in the inner segment pump Na out of the cell while Na/Ca exchange sites in the outer segment maintain low intracellular [Ca] (not illustrated). Light, by reducing Ca influx through light-sensitive channels, causes a decrease in intracellular Ca, which is thought to consititute a feedback signal that plays a role in receptor adaptation by influencing the recovery of cGMP levels and restoration of the dark current.

phorylating it. In the early 1980s, patch-clamp recording techniques (Chaps. 3 and 4) were applied to rods, and the basic properties of the light-sensitive channel were studied.[13] Among other things, it was shown to be a pore that conducted very little current.[22] Its apparent conductance in standard saline was about 0.1 pS, which is approximately 250 times smaller than the conductance of the end-plate acetylcholine receptor channel. The small conductance of the light-sensitive channel under physiological conditions is due to a blocking action of divalent cations. The conductance of the channel greatly increases to about 25 pS when Ca^{2+} and Mg^{2+} are removed from both the extracellular and intracellular solutions bathing the channel.[29] In the early part of 1985, several groups reported that cGMP opened a conductance in excised patches of outer segment membrane with the properties of the light-sensitive conductance.[19, 28] Since this occurs even in cell-free membrane patches, cGMP control of the channel is presently thought to be by a direct effect rather than through phosphorylation. Thus the light-sensitive channel may conceptually be very similar to many types of synaptic transmitter-activated channels; the major difference is that in the case of photoreceptors the agonist that operates the channel is an intracellular rather than extracellular messenger.

Although evidence presently seems to be solidly behind the cGMP hypothesis, a cautionary note is appropriate. Over the past 10 to 15 years, the controversy surrounding the identity of the intracellular transmitter has been characterized by unexpected discoveries causing abrupt changes in the field. In view of this history, one might predict that the story will undergo future changes. One of the loose ends is the role of calcium. There is good evidence that high amounts of calcium are associated with the disks, and there is no doubt that extracellular calcium reduces the dark current. Answers to the questions of how, where and why may contain several surprises. Photoreceptor adaptation, which we experience every day, arises through the combined effects of several independent events. One source of adaptation is the exhaustion of the supply of unbleached rhodopsin. There are also more subtle changes in receptor sensitivity (adaptation) that occur after the absorption of a single photon and have nothing to do with the availability of photopigment. The molecular basis of this adaptational signal is not known, but there is growing speculation that calcium is somehow involved. The fall in intracellular Ca during illumination (see earlier discussion) may

represent a feedback signal that influences receptor sensitivity by playing a role in the recovery of cGMP levels and restoration of the dark current.[44]

Cones. This discussion of photoreceptors has dealt almost exclusively with rods. A few comments about cones are in order. The principal structural difference between rods and cones has already been described. There are also photopigment differences that give rise to three classes of cones based on the wavelength of maximum light absorption. The spectroscopic differences arise from combining the same chromophore as in rods (11-cis retinal) with different membrane proteins, i.e., different opsins. The chromophore binding site of each class of opsin has a sufficiently different distribution of charge to shift the absorption spectrum of 11-cis retinal over the visible range. In spite of photopigment differences, the transduction process in cones is thought to be basically the same as in rods. Although it apparently uses cGMP as internal transmitter, there are undoubtedly differences in the details of the transduction mechanism that account for several functional differences between rods and cones. A primary example of these is the difference in single-photon sensitivity. Single-photon responses in rods are about 100 times larger than in cones. The origin of this difference is uncertain, perhaps arising in biochemical differences in the nucleotide cascade.

MECHANORECEPTORS

Cells that transduce mechanical stimuli into electrical signals are the most common type of sensory receptors. Vertebrates obtain information about a greater variety of events from these receptors than from any other type of sensory receptor. Their strategic association with hair follicles, diffuse distribution in skin, and scattered occurrence along the fascial planes of muscle allow them to detect the flow of air currents and touch and contact pressure along the body surface. Mechanoreceptors also monitor the development of pressure in teeth, blood vessels, bladder, and gut; the stretch of skeletal, smooth, and cardiac muscle; the extension of tendons and ligaments; the position of joints; the frequency and intensity of sounds; and the relative direction of gravity, head position, and acceleration.

The actual receptor cells responsible for this array of sensory inputs are structurally diverse but functionally similar. In all cases, it appears that accessory structures associated with the receptor transfer a mechanical disturbance in the local environment to a mechanosensitive region of the receptor. This transfer mechanism ranges from virtually direct coupling, as in touch, to elaborately involved linkages resembling Rube Goldberg devices. Consider the peripheral auditory system. The initial stimulus is an air-pressure wave (sound). It is directed by the external ear down the auditory canal, where it strikes the tympanic membrane, causing a displacement that sequentially moves three small bones, the last of which taps on the window of a fluid-filled chamber, creating a fluid wave that produces traveling waves in the basilar and tectorial membranes, which contain opposite ends of the mechanoreceptive hair cells. The basilar and tectoral membranes are hinged on different axes so that when displaced they move relative to each other, causing mechanical stimulation of the hair cells by unequal movement (shearing) of the two ends of the receptor cells. The ingenious features of the design of the accessory structures that couple sound to mechanical stimulation of the hair cell are discussed in Chapter 17, Audition. In this chapter, we are concerned principally with the question of how mechanoreceptors transduce mechanical stimuli into the electrical signals.

Pacinian Corpuscles. One of the first mechanoreceptors to be studied in detail was the Pacinian corpuscle,* an oval-shaped pressure receptor associated with the periosteum of bones, joint capsules, muscle fasciae, and subcutaneous tissues. Numerous, unusually large Pacinian corpuscles occur in the mesentery of the domestic cat. They look like overinflated footballs, with a myelinated nerve protruding from one end (Fig. 5–9). The nerve terminal within the capsule is unmyelinated and is surrounded by onionlike concentric layers of flat, lamellar cells separated by fluid-filled spaces.[57]

Since the nerve terminal of the Pacinian corpuscle is too small for intracellular recording, responses to mechanical stimuli have been studied using extracellular electrodes arranged as shown in Figure 5–9. An isolated corpuscle is placed with an electrode in a saline-filled chamber. Its axon is lifted by a second electrode into a layer of mineral oil floating on the surface of the aqueous solution. A glass stylus connected to a piezoelectric crystal

is positioned to deliver mechanical stimuli to the receptor capsule. A stimulus that causes a local change in the membrane potential of the axon terminal (a receptor potential) will produce a voltage difference between the terminal and unstimulated parts of the axon, causing current to flow in a loop that moves in opposite directions along the intra- and extracellular limbs of the circuit. The intensity of the current depends on the amplitude of the receptor potential and is thus a function of stimulus strength. If the saline/oil interface is positioned near the boundary between stimulated and unstimulated regions of the receptor, the oil will increase the resistance of the extracellular limb of the current loop. Current flowing in the extracellular resistance will produce a voltage difference between the two electrodes that, by Ohm's law ($V = IR$), is directly proportional to the size of the extracellular resistance. The sign of the voltage difference reflects the direction of extracellular current flow and depends on the polarity of the receptor potential.†

A few hundred microseconds after a brief mechanical displacement is presented to the receptor capsule, the electrode on the axon records a positive voltage relative to the electrode positioned in the saline. This indicates that following mechanical stimulation, extracellular current flows from the axon to the encapsulated ending and corresponds to the production of a depolarizing receptor potential. The amplitude of the receptor potential increases with the strength of mechanical stimulation (Fig. 5–9A). Detectable responses are evoked by mechanical displacements as small as 0.1 μm. Stronger stimuli produce progressively larger responses that reach maximum amplitude (saturation) with 2 to 5 μm displacements.[21] When a receptor potential spreads to the first node of Ranvier with sufficient amplitude to exceed the firing threshold, it generates one or two conducted action potentials in the axon.

The Pacinian corpuscle is a classic example of a rapidly adapting receptor; it is sensitive to the transient changes that occur when the stimulus is turned on or off but is insensitive to a maintained mechanical stimulus.[10] The response to a long-lasting mechanical stimulus reaches peak in 1 to 2 ms and then, in spite of the continued application of the mechanical stimulus, returns to the original

---

*Much of what is currently known about the physiology of the Pacinian corpuscle originated from experiments done in the late 1950s and early 1960s by Werner Loewenstein and colleagues. Their series of papers are classics and among the first examples of a systematic investigation of a sensory receptor.

---

†The basic principles discussed here apply whenever extracellular electrodes are used to record the electrical activity of biological tissue. The fundamental idea is to record the voltage difference between two external electrodes that is produced by current flowing in the extracellular resistance (Chap. 3).

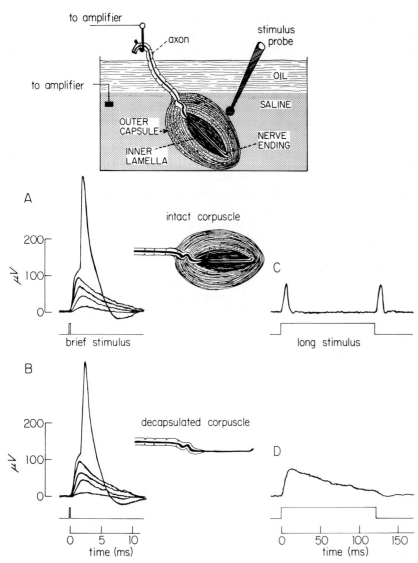

Figure 5–9 Responses of intact and decapsulated Pacinian corpuscles to brief and prolonged mechanical stimulation. The upper part of the figure illustrates the gross morphology of an intact Pacinian corpuscle and the method used to stimulate and record from it. The unmyelinated terminal of the receptor axon is surrounded by concentric layers of flat cells that form an inner lamellar sheath covered by an outer capsule. Mechanical stimuli are applied to the corpuscle by a piezoelectrically driven probe. An amplifier records the stimulus-evoked change in the voltage difference across an oil layer separating the axon from a saline solution bathing the receptor ending. (Adapted from Loewenstein, W.R. *Scientific American*, pp. 89–108, August, 1960). *A*, Responses evoked by progressively stronger brief mechanical stimuli applied to an intact corpuscle. The strongest stimulus evoked a response that exceeded threshold for generating an action potential. *B*, The same series of stimuli evoked similar responses from a decapsulated receptor. The response to long-duration stimulus, however, depends on whether or not the receptor has an intact capsule. *C*, A receptor with an intact corpuscle produces a transient response at the *on* and *off* of the long-duration stimulus. The strength of the stimulus used was not enough to evoke an action potential. *D*, A similar stimulus applied to a decapsulated receptor evoked a response that adapted slowly and failed to generate an *off* response. (After Gray, J.A.B.; Sato, M. *J. Physiol.* 122:610–639, 1953; Loewenstein, W.R.; Mendelson, M. *J. Physiol.* 177:377–397, 1965.)

baseline in 4 to 6 ms. The receptor is transiently reexcited when the stimulus is turned off (Fig. 5–9C). To investigate the origin of this behavior, Loewenstein and Rathkamp surgically removed the lamellar capsule and applied mechanical stimuli directly to the naked axon terminal.[47] These experiments gave two new insights. First, the response to a brief mechanical pulse is the same with or without the capsule (Figs. 5–9A, B), providing direct evidence that the nerve terminal rather than the lamellar cell capsule is the initial site of mechanotransduction. Second, removal of the capsule changes the adaptational properties of the receptor and abolishes the "off" response (Fig. 5–9D), providing evidence that the mechanics of

the capsule accounts for the rapid phase of adaptation and the production of the "off" response. Long-lasting stimuli applied directly to the denuded nerve terminal evoke a response that declines slowly over the duration of the stimulus.

As we have seen, removal of the lamellar capsule (the accessory structure) converts the Pacinian corpuscle into a slowly adapting receptor (Fig. 5–9D). In order to understand how this occurs, it first is necessary to know what physical attribute of the mechanical stimulus actually excites the receptor. The alternatives are pressure and membrane deformation. Experiments using a miniature hyperbaric chamber established that the receptor is not excited by uniform pressure. This ruled out the possibility that a pressure-sensitive chemical reaction underlies transduction. The results of these and other experiments indicate that the receptor is excited by stretching the membrane of the axon terminal.

Deformation of the capsule could be coupled to stretch of the terminal membrane in the following way.[48, 33] A steady mechanical displacement applied to the outer face of the capsule *initially* acts on a rigid column of lamellar cells and fluid spaces that effectively transfer the stimulus to the terminal, where it causes membrane deformation and receptor excitation. With time, fluid in the interlamellar spaces flows away from the region compressed by mechanical displacement; this distends other parts of the elastic capsule, equalizes the pressure on the terminal, and restores the membrane to its original shape, allowing the response to subside. When the mechanical stimulus is suddenly withdrawn, the pressure built up in the extended regions of the capsule deforms the terminal membrane for a second time and reexcites the receptor, producing an "off" response.

The receptor response is virtually abolished in sodium-free saline, suggesting that membrane stretch opens cation channels predominantly permeable to Na.[15] The presence of stretch-sensitive channels could explain the directional sensitivity of the Pacinian corpuscle.[38] Mechanical deformation directed along an axis perpendicular to that axis giving the maximum depolarizing response elicits a *hyperpolarizing* receptor potential. The explanation for this fact has to do with the elliptical cross-sectional geometry of the axon terminal. If internal volume is constant, the surface area will increase (membrane stretch) when the deforming stimulus is perpendicular to the major axis of the terminal, and it will decrease (membrane relaxation) when the stimulus is perpendicular to the minor axis. In one case, the stimulus would cause stretch-sensitive channels to open and in the other case would cause them to close. How stretch is coupled to the permeability change is not known. A possible mechanism is discussed at the end of the section on mechanoreceptors.

The distribution of stretch-sensitive channels was studied by mapping the mechanical sensitivity of different regions of the receptor. The node of Ranvier, but not myelinated portions of the axon, responds to mechanical stimuli. The threshold displacement needed to excite the node is about 100 times greater than the displacement needed to excite the terminal.[46] This suggests that both regions of the receptor contain stretch-sensitive channels. The difference in sensitivity may be due

in large part to differences between the geometry of the terminal and of the axon proper. A small displacement will produce a larger percentage of change in the surface area of a structure with an elliptical cross section than it does in one with a circular cross section (assuming a constant internal volume). There are other morphological specializations of the terminal that may contribute to its mechanosensitivity. It contains a prominent collection of closely packed mitochondria, intricate networks of 6-nm microfilaments, and scattered arrays of clear- and dense-core vesicles.[64] The relationship of these structures to the physiology of the Pacinian corpuscle is not known, but their absence from other parts of the receptor axon suggests that they may be functionally related to mechanotransduction.

Much research on mechanotransduction was done in the late 1950s to mid 1960s. During that time a favorite preparation in addition to the Pacinian corpuscle was the crayfish stretch receptor.[2] This is a primary sensory receptor that has a peripherally located cell body and an extensive dendritic arbor that twines among the connective tissue and myofibers of the muscle whose stretch it is designed to sense. Although the cell body is large enough to impale with microelectrodes and can even be voltage-clamped, biophysical analysis of the stretch receptor is difficult because transduction occurs at remote sites on the fine dendritic processes, and stretch applied to the muscle is difficult to quantitate in terms of dendritic membrane displacement.

Hair Cells. In the late 1970s and early 1980s, A.C. Crawford and R. Fettiplace in Cambridge, England[9, 10] and A.J. Hudspeth and colleagues in California led a renaissance in mechanoreceptor research. The two groups independently developed clever intracellular recording techniques and means of controlled mechanical stimulation to study the response properties of the quintessential mechanoreceptor—the vertebrate hair cell. Hair cells occur in the sensory organs that make up the "acousticolateralis system" and are responsible for the detection of sound, linear and angular acceleration, water motion, and gravity. This wide range of sensitivities is due to differences in the design of the various sensory organs and to their accessory structures, rather than to differences in the general morphology or function of the hair cells. Like many other kinds of sensory receptors—taste receptors, olfactory receptors, and photoreceptors, for example—hair cells are modified epithelial cells. They form with supporting cells a partition between two dissimilar environments.

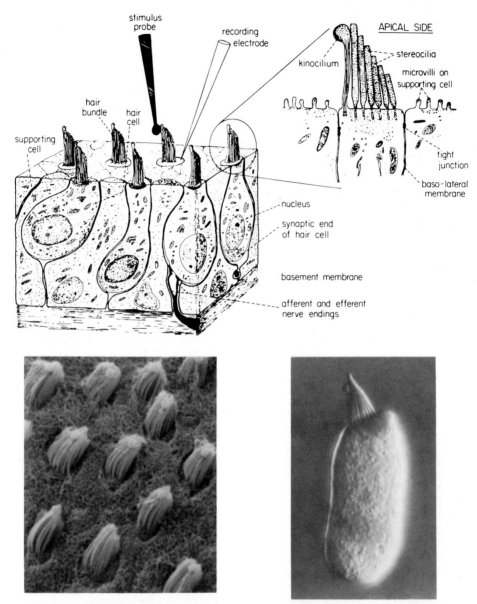

Figure 5–10 Hair cell structure. Cut-away schematic drawing of experimental preparation with an enlargement of the apical portion of a hair cell and neighboring epithelial cells. A stimulating probe is positioned on a hair bundle while an intracellular microelectrode records the membrane potential of the hair cell. (After Hudspeth, A.J.; Corey, D.P. *Proc. Nat. Acad. Sci. U.S.A.* 74:2407–2411, 1977.) *Lower left*: A scanning electron micrograph of bullfrog sacculus showing the ordered distribution of hair bundles in a matrix of supporting epithelial cells. The hair bundle of each hair cell is made up of many stereocilia and a single kinocilium with a bulbous distal swelling. (From Lewis, E.R.; Li, C.W. *Brain Res.* 83:35–50, 1975.) *Lower right*: Light micrograph of a living hair cell enzymatically isolated from the bullfrog sacculus. The hair bundle, including a single kinocilium, extends from the apical end of the cell. (Reproduced, with permission, from Hudspeth, A.J., *The Annual Review of Neuroscience*, Vol. 6, pp. 187–215, © 1983 by Annual Reviews, Inc.)

The basolateral surface of the cell is bathed in normal interstitial fluid and is the site of synaptic contacts with afferent, and in some cases efferent, axon terminals. The apical surface of the hair cell gives rise to a bundle of hairlike projections and is bathed in a special high-potassium, low-calcium solution called endolymph, a solution that has an ionic composition resembling that of cytoplasm. The structural details of the hair bundle vary greatly, depending on the species, the sense organ, the developmental stage of the animal, and sometimes, even the location of the hair cell.[37] In all cases, the bundle is mostly composed of stereocilia, which are modified microvilli 0.2 to 0.8 μm in diameter and filled with actin-containing microfilaments (Fig. 5–10). There are between 30 and 150 stereocilia per hair bundle. They extend from the surface in an ordered hexagonal array and are graded in length from one edge of the cell to the other. Most, but not all, hair bundles include a single true cilium with a 9 + 2 array of microtubules called the kinocilium. When present, it is the longest apical extension located at the edge of the cell adjacent to the row of longest stereocilia.

Mechanical stimulation of the receptor is accomplished by applying a force to the distal end of the hair bundle (Fig. 5–10). The absolute sensitivity of the receptor can be measured using a calibrated glass fiber to deliver precisely known forces to the hair bundle. Detectable responses are produced by displacements on the order of a few angstroms, whereas saturating (maximum) responses are produced by displacements of less than ± 1 μm. The cells show directional sensitivity; displacing the hair bundle toward the kinocilium depolarizes the cell; moving it in the opposite direction hyperpolarizes the cell; and displacements perpendicular to this axis have no effect. Responses are associated with a change in the ionic conductance of the apical cell membrane.[36] When the cell is in the resting (i.e., unstimulated) state, a few nonselective cation channels in the apical membrane are always open. Since potassium is the predominant cation in the endolymph, it enters the cell through the open cation channels, moving down an electrical, but not a chemical, gradient. There is little to no potassium gradient across the apical membrane, because the concentration of potassium in endolymph and cytoplasm is roughly the same (Fig. 5–11). There is, however, an electrical gradient across the apical membrane, because the intracellular potential of the hair cell is negative relative to the potential of the endolymphatic space. The influx of positively charged potassium ions produces a standing inward cur-

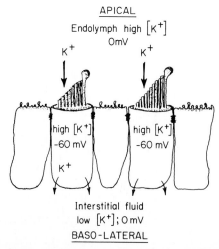

Figure 5–11 Current flow in hair cells. Schematic drawing showing the chemical and electrical gradients that support a steady flow of potassium ions into the cell across the apical membrane and out of the cell across the basolateral membrane. Supporting cells use metabolic energy to maintain a high potassium concentration in endolymph. The permeability of the intercellular connections between cells in the epithelium is nonselective and accounts for the fact that there is little-to-no potential difference between endolymphatic space and interstitial fluid. Potassium enters the hair cell through open mechanosensitive channels in the apical membrane. It moves down an electrical gradient—the potential inside of the cell is more negative than the potential in the endolymphatic space. There is no chemical gradient for K^+ across apical membrane. Potassium leaves the cell through K channels in the basolateral membrane. In doing so, it moves up an electrical gradient but down a larger chemical gradient—there is more K in cytoplasm than in interstitial fluid. Supporting cells complete the current loop by transporting potassium from the interstitial fluid back to the endolymph. Mechanical displacement of the hair bundle modulates the flow of current by changing the number of open channels in the apical membrane.

rent, called the receptor current, that keeps the cell relatively depolarized at −60 mV. Potassium ions that enter the cell through mechanosensitive channels in the apical membrane leave the cell through mechanoinsensitive potassium channels in the basolateral membrane. In order to leave the cell, potassium ions must move against (i.e., up) an electrical gradient. This is possible because they move down an even steeper chemical gradient. The basolateral portion of the hair cell is bathed in standard extracellular solution that has a much lower concentration of potassium than cytoplasm. The net result is that the difference in the potassium concentration of endolymph and extracellular fluid allows a steady current to circulate through the hair cell. Mechanical stimulation mod-

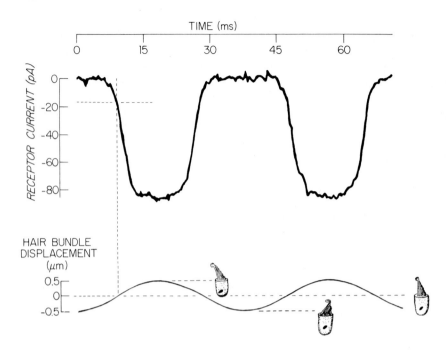

Figure 5–12 Hair cell receptor current. The upper trace shows the change in receptor current produced by displacement of the hair bundle (lower trace). In the resting state, there is a standing inward receptor current (intersection of dashed lines, upper trace) that is increased by displacement toward the kinocilium and decreased by displacement away from the kinocilium. Hair-bundle displacements are shown on the lower trace. (Adapted from Holton, T., and Hudspeth, A.J. *J. Physiol.* 375:195–227, 1986.)

ulates the current.* Moving the hair bundle towards the kinocilium opens more channels, the inward receptor current increases, and the cell depolarizes as more positive charges move into it (Fig. 5–12). Moving the bundle in the opposite direction closes channels causing hyperpolarization which reduces the inward receptor current.

The role of different parts of the hair bundle in the generation of the response was examined by removing selected portions of the bundle. Responses evoked by mechanical stimulation were recorded intracellularly while microsurgery was performed under a high-power light microscope, using an ultrafine dissecting needle.[35] Responses are unaffected by detachment of the kinocilium (Fig. 5–13) but decrease in proportion to the number of stereocilia removed. This shows that transduction is mediated by deflection of stereocilia, not of the kinocilium.

Patch-clamp studies of hair cells have provided information about the properties of channels operated by displacement of the stereocilia.[32] They have a reversal potential near zero millivolts and

an estimated single channel conductance at 37°C of about 25 pS, roughly the same as a single acetylcholine receptor channel (Chap. 6). Further analysis suggests that there are a small number of channels per stereocilium—possibly as few as four—and the open state of the average channel lasts for only a fraction of a millisecond, which could account for the ability of hair cells to respond to high-frequency stimuli.

The next question is, How does displacement control channels? There are at least two possibilities. Deflection of the stereocilia could pull directly on the channel to cause it to open, or it could modulate the channel by causing the release or generation of an intracellular second messenger. The delay between mechanical stimulation and hair cell response is 70 μs or less.[11] This leaves little time for the production and diffusion of a second messenger and argues against its involvement in mechanotransduction.

Stretch-Sensitive Channel. A stretch-activated ion channel with the properties of a mechanotransducer has recently been described in cultured embryonic skeletal muscle.[23]† The channel, like

---

*The steady current that circulates through the hair cell is analogous to the dark current in photoreceptors. In hair cells, however, it is the supporting cells in the epithelium that expend metabolic energy to maintain the ion gradient necessary for the standing current. Designing a receptor around the modulation of a standing current permits the receptor to respond both to increases and to decreases in the strength of a sensory stimulus.

†Since their initial discovery, stretch-sensitive channels have been reported to exist in a wide variety of cell types, including neurons, egg cells, red blood cells, bacteria, and yeast. One explanation for their widespread distribution is that they participate in volume regulation: When the cell swells, the membrane stretches and channels open, which leads to readjustment of cell volume.

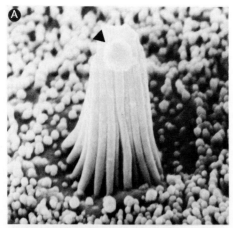

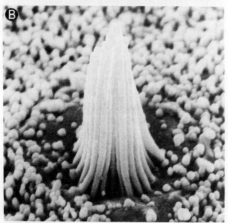

Figure 5–13 Scanning electron micrographs of hair bundles with *(A)* and without *(B)* an intact kinocilium from physiologically studied hair cells. *A,* Normal hair bundle consists of about 50 stereocilia and one kinocilium with a bulbous swelling at its distal tip (arrowhead). *B,* A hair cell whose kinocilium was removed during an experiment. Microdissection of the kinocilium did not change the receptor response to mechanical displacement of the hair bundle. (From Hudspeth, A.J., and Jacobs, R. *Proc. Natl. Acad. Sci. USA* 76:1506–1509, 1979.)

those described in various kinds of mechanoreceptors, has a reversal potential near zero millivolts, is cation-selective, and discriminates poorly between Na and K ions. It has been observed in intact cells as well as in membrane patches excised from the cell. This supports the conclusion that membrane stretch opens the channel directly, rather than through a mechanism that involves a second messenger. The mechanical linkage between channel and membrane is thought to be provided by cytoskeletal strings that pull the channel open when the membrane is stretched. Such cytoskeletal coupling may explain the prominent occurrence of microfilaments in the transducing

region of many types of mechanoreceptors. Increasing the number of "string" attachments between the channel and the membrane would allow force to be collected from a larger area of membrane and might be an important factor in accounting for the receptor's high sensitivity. The location of the attachment to the membrane could also be the basis for the directional sensitivity of certain mechanoreceptors.

CHEMORECEPTORS

Bacterial Chemotaxis. All living cells are chemoreceptive in that they are able to respond selectively to specific chemicals in the local environment. This point is well illustrated by the chemotactic behavior of bacteria.[49, 42] *Escherichia coli* is a unicellular prokaryote about 1 μm wide and 2 to 3 μm long with several flagella randomly distributed over its surface. Continuous rotation of the individual flagella cause them to coalesce into a hydrodynamically stable bundle that drives the bacterium as a propeller. *E. coli* swim towards some chemicals and away from others, depending on the importance of the chemical to the metabolism and survival of the bacterium. The presence of chemicals in the environment is monitored by specific receptor proteins (sensors) on the cell surface. A combination of biochemical, genetic, and behavioral studies have identified thirty different receptor proteins that are most commonly designed to detect carbohydrates and amino acids. Each type of sensor has a high affinity for a particular compound but few are specific for a single compound. (This is a common property of sensory receptors that detect chemical stimuli.) The occupancy of a sensor site triggers a transduction process that involves an additional ten proteins. The final output of transduction is control of the direction of flagella motor rotation. The mechanism is not fully understood, but the methylation/demethylation status of a membrane-associated protein appears to be an important factor in motor control. When rotation is reversed, the hydrodynamic forces that had caused individual flagella to coalesce into a neat bundle cease to exist, and the bundle flies apart causing the bacterium to tumble in a random direction. After a period, the flagellar bundle reforms and smooth swimming resumes. Bacteria migrate up or down chemical gradients by controlling the frequency of spontaneous tumbling. Bacteria moving in a "favorable" direction suppress tumbling and swim for a longer than normal distance in a straight line. When they are moving in an "unfavorable" direction they generate tumbling and randomly reorient the direction of their movement. Thus by controlling the number of chance movements (tumbling), the sensor-triggered transduction process biases the random walk of the bacterium and assures that it moves more in the right than in the wrong direction.

In *E. coli*, the generation of sensory information, its central processing, and the production of an appropriate behavioral response all take place within the confines of a single cell. In multicellular animals, these different tasks are assigned to separate cells. The job of providing information about the presence of specific chemicals in the environment is performed by chemoreceptors. They subserve the senses of taste and olfaction, and in addition may be used to measure the concentration of oxygen, CO_2, and other substances in the blood. There is not a lot of information about the cellular physiology of chemoreceptors. We review the present understanding of taste and olfaction by first considering the anatomy and then the physiology of each.

Taste. Cells specialized in detection of chemicals in food materials are classified as taste receptor cells. This definition provides a basis for distinguishing between taste and smell in aquatic animals. Taste receptors are commonly located near the mouth but may also occur at other locations on the body surface, such as on leg hairs of certain insects, the tentacles of cephalopods, and the barbels of catfish. Vertebrate taste receptors are mainly found within the oral cavity and most notably on the tongue. Small groups of receptor cells are arranged in ovoid cavities called taste buds embedded in blunt epithelial projections or papillae (Fig. 5–14). There is a small vestibule (taste pit) at the top of the bud that communicates with the epithelial surface through the taste pore. Both the pore and the pit are filled with a dense, amorphous substance characterized as neutral mucopolysaccharide. The wall and interior of the taste bud are lined with elongated epithelial cells that can be classified into four specific cell types[54] of, for the most part, uncertain function. Two cell types end in a brush of microvilli extending into the taste pore. One of these is light in color; the other is dark, with dense secretory granules that are thought to be the origin of the substance in the pit. The third cell type sends a narrow peglike process into the taste pore. It is the only cell in the bud that synapses with afferent nerve terminals, which makes it the leading candidate for the receptor cell. (The inability to identify unequivo-

cally which taste bud cell is the taste receptor cell is a major obstacle in taste research). At the bottom of the bud are undifferentiated basal cells that do not reach the taste pit and make up the fourth category of cell types. As generally true of epithelia, taste bud cells are joined together by tight junctions that form a continuous seal just below the taste pit. This divides the cell into two distinct regions; an apical region above the tight junction, which comes in contact with substances in the oral cavity, and a basal-lateral region below the tight junction, which is bathed in interstitial fluid. The other components of the taste bud are the afferent nerve fibers that form a highly branched network of unmyelinated terminal processes that coil and intertwine around all but the basal cells. The complex course and interrelationships of afferent terminals with three of the four cell types in the taste bud make it possible that light and dark cells, in addition to the presumed receptor cell, play a role in taste reception.

Another interesting morphological aspect is that taste cells are continually being renewed. This is fortunate, because they are frequently exposed to potentially damaging conditions, e.g., heat, cold, acids, mastication. They have a lifetime on the order of ten days. Tracer studies suggest that basal cells enter the bud from the surrounding epithelium and differentiate into new taste cells. Since the turnover of taste cells continues throughout adult life, there must be special mechanisms to remove old cells, insert new cells into the tight junction seal, form new synaptic contacts, and maintain the sensory specificity of neural coding.

Although it is estimated that the taste buds of the human tongue can distinguish between 4000 and 10,000 different chemicals, psychophysical studies since the mid 1800s have considered taste to be based on four primary qualities: salt, bitter, sour, and sweet. This led to the expectation that there were four corresponding categories of taste receptors. Beginning in the late 1960s, responses to chemical stimuli have been recorded intracellularly from different cells in the oral cavity of mammals and lower vertebrates. Although the details of the observations have varied greatly, the overall results are qualitatively similar and contain at least two surprises. The first surprise is that surface epithelial cells, regardless of their location, respond to taste stimuli.[18, 72] This has generated questions about the nature of the specialization that distinguishes taste cells from undifferentiated lingual epithelial cells and raises the possibility that surface cells play a role in gustation. The second surprise is that the same taste cell may be

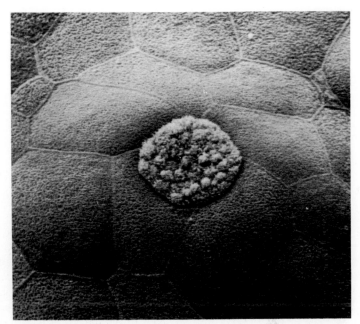

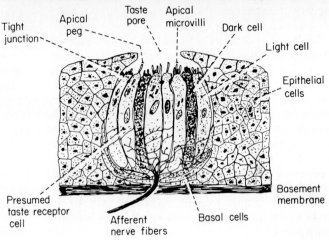

Figure 5–14 Taste cell morphology. *Above,* Scanning electron micrograph of a taste pore and surrounding epithelial cells in *Necturus* (courtesy of S. Roper). *Below,* Schematic drawing of cross section of a mammalian taste bud. Apical microvilli of light and dark cells and the apical pegs of the presumed taste receptor cells project into the taste pore, which contains an amorphous substance (not illustrated) secreted by supporting cells. Tight junctions between cells form a seal that prevents extracellular movement of fluids between the apical and basolateral portions of the cells.

sensitive to all four taste stimuli (Fig. 5–15A).[62, 63] The application of salt (NaCl), sour (acid), sweet (sucrose), and bitter (quinine) stimuli at concentrations above a critical level produces a depolarizing potential change in cells penetrated in the taste bud (a more precise identification of the cellular location of recording site has not been possible). The amplitude and time course of the depolarizing receptor potential vary from cell to cell according to both the specific stimulus and its concentration. Deionized water alone causes taste cells to hyperpolarize. Since experiments are done with the oral epithelium bathed in artificial saliva containing 0.1 M NaCl, the water response can be attributed to

dilution of NaCl and a reduction in the depolarizing response it produces.

Recent reports that taste cells are able to generate action potentials represents a third surprise about taste cell electrophysiology.[60] This is surprising because one would not expect that a small cell like the taste receptor would need action potentials to transmit information from the transductive region of the cell to the synaptic ending. It is possible, however, that the regenerative conductances underlying the generation of action potentials are present to shape the electrical response of the cell in a manner that plays a role in the ultimate coding of taste stimuli. This speculation

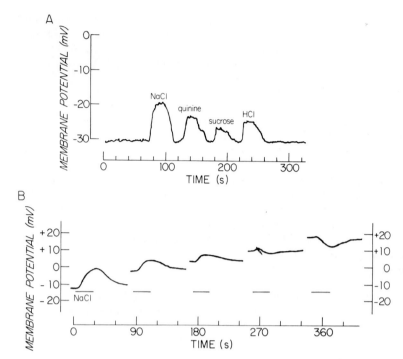

Figure 5–15 Taste cell responses. *A*, Intracellular recording from a hamster taste cell showing that all four classic taste modalities (salty, bitter, sweet, and sour) evoke the same polarity (depolarizing) receptor potential. Solutions were applied to the tongue surface with water rinses between stimuli containing 0.1 M NaCl, 0.02 M quinine hydrochloride, 0.5 M sucrose and 0.01 M HCl. (Data taken from Kimura, K., and Beidler, L.M. *J. Cell Comp. Physiol.* 58:131–139, 1961.) *B*, Effect of depolarization on frog taste cell responses. Membrane potential was depolarized by applying positive current steps (+0.41 to +1.13 nA) and the response to a solution containing 0.5 M NaCl was recorded. Water rinses were applied to the tongue between stimuli. The amplitude of the receptor potential decreased with positive shifts in membrane potential and reversed polarity at +7 mV. (Data from Sato, T., and Beidler, L.M. *J. Gen. Physiol.* 66:735–763, 1975.)

suggests that the electrophysiology of taste receptors could be more sophisticated and subtle than originally anticipated.

The basis of the receptor potential evoked by taste stimuli is not known, but in most cases it is thought to involve a change in the ionic permeability of the cell. An increase in membrane conductance is associated with the depolarizing responses produced by salt, whereas a large conductance decrease is associated with the response to acid and quinine. Sweet stimuli, on the other hand, have no apparent effect on membrane conductance.[72]

The amplitude of the receptor potential is graded with the concentration (intensity) of the stimulus in a manner that fits a Michaelis relationship (Equation 1).[4] This is consistent with the response being triggered by the binding (occupancy) of a stimulus molecule to a specific receptor site (receptor protein). Because responses may be associated with an increase, decrease, or no change in membrane conductance, different receptor sites must be used to detect different stimuli. This point is supported further by the identification of a specific sugar-binding protein in rat taste buds[31] and by human psychophysical experiments with taste modifiers. After the bitter tasting leaves of *Gymnema sylvestre* are chewed the sweet taste of sugar, but not the taste of bitter, salt, or sour substances, is suppressed, suggesting that Gym-

nema contains a compound that specifically blocks the sugar sensor.[51, *]

The occupancy of different types of receptor proteins must be able to trigger more than one type of transduction process. This follows from the observation that depending on the stimulus the receptor potential may or may not involve a conductance change. The nature of the different transduction processes is not known, nor is the identity of the ions involved in the receptor potential fully established. There is more information about the ionic basis of the depolarizing response evoked by NaCl than about the responses produced by other taste substances. The reversal potential has been measured in several animals and is typically in the vicinity of +10 to +30 mV,

---

*There are many examples of taste modifiers of plant origin. They are thought to be designed to change the behavioral response of animals to their leaves or fruits. Miraculin, a glycoprotein in the berries of an African shrub (*Synsepalum dulcificum*) causes sour substances to seem sweet. This tricks an animal into eating its sour fruit and dispersing its seeds which are stored in the fruit at an acid pH, possibly to reduce infestation by bacteria or insects. Since the miraculin-containing berries cause similar changes in human taste,[16] it may be argued that taste "perception" is similar in a wide range of animal types. (A further point: the pleasing scent of flowers was designed to attract insects rather than your local florist.) This suggests that there are basic similarities in the sensory systems of different animals (including insects) and that information obtained from one animal is relevant to others.

which is consistent with its being predominantly due to an increase in membrane permeability to Na^+ ions (Fig. 5–15B).[1, 63] The reversal potential, however, does not depend on the concentration of NaCl in the stimulus solution used to evoke the response. Similar results have been obtained using KCl as the stimulus.[55] This indicates that the conductance change occurs on a part of the cell that does not come in contact with the stimulus solution—perhaps the receptor proteins for salts and the site of the conductance change are on opposite sides of the tight junction seal. It is possible that receptor proteins on the apical surface of the cell trigger a transduction process that uses a diffusable intracellular second messenger to regulate the conductance of the basal-lateral membrane. The possibility that second messengers play a role in taste reception is speculative and based on very little information. This is a fundamental point that can be resolved only by additional experimental evidence.

Whatever the mechanism responsible for the taste receptor potential, it is thought to regulate the production of action potentials in the gustatory nerve by controlling the release of a neurotransmitter from the synaptic terminal of the receptor cell. The complexity of taste bud innervation and the marked chemical sensitivity of surface epithelium, however, raise the possibility that other events contribute to the transmission of gustatory information to the afferent nerve. It may be that electric fields produced by potential changes in surface cells, taste bud–supporting cells, or both influence afferent terminal excitability either by a direct action or through an effect on the efficacy of synaptic transmission.

Another mystery concerning taste involves the discrimination of taste quality. The observation that a single taste cell produces the same polarity of response to all four taste stimuli indicates that the quality of a taste sensation involves information other than that provided by the response of a single taste bud. In a classic study, Pfaffman[58] demonstrated that the chemical specificity of single gustatory nerve fibers is not absolute and suggested that taste quality did not depend on the all-or-none activation of a particular fiber group but on the relative activation of many neurons. This is still a viable explanation for taste discrimination (see Chap. 21).

Olfaction. Olfactory receptors are designed to detect specific molecules in the atmosphere or aquatic environment. Among the best studied olfactory stimulants are substances secreted by one member of a species and detected and responded to by another member. Such substances, called *pheromones*, play an important role in the behavior of many animals. In humans, the best known example involves the synchronization of the menstrual cycles of women living in close proximity. The olfactory basis for this was established by demonstrating that exposure only to the odor of the axillary secretions of a "dominant" woman was sufficient to entrain the menstrual cycles of a group of women.[61] The use of odors for purpose of intraspecies communication is well developed in lower animals. The sensitivity of the moth olfactory system to sexual pheromones is remarkable; a male can locate a female several miles away by following the pheromone odor plume she emits. The threshold concentration for evoking a behavioral response in the male is on the order of a few hundred pheromone molecules per cubic centimeter of air.[39]* This incredible sensitivity depends as much on the specialized properties of the receptor cell as on the accessory structures that are associated with it. In all animal species, olfactory cells are primary receptors. In insects, the cell bodies are located in the epithelium just below the cuticle of the antennae (Fig. 5–16). One end of the cell gives rise to an axon that projects into the central nervous system; the other end, to a dendrite that is separated into an inner and outer segment connected by a narrow neck with a ciliary structure. The outer segment is densely packed with neurofilaments (of unknown function) and extends into the lumen of a hollow, thin-walled cuticular (olfactory) hair. Each hair is perforated by several thousand pores 10 to 100-nm in diameter. To reach the dendrite, stimulant molecules must diffuse through the pores in the hair cuticle and then through a high-K^+ extracellular fluid (receptor lymph) that is secreted by supporting epithelial cells[67] and fills the free space inside the hair. Molecules that happen to be adsorbed between the pores in the hair are thought to diffuse laterally along the external lipid layer of the cuticle until they encounter a pore.

The antennae are covered with thousands of olfactory hairs and represent the insect's nose.† They sniff or, more correctly, filter the air stream for specific scent molecules. By virtue of the physical design of the antennae, it is estimated that

---

*Adding 0.5 μl of a 1 molar solution to a container 100 feet deep, 2 miles wide and 20 miles long containing 836 billion gallons of water will, with gentle shaking, yield a solution with a final concentration of 100 molecules/cc.

†Insect antennae are thus the objects referred to in the expression "cute as a bug's nose."

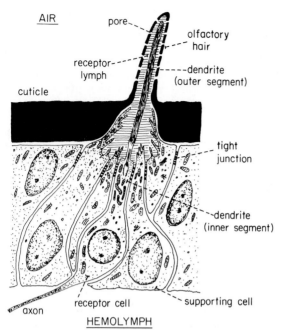

Figure 5-16 Insect olfactory receptor. One pole of the bipolar receptor cell body gives rise to an axon; the other, to a dendrite with an inner and outer segment joined by a ciliary neck. The outer segment extends into a cuticular hair and is bathed in receptor lymph, a high potassium solution secreted by supporting cells. Pores in the olfactory hair allow airborne odorant molecules to reach the dendrite membrane. Sensor proteins in the outer segment membrane bind odorant molecules and initiate a transduction process that culminates in the production of a depolarizing receptor potential and generation of action potentials in the receptor axon. (After Kaissling, K.E. In: Beidler, L.M., ed. *Handbook of Sensory Physiology,* Vol. 4, pp. 351–431, 1971.)

during a 1-second exposure to an odor stream, a moth flying at 60 cm/s is able to concentrate an odorant by about 10^5 times more than the concentration in the surrounding air.[40]

The concentration of odorants by the antennae, i.e., the accessory structure, is a major factor in accounting for the sensitivity of the moth's olfactory system. The other component is the receptor itself. To date, the small size of the olfactory cell and the technical difficulties posed by its intimate association with the cuticle have made studies using intracellular microelectrodes impractical. Information based on extracellular recording of spike activity generated by radio-labeled pheromones indicate, however, that a nerve impulse is evoked by the binding of a single pheromone molecule per olfactory receptor cell.[39, 41] This suggests that the binding step either gates a channel with a conductance that is large enough to trigger an action potential or it activates an intracellular proc-

ess that amplifies the signal produced by a single scent molecule so that it influences a large number of smaller conductance channels. At present, there is no experimental basis for choosing between these alternatives.

The olfactory system of terrestrial vertebrates shares several similarities with that of the moth. In analogy to the odor filtration system of the insect's antenna, the vertebrate nose has evolved elaborate passages to direct the flow of air over the receptive regions of the nasal cavity. The olfactory mucosa consists of a mosaic of receptor and supporting cells covered by a film of mucus and lying on a layer of basal cells (Fig. 5–17). The receptors are small (5 to 15 μm), flask-shaped bipolar cells.[59] One pole of the neuron gives rise to an axon of the olfactory nerve that projects to the olfactory bulb. The other pole gives rise to a cylindrical dendritic process that extends to the mucus-cell interface, where it expands to form a knoblike ending called the olfactory vesicle or olfactory knob. The olfactory vesicle is the only part of the dendrite that is in contact with the mucus layer and extends one to twenty cilia having the typical 9 + 2 array of microtubules.* The olfactory cilia of vertebrates vary in length and can be considerably complex. In the frog, the array of microtubules that make up the distal portion of the cilium split apart, forming enlarged regions that often contain vacuoles.[59] At least some of the cilia are motile, but unlike the cilia of the respiratory epithelium, they move in an uncoordinated and irregular manner. The olfactory receptors are interspersed in a matrix of columnar supporting cells with microvillar projections at their apical surface. The distal parts of these cells contain secretory granules that are released into the surface mucus; the chemical composition of the secretory product and its function is not known. One possibility is that it is involved in the removal (enzymatic?) of old odorant molecules. Supporting cells make tight junctions with each other and with the olfactory vesicles of neighboring receptor cells. These tight junctions form a seal between the external environment and the intercellular space of the epithelium, which is reminiscent of the tight-junction seal that exists in taste buds. Another similarity with taste is the continual re-

---

*One cannot help noting the parallel between the organization of the vertebrate's olfactory dendrite, the inner- and outer-segment arrangement of the insect's receptor dendrite, and the morphology of the vertebrate's photoreceptors, including the common occurrence of ciliary structures.

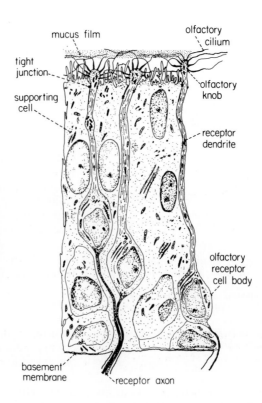

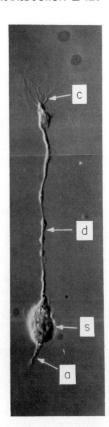

Figure 5–17 Vertebrate olfactory receptors. Drawing of the olfactory epithelium shows supporting cells and bipolar olfactory receptor. An axon and a dendrite project from opposite ends of the receptor cell body. The distal end of the dendrite swells to form an olfactory knob that gives rise to several olfactory cilia. The knob and the cilia are in contact with a mucous film that covers the epithelium. (Adapted from Andres, K.H. *Z. Zellfors.* 69:140–154, 1966.) Light micrograph of an enzymatically dissociated olfactory receptor from the tiger salamander. The axon, cell body, dendrite, and cilia are apparent and labeled. (Photograph courtesy of P.A.V. Anderson and K.A. Hamilton.)

placement of receptor cells. The basal cells of the olfactory mucosa are thought to be stem cells for the replacement of olfactory receptors, which have a life span of about two months. The intriguing mechanisms by which old receptors are identified and replaced with new ones making the correct synaptic connections in the olfactory bulb are not known.

Olfactory receptors respond to a wide variety of odorants, which suggests that each cell has receptor proteins for several types of odor molecules. Since removal of the cilium abolishes olfactory responses without disrupting the structural integrity of the epithelium, odor receptor proteins are thought to be located in the ciliary membrane.[45] This interpretation is consistent with the appearance of a dense collection ($10^3/\mu m^2$) of intramembranous particles in the surface membrane of freeze-fractured cilia from olfactory receptors but not from respiratory epithelial cells.

To reach receptor sites, scent molecules must first diffuse through the mucus film secreted by Bowman's glands distributed throughout the epithelium. Mucus covers the neuroepithelium, and

the ability of odorants to diffuse in it may play a role in determining what substances are able to reach receptor sites. In addition, the mucopolysaccharides contained in mucus retain water by a combination of polyionic and osmotic effects and thus provide a hydrated protective environment for the receptors to operate within.

In recent years there have been several reports of intracellular recordings from the soma of olfactory receptors. Appropriate odorants evoke a slow depolarizing receptor potential that gives rise to action potentials in the receptor cell's axon (Fig. 5–18). The amplitude of the receptor potential is graded according to the strength of the olfactory stimulus. It is associated with a conductance increase and has a reversal potential in the vicinity of 0 mV.[68] This suggests that the odorant-receptor interaction initiates an increased membrane permeability to sodium and potassium. The steps that couple the recognition of a scent molecule by a receptor protein to the permeability change, i.e., the transduction process, are not known.

Psychophysical studies of dogs and humans support the conclusion that one molecule is suffi-

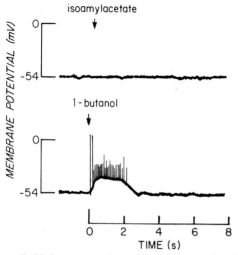

Figure 5–18 Response of an olfactory receptor to two different odorants. The traces show intracellular recordings from a vertebrate (salamander) olfactory receptor. The receptor is sensitive to 1-butanol (lower trace) but not isoamylacetate (upper trace). Brief exposure to 1-butanol evoked a depolarizing receptor potential with action potentials generated in the axon riding on top. (Data from Trotier, D., and McLeod, P. *Brain Res.* 268:225–237, 1983.)

cient to excite an olfactory receptor.[14, 53]* As in the earlier discussion of insects, this requires that odorant binding either directly gate a single large conductance channel or activate an amplification process expressed through many small channels. Amplification may involve either the release of an intracellularly sequestered second messenger such as Ca^{2+} or the activation of an enzyme cascade that changes the cytoplasmic concentration of a second messenger, e.g., cyclic nucleotide. A response generated through second-messenger control of many small channels would be capable of finer gradations than a response generated by the direct gating of a single large channel. Furthermore, the use of second messengers could provide a mechanism for separating the site of the ionic permeability change from the site of odorant binding. Such an arrangement would ensure that the receptor potential is independent of secondary effects owing to changes in the local environment

that are associated with the presence of an olfactory stimulus in the nasal cavity.

A report that isolated olfactory cilia contain adenylate cyclase at a much higher (15- to 100-fold) specific activity than respiratory cilia or membranes from brain has recently appeared. In a cell-free system, odorant molecules regulate the adenylate cyclase activity in a dose-dependent manner that requires the presence of GTP.[45, 56] This suggests that odorant binding is coupled to adenylate cyclase through a guanine nucleotide–binding protein (G protein) similar to that involved in the response produced by β-adrenergic receptor activation (Chap. 1). These results raise the possibility that changes in cAMP are involved in olfactory transduction.

THERMORECEPTORS

The popular press occasionally describes people who are able to "see" colors with their hands. Although such reports are rarely scientific and are often exaggerated, it is true that blindfolded people are able to choose the brighter of two colored metal plates more often than chance would predict.[50] This is apparently based on the ability of temperature receptors in the hands to detect subtle differences in reflected body heat. The cellular basis of thermal reception is one of the least understood areas of sensory neurobiology. One of the most remarkable and best studied temperature sensors is the rattlesnake infrared (IR) detector located in facial pits below each eye socket. The opening of the pit is guarded by a convex dry membrane of cornified epithelium 7 mm in diameter. The interior of the pit is highly vascularized (maximum distance between any point and a capillary is 30 μm). It is innervated by axons from the trigeminal nerve that taper to form tiny branchlets, which flatten out and intermingle with Schwann cell processes. The branchlets are filled with mitochondria that account for 50% of their cell volume. The combined innervation represents a neural mass that is approximately 10 μm thick and covers an area of 1500 μm². The trigeminal nerves project to the tectum in the brain, where they form a specific spatiotopic map of pit-organ space on the tectal surface. The similarity of this mapping to the way the amphibian retina is mapped on the tectum suggests that the pit organ could provide elaborate sensory information about heat-emitting objects in its receptive field.[27]

The firing rate of the infrared receptor increases transiently when the temperature of the mem-

---

*Human taste receptors are 10^5 to 10^6 times less sensitive than human olfactory receptors. The reason for this may be that highly sensitive taste has little evolutionary survival value. In terrestrial vertebrates, taste requires intimate contact with a substance and usually precedes ingestion. In the process of evolution, it is possible that animals appeared that had acquired by accident the ability to detect minute quantities of lethal substances but died before they discovered the substances were poisonous.

brane at the opening of the facial pit is changed by 0.003°C.[6] The effective stimulus is infrared radiant energy. Pit organs are used at night to "image" warm-blooded animals against the cold desert background. Rattlesnakes can detect a hand at a distance of 50 cm if it is at least 10° C warmer than the ambient temperature.[6] The pit organ is sensitive to electromagnetic radiation, having wave lengths that extend from 1.5 to at least 10.6 μm. Responses have also been obtained using microwaves with 2.8 cm wavelength.[26]* The energy carried by a photon is inversely proportional to its wavelength. Consequently an infrared photon carries little energy, which makes its detection a difficult problem. The energy transferred to a molecule by the absorption of an IR photon is enough to increase the vibrational energy of the molecule (heat) but is not enough to cause a conformational change like that produced by the absorption of a photon of visible (shorter wavelength) light by rhodopsin. This argues against a transduction process based on a photochemical reaction. An alternative possibility is that transduction is triggered by molecular vibration. How this could be engineered is not known. Whatever mechanism is responsible for thermotransduction, it must also have enough amplification to account for the high sensitivity needed to explain the infrared-based behavior of the rattlesnake or the ability of humans to "feel colors." Understanding how a heat detector with these specifications actually works is clearly a challenging problem.

SUMMARY

All living cells have the ability to translate physical events in their external environment into biologically relevant signals. Cells that exist to provide the nervous system with information about the detection and discrimination of different physical stimuli are called sensory receptors. They are input-output devices that transduce (convert) physical stimuli into biological signals. The input can be mechanical, chemical, electrical, thermal, or electromagnetic (light). The output in all vertebrate sensory receptors is a change in membrane potential called a receptor potential. The receptor potential, either directly or indirectly through the generation of action potentials, influences the re-

lease of neurotransmitter from the synaptic ending of the receptor, which relays the signal to the second cell in the afferent pathway.

The general sequence of events that couples a physical stimulus to the production of a receptor potential is similar in all types of sensory receptors. Accessory structures, which may be separate cells or part of the receptor, transform or modify the physical stimulus before transferring it to the transductive region of the receptor, where it is detected by specialized integral membrane proteins called sensors or receptor proteins. The combination of the physical stimulus with a sensor triggers a transduction process that either directly or indirectly through a second messenger gates an ion channel called a stimulus-sensitive channel. The stimulus-sensitive channel can be opened or closed by the transduction process. Either effect on the channel changes the ionic current flow across the cell membrane, resulting in a localized membrane potential change—a receptor potential. The amplitude of the receptor potential is graded according to the strength of the physical stimulus. The polarity of the potential change can be depolarizing or hyperpolarizing, depending on the ion selectivity of the stimulus-sensitive channel and on whether the physical stimulus opens or closes the channel. A maintained stimulus evokes a receptor potential that shows time-dependent changes referred to as adaptation.

More details about the general mechanism of sensory transduction are provided by specific examples.

In darkness, a steady current (dark current) circulates through vertebrate photoreceptors, keeping them partially depolarized. Cations—predominantly Na^+—enter the outer segment region of the receptor through light-sensitive channels that are maintained in the open state by a high intracellular concentration of cGMP. Current that flows into the outer segment leaves the inner segment by passive efflux of K^+ through light-insensitive K channels. Light-triggered isomerization of rhodopsin—the sensor, or receptor protein—initiates a long sequence of reactions that ultimately catalyzes the enzymatic destruction of cGMP. The influx of cations decreases as the cGMP concentration decreases and light-sensitive channels close. Potassium ions continue to leave the inner segment, causing the receptor to hyperpolarize.

In the resting state, a steady current also circulates through vertebrate hair cells. It is carried predominantly by K^+ ions, which enter the cell through open stretched-sensitive channels in the

*Electromagnetic radiation with a wavelength of 2.8 cm approaches the wavelength of radar. During World War II, the Armed Forces were interested in the possibility of using rattlesnakes as radar detectors.

membrane of the apical hair bundle and leave by stretch-insensitive channels in the basolateral membrane of the receptor. Depending on the direction, hair bundle displacement can either stretch or relax apical membrane. Membrane stretch directly opens more stretch-sensitive channels causing the receptor to depolarize. Membrane relaxation closes apical stretch-sensitive channels. Potassium continues to flow out basolateral K channels, and the receptor hyperpolarizes.

The events responsible for production of receptor potentials in chemoreceptors (taste and olfaction) are not completely known. Chemical stimuli combine with specialized receptor proteins in the apical membrane of the receptor. This triggers a transduction process that ultimately results in a change in ion permeability and a depolarizing receptor potential. It is possible, but not certain, that second messengers are involved in the transduction process.

ANNOTATED BIBLIOGRAPHY

Fesenko, E.E.; Kolesnikov, S.S.; Lyubarsky, A.L. Induction by cGMP of cationic conductance in plasma membrane of retinal rod outer segments. *Nature* 313:310–313, 1985.
By presenting the first direct evidence for a cGMP-operated conductance in outer segment surface membrane, this short paper had a dramatic and lasting effect on our understanding of phototransduction in vertebrate rods.

Hudspeth, A.J. Mechanoelectrical transduction by hair cells in the acousticolateralis sensory system. *Annu. Rev. Neurosci.* 6:187–215, 1983.
This review covers the recent advances in the field of mechanotransduction in vertebrate hair cells.

Kaissling, K.E. Insect olfaction. In Beidler, L.M., ed. *Handbook of Sensory Physiology,* 351–431. vol. 4/1, Chemical Senses. Springer-Verlag, Berlin, 1971.
This article provides a comprehensive review of the structure and function of insect olfactory receptors.

Lamb, T.D. Electrical response of photoreceptors. *Recent Adv. Physiol.* 10:29–66, 1984.
This review discusses the electrophysiology and biochemistry of vertebrate photoreception.

Lancet, D. Molecular view of olfactory reception. *Trends in Neurosci.* 7:35–36, 1984.
A short review of the data that addresses questions about involvement of second messenger cyclic nucleotides in olfaction.

REFERENCES

1. Akaike, N.; Noma, A.; Sato, M. Electrical responses of frog taste cells to chemical stimuli. *J. Physiol.* 254:87–107, 1976.
2. Alexandrowicz, J.S. Muscle receptor organs in the abdomen of *Homarus vulgaris* and *Palinurus vulgaris. Quant. J. Micro. Sci.* 92:163, 1951.
3. Baylor, D.A.; Lamb, T.D.; Yau, K.-W. Responses of retinal rods to single photons. *J. Physiol. (Lond.)* 288:613–634, 1979.
4. Beidler, L.M. Transductive coupling in the gustatory system. In Schmitt, F.O.; Crothers, D.M., eds. *Functional Linkage in Biomolecular Systems,* New York, Raven Press, 255–262, 1975.
5. Bodoia, R.D.; Detwiler, P.B. Patch-clamp recordings of the light-sensitive dark noise in retinal rods from the lizard and frog. *J. Physiol.* 365:183–216, 1985.
6. Bullock, T.H.; Diecke, F.D.J. Properties of an infrared receptor. *J. Physiol.* 134:47–87, 1956.
7. Burnside, B.; Nagle, B. Retinomotor movements of photoreceptors and retinal pigment epithelium: mechanisms and regulation. In Osborne, N.; Chader, G. eds. *Progress in Retinal Research 2.* New York, Pergamon Press, 1983.
8. Cervetto, L.; McNaughton, P.A.; Nunn, B.J. Calcium current and aequorin signals in isolated salamander rods. *Biophys. J.* 49:280a, 1986.
9. Crawford, A.C.; Fettiplace, R. The frequency selectivity of auditory nerve fibers and hair cells in the cochlea of the turtle. *J. Physiol.* 306:79–125, 1980.
10. Crawford, A.C.; Fettiplace, R. Non-linearities in the response of turtle hair cells. *J. Physiol.* 315:317–338, 1981.
11. Corey, D.P.; Hudspeth, A.J. Response latency of vertebrate hair cells. *Biophys. J.* 26:499–506, 1979.
12. Dartnall, H.J.A., ed. *Photochemistry of Vision,* Vol. 7, pt. 1, of *Handb. Sensory Physiol.* New York, Springer-Verlag, 1972.
13. Detwiler, P.B.; Conner, J.D.; Bodoia, R.D. Gigaseal patch-clamp recording from outer segment of intact retinal rods. *Nature* 300:59–61, 1982.
14. DeVries, H.; Stuiver, M. The absolute sensitivity of the human sense of smell. In Rosenblith, W.A., ed. *Sensory Communication,* 159–167. New York, Wiley & Sons, 1960.
15. Diamond, J.; Gray, J.A.B.; Inman, D.R. The relation between receptor potential and the concentration of sodium ions. *J. Physiol.* 142:383–394, 1958.
16. Eisner, T.; Halpern, B.P. Taste distortion and plant palatability. *Science* 172:1362, 1971.
17. Enoch, J.M.; Birch, D.G. Evidence for alteration in photoreceptor orientation. *Ophthalmology* 87:821–833, 1980.
18. Eyzaguirre, C.; Fidone, S.; Zapata, P. Membrane potentials recorded from the mucosa of the toad's tongue during chemical stimulation. *J. Physiol.* 221:515–532, 1972.
19. Fesenko, E.E.; Kolesnikov, S.S.; Lyubarsky, A.L. Induction by cGMP of cationic conductance in plasma membrane of retinal rod outer segments. *Nature* 313:310–313, 1985.
20. Gold, G. H.; Korenbrot, J.I. Light-induced calcium release by intact retinal rods. *Proc. Natl. Acad. Sci. USA* 77:5557–5561, 1980.
21. Gray J.A.B. Mechanical into electrical energy in certain mechanoreceptors. *Prog. Biophys.* 9:285–324, 1959.
22. Gray, P.; Attwell, D. Kinetics of light-sensitive channels in vertebrate photoreceptors. *Proc. R. Soc. Lond. [Biol.]* 223:379–388, 1985.
23. Guharay, F.; Sachs, F. Stretch-activated single ion channel currents in tissue-cultured embryonic chick skeletal muscle. *J. Physiol.* 352:685–701, 1984.
24. Hagins, W.A. The visual process: excitatory mechanisms in the primary receptor cells. *Annu. Rev. Biophys. Bioeng.* 1:131–158, 1972.
25. Hagins, W.A.; Yoshikami, S. A role for calcium in excitation of retinal rods and cones. *Exp. Eye Res.* 18:299–305, 1974.
26. Harris, J.F.; Gamow, R.I. Snake infrared receptors: thermal or photochemical mechanism. *Science* 172:1252–1253, 1971.
27. Hartline, P.H. Thermal reception in snakes, Chapter. In Fessard, A., ed. *Electroreceptors and Other Specialized Receptors in Lower Vertebrates,* Vol. 3, pt. 3, of *Handbk. Sensory Physiol.* New York, Springer-Verlag, 1974.

28. Haynes, L.; Yau, K.-W. cGMP-sensitive conductance in catfish cone outer segment membrane. *Nature* 317:252–255, 1985.

29. Haynes, L.W.; Kay, A.R.; Yau, K.-W. Single cGMP-activated channel activity in excised patches of rod outer segment membrane. *Nature* 321:66–70, 1986.

30. Hecht, S.; Shlaer, S.; Perenne, M.H. Energy, quanta, and vision. *J. Gen. Physiol.* 25:819–840, 1942.

31. Hiji, Y.; Sato, M. Isolation of the sugar-binding protein from rat taste buds. *Nature* 244:91–93, 1973.

32. Holton, T.; Hudspeth, A.J. The transduction channel of hair cells from the bullfrog characterized by noise analysis. *J. Physiol.* 375:195–227, 1986.

33. Hubbard, S.J. The study of rapid mechanical events in a mechanoreceptor. *J. Physiol.* 141:198–218, 1958.

34. Hubbell, W.L.; Bownds, M.D. Visual transduction in vertebrate photoreceptors. *Annu. Rev. Neurosci.* 2:17–34, 1979.

35. Hudspeth, A.J. Stereocilia mediate transduction in vertebrate hair cells. *Proc. Natl. Acad. Sci. USA* 76:1506–1509, 1979.

36. Hudspeth, A.J. Extracellular current flow in the site of transduction by vertebrate hair cells. *J. Neurosci.* 2:1–10, 1982.

37. Hudspeth, A.J. Mechanoelectrical transduction by hair cells in the acousticolateralis sensory system. *Ann. Rev. Neurosci.* 6:187–215, 1983.

38. Ilyinsky, O.B. Process of excitation in inhibition in single mechanoreceptors (Pacinian corpuscles). *Nature* 208:351–353, 1965.

39. Kaissling, K.E. Insect olfaction. In Beidler, L.M., ed. *Chemical Senses*, 351–431. Vol. 4, pt. 1, of *Handbk. Sensory Physiol.* New York, Springer-Verlag, 1971.

40. Kaissling, K.E. Sensory transduction in insect olfactory receptors. In *Biochemistry of Sensory Functions*, pp. 243–273. vol. 25 of Colloquium Gesellsch. Biolog. Chemie. L. Jaenicke, ed. Heidelberg, Springer-Verlag, 1974.

41. Kasang, G. Tritium-Markierung des Sexuallockstoffes Bombykol. *Z. Naturforsch. [B]* 23:1331–1335, 1968.

42. Koshland, D.E., Jr. Bacterial chemotaxis in relation to neurobiology. *Annu. Rev. Neurosci.* 3:43–75, 1980.

43. Lamb, T.D. Electrical response of photoreceptors. *Recent Adv. Physiol.* 10:29–66, 1984.

44. Lamb, T.D.; Matthews, H.R.; Torre, V. Incorporation of calcium buffers into salamander retinal rods: a rejection of the calcium hypothesis of phototransduction. *J. Physiol.* 372:315–340, 1986.

45. Lancet, D. Molecular view of olfactory reception. *Trends in Neurosci.* 7:35–36, 1984.

46. Loewenstein, W.R. On the specificity of a sensory receptor. *J. Neurophysiol.* 24:150–158, 1961.

47. Loewenstein, W.R.; Rathkamp, R. The sites for mechanoelectric conversion in a Pacinian corpuscle. *J. Gen. Physiol.* 41:1245–1265, 1958.

48. Loewenstein, WR.; Skalak, R. Mechanical transmission in a Pacinian corpuscle. *J. Physiol.* 182:346–378, 1966.

49. Macnab, R.M. Bacterial motility and chemotaxis: the molecular biology of a behavioral system. *Crit. Rev. Biochem.* 5:291–341, 1978.

50. Makous, W.L. Cutaneous color sensitivity: explanation and demonstration. *Psychol. Rev.* 73:280–294, 1966.

51. Meiselman, H.L.; Halpern, B.P. Effects of *Gymnema sylvestre* on complex tastes elicited by amino acids and sucrose. *Physiol. Behav.* 5:1379–1384, 1970.

52. Morton, R.A. The chemistry of visual pigments, Chap. 2. In Dartnall, H.J.A., ed. *Photochemistry of Vision*, Vol. 7, pt. 1, of *Handbk. Sensory Physiol.* New York, Springer-Verlag, 1972.

53. Moulton, D.G. Minimum odorant concentrations detectable by the dog and their implications for olfactory receptor sensitivity. In Muller-Schwartz, D.; Mozell, M.M., eds. *Chemical Signals in Vertebrates*, 455–464. New York, Plenum Press, 1977.

54. Murray, R.G. Ultrastructure of taste receptors. In Beidler, L.M., ed. *Chemical Senses.* Taste. pp 31–50. Vol. 4, pt. 2, of *Handbk. Sensory Physiol.* New York, Springer-Verlag, 1971.

55. Ozeki, M. Conductance change associated with receptor potentials of gustatory cells in rat. *J. Gen. Physiol.* 58:688–699, 1971.

56. Pace, U.; Hansk, E.; Salomon, Y.; Lancet, D. Odorant-sensitive adenylate cyclase may mediate olfactory reception. *Nature* 316:255–258, 1985.

57. Pease, D.C.; Quilliam, T.A. Electromicroscopy at the Pacinian corpuscle. *J. Biophys. Biochem. Cytol.* 3:331–357, 1957.

58. Pfaffman, C. Gustatory afferent impulses. *J. Cell. Comp. Physiol.* 17:243–258, 1941.

59. Reese, T.S. Olfactory cilia in the frog. *J. Cell Biol.* 25:209–230, 1965.

60. Roper, S. Regenerative impulses in taste cells. *Science* 220:1311–1312, 1983.

61. Russell, M.J.; Switz, G.M.; Thompson, K. Olfactory influences on the human menstrual cycle. *Pharmacol. Biochem. Behav.* 13:737–738, 1980.

62. Sato, T. Multiple sensitivity of single taste cells of the frog tongue to four basic taste stimuli. *J. Cell Physiol.* 80:207–218, 1972.

63. Sato, T.; Biedler, L.M. Membrane resistance changes of the frog taste cells in response to water and NaCl. *J. Gen. Physiol.* 66:735–763, 1975.

64. Spencer, P.S.; Schaumburg, H.H. An ultrastructural study of the inner core of the Pacinian corpuscle. *J. Neurocytol.* 2:217–235, 1973.

65. Stiles, W.S.; Crawford, B.H. The luminous efficiency of rays entering the eye pupil at different points. *Proc. R. Soc. Lond. [Biol.]* 112:428–450, 1933.

66. Stryer, L; Hurley, J.B.; Fung, B.K.-K. Transducin: an amplifier protein in vision. *TIBS* 6:245–247, 1981.

67. Thurm, U.; Wessel, G. Metabolism-dependent transepithelial potential differences at epidermal receptors of arthropods. *J. Comp. Physiol.* 134:119–130, 1979.

68. Trotier, D.; MacLeod, P. Intracellular recordings from salamander olfactory receptor cells. *Brain Res.* 268:225–237, 1983.

69. Yau, K.-W.; Nakatani, K. Electrogenetic Na-Ca exchange in retinal rod outer segment. *Nature* 311:661–663, 1984.

70. Yoshikami, S.; George, J.S.; Hagins, W.A. Light-induced calcium fluxes from outer segment layer of vertebrate retinas. *Nature* 286:395–398, 1980.

71. Young, R.W. Visual cells and the concept of renewal. *Invest. Ophthal.* 15:700–725, 1976.

72. West, C.H.K.; Bernard, R.A. Intracellular characteristics and responses of taste bud and lingual cells of the mud puppy. *J. Gen. Physiol.* 72:305–326, 1978.

Neuromuscular Transmission

INTRODUCTION

The contractile activities of skeletal muscles of the body are under direct nervous control. Commands are transmitted from axon to muscle fiber at a specialized region of close contact, the neuromuscular junction. This chapter concerns the anatomy of the neuromuscular junction and the sequence of events underlying transmission. Neuromuscular transmission is described as the best-studied example of chemical synaptic transmission.

Cell-to-Cell Transmission. The regulatory activities of the nervous system require that nerve cells communicate with one another and with their effector organs. A primitive type of communication, found in the endocrine system, uses chemical messages acting at a considerable distance from the cell that releases them. Often, these extracellular messenger molecules reach their targets through the circulatory system. For example, growth hormone, a peptide, is secreted by cells in the anterior pituitary and acts to promote fat catabolism and the deposition of protein in cells throughout the body. Such a scheme is suited for slow and diffuse control of the activities of entire tissues. It may be used to control metabolic rates or the rate of elaboration of certain gene products. However, the broadcasting of circulating hormones is quite unsuited for the quick and fine message handling that characterizes much nervous function, e.g., locomotion or visual perception. Speed and precision are instead achieved by specific cell-to-cell junctions designed to communicate a message from one prejunctional cell to one postjunctional cell.

Sherrington gave the name *synapse* to the specialized functional contacts of one nerve cell with another. At these junctions, the transmitting cell, called the *presynaptic cell*, and the receiving one, called the *postsynaptic cell*, come very close to each other, with no glial cells separating them in the region of synaptic contact. Traditionally, *synapse* has been reserved for neuron-to-neuron contacts, and *junction* is used more broadly for contacts between any excitable cells. However, with the realization that synapses and other junctions operate by similar mechanisms, the distinction in

Figure 6–1 Comparison of transmission at electrical and chemical synapses. *A,* In electrical transmission, the postsynaptic electrical change is due to local circuit current flowing through gap junction channels from the presynaptic cell. Presynaptic channels generate the current. *B,* In chemical transmission, no current flows from presynaptic to postsynaptic cell. Rather, transmission occurs when a liberated chemical transmitter (dots) opens channels in the postsynaptic membrane. The transmitter is released by a multistep process involving a presynaptic depolarization, calcium entry, and exocytosis of transmitter stored in presynaptic vesicles.

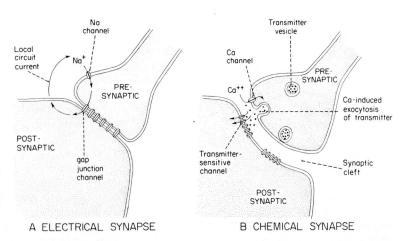

A ELECTRICAL SYNAPSE   B CHEMICAL SYNAPSE

terminology has seemed less important, and *synapse* is often used loosely to connote neuron–effector organ junctions as well as those between neurons. This relaxed definition will be used here. Thus the neuromuscular junction, where an axon terminal communicates messages to a muscle fiber, may be viewed as an exceptionally large peripheral synapse.

How are messages transmitted at synapses? Two fundamentally different modes, electrical and chemical transmission, are shown diagrammatically in Figure 6–1. Electrical transmission is conceptually the simpler. It occurs at structures that morphologists call *gap junctions.* Here the pre- and postsynaptic cell membranes are separated by a gap of only 2 nm, crossed by dodecameric protein tubes, connexons, that form aqueous channels bridging between the cytoplasmic compartments of the two cells. The connexon channels are tunnels large enough to pass molecules with a molecular weight up to 1500 daltons between the cells; and hence electric currents, carried by ions, flow easily through them from one cell to the other as do small metabolic intermediates and second messengers. The cells are said to be electrically coupled, since a voltage change in the presynaptic cell produces a voltage change in the postsynaptic cell via the low-resistance current pathway through the connecting channel. Thus electrical messages are communicated directly by local-circuit currents between the cells. Electrical coupling between smooth muscle cells and heart cells is discussed in Chapters 10 and 37, and electrical coupling as a synaptic mechanism is discussed in Chapter 11.

The second mode of synaptic transmission, chemical transmission, is both more complicated and more versatile than electrical transmission.

Most synapses in the nervous system are chemical synapses.* In this instance, the presynaptic terminal always contains numerous small, membrane-bounded vesicles (Fig. 6–1). The vesicles store neurotransmitter molecules in a readily releasable form. The steps of chemical synaptic transmission are (i) Presynaptic depolarization opens Ca channels in the presynaptic membrane. (ii) Entering Ca^{2+} ions promote the fusion of transmitter vesicle membranes with the surface membrane so that neurotransmitter is released into the synaptic cleft (exocytosis of transmitter). (iii) Neurotransmitter molecules bind to specific receptor molecules on the postsynaptic membrane. (iv) At this point, various mechanisms diverge: The transmitter-receptor complex may open postsynaptic ionic channels, which, depending on the ions involved, hyperpolarize or depolarize the postsynaptic membrane. The resulting localized slow wave, called the *postsynaptic potential,* is the electrical manifestation of transmission. Alternatively, the transmitter-receptor complex may act on membrane-bound enzymes or enzyme cascades that produce internal second messenger molecules. The second messengers in turn modulate activities of channels or other cellular processes. Often, this alternative mechanism is slow and does not require close, one-to-one contact of pre- and postsynaptic cells.

Chemical synapses with synaptic vesicles and electrical synapses with gap junctions are evolutionary innovations of the animals. They are present in all animals with nervous systems, first

---

*Counts are not available for higher animals, but in the nematode *Caenorhabditis elegans,* the 302 neurons of the adult hermaphrodite are joined by 7000 chemical synapses and 600 gap junctions.

appearing in some Cnidaria (coelenterates) with morphology very much like that in higher animals.

Neuromuscular Junction. By far the largest synapses in the vertebrate body are at the skeletal neuromuscular junctions. They are often called the *motor end-plates*. Here the presynaptic cell is a motoneuron, and the postsynaptic cell, a skeletal muscle fiber. Transmission is one-way and excitatory such that action potentials arising at the cell body of a motoneuron in the spinal cord propagate down the motor axon and ultimately cause muscle fibers in the periphery to contract (see Fig. 3–1). The neuromuscular junction is a popular subject in the study of physiology, both because of the importance of neuromuscular transmission *per se* and because much more is known about this junction than about any other synapse. Large size and easy access make it a favorite physiological and pharmacological tool. The successes of these many investigations make it the starting point for discussing all other, less-well-studied, synapses. Neuromuscular transmission is the subject of extensive reviews.[19, 24, 31, 32, 37, 49, 50]

The physiology of chemical synapses first became known through studies with exotic and poisonous natural products. In the last century, Claude Bernard, a giant in the history of physiology, showed that the Amazon dart poison, curare, blocks neuromuscular transmission without affecting either conduction in nerve or the contractile response of muscle to direct electric shock. Early in this century, J.N. Langley found that dilute solutions of nicotine evoke muscle twitches when applied directly to the neuromuscular junction, but not when applied to other parts of the muscle surface. The action of nicotine was antagonized by curare. These observations suggested that nicotine and curare compete for a receptive substance—now known as the acetylcholine receptor—in the neighborhood of the nerve terminals. Langley postulated that the nicotine-receptor complex has an excitatory effect on the muscle, whereas the curare-receptor complex ties up the receptor without excitatory effect.

The beat of the heart has long been known to be slowed by stimulation of the vagus nerve, an effect that is blocked by atropine (but not curare) and potentiated by eserine. In 1921, Otto Loewi found that the perfusion solution collected from a frog heart during vagal stimulation in the presence of eserine will slow the beating of a second heart. The action is blocked by atropine. He reasoned that the vagus nerve releases a labile, inhibitory substance that he called "Vagusstoff" and was later found to be the quaternary ammonium ion *acetylcholine* ($ACh, CH_3COOCH_2CH_2N(CH_3)_3^+$). Loewi and colleagues then showed that eserine prolongs the life of Vagusstoff by blocking an enzyme, cholinesterase, which actively hydrolyzes esters of choline, including ACh. He received the Nobel Prize for demonstrating chemical transmission of vagal inhibition. In 1936, Henry Dale, W. Feldberg, and colleagues proposed that ACh might also be the natural excitatory neurotransmitter at skeletal neuromuscular junctions. They showed that repetitive stimulation of the motor nerve to a cat skeletal muscle releases ACh into the solution bathing the muscle, and that ACh injected into an artery close to a muscle in a living animal causes a vigorous synchronous contraction of the muscle. To collect the small amount of ACh released by stimulation, Dale's group had to perfuse the muscle with solutions containing eserine. The perfusate was then assayed for ACh by a sensitive bioassay involving the contraction of strips of leech muscle.[50] Dale shared the Nobel prize with Otto Loewi for demonstrating chemical transmission at the neuromuscular junction.

These experiments led to the correct hypothesis that motor nerve terminals release the chemical transmitter ACh and that a complex formed between ACh and a postsynaptic receptor excites the muscle. Other substances, including nicotine, can also activate the receptor, although they are not the natural transmitter. They are called *agonists*. Conversely, curare blocks the receptor so that agonists cannot bind. It is called an *antagonist*. Because of these pharmacological properties, the neuromuscular junction was called by Dale a *nicotinic, cholinergic, chemical synapse*. The name *cholinergic* means that the presynaptic nerve fibers release ACh. The name *nicotinic* distinguishes the ACh receptor of skeletal muscle from those of heart, which Dale called *muscarinic* since they can be stimulated by the alkaloid muscarine (and not nicotine) and are blocked by atropine (and not curare). We now know that nicotinic and muscarinic ACh receptors are totally different classes of membrane proteins acting by unrelated mechanisms but both using ACh as an agonist.

In the period from 1939 to 1951, nerve stimulation and ACh applications were shown to evoke a localized depolarization of the muscle membrane, called the *end-plate potential*, that was attributed to chemical activation of a permeability increase to ions in the postsynaptic membrane. The alternative hypothesis, that neuromuscular transmission is accomplished by electrical coupling from the nerve terminal, was also considered possible until the early 1950s but then was clearly ruled out. A simple argument against it is the severe electrical mismatch between the tiny presynaptic nerve terminal and the giant postsynaptic membrane that needs to be depolarized to achieve effective transmission. The little axon terminal simply could not generate the current needed to depolarize the large muscle membrane without an interposed amplifying step.

With this brief historical introduction, we turn to a synthesis of our modern understanding of neuromuscular transmission. Most of this chapter is based on studies of cholinergic transmission in the leg, toe, and pectoral fast skeletal muscles of the frog; in the diaphragm muscle of the rat; and in the muscle-derived electric organs of electric eels and rays. In these popular and mechanistically quite similar preparations, every action potential arriving in the presynaptic nerve fiber elicits one postsynaptic action potential in the corresponding muscle fibers. Only at the end of the chapter is any comparison made with neuromuscular junctions in other cells, such as smooth muscle or invertebrate muscle, where some aspects of the transmission are different. Synaptic transmission in the central nervous system is described in Chapter 11.

ANATOMY AND CHEMISTRY OF THE NEUROMUSCULAR JUNCTION

Anatomy.[19] Vertebrate skeletal muscles contain many elongated, multinucleate contractile cells, the *muscle fibers*, each of which is innervated by a branch of an axon from a motoneuron whose cell body lies in the spinal cord. As the myelinated axon of a motoneuron reaches a muscle via a motor nerve trunk, it bifurcates many times at nodes of Ranvier to give short myelinated branches to several individual muscle fibers. These branches lose their myelin and branch several times again to give Schwann cell–covered nerve terminals that are applied closely to the muscle fiber in a localized region, the neuromuscular junction (Figs. 6–2 and 6–3). In the frog, the terminals are like extended fingers lying in grooves along the fiber surface, whereas in mammals, the terminal is often shaped more like a bunch of grapes pressed against cup-shaped depressions in the muscle fiber or like a foot pressing into the muscle surface. This latter geometry originally inspired the name *end-plate*, which is now used more generally to denote vertebrate neuromuscular junctions of any geometry. In all examples, the end-plate occupies only a small fraction of the total

muscle fiber surface, often near the center of the fiber. A muscle fiber many centimeters long may have an end-plate region only 100 to 500 μm long. A few vertebrate muscles have more than one end-plate region innervated by different motoneurons, but most muscle fibers have only a single endplate innervated by one axon.

The presynaptic terminals are filled with thousands of 50-nm-diameter synaptic vesicles and many mitochondria, reflecting a high level of metabolic activity (Figs. 6–2 and 6–3). In frog muscle, the terminals are only 1 to 1.4 μm in diameter, but within a total length of 500 μm, they may contain nearly half a million synaptic vesicles. Of all the vesicles seen in thin sections with the electron microscope, a small fraction lies lined up along so-called *active zones* of the presynaptic membrane. Each active zone may be regarded as a unitary presynaptic element with the Ca channels, transmitter vesicles, and releasing mechanism needed for chemical transmission. Electron microscope studies of the active zone show a band of electron-dense material underlying the prejunctional membrane in thin sections (Fig. 6–3) and a double row of intramembrane particles when the presynaptic membrane is viewed face-on after being split open by freeze-fracture methods (Fig. 6–2). Occasional profiles of synaptic vesicle membrane are seen fused with the cell membrane of the nerve terminal in a way that would permit the vesicle contents to enter the synaptic cleft by exocytosis.[19, 20] This is the general method for secreting vesicular packets of materials from a wide variety of secretory cells. Evidence that ACh is released by exocytosis at the end-plate is given in a later section. Membrane is recycled by being brought back into the nerve terminal by endocytosis in regions between the active zones.

The axon terminal and muscle cell are separated by a 60-nm synaptic cleft filled with an amorphous connective tissue meshwork, a *basal lamina* not very different in appearance from the basement membrane that envelops the entire muscle fiber. The basal lamina contains collagen fibers, proteoglycans, and enzymes derived from the pre- and postsynaptic cells, and its components help specifically to organize the pre- and postsynaptic specializations of these cells. On the postsynaptic side, the muscle membrane is subtended by dense cytoplasmic material and opposite the active zones is invaginated in junctional folds. Freeze-fracture electron micrographs of this acetylcholine-sensitive region of membrane reveal densely packed, large, intramembrane particles presumed to be the ACh receptor. Autoradiographic studies with ra-

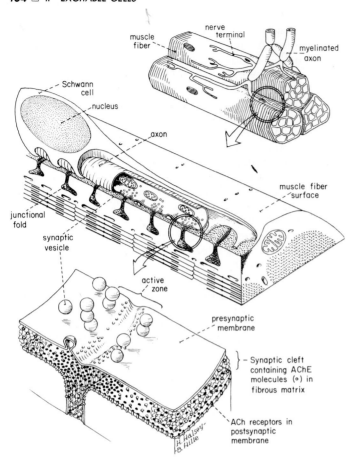

Figure 6–2 Diagram of an amphibian, skeletal neuromuscular junction. *Above,* Branches of myelinated nerve fibers lose their myelin and terminate on three striated muscle fibers. *Center,* The nerve terminal lies on a groove in the muscle surface and is covered by a thin sheath of Schwann cell cytoplasm. *Below,* At the active zone, synaptic vesicles lie nearly touching the presynaptic membrane. Two vesicles are drawn in the process of exocytosis. ACh molecules released into the snyaptic cleft encounter acetylcholinesterase molecules in the cleft and ACh receptors on the folded postsynaptic membrane. (After Peper, K.; Dreyer, F.; Sandri, C.; Ackert, K.; Moor, H. *Cell Tiss. Res.* 149:437–450, 1974, and Lester, H. *Scientific American* 236:106–118, 1977.)

Figure 6–3 Electron micrograph showing two active zones on the neuromuscular junction of a frog cutaneous pectoris muscle cut parallel to the long axis of the muscle. The calibration bar corresponds to 500 nm. The three cell types seen are: *A,* axon; *M,* muscle fiber; *S,* Schwann cell process. The muscle surface membrane is infolded opposite the darkened active zones of the presynaptic terminal. One active zone is marked by an asterisk in the synaptic cleft. The many 50-nm circles with gray centers are synaptic vesicles in the axon. Compare with the diagrams in Figure 6–2. (Electron micrograph courtesy of J. Heuser and L. Evans, University of California, San Francisco.)

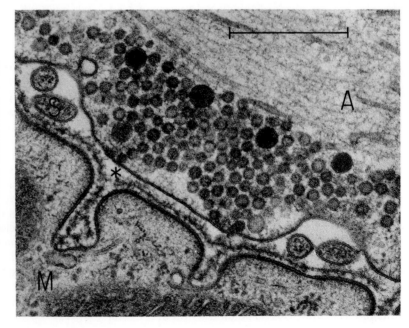

dioactively labeled, receptor-marking toxins show a nearly crystalline packing density of receptor molecules in the postsynaptic membrane.

The Acetylcholine Cycle. The natural transmitter, ACh, is synthesized in the presynaptic terminal by the enzyme choline acetyltransferase in a single step requiring choline and the ubiquitous metabolic intermediate, acetylcoenzyme A:

$$CH_3-\overset{\overset{O}{\|}}{C}\sim SCoA + HOCH_2CH_2-\overset{\overset{CH_3}{|}}{\underset{\underset{CH_3}{|}}{N^+}}-CH_3 \qquad \text{(Choline)}$$

Choline
acetyltransferase

$$HS-CoA + CH_3COCH_2CH_2\overset{\overset{CH_3}{|}}{\underset{\underset{CH_3}{|}}{N^+}}-CH_3 \qquad \text{(Acetylcholine)}$$

It is generally supposed that at the neuromuscular junction much of the transmitter is stored within the synaptic vesicles, as it is in a wide variety of other chemical synapses. When synaptic vesicles are isolated from the particularly extensive cholinergic synapses of the electric organ of the electric ray (*Torpedo*), the fractions obtained contain high concentrations of ACh together with ATP as the anion.[30] Since choline acetyltransferase seems to be free in the cytoplasm rather than associated with vesicles, ACh is made in the cytoplasm and must be packaged subsequent to its synthesis. The mechanism of transport into the vesicle is not definitively known, but a working hypothesis is that the packaging might involve the combined actions of a proton ATPase that creates a pH gradient across the vesicle membrane (acid inside) and an ACh-H$^+$ countertransport carrier system in the vesicle membrane.[2] Such a mechanism is known for packaging of amine neurotransmitters (Chap. 2).

Once released into the synaptic cleft by exocytosis, ACh molecules can interact with the postsynaptic receptor, a process described in a later section. All the ACh is eventually inactivated by being hydrolyzed back to choline and acetate by the extracellular enzyme *acetylcholinesterase* (AChE), a member of the large class of "serine protease" enzymes. Large numbers of AChE molecules with collagenlike tails are associated with collagen threads throughout the basal lamina meshwork in the synaptic cleft. Cholinesterases are also found in most nervous tissue and even on the extracellular surface of red blood cells. The hydrolysis is rapid, indeed it is among the fastest known enzymatic reactions. At maximum velocity, this enzyme cleaves an ACh molecule every 100 μs, and the velocity is half-maximal at 90 μM ACh.

A wide variety of chemical agents designed to inactivate acetylcholinesterase are used as insecticides, and some are used in the clinic. When used as insecticides they are applied at a dose high enough to disrupt cholinergic transmission, and when used in the clinic they are used at a low concentration to potentiate neuromuscular transmission when it is failing as a consequence of some diseases.[53] These anticholinesterases range from the long-lasting toxic "organophosphates," like the "nerve gas" diisopropylfluorophosphate, DFP, to the briefly acting and clinically useful quaternary ammonium compound endrophonium. Other anticholinesterase drugs useful in medicine and research are eserine (physostigmine) and neostigmine (Prostigmin). All of these compounds act as poor *substrates* of the AChE enzyme and lead to the formation of a covalent intermediate in which a part of the drug molecule remains attached for a long time to the essential serine group in the active site of the enzyme. The inhibition of AChE activity lasts seconds to days, depending on the stability of this acyl intermediate. A practical antidote, pyridine-2-aldoxime methylchloride, 2-PAM, has been designed to speed the breakdown of the acyl intermediate in the event of accidental anticholinesterase poisoning.[53]

Choline produced by ACh hydrolysis can be taken up again into the presynaptic terminal by a high-affinity choline uptake mechanism in the membrane. This transport mechanism is half-saturated by 1 μM of extracellular choline and can be blocked by a quaternary ammonium drug called *hemicholinium No. 3*. A neuromuscular junction stimulated repetitively while treated with hemicholinium 3 eventually runs out of transmitter.[17] The choline uptake mechanism has a requirement for external Na$^+$ ions and is, like many transmitter-uptake mechanisms, a cotransport device that uses the Na$^+$ ion gradient as a power source (Chap. 2).

THE POSTSYNAPTIC RESPONSE

We now turn to primarily electrophysiological studies of neuromuscular transmission. As the nerve impulse always proceeds from the presynaptic cell to the postsynaptic cell, it would be logical to begin the story with investigations of presynaptic events. However, for the electrophysiologist, the presynaptic terminal is too small (< 1 μm diameter) for penetration by a microelectrode, and the postsynaptic electrical response is almost the only signal accessible to study. Deductions about presynaptic events were therefore made by clever use of the postsynaptic response. The three major questions to be considered are (i) What permeability changes occur in the postsynaptic membrane? (ii) What is the nature of the acetylcholine receptor? (iii) How is the time course of the endplate response accounted for?

End-plate Potential. Our basic understanding of the nature of the end-plate potential (EPP) and of the mechanism of ACh release comes from the classical electrophysiological work of Bernard Katz and his collaborators, starting in the 1940s, work for which Katz received the Nobel Prize. Fatt and Katz[14] recorded the EPP from muscle fibers with an intracellular microelectrode. To avoid breaking the fine glass recording pipette with a muscle twitch, they had to depress the EPP amplitude below the firing threshold for propagated action potentials by adding sufficient curare to the bathing solution. Then electric shocks applied to the motor nerve evoked subthreshold EPPs in the end-plate region of the muscle (Fig. 6–4, top trace, and Fig. 6–5A). The nerve-evoked potential change follows the presynaptic shock with a delay, then rises to a peak in about 1 ms and falls slowly, with an exponential time constant of 15 to 30 ms. Potential records made at successive 0.5-mm distances from the junction show a more slowly rising and smaller potential change with increasing distance (Fig. 6–4).

Arguing from cable theory (Chap. 3), Fatt and Katz showed that these records may be quantitatively understood by supposing that during transmission a burst of inward current is injected into the muscle fiber at the junctional membrane for less than 1 ms. In the curarized muscle, all the subsequent spread of potential in distance and the decay in time simply reflect the passive electrical cable properties of the muscle fiber. Evidently, the EPP is generated by current localized at the junctional membrane, and when reduced by curare, the subthreshold depolarization spreads passively only a few millimeters in either direction along the muscle fiber. These properties are characteristic of any local, graded *slow potential*. Recall, however, that the muscle membrane just outside the end-plate has a specially high density of volt-

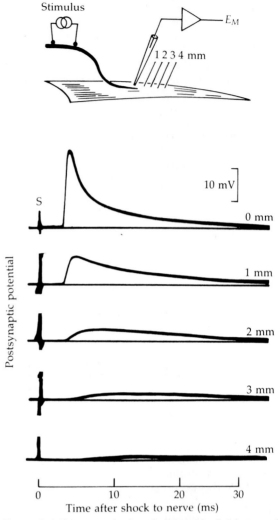

Figure 6–4 The spread of end-plate potential in a partly curarized frog muscle fiber. Membrane potential changes are recorded with an intracellular electrode while the motor nerve to the muscle is stimulated by brief electric shocks. *S,* Stimulus artifact, signaling the time of stimulus to the nerve. Number by each curve is distance in millimeters of the recording electrode from the end-plate region. As distance increases, the recorded potential becomes smaller and slower. (After Fatt, P.; Katz, B. *J. Physiol. [Lond.]* 115:320–370, 1951.)

age-gated Na channels. Therefore, in a normal, uncurarized twitch muscle, the membrane adjacent to the end-plate would reach its firing threshold in a fraction of a millisecond, and an action potential based on voltage-gated Na and K channels would propagate rapidly to both ends of the fiber in an all-or-nothing manner. Thus normal neuromuscular transmission involves the transmitter-induced depolarization of the chemically excitable membrane under the nerve terminal,

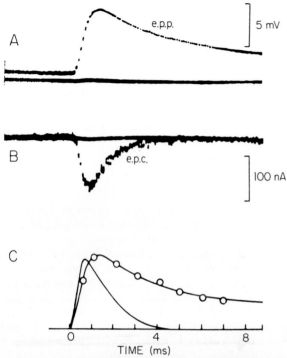

Figure 6–5 Nerve-evoked end-plate potential and end-plate current of a frog sartorius muscle fiber. The bath contains 6 μM d-tubocurarine so that the EPP remains subthreshold for the muscle action potential mechanism. *A,* End-plate potential recorded with an intracellular pipette. *B,* End-plate current recorded while the membrane potential is held at the resting potential by a two-microelectrode voltage clamp. *C,* Superposition of EPP and EPC records (solid lines), showing that the current is briefer than the resulting potential change. Circles show time course of the EPP, predicted by assuming that a current equal to the recorded EPC is injected into a passive cable whose membrane resistance and capacitance are chosen to match those of muscle. T = 17°C. (After Takeuchi, A; Takeuchi, N. *J. Neurophysiol.* 22:395–411, 1959.)

which in turn triggers the electrically excitable membrane that completely surrounds it.

Consider now the current flow that underlies the EPP. According to the early analysis by Fatt and Katz, the end-plate current, EPC or I_{ep}, lasts only about 1 ms at room temperature. Their conclusion was soon confirmed by voltage-clamp measurements done with two intracellular microelectrodes in the muscle. One electrode records the membrane potential, and the other is used to pass current. When the electrodes are connected through a feedback amplifier to make a voltage clamp, as in Figure 3–8, the postsynaptic membrane potential can be kept artificially constant while the ionic *current* flowing through ionic channels at the end-plate is measured directly. The end-plate current, first measured by Takeuchi and

Takeuchi,[51] is inward and transient (Fig. 6–5B). When compared with the EPP (Fig. 6–5C), the EPC rises and falls faster—with a 1-ms exponential time constant of decay at the resting potential, as expected from the work of Fatt and Katz.

Transmitter action probably needs to be kept short to avoid repetitive firing of muscle action potentials in response to one presynaptic impulse and to permit junctional transmission to follow closely spaced, presynaptic impulses. Some of the muscles controlling our eyeball movements need to respond to bursts of action potentials at frequencies up to 1000 Hz. How important is the rapid hydrolysis of ACh for keeping transmitter action brief? Fatt and Katz found that EPP amplitude and duration can be increased in curarized muscle by pretreatment of the preparation with neostigmine or other anticholinesterase drugs (Fig. 6–6). These augmented responses could again be analyzed in terms of muscle cable properties or by voltage clamp to show that the amplitude and duration of I_{ep} are increased 2- to 3-fold when the enzyme AChE is inhibited (Fig. 6–6). Hence, the esterase normally plays a significant role in limiting the duration of transmitter action, but in its absence, the process of diffusion still removes transmitter from the junctional region within a few milliseconds.

Conductance Increase

Properties of the Open Channel.[50] The voltage clamp is a convenient tool for studying channels opened by ACh at the neuromuscular junction. From voltage clamp currents measured at different membrane potentials one can determine the size of the underlying conductance increase and the electromotive force of the transmitter-gated channel. In the experiment of Figure 6–7, the muscle membrane potential is varied from −120 to +38 mV as the motor nerve is stimulated by brief shocks. At −120 mV, I_{ep} is −450 nA, a large inward current, while at +38 mV, it is +200 nA, an outward current. Closely spaced voltage steps show that I_{ep} reverses sign at −5 mV, the *reversal potential* of the end-plate response. From this kind of experiment, Takeuchi and Takeuchi[51] concluded that end-plate current may be described by Ohm's law, $I_{ep} = g_{ep} (E_M - E_{ep})$, where E_M is the membrane potential, E_{ep} is the reversal potential, and g_{ep} is the transmitter-induced conductance. In an uncurarized frog sartorius muscle fiber, g_{ep} is typically 10 to 15 μS at the peak of the EPP.

The end-plate channel cannot be perfectly selective for any single ion since its reversal potential, E_{ep}, is far from the equilibrium potential of any of the common ions. Takeuchi and Takeuchi[52] stud-

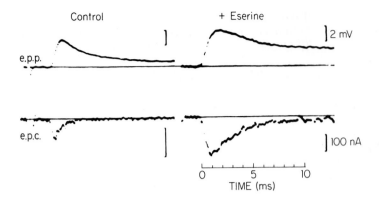

Control

+ Eserine

e.p.p.

2 mV

e.p.c.

100 nA

0   5   10
TIME (ms)

Figure 6–6 Enhancement of transmitter action by the anticholinesterase agent eserine (physostigmine). *Left,* The control EPP and EPC recorded from a curare-treated sartorius fiber, as in Figure 6–5. *Right,* Same responses after adding 36 μM eserine to the bath. Both current and voltage have higher peak values and last longer. (After Takeuchi, A.; Takeuchi, N. *J. Neurophysiol.* 22:395–411, 1959.)

ied the effect of changing the external ions on the reversal potential. They concluded that Na$^+$ and K$^+$ ions make about equal contributions, whereas Cl$^-$ ions make none. These results explain why E_{ep} is not far from 0 mV. Recall that the Goldman-Hodgkin-Katz voltage equation (Chap. 1) given the relationship among permeabilities, concentrations, and reversal potential. For a channel where Na$^+$ and K$^+$ ions are equally permeant ($P_{Na} = P_K$), the equation becomes

$$E = \frac{RT}{F} \ln \frac{[Na]_o + [K]_o}{[Na]_i + [K]_i}$$

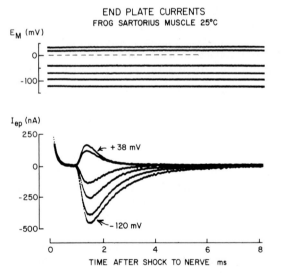

END PLATE CURRENTS
FROG SARTORIUS MUSCLE 25°C

E_M (mV)

0

-100

I_{ep} (nA)

250

+ 38 mV

0

-250

-500

- 120 mV

0   2   4   6   8
TIME AFTER SHOCK TO NERVE   ms

Figure 6–7 End-plate currents in a frog skeletal muscle fiber under voltage clamp. The membrane potential was held at six different potentials (E_M) as the motor nerve to the muscle was stimulated with a shock. The six different current responses (I_{ep}), each preceded by a transient shock artifact, are superimposed oscillographically. At negative voltage, I_{ep} is inward and slowly decaying. At a positive voltage, I_{ep} is outward and more rapidly decaying. (After Magleby, K.; Stevens, C. F. *J. Physiol. [Lond.]* 223:151–171, 1972.)

With the ionic concentrations given in Table 1–4, the predicted reversal potential is −3 mV. Modern experiments with a wide variety of test ions confirm that the end-plate channel is impermeable to anions and not very selective among the major cations. The permeability is roughly equal for Li$^+$, Na$^+$, K$^+$, Rb$^+$, and Cs$^+$, and even divalent ions like Ca^{2+} and Mg^{2+} are quite permeant. At least 60 different small organic cations can pass through the open channel, including triethylammonium and histidine, molecules requiring a pore diameter of at least 0.65 nm (see Fig. 4–2B).[12] Thus the end-plate channel is a relatively wide and poorly selective pore, quite unlike the voltage-gated Na or K channels in this respect. Of all the cations present under physiological conditions, however, only Na$^+$ and K$^+$ are in high enough concentration to contribute in an important way to the end-plate current. When the channels open, many Na$^+$ ions flow in down their steep electrochemical gradient, fewer K$^+$ ions flow out down their shallower electrochemical gradient, and the membrane potential moves towards −5 mV.

The ACh-activated channels of skeletal muscle end-plates were the first biological ionic channels for which single-channel properties were measured. Two powerful methods have been used—first, fluctuation ("noise") analysis and, later, direct patch–clamp recording of discrete current steps. In fluctuation analysis, a low concentration of ACh or some other agonist is applied extracellularly to the end-plate region to open a small fraction of the postsynaptic channels. As ACh molecules bind to and dissociate from their receptor hundreds of times a second, the instantaneous number of channels open fluctuates at random about its mean value. At one moment, 10,007 channels may be open and in the next, 9,963, and so forth. These fluctuations, seen as fluctuations of the current recorded by a voltage clamp, can be analyzed statistically to give both the amplitude of the unit conductance, γ, and the mean lifetime of the open state.[1, 21, 27] Probability theory tells us that when only a small fraction of the channels are open, the amplitude of the single-channel current will be equal to the variance of the current signal about the mean divided by the mean current. This technique, developed in the laboratories of B. Katz and C.F. Stevens in the early 1970s, has been especially useful for studying many synaptic channels. However, within five years, the patch-clamp method of Neher and Sakmann[18] appeared, and by now it is the preferred method for most single-channel work, since it resolves unitary currents directly. For synaptic work, the firepolished patch-clamp pipette is filled with a solution containing a minute concentration of agonist, e.g., 100 nM ACh, and it will activate an occasional channel in the membrane area where it is applied. The presynaptic terminal, which would normally prevent the pipette from reaching the subsynaptic membrane, must first be removed. The nerve-muscle preparation is treated with collagenase and a little nonspecific protease until the enveloping basal lamina of the muscle is digested away and the nerve fibers float free of the muscle fibers. Then the pipette can be pressed into the synaptic groove to make a seal against the postsynaptic membrane.

As with voltage-gated channels, the patch-clamp records of transmitter gated channels show sequences of rectangular unitary current steps (Fig. 6–8), indicating that opening and closing of individual channels in the patch are abrupt, all-or-nothing events. Channel opening requires agonist in the pipette and is blocked if an antagonist is added as well. Fluctuation and patch-clamp methods agree that a single open end-plate channel has a conductance of about 29 pS in amphibia and 45 pS in mammals, values two to three times larger than the single-channel conductance of the voltage-gated Na and K channels that generate action potentials.

Gating Kinetics.[34, 35, 49] What factors determine the kinetics of opening and closing of transmitter-gated channels during the EPC? As one hypothesis, the time course of the postsynaptic conductance might accurately parallel the rise and fall of available transmitter molecules in the cleft. Alternatively, it might be dominated by elementary opening and closing times intrinsic to the channel macromolecule once the transmitter arrives. Actually, the full explanation is a combination of these two ideas: As we shall see, the rise of the EPC is limited by the arrival of ACh molecules at ACh receptors, and the subsequent fall of the EPC

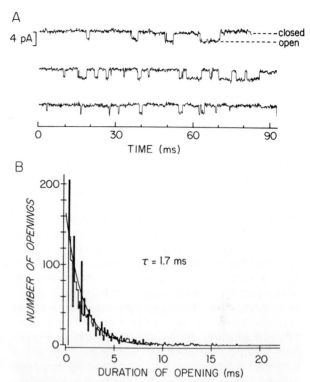

Figure 6–8 Patch-clamp recording of single acetylcholine receptor channels in the end-plate membrane of garter snake costocutaneous muscle. The presynaptic nerve fiber has been removed after enzymatic treatment of the neuromuscular junction, and a patch pipette containing 500 nM ACh was sealed against the receptor-rich membrane to record from individual channels. *A*, Single-channel currents (opening is downward). (From Leibowitz, M.D., and Dionne, V.E. *Biophys. J.* 45:153–163, 1984.) *B*, Open-time distribution showing the number of channel openings that last for the given time. The smooth curve is a decaying exponential function, $\exp(-t/\tau)$, with a time constant τ of 1.7 ms, the mean open time of these channels. T = 20°C. (From Dionne, V.E., Leibowitz, M.D. *Biophys. J.* 39:253–261, 1982.)

is limited by the lifetime of the transmitter-receptor complex and the open time of single channels.

The voltage-clamp traces in Figure 6–7 show that at each membrane potential, the evoked EPCs decay approximately exponentially after the peak. Surprisingly, however, the rate of decay is somewhat faster at +38 mV than at −120mV. The decay time constant in this frog muscle is 1.2 ms at −120 mV and only 0.6 ms at +38 mV. Although voltage steps by themselves do not open channels in the absence of agonist, depolarization clearly speeds the decay of the EPC. The rate of decay of the EPC is also speeded by warming, about 3-fold per 10°C ($Q_{10} = 3$).

The question is whether these kinetic properties of the decay are dictated by intrinsic properties of the channels or by the time course of the available ACh in the cleft. The distinction can be made from single-channel studies—either fluctuation analysis or patch-clamp recording—which give the open lifetime, and hence the gating kinetics, for individual channels bathed in a constant concentration of ACh. With the patch clamp, one sees that unitary channel openings have no fixed duration (Fig. 6–8A). They fluctuate from event to event. Nevertheless, there is a regularity in this random open time. If several hundred open times are individually measured and a histogram is formed of the number of open events versus their duration, the histogram describes a decaying exponential curve with a time constant in this snake muscle of 2 ms (Fig. 6–8B). In the frog, the corresponding single-channel time constant is 1 ms. Hence, if several hundred channels opened simultaneously, although each one contributes only a square pulse of current, the sum of all the current would appear to be a smoothly decaying exponential with a time constant of 1 ms, just as is seen in the natural EPC. In addition, the average single-channel open time can be shortened by membrane depolarization and lengthened by membrane hyperpolarization. Finally, the rate of closing is speeded by raising the temperature, with a Q_{10} of ~3. Thus we conclude that the duration of the end-plate current during normal transmission is set by the open-channel lifetime rather than by the lifetime of free ACh molecules in the cleft. The free ACh concentration actually falls much more rapidly, reaching a low concentration before I_{ep} reaches its peak.

Magleby and Stevens[34, 35] proposed a quantitative model for the time course of end-plate currents. A now widely accepted feature of their model is a separation of the drug-binding step from the activation step. They observed that the receptor is a protein whose conformation (seen as conductance) changes after its "substrate" (ACh) binds. By analogy with the well-studied cases of enzyme-substrate reactions, the binding of ACh might be a very rapid, diffusion-limited first step, and the channel opening and closing, a slower second step. With these assumptions, but allowing for the more recent additional observation that two bound ACh molecules are needed to activate

a channel, the following state diagram can be drawn. Here $R \cdot (ACh)_2$ is the closed conformation and $R^* \cdot (ACh)_2$, the open conformation of the receptor complex; β is the rate constant for opening channels, and α is the rate constant for closing, the step that is thought to be speeded by depolarization. In this model, the time course of the EPC would be identical to the time course of the open $R^* \cdot (ACh)_2$ complex. The model permits one to calculate back from the time course of the EPC first to get the transient time course of the closed $R \cdot (ACh)_2$ complex and then to estimate the time course of the concentration of free ACh. When this is done, the calculated time course of cleft ACh rises sharply and falls again to low values already by 400 μs after the beginning of the EPC.

Let us summarize the kinetic conclusions so far. During normal transmission, free ACh in the synaptic cleft rises and falls in a few hundred microseconds. End-plate current lasts somewhat longer because once channels open, they take, on the average, 1 ms to close. Then the end-plate potential in the absence of a muscle action potential would last more than 20 ms, because of the long electrical time constant, τ_M, of the muscle membrane. However, an action potential would normally be triggered and the end-plate region would be briefly depolarized further and then repolarized quickly by the successive activation of Na channels and then K channels, as in other electrically excitable cells.

Nicotinic ACh Receptor

Pharmacology and Agonist Action. The nicotinic ACh receptor (AChR) has long been defined operationally by pharmacological experiments as the hypothetical site of competitive binding of agonists and antagonists of neuromuscular transmission. As we have seen, acetylcholine receptors of vertebrate cells were classified by Dale as nicotinic or muscarinic, depending on which of the two diagnostic alkaloids, nicotine or muscarine, has a stimulatory effect. The receptor at the neuromuscular junction, like the primary one in sympathetic ganglia, is nicotinic. Either can be activated by a wide range of quaternary ammonium ions from the simple tetramethylammonium ion to ACh, carbachol (carbamylcholine), succinylcholine, suberyldicholine, decamethonium, and nicotine. Some

$$R + 2ACh \xrightarrow[\text{equilibrium}]{\text{rapid}} R \cdot (ACh)_2 \underset{\alpha}{\overset{\beta}{\rightleftharpoons}} R^* \cdot (ACh)_2$$

$$\text{closed} \qquad\qquad \text{closed} \qquad\qquad \text{open}$$
$$\text{channel} \qquad\qquad \text{channel} \qquad\qquad \text{channel}$$

of these compounds are called *partial agonists*, since even at optimal doses they increase the postsynaptic conductance far less than ACh does.

The first evidence that more than one agonist molecule must be bound to open a channel came from measurements of the relationship between free agonist concentration and the number of open channels—the *dose-response* curve.[9] From standard chemical kinetics, the end-plate conductance should rise as a linear function of the agonist concentration (in the low-concentration range) if just one ACh molecule is needed to activate one channel. If two ACh molecules are needed, conductance should rise as the square of the concentration and so forth.* Typically, the end-plate conductance g_{ep} is measured by a voltage clamp while defined quantities of agonist molecules are applied locally to the end-plate by the technique of *iontophoresis*. Here a concentrated solution of a salt of the agonist is placed in a fine-tipped glass micropipette that is held close to the end-plate region. The charged drug molecules are then expelled electrophoretically by passing a current pulse out of the tip of the pipette (like any dissolved ion, the drug molecules move in response to an applied electric field). The technique permits a steady delivery of drug for many seconds or a brief pulse of drug as short as a millisecond. When low concentrations of ACh are delivered iontophoretically, the end-plate conductance rises, not linearly but as a power function of ACh concentration, $[ACh]^n$, where n ranges from 1.5 to 3.0 in different experiments. Hence more than one ACh molecule binds to the receptor. The electrophysiological evidence suggests that the highest probability of opening a channel occurs when two ACh molecules bind. Structural evidence given later shows that there are two ACh binding sites per receptor protein.

How do channel openings induced by other agonists compare with those induced by ACh? Two major conclusions can be drawn by fluctuation and patch-clamp experiments. First, the single-channel conductance is independent of the agonist used to open the channel. Second, the lifetime of the open state does depend on the agonist. Compared with ACh, the channel open time with suberyldicholine is 2 to 3 times as long, that with carbachol is half as long, and that with

tetramethylammonium ion is less than a fifth as long. These revealing observations show that the open channel "remembers" what kind of agonist molecule opened it, arguing that the agonist molecules remain bound to the receptor until after the channel closes (as is implicit in the scheme of Magleby and Stevens). Once the channel closes, the agonist may finally dissociate, or the channel might open a second time if the agonist does not leave soon enough.

Many common experimental and clinical drugs block the postsynaptic response to ACh by binding to the ACh receptor. They are often classified as *nondepolarizing* or *depolarizing* blocking agents.[54] The distinction is actually more like a quantitative difference spreading across a continuum and varies depending on the organism being studied. The first class blocks competitively without depolarizing and includes, in humans, gallamine, β-erythroidine, d-tubocurarine (a purified component of curare), and pancuronium. These drugs bind to the receptor rapidly and reversibly, without inducing frequent channel opening. For them, the opening rate constant β is much smaller than the closing rate constant α. Other drugs, such as succinylcholine and decamethonium, are partial agonists. They somewhat depolarize the end-plate, but less effectively than ACh, because the open lifetime is short and the probability of opening even when the agonist is bound is less. They also block neuromuscular transmission, possibly by inducing a desensitized state of the postsynaptic receptors (see later) and also because the small depolarization they induce will inactivate the voltage-gated Na channels in the vicinity of the junction.

The common neuromuscular blocking agents all have one or more charged or quaternary nitrogen groups and are probably recognized by the ACh binding sites of the receptor because of their chemical resemblance to ACh. This resemblance gives most anticholinergic drugs a multiplicity of actions, since, for example, the ACh receptor, the acetylcholinesterase, and the presynaptic choline uptake mechanism are all pharmacological targets of quaternary ammonium drugs. Even the end-plate *channel* is blocked by many of these drugs, which evidently can enter the outer mouth of the channel and bind within the pore. Hence mechanistic conclusions reached by pharmacological studies at the end-plate must be drawn with more care than in other, less complex systems.

A valuable class of receptor-specific, nondepolarizing blocking agents has been obtained from the venoms of elapid snakes, including kraits and

---

*The argument is the same as that used in Chapter 2 for the Na^+-K^+ pump, where an upward curving relationship between pump rate and K^+ or Na^+ concentration showed that several ions must be bound to activate the pump.

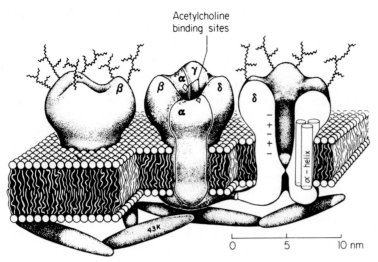

Acetylcholine
binding sites

Figure 6–9 Reconstruction of the orientation of the nicotinic acetylcholine receptor molecule in the lipid bilayer membrane. The drawing is based on electron microscopy and low-angle x-ray diffraction studies of crystalline arrays of receptors. Most of the mass extends into the extracellular medium (upward). Chains of extracellular glycosylation are represented as zigzag lines. The total vertical height of the molecule is 11 nm. On the intracellular side, the receptor pentamers are shown linked to each other by another rodlike protein. They could also normally be tied to the basal lamina in the extracellular space. (From Anholt, R.; Lindstrom, J.; Montal, M. In Martonosi, A., ed. *Enzymes of Biological Membranes*, 335–401. New York, Plenum Press, 1985.)

cobras. These venoms contain a paralytic "cocktail" of more than a dozen potent blocking agents and enzymes with pre- or postsynaptic actions at cholinergic synapses. The most useful have been the α-neurotoxins, such as α-bungarotoxin (from *Bungarus multicinctus*), which are paralytic peptides with from 61 to 74 amino acid residues. They have a potent, curarelike action at nanomolar concentrations and dissociate from the receptor exceedingly slowly (> 24 hours). These potent peptide agents are too toxic to be used in the clinic, but their use as specific labels of the ACh receptor is discussed in the next section.

Receptor Molecule.[3, 41] There has long been interest in identifying the receptors as physical molecules rather than as concepts needed to explain phenomena. However, to develop a chemical purification procedure, one needs a sensitive and reliable assay for the molecule in a test tube; the molecule must carry a radioactive label or have a measurable function. The snake α-neurotoxins provided the solution in this case, and nicotinic ACh receptor (AChR) molecules have now been purified from numerous sources. One of the simpler methods is to solubilize in detergent a rich AChR source, such as electric ray electric organ, and to pass it through a chromatography column containing α-neurotoxin covalently linked to the column—affinity chromatography. The receptor molecules remain specifically bound to the neurotoxin on the column and after a wash may be specifically eluted by rinsing the column with another strong receptor-binding agent such as carbachol.

The nicotinic AChR molecule obtained from a variety of vertebrates is a multisubunit glycoprotein with a protein molecular weight of 268,000 and five separate polypeptide chains. Two of the chains are identical, giving a subunit composition of $\alpha_2\beta\gamma\delta$. The purified molecule plus phospholipid can be induced to form two-dimensional crystalline arrays suitable for structural study by low-angle x-ray diffraction and electron-microscope image reconstruction. These techniques do not give atomic resolution, but they do give a low-resolution picture of the outline of the protein and its position in the membrane bilayer (Fig. 6–9). The five subunits are arrayed pentagonally, with much of the protein mass extending into the extracellular space and forming a pore in the space between them. Agonist binding sites are believed to lie near the outer tip of the molecule, surprisingly far—5 nm—away from the narrow gating region in the bilayer. In agreement with the physiological evidence that two ACh molecules must bind to open one channel, each of the two α chains carries an agonist binding site. Each α chain also carries an α-bungarotoxin binding site.

The purified AChR protein can be incorporated into closed lipid vesicles or planar bilayer membranes to reconstitute functional ionic channels that are opened and desensitized by solutions containing nicotinic agonists. Agonist action is blocked by curare. Such experiments prove that activation and desensitization of the nicotinic channel requires only the agonist and the pentameric receptor protein, not other cofactors, cytoplasmic messengers, or high-energy compounds.

One of the more powerful new approaches to understanding membrane proteins is through genetic techniques of molecular biology. Major success has already been achieved with the nicotinic

AChR. Messenger RNAs for the four peptide subunits from electric rays (*Torpedo*) were copied to make cDNA and then cloned. These cloned cDNAs have been used as probes to identify mammalian genes for the AChR.[40] Given this collection of genetic templates, one can now sequence, express, and mutate AChR proteins from various species.

The amino-acid sequences deduced from the cloned nucleic acids show that the four types of subunits have 30% to 50% sequence identity. Evidently, they evolved by gene duplication from a single ancestral subunit type, perhaps before the evolution of the chordates. Each shows four runs of strongly hydrophobic amino acids that are probably membrane-crossing helical sections. Another run of appropriately alternating hydrophobic and hydrophilic character could form a transmembrane helix with one side hydrophobic and the other hydrophilic. This *amphipathic* helix may form the pore lining, with the hydrophilic side contributed by each subunit forming one fifth of the wall of the ion-permeable pore. Thus we are beginning to develop a sketch of how this complex macromolecule is put together.

Genetic techniques not only reveal structure but also offer ways to test functional hypotheses. The genetic message can be specifically altered and then expressed to assay how function is affected. Acetylcholine receptor message can be expressed by injecting it through a micropipette into isolated oocytes of the African clawed toad *Xenopus laevis*.[39] Even though the message comes from other species, the toad oocyte will synthesize, assemble, and insert finished receptors into its plasma membrane. Then function can be assayed by voltage clamping the oocyte a day or two later while cholinergic agonists are applied in the bath. A simple experiment is to inject inappropriate amounts of the messages for the four different subunits. If a message for α or for β is omitted from the mixture, no response to ACh is found. On the other hand, when a message for γ or δ is omitted, a tiny response remains. Presumably when γ or δ is absent, a pentameric and functioning channel molecule still forms, but now containing some other subunit in place of the missing one. Another set of experiments exploits the possibility of selectively modifying the messages to cause peptide chains of altered sequence to be synthesized—site-directed mutagenesis. One result shows that small deletions in any of the five putative membrane-crossing helices do eliminate acetylcholine responses. Small deletions in some putative cytoplasmic domains do not. The experiment confirms the importance of hydrophobic and amphipathic sequences for final function. Such experiments are only the beginning of a new molecular-engineering approach to analyzing channel function.

The receptor molecule is the major protein component of the subsynaptic membrane. Cell fractionation of electric organs yields a membrane fraction presumed to be of subsynaptic origin that contains the receptor protein as 30% to 50% of the total protein. Autoradiography with radioactively labeled α-bungarotoxin in the rat and frog neuromuscular junctions reveals 20,000 toxin-binding sites per square micrometer of the thickened postsynaptic membrane opposite the presynaptic nerve terminal. The deeper parts of the junctional folds have a much lower site density. Altogether, there may be 10^7 to 10^8 binding sites in the whole neuromuscular junction.[38] This unusually high density of ionic channels combined with the high single-channel conductance of end-plate channels makes the neuromuscular junction an extraordinarily powerful current generator, capable of depolarizing a few millimeters of a large muscle fiber in less than a millisecond.

Desensitization.[29, 37, 49] If the end-plate is bathed in an agonist solution continuously for several seconds, the ability to depolarize in response to agonist is gradually lost; only after the agonist is completely washed away does the responsiveness slowly return. This phenomenon, called *desensitization* of the ACh receptor, is presumed to reflect temporary conversion of the receptor into a form that binds ACh strongly but does not open the gate of the channel. Katz and Thesleff[29] proposed a four-state model to describe the kinetics of desensitization:

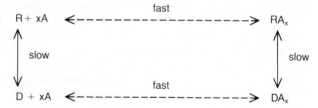

Here A is the agonist, x is the number of agonist molecules that must bind to induce desensitization (assumed to be one in the original model), R is the activatable form of the receptor, and D is the desensitized form. In this hypothesis, the complex RA_x *might* be the normal intermediate (called $R \cdot ACh_2$ in our previous diagram) leading to an open channel. D binds agonist even more strongly than does R, but the channel does not open from the complex DA_x. R and D are interconverted by slow steps. Newer work shows that D actually is a collection of desensitized states, some of which are reached in milliseconds from R and can revert to R in hundreds of milliseconds, and others of which form only after hundreds of milliseconds and revert to R only after seconds.

The significance of desensitization in normal physiology is uncertain; it might protect against excessive activation of muscle in pathological states. Desensitization needs to be considered in all experiments in which the exposure to agonist is anything but a very brief pulse. For example, in chemical experiments with purified ACh receptor, the desensitized receptor form is thought to be generated whenever agonist is added to the medium, making it difficult to study the drug-binding properties of the active form, R, in isolation. Finally, desensitized states are probably produced when in surgical procedures "depolarizing blocking agents" (succinylcholine) are used to induce muscle relaxation.

Denervation Supersensitivity.[32, 42, 43, 46] In addition to the rapid transmission of electrical signals, cells entering into synaptic contact have long-term influences on each other. Normal muscle fibers have acetylcholine receptors (α-bungarotoxin binding sites) and ACh sensitivity (electrical response to iontophoretic puffs of ACh) highly concentrated in subjunctional membrane by a factor of at least 1000:1 as compared with the rest of the muscle membrane. This extreme restriction on the placement of receptors is lost if the presynaptic nerve is cut or crushed; a vast acceleration of receptor synthesis follows, with new receptors appearing all over the muscle fiber membrane, "extrajunctional" receptors. The total number of toxin-binding sites on a muscle fiber increases 10- to 20-fold, and the membrane becomes sensitive to ACh everywhere, a phenomenon called *denervation supersensitivity* (sometimes, denervation hypersensitivity).

The effects of denervation appear within a few days in mammals and within a few weeks in cold-blooded animals like the frog. Since the extrajunctional ACh receptor molecules have a turnover time of less than a day,[42] the effects of denervation reverse quickly if synaptic contact is reestablished by regrowth of axons to the previous, still-remaining endplate region or to a new one. In mammalian muscle, the mean resting potential and the resting Cl permeability fall and the voltage-gated Na channels undergo changes, including loss of tetrodotoxin sensitivity, after denervation. During this time, the denervated muscle membrane re-

sembles embryonic muscle, which also has a diffuse sensitivity to ACh, rapid turnover of receptors, and TTX-insensitive Na channels.

Evidently, the "signs of denervation" actually require neither removal nor death of the presynaptic axon. Rather, they seem to arise when the normal pattern of electrical activity in the muscle is prevented. They can be induced by chronic application of local anesthetic or tetrodotoxin to the motor nerve, by nicotinic antagonists (curare), and by agents that prevent transmitter release. In a muscle receiving no impulses from its axons, the signs of denervation may be reversed simply by applying direct electrical stimuli to the muscle for a few days, imitating normal activity. Such changes illustrate some of the many effects of nerve and nervous activity on muscle and muscle on nerve. Similar interactions between pre- and postsynaptic cells must have an important role in the embryonic development of the nervous system and the neuromuscular system and in learning.

Summary of Postsynaptic Events. This concludes our discussion of the postsynaptic events of neuromuscular transmission. To review: In the immediate subjunctional region, the postsynaptic membrane contains a high density of an oligomeric protein molecule, the nicotinic acetylcholine receptor. The natural transmitter, acetylcholine, binds to receptor molecules and induces their ionic channels to open in the postsynaptic membrane. These channels have a high unit conductance and a high permeability to Na^+ and K^+ cations, so their physiological reversal potential is near -5 mV. In normal transmission at 20°C, the end-plate channels are open for about 1 ms, letting in Na^+ ions and depolarizing the postsynaptic membrane beyond the firing threshold for action potentials in the neighboring electrically excitable membrane. Any transmitter molecules remaining free in the synaptic cleft are quickly hydrolysed by the enzyme acetylcholinesterase. Once the postsynaptic action potential is elicited, the job of transmission is accomplished.

TRANSMITTER RELEASE

To complete the discussion of neuromuscular transmission, we need to consider how transmitter is released into the synaptic cleft. The major events, outlined in Figure 6–1, are (i) the propagation of an action potential down the axon of a motoneuron and into the nerve terminal, (ii) the opening of voltage-gated Ca channels and entry of Ca^{2+} ions into the terminal, and (iii) the release of ACh by exocytosis of the contents of some synaptic vesicles into the synaptic cleft. As we have mentioned, the interior of the axon terminal is too small to record from with an intracellular electrode, and most conclusions must be drawn by inference from extracellular recordings from the muscle. Hence while information on the postsynaptic side of the vertebrate neuromuscular junction is far more detailed than for any other chemical synapse, information on the presynaptic side is in several respects less complete than for some other systems with a large presynaptic cell, e.g., the squid giant synapse (see Chap. 11). Fortunately, the presynaptic mechanism, especially parts ii and iii above, seems to be the same in all chemical synapses, and one can generalize with some confidence.

Presynaptic Action Potential.[26] In any axon, the propagation of action potentials requires local-circuit currents to flow through the intracellular and extracellular fluids to depolarize unexcited membrane ahead of the advancing wave. The extracellular currents produce a small local triphasic voltage signal that can be detected with extracellular recording electrodes. Such measurements along the length of the branching nerve terminals of frog muscle (Fig. 6–10) reveal that each action potential continues beyond the myelinated portion of the motor axon, propagating to the tip of each branch with a low conduction velocity of 0.3 m/s (300 μm/ms) in the terminal (at 20° to 24°C). Conversely, antidromic action potentials, which propagate back to the myelinated axon, can be evoked in the terminal branch with a fine-tipped electrode. If tetrodotoxin is applied by an iontophoretic pipette at a point halfway to the end of a terminal branch, invading action potentials propagate up to that point and not beyond. Hence the presynaptic terminal has the properties of a typical, electrically excitable, unmyelinated axon with a Na channel–spiking mechanism. The presence of K channels in the terminal can be inferred from a vast increase in the postsynaptic response to a presynaptic impulse when K-channel blocking agents such as 4-aminopyridine are added to the medium. The presumption is that the increase in transmitter release can be attributed to a much prolonged presynaptic action potential.

Miniature End-plate Potential.[5, 15] A major step in understanding transmitter release was the discovery and analysis by Katz, Fatt, and del Castillo of the miniature end-plate potential (MEPP). Careful recording with an intracellular microelectrode inserted into a muscle at the junctional region

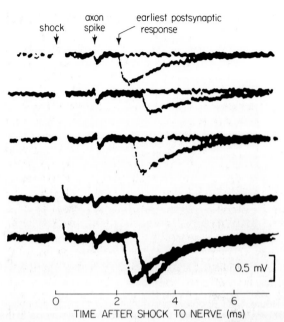

Figure 6–10 Focal extracellular recording at a frog sartorius muscle neuromuscular junction during stimulation of the motor nerve. The repeated traces show a large artifact at the moment of nerve stimulation, followed by a signal showing when the presynaptic action potential passes by the point of recording, and finally—about half the time—a large negative-going signal reflecting the current from one transmitter quantum acting on the muscle membrane. With such "focal" recording, the signals reflect events happening within 20 μm of the pipette since the extracellular space has no cable properties. Transmitter release occurs only locally in this experiment, since the bathing medium is calcium-free and the only available extracellular calcium comes from a calcium solution that has been placed in the recording pipette. T = 17°C. (From Katz, B.; Miledi, R. *Proc. R. Soc. Biol.* 161:483–495, 1965.)

reveals a spontaneous and random patter of tiny depolarizing postsynaptic potentials even when the motor nerve is quiescent (Fig. 6–11). These spontaneous MEPPs, occurring several times a second, are only 0.1 to 4 mV high without curare but otherwise have the properties of full-sized EPPs. They have the same time course, reversal potential, and sensitivity to curare; they have a similar localization to the junctional region; and their amplitude and duration can be increased by anticholinesterase agents such as eserine or edrophonium. Within any muscle fiber, the MEPPs tend to have a fairly uniform amplitude. Hence a MEPP is the response of a small number of end-plate channels to a spontaneously released and relatively uniform small amount of ACh. As we shall see, the MEPP arises when the contents of one presynaptic ACh-containing vesicle are re-

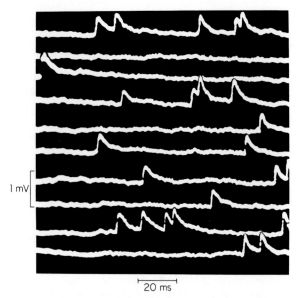

1 mV

20 ms

Figure 6–11 Spontaneous miniature end-plate potentials. Records are made with an intracellular microelectrode at the neuromuscular junction of frog toe muscle fiber. The average amplitude of MEPPs seen here is 0.6 mV. (From Fatt, P.; Katz, B. *J. Physiol. [Lond.]* 117:109–128, 1952.)

leased into the synaptic cleft. The peak conductance increase, g_{MEPP}, measured with a voltage clamp, is about 50 nS, which corresponds to the simultaneous opening of about 2000 end-plate channels if we take the single-channel conductance, γ, to be 25 pS.

Each MEPP can be recorded throughout the junctional region in the muscle fiber with an intracellular electrode, because intracellular signals spread electrotonically by the cable properties of the muscle. An extracellular micropipette electrode probing delicately along the surface of the neuromuscular junction also can detect tiny *extracellular* signals coincident with MEPPs, although these other signals have somewhat different properties. They are briefer than MEPPs and can be seen only at discrete, localized points along the terminal. At any one point, they are less frequent than MEPPs. These extracellular negative-going signals reflect the flow of a brief but intense inward current, MEPC, locally into the muscle during the rising phase of a MEPP, and show that each MEPP is generated by several thousand channels opening in a very restricted patch of the membrane rather than throughout the length of the junction.

Quantal Transmitter Release. [5, 6, 37, 48] Del Castillo and Katz used the miniature end-plate potential as the basis for their quantal hypothesis of transmitter release, which states that (i) ACh is released

from the presynaptic terminal in multimolecular packets called *quanta*, each sufficient to produce an MEPP and (ii) depolarization of the terminal by an action potential causes the nearly synchronous release of many quanta from different parts of the terminal, producing the full-sized EPP. Hence release of quanta must be voltage dependent.

Several simple experiments support the quantal hypothesis: (i) The rate of occurrence of MEPPs can be increased vastly by experimental depolarization of the presynaptic terminal, either by elevation of the external K concentration or by application of current to the nerve terminal from a suction electrode. In the latter experiment, tetrodotoxin is applied to block propagated activity; a stub of motor nerve, cut a few millimeters from the junction, is sucked into a pipette of just-matching inner diameter; and a current is applied between the pipette and the bath to depolarize the axon terminals electrotonically. (ii) Focal extracellular recording during nerve stimulation reveals highly localized regions along the neuromuscular junction (Fig. 6–10), where inward currents identical in size to spontaneous MEPCs are evoked by nerve stimulation. Presumably, the extracellular electrode happens to be placed at the site of release of one quantum, an active zone, and many other quanta are released at other active zones along the terminal. (iii) It is also possible, by manipulation of the bathing divalent ion (Ca^{2+} and Mg^{2+}) concentrations, to reduce the size of the intracellularly recorded EPP evoked by a normal nerve stimulus. This altered bathing solution does not change the size of the spontaneous MEPP. Evidently, the release process has been depressed to the point at which only a few quanta are released by a nerve impulse. When the EPP has been strongly depressed, its amplitude fluctuates from stimulus to stimulus in a stepwise fashion. The step size of this fluctuation is the same as the size of the spontaneous MEPP. Conditions needed to see such *quantal fluctuations* of the EPP include a lower external Ca^{2+} ion concentration, addition of Mg^{2+} ion or transition metal ions, or both; thus in 0.45 mM Ca^{2+} and 6 mM Mg^{2+}, successive EPPs fluctuate among quantal values equivalent to 0, 1, 2, 3, and 4 times the size of an MEPP. Hence the normal EPP is in effect the summation of many simultaneous MEPPs. Note that curare also reduces the EPP size, but its action is postsynaptic. Therefore it reduces the size of MEPPs and does *not* induce quantal fluctuations.

Many quantitative tests have been done for determination of the statistics of transmitter release. The number of quantal units making up an evoked EPP is termed its *quantal content*, often designated by the symbol m. Physiological EPPs have an average quantal content of 100 to 300 at the frog neuromuscular junction, meaning that 100 to 300 quanta are normally released by one action potential as it propagates the length of the presynaptic nerve terminal. Del Castillo and Katz[5, 6] reasoned that if quanta are released independently from a large available reservoir, the quantal content of EPPs should fluctuate randomly in a long train of stimuli according to Poisson statistics. Thus if the mean quantal content is m, then the probability, p_x, that an EPP will have a quantal content of x is

$$p_x = \frac{m^x e^{-m}}{x!}$$

so p_0 is e^{-m}, p_1 is me^{-m}, p_2 is $0.5\ m^2 e^{-m}$, and so forth. By lowering the extracellular Ca^{2+} concentration, del Castillo and Katz reduced the quantal content of nerve-evoked EPPs to such a low level that many evoked EPPs had quantal contents of zero (a "failure"), one, or two. Under these conditions, the Poisson formula was well obeyed. It is now common to view the whole neuromuscular junction as an array of several hundred unitary synapses—the periodically spaced active zones of the terminal—and to think of statistical independence of the whole release process as a statement of the independent operation of the spatially separate active zones. In a normal EPP, the quantal content *per active zone* averages out to close to one (but some active zones are more active than others).

Synaptic Vesicles and Exocytosis.[19, 30, 44] The year after the quantal hypothesis was formulated, synaptic vesicles were discovered in the presynaptic terminal by electron microscopy. Subsequent chemical analysis showed that the synaptic vesicles of cholinergic axons store ACh, and it is widely accepted that a quantal event (one MEPP) corresponds to the release of the contents of one synaptic vesicle by exocytosis. Thus the phrase "quantal content" could be taken as synonymous with "number of transmitter vesicles released."

Since some scientists have continued to question the vesicle hypothesis for release, favoring instead the idea that all stimulated release comes directly from ACh in the cytoplasmic compartment rather than from vesicles, it is worthwhile reviewing some experimental observations. The tests have been done on chemical synapses as well as on neurosecretory and other secretory cells in which

the secretory product is stored in membrane-bounded "secretory granules" analogous to synaptic vesicles. At least five lines of evidence exist for vesicular secretion: (i) Vesicles and granules disappear during secretion. (ii) New membrane is added to the cell surface during secretion and taken back into the cell later. (iii) The number of vesicles fused to the surface membrane is equal to the quantal content. (iv) The quantal response has a size that is consistent with the amount of transmitter that could be put into one vesicle. (v) The material secreted has the same composition as the contents of secretory granules. These ideas are elaborated in the following paragraphs.

In cells with granules large enough to be seen under the light microscope, such as mast cells that secrete histamine, the granules vanish from sight, like a balloon popping, during secretion. At the same moment, the effective surface area of the cell membrane increases stepwise, as can readily be measured electrically by recording tiny step increases in the membrane capacity (C_M, Chap. 3) of the whole cell.[16] Later, the membrane capacitance shows step decreases as membrane is brought back in by endocytosis.

At the neuromuscular junction, a single vesicle is too small to observe optically or to measure in terms of membrane capacity, but net addition of vesicular membrane can be seen after periods of stimulation. Antigens that are specific for vesicle membrane proteins can be demonstrated on the external membrane of the nerve terminal after intense chemical stimulation of transmitter release.[56] After ordinary electrical stimulation, the number of vesicle-specific proteins is small, however, so presumably their exposure is usually only transient. Endocytic reuptake of membrane can be demonstrated in a nerve terminal that is bathed with a high molecular–weight marker of the extracellular space. After a period of repetitive nerve stimulation, electron microscopy reveals that the marker molecule has been picked up by the terminal and is inside intracellular vesicles—after entering, presumably by endocytosis, at "coated pits."[19]

Additional evidence comes from the experiments of Heuser and colleagues,[20] who demonstrated a correspondence between the number of quanta observed electrophysiologically and the number of synaptic vesicles caught in the act of exocytosis by a combination of quick freezing and electron microscopic methods. Because the normal quantal content of 100 to 300 is expected to correspond to fusion of only one vesicle per active zone, the nerve-muscle preparation was placed in a medium containing 4-aminopyridine, increasing the quantal content of a single EPP to 5000. The tissue was mounted on a quick-freezing machine with the nerve attached to a stimulator; a single shock was applied to the nerve, and then within 5 ms, the tissue was driven against a copper block precooled to 4°K. Replicas of the freeze-fracture surfaces of the nerve terminals revealed about 13 pocketlike vesicle openings per active zone, a value agreeing exactly with the quantal content per active zone estimated electrophysiologically in these 4-aminopyridine treated fibers.

The number of ACh molecules required to make one MEPP is still not accurately known, but it may be in the range of 4000 to 10,000. Thus the conductance of one MEPP corresponds to the opening of 2000 channels, and each channel probably is opened by two ACh molecules. This would set a minimum requirement of 4000 molecules for a MEPP. In a difficult series of experiments on snake neuromuscular junctions, Kuffler and Yoshikami[33] were able to imitate the time course and amplitude of the spontaneous physiological MEPP with iontophoretic puffs of 10,000 ACh molecules from an ACh-containing pipette pressed carefully against the terminal. This number may be an upper limit to the number needed for a MEPP, since the presynaptic terminal is presumably in a position to deliver transmitter more efficiently than the glass micropipette. If ACh-containing vesicles are like epinephrine-containing ones that have been studied more thoroughly, the concentration of transmitter within the vesicle is high. For example, if the inner diameter of the vesicle is 37 nm and the ACh cation is concentrated as an isotonic solution, with ATP^{3-} and the phospholipids of the vesicle as counter ions, there could be 6000 ACh molecules per vesicle, not very different from the number needed for a MEPP.

Another line of evidence for vesicular release is comparison of the composition of the contents of a secretory granule with the composition of the secreted material. This has been done carefully for the epinephrine-containing chromaffin granules of chromaffin cells of the adrenal medulla. Isolated granules contain epinephrine; norepinephrine; ATP; a peptide transmitter, enkephalin; an enzyme, dopamine β-hydroxylase; and several "chromogranin" proteins that bind epinephrine. When chromaffin cells are induced to secrete, all of these substances appear simultaneously in the bathing medium and in appropriate concentration ratios for release from granules.[4] Other small molecules and enzymes known to be in the cytoplasm but not in granules are not released. Likewise, the

"zymogen granules" of pancreatic acinar cells contain the "pro" form of the digestive enzymes chymotrypsin, trypsin, and ribonuclease. Being secretory proteins, their messenger RNA has a signal sequence, and the proteins are synthesized directly on membrane-bound ribosomes on the rough endoplasmic reticulum into a vesicular compartment. When the acinar cell is stimulated to secrete, the digestive enzymes must be released by exocytosis without ever passing through the cytoplasm, since it would be lethal for the cell if they were to do so. Cholinergic vesicles contain ATP and ACh together,[30] and some ATP can be collected when ACh is released during nerve stimulation. However, the experiment has not been done quantitatively, as would be required to argue that ATP and ACh are co-released from vesicles.

In all experiments in which sensitive assays for ACh are used, it is agreed that much more ACh is released from unstimulated tissue than could be accounted for by the known average frequency of spontaneous MEPPs.[17, 30] This release of ACh must be from the nerve terminal, where because it is synthesized in the cytoplasm and accumulates there perhaps to nearly millimolar concentration, it constantly *leaks* out through the axon membrane at a low rate. Such constant dribbling from the terminal is called nonquantal or nonvesicular release.

Role of Calcium Ion.[24, 37, 44, 48] We now consider why the quantal content of an EPP is so remarkably sensitive to the concentrations of divalent ions in the bathing medium. In Ca^{2+}-free Ringer solution, nerve stimulation evokes no postsynaptic response. Extracellular electrodes show that the nerve action potential propagates in the normal manner to the tips of the nerve terminals but fails to evoke any release. Katz and Miledi[25] found that release can be restored locally in a Ca^{2+}-deprived terminal by brief extracellular puffs of Ca^{2+} ions delivered by a Ca^{2+} iontophoretic pipette just before or during the period when the nerve terminal is depolarized. However, a puff of Ca^{2+} applied just milliseconds *after* the depolarization is ineffectual. Thus transmitter release in response to presynaptic depolarization has an absolute requirement for extracellular Ca^{2+} to be present *during* the depolarization. Dodge and Rahaminoff[10] examined the relationship between quantal content of evoked EPPs and the concentration of Ca^{2+} ions in the medium, finding a steep power dependence, with quantal content being proportional to $[Ca^{2+}]^4$ (Fig. 6–12). They suggested that wherever the molecular site of action is, up to four Ca^{2+} ions need to bind to enable transmitter release.

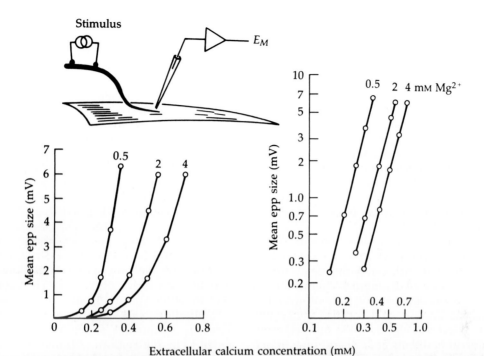

Figure 6-12 Calcium control of transmitter release at the neuromuscular junction. A frog nerve-muscle preparation is stimulated by shocks to the motor nerve, and the bathing Ca^{2+} and Mg^{2+} concentrations are varied. The EPP size has been averaged over many trials because of its fluctuating nature and is plotted on linear and log-log scales versus $(Ca^{2+})_o$. The slope of the lines in the log-log plot is 3.9. (After Dodge, F.A.; Rahaminoff, R. *J. Physiol. [Lond.]* 148:188–200, 1967.)

Katz and Miledi[24, 25] reasoned that Ca^{2+} ions might enter the terminal during a depolarization and then act intracellularly on specific divalent ion receptors. Their hypothesis has been amply confirmed in several chemical synapses. It is now known that presynaptic terminals have a high density of voltage-sensitive Ca channels, which open in response to depolarization, permitting Ca^{2+} ions to enter; and as is true of many cells that secrete by exocytosis, the intracellular Ca^{2+} concentration controls the probability of release of transmitter packets. Direct evidence obtained in the squid giant synapse is described in Chapter 11. It involves the demonstration that Ca channels exist at much higher density in the terminal than in the axon leading to it, that the free Ca^{2+} rises in the terminal during a presynaptic impulse, and that artificial injection of Ca^{2+}-containing buffers into the terminal will cause release of transmitter. After the terminal membrane repolarizes, the Ca channels shut quickly, and the intracellular Ca^{2+} concentration falls rapidly probably because of rapid binding to numerous cytoplasmic Ca-binding proteins and eventual extrusion by Na^+-Ca^{2+} exchange and the Ca ATPase. Although less effective, Sr^{2+} ions can substitute for extracellular Ca^{2+} in supporting release of transmitter.

Evoked transmitter release is depressed by such typical Ca channel–blocking ions as Mn^{2+}, Co^{2+}, or Ni^{2+}, as well as by elevated extracellular Mg^{2+}. For example, adding 10 mM Mg^{2+} to a Ringer solution containing 1.8 mM Ca^{2+} reduces the quantal content of EPPs at the neuromuscular junction to only 4% of normal (Fig. 6–12). The concentration dependence of this block suggests that Ca^{2+} and Mg^{2+} ions compete at some site.[10] As the typical Ca channels studied in membranes of neurons and crustacean muscle are not permeable to Mg^{2+} and are weakly blocked by it, the site of competition of Mg^{2+} and Ca^{2+} is probably at the point of Ca entry into the Ca channel. Calcium-dependent exocytosis is the most common biological mechanism for secretion of membrane-impermeant molecules in a rapidly controlled manner.[4, 11]

Because exocytosis at a synapse occurs within a few hundred microseconds after the impulse arrives, the released vesicles must have been nearly in contact with the membrane beforehand. There would not be sufficient time for a full cycle of, for example, an actin-myosin cross-bridge movement (Chap. 8) to move the vesicle to the membrane. The underlying mechanism—what Ca^{2+}-regulated proteins are used, how the vesicle fuses with the plasma membrane—remain a challenging unsolved problem today. There is some evidence that the Ca^{2+} detector is a calmodulinlike protein, and current thinking suggests that the fusion process could involve a protein conformational change that exposes hydrophobic surfaces and permits the two bilayers to blend together—a process that has been decribed for viral "fusion proteins."[44]

Compared with nerve-evoked transmitter release, the basal rate of spontaneous release (frequency of MEPPs) is relatively insensitive to the concentration of external divalent ions (Ca^{2+}, Mg^{2+}, or Mn^{2+}), except when the rate is already elevated by depolarization, as in solutions with elevated K^+ concentration.

Synaptic Delay and the Time Course of Transmitter Action. Chemical synaptic transmission is invariably associated with a synaptic delay that is used as one criterion to distinguish chemical from electrical synapses. Between the time of arrival of the presynaptic impulse and the first detectable postsynaptic electrical signal, there may elapse a "silent" interval of 0.5 ms at 22°C and of as much as 5 ms at 2.5°C (Fig. 6–10). Then the end-plate current takes an additional several hundred microseconds to rise to its peak value. Finally, the EPP rises more slowly than the underlying current (because of the membrane capacitance), and an action potential is elicited only after the membrane is depolarized to threshold. An obvious candidate for explaining the first part of the synaptic delay is the time taken for the released transmitter to reach receptors and activate postsynaptic channels. However, the calculations in the next paragraph show that these steps account for only a small fraction of the delay. Instead, it is now believed that the early, electrically silent period reflects the entirely presynaptic events of slow opening of Ca channels, buildup of intraterminal free calcium, and activation by Ca^{2+} ions of the steps of exocytosis.

The time course of the transmitter action that follows exocytosis of the contents of a vesicle has been reconstructed by computer simulation of the known steps.[28, 45, 55] Simultaneous processes of diffusion, binding to receptors, and hydrolysis of ACh are all taken into account. The detailed predictions depend on assumed parameters that are imperfectly known, but the outlines of the results are clear. Diffusion across the 0.05-μm synaptic cleft takes only 2 to 4 μs (consider the Einstein formula of Chap. 1). However, to reach enough ACh receptors, the approximately 10,000 ACh molecules released must also spread out laterally over a disk of membrane with an area of 0.2 to

0.4 μm.[2] This spread takes another 75 to 150 μs and could account for the time taken for a miniature end-plate current to rise to its peak. By this time, the free ACh concentration has fallen from the original 200-mM range down to 10 to 200 μM; about half the released molecules have been hydrolyzed, and most of the remainder are bound, two per channel, to receptors, causing channels to open. Then, as channels close randomly, ACh molecules dissociate from receptors and are hydrolyzed before they have a chance to activate another channel.* Transmitter action is thus terminated quickly. If the esterase is blocked, however, the calculations suggest that ACh molecules may rebind several times as they diffuse around the cleft, activating new channels even several milliseconds after the time of release. Binding and retention on receptor sites would slow the diffusion of transmitter out of the cleft considerably over the rate expected from free diffusion alone.[28]

Facilitation and Depression.[36, 37] When the nerve is stimulated repetitively at rates much above 1 Hz, the amount of transmitter released by each impulse is a function of the frequency and duration of stimulation, i.e., the quantal content depends on the pattern of stimulation. When the initial quantal content is low (as with low [Ca^{2+}_o]), repetitive stimulation leads to a progressive increase in the quantal content evoked by each nerve impulse. In one kinetic analysis, the increase has been subdivided into four components, with time constants of decay of 50 ms (first component of facilitation), 300 ms (second component of facilitation), 7 s (augmentation), and tens of seconds to minutes (potentiation).[36] The longer lasting potentiation is interesting as a possible analogue of short-term memory. The mechanisms of facilitation, augmentation, and potentiation are not clear but may involve Ca^{2+} accumulation or Ca^{2+}-activated changes in factors in the nerve terminal involved in release.

Under conditions of higher quantal content, repetitive stimulation often leads first to an increase and then a decrease in quantal content during the conditioning train; the increase, usually called *facilitation*, most likely results from increases in the above four components. The decrease in quantal content is called *depression* and may result from a depletion of the subpopulation of synaptic vesicles properly docked at the membrane of the active zones. Recovery from depression requires

---

*Acetylcholinesterase does not hydrolyze ACh molecules bound to the AChR. It acts only on ACh molecules that happen to collide with its active site as they diffuse in the cleft.

seconds following short conditioning trains and minutes following longer conditioning trains. During intense stimulation, e.g., at 100 Hz for 10 seconds, potentiation and depression both build up during the conditioning train. Following the conditioning train, depression recovers faster than potentiation. Thus, potentiation appears to come on with a delay leading to what has been termed *post-tetanic potentiation*. At the vertebrate neuromuscular junction, where one-to-one transmission is normally assured, these phenomena probably have little physiological importance. However, at central synapses, where the postsynaptic cell is summing small inputs from vast numbers of presynaptic inputs, facilitation and depression can have major effects (Chap. 11). Since such changes of quantal content sometimes last for minutes or more, they could contribute to short-term memory.

Botulism and Tetanus.[7, 30, 47] Several natural protein toxins have extreme effects on the release mechanism. Botulinum toxin, type A, produced by a common bacterium, *Clostridium botulinum*, is a dangerous agent of food poisoning. The toxin is produced when the bacterium grows anaerobically in a nourishing medium such as incompletely sterilized canned foods. The victim, after eating contaminated food, may experience progressive double vision, weakness, and eventual paralysis with effects that may persist for months. At the cellular level, botulinum toxin drastically depresses nerve-evoked EPPs. A simple description is that transmitter release is blocked at neuromuscular junctions, but in vitro studies show that the phenomenology is more complicated than that. The frequency of spontaneous MEPPs is reduced, and their size distribution is altered to include many giant MEPPs. Release can be partly restored by raising the external Ca^{2+} ion or by lengthening the presynaptic action potential with tetraethylammonium ion. Botulinum toxin may depress the sensitivity of the transmitter release mechanism to intracellular Ca^{2+} ions. Black widow spider venom (α-latrotoxin), an agent that normally stimulates intensive release, also stimulates release after botulinum toxin treatment.

Tetanus is an infectious disease that leads to muscle spasms, largely through effects on transmission in the central nervous system, but tetanus toxin, which is responsible, also acts on nerve-muscle preparations in the laboratory. Like botulinum toxin, tetanus toxin, secreted by *Clostridium tetani*, can stop neuromuscular transmission in less than 30 minutes by eliminating evoked release of transmitter. Presumably, in the central nervous system, it blocks transmitter release at inhibitory synapses, thus leading to spastic motor outputs.

The molecular targets of botulinum and tetanus toxins are not yet known. However, it is believed that the toxins act in a manner similar to that of diphtheria, cholera, and pertussis toxins. These well-characterized agents have a binding component that seeks a specific membrane ganglioside on the outside of the target cell and a second, *enzyme*, component that then enters the cytoplasm. The enzyme transfers an ADP-ribose moiety from the coenzyme NADP to a specific GTP-binding protein in the cell. Since they act catalytically as enzymes, one or two molecules of toxin would suffice eventually to poison every copy of their target protein in the cell. A different ganglioside preference and a different enzymatic component give each of these bacterial toxins a different tissue specificity and a different final action. Elucidation of the molecular action of botulinum and tetanus toxins should give us valuable new insight into molecules underlying transmitter release.

Myasthenia Gravis and Myasthenic Syndrome.[8, 13] *Myasthenia* means literally *muscle weakness*. In the disease myasthenia gravis, the patient experiences extreme muscular fatigability. Even a simple task such as gazing upwards brings on signs of weakness—the eyelids droop over the eyes within a few minutes. The symptoms are dramatically reversed by administering a short-acting anticholinesterase drug (edrophonium) to the patient, and when this test is positive, the patient may be prescribed a permanent regimen of, e.g., neostigmine to provide daily symptomatic relief.

The disease has a major autoimmune component. Most patients have circulating antibodies against their own acetylcholine receptors. These antibodies promote turnover and degradation of receptor molecules so that the neuromuscular junctions may contain only 20% of the normal number of ACh receptors; in addition, bound antibodies block the permeability response of the channel. Both factors reduce the responsiveness of the postsynaptic membrane. Symptoms of patients can be temporarily alleviated by removal of the immunoglobulin component of their serum (using plasmapheresis). Symptoms of the disease can temporarily be induced in animals by repeated injections of immunoglobulin G from patients, and symptoms can be induced permanently by immunizing animals with chemically purified ACh receptors. Perhaps in compensation for the low postsynaptic receptor numbers, the presynaptic motor terminals of these patients release more transmitter vesicles per impulse. This initial high quantal content may in turn lead to a quicker than normal depletion of available vesicles during continued activity, leading to failure of transmission during use.

Myasthenic syndrome is a rarer muscular weakness that improves with activity. Here, the quantal content is initially unusually low, and the frequency of MEPPs induced by depolarizing the terminal with KCl is lower than normal. The spontaneous MEPP is of normal size, suggesting no change of the amount of ACh per vesicle or of the postsynaptic receptor function. In myasthenic syndrome, strength may improve with activity because the low quantal content probably rises during repeated impulses through facilitation unimpaired by depletion.

SUMMARY AND COMPARISON WITH OTHER JUNCTIONS

The skeletal neuromuscular junction illustrates many of the general principles of chemical synaptic transmission and at the same time shows special adaptations to achieve a high safety factor in a one-to-one transmission of impulses from an axon to a large muscle fiber. Transmitter is packaged in vesicles and released in a quantal fashion from active zones when Ca^{2+} ions enter locally near docked vesicles. The long synaptic terminal releases several hundred vesicles per presynaptic impulse, guaranteeing that the postsynaptic response is sufficient to fire the action potential mechanism of the muscle fiber. The steps between Ca^{2+} entry and transmitter release are still only poorly characterized.

Most chemical synapses have a presynaptic terminal pressed close to the postsynaptic cell and filled with synaptic vesicles and mitochondria. Chemical transmission is always antagonized by external Mg^{2+} and enhanced by external Ca^{2+}, so it is presumed that neurotransmitter release is a quantal exocytosis controlled by Ca^{2+} ions entering when presynaptic Ca channels open. A wide variety of small molecules and peptides are used as transmitters.[31, 32] Therefore the synthetic machinery and blocking drugs differ for different chemical synapses. The transmitters interact with receptors on the postsynaptic membrane and often open ionic channels but may, instead, activate membrane enzymes through receptor coupling proteins. The ultimate effect may be excitatory or inhibitory. The ionic channels activated may be permeable to many cations, permeable to many anions, selective for Na^+ ions, or selective for K^+ ions. Often, the postsynaptic cell receives many synapses from many cells with the equivalent of only one or a few active zones in each synapse, and the quantal content of each synapse is therefore small. Then the activities of the postsynaptic cell depend on the *summation* of all excitatory and inhibitory inputs, rather than following automatically from the activity of a single presynaptic cell, as in skeletal muscle. The properties of nerve-nerve synapses are considered further in Chapter 11. Before finishing, however, we discuss briefly a few other neuromuscular junctions.

Autonomic Nerve-Effector Junctions.[22] In addition to skeletal neuromuscular junctions, the vertebrate nervous system makes many other nerve-muscle and nerve-effector junctions, including those with smooth and cardiac muscles, endocrine and exocrine glands, and adipose tissue. All of these chemical junctions are part of the autonomic nervous system described in Chapter 34. Their microphysiology is still not well characterized but is obviously varied.

The axons* innervating autonomic effectors are small-diameter unmyelinated fibers that branch profusely and often form nets of bundles of small fibers throughout the tissue. In some instances, a single axon may plunge directly into effector tissue, to be surrounded closely by effector cells, but frequently there are not obvious cell-to-cell junctions. Instead, the individual axons in a bundle may have widely spaced swellings (varicosities), like beads on a string, where vesicles of chemical transmitter are stored and released at some distance from any particular effector cell. Numerous transmitter substances are released from autonomic axons, including ACh (cholinergic transmission), norepinephrine (noradrenergic transmission), ATP (purinergic transmission), and many peptides (peptidergic transmission). The postsynaptic receptor for ACh on autonomic effectors is pharmacologically distinct from that in skeletal muscle. It is a *muscarinic cholinergic receptor*, since it can be stimulated by the alkaloid muscarine rather than by nicotine. Muscarinic receptors are blocked by atropine, not by curare, and couple to membrane enzymes and channels they control via GTP-binding coupling proteins. The postsynaptic receptors for norepinephrine are divided into two

---

*The axons described in this section are the "postganglionic" fibers of the autonomic nervous system (Chap. 34), and the junctions are those between postganglionic fibers and smooth muscle, cardiac muscle, glands, and so forth.

major pharmacological classes, α and β, again on the basis of their agonists, antagonists, and mechanism of action. The β receptors stimulate the membrane enzyme adenylate cyclase.

The physiology of autonomic nerve-effector transmission differs from tissue to tissue over a broad spectrum. Transmitter action can be excitatory or inhibitory. It can be accompanied by postsynaptic conductance increases *or decreases*, and it can stimulate enzymes that modulate cell function. Some junctions have properties that seem familiar after a study of skeletal neuromuscular transmission. For example, sympathetic nerve fibers produce an excitatory effect on the smooth muscle of the vas deferens of the guinea pig. One stimulus to the nerve bundle initiates one brisk contraction. When the nerve is quiescent, the muscle relaxes, but there still are spontaneous excitatory postjunctional potentials (SEPJP) of approximately 10-mV amplitude and 100-ms duration. Hence release appears to be quantal. Certain spider venoms greatly increase the frequency of SEPJPs for a while and deplete the axons of transmitter vesicles. Of the two neurotransmitters known to be released by these sympathetic nerves (noradrenalin and ATP), it seems to be ATP acting on P_2 purinoceptors that accounts for the fast component of the EPJP.[4a] By contrast, the heart and many smooth muscles beat rhythmically even without any active nervous input, and the autonomic nerves act only to slow or speed the rhythm. The primary stimulatory effect of noradrenergic input to the heart is not mediated directly by ionic channels opened by the transmitter. Instead, as we have discussed in earlier chapters, β receptor activation promotes the synthesis of cyclic 5'-adenosine monophosphate (cAMP) from intracellular ATP. This second messenger then activates cAMP-dependent protein kinase, which in turn may phosphorylate several kinds of K and Ca channels, thereby modifying their function. Further discussion of the variety of autonomic effectors is found in later chapters of this book.

Invertebrate Neuromuscular Junctions.[23, 32, 50] The muscles of invertebrates can also receive both excitatory and inhibitory effects from the nervous system. The best-studied example is the crustacean claw or leg muscle. Here, a single inhibitory axon and a single excitatory axon coming from the ventral ganglion chain may innervate an entire muscle. The excitatory axon releases the neurotransmitter, glutamic acid, which opens a cation-selective postsynaptic channel and depolarizes the muscle, much as at the vertebrate neuromuscular junction. The inhibitory axon releases γ-amino-

butyric acid (GABA), which opens an anion-selective postsynaptic channel and hyperpolarizes or at least opposes depolarization both of the muscle, at an inhibitory neuromuscular junction, and of the excitatory axon, at an inhibitory axon-to-axon synapse. Glutamate is an excitatory transmitter at molluscan as well as arthropod neuromuscular junctions, and ACh at nematode and annelid junctions (including the famous ACh-sensitive body-wall muscle of the leech). A survey of invertebrate neuromuscular junctions is given in Hoyle's monograph on muscle physiology.[23]

ANNOTATED BIBLIOGRAPHY

1. Katz, B. *The Release of Neural Transmitter Substances*. Springfield, Ill., Charles C Thomas, 1969.
   A short and clear outline of the experiments behind the quantal hypothesis from the hand of the master.
2. Kuffler, S.W.; Nicholls, J.G.; Martin, A.R. *From Neuron to Brain*. Sunderland, Mass., Sinauer Associates, 1984.
   An excellent textbook including several good chapters on chemical synaptic transmission.
3. Hall, Z.W.; Hildebrand, J.G.; Kravitz, E.A. *Chemistry of Synaptic Transmission: Essays and Sources*. Portland, Ore., Chiron Press, 1974.
   A stimulating collection of the classical papers concentrating on the transmitters themselves.
4. Heuser, J.E.; Reese, T.S.; Dennis, M.J.; Jan, Y.; Jan, L.; Evans, L. Synaptic vesicle exocytosis captured by quick freezing and correlated with quantal transmitter release. *J. Cell Biol.* 81:275–300, 1979.
   The ingenious research paper that captures pictures of vesicles during exocytosis.

REFERENCES

1. Anderson, C.R.; Stevens, C.F. Voltage clamp analysis of acetylcholine-produced end-plate current fluctuations at frog neuromuscular Junction. *J. Physiol. (Lond.)* 235: 655–691, 1973.
2. Anderson, D.C.; King, S.C.; Parsons, S.M. Proton gradient linkage to active uptake of [³H] acetylcholine by *Torpedo* electric organ synaptic vesicles. *Biochemistry* 21:3037–3043, 1982.
3. Anholt, R.; Lindstrom, J.; Montal, M. The molecular basis of neurotransmission: structure and function of the nicotinic acetylcholine receptor. In Martonosi, A.N., ed. *The Enzymes of Biological Membranes*, Vol. 3, 335–401, New York, Plenum Publishing Corp., 1985.
4. Baker, P.F.; Knight, D.E. Calcium control of exocytosis and endocytosis in bovine adrenal medullary cells. *Philos. Trans. R. Soc. Lond. [Biol.]* 296:83–103, 1981.
4a. Brock, J.A.; Cunnane, T.C. Relationship between the nerve action potential and transmitter release from sympathetic postganglionic nerve terminals. *Nature* 326:605–607, 1987.
5. Castillo, J. del; Katz, B. Quantal components of the end-plate potential. *J. Physiol. (Lond.)* 124:560–573, 1954.

6. Castillo, J. del; Katz, B. Statistical factors involved in neuromuscular facilitation and depression. *J. Physiol. (Lond.)* 124:574–585, 1954.

7. Cull-Candy, S.G.; Lundh, H.; Thesleff, S. Effects of botulinum toxin on neuromuscular transmission in the rat. *J. Physiol. (Lond.)* 260:177–203, 1976.

8. Cull-Candy, S.G.; Miledi, R.; Trautmann, A.; Uchitel, O.D. On the release of transmitter at normal, myasthenia-gravis– and myasthenic-syndrome–affected human end-plates. *J. Physiol. (Lond.)* 299:621–632, 1980.

9. Dionne, V.E.; Steinbach, J.H.; Stevens, C.F. An analysis of the dose-response relationship at voltage-clamped frog neuromuscular junctions. *J. Physiol. (Lond.)* 281:421–444, 1978.

10. Dodge, F.A., Jr.; Rahamimoff, R. Co-operative action of calcium ions in transmitter release at the neuromuscular junction. *J. Physiol. (Lond.)* 193:419–423, 1976.

11. Douglas, W.W. Stimulus-secretion coupling: the concept and clues from chromaffin and other cells. *Br. J. Pharmacol.* 34:451–474, 1968.

12. Dwyer, T.M.; Adams, D.J.; Hille, B. The permeability of the end-plate channel to organic cations in frog muscle. *J. Gen. Physiol.* 75:469–492, 1980.

13. Edwards, R.H.T.; Jones, D.A. Diseases of skeletal muscle. *Handbk. Physiol.*, Sec. 10:633–672, 1983.

14. Fatt, P.; Katz, B. An analysis of the end-plate potential recorded with an intra-cellular electrode. *J. Physiol. (Lond.)* 115:320–370, 1951.

15. Fatt, P.; Katz, B. Spontaneous subthreshold activity at motor nerve endings. *J. Physiol. (Lond.)* 117:109–128, 1952.

16. Fernandez, J.M.; Neher, E.; Gomperts, B.D. Capacitance measurements reveal stepwise fusion events in degranulating mast cells. *Nature* 312:453–455, 1984.

17. Gorio, A.; Hurlbut, W.P.; Ceccarelli, B. Acetylcholine compartments in mouse diaphragm: comparison of the effects of black widow spider venom, electrical stimulation, and high concentrations of potassium. *J. Cell Biol.* 78:716–733, 1978.

18. Hamill, O.P.; Marty, A.; Neher, E.; Sakmann, B.; Sigworth, F.J. Improved patch-clamp techniques for high-resolution current recording from cells and cell-free membrane patches. *Pflugers Arch.* 391:85–100, 1981.

19. Heuser, J.E.; Reese, T.S. Structure of the synapse. *Handbk. Physiol.*, Sec. 1, 1:194–261, 1977.

20. Heuser, J.E.; Reese, T.S.; Dennis M.J.; Jan, Y.; Jan, L.; Evans, L. Synaptic vesicle exocytosis captured by quick freezing and correlated with quantal transmitter release. *J. Cell Biol.* 81:275–300, 1979.

21. Hille, B. *Ionic Channels in Excitable Membranes.* Sunderland, Mass., Sinauer Associates, 1984.

22. Holman, M.E.; Hirst, G.D.S. Junctional transmission in smooth muscle and the autonomic nervous system. *Handbk. Physiol.*, Sec. 1, 1:417–461, 1977.

23. Hoyle, G. *Muscles and Their Neural Control*, 422–478. New York, John Wiley & Sons, 1983.

24. Katz, B. *The Release of Neural Transmitter Substances.* Springfield, Ill., Charles C Thomas, 1969.

25. Katz, B.; Miledi, R. The timing of calcium action during neuromuscular transmission. *J. Physiol. (Lond).* 189:535–544, 1967.

26. Katz, B.; Miledi, R. The effect of local blockage of motor nerve terminals. *J. Physiol. (Lond.)* 199:729–741, 1968.

27. Katz, B.; Miledi, R. The characteristics of end-plate noise produced by different depolarizing drugs. *J. Physiol. (Lond.)* 230:707–717, 1973.

28. Katz, B.; Miledi, R. The binding of acetylcholine to receptors and its removal from the synaptic cleft. *J. Physiol. (Lond).* 231:549–574, 1973.

29. Katz, B.; Thesleff, S. A study of the desensitization produced by acetylcholine at the motor end-plate. *J. Physiol. (Lond.)* 138:63–80, 1957.

30. Kelly, R.B.; Deutsch. J.W.; Carlson, S.S.; Wagner, J.A. Biochemistry of neurotransmitter release. *Annu. Rev. Neurosci.* 2:399–446, 1979.

31. Krnjevic, K. Transmitter in motor systems. *Handbk. Physiol.*, Sec. 1, 2(1):107–154, 1981.

32. Kuffler, S.W.; Nicholls, J.G.; Martin, A.R. *From Neuron to Brain.* Sunderland, Mass, Sinauer Associates, 1984.

33. Kuffler, S.W.; Yoshikami, D. The number of transmitter molecules in a quantum: an estimate from iontophoretic application of acetylcholine at the neuromuscular synapse. *J. Physiol. (Lond.)* 251:465–482, 1975.

34. Magleby, K.L.; Stevens, C.F. The effect of voltage on the time course of the endplate currents. *J. Physiol. (Lond.)* 223:151–171, 1972.

35. Magleby, K.L.; Stevens, C.F. A quantitative description of end-plate currents. *J. Physiol. (Lond.)* 223:173–197, 1972.

36. Magleby, K.L.; Zengel, J.E. A quantitative description of stimulation-induced changes in transmitter release at the frog neuromuscular junction. *J. Gen. Physiol.* 80:613–638, 1982.

37. Martin, A.R. Junctional transmission. II. Presynaptic mechanisms. *Handbk. Physiol.*, Sec. 1, 1:329–355, 1977.

38. Matthews-Bellinger, J.; Salpeter, M.M. Distribution of acetylcholine receptors at frog neuromuscular junctions with discussion of some physiological implications. *J. Physiol. (Lond.)* 279:19–213, 1978.

39. Mishina, M.; Tobimatsu, T.; Imoto, K.; Tanaka, K.; Fujita, Y.; Fukadu, K.; Kurasaki, M.; Takahashi, H.; Morimoto, Y.; Hirose, T.; Inayama, S.; Takahashi, T.; Kuno, M.; Numa, S. Location of functional regions of acetylcholine receptor alpha-subunit by site-directed mutagenesis. *Nature* 313:364–369, 1985.

40. Noda, M.; Furutani, Y.; Takahashi, H; Toyosato, M.; Tanabe, T.; Shimizu, S.; Kikyotani, S.; Kayano, T.; Hirose, T.; Inayama, S.; Numa, S. Cloning and sequence analysis of calf cDNA and human genomic DNA encoding alpha-subunit precursor of muscle acetylcholine receptor. *Nature* 305:818–823, 1983.

41. Popot, J.-L.; Changeux, J.-P. Nicotinic receptor of acetylcholine: structure of an oligomeric integral membrane protein. *Physiol. Rev.* 64:1162–1239, 1984.

42. Pumplin, D.W.; Fambrough, D.M. Turnover of acetylcholine receptors in skeletal muscle. *Annu. Rev. Physiol.* 44:319–335, 1982.

43. Purves, D.; Lichtman, J.W. *Principles of Neural Development.* Sunderland, Mass., Sinauer Associates, 1985.

44. Reichardt, L.F.; Kelley, R.B. A molecular description of nerve terminal function. *Annu. Rev. Biochem.* 52:871–926, 1983.

45. Rosenberry, T.L. Quantitative simulation of end-plate currents at neuromuscular junctions based on the reaction of acetylcholine with acetylcholinesterase. *Biophys. J.* 26:263–290, 1979.

46. Rosenthal, J. Trophic interactions of neurons. *Handbk. Physiol.*, Sec. 1, 1:775–801, 1977.

47. Schmitt, A.; Dreyer, F.; John, C. At least three sequential steps are involved in the tetanus toxin–induced block of neuromuscular transmission. *Naunyn Schmiedebergs Arch. Pharmacol.* 317:326–330, 1981.

48. Silinsky, E.M. The biophysical pharmacology of calcium-dependent acetylcholine secretion. *Pharmacol. Rev.* 37:81–132, 1985.
49. Steinbach, J.H. Activation of nicotinic acetylcholine receptors. In Cotman, C.W.; Poste, G.; Nicholson, G.L., eds. *The Cell Surface and Neuronal Function*, 120–156. Amsterdam, Elsevier North-Holland Biomedical Press, 1980.
50. Takeuchi, A. Junctional transmission. I. Postsynaptic mechanisms. *Handbk. Physiol.*, Sec. 1, 1:295–327, 1977.
51. Takeuchi, A.; Takeuchi, N. Active phase of frog's end-plate potential. *J. Neurophysiol.* 22:395–412, 1959.
52. Takeuchi, A.; Takeuchi, N. On the permeability of end-plate membrane during the action of transmitter. *J. Physiol. (Lond.)* 154:52–67, 1960.
53. Taylor, P. Anticholinesterase agents. In Goodman, L.S.; Gilman, A., eds. *The Pharmacological Basis of Therapeutics*, 6th ed. New York, Macmillan, 1980.
54. Taylor, P. Neuromuscular blocking agents. In Goodman, L.S.; Gilman, A., eds. *The Pharmacological Basis of Therapeutics*, 6th ed. New York, Macmillan, 1980.
55. Wathey, J.C.; Nass, M.M.; Lester, H.A. Numerical reconstruction of the quantal event at nicotinic synapses. *Biophys. J.* 27:145–164, 1979.
56. von Wedel, R.J.; Carlson, S.S.; Kelly, R.B. Transfer of synaptic vesicle antigens to the presynaptic plasma membrane during exocytosis. *Proc. Natl. Acad. Sci. USA* 78:1014–1018, 1981.

Excitation-Contraction Coupling in Skeletal Muscle

INTRODUCTION

Muscle consists of long cells that are specialized to provide controlled locomotion. They do so by virtue of their ability to contract, or shorten, against a load. In most animals, we distinguish *smooth* and *striated muscle*. In vertebrates, smooth muscle tissue is found in gastrointestinal tract, blood vessels, reproductive organs, glands, and eyes. There are many different types of smooth

muscle, described in Chapter 9. There are only two types of striated muscle: *skeletal* and *cardiac*. This chapter focuses on vertebrate skeletal muscle, because it is understood in the greatest detail. Before doing so, two generalizations about muscle cells can be made.

First, all muscle cells transform chemical energy into mechanical energy by means of the two so-called contractile proteins, actin and myosin. These protein molecules are water soluble as monomers but have a strong tendency to aggregate. Within a muscle cell, they aggregate into thin actin filaments and thicker myosin filaments. Motion is produced when actin and myosin filaments slide past each other to produce mechanical force. While doing so, they break high-energy bonds of ATP. Details about the biochemistry and mechanics of force production are given in Chapters 8 and 9.

Second, all muscle cells control their activity by varying the intracellular free [Ca²⁺]. Actin and myosin may interact, split ATP, and develop force only while $[Ca^{2+}]_i$ is above about 1 μM. Throughout, the regulation of actin and myosin interaction by internal [Ca²⁺] is achieved by a class of small proteins that includes troponin C and calmodulin. These proteins carry Ca binding sites with a high affinity to Ca^{2+} ($K_D = 1 - 10$ μM). Ca ions binding to these sites cause a large and reversible conformational change that ultimately results in the initiation of contraction, the detailed molecular mechanism depending on the muscle type. When Ca^{2+} dissociates from the Ca-binding protein, the conformational change is reversed, the interaction of actin and myosin is interrupted, and the muscle relaxes. Details about the control of contractile proteins by Ca^{2+} are given in Chapter 8.

Basic Anatomy and Physiology of Skeletal Muscle. The structure colloquially known as a skeletal muscle is a bundle of some 100 to 10,000 separate cells called muscle fibers. Within a muscle, each fiber, some 10 to 100 μm in diameter, runs all the way from one tendon to another. Next to neurons, muscle fibers are the longest cells known; in the human body, some are tens of centimeters long. Skeletal muscle cells have no cytoplasmic connections with each other; in this they are unlike the cells of most other tissues, which are coupled to each other by gap junctions. Hence the individual fibers within a muscle have little more in common than that they pull on the same tendon or are activated by the same motor neuron.

Figure 7–1A shows a light micrograph of a portion of a single skeletal muscle fiber dissected from a frog muscle. The cross-striations, which give this type of muscle its name, reflect the highly regular and periodic arrangement of actin and myosin filaments at the molecular level. Figure 7–1B is a scanning electron micrograph of a broken piece of a skeletal muscle fiber, showing its interior. The fiber is seen to be a tightly packed bundle of approximately 1 μm–thick striped rods called *myofibrils*. Each myofibril runs all the way from tendon to tendon. Figure 7–1C shows, at higher magnification, a few myofibrils sectioned longitu-

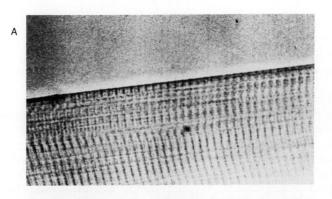

10 μm

Figure 7–1 *A,* Light micrograph under phase contrast optics; *B,* scanning electron micrograph, and *C,* transmission electron micrograph of a piece of frog skeletal muscle fiber. The fiber runs from left to right. In *C,* the A and I band are indicated for one myofibril, the Z line for another. (*A* from Weiss et al. *J. Gen. Physiol.* 87:955–983, 1986; *B* from Sawada et al. *Tissue Cell* 10:179–190, 1978; *C* from Huxley, H.E. In Brachet, J.; Mirsky, A.E., eds. *The Cell,* vol. 4, 365–481, 1960.)

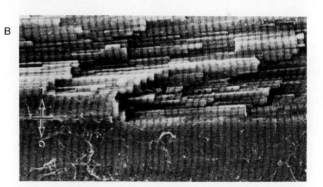

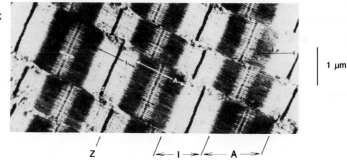

1 μm

dinally. The origin of the striations seen in Figure 7–1a is now apparent. Thin, dense lines, spaced about 2 μm apart, are seen to run periodically across the myofibrils; they are called Z *lines* and delimit the functional unit of a muscle cells, the *sarcomere*. (What appears as a Z line is actually an edge view of the Z disk, a planar network of filaments containing the proteins desmin and α-actinin.) Each sarcomere contains a dark electron-dense band, called the *A band*; it is sandwiched between two lighter regions of lesser electron density. Together, the light regions of two adjacent half-sarcomeres form the *I band*. The lighter regions contain only the 1.0 μm–long actin filaments, while the A band marks the location of the 1.5 μm–long myosin filaments. The region appears dark because it contains both myosin and actin filaments, and hence a relatively larger concentration of stained proteins. The striations visible in Figure 7–1A result from the alternating A and I bands. Other features of the sarcomere are described in Chapter 8.

Skeletal muscle in vivo is entirely under the control of the central nervous system. When stimulated by the motor nerve, and only then, the muscle fiber responds by giving a *twitch*, that is, a transient contraction lasting from ten to several hundreds of milliseconds, depending on the fiber type and temperature. A given motor neuron simultaneously innervates anywhere from 10 to 1000 fibers; together, the motor neuron and all fibers innervated by it are called the *motor unit*. An action potential propagating down the motor nerve axon of an intact muscle will elicit a twitch simultaneously from all fibers in the motor unit.

During continuous movements, muscles are stimulated repetitively via the motor nerve. For a single fiber in vivo, stimulation frequencies of 10 to 20 Hz are not unusual. At such frequencies, a fiber may not have enough time between stimuli to relax completely, so that the force of contraction summates to some degree from twitch to twitch. The resulting contraction is a series of rapidly repeating twitches superimposed on a maintained level of force. In vivo, the central nervous system regulates muscle force by varying both the frequency of impulses sent out by individual motor neurons and the fraction of motor units in a muscle to which nerve impulses are being sent. Under experimental conditions, muscle fibers can follow stimulation frequencies up to 100 Hz. Most muscles respond to such stimulation with a *fused tetanus*, in which, throughout the period of stimulation, the fiber maintains the maximal force of which it is capable.

Development and Regeneration of Skeletal Muscle. A muscle fiber develops by fusion of many small precursor cells called myoblasts. When myoblasts line up and fuse to form a muscle fiber, their nuclei survive in the adult cell; hence muscle fibers are multinucleated. Like adult neurons, mature skeletal muscle cells are unable to divide. No amount of training will increase the number of fibers in an athlete's muscle; training increases only the diameter of existing muscle cells, as well as improving their metabolism and the circulation to them. Skeletal muscle fibers surround themselves with a 20- to 100-nm thick layer of extracellular matrix called basal lamina or basement membrane. This layer carries special protein molecules marking the position of the neuromuscular junction and contains many small cells called *satellite cells*. If a muscle cell is damaged beyond repair but portions of the basement membrane remain intact, a new muscle fiber will develop from the satellite cells contained in the basement membrane. Satellite cells may be viewed as myoblasts that lie dormant within the basement membrane until they are awakened by an unknown chemical signal emitted by the dying muscle fiber. Once awakened, the satellite cells proliferate, fuse, and ultimately form a new muscle fiber within the sheath of basal lamina formerly occupied by the old. They inherit from the old fiber its connection with the central nervous system, forming neuromuscular junctions beneath the terminals of the old motor nerve.[24a]

Cardiac muscle cells have no satellite cells and cannot therefore be replaced. The only way the heart can compensate for cells lost, e.g., during an infarct, is by the enlargement of the remaining cells.

THE INITIATION OF THE TWITCH

The experiment in Figure 7–2 illustrates the major events during activation of vertebrate skeletal muscle. A single muscle fiber, perhaps 1 cm long, was dissected from a frog and mounted rigidly, with one end held by a tension transducer. The fiber was impaled by a microelectrode to measure its membrane potential, and was loaded with arsenazo III dye, which changes its light absorption when it binds a Ca ion. When either the muscle fiber or its motor nerve was stimulated electrically, an action potential was generated and traveled along the fiber (trace a). Nearly simultaneously, the myoplasmic Ca-concentration (measured as the light absorption of the dye) began to

Figure 7–2 *A*, Action potential; *B*, intracellular [Ca²⁺]; and *C*, tension in a frog muscle fiber at 16°C. In *D* and *E*, the action potential and time course of intracellular [Ca²⁺] were recorded at a faster time base and a lower temperature (7°C); the calibration of *B* does not necessarily apply to *E*. (*B* and *C* from Baylor, S.M., et al., *J. Physiol. [Lond.]* 344:625–666, 1983; *D* and *E* from Miledi, R., et al., *J. Physiol. [Lond.]* 333:655–679, 1982.)

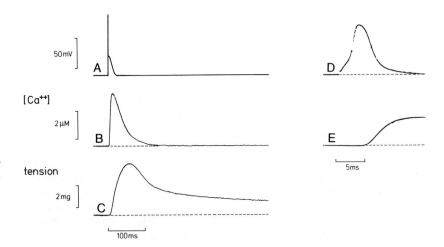

rise (trace b), transiently increasing ten- to a hundredfold above the low value of about 0.1 μM found at rest. After a brief delay, the muscle produced a twitch, visible as a transient increase in tension (trace c). In traces d and e, the relationship between action potential and Ca transient is shown in a high-speed recording. Even though the experimental temperature had been lowered to 7°C in order to slow the events preceding the twitch, Ca release is seen to follow the action potential within a few milliseconds. Similar events can also be observed in other muscle types, though nowhere do they happen as rapidly as in skeletal muscle. In all striated and many smooth muscles, activity begins with a cell membrane depolarization.

From the earliest days of muscle physiology, investigators have been impressed by the enormous speed at which twitches (and by inference, the changes in intracellular Ca concentration) follow the action potential. A humming bird, for example, can beat its wings at a frequency of up to 50 Hz, with only 20 ms allowed for the entire cycle of muscle contraction and relaxation. The increase and decline of myoplasmic [Ca²⁺] must occur at least as rapidly in these muscles. In such short times, Ca²⁺ does not have enough time to diffuse from the outside into and out of the depth of a muscle fiber, especially if one considers that it binds to Ca-binding proteins on the way. Hence the Ca²⁺ that appeared in the myoplasm in Figure 7–2 was probably already within the muscle fiber before the action potential started. There is conclusive proof for this view: a skeletal muscle fiber can give many tens of twitches even in a bathing solution made essentially Ca-free by the addition

of a Ca chelator.[5] Evidently, muscle fibers have intracellular Ca stores and are able to use the stored Ca²⁺ repeatedly to initiate contraction.

Two Intracellular Membrane Systems Controlling Myoplasmic [Ca²⁺]. The Ca store is readily visible under the electron microscope and has the appearance illustrated in the schematic drawing in Figure 7–3A. The figure shows five vertically oriented myofibrils. Each sarcomere of each myofibril is surrounded individually by a "cuff" of interconnected, membranous vesicles. Collectively, the system of vesicles is called the *sarcoplasmic reticulum* (SR). It is derived from the smooth endoplasmic reticulum and typically occupies 10% of the volume of a skeletal muscle fiber. Most of the Ca²⁺ inside a resting muscle fiber, very roughly 1 mmol/l fiber volume, is contained in the SR, where it can be detected by electron microprobe analysis. Most of the Ca contained in the SR (on the order of 10 mmol/l SR lumen) is loosely bound to *calsequestrin*, a highly anionic, 44,000-dalton protein that can bind 40 Ca ions per molecule. Nevertheless, the concentration of ionized Ca in the SR is probably still of the order of 1 mM, 10⁴ times higher than in the myoplasm. During activity, the SR allows Ca²⁺ to move down its electrochemical gradient into the myoplasm; after the action potential has passed, the Ca²⁺ thus released is reaccumulated.

If the SR is to function as a Ca store, it cannot be continuous with the extracellular space, since a muscle fiber continues to twitch even in Ca-free solutions. Lack of continuity is confirmed morphologically by the finding that electro-dense extracellular marker molecules cannot reach the lumen of the SR.

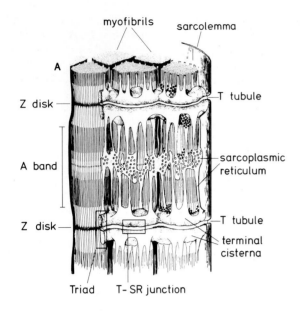

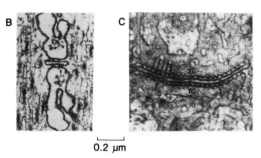

0.2 µm

Figure 7–3 A, Schematic drawing showing membrane systems in a frog skeletal muscle fiber. B and C, Transmission electron micrograph showing cross-sectional (B) and longitudinal (C) views of the triadic junction. (A, Reproduced from Peachey, L. D. *The Journal of Cell Biology*, 1970, Vol. 25, 209–231, by copyright permission of The Rockefeller University Press; B, after Franzini-Armstrong, C.; Nunzi, A. *J. Muscle Res. Cell Motil.* 4:233–252, 1983; C, courtesy of B. Eisenberg.)

Action potentials may propagate along the plasma membrane of the cell, also called *sarcolemma*. They are communicated to the SR by means of a second membrane system, the *transverse tubular system* (TTS). Transverse tubules are narrow invaginations of the sarcolemma (Fig. 7–3A) and form a network throughout the cytoplasm such that each mesh of the network encircles one myofibril. In frog skeletal muscle, there is one such network per sarcomere; it is located at the Z line. In reptilian and mammalian muscle, each sarcomere has two networks, located approximately in the planes defined by the tips of the myosin filaments. The transverse tubules act like minute

inside-out nerve fibers and conduct the action potential from the sarcolemma into the interior. Propagating along the transverse tubules, the action potential can travel around each sarcomere of each myofibril. The transverse tubular system possesses ten times less membrane than the SR and occupies only a fraction of 1% of the fiber volume. The lumen of the TTS is continuous with the extracellular space, since extracellular markers and membrane-impermeant fluorescent dyes may be seen to fill the TTS within seconds or minutes of external application. Not surprisingly, the membrane of the TTS is continuous with the sarcolemma, since the potential across it changes when the potential across the sarcolemma changes (see later). It is possible to obtain electron micrographs in which tubules are clearly seen to invaginate from the sarcolemma and form "mouths" that empty into the extracellular space.[13]

The functional importance of the transverse tubules was shown in a classic experiment[19] (see Fig. 7–4). Fire-polished glass micropipettes were

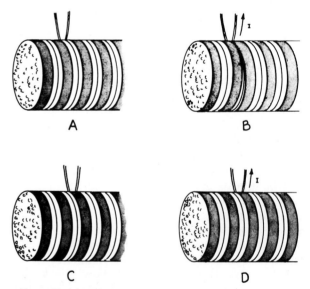

Figure 7–4 Local contraction in a frog muscle fiber where the transverse tubules occur at the Z lines. A and B, Micropipette applied to the muscle surface at a sensitive spot near Z line. When the patch membrane beneath the pipette tip is depolarized by making the inside of the pipette negative, the adjacent half of the sarcomere contracts locally (B). C and D are similar to A and B, but the pipette is applied to an A band. This time, a negative potential applied to the pipette does not cause contraction. The experiment shows that membrane depolarization is effective only at the Z line, where it can spread into the transverse tubular system. (After Fawcett, A.L. In Fishman, H.M., ed. *The Myocardium, Its Biochemistry and Biophysics.* New York, New York Heart Association, 1961. Illustrates an experiment by Huxley, A.F.; Taylor, R.E. *J. Physiol. [Lond.]* 144:426–441, 1958.)

pressed against the sarcolemma to isolate electrically a small patch beneath the about 1-μm wide pipette tip. The patch of sarcolemma was depolarized by applying an electric current to the pipette, and the fiber was observed under a high-power microscope to see whether the muscle would respond with a small local contraction. It was found that contractions could be elicited only from certain positions on the sarcolemma. In frog skeletal muscle, such "sensitive spots" were found when the tip of the micropipette was positioned over a Z-line; in lizards, they were found instead at the borders of the A-bands, which mark the position of the tips of the myosin filaments. The location of sensitive spots coincides with the locations of the TTS networks in the two preparations, suggesting that local stimulation is effective only when the pipette is above the mouth of a tubule. Further evidence for the importance of transverse tubules comes from findings on muscle fibers in which these tubules have been disconnected from the sarcolemma by osmotic shock treatment. In these fibers, action potentials no longer cause twitches.[13a] Evidently contraction is triggered by depolarization of the tubule membrane. Depolarization of the sarcolemma alone is not sufficient.

Inside the fiber, transverse tubules and SR come in close contact and form a structure called the "triad" or "T-SR junction." These structures are shown in Figure 7–3B,C, in which each tubule appears sandwiched between two large sacs of SR, one belonging to each sarcomere. The sacs are called *terminal cisternae*. Under high magnification, electron-dense spots (called "feet") may be seen to line up, regularly spaced, in two parallel rows along the zone of contact between tubules and SR membrane and to bridge the approximately 200-nm–wide gap between the two. It is natural to suggest that these electron-dense objects form not only the morphological connection but also the functional connection between transverse tubules and SR. Indeed, feet have been seen in the TTS-SR junction of all striated muscles in which they have been sought. A particularly interesting case are the muscle cells of *Amphioxus* (*Branchiostoma*), a primitive chordate. These cells are 1 μm–thick sheets, each containing a single layer of myofibrils. They possess no transverse tubules but do have numerous "subsurface cisternae," closed membranous sacs that lie immediately beneath the sarcolemma. Functioning as storage sites that release Ca^{2+} in response to action potentials, they are analogous to the SR in vertebrates. Interestingly, the narrow gap between sarcolemma and subsurface cisternae contains regularly spaced electron-

dense spots of the same appearance as those at the vertebrate TTS-SR junction.[17] Here, the feet connect not the SR and the TTS but instead the subsarcolemmal cisternae and the sarcolemma. It seems that in muscle, feet tend to be found wherever a functional connection is required between an intracellular Ca store and a portion of the cell membrane controlling the Ca movements into or out of that store. However, in cardiac muscle of mammals and birds one observes regions of SR with feet even though there are no tubules nearby.[20] The functional significance of feet without tubules is unknown.

In summary, three morphologic structures contribute to the control of myoplasmic $[Ca^{2+}]$: The sarcolemma, the transverse tubular system, and the sarcoplasmic reticulum. All three are discussed in the following section.

ELECTRIC EXCITATION OF SARCOLEMMA AND TRANSVERSE TUBULAR SYSTEM

The resting and action potentials in vertebrate skeletal muscle are generated by means of ion concentration gradients and selectively permeable ion channels in nearly the same way as in axons. The resting potential is due to two ion channels, one a Cl and the other a K channel. Though voltage-dependent, both tend to be open at negative membrane potentials such as the resting potential. The K channel is of a type found also in eggs and neurons and is called "inward rectifier," because it tends to shut on depolarization and to open on hyperpolarization.

Most skeletal muscle fibers are "twitch" fibers that are excited by the motor nerve via a compact neuromuscular junction located approximately midway between the tendons. When the motor nerve excitation causes an end-plate potential at the neuromuscular junction, an action potential is set up that propagates away from the neuromuscular junction in both directions. The role of the action potential is to provide a large electric signal to the transverse tubules and thereby cause Ca release from the SR. There also are so-called "tonic" fibers, e.g., the fibers within the extraocular muscles. These are relatively short fibers that have motor nerve endings over much of their length, that have only a poorly developed action potential mechanism and derive depolarization mainly from the end-plate potential. However, these occur relatively infrequently and are not considered further here.

Figure 7–5 shows action potentials in frog (A)

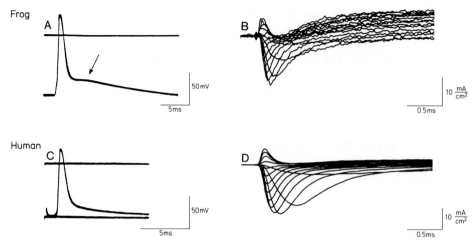

Figure 7–5 Action potentials *(A and C)* and membrane currents *(B and D)* under voltage clamp. *A* and *B* are from a frog muscle fiber and *C* and *D* from a human muscle fiber. Recordings were made at 20° to 23°C except for the human action potential *(D)*, recorded at 37°C. In *A* and *C* the horizontal lines indicate zero mV. Note afterdepolarizations that are thought to reflect the action potential in the transverse tubular system. Sometimes the depolarizing after potential rises to a distinct peak (arrow). *(A,* after Adrian, R.H., et al. *J. Physiol. [Lond.]* 208:607–644, 1970; *C,* after Ludin, P. *Europ. Neurol.* 2:340–347, 1969.) *B* and *D,* Membrane current traces, represent superimposed responses accompanying steps to potentials spaced at 10-mV intervals between −33 to 107 mV *(B)* and −53 to 77 mV *(D).* Membrane currents were recorded with a method that emphasizes currents across the sarcolemma; hence, there is no indication of delayed Na currents across the tubular membrane that might correspond to the afterdepolarizations in *A* and *B.* (After Roberts, W.M.; Almers, W. *Pflugers Arch.* 402:190–196, 1984.)

and human twitch-muscle fibers (C), along with families of ionic current traces recorded under a voltage clamp with step depolarizations of increasing amplitude (B,D). As in axons, there is a transient inward current through Na channels, followed by a delayed outward current through K channels. Human (and rat) fibers have far fewer voltage-dependent K channels than frog fibers, as judged by the comparatively small delayed outward currents. Having few K channels minimizes the amount of K lost from the muscles into the circulation and also limits the K accumulation in the transverse tubular system during activity (discussed later). Minimizing the loss of K during muscle action potentials is important, especially during vigorous exercise, because the heart may stop if the K concentration in the plasma rises above 8 mM.

Table 7–1 lists some of the ion channels that can be studied in intact skeletal muscle by electrophysiological recording and also compares the densities at which they populate sarcolemma and transverse tubular membranes. In the rat, the ion channels responsible for the resting potential, namely K and Cl channels, occur at approximately equal densities in TTS and sarcolemma. On the other hand, the two channels mediating the action potential, namely voltage-dependent Na and K channels, are concentrated on the sarcolemma.

The density of Na channels in the tubules of frog muscle is just high enough to allow a regenerative action potential there. Ca channels are discussed later.

Consequences of the Transverse Tubular System for the Electric Behavior of the Cell Membrane. The action potential propagates not only along but also radially into the fiber. It reaches the center of the TTS with a delay of about 1 to 2 ms after invading the sarcolemma.[14] The action potential in the tubule is thought to have nearly the

Table 7–1 The Distribution of Some Ion Transport Sites over the Sarcolemma and Tubular System of Vertebrate Skeletal Muscle[1]

Transport Site	Role	Density in Tubules ---- Density in Sarcolemma
Na channel	Action potential	0.7
Depolarization-activated K channel	Action potential	0.3
Cl channel (rat[2])	Resting potential	1
Inward rectifier K channel	Resting potential	1
Ca channel	Ca influx	>4
Na⁺-K⁺ pump	Na/K gradients	0.15

[1]After Almers & Stirling, *J. Membrane Biol.* 77:169–186, 1984.
[2]In frogs, Cl channels are reported to reside mainly in the sarcolemma.[11]

same amplitude as on the sarcolemma, and causes the *afterdepolarization* seen in Figure 7–5A (arrow). The comparatively small amplitude of the after potential illustrates that rapid electrical events across the tubule membrane are recorded across the sarcolemma only in attenuated form.

The action potential travels along a muscle fiber more slowly than it would along an unmyelinated nerve fiber of the same diameter. This is so because it takes time to depolarize the membrane belonging to the transverse tubular system. In electrophysiological terms, the transverse tubules act like a large capacitative load on the action potential mechanism. If referred to the cylindrical surface of a muscle fiber, the apparent *membrane capacity* is not the 1 $\mu F/cm^2$ characteristic for nonmyelinated nerve axons but 4 to 10 $\mu F/cm^2$ and proportional to the fiber diameter. The larger apparent membrane capacity is due to the additional membrane contributed by the TTS.

Some of the K ions leaving a muscle fiber during the falling phase of the action potential cause *K accumulation in the TTS* and temporarily make the K equilibrium potential there more positive. This causes an afterdepolarization that is slower and smaller than that shown in Figure 7–5A but lasts for hundreds of milliseconds and summates during repetitive activity. This *"late afterdepolarization"* is of physiological significance mainly because it carries the danger of making tetanic excitation of muscle fibers self-perpetuating. The danger of repetitive activity is probably the reason why skeletal muscle cells, unlike axons, need a large Cl conductance. Especially in mammalian muscle, the Cl conductance is ten times larger than the resting K conductance. Cl influx is expected to contribute prominently to repolarization after an action potential, thereby making a large K efflux unnecessary and lessening the K accumulation in the tubule lumen. Also, a large Cl conductance tends to diminish the depolarizing effects of K accumulation in the TTS.

Pathological Defects in Electric Excitation of Muscle. The importance of a large Cl conductance is illustrated by *congenital myotonia*,[25] a rare disease that afflicts certain mammals, including human beings. Myotonic patients cannot readily relax their muscles after a vigorous contraction; instead, their muscles may continue to contract involuntarily for tens of seconds. Electromyographic observation shows that this aftercontracture is accompanied by trains of muscle action potentials that continue even though activity in the motor nerve has ceased. The symptom has been investigated by intracellular recording from muscles taken from myotonic goats. Under tetanic stimulation, the late afterpotential of such muscles is seen to summate and increase until it is sufficient to sustain a train of action potentials without external stimulation. The effect is entirely due to a lack of Cl conductance in muscles of myotonic goats, since muscles from normal goats become myotonic when external Cl is removed.[1, 3] There are also other diseases with myotonic symptoms.[25]

The periodic paralyses are another group of rare muscle diseases that are probably related to deficiencies in the ion permeability of the cell membrane. Patients thus afflicted experience episodes of flaccid (i.e., relaxed) paralysis that are usually accompanied by K changes in the serum. In *hypokalemic periodic paralysis*,[15] K in the serum may fall from 5 to 2 mM; attacks are triggered by a heavy carbohydrate meal, prolonged rest, or emotional anxiety, and the flaccid paralysis may last for a few hours or up to 1 to 2 days. Oral or intravenous administration of K relieves the paralysis. The origin of these symptoms is still unclear. One hypothesis is based on the finding that, under certain experimental conditions, skeletal muscle fibers depolarize when external K is lowered. The result is attributed to a peculiarity of the inward rectifier channel responsible for the K permeability of the resting muscle. This ion channel requires external K in order to conduct K. When external $[K^+]$ drops, then so does the K permeability of the inward rectifier; it is hypothesized that this permits the cell to depolarize to potentials at which Na channels become inactivated, and the cell can no longer generate action potentials. Details of the mechanism are still lacking; for instance, it is not known which ion channels are affected by the disease.

During attacks of *hyperkalemic periodic paralysis*, muscle fibers are depolarized to potentials at which Na channels are inactivated. At the same time, serum K is increased, but the increase is only slight, perhaps from 5 to 7 mM, and is insufficient in itself to explain the strong depolarization. There is recent evidence[16] that in a normal muscle, increasing external $[K^+]$ from 4 to 10 mM causes only the relatively slight depolarization expected from the Nernst equation and leads to a decrease in intracellular $[Na^+]$ due to stimulation of the Na pump. In diseased muscle, however, the slight depolarization caused by external K leads into a second phase of depolarization, during which intracellular $[Na^+]$ rises dramatically. The rise in intracellular Na indicates an increased Na permeability, and indeed, when tetrodotoxin is added to block Na channels, a normal membrane potential is restored and Na influx stops. Probably, the strong depolarization in diseased muscle is due to a small population of altered voltage-sensitive Na channels that open aberrantly during the slight depolarization caused by increased $[K^+]_o$.

Biochemical Studies of Tubule Membranes. Membranous vesicles derived purely from transverse tubules can now be made. They are found to contain Na pumps (Na-K ATPase)[21] and Ca pumps that would enable an intact muscle cell to extrude Na and Ca^{2+} against an electrochemical gradient. They also contain tetrodotoxin binding sites that represent Na channels, as well as sites that specifically bind 1,4-dihydropyridines such as the drug nifedipine.[8, 12] Nifedipine is clinically used as a "Ca antagonist" to relax vascular smooth (though not skeletal) muscle, and blocks Ca channels in the cell membrane. It has been suggested that the 1,4-dihydropyridine binding sites are, in fact, Ca channels,[8, 12] even though, relative to the Ca currents recorded electrophysiologically, their number is surprisingly large.[27a]

Transverse tubule vesicles may be allowed to fuse with planar lipid bilayers so that the ion channels contained in them may be studied electrically at the single-channel level. In this way, recordings have been made from K and Ca channels.[2, 30] The K channels seen in this way appear to be not of the "inward rectifier" type but rather K(Ca) channels that are activated by membrane depolarization as well as intracellular Ca^{2+} (Chap. 4). Such recordings provide the most direct evidence that K(Ca) channels exist in the transverse tubules of adult muscle.

Ca Pumps and Ca Channels in the Muscle Cell Membrane. A storage system such as the SR can control only the distribution of Ca^{2+} within the cell, not the cellular Ca content. To keep intracellular Ca low in the long run, muscle cells depend, like all other eukaryotic cells, on a *Ca pump in the cell membrane* that is capable of extruding Ca^{2+} against a large electrochemical gradient. This Ca pump is a 130,000-dalton ATPase that requires both Ca^{2+} and Mg for activation, and is regulated by calmodulin, an intracellular Ca-binding protein that changes its conformation in response to changes in intracellular $[Ca^{2+}]$. Muscle cell membrane also possesses a Na^+-Ca^{2+} exchange system that catalyzes the electrogenic exchange of Na and Ca^{2+} across the cell membrane. The Na^+-Ca^{2+} exchanger can extrude Ca^{2+} without hydrolyzing ATP and derives the energy for this effort from the Na gradient established by the sodium pump. It, too, is found in the cell membranes of all types of muscle and other excitable cells (Chap. 2).

Given that Ca extrusion devices in the cell membrane are a necessity anyway, it may be thought that instead of having an SR and TTS to control intracellular $[Ca^{2+}]$, it would be simpler to keep intracellular $[Ca^{2+}]$ low at rest by extruding Ca^{2+} across the cell membrane and to raise $[Ca^{2+}]_i$ during activity by opening Ca channels in the cell membrane. In this way, Ca channels could serve the dual role of mediating cell membrane excitation and helping to supply the intracellular messenger needed to activate contraction. Indeed, Ca channels play such a role in nearly all types of muscle other than vertebrate skeletal muscle. In cardiac muscle, Ca channels determine the duration of the action potential and contribute to the increase of $[Ca^{2+}]_i$. In arthropod skeletal muscle, cell membrane depolarization is mediated jointly by neuromuscular transmission and Ca channels and requires no voltage-sensitive Na channels. Some types of smooth muscle contain only a little SR, and presumably much of the Ca^{2+} that activates contraction in them is supplied by Ca influx through Ca channels.

The importance of SR and TTS lies entirely in speeding up the supply and withdrawal of Ca^{2+} to and from the myofibrillar space, so that muscles may contract and relax more rapidly. Vertebrate skeletal muscle is optimized for speed, and here, Ca movements across the SR membrane have completely supplanted those across the cell membrane in regulating myoplasmic $[Ca^{2+}]$. Nevertheless, even vertebrate skeletal muscle possesses both voltage-dependent Ca channels[4, 26] and Ca pumps. The Ca channels are situated nearly exclusively in the transverse tubular system (Table 7–1) and respond so slowly to potential changes that only a few of them are opened during an action potential, far too few to contribute significantly to the $[Ca^{2+}]$ changes shown, e.g., in Figure 7–1.[4] Not surprisingly, the rapid activation of contraction in skeletal muscle is unimpaired when one blocks Ca channels by nifedipine. What is the physiological role of Ca channels in vertebrate skeletal muscle? Most likely, they are necessary for regulating cellular Ca content over times much longer than a single twitch. When myoplasmic $[Ca^{2+}]$ rises during activity, the Ca pump in the cell membrane extrudes more Ca, so that Ca efflux also rises. To maintain cellular Ca reserves, there must therefore be a pathway for Ca entry that is linked in some way to muscle activity. Depolarization-activated Ca channels would be suitable for providing such a pathway. Moreover, the fact that Ca channels reside nearly exclusively in the TTS guarantees that the SR throughout the fiber cross-section has equal access to the Ca^{2+} entering through Ca channels. This is clearly desirable but would not be the case for Ca^{2+} entering through channels in the sarcolemma.

THE SARCOPLASMIC RETICULUM

The SR functions not only as an internal store for Ca^{2+} but also, with its large surface area, as a "parking lot" for Ca pump sites. In a rat gastrocnemius, a fast-twitch muscle, there are 45,000 cm^2 of SR membrane per 1 ml of cytoplasm;[9] less is present in slower muscles. The SR membrane is tightly packed with a 100,000-dalton ATPase that is jointly activated by Ca^{2+} and Mg^{2+}. The ATPase functions as a *Ca pump in the SR* (Chap. 2). Constantly on the lookout for myoplasmic Ca ions, it captures and transfers them into the lumen of the SR. The SR Ca pump is unique to internal Ca stores derived from the endoplasmic reticulum and differs, for example, from the Ca-Mg ATPase found in plasma membranes of most cells. There are 20,000 pump molecules per square micrometer of SR membrane; the packing is so dense that adjacent pump molecules touch each other. The SR with its Ca pump sites pervades a muscle fiber so thoroughly that nowhere within the cytoplasm can a Ca^{2+} diffuse for a distance of more than 0.5 μm before colliding with SR membrane and being captured by a Ca-pump molecule. A pump molecule transfers two Ca ions in each cycle, and the total concentration of pump sites in rat gastrocnemius is estimated to be about 0.1 mM. If 0.4 mM Ca^{2+} are released into the myoplasm during a typical twitch, each pump molecule would have to turn over only twice for all the released Ca^{2+} to be reaccumulated.

Most of the Ca pump molecule protrudes out of the lipid bilayer into the cytoplasm. Pump molecules appear as approximately 4.5 nm particles in negatively stained SR membrane. They have a tendency to aggregate into oligomers and, under certain experimental conditions, even into two-dimensional crystals.

Biochemical Studies of Ca Uptake by the Sarcoplasmic Reticulum. Vesicular preparations of SR are readily obtained by differential centrifugation of homogenized muscle. In SR vesicles, 80% of the dry weight is protein, and 80% of this protein is the Ca-Mg ATPase. Most of the remaining protein is *calsequestrin*, a protein that is not embedded in the membrane but instead fills most of the lumen of the SR and functions as a Ca buffer, loosely binding the Ca^{2+} accumulated there. The SR of skeletal muscle is a naturally occurring, nearly pure preparation of a single membrane transport protein. It is not surprising that the Ca-Mg ATPase of SR is one of the most extensively studied transport proteins and also one of the first to be purified. The amino acid sequence of the protein is known.[23]

Ca transport is a multistep reaction that requires the hydrolysis of one ATP for every two Ca^{2+} transported. The first step is the reversible binding of two cytoplasmic Ca^{2+} to a pair of high-affinity binding sites on the cytoplasmic face of the pump, with half-saturation occurring at about 0.3 μM Ca^{2+}. In the next step, the pump accepts a phosphate from an ATP molecule; once this has occurred, the two bound Ca ions are "occluded," i.e. they can no longer dissociate into the cytoplasm. Next, the Ca ions are translocated by an as yet unknown mechanism to the luminal side of the membrane. The binding affinity is strongly reduced (dissociation constant in the millimolar range) and the two Ca ions leave the pump and diffuse into the SR lumen. The energy of the phosphorylated pump molecule is now spent, and a Mg-dependent dephosphorylation converts the enzyme into a form allowing it to once more capture a pair of Ca ions in the myoplasm. Each Ca pump can maximally transport about 20 Ca^{2+}/s, and half of this velocity is reached with 0.1 to 0.3 μM myoplasmic $[Ca^{2+}]$. The Ca pump can lower the myoplasmic free Ca^{2+} to or below 10 nM, and it can establish a Ca gradient of 1000-fold or greater across the SR membrane.

In cardiac muscle, the activity of the Ca pump is modulated by a 20,000-dalton regulatory peptide. This peptide, termed *phospholamban*, can be phosphorylated by a cAMP-dependent protein kinase. Phosphorylation of the peptide increases SR pump activity.

Ca Release. The rapid release of the Ca accumulated in the SR occurs through a Ca-permeable channel in the SR-membrane that connects the lumen of the SR with the cytosol and is controlled by the potential across the transverse tubule membrane. Details about the mechanism of Ca release have emerged only recently. The starting point was an older finding that Ca release from the SR can be stimulated by a rapid increase in cytosolic Ca^{2+} *(Ca-induced Ca release)*. A variety of adenine nucleotides, ATP among them, can also stimulate this release pathway, and Mg^{2+} is inhibitory. Later, SR vesicles incorporated into lipid bilayers were found to contain Ca-permeable channels that are activated by Ca and ATP and inhibited by Mg.[28]

SR vesicles were also found to contain binding sites of high affinity for the plant alkaloid ryanodine,[11a] a substance known to cause Ca-release from the SR under some conditions. The binding sites were localized to the portion of the SR participating in the "triad." Purified ryanodine receptors were found to be large molecules, each composed of probably four identical subunits of 400,000 MW and binding one molecule of ryanodine. When the protein was reconstituted into planar lipid bilayers, it was found to form Ca-permeable channels similar to those previously found in SR vesicles, that is, channels that were activated by Ca, ryanodine or ATP and inhibited by Mg.[19a, 20a] Under the electron microscope, the protein looked identical to the electron-dense "feet"[20a] that bridge the gap between transverse tubules and SR. Apparently, the SR-membrane Ca channel protein provides most of the material constituting the "feet." Probably only a small portion of this protein spans the SR membrane to form a channel, while the remaining large cytosolic domain may be involved in signal transduction between transverse tubule and channel. There is only about one "foot" for every 1000 Ca pump molecules, hence the Ca channel protein is present in only relatively small amounts.

"Ca-induced Ca release" suggests that Ca may be the key cytosolic messenger coupling the Ca-release channel to events in the cell membrane. Indeed, in the mammalian heart, Ca release from the SR is thought to be triggered by a sudden and small rise in cytosolic $[Ca^{2+}]$ caused by Ca influx through voltage-sensitive Ca channels in the plasma membrane.[11] In skeletal muscle, however, the high concentration of Mg^{2+} in the cytosol prevents cytosolic Ca^{2+} from being an effective activator of the Ca release channel, suggesting that this channel is regulated in a different way (see below).

A defective Ca release channel may be the cause of *malignant hyperthermia*, a hereditary disorder observed in humans and pigs. In individuals so afflicted, the general anesthetic *halothane* triggers Ca release from the SR in the entire skeletal musculature. The Ca in turn activates the contractile proteins, causing muscle rigidity and a massive heat development that results in high fever and frequently in death. The crisis can be reversed by giving *dantrolene*, a muscle relaxant known to attenuate Ca release during the twitch. Muscles from susceptible humans and pigs behave as though their SR contained Ca release channels that open aberrantly in response to halothane.

CONTROL OF Ca RELEASE BY THE MEMBRANE POTENTIAL IN THE TRANSVERSE TUBULES

Though the voltage dependence of contraction was already described in 1960,[18] the mechanism whereby the potential in the tubules controls Ca release from the SR has remained one of the major unsolved problems in muscle physiology. So far this topic has been studied mainly on intact fibers (or on pieces thereof), because it has been difficult to obtain useful cell-free preparations in which the functional coupling between tubules and SR remains intact. To control the membrane potential in the transverse tubules of intact fibers, one must prevent the large increases in Na and K conductances that normally accompany cell membrane depolarization. If these conductances are blocked pharmacologically, then voltage jumps imposed on the sarcolemma spread with little or no attenuation to the membranes of the TTS. Figure 7–6A, B shows an experiment on a short snake-muscle fiber. As the sarcolemma is depolarized with a voltage clamp, the fiber develops tension. The fiber contracts, but only while the sarcolemma (and hence the transverse tubule membrane) remains depolarized. The experiment suggests that Ca release remains at all times under the control of the tubule membrane potential, since repolarization is always and immediately followed by relaxation.

Ca release may be studied more directly by measuring the concentration of myoplasmic Ca by means of a Ca-sensitive dye. Figure 7–6C-E shows a voltage-clamp experiment on a fiber loaded with the dye Antipyrylazo III. Light absorption changes of this dye indicate that step depolarization (C) is accompanied by a maintained increase in myoplasmic $[Ca^{2+}]$ (D). On repolarization, $[Ca^{2+}]$ declines as release stops and as Ca^{2+} is reaccumulated by the SR. The Ca release mechanism must have remained active throughout the pulse, because myoplasmic $[Ca^{2+}]$ remains elevated even though Ca^{2+} is being taken from the myoplasm not only by the Ca pump of the SR but also by the various myoplasmic Ca-binding proteins. The rate at which Ca is taken up varies as a function of time and can be measured in separate experi-

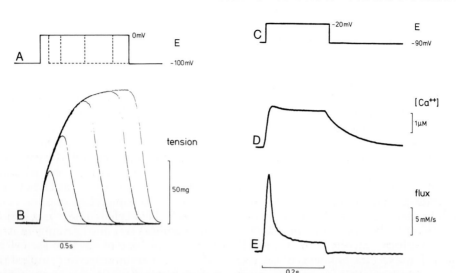

Figure 7–6 A, Membrane potential, and B, contraction tension, recorded from a voltage-clamped, short snake muscle fiber at about 20°C. Note that the fiber begins to relax immediately after repolarization. (After Heistracher, P.; Hunt, C.C. *J. Physiol.* 201:589–611, 1969.) C, Membrane potential; D, intracellular [Ca²⁺]; and E, time course of Ca flux out of the SR at 10°C in a voltage-clamped frog muscle fiber that had previously been loaded with a Ca-sensitive dye, Antipyrylazo-III. Trace E was calculated from trace D, and for the calibration it was assumed that 49 out of 50 Ca ions are bound immediately after being released and therefore do not contribute to raising myoplasmic [Ca²⁺]. Note that Ca release rises to a transient peak soon after depolarization but continues at a lower level for as long as the fiber is kept depolarized. Even the relatively low, maintained rate of Ca release is sufficient to sustain an essentially saturating myoplasmic [Ca²⁺]. (C, D, E, Modified from Melzer, W., et al. *The Biophysical Journal* 1984, vol. 45, pp. 637–641 by copyright permission of the Biophysical Society.)

ments. When the time derivative of trace D is formed and corrected for Ca uptake by the various components of the myoplasm, one obtains the rate at which Ca²⁺ leaves the SR through the release mechanism (Fig. 7–6E). Figure 7–6E represents the Ca current across the SR membrane that one might record if electrical studies of the intact SR were possible. Evidently, maintained depolarization causes an initial burst of Ca release that lasts tens of milliseconds. Then Ca release diminishes and is maintained at a level that is still sufficient to keep [Ca²⁺]ᵢ elevated for as long as the fiber stays depolarized. Similar observations were also made during trains of action potentials.[6] Evidently, the time course of Ca release may be complicated, but release may be terminated at any time by repolarization. In other experiments, release is found to grow in a graded fashion with the amplitude of depolarization, just as K and peak Na conductances do in a voltage-clamped axon. Ca release is under tight control of the tubule membrane potential and is evidently not a regenerative, all-or-none process in the way an action potential is.

How is the potential in the transverse tubular membrane sensed by the Ca-release mechanism? Early speculations have been influenced by the finding that each action potential is accompanied by a small Ca influx, perhaps through Ca channels in the tubule membrane. This influx is clearly much too small to cause significant activation of the contractile proteins, but it has been proposed that the influx may act as a "trigger," provoking a much larger Ca release from the SR. Indeed, an increase in myoplasmic [Ca²⁺] may, under certain conditions, stimulate Ca release (see the preceding discussion). A difficulty with this hypothesis is that skeletal muscle continues to twitch in Ca²⁺-free solutions in which no influx of "trigger Ca²⁺" can occur. Also, myoplasmic Ca²⁺ clearly cannot be the major variable controlling Ca release, since the process can be stopped by tubule repolarization even while myoplasmic [Ca²⁺] is high, much higher than it could ever rise as a consequence of the "trigger" influx. The Ca release that was terminated by repolarization in Figure 7–5E clearly could not have been under the control of myoplasmic Ca²⁺.

Another hypothesis was inspired by the discovery of gap junctions. Could the electron-dense spots in the tubule-SR junction be regions of electric contact between tubules and SR? If so, tubule depolarization could spread into the SR and open voltage-dependent Ca channels in the SR membrane, if such channels exist. The difficulty is that if the SR had electrical connections with the surface, the SR membrane should contribute to the total electrical capacitance of the cell membrane. Since in a typical fast-twitch fiber, there are roughly 100 cm^2 of SR membrane for every square centimeter of sarcolemma, all within 1 μm of the tubule-SR junctions, the membrane capacity should be 100 μF/cm^2 of sarcolemma, not the 4 to 10 μF/cm^2 observed experimentally. Clearly, electric coupling between tubules and SR is inconsistent with the electric behavior of the cell membrane.

"Charge Movements": a First Step in Tubule-SR Coupling? As discussed in Chapter 4, the opening and closing of voltage-dependent Na channels is accompanied by "gating currents," capacitive currents that result from the reorientation within the membrane of charged or dipolar portions of the Na-channel molecule. Similar gating currents are expected for any regulatory process that starts with a potential-dependent change of state in a cell membrane protein. Indeed, it is possible to record from skeletal muscle an electric signal, called "charge movement," that is similar to Na-channel gating currents except for having about a 100 times slower time course.[27] "Charge movements" are thought to represent the voltage-dependent rearrangements in a "voltage sensor," a tubule membrane macromolecule that controls the opening and closing of the Ca release channel in the SR.

Both charge movements and Ca release can be blocked by dihydropyridines and D-600, substances known to block voltage-sensitive Ca channels in cell membranes. It has been suggested that the voltage sensor is molecularly similar to a voltage-sensitive Ca channel in having the same drug binding sites.[25a] This would explain why there are many more high-affinity dihydropyridine receptors in skeletal muscle than there are Ca channels that can be activated to pass current. Dihydropyridine receptors in skeletal muscle are known to be voltage-sensitive,[27a] and may thus serve as voltage sensors.

Signaling Between the Cell Membrane and Intracellular Ca Stores. It is still unknown how the transverse tubules and the SR communicate. One hypothesis postulates that the electron-dense "feet" (Fig. 7–3B,C), now known to be the cyto-

plasmic domain of the Ca-release channel, directly link the tubule membrane to the release channel.[7] In reaching out to the tubule membrane, the "foot" might touch a dihydropyridine receptor, whose molecular rearrangements (visible as charge movements) are thus allosterically coupled to the Ca release channel. Others consider that one or more chemical messengers link the Ca channel in the SR to a voltage sensor in the tubule membrane.

Muscle is not the only tissue in which cell membrane events control the release of Ca from an internal store. In a variety of secretory cells (pancreatic acinar cells, parotid salivary gland, pancreatic beta cells), and in eggs, blood platelets, and lymphocytes, the binding of an agonist (or sperm in the case of eggs) to the cell membrane is followed by an increase in cytoplasmic [Ca^{2+}] that is due, at least partly, to release from an internal store. In many types of neurons, subsurface cisternae are found beneath the cell membrane; as in muscle, these are probably intracellular Ca stores controlled by receptive structures in the cell membrane. In sympathetic ganglion cells, Ca release from internal stores can be provoked by caffeine, a substance known to stimulate Ca release also from the SR of skeletal muscle.[22]

Wherever intracellular Ca stores are under the control of the cell membrane, a message must pass between them to stimulate the release of Ca^{2+}. In many cases the messenger is *inositol 1,4,5-triphosphate*, IP$_3$ (Chap. 1). Intracellular application of this compound in micromolar concentration has been reported to release Ca^{2+} from internal stores of pancreatic acinar cells, liver cells, macrophages, and even vascular smooth muscle cells. It is thought that in these cells, the formation of IP$_3$ is initiated by the binding of a "first messenger," e.g., a hormone, to a cell membrane receptor. The receptor in turn activates a membrane-bound phosphodiesterase. This enzyme hydrolyzes phosphatidylinositol 4,5-bisphosphate, a lipid found in the cell membrane bilayer, to form IP$_3$ and diacylglycerol, both thought to be intracellular messengers. IP$_3$ is water soluble and so may reach cytoplasmic Ca stores, whereas diacylglycerol probably remains in the plasma membrane to activate a membrane-associated protein kinase, C-kinase.

Recent evidence suggests that IP$_3$ may mediate Ca release when contractions of vascular smooth muscle are induced by norepinephrine rather than by cell membrane depolarization.[29] IP$_3$ has been reported to stimulate Ca-release even if applied to skinned skeletal muscle fibers. The finding has led to the suggestion that the tubule membrane of skeletal muscle contains a phospholipase that can generate IP_3, and that a voltage-dependent activation of this enzyme is an early step in initiating Ca release.[31] However Ca release from the SR is initiated within milliseconds, while in all known examples of IP_3-controlled Ca release, many seconds are required for the coupling between stimulus and Ca release. Hence it seems unlikely that IP_3 is the major mediator of Ca release in skeletal muscle.

SUMMARY

Muscle contraction is regulated by intracellular Ca^{2+}. Ca^{2+} is sequestered into an intracellular store, the sarcoplasmic reticulum (SR) by an ATP-driven Ca pump. During activity, Ca^{2+} is released from the SR by a Ca channel that is controlled by the potential across invaginated portions of the cell membrane. Jointly, Ca pump and Ca channel in the SR membrane control the distribution of intracellular Ca^{2+}.

In virtually all muscle cells, the cell membrane also contains Ca channels and ATP-driven Ca pumps. In the long run, these control the Ca content of the cell. They are biochemically and pharmacologically distinct from the Ca transport proteins of the SR membrane.

Vertebrate skeletal muscle fibers generate sodium-dependent action potentials in the same way as do nerve axons. In addition to the SR, they contain another intracellular membrane system, the transverse tubular system (TTS). Transverse tubules are invaginations of the cell membrane that conduct the action potential to the fiber interior. In the larger mammals, transverse tubules are found also in cardiac muscle.

In vertebrate skeletal muscle, Ca release from the SR is controlled by the membrane potential in the TTS. Depolarization of the transverse tubule membrane initiates Ca release, and repolarization stops Ca release. The functional coupling between TTS and SR membrane is not understood at the molecular level. It may involve an intracellular messenger.

ANNOTATED BIBLIOGRAPHY

Berridge, M.J.; Irvin, R.F. Inositol triphosphate, a novel second messenger in cellular signal transduction. *Nature* 312:315–321, 1984.
This article reviews the inositide metabolism and the evidence for inositol triphosphate and diacyl glycerol being second messengers. Emphasis is given to the hypothesis that inositol triphosphate is a

universal messenger that stimulates Ca release from the endoplasmic reticulum and the organelles derived from it.

Martonosi, A.N. Mechanisms of Ca^{2+} release from sarcoplasmic reticulum of skeletal muscle. *Physiol. Rev.* 64:1240–1320, 1984.
This review covers the biochemistry and function of the sarcoplasmic reticulum and also reviews several hypotheses concerning the mechanism of Ca release from this organelle.

Peachey, L.D.; Franzini-Armstrong, C. Structure and function of membrane systems of skeletal muscle cells. *Handbk. Physiol.*, Sec. 10:23–73, 1983.
This review covers the morphology of membrane systems in skeletal muscle and includes many electron micrographs.

REFERENCES

1. Adrian, R.H.; Bryant, S.H. On the repetitive discharge in myotonic muscle fibers. *J. Physiol. (Lond.)* 240:505–515, 1974.
2. Affolter, H.; Coronado, R. Agonists Bay-K8644 and CGP-28392 open calcium channels reconstituted from skeletal muscle transverse tubules. *Biophys. J.* 48:341–347, 1985.
3. Almers, W. Potassium concentration changes in the transverse tubules of vertebrate skeletal muscle. *Fed. Proc.* 39:1527–1532, 1980.
4. Almers, W.; Palade, P.T. Slow calcium and potassium currents across frog muscle membrane: Measurements with a vaseline-gap technique. *J. Physiol. (Lond.)* 312:159–176, 1981.
5. Armstrong, C.M.; Bezanilla, F.M.; Horowicz, P. Twitches in the presence of ethylene glycol bis(beta-aminoethyl ether)N,N'-tetraacetic acid. *Biochim. Biophys. Acta* 267:605–608, 1972.
6. Baylor, S.M.; Chandler, W.K.; Marshall, M.W. Sarcoplasmic reticulum calcium release in frog skeletal muscle fibres estimated from arsenazo III calcium transients. *J. Physiol. (Lond.)* 344:625–666, 1983.
7. Chandler, W.K.; Rakowski, R.F.; Schneider, M.F. Effects of glycerol treatment and maintained depolarization on charge movement in skeletal muscle. *J. Physiol (Lond.)* 254:285–316, 1976.
8. Curtis, R.M.; Catterall, W.A. Purification of the calcium antagonist receptor of the voltage sensitive calcium channel from skeletal muscle transverse tubules. *Biochemistry* 23:2113–2118, 1984.
9. Eisenberg, B. Quantitative ultrastructure of mammalian skeletal muscle. *Handbk. Physiol.* Sec. 10:73–112, 1983.
10. Eisenberg, R.S.; Gage, P.W. Ionic conductances of the surface and transverse tubular membrane of frog sartorius fibers. *J. Gen. Physiol.* 53:279–297, 1969.
11. Fabiato, A. Calcium-induced release of calcium from the cardiac sarcoplasmic reticulum. *Am. J. Physiol.* 245:C1–C14, 1983.
11a. Fleischer, S.; Ogunbunmi, E.M.; Dixon, M.C.; Fleer, E.A.M. Localization of Ca^{2+} release channels with ryanodine in junctional terminal cisternae of sarcoplasmic reticulum of fast skeletal muscle. *Proc. Natl. Acad. Sci. USA* 82:7256–7259, 1985.
12. Fosset, M.: Jaimovich, E.; Delpont, E.; Lazdunski, M. [³H]Nitrendipine receptors in skeletal muscle. *J. Biol. Chem.* 258:6086–6092, 1983.
13. Franzini-Armstrong, C.; Porter, K. Sarcolemmal invaginations constituting the T-system in fish muscle fibers. *J. Cell Biol.* 22:675–696, 1964.
13a. Gage, P. W.; Eisenberg, R. S. Action potentials, afterpotentials and excitation-contraction coupling in frog sartorius

fibers without transverse tubules. *J. Gen. Physiol.* 53:298–310, 1969.

14. Gonzales-Serratos, H. Inward spread of activation in vertebrate muscle fibers. *J. Physiol. (Lond.)* 212:777–799, 1971.

15. Ruff, R.L.; Gordon, A.M. Disorders of muscle. The periodic paralyses. In Andreoli, T.E.; Hoffman, J.F.; Fanestil, D.D.; Schultz, S.G.; eds. *Physiology of Membrane Disorders.* New York, Plenum Publishing Corp., 1986.

16. Grafe, P.; Ballanyi, K.; Küther, G. Intracellular Na^+ activity in normal human intercostal muscle and in one case of hyperkalemic periodic paralysis. *Pflugers Arch.* 403:817–829, 1985.

17. Grocki, K. The fine structure of the deep muscle lamellae and their sarcoplasmic reticulum in *Branchiostoma lanceolatum. Eur. J. Cell Biol.* 28:202–212, 1982.

18. Hodgkin, A.L.; Horowicz, P. Potassium contractures in single muscle fibres. *J. Physiol. (Lond.)* 153:386–403, 1960.

19. Huxley, A.F.; Taylor, R.E. Local activation of striated muscle fibres. *J. Physiol. (Lond.)* 144:426–441, 1958.

19a. Imagawa, T.; Smith, J.S.; Coronado, R.; Campbell, K.P. Purified ryanodine receptor from skeletal muscle sarcoplasmic reticulum is the Ca-permeable pore of the calcium release channel. *J. Biol. Chem.* 262:16636–16643, 1987.

20. Jewett, P.H.; Sommer, J.R.; Johnson, E.A. Cardiac muscle. Its ultrastructure in finch and hummingbird with special reference to the sarcoplasmic reticulum. *J. Cell Biol.* 49:50–65, 1971.

20a. Lai, F.A.; Erickson, H.P.; Rosseau, E.; Liu, Q.-Y.; Meissner, G. Purification and reconstitution of the calcium release channel from skeletal muscle. *Nature* 331:315–319, 1988.

21. Lau, Y.H.; Caswell, A.H.; Garcia, M.; Letellies, L. Ouabain binding and coupled sodium, potassium and chloride transport in isolated transverse tubules of skeletal muscle. *J. Gen. Physiol.* 74:335–349, 1979.

22. Lüttgau, H.C.; Oetliker, H. The action of caffeine on the activation of the contractile mechanism in striated muscle fibres. *J. Physiol. (Lond).* 194:51–74, 1968.

23. MacLennan, D.H.; Brandl, C.J.; Korczak, B.; Green, N.M. Amino-acid sequence of a Ca^{2+} + Mg^{2+}-dependent ATPase from rabbit muscle sarcoplasmic reticulum, deduced from its complementary DNA sequence. *Nature* 316:696–700, 1985.

24. Martonosi, A.N.; Beeler, T.J. Mechanism of Ca^{2+} release from sarcoplasmic reticulum of skeletal muscle. *Handbk. Physiol.*, Sec. 10:417–485, 1983.

24a. McMahan, U.J.; Edgington, D.R.; Kuffler, D.P. Factors that influence regeneration of the neuromuscular junction. *J. Exp. Biol.* 89:31–42, 1980.

25. Rüdel, R.; Lehmann-Horn, F. Membrane changes in cells from myotonia patients. *Physiol. Rev.* 85:310–356, 1985.

25a. Rios, E.; Brum, G. Involvement of dihydropyridine receptors in excitation-contraction coupling in skeletal muscle. *Nature* 325:717–720, 1987.

26. Sanchez, J.A.; Stefani, E. Inward calcium current in twitch muscle fibres of the frog. *J. Physiol. (Lond.)* 283:197–209, 1978.

27. Schneider, M.F.; Chandler, W.K. Voltage-dependent charge movement in skeletal muscle: a possible step in excitation-contraction coupling. *Nature* 242:244–246, 1973.

27a. Schwartz, L.M.; McCleskey, E.W.; Almers, W. Dihydropyridine receptors in muscle are voltage-dependent but most are not functional Ca channels. *Nature* 314:747–751, 1985.

28. Smith, J.S.; Coronado, R.; Meissner, G. Single channel measurements of the calcium release channel from skeletal muscle sarcoplasmic reticulum. *J. Gen. Physiol.* 88:573–588, 1986.

29. Somlyo, A.V.; Bond, M.; Somlyo, A.P.; Scarpa, A. Inositol trisphosphate–induced calcium release and contraction in vascular smooth muscle. *Proc. Natl. Acad. Sci. USA* 82:5231–5235, 1985.

30. Vergara, C.; Latorre, R. Kinetics of Ca-activated K^+ channels from rabbit muscle incorporated into planar bilayers. Evidence for Ca^{2+} and Ba^{2+} blockade. *J. Gen. Physiol.* 82:543–568, 1983.

31. Vergara, J.; Tsien, R.Y.; Delay, M. Inositol 1,3,4,-trisphosphate: a possible chemical link in excitation-contraction coupling in muscle. *Proc. Natl. Acad. Sci. USA* 82:6352–6356, 1985.

Molecular Basis of Contraction

INTRODUCTION

Muscle cells are specialized to change stored chemical energy into mechanical energy for movement or production of tension. This chapter discusses the molecular mechanism of contraction in striated skeletal muscles—muscles whose contractile proteins are packed into regularly organized arrays, giving each cell the striated or banded appearance described in Chapter 7. Contraction in other tissues lacking such organized arrangement (smooth muscle and nonmuscle contractile cells) is discussed in Chapter 10. The mechanism by which muscles contract has fascinated scientists for centuries and has been the subject of numerous theories, many of which seem fanciful indeed today. Galen, the Greek physician and writer who lived from 130 to 200 A.D., taught that animal

spirits pass down the nerve to the muscles, swelling the muscles and causing their ends to approximate. This theory was based on the observation that nerve section causes muscle paralysis. Later, this hydraulic theory was modified by the postulate that "nitro-aerial spirits" (oxygen) transverse the nerve to react in the muscle with "salino-sulfurous particles" brought there by circulation. The reaction was thought to produce effervescence and swelling of the muscle. The supporting evidence for this theory was that interruption of the circulation of a muscle leads to paralysis. This swelling theory of contraction was finally disproved by Jan Swammerdam from experiments done around 1700.

Later, springs, dashpots, folding filaments, and other elements formed the basis of theories of contraction. The modern view is that muscle contraction results from molecular interactions between the contractile proteins actin and myosin contained in filaments and that this interaction causes the longitudinally oriented filaments to slide by one another, thus shortening the muscle. This modern theory is based on a detailed understanding of the structure and molecular properties of muscle constituents. There are a number of recent reviews of the molecular basis of contraction.[4, 4a, 42]

Muscle contraction occurs in the sarcomeres, the repeating structures that constitute the alternating bands. The nomenclature of these bands, discussed in Chapter 7, is illustrated again in Figure 8–1. Electron micrographs (Fig. 8–2) reveal that the bands are created by the overlap of macromolecular filaments.

MYOFILAMENTS AND ATTACHMENTS

Electron micrographs of muscle fibers reveal intracellular thick and thin filaments (Figs. 8–2; 8–3; see also Figs. 7–1 and 8–11). Thick filaments,

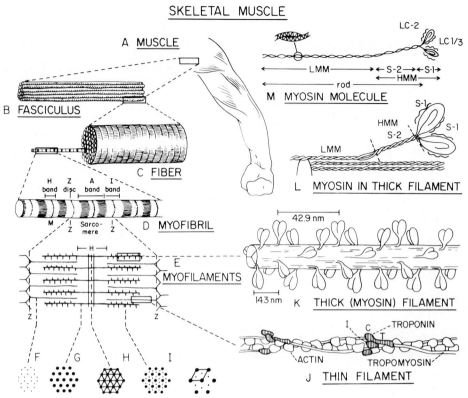

Figure 8–1 Diagram of muscle structure. *A*, the deltoid muscle; *B*, fasciculus, or bundle of muscle fibers, from this muscle. *C*, An individual, striated muscle fiber; *D*, a myofibril from this fiber showing the striated, or banded, pattern. *E*, A sarcomere showing overlapping thick and thin filaments; *F–I*, transverse cross sections through the sarcomere. Notice the hexagonal arrangement of filaments in G-I and the inset to the right of I, where the unit cell is outlined. *J*, The thin filament with the component regulatory proteins from fast, striated vertebrate muscle. *K*, Thick filament with proposed helical arrangement of myosin heads. *L*, Arrangement of the myosin molecules in a thick filament; the components of the molecule (HMM, S-1, S-2, and LMM) are indicated. *M*, diagram of single myosin molecule with the two light chains associated with each S-1 component. (Redrawn from Bloom, W.; Fawcett, D. W. *A Textbook of Histology*, 10th ed., Philadelphia: W. B. Saunders Co., 1975.)

up to 1.6 μm long, have lateral projections along much of their length (Figs. 8–3), except for a bare region in the center. The center of the filament is fatter at the site of the M-line; the ends are tapered. Thin filaments appear as twisted double strands resembling a two-stranded, right-hand rope. They measure up to 1 μm in length in isolated specimens or up to about 2 μm when the Z-line attachments are included (Figs. 8–1; 8–2).

Transverse sections through vertebrate muscle in the region of filament overlap reveal that the filaments are arranged in a hexagonal lattice (Fig. 8–1; also see Fig. 8–13). Each thin filament is surrounded by three thick filaments, and each thick filament by six thin filaments. This packing pattern gives a ratio of two thin filaments for every thick filament, a ratio that can be obtained by counting the number of filaments and partial filaments within the "unit cell" outlined near the bottom of Figure 8–1.

In other muscles, there is a different arrangement; for example, in insect muscle, the thin filaments occupy an intermediate position between two consecutive thick filaments. In other muscles, there may be more than six thin filaments surrounding each thick filament, or the thick filament may be thicker, indicating more molecules packed in per unit length.

Muscles fatten as they shorten. X-ray diffraction studies show that during muscle shortening the distance between the filaments changes inversely with sarcomere length (actually, the square root of the sarcomere length) so that the volume of the filament lattice remains constant. That is, as sarcomere length decreases, the separation between the interacting filaments increases. In mammalian muscle at a sarcomere length of about 2.5 μm, the center-to-center separation of neighboring thick filaments is about 40 nm and that of neighboring interacting thick and thin filaments, 24 nm. For

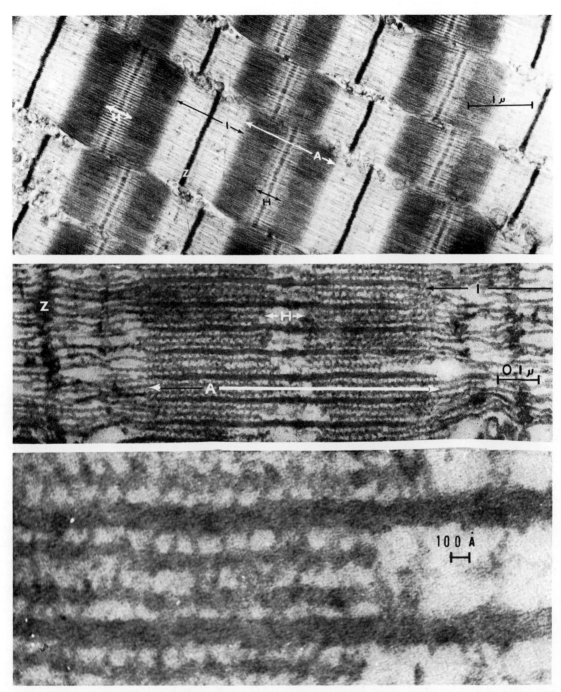

Figure 8–2 Electron micrographs of longitudinal sections of rabbit psoas muscle. *Top,* Low-power view of several sarcomeres, showing A, H, and I bands and Z line. *Center,* One sarcomere. Thick filaments run from one end of the A band to the other. Pairs of thin filaments are attached at the Z line, interdigitate with the thick filaments, and terminate at the edge of the H band. *Bottom,* High magnification of A band showing cross bridges between thick and thin filaments. (From Huxley, H. E. In Brachet, J.; Mirsky, A. E., eds. *The Cell,* vol. 4. New York, Academic Press, 1960.)

the filament dimensions given, the distance between surfaces of the backbone of the filaments is about 28 nm for neighboring thick filaments and 12 nm from a thick filament to the nearest thin filament at this sarcomere length. However, the important 20-nm lateral projections from the backbone of the thick filaments discussed below (Fig. 8–3) are long enough to bridge the gap to the thin filament.

The thick filament is composed of many molecules of the protein *myosin* along with some other minor proteins, such as the C protein and the M-line protein. The thin filament is made up of the major protein *actin* and the regulatory proteins *tropomyosin* and *troponin*. The location of the various proteins in the filament has been determined in several ways. One is by selective extraction; myosin and thick filaments are removed by soaking myofibrils in a concentrated salt solution and seeing what part of muscle cell remains. Another method is to treat muscle with antibodies against a contractile protein—antibodies that are labeled with either fluorescent or electron-dense markers so the antibody-antigen complex can be visualized within the fiber in light or electron micrographs respectively. A third method is to show that the purified isolated proteins reassemble *in vitro* to form structures identical to the native filament.

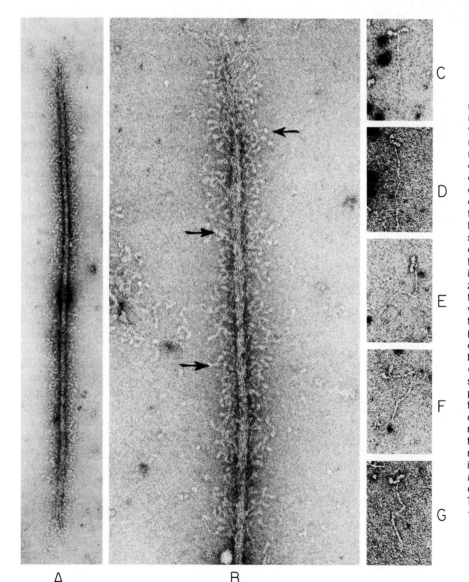

A   B   C   D   E   F   G

Figure 8–3 Thick filaments and myosin molecules as visualized in the electron microscope with negative stain. Thick filaments (*A* and *B*) and individual myosin molecules (*C–G*) were dried onto specially activated carbon films in the presence of a relaxing solution and uranyl acetate. The uranyl ions provide the dark staining around the lighter myosin molecules. *A* shows quite clearly the thick filament tapering at the ends and the projecting myosin molecules in all regions except the bare zone in the center (magnification: 104,000 ×). *B*, a higher power view of a thick filament with the two-headed myosin molecules (arrows) linked to the backbone of the thick filament clearly visible (magnification 215,000 ×). *C–G*, individual myosin molecules, showing the two heads with no preferred orientation and the two regions where sharp bends can occur in the tail (about 44 nm and 76 nm from the head-tail junction). (*A* and *B*, electron micrographs courtesy of J. Trinick, from Knight, P.; Trinick, J. *J. Mol. Biol.* 177:461–482, 1984; and *C–G* from Walker, M., Knight, P., Trinick, J. *J. Mol. Biol.* 184:535–542, 1985.)

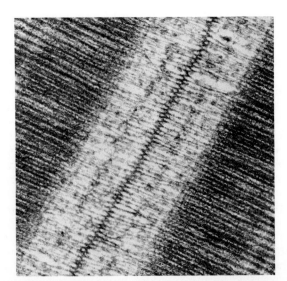

Figure 8–4 Electron micrographs of a longitudinal section through guppy tail muscle. The Z line goes diagonally across the micrograph in the center of the I band. The Z line is narrow, showing a distinct zigzag pattern. The thin filaments can be followed from one side to the Z line, where they terminate in contact with denser filaments; the latter in turn connect to thin filaments coming from the other side of the I band. Also visible in the upper left hand corner in the middle of the A band are the lines making up the M lines. (Courtesy of C. Franzini-Armstrong.)

The detailed organization of the proteins forming both the thin and thick filaments is considered later, after the discussion of the properties of the individual proteins.

The thick filaments are linked together in the center of the A band by the M line structures, which are radial projections that bind the central regions of thick filaments together. At least two proteins are found in the M line. One is an enzyme that is important for muscle energetics, *creatine phosphokinase* (CPK), (also called creatine phosphoryl transferase). The other protein, myomesin, is probably responsible for the structural attachment between thick filaments.

Several models of the structure of the Z line of muscle have been proposed; none appears to be applicable to all Z lines. The simplest form of Z line observed is the zig-zag pattern shown in Figure 8–4. It is supposed that this pattern is formed by thin filaments from adjacent sarcomeres that are cross-linked at the Z line by the protein α-actinin.

Filaments other than the classic thick and thin filaments can be detected, especially in muscle fibers extended to sarcomere lengths, at which the thick and thin filaments no longer overlap. Some of these run parallel to the thick and thin fila-

ments, attaching two Z lines to one another or attaching the thick filaments to the Z line. They contain the proteins titin and nebulin and probably function to hold the sarcomere and myofibrils together, providing additional elasticity to the sarcomere.

MUSCLE PROTEINS: PROPERTIES AND LOCALIZATION

The back muscles of rabbits have been a favorite preparation for muscle biochemistry. The proteins found are listed in Table 8–1 along with their location in the sarcomere, molecular weight, and approximate proportion by weight. The values should be taken only as examples; the concentration and properties of muscle proteins vary, depending on the muscle studied and the method of determination. The data in Table 8–1 were determined by selective protein extraction.

Muscle proteins may be subdivided into the contractile proteins, myosin and actin; the regulatory proteins, tropomyosin and troponin; and additional structural proteins, such as α-actinin, titin, nebulin, and myomesin. Other proteins, present in small quantities, are involved in cell

Table 8–1 Protein Composition of the Myofibril

Component	Location	Molecular Weight	%
Myosin	Thick filaments	485,000	44
Actin	Thin filaments	42,000	22
Tropomyosin	Thin filaments	68,000	5
Troponin	Thin filaments	70,000	5
TnC		18,000	
TnI		21,000	
TnT		31,000	
Titin	Gap filament	700,000–1,000,000	10
Nebulin	N line	600,000	5
C-Protein	Thick filament	140,000	2
Myomesin	M line	165,000*	2
α-Actinin	Z line	180,000–190,000	2
β-Actinin	Thin filament	71,000	<1
γ-Actinin	Thin filament	35,000*	<1
Eu-Actinin	Z line	42,000	<1
Creatine kinase	M line	84,000	<1
55 kDalton protein	Z line	55,000*	<1
H protein	Thick filament	74,000	<1
I protein	A band	50,000	<1
Filamin	Z band	480,000	<1
Desmin	Peripheral	55,000*	<1
Vimentin	filaments of	58,000	<1
Synemin	the Z line	280,000	<1

*Subunit weight.

metabolism, protein synthesis, membrane ion transport, and other general cellular processes.

The major contractile proteins are really multigene families. Each mammalian cell contains genes that code for at least 11 different myosins, and 6 different actins and numerous variations of the other contractile proteins. Different isoforms of myosin and actin appear at sequential times in developing muscle and during adaptation to various physiological and pathophysiological conditions. Table 8–2 is a partial list of the major isoforms of the myosin heavy chain (see below) from rat muscle. These different isoforms have slightly different properties, but there is great similarity in the amino acid sequence, with most substituted amino acids having similar structures. Consequently, the structural and molecular properties are similar as well. This section discusses the general properties of the protein families of adult fast skeletal muscle. Other isoforms are discussed later in connection with the specific tissues in which they predominate.

Myosin. The myosin molecule is a hexamer composed of 2 heavy chains with a molecular weight (M.W.) of approximately 200,000 each and 4 light chains of about 20,000 M.W. each. Complete or partial amino acid sequences have been determined for myosins derived from different tissues and different species. The myosin molecule resembles two golf clubs (heavy chains) with their shafts twisted helically around one another as in Figures 8–1M and 8–3. The tail of the molecule consists of the twisted, α-helical long chains (shafts) and the heads of the clubs are the globular ends of the two chains.[29] Two light chains are associated with each head. The shaft or tail of the molecule can be dissociated from the heads by partial digestion with the proteolytic enzyme trypsin; these digestion products are called light mero-

myosin (LMM) (tail only) and heavy meromyosin (HMM) (heads plus neck region). An essential biochemical property of myosin is its ability to catalyze the hydrolysis of ATP to form ADP and inorganic phosphate—an energy yielding reaction. The ATPase activity is confined to the heads, which also contain the actin binding sites crucial for contraction. Heavy meromyosin (HMM) can be further fractionated by digestion with the proteolytic enzyme papain, yielding two products called S-1 and S-2; each molecule of HMM contains two S-1 molecules and one S-2 molecule. The S-1 fraction is the globular portion or heads of the original myosin, while the S-2 fraction derives from the necks that connect the heads to the tails. Both the ATPase activity and the actin binding ability are associated with the S-1 fraction. Furthermore, each of the heads of the S-1 fraction has associated with it two low–molecular weight proteins or light chains (not to be confused with light meromyosin). By appropriate chemical manipulation, the light chains can be removed from the S-1 heads. The light chains in combination with the heavy chains control the ATPase activity and actin-binding characteristics of myosin. Properties of the light chains in different muscle types are summarized in Table 8–3.

Under appropriate ionic conditions, myosin molecules polymerize to form filamentous structures that closely resemble naturally occurring thick filaments (Figs. 8–1, 8–2, 8–3).[22] The basic structure of the thick filament is a bipolar one in which the myosin molecules point in opposite directions at each end of the filament. The tails (LMM portion) of myosin adhere side to side to form the backbone of the filament; the heads project laterally every 14.3 nm. The result is a bundle with knobby, lateral surfaces (Fig. 8–3). The molecules are staggered within the bundle so that the heads repeat at regular intervals of 42.9 nm (frog and rabbit muscle) and are helically arranged along the length of the structure so that heads appear every 14.3 nm along the helical path (Figs. 8–1, 8–3; see also Fig. 8–12).[26] The complete filament consists of two such bundles interlaced tail-to-tail within the central region, so that the laterally projecting heads, present everywhere else along the thick filament, are lacking in this central junctional region, the bare zone.

Actin. Actin is a globular, peanut-shaped protein. It is readily extractable not only from muscle but possibly from all other types of tissue; the amino acid sequences of actin from different sources are similar but not identical. Under appropriate ionic conditions, extracted globular actin (G-

Table 8–2 Distinct Mammalian Myosin Heavy-Chain Isoforms Determined from Proteins and mRNA

Source	Number	Type	Same as
Skeletal Muscle			
Fast Twitch	2	IIA, IIB	
Slow Twitch	1	I	Cardiac V_3
Embryonic	1	Embryonic	
Neonatal	1	Neonatal	
Tonic extraocular	1	Tonic	
Cardiac Muscle			
Ventricle	2	V_1, V_3 (α,β)	V_3 = slow twitch, I
Atrium	1	V_1 (α)	Ventricle V_1
Smooth Muscle	1		?
Nonmuscle	2		?

Table 8–3 Myosin Light Chains

Muscle	Light Chain	Molecular Weight (Approx)	Function	Number Per S–1
Skeletal fast				
DTNB*	LC_{2F}	18,000	(Ca²⁺ binding) (phosphorylatable but function?)	1
Alkali†	LC_{1F}	21,000		1
	LC_{3F}	17,000	Modify myosin ATPase	
Cardiac slow	LC_{1S}	27,000	Modify myosin ATPase	1
Skeletal	LC_{2S}	18,000–20,000	(Phosphorylatable, but function?) (Ca²⁺ binding)	1
Smooth	LC_P	20,000	Phosphorylated under Ca²⁺ control	1
	LC_1	17,000	?	1
Scallop skeletal				
EDTA‡	LC_R	17,000	Ca²⁺ sensitivity	1
SH§	LC_1	17,000	?	1

*Dissociated by 5,5′ dithiobis - (2-nitrobenzoic acid). †Dissociated by alkali solutions. ‡Dissociated by EDTA (ethylenedinitrilo-tetra acetic acid). §SH-containing. The initial nomenclature for the light chains was based on the specific extracting solution. The present nomenclature is according to the relatiive molecular weight and muscle of origin, whether it is phosphorylatable or not, and its function.

actin) polymerizes in vitro to form a filamentous (F-actin) structure very similar to the backbone of the thin filament: a two-stranded, twisted chain of peanuts. The periodicity of the globules along a single strand is one every 5.4 nm, but since the strands are staggered half-a-peanut with respect to each other, there is an actin every 2.7 nm along the thin filament. The distance occupied by one complete twist of the paired chains is about 36.5 nm, which is different from the 42.9-nm periodicity of the thick filament. Thus the projections from the thick filament that interact with the thin filament do not all "see" the same arrangement of actins[26] (see Fig. 8–12).

Actin binds strongly with myosin in vitro and in vivo and activates the ATPase of the myosin head. Both reactions are important in muscle contraction and are discussed in more detail later.

Regulatory Proteins Tropomyosin and Troponin. Tropomyosin is a rod-shaped molecule about 40 nm long, formed of two twisted α-helical subunits. In the thin filaments of vertebrate skeletal muscle, tropomyosin lies along but not packed into the depth of the groove formed by the two helical actin strands (see Fig. 7–8). One tropomyosin molecule spans about seven actins and overlaps with each neighboring tropomyosin at either end. In relaxed skeletal muscle, the tropomyosin molecule is thought to be strategically located to modify the binding of the myosin molecule to actin and thereby to suppress its ATPase activity. For contraction to occur, the tropomyosin must move deeper into the groove of the thin filament, a change that modifies the actin-myosin

binding and enhances the myosin ATPase activity (see Fig. 8–17).[25, 44]

Troponin is a globular calcium-binding protein composed of three subunits. In vertebrate thin filament, molecules of troponin are bound to the tropomyosin ribbon at intervals of about 40 nm, the binding being accomplished through the subunit called TnT. Two other troponin subunits, TnC (calcium troponin) and TnI (inhibitory troponin), bind calcium and inhibit actin activation of myosin ATPase, respectively. When Ca²⁺ is bound to troponin, the molecule changes its configuration, allowing the tropomyosin to move farther into the actin groove. Consequently, the myosin-actin interaction is modified such that the ATPase activity of the myosin heads is greatly enhanced (see Fig. 8–17).

TnC is a dumbbell-shaped molecule about 8 nm long. It binds 4 Ca²⁺ ions, 2 at either end of the dumbbell. The pairs of binding sites located at opposite ends of the molecule are not identical; those at one end have a higher affinity for Ca²⁺ and bind Mg²⁺ competitively with Ca²⁺, while those at the other have a lower affinity for Ca²⁺ and are more specific for Ca²⁺. The two ends of the molecule are connected by a nine-turn α-helix that allows some flexibility and bending. TnI is a globular subunit that binds TnC. When it is added in vitro to actin, even in the absence of tropomyosin, it suppresses the ability of actin to activate the myosin ATPase, an action suggesting that TnI can interfere with the ATPase-activating sites on actin. In this action, TnI is much less effective molecule for molecule than tropomyosin plus TnI, because one tropomyosin molecule stretches along the thin filament affecting a number of binding sites, whereas TnI affects only one site.

The controlling factor in triggering muscle contraction is the intracellular free calcium concentration. In resting muscle, the concentration is about 10^{-7}M. When the concentration rises (e.g., to 10^{-5}M) in vertebrate striated muscle, calcium combines with troponin; the troponin reorients, pulling tropomyosin into the actin groove, altering the myosin binding to actin, and allowing the myosin ATPase to become active (see Fig. 8–17). The mechanism underlying the intracellular Ca^{2+} increase during a muscle action potential has been described in Chapter 7.

Structural Proteins. In addition to the four proteins discussed above, several others have been isolated from muscle (Table 8–1). They include alpha- and beta-actinin. Alpha-actinin is an architectural constituent of the Z line. Beta-actinin appears to regulate the polymerization of actin and thus determines thin filament length. The M-line proteins (myomesin and creatine phosphokinase) reside in the middle of the A band and tie together the thick filaments and provide for the enzymatic phosphorylation of creatine. C protein, localized in the A band, appears to form bands along the thick filament.

Titin and nebulin, very large structural proteins, form the elastic filaments that probably stabilize sarcomeres and contribute to the internal elasticity of muscle cells.

SLIDING FILAMENT THEORY

When a muscle shortens, the thick and thin filaments slide by one another, the length of each filament remaining fixed. This shortening is hypothesized to be due to the sequential formation and breakage of cross bridges between thick and thin filaments. The bridges pull the thin filaments toward the center of the sarcomere through a sequence of events that couples the hydrolysis of ATP with movement. Although each sarcomere can at most shorten by only one micrometer from its rest length, a fiber containing many sarcomeres in series can shorten appreciably.

The sliding filament hypothesis was independently proposed by A.F. Huxley and Niedergerke[18] and by H.E. Huxley and Hanson[27] over a quarter of a century ago. Their papers created a revolution in muscle physiology, for most previous models of muscle contraction were built on the concept of folding or coiling of the protein molecules composing the filaments. The Huxleys and their collaborators showed that when a muscle fiber contracts, the length of its filaments remains almost constant, and they therefore concluded that the interdigitating thick and thin filaments must slide between one another as the sarcomere shortens.

Some of the important evidence for the sliding of filaments is diagrammed in Figure 8–5. High-power electron micrographs, taken with special precautions against shrinkage artifacts, show no significant difference in filament length in muscle specimens fixed during varying degrees of contraction within the physiological range of sarcomere shortening.[21] Light micrographs also show that as a muscle contracts (shortens), the A band (thick filament length) remains constant while the I and H bands decrease in length in proportion to the sarcomere shortening. These data show clearly that muscle shortening is accompanied by an increased overlap of thick and thin filaments.[19]

Further support for the constancy of filament

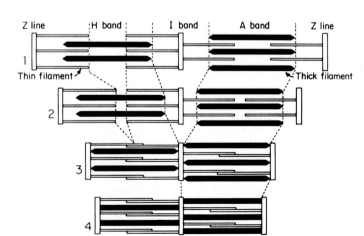

Figure 8–5 Diagrammatic representation of filament positions in muscle at different sarcomere lengths. As the sarcomere shortens, the thick and thin filaments remain at constant length; all the length change occurs by a relative sliding of the two sets of filaments. Both the *I* and the *H* bands decrease in width until they are indistinguishable as bands. The diagrams are constructed from electron micrographs taken at various sarcomere lengths. (From Huxley, H. E. Muscle cells. In Brachet, J.; Mirsky, A. E., eds. *The Cell*, vol. 4. New York, Academic Press, 1960.)

length comes from studies employing x-ray diffraction, a technique that allows one to measure periodicities or repeat spacings of structures along the filaments. The results indicate little or no change in the periodicities of filament structures as the sarcomeres shorten during activation, a finding inconsistent with any hypothesis of contraction that features shortening of filaments.

The described changes in striation pattern refer to fibers with sarcomere lengths greater than 2.0 μm. At shorter lengths, the microscopic picture is complicated by overlap or collision between the thin filaments from the two ends of the sarcomere. If the sarcomere actively shortens to less than 1.6 μm, the ends of the thick filaments butt against the Z lines. Under these conditions, accurate measurements of filament length are difficult, but there is no reason to suspect that even then filament lengths are significantly altered.

FILAMENT INTERACTION[2, 4a, 11]

Although the sliding-filament theory provides a reasonable picture of the mechanics of muscle contraction, it leaves unanswered the question of how the filaments interact to generate the force that makes them slide. This riddle has spawned many ingenious and imaginative theories; the most widely accepted and best documented is the cross-bridge theory. According to this theory, movement of thin filaments between the thick filaments is accomplished by the sequential formation and dissolution of cross bridges between thick and thin filaments. In effect, myosin molecules crawl ratchetlike along the chain of actin molecules in the thin filament. Each cross bridge (or bound myosin head) is postulated to undergo a conformational change in concert with the other cross bridges, creating a force that pulls the thin filaments toward the center of the sarcomere. The bridges then break, and in a subsequent cycle, the myosin head forms a new bridge with a site on the actin chain farther from the center, and again pulls to generate further sliding. The energy for this translation is supplied by the hydrolysis of ATP. Because of the small size of the structures involved, cross-bridge cycling cannot be observed directly, but much experimental evidence supports the theory. We now consider in detail how one can demonstrate and describe interactions of myosin with actin like those postulated in the cross-bridge theory. The evidence comes from a wide range of techniques including biochemistry, electron microscopy, x-ray diffraction, and muscle mechanics.

Actin and Myosin Interaction in Solution. Our search for cross-bridgelike interactions starts in the test tube. We have already noted that the myosin heads (S-1 fraction) display ATPase activity. Heavy meromyosin or S-1 fractions* split ATP, producing ADP and inorganic phosphate (P_i). The reaction rate, however, is far too slow to account for the ATP consumption that occurs during muscle contraction; one HMM or S-1 molecule can split one molecule of ATP only once every 20 seconds or so. More detailed observations reveal that the instantaneous turnover rate is a function of the time the reactants are allowed to interact (Fig. 8–6). Thus, when the reactants are first mixed, hydrolysis (as measured by the production of inorganic phosphate) is rapid, but then the machinery seems to bog down, and hydrolysis becomes very slow indeed. Lymm and Taylor[30] discovered that the ADP and P_i formed in the early burst remain attached to the myosin for some time. They postulated that hydrolysis itself is fast but that the release of the reaction products, ADP and P_i, from myosin is the slow, rate-limiting step in the reaction. Later, Bagshaw and Trentham[3]

---

*Either HMM or S-1 fraction, rather than myosin, is usually used for in vitro experiments because both are much more soluble at physiological ionic strength than is intact myosin.

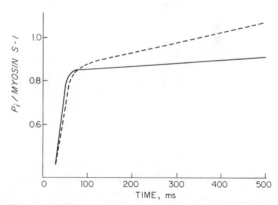

Figure 8–6 Rate of production of phosphate by myosin ATPase in the presence (dotted line) and absence (solid line) of actin. Phosphate formation is expressed per S-1 head of myosin and includes phosphate both bound to the head and in solution. The reaction begins at time zero. At the outset, phosphate formation is rapid (the early burst) in both experiments, amounting to formation of about one phosphate per myosin head. Subsequently, in the absence of actin, the rate slows dramatically to a steady state. In the presence of actin, this steady-state rate is increased dramatically (up to 100 times). (Data reprinted with permission from Lymm, R. W.; Taylor, E. W. *Biochem.*, 10:4617–4624, copyright [1971], and from Taylor, E. W. *Biochem.*, 16:732–740, copyright [1977] American Chemical Society.)

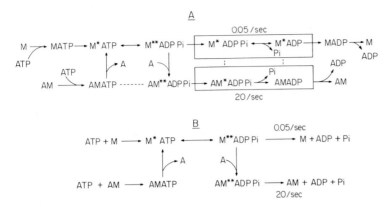

Figure 8–7 Steps in myosin-actomyosin ATPase reactions. The upper line diagrams the steps for myosin *(M)*. Distinct states identified by their fluorescence are shown with superscripts, M^* and M^{**}. The lower line shows the steps in the actin *(A)* activation of the myosin ATPase. Numbers indicate the approximate forward rates for the myosin steps, and in the lower line, for the actomyosin steps. Actin increases the rate at which the hydrolysis products, once formed, can dissociate from myosin thus allowing more rapid cycling of the ATPase. Compare with cross-bridge cycle (Fig. 8–15).

showed that the rate-limiting step actually occurs between hydrolysis and product release, but in any case, a myosin molecule encumbered by ADP and P_i from a previous reaction is unable to split another ATP. However, the block can be overcome if actin is added to the reaction mixture. Lymm and Taylor[31] found that actin greatly hastens the release of reaction products and so speeds the turnover that one molecule of ATP is hydrolyzed every 50 ms— a 400-fold increase over the stable rate in the actin-free mixture (Fig. 8–6). Thus actin makes myosin a more effective ATPase by accelerating the release of reaction products from the enzyme, making it available to split additional ATP. Actually, it is polymerized actin (not monomeric actin) that stimulates the myosin ATPase. Figure 8–7 diagrams the steps in the myosin- and actomyosin-ATPase reactions.

In the reaction that releases the ADP and P_i, actin and myosin associate to form actomyosin, a reaction that can also be followed as an increased viscosity and decreased transmission of light through the solution. The actin-myosin bond is strong and can be broken under physiological conditions only at the expense of the energy associated with the binding of ATP to myosin. After ATP is bound and myosin-ATP dissociates from actin, ATP can be split; then the myosin-ADP-P_i complex can again bind to the actin. The reaction thus proceeds by a cyclical or periodic formation and dissolution of actomyosin at the expense of ATP.

Within this scheme, the hydrolysis of ATP by myosin takes place when actin and myosin are dissociated. However, the ATPase rate is high even in the presence of high concentrations of actin in solution or when actin is chemically cross-linked to myosin, conditions under which full dissociation of actomyosin cannot occur. Thus, although total dissociation of actomyosin is not required for hydrolysis to proceed, the hydrolysis of ATP while actin and myosin are detached is strongly favored over that occurring while they are attached both in vitro and in vivo.

In in vitro experiments, the cyclical interaction between actin and myosin at the expense of ATP causes progressive depletion of ATP and the buildup of ADP and P_i. If ATP becomes completely depleted, the reaction ends with actin and myosin firmly bound together, and the actomyosin precipitates from solution as a dense, contracted plug, a phenomenon known as *superprecipitation*.* The parallel of this state in intact muscle is rigor mortis, the stiffly contracted state of muscle after death when ATP synthesis ceases. In resting muscle, many of the myosin heads are loaded with ADP and P_i. As we see later, what is required to activate resting muscle is elevation of sarcoplasmic Ca^{2+} to levels sufficient to allow actin and myosin to interact; the cycling process that underlies contraction then begins.

The free energy released through the hydrolysis of ATP is stored in a series of states in which the affinity of myosin for actin increases. The lowest free energy state is attained following ADP and P_i release, when myosin and actin are strongly bound together. The binding of ATP then dissociates the complex since M·ATP binds only weakly to actin. The chemical energy released by the

_____

*The time of occurrence of superprecipitation after mixing together actin, myosin, and ATP has been used as a measure of ATPase activity. In the presence of ATP, the actin and myosin are optically clearer in its absence (thus this is called the "clearing phase"). When the ATP is used up by the actomyosin in ATPase, actin and myosin form "tight bonds" and the whole mixture superprecipitates into a dense plug. Thus, the time to superprecipitation is the time taken to use up the added ATP, a measure of the ATPase rate.

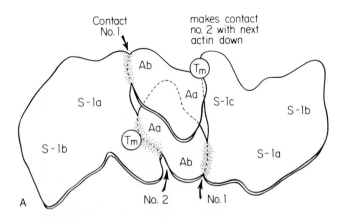

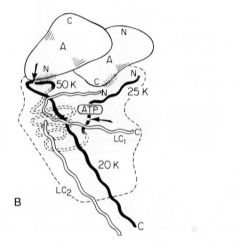

Figure 8–8 *A*, View down the axis (toward the Z line) of a thin filament decorated by added myosin S-1 heads. (Fig. 8–9 is a longitudinal view of the decorated thin filament.) The lines represent the most probable boundary of the individual monomeres (S-1 and A-actin). *Tm* is the probable position of the tropomyosin molecule that runs along the thin filament (see Fig. 8–1J). The possible domains of each molecule are indicated as are the possible areas of contact between myosin and actin. Note that there may be two actins contacted by each myosin S-1. The S-1 on the right has its second contact with the next actin down the thin filament (not shown).

B, In a possible model of the actin S-1 interaction, an attempt to show the sequential regions of the myosin S-1 heavy chains (dark coils) and light chains (light coils). *N* refers to amino terminal and *C* to the carboxyl terminal amino acids in each sequence. The two cleavage sites on the heavy chain are indicated by arrows producing 25, 50, and 20 kD regions (labeled 25 K, 50 K, and 20 K) extending from the N terminal to C terminal. Note the region of the 50 kD heavy chain, which probably is responsible for the one actin-binding site (plus some 20 kD cooperation), and the 25 kD region, which makes up the second possible binding site. The possible ATP-binding site is also shown. (Modified, with permission, from Amos, L.A., The Annual Review of Biophysics and Biophysical Chem., Vol. 14, © 1985, by Annual Reviews, Inc.)

various steps in hydrolysis, actin binding, and release of products is converted to muscle work and heat.

The roles of different domains of the myosin head can be explored by cutting the molecule into smaller pieces. S-1 can be split by means of proteolytic enzymes into three portions, having molecular weights of about 25,000, 50,000, and 20,000 (moving from the NH_2 terminal end to the COOH terminal that is connected to S-2 and the rest of the myosin molecule). As shown in Figure 8-8, both the 20,000 and 50,000 portions probably participate in the binding to actin. The binding is not covalent but may involve an electrostatic interaction at one site and a hydrophobic interaction at the other. ATP binding and hydrolysis occur at a distance from the actin binding sites, mainly on the 25-kD fragment.

Interaction of Filament Components in Vitro.[11, 2, 37] Thin filaments for test-tube experiments can be isolated from muscle or reconstructed by allowing extracted globular actin to polymerize. When isolated thin filaments are mixed with soluble forms of myosin (HMM or S-1) *in the absence of ATP*, the filaments and myosin bind strongly to form actomyosin.[22] An electron micrograph of the reaction product is shown in Figure 8–9A, and Figure 8–9C shows a model of it reconstructed from EM and optical diffraction data. Myosin heads attach to actin globules of the twisted chain; because of a hooklike configuration of the heads, they resemble arrowheads or chevrons (Fig. 8–9A). The tips of the arrowheads attach to the thin filaments pointing away from its Z-line end (see Figs. 8–8A and 8–11).[36, 46] The chevron pattern of actomyosin is so characteristic that its

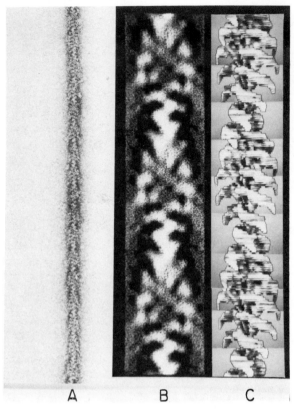

A   B   C

Figure 8–9 "Decoration" of thin filaments with myosin S-1 fragments. A, Electron micrograph of a negatively stained preparation of thin filaments that have been reacted with myosin S-1. Note the arrowhead-like chevron appearance. The image is equivalent to a projection.

B, Computer-processed image of the electron micrograph projection shown in A, constructed from diffraction and electron micrographic data (described in Taylor, K.A.; Amos, L.A. J. Mol. Biol. 147:297–324, 1981). Notice that the individual S-1 molecules are tilted less steeply with respect to the long axis of the filament than the "arrowheads" which are a pattern of several superpositioned S–1s. The tilted mode of attachment gives rise to the characteristic arrowhead-like appearance.

C, Montage of several different views of a balsa-wood model constructed from the computer processing of the electron micrograph shown in A and described in Taylor and Amos (B, above). These show the surface of the decorated filament and the individual S–1s. (A, B, and C, photographs courtesy of L.A. Amos.)

occurrence in specimens reacted with soluble myosin is taken as evidence of the presence of actin filaments. This technique is especially useful in identifying actin filaments in nonmuscle tissues.

We presume that the chevron structures are the cross bridges that, in the presence of ATP, would pull the thin filaments toward the center of the thick filament. When muscle S-1 is chemically cross-linked to actin in the presence of ATP, hy-

drolysis occurs at a maximum rate, and the electron micrographs of the resultant product show no regular pattern of arrowheads.[6]

Another method to observe the interaction of S-1 with thin filaments (or polymerized actin) employs optical or magnetic probes of molecular structure. The S-1 head of myosin has intrinsic fluorescence due to its tryptophan content. However, more detailed information can be obtained by use of a specific extrinsic molecular probe attached covalently to an amino acid in S-1 near the active ATPase site (see Thomas[47] for a description of these techniques). The depolarization of intrinsic or extrinsic fluorescence (change in the angle of polarization) due to the rotation of the S-1 head shows that it has a shape similar to that shown in Figures 8–1 and 8–3, that it can move relatively freely with respect to the tail of the myosin molecule (see Fig. 8–3), and that when it attaches to the thin filament in the absence of ATP, it becomes relatively immobilized.[35]

The sliding-filament theory postulates that, after attaching to actin, the myosin heads swivel with respect to their tails, thereby generating a movement that propels the thin filament toward the center of the sarcomere; the bridge then breaks and after hydrolysis of another molecule of ATP, the heads reattach at a new site and the cycle is repeated. The repetitive, asynchronous ratchetlike action of many myosin heads provides the force for contraction. Attempts to measure the movement of myosin heads employ myosin covalently labeled with a fluorescent or phosphorescent group that can be excited with polarized light. Changes in the angle of polarization indicate movements of the labeled head, and the time taken for the change gives an estimate of the rate of movement (Fig. 8–10). The measurable time scale is limited to the lifetime of the excited state—microseconds for fluorescent labels but much longer for phosphorescent ones.

Such optical studies[47] show relatively rapid motion of the S-1 heads of myosin in the absence of actin (only 2 to 3 times slower than for isolated S-1) and a somewhat slower movement of the whole HMM (S-1 plus S-2). In the presence of actin, S-1 remains rigidly bound to actin in the absence of ATP but is very mobile in the presence of ATP. This suggests that the S-1 heads do not remain fixed in orientation during the hydrolysis of ATP but move, presumably detaching and reattaching to actin.

Electron paramagnetic resonance can also be used to measure motion of S-1 (EPR, or electron spin resonance, ESR) in conjunction with a "spin"

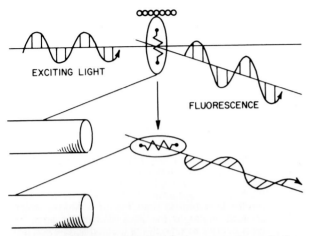

Figure 8–10 Diagram illustrating fluorescence depolarization. Fluorescent probe label attached to myosin S-1 head with absorption and emission dipole aligned along S-1 axis. Exciting light polarized as indicated will stimulate the fluorescent probe (upper diagram). If the myosin head and, therefore, the label stay bound in the vertical direction, the emitted fluorescence will also be polarized as indicated in the upper diagram. If the label or head rotate during the lifetime of the fluorescent state to a position such as that in the lower diagram, the emitted fluorescence will be depolarized, with plane of polarization at right angles to the exciting polarization. Thus, the change in polarization of the emitted fluorescence (the depolarization) indicates head movement in conditions in which it is attached or detached. (After Bagshaw, C.R. *Muscle Contraction*, p. 35. London, Chapman and Hall, 1982.)

label molecule, i.e., one having an unpaired electron (such as in the nitroxide group). With conventional techniques, rotations in the microsecond range can be measured. New "saturation transfer" methods[47] allow this technique to detect rotations that take milliseconds to achieve. Rotation (correlation) times can be calibrated using well-defined molecules rotating in solutions of different viscosities. When this EPR technique is applied to labels on S-1, electron paramagnetic resonance studies also show that S-1 can move rapidly unless actin is present. In the presence of actin, S-1 motion is relatively slow in the absence of ATP, but is accelerated when ATP is present.

Interaction of Filaments in Vivo. The interaction of myosin and actin so clearly demonstrated in vitro is less easily demonstrated in vivo. Evidence comes from four sources.

Electron Microscopy

Figure 8–11B shows an electron micrograph of muscle deprived of ATP and hence in a state of stiff rigor. The myosin heads form chevrons the directions of which are opposite at the two ends

of the thick filament; the arrowheads always converge on the thin filament and point away from the Z line. In electron micrographs of muscles at rest, the bridges are mainly perpendicular to the thick filament (Figure 8–11A).[40] The data are consistent with the postulate that the bridges when first formed are nearly perpendicular to the two filaments. During their reaction however, they undergo an angular conformational change, pulling the thin filaments toward the center of the sarcomere. The muscle in rigor thus shows the end stage of a cycle frozen in place, because in the absence of ATP, the bond cannot be broken. In the presence of ATP, the bridges break (detach) at the end of the pull; when ATP is hydrolyzed, the heads spring back and attach to a binding site farther from the center. Repeated cycles of this sort pull the thin filaments along between the stationary thick filaments and drag the Z lines toward one another.[24]

Electron Paramagnetic Resonance and Fluorescent Probes

The "spin label" and fluorescent probe label techniques have also been used to study cross-bridge orientation in relaxed and contracted muscle. Because the methods require chemical modification of a protein, they cannot be used on intact muscle but are feasible on "skinned" muscle fibers in which the sarcolemma has been disrupted but the filament structure left intact. As described earlier, both of these techniques provide estimates of the orientation of the molecular probes and thus of the molecules to which they are attached. The spin label technique allows measurement of orientation of S-1 in muscle at rest, in rigor, and in contraction. In relaxed muscle, Cooke and Thomas[48] found that the label (S-1 head) shows relatively free movement. In rigor, nearly all of the labels are oriented, with the heads attached at an angle of approximately 70 degrees to the fiber axis. During isometric contraction, most of the labels move relatively rapidly, but 20% were attached in the rigor angle configuration.[5] Cooke and Thomas's studies suggest that during contraction the cross bridges are at the rigor angle, or in motion. In other words, the bridge appears to step quickly from the perpendicular attachment configuration to the angled (rigor) configuration.

Fluorescent probe studies provide another measure of cross-bridge (label) orientation but yield data less clear than those obtained by the spin label technique. Results from these studies agree with the spin label studies that the cross

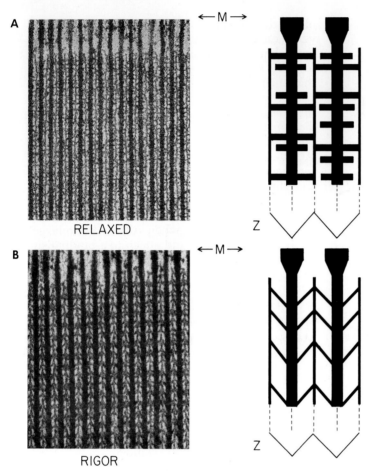

Figure 8–11 Cross-bridge position in electron micrographs of insect muscles fixed in the relaxed and rigor states. On the left are electron micrographs, on the right are diagrams constructed from the micrographs. In a relaxed state, the cross bridges project at right angles to the thick filament. In rigor, the cross bridges assume the characteristic chevron pattern. Since the M line is at the top of the picture, the arrowheads all point away from the Z line when referred to the thin filaments. These pictures are strong evidence that cross-bridge orientation changes during contraction. (Courtesy of M. C. Reedy and M. F. Reedy.)

bridges are in motion in the relaxed muscle, are attached in rigor, and are both attached and moving during contraction. However, in contrast with the spin label studies, fluorescence studies indicate different angles of attachment during contraction and rigor.

In summary, cross-bridge attachment during rigor and contraction is well documented, but the angles of attachment in contraction remain uncertain, and there is little evidence of slow rotation of attached heads during contraction.

X-Ray Diffraction

X-ray diffraction studies of muscle[14] provide another line of evidence that filaments interact in vivo. We have already mentioned that x-ray diffraction permits the investigator to measure with precision the spacing of structures along the filament. The technique is complex. For the interested reader, the following section presents a simplified description that can be omitted by those not in-

trigued with methodology. The major findings of x-ray diffraction are listed here: (i) Filament lengths do not change significantly during contractions. (ii) During contraction, slightly preceding force development, portions of the thick filaments move toward the thin filaments, specifically toward actins, consistent with the hypothesis that myosin heads combine with actin. (iii) Myosin heads can and do move on the thick filaments and are thus capable of some degree of "seeking" for attachment sites on actin. (iv) The spacing between filaments changes little with changes in muscle length during isometric contraction. The major change in filament spacing occurs in such a way that the lattice volume remains constant, i.e., as the sarcomere shortens, filament separation increases.

X-ray diffraction can be employed to study the structure of living muscle either at rest or during contraction. The special virtue of x-rays is their short wavelength, a property that allows much greater resolving power than can be obtained using electromagnetic radiation in the visible range. The data obtained are interference patterns of x-rays scattered by atoms in the path of the x-ray beam and are quite different from the images formed in the light and electron microscope. Structure is deduced, not visualized, and such deduction is a highly specialized art that requires skill, imagination, and perseverance.

The principle behind the technique is most clearly exemplified by its use to analyze the spacing of the atomic lattice of crystals. If a parallel beam of "monochromatic" x-rays strikes such a regular structure, the orbital electrons in its constituent atoms are set into vibration and emit radiations having the same wavelength as the exciting x-rays. Each atom thus becomes a point source of radiation that spreads radially from the source. The original parallel beam is scattered. The degree to which an atom scatters x-rays varies with the number and distribution of its orbital electrons. X-rays emanating from multiple coherent point sources (coherent sources are sources from which waves of like phase emanate) produce interference patterns similar to the interference patterns observed when visible light is scattered by a diffraction grating. In fact, for the short-wavelength x-rays, the regularly spaced atoms in a crystal act like a grating, each atom serving as a coherent source, analogous to the slits of the grating, for visible light. The student should refer to a physics text to review how waves emerging from a diffraction grating interact and produce a series of light and dark lines, the spacing of which can be related to the distance between the lines on the diffraction grid if the wavelength and distance of the screen from the grid are known.

The distance (d) between the lines on the diffraction grating can be computed from the angle at which the x-rays are scattered (θ), the wave length (λ) of the x-rays, and the order of the diffraction line (n), counting from the undiffracted line, by $d = n\lambda/\sin\theta$ (see Figure 8–12). An important feature of this relationship is that d and θ are inversely related; a closer spacing of the grid lines gives a wider separation of the diffraction lines. Paradoxically, close spacings are more easily resolved than broader ones. Usually, the x-ray source is a copper target with $\lambda = 0.154$ nm. Since the distances involved in periodic structures in muscle are of the order of 10–40 nm, the ratio λ/d is very small and θ is a small angle.* With this

small angle, the relationship between the spacing (d) between lines of the grating, the distance (x) of the diffracted line from the undiffracted position, and the distance from the grating to the detection screen (L) becomes $d = Ln\lambda/x$.

In considering the diffraction of x-rays by a crystal rather than the simple grating described, above, Bragg introduced a simplifying formulation in which the rays are viewed as reflecting off planes formed by the ordered atoms in the crystal, much as light reflects off a mirror. This gives similar results to the above, but the spots in the pattern due to constructive interference are referred to as *reflections*, a term which is used frequently.

In addition to giving information about spacing of atoms or molecules in the structure being examined, the *intensities* in the x-ray diffraction pattern give information about the basic scattering structures. (In muscles, the scattering structures are molecules or collections of molecules.) A higher density of x-ray scattering structures causes the reflections to be more intense than sparser scattering structures. Also, the repeating structure (atom, molecule, or collection of molecules) has its own intrinsic x-ray diffraction pattern that influences the overall pattern by enhancing or diminishing the reflections in the pattern formed by the basic repeat. The diffraction pattern produced by a grating with a series of slits depends not only on the spacing between the slits but also on the size and shape of each slit. In a crystalline array of molecules, the diffraction pattern depends both on the arrangement of molecules and on the arrangement of atoms in each molecule.

Because of its regular structure, skeletal muscle diffracts light and x-rays, although the complexity of its structure, compared with that of a simple crystal, correspondingly increases the complexity of the patterns. The recurring structures in muscle are molecules rather than atoms. Moreover, the molecules are arranged in complex helical configurations that present formidable problems to the investigator seeking to work from the diffraction patterns to the underlying molecular structure.

Nevertheless, a great deal of valuable information can be obtained using the diffraction technique, especially when the data are combined with those obtained through light and electron microscopy. Figure 8–12 shows diagrammatically how the technique is applied to muscle. Conventionally, x-rays are produced by bombarding a metal target with high-speed electrons. More recently, x-ray radiation from high-energy electron accelerators, synchrotrons, have been used because they

---

*Thus, with these dimensions, the type of diffraction is called "low angle", in contrast with the case in molecular structure in which d is much smaller and θ is larger.

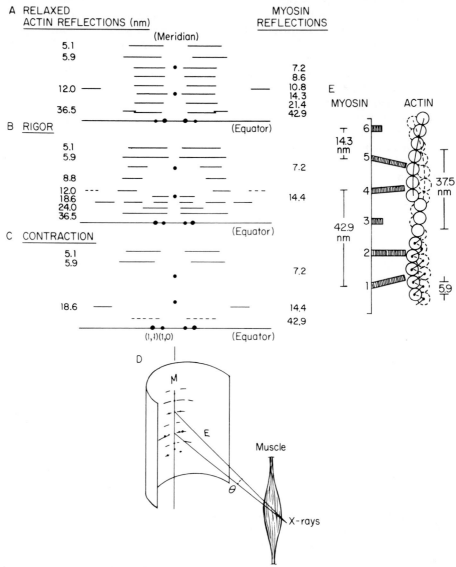

Figure 8–12 X-ray diffraction studies on muscle. The lower section *(D)* shows the experimental arrangement. The upper portion diagrams the diffraction pattern, indicating the intensities (reflections) associated with periodicities in the thin filament, actin (left) and the thick filament, myosin (right), for muscle in the relaxed state *(A)*, in rigor *(B)*, or during contraction *(C)*. Along the equator, reflections due to the specific (1,0) and (1,1) planes are indicated. On the right are shown the helical arrangement of cross bridges compared with the helical arrangement of actins in the thin filament. Notice the differences in periodicity between the thick and thin filaments and how in rigor the reflections from the cross bridges can take on a periodicity more closely related to those of the thin, actin filament. (After Haselgrove, J. C. *Handbk. Physiol.*, Sec. 10; pp. 143–172, 1983.)

are more intense sources. The x-ray beam is first made parallel by passing it through a slit or focusing it with a crystal. The beam is then incident on a vertically suspended muscle, and the diffraction pattern is recorded on a curved photographic film or by radiation detectors and electronic counters.

The diffraction pattern can be sampled with the muscle at rest, in rigor, or during contraction. During contraction, a lead shutter can be triggered to open when the force reaches a certain level, restricting exposure of the muscle and the recording photographic film to x-rays. If electronic counters are used, the time "window" of sampling can be set electronically to occur when the force is at a given level or at a given time after stimulation. Two dimensions of the pattern are distinguished: meridional and equatorial. The intensity of diffraction along the meridian gives information about periodic structures (e.g., myosin heads) occurring along the long axis of the muscle fibers, whereas diffraction along the equator provides clues concerning the arrangement of elements lying side by side (e.g., interfilament distances).

Figure 8–12 shows a schematic representation of diffraction observed for whole muscle in the relaxed, rigor, and contracted states. (Of course for whole muscle the pattern is repeated below the equator.) The diffraction pattern along the equator can be considered as reflections off lattice planes formed by the regular arrangement of filaments seen in a transverse section through the muscle. As described earlier and as shown in Figure 8–13, the filaments form a hexagonal pattern. Longitudinal planes parallel to the muscle axis can be drawn to include either thick filaments only (the [1,0] planes in Figure 8–13) or to include both thick and thin filaments (the [1,1] planes in Figure 8–13). Reflections off these planes form diffraction intensities along the equator. Because of the inverse relationship between interplanar distance and the deviation of the diffracted spot (line) from the undiffracted position, x-rays reflected from the (1,0) plane are closer to the central spot than those reflected from the (1,1) plane. The distance between these spots is used to compute the separation between the filaments.

Measurement of the distance to the (1,0) and (1,1) reflections as a function of the sarcomere length has established that in unstimulated muscle, the interfilament distances change to maintain the constant volume referred to earlier. As the muscle is shortened, the (1,0) and (1,1) reflections move in along the equator, indicating that the filaments interacting in contraction move further apart by an amount required to keep the volume

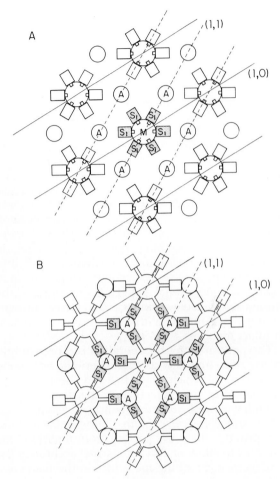

Figure 8–13 Filament arrangement in a transverse section. The thin filaments (A) and thick filaments (M) are shown in their hexagonal arrangement in vertebrate striated muscle. The two planes (1,0; 1,1) are identified as the planes containing only thick filaments or both thick and thin filaments, respectively. Diagrams *A* and *B* are consistent with the x-ray diffraction data discussed in the text. Myosin projections are represented as single S-1 heads even though the heads are paired and in rigor attach to the thin filament in the manner shown in Figure 8–8A. *A,* The relaxed state. The S-1 projections are associated with the thick filaments, consistent with the greater intensity of the reflections from the (1,0) planes. *B,* Rigor, or the contracted, state. The S-1 fragments are more closely associated with the thin filaments, consistent with the increase in the intensity of reflections from the (1,1) planes.

constant.[21] This constant lattice-volume relationship does not hold for "skinned" muscle fibers, suggesting that it depends more on the osmotic properties of the muscle membrane than on any inherent property of the filaments.[34] In contrast, measurement of the (1,0) and (1,1) reflections during isometric contraction also indicates that interfilament distances change little, thus cross-bridge forces do not have a strong radial component at these filament separations.[13]

The relative intensities of the two reflections on the equator give information about the relative amount of x-ray scattering material contained in the (1,0) and (1,1) planes. Because the thick filaments have more mass than thin filaments, the (1,0) plane (thick filament only) of resting muscle scatters x-rays better than (1,1) planes that contain both thin and thick filaments. However, the ratio of ([1,1] intensity):([1,0] intensity) increases as muscle contracts and decreases as it relaxes.[23] During the increase in force, the (1,1) increase slightly precedes the force. The largest increase occurs when the muscle is in rigor. The implication is a significant one: some parts of the thick filament (such as the bridges) have moved out to the region of the thin filament and attached to the thin filament to produce contraction. Changes of equatorial and meridional reflections constitute the strongest available evidence that cross bridges from myosin interact with actin during contraction in intact muscles.

The pattern along the meridian is more complicated, reflecting the complex periodicities along both thin and thick filaments and the helical arrangement of the molecules in the filaments. Although work done on other helical molecules (e.g., DNA and α-helical proteins) provides a helpful background, the x-ray pattern from muscle is too complex to allow one to reconstruct accurately the arrangement of all the molecules in the filaments. Nevertheless, one can make deductions about the resting arrangement in both thin and thick filaments and about the changes that occur in the myosin heads accompanying contraction.

Figure 8–12 shows a simplified diagram of the diffraction pattern of muscle at rest, in rigor, and in contraction referred to the thick (right) and thin (left) filaments. The simplified pattern shows reflections both on and off the meridian, a characteristic of helically arranged molecules. The reflections on the lines parallel to the equator are called layer lines. The important dimensions of helically arranged molecules are diagrammed in Figure 8–12. For the thick filament, these are 14.3 nm and 42.9 nm. The distance along the filament axis between occurrences of myosin heads is 14.3 nm. This can be obtained from the reciprocal of the distance between the diffraction equator and the first intense spot on the meridian. The axial distance for repeat of the helical pattern in thick filaments, with 3 subunits per repeat, is 42.9 nm. In the diffraction pattern, this is shown by the off-meridional spot closer to the equator. Because of their large mass, it is presumed that the diffracting subunits are myosin heads projecting laterally from the backbone (see Fig. 8–3); electron micrographic measurement of spacing of cross bridges satisfactorily agrees with this interpretation.

During contraction, the 14.3-nm meridional spot moves to a position indicative of a 14.4-nm repeat without change of intensity. This change suggests that the thick filament does not shorten but rather lengthens slightly during contraction and that the cross bridges have not moved so much along the thick filament during contraction that the regular repeat is disrupted. In contrast, the intensity of the 42.9-nm layer line intensity decreases, a finding that suggests that this major repeat along the thick filament is disrupted and that, in the absence of movement of bridges along the filament, an azimuthal movement around the thick filament occurs as if the bridges were seeking the nearest thin filament to hook onto. In rigor, the 42.9 pattern disappears and is replaced by one that has the periodicity of the thin filament, an indication that in the absence of ATP the bridges attach irreversibly to thin filaments (Figure 8–12).

Other intensities in the x-ray diffraction pattern (Figure 8–12) are contributed by the thin filament. Some are used to deduce the *double helical* arrangement of the actin monomers as shown in Figure 8–1 and 8–12. During rigor and contraction, the intensities of these spots increase, which is consistent with myosin attachment to actin. Other lines give information about the shape of the thin filament. Changes in the relative intensity of these lines support the argument that the shape of the thin filament changes during calcium activation (see Fig. 8–17 and the article by Haselgrove[14] for discussion of both this and the relevant points in the thick filament reflections).

Length-Tension Relation In Single Muscle Fibers

Length-tension studies provide another source of evidence for the in vivo interaction of thick and thin filaments through the cross bridges.[17] If the cross bridges are the site of interaction between the two sets of filaments, one would expect that the contractile force would vary with the number of sites that can interact, ranging from zero at long sarcomere lengths at which thin and thick filaments do not overlap (and thus cannot interact) to a maximal force at the sarcomere length allowing maximal overlap of filaments. Figure 8–14 shows a length-tension relation determined by

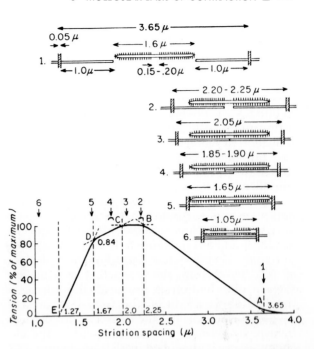

Figure 8–14 The relationship between the length-tension curve and striation spacing for single frog muscle fibers. Diagrams show filament relations at different sarcomere lengths; the numbers to the left correspond to numbers marked with arrows on the length-tension curve below. A = the long sarcomere length at which active tension is zero; B = the beginning of the plateau phase of tension; C = the shorter sarcomere length limit of the plateau; D = the knee or substantial change in slope at the short sarcomere length side; E = the short sarcomere length limit at which active tension is zero. (After Gordon, A.M., *et al. J. Physiol. [Lond.]* 184:170–192, 1966.)

Gordon, Huxley, and Julian,[12] for a single muscle fiber in which the sarcomere length was controlled so that force could be related to sarcomere length and therefore to filament overlap. The diagrams in the upper part of the figure show filament positions at various key points in the length-tension relation. At long sarcomere lengths in which there is no overlap, the force is zero, but it increases as greater overlap is allowed. Because the middle of the thick filament lacks cross bridges (myosin heads), shortening of the sarcomere from 2.2 to 2.0 μm generates no additional opportunity for bridge formation, and over this length of range, force remains constant. At still shorter lengths,

force drops again, perhaps because of mechanical collisions of thin filaments with one another and of thick filaments with Z lines. The data taken together thus give strong support for the hypothesis that cross bridges are the site of interaction among the filaments and that they generate the force to produce the sliding of these two sets of filaments and muscle contraction.

Coupling of Mechanical and Chemical Events at the Cross Bridge.[49, 11, 10] Cross bridges are the myosin projections from the thick filament (Fig. 8–14; see also Figs. 8–1, 8–2, 8–3, and 8–11). Their linkage appears to be flexible between the S-1 and S-2 portions and between the S-2 and LMM por-

Figure 8–15 A hypothetical cross-bridge cycle relating changes in attachment to the thin filament and tilting of the cross-bridge head to the steps of actin activation of myosin ATPase activity. The myosin S-1 heads are stippled. The circles on the thin filament represent actin molecules. *1*, at rest; *2*, initial attachment; *3*, after release of inorganic phosphate, beginning of force production; *4*, during force generation causing filament sliding and muscle shortening; *5*, after release of ADP (rigor-like configuration); *6*, ATP binds to the attached S-1. The addition of ATP causes *7*, the dissociation of actin and myosin; *8*, the subsequent hydrolysis of ATP by myosin and the return to state *1*. This scheme couples the hypothesized mechanical attachments with the biochemical kinetics of the myosin ATPase. (Modified, with permission from Lymm, R.W.; Taylor, E.W. *Biochemistry* 10:4617–4624. Copyright [1971] American Chemical Society.)

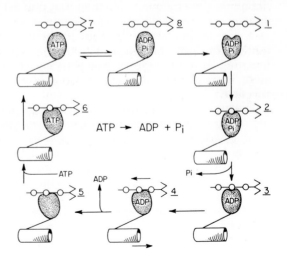

tions, so that S-1 heads can move away from the thick filament. In addition, the S-1 portion is long enough to reach around the neighboring thin filament to make contact, as in Figures 8–8 and 8–9. Each thick filament is surrounded by six thin filaments (Figure 8–1). With the helical projection of bridges, the possible sites of interaction between actin and myosin are multiple, an arrangement that allows cyclical interaction of actin and myosin and continuous generation of force in shortening muscle as the myosin heads "crawl" along the thin filament.

A possible model of interaction between actin and myosin is suggested by the chevron appearance when S-1 reacts with thin filaments (see Figs. 8–8, 8–9, and 8–11). Figure 8–15 traces a full cycle of actomyosin interaction, including the hydrolysis of ATP by myosin, the interaction of actin and myosin plus hydrolysis products, the release of the hydrolysis products, the formation of actomyosin, and, finally, the dissociation of actomyosin by ATP. This cycle, identical to the biochemical cycle described earlier in Figure 8–7, includes a postulated rotation of the myosin head to generate a relative force. Such a model was first proposed by H.E. Huxley,[24] with the biochemical correlates by Lymm and Taylor,[31] and is supported by mechanical studies of A.F. Huxley and R.M. Simmons[20] (see Chap. 9). During isometric contraction, the heads attach, generate force, and detach, cycling on and off slowly at the same or a neighboring actin site on the thin filament and generating a relative force between the two sets of filaments during each cycle. Asynchronous activation of many sites along a single thin filament generates a sustained force that at any time is a function of the number of interacting sites and the force generated at each site. During shortening, filaments slide by one another so that one S-1 site attaches, generates force, and detaches to reattach at a site farther along the thin filament. In the model, bridges interact cyclically making and breaking in a continuous, asynchronous manner along each filament throughout the muscle while it is active.

The exact modes of cross-bridge attachment during contraction of living muscle remain to be established. The chevron pattern is seen in rigor and is hinted at by the magnetic spin label and fluorescent probe experiments described earlier. However, possible intermediate attached states of myosin heads on actin as well as large movements of attached myosin heads remain to be shown. Magnetic and fluorescent probe studies are ambiguous on this point. Experiments using an ATP analog such as β,γ-imido ATP that binds to but is not hydrolyzed by myosin have been done to attempt to identify another attachment position.[33] In the presence of this substance, myosin may bind to the thin filament at angles other than in the more oblique chevron attachment seen in the absence of ATP. In all the foregoing discussions, we have ignored a possible interaction between the two S-1 heads on each myosin molecule during contraction and the role of the two possible actin binding-sites on each S-1. In summary, it is clear that myosin interacts with the thin filament during contraction, but the specific molecular configurations occurring during contraction remain to be demonstrated.

CALCIUM REGULATION OF FILAMENT INTERACTION[8, 1]

Since 1883, it has been known that calcium plays an important role in the regulation of muscle contraction. In that year, Sydney Ringer noted that frog hearts immersed in solutions made with London tap water continued to beat but that they came to a standstill if the bathing solutions were prepared with other water. The critical ingredient of London water proved to be dissolved calcium ions. Over half a century later, Heilbrunn and Wiercinski[15] found that muscle fibers contract when injected intracellularly with solutions of calcium salts but not with solutions of sodium, potassium, or magnesium salts.

Our understanding of the actions of calcium was greatly advanced after the introduction of several simplified muscle preparations that allowed better control of the chemical environment than can be achieved in whole muscle. One of these was Albert Szent-Györgyi's glycerol-extracted preparation,[45] produced by storing muscles at a low temperature for long periods in a 50% aqueous glycerol solution. This and other chemical treatments destroy the sarcolemma, as well as the sarcoplasmic reticulum, exposing the contractile proteins directly to constituents of the bathing medium by eliminating the diffusion barrier of the surface membrane. Single muscle fibers can also be "skinned" mechanically, a procedure that tears or dissects away the sarcolemma.[38] "Skinned" fibers contract when ATP and calcium ions are applied. An even more radical but still useful procedure is to isolate myofibrils (see Fig. 8–1) from muscle homogenized with a blender. Isolated myofibrils hydrolyze ATP in the presence of low concentrations of free calcium. Finally, purified contractile proteins can be combined and will react in vitro in the presence of ATP and calcium and contract into a dense plug (superprecipitation or syneresis, described on page 180).

The importance of calcium ions was overlooked in early studies because calcium was usually present as an unsuspected contaminant of the ATP and other reagents used. Attention was focused on calcium following the work of Marsh,[32] who isolated from muscle homogenates a factor that relaxed myofibrils previously contracted by ATP. Ebashi[7] showed that this relaxing factor acted by removing calcium ions from the myofibrils; it was subsequently shown to consist mainly of fragmented sarcoplasmic reticulum membranes whose action to take up and release calcium is described in Chapter 7. Ebashi and colleagues[9] later showed that calcium regulation of the actin and myosin interaction occurred through previously unsuspected proteins (troponin and tropomyosin), which were contaminants of the "actin" and "myosin" preparations.

Because the concentrations at which intracellular calcium is effective in regulating contraction are minuscule, special precautions must be taken to control the free calcium concentration in experimental solutions. This control is usually accomplished by the use of chelating agents that bind calcium stoichiometrically with high affinity and thus can be used as calcium buffers. The ATPase activity of isolated myofibrils and the force in active "skinned" muscle fibers of both skeletal and cardiac muscle vary with calcium concentration over the range of 10^{-6} to 10^{-4}M (Fig. 8–16), which is also the range over which the sarcoplasmic reticulum is capable of controlling the calcium ion concentration. Ebashi and colleagues[9] showed that in rabbit fast-twitch muscles, the binding of calcium to troponin initiates activation; in the absence of the troponin-tropomyosin complex, actomyosin ATPase is always active and not regulated by calcium.

As discussed on page 177, troponin is made up of a number of subunits, only one of which, TnC, binds calcium. TnC is structurally similar to a number of other calcium-binding proteins in muscle, including calmodulin, parvalbumin and a light chain from myosin. In skeletal muscle, TnC has four calcium binding sites, two of which have a much higher affinity for calcium than for magnesium; the other two have sufficiently similar affinities for calcium and magnesium that at their physiological concentrations, magnesium competes successfully with calcium.[39] Much data suggests that the calcium-specific site is the critical one in regulation of muscle.[39] The other site may stabilize the TnC interaction with the other troponin subunits. Figure 8–17 shows a simple model

prepared to explain calcium activation of contraction in a fast mammalian muscle fiber. At rest, tropomyosin is in a position to "block" the interaction of actin with myosin. When calcium binds to troponin, tropomyosin rolls toward the "groove" of the double-stranded thin filament and

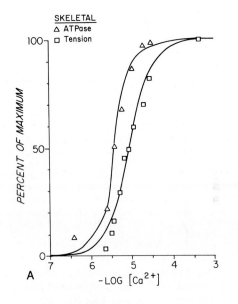

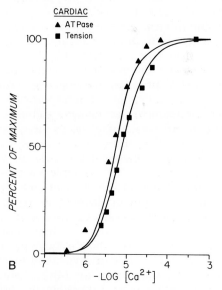

Figure 8–16 Calcium activation of contraction. The percent of maximum activation of myofibrillar ATPase (triangles) or tension (squares) produced by mechanically skinned fibers is plotted as a function of $-\log_{10}[Ca^{2+}]$ (pCa). A, data from rabbit adductor magnus skeletal muscle fibers (fast-twitch). B, data from rabbit cardiac muscle. Tension and ATPase activity are activated over approximately the same narrow range of calcium concentrations. (Data from Kerrick, W.G.L. et al., Pflugers Arch. 386:207–213, 1980.)

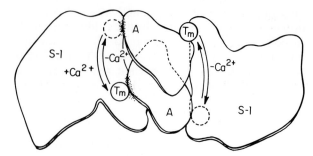

Figure 8–17 Model for activation of the thin filament by calcium binding to troponin. Cross section of the decorated filament showing the position of the actin (A) and myosin S-1 molecules, attached in a rigor link, and the probable position of the rod-shaped tropomyosin molecule (Tm) in the activated state (solid line) and inactivated state (dotted line). In the relaxed state ($-Ca^{2+}$), the tropomyosin (Tm) modifies the binding of the S-1 head to actin so that the ATPase is not stimulated. When calcium binds to TnC, tropomyosin (Tm) rolls farther into the groove of the thin filament, modifying the binding of the S-1 head to actin, activating the myosin ATPase. In the presence of ATP, the S-1 head can then interact cyclically with actin. When myofibrillar calcium drops, calcium dissociates from troponin, and the controlling molecules return to the initial state. (Modifed, with permission from Amos, L.A. *Annual Review of Biophysics and Biophysical Chemistry*, vol. 14. © 1985 by Annual Reviews, Inc.)

allows the myosin to interact with actin. Measurements on changes in the x-ray diffraction pattern from the thin filament support this movement of tropomyosin and show that it precedes the increase in the intensity of the (1,1) x-ray reflection (myosin heads approaching actin) and the production of force. Biochemical data indicate that tropomyosin controls the interaction of myosin with seven or more actin units, in agreement with the molecular dimensions; the 40-nm tropomyosin rod can cover approximately seven actin globules with diameters of 5.5 nm and overlaps neighboring tropomyosin at either end.

The model diagrammed in Figure 8–17, in which calcium appears to act as a simple switch that alters the thin filament to allow actin and myosin interaction, is an oversimplification. Sarcoplasmic free calcium concentrations affect not only force production but may also affect the maximum velocity of shortening. Also, the absence of calcium does not prevent actin and myosin interaction, particularly that occurring at low MgATP concentrations and low ionic strength.[10] Finally, the relationship between force and calcium is steeper than could be explained by simple calcium binding and is also shifted by many factors such as pH, magnesium, fiber type, and sarcomere length that do not appear to affect calcium binding to TnC in a simple manner.

Troponin is not the site of calcium regulation in all muscles. While investigating calcium regulation in scallop muscle, Andrew Szent-Györgyi and co-workers[28] discovered that calcium activates by binding to myosin rather than to the thin filaments, which in the scallop contain little troponin. This muscle was then said to show *thick-filament regulation* since the Ca-binding component was part of that filament in contrast to the troponin–thin filament system. One of the light chains of scallop myosin (similar to the P-light chain or LC-2 of vertebrate muscle, Table 8–3) binds calcium and allows actin to activate the myosin-ATP hydrolysis. The regulatory light chain from scallop is so similar to light chains from other muscles that when it is dissociated from myosin with EDTA, it can be at least partially replaced as a regulator of actin-myosin interaction by a light chain from rabbit muscle and completely replaced by light chains from muscles (such as smooth muscle) that show thick filament regulation.

Szent-Györgyi and his co-workers[28] have developed ingenious criteria for distinguishing muscles regulated by their thin filaments (e.g., rabbit skeletal muscle) from those, like the scallop muscle, that are regulated through myosin light chains. Thin-filament regulated muscles are identified by mixing their thin filaments with rabbit myosin and measuring the ATPase activity in the presence of 10^{-4}M of a Ca^{2+} chelator (thus low free Ca^{2+}) or in the presence of 10^{-4} Ca. Thin filaments from thin-filament regulated muscles require calcium to activate rabbit myosin ATPase, whereas thin filaments from thick-filament regulated muscles (e.g., scallop) activate rabbit myosin ATPase in the absence of calcium.

Another test is to mix *pure* rabbit actin (i.e., actin devoid of troponin and tropomyosin), with myosin from the test muscle and to determine whether the ATPase activity is calcium sensitive. Pure actin from thin-filament regulated muscle readily activates rabbit myosin ATPase in the absence of calcium because the "shielding" actions of troponin and tropomyosin are lacking. The same actin mixed with myosin from a thick-filament–regulated muscle activates the ATPase only if calcium is present.

To simplify the procedures further, Szent-Györgyi and colleagues[28] isolated myofibrils and measured the ATPase activity of these myofibrils with the addition of either rabbit actin or rabbit myosin both in the absence and presence of Ca^{2+}. The rationale for these experiments is as follows. Pure rabbit actin plus rabbit myosin does not require Ca^{2+} to hydrolyze ATP at a high rate. If the system

Table 8–4 Thin and Thick Filament Regulation*

Phylum, Class	Animal	Muscles Tested	Ca^{2+} Sensitivity with Rabbit Actin	Ca^{2+} Sensitivity with Rabbit Myosin	Thick or Thin Regulation
Chordata: vertebrata					
Mammalia	Rabbit	Back	0	High	Thin
Mammalia	Mouse	Leg	0	High	Thin
Amphibia	Frog	Leg	Low	High	Thin
Aves	Chicken	Pectoral	0	High	Thin
Arthropoda					
Crustacea	Hermit crab	Claw	0	High	Thin
Crustacea	Sand crab	Leg	0	High	Thin
Crustacea	Lobster	Claw	High	High	Both
Insecta	Giant waterbug	Leg	High	High	Both
Insecta	Cockroach	Leg	High	High	Both
Annelida	Featherduster worm	Body wall	High	High	Both
Mollusca					
Pelecypoda	Scallop	Striated adductor	High	0	Thick
Pelecypoda	Razor clam	Foot	High	0	Thick
Cephalopoda	Squid	Ventral pharynx retractor	High	0	Thick
Brachiopoda	Lampshell	Pedunculus	High	0	Thick
Echinoderma					
Holothuroidea	Sea cucumber	Lantern retractor	High	0	Thick
Nemertinea	Ribbon worm	Oral region	High	0	Thick

*Modified from Lehman, W.; Szent-Györgyi, A. G.: The Journal of General Physiology, 1975, vol. 66, pp. 1–30, by copyright permission of The Rockefeller University Press.

of rabbit actin or rabbit myosin and other animal myofibrils requires Ca^{2+} for activity, then the calcium must be activating the appropriate filament of the other animal's myofilaments; thick filaments, in the case of rabbit actin; or thin filaments, for rabbit myosin.

Szent-Györgyi and colleagues[28] were able to screen and classify muscles from a wide variety of animal species. A surprising finding was that in many species some muscles appeared to require calcium in both tests, a finding that suggests that these muscles have a dual regulatory system in which calcium activates both thick and thin filaments. A partial listing of their findings is given in Table 8–4.

Another variation of thick-filament regulation occurs in some smooth muscles and nonmuscle cells in which calcium appears to activate the myosin ATPase by first binding to the protein calmodulin. The resulting complex activates an enzyme that phosphorylates a myosin light chain (the P-light chain); the myosin with the phosphorylated light chain is an effective ATPase and reacts with actin.[41, 43] In this instance, then, calcium activates myosin not by binding to it directly but by activating an enzyme that phosphorylates a myosin light chain.

It is thus obvious that the machinery of calcium regulation in different species and in different muscles of the same species is varied; it may be that these variations are related to the way in which calcium influences other cellular processes (glycolysis, secretion, cell motility, conductance of gap junctions) in different tissues. It should be noted that there is close similarity of the proteins involved in Ca^{2+} regulation of many different cellular processes. Troponin, the myosin light-chain 2 (phosphorylatable, P-type, Ca-binding light chains), calmodulin, and parvalbumin (a Ca^{2+}-binding protein present in high concentrations in some muscles) have strikingly similar amino-acid sequences and properties. They may all have evolved from the same ancestral gene.

SUMMARY

Skeletal muscle contraction occurs in the sarcomeres, the repeating structures of striated muscle. Aligned arrays of thick and thin filaments produce these striations and slide by one another as the muscle shortens. Force is generated when the S-1 portions of the myosin molecules (the cross bridges) that project out from the thick myosin filament interact with actin, which resides in the thin filament. This interaction with actin stimulates the dissociation of ADP and P_i formed on the myosin heads by ATP hydrolysis. ATP then binds and dissociates the actomyosin complex so that ATP is hydrolyzed in a cyclical interaction of the cross bridges with actin. This cyclical interaction propels filament sliding and muscle contraction. Evidence for cross-bridge interaction comes from biochemistry, electron microscopy, x-ray diffraction, and molecular-probe studies of a variety of muscles.

Intracellular Ca ions regulate cross-bridge interactions by several mechanisms. In vertebrate striated muscle, Ca^{2+} binds to a troponin on the thin filament. Together with tropomyosin, troponin regulates the actin-myosin interaction so that in the absence of Ca^{2+}, the ability to hydrolyze ATP and produce force is low, and in its presence the activity is high. In other muscles, Ca^{2+} binds to myosin, and in still others, Ca^{2+} regulates phosphorylation of myosin. Thus elevated internal Ca^{2+} allows productive cross-bridge interaction and initiates filament sliding, muscle shortening, force production, and a rapid transformation of stored chemical energy into mechanical energy. This contractile mechanism regulates motility in cardiac, smooth, and nonmuscle cells as well as skeletal muscle. The mechanical consequences of those molecular interactions are considered in the next chapter, which describes the properties of skeletal muscles and how they can be understood from the molecular basis of contraction.

ANNOTATED BIBLIOGRAPHY

Bagshaw, C.R. *Muscle Contraction (Outline Studies in Biology)*. London, Chapman and Hall, Ltd., 1982.
An excellent, short, readable review with a stress on molecular aspects of muscle contraction. The author studies the biochemistry of myosin and actin and writes clearly, using good illustrations and examples.
Huxley, A.F.; Niedergerke, R. Structural changes in muscle during contraction. *Nature* 173:971-977, 1954.
Huxley, H.E.; and Hanson, J. Changes in the cross-structure of muscle during contraction and stretch and their structural interpretation. *Nature* 173:978-987, 1954.
These two back-to-back papers propose the sliding filament mechanisms of muscle contraction. They are important historically because their carefully executed and interpreted experiments altered the direction of scientific thought.
Huxley, A.F. *Reflections on Muscle*. Princeton, N.J., Princeton University Press, 1980.
A stimulating little book by a central contributor to our understanding of muscle contraction that reviews the early work on muscle structure and contraction and the studies leading up to the sliding filament–cross-bridge theory. Later chapters discuss "certainties" and "uncertainties" about cross bridges and how they may work.

Squire, John. *The Structural Basis of Muscular Contraction*. New York, Plenum Press, 1981.
A long, comprehensive discussion of the structural—molecular and macromolecular—basis of contraction, stressing the interrelationships among techniques, observation, and interpretation, written by one who has contributed significantly to this literature. Muscles of many species are considered.
Sheterline, D. *Mechanisms of Cell Motility: Molecular Aspects of Contractility*. New York, Academic Press, 1983.
Although this book's main strength is an excellent comprehensive treatment of cell motility, in addition, there are up-to-date, well-written chapters on the muscle contractile proteins and their control. It provides good reading on the molecular basis of contraction and cell motility.

REFERENCES

1. Adelstein, R.S.; Eisenberg, E. Regulation and kinetics of the actin-myosin-ATP interaction. *Annu. Rev. Biochem.* 49:921-956, 1980.
2. Amos, L.A. Structure of muscle filaments studied by electron microscopy. *Annu. Rev. Biophys. Biophys. Chem.* 14:291-313, 1985.
3. Bagshaw, C.R.; Trentham, D.R. The characterization of myosin-product complexes and of product-release steps during the magnesium ion-dependent adenosine triphosphatase reaction. *Biochem. J.* 141:331-349, 1974.
4. Bagshaw, C.R. *Muscle Contraction; Outline Studies in Biology*, Chapman and Hall Ltd., London, 1982.
4a. Cooke, R. The mechanism of muscle contraction. *CRC Critical Reviews in Biochemistry* 21:53-118, 1986.
5. Cooke, R; Crowder, M.S.; Thomas, D.D. Orientation of spin labels attached to cross-bridges in contracting muscle fibres. *Nature* 300:776-778, 1982.
6. Craig, R; Greene, L.E.; Eisenberg, E. Structure of the actin-myosin complex in the presence of ATP. *Proc. Natl. Acad. Sci. USA* 82:3247-3251, 1985.
7. Ebashi, S. Calcium binding and relaxation in the actomyosin system. *J. Biochem.* 48:150-151, 1960.
8. Ebashi, S.; Endo, M. Calcium ion and muscle contraction. *Prog. Biophys. Mol. Biol.* 18:123-183, 1968.
9. Ebashi, S.; Kodama, A.; and Ebashi, F. Troponin. I. Preparation and physiological function. *J. Biochem.* 64:465-477, 1968.
10. Eisenberg, E.; Hill, T.L. Muscle contraction and free energy transduction in biological systems. *Science* 227:999-1006, 1985.
11. Gergely, J.; Seidel, J.C. Conformational changes and molecular dynamics of myosin. In *Skeletal Muscle. Hand. Physiol., Sec. 10*, 257-274, 1983.
12. Gordon, A.M.; Huxley, A.F.; Julian, F.J. The variation in isometric tension with sarcomere length in vertebrate muscle fibres. *J. Physiol. (Lond.)* 184:170-192, 1966.
13. Haselgrove, J.C.; Huxley, H.E. X-ray evidence for radial cross-bridge movement and for the sliding filament model in actively contracting skeletal muscle. *J. Mol. Biol.* 77:549-568, 1973.
14. Haselgrove, J.C. Structure of vertebrate striated muscle as determined by X-ray diffraction studies. In *Hand. Physiol. Sec. 10*, 143-172, 1983.
15. Heilbrunn, L.V.; Wiercinski, F.J. The action of various cations on muscle protoplasm. *J. Cell Comp. Physiol.* 29:15-32, 1947.
16. Herzberg, O.; James, M.N.G. Structure of the calcium regulatory muscle protein troponin-C at 2.8A resolution. *Nature* 313:653-659, 1985.

17. Huxley, A.F. Review lecture. Muscular Contraction. *J. Physiol. (Lond.)* 243:1-43, 1974.
18. Huxley, A.F.; Niedergerke, R. Structural changes in muscle during contraction. *Nature* 173:971-977, 1954.
19. Huxley, A.F.; Niedergerke, R. Measurement of the striations of isolated muscle fibres with the interference microscope. *J. Physiol. (Lond.)* 144:403-425, 1958.
20. Huxley, A.F.; Simmons, R.M. Proposed mechanism of force generation in striated muscle. *Nature* 233:533-538, 1971.
21. Huxley, H.E. The double array of filaments in cross-striated muscle. *J. Biophys. Biochem. Cytol.* 3:631-647, 1957.
22. Huxley, H.E. Electron microscope studies on the structure of natural and synthetic protein filaments from striated muscle. *J. Mol. Biol.* 7:281-308, 1963.
23. Huxley, H.E. Structural difference between resting and rigor muscle; evidence from intensity changes in the low-angle x-ray diagram. *J. Mol. Biol.* 37:507-520, 1968.
24. Huxley, H.E. The mechanism of muscular contraction. *Science* 164:1356-1366, 1969.
25. Huxley, H.E. Structural changes in the actin- and myosin-containing filaments during contraction. *Cold Spring Harbor Symp. Quant. Biol.* 73:361-376, 1973.
26. Huxley, H.E.; Brown, W. The low-angle x-ray diagram of vertebrate striated muscle and its behavior during contraction and rigor. *J. Mol. Biol.* 30:383-434, 1967.
27. Huxley, H.E.; Hanson, J. Changes in the cross-striations of muscle during contraction and stretch and their structural interpretation. *Nature* 173:978-987, 1954.
28. Lehman, W.; Szent-Györgyi, A.G. Regulation of muscular contraction. Distribution of actin control and myosin control in the animal kingdom. *J. Gen. Physiol.* 66:1-30, 1975.
29. Lowey, S.; Slayter, H.S.; Weeds, A.G.; Baker, H. Substructure of the myosin molecule. *J. Mol. Biol.* 42:1-29, 1969.
30. Lymm, R.W.; Taylor, E.W. Transient-state phosphate production in the hydrolysis of nucleotide triphosphate by myosin. *Biochem.* 9:2975–2983, 1970.
31. Lymm, R.W.; Taylor, E.W. Mechanism of adenosine triphosphate hydrolysis by actomyosin. *Biochem.* 10:4617-4624, 1971.
32. Marsh, B.B. The effects of adenosine triphosphate on the fibre volume of a muscle homogenate. *Biochem. Biophys. Acta* 9:247-260, 1952.
33. Marston, S.B.; Rodger, C.D.; Tregear, R.T. Changes in muscle cross bridges when β,γ-imido-ATP binds to myosin. *J. Mol. Biol.* 104:263-276, 1976.
34. Matsubara, I.; Elliott, G.F. X-ray diffraction studies on skinned single fibres of frog skeletal muscle. *J. Mol. Biol.* 72:657-669, 1972.
35. Mendelson, R.A.; Morales, M.F.; Botts, J. Segmental flexibility of the S-1 moiety of myosin. *Biochem.* 12:2250-2255, 1973.
36. Moore, P.B.; Huxley, H.E.; DeRosier, D.J. Three-dimensional reconstruction of F-actin, thin filaments, and decorated thin filaments. *J. Mol. Biol.* 50:279-295, 1970.
37. Morales, M.G.; Borejdo, J.; Botts, J.; Cooke, R.; Mendelson, R.A.; Takashi, R. Some physical studies of the contractile mechanism in muscle. *Annu. Rev. Phys. Chem.* 33:319-351, 1982.
38. Natori, R. The property and contraction process of isolated myofibrils. *Jikei Med. J.* 1:119-126, 1954.
39. Potter, J.D.; Gergely, J. The calcium and magnesium binding sites on troponin and their role in the regulation of myofibrillar adenosine triphosphatase. *J. Biol. Chem.* 250:4628-4633, 1975.
40. Reedy, M.K. Cross-bridges and periods in insect flight muscle. *Am. Zool.* 7:465-481, 1967.
41. Sherry, J.M.F.; Gorecka, A.; Adsoy, M.O.; Dabrowska, R.; Hartshorne, D.H. Roles of calcium and phosphorylation in the regulation of the activity of gizzard myosin. *Biochem.* 17:4411-4418, 1978.
42. Sheterline, P. *Mechanisms of Cell Motility: Molecular Aspects of Contractility*, New York, Academic Press, 1983.
43. Sobieszek, A.; Small, J.V. Myosin-linked calcium regulation in vertebrate smooth muscle. *J. Mol. Biol.* 101:75-92, 1976.
44. Squire, J.M. Muscle filament structure and muscle contraction. *Annu. Rev. Biophys. Bioeng.* 4:137-163, 1975.
45. Szent-Györgyi, A. Free-energy relations and contraction of actomyosin. *Biol. Bull.* 96:140-161, 1949.
46. Taylor, K.A.; Amos, L.A. A new model for the geometry of the binding of myosin cross-bridges to muscle thin filaments. *J. Mol. Biol.* 147:297-324, 1981.
47. Thomas, D.D. Large-scale rotational motions of proteins detected by electron paramagnetic resonance and fluorescence. *Biophys. J.* 24:439–462, 1978.
48. Thomas, D.; Cooke, R. Orientation of spin-labeled myosin heads in glycerinated muscle fibers. *Biophys. J.* 32:891-906, 1980.
49. Tregear, R.T.; Marston, S.B. The cross-bridge theory. *Annu. Rev. Physiol.* 41:723-737, 1979.

Contraction of Skeletal Muscle

INTRODUCTION

Skeletal muscle is called voluntary muscle because it is under direct neural control. Flexion of the elbow requires contraction of fibers in the biceps muscle. These muscle fibers are innervated by motor nerve fibers whose cell bodies lie in the spinal cord and from which axons project through the ventral roots to innervate the biceps muscle fibers. Each nerve fiber or axon forms a functional neuromuscular junction with several muscle fibers. Activation of the motoneuron activates all the muscle fibers it innervates. The motoneuron plus the muscle fibers that it innervates thus form a functional unit of contraction called the *motor unit*. The number of muscle fibers in a motor unit varies from muscle to muscle, ranging from 200 for leg muscles to 3 to 6 for extrinsic eye muscles. Not surprisingly, the tension developed by a motor unit varies with the number of muscle fibers in the unit, for they all contract together when the motoneuron discharges. Because the discharge of a motor unit provides the minimal increment of tension in a muscle, the innervation ratio deter-

mines fineness of gradation of contraction. Properties of motor units are discussed in Chapter 24.

The events leading to contraction are diagrammed in Figure 9–1A. Activation of a motoneuron (I) causes an action potential to propagate down the axon (II) to the neuromuscular junctions on the muscle fibers (III). Neuromuscular (III) transmission depolarizes the muscle fiber membrane at the end-plate (Chap. 6) and initiates an action potential that propagates in both directions along the muscle fiber (IV). This action potential leads to an elevation in the intracellular calcium concentration (V) in the muscle (Chap. 7). Calcium allows the contractile proteins to interact and generate force and muscle shortening (VI). The phasic contraction (Fig. 9–1B), which is the muscle response to a single stimulus, is called a *twitch*. The time course of the twitch varies in different muscles, being brief (50 ms) for fast muscles and more prolonged (several hundred milliseconds) in slower muscle fibers.

The central nervous system grades contractile force of a muscle by two mechanisms, *recruitment* of additional motor units and *summation* of contraction due to repetitive discharge in individual motor units (Chap. 24). An increased number of active motor units produces increased contraction by activation of more muscle fibers. Repetitive discharge of a single motor unit increases tension by summation of successive twitches as illustrated diagrammatically in Figure 9–2A. When the fiber is stimulated twice at an interval exceeding the twitch duration, two distinct, phasic twitches occur. At shortened interstimulus intervals, the twitches sum, producing more force than that of a single twitch. In contrast, the action potentials in the muscle fiber membranes remain discrete because of the all-or-none property and the brevity of the action potential.

The response of the muscle of repetitive stimulation of either the motor nerve or the muscle fiber

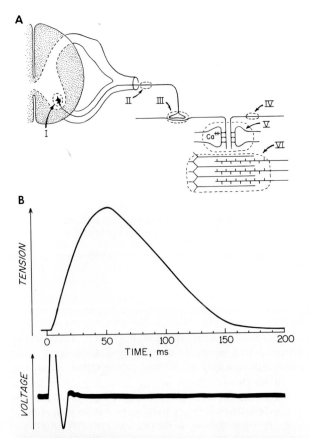

A

B

Figure 9–1 Events leading to contraction in a muscle. *A*, The structures responsible for the control of contraction in skeletal muscle and the various processes responsible are diagrammed and numbered. I, The motor neuron in the spinal cord. II, the motor axon; III, the neuromuscular junction; IV, the muscle cell membrane; V, the sarcoplasmic reticulum responsible for calcium storage and release in the muscle; and VI, the contractile filaments responsible for generation of muscle force and shortening. *B*, Tension (above) and action potential (below) recorded in cat tibialis muscle. Abscissa represents time; upper ordinate, tension in arbitrary units; lower ordinate, voltage in arbitrary units (upper deflection incomplete). Note that electrical activity starts several ms before contraction, but contraction far outlasts electrical activity. Action potential is diphasic because both electrodes are on active tissue. (After Creed, *et al. Reflex Activity of the Spinal Cord.* Oxford, Clarendon Press, 1932.)

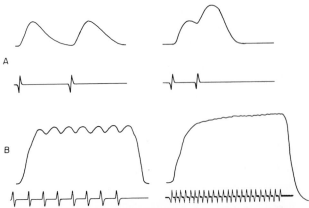

A

B

Figure 9–2 Responses of muscle to dual and repetitive electrical stimulation of the motor nerve. Upper traces, force; lower traces, electrical activity. *A*, Responses to two brief shocks at long (left) and short (right) intervals. With the short but not the long interval, the force sums, producing a greater total tension than that of a single twitch. Electrical responses do not sum. *B*, Responses to repetitive stimulation at low (left) and higher (right) frequency. In both, the force sums, creating a tension greater than that of a twitch. At the lower frequency, the force oscillates at the frequency of the stimulus; with the higher frequency, the force plateau is relatively smooth.

tracting muscles, down to about 40/s for slow muscles. The molecular mechanism responsible for summation of contraction is discussed later. The force that a muscle generates also depends on the muscle length and on its speed of shortening. Figure 9–4 shows an experiment demonstrating the dependence of force on length. As the muscle is stretched without stimulation, force increases as one expects for the passive length-force relation

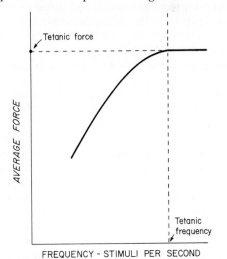

Figure 9–3 Relationship between average force and frequency of stimulation of motor nerve or muscle. With increasing frequency of stimulus, force increases to a maximum, the tetanic force. Further increases in stimulation frequency do not increase the force.

membrane is illustrated in Figure 9–2B. The force produced is much greater than that of a single twitch. An increase of the stimulus frequency leads to increased force up to a critical frequency called the *tetanic* frequency, after which further increase results in no further increase of force (Fig. 9–3). This maximal contraction at stimulation frequencies at or above the tetanic frequency is called a *tetanus*, or *tetanic contraction.* The tetanic frequency depends on the contraction speed of the muscle, varying from about 300/s for rapidly con-

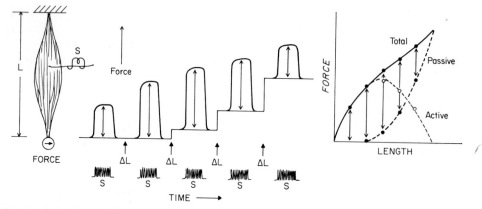

Figure 9–4 The length-tension relationship in muscle. Left, diagram of experimental arrangement; muscle length is varied, and tetanic tension is measured isometrically at each length. Resting and tetanic forces are shown in the traces in the middle of the figure; ΔL, length increases of 1 mm; S, stimulus. Plotted on the right are the relationships between passive (resting) and active (tetanic) and between total force and resting length. Passive resting tension increases continuously with length, whereas active force increases to a maximum and then declines.

of any elastic body. If the muscle is stimulated and its length is held constant, the resulting active force is termed *isometric*. If the stimulus is repetitive at a frequency above that required to produce tetanus, maximum active force is achieved. The active force added to the passive force yields a total force that in most muscles is low at short lengths and increases with increasing muscle length. The active tetanic force (total force minus passive force) also varies with length. It is low at short muscle lengths, increases to a maximum at a muscle length that is close to the length of the resting muscle in the body, and declines at longer muscle lengths. The molecular basis for this relationship involves the overlap of thin and thick filaments, as discussed in Chapter 8.

Maximum active force of muscle is usually expressed in terms of the cross-sectional area of the muscle, because muscles with larger cross sections generate higher forces. A single muscle fiber can generate a force of about 3 kg/cm² (the same number expressed in SI units is 0.003 N/m²). If all muscle fibers in the body were lined up in parallel, the resultant "muscle" would generate a force sufficient to support a 25-ton weight. During running, the gastrocnemius generates forces equal to approximately six times the body weight.

Contracting muscles shorten, lift loads, and perform work. During sustained shortening, the force equals the load being lifted. Shortening of a muscle against a constant load (as diagrammed on Fig. 9–5A) is termed an *isotonic* contraction. The speed of shortening varies with the load and with muscle type. Larger loads are lifted more slowly than

lighter loads (Fig. 9–5B). The maximum speed of shortening of fast muscle is nearly 20 muscle-lengths per second; a muscle 1 foot long thus shortens at a rate of 20 feet per second, or nearly 14 miles per hour.

Since the time of Helmholtz, biophysicists have regarded muscle as a machine. From the first, it was recognized as a device to convert chemical energy into mechanical work, with some inefficiency that produces heat. Indeed, it was in muscle that many of the energy-supplying reactions of intermediary metabolism were worked out. In this chapter, we outline the modern theory of how muscles pull. Having already discussed the major contractile proteins in the previous chapter, the emphasis here is first on the supply and utilization of metabolic energy and then on what can be learned by careful measurements of length, tension, and velocity of active muscle. These experiments provide the additional clues that inspired a provisional molecular description of contraction.

MUSCLE ENERGETICS

As we explained in Chapter 8, the immediate source of energy for muscle contraction is the high-energy phosphate compound adenosine triphosphate (ATP), whose hydrolysis yields energy, adenosine diphosphate (ADP), and inorganic phosphate. During isometric contraction in skeletal muscle, ATP is hydrolyzed at the rate of about 0.3 mM/twitch. The amount of ATP hydrolyzed increases in proportion to the work done by the

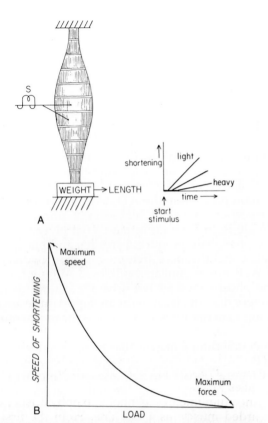

Figure 9–5 Relationship between the muscle speed of shortening and the load. *A,* Diagram of experiment to measure speed of shortening under varying loads. The muscle is stimulated tetanically through its motor nerve. The weight, initially supported, is not lifted until the muscle force equals the load. Shortening as a function of time for three different loads, light to heavy (from above downward), is shown on the right. The slope is the speed of shortening. *B,* Plot of data obtained from *A* showing an inverse relationship between speed and load. The load the muscle cannot lift is the maximum isometric force.

muscle. On a short-term basis, ATP levels are maintained by enzymatic transfer of phosphate from another high-energy phosphate compound, creatine phosphate (CP), to ADP. Eventually, the muscles must restore depleted ATP and CP by slower metabolic processes.

Heat Production in Muscle. Muscles do work by converting stored chemical energy into mechanical energy. In the process, heat is evolved. During an isometric twitch, a frog muscle generates about 0.003 calorie (0.013 J) per gram of tissue, hence its temperature rises about 0.003°C. During this process, the muscle hydrolyzes about 0.3 mM of ATP, which is rapidly restored by rephosphorylation of ADP from creatine phosphate and eventually by anaerobic and aerobic metabolism. In this section, we consider heat production of mus-

cle and the changes in high energy phosphate compounds that occur during contraction. The review by Kushmerick[22] and the book by Woledge, Curtin, and Homsher[37] are the most recent comprehensive discussions of this subject.

Measurement of the minute temperature changes accompanying muscle contraction requires extremely sensitive multijunction metal thermocouples or thermopiles and careful corrections for heat losses, techniques developed primarily by A.V. Hill. (For an engaging discussion of this subject, the reader is referred to Hill's book.[13]) Heat liberated during contraction and relaxation is called initial heat; the increased heat production that persists for minutes following contraction is called recovery heat. Recovery heat is associated with metabolic reactions required to return the muscle to its initial state; it requires oxygen.

Attempts have been made to associate initial heat with the various aspects of contraction such as activation, force production, the work done, and muscle shortening.[22, 37] Because these multiple exothermic processes occur simultaneously along with endothermic ones during contraction, assignment of their individual roles in heat production is difficult and requires special experimental procedures. An example is the procedure used to measure the heat associated with activation. Heat should be measured when the muscle is activated but not producing force, a condition that can be achieved by measuring heat in a muscle stretched so far that it cannot generate active force (Fig. 9–4; see also Fig. 8–14). Under these conditions, heat production is approximately 30% of that liberated during maximal isometric contraction.[14] Assuming that activation is not affected by stretch, we could thus conclude that about 30% of the total energy is consumed in "activation." Careful analysis has actually shown that most of this energy consumption is associated with relaxation—calcium transport back into the sarcoplasmic reticulum—although other steps in the complex activation process contribute.

Fenn[8] investigated the extra energy liberated when work is being done. He found that muscle mobilizes additional energy in proportion to the work performed. This so-called Fenn effect disproved the then current viscoelastic theory of muscle contraction, a theory postulating that activation converts muscle into an elastic body with an equilibrium length less than the existing muscle length, so that it shortens like a stretched rubber band at a rate depending on muscle viscosity and does work in the process. This theory assumes that the muscle when stimulated is endowed with all of the energy needed for the subsequent contraction. Fenn's work showed that in contrast, the energy production varies in accordance with the amount of work the muscle is required to perform and that the muscle can thus modify its energy output while contraction proceeds. This circumstance demonstrates a coupling between the mechanical behavior of the muscle and the rate constants of energy turnover, the ATPase cycle. Muscles show reasonable efficiencies, the ratio of work to heat-plus-work being about 0.4.

Metabolic Supply of Free Energy. Muscle derives its energy from the free energy stored in chemical bonds of muscle metabolites. The immediate energy source is free energy derived from the hydrolysis of the terminal phosphate group of ATP, an energy-yielding reaction that is used in each twitch to drive the cross-bridge interactions, the calcium transport system of the sarcoplasmic reticulum, and the membrane Na-K pump system. The ATP consumed must eventually be resynthesized by cellular metabolism of glucose or glycogen, fatty acids, or other metabolic intermediates.

The metabolic pathways for glycolysis and oxidative metabolism in muscle are like those of many cells. Glycolysis occurs in the cytoplasm. It splits glycogen into its constituent glucose units and metabolizes them to pyruvate or lactate with the net generation of two molecules of ATP per glucose molecule (in the absence of oxygen). Like contraction, glycolysis can be triggered by the elevation of sarcoplasmic free calcium during activity. The raised $[Ca^{2+}]_i$ activates phosphorylase kinase, which in turn activates phosphorylase, the enzyme that initiates the breakdown of glycogen to glucose-1-phosphate to start glycolysis. Glycolysis is also stimulated by extracellular epinephrine and by decreased intracellular ATP and increased ADP concentrations. It is generally synchronized with cellular oxidative metabolism.

Oxidative metabolism or oxidative phosphorylation coupled with the citric acid cycle requires oxygen and converts the products of glycolysis and other metabolites to carbon dioxide and water plus about 30 molecules of ATP per molecule of glucose. Glycolysis, although operative anaerobically, is more efficient in the presence of oxygen and can then produce six, rather than two, molecules of ATP per molecule of glucose. Thus under optimal circumstances, the metabolism of a molecule of glucose can yield almost 36 molecules of ATP. Together, glycolysis and oxidative phosphorylation regenerate the ATP used in contraction.

Another source of high-energy phosphate in muscle is creatine phosphate. Hydrolysis of creatine phosphate producing creatine and phosphate yields slightly more free energy than hydrolysis of ATP under similar conditions. Although creatine phosphate does not drive the contractile machinery directly, it serves as an important reservoir of high energy phosphate groups for regeneration of ATP. The reaction ADP + CP ↔ ATP + C has an equilibrium constant that favors production of ATP.

Creatine phosphate may also serve another role: to shuttle energy from the mitochondria to the myofilaments. ATP is generated by oxidative phosphorylation in the mitochondria and is used by the myosin ATPase at the myofilaments. Creatine phosphokinase (CPK) occurs in both the mitochondrial outer membrane and in a different form at the myofilament (in the M line) and free in the cytoplasm. Thus the enzymatic machinery is present at the mitochondria to transfer the high-energy phosphate from ATP to CP and again at the myofilament, where the reverse reaction would yield ATP (Fig. 9–6). This hypothesis is supported by some experimental evidence but remains speculative.[3]

Finally, muscle tissue contains the enzyme myokinase, which catalyzes the conversion of two

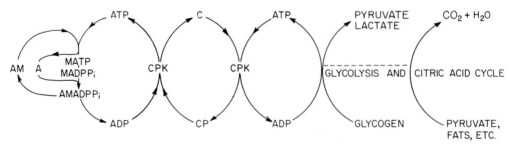

Figure 9–6 Summary diagram showing the breakdown of ATP by myosin ATPase, on the left, and its resynthesis via enzymatic transfer of high-energy phosphate from creatine phosphate [CP] and, eventually, from glycolysis and the citric acid cycle through oxidative phosphorylation. CPK is creatine phosphokinase. There may be two distinct sites of CPK action. The site on the left could be at the M line; the one at the right, on the mitochondrial external membranes, as discussed in the text.

ADP molecules to one ATP and one AMP (adenosine monophosphate). Myokinase thus provides another back-up route for regenerating ATP. It probably functions only in extreme circumstances, since it operates only at unphysiologically high ADP concentrations.

Taken together, the multiple synthetic pathways for ATP provide muscles with latitude to function for limited periods under conditions in which ATP hydrolysis exceeds synthetic replacement via the glycolytic and oxidative pathways. The debts thus established are eventually repaid through the combined operation of the synthetic pathways (Fig. 9–6).

Cellular concentrations of ATP, CP, ADP, and P_i can be measured in several ways. One is by direct chemical analysis of quick-frozen tissue. Such analyses are satisfactorily sensitive but subject to error from possible chemical changes during freezing. Also, direct analysis allows only one measurement for each tissue, so that the tissue cannot serve as its own control. A more recently introduced technique is phosphorus nuclear magnetic resonance (NMR).[22] The ^{31}P nucleus has a magnetic moment. In the presence of a magnetic field, it takes on quantized energy levels whose values depend on both the applied magnetic field and the local magnetic field due to the local environment of each phosphorus atom. Phosphorus ^{31}P present in CP, in the α, β, and γ positions of ATP, and in P_i can be distinguished by NMR. Figure 9–7 shows an NMR spectrum for muscle. The position of the peak is characteristic of the particular phosphorus species (CP, P_i, etc.). The area under each peak measures the concentration of that species. This technique allows for nondestructive measurement, but the signal from phosphorus is small, so that a number of spectra must be averaged to measure and detect ATP, CP, and P_i in physiological concentrations. Even with these limitations, changes in concentrations can be followed in time with each muscle serving as its own control. These measurements are now done not only in isolated tissues but also in situ on living subjects. The ^{31}P NMR spectrum can also be used to calculate intracellular pH since the position of the inorganic phosphate peak depends on ratio between HPO_4^{2-} and $H_2PO_4^-$, which in turn depends on the pH.

Study of high-energy phosphate metabolism in muscle is facilitated by use of metabolic poisons that selectively block one or another synthetic pathway. Glycolysis is readily blocked by iodoacetic acid, which inactivates the essential enzyme glyceraldehyde dehydrogenase, and oxidative phosphorylation can be selectively blocked either by depriving the tissue of oxygen or by lowering its temperature. Even when the glycolytic and oxidative pathways are both blocked, muscle can still contract many times, because phosphate groups are enzymatically transferred from stores of creatine phosphate to ADP to re-form ATP. In anoxic, iodoacetic acid–poisoned muscle, creatine phosphate levels drop, and free creatine increases in direct proportion to the number of twitches elicited. When in addition, creatine phosphokinase is inactivated by the poison fluorodinitrobenzene (FDNB), ATP concentrations fall,[4] and contractions soon fail even when creatine phosphate levels are high. The failure indicates both that ATP is broken down directly in contraction and that creatine phosphate is not in itself available for hydrolysis by myosin.

When the ATP-synthetic pathways and creatine phosphokinase are blocked in frog muscle, contractions deplete ATP at the rate of 0.3 mM of ATP per twitch.[24] During a tetanus, about 1.5 mM of ATP is used per second.[24] In mouse rapidly contracting muscles (extensor digitorum longus, EDL) and slower contracting muscles (soleus) these numbers are 4 and 1.3 mM/s during a tetanus.[22] Hydrolysis of one mole of ATP liberates about 65 kJ (Chap. 1), which supports activation and generation of force and mechanical work done. Attempts to partition the energy between these processes and to correlate them with breakdown of high-energy compounder are complicated, because many of the processes occur simultaneously, but some results are available.[22, 37]

Table 9–1 lists the energy stores in two different muscles. Surprisingly, the store of ATP is small. The creatine phosphate stock is more generous

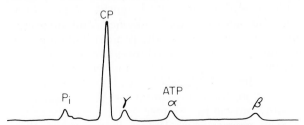

Figure 9–7 ^{31}P nuclear magnetic resonance (NMR) spectrum from frog sartorius muscle. The absorption of the applied radio waves by the ^{31}P atoms in the muscle placed in a magnetic field depends on the local environment of the phosphorus atoms. The absorption for ^{31}P present in inorganic phosphate (P₁), creatine phosphate (CP), and the various positions of ATP (α, β, and γ) are identified in the spectrum above. The area of each peak gives a measure of the relative quantity of ^{31}P present in each species. (From Dawson, Godian, and Wilkie. *J. Physiol.* 267:703 735, 1977.)

Table 9–1 Stored Energy in Muscle

Muscle	Metabolite	Utilized Through	Contents (mmol/kg)	ATP Used/Tetanic Rate	Provides Energy for Contraction (Seconds)
Mouse EDL	ATP	ATP → ADP + P_i	6	4 mmol/kg-s	1.5
	CP	CP + ADP → ATP + C	24		6
	Glycogen	Glycolysis (no O_2)	41 (hexoses)		20.5
		Glycolysis (+ O_2)	41 (hexoses)		41
	Pyruvate from glycogen only	Citric acid cycle plus oxidative phosphorylation	82		308
					377
Mouse soleus	ATP	ATP → ADP + P_i	5	1.3 mmol/kg-s	4
	CP	CP + ADP → ATP + C	13		10
	Glycogen	Glycolysis (no O_2)	30 (hexoses)		46
		Glycolysis (+ O_2)	30 (hexoses)		92
	Pyruvate from glycogen only	Citric acid cycle plus oxidative phosphorylation	60		696
					844

but even so is less than abundant, especially when one considers the phosphate requirements of repeated muscle contractions. The glycolytic pathway, acting as an anaerobic back-up system, can generate substantially more energy. Finally, however, in sustained muscular effort, the muscle must be supported by oxidative metabolism.*

Table 9–1 shows that different muscle types manifest only small differences in amounts of metabolites but show larger differences in rates of ATP utilization. Mouse EDL fibers have high shortening velocities and contain a myosin isozyme with a high intrinsic actin-activated ATPase rate. In contrast, most soleus muscle fibers have slower shortening velocities and contain a myosin isozyme with a relatively low actin-activated ATPase rate (see below). The more slowly contracting soleus muscle utilizes its energy stores more efficiently for force production than does the "fast" muscle.

Table 9–1 shows the amount of energy available from different sources but gives no information about the rate at which energy can be supplied from each source. The rate of supply depends on the enzymes involved, both the quantity and specific rate of catalysis, and the availability of O_2 for oxidative phosphorylation or other required factors. Thus the activity of enzymes for glycolysis, citric acid cycle, and oxidative metabolism; the

*The data in Table 9–1 understate the resources that can be drawn on by listing cellular glycogen as the only energy source. In muscle active in vivo, cellular carbohydrate supplies are abundantly supplemented by blood-borne glucose—a source that homeostatically increases during muscular exercise. Metabolism of fatty acids will also increase the energy available and may serve as the primary source of energy under aerobic conditions in many muscles.

relative storage of O_2; and the rapid exchange of metabolites and O_2 via the normal blood circulation all affect the rate of utilization of the "available" energy supplies.

Discrepancy between demand and supply rates leads to muscle fatigue and a decline in muscle performance due to either depletion of the energy supplies or accumulation of P_i. The molecular mechanism of muscle fatigue is not well understood. During sustained exercise, creatine phosphate levels decline while ATP levels remain constant, changes that are consistent with CP serving as a buffer for ATP. Intracellular pH initially rises as CP is utilized (a proton is absorbed) and finally decreases as ATP is utilized (a proton is liberated). Sustained anaerobic metabolism decreases the pH further as a lactic acid is produced.

As would be expected, factors that enhance the oxidative capability of muscle increase its resistance to fatigue. These factors include an increased capillary supply, increased density of mitochondria and mitochondrial enzymes, and increased amounts of myoglobin, the protein that stores O_2 in muscle. All of these features are prominent in slowly contracting muscles and enable them to sustain their ATP levels by oxidative metabolism. In addition, the requirement for ATP is decreased in slow muscle by a lower ATPase activity of the myosin isoenzymes. Factors that regulate gene expression controlling myosin isoenzymes and cellular metabolism are discussed in Chapter 24.

MUSCLE MECHANICS

Muscles are mechanical structures that generate force, shorten, lift weights and perform work both on external objects and on body structures. The classic mechanical models of Fenn and Marsh[10] and Hill[12, 13] conceived of muscles as composed of two kinds of elements: elastic elements that do not change with stimulation and a contractile element that does. The model comprised at least two elastic elements, one in series and one in parallel with the contractile element. The Fenn-Marsh-Hill model satisfactorily described many properties of muscle but is too simple to explain others. We now show how our modern understanding of the molecular basis of contraction can be combined with the measured mechanical properties to give a unified model. Excellent reviews of the subject are those of Podolsky and Schoenberg[28] and Simmons and Jewell.[34] In most of the studies to be described, the muscle is placed in a steady state of activation by tetanic stimulation. In this case, time after stimulation is not a variable. Phasic twitch contractions are used in investigations of the effects of activation.

Length-Tension Relationship. As we have seen, force in an unstimulated muscle increases as it is passively stretched (Figs. 9–4 and 9–8). When the muscle is stimulated tetanically with its ends fixed, an isometric contraction, it produces additional

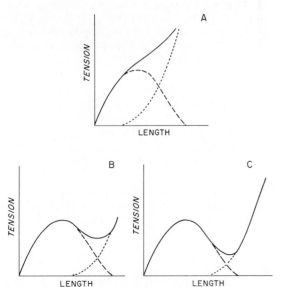

Figure 9–8 Effect of connective tissue content on length-tension curves of muscle. Dotted lines are passive or unstimulated tension, solid lines are the total muscle tetanic tension and the dashed lines, the difference representing the active muscle tetanic tension. The muscle in *A* has more connective tissue and, hence, much more resting or passive tension than those in *B* or *C*. The amount of connective tissue greatly influences the total tension maximum. (After Carlson, F. D.; Wilkie, D. R. *Muscle Physiology.* Englewood Cliffs, N.J., Prentice-Hall, 1974.)

active force that is also length-dependent. In the early models, the passive force curve was ascribed to elastic elements that are unaltered by muscle activation and are arranged in parallel with the contractile elements. The total "length-tension" curve* of the active muscle was ascribed to the influence of the contractile elements superimposed on the passive (elastic) force curve. Subtraction of the passive from the total force was regarded as giving the contribution of the contractile elements (Fig. 9–8). Ramsey and Street[29] showed that the total force is the sum of an active and a passive component provided by two elements in parallel. The elastic element is mainly the connective tissue surrounding the muscle fibers but also has an intracellular component, since even "skinned" muscle fibers denuded of sarcolemmal and external connective tissue component still display elasticity.[26] As shown in Figure 9–8, the passive tension curve varies from muscle to muscle and correlates well with connective tissue content, being high in gastrocnemius, lower for sartorius, and even less for a single fiber, which would be virtually free of connective tissue. In contrast, the active length-tension curves of different muscles are more nearly uniform.

The relationship between the active length-tension curve and the sliding filament model of contraction has already been discussed and illustrated in Figure 8–14 with observations on a single muscle fiber. Here, the passive tension was insignificant, except at very long sarcomere lengths, and could be ignored. In the discussion in Chapter 8, we showed that the shape of the curve can be understood if cross bridges are the site of force production and the force per cross bridge is a constant. At short sarcomere lengths, we argued that force declines because of collisions of filaments and other parts of the sarcomere structure. In addition, because muscles contract at a constant volume, shortening is accompanied by increased cross-sectional area, a change that may be resisted by the connective tissue surrounding the muscle fibers. Finally, there is some evidence that activation by Ca is compromised at short sarcomere lengths.[32] In any case, a restoring force clearly exists at short sarcomere lengths; a muscle stimulated to contract to a short sarcomere length reextends to a sarcomere length of about 2.2 μm when the stimulus is withdrawn.

If steady-state activation is not achieved, the shape of the active length-tension curve is differ-

---

*The word tension has traditionally been used in studies of isometric contraction. It is synonymous with force.

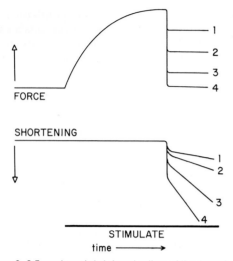

FORCE

SHORTENING

STIMULATE

time ⟶

Figure 9–9 Experimental determination of the force/velocity curve for contracting muscle. The muscle is stimulated tetanically during the time indicated, and the force is measured isometrically, as shown in the upper trace. Force increases to a plateau, at which point it is released to shorten, lifting a load indicated by the forces in 1, 2, 3, and 4. The lower trace shows the muscle shortening. Since the contraction is initially isometric, there is no shortening until the muscle is released and allowed to lift a load. After the initial rapid shortening, the muscle attains a steady shortening velocity that is greater for lighter loads (4) than for heavier loads (1).

ent. In frog muscles, the sarcomere length corresponding to maximal active force is greater for twitches than for tetani.[6] In cardiac muscle, where almost any contraction is a twitch, maximal active force falls off more rapidly with decreasing sarcomere lengths than in skeletal muscle.

Force-Velocity Relations. When a muscle lifts a load, the rate of shortening (speed) soon becomes constant at a value that varies inversely with the load. Figure 9–9 shows the length and force changes in a contracting muscle that is first held isometric but is then suddenly permitted to shorten and lift various loads. The loads are less than the isometric tension established prior to release. Shortening occurs in three phases: a brief high-speed shortening followed by a transition phase to a more prolonged shortening at a lower but constant speed. The stable speed (as indicated by the slope of the length-time record) varies inversely with the load, large loads being lifted more slowly than light loads.

The initial rapid change in length is consistent with a model in which a passive elastic element is arranged in series with a contractile force-generating element. A tentative length-tension curve can be plotted for the series-elastic element by

measuring the muscle length change as a function of the change in force (difference between isometric tension and the imposed load). As the curve is not linear, the stiffness of the elastic element is often given as the change of muscle length required to unload the muscle and bring the force down to zero. This has a value between 1% and 10% of the muscle length, depending on the muscle and the elasticity in the mechanical coupling of the recorder to the muscle.

Following the rapid length change, there ensues a brief period of transition before a steady shortening velocity is attained (see Fig. 9–14). Once thought to be a mechanical artifact, this transient was shown by Podolsky[5] to be real. The information it gives about the contractile mechanism is discussed later.

The relationship between the load and the steady velocity of shortening (the force-velocity relation) diagrammed in Figure 9–5 is drawn in detail in Figure 9–10A. Several equations have been fitted to this curve; the best known is Hill's characteristic equation:[12] $(P + a)(V + b) = c$, where P is the load (muscle force), V the velocity, and a, b, c are constants. Hill initially derived this equation from measurements of heat and work in muscles;[12] although it was later shown that his assumptions were incorrect, the equation nevertheless describes the data fairly well. At higher loads, the relationship breaks down, and the velocity becomes somewhat independent of load as can be seen in Figure 9–10A. The values of a, especially, but also b, which together determine the shape of the curve, differ somewhat from muscle to muscle. Since V_{max} (the maximum shortening velocity at zero load) is given by the expression bP_o/a, where P_o is the maximum isometric force, b is related to the maximum shortening velocity and varies greatly as does V_{max} with different muscles.

As discussed in the previous section on energetics, variation in the velocity of shortening is correlated with the myosin ATPase activity. The correlation holds both for muscles from one animal (Fig. 9–11) and for muscles taken from different animals. Barany[2] found that for shortening velocities varying over about 4 orders of magnitude, the actin-activated myosin ATPase varies in almost the same proportion. It has been speculated that the specific rate constant underlying this correlation is the rate of dissociation of ADP from the actin-myosin ADP complex (see Chapter 8).[33]

If the load applied to the stimulated muscle is greater rather than less than the isometric tension, the muscle stretches but resists the stretch. After

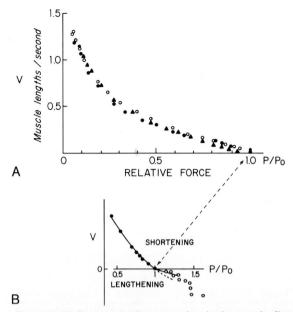

A

B

Figure 9–10 Force velocity curves for single muscle fibers and for whole muscles. *A* shows the steady shortening velocity plotted against the muscle load, expressed as a fraction of the total isometric tension. Data from three different sarcomere lengths, open circles 2.03 μm, closed circles, 2.23 μm, and triangles, 2.43 μm. (Data from Edman and Hwang. *J. Physiol.* [Lond.] 269:255–272, 1977.) *B* shows the shortening and lengthening velocities in a whole muscle as the load is changed from less than isometric to greater than isometric. The closed circles are shortening velocities and the open circles, lengthening velocities. The solid line shows the shortening velocity with the dotted portion extrapolated for lengthening through the isometric tension point. The actual velocity of lengthening is less than predicted by extrapolation, thus the muscle is stiffer in lengthening than in shortening. When the force reaches about 1.5 times isometric tension, P_o, the muscle gives, and the lengthening velocity increases rapidly. (After Aubert, X. *Le couplage énergétique de la contraction musculaire.* Thése d'agregation, Université Catholique de Louvain. Editions Arscia, 1956.)

because it produces too little force suddenly becomes stiffer, requiring forces substantially higher than isometric to stretch it. This gives the muscle greater stability. At forces above about 1.6 times the isometric force, however, muscles tend to give and elongate rapidly.[21]

Mechanical Transients. In the preceding section, we considered the response of muscle to a step change in *load* applied during steady isometric contraction. We now consider the effect of a step decrease in *length* applied to a contracting muscle. The focus will be on the force transients as the muscle adjusts to its new length. In response to a large step decrease, the force drops to nearly zero and then redevelops to a value consistent with the force corresponding to the new muscle length as defined by the length-tension relationship. The

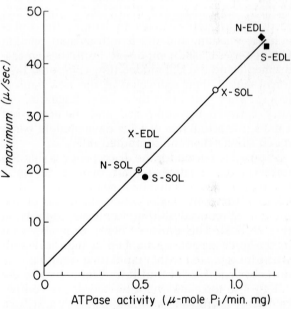

Figure 9–11 Relationship between maximum shortening velocity (V_{max}) and the ATPase activity of myosin activated by actin. Different points are different muscles under various innervation conditions. All muscles are from the rat. SOL is the soleus muscle, predominantly composed of slow-twitch muscle fibers. EDL, the extensor digitorum longus muscle, is predominantly fast twitch muscle fibers. N is normally innervated, S is self-innervated (reinnervated following section of the motor nerve), and X is cross innervated (reinnervated by a nerve normally supplying a muscle of different composition; see Chap. 23). Maximum shortening velocity is correlated with ATPase activity. Soleus muscle is normally slower and has less ATPase activity than EDL muscle. Cross innervation changes both the speed and ATPase activity of the soleus toward that for a normal EDL muscle and changes speed and ATPase activity of the EDL muscle toward that expected for the soleus. (After Barany and Close. *J. Physiol.* [Lond.] 213:455–475, 1971.)

an initial increase in length, it lengthens at an approximately constant rate. The force-velocity relationship of lengthening is shown in Figure 9–10B. The scale is expanded to show clearly the change in slope of the curve at the transition from shortening to lengthening at the isometric tension point. For forces greater than the isometric tension, the velocity of lengthening is much less than the velocity of shortening at the corresponding increment below isometric force. Thus, the muscle appears stiffer when it is forcibly lengthened than when it is shortened, since higher forces are required to achieve a lengthening velocity than to achieve a comparable shortening velocity.[21] A muscle (or a sarcomere) that is being stretched redevelopment of force is more rapid than the initial rise in force at the beginning of the tetanus (see Fig. 9–15).

A.V. Hill ascribed this rapid fall and subsequent redevelopment of force to the response of a contractile and an elastic component in series. The elastic component can shorten much more rapidly than the contractile component and unloads it. The contractile component then shortens more slowly, restretching the elastic component and redeveloping the force. Hill hypothesized that this latter shortening took place at a velocity set by the force-velocity relationship for the contractile component. The properties of the elastic components vary substantially in different muscles and different experimental conditions; in whole muscle, a 4% or 5% change in length usually causes the force to drop from the isometric value to zero, but in a single fiber, a 1% change may suffice (see later). The Hill model was tested by Jewell and Wilkie,[19] who measured in the same muscle the response to step changes in length, the force-velocity relation, and the length-tension relation for the series elastic component. From these measurements and the Hill model, they computed how force should then redevelop at the new muscle length. They found that there was a large discrepancy between the computed and the measured values, the actual rate of force development being much slower than the computed rate.

A possible reason for the discrepancy is that the force-velocity relation is not obeyed instantaneously, as the calculations assume.[5] Instead, a velocity transient occurs when the force of the muscle is reduced during isometric contraction (Fig. 9–12A). This transient behavior is also seen in the force response to a step change in length when investigated with a rapid time resolution, as done by A.F. Huxley and R.M. Simmons (Fig. 9–12B).[16] The force and velocity records are complicated but mirror one another, displaying at least four phases rather than the simple two-phase sequence described earlier. Huxley and Simmons[17] concluded that each cross bridge includes a series elastic element and that the transient changes in force may be attributed to transient changes in attached cross-bridge force generators. In this concept, a simple two-component model is inadequate to describe muscle characteristics, because both elastic and contractile elements are contained in one and the same intracellular structure.

The foregoing should not be taken to mean that a muscle in situ has no series elasticity contributed by elements such as the tendon. The muscles in the Huxley and Simmons experiment were pre-

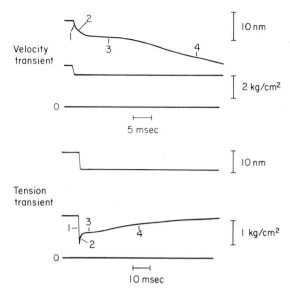

Figure 9–12 Mechanical transients in frog single muscle fibers during tetanic stimulation. The upper record shows the velocity transients, i.e., the length change (upper trace) in response to a step change in force (lower trace). Four phases of shortening are labeled: (1) a rapid drop accompanying the step change in force, (2) a rapid shortening phase, (3) a slowing or reversal of shortening, and (4) a final steady shortening, the steady velocity of shortening described previously. The lower record shows the tension transients (lower trace) in response to the step change in length (upper trace) that occur during the plateau of isometric tetanic tension. Again, four phases of the transient are labeled: (1) force accompanying the length step, (2) the rapid redevelopment of force, (3) a slowing down and reversal of this redevelopment, and (4) a final recovery of force back to the maximum level. The numbered phases of the two types of transients correspond to the same processes. (After Huxley, A. F. *J. Physiol. [Lond.]* 243:1–43, 1974.)

pared in a special way to minimize series elasticity so that a length change of less than 1% was needed to reduce the full isometric force to zero. In situ, the series elasticity is greater because of the contributions of the muscle tendons.

Every twitch of a muscle produces a force transient. The build-up of force at the beginning of the twitch depends on at least three processes. First, activation of the muscle causes release of stored calcium, which binds to the control proteins. The second step is the attachment of force-generating bridges within the muscle and the development of force. The third step is the stretching of any series elastic elements. At every step there is some additional delay, so as shown earlier, the sarcoplasmic calcium concentration rises more rapidly than does the final force production (Fig. 9–13; see also Fig. 7–2). Calcium binding to produce activation presumably follows the rise in intracellular calcium, and dissociation proceeds after the free calcium declines. Activation permits cross-bridge interaction, and subsequent delays in force production include the time to generate force by the bridges and to stretch whatever series elasticity lies external to the cross bridges in the tendons or other elements, such as the Z lines. Only then does force appear externally. The many steps preceding appearance of force account for the variable time courses of twitches in different muscles, fast muscles with high maximum velocity of shortening showing more rapid force development than slow muscles. Variations in the stiffness of the elastic elements and in the time course of calcium release may also affect the rising phase of twitch tension.

The declining phase of twitch contraction depends on calcium uptake by the sarcoplasmic reticulum. As free calcium declines, calcium leaves the activating sites, but force is sustained until the force-generating cross bridges break; there is good evidence that force production outlasts the calcium activation (Fig. 9–13). Thus, the rate of relaxation depends on the rate of sequestration of calcium by the sarcoplasmic reticulum, the rate at which bound calcium dissociates, and the speed with which cross bridges once made can be broken, bringing about ultimate relaxation.[20]

The intracellular Ca and the force transient last much longer than the membrane action potential that initiates them. Therefore, the Ca transients from two closely spaced action potentials will sum, and the resulting force may be greater. This is illustrated in Figure 9–13 for a barnacle single muscle fiber in which free calcium and force are simultaneously measured during a twitch. A second stimulus delivered shortly after the first produces extra force that adds onto or sums with the first as illustrated in Figure 9–13 B, C, and D (see also Fig. 9–2A). What is clear from this figure is that the additional force derives from recruitment of more force-generating elements that add to those persisting from the first stimulus. Even though less calcium is released by the second depolarization (the calcium-releasing process takes time to recover) the force produced by the second stimulus exceeds that from the first. It is presumed that the time course of Ca bound to troponin is

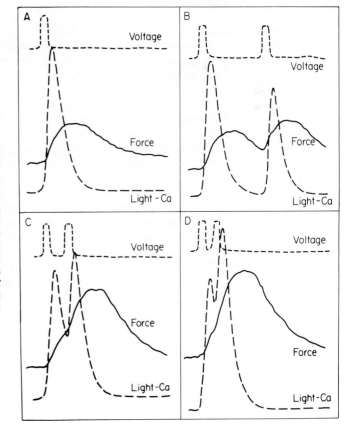

Figure 9–13 Summation of tension in a barnacle single muscle fiber. A, Response to a single stimulus. Membrane voltage, light (from the Ca-sensitive photoprotein aequorin, microinjected into the fiber, which luminesces after it binds Ca), which is a measure of free Ca concentration, and muscle isometric force. Note that the rapid voltage change produces a somewhat slower, prolonged rise in intracellular free Ca, which in turn produces a more slowly rising, longer lasting isometric force. B, C, and D show the measured response to two stimuli (voltage pulses) of equal height separated by a decreasing interval. In B note that the Ca released by the second pulse is actually less than the Ca released by the first pulse, but the force is increased by the second stimulus; the force sums. In C and D, the interval between the two stimuli is decreased so that the Ca concentration sums and force is elevated even further.

intermediate between free calcium concentration and force development, since the presumed steps in calcium regulation are $2Ca + Tn \rightarrow Ca_2Tn \rightarrow$ thin filament activation $\rightarrow$ cross-bridge attachment $\rightarrow$ force and shortening. Thus during force rise after the initial stimulation, Ca binding, thin filament activation, and cross-bridge attachment each precede force development. Force summation can occur through increases in each of the terms in the sequence but eventually is displayed as additional, attached cross bridges.

Tetanus is a similar fusion of Ca transients and force transients occurring during repetitive stimulation. Here, the free Ca builds up to a level at which release and uptake balance (see Fig. 8–16). The plateau value depends on the relative rates of these two processes. In some cells, such as heart muscle, there can be hardly any fusion and no tetanus. In these cells, each action potential is followed by such a long period of refractoriness that the Ca transient and the force transient have fully died away before the next action potential can be initiated. Hence the conditions needed for summation do not exist.

THEORIES OF CONTRACTION

We have argued earlier that contraction takes place through interaction between actin-containing thin filaments and myosin-containing thick filaments, so that the filaments move with respect to one another. Fenn[9] concluded from his studies of the extra energy utilization accompanying muscle work that muscle contraction, or shortening, takes place in a manner analogous to a "chain and windlass" such that "every link of the chain which is wound up involves the expenditure of so much energy at the moment of winding."

In 1957, soon after the birth of the sliding filament theory, A.F. Huxley[15] proposed a model of interaction between actin and myosin to account for filament sliding; it is diagrammed in Figure 9–14A. M is a reactive site on the myosin that oscillates about its equilibrium position and may attach to site A. When attached to A, M tends to return to the equilibrium position because of the force generated in the stretched spring. Huxley selected a linear spring so that the force produced by the bridge would be proportional to the displacement of M and would either aid or impede muscle shortening, depending on which side of the equilibrium point the attachment occurred. Huxley proposed that the rate constants for attachment and detachment depended on the dis-

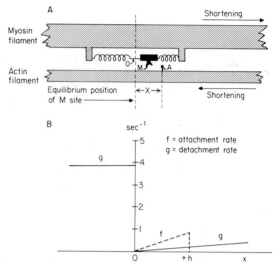

Figure 9–14 *A,* Model of contraction proposed by A. F. Huxley. The M site on the thick filament oscillates about an equilibrium position; it can attach to the A site on the thin filament. In *B,* the rate constants for attachment (f) and detachment (g) are shown as a function of the distance x between A and M. The attachment rate is chosen so that attachment occurs predominantly to the right as shown and results in muscle shortening when the attached sites moved toward the equilibrium position. (Modified with permission from *Progress in Biophysics* after Huxley, A. F. Muscle structure and theories of contraction, copyright 1957, Pergamon Journals Ltd.)

placement of the site M from its equilibrium position, as illustrated in Figure 9–14B. He selected the rate constants so that attachment would most likely occur when the sites were displaced in a direction from which return to the equilibrium position would cause the filaments to slide and shorten the muscle, whereas the detachment rate was set so that bridges in a position to oppose shortening would break. Different M sites were assumed to act independently, so that the total force equalled the sum of the forces generated by each site and depended both on the number of sites and on the displacements of the sites from their equilibrium positions. The assumption of position-sensitive attachment and detachment rate constants is reasonable, considering what we now know of the structures of the thick and thin filaments.

Huxley also assumed that in the attachment-detachment cycle, the energy, released by the hydrolysis of ATP, was stored in the spring and dissipated in returning the actin-myosin complex to the equilibrium position. Huxley solved the equations for the number of attached bridges under conditions of constant shortening velocity for different attachment and detachment rate con-

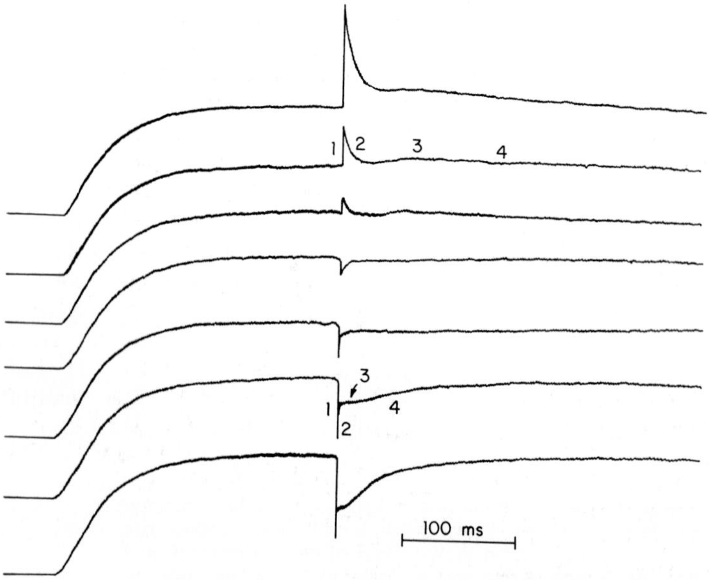

Figure 9–15 Family of tension transients to step changes in length, shown for different amounts of shortening and stretch of muscle. The bottom trace is the largest shortening; the top, the largest stretch. The baseline of tension has been adjusted for each trace. Note that the time course of phase 2 varies appreciably, being very rapid for shortening steps shown below and much slower for lengthening steps shown above. (From Huxley, A. F., and Simmons, R. M. *Cold Spring Harbor Symp. Quant. Biol.* 37:669–680, 1972.)

stants as a function of position and was able to fit the known mechanical and energetic results well. The model was proposed prior to H.E. Huxley's demonstration[18] of cross-bridges between thick and thin filaments. Ingenious modifications of the original theory have been attempted to account for twitches and mechanical transients, but without complete success.[20, 27]

A new explanation of the rapid transients has been proposed by A.F. Huxley and R.M. Simmons.[16] It is based on measurements of the transient changes in force accompanying step changes in length and assumes that cross bridges already attached rotate to regenerate the force. Figure 9–15 shows some of their records obtained from muscles subjected to various abrupt stretches and releases. The force response has four phases, described here for a shortening step. The first phase is an abrupt drop in force simultaneous with the shortening of the muscle; the behavior resembles that of a pure elastic element in series with the force generator. The second phase consists of a very rapid redevelopment of force that merges with the third phase, in which force development pauses prior to the fourth phase, during which tension gradually recovers back toward the initial value. These four phases have homologs in the velocity transient (compare Fig. 9–12, suggesting that similar processes are operating for force and velocity). Figure 9–16 shows a plot of length change against the forces attained in the rapid shortening (elastic) phase (T_1) and during the rapid

recovery phase (T_2). The T_1 curve is almost linear, as expected from a simple elastic element; its slope gives the stiffness of the element; the intercept with the abscissa, the release required to reduce the force to zero. The latter value is only about 6 nm per half sarcomere (sarcomere length is about 2.2 μm), or less than a 1% change in muscle length. Thus, the elastic element is very stiff. Huxley and Simmons[16] argued that this elastic element is in series, with each force-generating site as an integral part of the bridge, because, as they varied sarcomere length (and hence the number of bridges between overlapping thick and thin filaments), the muscle stiffness (slope of the curve) varied precisely in proportion to the amount of filament overlap and maximum force. Such behavior would not be expected if this elastic element were in series with all the bridges, e.g., if it were located at the end of the muscle or in the Z line (in which case it would have a fixed value), or if the elastic element were the whole thick or thin filament or both.

The position of the elastic element in series with the cross-bridge force generator means that it contributes to muscle stiffness only when the cross bridge is attached; hence measurements of stiffness give an estimate of the number of attached cross bridges. Stiffness (change in force for a rapid change in length) has been assessed during the rise of tension at the beginning of stimulation and during steady muscle shortening as well as under other critical conditions. The results indicate that

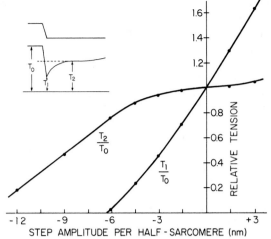

Figure 9–16 Tension during length transients as a function of step amplitude. Insert shows length step, above, and tension, below. Graph shows effect of step amplitude on T_1 and on T_2, both expressed as ratios to T. Step releases are negative and stretches are positive. (After Ford, L. E. et al. *J. Physiol. [Lond.]* 269:441–515, 1977.)

stiffness (cross-bridge attachment) precedes by a few milliseconds the rise in force after stimulation and that during steady shortening, there are fewer cross bridges attached at any time than in the isometric condition. Both results are important in relating theory to molecular mechanisms. To explain the rapid restoration of force during phase 2, Huxley and Simmons[16] proposed that an attached cross-bridge rotates to a new position as illustrated in Figure 9–17. They suggested that the S-1 portion of the myosin molecule was free to swivel about its attachment to the S-2 portion, which in turn was hinged to the (LMM) portion of the backbone. These locations for the force-producing and elastic elements are most consistent with x-ray, EM, and biochemical data, although other locations would be possible. Once the head attaches to the thin filament, it rotates to a low-energy state, stretching the spring that attaches it to the thick filament and thus produces the force that makes thin filaments slide. The response of a muscle to step decreases in length is a rapid shortening of the spring (illustrated as the S-2 myosin segment) corresponding to the drop in force to the T_1 value. Recovery then occurs by a rapid rotation of the S-1 head to a lower energy state, stretching the spring S-2 to a new equilibrium value.* The total energy in the system of

head plus spring is equal to the sum of the energy stored in the head as it assumes different energy levels corresponding to a new attachment state on the thin filament, plus the energy stored in the stretched spring that attaches to the backbone of the thick filament. The force in this new state is given by the T_2 curve. For a more complete discussion of how this quantitative theory fits the data, the reader is referred to the paper by Huxley and Simmons[17] and reviews by Podolsky and Schoenberg[28] and Simmons and Jewell.[34]

An attractive feature of the rotating head model is that it correlates mechanical and biochemical events. For example, various energy states of the

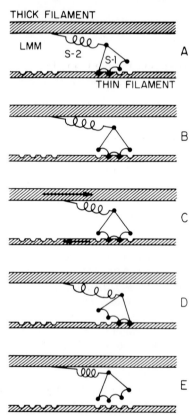

Figure 9–17 Model proposed by Huxley and Simmons[16] to account for tension transients. It is one of several models that fit the data and are reasonable on the basis of the known structure of the thick and thin filaments and cross bridges. The S-1 portion of myosin attaches to the thin filament at several sites, and the energy decreases as the head rotates in clockwise direction, A to B. This rotation of the head stretches the spring (here drawn as the S-2 element). Force is transmitted to the thick filament through the S-2 portion, pulling the thin filament to the left. Rapid release of the muscle causes the shortening of the spring in C. The subsequent rapid rotation of the S-1 head (in D) is hypothesized to account for the rapid redevelopment of force during phase 2 of tension redevelopment (see Figs. 9–12, 9–15, and 9–16). However, the role of the defective protein dystrophin is still to be identified. (After Huxley, A. F. *J. Physiol.* 243:1–43, 1974.)

---

*Since S-2 is a coiled-coil structure, it may not be very elastic (unless it undergoes a change of state) so that it may be better to model the elasticity elsewhere, such as in the actin-S-1 or S-1–S-2 attachment.

myosin-actin attachment may have as their counterparts the various biochemical states associated with the release of the hydrolysis products. Release of inorganic phosphate may be associated with initial binding, with ADP release occurring later.

As discussed in the previous chapter, EPR studies of cross-bridge orientation during contraction support the notion of rapid rotation of attached cross bridges but do not provide evidence for a series of well defined attached orientations or states. During maximal calcium activation, about 20% of the bridges have an orientation characteristic of rigor.

PATHOPHYSIOLOGY OF MUSCLE CONTRACTION

This chapter has focused on the normal physiology of muscle cells. Such research provides information concerning normal processes and also the tools to understand the pathophysiology of muscle diseases. Study of muscle diseases in turn helps clarify normal function. In this section, we will give examples of diseases affecting the muscle contractile mechanism and its regulation.

Muscular Dystrophy. The most common muscle disease affecting the contractile mechanism is muscular dystrophy, a disease in which muscle fibers progressively atrophy, producing weakness and often death. Unfortunately, despite much research, the cause of muscular dystrophy is obscure. The disorder, which has many forms, may have multiple causes. An exciting new discovery is the isolation of the defective gene in one form, Duchenne's dystrophy, a very large gene presumably coding for a protein with molecular weight near one million Daltons.[23a] However, the role of the defective protein dystrophin is still to be identified.

McArdle's Disease. McArdle's disease in characterized by muscular stiffness that may, during exercise, lead to severe and painful cramps. A diagnostic procedure is to have the patient exercise muscles of the forearm while the blood supply is restricted with a tourniquet. The result would be cramps that are not accompanied by muscle electrical activity as measured by the electromyogram (EMG). Such electrically silent contractions are termed *contractures*. Muscle biopsy specimens from patients with McArdle's disease are deficient in phosphorylase,[25] the enzyme that catalyzes the initial step in glycolysis, the production of glucose-1-phosphate from glycogen. Normally, glucose-1-phosphate enters the glycolytic pathway, being first converted to glucose-6-phosphate, which

eventually produces ATP and pyruvate or lactate. In McArdle's disease, exercise fails to increase blood lactate, and the cramps can be relieved by intravenous injection of glucose (which enters glycolysis via glucose-6-phosphate, beyond the step catalyzed by phosphorylase).

At first, it was postulated that blocked glycolysis limits ATP production and leads to a rigor-like condition in which the muscle is unable to dissociate actin and myosin (see Fig. 8–15). But measurements of total muscle ATP concentrations in McArdle's disease, either at rest or during contractures, failed to show any significant decrease in total muscle ATP.[31] New information comes from the phosphorus nuclear magnetic resonance (NMR) technique (see p. 201), which permits measurements of phosphorylated metabolites in vivo. NMR showed that although the total ATP levels did not fall during exercise, the muscle creatine phosphate level in a patient with McArdle's disease fell much more than in a normal subject during ischemic exercise.[30] Moreover, in sharp contrast to normals, the patient's muscle intracellular pH did not fall but actually rose. In normal subjects, the intracellular pH falls substantially due to production of lactic acid under anaerobic conditions. Thus the contractures seem to be associated with relatively normal ATP levels but decreased creatine phosphate levels and high intracellular pH. How these conditions lead to contraction is not certain, but high pH does enhance Ca stimulation of muscle contraction, increasing myofilament sensitivity to Ca and decreasing Ca uptake by the sarcoplasmic reticulum.

Hypothyroidism. In *hypothyroidism*, reflexes are abnormally slow, whereas patients with hyperthyroidism show overly brisk reflexes. These findings are partly due to changes in muscle.[35] Twitches of muscle from animals with experimentally induced hypo- or hyperthyroidism show changes in both total duration and time-to-peak tension. In addition, thyroid disease produces changes in muscle mass. Muscles in hypothyroidism show a pseudohypertrophy with proliferation of connective tissue but little change in muscle mass. The muscles of hyperthyroid patients, on the other hand, are reduced in mass. Mass changes are not directly related to change in contraction kinetics.

Changes in contraction kinetics with thyroid dysfunction are apparently linked to the role of the thyroid hormone in controlling the synthesis of proteins of both the myofibrils and the sarcoplasmic reticulum. In cardiac ventricular muscle, the presence of thyroxine leads to the preferential synthesis of a myosin heavy chain with high intrinsic ATPase activity, producing muscle with high shortening velocities. The same is true for some skeletal muscles investigated, particularly the more slowly contracting ones. In addition, thyroxine has been shown to stimulate the synthesis of more sarcoplasmic reticular Ca^{2+}-ATPase of a type that has a higher intrinsic calcium pumping ability.[1] Thus, thyroxine apparently aids the muscle cell in switching expression from one set of genes for a more slowly contracting fiber to another set for a more rapidly contracting fiber. This demonstrates the plasticity of adult muscle cells.

SUMMARY

Muscles shorten, develop force, and turn stored chemical energy into mechanical energy. These properties can be understood from the interaction of sliding filaments presented in the previous chapter. Since skeletal muscles are voluntary muscles made up of many fibers controlled by the nervous system, their contraction is regulated by the recruitment of motor units by the nervous system. In addition, contraction is regulated through temporal summation of contraction within each fiber, determined by the frequency of stimulation of each unit. Summation of contraction (of interacting cross bridges) can be understood in terms of the kinetics of cross-bridge interaction, the sarcoplasmic Ca signal, and Ca binding to activating sites.

Muscles hydrolyze ATP through the cyclical interaction of actin and myosin as they contract. This ATP must be resynthesized via glycolysis, the citric acid cycle, and oxidative phosphorylation. However, the immediate back-up buffer for ATP is creatine phosphate, the high-energy phosphate of which can be transferred to ADP to resynthesize ATP. Because ATP and CP provide only a small reservoir of free energy, glycolysis and oxidative metabolism must be well developed to keep a muscle from fatiguing during activity from a decline in available free energy.

Muscle isometric force is a function of muscle length, passive force increasing as the muscle is stretched, active force going through a maximum, with increasing lengths peaking about the in situ length. The passive force is borne by connective tissue elements and also by intracellular elastic filament structures. The active force is generated by the cross bridges. Its decrease at long sarcomere lengths occurs as the overlap between thick and thin filaments, and thus the number of active cross bridges declines. Its decrease at short muscle

lengths may be due both to mechanical factors and to a deficit in the Ca activation of contraction.

Muscles shorten faster isotonically when lifting light loads than when lifting heavy ones. The maximum speed of shortening differs between muscle but depends on the molecular properties of the myosins, increasing with increasing myosin ATPase rates. More rapidly cycling myosin cross bridges lead to higher maximum shortening velocities.

The complex response of muscles to step changes in length or load leads to a theory of contraction involving the attachment and rotation of cross bridges. This theory ties together biochemical, structural, mechanical, and energetic data in a theory of muscle contraction. Thus the diverse properties of muscle can be understood on the basis of the molecular interactions within the cell.

ANNOTATED REFERENCES

Carlson, F.D.; Wilkie, D.R. *Muscle Physiology.*, Englewood Cliffs, N.J., Prentice-Hall, 1974.
This book, now out of print, provides a clear discussion of classic muscle mechanics and energetics by two outstanding researchers and writers.

Hoyle, G. *Muscles and their Neural Control.* New York, Wiley and Sons, 1983.
A comprehensive treatment of muscle and neural control covering virtually the entire animal kingdom and discussing topics from the molecular to the organismic. The prospect propounded by the late Graham Hoyle is unique; the range of topics is astounding.

Huxley, A.F. Muscle structure and theories of contraction. *Prog. Biophys.* 7:279-318, 1957.
Huxley, A.F.; Simmons, R.M. Proposed mechanisms of force generation in striated muscle. Nature 233:533-538, 1971.
Quantitative treatments of the cross-bridge model of muscle contraction at two major stages in its development.

Treherne, J.E., ed. Design and performance of muscular systems. *J. Exper. Biol.* 15:1-412, 1985.
This review volume contains many fascinating articles, each written by an expert in the field, about muscles: from the basic molecular level to neural control and energetics at the cellular level to specialized operations, limitations, and changes in performance of whole muscle systems.

Woledge, R.C.; Curtin, N.A.; Homsher, E. *Energetic Aspects of Muscle Contraction*, New York, Academic Press, 1985.
A detailed treatment of muscle mechanics and energetics integrated with theories of contraction. The unique treatment combines muscle energetics and molecular energetics, the research expertise of the authors.

REFERENCES

1. Ash, A.S.F.; Besch, H.R.; Harigaya, S.; Zaimis, E. Changes in the activity of sarcoplasmic reticulum fragments and actomyosin isolated from skeletal muscle of thyroxine-treated cats. *J. Physiol. (Lond.)* 224:1-19, 1972.

2. Barany, M. ATPase activity of myosin correlated with speed of muscle shortening. *J. Gen. Physiol.* 50(Suppl.):197-216, 1967.

3. Bessman, S.P.; Geiger, P.J. Transport of energy in muscle: the phosphorylcreatine shuttle. *Science* 211:448-452, 1981.

4. Cain, D.F.; Davies, R.E. Breakdown of adenosine triphosphate during a single contraction of working muscle. *Biochem. Biophys. Res. Commun.* 8:361-366, 1962.

5. Civan, M.M.; Podolsky, R.J. Contraction of kinetics of striated muscle fibres following quick changes in load. *J. Physiol. (Lond.)* 184:511-534, 1966.

6. Close, R.I. The relations between sarcomere length and characteristics of isometric twitch contractions of frog sartorius muscle. *J. Physiol. (Lond.)* 220:745-762, 1972.

7. Eisenberg, E.; Greene, L.E. The relation of muscle biochemistry to muscle physiology. *Annu. Rev. Physiol.* 42:293-309, 1980.

8. Fenn, W.O. A quantitative comparison between the energy liberated and the work performed by the isolated sartorius muscle of the frog. *J. Physiol. (Lond.)* 58:175-203, 1923.

9. Fenn, W.O. The relation between the work performed and the energy liberated in muscular contraction. *J. Physiol. (Lond.)* 58:373-395, 1924.

10. Fenn, W.O.; Marsh, B.S. Muscular force at different speeds of shortening. *J. Physiol. (Lond.)* 85:227-297, 1935.

11. Gordon, A.M.; Huxley, A.F.; Julian, F.J. The variation in isometric tension with sarcomere length in vertebrate muscle fibres. *J. Physiol. (Lond.)* 184:170-192, 1966.

12. Hill, A.V. The heat of shortening and the dynamic constants of muscle. *Proc. Roy. Soc., London, S.B.* 126:136-195, 1938.

13. Hill, A.V. *Trails and Trials in Physiology.* Baltimore, Williams & Wilkins, 1965.

14. Homsher, E.; Mommaerts, W.F.H.M.; Ricchiuti, N.V.; Wallner, A. Activation heat, activation metabolism and tension-related heat in frog semitendinosus muscles. *J. Physiol. (Lond.)* 220:601-625, 1972.

15. Huxley, A.F. Muscle structure and theories of contraction. *Prog. Biophys.* 7:255-318, 1957.

16. Huxley, A.F.; Simmons, R.M. Proposed mechanism of force generation in striated muscle. *Nature* 233:533-538, 1971.

17. Huxley, A.F.; Simmons, R.M. Mechanical transients and the origin of muscular force. *Cold Spring Harbor Symp.* 37:669-680, 1972.

18. Huxley, H.E. The double array of filaments in cross-striated muscle. *J. Biophys. Biochem. Cytol.* 3:631-647, 1957.

19. Jewell, B.R.; Wilkie, D.R. An analysis of the mechanical components in frog's striated muscle. *J. Physiol. (Lond.)* 143:515-540, 1958.

20. Julian, F.J. Activation in skeletal muscle contraction model with a modification for insect fibrillar muscle. *Biophys. J.* 9:547-570, 1969.

21. Katz, B. The relation between force and speed in muscular contraction. *J. Physiol. (Lond.)* 96:45-64, 1939.

22. Kushmerick, M.H. Energetics of muscle contraction. *Handbk. Physiol.* Sec. 10, 189-230, 1983.

23. Layzer, R.B.; Rowland, L.P. Cramps. *N. Engl. J. Med.* 285:31-40, 1971.

23a. Monaco, A.D.; Neve, R.L.; Colletti-Feener, C.: Bertelson, C.J.; Kurnit, D.M.; Kunkel, L.M. Isolation of candidate cDNAs for portions of the Duchenne muscular dystrophy gene. *Nature* 323:646-650, 1986.

24. Mommaerts, W.F.H.M. Energetics of muscular contraction. *Physiol. Rev.* 49:427-508, 1969.

25. Mommaerts, W.F.H.M.; Illingworth, B.; Pearson, C.M.; Guillory, R.J.; Seraydarian, K. A functional disorder of muscle associated with the absence of phosphorylase. *Proc. Nat. Acad. Sci. USA* 45:791-797, 1959.

26. Podolsky, R.J. The maximum sarcomere length for contraction of isolated myofibrils. *J. Physiol. (Lond.)* 170:110-123, 1964.

27. Podolsky, R.J.; Nolan, A.C.; Zaveler, S.A. Cross-bridge properties derived from muscle isotonic velocity transients. *Proc. Nat. Acad. Sci. USA* 64:504-511, 1969.

28. Podolsky, R.H.; Schoenberg, M. Force generation and shortening in skeletal muscle. *Hand. Physiol.* Sec. 10, 257-274, 1983.

29. Ramsey, R.W.; Street, S.F. The isometric length-tension diagram of isolated skeletal muscle fibers of the frog. *J. Cell. Comp. Physiol.* 15:11-34, 1940.

30. Ross, B.D.; Radda, G.K.; Gadian, D.G.; Rocker, G.; Esiri, M.; Falconer-Smith, I. Examination of a case of suspected McArdle's syndrome by 31P nuclear magnetic resonance. *N. Engl. J. Med.* 304:1338-1342, 1981.

31. Rowland, L.P.; Araki, S.; Carmel, P. Contracture in McArdle's disease. *Arch. Neurol.* 13:541-544, 1965.

32. Rudel, R.; Taylor, S.R. Striated muscle fibers: facilitation of contraction at short lengths by caffeine. *Science* 172:387-388, 1971.

33. Siemankowski, R.F.; Wiseman, M.O.; White, H.D. ADP dissociation from actomyosin subfragment is sufficiently slow to limit the unloaded shortening velocity in vertebrate muscle. *Proc. Nat. Acad. Sci. USA* 82:658-662, 1985.

34. Simmons, R.M.; Jewell, B.R. Mechanics and models of muscular contraction. *Recent Adv. Physiol.* 9:87-147, 1974.

35. Takamori, M.; Gutmann, L.; Shane, S.R. Contractile properties of human skeletal muscle. *Arch. Neurol.* 25:535-546, 1971.

36. Wilkie, D.R. The mechanical properties of muscle. *Br. Med. Bull.* 12:177-182, 1956.

37. Woledge, R.C.; Curtin, N.A.; Homsher, E. *Energetic Aspects of Muscle Contraction.* New York, Academic Press, 1985.

Chapter 10

Albert M. Gordon

Contraction in Smooth Muscle and Nonmuscle Cells

INTRODUCTION

We have now discussed excitation and contraction in skeletal muscle in detail. Similar mechanisms account for the behavior of cardiac muscle, smooth muscle, and some motile nonmuscle tissues. In this chapter, we consider excitation and contraction in smooth muscle and motility of some nonmuscle systems. Cardiac muscle is described in Chapters 37 and 39.

SMOOTH MUSCLE

Many organs and glands in the body contain muscles that differ from skeletal muscle in not being under voluntary control and in not having a striated or banded appearance under the light microscope. These smooth muscles lack striations because their contractile filaments are not organized in regular arrays of aligned sarcomeres. Smooth muscles are usually specialized for long, slow contractions. Rather than taking instructions from motoneurons with cell bodies in the central

nervous system, they are controlled by nerve fibers of the autonomic nervous system and by circulating hormones.

Smooth muscles perform a wide variety of functions. They control the diameter of arterioles and thus help regulate blood pressure. They form muscular sphincters at branches of the vascular tree and thus determine the distribution of blood to different capillary beds. Gastrointestinal sphincters control passage of intestinal contents from one section of the gut to the next. Smooth muscles regulate the size and internal pressure in hollow organs such as the urinary bladder and the uterus. Smooth muscles vary greatly in electrical activity, degree of automaticity, innervation, response to circulating hormones and drugs, and extent of cell-to-cell coupling. Several good reviews are available.[31, 3, 2]

Structure. Smooth muscle cells range from 2 to 10 μm in diameter and from 20 to 600 μm in length; the smallest are from arterioles and the largest from pregnant uterus. The cells are irregularly shaped and mechanically joined together into bundles or sheets by structures known as attachment plaques, intermediate junctions, or adherens junctions. At these desmosomelike junctions, the cell membranes are thickened and separated by a gap of about 60 nm containing a dense, granular material. Smooth muscle cells are also electrically intercoupled at gap junctions (often called *nexuses* in the literature) in which the membranes of adjacent cells come into close apposition, leaving a gap of only 2 to 3 nm. The gap is traversed by connexons, proteinaceous tubes connecting adjacent cells and allowing passage of ions and small molecules (molecular weights less than 1200 to 1400) from cell to cell (Chap. 11). The gap junction provides a path of low electrical resistance between cells, and tissues having many such junctions behave electrically like a syncytium. The number of functional junctions varies greatly from tissue to tissue and from time to time in the same tissue. Thus in the guinea pig ileum, they are virtually absent in the longitudinal smooth muscle but abundant in the circular smooth muscle, having a density of about 48/1000 μm^2 of cell surface and occupying about 0.2% of the cell surface area. In the rat myometrium, nexuses are present when young are delivered but are difficult to detect a day or two later. Electrical coupling synchronizes electrical activity in adjacent cells. Also, because the connexon tubes are large enough to pass small metabolites and second messengers, they couple cells metabolically.

Most smooth muscle plasma membranes contain numerous, flask-shaped invaginations called *caveolae*. In guinea pig taenia coli, they cover almost 50% of the cell surface area. The function of caveolae is unknown, but they greatly expand the surface area available to the cell. Smooth muscle cells have a well-defined sarcoplasmic reticulum (SR) but no transverse tubules. The amount of SR varies greatly, ranging from 2% to 7.5% by volume. The surface area of the SR can be approximately equal to that of the cell membrane. Often it occurs as sacks near the cell membrane with which it forms peripheral couplings with electron-dense material (feet) between the apposed membranes (see Fig. 7–4). Throughout the cytoplasm, SR may be oriented radially or scattered in closer association with the myofilaments or mitochondria. The sarcoplasmic reticulum contains calcium and probably plays the same role in calcium release and uptake that it does in skeletal muscle.

Single functioning smooth muscle cells can be isolated by treating whole muscles with proteolytic enzymes such as trypsin and collagenase. Enzymatically dissociated preparations are useful for mechanical, electrical, and pharmacological studies. Because isolated cells are not innervated, comparison of their properties with those of intact tissue cells allows some evaluation of the neural influence on smooth muscle.

Types of Smooth Muscle. Almost half a century ago, Bozler[2a] classified smooth muscle into two types, viz., visceral, or unitary, smooth muscle and multiunit smooth muscle.

In *unitary muscle*, contraction is synchronized so that the muscle acts as a unit, all fibers either contracting or resting. Unitary muscles are spontaneously active, display electrical pacemaker activity, and respond to stretch with increased activity. The innervation density is low, and the cells are tightly coupled electrically through gap junctions so that activity, once initiated, spreads promptly from cell to cell—the multicellular muscle acts as a single unit (Figure 10–1B). Unitary smooth muscle thus resembles cardiac muscle more closely than striated muscle. Examples are the gut and uterus, organs that generate and propagate their own slow, rhythmic movements.

Multiunit smooth muscle has motor units that resemble those of skeletal muscle and contractile tension is graded by variation in the number of active units. The innervation density is high and there is little or no spread of activity from cell to cell (Fig. 10–1A). Examples of multiunit smooth muscle are the cat nictitating membrane, the ciliary body (the muscle that focuses the eye) and the

Figure 10–1 Diagram of two extreme cases of autonomic innervation and control in smooth muscle. In *A* the parasympathetic and sympathetic neurons have varicosities and sites of transmitter release near virtually every smooth muscle cell, and the smooth muscle cells have few gap junctions connecting cells. This is characteristic of multiunit smooth muscle. In *B* the parasympathetic neurons release transmitters near very few cells, but the cells are electrically coupled through gap junctions. This is characteristic of unitary (visceral) smooth muscle.

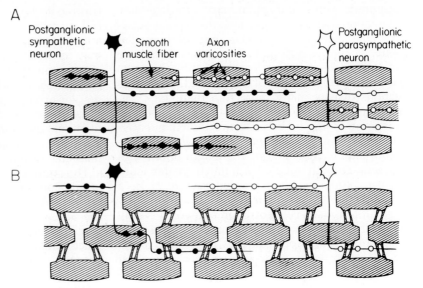

A

Postganglionic sympathetic neuron
Smooth muscle fiber
Axon varicosities
Postganglionic parasympathetic neuron

B

Table 10–1 Comparison of Characteristics of Different Muscle Types

Characteristics	Skeletal	Smooth		Cardiac
		Multiunit	*Unitary*	
Thick and thin filaments	Yes	Yes	Yes	Yes
Striations—distinct sarcomeres	Yes	No	No	Yes
Speed of contractions	Fast and slow	Very slow	Very slow	Slow
Transverse tubules	Yes	No	No	Yes
Sarcoplasmic reticulum: % cell vol.	1–9%	2–5%	2–5%	1–4%
Source of activating Ca	SR	SR and some extracellular	SR and some extracellular	SR and some extracellular
Site of Ca regulation	Troponin	Myosin + ?	Myosin + ?	Troponin
Spontaneous electrical activity	No	No	Yes	Yes
Gap junctions between cells	None	Few	Many	Many
Activity spreads between cells	No	Some	Yes	Yes
Stretch increases muscle activity	No	Some	Yes	Some
Extent of innervation	Each cell	Dense	Sparse	Variable
Effect of nerve stimulation	Excitation	Excitation or inhibition	Excitation or inhibition	Excitation or inhibition
Effects of hormones on contraction	Some	Large	Large	Large

pilomotor muscles that erect body hair. Table 10–1 compares the properties of these two types of smooth muscle with those of skeletal and cardiac muscle.

Contractile Mechanism.[19, 30, 9] Smooth muscle cells contain thin and thick filaments (as well as intermediate diameter 10 nm, filaments of unknown function). Thin filaments contain actin, which can be extracted chemically and repolymerized, in vitro into thin filaments. In the presence of heavy meromyosin, these repolymerized actin filaments show the arrowhead pattern similar to that shown in Figure 8–9. The thick filaments contain myosin. The intermediate filaments are composed of proteins called desmin and vimentin. Thick filaments have lengths of 2.2 μm in rabbit mesenteric veins and pulmonary arteries; they have tapered ends but may lack the bipolar structure characteristic of skeletal muscle thick filaments as central regions bare of projections have not been detected. The length of thin filaments has not been satisfactorily measured, but cross sections display many more thin than thick filaments (15 to 1). Hence either thin filaments are more numerous, or because the thin filament is longer, a number of short individual thick filaments can be aligned along the length of one thin filament. Smooth muscle has a much greater actin-myosin ratio (3:1) than striated muscle (1:3), a datum consistent with either of the two possibilities. Although the concentration of myosin is only one third that of skeletal muscle, smooth muscle cells can produce approximately equivalent forces and can shorten to a small fraction of their rest length, sending their cell membranes into prominent folds (Fig. 10–2).

Within the smooth muscle cell, thin filaments are organized into bundles inserting into dense bodies. They are distributed at periodic intervals through the cell and some are attached to the cell membrane. Dense bodies appear to function as do Z lines in striated muscle cells and contain a similar protein, α-actinin. The thin filaments have opposite polarities on either side of the dense bodies, as indicated by the arrowheads formed by reaction with myosin, which point in opposite directions. As the cell contracts, the lines of dense bodies become more oblique to the cell axis (Fig. 10–2) and also get closer together, as if "sarcomeres" are shortening. In the shortened cells, the thin filaments–dense body formation occurs in the troughs of the folds that form the membrane. The contractile force appears to be transmitted to the cell membrane here through an intervening protein called vinculin. The intermediate filaments may connect the dense bodies of neighboring strings much as similar proteins connect Z lines in striated muscle. Thick filaments have a less regular arrangement than thin filaments but as in skeletal muscle, show more overlap with thin filaments at shorter muscle lengths. Thus the filament arrangement is consistent with a sliding filament contractile mechanism.

The contractile mechanism presumably involves interaction of actin and myosin as in striated muscle. Although there are small differences in the amino acid sequences of actin from smooth and striated muscle, the two proteins appear to react similarly with myosin from skeletal muscle. Smooth muscle myosin, however, has a lower ATPase activity consistent with the slower contractions. Of particular significance are the unique

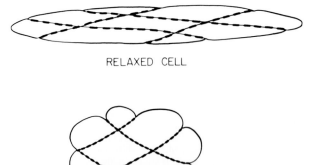

RELAXED CELL

CONTRACTED CELL

Figure 10–2 Schematic representation showing contraction in a smooth muscle cell. Attachment sites for actin on both surface and aligned dense bodies are shown (localization of α-actinin). Actin and myosin are thought to be located along these lines in thin and thick filaments that slide to decrease the distance between the dense bodies and shorten the cell during contraction. During contraction, the cell balloons out between anchor points. (After Fay, F. S. and Delise, C. M. *Proc. Nat. Acad. Sci. USA* 70:641–645, 1973; and Fay, F. S.; Fujiwara, K; Rees, D. D.; Fogarty, K. E. *J. Cell. Biol.* 96:783–795, 1983.)

properties of myosin light chains and their calcium-dependent role in regulation of contraction. As indicated in Table 8–3, smooth muscles contain two types of light chains a P or phosphorylatable type and another type that is similar to the so-called alkali light chains of striated muscle.[27] Phosphorylated P-type light chains are required before actin can activate the smooth muscle myosin ATPase. The light chains may also participate in the regulation of the polymerization of myosin into thick filaments. Phosphorylation of the regulatory light chains promotes polymerization of smooth muscle myosin by shifting its structure from a folded to a linear configuration that is better suited for association with other myosins and for polymerization. Physiological intracellular concentrations of K^+ and MgATP also promote this transition and polymerization and indeed may be the physiologically more important factors since polymerized myosin from relaxed smooth muscle is little phosphorylated.

Studies of rapid length changes are difficult in intact smooth muscle because of the many cells in series, but they can be done on isolated cells. Such studies show that smooth muscle behaves much like skeletal muscle (Fig. 9–12) except that the elastic element is more compliant and the redevelopment of force after a step decrease in length is slower. The similarities imply strongly that the cross-bridge contractile mechanism also applies to smooth muscle.

Contraction of vertebrate skeletal muscle is triggered by the binding of calcium to troponin, a constituent of the thin filaments. Thin filaments of smooth muscle contain tropomyosin but not troponin. There are small concentrations of a troponinlike calcium-binding protein called leiotonin, but its role is not clear.[22] In any case, the thick filaments seem to play a major role in initiating contraction.[29] Smooth muscle actin and myosin interact weakly in the absence of the regulatory light chains and tropomyosin and calcium, whereas skeletal muscle actin and myosin interact strongly in the absence of troponin and tropomyosin, a reaction that is inhibited by these substances in the absence of calcium.

How then is contraction triggered in smooth muscle? Two models have been proposed, the phosphorylation model and the calcium direct-control model[14] (Fig. 10–3). One postulates a series of reactions that lead to the phosphorylation of the myosin P-type light chain. Myoplasmic calcium binds to the soluble cytoplasmic protein calmodulin and the Ca-calmodulin complex activates an enzyme called myosin light-chain kinase. The activated enzyme catalyzes phosphorylation of the myosin-P light chain, a reaction that allows actin to activate the myosin ATPase. From this point on, the process presumably proceeds as in skeletal muscle, with the formation of cross bridges and sliding of filaments. Phosphorylation of myosin in contraction of smooth muscle is clearly established. First, calcium-dependent activation and inactivation of myosin ATPase is accompanied by phosphorylation and dephosphorylation of myosin, respectively.[27, 29] Second, in "skinned" smooth muscle cells, irreversible phosphorylation of the myosin initiates a sustained contraction that is insensitive to removal of calcium.[6] However, during a sustained contraction of smooth muscle, the amount of phosphorylation increases early in contraction when the muscle can shorten rapidly but declines later in the contraction, even though force is sustained. This sustained force depends on the external calcium in intact fibers and on filament calcium in "skinned" fibers.[21] These observations have led to the hypothesis of a second type of calcium regulation, with calcium serving a more direct role than that of activating a kinase: calcium binding to thin or thick filaments. There is also evidence that the two types of calcium regulation (phosphorylation and binding) may govern different properties of the actomyosin interaction—phosphorylation regulating the velocity of shortening while binding regulates the production of sustained force.[21] Indeed, the reactions may occur sequentially: first

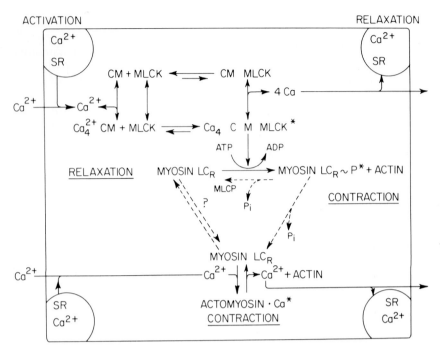

ACTIVATION   **RELAXATION**

Figure 10–3 Regulation of smooth muscle contraction by Ca^{2+}. The major mechanism through Ca^{2+}, calmodulin (CM) activation of myosin light-chain kinase (MLCK) to phosphorylate the regulatory light chain of myosin (LC_R) (with subsequent dephosphorylation via myosin light-chain phosphatase—MLCP) is shown in the upper part of the figure. In the lower part of figure is a hypothesized activation by direct binding of Ca^{2+} to activate actomyosin. Ca^{2+} is shown as coming from storage sites in the sarcoplasmic reticulum (SR) or crossing the cell membrane. Ca^{2+} removal from the myoplasm is into the SR or across the cell membrane. (Modified from Kamm, K. E.; Stull, J. T., with permission from the *Annual Review of Pharmacology & Toxicology*, Vol. 25, © 1985 by Annual Reviews, Inc.)

phosphorylation, followed by direct calcium activation by binding (see Fig. 10–3).

Excitation and Excitation-Contraction Coupling. Resting smooth muscle cells have a membrane potential in the range of -20 to -65 mV. The resting membrane resistance is tenfold higher than in skeletal muscle, so fewer ionic channels must be open. The resting potential is partly set by the relative permeabilities of the cell membrane to the permeant ionic species as well as by action of electrogenic ion pumps (Chap. 2) such as the Na^+- K^+ pump.

The identification of voltage-sensitive ionic channels in intact smooth muscle is complicated by the interconnecting gap junctions, which defeat efforts at voltage clamping one cell. The newer patch-clamp technique and the use of dissociated cells satisfactorily circumvent this difficulty. The most important channel for excitation is the voltage-gated Ca channel. Tetrodotoxin-sensitive Na channels are not found. The Ca channels are selective for calcium and are readily blocked by Mn^{2+}, Co^{2+}, and the drugs verapamil and nifedipine (Chap. 4).* Smooth muscle membranes also

have potassium channels, both the delayed-rectifier type characteristic of axons and those activated by elevated intracellular calcium. Smooth muscle membranes also have the Na^+-Ca^{2+} exchange mechanism discussed in Chapter 2.

Smooth muscles show a variety of electrical activities (see Fig. 10–4). Some muscles, such as rabbit pulmonary artery, small arteries from rat and guinea pig, and canine stomach antrum are electrically inexcitable. Others produce a variety of action potentials (Fig. 10–4), ranging from brief spikes to long-lasting, plateaulike responses or plateaus with superimposed spikes. In some tissues, primarily in unitary smooth muscle, the cells show slow spontaneous oscillations in membrane potential, variously termed slow waves, pacesetter potentials, or "basic electric rhythm." These can lead to spontaneous contractions and intrinsic regulation of those tissues. In the intestine, the frequency of slow waves is from 9 to 11/min, which provides the rhythmic contractions important to gastrointestinal motility.

The origin of the slow waves is debated. In the exceptional case of the longitudinal smooth muscle of the guinea pig ileum or gastric antrum, slow-wave depolarizations appear to be generated through neural influences, since they can be produced by the application of the neurotransmitter acetylcholine (ACh), blocked by tetrodotoxin and altered by cholinergic drugs. However, in most tissues they seem to be intrinsic to the muscle, for

---

*Ca channels have been separated into different types on the basis of a differential sensitivity to blocking drugs. However, since the affinity of the Ca channel for the drug may be increased when the channel is open or refractory, states promoted by membrane depolarization, caution must be exercised in inferring from differing drug sensitivities in cells with different resting potentials the presence of different channels.

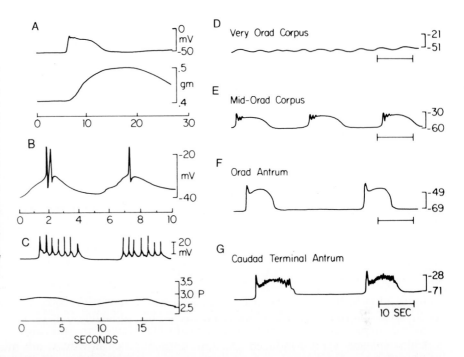

Figure 10–4 Example of intracellularly recorded action potentials and slow-wave potentials from a variety of smooth muscle cells showing the diversity of wave forms. In some cases, associated contractile responses are included. *A,* turtle aorta; *B,* rat small mesenteric artery; *C,* guinea pig portal vein; *D, E, F,* and *G* spontaneous potentials from various parts of the canine stomach. (*A, B,* and *C* after Johansson, B. and Somlyo A. P. *Handbk. Physiol.,* Sec. 2, 2:301–324; *D, E, F,* and *G* after Szurszewski, J. In Johnson, L. R., ed. *Physiology of the Gastrointestinal Tract,* vol. 2. New York, Raven Press, 1981.)

they are not affected by drugs that block nerve activity (e.g. tetrodotoxin) or by autonomic ganglionic blocking agents. Moreover, slow waves occur in isolated smooth muscle cells. One theory of slow-wave genesis postulates that slow oscillations in metabolic activity affect the activity of ion pumps such as the electrogenic Na^+-K^+ pump; increased activity causes hyperpolarization and decreased pumping, depolarization. In support of this theory is the demonstration that conditions that slow pumping, e.g., treatment with cardiac glycosides (Chapter 2), decreases in extracellular $[K^+]$, hypoxia, and lowered temperature, also depress slow-wave activity. Conversely, slow waves are enhanced by conditions that stimulate sodium pumping, such as elevated internal $[Na^+]$ and elevated temperature. Membrane conductance changes little during slow waves. A difficulty for the theory is that the changes in membrane potential during slow waves are larger than pump variations can account for. Also, the lag in the development depolarization following administration of pump inhibitors is longer than expected. A variant of the theory postulates that alteration of Na pumping in turn affects the electrogenic Na^+-Ca^{2+} exchange, which in turn changes $[Ca^{2+}]_i$.

Some unitary smooth muscles have a pacemaker region, i.e., a group of cells with a high rate of generation of slow waves. Because the cells are coupled electrically, activity can spread from the pacemaker region by current flow from the slow waves themselves or by action potentials they produce. In the ureter, slow waves start at a pacemaker in the pelviureteral junction and spread toward the bladder, moving urine distally.

Contraction in smooth muscle normally is produced by electrical activity (Fig. 10–4), as in skeletal muscle, but it can also be produced by the direct action of neurotransmitters, such as ACh on the guinea pig ileum or norepinephrine on vascular smooth muscle cells of the portal vein, without any change in membrane potential. As in skeletal and cardiac muscle, contraction of smooth muscle is triggered by elevated internal Ca^{2+}, a correlation that is well documented by studies employing simultaneous measures of Ca^{2+} and force. The source of the activating Ca^{2+} is mainly the sarcoplasmic reticulum, but in addition, substantial Ca^{2+} can enter across the surface membrane via voltage-gated calcium channels. The SR of smooth muscle cells has a calcium pump that accumulates calcium when the cell is at rest. The SR stores enough calcium to allow the cells to contract to chemical stimuli for some time in the absence of external calcium. Normally, transmembrane calcium currents during excitation provide additional calcium. Figure 10–5 summarizes the factors involved in the elevation of intracellular calcium during contraction.

The mechanism of calcium release from the sarcoplasmic reticulum is no better understood in

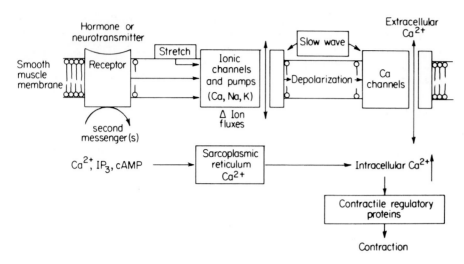

Figure 10–5 Summary of membrane and intracellular mechanisms for the regulation of intracellular calcium in contracting smooth muscle.

smooth muscle than in skeletal muscle. One hypothesis is that transmitter binding to a membrane receptor opens calcium channels that activate SR calcium release via Ca-induced calcium release (Chap. 7). Another hypothesis was initially suggested as the mechanism for calcium release from the endoplasmic reticulum in nonmuscle tissues such as pancreatic acinar cells, liver cells, macrophages, and many cultured cell lines. This is that the binding of hormones or transmitters to a membrane receptor activates a phosphodiesterase (phospholipase C) to form diacylglycerol (DG) and inositol trisphosphate (IP_3) from the membrane phospholipid, phosphatidyl inositol bisphosphate.[3] IP_3 causes release of calcium from the sarcoplasmic reticulum in smooth muscle cells. DG with calcium activates a protein kinase C that may phosphorylate cell regulatory proteins.

Mechanical Properties and Energetics.[20, 5] Contraction of smooth muscle is characteristically slow, lasting many seconds to minutes. In muscles whose contraction is triggered by spikelike electrical activity, summation and tetanus occur readily (Fig. 10–4). In cells having plateaulike action potentials (e.g., gastric smooth muscle), each action potential produces a long-lasting contraction resembling a fused tetanus.

Smooth muscle can operate at muscle lengths ranging from a small percentage of the initial length up to substantial stretches. The length-tension relationship is similar to that of skeletal muscle (Fig. 10–6). Active force is maximal at a characteristic length. Smooth muscle is more plastic than skeletal muscle; if it is stretched substantially, the characteristic length for maximal active force moves reversibly to longer lengths.[24] The

maximum force per unit cross-sectional area approaches that of striated muscle, despite a much lower myosin concentration; the shortening velocities however, are much lower,[24] in the range of 0.1 to 0.7 muscle lengths/second. The broad working range of smooth muscle enables it to serve in organs that must close down tightly while at the same time being capable of vast distention.

Smooth muscles generate force efficiently.[5] They have 300 to 500 times lower ATPase rates associated with a given force level than skeletal muscle. This is due both to the low myosin-ATPase rate (a

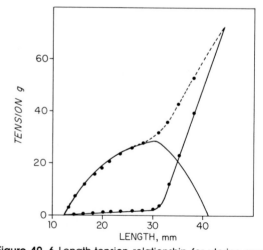

Figure 10–6 Length-tension relationship for uterine smooth muscle (rabbit). Dots and solid line, passive tension; dots and dashed line, total tension; solid line, active tension. The curves resemble those of skeletal muscle (Figs. 9–4 and 9–8). (After Csapo. In Bourne, G. H., ed. *Structure and Function of Muscle*, New York, Academic Press, 1960.)

property of the myosin isoenzyme) and to the low concentration of myosin. Energy utilization by the contractile apparatus can be measured as a change in creatine phosphate, even though ATP is the true energy source, since (as in skeletal muscle) the ATP used is rapidly resynthesized from creatine phosphate via the enzyme creatine phosphokinase (Chap. 9). During a sustained activation, the force remains elevated, but the rate of hydrolysis of ATP declines. Thus the efficiency for producing force increases during continued activation. Cross bridges that generate force so efficiently are termed "latch" cross bridges.[21] Smooth muscles show a Fenn effect, that is, additional energy is expended when they shorten and perform work (Chap. 9). Although smooth muscles generate force efficiently, they do not shorten efficiently.

Many smooth muscle cells (cat intestine, pig ureter, rat portal vein), primarily of the unitary type, become depolarized and contract when they are stretched, a phenomenon termed the *myogenic response* (Fig. 10–7). In cells that are spontaneously active, stretch increases the frequency of spontaneous action potentials and augments the contractile tension. The stretch-induced depolarization is clearly an intrinsic property of the smooth muscle, for it occurs in isolated smooth muscle cells and in tissues treated with TTX and with ganglionic blocking drugs and may involve the opening of stretch–sensitive ionic channels. This property is important to the regulation of smooth muscle function such as intestinal motility—a contractile response to the presence of food in the intestine.

In other smooth muscle cells (pig renal artery, cat nictitating membrane), stretch leads to increased tension that eventually subsides in time (Fig. 10–7). This drop in tension is called *stress relaxation*. Stress relaxation is an important property of organs such as the bladder, uterus, and stomach that in normal function are required to adapt to large changes in internal volume without inordinate increase in pressure. Stress relaxation is partly due to neural regulation, but much of it is due simply to the viscoelastic properties of the muscle.

Innervation and Transmitter Actions.[3] The innervation of smooth muscle differs in many respects from that of skeletal muscle, discussed in Chapter 6. Skeletal muscle, largely under voluntary control, is always innervated by a cholinergic motoneuron whose cell body lies in the spinal cord or in the higher central nervous system. The transmitter always acts on highly localized nicotine acetylcholine (ACh) receptors to open their ionic channels and initiate a depolarizing postsynaptic response. Smooth muscle, under involuntary control, is innervated by a variety of axons of the autonomic and enteric (in the gut) nervous systems, whose cell bodies lie in ganglia and plexuses outside of the central nervous system. The transmitters released, ACh, norepinephrine, and a range of peptides, act on receptors dispersed widely over the muscle cell surfaces. The actions

Figure 10–7 Response of smooth muscle to stretch. Muscle force is plotted against times for stretch in 6 smooth muscle preparations. The initial rapid rise of force indicates the time of stretch. *A, B,* and *C,* demonstrate stretch activation in 3 smooth muscle preparations as an increase in the active force that occurs with a delay after the stretch. *D* and *E,* demonstrate stress relaxation (not stretch activation), a decline in force with sustained stretch and no delayed increase in force, in 2 other smooth muscle preparations. *F* demonstrates that stretch activation involves opening of Ca channels, because the addition of a Ca channel blocker, D-600, turns a stretch activation response [seen in the dog trachea preparation when the K channels are blocked with tetraethylammonium (TEA)] into stress relaxation. (*A, B, C, D* and *E* after Burnstock and Prosser. *Am. J. Physiol.* 199:553–559, 1960; *F* modified from Kroeger, E. A. In Stephens, N. L., ed. *Smooth Muscle Contraction,* by courtesy of Marcel Dekker, Inc., 1984.)

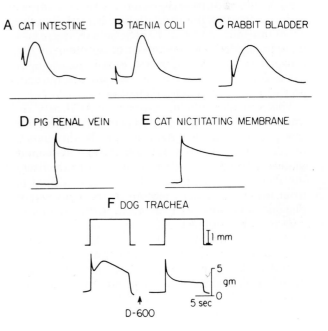

may be excitatory or inhibitory. Perhaps none of the receptors follow the design of the nicotinic ACh receptors, with channel and agonist sites part of the same molecule. They may all couple to membrane enzymes and channels through additional proteins and second messengers.

The nerve fibers supplying smooth muscle form vesicle-laden bulges or *varicosities* in the vicinity of the muscle fibers; there is little specialization of pre- and postjunctional membranes. The distance between these membranes ranges widely from 20 nm to several micrometers. The density of innervation of muscle cells varies widely in different tissues. In some multiunit muscles, such as the vas deferens, almost every muscle cell receives innervation (Fig. 10–1A). In other unitary muscles (uterus, intestine, ureter), only a few cells are innervated (see Fig. 10–1B). In such tissues, excitation presumably spreads from innervated to noninnervated cells through gap junctions. In the arterial wall, where only those cells around the periphery of the muscle layer are innervated, those near the lumen are ideally positioned to be effected by transmitter substances circulating in the blood.

Parasympathetic and sympathetic nerves usually have antagonistic effects on smooth muscle. As explained in Chapter 34 (Autonomic Nervous System), the effects of parasympathetic and sympathetic stimulation differ from tissue to tissue. The former prepares the body for rest and vegetative functions, whereas the latter sounds the alarm and prepares the body for "fight or flight." In these terms, it is understandable that for intestinal motility during digestion, parasympathetic activity increases the strength of contraction of the small intestine, and sympathetic activity decreases it; whereas for the regulation of air flow in the lungs necessary to support exercise, parasympathetic activity constricts bronchial smooth muscle, and sympathetic activity relaxes it.

The parasympathetic transmitter ACh acts on muscarinic ACh receptors and is mostly excitatory. Electrically excitable cells increase the frequency of action potentials or develop a more sustained plateau depolarization. Intracellular calcium rises, but it is not always clear whether this calcium is from the extracellular medium or from the sarcoplasmic reticulum. However, in smooth muscle cells from the stomach, ACh can cause contraction in the absence of external Ca^{2+}. The muscarinic ACh receptor can activate phospholipase C to stimulate the production of IP_3 in iris smooth muscle cells, which in turn promotes Ca release from internal stores.

As described in Chapter 1, β-adrenergic effects are mediated through β-receptors that activate adenylate cyclase, leading to production of cAMP, activation of a protein kinase, and phosphorylation in one or more proteins. The result in smooth muscle is usually inhibitory. The phosphorylated proteins could lead to membrane hyperpolarization and to a decrease in intracellular calcium. There is evidence that cAMP decreases intracellular calcium by two action: stimulation of the calcium pump of the SR (and possibly the surface membrane) and stimulation of the Na^+-K^+ pump, thus promoting more Na^+/Ca^{2+} exchange.[25] β-Adrenergic stimulation also increases the phosphorylation of myosin light-chain kinase, thereby decreasing its effectiveness in phosphorylating the myosin P light chain, which along with the Ca and calmodulin activates contraction (see Fig. 10–3).

α-Adrenergic effects on smooth muscle may be excitatory (as in vascular, urinary, genital smooth muscle) or inhibitory (as in intestinal smooth muscle). Excitatory action is correlated with increased intracellular calcium. Because contraction can take place without change in membrane potential and in a calcium-free extracellular medium, it is believed that the alpha excitatory effect is mediated by calcium release from intracellular stores, again possibly through IP_3. In those cases in which α-adrenergic action is inhibitory, the smooth muscle cells hyperpolarize, probably due to an increased potassium conductance. However, this, too, can be mediated through elevated intracellular calcium, since elevated intracellular Ca in the immediate vicinity of the membrane can open calcium-activated potassium channels (Chap. 4). Thus, elevated calcium can produce contraction if it is elevated sufficiently throughout the cell, or it can produce relaxation through hyperpolarization and decreased excitation if it is elevated primarily near the membrane Ca-activated K channels. This could explain the excitatory or inhibitory effects of α-adrenergic stimulation in different tissues.*

In addition to the neurotransmitters, smooth muscle is sensitive to circulating hormones. Consider the profoundly excitatory effect of gastrin (a peptide) on the smooth muscle of the stomach.[32]

---

*Another consideration is that the release of ACh varicosities of parasympathetic neurons can be inhibited through an α-adrenergic effect, presynaptic inhibition. Thus in tissues in which ACh activates, α-adrenergic inhibition could occur via either action on the nerve terminals or action on the muscle directly.

It increases the amplitude of the plateau of action potentials. The increased depolarization presumably leads to increased contraction of cells and accounts for the increased gastric motility produced by gastrin. There are many other examples of hormone action on smooth muscle, including the actions of histamine, prostaglandins, and a number of peptides. The hormones progesterone and estrogen cause the uterus to hypertrophy and to stay quiescent during gestation. At the time of parturition, uterine muscles respond dramatically to oxytocin with increased electrical activity and synchronized contraction. The reader is referred to later chapters in which the hormonal regulation of specific organs is discussed.

CONTRACTION IN NONMUSCULAR CELLS[11, 15, 28]

In this section, we will consider cells that are not muscle cells but nevertheless exhibit motility either of the whole cell or of its parts, e.g., cell organelles. Motility in these cells depends on cytoplasmic filaments that resemble those in muscle cells and include thin, actin-containing filaments; occasional short, thick myosin filaments; and also intermediate-size filaments and microtubules.

Actomyosin Systems. Thin actin-containing contractile filaments in conjunction with myosin and other contractile proteins are mainly responsible for cell surface motility, giving rise to such varied phenomena as amoeboid movement, movement of microvilli, ruffling of cell surfaces, movement of antigen antibody complexes on cell surfaces, cytokinesis (pinching off of daughter cells in mitosis),[26] and contraction of retinal rods and cones. In some examples, e.g., the microvilli, motility depends on the actin filaments that may be permanently present as structural elements, much as in striated and smooth muscle. In other instances, e.g., amoeboid movement, the thin-filament structures are disassembled and reassembled in a different part of the cell to change the direction of movement and force; thus control involves not only regulation of actin-myosin interaction but the selective formation and dissolution of the contractile structure itself.

Actin is an almost universal constituent of cells and has been identified in virtually all cell types carefully examined. It is readily identified in filamentous form by its reaction with heavy meromyosin to produce the chevron pattern previously described (see Fig. 8–9). It accounts for 30% of the total protein of motile cells, such as amoeba and

human blood platelets, but only 1% to 2% of sessile cells, such as those of the liver. Actins, regardless of source, have about the same molecular weight and amino-acid sequences, only about 6% of the actin amino-acid residues varying among different sources. Like muscle actin, nonmuscle actin can polymerize from the G actin (globular) form to F (fibrous) actin filaments; the fibrous form activates myosin ATPase.

In nonmuscle cells, cytoplasmic F actin occurs both in microfilaments and in a free, unpolymerized form. Filaments are formed by polymerization of actin with actins binding at either end at binding sites in equilibrium with the free actin. In the presence of ATP, each attachment of actin resulting in hydrolysis of one molecule of ATP. The reaction product (ADP) remains attached to the actin with the result that the two ends of the thin filament have different equilibrium constants for additional actin attachment, with the barbed end (as defined by the polarity of myosin decoration—see Fig. 8–9) being favored. Filament growth is thus controlled by the concentration of free actin monomers and by the availability of these barbed ends, which can be blocked by attachment to other proteins. In the G form, actin interacts one-to-one with a small (16,000 molecular weight) protein, profilin, and also with a seemingly unlikely companion, DNase I, reactions that prevent it from polymerizing to form structural or "contractile" filaments. Filamentous F actin interacts with a variety of proteins to form different cytoskeletal structures. The functions and names of these proteins include: (i) stabilization of thin filaments (tropomyosin); (ii) cross-linkage of actin filaments into networks that can be disrupted by calcium (villin, present in many cells) or can remain stable in tightly bundled structures (fimbrin and α-actinin) or more loosely bound structures (filamin, found in muscle and nonmuscle cells, and α-actinin); (iii) termination and cross linkage together of thin filaments at the barbed ends in transverse structures, either within or at the surface of cells (α-actinin, as in skeletal, cardiac, and smooth muscle); or (iv) termination in the region near cell surfaces (spectrin in erythrocytes and fodrin and other proteins in intestinal microvilli and other cells). These proteins provide cells with the ability to form and dismantle isolated thin filaments and whole networks of filaments in response to the need to move in a specific direction, to keep thin filaments bundled together in order to provide necessary rigidity, or to anchor intracellular networks to surfaces for support

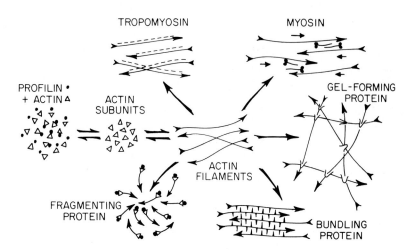

TROPOMYOSIN   MYOSIN

PROFILIN •
+ ACTIN △

ACTIN
SUBUNITS

GEL-FORMING
PROTEIN

ACTIN
FILAMENTS

FRAGMENTING
PROTEIN

BUNDLING
PROTEIN

Figure 10–8 Actin-binding proteins controlling actin polymerization to form thin filaments and networks of filaments. Profilin binds actin to prevent polymerization. Tropomyosin binds to actin filaments, stabilizing them. Myosin attaches driving filament movement. Gel-forming proteins (filamin and α-actinin) aid in network formation. Bundling proteins (fimbrin and villin) bridge the filaments forming bundles. Fragmenting proteins (villin and gelsolin) plus Ca promote the dissociation of filaments. (After Sheterline, P. *Mechanisms of Cell Motility: Molecular Aspects of Contractility.* New York, Academic Press, 1983; and Alberts, B., et al. *The Molecular Biology of the Cell.* New York, Garland, 1983.)

or structural rigidity. Examples of these various functions and arrangements are illustrated in Fig. 10–8.

Myosin or myosinlike proteins are also found widely distributed in a variety of nonmuscle cells (brain, liver, blood platelets, leukocytes, adrenal medulla), usually in a soluble, cytoplasmic form and in molar concentrations much lower than in muscle cells. These myosins are mostly similar to muscle myosin in shape and molecular weight. Nonmuscle myosin, like that from muscle, forms complexes with actin that can be dissociated by magnesium ATP, has ATPase activity (although less than that of muscle myosin), and can polymerize to form thick filaments. As with smooth muscle myosin, the polymerization can be regulated by the phosphorylation of myosin, with the phosphorylated form polymerizing more readily. The thick filaments formed from nonmuscle myosin have the same 14.3-nm periodicity of heads found in skeletal muscle thick filaments but often lack the bare central region, the resulting filament being mono- rather than bipolar.

Interaction of myosin and actin in nonmuscle cells is controlled by calcium, although the way in which calcium exerts its activating effects is less well understood than in skeletal and cardiac muscle and varies in different species. Although analysis of human blood platelets and brain cells reveals troponin and tropomyosin,[7] it is thought that the predominant activation in mammalian nonmuscle cells is through the thick rather than thin filaments. In macrophages and blood platelets, the process of calcium activation involves phosphorylation of myosin light chains through the intermediation of calmodulin; here, the process is similar to that found in smooth muscle cells.[15]

The movement of intestinal microvilli, illustrated in Figure 10–9, demonstrates two roles for actin filaments: providing structural rigidity of the microvilli and, in interaction with myosin, moving the microvilli through contraction of the ring surrounding each cell.[16] Bundles of actin filaments that traverse the long axis of the villus attach to the membrane at the distal tip of the villus. The attachment is accomplished by α-actinin and is reminiscent of the attachment of thin filaments to the Z line in muscle. The actin filaments are bundled together by fimbrin and villin and are attached to the lateral membranes by still another protein. The proximal ends of some of the filaments insert into a terminal web containing myosin, actin, and other filaments below the troughs of the villi, where the actin filaments interact with horizontally arranged thick myosin filaments. The microvilli increase the surface area for absorption in these intestinal epithelial cells. The role of these thick myosin filaments is not certain but they may serve to stabilize the microvilli in an upright orientation. Bundles of actin filaments also wrap circumferentially around each intestinal epithelial cell attaching to the lateral membranes at the zonula adherens, the dense, belt-like adhesion plaque that sticks epithelial cells together. The actin filament attachment also involves α-actinin and vinculin. Myosin is also located along with these circumferentially oriented actin filaments. The polarity of these filaments, assessed by myosin decoration, can be opposite in adjacent filaments allowing for a possible sliding mechanism of contraction. These thin filaments

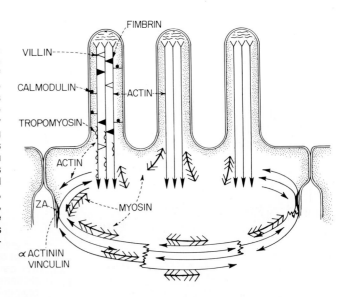

Figure 10–9 Diagram of the filament arrangement in intestinal microvilli. Actin filaments are present in both the microvilli and in the circumferential ring [at the level of the zonula adherens (ZA)]. In the microvilli, they attach to α-actinin at the top and overlap the short thick filaments in the valleys between microvilli. The arrowheads indicate the polarity of the actin filaments to decoration by myosin. Other proteins, fimbrin and villin, aid in bundling the actin filaments together. Tropomyosin also exists along the thin filaments. Clamodulin is important in Ca regulation. The circumferentially oriented actin filaments attach to the zonula adherens through α-actinin and vinculin. They also are bundled together with myosin, which when used to decorate these actin filaments, shows that neighboring filaments can have opposite polarities. Contraction of this circumferential ring is thought to cause movement of the microvilli. (After Mooseker, M. S. *Cell* 35:11–13, 1983. Copyright by Cell Press.)

are probably involved in microvilli motility since addition of ATP and low concentrations of calcium to glycerinated preparations with their surface membrane disrupted causes contraction of this ring, squeezing the cell circumferentially and splaying out the microvilli. This calcium activation involves phosphorylation of myosin light chains presumably through the same mechanism as in smooth muscle. The microvilli actin filaments are probably more structural. Higher concentrations of calcium can cause the microvilli to shorten, but this is because calcium causes the disruption of these bundled actin filaments in the microvilli. This phenomenon is mediated by calcium activation of villin (See Fig. 10–8). Thus actin filaments are used here for both structural elements (microvilli) and contractile elements (the circumferential ring) to bring about the movements of microvilli, presumably increasing intestinal absorption.

The retinal photoreceptors of bony fish change shape in response to the level of illumination. During light adaptation, the cones slowly contract (2 μm/min), shortening by as much as 80 μm, while the rods elongate so that their outer segments (which contain the photopigment) become buried in the overlying pigment epithelium. During dark adaptation, reverse changes occur. The result is to bring the photosensitive portion of the appropriate receptor closer to the incoming light. Burnside[4] has studied the changes in light adap-

tation by electron microscopy. Fish cones contain bundles of thin filaments that are attached to dense distal structures below the outer segment and run the length of the elongated cone (Fig. 10–10). Other thin filaments originate from the proximal portion of the cone and end blindly, overlapping with the "upper" set of filaments and with thick filaments. It is proposed that in light adaptation cycling cross bridges ratchet the two sets of thin filaments past the thick filaments and thus shorten the receptor, pulling its outer segment closer to the incoming light. Calcium appears to activate the actin-myosin interaction through myosin light chain phosphorylation, but the intriguing question of how the level of illumination affects the calcium level remains to be explored.

Amoeboid movement and pinocytosis, the engulfment of particles by cell membranes, have been ascribed to actin-myosin interaction. Electron micrographs reveal thin filaments in pseudopodia and especially in the cytoplasm underlying the membrane near incipient pinocytosis. Studies with intracellular aequorin reveal increases in cytoplasmic calcium associated with pseudopod formation and pinocytosis.[33, 34] In addition to its role of facilitating actomyosin interaction, calcium may cause the breakdown of the cytoplasmic thin filament network at the site of movement, allowing the movement of material, polymerization of thin filaments, and redirection of the filaments in the

DARK ADAPTED

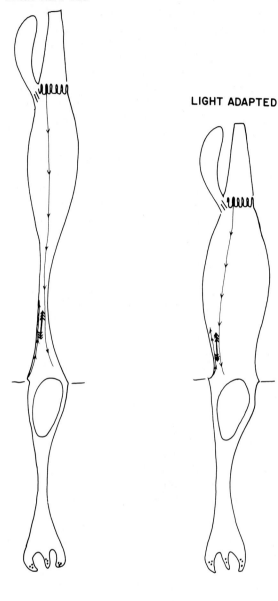

LIGHT ADAPTED

Figure 10–10 Diagram of the thin filament arrangement in the cones of fish. The cone on the left is dark-adapted and extended. The cone on the right has been exposed to light and has shortened, pulling the light-detecting outer segment out of the pigment layer of the retina in which it is buried when in the dark. Some thin filaments attach to the outer segment at the top of the figure and project to overlap short, thick filaments. Others attach to the base and project upward to overlap these thick filaments. Light activates interaction of actin and myosin, and the thick filaments reel in the long, thin filaments from above and below, thus shortening the cone. (From Burnside, B. *J. Cell Biol.* 78:227–246, 1978.)

new direction of movement. The role of filament assembly in cell movement is suggested by the blocking of amoeboid movements and cytokinesis by cytochalasin B, an agent that binds to the barbed end of actin filaments and prevents further polymerization.

Bundles of actin filaments are prominent in the so-called stress fibers seen particularly in cultured cells grown on a flat surface. Stress fibers connect the points at which a cell adheres to the surface with intermediate filaments surrounding the nucleus. The actin filaments, some with opposite polarity, are bundled together with the protein filamin. Periodocities of protein density occur every 0.4 μm or so along the stress fiber with α-actinin and myosin localized in alternating regions. These have been likened to "little" sarcomeres; however, they may alternatively be anchoring structures, unique to immobile cultured cells.

Finally, in some nonmuscle cells, actin appears to play a structural role unrelated to movement. For example, actin filaments are found in the stereocilia of cochlear hair cells (Chap. 5). The sterocilia are pencil-like structures made of bundles of actin filaments with several thousand filaments near the tip, tapering sharply down to several dozen near the base, in a narrow attachment of each cilium to the receptor cell.[10, 35] The filaments are actin, because the S-1 portion of myosin forms arrowhead structures on them pointing toward the base of the stereocilia, away from the tip. In this way, stereocilia resemble the microvilli of intestinal epithelium. There is no evidence that these stereocilia contract with or without calcium and ATP;[35] however, hair cells have been shown to change shape when stimulated, so as in the intestinal epithelial cells some component is contractile.

Microtubule-Based Motility. Other forms of motility are based on microtubules. Microtubules are long hollow cylinders with internal and external diameters of about 14 and 24 nm, respectively, formed from helical arrays of the protein units of *tubulin*. These microtubules are assembled from α and β tubulin dimers polymerized into protofilaments by a process in which the energy is provided by the hydrolysis of GTP (instead of ATP). Polymerization is similar to that of actin in the thin filament. Tubulin will polymerize in the absence of GTP but grows in either direction. With GTP, the microtubules selectively grow at one end. Microtubules play a dynamic role and assemble and disassemble in response to physiological needs. Many originate in the region of the cell close to the centrioles and during mitosis function in positioning cell structures and moving chromosomes.

Microtubules play a direct role in the force generation and movement in cilia and flagella in which the microtubules are arranged as doublets in a characteristic 9 + 2 array: nine circumferentially arranged doublets surrounding a central doublet to which they are attached by spokes. If the spokes are disrupted, the addition of ATP initiates a sliding of the doublets with respect to one another. If the spokes are intact, the flagellum or the cilium bends. The force underlying these movements is produced by a protein, dynein, that forms radiating arms from one of the doublets and projects out to interact with a microtubule of the adjacent doublet. Dynein is an ATPase.[13] There is much similarity with myosin. Both are large molecules with more than one head. As with the kinetics of the myosin ATPase, when dynein reacts with ATP, it shows an early burst of phosphate production that quickly tapers off to a slower rate of product release (see Fig. 8–6). However, product release from dynein is much faster than from

myosin. As with actin-myosin, ATP dissociates the dynein-microtubule complex. Unlike for myosin, there is some disagreement about the dynein structure and contractile mechanism. In some cilia, dynein appears to have three heads, in others two. There is no dynein filament. Dynein connects two microtubules and propels them along. The role of calcium is not clear. It may be important in regulating the frequency of beating and may not be needed for activating the dynein ATPase.

Another important role for microtubules occurs in the movement of organelles and proteins within a cell. This has been investigated most extensively in nerve cells, which must transport organelles such as synaptic vesicles and cell proteins long distances, down axons from the cell bodies where they are synthesized to their site of action, synaptic terminals up to a meter away. Movement can be at rates varying from a few mm/day for some proteins to 200–400 mm/day for membrane-bound organelles—rapid oxoplasmic transport. This fast transport is saltatory, not smooth. Axons contain many microtubules aligned along the axon, as well as intermediate neurofilaments (see the next section) and actin. The microtubules mediate this transport, since disrupting them with colchicine impedes transport, whereas disrupting actin filaments with actin binding proteins (e.g., DNase I, see page 223) or cytochalasin B does not. Particles can move along the microtubules in both directions, down the axon (anterograde) or back (retrograde), even though the microtubules themselves have a single polarity. A recently discovered protein called kinesin binds particles to the microtubules and provides the propelling force for their anterograde movement. Kinesin purified from squid axon axoplasm and optic lobes[35a] can be shown to catalyse movement in vitro. It even induces movement of purified microtubules along a glass coverslip and of latex beads along microtubules! Kinesin, like dynein, is an ATPase and binds to microtubules, but it is different from dynein in its molecular weight and enzymatic properties. Thus the kinesin-microtubule system seems to be the motor for anterograde rapid axoplasmic flow. Probably a dyneinlike molecule catalyzes retrograde movements.

Intermediate Filaments. Although they may not play a direct role in motility, no discussion of cytoplasmic filaments would be complete without mention of intermediate filaments, so named because their size falls between those of actin filaments and microtubules. Although intermediate filaments from different cells have similar diameters, the estimated weights of their constituent proteins vary greatly: from 40,000 to 200,000. Each cell type usually contains intermediate filaments made from a different constituent protein, e.g., neurofilaments in nerve cells, keratin in epithelia, and vimentin and desmin in muscle cells. The function of intermediate filaments is not known, but they are believed to play a major structural role in giving cells their resting shape and stiffness and in anchoring intracellular structures.

ANNOTATED REFERENCES

Alberts, B.; Bray, D.; Lewis, J.; Raff, M.; Roberts, K.; Watson, J.D. *Molecular Biology of the Cells.* New York, Garland Publishing, 1983.
This exceptional text, in addition to its clearly presented and illustrated material on other aspects of molecular biology, includes a readable summary chapter on "the cytoskeleton," discussing cell motility from muscle to nonmuscle, myosin to microtubule.

Bagby, R.M.; Young, A.M.; Dotson, R.S.; Fisher, B.A.; McKinnen, K. Contraction of single isolated smooth muscle cells from *Bufo marinus* stomach. *Nature* 234:351-352, 1971.

Fay, F.S.; Fujiwana, K.; Rees, D.D.; Fogarty, K.E. Distribution of α-actinin in single isolated smooth muscle cells. *J. Cell Biol.* 96:783-795, 1983.

Singer, J.J.; Walsh, J.V. Large conductance Ca^{2+}-activated K^+ channels in smooth muscle cell membrane. *Biophys. J.* 45:68-70, 1984.
These and several articles in the Stephens book Smooth Muscle Contraction discuss experiments on the isolated smooth muscle cell, a preparation offering great promise.

Bulbring, E.; Brading, A.F.; Jones, A.W.; Tomita, T., eds. *Smooth Muscle: An Assessment of Current Knowledge,* Austin, University of Texas Press, 1981.
A collection of review papers by experts in the field on structure, electrical properties, and regulation of various smooth muscle.

Johnson, K.A. Pathway of the microtubule-dynein ATPase and the structure of dynein: a comparison with actomyosin. *Annu. Rev. Biophys. Biophys. Chem.* 14:161-188, 1985.
An excellent review of the data, much of it produced by the author, on the dynein structure and ATPase plus a detailed comparison with the actomyosin system, showing the extensive similarities.

Sheterline, P. *Mechanisms of Cell Motility: Molecular Aspects of Contractility,* New York, Academic Press, 1983.
An excellent comprehensive treatment of myosin-based cell motility with particular stress on smooth muscle and nonmuscle motility. It also contains up-to-date information on the contractile proteins and their function in muscle and nonmuscle motility.

Stephens, N.L., ed. *Smooth Muscle Contraction.* New York, M. Dekker, 1984.
A collection of short papers by experts, combining reviews and original experiments on the ultrastructure, mechanics, contractile mechanisms, regulation, functional properties, and pathophysiology of smooth muscle. The treatment is not integrated, but the papers provide interesting reading and show the trends in the field, including measurements using isolated smooth muscle cells.

REFERENCES

1. Berridge, M.J.; Irvine, R.F. Inositol trisphosphate, a novel second messenger in cellular signal transmission. *Nature* 312:315-321, 1984.

2. Bohr, D.F.; Somlyo, A.P.; Sparks, H.V.; eds. *Vascular Smooth Muscle. Handbk. Physiol.* Sec. 2, 2:1-671, 1980.

2a. Bozler, E. Actin potentials and conduction of excitation in muscle. *Biol. Symp.* 3:95-110, 1941.

3. Bulbring, E.; Brading, A.F.; Jones, A.W.; Tomita, T.; eds. *Smooth Muscle: An Assessment of Current Knowledge.* Austin, University of Texas Press, 1981.

4. Burnside, B. Thin (actin) and thick (myosinlike) filaments in cone contraction in the teleost retina. *J. Cell Biol.* 78:227-246, 1978.

5. Butler, T.M.; Siegman, M.J. High-energy phosphate metabolism in cardiovascular smooth muscle. *Annu. Rev. Physiol.* 47:629-643, 1985.

6. Cassidy, P.; Hoar, P.E.; Kerrick, W.G.L. Irreversible thiophosphorylation and activation of tension in functionally skinned rabbit ileum strips by [^{35}S]ATP S. *J. Biol. Chem.* 254-11148-11153, 1979.

7. Cohen, I.; Kaminski, E.; De Vries, A. Actin-linked regulation of the human platelet contractile system. *FEBS Letters* 34:315-317, 1973.

8. Collins, J.H.; Dron, E.D. Actin activation of Ca^{2+}-sensitive Mg^{2+}-ATPase activity of acanthameba myosin II is enhanced by dephosphorylation of its heavy chains. *J. Biol. Chem.* 255:8011-8014, 1980.

9. Fay, F.S.; Fogarty, K.; Fujiwara, K. The organization of the contractile apparatus in single isolated smooth muscle cells. In Stephens, N.L., ed. New York, M. Dekker, *Smooth Muscle Contraction,* 75-90, 1984.

10. Flock, A.; Cheung, H.C. Actin filaments in sensory hairs of inner ear receptor cells. *J. Cell Biol.* 75:339-343, 1977.

11. Goldman, R.D.; Milsted, A.; Schloss, J.A.; Starger, J.; Yerna, M.J. Cytoplasmic fibers in mammalian cells: cytoskeletal and contractile elements. *Annu. Rev. Physiol.* 41:703-723, 1979.

12. Hinssen, H.; D'Haese, J.; Small, J.V.; Sobieszek, A. Mode of filament assembly of myosins from muscle and nonmuscle cells. *J. Ultrastruct. Res.* 64:282-302, 1978.

13. Johnson, K.A. Pathway of the microtubule-dynein ATPase and the structure of dynein: a comparison with actomyosin. *Annu. Rev. Biophys. Biophys. Chem.* 14:161-188, 1985.

14. Kamm, K.E.; Stull, J.T. The function of myosin and myosin light chain kinase. Phosphorylation in smooth muscle. *Annu. Rev. Pharmacol. Toxicol.* 25:593-620, 1985.

15. Korn, E.D. Biochemistry of actomyosin-dependent cell motility (a review). *Proc. Natl. Acad. Sci. USA* 75:588-599, 1978.

16. Mooseker, M.S.; Tilney, L.G. The organization of an actin filament-membrane complex: filament polarity and membrane attachment in the microvilli of intestinal epithelial cells. *J. Cell Biol.* 67:725-743, 1975.

17. Morgan, J.P.; Morgan, K.G. Vascular smooth muscle: the first recorded Ca^{2+} transients. *Pflugers Arch.* 395:75-77, 1982.

18. Morgan, J.P.; Morgan, K.G. Alteration of cytoplasmic ionized calcium levels in smooth muscle by vasodilators in the ferret. *J. Physiol.* 357:539-551, 1984.

19. Murphy, R.A. Filament organization and contractile function in vertebrate smooth muscle. *Annu. Rev. Physiol.* 41:737-748, 1979.

20. Murphy, R.A. Mechanics of vascular smooth muscle. *Handbk. Physiol.* Sec. 2, 2:325-442, 1980.

21. Murphy, R.A.; Aksoy, M.O.; Dillon, P.F.; Gerthoffer, W.T.; Kamm, K.E. The role of myosin light chain phosphorylation in regulation of the cross-bridge cycle. *Fed. Proc.* 42:51-56, 1983.

22. Nonomura, Y.; Ebashi, S. Calcium regulatory mechanism in vertebrate smooth muscle. *Biomed. Res.* 1:1-14, 1980.

23. Pollard, T.D.; Korn, E.D. Acanthameba myosin. II. Inter-

action with actin and with a new co-factor protein required for actin activation of Mg^{2+} adenosine triphosphatase activity. *J. Biol. Chem.* 248:4682-4690, 1973.

24. Prosser, C.L. Smooth muscle. *Annu. Rev. Physiol.* 36:503-535, 1974.
25. Scheid, C.R.; Fay, F.S. α-Adrenergic stimulation of ^{42}K influx in isolated smooth muscle cells. *Am. J. Physiol.* 240:C415–421, 1984.
26. Schroeder, T.E. Dynamics of the contractile ring. In Inoue, S.; Stephens, R.E., eds. *Molecules and Cell Movement.* New York, Raven Press, 305–334, 1975.
27. Sherry, J.M.F.; Gorecka, A.; Adsoy, M.O.; Dabrowska, R.; Hartshorne, D.J. Roles of calcium and phosphorylation in the regulation of the activity of gizzard myosin. *Biochemistry* 17:4411-4418, 1978.
28. Sheterline, P. *Mechanisms of Cell Motility: Molecular Aspects of Contractility.* New York, Academic Press, 1983.
29. Sobieszek, A.; Small, J.V. Myosin-linked calcium regulation in vertebrate smooth muscle. *J. Mol. Biol.* 101:75-92, 1976.
30. Somlyo, A.V.; Bond, M.; Butler, T.M.; Berner, P.F.; Ashton, F.T.; Holtzer, H.; Somlyo, A.P. The contractile apparatus of smooth muscle: an update. In Stephens, N.L., ed. *Smooth Muscle Contraction,* New York, M. Dekker, 1-20, 1984.
31. Stephens, N.L., *Smooth Muscle Contraction,* New York, M. Dekker, 1984.
32. Szurszewski, J.H. Mechanism of action of pentagastrin and acetylcholine on the longitudinal muscle of the canine antrum. *J. Physiol. (Lond.)* 252:335-361, 1975.
33. Taylor, D.L.; Blinks, J.R.; Reynolds, G. Contractile basis of ameboid movement. VIII. Aequorin luminescence during ameboid movement, endocytosis, and capping. *J. Cell Biol.* 86:599-607, 1980.
34. Taylor, D.L.; Wang, Y.-L.; Heiple, J.M. Contractile basis of ameboid movement. VII. The distribution of fluorescently labeled actin in living amebas. *J. Cell Biol.* 86:590-598, 1980.
35. Tilney, L.G.; DeRosier, D.J.; Mulroy, M.J. The organization of actin filaments in the stereocilia of cochlear hair cells. *J. Cell Biol.* 86:244-259, 1980.
36. Warshaw, D.M.; Fay, F.S. Cross-bridge elasticity in single smooth muscle cells. *J. Gen. Physiol.* 82:157-199, 1983.
37. Williams, D.A.; Fay, F.S. Calcium transients and resting levels in isolated smooth muscle cells as monitored by Quin2. *Am. J. Physiol.,* 250:C779–C791, 1986.
A. Vale, R.D.; T.S. Reese; Sheetz, M.P. Identification of a novel force-generating protein, kinesin involved in microtubule-based motility. *Cell* 42:39-50, 1985.

ALBERT F. FUCHS
Section Editor

Section III

Processing by Central Neurons

Peter B. Detwiler
Wayne E. Crill

Chapter 11

Synaptic Transmission

INTRODUCTION

A jellyfish phosphoresces, a snake rattles, a bee dances, a bird sings, and a mother and child exchange emotions through a smile—these are a few examples of the diverse repertoire of animal behaviors. Each response is orchestrated by the nervous system and requires communication at many levels and over distances ranging from micrometers to meters. Surprisingly, our large catalogue of behavioral responses uses only two types of signals: electrical and chemical. Previous chapters have presented the modern view of transmission of electrical signals over distance by axons and chemical transmission at the neuromuscular junction. This chapter is about the special adaptation of these two processes for communication between nerve cells.

Neuron Morphology and Interconnections. The fundamental unit for the transfer and integration of information in the nervous system is the neuron. Both the surface area and physical extent of neurons are larger than those of any other cell type in the body. While neurons come in an enormous array of shapes and sizes (Fig. 11–1), the neuron is basically a cell body that has undergone morphological specialization to receive input at one end and to send output from the other end. The cell body, or soma, which contains the nucleus and is the metabolic center of the cell, ranges in diameter from only a few micrometers to one tenth

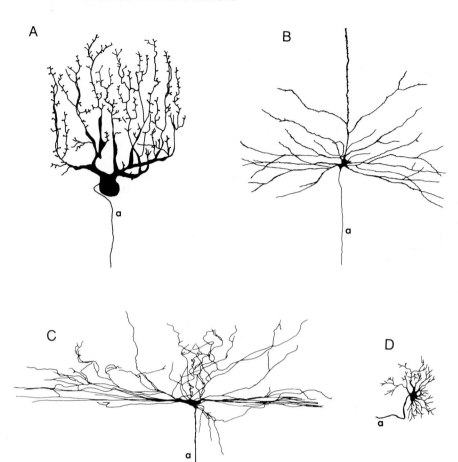

A

B

C

D

Figure 11–1 A collection representative of isolated neurons. Drawings show the dendritic arbor, cell body, and axon (a) of a cerebellar Purkinje cell *(A)*, a pyramidal cell in the cerebral cortex *(B)*, a phrenic motoneuron *(C)*, and an interneuron in the spinal cord *(D)*. (A, B, and D, after Cajal, S. R. *Histology du systeme nerveux de l'homme et des vertébrés*, vol. I. Paris, Maloine, 1909. C, from Cameron, W. E.; Averill, D. B.; Berger, A. J. *J. Comp. Neurol.* 219:70–80, 1983.)

of a millimeter. A typical neuron receives input on its cell body and on 10 to 20 highly branched processes called dendrites. It is the many branching dendrites, often extending for more than a millimeter, that are responsible for a neuron's extensive surface area. In spinal motoneurons, for example, the dendritic surface area is 10 to 20 times greater than the area of the soma membrane.[3]

The dendrites and cell body of a typical central neuron receive thousands of synaptic contacts from hundreds of different neurons. This is an example of *convergence* of activity; inputs from many cells converge onto a single cell. On the output side of a typical neuron there is an axon. It projects from the soma and extends for distances that can vary from a few micrometers in some cells to more than a meter in others. The parent axon frequently divides along its course to form one or more axon collaterals that branch profusely as they near their terminations. The arbor of terminal branches allows a single neuron via its axon to influence hundreds of other neurons. This is

an example of *divergence* of activity; the output of a single neuron diverges onto many neurons.

Divergence and convergence are general statements about the fact that a single neuron tells many neurons about information it has received from many other neurons; neurons gossip. The precise anatomy or "wiring diagram" of neuronal circuitry is another piece in the puzzle of how the nervous system is able to use relatively simple chemical and electrical signals to receive, translate, and process complex information. This aspect of nervous system function is dealt with in Chapters 13 through 21 on the processing of sensory information by higher neural centers. The focus of the present chapter is at a more basic level; rather than being concerned with where information is coming from and where it is going to, we will be concerned only with how chemical and/or electrical signals are used to pass information from one cell to another.

Interneuronal Communication: General Properties. To appreciate the full range of communication that may take place between nerve cells, it is important to have a feeling for the environment in which they live. While the specific nature of their surroundings may vary depending on the exact cell type considered, neurons in general exist in very similar environments. They live in crowded spaces with many other nerve cells and a relatively large number of supporting cells in a sea of gently circulating interstitial fluid having a carefully controlled ionic composition. Fanciful illustrations of the neuronal neighborhood (Fig. 11–2) succeed in conveying a vivid feeling for the structural jungle of the nervous system, but they fail to portray the extraordinary complexity of the extracellular space. The nervous system floats in an aqueous environment that provides an ideal medium for the dispersal and distribution of a variety of chemical messages. As chemical signals circulate in the extracellular fluid, they also percolate through a matrix of carbohydrate and protein molecules that comprises the packing material that surrounds the cells and fills the spaces between them. The functional significance of these extracellular molecules is only beginning to be appreciated. There is growing evidence, for example, that the matrix provides scaffolding as well as directional instructions for the growth of neuronal processes during development. There are fewer clues, however, about the role of the matrix

in the adult nervous system. It may act in the maintenance of cells and in the regulation of the chemical composition of the extracellular environment. The latter function may involve the removal of some chemical messages and possibly the enzymatic amplification of others. Superimposed on the extracellular traffic of chemical signals there are also extracellular electrical signals. These are generated by the electrical activity of localized groups of nerve cells, which create small extracellular voltage gradients (called field potentials) between active and inactive regions of the tissue. The importance of field potentials in the nervous system is not known. They may be a minor but significant form of communication or they may be an insignificant byproduct of the normal electrical activity of neurons.

There are two general kinds of interneuronal communication: public announcements and private conversations. Communications that are mediated by the broad dispersal of signals into the extracellular fluid are nonspecific and analogous to public announcements. This kind of communication is difficult to study and little is known about it. There is growing recognition, however, that it plays an important role in the modulation of central nervous system activity. Recent studies suggest that the principal chemical mediators of

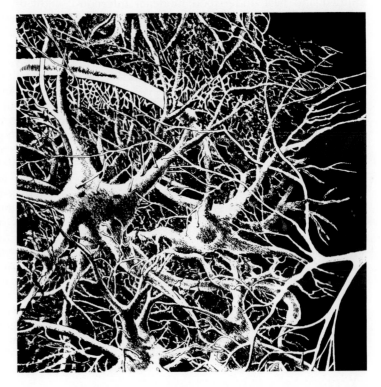

Figure 11–2 The neuronal neighborhood. This illustration showing the intermingled dendrites, cell bodies, and axons of a small group of neurons is a fanciful attempt to portray the structural complexity of the nervous system. Presynaptic processes swell slightly as they terminate to make synaptic contact with the dendrite, cell body, or axon of a second cell. The drawing is an oversimplification; it underestimates the true neuron density and does not show the presence of non-neural supporting cells. (From Ornstein, R.; Thompson, R. F. *The Amazing Brain*. Boston, Houghton Mifflin Co., 1984.)

these modulatory effects are peptides.[9] Although the effects of peptides on target cells have been studied most thoroughly in invertebrate nervous systems, immunocytochemistry has localized similar and in some cases identical neuropeptides in vertebrate neurons. It is thought that most neurons contain classic neurotransmitters as well as neuropeptides.[9] While the classic neurotransmitters are released at discrete, morphologically specialized sites (see below), granules containing neuropeptides may be released from less specialized regions of the plasmalemma. The wide distribution of neuropeptide-releasing sites provides a means of dispersing peptides into the general circulation of extracellular fluid and thus promotes their distribution and access to a large number of neurons. The mechanisms by which neuropeptides and other neuromodulators exert their effects on the nervous system are not well defined. Most commonly they are thought to influence target cells by causing changes in intracellular second messengers, an action that we will discuss more fully in a later section of this chapter.

Nerve cells are also capable of having private conversations with each other. Communications that are mediated by the focused delivery of a signal from one cell to another have a very precise or specific target. This type of communication is called *synaptic transmission*. It is the most widely studied and best understood mode of interneuronal communication and is the principal subject of this chapter. Synaptic transmission takes place at a morphologically specialized site called a synapse,* where two cells are in close contact. The signal that is exchanged at a synapse can be either *chemical* or *electrical*.

Overview: Chemical Synaptic Transmission. The basic chain of events in chemical transmission is shown in Figure 11–3. A membrane potential change reaches the synaptic ending of the first cell (the presynaptic cell), where it causes voltage-sensitive calcium channels in the surface membrane to open. Calcium enters the synaptic ending and promotes fusion of synaptic vesicles with the

plasma membrane, resulting in exocytosis of the neurotransmitter contained in the vesicles. Classic neurotransmitters are simple organic molecules of low molecular weight. In the central nervous system they include acetylcholine, adenosine, amino acids (glycine, gamma aminobutyric acid [GABA], and glutamate), and amines (norepinephrine, dopamine, serotonin, and histamine). Released neurotransmitter diffuses across the synaptic cleft separating the two cells and combines with receptor proteins on the surface of the second cell (the postsynaptic cell). The binding of transmitter to receptor triggers a conformational change in the protein, which initiates one of two possible processes in the postsynaptic cell. The first possible process resembles that evoked at the neuromuscular junction by acetylcholine (see Chapter 6). The conformational change triggered by transmitter binding forms an aqueous pore or channel in the receptor protein that extends across the membrane and is permeable to specific ions. The flow of ions through the open pore changes the membrane potential of the postsynaptic cell. The postsynaptic potential change is the signal that contains the information transmitted from the presynaptic cell. We will refer to postsynaptic responses generated directly by neurotransmitter gated channels as *first messenger transmission*. The second possible process resembles the action of light on a vertebrate photoreceptor (see Chap. 5) more than the action of acetylcholine on the neuromuscular junction. Here the conformational change in the receptor protein initiates a chemical chain reaction that leads to the production of intracellular *second messengers*. The second messengers then bring about profound changes in the postsynaptic cell that represent the information synaptically transmitted from the presynaptic cell. The changes triggered by the second messenger are not restricted to changes in the membrane potential of the postsynaptic cell. However, since most of what we know about synaptic transmission has been gleaned from electrical recordings we will not consider the "electrically silent" components of synaptic transmission and will focus instead on synaptic events that are electrically demonstrable. The principal way that second messengers influence the electrical activity of a postsynaptic cell is by opening or closing ion channels. We will refer to postsynaptic responses generated indirectly by the action of intracellular second messenger as *second messenger transmission*.

Chemical synaptic transmission has several additional distinguishing characteristics. First, chemical transmission takes time; it is slower than action

---

*Strictly speaking, a contact specialized for purposes of communication between *two nerve cells* is called a synapse, whereas any other similar specialized contact, whether it is between a nerve cell and a non-nerve cell or between two non-nerve cells, is called a *junction*. Although this nomenclature points out a clear morphological distinction between synaptic and junctional transmission, it is commonly ignored in informal (and sometimes formal) discussions because there is no essential difference in the cellular steps responsible for the transmission of the chemical signal at a synapse or a junction.

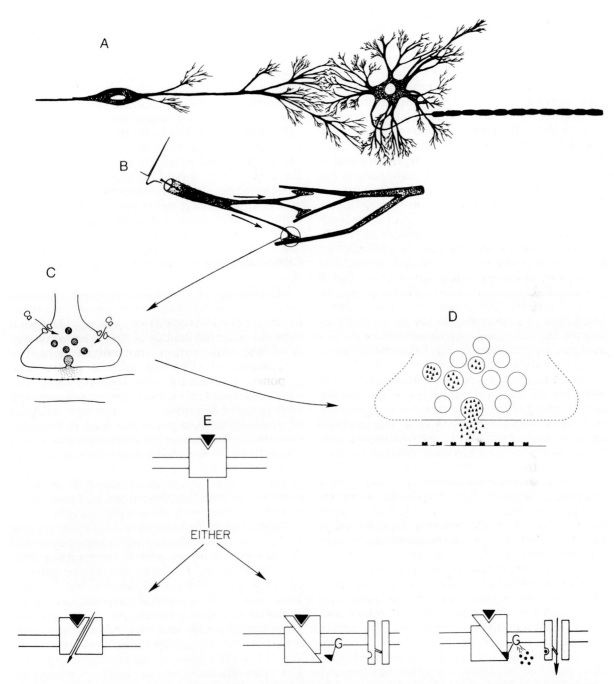

Figure 11–3 The principal steps in synaptic transmission. Proceeding from top to bottom, illustrated are progressively finer details about the underlying events of chemical synaptic transmission. *A*, The presynaptic *(left)* and postsynaptic *(right)* neurons. *B* and *C* show the propagation of an action potential into the fine presynaptic terminals of the first cell, where membrane depolarization opens voltage-gated calcium channels. Calcium influx increases intracellular calcium and triggers exocytosis of neurotransmitter molecules (solid triangles) from synaptic vesicles *(D)*. *E* shows a transmitter molecule bound to a postsynaptic receptor protein. The binding of transmitter to receptor either opens an ion channel directly *(left)* or activates a G protein *(right)* to generate an intracellular second messenger (solid circles) which in turn opens an ion channel.

potential propagation along an axon. During first messenger transmission a *synaptic delay* of 0.3 to 3 ms occurs between the invasion of the presynaptic terminal by an action potential and the production of a detectable electrical change in the postsynaptic neuron.[14] When second messengers are involved, the synaptic delay may be hundreds of milliseconds. Second, since the neurotransmitter is released from the presynaptic terminal and detected by receptors in the postsynaptic membrane, transmission at a chemical synapse is *unidirectional*. It always proceeds from the presynaptic terminal to the postsynaptic cell. The presynaptic cell is not influenced by membrane potential changes, including action potentials, in the postsynaptic cell. Third, chemical synaptic activity can either depolarize (*excite*) or hyperpolarize (*inhibit*) postsynaptic neurons. Whether a transmitter is excitatory or inhibitory depends on the ionic permeability of the channel in the postsynaptic membrane that is gated by the neurotransmitter rather than on the chemical identity of the transmitter. The common description of a transmitter as excitatory or inhibitory is imprecise and can cause confusion. For example, acetylcholine causes inhibition at some synapses and excitation at others.

Overview: Electrical Synaptic Transmission. In order for an electrical signal to be transmitted directly from one cell to another it is necessary for the ionic currents that give rise to the electrical signal in the first cell to flow into the second cell. This is possible at an electrical synapse because the short distance between the cells is spanned by a unique class of ion channels that provide a direct connection between the cytoplasmic compartments of the two cells. At an electrical synapse many of these *intercellular channels* are collected in a localized region to form a morphologically specialized contact called a *gap junction*.

Since the intercellular channels provide cytoplasmic continuity between the cells, the gap junction allows not only the transmission of electrical signals but also the direct exchange of ions and other small molecules. Although the pores through the intercellular channels are not large enough to permit passage of intracellular organelles, proteins, or nucleic acids, the fact that the cytoplasm of the two cells is connected places cells that are joined by gap junctions in a special class. They are a little like Siamese twins in that they are not exactly separate cells but neither are they exactly a single cell. Consequently, *electrical synaptic transmission* may be considered analogous to the spread of an electrical signal from one part of a single cell to another.

This analogy emphasizes a clear distinction between electrical and chemical synaptic transmission and suggests comparison on the basis of three properties. First, electrical transmission has fewer intervening steps and thus is faster than chemical transmission. Electrical synapses have no delays. Second, most but not all gap junctions allow electrical signals to spread in either direction; consequently electrical synaptic transmission is typically bidirectional. The factor that determines if an electrical synapse is bidirectional or unidirectional will be discussed in a later section. Third, while the polarity of a signal may be inverted by transmission across a chemical synapse, the polarity of an electrical signal is never reversed by transmission across an electrical synapse.

CHEMICAL SYNAPTIC TRANSMISSION

Morphology of Chemical Synapses. Chemical synapses occur between specific parts of neurons. In the central nervous system most presynaptic elements are axon terminals that synapse on the dendrites and the soma of the postsynaptic neuron to form what are called *axo-dendritic* and *axo-somatic* synapses, respectively. The large dendritic and somatic surfaces are almost completely covered with synaptic terminals. Less common are axon terminals that synapse with the axon of the postsynaptic cell to form an *axo-axonic* synapse. These connections are the anatomical substrate for presynaptic inhibition (see below).[14] In a few regions of the nervous system, *dendro-dendritic* synapses occur where both the presynaptic and postsynaptic elements are dendrites of different cells.[39]

Regardless of the location of a chemical synapse or the anatomical identity of its presynaptic and postsynaptic elements, each is designed to release a chemical messenger onto a receptive patch of specialized postsynaptic membrane, and hence all have similar ultrastructural morphologies. In the mammalian nervous system the presynaptic element typically ends in a small (about 1-μm diameter) bulbous enlargement called a *synaptic bouton* (Fig. 11–4). The bouton contains two morphological specializations: one is tufts of amorphous material that appear dense in tissue stained for the electron microscope, and the other is an accumulation of synaptic vesicles. The dense tufts are thought to be made up of filamentous proteins (Fig. 11–4A, B) that extend from synaptic vesicles and contact one another or larger filaments in the axoplasm.[29] In many preparations the tufts have a triangular profile extending into the cytoplasm

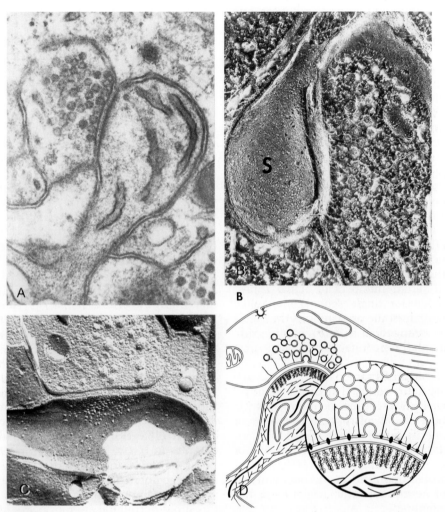

Figure 11–4 Four views of a chemical synapse in the mammalian nervous system. *A*, Transmission electronmicrograph of the synaptic junction between the terminal of a parallel fiber axon and the dendritic spine on a cerebellar Purkinje cell. (Magnification 61,500 ×.) The presynaptic element *(upper left)* swells slightly as it terminates, forming a synaptic bouton that contains a collection of synaptic vesicles. The space between vesicles contains an amorphous electron-dense material that is most apparent near the presynaptic membrane. The structure of the amorphous material is better seen in shallow-etched, rapidly frozen material. *B*, The dendritic spine (S) originates from a dendrite *(top)*. The neck of the postsynaptic spine has been cross-fractured, revealing fine cytoskeletal filaments. The axoplasm of the presynaptic element contains synaptic vesicles and a variety of cytoskeletal filaments. A larger, spectrin-like filament, *(horizontal arrow)* can be seen arising from the presynaptic membrane and extending into the axoplasm in the vicinity of synaptic vesicles. Outside of the domain occupied by synaptic vesicles, the cytoplasm is more coarsely granular and contains actin-like filaments *(oblique arrows)*. (Magnification 64,000 ×.) *C*, In freeze fracture the synaptic bouton (entering from *top*) and its complement of vesicles are shown in relief. The postsynaptic membrane of the dendritic spine that enters from the left contains a collection of membrane particles that are assumed to be a component of the electron-dense fuzz that lines the postsynaptic membrane shown in *A*. (Magnification 65,000 ×.) *D*, Diagram summarizing the principal morphological elements at a chemical synapse. The presynaptic terminal with its complement of synaptic vesicles is shown at the top. Fine filaments with the shape and dimensions of synapsin I extend from synaptic vesicles and contact one another or larger filaments. The larger filaments, which have the dimensions of spectrin, extend from the active zones in the presynaptic membrane. In order for a vesicle to fuse with the presynaptic membrane, it is thought that it must first dissociate from the presynaptic protein mesh. It is possible that a rise in intracellular calcium promotes this dissociation. On the postsynaptic side of the synapse the most salient morphological feature is a lining of unidentified material that represents the electron-dense fuzz shown in *A* and the postsynaptic particles shown in *C*. (Reproduced with permission from Landis, D. M. D.; Hall, A. K.; Weinstein, L. A.; Reese, T. S. Neuron 1:201–209, 1988 and *J. Electronmic. Tech.*, in press.)

from a base on the presynaptic membrane. The origin of the triangular structures is not clear, and it is uncertain whether such profiles exist in unfixed terminals or are formed artifactually from the collapse of cytoskeletal elements during fixation. In any case synaptic vesicles typically surround the filamentous projections and, because of this close relationship, they are thought to play a role in vesicle exocytosis (Fig. 11–4D). This interpretation is supported by studies on the neuromuscular junction where a few milliseconds after the arrival of an action potential in the motor nerve terminal, omega-shaped profiles of vesicles in the act of exocytosis appear in the presynaptic membrane along the boundaries of the dense projections. For this reason the dense projections and the associated vesicles are called the active zone and are thought to represent the region of the presynaptic terminal where transmitter is released. Synaptic terminals in the central nervous system have different numbers of active zones. Some boutons have a single active zone, while others have many. In some preparations the dense projections are arranged in a "presynaptic grid" with fine filaments interconnecting adjacent projections.

The postsynaptic membrane opposite the active zone is the target for released neurotransmitter. The only specialization of this membrane that is apparent in the electron microscope is a layer of dense fuzz along its inner surface (Fig. 11–4A, C, D). The function of this fuzzy material is not known, but it is tempting to speculate that it represents a protein involved in the maintenance and renewal of neurotransmitter receptors in the postsynaptic membrane.

Postsynaptic Mechanisms. For over four decades physiologists have asked questions and tested hypotheses about the mechanisms of chemical synaptic transmission at both a cellular and a molecular level. Although the signal under examination is chemically mediated, most advances have used electrical recording to study synaptic physiology. The continued presence of electrical methods in the forefront of synaptic research is a testimony to their exquisite sensitivity and reliability. In most neurons direct electrical recording from axon terminals is not practical because they are too small; recording from the large postsynaptic neuron, however, is relatively easy. Consequently, presynaptic mechanisms often are deduced from direct and precise postsynaptic measurements. Since it is necessary to understand postsynaptic responses in order to appreciate what they reveal about presynaptic events

we will discuss the physiology of postsynaptic mechanisms before we consider presynaptic mechanisms.

The steps that ultimately generate the postsynaptic response are initiated by the binding of neurotransmitter to receptors in the subsynaptic membrane of the postsynaptic cell. The resulting conformational change in the receptor either opens an ion channel through the protein or triggers a cascade of biochemical reactions that generate an intracellular second messenger, which in turn changes the ionic permeability of the cell. In either case there is a change in the ionic current flow across the cell membrane, which results in a postsynaptic potential change. First, we will discuss the ionic basis of postsynaptic potentials (PSPs) and their general properties and then we will consider the mechanisms that operate the ion channels involved.

The synaptic responses recorded from spinal motoneurons serve as a model for chemical transmission in the central nervous system (CNS). The first direct measurements of postsynaptic potentials were made in spinal motoneurons of the anesthetized cat by Sir John Eccles and his colleagues.[8, 13] For this work he shared the 1963 Nobel Prize for Physiology and Medicine with A. L. Hodgkin and A. F. Huxley. Eccles obtained electrical contact with the interior of a motoneuron by impaling it with a fine glass electrode filled with an electrolyte (KCl). As shown in the schematic diagram in Figure 11–5, the spinal cord offers the experimentalist access to different classes of afferent nerve fibers that have different synaptic effects on the motoneuron under study. The lowest threshold peripheral afferent nerve fibers are from muscle stretch receptors (see Chap. 24) and they make direct synaptic contact with spinal motoneurons innervating the same muscle or its synergists. Since no neurons are intercalated between the primary afferent fibers from stretch receptors and the motoneuron, the connection is called *monosynaptic*. Synchronous stimulation of stretch receptor afferents (e.g., in the tibialis anterior muscle) evokes a subthreshold depolarizing potential change that is called an *excitatory postsynaptic potential (EPSP)* because it is depolarizing and increases the excitability of the motoneuron by moving the membrane potential closer to the threshold for action potential initiation (upper trace, Fig. 11–5A). By contrast, stimulation of afferent nerve fibers from stretch receptors in an opposing, or *antagonist*, muscle (e.g., the gastrocnemius muscle) hyperpolarizes the motoneuron, causing inhibition. This graded hyperpolarizing PSP is called an

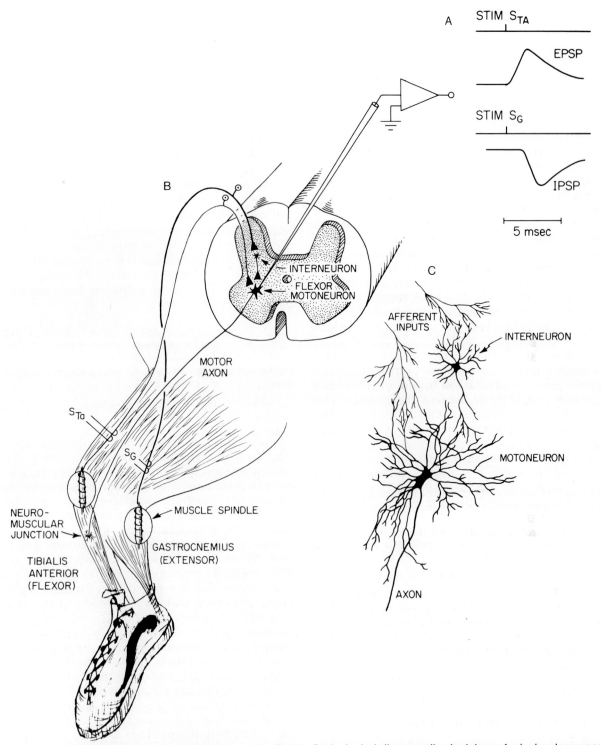

Figure 11–5 Diagram showing the preparation used by Sir John Eccles to study the synaptic physiology of spinal motoneurons, in which a flexor motoneuron in the ventral horn of the spinal cord is impaled with an intracellular microelectrode. *A,* The electrode records excitatory or inhibitory postsynaptic potentials evoked by stimulation of muscle spindle (stretch receptor) afferents from opposing (antagonistic) muscle groups. In *B, upper right,* an excitatory postsynaptic potential is recorded when afferents from the tibialis anterior (the muscle innervated by the impaled motoneuron) are excited by stimulating electrode (Sta). An inhibitory postsynaptic potential is recorded when afferents from the gastrocnemius (an antagonistic muscle) are excited by stimulating electrodes (S_G). C shows that excitatory afferents make direct (monosynaptic) contact with the motoneuron, whereas the afferents that cause inhibition do so indirectly via an interposed interneuron. The gastrocnemius afferents excite the interneuron, which in turn causes inhibition of the motoneuron.

inhibitory postsynaptic potential (IPSP) (lower trace, Fig. 11–5A). Unlike the excitatory pathway, inhibitory input from muscle stretch receptors in antagonist muscles influences motoneurons through an intercalated neuron called an interneuron (see Chap. 24). The primary afferent fibers from stretch receptors in the antagonist muscles evoke EPSPs in the inhibitory interneurons, which in turn release a neurotransmitter causing inhibition in the motoneuron. Interneurons that inhibit other neurons often have short axons and occur throughout the nervous system.

Ionic Basis of Postsynaptic Excitation. The ionic basis of the EPSP is similar to that of the end-plate potential (see Chap. 6). At an excitatory synapse, neurotransmitter opens channels in the postsynaptic membrane that are permeable to cations, principally Na and K. The experimental evidence for this is based on measurements of the EPSP reversal potential. You will recall from Chapter 3 that the reversal potential is the membrane potential (E_m) at which the net ionic current through a channel is zero. The net ionic current (I_{EPSP}) that gives rise to an EPSP is the sum of the individual currents carried by all the ions that permeate the channel. If we consider only the principal ions, Na and K, we can write

$$I_{EPSP} = I_{Na} + I_K \tag{1}$$

The expressions for the current carried by Na and K, I_{Na} and I_K, were derived in Chapter 3 and are written below:

$$I_{Na} = g_{Na} (E_m - E_{Na}) \tag{2}$$
$$I_K = g_K (E_m - E_K) \tag{3}$$

where g_{Na} and g_K are conductances proportional to the permeability of the channel to Na and K, and E_{Na} and E_K are the Nernst potentials for these ions. Substituting equations 2 and 3 into 1 we obtain the following expression for the net ionic current flowing through an excitatory synapse channel:

$$I_{EPSP} = g_{Na} (E_m - E_{Na}) + g_K (E_m - E_K) \tag{4}$$

The value of the membrane potential when $I_{EPSP} = 0$ is the reversal potential (E_{rev}). Solving equation 4 for E_M at $I_{EPSP} = 0$ gives an expression for E_{rev}:

$$E_{rev} = \frac{g_{Na} E_{Na} + g_K E_K}{g_{Na} + g_K} \tag{5}$$

If we use the values for E_{Na} and E_K given in Table 1–4 of Chapter 1—i.e., +67 mV and −98 mV, respectively, and assume that the Na and K conductance of the EPSP channel are equal ($g_{Na} = g_K$) the predicted reversal potential would be

$$E_{rev} = \frac{67 \text{ mV} - 98 \text{ mV}}{2} = -16 \text{ mV} \tag{6}$$

The experimentally observed reversal potential for a motoneuron EPSP, however, is not −16 mV; it is approximately 0 mV (Fig. 11–6). This tells us that our assumption that g_{Na} and g_K are equal is a little off. In order to make the E_{rev} more positive (i.e., closer to 0 mV), we must postulate that the excitatory synapse channels have greater conductance to sodium than potassium. Rearrangement of equation 5 gives the following expression for the $g_{Na} \cdot g_K$ ratio:

$$\frac{g_{Na}}{g_K} = \frac{E_K - E_{rev}}{E_{rev} - E_{Na}} \tag{7}$$

In the case of the excitatory synapse we calculate that

$$\frac{g_{Na}}{g_K} = \frac{-98 \text{ mV}}{-67 \text{ mV}} = 1.5 \tag{8}$$

Therefore the Na conductance of the channel is 1.5 times the K conductance. In drawing this conclusion we have used the measurement of reversal potential—determined in experiments like the one illustrated in Figure 11–6—and have considered any current carried by ions other than Na and K to be insignificant.

Ionic Basis of Postsynaptic Inhibition. Let us now try to infer the ionic mechanisms for the IPSP, using its reversal potential as our principal clue as we did for the EPSP. Experimentally, Eccles[8] observed that the amplitude of the IPSP increased when the motoneuron membrane potential was depolarized, and that the amplitude decreased when the potential was hyperpolarized (Fig. 11–7A). At about −80 mV the IPSP can no longer be detected, and it becomes depolarizing at more negative membrane potentials. Thus the reversal potential for the IPSP is about −80 mV. Which ions might be flowing through the channels opened by neurotransmitters at an inhibitory synapse? The primary suspect is chloride because its Nernst potential (E_{Cl}) is −80 mV, close to the IPSP reversal potential. If the inhibitory synapse channels are permeable exclusively to Cl the inhibitory synaptic current is

$$I_{IPSP} = g_{Cl} (E_m - E_{Cl}) \tag{9}$$

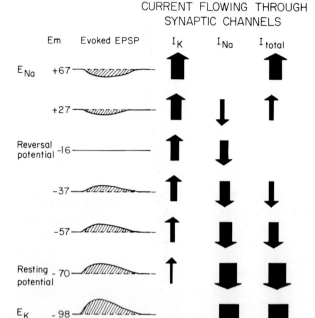

CURRENT FLOWING THROUGH
SYNAPTIC CHANNELS

Figure 11–6 Ionic mechanism of the excitatory postsynaptic potential (EPSP). The amplitude and polarity of the EPSP *(shaded area)* depends on the membrane potential (Em) of the postsynaptic cell. The amount of transmitter released at each potential is assumed to remain constant. The width of the arrows represents the magnitude of the synaptic current carried by Na^+ (I_{Na}) and K^+ (I_K) and, in the last column, the magnitude of the net synaptic current (I_{EPSP}) (i.e., the sum of I_{Na} + I_K). Outward current is upward (positive) and inward current downward (negative). Note that at the Nernst potential for Na (+67 mV) there is no inward Na current and at the Nernst potential for K (−98 mV) there is no outward K^+ current. By definition the membrane potential at which there is no *net* synaptic current (E_m = −16 mV), and thus no apparent potential change, is the reversal potential of the EPSP.

When the resting membrane potential is −70 mV, the direction of the I_{IPSP} is outward; an outward Cl^- *current* represents an influx of the negatively charged Cl^-, which will hyperpolarize the cell. The outward synaptic current will increase as the membrane potential is made less negative (depolarized) and decrease as the potential is made more negative (hyperpolarized). While this is consistent with the experimental observations, another possibility is that the inhibitory synapse channel is permeable to K^+, with a Nernst potential of −90 mV, and a second ion, with a more positive Nernst potential (either Na^+ or Cl^-, would qualify).

In a separate series of experiments Eccles demonstrated directly that chloride is involved in the IPSP (Fig. 11–7*B*). Chloride ions were injected from the KCl-filled recording microelectrode into the impaled motoneuron to raise the intracellular Cl^- concentration and shift E_{Cl} to a value positive to the resting potential. Opening the inhibitory synaptic channels at the resting potential now causes chloride ions to leave the cell. The chloride current has become negative (inward) and depolarizing. The observation that the hyperpolarizing IPSP becomes a depolarizing response after the injection of chloride shows that inhibitory synapse channels are permeable to chloride ions but it does not prove that the channels are not also permeable to K^+. This issue was not resolved until the advent of single-channel recording techniques, which

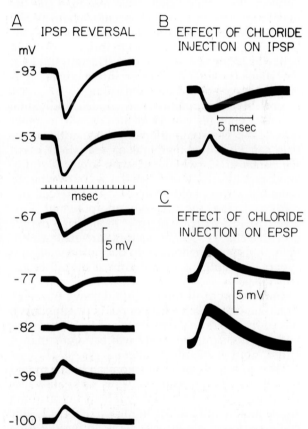

Figure 11–7 Identification of the ionic mechanism of the inhibitory postsynaptic potential (IPSP). *A* shows how the IPSP recorded from a motoneuron changes as the membrane potential of the motoneuron is polarized to the values shown on the left. The reversal potential (i.e., the potential at which there is no apparent IPSP) is about −80 mV. *B*, Intracellular injection of Cl^- ions changes the reversal potential of the IPSP. The upper trace shows an IPSP recorded at −70 mV before Cl^- injection. The lower trace shows that the polarity of the IPSP recorded at the same membrane potential is reversed after Cl^- injection. Resting potential does not change because the motoneuron has a low resting chloride permeability. *C*, Chloride injection has no effect on the excitatory postsynaptic potential (EPSP). (After Coombs, J. S.; Eccles, J. C.; Fatt, P. *J. Physiol.* 130:326–373, 1955.)

showed that glycine—the neurotransmitter at the inhibitory synapse studied by Eccles—opens channels that are permeable to only chloride ions (see Fig. 11–10).

General Properties of Postsynaptic Potentials. A postsynaptic potential, like the end-plate potential (Chap. 6) and the receptor potential (Chap. 5), is a graded potential change. The amplitude of the postsynaptic potential increases with the amount of neurotransmitter released; that is, it increases with the number of receptors bound by neurotransmitter. As a consequence, the amplitude of the postsynaptic potential may have any value between the smallest detectable response and the maximum response, which occurs when all the receptors are bound by neurotransmitter. This behavior is in marked contrast to the all-or-nothing behavior of the action potential. The amplitude of the action potential is never intermediate, it is either zero or some maximum value.

Since synaptic potentials are graded, individual potential changes can summate, or add together (Fig. 11–8). The summation of postsynaptic potentials can be either temporal or spatial. *Temporal summation* arises when the same input is stimulated repetitively, such that the second synaptic potential arrives before the postsynaptic cell has recovered from the first synaptic potential. This gives rise to a staircase increase in the postsynaptic potential change. *Spatial summation* refers to essentially the same phenomenon, but now the second synaptic potential is provided by a second input, which makes synaptic contact in close proximity to the first input. If the two inputs have the same sign (i.e., either both excitatory or both inhibitory), the postsynaptic response evoked by stimulating both inputs simultaneously will be larger than if either input is stimulated alone.

The importance of summation can be illustrated by considering an example. It is estimated that there are 20,000 individual synaptic boutons on a single motoneuron in the lumbar spinal cord. This could represent about 6600 separate inputs, each of which contributes an average of three synaptic boutons. Stimulation of a single input evokes a 100-μV EPSP, which falls far beneath the 8- to 10-mV threshold depolarization needed to evoke an action potential in the motoneuron. The only way that the amplitude of the EPSP in a motoneuron can ever be large enough to evoke an action potential is if a single input is stimulated repeatedly (temporal summation) or if many inputs are stimulated simultaneously (spatial summation).

Thus far our discussion of summation has considered only the summation of postsynaptic po-

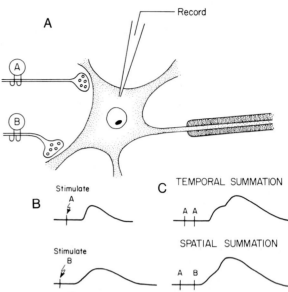

TEMPORAL AND SPATIAL SUMMATION

Figure 11–8 Temporal and spatial summation of synaptic inputs to a central neuron. *A* shows the experimental arrangement in which a neuron is impaled by an intracellular recording electrode, and stimulating electrodes A and B activate two separate inputs that make synaptic contact near each other. *B,* Stimulation of either synaptic input alone evokes an excitatory postsynaptic potential (EPSP) *C,* The amplitude of the EPSP can be increased either by stimulating the same synaptic contact twice at a short interval (temporal summation) or by stimulating one contact shortly after the other (spatial summation).

tentials having the same sign (excitation or inhibition), but in fact neurons like the spinal motoneuron are under a constant barrage of both inhibitory and excitatory synaptic inputs. At any moment the membrane potential of the neuron reflects the sum of all these inputs. A decrease in the level of tonic excitatory input will hyperpolarize the cell, whereas a decrease in the level of tonic inhibitory input will depolarize the cell. These changes are referred to as *disfacilitation* and *disinhibition*, respectively, and are discussed in more detail in Chapter 29 on the cerebellum.

Molecular Mechanisms of Channel Gating. The binding reaction between neurotransmitter and postsynaptic receptor protein causes a conformational change that opens (or closes) ion channels in the postsynaptic membrane either by directly forming a pore through the protein molecule or by triggering the generation of a second messenger, which subsequently changes the ionic permeability of the cell. The former mechanism needs little explanation. Chapter 6 discusses the prototype of a ligand-gated channel, the nicotinic acetylcholine receptor channel at the neuromuscular junction. For this channel the transition between closed and open states occurs in microseconds.[7] Ligand-gated channels are assumed therefore to be responsible for the generation of fast synaptic potentials, whereas second messenger–operated channels, which involve more steps and are slower, are assumed to be responsible for slow synaptic potentials.

What are the underlying steps in second messenger–mediated synaptic transmission? In all the best studied examples the first step following the formation of a neurotransmitter receptor complex is the activation of a GTP binding protein (G protein). G proteins are intracellular peripheral membrane proteins that are believed to be present in every eukaryotic cell (Chap. 1). They act as a "go-between," which couples extracellular membrane receptors to the production of an intracellular response. They play a key role in mechanisms as different as transduction of sensory stimuli (vision and olfaction), control of cell proliferation, regulation of ribosomal protein synthesis, and the cellular action of hormones and transmitters. From this it is obvious that there are several different kinds of G proteins—for example G_s, G_i, and G_o. These, like all G proteins, are heterotrimers consisting of alpha, beta, and gamma subunits.[19] They are activated when GDP bound to the alpha subunit is replaced by GTP (Fig. 11–9). GTP-GDP exchange is catalyzed by a transient interaction with the neurotransmitter-receptor complex. A single neurotransmitter-receptor molecule can activate many molecules of G protein and thus amplify the effect of the neurotransmitter. The exchange of GTP for GDP causes the alpha subunit to dissociate from the beta-gamma subunits. The next step in the second messenger cascade depends on the specific system being considered. There are two possibilities. Either an activated subunit of the G protein can directly gate an ion channel (Fig. 11–9C) as, for example, in the case of muscarinically activated potassium channels in the heart.[38] Or the activated alpha subunit (G_α-GTP) can change the activity of an enzyme* such as adenylate cyclase (which is turned on by $G_{s\alpha}$-GTP and off by $G_{i\alpha}$-GTP) or phospholipase C (which is thought to be turned on by $G_{o\alpha}$-GTP) (Fig. 11–9A, B). Turning on adenylate cyclase causes synthesis of cyclic adenosine monophos-

phate (cAMP), an intracellular second messenger. Turning on phospholipase C causes hydrolysis of membrane phospholipids and the formation of two second messengers, diacylglycerol (DAG) and inositol triphosphate (IP_3). DAG is a lipid and stays in the lipid phase of the membrane; IP_3 is water soluble. It enters the cytoplasm where it triggers an increase in intracellular Ca^{2+} by releasing Ca from intracellular stores. The increase in intracellular Ca represents a third intracellular signal produced by the activation of phospholipase C.

The α subunit of the G protein has intrinsic GTPase activity which hydrolyzes GTP to GDP in a few seconds, causing G_α-GTP to revert to its inactive form (G_α-GDP) (Fig. 11–9A). This form of the α subunit reassociates with the β-γ subunits and returns the G protein to its resting state, causing inactivation of the subsequent steps of the second messenger cascade.

How do second messengers (cAMP, IP_3/Ca^{2+}, and DAG) affect the ionic permeability of the cell? There are two established mechanisms. First, second messengers can operate ion channels directly, similar to the way that acetylcholine operates the ACh receptor channel; the only difference is that a second messenger is an internal transmitter that binds to the intracellular side of the ion channel. Channels that are gated by Ca^{2+}, cAMP, and IP_3 have been described in a variety of cell types (see Chap. 12). Second, and possibly more commonly, the second messenger can activate a protein kinase that phosphorylates a channel and changes its functional state.[6] Kinases that are specifically activated by cAMP (cAMP-dependent protein kinase), Ca^{2+} (Ca-calmodulin kinase), and DAG (protein kinase C) have been identified (see Chap. 33). The effect of phosphorylation on an ion channel varies according to the channel in question. Phosphorylation has been shown to open some channels, and is thought to close others.

Neurotransmitters. Ideally there are several logical criteria that should be satisfied before a proposed substance is accepted to be a neurotransmitter. The putative neurotransmitter must be shown to be present in the presynaptic terminal and released by stimulation. This is much easier said than done. The high density of cells in the central nervous system and the small size of their fine terminal processes make it difficult to be certain that the right cells are sampled. Some of the resolution problems have been circumvented by using various histological techniques, like autoradiography and immunocytochemistry. Here tissues incubated with radioactive precursors or stained with specific antibodies are sectioned for

---

*The activation of an enzyme represents a second stage of amplification; a single enzyme molecule will generate many molecules of product.

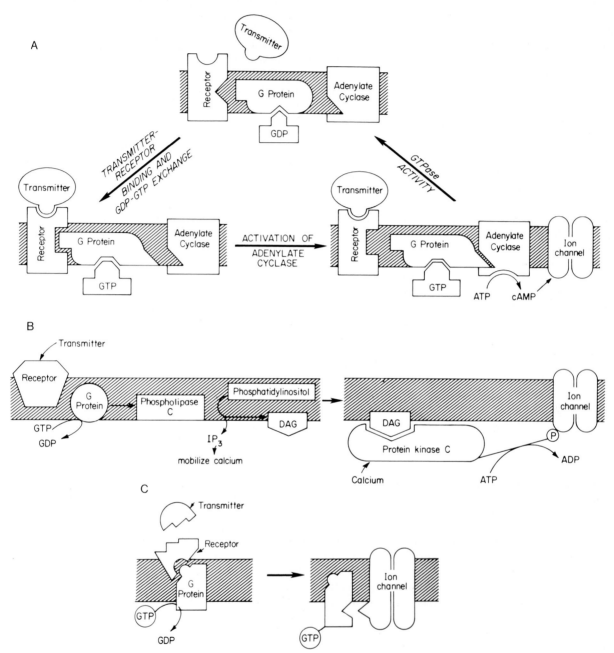

Figure 11–9 Diagram summarizing three ways that transmitter-activated G proteins influence ion channels. (ADP = Adenosine diphosphate; ATP = adenosine triphosphate; cAMP = cyclic adenosine monophosphate; DAG = diacylglyceride; GDP = guanosine diphosphate; GTP = guanosine triphosphate; IP$_3$ = inositol trisphosphate.) In each case transmitter binding to a surface receptor initiates a conformational change in the receptor that subsequently activates a G protein by catalyzing GDP-GTP exchange. *A,* Activated G protein stimulates adenylate cyclase, causing production of cAMP which, in turn, influences ion channels either directly or through phosphorylation by an activated protein kinase. *B,* Activated G protein stimulates phospholipase C, causing the formation of IP$_3$ and DAG. IP$_3$ causes calcium mobilization while DAG activates protein kinase C causing phosphorylation of ion channels and hence changes in their properties. *C,* Activated G protein directly interacts with an ion channel. Each of these cases involves a different G protein. The specific G protein responsible has not been identified in all cases. In *A* the G protein is G$_s$, in *B* it may be G$_o$, and in *C* it is unclassified.

histology and examined microscopically. With these techniques it has been possible to localize transmitters and/or enzymes that synthesize them in specific cells. It has been much more difficult to obtain convincing evidence that stimulation causes an identified cell to release a putative transmitter. The typical approach is to preload cells with radiolabeled transmitter and then demonstrate that stimulation of the loaded cells causes redistribution of the label. While there are many potential pitfalls in these kinds of experiments, some have succeeded in making a strong case for the release of specific transmitters from selected cells.

The existence of a mechanism for inactivating released transmitter is another criterion used to identify a functional neurotransmitter. In the central nervous system, as in the peripheral nervous system, there are two mechanisms of inactivation. The released transmitter can be either transported back into the presynaptic terminal (re-uptake) or destroyed by extracellular enzymes. Re-uptake is the more prevalent method of transmitter inactivation in the central nervous system; this is particularly true for the amino acid transmitters. There are also examples of central nervous system transmitters that are enzymatically inactivated (e.g., acetylcholine and peptides) or that use a combination of re-uptake and enzymatic destruction (e.g., norepinephrine).

No putative neurotransmitter can be taken seriously until it is shown that exogenous application of "reasonable" concentrations of the substance have the same effect on the postsynaptic cell as stimulation of the presynaptic neuron. Furthermore, chemical agents that block or potentiate transmission at the synapse in question must have a similar effect on the action of exogenously applied transmitter. These pharmacological experiments are the acid test for a putative neurotransmitter. While one may dismiss inconclusive evidence about the localization and release of a proposed transmitter because of technical difficulties, one cannot dismiss negative evidence about the pharmacology; a transmitter candidate must be rejected if its action is not blocked by a drug that blocks transmission at the synapse under study. For this reason most of what we know about CNS neurotransmitters is based on pharmacological evidence gathered from experiments on intact nervous system, in vitro slices of functional nervous system, or tissue-cultured neurons.

In no case has it been possible to completely satisfy all the criteria for a neurotransmitter in the central nervous system. The persistent "doubting Thomas" can find some flaw in the detailed argument for just about any of the putative transmitters. Although the counter arguments grow more and more implausible in the face of an expanding body of circumstantial evidence, it is still best to place many central nervous system neurotransmitters in the category of being almost but not 100% certain. We will now briefly survey some of the best neurotransmitter candidates concentrating on transmitters that are thought to mediate excitatory and inhibitory synaptic transmission.

Excitatory Transmission. The acidic amino acids, especially glutamate, are thought to play the same role in the central nervous system as acetylcholine plays at the neuromuscular junction. Acetylcholine mediates excitatory synaptic transmission between nerve and muscle, whereas glutamate mediates fast excitatory synaptic transmission between nerve cells in the brain.[49] It is estimated that more than half of the neurons in the central nervous system use glutamate as their neurotransmitter. It has been thought that neurons that responded to glutamate or to aspartate, another acidic amino acid with similar actions, have several distinct types of glutamate receptors. They were classified according to the agonist that excited them into NMDA (N-methyl-D-aspartate) and non-NMDA types; the latter could be excited by kainic and quisqualic acids. Both receptor types operated channels directly, as opposed to indirectly via second messengers, and the channels they gated appeared to have different properties. The conductance activated by NMDA is permeable to sodium, potassium, and calcium and is blocked in a voltage-dependent manner by magnesium.[36] The channel activated by kainic acid, on the other hand, is not blocked by magnesium, and is permeable to sodium and potassium but not calcium.

These different conductance mechanisms, often thought to be due to different channels, have been proposed alternatively to represent different functional states of the same ion channel.[10] This new view comes from recent work using single-channel recording from cultured hippocampal and cerebellar neurons.[10, 24] In most membrane patches excitatory amino acids activate ion channels with multiple conductance states. There appear to be five main open states ranging from 10 to 50 pS spaced at approximately equal 10-pS intervals. Different agonists have different probabilities of opening the channel to the various conductance states. Kainic acid, for example, most frequently opens the channel to a low conductance level, whereas NMDA most often opens the channel to its highest conductance. If a single channel can have multiple conductance states then it is possible that there is a single glutamate receptor/ion channel complex with two ligand binding sites; each site would bind glutamate, but one would have a high affinity for NMDA while the other would have a high affinity for kainic acid. According to this model the channel would have several conformational states corresponding to the different channel conductance levels. The probability of the channel having a particular conformational state and hence conductance would depend on the occupancy of the two binding sites. Presumably the concentration of glutamate would determine which sites are occupied and thus the conductance of the channel. These observations can still be interpreted by a several-channel model, namely that there are at least two distinct glutamate receptor/ion channel complexes with different affinities for NMDA and kainic acid. According to this model the two channels activated would have similar substates of conductance but their predominant conductance would be different. More work is needed on this interesting question.

Regardless of whether all of the conductance levels observed in single channel recordings are due to a single membrane complex or many protein complexes, it is clear that the conductance mechanism activated by excitatory amino acids is a lot more complicated than the conductance mechanism activated by acetylcholine at the neuromuscular junction. The greater complexity of excitatory transmission in the brain is not surprising when you consider the greater complexity of brain function. Learning and memory represent examples of two such complex functions (Chap. 33). In 1949 D.O. Hebb[22] suggested that one expression of learning at the cellular level would be a long-term change in a cell contingent on the coincidence of a presynaptic input with a postsynaptic action potential. The large conductance activated by glutamate binding to the NMDA-type receptor could provide the molecular mechanism for such a long term associative change. This might work in the following way. At the normal resting potential of approximately -60 mV, extracellular magnesium ions are known to keep the glutamate channel in its low conductance state. Under these conditions glutamate released from an excitatory synaptic terminal would bind to postsynaptic receptors causing low conductance channels to open resulting in a modest depolarizing potential. If this presynaptic excitatory terminal were activated at the same time as the postsynaptic cell is strongly depolarized by another synaptic input eliciting an action potential, the depolarization would remove the block by extracellular magnesium ions causing the channels opened by release of glutamate to be in the large conductance state. In this large, but not in the small, conductance state the glutamate channel is permeable to calcium. The influx of calcium might act as a second messenger to trigger biochemical processes that would cause long term changes in the cell. Whether this sequence of events is actually involved in learning and memory is not known, but there is currently much interest in the glutamate receptor because its complexity is sufficiently great that it has the potential of playing a role in many brain functions.

Inhibitory Transmission. Inhibitory postsynaptic potentials are generated by transmitters that cause flow of outward current carried by chloride and in some cases potassium. The postsynaptic channels responsible are generally thought to be gated by two different transmitters: gamma aminobutyric acid (GABA) and glycine. There is a single type of postsynaptic glycine receptor (blocked by strychnine) and two types of GABA receptors (type A and B) that have different pharmacologies and different mechanisms of action. Muscimol is a specific agonist for GABA$_A$ receptors which are selectively blocked by picrotoxin and bicuculine. GABA$_B$ receptors are activated by baclofen and blocked by 5-aminovaleric acid. The GABA$_A$ receptor and the glycine receptor are protein macromolecules that are a receptor/chloride channel complex. The chloride channels gated by both types of receptors have similar biophysical properties, which include identical ion selectivities and four major conductance states ranging from about 10 to 50 pS (Fig. 11–10).[5] Evidently these striking functional similarities carry over into the structure as well, since the amino acid sequences of subunits of the two receptor types show some regions of identity and many regions of similarity.[20a, 43a] Surprisingly these subunits have struc-

---

Figure 11–10 Single-channel currents activated by glycine and gamma aminobutyric acid (GABA). *A,* Glycine-activated currents at the different membrane potentials indicated at the right of each trace. Inward current is downward. *B,* GABA-activated single-channel currents at the same membrane potentials as in *A.* The distribution of the single-channel currents activated by glycine and GABA at +50 mV and −90 mV are shown in *C* and *D,* respectively. (Data from Bormann, J.; Hamill, O. P.; Sakmann, B. *J. Physiol.* 385:243–286, 1987.)

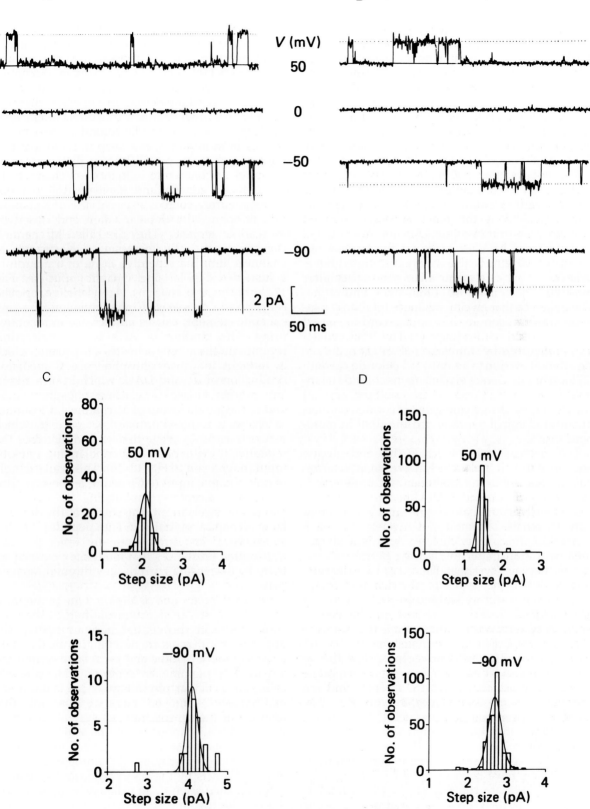

Figure 11–10 *See legend on opposite page*

tural homology with those of the nicotinic acetyl-choline receptor, meaning that all of these ligand-gated channels belong to a supergene family and derive from some common ancestral protein.

The GABA$_B$ receptor, unlike the GABA$_A$ and glycine receptors, is not part of a single receptor/ion channel complex. The GABA$_B$ receptor acts via a G protein to open potassium channels.[1, 18] It is thought that this occurs through a direct action of the G protein on the potassium channel.

The specific transmitter and/or receptor type that is used to mediate inhibitory synaptic transmission varies in different parts of the nervous system. GABA is the most widely distributed inhibitory neurotransmitter. All neurons studied so far have GABA receptors on their soma and dendrites. Glycine is also widely distributed but is believed to act as a mammalian neurotransmitter mostly in the spinal cord, brain stem, and retina. The evidence that glycine mediates inhibitory synaptic transmission in the spinal cord is based primarily on three findings: (1) The IPSP evoked in motoneurons by stimulation of antagonist muscle afferents is due to an outward chloride current; (2) local application of glycine opens chloride channels on tissue-cultured spinal neurons; and (3) strychnine blocks glycine-gated chloride channels in cultured spinal neurons and the IPSP in motoneurons.

The mechanism by which GABA and glycine are concentrated and stored in presynaptic terminals as releasable transmitter is not understood. The action of released GABA, and probably glycine as well, is terminated by re-uptake by presynaptic terminals.

Acetylcholine. Acetylcholine (Ach) is a ubiquitous neurotransmitter in both the peripheral and central nervous systems. There are two pharmacologically distinct types of acetylcholine receptors: nicotinic and muscarinic (see Chaps. 6 and 34). Nicotinic receptors are rare in the central nervous system, whereas muscarinic receptors are common, especially in the cerebral cortex and basal ganglia. The nicotinic receptor, like that at the neuromuscular junction, is a single membrane protein complex that is both a receptor and ion channel. The muscarinic receptor, on the other hand, and the various ion channels that it controls are separate membrane proteins. Muscarinic receptors control ion channels through the generation of intracellular second messengers, and three mechanisms have been identified[21, 38]: (1) Muscarinic receptors are coupled to G$_i$, which inhibits the action of adenylate cyclase and decreases the production of cAMP; (2) muscarinic receptors ac-

tivate phospholipase C through a G protein–mediated step causing the breakdown of inositol phospholipids and the formation of the second messengers DAG and IP$_3$; (3) muscarinic receptors are also known to activate a G protein that acts by itself as a second messenger and opens K channels and perhaps other channels by a direct interaction.

We will now consider the second of these mechanisms in more detail, but keep in mind that it is only one of three possible ways that muscarinic receptors are known to influence ion channels. In sympathetic ganglia and some central nervous system neurons there is a special class of K channel that is opened by depolarization and closed by muscarinic agonists. They are called M channels (for muscarinic). By increasing a K current, M channels limit the repetitive firing of a neuron to a sustained depolarizing synaptic input (see Fig. 12–16 in the next chapter). Acetylcholine, applied exogenously or released by stimulation of a presynaptic element, causes an increase in repetitive firing. The binding of ACh to the muscarinic receptor on these cells activates a G protein, which is thought to cause phospholipase C–mediated production of IP$_3$ and DAG, which in turn stimulate protein kinase C, causing phosphorylation and subsequent closure of M channels. Closure of M channels reduces outward K$^+$ current, which makes it easier to excite the cell and increases the repetitive firing evoked by a depolarizing synaptic input (refer again to Fig. 12–16). This is an example of one synaptic input (ACh acting on postsynaptic muscarinic receptors) modulating the action of another synaptic input (the source of the depolarizing synaptic potential). The peptides LHRH, substance P and bradykinin are other putative transmitter substances that modulate neuronal activity by decreasing current flow through M channels.

Catecholamines and Serotonin. One of the most common groups of neurotransmitters is the catecholamines. In the central nervous system the important candidates are norepinephrine (NE) and its precursor dopamine and the related compound 5-hydroxytryptamine, or serotonin. The pharmacology and cellular physiology of NE is discussed in Chapters 34 and 65. Here we note only that neurons of the central nervous system have both the α and β adrenergic receptors, which influence ion channels by second messenger mechanisms that involve primarily cAMP, although one class of receptor acts by the phosphoinositol system.

Rather than attempt an exhaustive discussion of the bewildering spectrum of action of these compounds on the nervous system, we will focus on one well-studied example. In *Aplysia,* an invertebrate sea slug that has been used for pioneering studies on synaptic physiology, stimulating one neural pathway causes an increase in the transmitter released by stimulating a second pathway. Serotonin is the transmitter released by the first pathway. It binds to receptors on the postsynaptic terminals of the second pathway causing a special class of potassium channels (called S channels) to close.[44] S channels normally function to promote the repolarization of the action potential. Their closure delays repolarization and prolongs the action potential in the presynaptic terminal, causing a greater influx of calcium and enhanced transmitter release (see section on Presynaptic Mechanisms). Serotonin has been shown to close S channels by activating a G protein (presumably G_s) that stimulates adenylate cyclase, resulting in an increase in cAMP. The rise in cAMP, in turn, activates a kinase that phosphorylates S channels, causing them to close. The presynaptic changes in transmitter release caused by serotonin action on S channels is a classic example of synaptic plasticity in a simple nervous system. It is thought that changes of this sort may possibly represent a primitive expression of one of the basic mechanisms for learning and memory in more advanced nervous systems[20] (see Chap. 33).

Peptides. Our concepts about intercellular communication are changing rapidly.[25] The turnabout in our thinking in part stems from the discovery of a new type of neuroactive compound in nerve terminals: the neuropeptides (see Chap. 62). Several dozen peptides have been found that affect central neurons. Peptides can carry messages to neurons, glia, and smooth muscle cells in the central nervous system. As proteins, their synthesis must occur in the cell body and is regulated by messenger RNA. The peptide or its precursor is transferred to the synaptic terminal of the neuron, where it is released by a calcium-dependent process (see below). A decade ago we thought that a single neuron released only one type of transmitter. This view has been called Dale's principle. We now know that several putative neurotransmitter substances can coexist in a single synaptic terminal. Often one of the additional compounds is a neuroactive peptide. For example, somatostatin coexists with GABA in cortical neurons; vasoactive intestinal peptide (VIP) and ACh are present in neocortical nerve terminals, and substance P colocalizes with ACh in pontine nerve terminals. Although we do not know the precise physiological action of peptides, many think that they modulate the action of traditional transmitters. For

example, VIP increases the blood flow in salivary glands and enhances parasympathetic control of saliva secretion, whereas neuropeptide Y sensitizes smooth muscle to adrenergic transmitters. An emerging view is that synaptic terminals contain a menu of transmitters that allow rapid private conversation with immediate neighbors as well as modulators that act more slowly to regulate the communications between members of a wider and more distant audience.

This concludes our discussion of central nervous system neurotransmitters. One thing that should be clear about interneuronal communication, whether it is mediated by neurotransmitters that operate channels directly or via second messengers or in concert with a modulatory peptide, is that synaptic events in the central nervous system include many mechanisms that allow the transmission and storage of information to be modified. Synaptic transmission between central neurons is multi-dimensional; it is a finely tuned and adjustable process. In this regard it differs distinctly from transmission at the neuromuscular junction, which in comparison should be viewed as an unsophisticated relative.

Presynaptic Mechanisms. Behavioral development such as learning and the progressive modification of motor skills requires plastic changes within the nervous system (see Chap. 33). Many of these processes have been described by behavioral scientists but we are only beginning to understand their molecular basis. Most neurobiologists feel that these complex expressions of neurological function involve modification of the biological mechanisms for communication between cells. The presynaptic terminal presents a prime site for modulating intracellular communication. For example, the number of synaptic boutons formed by an afferent fiber can change by sprouting, or the synthesis and release of neurotransmitter can be modified. These effects will alter the weight of particular inputs and modify the information transferred through a neural pathway. Before we can discuss modulation of neurotransmitter release, however, we must address the elementary principles of the process.

The machinery for packaging and release of neurotransmitter is located in the presynaptic terminals, but these structures are so small that in most preparations they cannot be examined directly by electrophysiological techniques. Nonetheless, ingenious experiments have given us some insight about the rules for transmitter release. Many of our ideas have been generalized from the work at the neuromuscular junction.

Although the motor nerve terminals at the neuromuscular junction are tiny, they can be visualized in the isolated nerve-muscle preparation and the mechanism of transmitter release can be investigated by applying pharmacological agents either to the entire junction or topically to different parts of the synapse. The giant stellate synapse of the squid is even more accessible; both the presynaptic and postsynaptic regions of this synapse are large and easily impaled with one or more microelectrodes. Many quantitative studies on synaptic release have used this preparation.

ACTION POTENTIAL INVASION OF THE PRESYNAPTIC TERMINAL. In spiking neurons, the presynaptic action potential triggers transmitter release. Most axons branch extensively near their termination and form synapses with many different neurons (i.e., divergence). At each branch point the axon decreases in diameter, and in myelinated fibers the myelin sheath disappears from the terminal branch near the synaptic bouton. Direct intracellular recording from the presynaptic terminal of the squid giant synapse and extracellular recording of presynaptic terminal responses in the central nervous system indicates that action potentials invade the terminal. Hence, the terminals,

like the axon proper, must contain the voltage-dependent Na and K channels necessary for action potential generation.

COUPLING THE NERVE IMPULSE TO TRANSMITTER RELEASE. In a series of landmark experiments using the squid giant synapse, B. Katz and R. Miledi asked which parameter of the presynaptic action potential is the signal for activating the release mechanism.[26, 27] It could be the influx of sodium ions, the efflux of potassium ions, or membrane potential itself. They placed two intracellular electrodes in the presynaptic terminal of the giant squid synapse—one to pass current and one to record the presynaptic membrane potential. Another electrode was placed in the postsynaptic cell to record the postsynaptic response to transmitter released from the presynaptic terminal. With the application of tetrodotoxin (TTX), the Na channels underlying the presynaptic and postsynaptic action potentials slowly became blocked, causing a gradual decline in the size of the presynaptic potential change. During the progressive decrement of the presynaptic action potential the size of the postsynaptic potential (not itself sensitive to TTX) also decreased (Fig. 11–11A). Presynaptic potential changes smaller than about 40 mV

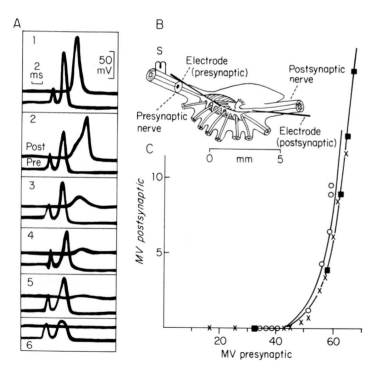

Figure 11–11 The effects of presynaptic depolarization on the size of the postsynaptic potential in the squid giant synapse. Separate intracellular electrodes were used to obtain simultaneous recordings from the presynaptic terminal and the postsynaptic neuron (see *B*). In response to stimulation of the presynaptic nerve the electrode in the presynaptic terminal records a stimulus artifact followed by an action potential (lower trace of each panel in *A*). After a synaptic delay the presynaptic action potential evokes an excitatory postsynaptic potential (EPSP) which triggers an action potential in the postsynaptic neuron (upper trace, *A1*). The EPSP becomes more apparent as the amplitude of the presynaptic action potential is gradually reduced by the blocking action of tetrodotoxin (TTX) (*A2–A5*). The presynaptic depolarization ultimately becomes too small to evoke a postsynaptic response (*A6*). The relationship between the amplitude of the presynaptic depolarization and the resulting postsynaptic response is plotted in *C*. The solid squares graph the postsynaptic potential as a function of the amplitude of the presynaptic action potential. The ×s and circles plot the postsynaptic potential evoked by 1-ms and 2-ms depolarizing pulses applied directly to the presynaptic terminal in the presence of TTX to prevent action-potential generation. (Data from Katz, B.; Miledi, R., *J. Physiol.* 192:407–436, 1967.)

evoked no detectable postsynaptic response. For presynaptic potential changes larger than 40 mV, the relationship between the size of the presynaptic potential and the postsynaptic response was exponential (Fig. 11–11C). There was a tenfold change in the size of the postsynaptic potential for a 13.5-mV change in the presynaptic potential. To test if the signal for release depended upon depolarization of the terminal rather than on the influx of sodium ions or the efflux of potassium ions, Katz and Miledi depolarized the presynaptic terminal to different levels in the presence of TTX and tetraethylammonium (TEA) to block the sodium and potassium currents underlying the action potential. The postsynaptic response to experimentally evoked steady depolarizations in the presynaptic terminal was the same as the responses to variably sized action potentials during the onset of TTX blockade. These experiments showed that depolarization of the presynaptic membrane per se initiates the release process.

The next step was to identify the messenger that coupled depolarization of the presynaptic terminal to transmitter release. At the neuromuscular junction external calcium ions are necessary for the release of acetylcholine. Junctional transmitter release is inhibited by decreasing $[Ca^{2+}]_0$ or increasing $[Mg^{2+}]_0$ (see Chap. 6). Katz and Miledi postulated that presynaptic depolarization activated voltage-dependent calcium channels in the presynaptic membrane. Other experiments on the squid giant synapse supported the calcium hypothesis.

Llinas and Heuser were the first to demonstrate directly a relationship between the influx of presynaptic calcium and the size of the transmitter-induced postsynaptic response.[30] They voltage-clamped the presynaptic membrane of the squid giant synapse in the presence of TTX and 3-aminopyridine, another K channel blocker, leaving only voltage-dependent Ca channels unaffected. Presynaptic inward calcium currents evoked by a family of depolarizing voltage steps preceded the graded depolarizing potential changes recorded in the postsynaptic cell. If arsenazo III, a calcium-sensitive dye, is injected into the presynaptic terminal, changes in $[Ca^{2+}]_i$ can be measured directly. Such experiments show a power-law relationship between intracellular calcium concentration and the release of neurotransmitter, which indicates that a small increase in calcium concentration will cause a large increase in transmitter release.

NEUROTRANSMITTER IS RELEASED IN PACKETS. When we consider the release process itself, the principles first described at the neuromuscular junction again apply. (The fundamental features of quantal transmission at the neuromuscular junction were presented in Chapter 6.) Briefly, transmitter is released from the terminal in unitary packets or quanta. The anatomical correlate of a single packet, or quantum, is a single presynaptic vesicle. At the neuromuscular junction, ACh is concentrated in the presynaptic vesicles such that each vesicle contains from 4000 to 10,000 molecules. The release of vesicles is probabilistic and the probability of release increases with increases in intracellular calcium concentration. At rest, the probability of release is low and the response to the release of individual vesicles can be recorded from the postsynaptic cells as spontaneous miniature end-plate potentials (MEPPs). During a presynaptic action potential, the entrance of calcium markedly increases the probability of packet release. As expected with this mechanism of release, increasing the influx of calcium into the presynaptic terminal does not change the size of a quantum but only increases the number of quanta (quantal content) released.

In all chemical synapses the effectiveness of the release process is assessed by measuring the response of postsynaptic neurons. At the neuromuscular junction the single innervating motor axon is easily isolated for selective stimulation, allowing the measurement of the response evoked by a single synaptic input. Such isolation is not so easy in the central nervous system, where a single cell often receives thousands of synaptic contacts from almost as many different afferent nerve fibers. With so many synaptic inputs, the cell is never "at rest"; it is under a continuous bombardment of excitatory and inhibitory inputs. In order to examine basic mechanisms at central synapses, it is necessary to study the response evoked by a single afferent axon. Fortunately, neuroscientists have been able to combine computer methods with signal analysis techniques to identify the postsynaptic responses evoked by an impulse in a single afferent axon. In Figure 11–12 a single afferent fiber making monosynaptic connections with a motoneuron and the motoneuron itself were studied at the same time using microelectrodes. The EPSP in Figure 11–12B is the average postsynaptic response evoked by stimulation of the afferent axon. The EPSP evoked by a single stimulus, however, varied in amplitude from one stimulus trial to the next.[41] The amplitude distribution in Figure 11–12C shows that 65% of the EPSPs had a peak amplitude of 320 μV; all the other EPSPs (the remaining 35%) had an amplitude of 420 μV. Injection of both the afferent axon and the post-

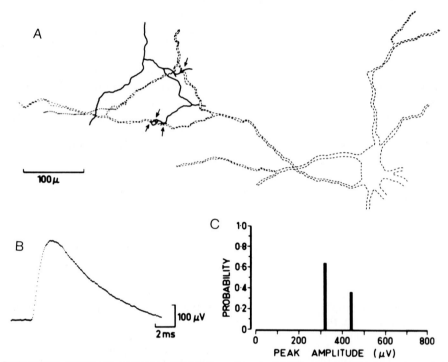

Figure 11–12 Postsynaptic responses evoked by stimulation of a single afferent fiber fluctuate in amplitude. *A,* A single afferent fiber *(solid line)* has four synaptic boutons *(arrows)* on the dendrites of this motoneuron *(dashed line). B,* The average excitatory postsynaptic potential (EPSP) recorded from the motoneuron in response to many stimuli applied to the single afferent fiber. The amplitude of the EPSP evoked by each stimulus fluctuated between two values. *C* illustrates that 65% of the EPSPs were 320 μV and 35% were 420 μV. (Reproduced with permission from Redman, S; Walmsley, B. *Trends Neurosci.* 4:248–251, 1981.)

synaptic cell with different intracellular stains revealed that this presynaptic terminal had only four synaptic boutons with the postsynaptic cell (Fig. 11–12A). This suggests that transmission always occurred at the same three boutons, generating a 320-μV synaptic potential, whereas transmission at the fourth bouton occurred only 35% of the time, adding an additional 120 μV to the peak of the EPSP. These observations supported an old hypothesis that in many neurons the unitary nature of the postsynaptic response reflects the release of neurotransmitter from individual boutons and that the response produced by activation of an individual bouton has an invariant size.

The question, then, is how can the size of the postsynaptic response evoked by a single bouton be invariant. One would expect that the number of vesicles, and hence the amount of transmitter, released by a presynaptic action potential, would vary in a quantal manner from one stimulus to the next as it does at the neuromuscular junction (see Chap. 6). There are two possible explanations. One is that while the amount of transmitter release may vary, it is always enough to saturate the postsynaptic receptor and thus evokes a constant amplitude response. The other is that a single bouton can release only a single vesicle. By studying the postsynaptic conductance change evoked by activation of a single inhibitory bouton, Farber, Korn, and Triller have obtained evidence that supports the latter explanation.[50] They found that activation of a single bouton released in an all-or-none manner enough transmitter to open 1000 to 10,000 chloride channels in the postsynaptic membrane. The number of channels activated was similar to the number activated at the neuromuscular junction by the release of a single vesicle. This finding, combined with ultrastructural data showing a single active zone per inhibitory synaptic bouton, suggests that an individual bouton can release no more than one vesicle in response to a presynaptic action potential. According to this, the unitary components in the postsynaptic response of a neuron would represent fluctuations in the probability of exocytosis from single active zones (i.e., individual boutons).

While the experiments described above focus on the probabilistic nature of synaptic mechanisms, there is another potential source of variability. It is possible that some of the variations in postsynaptic responses are due to changes in the probability of action potential conduction in the different branches of a presynaptic terminal. This is an appealing idea because it suggests another mechanism by which the strength of communication between nerve cells could be regulated. Unfortunately there is no direct means of monitoring the spread of an impulse into the fine terminals of the presynaptic arbor. As a result, at the present time, the role of branch point failure in synaptic transmission is uncertain.

MOLECULAR MECHANISMS OF TRANSMITTER RELEASE. Invasion of the presynaptic terminal causes calcium entry, which initiates neurotransmitter release. How increases in terminal calcium concentration lead to exocytosis is largely unknown, but recent experiments have given us some important clues as to the types of processes involved. Vesicles ready for release appear to be close by or attached to dense filamentous projections in the presynaptic membrane (see Fig. 11–4). A signal initiated by calcium entry allows the vesicle to contact and fuse with the presynaptic membrane. The contents of the vesicle are released into the synaptic cleft, and the vesicle membrane is incorporated into the presynaptic membrane for later retrieval and manufacture of new vesicles (see Chap. 6).

Recently Paul Greengard has identified a protein called synapsin I that is part of the cytoskeleton in nerve terminals.[34] Presynaptic vesicles are bound to synapsin; it binds to the surface of small presynaptic vesicles. Phosphorylation of synapsin I by calcium/calmodulin-dependent protein kinase II causes a marked decrease in vesicle binding. Greengard and Llinas have postulated that synapsin I holds the vesicles until phosphorylation, triggered by an increase in intraterminal calcium, frees them to move to the membrane for exocytosis.[31] In the squid giant synapse the presynaptic terminal contains synapsin I and calcium/calmodulin-dependent protein kinase II. Injection of dephosphorylated synapsin I into the terminal decreases the measured postsynaptic potential, presumably by increasing vesicle binding. Injection of kinase II frees vesicles and increases the postsynaptic potential.

PRESYNAPTIC INHIBITION. In 1957 Frank and Fuortes discovered a new inhibitory mechanism in the mammalian central nervous system.[16] They impaled a motoneuron with an intracellular electrode and recorded an EPSP in response to stimulation of an afferent (Fig. 11–13). The amplitude of the EPSP was reduced by stimulation of a second afferent axon that will be referred to as the presynaptic axon. Stimulation of the presynaptic axon alone, however, had no detectable effect on

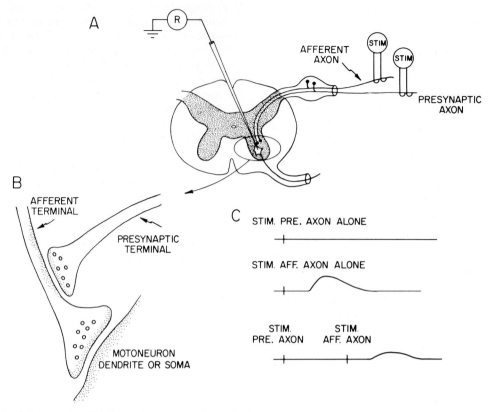

Figure 11–13 Presynaptic inhibition in the mammalian spinal cord. (AFF. = afferent; PRE. = presynaptic; STIM. = stimulation.) *A,* An intracellular electrode records from a motoneuron that receives monosynaptic excitatory input from an afferent axon which is prepared for stimulation. A stimulating electrode is also placed on a presynaptic axon that forms an axo-axonic synapse with the terminals of the afferent axon. *B,* Enlarged drawing of the relationship of the afferent axon terminal, presynaptic terminal, and motoneuron. *C, Top trace:* Stimulation of the presynaptic axon causes no detectable change in motoneuron membrane potential since it does not synapse on the motoneuron. *Middle trace:* Stimulation of the afferent axon evokes an excitatory postsynaptic potential (EPSP) in the motoneuron. *Botton trace:* The EPSP evoked by stimulation of the afferent axon is reduced by preceding stimulation of the presynaptic axon.

the motoneuron (Fig. 11–13C). This meant that the presynaptic axon was able to inhibit the EPSP without directly affecting the postsynaptic cell. For this reason this type of inhibition is called *presynaptic inhibition.* Subsequent studies demonstrated that the terminal of the presynaptic axon makes an axo-axonic synapse with the terminal of the afferent axon (Fig. 11–13B). Activation of this synapse by stimulation of the presynaptic axon reduces the amount of transmitter released by the arrival of an action potential in the terminal of the afferent axon; the less transmitter released, the smaller the EPSP evoked in the motoneuron.

The mechanism of presynaptic inhibition has been studied most directly at the crayfish neuromuscular junction, which receives both excitatory

and inhibitory inputs.[12] If stimulation of the inhibitory axon precedes stimulation of the excitatory axon by 1 to 3 ms, the synaptic potential in the muscle is markedly reduced. This reduction following stimulation of the inhibitory input is due to a decrease in the number of quanta released from the excitatory axon. The postsynaptic response to a single quantum is not changed. This shows that the inhibitory axon has no direct effect on the postsynaptic cell and indicates that it exerts its inhibitory effect by acting presynaptically on the terminals of the excitatory axon. At the crayfish neuromuscular junction the inhibitory axon releases GABA, which opens chloride channels in the terminal of the excitatory axon. The increased chloride conductance decreases the amplitude of the action potential in the excitatory axon terminal. The smaller action potential causes fewer calcium channels to open, resulting in less calcium entry and thus less transmitter release.

Indirect evidence suggests that a similar mechanism may be responsible for presynaptic inhibition in the mammalian central nervous system. During stimulation of the presynaptic axon (Fig. 11-13) extracellular currents are recorded from the region of the afferent fibers, which is consistent with subthreshold depolarization of the afferent axon terminal.[14] Although the ionic mechanism is not known, it has been postulated that the subthreshold depolarization of the afferent terminal partially inactivates the sodium conductance mechanism leading to a smaller presynaptic action potential and hence reduced transmitter release. The similarity of this mechanism with that proposed for the crayfish neuromuscular junction apparently includes the identity of the neurotransmitter substance that mediates inhibition. In both the mammalian spinal cord and the crayfish neuromuscular junction presynaptic inhibition is blocked by the $GABA_A$ antagonists picrotoxin and bicuculline.

ELECTRICAL TRANSMISSION

Conceptually, electrical synaptic transmission is straightforward.[4] Current carried by ions flows directly from one cell to another just as current carried by electrons flows along a copper wire. Electrical synapses fall into one of two categories, depending upon the electrical properties of the synaptic element or junction that connects the cells. Junctions that behave like a simple resistance are classified as *nonrectifying*; like a copper wire, they pass positive and negative current equally well in either direction. Junctions that pass current in one direction more readily than the other are classified as *rectifying*. In this case the resistance offered by the junctional element depends on the direction of current flow.

The two categories of electrical synapse were first described in experiments on the crayfish nervous system. In 1959 Furshpan and Potter[17] reported that the crayfish giant motor synapse—a synapse between a giant axon in the abdominal nerve cord (the presynaptic fiber) and a postsynaptic motor axon that innervates flexor muscles in the tail—is an electrical synapse.[17] This was demonstrated by simultaneously recording the responses of the pre- and postsynaptic fibers to currents applied to either side of the synapse (Fig. 11-14). They found that subthreshold depolariza-

tion of the presynaptic fiber was accompanied by significant although smaller depolarization of the postsynaptic fiber (Fig. 11-14A). Depolarizing currents applied to the postsynaptic axon, however, produced an insignificant change in the membrane potential of the presynaptic cell (Fig. 11-14B). The opposite was true for hyperpolarizing potential changes. Hyperpolarization of the presynaptic fiber had a negligible effect on the postsynaptic fiber, whereas hyperpolarization of the postsynaptic fiber was accompanied by an appreciable hyperpolarizing potential change in the presynaptic fiber. Evidently this synapse transmitted depolarizing potential changes only in the pre- to postsynaptic direction (Fig. 11-14C), just like a proper chemical synapse, but, unlike a standard chemical synapse, it also transmitted hyperpolarizing potential changes in the opposite direction (postsynaptic to presynaptic) (Fig. 11-14D). These and other results indicated that the two cells were electrically coupled. The synaptic connection responsible for this behavior was studied by clamping the voltage of the two cells to measure the current flowing across the synapse (I_s) as a function of the trans-synaptic voltage ($V_s = V_{pre} - V_{post}$).[46] The conductance of the synaptic element ($g_s = I_s/V_s$) depends on the value of V_s; it is *voltage-dependent*. It increases when the potential of the presynaptic fiber is more positive than the potential of the postsynaptic fiber and decreases when the potential of the presynaptic fiber is more negative than the potential of the postsynaptic fiber. As a result, the synaptic conductance behaves like a rectifier that allows positive current to flow in one direction only.

The unidirectional transmission of positive current (i.e., depolarizing potential changes) at the giant motor synapse is markedly different from what occurs at a second type of electrical synapse in the crayfish nervous system, the segmental synapse. The abdominal nerve cord contains two pairs of giant axons. The medial pair is typical of giant axons; each is a continuous, uninterrupted process of a single nerve cell. The lateral pair of giant axons, however, is unusual. Each is a segmental or septate axon made up of a longitudinal chain of short individual axon segments. Adjacent segments are connected by a specialized junction called a segmental synapse. In 1961 Watanabe and Grundfest showed that an electrical synapse was responsible for the transmission of signals from one segment to another.[54] They used the same experimental protocol as Furshpan and Potter but found that the segmental synapse, unlike the giant motor synapse, conducted positive current equally well in either direction (Fig. 11-15). The conduc-

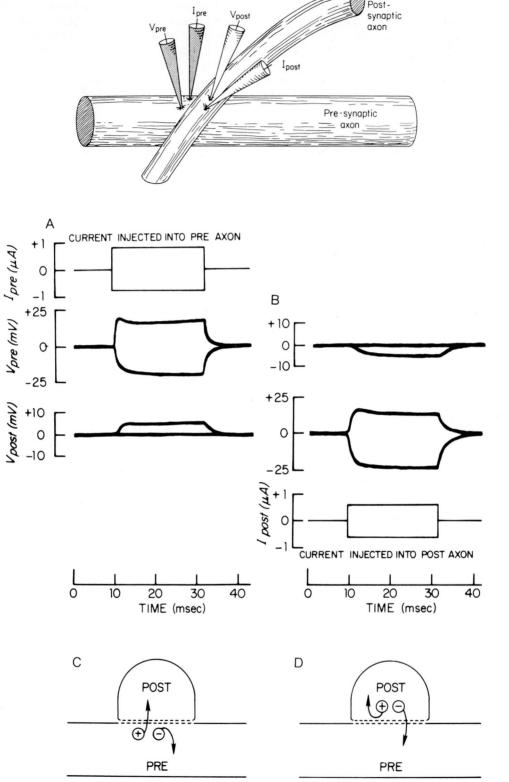

Figure 11–14 *See legend on opposite page*

tance of the synaptic connection was independent of the trans-synaptic voltage over a range of ±25 mV. The segmental synapse is nonrectifying. In this electrical synapse the junctional element is *voltage-insensitive*; it behaves like a simple resistor that passes current equally well in either direction.

The classic description of electrical transmission in the crayfish had a major impact on neurobiology. It provided unequivocal evidence for the existence of electrical coupling between nerve cells and justified a search for electrical transmission in other neural and non-neural tissues. In time, electrical junctions were found to occur between representatives of nearly every major cell type: epithelia, glands, and smooth and cardiac muscle. Notable exceptions are circulating blood cells and skeletal muscle. The various electrical junctions showed either rectifying (voltage-dependent) or nonrectifying (voltage-independent) transmission and were classified as such, using the crayfish giant motor synapse and septate synapse, respectively, as prototypic examples.

The Gap Junction. Like many topics in biology, advances in understanding the mechanism of electrical transmission required information about structure and function. It is surprising, however, that before there was experimental evidence for the existence of electrical transmission there were discussions about its morphological basis. The issue was whether coupled cells were in cytoplasmic continuity or only very close together. An experiment done in the 1940s by one of the first women electrophysiologists, Anna Marie Arvani-

taki, supported the latter alternative.[2] Two squid giant axons were isolated and positioned so they were touching each other over a limited region. Electrical stimulation of one axon produced a subthreshold depolarization of the other axon. The interpretation was that extracellular currents generated by the action potential in the stimulated axon spread directly to the unstimulated axon through an artificial synapse (called a *ephapse*) created by the region of close contact. The coupling efficiency of the ephapse was exceptionally poor, so it was argued that naturally occurring electrical synapses had structures designed to shunt the flow of extracellular current from one cell to another. The possible existence of such structures could not be properly investigated until the advent of the electron microscope. Subsequent ultrastructural studies of electrical junctions, however, revealed morphological specializations that appeared to be designed to direct the flow of intracellular, rather than extracellular, current between coupled cells.

The morphological specialization associated with electrical transmission is the *gap junction*, which is also called a nexus, or macula communicans. In cross-section it is identified as a region where two cells are in extra-close contact (20 nm) and bridged by a number of electron-dense bars. Freeze-fracture studies indicate that each bar is made up of two identical transmembrane proteins, one from each cell, that meet in the middle of the gap. There is a slight depression at the tip of each transmembrane protein that is thought to be the

---

Figure 11-14 Rectifying electrical transmission at the crayfish giant motor synapse. The upper drawing shows the position of current-passing and voltage-recording intracellular microelectrodes. The traces in *A* and *B* illustrate the voltage changes in the presynaptic axon (upper panel) and postsynaptic axon *(lower panel)* evoked by current injected into either the presynaptic axon *(A)* or postsynaptic axon *(B)*. *A,* Depolarizing current injected into the presynaptic axon *(top panel)* where it caused a large depolarizing potential change *(middle panel)*, spread into the postsynaptic axon, where it caused a small depolarizing potential change *(bottom panel)*. In contrast, hyperpolarizing current injected into the presynaptic axon did not spread into the postsynaptic axon. This is shown by the fact that a large hyperpolarizing potential change in the presynaptic cell *(top panel)* is not associated with a hyperpolarizing potential change in the postsynaptic cell (flat trace, *bottom panel)*. *C* illustrates that the synapse passes positive current from the presynaptic to the postsynaptic cell, but not negative current. *B,* Depolarizing current injected into the postsynaptic axon (current monitor bottom) does not spread into the presynaptic axon. The large depolarizing potential change in the postsynaptic cell *(bottom panel)* is not associated with a depolarizing potential change in the presynaptic cell (flat trace, *top panel)*. Hyperpolarizing current, on the other hand, injected into the postsynaptic axon spreads from the presynaptic cell, where it produces a large hyperpolarizing potential change *(bottom panel)* into the postsynaptic cell where it produces a small hyperpolarizing potential change *(top panel)*. *D* illustrates that the synapse passes negative current from the postsynaptic to the presynaptic cell, but not positive current. (After Furshpan, E. J.; Potter, D. D. *J. Physiol.* 145:289–325, 1959.)

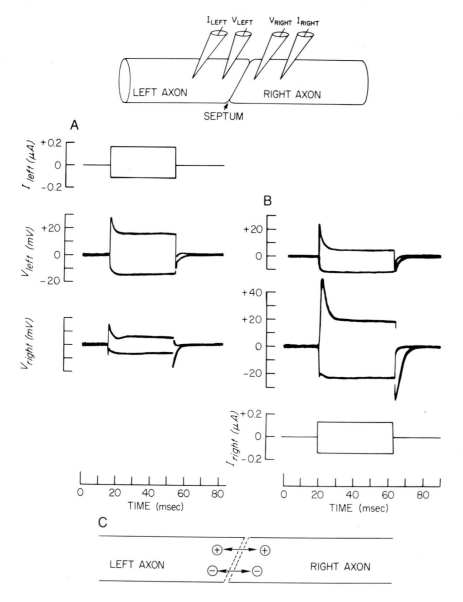

Figure 11–15 Non-rectifying electrical transmission at the crayfish septate axon. The upper drawing shows the position of current passing and voltage recording intracellular microelectrodes. The traces in *A* and *B* show the voltage response evoked by passing current into either the left *(A)* or right *(B)* axon. *A,* Depolarizing and hyperpolarizing currents (current monitor at top) injected into the left axon spread from the left cell, where they caused depolarizing and hyperpolarizing potential changes *(top panel),* into the right cell, where they cause smaller depolarizing and hyperpolarizing potential changes *(bottom panel).* *B,* Depolarizing and hyperpolarizing current (current monitor at bottom) injected into the right axon spread from the right cell, where they cause large depolarizing and hyperpolarizing potential changes *(bottom panel),* into the left cell, where they cause smaller depolarizing and hyperpolarizing potential changes *(top panel).* *C* shows that positive and negative currents spread equally well between the left and right axons. (After Watanabe, A.; Grundfest, H. *J. Gen. Physiol.* 45:267–308, 1961.)

Figure 11–16 Schematic drawing of a gap junction. The panel to the right is a tracing of an electron micrograph of a gap junction, showing electron dense bars spanning a region of close membrane apposition (scale = 75 nm). The larger drawing illustrates a sectional view of a gap junction. Intercellular channels formed by pairs of membrane proteins, hemichannels one from each cell, connect the cytoplasmic compartments of the two cells. Each hemichannel is made up of six connexin subunits. The intercellular channels are large enough to permit passage of ions, cAMP (an intracellular second messenger), fluorescein (a fluorescent dye) and other small molecules, but not proteins, nucleic acids, or intracellular organelles. After Loewenstein, W.R. (Chapter 11). In Weissmann, G.; Cailborne, R., eds. *Cell Membranes: Biochemistry, Cell Biology and Pathology*, pp. 105–114. New York, H. P. Pub. Co., 1974.

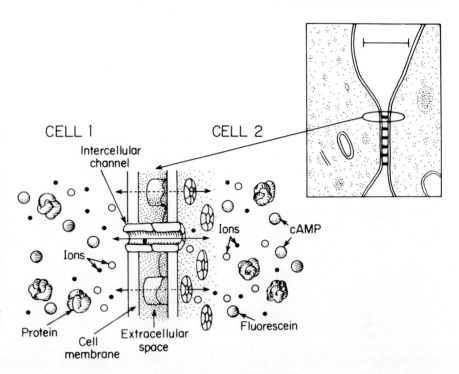

mouth of a pore that passes through the center of the protein molecule. On the strength of such structural evidence, morphologists proposed that electrically coupled cells were in cytoplasmic continuity and that the dense bars in the gap junction were *intercellular channels* made up of two transmembrane proteins or hemichannels joined together in the middle of the gap (Fig. 11–16). Electron diffraction studies of gap junctions indicate that each hemichannel is made up of six rod-shaped subunits that aggregate to form a cylindrical structure 7 nm in diameter and 7.5 nm long. The six subunits are thought to rotate relative to each other like the elements in an iris diaphragm to open a 1- to 2-nm diameter central pore.[52] Biochemical studies of isolated gap junctions from liver cells have identified a 27-kilodalton protein, called *connexin*, which is thought to represent a single subunit.[15] The protein has been cloned[11] and more detailed information about the molecular properties of the gap junction should be available in the near future.

Cytoplasmic Continuity. Tracer studies provided the first direct evidence for cytoplasmic continuity between electrically coupled cells. Fluorescent dye molecules injected into the cytoplasm of one cell diffuse across the gap junction into the cytoplasm of a second cell. The intercellular channel connecting the two cells is large enough to

accommodate molecules with a molecular weight of up to 1200 daltons.[45] Thus a large variety of intracellular molecules are small enough to be exchanged between coupled cells. These include nucleotides, sugars, amino acids, vitamins, and many products of intermediate metabolism. The capacity for metabolic coupling makes it possible for cells to communicate by other than electrical means. Gap junctions allow exchange of chemical messengers (second messengers), nutrients, and metabolites (see Fig. 11–16).

There appears to be no morphological difference in the structure of gap junctions at rectifying and nonrectifying electrical synapses. Fluorescent dyes are able to diffuse through the intercellular channels at both junctions.[33] The only difference is that dye permeability at a rectifying synapse depends on the trans-synaptic voltage. When junctional conductance is lowered by imposing the appropriate trans-synaptic voltage, dye permeability is reduced. This indicates that the principal difference between the intercellular channels that connect cells at rectifying and nonrectifying synapses is that one is gated by relatively small voltage changes and the other is not.

Gap Junction Blockers. Gap junction channels are blocked by the binding of specific ions to the cytoplasmic face of the channel. Intracellular injection of calcium or hydrogen ions rapidly and reversibly uncouples cells.[42, 51] Since Ca^{2+}/H^+ exchange is a source of cytoplasmic buffering it is possible that H^+ ions affect junctional conductance indirectly by causing changes in intracellular $[Ca^{2+}]$. This suggestion was tested by M. V. L. Bennett and colleagues who used a combination of intracellular recording and whole-cell gigaseal recording to record from a coupled cell pair while the cytoplasmic aspect of the junction was perfused with buffered solutions of known compositions.[48] By changing the concentration of H^+ or Ca^{2+} while keeping the concentration of the other ions constant, they were able to show that nanomolar changes in H ion concentration produced the same change in synaptic conductance as millimolar changes in $[Ca^{2+}]$; hydrogen ions were about 10,000 times more effective than Ca ions in controlling junctional conductance. The gating mechanism is not known, but the reversibility and rate of uncoupling suggests that both ions bind directly to sites on the cytoplasmic surface of the intercellular channel. The relation between junctional conductance and Ca^{2+} or H^+ ion concentration suggests that both ions act on the same site and that a number of ions must be bound simultaneously in order to operate the channel-gating mechanism.[47] It has been hypothesized that there is a single negative binding site per connexin subunit and that the channel closes when all six binding sites on a hemichannel are neutralized by binding six H^+ ions or three Ca^{2+} ions.

There are several reports that the gating of the channel is associated with changes in the fine structure of the gap junction.[37, 40] The principal difference is that freeze-fractured images of uncoupled junctions show an ordered array of intramembranous particles, whereas the particles at coupled junctions are more randomly distributed. There is still conjecture about whether the regular structure associated with the low conductance state is a primary or slow secondary effect of uncoupling.

Gap Junction Formation. The conditions that uncouple gap junctions are understood more fully than are the events responsible for their formation. The electrical changes associated with the formation of intercellular channels have been studied most extensively in embryonic or tissue-cultured cells. Under these conditions channel formation is a capacity of most, if not all parts of the plasma membrane. The only requirement is intimate contact (adhesion) between the membranes of the two cells. The cells do not have to be the same type; in fact, electrical junctions will form between cultured cells from different organs and different species. For example, a perfectly viable junction develops between a rabbit lens cell and human skin fibroblast.

The development of intercellular channels can be followed by measuring the electrical coupling between two tissue-cultured or embryonic cells that are brought together mechanically.[23] After a latency of 1 to 2 min when no current flows between the cells, there is a gradual increase in coupling, which reaches a maximum in 10 to 15 min. This phenomenon does not occur between adult cells and is not the basis of transmission at the contact point between two axons described by Arvanitaki.[2] The gradual increase in coupling may indicate that the entire gap junction, with all its intercellular channels in place, is formed at once and the conductance of the whole junction is gradually increased. Another possibility is that each intercellular channel contributes an all-or-none unit of coupling such that the total amount of coupling increases as the number of intercellular channels increases. The latter interpretation was shown to be correct by using highly sensitive recording electronics to detect minute changes in the potential of one cell produced by the steady application of current to a second cell.[32] After the cells were physically brought together and junction formation began, small potential steps were recorded from the first cell while constant current was applied to the other cell. The potential steps were quantal in that their amplitudes were discrete multiples of the amplitude of the smallest potential jump. The conclusion of these findings is that each time an intercellular channel appears, it contributes a fixed amount of junctional conductance. This allows a fixed amount of current to flow between the cells, causing a discrete change in the potential of the recorded cell. Estimates of the conductance of a single intercellular channel have been obtained by using patch-clamp recording methods to monitor whole-cell currents in pairs of coupled cells. The specific properties of the channel vary with cell type, but in general the intercellular channel has a conductance of about 160 pS and opens and closes spontaneously.[53] There is some evidence that the transition between the closed and open states occurs gradually, like the opening of an iris diaphragm.

Regulation of Junctional Transmission. It was once thought that regulation was a principal difference between electrical and chemical synaptic transmission. Transmission at an electrical synapse was thought to be fixed and nonadjustable, whereas transmission at a chemical synapse could be adjusted by any one of a number of mechanisms. This view is clearly an oversimplification.

Since gap junctional conductance is highly sensitive to H^+ ions, it is possible that physiological changes in intracellular pH could regulate intercellular communication. For example, during early development embryonic cells become progressively more alkaline; this, by increasing junctional conductance, could play an important role in the orchestration of normal growth. In the vertebrate retina, electrical coupling between horizontal cells is regulated by dopamine,[35] a retinal neurotransmitter that acts through a second messenger mechanism. In liver cells cAMP increases gap junctional conductance and stimulates phosphorylation of connexin subunits.[41] Thus it is clear that intercellular communication can be controlled by a number of intracellular biochemical events, second messenger mechanisms in particular. What about regulation by calcium? As discussed earlier, junctional conductance is not very sensitive to calcium. When pH is held constant, intercellular channels are not blocked until calcium reaches millimolar levels. The only time cytoplasmic calcium would reach millimolar concentrations is during cell death or membrane disruption. This catastrophic situation, nevertheless, provides a very important protection mechanism. A large increase in intracellular calcium would isolate a damaged or diseased cell from its coupled neighbors and thus prevent the cytoplasmic constituents in healthy cells from leaking into the extracellular space by diffusing through open intercellular channels into the cell with a broken membrane.

SUMMARY

This chapter is about communication between nerve cells. In most cases communications take place at a morphologically specialized site called a synapse where the two cells are in close contact. Exchange of information at a synapse is mediated either by the direct flow of ionic current from one cell to the other (electrical synaptic transmission) or by the release of an extracellular chemical messenger (chemical synaptic transmission).

Cells that exchange information by electrical synaptic transmission are joined together by a gap junction made up of intercellular channels that connect the cytoplasmic spaces of the two cells. Intercellular channels are large enough to allow ions and other small molecules (<1200 daltons) to pass between coupled cells. Hence, cells joined by gap junctions are electrically and metabolically coupled together. Based on electrical behavior there are two classes of gap junctions: those that

pass positive current equally well in either direction (nonrectifying) and those that pass positive current more easily in one direction than the other. The conductance of both kinds of gap junction is sensitive to changes in intracellular pH and Ca concentration. Gap junctions may also be regulated by extracellular signals (e.g., neurotransmitters) that cause changes in intracellular second messengers.

In chemical synaptic transmission, depolarization of the presynaptic terminal opens voltage-gated calcium channels, causing calcium influx and an elevation of intracellular calcium that triggers exocytosis of vesicles filled with any one of a number of possible neurotransmitter substances. Released neurotransmitter binds to receptor proteins in the postsynaptic membrane and evokes either a depolarizing or a hyperpolarizing potential change. A depolarizing potential change is called an excitatory postsynaptic potential (EPSP) because it moves the membrane potential of the postsynaptic cell toward the threshold for action potential production. A hyperpolarizing potential change is called an inhibitory postsynaptic potential (IPSP) because it moves the membrane potential away from the threshold for action potential production. The binding of neurotransmitter to receptor protein causes a potential change either by directly opening an ion channel in the receptor protein or by triggering the production of an intracellular second messenger which changes the ion conductance of the postsynaptic cell. One of the principal characteristics of chemical synaptic transmission in the central nervous system is modifiability. Changes in the strength of synaptic transmission can be modified by a presynaptic action (e.g., presynaptic inhibition) or by a postsynaptic action that influences the efficacy of the mechanism responsible for the generation of the postsynaptic potential.

ANNOTATED BIBLIOGRAPHY

Cotman, C.W.; Iversen, L.L. Excitatory amino acids in the brain—focus on NMDA receptors. *Trends Neurosci.* 10:263–265, 1987.
This paper introduces a special issue on trends in neuroscience that is entirely devoted to the topic of excitatory amino acids. Short review articles by various authors provide a clear and concise overview of this topic.

Dunlap, K.; Holz, G.G.; Rane, S.G. G proteins as regulators of ion channel function. *Trends Neurosci.* 10:241–244, 1987.
A short review that discusses the various mechanisms by which G proteins regulate ion channels in synaptic transmission.

Eccles, J.C. *The Physiology of Synapses.* Berlin, Springer-Verlag, 1964.

This book, written by one of the founding fathers of synaptic physiology, presents an excellent summary of early work by the author that established many of the basic principles of chemical synaptic transmission.

Katz, B. *The Release of Neurotransmitter Substances.* Liverpool, Liverpool University Press, 1969.

An excellent book that describes classic experiments on the release of neurotransmitter.

Spray, D.C.; Harris, A.L.; Bennett, M.V.L. Comparison of pH and calcium dependence of gap junctional conductance. In Nuccitelli, R.; Deamer, D., eds., *Intracellular pH: Its Measurement, Regulation and Utilization in Cellular Functions.* New York, Alan R. Liss, pp. 445–461, 1982.

A review paper that specifically addresses the regulation of gap junctional conductance by pH and calcium but also contains many references to earlier work on the physiology of electrical synaptic transmission.

REFERENCES

1. Andrade, R.; Malenka, R.C.; Nicoll, R.A. A G protein couples serotonin and GABA_b receptors to the same channels in hippocampus. *Science* 234:1261–1265, 1986.
2. Arvanitaki, A. Interactions electriques entre deuz cellules nerveuses contigues. *Arch. Int. Physiol.* 52:381–407, 1942.
3. Barrett, J.N.; Crill, W.E. Specific membrane properties of cat motoneurons. *J. Physiol. (Lond.)* 239:301–324, 1974.
4. Bennett, M.V.L. Electrical transmission: A functional analysis and comparison to chemical transmission. *Handbk. Physiol.* 1(1):357–416, 1977.
5. Bormann, J.; Hamill, O.P.; Sakmann, B. Mechanism of anion permeation through channels gated by glycine and gamma-aminobutyric acid (GABA) in mouse cultured spinal neurons. *J. Physiol.* 385:243–286, 1987.
6. Browning, M.D.; Huganir, R.; Greengard, P. Protein phosphorylation and neuronal function. *J. Neurochem.* 45:11–22, 1985.
7. Changeux, J.P. The acetylcholine receptor. An allosteric membrane protein. *Harvey Lecture Series* 75:85–255, 1981.
8. Coombs, J.S.; Eccles, J.C.; Fatt, P. The specific ionic conductances and the ionic movements across the motoneuronal membrane that produce the inhibitory post-synaptic potential. *J. Physiol. (Lond.)* 130:326–373, 1955.
9. Costa, E.; Trabucchi, M. *Neural Peptides and Neural Communication.* New York, Raven Press, 1980.
10. Cull-Candy, S.G.; Usowicz, M.M. Multiple-conductance channels activated by excitatory amino acids in cerebellar neurons. *Nature* 325:525–528, 1987.
11. Dahl, G.; Miller, T.; Paul, D.; Voellmy, R.; Werner, R. Expression of functional cell-cell channels from cloned rat liver gap junction complementary DNA. *Science* 236:1290–1293, 1987.
12. Dudel, J.; Kuffler, S.W. Presynaptic inhibition at the crayfish neuromuscular junction. *J. Physiol. (Lond.)* 155:543–562, 1961.
13. Eccles, J.C. *The Physiology of Nerve Cells.* Baltimore, Md., Johns Hopkins Press, 1957.
14. Eccles, J.C. *The Physiology of Synapses.* Berlin, Springer-Verlag, 1964.
15. Finbow, M.; Yancey, S.B.; Johnson, R.; Revel, J.P. Independent lines of evidence suggesting a major gap junctional protein with a molecular weight of 26,000. *Proc. Natl. Acad. Sci. USA* 77:970–974, 1980.
16. Frank, K.; Fuortes, M.G.F. Presynaptic and postsynaptic inhibition of monosynaptic reflexes. *Fed. Proc.* 16:39–40, 1957.
17. Furshpan, E.J.; Potter, D.D. Transmission of the giant motor synapses of the crayfish. *J. Physiol.* 145:289–325, 1959.
18. Gähwiler, B.H.; Brown, D.A. GABA_B-receptor-activated K$^+$ current in voltage-clamped C_{A3} pyramidal cells in hippocampal cultures. *Proc. Natl. Acad. Sci. USA* 82:1558–1562, 1985.
19. Gilman, A. G proteins: Transducers of receptor-generated signals. *Ann. Rev. Biochem.* 56:615–649, 1987.
20. Goelet, P.; Castellucci, V.F.; Schacher, S.; Kandel, E.R. The long and short of long-term memory—a molecular framework. *Nature* 322:419–422, 1986.
20a. Grenningloh, G.; Rienitz, A.; Schmitt, B.; Methfessel, C.; Zensen, M.; Beyreuther, K.; Gundelfinger, E.D.; Betz, H. The strychnine-binding subunit of the glycine receptor shows homology with nicotinic acetylcholine receptors. *Nature* 328:215–220, 1987.
21. Harden, T.K.; Tanner, L.I.; Martin, M.W.; Nakahata, N.; Hughes, A.R.; Hepler, J.R.; Evans, T.; Master, S.B.; Brown, J.H. Characteristics of two biochemical responses to stimulation of muscarinic cholinergic receptors. *Trends Pharm. Sci. Suppl.* Feb.:14–18, 1986.
22. Hebb, D.O. *The Organization of Behaviour.* New York, John Wiley & Sons, 1949.
23. Ito, S.; Sato, E.; Loewenstein, W.R. Studies on the formation of a permeable cell membrane junction. *J. Mem. Biol.* 19:205–337, 1974.
24. Jahr, C.E.; Stevens, C.F. Glutamate activates multiple single-channel conductances in hippocampal neurons. *Nature* 225:522–525, 1987.
25. Kaczmarek, L.K.; Levitan, I.B. Neuromodulation. In Kaczmarek, L.K.; Levitan, I.B., eds. *The Biochemical Control of Neuronal Excitability.* New York, Oxford University Press, 1987.
26. Katz, B.; Miledi, R. A study of synaptic transmission in the absence of nerve impulses. *J. Physiol. (Lond.)* 192:407–436, 1967.
27. Katz, B.; Miledi, R. The timing of calcium action during neuromuscular transmission. *J. Physiol. (Lond.)* 189:535–544, 1967.
28. Katz, B.; Schmitt, O.H. Electrical interaction between two adjacent nerve fibers. *J. Physiol.* 97:471–488, 1940.
29. Landis, D.M.D.; Hall, A.K.; Weinstein, L.A.; Reese, T.S. The organization of cytoplasm at the presynaptic active zone of a central nervous system synapse. *Neuron* 1:201–209, 1988.
30. Llinas, R.; Heuser, J.E. Depolarization-release coupling systems in neurons. *Neurosci. Res. Program Bull.* 15:555–687, 1977.
31. Llinas, R.; McGuinness, T.L.; Leonard, C.S.; Sugimori, M.; Greengard, P. Intraterminal injection of synapsin I or calcium/calmodulin-dependent protein kinase II alters neurotransmitter release at the squid giant synapse. *Proc. Natl. Acad. Sci. USA* 82:3035–3059, 1985.
32. Loewenstein, W.R.; Kanno, Y.; Socolar, S.J. Quantum jumps of conductance during formation of membrane channels at cell-cell junction. *Nature* 274:133–136, 1978.
33. Margiotta, J.F.; Walcott, B. Conductance and dye permeability of a rectifying electrical synapse. *Nature* 305:52–55, 1983.
34. Miller, R.J. Second messengers, phosphorylation and neurotransmitter release. *Trends Neurosci.* 8:463–465, 1985.

35. Negishi, K.; Drujan, B.O. Reciprocal changes in center and surrounding S potentials of fish retina in response to dopamine. *Neurochem. Res.* 4:313–318, 1979.

36. Nowak, L.; Bregestovski, P.; Ascher, P.; Herbet, A.; Prochiantz, A. Magnesium gates glutamate-activated channels in mouse central neurons. *Nature* 307:462–465, 1984.

37. Peracahia, C. Gap junctions: Structural changes after uncoupling procedures. *J. Cell Biol.* 72:628–641, 1977.

38. Pfaffinger, P.S.; Martin, J.M.; Hunter, D.D.; Nathanson, N.M.; Hille, B. GTP-binding proteins couple cardiac muscarinic receptors to a K channel. *Nature* 317:536–538, 1985.

39. Rakic, P. Local circuit neurons. *Neurosci. Res. Program Bull.* 13:289–446, 1975.

40. Raviola, E.; Goodenough, D.G.; Raviola, G. Structure of rapidly frozen gap junctions. *J. Cell Biol.* 87:273–279, 1981.

41. Redman, S.; Walmsley, B. The synaptic basis of the monosynaptic stretch reflex. *Trends Neurosci.* 4:248–251, 1981.

42. Rose, B.; Loewenstein, W.R. Permeability of cell junction depends on local cytoplasmic calcium activity. *Nature* 254:250–252, 1975.

43. Saez, J.; Spray, D.C.; Nairn, A.; Hertzberg, E.L.; Greengard, P.; Bennett, M.V.L. cAMP increases junctional conductance and stimulates phosphorylation of the 27-KDa principal gap junction polypeptide. *Proc. Natl. Acad. Sci. USA* 83:2473–2477, 1986.

43a. Schofield, P.R.; Darlison, M.G.; Fujita, N.; Burt, D.R.; Stephenson, F.A.; Rodriguez, H.; Rhee, L.M.; Ramachandran, J.; Reale, V.; Glencorse, T.A.; Seeburg, P.H.; Barnard, E.A. Sequence and functional expression of the $GABA_A$ receptor shows a ligand-gated receptor superfamily. *Nature* 328:221–227, 1987.

44. Siegelbaum, S.A.; Camardo, J.S.; Kandel, E.R. Serotonin and cyclic AMP close single K^+ channels in Aplysia sensory neurons. *Nature* 299:413–417, 1982.

45. Simpson, I.; Rose, B.; Loewenstein, W.R. Size limit of molecules permeating the junctional membrane channels. *Science* 195:294–296, 1977.

46. Spray, D.C.; Harris, A.L.; Bennett, M.V.L. Voltage-dependence of junctional conductance in early amphibian embryos. *Science* 204:432–434, 1979.

47. Spray, D.C.; Harris, A.L.; Bennett, M.V.L. Comparison of pH and calcium dependence of gap junctional conductance. In Nuccitelli, R.; Deamer, D., eds. *Intracellular pH: Its Measurement, Regulation and Utilization in Cellular Functions.* New York, Alan R. Liss, pp. 445–461, 1982.

48. Spray, D.C.; Stern, J.H.; Harris, A.L.; Bennett, M.V.L. Gap junctional conductance: comparison of sensitivities to H and Ca ions. *Proc. Natl. Acad. Sci. USA* 79:441–445, 1982.

49. Stevens, C.F. Are there two functional classes of glutamate receptors? *Nature* 322:210–211, 1986.

50. Triller, A.; Korn, H. Transmission at a central inhibitory synapse. III. Ultrastructure of physiologically identified and stained terminals. *J. Neurophysiol.* 48:708–736, 1982.

51. Turin, L.; Warner, A.E. Carbon dioxide reversibly abolishes ionic communication between cells of early amphibian embryo. *Nature* 207:56–57, 1978.

52. Unwin, P.N.T.; Zanpighi, G. Structure of the junction between communicating cells. *Nature* 283:545–549, 1980.

53. Veenstra, R.D.; DeHan, R.L. Measurement of single channel currents from cardiac gap junctions. *Science* 233:972–974, 1986.

54. Watanabe, A.; Grundfest, H. Impulse propagation at the septal and commissural junctions of crayfish lateral giant axons. *J. Gen. Physiol.* 45:267–308, 1961.

Transformation of Synaptic Input into Spike Trains in Central Mammalian Neurons

INTRODUCTION

The nervous system is a remarkable structure that is responsible for such diverse phenomena as sensation, movement, memory, cognition, and language. Although the specific neural mechanisms underlying these phenomena are poorly understood, they clearly are the result of interactions between individual nerve cells or *neurons*. Most neurons are electrically excitable and generate all-or-nothing action potentials (also called *spikes* or *impulses*) in response to depolarization of their membranes. The action potentials propagate without decrement along a neuron's axon to its synaptic terminals, where transmission to another cell occurs, usually by the release of a chemical neurotransmitter (see Chap. 11). Each neuron must transform graded synaptic input into physiologically meaningful patterns of all-or-none action potentials for propagation down its axon. This encoding occurs in the cell body, in the dendrites, and in the proximal axon of the neuron and requires ionic channels in addition to those responsible for spike propagation in axons. The mechanisms underlying the transformation of synaptic current into spike trains are the main topic of this chapter.

MORPHOLOGY OF CENTRAL NEURONS

The vast majority of neurons have a cell body and a number of radiating processes. The cell body or *soma* of a neuron (Fig. 12–1) contains the nucleus and the subcellular machinery used for metabolic and synthetic functions. Arising from the soma are a number of radiating, branching processes, or *dendrites*, which form the *dendritic tree* of the neuron. The dendrites often extend for several millimeters, and the surface area of the dendritic tree may be 100 times that of the soma.

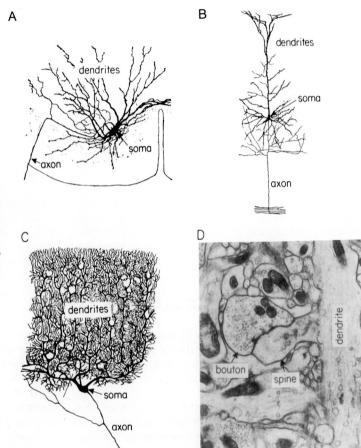

Figure 12–1 Golgi silver stains of mammalian central nervous system neurons. *A*, Cat spinal motoneuron; *B*, rabbit neocortical pyramidal cell; *C*, human cerebellar Purkinje cell; and *D*, an electron micrograph of a synaptic connection on a spine of a neocortical neuron. (*A*, *B*, and *C* are adapted from Ramon Y Cajal, S. *Histologie systeme du nerveux de l'homme et des vertébrés*, Vol. 1. Paris, Maloine, 1909. *D*, Courtesy of Dr. L. Westrum).

The dendrites receive the great majority of synaptic inputs to the neuron. Some dendrites have small, lateral outpouchings called *spines* (Fig. 12–1D), which are special structures for receiving synaptic connections; synapses occur on the smooth portion of the dendrites as well.

In most neurons, a single axon extends from the soma. As we shall see later, the axon-soma junction, called the *initial segment*, has a specialized role in spike generation. Axons of some cells are only a few micrometers long, whereas others may extend over a meter. Axons usually branch extensively, so that a single neuron may make synaptic contact with many other neurons. The enlarged knoblike synaptic terminals of the axon are called *boutons*. The dendrites, soma, and initial segment are covered with boutons from thousands of axons.

As seen in Fig. 12–1, neurons come in many different sizes and shapes. The morphology of a neuron often is correlated with its function. Cell size influences function in several ways. For ex-

ample, the largest cell bodies have the largest diameter (and usually very long) axons and conduct impulses at the fastest velocity. They also require more depolarizing synaptic current to evoke action potentials than small cells. The branching patterns of their different dendritic trees give certain neuron types a distinctive structure that can also be correlated with function. For example, the size of the dendritic tree limits how many synaptic inputs the neuron can receive. In addition, the orientation of the dendritic tree determines the types and number of sources from which it can receive synaptic connections. Some typical dendritic patterns are shown in Fig. 12–1. Most neurons have 10 to 20 major dendritic trunks attached to the soma from which the dendrites radiate in a spherically symmetrical pattern. The large neurons in the spinal cord that innervate skeletal muscle (spinal motoneurons) have this shape (Fig. 12–1A), as do many small neurons throughout the central nervous system. A different dendritic pattern, consisting of radially symmetric dendritic trees, is found in many neurons in regions of the nervous system where the cell bodies are organized into layers. For example, the soma of a large pyramidal cell of the cerebral cortex (Fig. 12–1B) lies in a deep cortical layer. A single primary apical dendrite extends from the soma toward the cortical surface, and secondary and higher-order branches emerge from the primary apical dendrite. These higher-order dendritic branches extend through (and receive synaptic input from) all layers of the cortex superficial to the cell body. In contrast, Purkinje cells of the cerebellum have a massive, flat dendritic tree occupying a single plane (Fig. 12–1C), a morphology that allows them to be closely packed. Axons running orthogonal to the plane of the dendrites make synapses en passant with thousands of Purkinje cells (see Chap. 29).

REGIONAL EXCITABILITY AND SPIKE INITIATION

The excitable components of the neuron reside in its surface membrane. As discussed in Chapter 1, the membrane lipid bilayer contains specialized macromolecules that serve as selective channels for ion flow into and out of the cell. In addition, the membrane contains receptors for various neurotransmitters and other chemicals and pumps for the active transport of various substances into and out of the cell against concentration gradients. In Chapter 3, we learned that the density of ionic channels in the membrane of myelinated axons depends upon their location along the axon, and thus different regions have different excitable properties. For example, in the myelinated axon, the sodium channels responsible for regenerative depolarization during the action potential are restricted to the nodal regions, whereas (in mammalian axons) the voltage-gated potassium channels appear to be present only in the internodal regions. A regional variation in excitability also underlies the mechanism of spike initiation in the soma/dendritic region of neurons. Spikes are normally initiated at a low-threshold ''trigger zone'' that is more excitable than any other part of the soma or dendrites. The trigger zone is thought to be located at the axon initial segment. If depolarizing synaptic current is large enough, it brings the membrane potential of the initial segment to threshold, and a spike is initiated there. The spike arising in the initial segment propagates down the axon; however, it also propagates ''backwards'' into the soma/dendritic region and usually provides enough depolarization to trigger a spike in this less-excitable region.

The spatial variation of excitability in a neuron is perhaps the most important single factor governing the transformation of synaptic input into spike output. Since the trigger zone is located distant from most synaptic inputs, it sees a weighted average of the synaptic currents flowing from the entire dendritic tree and the soma. Synaptic inputs located close to the trigger zone suffer less electrotonic decrement and are more effective in depolarizing or hyperpolarizing it than synaptic inputs of equal strength located farther away. The trigger zone therefore acts as a summing point for all synaptic current, and it is the site where the decision is made to send a spike down the axon.

It is worth examining some of the evidence that led to the trigger zone concept, because these examples illustrate techniques and problems associated with recording in the central nervous system, where one can usually record only from the soma of a neuron that cannot be seen. The mechanism of spike initiation has been studied in the most detail in motoneurons of the lumbar part of the spinal cord (lumbar motoneurons).[14, 15, 17]

The regional variation in excitability was first surmised from the shape of the action potential recorded in the soma. An action potential may be evoked in a neuron experimentally by the three methods indicated diagrammatically in the top row of Figure 12–2. During normal function, excitatory synaptic activity depolarizes the cell to spike threshold and an *orthodromic* action potential is propagated down the axon (Fig. 12–2A1). The experimenter can mimic this natural activation by injecting depolarizing current through an intracellular recording microelectrode placed in the soma (Fig. 12–2B1). This *direct* activation also evokes an action potential that propagates orthodromically down the axon. Sometimes it is desirable to generate an action potential without previously depolarizing the soma. This can be accomplished by electrically stimulating the axon of the recorded cell to generate an *antidromic* spike, which travels backward along the axon, opposite the normal direction of propagation, and invades the soma (Fig. 12–2C1). The lower portion of Figure 12–2 (A2 to C2) shows action potentials (V) recorded in a motoneuron soma during each type of stimulation. Below each action potential is its time derivative (dV/dt). Notice that there is an inflection on the rising phase of each spike (at the arrow), and this *A-B inflection* is revealed as a distinct notch on the dV/dt records. This inflection was recognized as unusual by early investigators since it did not occur on axon spikes. Experiments seeking the cause of the inflection led to the spike trigger zone concept, as we now explain.

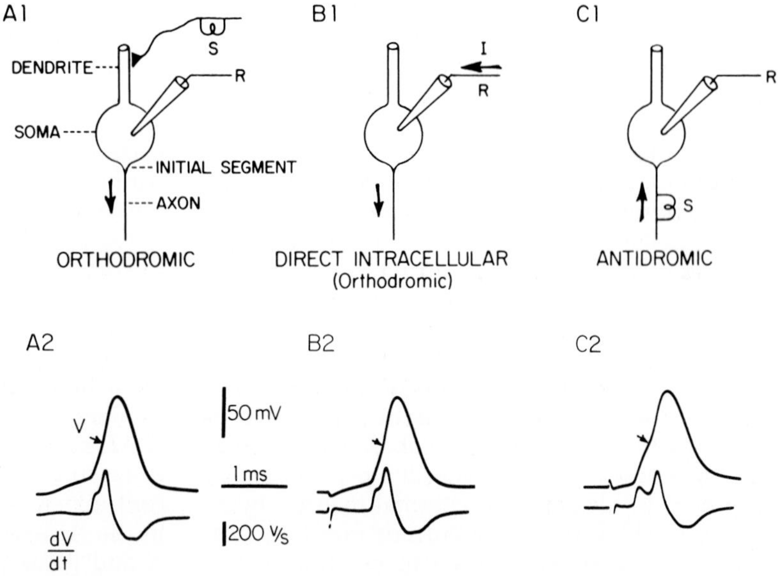

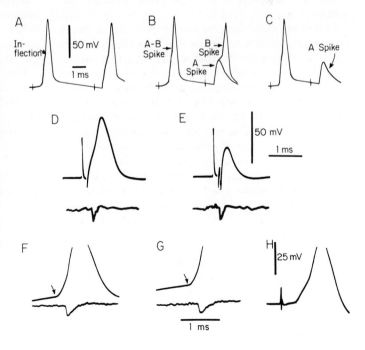

Figure 12-2 Three experimental methods of evoking spikes in a neuron. The schematic neurons (above) are activated orthodromically either by a synaptic input *(A1)* or intracellular current (I in *B1*) or antidromically by axonal stimulation (S in *C1*). *A2, B2,* and *C2* show actual recordings (R) of lumbar motoneuron action potentials (V) evoked by the corresponding methods of the top row and their time derivative (dV/dt). See text for further explanation.

Action Potential Components. Unlike the spike in an axon, the spike recorded in the soma of a neuron is not a unitary event. Experimental manipulations can decompose it into components reflecting activation of spatially distinct regions of the neuron. Since the components of the spike have different refractory periods, one way to perform the decomposition is to evoke antidromic spikes at progressively shorter intervals, as shown in Figure 12-3,A-C. With a sufficiently long interval between stimuli, the second antidromic action potential is identical with the first (Fig. 12-3A).

As the interval between stimuli is decreased, the inflection on the rising phase of the second spike becomes more prominent (Fig. 12-3B). When the stimuli are applied at a sufficiently short interval, a portion of the second spike fails in an all-or-nothing manner, revealing a smaller component called the *A spike* (Fig. 12-3B,C). The larger spike that fails during stimulation at short intervals is called the *B spike*; it has a longer refractory period. The spike recorded in the soma usually is composed of both components and is called the *A-B spike*.

Figure 12-3 Evidence that spike initiation occurs on the proximal axon. *A, B,* and *C* show two antidromic spikes in lumbar motoneurons evoked at increasingly shorter intervals. *D, E, F,* and *G:* Intracellular recordings from a motoneuron soma (upper records) and extracellular records from its axon (lower records) showing spikes at slow *(D, E)* and fast *(F, G)* sweep speed produced by brief depolarizing current pulses injected through the recording microelectrode. In *E,* the depolarizing current pulse is followed by a hyperpolarizing pulse sufficient to block the B spike component. *H* shows the antidromic spike in the same neuron evoked by stimulation of its axon at the same point at which the axon spike is recorded in *F* and *G.* Arrows in *F* and *G* indicate the time at which the spike recorded in the axon was initiated, based on the conduction time from axon to soma measured in *H.* *(D-H* are adapted from Coombs, J. S., et al. *J. Physiol. [Lond.]* 139:232–249, 1957.)

A similar decomposition of an antidromic spike can be produced by artificially hyperpolarizing the soma. As the soma is hyperpolarized, the A-B inflection becomes more prominent until the B spike fails, all-or-none, leaving the A spike alone. The decomposition of the A-B spike by hyperpolarization suggests that the A and B spikes are generated in different regions of the cell. Since the A spike is both smaller and more difficult to block than the B spike, the region where the A spike is generated (the A region) must be at some distance from the soma.

The most important inference that has been drawn from the decomposition of the antidromic spike is that the inflection on the A-B spike is caused by the A spike occurring before the B spike and triggering it. An A-B inflection is also seen on the spikes produced in the normal manner by an EPSP (Fig. 12–2A) and by current injection (Fig. 12–2B). These observations suggest that the A spike also precedes and triggers the B spike in normal neuronal function.

Site of the A Region. Where is the A region of the neuron? The A spike may originate either on the dendrites or on the proximal axon. If it is initiated on the proximal axon, the A spike may be recorded in the cell's axon before the B spike appears in the soma, and the B spike is not required for occurrence of a propagated spike in the axon. On the other hand, if the A spike is initiated in the dendrites, the axon spike cannot occur before the A or B spikes are recorded in the soma, because a spike propagating from the dendrites to the axon must first pass through the soma.

Experiments designed to locate the site of the A region are shown in Figure 12–3, D-H. In Figure 12–3, D and E, the top trace is the action potential recorded intracellularly (evoked by a very brief depolarizing current pulse), and the bottom trace is an extracellular recording of the spike in the axon. As seen in Figure 12–3D, the axon spike is recorded before the B spike is recorded in the soma. In Figure 12–3E, the soma is first depolarized briefly (to evoke the A spike), then hyperpolarized by applying current through the recording electrode to block the B spike. Since an action potential still is recorded in the axon, the A spike alone is sufficient to evoke a propagating spike. In Figure 12–3, F-H show an experiment designed to estimate the exact time at which the axon spike is initiated relative to the A and B spikes. The records in 12–3F and 12–3G are at higher gain and faster sweep from another neuron in which a longer, smaller current pulse was used to trigger

the spikes (so that the "on" and "off" artifacts resulting from current injection are off-screen). Figure 12–3H shows the antidromic spike recorded in the soma of the neuron of 12–3F and 12–3G when its axon was stimulated at the same distant, extracellular site where the axon spike was recorded. By measuring the time between stimulation of the axon and arrival of the antidromic spike in the soma in Fig. 12–3H, it can be calculated that the orthodromic axon spike recorded in 12–3F and 12–3G must have been initiated slightly before the A spike was observed in the soma (i.e., at the arrows in Fig. 12–3, F and G). Thus, the A region must be on the axon side of the soma, and it is the first region to be excited when the soma/dendritic region is depolarized. Although it cannot be assumed that every type of neuron in the central nervous system is first activated by a spike in the A region, the decomposition of spikes into A and B components or A-B inflections has been observed in all the mammalian neurons that have been tested thus far.

The A spike is also called the *IS spike*, because the A region is thought to be the axon initial segment. Recordings from invertebrate neurons in vitro, where the recording site can be visualized, have shown that the normal spike trigger zone occurs on a proximal portion of the axon analogous to the initial segment of a central mammalian neuron. It is interesting that full-size spikes are recorded from sites in motoneurons thought to be the A region.[14] Apparently, the A spike appears small when recorded at the soma both because the initial segment is electrically remote from the soma and because the soma/dendritic region provides a low-impedance shunt to the limited amount of ionic current that can be generated by the initial segment membrane.

The low threshold of the initial segment has been attributed to a higher density of sodium channels compared with the soma/dendritic region. Catterall[11] has used immunohistochemical techniques in spinal neurons grown in tissue culture to show that one neurite (i.e., an outgrowth from the soma) has the highest density of sodium channels and presumably corresponds to the axon initial segment.

The B Region. The B spike is also called the *SD spike*, because it is thought to be generated in the soma/dendritic region. Although the B spike is not propagated down the axon, it still plays an important role in neuronal function, since the depolarization provided by this component activates the many voltage-sensitive conductances that occur only in the soma/dendritic region. One such conductance is a potassium conductance responsible for the long-lasting afterhyperpolarization (AHP) of central neurons. The AHP is associated only with the B spike, since the A spike alone produces no AHP.

The size of the B region is somewhat uncertain, because the excitability of the dendrites in central mammalian neurons is difficult to examine experimentally. Action potentials have however been recorded from both the soma and the dendrites of pyramidal cells of the hippocampus[48] and Purkinje cells of the cerebellum.[30] As seen in Figure 12–4, two types of action potentials can be recorded in Purkinje cells in vitro.[29, 30] The action potentials labeled Ca in the figure can be recorded from points all along the dendritic tree and are calcium-dependent. The spikes that are more prominent during somatic recording (labeled Na) are sodium-dependent and correspond to the A-B spikes discussed above. As indicated in Figure 12–4, the sodium-dependent spikes recorded in the most proximal dendrites are very small and are not detectable at all in the distal dendrites. This observation suggests that the B component of the A-B spike is generated in the soma and in only the most proximal dendrites in this type of cell. Notice that the dendritic calcium spikes themselves do not propagate through the soma and down the axon. Rather, their action is like that of a large EPSP; they tend to depolarize the initial segment and trigger an A-B spike. In the Purkinje cells, the size of the dendritic calcium spike reaching the initial segment is large enough to trigger several A-B spikes at a fast rate.

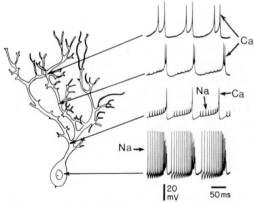

Figure 12–4 Recordings obtained from different points in a cerebellar Purkinje cell (in vitro), showing sodium-dependent A-B spikes (Na) and calcium-dependent, dendritic spikes (Ca). (Adapted from Llinas, R.; Sugimori, M. *J. Physiol. [Lond.]* 305:197–213, 1980.)

CABLE PROPERTIES OF CENTRAL NEURONS

The most striking anatomical feature of a central mammalian neuron is its dendritic tree (see Fig. 12–1). Because dendrites are long, narrow, branching structures, the synaptic signal produced in the dendrites is likely to be significantly attenuated by the time it reaches the soma. How effective, then, are synapses on the distal dendrites in depolarizing the initial segment? Since both the dendrites and somata of many central neurons contain voltage-gated conductances, how does activation of voltage-gated conductances in one part of the neuron affect membrane potential (and thus other voltage-gated channels) in another part?

Dendrites as Leaky Cables.[23, 35] A theoretical framework for answering the questions raised above is provided by the same core-conductor theory used to analyze the passive cable properties of axons (see Chapter 3). A dendrite may be considered to be an electrically leaky cable having a relatively low-resistance cytoplasm surrounded by a membrane consisting of resistive and capacitive elements in parallel. If a steady electrical signal is applied to the end of a dendrite, the attenuation of the signal with distance will depend critically on the specific membrane resistance, R_m, of the dendrite. The value of R_m determines how much current can leak out of the membrane before the signal reaches its destination. If R_m is infinite, no attenuation will occur even if the dendrite is very long, because no current can leak through the membrane. If R_m is sufficiently small, an electrical signal applied to one end of the dendrite will not be seen at the other end, even if the dendrite were physically short. Therefore, the amount of signal attenuation along a dendrite cannot be determined from knowledge of its physical length alone. Some other measure of "effective" length that reflects the amount of signal attenuation with distance is required.

If a constant current is injected into one end of a cable that has a constant diameter and extends to infinity, the membrane potential declines exponentially with distance from the point of current injection. This relation is represented by the dashed line in the semilogarithmic plot of Figure 12–7C. A measure of the "leakiness" of such a cable is the distance (e.g., in millimeters) over which the voltage decays by 1/e, in which e is the base of the natural logarithms. This characteristic distance, the space constant (λ), is determined from properties of the cable by the equation[23, 35]

$$\lambda = (rR_m/2R_a)^{1/2} \qquad (1)$$

in which r is the radius of the cable and R_a is the specific resistance of the cytoplasm. It is assumed that R_a is rather constant (at about 70 Ω-cm) among neurons. Thus for a given r, λ will depend on R_m. As R_m becomes larger, voltage is attenuated less rapidly with distance along the cable.

Three simplifying assumptions were used to derive Equation 1. The cable must have a uniform diameter, must be infinitely long, and must have a constant membrane resistance. Real dendrites do not satisfy these assumptions, so the space constant of a dendrite cannot be computed directly from Equation 1. We must therefore consider how the concept of the space constant can be applied to a more realistic case that approximates an actual dendrite. We will consider each of the three assumptions in turn.

Effects of Dendritic Length. First, we assume that dendritic diameter and membrane resistance are constant and consider the effect of dendritic length. In the infinitely long cable used to derive Equation 1, membrane potential decays exponentially with distance. For practical purposes, an exponential decay is 98% complete in 4 space constants. Thus, a cable is effectively infinitely long as far as electrical signals are concerned if it extends at least 4 space constants from the point where the electrical signal is applied. Axons or muscle fibers may satisfy this criterion since their physical length is often many times greater than their space constants (see Chap. 3). However, the physical length of a particular dendrite of a central neuron may correspond to only 1 to 2 space constants, and for such short cables, the decay of voltage with distance is not even exponential. Thus, the space constant of a short dendrite calculated from Equation 1 is only a *virtual* quantity, determined as if the dendrite extended to infinity. Nevertheless, the space constant concept is useful, because a comparison of space constants among dendrites can give a relative measure of attenuation, and because the exact, nonexponential, voltage attenuation can be calculated once the space constant is known. For example, Curve 1 in Figure 12–7C shows such a calculation of membrane potential along a short cable whose "length" is one space constant (i.e., the cable ends at the physical length that would correspond to one space constant if the cable were, in fact, infinitely long). By comparing Curve 1 with the dashed line in Figure 12–7C, it can be appreciated that far less attenuation with distance occurs in the short cable than in one that is infinitely long.

Effect of Dendritic Diameter. The second major assumption used to derive Equation 1 is that the diameter of the dendrite does not change with distance. Although most dendrites are, in fact, tapered, the dendrite may be approximated by a series of cable segments, each of which has a constant diameter equal to its average diameter. In this more realistic case, it is possible to calculate the space constant of each segment, and, using a weighted average of these values, to determine an equivalent space constant for the whole dendrite. The length of the dendrite then could be given in terms of its equivalent space constant rather than its physical length; e.g., one could say the dendrite was 2.5 space constants long rather than 1.8 mm long. This designation of electrical length is called *the equivalent electrotonic length* (L). Finally, if the equivalent electrotonic lengths of all a neuron's dendrites are known, one can compute the equivalent electrotonic length of the whole dendritic tree. The utility of reducing the dendritic tree to a single uniform-diameter cable will now be discussed.

Dendritic Trees as Equivalent Cylinders.[23, 35] To evaluate the degree of attenuation of a distal synaptic input it is necessary to determine R_m, λ, or L for the dendrites. Rall[35] developed a mathematical model of a neuron by representing the entire dendritic tree as a single, electrotonically equivalent cylinder (as defined above) attached to an isopotential soma. Although many of its assumptions are not strictly obeyed in real neurons, it has proved useful in many ways. For example, the Rall model allows properties such as the L value of a dendritic tree to be deduced from measurements of the time course of the membrane potential produced by current injected into the neuron's soma.

The Rall model also has been used to predict how the shape and amplitude of postsynaptic potentials recorded at the soma are affected by varying the location of the synaptic input, the dendritic cable properties, and the properties and time course of synaptic conductance changes. Figure 12–5 shows examples of the theoretical shapes of EPSPs recorded in the soma owing to synaptic input at different dendritic loci. Not only are the EPSPs generated by distal inputs (e.g., to compartments 8 and 9) smaller, but they also rise and fall more slowly than EPSPs resulting from more proximal inputs (e.g., to compartments 2 and 3). Rall has shown that it is possible to determine the relative location (as a fraction of L) of a synaptic input on the dendritic tree based on the shape of the EPSP recorded at the soma, and his "shape index" is widely used to estimate the dendritic locus of a synaptic input.[35]

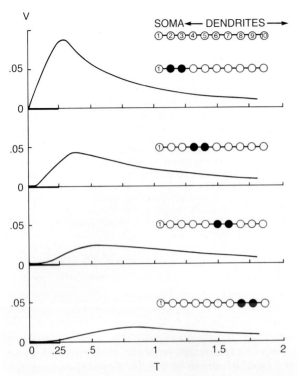

Figure 12–5 Theoretical shapes of EPSPs in the soma, produced by an excitatory increase in membrane conductance at different locations on a dendritic tree. The dendritic tree is idealized as a uniform cable 1.8 space-constants long and divided into 9 isopotential compartments (2–10). Compartment 1 represents the soma. The conductance increase occurs in the two black compartments for the duration indicated by the thick line on the abscissa. Time (T) is plotted as a fraction of the membrane time constant, and depolarization (V) as a fraction of the driving potential for the EPSP. (Adapted from Rall. In Reiss, R., ed. *Neural Theory and Modeling.* Stanford, CA: Stanford University Press, pp. 73–96, 1964.)

Determination of Cable Properties from Anatomy.
An alternative method of determining neuron cable properties is based on the anatomical reconstruction of a recorded neuron.[5, 46] In this method, the input resistance at the soma (R_N) is measured by current injection; the neuron is then stained by intracellular injection of an appropriate substance and recovered later in histological sections to examine its precise anatomical structure. A value for the input resistance of each dendrite can be computed by (1) assuming a value for R_m, (2) approximating the actual dendritic geometry of a series of short cable segments of appropriate length and diameter, and (3) realizing that the input conductance of a distal cable segment is the termination conductance of the next most proximal cable segment. After computing the tentative input resistance for one dendrite, the process is

repeated for all the dendrites and for the soma to obtain the input resistance of the whole neuron. The process is repeated, assuming various values of R_m, until the computed R_N equals the measured R_N. Once the appropriate value of R_m is obtained, the equivalent values of λ and L for each dendrite can be calculated, and the average value of these parameters for the whole neuron can be obtained.

Dendrites Are Electrically Short. Results obtained from both the Rall model and the neuronal reconstruction suggest that the dendritic trees of central mammalian neurons have relatively short equivalent electrotonic lengths. Furthermore, the Rall model and the anatomic reconstruction method yield similar values of L for a given neuron type. By either method, the average value of L is about 1.5 space constants for lumbar motoneurons and about 1 space constant for hippocampal neurons in vitro. Values of R_m derived from an analysis of cable properties are generally greater than 2000 Ω-cm.[2]

Since all neurons have a small value for L, even the most distant synaptic input will produce some effect on the spike trigger zone. The anatomic reconstruction method, although providing a value of L similar to the Rall model, has revealed that real neurons are even more efficient in transferring synaptic current to the soma than a uniform cable of the same electrotonic length. This point is illustrated in Figure 12–6, which is based on the anatomical reconstruction of a lumbar motoneuron.[5, 6] As illustrated in the lower portion of the figure, a real dendrite is more like a tapered cable, with the smallest diameter at the tip of the dendrite. The input resistance of a cable is a function of its diameter. The input resistance at the tip of the dendrite may be 100 times that of the soma, and it decreases with proximity to the soma. Consequently, synaptic current flows toward the soma more easily than it would in a uniform cylinder, and less current is lost through the membrane. This point is seen by comparing the calculated decay of current along an actual dendrite (I_D in Fig. 12–6) with the calculated decay of both current and voltage in a uniform equivalent cylinder of the same electrotonic length (V_C, I_C in Fig. 12–6). About half the current applied at the tip of the actual dendrite reaches the soma (at zero distance in Fig. 12–6), compared with about 20% for a uniform cylinder of the same electrotonic length. This result is significant physiologically, because as discussed in the next section, the strength of the steady, summed, synaptic current reaching the soma provides the drive for repetitive firing. The decrement of voltage along the actual dendrite

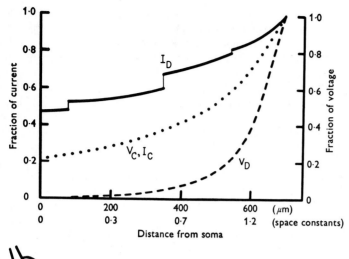

Figure 12–6 Decrement of current and voltage along a motoneuron dendrite that was anatomically reconstructed as described in the text. A constant current applied to its tip results in the indicated variation of normalized current (I_D) and normalized voltage (V_D), with distance along an actual motoneuron dendrite, shown schematically below. Discontinuities in the current plot are caused by loss of current to side branches. Dotted line indicates the identical variation of both current and voltage (V_C, I_C) in a uniform-diameter cable having the same electrotonic length as the dendrite. (Adapted from Barrett, J. N.; Crill, W. E. *J. Physiol. [Lond.]* 293:325–345, 1974.)

(V_D in Fig. 12–6) is greater than that in a uniform cylinder, because the current produces much less of a voltage drop at the base of the dendrite (owing to its low input resistance) than it would in a uniform cylinder. Because of the variation of input resistance along the dendrite of a motoneuron, a synaptic current producing a 20 mV EPSP at the tip of the dendrite produces an EPSP of only about 0.1 mV at the soma.[6]

The cable properties described above apply to situations in which the summed synaptic current is steady or varies slowly with time and those in which background synaptic input to the neuron is minimal. In other situations, the amount of signal attenuation may be quite different. For example, if the synaptic input varies with time, current is lost not only through the membrane resistance but through the membrane capacitance as well. Thus the more rapidly the signal changes with time, the faster it is attenuated with distance.

Effects of Varying R_m. Both the Rall model and the anatomical reconstruction method assume that R_m is constant over the length of the dendrite. In fact, it is more likely that R_m is labile and influenced by a variety of conductance changes. For example, a tonic increase of a transmitter-sensitive conductance along the dendritic tree would tend to decrease the effective R_m, allow more membrane current to be lost through the membrane resistance, and result in a greater decrement of the synaptic potential with distance.[6] The effective R_m may also be altered by activation of voltage-de-

pendent conductances in the dendrites. Depending on the nature of the voltage-dependent conductances, the attenuation of a signal applied to one end of a cable with such nonlinear, "active" conductances can be much worse or much better than in a linear, "passive" cable.[23]

The effect of voltage-dependent dendritic conductances on the efficacy of signal transmission along a dendrite is illustrated by examining two hypothetical cables whose membranes have the current-voltage (I-V) relations shown in Figure 12–7, A and B. The shape of the I-V relation in Figure 12–7A is typical of a membrane with a voltage-gated potassium conductance that increases with depolarization. An I-V curve similar to that of Figure 12–7B is found in lumbar motoneurons.[40] The negative slope is caused by an inward calcium current that opposes the outward potassium current over a significant voltage range, so that membrane conductance actually decreases as the membrane is depolarized from rest (the origin of the I-V curve) toward point 3 in Figure 12–7B.

If the membrane of a cable has the I-V curve of Figure 12–7A, and one end of the cable is depolarized by only a small amount (e.g., to arrow 1), the attenuation of voltage along the cable is shown by Figure 12–7C, Curve 1. If the end of the cable undergoes a large depolarization (e.g., to arrow 2) the attenuation of voltage (Fig. 12–7C, Curve 2) is much greater, since the membrane conductance (the slope of the I-V curve) at the depolarized end

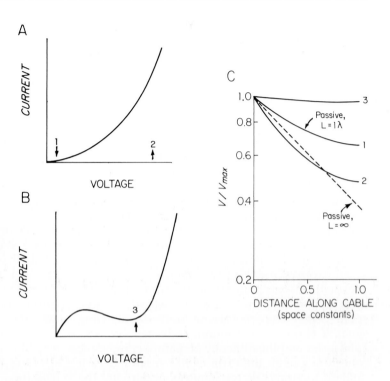

Figure 12–7 Effects of cable length and voltage-dependent conductances on the attenuation of membrane potential with distance along a cable. Curves in *C* show the calculated decrement of normalized membrane potential with distance in four different cables. The dashed line is for a passive cable of infinite length. Curve 1 is for a passive cable 1 space-constant long, the membrane of which has a linear current-voltage relation. Curves 2 and 3 are for two cables 1 space-constant long, membranes of which have the nonlinear current-voltage relations shown in *A* and *B*, respectively. Calculations for the short cables assume that only one end of each cable is polarized and that no current flows out of the other end. (Adapted from Jack, J. J. B., et al. *Electrical Current Flow in Excitable Cells.* Oxford, Oxford University Press, 1975.)

is much greater than during the small depolarization. Consequently, much of the current required to depolarize the end of the cable leaks out of the membrane at and near the point of depolarization, and relatively less current is available to depolarize more distal points. In contrast, if a cable whose membrane has the I-V relation of Figure 12–7B is depolarized to point 3, very little decrement of membrane potential occurs along the cable, as seen in Curve 3 of Figure 12–7C. Better signal transmission is attained because the membrane conductance near point 3 in Figure 12–7B is actually less than during small depolarizations. Thus, more of the current applied to the depolarized end of the cable is available to depolarize the distal membrane than in a "passive" cable whose membrane conductance doesn't change from its value at resting potential. The improved electrotonic transmission resulting from a membrane conductance that decreases with depolarization has been observed experimentally in cerebellar Purkinje cells in vitro. In these cells, electrotonic coupling between soma and dendrites is poor when the cell is hyperpolarized but becomes increasingly better as the cell is depolarized and the effective dendritic membrane conductance decreases. Functionally, this effect aids transmission of dendritic depolarizations (resulting from EPSPs or calcium spikes) to the soma.

In summary, a neuron's cable properties are not fixed quantities. Efficient transmission of potentials along the dendrites depends on the rate of change of the applied potential, the magnitude of background synaptic conductance changes, and the type of voltage-gated channels, if any, in the dendritic membrane. These factors, together with the location of the synaptic inputs on the dendritic tree, determine the shape and amplitude of synaptic potentials reaching the soma and the initial segment. We have seen, however, that even the most distal synaptic input probably has some influence on the membrane potential of the soma. Next, we will discuss how the filtered, weighted, synaptic current that reaches the soma is encoded into repetitive spikes.

REPETITIVE FIRING

Extracellular recordings in alert, behaving animals reveal that many neurons throughout the central nervous system fire action potentials more-or-less continuously. As will become apparent in subsequent chapters, the firing rate varies from a few to hundreds of spikes per second, depending on the type of neuron and how it is activated. Since the spikes themselves are invariant in size, the information the neuron delivers to the end organ, muscle, or other neurons that it innervates is coded as interspike interval or spike frequency. A neuron can be considered a transducer that transforms sustained, graded, synaptic input into an output consisting of repetitive, all-or-none spikes. In this section, we examine the relation between sustained synaptic input and spike frequency.

The Neuron at Rest. When an intracellular recording is made from a deeply anesthetized animal or from an in vitro preparation, the recorded neuron usually fires no action potentials in the absence of an applied stimulus. The membrane potential of most quiescent neurons in the central nervous system ranges between −55 and −80 mV. However, the soma/dendritic region of a central neuron is never at rest in the same sense as an axon or muscle fiber. Even in the deeply anesthetized, intact animal or in an in vitro brain slice preparation, the neurons constantly receive spontaneous hyperpolarizing and depolarizing synaptic input. This continuous barrage of synaptic input and the difficulty of manipulating ionic concentrations in the extracellular space makes it difficult to study the mechanisms underlying the resting potential of central neurons. Such evidence as exists suggests that when the spontaneous synaptic input is minimal the resting potential is determined by mechanisms similar to those found in axons (see Chap. 1), namely, a greater membrane permeability to potassium than to other ions.

The Repetitive Response to Steady Current. Our ideas about the events underlying neuronal activation during natural stimulation in the mammalian central nervous system come mainly from intracellular recordings from cat lumbar motoneurons. As seen in Figure 12–8, the membrane potential of a quiescent motoneuron depolarizes more-or-less smoothly to spike threshold as the intensity of a steady, excitatory stimulus is increased. The slow depolarization is caused by the summation of individual EPSPs from hundreds of afferent fibers. The EPSP recorded in the motoneuron soma by stimulation of only one of these afferent fibers is on the order of 0.1 mV, but the summation of the individual EPSPs from many fibers (each capable of delivering hundreds of EPSPs per second) results in a steady, continuous depolarization in which individual EPSPs cannot be seen. However, the summation of the numerous, small EPSPs produces small, rapid membrane potential fluctuations (called *synaptic noise*) that are superimposed on the steady depolarization (Fig. 12–8).

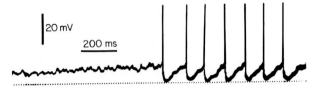

Figure 12–8 Intracellular recording from a lumbar motoneuron showing slow depolarization with superimposed synaptic noise and repetitive firing during natural stimulation. Spikes are truncated. Dotted line indicates resting potential (−65 mV) in the absence of stimulation. (Adapted from Kolmodin, G. M.; Skoglund, C. R. *Acta. Physiol. Scand.* 44:11–54, 1958.)

Once repetitive firing begins in Figure 12–8, the steady synaptic depolarization itself is no longer identifiable. The membrane potential trajectory between repetitive spikes (called the *pacemaker potential*) is the result of the interaction of synaptic current flowing through transmitter-gated channels and ionic currents flowing through voltage-gated channels. When the membrane potential is negative to spike threshold, it is convenient to speak of excitatory or inhibitory synaptic potentials (i.e., EPSPs and IPSPs), but once repetitive firing begins, synaptic *current* is the only possible measure of the strength of the synaptic drive.

If the stimulus intensity were increased in the foregoing example, the synaptic current would increase, the pacemaker potential would become more shallow, and the motoneuron would fire faster. A first step in understanding how a neuron responds to sustained synaptic input is to determine the relation between synaptic current and firing rate. It is difficult, however, to measure the synaptic current directly and to control its amplitude and time course using natural stimulation. Therefore, the effect of sustained synaptic current is mimicked by injecting current through the recording microelectrode.

When a long-lasting, constant current of sufficient magnitude is injected into a lumbar motoneuron, the neuron fires repetitively (Fig. 12–9A).[24, 25] The current strength required to initiate steady firing is slightly greater than that required to produce a single spike. The minimum firing rate that can be sustained varies from about 5 to 20 spikes per second in different motoneurons. A plot of the reciprocal of each interspike interval (the "instantaneous" firing rate) versus time during the current step reveals that the firing rate declines to a finally steady rate (Fig. 12–9B). The decline of firing rate over time is called *adaptation* by analogy with the similar decline of the firing of sensory receptors (see Chap. 5).

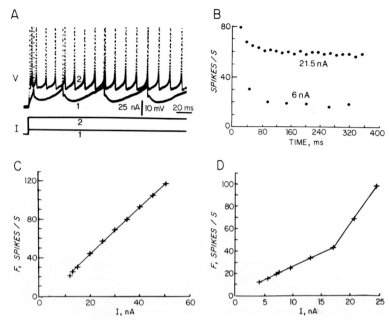

Figure 12-9 Repetitive firing properties of lumbar motoneurons during intracellular injection of constant current steps. *A* (above), two superimposed records of repetitive firing (spikes truncated) in response to a smaller (trace 1) and larger (trace 2) current step (below). *B*, Reciprocal of the interspike interval versus time during the two indicated injected current strengths for the same cell. *C* and *D*, Plots of final, steady firing rate (F) versus injected current (I) in two different motoneurons.

For some motoneurons, the plot of steady firing rate versus injected current (the *f-I relation*) is linear over a substantial range of current strengths (Fig. 12–9C). Thus, the motoneuron in Figure 12–9C behaves like a linear transducer in which steady current is the input and spike rate is the output. On the other hand, many lumbar motoneurons (and other cell types as well) show a steady f-I relation as in Figure 12–9D that is best described by two straight lines.[25] The initial linear portion of such an f-I relation is called the *primary range* of firing, and the steeper segment is called the *secondary range*. The slope of the f-I relation is the gain of the transducer, and it indicates by how many spikes per second the firing rate will increase for a unit increase of steady synaptic current. If firing progresses from the primary to the secondary range, a sudden change in the gain of the neuron's input/output relation occurs. An initial slope of 1 to 3 spikes/s/nA is typical for lumbar motoneurons, whereas the slope of the secondary range can be 3 to 6 times greater.[25] In motoneurons, the secondary range is steeper, but the thalamic neurons (see Fig. 12–15C) or neocortical neurons[43] in vitro exhibit a secondary range that is less steep than their primary range.

Our discussion has focused on lumbar motoneurons because their firing properties have been studied most extensively. Lumbar motoneurons are unusual in comparison with most other mammalian central neurons, however, because they

are capable of slow sustained firing rates, and because the primary range of their f-I relation has a shallow slope. Nevertheless, many different types of central neurons have firing properties that are *qualitatively* similar to lumbar motoneurons. Such similarities include an initial period of adaptation, a substantial linear portion of the f-I relation, a characteristic minimum current strength required to initiate repetitive firing, and characteristic minimum and maximum firing rates in response to injected current. The *quantitative* details of each of these properties, however, vary markedly among the different neuron types. For example, the slopes of the primary range f-I relations in various neurons in the brainstem[3, 19, 20, 45] and the cerebral cortex[26, 43] may be 5 to 20 times those of lumbar motoneurons, and their maximum steady firing rates may approach 1000 spikes/s, rather than only 100 to 200 spikes/s for lumbar motoneurons.

Cells That Discharge Bursts of Action Potentials. Although most central neurons respond to prolonged injected current with repetitive firing patterns that are qualitatively similar to those of motoneurons, some mammalian neurons respond with clusters of closely spaced spikes. These clusters of spikes are commonly called *bursts*. An example of such a burst, produced all-or-none by a brief current pulse in a pyramidal cell from the CA3 region of the hippocampus, is shown in the inset of Figure 12–10. The bursts are known to

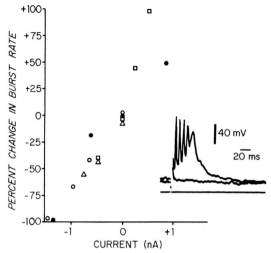

Figure 12–10. The percentage of change in the occurrence rate of successive bursts as a function of steady injected current in four pyramidal neurons (indicated by different symbols) from the CA3 region of the hippocampus (in vitro). Inset shows a single burst, which is evoked all-or-none by a just-threshold, brief, injected current pulse. (Adapted from Hablitz, J. J.; Johnston, D. *Cell. Mol. Neurobiol.* 1:325–334, 1981.)

result from intrinsic membrane properties rather than from underlying synaptic potentials,[22] and it is the average rate of successive bursts, rather than the frequency of individual spikes within a burst, that is a linear function of injected current (Fig. 12–10). Other central mammalian neurons that can discharge in a bursting pattern include Purkinje cells of the cerebellum[29] (see Fig. 12–4), thalamic

neurons[28] under certain conditions (see Fig. 12–15), and certain neurons in the hypothalamus.[2]

The Response to Time-Varying Injected Current. The firing rates of neurons in alert animals is rarely steady for prolonged periods because the underlying synaptic drive changes with time. Intracellular studies using injected current to mimic a varying synaptic drive have shown that the spike encoding process is sensitive to the rate of change of injected current as well as its magnitude.[43]

If a steady injected current is attained by applying a current that increases linearly with time (i.e., a ramp of current), as in Figure 12–11, A and C, the firing rate reaches a higher rate during the current ramp than during the subsequent steady current (Fig. 12–11, B and D). Thus the cell responds to the rate of change of current during a ramp in addition to the final current amplitude. As also seen in Fig. 12–11, B and D, the maximum firing rate is faster for the current ramp with the steeper slope (i.e., the faster rate of change of current). Although firing rate increases monotonically with time during the current ramp, the instantaneous firing rate is a complex function of current amplitude, rate of change of current, and time.

If the rate of change of the injected current (i.e., the slope of the ramp) is kept very slow, the threshold for spike initiation increases, a process called *accommodation*. When a very slowly rising current ramp is applied to an axon, no spikes at all are generated, regardless of the amplitude of the applied current. In contrast, injection of a

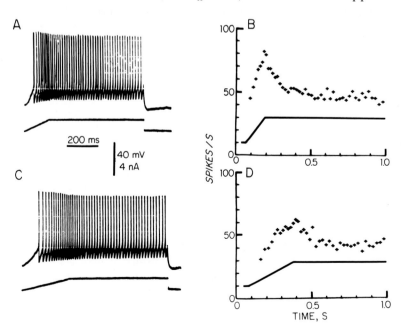

Figure 12–11 Repetitive response of a neuron from layer V of cat sensorimotor cortex (in vitro) to time-varying injected current. *A* and *C* show the repetitive response (above) during injection of ramps of current having different slopes but the same final value (below). *B* and *D* show the time course of instantaneous firing rates during the injected current. (Adapted from Stafstrom, C. E., et al. *J. Neurophysiol.* 52:264–277, 1984.)

current ramp into the soma of a healthy neuron always results in spike generation, no matter how slowly the current rises.[39] However, spike threshold does increase during slowly rising currents, so some accommodation does occur. There is evidence that the accommodation produced by somatic current injection reflects the accommodative properties of the initial segment.[41]

IONIC CONDUCTANCES

The repetitive firing described in the preceding section depends on neuronal cable properties, on synaptic current, and on ionic current flowing through voltage-gated channels. Compared with axons, the soma and dendrites of mammalian neurons have a large variety of voltage-gated membrane channels. At the present time, at least 14 apparently different types of voltage-gated membrane conductances have been distinguished in mammalian neurons based on criteria such as ionic selectivity, voltage- and time-dependent activation and inactivation properties and pharmacological sensitivity to various agents. At least 9 such conductances have been distinguished in hippocampal pyramidal neurons alone. In this section we describe only those conductances which are presently thought to play a major role in repetitive firing in several types of central mammalian neurons.

Action Potential Mechanisms. All available evidence supports the view that the upstroke of the action potential in mammalian central neurons is caused by the same mechanism as in axons, namely, a voltage-dependent increase in sodium conductance. In each of the many types of central neurons that have been studied in vitro, tetrodotoxin (TTX), a specific blocker of the sodium current, I_{Na}, blocks the action potential. However, the I_{Na} measured in the soma of lumbar motoneurons[7, 41] differs in two ways from that of axons. First, I_{Na} is not activated until the soma is depolarized by 25 to 40 mV positive to resting potential. The high threshold for activation of I_{Na} in the soma is consistent with the idea described earlier that the B spike is normally triggered by the low-threshold A spike. Second, the somatic sodium channels exhibit little inactivation at resting potential, and significant sodium inactivation does not occur until the soma is depolarized about 20 mV positive to rest. Because of this property, sustained synaptic depolarization of the soma can produce repetitive firing without spike inactivation.

In addition to I_{Na} inactivation, action potential repolarization is assisted by activation of a fast, outward, potassium current similar to the "delayed rectifier" described in Chapter 4. Although this current has been studied directly only in lumbar motoneurons,[8] a similar current probably exists in the soma of all types of central mammalian neurons. Remarkably, the delayed rectifier is present and large in the motoneuron soma, even though it is absent or small at the node of Ranvier of mammalian axons. Because of its fast kinetics, the delayed rectifier is activated much more fully during a spike than other, slower, potassium currents (see the following section) and therefore plays a major role in spike repolarization. Rapid repolarization is much more critical in the soma than in the axon because the depolarization provided by the A-B spike can activate one or more regenerative, persistent, inward currents (i.e., a calcium current, a persistent sodium current, or both—see later) that tend to maintain depolarization of the soma. If the repolarizing currents are blocked pharmacologically, a prolonged action potential lasting several seconds can be triggered in central neurons. The delayed rectifier aids in preventing this abnormal behavior.

A Calcium-Sensitive Potassium Conductance. Spikes recorded in the soma of mammalian central neurons are followed by a long-lasting afterhyperpolarization (AHP). Figure 12–12 shows that the AHP in hippocampal pyramidal cells may last several seconds. Membrane conductance increases during the AHP, and the AHP decreases in size as the soma is hyperpolarized towards the presumed Nernst equilibrium potential for potassium ions. For these reasons, it is assumed that the AHP reflects a residual increase in a potassium conductance that is activated during the B spike and subsequently decays very slowly. The AHP reduces excitability for an extended period of time, both by repolarizing membrane potential and by increasing membrane conductance. Consequently, the conductance increase underlying the AHP is thought to be the main determinant of the pacemaker potential during repetitive firing, especially at slow rates.[33] The reconstruction of motoneuron firing properties based on voltage-clamp data indicates that a slow potassium conductance is indeed responsible for their AHP, their ability to fire repetitively at slow rates, and the adaptation of firing rate during constant current injection.[8]

The conductance underlying the AHP is thought to be a calcium-sensitive potassium conductance, $G_{K(Ca)}$, of the type discussed in Chapter 4. As is discussed in more detail later, the depolarization caused by the action potential, or even subthres-

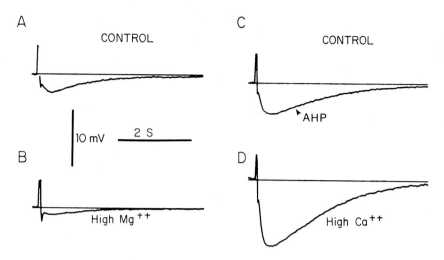

Figure 12–12 The slow afterhyperpolarization (AHP) of a hippocampal pyramidal neuron (in vitro) and its response to manipulations that affect the entry of calcium ions. *A* and *C* show control recordings of the AHP in two different cells at a very slow time base. Bathing the cell with a higher-than-normal magnesium concentration results in a depressed AHP *(B)*, while bathing the cell with a higher-than-normal calcium concentration results in a larger AHP *(D)*. See text for further details. (Adapted from Gustafsson, B.; Wigstrom, H. *Brain Res.* 206:462–468, 1981, with permission from Elsevier Science Publishers BV.)

hold depolarization in some cases, opens voltage-gated calcium channels. The resulting temporary rise in intracellular calcium concentration opens $G_{K (Ca)}$, and it may then be opened still further by depolarization. In mammalian central neurons, the idea that the AHP is caused by $G_{K(Ca)}$ activation is usually based on indirect evidence. For example, the slow AHP of the hippocampal pyramidal neuron in Figure 12–12 is enhanced by increasing the extracellular calcium concentration (a manipulation expected to increase calcium entry during depolarization) and decreased by increasing extracellular magnesium concentration (a manipulation expected to block calcium channels).[21] These results are consistent with the idea that the conductance underlying the AHP is calcium-mediated.

The exceedingly long duration of the AHP of the hippocampal pyramidal cell (see Fig. 12–12) raises the question of how the underlying conductance change is implemented. Most voltage-gated channels have time constants of opening and closing ranging from tenths to tens of milliseconds rather than the several hundred milliseconds required for an AHP that decays over seconds. Studies in molluscan neurons[4] suggest that the slow decay of $G_{K(Ca)}$ (and thus the AHP) results from the slow removal of calcium ions via diffusion and the intracellular uptake of calcium from the inner surface of the membrane rather than from the kinetics of $G_{K(Ca)}$ itself.

The control of AHP duration by intracellular calcium ions allows neuronal firing rates to be modified by several different mechanisms. Intracellular calcium concentration can be influenced by calcium entering through voltage- or transmitter-sensitive channels, by the activity of pumps and other ion-exchange mechanisms (see Chap. 2), by release from intracellular stores, and by the influence of other intracellular second messengers (see Chap. 1) on each of these processes.

The A Current. Another potassium current, called the A current, also has been implicated in the control of repetitive firing. The A current (I_A) as been studied in most detail in the soma of central molluscan neurons,[12, 13] but an analogous current presumably having a similar function has been detected during voltage-clamp studies of many other neurons, including mammalian central neurons.[38] I_A exhibits a transient time course upon depolarization (Fig. 12–13A), and its underlying conductance, G_A, undergoes voltage-dependent inactivation similar to the sodium conductance. The variation of I_A during the first interspike interval in a molluscan neuron is shown in Figure 12–13B, where the ionic currents have been calculated during reconstruction of the firing properties of the cell from voltage-clamp data. I_A is partially inactivated at the relatively low resting potential of the molluscan neurons (about -40 mV). Inactivation is removed, however, during the AHP following a spike (see Fig. 12–13B), and I_A remains activated to some degree during most of the interspike interval. I_A activation tends to repolarize membrane potential; it thus delays the slow depolarization to spike threshold and prolongs the interspike interval. Since I_A is the largest current present during the interspike interval, its activation is critical for the control of low-rate firing in molluscan neurons, even though it is very much smaller than the large inward (I_I) and outward (I_K)

Figure 12–13 The A current (I_A) in a molluscan neuron. *A,* Superimposed records of the transient A current during voltage clamp steps from -100 mV to the potentials indicated. *B,* Variation of both the actual membrane potential (E, above) and that reconstructed from voltage clamp data (arrows) and the ionic current (I, below) during the first interspike interval of repetitive firing evoked by a steady, depolarizing, injected current step. I_i and I_K are the (truncated) inward and outward currents, respectively, which are involved in action-potential generation. See text for further explanation. (Adapted from Connor, J. A.; Stevens, C. F. *J. Physiol. [Lond.]* 213:21–30 and 213:31–53, 1971.)

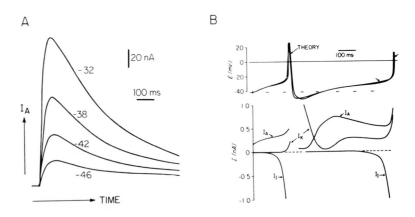

currents associated with the action potential (Fig. 12–13B).

Calcium Conductances. Potassium conductances such as $G_{K(Ca)}$ and G_A provide a basis for repetitive firing because they repolarize membrane potential and retard depolarization for an extended period of time following a spike. In addition to those conductances, however, neuronal firing patterns also may be strongly influenced by an inward calcium current, I_{Ca}, that flows through voltage-gated calcium channels. Although it is likely that the soma/dendritic region of every neuron type contains such channels, the apparent voltage dependence and kinetics of the calcium conductance and the contribution of the calcium current to membrane depolarization varies widely among neuron types. We will consider three types of calcium channels that have been detected in central mammalian neurons and their influence on neuronal firing patterns.

Low-Threshold, Persistent I_{Ca}. When a voltage clamp is applied to lumbar motoneurons,[40] a slow, persistent, calcium current is activated by depolarizations about 10 mV positive to resting potential. Although I_{Ca} is inward, it is superimposed on larger outward ionic currents that are activated at the same potentials. Thus, this low-threshold I_{Ca} is activated in the voltage range traversed by the pacemaker potentials during repetitive firing, and its activation reduces the net outward ionic current existing in this voltage range. This effect may be important in helping to sustain the faster firing rates associated with larger synaptic current, especially those that produce secondary range firing.[41]

High-Threshold I_{Ca}. In contrast with the low-threshold, persistent I_{Ca} of lumbar motoneurons other central mammalian neurons have a high

threshold calcium conductance that is apparent only when the soma is subjected to large depolarizations. In most instances, evidence for the presence of this calcium conductance has been obtained by showing the existence of an all-or-none calcium spike under appropriate conditions rather than by direct observation of I_{Ca} during voltage-clamp. Figure 12–14A shows the large difference in threshold between the sodium-dependent spikes (labeled Na) and a calcium-dependent spike (labeled Ca) evoked by injected current in a neocortical neuron in vitro. The calcium spike was evoked after pharmacological blockade of sodium and potassium conductances in this cell. Figure 12–14A also illustrates the typical finding that calcium spikes are smaller and rise and fall more slowly than sodium-dependent spikes. Extensive evidence suggests that the calcium channels contributing to the high-threshold calcium spike in neocortical and other neurons are located mainly in the dendrites.

The amount of depolarization produced by the high-threshold I_{Ca} varies widely among different types of neurons. For example, the calcium spike cannot be seen at all in the large neocortical neurons unless potassium currents are first suppressed pharmacologically (and I_{Na} is abolished by TTX) since the potassium currents are normally much larger than the calcium current at the large depolarizations where the calcium conductance is activated.[44] Although there is calcium entry, it does not normally result in depolarization. In hippocampal pyramidal cells, however, a calcium spike can be distinguished without suppression of potassium conductances provided that I_{Na} is first abolished by TTX application.[47] In these neurons, calcium entry is postulated to provide the slow underlying depolarization that results in a burst

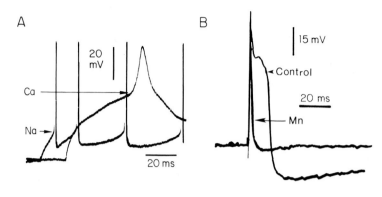

Figure 12–14 Calcium-dependent responses in mammalian central neurons. *A,* Comparison of thresholds for regenerative sodium spikes (Na) and calcium spikes (Ca) evoked by depolarizing current injected in a neocortical neuron (in vitro). The calcium spike could be evoked only after pharmacological blockade of the sodium conductance and depression of potassium conductances. *B,* The action potential (evoked by a brief, injected current pulse) in a neuron from the inferior olive nucleus (in vitro) before (control) and after (Mn) addition of manganese (a calcium channel blocker) to the perfusate. (*A* and *B* are adapted from Stafstrom, C. E., et al. *J. Neurophysiol.* 53:221–226, 1985; and Llinas, R.; Yarom, Y. *J. Physiol. [Lond.]* 315:549–567, 1981, respectively.)

of normal, sodium-dependent spikes[22] (see Fig. 12–10). On the other hand, cells from the inferior olive nucleus of the brain stem have a complex spike consisting of a rapid rise and peak followed by a prolonged depolarization, as seen in the control record of Fig. 12–14B. Application of manganese (to block voltage-gated calcium channels) abolishes the prolonged depolarization, leaving a fast spike (Fig. 12–14B, Mn) which can be abolished by TTX. This and other evidence suggests that the depolarization provided by the fast, sodium-dependent spike activates the high-threshold calcium conductance that in turn is responsible for producing the prolonged depolarization.[31, 32] Thus I_{Ca} is powerful enough in olivary neurons to produce a depolarization that is part of their normal, all-or-none response.

Low-Threshold, Transient I_{Ca}. A third type of calcium conductance actually can change the nature of the neuronal firing pattern.[28, 31, 32] In the thalamic neuron of Figure 12–15A, an injected current pulse produces a subthreshold voltage response when the cell is at rest, and it evokes steady repetitive firing when the soma is depolarized by steady injected current. The repetitive firing, like that of motoneurons, can be characterized by an f-I curve (Fig. 12–15C). However, when the soma is first hyperpolarized by steady injected current, as in Figure 12–15B, the same depolarizing current pulse results in a burst of high-frequency spikes atop a slow, regenerative depolarization. The explanation for this unusual behavior is shown in Fig. 12–15, D-F. In Figure 12–15D the membrane is more polarized, so that application

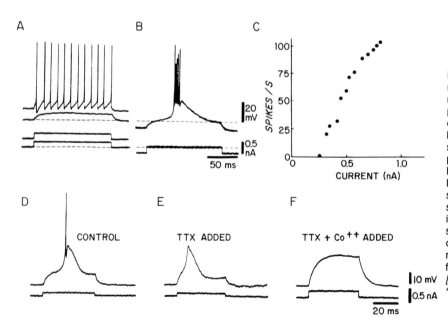

Figure 12–15 Properties of thalamic neurons in vitro. *A,* Voltage responses (above) of a thalamic neuron to the same injected current pulse (below) at resting potential and when the soma is depolarized by a steady injected current. *B,* Response of same cell to same current pulse as in *A* but during maintained hyperpolarization of the soma. *C,* f-I relation of same cell under the stimulating condition of *A. D–F:* Responses in another cell hyperpolarized by steady, injected current and subjected to various blocking agents. See text for further explanation. (Adapted by permission from Llinas, R.; Jahnsen, H. *Nature [Lond.]* 297:406–408. Copyright © 1982 Macmillan Journals Limited.)

of the current pulse produces only a single fast spike superimposed upon a slower, transient depolarization. After the fast spike is abolished by TTX (Fig. 12–15E), only the slow spike remains. Finally, in Figure 12–15F, the application of cobalt (to block voltage-gated calcium channels) abolishes the slow spike, and only the passive response of the cell to the current pulse remains. These and other data suggest that thalamic neurons possess a voltage-dependent calcium conductance that is inactivated at the normal resting potential. Hyperpolarization removes the inactivation, and subsequent depolarization activates the conductance transiently (like I_{Na}), to produce a slow, regenerative calcium spike. The slow depolarization provided by the calcium spike triggers the burst of fast, sodium-dependent action potentials in Fig. 12–15B. Thus, the low-threshold, transient calcium conductance allows these cells to have two possible modes of repetitive firing: a normal, "motoneuronlike" mode and a bursting mode. Which mode a particular synaptic depolarization will evoke depends on the level of membrane potential preceding the synaptic depolarization.

The above examples show how the presence, absence, and combination of different calcium conductances can result in quite different firing patterns among neuron types. The firing patterns are influenced by the amount of depolarization produced by the calcium entry, and as we have seen, this varies greatly among neuron types. Even if it does not produce depolarization, the entry of calcium is still important in neuronal function because it can activate $G_{K(Ca)}$. This latter function is similar to the well-known role of calcium as an intracellular second messenger of hormone action (see Chap. 1). In its role as a second messenger, calcium can regulate a wide variety of cellular functions.

Persistent Sodium Current. Although the low-threshold, persistent calcium current of motoneurons has not been detected in other types of mammalian central neurons, another low-threshold, persistent inward current carried by sodium ions has.[18, 29, 44] Unlike the I_{Na} responsible for the upstroke of the action potential, this persistent sodium current (I_{NaP}) shows no voltage-dependent inactivation. Like the low-threshold, persistent calcium current of motoneurons, I_{NaP} is activated in the voltage range traversed by pacemaker potentials during repetitive firing,[43] but I_{NaP} turns on much faster than the calcium current. Because of its similar voltage dependence, I_{NaP} is thought to act like the calcium current, boosting depolarization in the voltage range of the pacemaker poten-

tials and influencing the slope of the f-I curve. Because of its faster time course and relatively greater magnitude, I_{NaP} can perform these functions more effectively than the slow and relatively weak calcium current of motoneurons.

Summary. Table 12–1 lists the conductances described above and summarizes their putative functions in repetitive firing. The potassium conductances provide the basis of repetitive firing. $G_{K(Ca)}$ provides the very slow, repolarizing, conductance increase required for slow-rate firing and adaptation. G_A provides fine control of the interspike interval duration during slow-rate firing. The low-threshold, persistent I_{Ca} of motoneurons and the I_{NaP} found in other neurons help to maintain firing at high rates. The direct influence of the high-threshold I_{Ca} on firing patterns is variable because of its variable efficacy in producing depolarization in the different types of neurons. Cells with the low-threshold, transient calcium conductance can exhibit both normal and bursting modes of repetitive firing. Calcium entry through any of the calcium channels also activates $G_{K(Ca)}$ and probably regulates other cellular functions as well.

NEUROTRANSMITTER EFFECTS

In the previous section we learned that most of the membrane conductances that control repetitive firing are voltage dependent. The voltage-depend-

Table 12–1 Functions of Major Ionic Currents of Central Mammalian Neurons

Type of Current	Role in Repetitive Firing
Sodium	
Transient	Spike upstroke (normally activated by A spike)
Persistent	Amplifies synaptic depolarization; helps maintain fast firing
Calcium	
Persistent, low threshold	Amplifies synaptic depolarization; helps maintain fast firing
High threshold	Variable: produces zero to large depolarization in different cells
Transient, low threshold	Enables burst firing following prolonged hyperpolarization
Potassium	
Fast, persistent ("Delayed Rectifier")	Rapid spike repolarization
Fast, transient ("A-Current")	Delays rise to threshold during slow-rate firing
Calcium-mediated	Governs interspike interval & adaptation during slow-rate firing

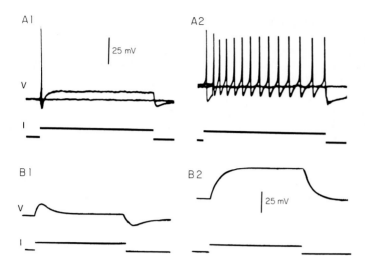

Figure 12–16 Role of the muscarine-sensitive potassium conductance (G_M) in the excitability of a rat sympathetic neuron. *A*, The voltage response (V) to a 1/2-nA, depolarizing, injected current pulse (I) in normal perfusate *(A1)* and after application of a muscarinic agonist *(A2)*. *B*, Time course of membrane potential (V) in the absence of spikes during injection of a 1/2-nA, depolarizing, constant current pulse (I) with G_M present *(B1)* and absent *(B2)*. The variation of membrane potential in *B* was calculated from voltage-clamp data. (Adapted from Brown, D. A.: *Trends Neurosci.* 6:302–307, 1983.)

ent conductances can be affected by neurotransmitters in several ways. In the traditional view, the transmitter-sensitive conductances and the voltage-sensitive conductances that underlie cell excitability are separate entities. In this view, the transmitter-sensitive conductances influence the voltage-sensitive conductances only indirectly, i.e., by altering membrane potential and thereby turning the voltage-sensitive conductances on or off. Although the traditional view is correct in many cases, there is now much evidence that some neurotransmitters also alter the voltage-sensitive conductances themselves,[42] and that some transmitter-gated channels are highly voltage sensitive.

Some of the neurotransmitters that alter voltage-sensitive conductances include acetylcholine, 5-hydroxytryptamine (serotonin), dopamine, norepinephrine, γ-aminobutyric acid (GABA), and substance P. The dominant effect of a particular neurotransmitter may vary from neuron to neuron. For example, serotonin depresses a potassium conductance to enhance excitability in some neurons, and enhances a potassium conductance to depress excitability in others. The effects of several of these transmitters are thought to be mediated through intracellular second messengers (see Chap. 1). Evidence for these more complicated actions of neurotransmitters has been obtained mainly in invertebrate neurons and in the peripheral neurons of vertebrates.

A well-studied example of a voltage-sensitive conductance that is altered by a neurotransmitter is the M conductance, G_M. This conductance, which has been studied most extensively in sympathetic neurons,[1] is a voltage-dependent, persistent potassium conductance that is depressed by acetylcholine and a variety of other muscarinic agonists. In the rat sympathetic neuron of Figure 12–16A1, G_M is already activated at the resting potential of −40 mV. Application of a muscarinic agonist suppresses G_M in Figure 12–16A2. This causes depolarization by closing the M channels; furthermore, a current pulse which initially evoked only a single spike in Figure 12–16A1 now evokes repetitive firing. If action potentials are prevented from occurring, as in Figure 12–16B, it can be seen that suppressing G_M raises excitability in two ways: not only does resting potential become more depolarized (a common effect of excitatory neurotransmitters) but the injected current pulse actually produces a larger depolarization (compare Fig. 12–16B1 and 12–16B2). It is the later effect, equivalent to an increase of the cell's input resistance, that is unusual. In Figure 12–16B1, before G_M blockade, the depolarization produced by the current pulse causes G_M to increase above its value at resting potential, and this additional activation of G_M causes the "sag" of membrane potential towards rest during the pulse and the AHP following the pulse. When G_M is blocked in B2, however, the same current pulse evokes the larger, passive depolarization. Hippocampal and neocortical neurons also receive extensive cholinergic innervation, and muscarinic agonists also depress a slow outward current and markedly increase excitability in these cells.[9]

Conductances like G_M are activated by voltage

and suppressed or augmented following application of a neurotransmitter. In contrast, activation of the "NMDA receptor" opens transmitter-gated channels that are opened further by depolarization.[34] Although this receptor is thought to be activated normally by endogenous excitatory amino acids such as L-glutamate or L-aspartate, it is most sensitive to the artificial excitatory amino acid, N-methyl-D-aspartate (NMDA). When the NMDA receptors are activated, a "new" voltage-gated, depolarizing conductance is suddenly added to the neuron. In neocortical neurons, the addition of these "new," voltage-gated channels by NMDA application can cause these cells suddenly to fire in a rhythmic, bursting mode instead of their usual "motoneuronlike" repetitive firing mode.[16]

The discovery that some neurotransmitters may alter or induce voltage-sensitive conductances adds another dimension to the complexity of neuronal function. These findings suggest that a postsynaptic neuron could have quite different membrane properties after the presynaptic liberation of such a transmitter than existed a few moments before. A related complication is the possible modulation of transmitter release or transmitter action by another neurotransmitter (see Chaps. 11 and 34). Consequently, a whole spectrum of behavior may be obtained from the same neuron, depending on the interaction of neurotransmitters with each other and with voltage-sensitive conductances.

ANNOTATED BIBLIOGRAPHY

Adams, D.J.; Smith, S.J.; Thompson, S.H. Ionic currents in molluscan soma. *Annu. Rev. Neurosci.* 3:141–167, 1980.
A review of the major conductance systems in the molluscan neuron soma. These are similar to the conductances found in mammalian neurons and are understood in more detail.
Dingledine, R., ed. *Brain Slices.* New York, Plenum Press, 1983.
Contains summaries of the properties of the many mammalian central neurons that have been studied in brain slice preparations.
Schwartzkroin, P.A.; Mueller, A.L. Electrophysiology of hippocampal neurons. In Jones, E.G.; Peters, A., eds. *The Cerebral Cortex, vol. 6. Further Aspects of Cortical Function, Including Hippocampus.* pp 295–343. New York, Plenum Press, 1987.
A summary of the extensive literature on hippocampal pyramidal neuron properties, mainly based on in vitro studies.
Schwindt, P.C.; Crill, W.E. Membrane properties of cat spinal motoneurons. In Davidoff, R.A., ed. *Handbook of the Spinal Cord,* vols. 2 & 3, 199–242. New York, M. Dekker, 1984.
A more detailed outline of motoneuron properties and the role of their conductances in repetitive firing.

REFERENCES

1. Adams, P.R.: Brown, D.A.; Constanti, A. Pharmacological inhibition of the M-current. *J. Physiol. (Lond.)* 332:223–262, 1982.
2. Andrew, R.D.; Dudek, F.E. Analysis of intracellularly recorded phasic bursting by mammalian neuroendocrine cells. *J. Neurophysiol.* 51:552–566, 1984.
3. Baker, R.; Precht, W. Electrophysiological properties of trochlear motoneurons as revealed by IVth nerve stimulation. *Exp. Brain Res.* 14:127–157, 1972.
4. Barish, M.E.; Thompson, S.H. Calcium buffering and slow recovery kinetics of calcium-dependent outward current in molluscan neurones. *J. Physiol. (Lond.)* 337:201–219, 1983.
5. Barrett, J.N.; Crill, W.E. Specific membrane properties of cat motoneurones. *J. Physiol. (Lond.)* 293:301–324, 1974a.
6. Barrett, J.N.; Crill, W.E. Influence of dendritic location and membrane properties on the effectiveness of synapses on cat motoneurones. *J. Physiol. (Lond.)* 293:325–345, 1974b.
7. Barrett, J.N.; Crill, W.E. Voltage clamp of cat motoneurone somata: properties of the fast inward current. *J. Physiol. (Lond.)* 304:231–249, 1980.
8. Barrett, E.F.; Barrett, J.N.; Crill, W.E. Voltage sensitive outward currents in cat motoneurones. *J. Physiol. (Lond.)* 304:251–276, 1980.
9. Benardo, L.S.; Prince, D.A. Cholinergic excitation of mammalian hippocampal pyramidal cells. *Brain Res.* 249:315–331, 1982.
10. Brown, D.A.; Slow cholinergic excitation—a mechanism for increasing neuronal excitability. *TINS* 6:302–307, 1983.
11. Catterall, W.A. Localization of sodium channels in cultured neural cells. *J. Neurosci.* 1:777–783, 1981.
12. Connor, J.A.; Stevens, C.F. Voltage clamp studies of a transient outward membrane current in gastropod neural somata. *J. Physiol. (Lond.)* 213:21–30, 1971a.
13. Connor, J.A.; Stevens, C.F. Prediction of repetitive firing behaviour from voltage clamp data on an isolated neurone soma. *J. Physiol. (Lond.)* 213:31–53, 1971b.
14. Coombs, J.S.; Curtis, D.R.; Eccles, J.C. The interpretation of spike potentials of motoneurones. *J. Physiol. (Lond.)* 139:198–231, 1957a.
15. Coombs, J.S.; Curtis, D.R.; Eccles, J.C. The generation of impulses in motoneurones. *J. Physiol. (Lond.)* 139:232–249, 1957b.
16. Flatman, J.A.; Schwindt, P.C.; Crill, W.E. The induction and modification of voltage-sensitive responses in cat neocortical neurons by N-methyl-D-aspartate. *Brain Res.* 363:62–77, 1986.
17. Fourtes, M.G.F.; Frank, K.; Becker, M.D. Steps in the production of motoneuron spikes. *J. Gen. Physiol.* 40:735–752, 1957.
18. French, C.R.; Gage, P.W. A threshold sodium current in pyramidal cells in rat hippocampus. *Neurosci. Lett.* 56:289–293, 1986.
19. Grantyn, R.; Grantyn, A. Morphological and electrophysiological properties of cat abducens motoneurons. *Exp. Brain Res* 31:249–274, 1978.
20. Grantyn R.; Grantyn, A.; Schierwagen, A. Passive membrane properties, afterpotentials and repetitive firing of superior colliculus neurons studied in the anesthetized cat. *Exp. Brain Res.* 50:377–391, 1983.
21. Gustafsson, B.; Wigstrom, H. Evidence for two types of afterhyperpolarization in CA1 pyramidal cells in the hippocampus. *Brain Res.* 206:462–468, 1981.

22. Hablitz, J.J; Johnston, D. Endogenous nature of spontaneous bursting in hippocampal pyramidal neurons. *Cell. Mol. Neurobiol.* 1:325–334, 1981.

23. Jack, J.J.B.; Noble, D.; Tsien, R.W. *Electrical Current Flow in Excitable Cells.* Oxford, Oxford University Press, 1975.

24. Kernell, D. The adaptation and the relation between discharge frequency and current strength of cat lumbosacral motoneurones stimulated by long-lasting injected currents. *Acta Physiol. Scand.* 65:65–73, 1965a.

25. Kernell, D. High-frequency repetitive firing of cat lumbosacral motoneurones stimulated by long-lasting injected currents. *Acta Physiol. Scand.* 65:74–86, 1965b.

26. Koike, H.; Mano, N.; Okada, Y.; Oshima, T. Repetitive impulses generated in fast and slow pyramidal tract cells by intracellularly applied current steps. *Exp. Brain Res.* 11:263–281, 1970.

27. Kolmodin, G.M.; Skoglund, C.R. Slow membrane potential changes accompanying excitation and inhibition in spinal moto- and interneurons in the cat during natural activation. *Acta Physiol. Scand.* 44:11–54, 1958.

28. Llinas, R.; Jahnsen, H. Electrophysiology of mammalian thalamic neurones in vitro. *Nature* 297:406–408, 1982.

29. Llinas, R.; Sugimori, M. Electrophysiological properties of in vitro Purkinje cell somata in mammalian cerebellar slice. *J. Physiol. (Lond.)* 305:171–195, 1980a.

30. Llinas, R.; Sugimori, M. Electrophysiological properties of in vitro Purkinje cell dendrites in mammalian cerebellar slices. *J. Physiol. (Lond.)* 305:197–213, 1980b.

31. Llinas, R.; Yarom, Y. Electrophysiology of mammalian inferior olivary neurones in vitro. Different types of voltage-dependent ionic conductances. *J. Physiol. (Lond.)* 315:549–567, 1981a.

32. Llinas, R.; Yarom, Y. Properties and distribution of ionic conductances generating electroresponsiveness of mammalian inferior olivary neurones in vitro. *J. Physiol. (Lond.)* 315:569–584, 1981b.

33. Madison, D.V.; Nicoll, R.A. Control of the repetitive discharge of rat CA1 pyramidal neurones in vitro. *J. Physiol. (Lond.)* 354:319–331, 1984.

34. Mayer, M.L.; Westbrook, G.L. The actions of N-methyl-D-aspartic acid on mouse neurones in culture. *J. Physiol (Lond.)* 361:65–90, 1985.

35. Rall, W. Core conductor theory and cable properties of neurons. In *Handbk. Physiol.* Sec. 1, 1:877–944, 1977.

36. Rall, W. Theoretical significance of dendritic trees for neuronal input-output relations. In Reiss, R., ed. *Neural Theory and Modeling*, 73–96. Stanford, Stanford University Press, 1964.

37. Ramon y Cajal, S. *Histologie systeme du nerveux de l'homme et des vertébrés*, vol. 1. Paris, Maloine, 1909.

38. Rogawski, M.A. The A-current: how ubiquitous a feature of excitable cells is it? *TINS* 9:214–219, 1985.

39. Schule, W.R.; Richter, D.W.; Mauritz, K.-H.; Nacimiento, A.C. Responses of cat spinal motoneuron somata and axons to linearly rising currents. *J. Neurophysiol.* 37:303–309, 1974.

40. Schwindt, P.C.; Crill, W.E. Properties of a persistent inward current in normal and TEA-injected motoneurons. *J. Neurophysiol.* 43:1700–1724, 1980.

41. Schwindt, P.C.; Crill, W.E. Factors influencing motoneuron rhythmic firing: results from a voltage clamp study. *J. Neurophysiol.* 48:875–890, 1982.

42. Siegelbaum, S.A.; Tsien, R.W. Modulation of voltage-gated ion channels as a mode of transmitter action. *TINS* 6:307–313, 1983.

43. Stafstrom, C.E.; Schwindt, P.C.; Crill, W.E. Repetitive firing in layer V neurons from cat neocortex in vitro. *J. Neurophysiol.* 52:264–277, 1984.

44. Stafstrom, C.E.; Schwindt, P.C.; Chubb, M.C.; Crill, W.E. Properties of persistent sodium conductance and calcium conductance of layer V neurons from cat sensorimotor cortex in vitro. *J. Neurophysiol.* 53:153–170, 1985.

45. Takata, M.; Fujita, S.; Kanamori, N. Repetitive firing in trigeminal mesencephalic tract neurons and trigeminal motoneurons. *J. Neurophysiol.* 47:23–30, 1982.

46. Turner, D.A.; Schwartzkroin, P.A. Steady-state electrotonic analysis of intracellularly stained hippocampal neurons. *J. Neurophysiol.* 44:184–199, 1980.

47. Wong, R.K.S.; Prince, D.A. Participation of calcium spikes during intrinsic burst firing in hippocampal neurons. *Brain Res.* 159:385–390, 1978.

48. Wong, R,K.S.; Prince, D.A.; Basbaum, A.I. Intradendritic recordings from hippocampal neurons. *Proc. Natl. Acad. Sci. USA.* 76:986–990, 1979.

Sensory Processes

Chapter 13

Arthur C. Brown

Introduction to Sensory Mechanisms

INTRODUCTION

Sensation and Perception

All of our knowledge about the external world in which we are immersed and all of our knowledge about the internal world of our own bodies begins with the generation of action potentials at afferent nerve endings. The process through which we become aware of our internal and external environments normally is initiated by the application of some form of physical energy or *stimulus* in the vicinity of an afferent ending (i.e., at the periphery). Conversion of the stimulus into action potentials at the nerve ending is called *transduction*, the general aspects of which are discussed in Chapter 5. Once initiated at primary (or first-order) endings, action potentials are conducted into the central nervous system (CNS) along afferent axons. Upon entering the CNS, they synapse with second-order neurons, which, in turn, excite higher-order neurons. Since the general direction of the conduction of information is cephalad and humans are classified as upright, these pathways are designated as *ascending*. Eventually the ascending excitation reaches the regions of the CNS associated with the conscious appreciation of sensory stimuli, usually considered to be the sensory regions of the cerebral cortex.[1, 9] The elements of this "classical" ascending sensory system, depicted in Figure 13–1, have long been well known. More recently, it has been demonstrated that descending signals also can influence (or modulate) the flow of sensory information at virtually every stage of the ascending pathway. The *sensory system* comprises all of the afferent nerve endings, the afferent pathways, the neurons involved in the modulation of sensory information, and the CNS areas responsible for conscious recognition of applied stimuli.

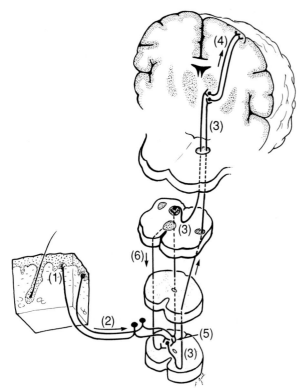

Figure 13–1 Schematic of the elements of a sensory system. (1) The afferent ending to which a stimulus is applied and at which neuronal activity originates; (2) the peripheral axon that conducts action potentials initiated at the afferent ending into the CNS; (3) CNS axons and synapses that conduct sensory information to "higher" regions responsible for conscious sensation; (4) afferents to the cerebral cortex associated with conscious sensation of the stimulus; (5) other peripheral inputs; and (6) CNS descending pathways that modify (modulate) the excitability of the ascending sensory pathway.

The basic recognition of a stimulus is defined as *sensation*. The interpretation of sensations, sometimes called *perception*, involves the appreciation of patterns of sensation. Sensation is the elementary process; perception involves the comparison, discrimination, and integration of several sensations. Being aware that a stimulus is being applied to the skin is a sensation. Recognizing the stimulating agent as a coin is a perception.

The sensation evoked by a particular stimulus can vary in quality and intensity, depending on activity in other afferent pathways and activity originating within the CNS itself. The result might be a selective augmentation of some sensations and an attenuation of others. This control of ascending information is called sensory *modulation*.

The appropriate stimulation of an afferent ending does not always ultimately result in a con-scious sensation. For example, the excitation of afferent receptors responding to blood pressure or to the chemical composition of blood is important in cardiovascular and respiratory regulation, but such activity never reaches the level of consciousness. Thus, the designator "afferent" is broader than "sensory," since the former refers to all information entering the CNS, while the latter is limited to information that is consciously appreciated.

The clinical term for sensation is *esthesia* (from the Greek word for "feeling"). This term is often incorporated into words designating normal and abnormal sensations. For example, "anesthesia" denotes a lack of sensation, "parasthesia" denotes abnormal spontaneous sensation, and "kinesthesia" denotes sensation of movement.

Dimensions of Sensation

When a sensory stimulus intrudes on our mental affairs, we generally can appreciate that it has several aspects: modality, location, intensity, and affect. These elements are the dimensions of the sensory experience.

Modality. The quality or type of a sensation is its modality. Examples of modalities are tactile sensations such as touch, thermal sensations such as warmth and cold, visual sensations, and pain.

Location. Most sensations convey a sense of the location from which the stimulus originated, whether on the body surface, within the body, or in the external environment.

Intensity. Intensity is the perceived strength of the stimulus. Some sensations are characterized by constant intensity as long as the stimulus is applied; others seem to fade away with time, even though the stimulus remains steady. Sensations that spontaneously decrease in intensity are said to *adapt*; sensations whose intensity remains relatively constant as long as the stimulus is applied are *non-adapting*. For example, sitting down initially generates a distinct sensation of pressure originating in the anatomical regions that contact the chair, but the sensation of pressure adapts, or declines, with time. In contrast, toothaches are generally non-adapting.

Affect. Some sensations have an emotional component or *affect*; others are neutral. For example, warmth and sexual sensations are pleasant and have a positive affect. Pain and itch are characterized by a negative affect, since they are inherently unpleasant. Presumably vision is neutral at birth, since the emotional content of visual information is not an inherent property of visual pathways. But visual sensations may acquire affect based on learned associations—for example, the positive affect induced by seeing one's neurophysiology instructor who has consistently delivered stimulating lectures.

Methods of Study

Human Psychophysics. The clinical evaluation of a sensory response typically involves application of a known stimulus, after which the subject is queried about the resulting sensation. By incorporating proper experimental controls, sophisticated instrumentation, and a systematic analysis of the responses, this technique has been elaborated into the scientific study of the quantitative relation between stimulus and sensation, a field called *psychophysics*.[3, 17]

Animal Operant Conditioning. Human psychophysical experiments are useful for describing the input-output relations of sensory processes but can give little insight into the particular role of the many neurons and synapses involved in the elaboration of sensation. Although electrical potentials arising from sensory neurons following stimulation can be recorded in humans using surface electrodes on the head, such *evoked potentials* result from the summed activity of many simultaneously active cells (see Chap. 32 for a discussion of evoked potentials). Thus, it is impossible from external recording to attribute specific functions to particular groups of neurons.

The elucidation of specific function requires invasion, and perhaps disruption, of the nervous system. In civilized societies, studies of such disruptions of function in humans are acceptable only if they are a by-product of automobile collisions, wars, disease, and the like. Although a clever observer can learn much from such calamities, chance events do not lend themselves to a systematic evaluation of function. As in other areas of biology, experiments that may not be done on humans may use experimental animals.

Animals are incapable of reporting their sensory experiences verbally. Therefore, they must be given a "language" by which to express their opinions of the stimuli to which they are subjected. The nonverbal language is generated by shaping the animal's behavior with positive reinforcements (rewards) and/or negative reinforcements (punishments) until it reliably performs a particular task in response to a particular stimulus. This behavior can then be used as a measure of sensation and may be correlated with experimental interventions, such as the destruction of certain sensory areas or the discharge of sensory neurons.[15] Figure 13–2 shows such an operant conditioning apparatus. An experimental animal (in this case, a monkey) is restrained in a chair to direct its attention, constrain it from wandering off, and prevent it from destroying the experimental apparatus. Mechanical stimuli of controlled strength and duration are applied to the immobilized left hand. Positive reinforcement is provided by a food dispenser, which delivers a squirt of orange juice or a banana pellet when the animal makes the correct response.

Invasive Recording and Stimulation. The reporting, in words or behavior, of the sensation produced by an external stimulus cannot reveal the details of the neural mechanisms responsible for mediating the sensation. To understand the detailed neural processes, invasive recording, invasive electrical stimulation, and/or internal application of interesting chemicals must be employed.

Until recently, recording from single neurons was limited to acute experiments in anesthetized animals. Recent advances in experimental surgery and neurophysiological instrumentation now

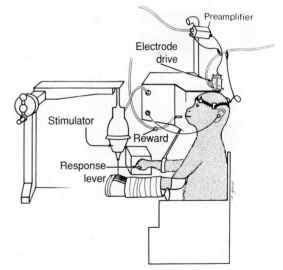

Figure 13–2 An apparatus in which the activity of single neurons may be recorded in a monkey trained to make a somatosensory discrimination. Mechanical stimuli applied to the restrained left hand must be properly discriminated so that an appropriate response can be made by the right hand. The appropriate response is rewarded by a squirt of orange juice through the tube positioned in front of the monkey's mouth. An electrode drive allows the investigator to position a recording electrode by remote control in different brain locations. The recorded electrical activity is processed by a nearby preamplifier.

make possible humane experiments involving direct recording from and stimulation of CNS structures in conscious, unanesthetized animals. Indeed, Figure 13–2 shows such an experiment, in which a recording apparatus has been mounted over a hole in the skull, through which the apparatus, driven by the investigator, can advance a microelectrode into the cerebral cortex. Similar recordings can be obtained from human peripheral nerves by the transcutaneous insertion of fine tungsten microelectrodes, a technique termed *microneurography* (illustrated in Fig. 13–3 and described in Chap. 15), and from superficial CNS neurons during human brain surgery.[7, 19]

Human psychophysics, animal operant conditioning, and invasive recording, lesions, and stimulation have all contributed to the findings de-

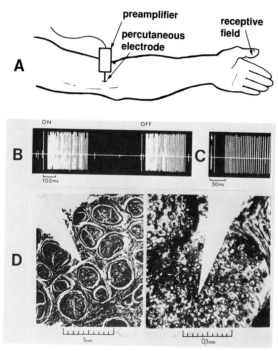

Figure 13–3 Arrangement for recording from and stimulating single peripheral axons in human subjects using fine tungsten electrodes. Shown are electrode placement and axon receptive field *(A)* and responses to mechanical *(B)* and electrical *(C)* stimuli applied to the peripheral receptive field. Photomicrographs of the median nerve with a superimposed silhouette of an electrode are shown in *D*. Fascicles containing densely packed axons are shown in the lower magnification micrographs (left); individual axons can be discerned at higher magnification (right). The electrode is insulated except at its tip (uninsulated length <30 μm). (*A*, *B*, and *C* after Torebjork, H.E.; Ochoa, J.I. *Acta Physiol. Scand.* 110:446–447, 1983; *D* from Hagbarth, K.E.; Hongell, H.; Gallin, R.G.; Torebjork, H.E. *Brain Res.* 24:423–442, 1970.)

scribed in this and subsequent chapters on sensory physiology.

MODALITY

Basis of Modality Discrimination

Two mechanisms are potentially available to the nervous system for conveying information on the modality or quality of a sensation: receptor-pathway specificity and pattern encoding. The first requires that different afferent endings respond to different modalities and that the action potentials from these different endings be processed by separate pathways. The second potential mechanism suggests that different forms of stimulus energies cause an afferent ending to respond with different patterns or sequences of action potentials, which are interpreted by CNS structures as different modalities.

The separate receptor-pathway theory, now called the *specificity theory*, was proposed in 1826 by Johannes Mueller, who termed it the "doctrine of specific nerve energies." In its modern form, this theory states that "the quality of evoked sensation is specific for each type of sensory unit."[12] That is, when the skin is touched with a probe, we are aware of the stimulus pressure because skin distortion activates specific, mechanically sensitive cutaneous afferent endings. If the probe is warm, we identify the thermal quality of the stimulus through the discharge of a different group of thermally sensitive cutaneous endings, which are separate from the mechanoreceptors. Not only do different groups of endings respond to different forms of stimulus energies, but these different sensations reach consciousness via separate ascending pathways. Thus, the specificity theory assumes that modality depends only on which group of afferent endings and associated ascending neurons are activated by the stimulus. Because each axon can convey only a single modality, this theory is also called the *labeled line* theory.

The competing theory of modality discrimination is called the *pattern theory* because it postulates that a single ascending pathway conveys information concerning several sensory modalities by the temporal pattern of action potentials.[8, 14] Thus, a train of uniformly spaced action potentials may signal one modality, whereas a train of action potentials occurring in bursts may signal a different modality (Fig. 13–4).

Modern techniques for recording from single

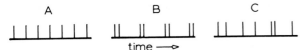

Figure 13–4 Three patterns of action potential discharge, all with the same mean frequency, by which a neuron theoretically could represent different modalities. *A*, Constant frequency; *B*, discharge of pairs of action potentials; *C*, random pattern.

axons in conscious humans and animals allow the simultaneous monitoring of action potential discharge and the sensation produced by the discharge.[18] Such experiments have resulted in overwhelming evidence for the specificity theory (see Chap. 15). However, a few afferent endings do appear to be *multimodal* because, depending on the stimulus, more than one sensation can be evoked by their activation.[20] In general, however, modality depends on which population of afferent endings is activated and the adequate stimulus for that population, and not on the pattern of transmitted action potentials.

Adequate Stimulus

The form of energy to which an afferent ending responds in normal function is called its *adequate stimulus*. In unusual circumstances, however, the ending also may be caused to discharge by other forms of energy.

A corollary of the specificity theory is that the perceived sensation is that of the receptor's adequate stimulus, irrespective of the actual form of energy that initiates the action potential discharge at the ending or along the pathway. For example, the adequate stimulus for the visual system is a photon impinging on a retinal photoreceptor in the eye, and the resulting sensation is visual luminosity. Mechanical pressure on the retina also can activate photoreceptors, but the resulting sensation is visual, not mechanical. The adventurous reader can verify this phenomenon by closing his eyes, looking toward the left (to bring the edge of the retina forward), and then pressing gently with a fingertip on the lid overlying the right corner of the right eye; the resulting sensation will be a semicircle of light in the left visual field, corresponding to the site of maximum mechanical distortion of the retina. Similarly, stimulation of cochlear eighth nerve afferents by drugs (e.g., aspirin) or mechanical irritation produced by a tumor may result in the sensation of ringing in the ears (tinnitus) because the adequate (normal) stimulus for cochlear afferent endings is sound waves.

However, if the drug or tumor stimulates other eighth nerve axons that innervate the vestibular organs of the inner ear, the resulting sensation is turning relative to one's surroundings (i.e., vertigo), because the adequate stimulus of the vestibular organs is linear or angular acceleration of the head (see Chap. 27).

Classification of Afferent Receptors

Sensation and sensory receptors are classified in several ways (Table 13–1). The most common classification is by modality and adequate stimulus: that is, afferent endings that normally respond to mechanical distortion are called mechanoreceptors, endings excited by temperature are called thermoreceptors, chemically sensitive endings are chemoreceptors, and so forth. Often different receptor subpopulations of the same modality respond to different aspects of the stimulus energy spectrum. Thus, thermoreceptors can be divided into warm receptors, which respond maximally at temperatures sensed as warm, and cold receptors, whose maximum response occurs at cool temperatures. The sensations associated with subpopulations are called *submodalities*. For example, the submodalities of the taste receptors—oral chemoreceptors that respond to dissolved substances—are sweet, sour, bitter, and salty (see Chap. 21).

A second classification, proposed by Sherrington in 1906,[13] divides receptors into four groups, based on function. *Exteroceptors* are located in the skin and provide information concerning the immediate environment. *Teleceptors* sense events that originate at some distance from the body; these receptors include the sensory endings of the eyes, ears, and nose. *Interoceptors* respond to stimuli originating in the visceral organs. *Proprioceptors* convey information about the relative positions of the body parts and are located in skeletal muscles, tendons, and joints.

The common clinical classification is derived from the pathway by which the primary afferent axons enter the CNS. Thus, *special* sensory receptors are served by certain cranial nerves; *cutaneous* (superficial) receptors are innervated by superficial branches of spinal and analogous cranial nerves; *deep* receptors, located in muscles, tendons, and joints, receive their axons from deeper branches of spinal nerves and analogous cranial nerves; and *visceral* receptors are served by afferent axons associated with the autonomic nervous system.

Finally, receptors are classified as *somatic* or

Table 13–1 Receptor Classification

Classification by Adequate Stimulus		
Class	*Stimulus*	*Examples*
Mechanoreceptors	Mechanical distortion	Cutaneous touch receptors, pacinian corpuscles, hair cells of the ear
Chemoreceptors	Chemical	Taste receptors, olfactory receptors
Thermal receptors	Temperature	Warm cutaneous receptors, cold cutaneous receptors
Photoreceptors	Photons	Rods and cones of the retina
Classification by Function		
Class	*Definition*	*Examples*
Exteroceptors	Respond to cutaneous stimuli	Cutaneous afferent endings
Proprioceptors	Signal position of body parts	Muscle spindle endings
Interoceptors	Respond to internal body stimuli	Alimentary tract endings
Teleceptors	Respond to stimuli that originate remotely	Rods and cones of the retina
Clinical Classification		
Class	*Location*	*Examples*
Superficial	Skin	Touch and pressure
Special	Head; complex organs served by special cranial nerves	Vision, audition, equilibrium (vestibular), olfaction, taste
Deep	Skeletal muscles, tendons, joints, bone	Proprioception
Visceral	Viscera	Visceral pain
Classification by Embryological Origin		
Class	*Derivation*	*Examples*
Somatic	Somatopleura	Cutaneous and muscle receptors
Visceral	Visceropleura	Alimentary tract receptors

visceral according to whether their embryological origin is the somato- or visceropleura.

LOCALIZATION

Projection

Knowledge of the site to which a stimulus has been applied is termed *topognosis*, or localization. For cutaneous senses, stimulus location is sensed as the site of the stimulated primary afferent ending. In other words, CNS neuronal activity resulting in conscious sensation is mentally "projected" backward, down the ascending pathway, and out to the peripheral site where the stimulus was applied, and that site is identified as the location of the stimulus.[6] In order for the *projection* to the assumed site of the stimulus origin to be accurate, ascending pathways must preserve information about location. That is, ascending pathways must be *topographically* organized to convey an orderly map of the spatial pattern of peripheral stimulation, so that a particular CNS sensory neuron responds only when the afferent endings in a circumscribed peripheral area are stimulated. Occasionally, a given peripheral region may be the exclusive domain of a single primary afferent neuron; more commonly, the receptive domains of adjacent endings overlap (as in Fig. 13–5A), so that stimulation of a particular point on the body surface activates several afferents.

The spatial region throughout which application of a stimulus causes an afferent neuron to respond is defined as that neuron's *receptive field*. The receptive field concept is also applicable to a higher-order sensory neuron, for which the receptive field is that region of the periphery within which application of a suprathreshold, adequate stimulus causes an alteration of the discharge rate. A somatosensory neuron is generally most sensitive to a stimulus applied in the center of its receptive field; the response decreases as the stimulus is moved to the outer edge of the field (see Fig. 13–5B). As the stimulus intensity increases, each receptor responds more to a central stimulus and, because of the greater skin distortion, the stimulus can be moved farther from the center of the receptive field before the receptor ceases to discharge. Consequently, stronger stimuli cause an effective receptive field expansion and an increasing overlap of adjacent receptive fields (Fig. 13–5B). As we shall see in the next section, both of these factors reduce somatosensory acuity.

A

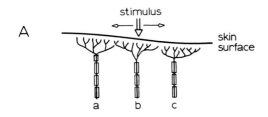

B

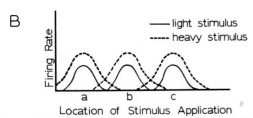

Figure 13–5 A schematic illustration of the distribution of the cutaneous terminations of three adjacent touch receptors, a, b, and c *(A)* and their associated action potential firing rates *(B)*. B shows that increasing the stimulus intensity causes an effective increase in receptive field size (location over which a stimulus elicits a response), leading to an overlap of previously separate but adjacent receptive fields (see text for details).

Acuity

The precision of stimulus localization is called *acuity*. Acuity varies widely between different modalities and over different body regions. For example, sensation of a pin prick on the finger tip is characterized by high acuity, whereas a stomach ache, which is generally felt diffusely over the entire abdominal region, is characterized by low acuity. One measure of acuity is *two-point threshold*, which is determined by applying two similar stimuli at separate sites and then bringing them progressively closer together until they are felt as one. Two-point threshold is defined as the smallest separation at which the two stimuli are still felt as distinct. The two-point threshold for cutaneous tactile sensation varies over the body surface (as shown in Fig. 14–5 in Chap. 14).

Acuity is determined by several neurophysiological factors, including receptive field size, innervation density, central convergence, and lateral inhibition. Clearly, high acuity must be associated with small receptive field size, and low acuity with large receptive field size, as a subject cannot distinguish where, within a receptive field, a stimulus has been applied to excite a given afferent receptor. Also, the more closely the receptors are spaced (called the *innervation density*), the higher the acuity, since a small shift in stimulus location will

alter the population of excited afferent endings when the endings are closely spaced, whereas a greater displacement is required to change the stimulated population for more widely separated endings.

The receptive field of a higher-order central sensory neuron is determined by the summed receptive fields of the lower-order neurons that converge on it. If there is little convergence of lower-order neurons and the receptive fields of the lower-order neurons are themselves small, then acuity is high, and vice versa. For example, visual acuity is maximal for those cells in the primary visual cortex that receive their projections from the foveal (central) region of the retina, since pathways from the fovea are essentially "private lines"; in contrast, visual acuity is much lower in cells with input from the peripheral retina, since peripheral pathways are characterized by considerable convergence (see Chap. 19 for further discussion).

Finally, acuity can be enhanced in higher-order neurons by the mechanism of *lateral inhibition*, sometimes called *surround inhibition*.[10, 11] In some higher-order sensory neurons, application of a stimulus to the center of the receptive field excites the neuron, whereas application of the stimulus near the edge of the receptive field inhibits it. In other words, the excitatory central region is surrounded by a zone of inhibition. Figure 13–6 shows such lateral inhibition in the receptive field of a cell in monkey somatosensory cortex. Tactile stimuli applied to the skin surface labeled "excitatory" (Fig. 13–6A) cause an increase in firing rate, but, as seen in the discharge pattern in Figure 13–6B, stimulation applied 3 seconds later to the "inhibitory" surface causes a complete inhibition of neuronal activity.

Lateral inhibition enhances acuity by increasing the precision of localization, decreasing two-point threshold, and emphasizing edge contrast. Figure 13–7, which depicts hypothetical interconnections between first-order cutaneous mechanoreceptors and second-order neurons, illustrates the mechanism of lateral inhibition; a similar arrangement could exist at every level of ascending synapses. In this figure, stimulation of primary afferent endings A, B, and C leads to excitation of the corresponding second-order neurons a, b, and c, respectively. In addition, the second-order neurons inhibit synaptic transmission in their neighbors with adjacent receptive fields through inhibitory interneurons (shown as solid ovals in Fig. 13–7). Figure 13–7A shows the consequences of this arrangement when a blunt stimulus is applied.

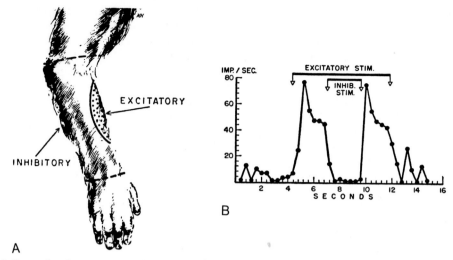

Figure 13–6 *A,* Recording from a neuron in the sensory cortex of a monkey showing experimental evidence for lateral inhibition. The cell was excited (see *B*) by stimuli applied within the receptive field, labeled *excitatory* in *A,* and inhibited (see *B*) by stimuli applied to the large surrounding area, labeled *inhibitory* in *A.* (From Mountcastle, V.B.; Powell, T.P.S. *Johns Hopkins Med. J.* 105:201–232, 1959.)

The blunt stimulus excites all three receptors A, B, and C. In the absence of lateral inhibition, neurons a, b, and c would discharge in accordance with the stimulation (indentation) of their corresponding excitatory primary endings. Thus the second-order neurons would register the blunt probe as a similarly "blunt" spatial profile of firing frequency, shown as a dashed line in the top panel. With lateral inhibition, the firing rates of a, b, and c are reduced, but they are not reduced equally. Since B is excited more than A and C, the interneurons driven by B inhibit a and c more than those driven by A and C inhibit b, leading to the firing frequency spatial profile shown as a solid line. The effect of lateral inhibition, therefore, is a "sharpening" or selective spatial emphasis of

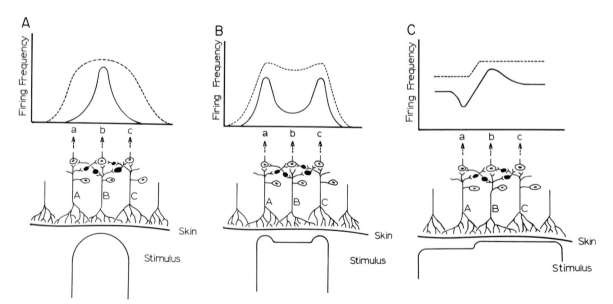

Figure 13–7 Effect of lateral inhibition on acuity *(A),* two-point threshold *(B)* and edge-contrast enhancement *(C).* In each example, the stimulus probe is shown below; three primary afferents (A, B, C) and their connections with three second-order neurons (a, b, c) are schematized in the middle panel; in the top panel, the spatial profile of neural activity in higher-order neurons with (solid line) and without (dashed line) lateral inhibition is indicated (see text for detailed discussion).

the higher-order neurons corresponding to the center of the stimulated region, enabling more precise localization of the site of stimulation.

Lateral inhibition also serves to decrease two-point threshold, as illustrated in Figure 13-7B. To recognize that two stimuli have been applied, the spatial profile of activity representing the stimulated areas must be separated by an unstimulated or less stimulated area. Because of lateral inhibition, the more strongly stimulated afferents A and C, subserving the two "points," combine to inhibit transmission of ascending activity from neuron B, which subserves the region between the points. Therefore, rather than the dashed spatial profile of activity indicated by the curve in the top of panel Figure 13-7B, lateral inhibition produces the solid curve of spatial activity, thereby enhancing the ability to recognize the two points as distinct.

Finally, lateral inhibition leads to contrast or edge enhancement. Consider the situation represented by Figure 13-7C, in which the mechanical stimulus contains a "step" (i.e., a hard-soft interface), which produces a strongly stimulated area adjacent to a weakly stimulated one. Owing to lateral inhibition, the second-order sensory neurons with receptive fields subserving the strongly stimulated cutaneous areas away from the edge will be attenuated by activity from the strongly stimulated afferents with neighboring receptive fields. However, transmission from strongly stimulated afferents at the step interface will be less inhibited because afferents subserving the weakly stimulated adjacent area provide less effective lateral inhibition. Similarly, afferent transmission from the weakly stimulated area near the boundary will be more strongly inhibited than will transmission from the more distant weakly stimulated regions. The result is an increased spatial gradient of activity in the neurons with receptive fields representing the boundary between the two regions (i.e., the solid lines in the spatial activity profile of Fig. 13-7C).

In summary, the receptive field is the unit of localization, with excitation of an afferent pathway being localized by projection to the afferent's receptive field. The ability to discern small changes in location depends on the peripheral factors of receptive field size and innervation density and the central factors of convergence and lateral inhibition.

Errors of Localization

Since localization occurs by projection no matter where the actual excitation originates, errors in localization can and do occur. For example, when axons are stimulated in mid-course by pressure, the victim generally reports sensation arising from the peripheral receptive fields of the axons that were excited. Exciting the ulnar nerve by striking an elbow (the "funny bone") is reported as tingling in the hand; pinching a lumbar dorsal root by displacing or rupturing a vertebral disc leads to pain sensed as originating in the leg or foot. Such "errors of projection" give the neurologist valuable clues concerning the actual site of the nervous system lesion that gives rise to the sensation.

Among the more bizarre localization errors is the "phantom limb" phenomenon. After an extremity has been amputated, the severed central end of the peripheral nerve may be sensitive to external irritation, or postsynaptic neurons along the afferent pathway may begin to discharge spontaneously. In either case, the patient reports sensation in the amputated part—e.g., "My leg hurts" or "My foot is cold"—projecting the sensation to a receptive field with which he or she is no longer associated.

INTENSITY

Coding of Intensity

When an adequate stimulus is applied in the vicinity of a population of afferent endings, some endings will discharge and some will not, depending both on their proximity to the stimulus and their relative sensitivities. As the stimulus strength increases, the less responsive neurons also will begin to discharge, and the originally excited neurons will fire at more rapid rates. These two mechanisms, *recruitment* from a receptor population with various sensitivities and *frequency modulation* of individual receptors, form the basis for sensory intensity discrimination.

If a stimulus is applied to a sentient subject who is asked to report the sensed stimulus intensity, a plot like that of Figure 13-8 typically results. As stimulus strength increases, the subject eventually reaches a sensation *threshold* (i.e., the lowest stimulus intensity that he can sense). Above threshold, the magnitude of the sensation increases monotonically but not linearly with stimulus intensity.

The various sensory modalities differ in their threshold energies, in the range of stimulus intensities they can accommodate, and in their sensitivities to a change in stimulus magnitude. For example, the threshold of the visual system is a single photon of light, but vision can operate over

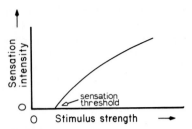

Figure 13–8 General relation of sensation intensity as a function of stimulus strength.

a 10^{10}-fold range of light intensities. In contrast, primitive sensations, such as pain and coarse touch, have high thresholds and can resolve a relatively modest range of stimulus intensities.

Stimulus Strength and Perceived Intensity

Psychophysicists have expended much effort and argument in attempting to characterize the relation between stimulus strength and perceived magnitude for each sensory modality. One of the first of such investigations was conducted in 1834 by Ernst Weber, who tested the ability of human subjects to distinguish differences in weight. The smallest difference that could be detected was called the "just noticeable difference," often abbreviated j.n.d. or ΔS (for change in stimulus strength). Weber concluded that the j.n.d. for weight estimation was not constant, but that it varied in direct proportion to the absolute magnitude of the stimulus being tested. For example, a normal subject could tell the difference between weights of 30 and 31 grams but could not distinguish between more closely spaced weights. Thus at 30 grams, the j.n.d. for weight discrimination was 1 gram. But at 300 grams, a difference of at least 10 grams was required for the subject to notice that the weights were not equal. These results have been expressed as an equation known as *Weber's ratio*:

$$\Delta S/S = c \qquad (1)$$

where c is a constant that characterizes sensory magnitude discrimination for a given modality. For weight discrimination, c averaged 1/30; that is, two weights must differ by at least 3.3% for subjects to determine that they are not the same.

Some years later, Gustav Fechner repeated and reanalyzed Weber's experiments.[2] Fechner assumed that the smallest quantal change in sensa-

tion intensity, ΔI, was directly proportional to Weber's ratio, or

$$\Delta I = k \cdot \Delta S/S \qquad (2)$$

This equation can be integrated to give the relation

$$I = k \cdot \log(S/So) \qquad (3)$$

where I is sensation intensity, S is stimulus strength, So is the threshold stimulus, and k is a constant. Equation 3 is known as the *Weber-Fechner law*.

In the past century, the Weber-Fechner Law has generated considerable debate, and several alternative relations have been proposed. The most widely accepted is the formulation by Stevens,[17] who noted that when the log of sensation intensity was plotted against the log of stimulus magnitude, the resulting relation could be best fit by a straight line. This observation implies a power relationship, which can be expressed as *Stevens' law*:

$$I = k \cdot (S-So)^n \qquad (4)$$

where k and n are constants characteristic of a particular sensory modality. For example, for the estimate of warmth sensation in the experiment illustrated in Figure 13–9, n = 1.5 and k = 0.5.[16]

In any case, whatever the exact equation, the general relation of stimulus magnitude to sensation intensity is clear. The stimulus must be of at least threshold amplitude to be sensed. Above

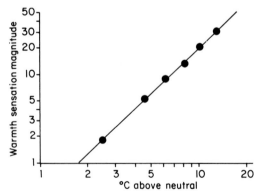

Figure 13–9 Thermal sensation intensity, using a scale of 0 to 50, as a function of skin temperature above 32.5° C ("neutral") plotted with logarithmic scaling on both axes. A straight-line log-log plot represents a power relation, and the slope of the line determines the exponent. The data here can be approximated by the equation $I = 0.5 \cdot T^{1.5}$, where I is the sensation intensity and T is the degrees above neutral. (After Stevens, J.C.; Stevens, S.S. *J. Exper. Psychol.* 60:183–192, 1960.)

threshold, sensation increases with increasing stimulus strength, due both to an increased firing frequency of individual primary afferent endings and to recruitment. As stimulus strength increases, a larger alteration in stimulus magnitude is required before a change can be detected—a relation that has been approximated by both a logarithmic function and a power function.

Adaptation

To complicate matters still further, the relation between stimulus strength and sensation intensity generally changes if a stimulus is applied for an extended period of time. For long-lasting stimuli, sensation reaches a maximum shortly after stimulation begins, and then declines with time, eventually reaching a steady level, as illustrated in Figure 13–10. The component that declines with time is called the phasic or dynamic or *adapting* component, and the phenomenon is termed *adaptation*. The component that remains after the phasic part of the response has run its course is called the *tonic* or static component.

The sense of temperature provides a typical example of these two components. On first entering a cool swimming pool, one has the sensation of intense cold. After a few minutes, the water feels less cold, although in fact neither the water temperature nor the temperature of the skin thermoreceptors has changed. Some thermal sensation remains, since the water still feels cool, but the intensity of the sensation has decreased. The initial sensation intensity is the sum of the phasic and

tonic components; the final intensity is the tonic component alone.

Sensations can be characterized according to how much and how fast they decrease. The first characteristic is measured by the ratio of the phasic to the tonic component. The measure of the second characteristic is the time constant of the phasic component, i.e., the time required for the maximum response to decrease by 63%. Based on these characteristics, sensations generally fall into one of three classes: rapidly adapting, slowly adapting, and non-adapting.

A *rapidly adapting* sensation has a large phasic component, fades away rapidly (short time constant), and has little, if any, tonic component. A *slowly adapting* sensation has a longer time constant and both phasic and tonic components. A *nonadapting* sensation has only a tonic component. An example of a rapidly adapting sensation is light touch applied to the skin. More intense or coarse pressure is characterized by slow adaptation. The sensation that most closely approximates the non-adapting pattern is pain.

An associated characteristic of adapting sensations is the rise in sensation threshold produced by a maintained stimulus. For example, if a stimulus for a rapidly adapting sensation is applied sufficiently slowly, it will never be felt, since the threshold will rise faster than the stimulus intensity. In other words, sensations characterized by rapid adaptation are evoked only by stimulus change and may be viewed to result from stimulus velocity rather than stimulus magnitude. Rapidly adapting sensory pathways, therefore, function as high-pass filters, emphasizing the changing aspects of the external or internal environment. On the other hand, slowly adapting and non-adapting pathways provide information concerning maintained or slowly changing stimuli.

The property of adaptation reflects, in part, the characteristics of individual afferent endings. For example, Figure 13–11 demonstrates that muscle spindle endings have both a phasic and tonic component, and that the phasic component decreases as the muscle is stretched more slowly.[5] Different populations of primary afferents differ in their phasic and tonic responses and their rates of adaptation. These differences are catalogued in Table 13–2. However, the time course and the phasic/tonic ratio of the adaptation of sensation often differ from the time course and relative size of the phasic and tonic components of the discharge of the associated primary afferent endings. Thus, central structures must also contribute to adaptation.

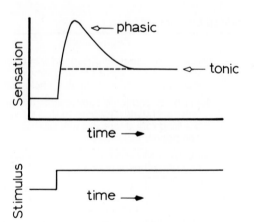

Figure 13–10 The general time course of the intensity of sensation following application of a step change in the stimulus, showing both a phasic adapting response and a tonic nonadapting response.

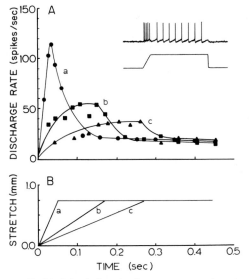

Figure 13–11 Adaptation in the response of a muscle spindle primary (annulospiral) ending. Inset shows an actual phasic-tonic discharge pattern produced by stretching the muscle at a constant velocity; *A*, shows the discharge patterns produced by the three different velocities of stretch in B to show that the phasic component decreases as the stretch becomes slower. (After Hunt, C.C.; Ottoson, D. *J. Physiol.* 252:259–281, 1975.)

Modulation

Sensation intensity depends not only on stimulus strength but also on the ease with which sensory information is passed from neuron to neuron across the synapses of ascending pathways. A change in the efficiency of this ascending transmission is called sensory *modulation*. Synaptic

excitability can be influenced by collaterals from peripheral neurons associated with other sensory modalities (*peripheral modulation*) and by neurons descending from higher CNS structures (*central* or *descending modulation*). An example of peripheral modulation is the reduction of painful stimuli produced by rubbing the uncomfortable area with your hand to activate tactile afferents. An example of central modulation is the suppression of pain produced in a fakir lying on a bed of nails.

AFFECT

The general meaning of the term *affect* when used as a noun (and pronounced with the accent on the first syllable) is "mood" or "feeling." In the context of sensory physiology, affect refers to the pleasantness or unpleasantness of a sensation. For example, warmth and satiety have a pleasant or positive affect, whereas cold and pain have an unpleasant or negative affect. Some sensations do not influence mood and so have a neutral affect.

Affect is a distinct dimension of sensation rather than a vague concomitant of modality and intensity. This is demonstrated by two lines of evidence. First, affect can be selectively diminished or enhanced without affecting the other dimensions of sensation. For example, certain drugs (e.g., morphine) and certain lesions of the CNS can reduce the unpleasantness associated with noxious stimuli without attenuating other aspects of the pain response; the subject still recognizes the pain and can identify its location and intensity but it doesn't "hurt." Other drugs can selectively increase affect.

Table 13–2 Classification of Afferent Endings by Adaptation

Type	Name	Characteristics
RA	Rapidly adapting	Phasic component only; respond with a single spike or burst of spikes to step stimulus; rapidly cease firing if stimulus is held constant
SA	Slowly adapting	Both phasic and tonic components; respond with continuous discharge to step stimulus, but firing rate decreases with time as phasic component fades away
SAI	Slowly adapting Type I	Subclass of SA receptors for which the phasic component is large compared to the tonic component
SAII	Slowly adapting Type II	Subclass of SA receptors for which the tonic component is large compared to the phasic component
NA	Non-adapting	Tonic component only; firing rate determined only by stimulus magnitude, not stimulus rate of change

The second type of evidence is based on self-stimulation experiments, in which an animal controls an electrical stimulator attached to electrodes implanted in specific regions of its own CNS. When the electrodes are located in certain sites in the hypothalamus and limbic system, the animal will work to activate the stimulator just as it would work for food or other rewards. In fact, given the choice between food and electrical stimulation, the animal often prefers the latter. Since these brain structures are not otherwise associated with conscious sensation, it is concluded that the stimulated sites are the "pleasure centers" of the brain—that is, the centers responsible for the positive affect of sensation. Electrodes at other sites in the hypothalamus and limbic system cause the animal to work to avoid stimulation and are thus identified as the centers responsible for negative affect (see Chap. 32).

SUMMARY

A sensation, which is the conscious realization that a stimulus has been applied, conveys four distinct types of information, each of which has a specific neurophysiological basis. The quality or modality of the sensation is determined by which afferent endings are excited and which ascending pathways are activated. Location is based on projection from the anatomical mapping of peripheral receptors on central neural structures. Intensity depends on the firing frequency of individual receptors and recruitment from a population of receptors; intensity can be modified by adaptation and modulation. The basis for affect is much more obscure but may involve ascending collaterals that excite specific "affect centers" that are separate from, but influenced by, traditional sensory pathways.

REFERENCES

1. Adrian, E.D. *The Basis of Sensation.* London, Christophers, 1928.
2. Fechner, G. In Howes, D.H. and Boring, M.G., eds. *Elements of Psychophysics.* Vol. 1. H.E. Adler, trans. New York, Holt, Rinehart & Winston, 1966.
3. Galanter, E. Detection and discrimination of environmental change. *Handbk. Physiol.,* Sec. 1, 3:103–121, 1984.
4. Hagbarth, K.E.; Hongell, H.; Gallin, R.G.; Torebjork, H.E. Afferent impulses in median nerve fascicles evoked by tactile stimuli of the human hand. *Brain Res.* 24:423–442, 1970.
5. Hunt, C.C.; Ottoson, D. Impulse activity and receptor potential of primary and secondary endings of isolated mammalian muscle spindles. *J. Physiol.* 252:259–281, 1975.
6. Knibestöl, M.; Vallbo, A.B. Single unit analysis of mechanoreceptor activity from human glabrous skin. *Acta Physiol. Scand.* 80:178–195, 1970.
7. Johansson, R.S.; Vallbo, A.B. Tactile sensory coding in the glabrous skin of the human hand. *Trends Neurosci.* 6:27–32, 1983.
8. Melzack, R.; Wall, P.D. On the nature of cutaneous sensory mechanisms. *Brain* 85:331–356, 1962.
9. Mountcastle, V.B. Neural mechanisms in somesthesis: Recent progress and future problems. In von Euler, C.; Franzen, O.; Lindblom, U.; Otteson, D., eds., *Somatosensory Mechanisms.* New York, Plenum Press, 1984.
10. Mountcastle, V.B.; Powell, T.P.S. Neural mechanisms subserving cutaneous sensibility, with special reference to the role of afferent inhibition in sensory perception and discrimination. *Johns Hopkins Med. J.* 105:201–232, 1959.
11. Ratliff, F.; Hartline, H.K. The response of limulus optic nerve fibers to patterns of illumination on the receptor mosaic. *J. Gen. Physiol.* 42:1241–1255, 1959.
12. Schady, W.J.L.; Torebjork, H.E.; Ochoa, J.L. Peripheral projections of nerve fibers in the human median nerve. *Brain Res.* 277:249–261, 1983.
13. Sherrington, C.S. *The Integrative Function of the Nervous System.* New Haven, Yale University Press, 1906.
14. Sinclair, D.C. Cutaneous sensation and the doctrine of specific energy. *Brain* 78:584–614, 1955.
15. Stebbens, W.C.; Brown, C.H.; Peterson, M.R. Sensory function in animals. In *Handbk. Physiol.,* Sec. 1, 3:123–148, 1984.
16. Stevens J.C.; Stevens, S.S. Warmth and cold: dynamics of sensory intensity. *J. Exper. Psychol.* 60:183–192, 1960.
17. Stevens, S.S. Sensory power functions and neural events. In Lowenstein, W.R., ed. *Principles of Receptor Physiology,* 226–242. Berlin, Springer-Verlag, 1971.
18. Torebjork, H.E.; Ochoa, J.L. Specific sensations evoked by activity in single identified sensory units in man. *Acta Physiol. Scand.* 110:445–447, 1980.
19. Vallbo, A.B.; Hagbarth, K.E. Activity from mechanoreceptors recorded percutaneously in awake human subjects. *Exper. Neurol.* 21:270–289, 1968.
20. Wall, P.D.; McMahon, S.B. Microneurography and its relation to perceived sensation. A critical review. *Pain* 21:209–229, 1985.

Somatic Sensation: Peripheral Aspects

INTRODUCTION

Sensations are broadly classified as *special, visceral,* or *somatic*. Special senses arise from complex cephalic organs innervated by special cranial sensory nerves. These senses include vision, audition, olfaction, taste, and the vestibular senses. Visceral sensation originates in the viscera of the thorax and abdomen, such as the lungs, alimentary tract, and bladder. Somatic sensation, also called *somesthesis*, arises from afferent endings on the body surface and in the body wall, muscles, tendons, bones, joints, and connective tissue. Somatic sensory receptors have the following general characteristics.[12]

1. Somatosensory endings are less complex than those of special sensory organs and are widely distributed throughout the body.

2. Somatosensory endings exhibit morphological differentiation and functional specificity. In some cases, endings are characteristically associated with non-neural tissue (i.e., the accessory structure; see Chap. 5), which may participate in the transduction process. For the majority of receptors, however, the nerve ending itself is the transducer.

3. Somatic sensibility is conveyed into the central nervous system (CNS) by a wide variety of peripheral axons, ranging from the largest myelinated peripheral fibers to the small unmyelinated C fibers. However, a specific type of ending is innervated by a specific type of axon.

4. Afferent endings innervated by myelinated axons are not themselves myelinated. The myelin sheath is lost before the axon terminates in the ending.

5. Axons enter the CNS via spinal nerve roots or as general somatic afferent fibers in cranial nerves.

In this chapter we consider the somatic sensations, particularly mechanical sensation (touch, pressure, vibration), thermal sensation, and proprioception and the peripheral endings from which they arise. Visceral sensation is assumed to have many of the characteristics of somesthesis and therefore is treated in this chapter also. Central somatosensory processing is covered in Chapter 15. The senses of pain and itch are discussed separately in Chapter 16.

SOMATIC MECHANORECEPTORS

Somatosensory information originates in a variety of mechanoreceptors. All share the defining characteristic that mechanical distortion causes an alteration of their firing frequency, but otherwise they differ in morphology, distribution, density, sensitivity, adaptation rate, and (presumably) sensory submodality (see Chap. 13 for definition of submodality).[12]

Merkel disks and their associated Merkel cells in the epidermis of both glabrous (hairless) and hairy skin are the most superficial mammalian mechanoreceptors. The disks are formed by the flattened terminations of primary afferent axons that lose their myelin sheaths before entering the epidermis. The diameters and conduction velocities of their axons lie in the Aβ range (Table 14–1).

The function of the associated Merkel cells is unclear. Because they make synapselike junctions with the disk, Merkel cells are sometimes assigned the role of primary electromechanical transduction, a speculation that awaits experimental verification. The entire Merkel apparatus functions as a slowly adapting mechanoreceptor of the SA-I type[13] (see Chap. 13 for the classification of receptors according to their rates of adaptation).

In the cutaneous dermal layer are several types of encapsulated mechanoreceptors: Meissner corpuscles, Krause end-bulbs, Ruffini endings, Golgi corpuscles, and pacinian corpuscles (Fig. 14–1).[34] Meissner corpuscles are relatively large receptors (150 μm $\times$ 50 μm) that are located in dermal ridges of glabrous skin. They are characterized by rapid adaptation and therefore are classified as rapidly adapting (RA) receptors. Krause cylindrical end-bulbs lie in the dermis, often close to the epidermis, and are widely distributed throughout the skin. Although designated by von Frey in 1895

Table 14–1 Afferent Axon Classification*

Cutaneous Axons	Muscle/Tendon Axons	Axon Diameter (μm)	Conduction Velocity (m/s)
Aα	Group I	13–20	80–120
Aβ	Group II	6–12	35–75
Aδ	Group III	1–5	5–30
C	Gourp IV	0.2–1.5	0.5–2

*Myelinated axons include Aα, Aβ, Aδ, and Groups I, II, and III. Axons of C fibers and Group IV neurons are unmyelinated.

as cold receptors, a classification promulgated for many years in textbooks, these endings are now considered to be RA mechanoreceptors.[14]

Ruffini, Golgi, and pacinian endings represent a range of laminated endings with a progressively increasing number of lamellae. Ruffini endings are the smallest and have a distinct outer capsule of 3 to 5 layers and are located in both glabrous and hairy skin. These receptors function as slowly adapting (SA) type II (SA II) mechanoreceptors. Golgi (or Golgi-Mazzoni) corpuscles were first identified in tendon and muscle. Their endings are encapsulated by 10 to 15 lamellae and are 150 to 250 μm in diameter. Their function remains to be established, but they are assumed to be RA mechanoreceptors. The largest mechanoreceptor is the pacinian corpuscle. It occurs not only in skin but also in the mesentery, where it is readily isolated and studied. The pacinian corpuscle, de-

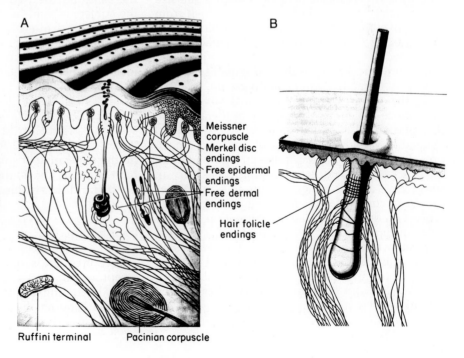

Figure 14–1 *A,* Schematic illustration of the distribution of cutaneous mechanoreceptors in glabrous skin. *B,* Hair follicle receptors. Note that some endings encircle while others lie parallel to the follicle. (After Williams, P. L.; Warwick, R., eds. *Gray's Anatomy,* 36th British Ed. London, Churchill Livingstone, 1980.)

Meissner corpuscle
Merkel disc endings
Free epidermal endings
Free dermal endings
Hair follicle endings

Ruffini terminal   Pacinian corpuscle

scribed in Chap. 5, has 20 to 70 lamellae and its dimensions are 0.5 to 2 mm long by 0.7 mm wide in diameter. The pacinian corpuscle functions as a very rapidly adapting mechanoreceptor that is capable of following high frequencies of oscillating stimulation.[10]

Afferent endings are often present adjacent to hair follicles. Pressure on the hair causes it to bend; the resulting torque is transmitted to the follicle, which distorts and stimulates the associated ending. There are three types of hair follicles which are innervated by morphologically distinct afferent endings.

Simple hair follicles (follicles without erectile tissue) are surrounded by an array of unmyelinated nerve terminals derived from myelinated axons. Functionally, these comprise several subclasses of RA endings.[2, 33] The second type of hair follicle, the nonsinus facial hair, is associated with spraylike terminals, resembling glabrous Ruffini endings both in morphology and in electrophysiological characteristics (i.e., SA II). The third type, the sinus hair follicle, has a hair of large diameter, a vascular sinus, associated erectile muscle, and a rich nerve supply. In various regions of the body, these hairs are known as whiskers, vibrissae, or tactile hairs. They are invested with several types of endings, some of which are rapidly adapting whereas others adapt only slowly.[6, 32]

Some mechanoreceptors seem to be innervated by unmyelinated C fibers. They are difficult to identify and investigate because of their small size and because they are difficult to distinguish from the unmyelinated terminations of myelinated afferent axons. Their characteristics remain to be elucidated.

In sum, mechanosensation originates from a variety of somatic afferent endings, an arrangement that affords several functional advantages. First, since the ending types vary in the aspect of the mechanical stimulus to which they respond (e.g., hair bending, superficial pressure, deep pressure), discrimination between submodalities is enhanced. Second, since ending types vary in sensitivity, a wide range of stimulus intensities can be accommodated. Third, since the ending types vary in adaption rate, information is available concerning the time course of the stimulus. In other words, a mechanical stimulus causes an encoded afferent barrage from the several receptor populations in which much of the work of sensory classification and discrimination has already been accomplished (Table 14–2).

TOUCH-PRESSURE

Adequate Stimulus and Sensitivity

The adequate stimulus for mechanical somesthesia is physical distortion of the afferent terminal. The stimulus strength may be measured in units of length or angle (e.g., indentation of the skin in millimeters, bending of a hair in degrees), in units of force necessary to achieve distortion (e.g., grams), or in units of pressure applied to the tissue containing the afferent ending (e.g., grams per square millimeter).

The most sensitive somatic mechanoreceptors have a remarkably low threshold, enabling humans to recognize stimuli that produce only 10 μm of distortion or 5 mg of force. For example, we can discern the bending of a single hair. Rapidly adapting receptors tend to have the lowest thresholds; thus a brief or oscillating mechanical stimulus, such as a vibrating object in contact with the skin, is the stimulus most effective in evoking sensation.

Sensitivity is not uniform over the body surface but varies with the number of afferent endings per unit area (called *innervation density*) and the types of endings present. The lowest thresholds are found on the face and hands; the back, legs, and other more sparsely innervated areas have thresholds that are an order of magnitude higher (Fig. 14–2).[35] Mechanical sensitivity is more or less the same on the two sides of the body whether the subject is right- or left-handed.

The sensed intensity of a mechanical stimulus is a linear function of stimulus strength (Fig. 14–3A),[22] a relation that presumably reflects the stimulus-response characteristics of the primary afferent receptors (Fig. 14–3B) as determined from animal studies.[27, 28] Intensity discrimination, as measured by the just noticeable difference (j.n.d.; see Chap. 13 for further discussion of the j.n.d.), varies from about 3% to 10% of stimulus strength (in the mid-range of stimulus strength); at low and high stimulus levels, discrimination is poorer; that is, the j.n.d. is greater.

Localization

Receptive fields of primary afferent mechanoreceptors vary in size from a few mm² to several cm², depending on the type of ending and its location. The smallest receptive fields are found

Table 14–2 Somatic Sensory Endings

Structure	Afferent Fibers	Adequate Stimulus	Adaptation	Associated Sensation
Mechanosensation				
Merkel cell	Aβ	Skin distortion	Slow	Touch-pressure
Meissner corpuscle	Aβ	Vibration	Rapid	Flutter, contact
Ruffini ending	Aβ	Skin distortion	Slow	Touch
Pacinian corpuscle	Aβ	Vibration	Very rapid	High-frequency vibration
Hair follicle	Aβ	Hair movement	Rapid	Contact, touch
Free ending	Aδ	Distortion	Rapid	Contact (coarse)
Thermal Sensation				
Free ending	Aδ, C	15–30°C	Intermediate	Cold
Free ending	C	30–42°C	Intermediate	Warm
Proprioception				
Muscle spindle—annulospiral	Group I	Spindle stretch	Slow	Proprioception
Muscle spindle—flower spray	Group II	Spindle stretch	Slow	?
Golgi tendon ending	Group I	Tendon tension	Slow	Muscle force (?)
Joint receptor	Group II, IV	Joint movement and pressure	Slow	Proprioception
Pain				
Free ending	Aδ	Noxious	Slow	Pricking pain
Free ending	C	Noxious	Slow	Burning pain
Itch				
Free ending	C	Pruritogenic	Slow	Itch

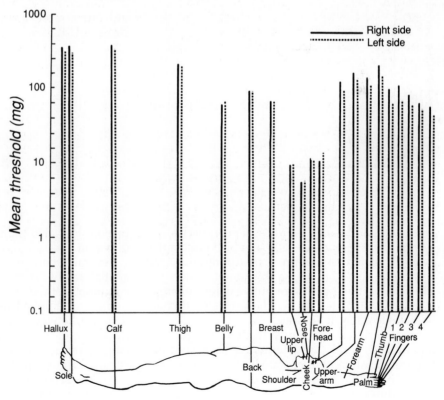

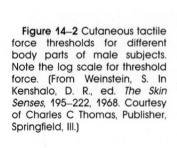

Figure 14–2 Cutaneous tactile force thresholds for different body parts of male subjects. Note the log scale for threshold force. (From Weinstein, S. In Kenshalo, D. R., ed. *The Skin Senses,* 195–222, 1968. Courtesy of Charles C Thomas, Publisher, Springfield, Ill.)

A

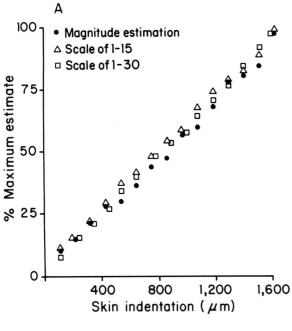

B

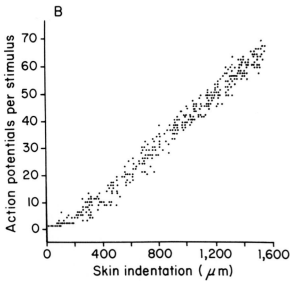

Figure 14–3 Psychophysical and neural responses to step indentations 0.6 to 0.9 s in duration, delivered at intervals of 8 to 12 s, using a rounded probe with a 2-mm diameter tip. *A,* Estimation of stimulus magnitude by human subjects. Three different rating scales were used. (1) Scale 1–15: subjects reported magnitude as a number between 1 and 15. (2) Scale 1–30: subjects reported magnitude as a number between 1 and 30. (3) Magnitude estimation: subjects used numbers of their own choice, based on the ratio of one stimulus to another. *B,* Number of action potentials per stimulus recorded from a myelinated axon innervating the glabrous skin of the monkey hand. (*A* from LaMotte, R. H. In Norman, R. S.; Hollies, R. S.; Goldman, R. F., eds. *Clothing Comfort: Interaction of Thermal, Ventilation, Construction and Assessment Factors,* pp. 83–105. Ann Arbor, Mich., Ann Arbor Science Pub., 1977. *B* from Mountcastle, V. B.; Talbot, W. H.; Kornhuber, H. H. In de Reuck, A.V.S.; Knight, B. A., eds. *Ciba Foundation Symposium on Touch, Heat and Pain,* 325–351. Boston, Little, Brown & Co., 1966.)

on the face and hands for Meissner corpuscles and Merkel disks. Generally, all parts of the receptive field of an afferent neuron are contiguous, but because of axon branching this is not always the case.[21] Also, neighboring receptive fields generally overlap, so that no area is the exclusive domain of a single afferent ending.

Examples of the receptive fields of receptors in the human hand, as determined by recordings from single afferent fibers, are illustrated in Figure 14–4.[16] Note that elicitation of afferent activity from the Ruffini corpuscles required not only skin dis-

tortion but also skin stretching in a particular direction.

Several techniques have been used to assess spatial acuity for mechanosensation. In the most common method, two sharp stimuli (e.g., caliper points) are applied repeatedly to the skin as the distance between the stimuli is progressively reduced, until the subject reports the sensation as a single point. The *two-point threshold* is the smallest separation for which the two stimuli are still perceived as separate. Figure 14–5 maps the regional variation in two-point threshold over the body

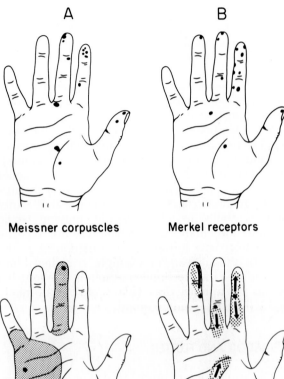

A
B

Meissner corpuscles   Merkel receptors

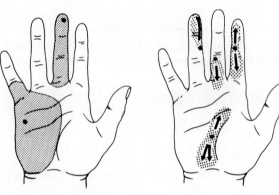

Figure 14–4 Receptive fields of several afferent receptors of the human hand. *A,* Rapidly adapting (RA) mechanoreceptors: Meissner corpuscles and pacinian corpuscles. For the pacinian corpuscles, the shaded area represents the total receptive field, and maximum sensitivity is denoted by the dark circle. *B,* Slowly adapting (SA) receptors: Merkel receptors and Ruffini corpuscles. The arrows on the latter indicate the direction the skin must be stretched to cause discharge. (From Johanson, R. S.; Vallbo, A. B. *Trends Neurosci.* 6:27–32, 1983.)

Pacinian corpuscles   Ruffini corpuscles

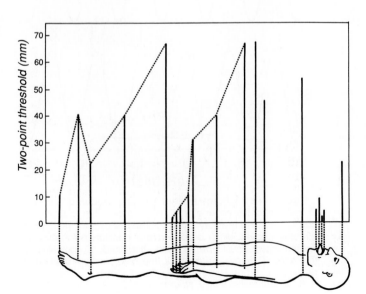

Figure 14–5 Distribution of two-point thresholds over the body surface. (From Patton, H. D.; Sundsten, J. W.; Crill, W. E.; Swanson, P. D., eds. *Introduction to Basic Neurology,* 160. Philadelphia, W. B. Saunders, 1976.)

surface.[30] Acuity is best in the facial region (particularly on the tongue), where the threshold separation is a few millimeters. The next best resolution is found on the hands and toes. The trunk, legs, and arms are characterized by surprisingly large threshold separations. For example, on the back, stimuli must be more than 60 mm apart in order to be judged as distinct.

More sophisticated psychophysical tests, such as detecting a narrow groove in a smooth surface or identifying the orientation of a grating consisting of a series of narrow grooves, give smaller values for spatial acuity: 1 mm or less for the tips of the fingers.[31] Because of its high spatial acuity and its ability to follow rapidly moving tactile stimuli, the cutaneous mechanoreceptor has the highest capacity to transfer information of all the somatic modalities. For example, a skilled blind individual can read Braille at the rate of about 100 words per minute, close to the rate of a normally sighted person reading aloud.

VIBRATION AND FLUTTER

The application of mechanical stimuli oscillating rapidly at from 2 to 400 Hz elicits a unique sensation described by terms such as "buzzing," "trembling," or "vibrating." The sensation evoked by oscillations at the lower part of this frequency range is called *flutter*; the higher frequency sensation is called *vibration*.

The afferent endings responsible for vibration and flutter differ in several ways from touch-pressure endings. First, vibration and flutter endings are very rapidly adapting, often ceasing to discharge a few milliseconds after the application of a constant mechanical stimulus. Second, each ending is most sensitive to a narrow range of frequencies, and is progressively less responsive to higher- and lower-frequency stimuli; that is, the ending is "tuned" to a particular frequency range. Third, these receptors code mechanical information in two ways: by mean discharge frequency (similar to other receptors) and by discharge pattern.

Figure 14–6 illustrates the discharge pattern recorded from a vibration neuron.[32a] At threshold, the neuron responds with occasional action potentials that always occur near the peak of vibration (Fig. 14–6B). As the size of the displacement increases further (Fig. 14–6C), an action potential occurs near the peak of *every* stimulus cycle, producing a so-called phase-locked response. Still larger displacements also produce one action po-

tential per cycle, but the action potential occurs progressively earlier in the stimulus cycle (Fig. 14–6D). At the largest displacements, the neuron discharges continuously throughout the stimulus cycle (Fig. 14–6E). Thus the CNS receives two types of information from vibration neurons. Stimulus intensity is conveyed by average discharge rate and by recruitment. Stimulus frequency is coded by the interval between the bursts of action potentials from individual endings and, as discussed above, by the particular frequency range for which the endings are tuned.[18] These psychophysical dimensions of frequency and intensity anticipate the response of receptors responsible for the sense of hearing in which the mechanisms

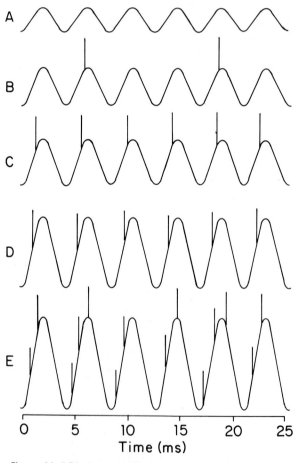

Figure 14–6 Discharge pattern, superimposed upon stimulus amplitude, of first order afferent axons innervating a pacinian corpuscle in the monkey. The response is shown for increasing stimulus amplitudes for a sine wave of constant frequency. *A*, Subthreshold amplitude; *B*, near threshold; *C*, phase-lock threshold; *D*, Supra–phase lock threshold; *E*, highest stimulus amplitude. (After Talbot, W. H.; Dorian-Smith, I.; Kornhuber, H. H.; Mountcastle, V. B. *J. Neurophysiol.* 32:301–334, 1968.)

of tuning and action potential entrainment also are in evidence (see Chapter 17).

The comparison of human psychophysical measurements with the responses of somatosensory afferents in the monkey has permitted the tentative identification of the receptors associated with vibratory submodalities.[23] At frequencies of periodic mechanical stimulation below about 40 Hz, the sensation described by human subjects is one of a "fluttering vibration" distinct from ordinary vibration. Mechanical stimuli occurring at less than 40 Hz correspond to the response frequency of the Meissner ending, whose tuning curve is maximum near 30 Hz (Fig. 14–7); therefore, the Meissner ending is assumed to be responsible for flutter. As frequency increases through the range of 40 to 80 Hz, sensation changes from flutter to vibration ("vibratory hum"). From 80 to 300 Hz, only vibration is felt. Since pacinian corpuscles are maximally sensitive at higher frequencies, with peak responses at about 250 Hz (see Fig. 14–7), they are assumed to be the end organs of the vibratory sense. Flutter-vibration localization supports these structure-function correlations. The flutter sensation is localized to the skin where the Meissner corpuscles lie; vibration, on the other hand, is felt as diffuse and deeper, consistent with the distribution of pacinian endings.

In summary, application to the body surface of rapidly oscillating mechanical force in the low audio frequency range evokes the sensations of flutter, vibration, or both, distinct submodalities encoded according to the tuning curves of the Meissner and pacinian corpuscles from which they originate.

THERMAL SENSATION

Temperature is a continuous quantity beginning at absolute zero (0° Kelvin, a temperature scale, like Centigrade) and increasing without limit. Within this infinite range, our thermally sensitive afferent endings can provide information from a low of approximately 10°C to a high of about 45°C. Below 10°C, the molecular processes that underlie sensory transduction and impulse propagation are slowed to such an extent that action potentials cannot be generated or conducted from either thermal fibers or any other fibers. Therefore, cold is an excellent anesthetic. At the upper limit, a maintained tissue temperature above 45°C is inconsistent with life for most multicellular terrestrial organisms, including humans, and thus an extreme body temperature may also be viewed as an anesthetic, although in a rather macabre sense.

Because a thermal stimulus is continuously variable, it may be speculated that a single receptor type that responds with increasing frequency as temperature increases throughout the sensible range would be utilized to encode thermal sensation. The design of our temperature sensory system, however, is based on a different plan. Instead, we possess two groups of receptors, each generating action potentials within a limited temperature range. One group is called *cold receptors* because the temperature pattern most effective in activating them evokes the subjective sensation of cold. For a similar reason, the other group is called *warm receptors*.

Thermoreceptor Morphology and Location

Thermoreceptors can be readily mapped by exploring the body with a fine wire or other probe of controlled temperature. Such a probe is sometimes called a *thermode*. Placing a heated wire over certain sites causes the subject to report the sensation of "warm"; at intermediate locations, no thermal sensation is reported. Cooling the wire and repeating the exploration results in the identification of "cold" spots. That the warm and cold spots have distinct and separate distributions has long been known, and this fact provided the first

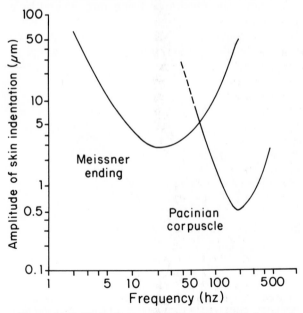

Figure 14–7 "Tuning" curves relating threshold stimulus intensity to stimulus frequency for Meissner and pacinian afferent endings.

evidence that thermal sensation consists of two separate submodalities.

Such mapping explorations have demonstrated that thermal receptors are distributed widely over the body surface and also are present, although more sparsely, in the mouth and nasal cavity. In body regions with high thermal sensitivity, receptor density is high. For example, the glabrous skin of the fingers and palm is estimated to contain 50 to 70 cold fibers/cm² or about one third of all Aδ fibers from the region. The same region gives rise to about an equal number of warm fibers.[5, 17]

The receptive field of individual thermal fibers depends not only on receptor location and arborization but also on the thermal conduction properties of the surrounding tissue. Thus the size of the receptive field is a function of probe temperature, probe size, and the duration of stimulus application. The brief application to the skin of fine probes at 20°C for cold fiber stimulation or 40°C for warm fibers reveals receptive fields of 1 mm or less. Larger probes applied for a longer period elicit responses from an area of about 3 to 5 mm in diameter.

Conduction velocity studies on fibers excited by cutaneous thermal stimulation indicate that the axons innervating cold endings are small myelinated Aδ fibers and those innervating warm endings are unmyelinated C fibers. Progress in identifying the morphology of the endings transducing thermal stimuli has been disappointingly slow. The technique is clear enough: while recording from afferent fibers, identify the thermal ending location, using a fine thermal probe; mark and later excise the tissue; and examine the ultrastructure of nerve endings from the excised section. The technical difficulties in studying small unmyelinated terminals of Aδ and C fibers, however, have not yet been overcome. The best data are for cat facial cold receptors, whose small myelinated axons lose their sheaths but not their Schwann cell envelopes in the dermis and then branch and terminate near the dermal-epidermal border.[8] Warm fiber endings and cold fiber endings in glabrous skin remain to be identified.

In addition to cutaneous thermoreceptors, temperature-sensitive endings are present elsewhere in the body, particularly in the hypothalamus and perhaps in the spinal cord. While these deeper endings are extremely important in the unconscious reflexes that regulate body temperature, they do not appear to contribute to the conscious sensation of temperature. Our knowledge of temperature seems to be derived exclusively from the discharge of thermally sensitive endings located on the body surface.

Adequate Stimulus

When held at a constant temperature within their sensitive range, primary afferent thermal receptors soon reach a steady rate at which they discharge indefinitely. The maximum steady-state firing rates are in the order of 5 to 10 spikes/s. The temperature at which the maximum rate is achieved differs for each ending; in general, cold fibers fire maximally in the range of 23 to 28°C and warm fibers fire maximally in the range of 38 to 43°C. The discharge rate falls off on either side of the "best" temperature. As Figure 14–8 illustrates, the cold fiber response extends over 10 to 40°C, and the warm fiber response over 30 to 45°C.[17]

As shown in Figure 14–8,[5, 18a, 20a] some cold endings unexpectedly begin to discharge again as temperature is raised above 40°C and increase their firing as temperature increases still further. This has been termed the *paradoxical cold response*. The cause of such behavior remains to be explained, but it appears to be related to changes in local blood flow induced by thermal or other influences.[24]

As with other adapting endings, thermal receptors are sensitive to the rate of change of temperature as well as to temperature per se. Consequently, a step change in temperature leads to a transient overshoot in discharge rate, followed by an asymptotic decrease to a new steady-state firing frequency. The maximum firing rate elicited by a rapid temperature change is several times greater than the maximum steady-state discharge. Cold and warm fibers differ not only in the range of

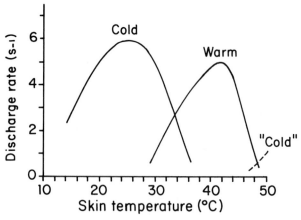

Figure 14–8 Steady-state discharge rate as a function of temperature for typical primate thermoreceptors. (After Kenshalo, D. R.; Duclaux, R. *J. Neurophysiol.* 40:319–332, 1977. Darian-Smith, I.; Johnson, K. O.; LaMotte, C.; Sigenaga, Y.; Kenin, P.; Champness, P. *J. Neurophysiol.* 42:1297–1315, 1979; Konietzny, F. & Hensel, H. *Pflugers Arch.* 370:111–114, 1977.)

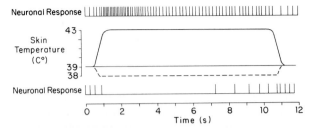

Figure 14–9 Responses of a warm receptor to rapid increases (top trace) and decreases (bottom trace) in temperature. The solid line in the middle represents the time course of the 10-s increase of temperature from 39 to 43 deg, whereas the dashed line shows the 10 s decrease from 39 to 38 deg. Note that (1) the temperature increase produces a transient overshoot that subsides to the steady-state level in about 8 to 10 seconds and (2) the rapid temperature decrease leads to a transient silent period. For cold receptors (not shown), the response is the mirror image of this response, with a transient overshoot in discharge rate following a temperature decrease and a silent period following a rapid temperature rise. (After Darian-Smith, I.; Johnson, K. O.; LaMotte, C.; Shigenaga, Y.; Kenin, P.; Champness, P. *J. Neurophysiol.* 42:1297–1315, 1979; and after Darian-Smith, I.; Dykes, R. W. In Dubner, R.; Kawamura, Y., eds. *Oral-Facial Sensory and Motor Mechanisms*, 7–22, 1971. Courtesy of Appleton-Century-Crofts, Publishing Division of Prentice-Hall, Inc., Englewood Cliffs, N.J.)

temperatures over which they discharge but also in the direction of temperature change that elicits the overshoot in discharge. Cold receptors are stimulated by decreasing temperature (negative rate of change) while warm receptors overshoot in response to a temperature increase (positive rate of change), as shown in Figure 14–9.[4, 5]

The curious reader can personally demonstrate the adapting properties of cutaneous thermal endings by the following procedure: (1) Fill three beakers, one with hot water (40–45°C), one with cold water (15–20°C), and the third with water of a neutral temperature (30–35°C). (2) Place one index finger in the hot water, the other index finger in the cold water, and then wait a minute or so for adaptation to be complete. (3) Then place both fingers simultaneously in the neutral temperature water. For several seconds the formerly hot finger will feel cold and the formerly cold finger will feel hot, even though both fingers are at the same "neutral" temperature. The explanation lies in the responses of warm and cold receptors when subjected to rapid temperature changes. In the finger moved from hot to neutral, the cold receptors respond with a burst of impulses because of a decrease in temperature while the warm receptors are temporarily silent; hence, the sensation is one of cold. The thermal receptors in the finger moved from cold to neutral respond inversely.

Temperature Psychophysics

The temperature reported by a human subject as arising from a given area of skin depends on two factors. The first factor is the firing rate of the thermal receptors innervating the area—or rather the relative firing rate of warm and cold receptors. Humans report thermal neutrality ("thermal comfort") when skin temperature is about 32 to 33°C. As noted in Figure 14–8, in this range, warm and cold receptors are discharging at about the same rate. Temperatures at which the discharge of warm receptors is appreciably greater than that of cold receptors are reported as warm or hot, depending on the firing rate of the warm receptors and the degree to which warm receptor discharge exceeds that from cold receptors. Conversely, cold is the sensation reported when cold receptors discharge more rapidly than warm receptors.

As mentioned above, the firing rate of a thermoreceptor depends on the rate of change of temperature as well as the absolute temperature level. The j.n.d. for thermal stimulation reflects this sensitivity to different rates of change. For very slow temperature changes in the neutral range, the j.n.d. can be as large as 7°C. For rapid temperature changes, the j.n.d. decreases to about 0.1°C.[19] Following a rapid temperature change in the neutral range, sensation intensity decreases with time. However, above 40°C and below 22°C, sensation persists indefinitely even at constant temperature, suggesting that at temperature extremes adaptation is less of a factor.

The second factor determining the intensity of thermal sensation is the number of receptors stimulated, which in turn depends upon the size of the area stimulated. Because of recruitment and convergence, a larger temperature change is required for a j.n.d. when a small area is stimulated than when a large area is involved. Similarly, a suprathreshold stimulus applied to a large area elicits a greater psychophysical response than a stimulus of equal temperature applied to a small area (Fig. 14–10).[20]

When a warm or cool object is touched, identification of the site of stimulation is generally accurate, but this acuity is due to stimulation of tactile mechanoreceptors rather than to activation of thermal receptors. When thermal endings alone are stimulated—for example, by radiant heat—localization is poor. Radiant warming of one finger is often reported as originating from an adjacent finger; warming the back may be sensed as arising from the chest.

At the extremes of the range of thermal sensibility, judgment of temperature is vague. At high temperatures, above 45°C, discharge of pain receptors and perhaps thermoreceptors responsible for the paradoxical cold response add confusion to the afferent input. At temperatures below 15°C,

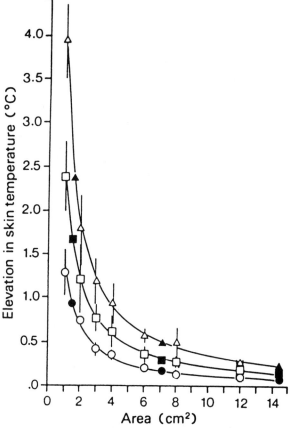

Figure 14–10 Relationship between the temperature increase required for detection (j.n.d.) and the area warmed. The curves can be fit by the equation j.n.d. = kA + c, where k is a constant varying with body region and c is the threshold for detection of warmth for a very large skin area (200 cm²). (After Kenshalo, D. R.; Decker, H. T.; Hamilton, A. J. *Comp. Physiol. Psychol.* 63:510–515. Copyright 1967 by the American Psychological Association. Adapted by permission of the author.)

pain endings may discharge. Thus, very cold stimuli may be reported as hot,—for example, "burning" by dry ice (solid carbon dioxide)—or very hot stimuli as cold. As the temperature falls still farther all afferent firing ceases (local anesthesia).

To summarize, a rapid temperature change of as little as 0.1°C is readily perceived. In contrast, a slow thermal change requires a large temperature increase or decrease to evoke a conscious sensation. Because there is an inverse relation between the thermal j.n.d. and the area involved, stimulation of a large area can compensate to some extent for a small or slow temperature change. When accompanied by tactile activation, thermal stimuli are accurately localized; when unaccompanied by additional sensory input, thermal localization is poor.

The temperature that is sensed is that at the site of the primary afferent endings, which are a fraction of a millimeter beneath the skin surface. Skin temperature depends upon (1) ambient temperature via conduction, convection, and radiant heat exchange with the environment; (2) deep body temperature via circulation of warm blood to the body surface; and (3) local influences, such as cooling through the evaporation of sweat.

In a neutral thermal environment and with normal thermal balance, skin temperature is the weighted mean of deep and ambient temperature. Since deep body temperature is constant, changes in skin temperature are due mainly to changes in local ambient temperature. Thus our conscious information reflects the thermal state of the immediate environment.

In conditions of cutaneous vasodilation, however, circulation becomes the dominant influence and skin temperature approaches deep body temperature. In contrast, during vasoconstriction the skin is approximately at the ambient temperature. The victim of Raynaud's disease, an inappropriate vasoconstriction in the extremities, may complain bitterly of cold even on a mild day because his skin really is colder than normal. The warm glow suffusing the imbiber of an alcoholic beverage is due not only to ethanol's effect on the CNS but also to its action as a peripheral vasodilator.

KINESTHESIA AND PROPRIOCEPTION[3]

Proprioception is defined as the afferent input arising from stimulation of muscle, tendon, and joint mechanoreceptors. Kinesthesia is defined as the sensation arising from movement of the body parts in relation to one another. These terms also are often used to designate sensory information concerned with both dynamic movement and the static position of body parts generated by somatosensory position-sensitive and force-sensitive mechanoreceptors. Note that both definitions exclude the position and orientation information obtained from vision and the vestibular apparatus (see Chap. 27).

Afferent Ending Morphology

The majority of joint mechanoreceptors are SA endings that fire continuously when the joint is positioned within a specific angular range of flexion-extension, as illustrated in Figure 14–11. They are innervated by large myelinated axons (the

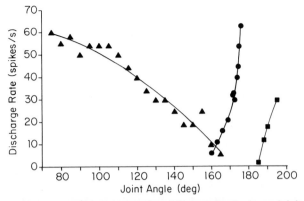

Figure 14-11 Discharge rate of three primate knee joint neurons as a function of angle. Two neurons (circle and square symbols) are typical of most knee-joint receptors in that they discharge over a relatively narrow range of joint angles only at the extremes of joint movement. Occasional neurons (triangle symbol) are responsive over a wide angular range. (After Griggs, P.; Greenspan, B. J. *J. Neurophysiol.* 40:1–8, 1977.)

majority in groups II and III) that are assumed to arise from Ruffini receptors. Joints also contain RA endings, presumed to be paciniform joint receptors and pacinian corpuscles, innervated by axons of groups II and III. The RA fibers discharge during joint extension and flexion but fall silent when the joint is stationary.[7]

Skeletal muscles contain two classes of afferent endings associated with the intrafusal fibers that compose muscle spindle organs. Primary endings (also called *annulospiral endings*) are located on both nuclear bag and nuclear chain intrafusal muscle fibers and are innervated by group I afferents. Secondary endings (also called *flower-spray endings*) are located almost exclusively on nuclear chain intrafusal fibers and are innervated by group II afferents. Because the muscle spindle lies in parallel with the regular (extrafusal) muscle fibers, muscle spindle afferent discharge is determined by muscle length and velocity. The sensitive range of muscle spindle afferents is controlled by contraction of the intrafusal fiber, which is innervated by gamma motoneurons. The detailed properties of the muscle spindle and its role in reflexes contributing to posture and movement are discussed in Chapter 24.

Also associated with skeletal muscle are the afferent endings that innervate the Golgi tendon organ. These receptors are innervated by group I axons. As discussed in Chapter 24, the adequate stimulus for the tendon organ is muscle contraction. The cutaneous mechanoreceptors in the skin overlying the joint may also contribute to kinesthesia, since they are distorted as the skin is compressed or stretched during movement.

Psychophysics

To reach conscious sensation, joint movement must exceed a minimum magnitude and velocity. Proximal joints tend to have a lower j.n.d. than distal joints. For example, passive movement of the shoulder of 0.2 degrees at a rate of at least 0.3 degrees/s is detected by human subjects. On the other hand, a finger displacement of 1.2 degrees at a rate of 10 degrees/s is required to be detected.

Information concerning joint position is available to the CNS even when joint movement is too small to be consciously sensed. For example, the position of one knee, reached by passive flexion or extension at a rate so slow that no movement is detected, can be matched within 2 to 3 degrees by the other knee.[9]

Origin of Kinesthesis

The relative contribution of the several afferent endings to the sense of position and movement has been a subject of controversy for over a century. Candidates for the office of "primary kinesthetic sensor" include joint receptors, muscle spindle and tendon organ afferents, and cutaneous mechanoreceptors. Also, as first noted by Helmholtz, descending and spinal somatic motor discharge provides an additional potential source of information, since most joint movement is accomplished through active contraction of one or more skeletal muscles. Accordingly, Helmholtz postulated that the motor command to efferent pathways innervating skeletal muscles was accompanied by a "corollary discharge" to the sensory system, thus informing the brain of expected limb movement and position[1] (see Chap. 22).

From the 1960s to the mid-1970s, joint receptors were considered to be the most likely primary source of kinesthetic information. Recent findings, however, have all but demolished this view. First, recordings from single neurons have demonstrated that most joint receptors fire rapidly only at *extreme* joint angles and are relatively insensitive in the mid-range where position information is most required; also, many joint receptors fire tonically in both flexion *and* extension, thus providing ambiguous information. Second, elimination of joint afferent input at the knee by intracapsular injection of a local anesthetic leaves position sense at this joint unaffected. Third, although replacement of a joint such as the hip joint with an artificial prosthesis destroys joint afferents, position sense is retained. Fourth, when input from nonjoint sources (such as muscle spindles) is minimized so that most input is from joint receptors, human subjects show a much reduced position acuity. Thus joint receptors seem neither sufficient nor necessary for kinesthetic sensibility. Instead, they seem to subserve the sensation of deep pressure. (See Matthews[25] for a detailed review of the experiments described above.)

Cutaneous afferents also have been candidates because joint movement distorts the overlying skin, causing endings in the skin to discharge. In fact, the majority of mechanoreceptors in the glabrous skin of the hand fire during voluntary finger movement. Also, the sense of position is enhanced when the sense of touch is involved—as, for example, when estimating interphalangeal angle while touching the tips of the fingers against a solid surface. On the other hand, interdiction of knee joint skin afferents by infiltration of a local anesthetic has no detectable effect on position detection. Therefore, information from tactile afferents probably serves more to complete or enrich the mental "body image" than to contribute to the sensation of joint position. Thus cutaneous mechanoreceptors contribute to kinesthesis but are not the primary endings for this modality.

Corollary input from motor discharge may contribute at least indirectly to kinesthesia, as explained below. But such discharge cannot be the primary source of position information, since angle is reported as accurately when a joint is passively rotated with no active contraction as when rotation is due to voluntary movement. However, corollary activity does appear to contribute to the subjective sense of the weight of lifted objects.[26]

By elimination, then, muscle and perhaps tendon organ afferents must be considered the primary receptors for proprioception. This conclusion is buttressed by several observations. First, during passive joint movement, the discharge of muscle receptors signals joint position over the entire range of rotation, with receptors from flexors and extensors exhibiting inverse discharge patterns (Fig. 14–12).[2] Second, stimulation of muscle spindles by application of a vibrator to the skin results in the illusion of joint movement.

If muscle receptors are responsible for position sense, then pulling on a tendon severed from its bony attachment at the joint should convey the sensation of joint movement, even though the joint is stationary. Such experiments have been tried during surgery when the transection of a tendon was required for medical reasons. Unfortunately, the results were ambiguous, with some patients reporting sensation and others not. To

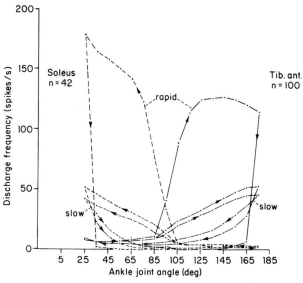

Figure 14–12 Mean discharge frequency of muscle spindle afferents from soleus (an ankle extensor) and tibialis anterior (an ankle flexor) as the ankle joint is rotated through rapid and slow cycles. Note that (1) receptors from each muscle serve about half the range of ankle rotation, and (2) discharge frequency depends strongly on rotation rate. (After Burgess, P. R.; Perl, E. R. In Iggo, A., ed. *Handbook. Sensory Physiology*, vol. 2, pp. 29–78. New York, Springer-Verlag, 1973.)

settle the matter, McCloskey had the tendon of his own big toe sectioned and then pulled without his knowledge. Upon this heroic maneuver, McCloskey experienced the sensation of toe movement.[25] This experimental design has not been repeated by other investigators.

The output of muscle receptors now seems firmly established as the source of the primary information in kinesthesis; however, problems remain. First, the stimulus to muscle spindles depends not only on muscle length (and thus joint angle) but also on the tone or state of contraction of the intrafusal fibers. Second, spindle afferent discharge is characterized by partial adaptation so that the firing frequency depends on the angular velocity of a joint as well as its position. Thus, a given spindle discharge rate does not indicate a unique joint angle. It is assumed, therefore, that the central nervous system must be capable (1) of decoding spindle output by integrating the signal (in the mathematical sense) to derive position information from the velocity component of the discharge and (2) of allowing for the shift in spindle sensitivity caused by gamma efferent activity. Verification of these assumptions would be welcome.

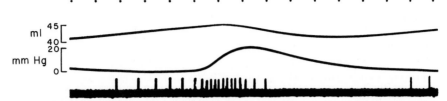

Figure 14–13 Single afferent recording from an "in-series" mechanoreceptor located in the urinary bladder of the cat. The upper trace is a recording of bladder volume; the middle trace, pressure within the bladder; and the lower trace, the discharge pattern of an afferent fiber. Bladder distention leads to a modest increase in firing frequency until the micturition reflex causes contraction of the bladder wall muscle, resulting in a rise in pressure, fluid expulsion and a great increase in firing frequency—even though the bladder is contracting. (From Iggo, A. *J. Physiol.* 128:593–607, 1955.)

VISCERAL SENSATION

Conscious sensation originates not only from special sensory organs and somatic structures but also from viscera such as the lungs and airways, the alimentary tract organs, the heart and blood vessels, the kidney, the bladder and urinary tract, and the sex organs. Examples of sensations that arise in whole or in part from viscera are dyspnea (breathlessness), satiety, heartburn, stomachache, angina, and bladder fullness.

In contrast to the intensive efforts that have been lavished on the study of the special and somatic senses, visceral sensory mechanisms have received relatively little attention. The reasons for this neglect involve both technique and fashion. In the past, it was technically difficult to record from the small axons that convey afferent impulses from visceral organs into the CNS. Also, visceral sensations command a person's attention relatively rarely and then are judged as primitive and vague; therefore, they have generated little excitement among sensory physiologists. Even with the recent advances in technique and a broadening of interest,[29] our knowledge of visceral sensation remains limited.

The viscera are a source of a variety of afferent inputs that are conveyed into the CNS via both small myelinated and unmyelinated axons. Visceral afferent fibers travel with sympathetic and parasympathetic efferents in visceral nerves. The innervation density of visceral afferents is low. The thoracic, abdominal, and pelvic cavities are subserved by 22,000 to 25,000 neurons, which represent only about 2% of all spinal afferent neurons.[15]

After pain (see Chap. 16), the best established visceral modality is mechanosensation. Mechanosensitive endings are stimulated by tissue distortion, which in a hollow organ (e.g., colon, urinary

bladder, gallbladder) is a function of luminal pressure. Many neurons discharge during smooth muscle contraction as well as organ distention and are thus assumed to lie in series with muscle fibers (Fig. 14–13).[11] Visceral afferents display a spectrum of adaptation times, ranging from rapidly adapting to slowly adapting. Sensory thresholds range from a few to 80 mm Hg (Fig. 14–14).[15] The spatial resolution of mechanical sensations originating from the viscera is poor.

Other stimuli that lead to discharge of visceral afferents include temperature (particularly heat), ischemia, a variety of chemicals (including bradykinin, serotonin, H^+, and K^+), and application of hypertonic solutions.

A puzzling feature of visceral afferents is their functional homogeneity. Visceral afferents cannot be subdivided into functionally distinct types (e.g., "pure" nociceptors, mechanoreceptors, chemoreceptors) based either on discharge patterns, stimulus-response relations, thresholds to natural stimuli, resting (unstimulated) activity, or conduction velocity. If no functional subclassification can

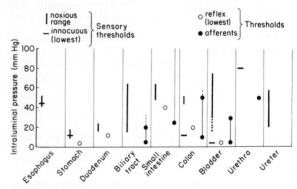

Figure 14–14 Summary of intraluminal pressures that elicit innocuous sensation and pain in humans, and pupillary and visceral reflexes and afferent fiber discharge in animals. (After Jänig, W.; Morrison, J. F. B. *Prog. Brain Res.* 67:87–114, 1986.)

be established, it must be concluded that the various sensations and reflexes that arise from visceral organs originate from common afferent endings and that sensory coding is based on discharge rate rather than on endings with different specificities. Future experiments will determine whether this challenge to the specificity theory is sustained.

ANNOTATED BIBLIOGRAPHY

Darian-Smith, I. The sense of touch: performance and peripheral neural processes. *Handbook of Physiology*, Vol. 3, Pt. 1, pp. 739–788, 1984.
A comprehensive, well-written review of peripheral somatic sensation in which the results of primate psychophysical experiments are compared with the discharge patterns of primary afferents; includes 250 references.

REFERENCES

1. Boring, E.G. *Sensation and Perception in the History of Experimental Psychology*. New York, Appleton-Century, 1942.
2. Burgess, P.R.; Perl, E.R. Cutaneous mechanoreceptors and nociceptors. In Iggo, A., ed. *Handbook of Sensory Physiology*, Vol. 2, 29–78. New York, Springer-Verlag, 1973.
3. Burgess, P.R.; Wei, J.Y., Clark, F.J., Simon, J. Signaling of kinesthetic information by peripheral sensory receptors. *Ann Rev Neuroscience* 5:171–187, 1982.
4. Darian-Smith, I.; Dykes, R.W. Peripheral neural mechanisms of thermal sensation. In Dubner, R.; Kawamura, Y., eds. *Oral-Facial Sensory and Motor Mechanisms*, 7–22. New York, Appleton-Century-Crofts, 1971.
5. Darian-Smith, I.; Johnson, K.O.; LaMotte, C.; Shigenaga, Y.; Kenin, P.; Champness, P. Warm fibers innervating palmar and digital skin in monkey: responses to thermal stimuli. *J. Neurophysiol.* 42:1297–1315, 1979.
6. Gottschaldt, K.M.; Iggo, A.; Young, D.W. Functional characteristics of mechanoreceptors in sinus hair follicles of the cat. *J. Physiol.* 235:287–315, 1973.
7. Griggs, P.; Greenspan, B.J. Response of primate joint afferent neurons to mechanical stimulation of knee joint. *J. Neurophysiol.* 40:1–8, 1977.
8. Hensel, H.; Andres, K.H.; von Duering, M. Structure and function of cold receptors. *Pflugers Arch.* 329:1–8, 1974.
9. Horsch, K.W.; Clark, F.J.; Burgess, P.R. Awareness of knee joint angle under static conditions. *J. Neurophysiol.* 38:1436–1447, 1975.
10. Hunt, C.C. The pacinian corpuscle. In Hubbard J.I., ed. *The Peripheral Nervous System*, 405–420. New York, Plenum Press, 1974.
11. Iggo, A. Tension receptors in the stomach and urinary bladder. *J. Physiol.* 128:593–607, 1955.
12. Iggo, A.; Andres, K.H. Morphology of cutaneous receptors. *Ann. Rev. Neurosci.* 5:1–31, 1982.
13. Iggo, A.; Muir, A.R. The structure and function of a slowly-adapting touch corpuscle in hairy skin. *J. Physiol.* 20:763–796, 1969.
14. Iggo, A.; Ogawa, H. Correlative physiological and morphological studies of rapidly adapting mechanoreceptors in cat's glabrous skin. *J. Physiol.* 266:275–296, 1977.
15. Jänig, W.; Morrison, J.F.B. Functional properties of spinal visceral afferents supplying abdominal and pelvic organs, with special emphasis on visceral nociception. *Prog. Brain Res.* 67:87–114, 1986.
16. Johanson, R.S.; Vallbo, A.B. Tactile sensory coding in glabrous skin of the human hand. *Trends Neurosci.* 6:27–32, 1983.
17. Johnson, K.O.; Darian-Smith, I.; LaMotte, C. Peripheral neural determinants of temperature discrimination in man: a correlative study of responses to cooling skin. *J. Neurophysiol.* 36:347–370, 1973.
18. Keidel, W.D. The sensory detection of vibrations. In Dawson, W.W.; Enoch, J.M., eds. *Foundations of Sensory Science*, 465–512. New York, Springer-Verlag, 1984.
18a. Kenshalo, D.R.; Duclaux, R. Response characteristics of cutaneous cold receptors in the monkey. *J. Neurophysiol.* 40:319–332, 1977.
19. Kenshalo, D.R.; Scott, H.A. Temporal course of thermal adaptation *Science* 151:1095–1096, 1966.
20. Kenshalo, D.R.; Decker, H.T.; Hamilton, A. Spatial summation on the forehead, forearm, and back produced by radiant and conducted heat. *J. Comp. Physiol. Psychol.* 63:510–515, 1967.
20a. Konietzny, F.; Hensel, H. The dynamic response of warm units in human skin nerves. *Pflugers Arch.* 370:111–114, 1977.
21. Kruger, L.E.; Perl, E.R.; Sedivic, M.J. Fine structure of myelinated mechanical nociceptor endings in cat hairy skin. *J. Comp. Neurol.* 198:137–154, 1981.
22. LaMotte, R.H. Psychophysical and neurophysiological studies in tactile sensibility. In Norman, R.S.; Hollies, R.S.; Goldman, R.F., eds. *Clothing Comfort: Interaction of Thermal, Ventilation, Construction and Assessment Factors*, 83–105. Ann Arbor, Mich. Science Pub., 1977.
23. LaMotte, R.H.; Montcastle, V.B. Capacities of humans and monkeys to discriminate between vibratory stimuli of different frequency and amplitude: a correlation between neuronal events and psychophysical measurements. *J. Neurophysiol.* 38:539–559, 1975.
24. Long, R.R. Sensitivity of cutaneous cold fibers to noxious heat: paradoxical cold discharge. *J. Neurophysiol.* 40:489–502, 1977.
25. Matthews, P.C.B. Where does Sherrington's "muscular sense" originate? Muscles, joints, corollary discharges? *Ann. Rev. Neurosci.* 5:189–218, 1982.
26. McCloskey, D.I. Corollary discharges and motor commands. *Handbk. Physiol.* 2:1415–1447, 1981.
27. Mountcastle, V.B.; LaMotte, R.H.; Carli, G. Detection thresholds for stimuli in humans and monkeys: comparison with threshold events in mechanoreceptive afferent nerve fibers innervating monkey hand. *J. Neurophysiol.* 35:122–136, 1972.
28. Mountcastle, V.B.; Talbot, W.H.; Kornhuber, H.H. The neural transformation of mechanical stimuli delivered to the monkey's hand. In de Reuck, A.V.S. and Knight B.A. ed. *Ciba Foundation Symposium on Touch, Heat and Pain*, 325–351, Boston, Little, Brown & Co., 1966.
29. Paintal, A.S. The visceral sensations—some basic mechanisms. *Prog. Brain Res.* 67:3–19, 1986.
30. Patton, H.D. Introduction to sensory physiology. In Patton, H.D.; Sundsten, J.W.; Crill, W.E.; Swanson, P.D.,

eds. *Introduction to Basic Neurology*, 160. Philadelphia, W.B. Saunders, 1976.

31. Phillips, J.R.; Johnson, K.O. Tactile spatial resolution. III. A continuum mechanics model of skin predicting mechanoreceptor responses to bars, edges, and gratings. *J. Neurophysiol.* 46:1204–1225, 1981.

32. Pubols, B.H.; Donovick, P.J.; Pubols, L.M. Opossum trigeminal afferents associated with vibrissa and rhinarial mechanoreceptors. *Brain Behav. Evol.* 7:360–381, 1973.

32a. Talbot, W.H.; Darian-Smith, I.; Kornhuber, H.H. and Mountcastle, V.B. The sense of flutter-vibration: comparison of the human capacity with response patterns of mechanoreceptive afferents from the monkey hand. *J. Neurophysiol.* 32:301–334, 1968.

33. Tuckett, R.P.; Horch, K.W.; Burgess, P.R. Response of cutaneous hair and field mechanoreceptors in cat to threshold stimuli. *J. Neurophysiol.* 41:138–149, 1978.

34. Weinstein, S. Intensive and extensive aspects of tactile sensitivity as a function of body part, sex and laterality. In Kenshalo, D.R., ed. *The Skin Senses*, 195–222. Springfield, Ill., Charles C Thomas, 1968.

35. Williams, P.L.; Warwick, R., eds. *Gray's Anatomy, 36th British Edition*. London, Churchill Livingstone, 1980.

Chapter 15

James O. Phillips
Albert F. Fuchs

Somatic Sensation: Central Processing

INTRODUCTION

Somatic sensation encompasses a variety of sensory experiences. Some, such as the sensations of contact, flutter, vibration, and pressure can be explained largely by the behavior of primary peripheral afferent fibers. Others, such as tactile recognition of an unseen object or the ability to sense the direction of a moving cutaneous stimulus, require the integration of information originating in many afferent fibers. In this chapter, we will examine what is known about the role of the central somatosensory system in producing somatic sensation and ultimately the perceptions that define our somesthetic experience. In so doing, we will travel from peripheral nerves, which report the modality and position of stimuli, to the complex integrative elements of somatosensory cortex. We will review both the functional anatomy and the physiology of the ascending and descending systems that serve the somatic senses and will describe what is currently known about the contribution of each to the coding of somatic sensation.

PERIPHERAL NERVE

Organization

Somatosensory signals travel to the central nervous system (CNS) over a network of peripheral nerves, which contain the distal axons of pseudounipolar sensory neurons whose somata lie in the dorsal root ganglia. Sensory fibers range from the smallest unmyelinated C or group IV fibers (0.2–1.5 μm in diameter) to the largest myelinated Aα or group I fibers (13–20 μm). Within the peripheral nerve, sensory fibers are organized randomly. Large fibers may be adjacent to small fibers, and fibers adjacent at one point may be separated only a few centimeters further on. Adjacent nerve fibers often have widely separated receptive fields (defined in Chapter 13) and stimulation of adjacent fibers can produce projected sensations in widely separated loci (see below).

As can be seen in the right half of Figure 15–1, which shows the parts of the body innervated by individual peripheral nerves, each nerve serves a continuous patch or region of skin. Although not depicted in Figure 15–1, the cutaneous distributions of individual peripheral nerves overlap, such that adjacent cutaneous regions are innervated by two or more nerve trunks. This overlap is more extensive for afferent fibers reporting pain and extreme temperature than for those signaling discriminative touch and innocuous temperature. Injury to a peripheral nerve, therefore, may produce limited regions that are anesthetic surrounded by regions with preserved pain sensibility and reduced, or possibly absent, touch. These latter regions are called *intermediate zones* because they lie between normal and anesthetic skin.

Neuronal Responses and Sensation

As discussed in Chapter 14, the response characteristics of individual nerve fibers are determined by the receptors that excite them. Thus, they have both an adequate, or best, stimulus and an excitatory receptive field, which is determined by the location of their receptors or free nerve endings. As mentioned in Chapter 13, investigators in the nineteenth century thought that individual afferent fibers carried information about a

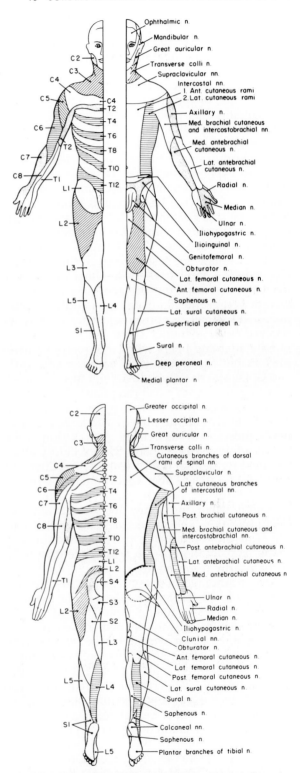

Figure 15–1 Anterior and posterior view of the dermatomes (on left side of figures) and the cutaneous areas supplied by individual peripheral nerves (on right side). (From Carpenter, M. B.; Sutin, J., *Human Neuroanatomy.* © 1983, The Williams & Wilkins Co., Baltimore.)

specific modality (e.g., pressure), and that activation of such a "labeled line" produced a discrete sensation. Others suggested that the temporal and spatial pattern of afferent fiber input defined a modality (e.g., touch), and that individual fibers could signal, or contribute to, different modalities by changing their patterns of activity.

In the early 1980s, two groups independently claimed that activation of individual fibers can indeed lead to simple human somatic sensations.[38] In these experiments, small tungsten microelectrodes were driven into nerves innervating the glabrous skin of the hand in alert human volunteers. First, the responses of a single large peripheral fiber to natural stimuli were recorded (microneurography; see Fig. 13–3, Chap. 13), and then small currents were passed through the recording electrode to stimulate the same fiber (intraneural microstimulation). Four different fiber types were recorded from and stimulated: (1) rapidly adapting fibers with small receptive fields presumably associated with Meissner (type RAI) endings, (2) rapidly adapting fibers with large receptive fields presumably associated with pacinian corpuscles and Golgi-Mazzoni (type RAII) endings, (3) slowly adapting fibers with small receptive fields presumably associated with Merkel (type SAI) endings, and (4) slowly adapting fibers with large receptive fields presumably associated with Ruffini (type SAII) endings (see Chap. 14). The excitatory receptive field for natural (i.e., adequate; see Chap. 13) stimulation of each fiber was compared with the area of skin over which the subject reported a sensation when the same fiber was stimulated electrically (i.e., the projected field; recall Chap. 13). The projected fields corresponded in size, shape, and location to the actual receptive fields of the nearby recorded fibers.

Several findings in these experiments are consistent with the labeled line theory. For example, for all but the SAII fibers, stimulation of a single fiber produced a clearly defined sensation. Also, the modality of that sensation was *not* altered by changing the pattern of stimulation. For example, the tapping (flutter) sensation elicited by stimulation of FAI fibers increased in frequency with increases in the frequency of stimulation but did not change to another sensation (e.g., pressure). Similarly, the sensation of sustained pressure elicited by stimulation of SAI fibers became more intense with increases in stimulation frequency but did not change its modality. Because the sensation produced by stimulation of a single fiber was localized to the receptive field of that fiber, activation of a single fiber was sufficient to define both the modality and the location of the sensation.

Whereas the experiments just described suggest the existence of specific labeled lines related to some cutaneous fiber types for some sensations, some fibers might participate in multiple sensations. Indeed, convergence of many fibers and fiber types is required to produce many common somatic sensations. For example, the sensation of proprioception results from convergence of skin, muscle, and joint afferent fibers. Also, the sensation of texture results from a complex spatial and temporal pattern across cutaneous afferent fibers.[11] Indeed, subjects never report the sensation of "touch" in response to the stimulation of an individual nerve fiber. Even touch, therefore, appears to be a complex sensation requiring the summation of responses originating in a variety of afferent fibers. Finally, the modality coding of the SAII fibers, whose stimulation fails to elicit consistent sensations, and of the small unmyelinated fibers, remains in doubt.

Pathophysiology

Reversible or irreversible impairment of peripheral nerve function may result from a variety of pathologies, including axotomy, demyelination, and ischemia.

Axotomy and Regeneration

Transection of a nerve fiber (i.e., axotomy) results in degeneration of the axon distal to the site of injury, subsequent degeneration of the myelin sheath, and finally digestion of the sheath by proliferating Schwann cells. These distal changes are accompanied by several proximal changes (retrograde degeneration), including retraction of the axon from the point of injury, chromatolysis, and swelling of the neuron soma. The proximal changes are thought to help prepare the neuron for subsequent regeneration. If regeneration occurs, new fibers sprout from the tips of the proximal axons as growth cones. After sprouting, the fibers reach the site of injury in about one day, then require many days to grow through the scar tissue. After penetrating the scar at a rate of 0.25 mm per day, the fibers grow 4 mm per day. Although the remaining endoneurial tube and Schwann cell bodies may provide a guide for regenerating nerve fibers, unmyelinated peripheral nerve fibers also regenerate.

Central axons seem less able than peripheral axons to regenerate. After spinal cord injury, for example, regenerating tract fibers usually fail to cross the site of the insult, mainly because the dense scar tissue forms a barrier to the advancing nerve processes. One experimental solution is to bypass the scarred cord with a peripheral nerve bridge through which the central axons can grow (see Chap. 33).

The peripheral nerves of lower vertebrates, especially immature animals, display an extraordinary capacity for regeneration. For example, if the skin of a tadpole is transplanted from its belly to its back, it is reinnervated by the original nerves, and tactile stimuli delivered to the back result in a ventral wiping behavior. In contrast, mammalian regeneration is often abnormal. Human peripheral nerve fibers have unusual receptive field shapes and locations following regeneration. Even in mammals, however, regeneration can result in a normal distribution of peripheral receptors if the target of reinnervation is relatively close to the point of injury.[7]

Demyelination

Demyelination of peripheral nerve fibers, a common form of human neuropathy, may lead to slowing or failure of spike conduction, hyperexcitability, and interaction between nerve fibers. Slowed conduction occurs only at the site of demyelination, whereas the rest of the fiber has a normal conduction velocity. Unequal changes in conduction velocities in different fibers may contribute to altered sensation if the sensation requires the synchronous activity of many fibers in the nerve. For example, the tendon jerk reflex and the sense of vibration are often impaired in peripheral neuropathy. Either spontaneous activity due to hyperexcitability or the cross-excitation of one demyelinated fiber by another could contribute to *paresthesias,* in which false peripheral events are signaled by inappropriately active nerves.

The polyneuropathies related to chronic alcoholism, diabetes, and a variety of toxins (e.g., acrylamide and heavy metals) are associated with a degeneration of the distal axons of peripheral neurons (distal axonopathy). Whereas these neurons have normal conduction velocities, somatic metabolism, and axonal transport, they fail to support their distal axons, producing sensory deficits in distal body parts. Studies on regenerating acrylamide-poisoned neurons suggest that normally transported materials cannot be utilized by the distal processes and simply accumulate in swellings in the degenerate distal axon.

Ischemia

A 15-minute period of ischemia produced by an inflated pressure cuff around the arm leads to the loss of touch, temperature, rapidly conducted pain (see Chap. 16), and sensations of passive movement. These sensations are all mediated by myelinated fibers that are more sensitive to anoxia than are unmyelinated C fibers, which are involved in sensations of slowly conducted pain. In animals, ischemic anoxias lasting longer than 7 hours produce a lasting impairment of function and result in degeneration of nerve fibers. Ischemia of shorter duration blocks fast axoplasmic transport and impulse propagation (called conduction block).

Compression of a nerve causes distortion of the myelin of large fibers (>5 μm in diameter) at the nodes of Ranvier but spares the small fibers; thus direct mechanical damage often eliminates sensations that are communicated by large fibers (e.g., touch), while preserving those that are carried over unmyelinated fibers (e.g., slow pain). We all have experienced acute compression when, after sitting in an awkward position for some time, a limb "falls asleep." Chronic compression can occur when a nerve becomes entrapped within a fibrous tunnel, often against a bone. Such chronic compression can produce hyperexcitability in which the nerve becomes either abnormally susceptible to mechanical excitation or spontaneously active, leading to paresthesias. The relief of either acute or chronic compression eliminates the symptoms.

Whereas ischemia and compression eliminate touch first and then pain, local anesthesia eliminates pain first and then touch. Recording and electrical stimulation during application of a local anesthetic demonstrates that C fiber activity is blocked initially, and then A fiber activity.[37] Thus, the small unmyelinated fibers are more sensitive to local anesthetics but less sensitive to anoxia and direct mechanical damage than are large myelinated fibers.

DORSAL ROOT

Organization

Peripheral afferent fibers enter the spinal cord via the dorsal root ganglia, which are collections of pseudounipolar neurons lying just outside the spinal cord. The distal axons of dorsal root ganglion cells are peripheral nerve fibers; the proximal axons (called dorsal root fibers) enter the spinal cord. The region of skin innervated by a single dorsal root ganglion is called a *dermatome.* The general location of the dermatomes on the human body is displayed on the left side of Figure 15–1. Although these dermatomes are displayed as nonoverlapping regions, the innervation provided by adjacent ganglia has so much actual overlap that a lesion to any individual ganglion does not produce anesthesia in the dermatome of that root. Therefore, dermatomes can be revealed only by destroying a series of adjacent dorsal roots and reconstructing the borders of those remaining (constructive method), or by damaging contiguous roots above and below an intact root (method of residual sensitivity).

The locations of dermatomes differ according to the means by which they are revealed. Some investigators map dermatomes by charting the zones of hypalgesia (diminished sensitivity to pain) created by lesions of individual dorsal roots. Hypalgesia results because, in dermatomes as in peripheral nerve, the overlap for touch is greater than that for pain. This procedure produces dermatomes that are smaller and somewhat inconsistent with those presented in Figure 15–1. Similar reduced dermatomes have been constructed by measuring the vasodilation produced by stimulation of dorsal roots or by mapping the distribution of herpes zoster, which infects dorsal root ganglia and peripheral nerves.

Figure 15–1 shows that in the limbs there is a loose correspondence at best between the dermatomal organization and the distribution of individual peripheral nerves. Individual peripheral nerves contain fibers that have cell bodies in several dorsal root ganglia, because of the complex interweaving of axons in the brachial and lumbosacral plexi. For example, fibers of the superficial peroneal nerve originate in three dorsal root ganglia (L4 to S1). There is also little correspondence between the dermatomes and the groups of muscles or bones innervated by individual dorsal root ganglia (called myotomes and sclerotomes, respectively). In the trunk, there is a closer correspondence between the dermatomal organization and innervation of individual peripheral nerves (not shown in Fig. 15–1) because there is no plexus subserving the trunk, and hence no mixing of fibers.

Pathophysiology

Like peripheral nerves, dorsal roots can become damaged or irritated due to compression, transection, or disease. The effects generally are similar to those produced by peripheral nerve lesions but differ in their severity and distribution. A common form of dorsal root injury is compression from herniation of an intervertebral disc. As many of us know, this injury can produce irritation of the dorsal root, resulting in pain over all or part of the affected myotome, sclerotome, or dermatome. Irritation can also produce paresthesias (e.g., a tingling sensation) or hyperesthetic regions (i.e., areas of increased sensitivity) within the dermatome. If several dorsal roots receive sufficient damage, frank anesthesia of one or several dermatomes can result (called segmental anesthesia). Other clinically relevant forms of dorsal root impairment result from direct mechanical effects such as traction, inflammatory processes and diseases such as tabes dorsalis (tertiary syphilis), and ischemia.

SPINAL CORD

Segmental Synaptic Connections

The vast majority of proximal axons from the dorsal root ganglia enter the dorsal horn of the spinal cord via the dorsal roots and (1) contact secondary neurons within the same segment of the spinal cord, or (2) ascend or descend a few spinal segments before contacting secondary neurons, and/or (3) enter a major ascending spinal tract. Dorsal root fibers in each of these categories have a characteristic distribution within a spinal cord segment.

Not all afferent fibers enter via the dorsal roots. If the dorsal roots are cut and horseradish peroxidase (HRP) is injected into the spinal cord, many dorsal root ganglion cells are labeled.[23] If, however, the ventral roots are also cut, no labeled cells are found, suggesting that the ventral roots contain many afferent fibers. Most ventral root afferent fibers seem to be small, unmyelinated nociceptors, and many may be visceral rather than somatic.

As dorsal root fibers enter the spinal cord, they divide roughly into a lateral aggregate of small, unmyelinated fibers and a more medial aggregate of larger myelinated fibers. This segregation is the beginning of a sorting process that separates the fibers by size, and therefore by modality, since small fibers are largely nociceptive and thermoceptive, whereas the large fibers convey discriminative touch and proprioceptive information (recall Chap. 14).

Fibers of different diameters terminate in different spinal cord laminae; Rexed[31a] first identified these laminae by their cytological features in the cat. The cat spinal cord gray matter was divided from dorsal to ventral into lamina I through IX, and area X, which is the gray substance surrounding the spinal canal.[9] Rexed's laminae for a human spinal segment (C6) are shown in Figure 15–2A.

descend only a few segments to terminate on second-order cells in the outer portions of lamina II, and in laminae III through V. A representative large-diameter fiber also is shown in Figure 15–2B (thick line).

The injection of HRP into single, functionally identified cat dorsal root fibers demonstrates that large fibers innervating different receptors have different characteristic terminations (Fig. 15–3).[5] Hair follicle afferent fibers terminate in flame-shaped arborizations primarily in lamina III, but also in laminae II and IV. Since individual hair follicle fibers also arborize extensively in the rostrocaudal dimension, their arborizations form long (7 mm) rostrocaudal sheets within the spinal cord (Fig. 15–3A). Like hair follicle afferent fibers, rapidly adapting mechanoreceptor fibers have flame-shaped arbors confined largely to lamina III, but they do not arborize in continuous rostrocaudal sheets (Fig. 15–3D). In contrast, type I slowly adapting afferent fibers form a rostrocaudal series of spherical arborizations in laminae III through V (Fig. 15–3B), whereas type II fibers form a rostrocaudal series of plate-like arborizations in laminae III through VI (Fig. 15–3C). Finally, fibers innervating pacinian corpuscles arborize in laminae III through V (Fig. 15–3E). Therefore, all of the cutaneous modalities subserved by large fibers project to the dorsal horn in sagittal strips or columns. Also, each type of afferent fiber has a unique, *but often overlapping,* termination in different laminae.

Somatotopic Organization and Sprouting

Normally, afferent fibers from a single dorsal root terminate over a number of segments. For example, the L6 dorsal root of the cat has intersegmental terminations that extend rostrally to T9. In addition, fibers with receptive fields on the distal extremities and/or on the ventral surface of the body terminate in the medial part of the cord, whereas those with receptive fields on the proximal and/or dorsal body surface terminate laterally.[5] For example, fibers innervating pacinian corpuscles that are located in the glabrous skin of the foot and toe pads arborize on the medial edge of the dorsal horn (see Fig. 15–3E). Thus, dorsal root afferent fibers terminate in such a way that the skin surface is mapped in axial columns across the medial-lateral extent of the cord, with individual dermatomes represented as columns several segments in length.

Damage to the dorsal root or peripheral nerve may alter this pattern of termination. If all dorsal roots but one are cut on one side in cats, the intact root on the damaged side establishes far more extensive terminations within the cord than does the homologous root on the intact side. Therefore, sprouting collaterals from intact dorsal root fibers

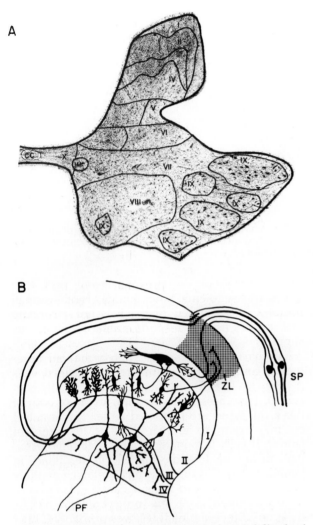

Figure 15–2 Rexed's laminae in the spinal cord. *A,* Structural lamination indicated on a thick section of the human cord in the sixth cervical segment. *B,* Schematic diagram of cutaneous input to the dorsal horn of the cat showing laminae I through IV. The thin line represents a small-diameter afferent fiber; the thick line represents a large-diameter myelinated afferent fiber. PF = projection fibers; SP = dorsal root (spinal) ganglion; ZL = zone of Lissauer (shaded). (From Carpenter, M. B.; Sutin, J. *Human Neuroanatomy.* © 1983, The Williams & Wilkins Co., Baltimore.)

Small fibers, which branch and ascend or descend 1 or 2 segments in the dorsolateral funiculus (Lissauer's tract) synapse on second-order cells in Rexed's lamina I (the posteromarginal nucleus) and the outer zone of lamina II. A representative fiber of this type is displayed in Figure 15–2B (thin line). The larger fibers also branch, but many ascend in the dorsal funiculi (dorsal columns) all the way to the lower medulla, where they terminate in the nucleus gracilis and nucleus cuneatus (the dorsal column nuclei). Others may ascend or

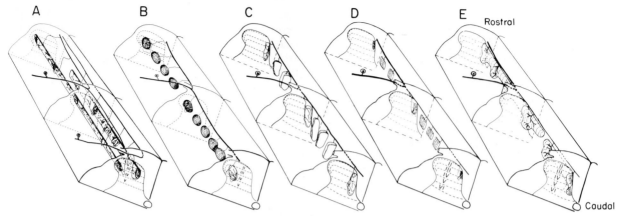

Figure 15–3 The pattern of termination in the cat spinal cord of five different large-diameter somatosensory afferent fiber types. *A*, hair follicle; *B*, type I slowly adapting; *C*, type II; *D*, rapidly adapting mechanoreceptor; *E*, pacinian corpuscle. (After Brown, A. G. *Organization of the Spinal Cord.* New York, Springer-Verlag, 1981.)

may invade denervated regions of the cord. This sprouting, however, seems restricted to regions previously innervated, albeit sparsely, by the intact root. Thus, the rostrocaudal extent of the termination of a particular root remains unchanged.

Lesions of peripheral nerves can produce expansions in the receptive field dimensions of spinal cord neurons, shifts in receptive field locations, and changes in the distribution of modalities to which they respond (e.g., the damage may produce a higher percentage of cells responsive to high-threshold nociceptive stimuli). It is unclear whether these changes result from a sprouting of afferent fiber terminals or from an unmasking of pre-existing but weak (subliminal) inputs (see Chap. 33 for additional discussion).

SECONDARY SPINAL CORD NEURONS

Organization

A neuron in the spinal cord can be characterized by its location, morphology, and response to natural stimuli. While the somata of these cells are restricted to individual laminae, their dendritic arbors can extend over many laminae (Fig. 15–4A) and therefore can potentially sample a number of different inputs.

Lamina I (the marginal zone) contains marginal cells (M, Fig. 15–4), which have large, flattened disk-shaped dendritic arbors confined to lamina I. Many marginal cells project axons great distances to a variety of structures (e.g., the lateral cervical nucleus, dorsal column nuclei, thalamus), whereas others are interneurons whose axons reach only a few nearby segments. Since lamina I receives small nociceptive Aδ and C fibers, it is not surprising that most of its cells respond primarily to noxious stimuli.

Lamina II (substantia gelatinosa) has an outer region composed predominately of stalked cells (S, Fig. 15–4) and an inner region of mostly islet cells (I, Fig. 15–4). The dendritic arbors of both of these cell types are restricted largely to lamina II but may extend to laminae III or I. Most axons of lamina II cells terminate locally; islet cells terminate in lamina II and stalked cells terminate in lamina I (Fig. 15–4B) or project intersegmentally via Lissauer's tract to the substantia gelatinosa of nearby segments. A few project as far as the thalamus. Since lamina II receives primarily C fiber inputs, most lamina II cells respond to nociceptive or strong mechanical stimuli. A small number, however (perhaps 20%), respond both to high-threshold or nociceptive stimuli *and* to low-threshold stimuli such as light touch, presumably because their dendrites sample C fiber inputs to lamina II as well as myelinated afferent fiber inputs to lamina III. Unlike those of primary afferent fibers, the responses of lamina II cells habituate to repeated stimuli.

Lamina II also contains three classes of "inverse" neurons, which are inhibited rather than excited by natural peripheral stimuli; one class is inhibited by nociceptors, one by mechanoreceptors and nociceptors, and one only by sensitive cutaneous mechanoreceptors. These cells are thought to be involved in the modulation of nociceptive input (see Chap. 16).

In contrast to laminae I and II, the remaining dorsal horn laminae (III–VI) contain most of the

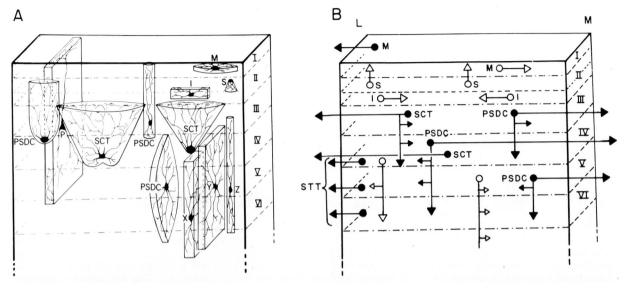

Figure 15–4 The morphology and connections of various cells in the dorsal horn of the cat spinal cord. *A*, Diagrammatic representation of the organization of dendritic trees viewed in a parasagittal block of tissue. *B*, Schematic illustration of the connections of the same neuron types depicted in *A*; solid circles are projection neurons; open circles are interneurons. I = islet cell (lamina II); M = marginal cell (lamina I); P = pyramidal cell (lamina III); PSDC = three main types of cells projecting to the dorsal columns; S = stalked cell (lamina II); SCT = spinocervical tract cell; STT = spinothalamic tract cell; X, Y, and Z = interneurons (laminae V & VI). (After Brown, A. G. *Organization of the Spinal Cord.* New York, Springer-Verlag, 1981.)

cells whose axons ascend in (or project into) the main ascending somatosensory pathways. Although these more ventral laminae receive myelinated A fiber inputs (see above), both projection and intrinsic neurons in laminae III through VI can have large dendritic arbors (see Fig. 15–4A) that may extend across several laminae (including lamina II) and therefore can have access to a variety of other inputs. For example, many laminae V and VI neurons respond both to light touch and pinch. Other lamina VI cells respond to muscle and joint inputs. On the other hand, many lamina IV cells respond exclusively to light mechanical stimuli, presumably because they have more restricted dendritic arbors. For example, many of the lamina IV cells that project into the spinocervical tract respond primarily to hair movement and have dendrites that are restricted to laminae containing hair follicle afferent fiber terminals (laminae III and IV). Thus, as one moves ventrally from lamina III to VI, cells progress from favoring light mechanical stimuli to responding to several modalities (*i.e., they become multimodal*).

Consistent with the increase in their dendritic arborizations, cells more ventral in the dorsal horn have larger, more complex receptive fields. For example, the majority of cells in lamina IV have simple excitatory receptive fields. In contrast, cells in lamina V have receptive fields with both excitatory and inhibitory subregions. For some cells whose axons ascend in the dorsal columns, the inhibitory field may surround the excitatory one. As in the visual system (see Chap. 19), this *surround inhibition* may help sharpen the spatial acuity of our somatic sensations (see Chap. 13). Another, more diffuse, type of inhibitory field is shown in Figure 15–5A. This cell, which projects into the spinothalamic tract, is excited only by noxious stimuli delivered to the right foot and is inhibited by noxious stimuli presented over the rest of the body.

Stimulus location alone does not determine the response of all spinothalamic tract cells.[48] Often, cells that are excited by one modality are inhibited by another presented to the same receptive field location. For example, the cell in Figure 15–5B is excited by pinching and squeezing of the tail. Other, less noxious stimuli (e.g., a brush), delivered to the same skin location inhibit spontaneous activity or (for pressure) show a mixture of early excitation and later inhibition. This type of interaction may explain the decrease in subjective pain sensation associated with rubbing an injury.

As expected from the axonal terminations described earlier, neurons in the medial part of the dorsal horn have distal receptive fields, whereas those in the lateral part of the dorsal horn have proximal fields.[43] In addition, cells in the medial part of the dorsal horn usually are more responsive to stimuli on the ventral body surface, and cells

A

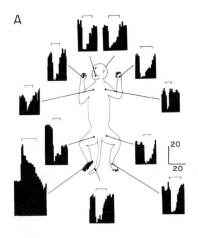

B

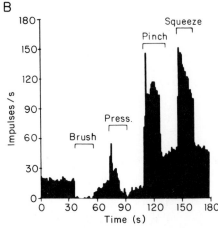

Figure 15–5 Responses of two cells in the monkey with axons that ascend in the spinothalamic tract. *A,* Neuron that is excited by squeezing the skin of the right foot with a forceps and is inhibited by squeezing a wide expanse of the rest of the body and face. *B,* A high-threshold neuron with a receptive field on the tail. Brushing the hair of the tail inhibited the resting rate; pressure produced by an arterial clip produced an initial excitation and then inhibition; pinching with greater force and squeezing the skin with a forceps produced strong excitatory responses. Brackets in *A* and *B* indicate the time of stimulus application. In *A,* the calibration "L" above the lower right histogram indicates 20 spikes/sec (ordinate) and 20 sec (abscissa). (After Willis, W.D. In Rowe, M.; Willis, W.D., eds. *Development, Organization and Processing in Somatosensory Pathways,* pp. 333–346. New York, Alan R. Liss, 1985.)

in the lateral parts are more responsive to dorsal body stimulation. As we shall see, this organization is *refined* in the dorsal columns as a first step toward a complete, orderly representation of the body surface.

Descending Influences

Descending inhibition can modulate the flow of sensory information through the spinal cord. These effects can be demonstrated by comparing somatosensory responses in dorsal horn cells before and during stimulation of central structures or before and after transection of descending tracts.

Descending influences come primarily from the cortex via the corticospinal tract (terminating in laminae III–VI) and from the brain stem via both the raphespinal tract (so-named because it originates in the raphe nuclei on the midline of the brain stem reticular formation and terminates in laminae I, II, and V–VII of the spinal cord) and the dorsal and ventral reticulospinal tracts, which terminate in laminae I, II, and V–VII, and in laminae V, VII, and VIII, respectively; see also Chap. 26). Electrical stimulation of the cortex or the corticospinal tract can produce either inhibition or excitation of many dorsal horn neurons.[12] Electrical stimulation of the nucleus raphe magnus

inhibits many dorsal horn interneurons and spinothalamic tract cells (see Chap. 16). Hemisection of the spinal cord, which severs these descending pathways, increases both the receptive field size of dorsal column cells and the magnitude of their response to sural nerve stimulation.

ASCENDING PATHWAYS

As is the case for all sensory modalities, somatosensory information is communicated to the somatosensory regions of the cerebral cortex via thalamic nuclei. Somatosensory signals from the spinal cord reach the thalamus over several separate routes, including the dorsal column pathway (via the dorsal funiculi) and the spinothalamic, spinocervicothalamic, and spinoreticulothalamic pathways (Fig. 15–6). The dorsal column system is by far the most extensive ascending somatosensory system, containing in carnivores at least 10- to 100-fold more fibers than either the spinothalamic or the spinocervicothalamic pathway.[25]

Dorsal Column Pathway

As mentioned above, many of the larger dorsal root fibers branch as they enter the spinal cord, and a main branch ascends in the dorsal columns

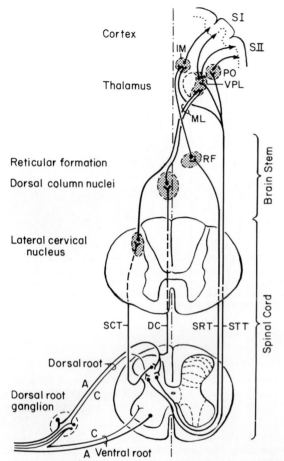

Figure 15–6 Primary afferent fiber and ascending pathways involved in somesthesis. All large myelinated (A) and most small unmyelinated (C) fibers enter through the dorsal roots; a few C fibers enter via the ventral root. Ascending sensory pathways include the spinocervical tract (SCT), dorsal columns (DC), spinoreticular tract (SRT), and spinothalamic tract (STT). Sensory relay nuclei in the thalamus include the ventral posterolateral nucleus (VPL), posterior nuclei (PO), and intralaminar nuclei (IM). SI and SII are the two primary somatosensory areas in cerebral cortex. (After Iggo, A. In Barlow, H.B.; Mollon, J.D., eds. *The Senses*, p. 390. Cambridge, Cambridge University Press, 1982.)

ment in primates; in humans they occupy about 40% of the total cross section of the high cervical cord.

The cortical representation of many sensory modalities is characterized by a *topographic organization* in which neighboring cortical neurons are influenced by stimuli delivered to adjacent points in the sensory world. Visual cortex has an excellent retinotopic map of visual space, auditory cortex a fairly discrete tonotopic map of sound frequency, and somatosensory cortex a rather gross somatotopic map of the body surface.

The topographic organization that characterizes higher levels of the somatosensory system appears clearly for the first time in the dorsal columns. If one cuts selective dorsal roots and traces the resulting degeneration in the dorsal columns, fibers from successively more rostral dorsal roots are seen to occupy progressively more lateral positions. For example, the dorsal columns in the caudal (lumbar) cord are organized according to the dermatomes of the successively applied dorsal roots. However, the receptive field positions of individual fibers show a less orderly arrangement, as can be seen in Figure 15–7A. For example, fibers with receptive fields on the foot are widely dispersed throughout the two electrode penetrations shown and consequently are extensively intermingled with fibers innervating receptive fields on other parts of the limb.[45] Similar penetrations through the more rostral (i.e., cervical) spinal cord, however, reveal that *at that level* the hindlimb is represented by an orderly series of mediolateral fiber bands, each containing axons innervating a local region of the ipsilateral body surface. For example, the shaded area of Figure 15–7B shows the mediolateral band of fibers innervating the foot. This re-sorting of hindlimb axons to produce a sequence of fibers whose receptive fields map a continuous path on the lower body is a necessary step in establishing a somatotopic organization. Much less is known about the distribution of dorsal column axons from the forelimb.

As suggested earlier, the vast majority of the dorsal root fibers that enter the dorsal columns are specifically mechanoreceptive. These include the full complement of larger (3–4 μm) dorsal root fibers, such as those from muscles, and the Aβ fibers from the skin, joints, and other deep structures. Few Aδ or C afferent fibers enter the dorsal columns. Figure 15–7A shows that, as expected, fibers with both deep and cutaneous receptive fields on the hindlimb are found in the dorsal columns near the level where dorsal root fibers enter the spinal cord. However, electrode penetra-

to terminate in the dorsal column nuclei of the medulla (see Fig. 15–6). Although the vast majority of dorsal column fibers are primary afferent fiber branches destined for the medulla, other axons traveling in the dorsal columns include (1) primary axons that leave at higher segmental levels, (2) axons of intraspinal neurons that either leave at higher segmental levels or project to the dorsal column nuclei, and (3) descending axons of forebrain and brain-stem origin. Although they exist in reptiles, birds, and all orders of mammals, the dorsal columns reach their greatest develop-

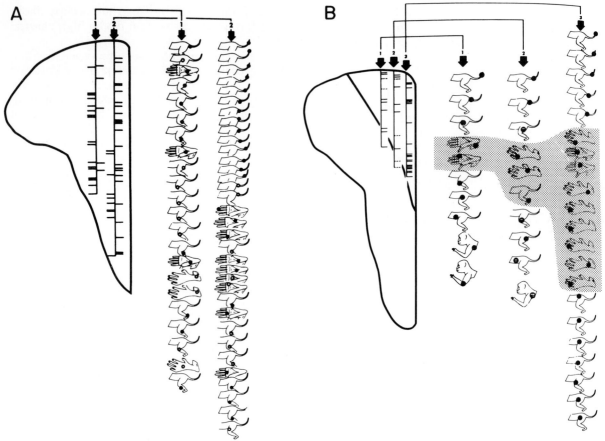

Figure 15–7 Modality, receptive field location, and position of single fibers within the lumbar *(A)* and cervical *(B)* fasciculus gracilis of the squirrel monkey. In the two penetrations in *A* and the three in *B*, the locations of single recorded cutaneous fibers are indicated by the horizontal lines to the right of the penetration, and deep fibers by the horizontal lines to the left; dotted lines indicate multineuron recordings. The receptive field center of each skin fiber is indicated by a solid circle, and each deep fiber by an open circle; the "striped" circle indicates the position of multineuron recordings. Shaded area in *B* shows the mediolateral band devoted to the foot. (After Whitsel, B.L.; Petrucelli, L.M.; Sapiro, G. *Exp. Neurol.* 29:227–242, 1970.)

tions though the dorsal columns at cervical levels encounter mostly rapidly adapting fibers with cutaneous receptive fields, suggesting that a re-sorting for modality has occurred.[46] Apparently, the re-sorting of some axons and the exit of others shapes the dorsal columns (in this case, the part called the fasciculus gracilis) into a system specialized for transmitting signals about dynamic changes in the patterns of cutaneous stimuli.

About 10% of the dorsal column fibers in the cat and less than 10% in the monkey arise from cells of the dorsal horn. These neurons are located mainly in lamina IV and the medial part of lamina V, with some additional contribution from laminae III in the cat (see Fig. 15–4B). In the cat, over three fourths of these cells receive convergent input from cutaneous and muscular afferent fibers, and a significant number can be activated by noxious stimulation of the skin. The information that

these fibers deliver, therefore, differs from that communicated by first-order dorsal column afferent fibers; their role in somesthesis is unknown.

Dorsal Column Nuclei

Fibers from the dorsal columns terminate in the gracile and cuneate nuclei (i.e., dorsal column nuclei), which lie at the junction of the spinal cord and medulla (see Fig. 15–6). In the cat, the animal used in most experiments, the dorsomedially placed gracile nucleus receives information from the hindlimbs and trunk, whereas the laterally placed cuneate nucleus receives afferent fibers from the forelimbs. Penetrations through the dorsal column nuclei successively encounter cells whose receptive fields are located sequentially along the ipsilateral body surface. For example, the most dorsal cells in the medial penetration through the cuneate nucleus shown in Figure 15–8 have receptive fields at the tips of the toes, and the receptive fields of deeper cells occupy more and more proximal sites along the leg. The distribution of receptive fields on the body surface encountered in such penetrations is often depicted in a cartoon like that at the top of Figure 15–8. This representation demonstrates that more cells (a greater amount of neural tissue) in the dorsal column nuclei are devoted to the parts of the body with the highest tactile acuity (i.e., the toes and paws) and that fewer cells respond to stimuli delivered to the trunk. Also, receptive fields on the toes and paws are usually considerably smaller than those on the legs and trunk (see the track through the feline gracile nucleus in Fig. 15–8). The head of the figurine is provided by somatosensory input to neurons of the trigeminal complex (Fig. 15–8, left-most track), which will be considered separately at the end of this chapter.

Within the dorsal column nuclear complex there are two cytoarchitectonically differentiable zones with possibly different functional roles. Relay neurons that project to the thalamus via the medial lemniscus (see Fig. 15–6) are largely but not solely concentrated in clusters in the caudal half of the core of the nuclei. Most of these "cluster" neurons are medium-sized to large spherical cells that receive direct dorsal column inputs. The surrounding rostral and deeper portions of the nuclei contain a mixture of large and small cells that receive (1) mostly non–dorsal column inputs from laminae I and II and (2) dense descending projections from the cortex. These noncluster regions contain more interneurons and also contain most of the cells that project to non-thalamic areas such as the spinal cord, inferior olive, cerebellum, pretectum, tectum, and red nucleus.[4]

Two thirds of a large sample of neurons in the dorsal column nuclei of the cat responded either to light touch or the displacement of a few hairs.[21] Twelve per cent responded only to joint movement, and the remainder responded only to pressure, but it was often difficult to determine whether deep or cutaneous pressure was the adequate stimulus. None of the cells in this anesthetized preparation responded to noxious stimuli. Almost three fourths of the cells were rapidly adapting.

In summary, the direct pathway from the dorsal columns through the cluster neurons of the dorsal column nuclei and then to the thalamus seems uniquely suited to convey information related to rapidly changing spatial patterns of tactile stimu-

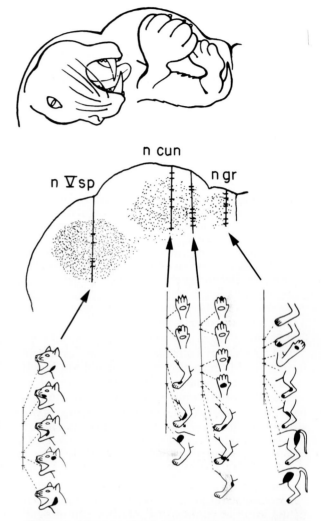

Figure 15–8 The receptive field size and location of neurons recorded at different depths in the nucleus cuneatus (n cun), nucleus gracilis (n gr), and the spinal nucleus of cranial nerve V (n V sp) in the cat. The hindlimb is represented in n gr, the forelimb in two penetrations through n cun, and the head in n V sp. The figure at the top shows the amount of gray matter devoted to different body parts. (After Kruger, L.; Siminoff, R.; Witkovsky, P. *J. Neurophysiol.* 24:333–349, 1961.)

lation. This information, which is kept largely separate according to modality, may be intimately involved in form discrimination. On the other hand, extra-cluster cells are of mixed modality, have larger receptive fields, and project to areas that are traditionally thought to have a role in movement. The dorsal column nuclei should not be thought of as passive relays, however, since the large majority of terminals on dorsal column nucleus cells originate from sources other than the dorsal columns, such as interneurons, recurrent collaterals of relay neuron axons, and descending axons from the cerebral cortex. The role of these other influences is yet to be elucidated.

The direct and indirect (via the reticular formation) cortical projection to the dorsal column nuclei may modulate the transmission of sensory information. Stimulation of somatosensory area SI (see later) or motor cortex (see Chap. 28) can either inhibit or facilitate the activity of cells in the dorsal column nuclei.[40] These effects, which are mediated primarily by pyramidal tract fibers and bulbar reticular neurons or local interneurons of the dorsal column nuclei, are strongest when the stimulation is in that part of the contralateral somatosensory cortex that contains neurons with receptive field locations similar to that of the recorded dorsal column nucleus neuron. Thus, the cortical areas receiving somatosensory information are reciprocally connected with those regions of the dorsal column nuclei that relay the same information. These point-to-point connections may allow somatosensory input to be shaped according to selective attention or arousal (see Chap. 31).

Spinocervicothalamic Pathway

The spinocervicothalamic system originates from dorsal horn neurons lying mainly in lamina IV in the cat (with a lesser contribution from lamina III) and laminae IV and V in the monkey[6] (see Fig. 15–4B). Axons from these neurons constitute the spinocervical tract, which ascends in the ipsilateral dorsolateral column to the C1-C2 levels of the spinal cord to terminate on cells of the lateral cervical nucleus (see Fig. 15–6). Axons of the cells in the lateral cervical nucleus, which lies ventrolateral to the dorsal horn, cross the cord in the anterior commissure, traverse the medulla lateral to the inferior olive, pass through the pons, and join the medial lemniscus. The spinocervicothalamic pathway is more highly developed in carnivores than in primates.

The cells of origin of the spinocervical tract can be identified by antidromic activation from stimulating electrodes in the lateral cervical nucleus. In both the anesthetized cat and monkey, the vast majority of such cells respond to hair movement with rapidly adapting discharges.[5, 6] In addition, some also respond to pressure or pinch of the skin with slowly adapting discharges. Only the occasional cell is excited by noxious stimuli. Few if any responses attributable to stimulation of deep tissues have been observed.

Most spinocervical neurons are excited by their adequate stimulus; in addition, some are inhibited by stimuli placed quite distant (often on the contralateral limb) to the excitatory field. This separation of the excitatory and inhibitory portions of a somatosensory receptive field, which is quite different from the contiguous receptive fields seen in peripheral somatic afferent fibers, indicates a considerable convergence of peripheral information onto these cells.

Similar to the organization demonstrated above for the dorsal columns, there is a somatotopic map of the body surface of the cat on the sheet of spinocervical neurons. For the hindlimb, the toes are represented medially, the foot somewhat more laterally, and the leg more laterally still. Again, there is a distorting enlargement of the representation for the toes.

Lateral Cervical Nucleus

As might be expected from the behavior of their input cells, the majority of cells in the lateral cervical nucleus can be activated by small displacements of one or a few hairs on the ipsilateral body surface. Most cells have single excitatory receptive fields with no evidence of a surrounding inhibitory region. The sizes of the receptive fields range from a single toe to a whole limb, or even to much of the ipsilateral body surface.[10] Distal receptive fields are smaller than proximal ones. A few cells have either multiple excitatory receptive fields or an inhibitory receptive field at some distance from the excitatory one; that is, inhibition can often be elicited by the pinch of a distant body part.

Thus, the spinocervicothalamic pathway is dominated by low-threshold, velocity-sensitive information that could be useful for signaling small, rapid changes in the tactile environment. This information differs from that conveyed by the dorsal columns in that (1) it carries a balanced representation of both the fore- and hindlimbs (the dorsal column nuclei collectively contain a much greater forelimb than hindlimb representation; see Fig. 15–8); (2) it carries information primarily from hair follicle receptors and not from receptors in the glabrous skin; and (3) it is most highly developed in quadrupedal carnivores (e.g., the cat). Thus, unlike the dorsal columns, it is not best suited for fine tactile discrimination with the forepaws; rather, it is a system for rapidly and reliably relaying information about hair movement. It has been suggested that such information could be useful in the regulation of locomotion.

Spinothalamic Pathway

Like the dorsal column pathway, the spinothalamic pathway includes one relay neuron between the primary afferent fibers and the thalamus. These relay neurons, which are located primarily in laminae IV through VI (in primates, the highest concentration is found in lamina V; see Fig. 15–4B), give rise to axons that cross in the anterior commissure and ascend in the contralateral anterolateral and ventral funiculi as the *spinothalamic tract.* In contrast to the dorsal columns, spinothalamic axons from successively more rostral spinal cord segments are applied more medially. This arrangement forms a loose topographic pattern with distal body parts more lateral and proximal body parts more medial. Based largely on studies with humans, it is thought that the spinothalamic tract has functionally separate ventral and lateral components that communicate different types of sensory information. The thalamic destinations of the spinothalamic fibers are the ventral posterolateral nucleus (VPL), the ill-defined posterior group of nuclei (PO), and the intralaminar nuclei (IM).

About one third of spinothalamic cells, which are identified by antidromic activation from electrodes placed near VPL, respond only to noxious stimuli such as excess temperature and pinching of the skin with a clip.[30] The remaining cells respond to several stimulus modalities, including ones with noxious qualities.[49] For example, some cells respond transiently to gentle stimulation of hairs or weak stimulation of the skin but respond more robustly to noxious mechanical or thermal stimuli. Furthermore, the maintained firing rates in response to noxious stimuli increase with increases in stimulus magnitude (e.g., pressure or temperature). Only a few spinothalamic cells respond selectively to low intensity tactile stimuli. Consequently, the spinothalamic pathway is thought to deliver information primarily about noxious stimuli. The processing of pain information in this pathway is considered separately and in detail in Chapter 16.

A fourth ascending pathway constitutes a spinoreticulothalamic system. Most fibers in this pathway are intermingled with those of the spinothalamic pathway. They terminate in the medial two thirds of the reticular formation, largely in the nucleus reticularis gigantocellularis and the reticular nuclei pontis oralis and caudalis. These nuclei contain large cells with long axons that project to the thalamus. Antidromically activated spinoreticular tract cells give complex responses to both noxious and non-noxious cutaneous stimuli and reside primarily in laminae VIII and IX.[22] (The association of this system with slowly conducted pain sensation is discussed more thoroughly in Chapter 16.)

Pathophysiology

Deficits produced by transections of either the dorsal columns or the anterolateral pathways in monkeys and humans enable us to contrast the roles of these two major pathways in somesthesis. Primates with lesions of the dorsal columns consistently fail to perform sophisticated somatosensory discriminations. For example, monkeys with dorsal column lesions fail to discriminate among contoured, flat, and three-dimensional forms by either active palpation with the hand or movement of the form over a stationary hand.[2] Monkeys with dorsal column damage can still detect the motion of a cutaneous stimulus but fail to specify its direction. Also, they have difficulty with other discriminations that might aid them in identifying a form, such as appreciation of the relative positions and distances between point or edge contours and discrimination between different forces produced by stationary indentations of the skin.[42] Human patients with dorsal columns damaged by stab wounds or tumors[44] fail to detect the direction or speed of moving cutaneous stimuli; they also lose the capacity to identify numbers written on the skin and exhibit modest deficits in two-point discrimination, point localization, and vibratory sensation (see Chap. 13). In addition to deficits in somatosensory discrimination, monkeys with dorsal column lesions show long-lasting deficits in the control of hand movements that require sensory guidance such as the catching of a falling bait and the execution of small skilled movements of the fingers. Such monkeys scrape rather than pick food out of depressions and often drop grasped food pellets.[42]

In contrast to the devastating effects of dorsal column lesions on somatosensory sensibility, damage to the anterior or lateral columns in humans produces analgesia but only a mild elevation in tactile thresholds. This latter modest deficit suggests that the spinothalamic pathway delivers some information about touch. The major role of the spinothalamic system, however, is clearly to relay pain and temperature information. Indeed (as will be discussed in Chap. 16), anterolateral cordotomy is a treatment for intractable pain. During such procedures, local stimulation of the intact ventral quadrants elicits reports of pain, heat, or cold but never of touch or vibration.[26] Of course, pain and temperature sensations are extremely compelling and may reduce or obscure other sensations.

THALAMIC NUCLEI

The thalami, bilateral egg-shaped nuclear structures sitting atop the brain stem, are composed of a number of discrete nuclei, which are schematized in Figure 15–9. The nuclei by which most afferent information reaches the cerebral cortex are called the specific relay nuclei. Other thalamic nuclei (called association and non-specific nuclei) contain thalamic interneurons or project diffusely to the cortex or to subcortical structures. As we shall see, somatosensory information reaches all of these different types of thalamic nuclei.

Organization

The four ascending somatosensory pathways terminate primarily in certain thalamic nuclei (the shaded portions of Fig. 15–9). Most of these afferent signals arise from contralateral peripheral receptors. Axons from cells in the dorsal column nuclei cross the brain stem immediately at the level of the medulla, and travel in the contralateral medial lemniscus to terminate primarily in the ventral posterolateral nucleus (VPL) of the thalamus; a few fibers reach the medial portion of the posterior thalamus (PO). Axons from the lateral cervical nucleus of the spinocervicothalamic pathway cross at a high cervical level to join the contralateral medial lemniscus. Most fibers of the spinothalamic pathway travel in the contralateral

anterolateral funiculus and spinothalamic tract to terminate in the VPL, PO, and IM. Finally, reticulothalamic fibers, which are associated with the spinoreticulothalamic pathway, are largely uncrossed and terminate in IM, including the center median (CM) and parafascicular (Pf) nuclei, and in the thalamic reticular nuclei (RET). The VPL and PO project to a limited number of cortical regions. IM has a diffuse cortical projection and strong subcortical projections, and RET is reciprocally connected with other thalamic nuclei.

The ventral intermediate nucleus (Vim), and portions of the ventral lateral (VL) and ventral anterior (VA) nuclei receive some somatosensory information in some species,[1] but are primarily related to other systems (e.g., VL receives heavy projections from the cerebellum)[11] and will not be discussed here.

Ventral Posterolateral Nucleus (VPL)

The VPL receives projections from a variety of ascending somatic pathways, and each has a characteristic pattern of termination. Although medial lemniscus fibers from the dorsal column nuclei terminate throughout the VPL, its central core receives largely cutaneous inputs, its dorsal aspect mostly deep inputs, and its anterior surface predominately muscle inputs. Thus, the VPL sometimes is considered to be composed of a core of cutaneous inputs and a shell of deep inputs.

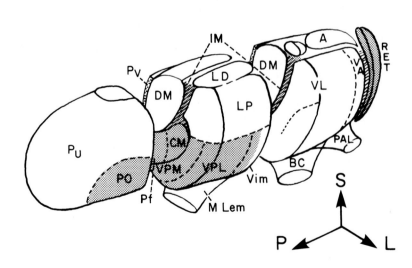

Figure 15–9 Lateral schematic view of the right thalamus as viewed from above and behind. Shaded regions receive input from the ascending somatosensory pathways and are discussed in detail in the text. A = anterior; BC = brachium conjunctivum; CM = nucleus "center median"; DM = dorsalis medialis; IM = intralaminar nuclei; LD = nucleus lateralis dorsalis; M Lem = medial lemniscus; LP = nucleus lateralis posterior; PAL = pallidal afferences; Pf = nucleus parafascicularis; PO = posterior nuclei of the thalamus; P_u = pulvinar; P_v = nuclei paraventriculares; RET = thalamic reticular nuclei; VA = ventral anterior nucleus; Vim = ventral intermediate nucleus ; VL = ventral lateral nucleus; VPL = ventral posterolateral nucleus; VPM = ventral posteromedial nucleus. P, S, and L in the inset indicate posterior, superior, and lateral, respectively. (After Albe-Fessard, D; Besson, J.M. In Iggo, A., ed. *Handbook of Sensory Physiology*, vol 2. *Somatosensory System*, pp. 489–560. New York, Springer-Verlag, 1973.)

In primates, spinothalamic afferent fibers terminate throughout the VPL, whereas in cats and other lower mammals the terminations are more restricted to its anterior border. The VPL receives inputs from both lamina I neurons (presumably nociceptive) and lamina V neurons (presumably multimodal cells). Recordings suggest that the majority of spinal neurons projecting to the VPL are multimodal, with high-threshold nociceptive cells being the next most common class; none are low-threshold mechanoreceptive neurons. Nearly all of these spinothalamic cells have cutaneous receptive fields.

Receptive Field Properties

Individual VPL neurons in the monkey respond to stimuli of a specific modality, delivered to a particular receptive field location.[27, 28] Therefore, although the VPL receives a considerable spinothalamic input, the discharge of VPL neurons seems to be dominated by their lemniscal inputs. VPL neurons are activated only by mechanical stimuli—specifically, light cutaneous stimuli (42%), stimulation of deep tissue (32%), or joint movement (26%). Most cells have relatively small contralateral receptive fields and, as in the dorsal column nuclei, the smallest receptive fields are located in the distal extremities. Strictly high-threshold nociceptive neurons and neurons with bilateral receptive fields, features characteristic of presumed spinothalamic inputs, are only rarely encountered.

The receptive field locations of VPL neurons are somatotopically organized, such that adjacent regions of the body are represented in adjacent layers, or lamellae. Electrode penetrations driven perpendicular to these layers (Fig. 15–10A) encounter cells with receptive field locations that progress across the body surface, while those

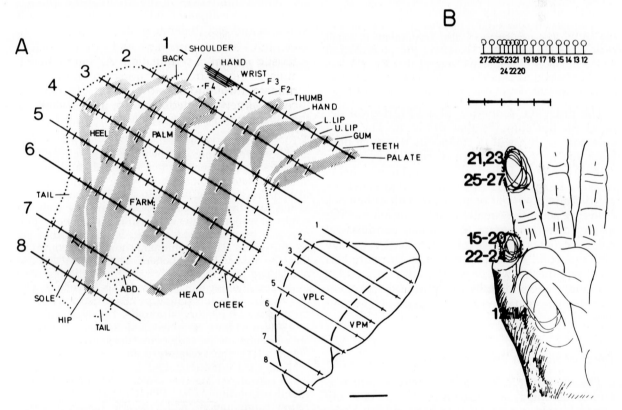

Figure 15–10 Topographic representation of the body surface in the ventral posterolateral nucleus (VPL) and ventral posteromedial nucleus (VPM) of the barbiturate-anesthetized monkey. *A,* Eight electrode penetrations made from anterolateral to posteromedial. Clear and stippled strips represent regions in which multineuron activity was recorded in response to cutaneous stimulation of the indicated body parts; bars designate the location of isolated neurons. *B,* Part of a single horizontal penetration in VPL, parallel to a lamella, with the location of individually numbered isolated cells along the track indicated above and the receptive fields of the numbered cells displayed on the hand below. The numbers indicate the sequential position of the cell in the track; cell No. 12 is the first cell in the functionally identified lamella. In both *A* and *B,* the calibration bar is 1 mm. (After Jones, E.G. *The Thalamus.* New York, Plenum Press, 1985.)

running parallel to the layers (Fig. 15–10B) reveal cells with similar receptive field locations.[18] The layers representing caudal body parts (e.g., the tail and foot) lie anterolaterally, whereas those representing rostral body parts (e.g., the hand) lie posteromedially. The face and head are represented in the ventral posteromedial nucleus (VPM), which is posteromedial to the VPL. The VPM, which is often studied in conjunction with the VPL, receives trigeminal inputs and will be discussed at the end of this chapter.

This lamellar organization has been attributed to the termination patterns of medial lemniscus fibers. As the medial lemniscus enters the thalamus, its fibers are laid down sequentially, in parallel with the physiologically defined lamellae.[18] The lamellae also appear to have a suborganization, since small clusters of cells within a lamella have similar discharge properties. Focal injections of marker substances, which are transported in the axons of nerve cells and can be subsequently visualized, have been made in the dorsal column nuclei (tritiated amino acids for orthograde transport to the VPL) and in the cortex (HRP for retrograde transport to the VPL). These substances also accumulate in small clusters within the VPL. Therefore, the physiological and anatomical observations suggest that the clusters relay a labeled line of modality and place specific information from the dorsal column nuclei to the cortex.

The discharge properties of a VPL neuron depend on the stimulus that drives it. The most common neuron with a cutaneous receptive field responds to the movement of body hairs and is rapidly adapting. Other cells with cutaneous fields respond to light touch and can be either slowly or rapidly adapting. All of the cells with cutaneous fields are activated equally well by stimuli moving in any direction across their small, contralateral receptive fields.[13] Some cells with deep receptive fields respond to light or moderate pressure and may be either slowly or rapidly adapting; others respond either tonically or phasically to joint movement, often over a broad range of joint angles.

Effect of State on VPL Responses

The view of the VPL just presented, which is widely accepted, was established from data obtained largely in animals anesthetized with barbiturates. Data obtained under different types of anesthesia (e.g., alpha-chloralose) or in awake and unparalyzed animals provide a different picture. For example, in unparalyzed and unanesthetized cats, VPL neurons have large contralateral receptive fields that often encompass whole limbs and the adjacent trunk.[3] In alpha-chloralose–anesthetized cats, cells with even larger receptive fields that are sometimes discontinuous, and often ipsilateral, may be found. Many of these cells also can be driven by more than one stimulus modality. Under some conditions, even small numbers of VPL neurons with small nociceptive receptive fields can be recorded. Taken together, these data suggest that cells in the VPL receive convergent inputs that can be revealed under the appropriate conditions. Indeed, such multimodal cells with large receptive fields are present, albeit in small numbers, even in awake primates. Therefore, the organization of the VPL seems to be state dependent; it can range from the somatotopically organized modality and place specific responses that are commonly observed, to the large, often ipsilateral and occasionally discontinuous receptive fields, and multimodal responses seen under alpha-chloralose anesthesia. These disparate findings may also reflect differences in the theoretical approach of various investigators, who use different stimuli to reveal and test thalamic units.

Surround Inhibition

A small percentage of the cells in the VPL are inhibited by stimulation of areas just outside of their excitatory receptive fields.

To reveal the spatial extent of the excitatory and inhibitory areas, Jänig and coworkers[17] used a condition-test paradigm on neurons that exhibited an early excitation and late inhibition in response to tactile stimuli. Maximal excitatory responses to single stimuli delivered to the receptive field center were compared with the same test responses after a preceding conditioning stimulus had been delivered at other locations within the receptive field. The conditioning stimulus was timed so that the excitatory component of the test response overlapped the inhibitory component of the response to the conditioning stimulus. With this procedure, 89% of all neurons in the ventral posterior thalamus of the cat displayed inhibitory surrounds. The average excitatory and inhibitory receptive field dimensions and relative locations of the center and surround are shown in Figure 15–11; the excitatory center (Fig. 15–11A) and inhibitory surround (Fig. 15–11B) are coextensive and, when considered together, account for the center-surround organization (Fig. 15–11C). Such a center-surround configuration, which is also found in dorsal column and cortical cells, could increase the spatial resolution of the somatosensory system and facilitate two-point discrimination (see Chap. 13).

It is difficult to believe that lateral inhibition is the sole mechanism for enhanced somatic acuity, however, since the condition-test studies are open to criticism (e.g., inhibition may be confused with occlusion[39]), and

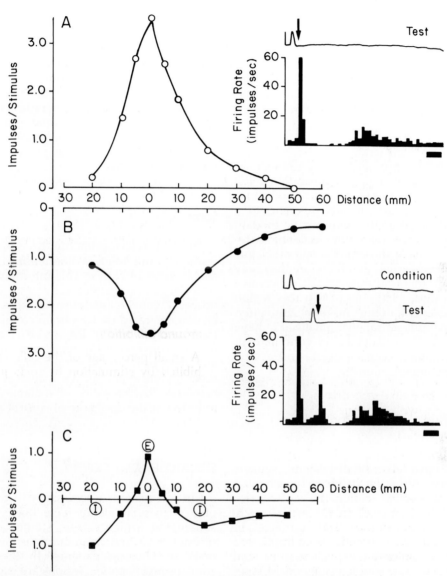

Figure 15–11 Surround inhibition in the cat thalamus as revealed by a condition-test paradigm. *A,* Average initial excitatory response (see arrow on accompanying histogram for a typical response) of 8 cortically projecting ventral posterolateral nucleus (VPL) neurons to air-jet stimuli delivered to their receptive fields, located on the forepaw and lower foreleg. The receptive field centers are at 0; other stimulus locations are indicated by their distance from the receptive field centers. *B,* Inhibition produced by conditioning a maximally excitatory test response produced by stimulation of the receptive field center, with a conditioning stimulus delivered earlier at other locations (indicated on the abscissa) in the receptive field. The two stimuli are timed so that the initial excitatory response to the test stimulus overlaps the late inhibitory response produced by the conditioning stimulus (as simulated in the accompanying inset). Inhibition is displayed as the difference (impulses/ stimulus) between the maximal response with (arrow, inset in *B*) and without (arrow, inset in *A*) conditioning. *C,* Summation of curves in *A* and *B* yields the overall spatial response profile of the cells to both direct excitation and lateral inhibition. Note how the excitatory center (E) and the inhibitory surround (I) can be accounted for by a summation of the individual E and I processes. (After Jänig, W.; Spencer, W.A; Younkin, S.G. *J. Neurophysiol.* 42:1450–1460, 1979.)

other paradigms only rarely reveal inhibitory surrounds. Also, condition-test studies in somatosensory cortex reveal inhibitory fields that are similar in size to the excitatory fields and thus, cannot provide surround inhibition. Nevertheless, the concept, which is perhaps best demonstrated in the visual system, illustrates how lateral inhibition in the CNS *could* sharpen sensory acuity.

Posterior Nuclei (PO)

In PO, neurons with large receptive fields and multimodal convergence are the rule rather than the exception. In a variety of preparations, PO cells respond to nociceptive stimuli, and many, especially those in alert animals, respond to a variety of somatic and even non-somatic modalities (e.g., audition[27]). Many PO cells have responses that resemble those of the spinothalamic neurons described earlier, and therefore they may be involved in the transmission of nociceptive information. Indeed, large lesions in this region can produce analgesia in humans, and stimulation can produce painful sensations. Finally, PO have neurons with both ipsilateral and contralateral, often discontinuous, receptive fields, and exhibit no somatotopic organization.

Intralaminar Nuclei (IM) and Thalamic Reticular Nuclei (RET)

Cells in IM have large, often bilateral, somatic receptive fields.[1] Although they can be activated by punctate nociceptive stimuli, they often also are multimodal. Responses in IM are strongly affected by both anesthesia and the level of arousal. Although this region has been implicated in nociceptive transmission, as well as in general arousal, attention, and affect, lesions of IM (specifically CM) to control intractable pain in human patients have produced equivocal results. Indeed, ironically, one potential result of thalamic lesions is the thalamic pain syndrome, in which normally innocuous stimuli produce profoundly debilitating pain. Like IM, RET contains cells that respond at long latencies to a variety of somatosensory stimulus modalities. In contrast to the specific connections of the VPL to cortex, IM has a diffuse projection to the cortex, but a strong projection to the striatum of the basal ganglia (see Chap. 30). RET does not project to cortex at all but is reciprocally connected to IM, VPL, and many other thalamic nuclei; consequently it is thought to be involved in the modulation of thalamic activity.

Pathophysiology

Lesions of the VPL due to vascular accidents produce a severe impairment of the sensations of contralateral discriminative touch and pressure, whereas diffuse touch, temperature, and pain sensations are often less impaired. Vascular lesions often spare the medial thalamus, including the VPM, and hence sensations in the face and head often remain intact.

Summary

The VPL relays primarily discriminative touch, pressure, and joint position information to the somatosensory cortex. This information is organized largely somatotopically and by modality; however, convergence of somatosensory input can be demonstrated under certain conditions. In contrast, neurons in PO, IM, and RET are characterized by convergent inputs from many modalities, including pain, and show no evidence of somatotopic organization. As we shall see, many of the response characteristics of thalamic neurons are reflected in the discharge of cortical neurons.

SOMATOSENSORY CORTEX

Cortical regions associated with the processing of a sensory modality have been identified by converging evidence from studies using electrophysiological and anatomical techniques in animals and human patients. In the case of the somatosensory cortex, lesions are expected to produce an impairment of sensation in contralateral body parts; stimulation in humans should produce somatic sensory experiences; heavy projections should exist from the ventrobasal thalamus; and cortical neurons should respond to touch, pressure, pain and proprioceptive stimuli, but not to other modalities. All of these expectations are met in the primary somatosensory area (SI), which resides in the anterior parietal lobe on the posterior bank and floor of the central sulcus (i.e., Brodmann's areas 3, 1, and 2; see Fig. 15–12). A second, less well studied primary somatosensory area (SII) lies on the anterior wall of the lateral (Sylvian)

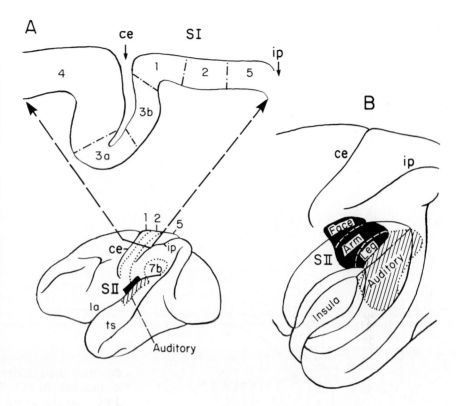

Figure 15–12 Anatomical location of SI and SII in the rhesus monkey. SI is divided into subregions 1, 2, 3a, 3b, and 5 according to Brodmann's classification; areas 3a and 3b are seen only after a cut through SI reveals the depths of the central sulcus (ce) in *A.* Also displayed are locations of the lateral (la), intraparietal (ip), and superior temporal (ts) sulci. Similarly, SII, which lies on the anterior bank of the lateral sulcus, is revealed only after the temporal lobe is pulled aside, as shown schematically in *B* (not to scale). The locations of the face, arm, and leg area of SII, as defined by evoked potential recordings, are shown, as is the location of the adjacent auditory cortex. (After Burton, H.; Robinson, C.J. In Woolsey, C.N., ed. *Cortical Sensory Organization,* vol. 1. *Multiple Sensory Maps,* pp. 67–119. Clifton, New Jersey, Humana, 1981.)

sulcus in primates. SII is best revealed by pulling back the temporal lobe to expose the inner bank of the lateral sulcus and the deeper insular cortex (Fig. 15–12). The cortex surrounding SII (especially the dorsally lying area 7b) also contains cells that respond to somatosensory stimuli, a fact that apparently confounded early descriptions of the discharge properties of SII neurons.

Somatosensory signals also reach the precentral cortex (Brodmann's area 4) and the somatosensory association cortex in the posterior parietal lobe (Brodmann's area 5). Precentral cortex is most logically considered as "motor" cortex, since lesions there produce motor disorders but no clear somatosensory deficits (see Chap. 28). The posterior parietal association area is considered to be a "higher" somatosensory area because its input is derived largely from SI and SII, and lesions of Brodmann's area 5 produce more subtle disorders in somatic sensation. The portion of the posterior parietal association area on the medial wall of the macaque parietal lobe is called the supplementary sensory area (SSA).

Figure 15–13 summarizes the major pathways flowing to and from the cortical somatosensory areas. Several general conclusions can be drawn. First, all subareas of SI make reciprocal, specific point-to-point connections (not shown in Fig. 15–13) with the caudal portion of the VPL in the thalamus; the heaviest thalamic input is to area 3a. SII also receives direct somatosensory input from the ventrobasal thalamus. Second, there is a serial flow of information from area 3b to area 5, but area 3a is reciprocally connected with areas 1 and 2. Third, SI and SII have extensive, reciprocal connections. Fourth, area 5 receives its only thalamic input from the lateral posterior nucleus (LP), a structure that receives no direct ascending somatosensory input (recall Fig. 15–9). Instead, area 5 receives its somatosensory input from areas 1 and 2. Fifth, SI and SII are each linked via the corpus callosum with their homologous areas in the opposite hemisphere (not shown). Finally, SI, SII, area 5, and SSA all project to precentral areas involved in the regulation of movement (i.e., Brodmann's areas 4 and 6). Consequently, like that of other sensory modalities (see Chaps. 18 and 20), the cortical processing of somatosensory information involves a complicated network of serial and parallel pathways.

Both SI and SII also receive weak inputs from other structures, which include the monoaminergic nuclei of the brain stem core, e.g., the locus ceruleus and raphe

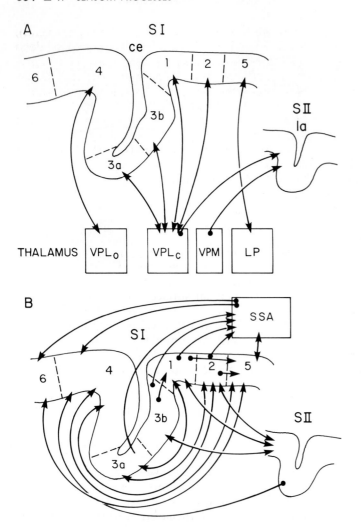

Figure 15–13 Thalamic *(A)* and intracortical *(B)* connections of areas SI and SII. Areas are designated numerically according to Brodmann's nomenclature. Thalamic nuclei include the ventral posterolateral nucleus pars oralis and caudalis (VPL$_o$ and VPL$_c$, respectively), the ventral posteromedial nucleus (VPM), and the lateral posterior nucleus (LP). ce = Central sulcus; la = lateral sulcus; SSA = supplementary sensory area.

nuclei. These nuclei are thought to be involved in regulating cortical excitability. In this regard, it is interesting that the discharge of some postcentral neurons in the monkey is affected by the animal's state of arousal. We have already seen that "state" effects can be detected at the thalamic level.

Somatosensory Area SI

Most of the information that we have about somatosensory cortex concerns its functional neuroanatomy and addresses questions such as (1) Which somatic signals reach the cortex? (2) Are these signals localized to cortical subregions? and (3) Do these subregions have a functional organization? To answer these questions, investigators have recorded neuronal activity from single or multiple neurons while probing the body surface and its deep structures (muscle and joints) with various stimuli. Unfortunately, each investigator usually has asked different questions, so that (1) dissimilar batteries of stimuli are applied, (2) frequently only the representations of specific body parts (e.g., the hand or hind limb) are explored, and (3) different species in different states of anesthesia (ranging from deeply anesthetized to fully alert) are used. Nevertheless, these various approaches have yielded some common insights into the organization of SI, which we will discuss for the monkey.

Modality Specific Subregions

Cortical neurons have been identified according to whether they respond to cutaneous or deep stimuli.[29, 35] The distinction between the activation of cutaneous and deep receptors is often a qualitative one. Stimuli believed to elicit cutaneous responses include air puffs, stroking with a camel's hair brush, touching with von Frey hairs, and light application of a small rod. In contrast, deep receptors respond to joint movement or the tapping of a muscle. The occasional investigator differentiates cutaneous from deep stimuli according to the weight of the stimulus necessary to elicit a response.

Most studies agree that the majority of neurons in area 3a are driven by stimulation of deep receptors, many of which lie in muscle or joints. Similarly, the majority of neurons in area 2 also prefer deep stimuli and respond poorly to cutaneous stimulation. In contrast, the majority of cells in area 3b prefer cutaneous stimulation. Area 1 neurons may be driven by stimulation of either cutaneous or deep receptors. Individual SI neurons are rarely excited by *both* cutaneous and deep stimuli, although occasionally neurons may be excited by one modality and inhibited by another (usually in a distant part of the receptive field). Neurons responding to cutaneous stimuli can be either rapidly or slowly adapting. In areas 3b and 1, rapidly adapting neurons greatly outnumber slowly adapting ones, although there are relatively more slowly adapting neurons in area 3b. As pointed out by Powell and Mountcastle,[29] it is inaccurate to attribute specific submodalities to specific sub areas of SI, since the proportion of deep and cutaneous cells varies *continuously* with distance from the central sulcus rather than changing precipitously at subregion boundaries.

Topographic Organization

The positions on the body surface of the somatosensory receptive fields of single cortical neurons are mapped in an orderly sequence along the cortical surface. All receptive fields are on the body surface contralateral to the recording site. Figure 15–14A shows the receptive field locations of 6 SI neurons.[19] As the locations of the neurons shift from medial to lateral on the cortical surface, the receptive fields move progressively from the foot to the trunk, hand, and head. A comprehensive mapping of the locations of the receptive fields of many postcentral neurons, as summarized in Figure 15–14B, shows that the entire body surface is represented at least twice in SI, once in area 3b, and again in area 1.[26] The tail and foot regions are represented on the medial bank of SI with an orderly progression of receptive field locations from the leg to the face along the lateral hemisphere. Although area 2 neurons with cutaneous receptive fields are sparse, they allow demonstration in area 2 of a partial representation of

parts of the body surface (at least the hand), together with a complete representation of deep body structures.[25] As with the topographic representation of the body surface in other somatosensory nuclei (recall Fig. 15–8), the cortical representation for the foot, hand, and face greatly exceeds that devoted to other body regions with poor somatosensory acuity (e.g., the thigh and trunk). Also, as shown in Figure 15–14A, receptive fields in regions with high somatosensory acuity (e.g., the perioral region and fingers) are much smaller than those elsewhere on the body.

Columnar Organization

As mentioned earlier, most SI neurons respond to either deep or cutaneous stimuli but not to both. If an electrode penetration is run perpendicular to the cortical surface (as in tracks A, B, C, and F in Figure 15–15, all the cells encountered throughout the entire cortical thickness tend to respond to the same modality, i.e., either to cutaneous or deep stimuli. Furthermore, as expected from topographic maps like those of Figure 15–14B, all of the neurons encountered in such a perpendicular penetration tend to have very similar receptive field locations. For example, all of the receptive fields encountered on track F lie on the ulnar side of the wrist and many of them overlap (neurons 1, 2, 5, 6, and 7). Based on his original and similar observations in cat SI, Mountcastle suggested that somatosensory cortex is organized in vertical, "modality-specific" columns. Indeed, when penetrations are run across the putative vertical columns (tracks D, E, and G in Fig. 15–15), the recording electrode encounters successive groups of neurons that are driven alternately by deep and cutaneous stimuli.

The notion of a columnar organization for different somatosensory modalities has had a great influence on models for cortical organization in a variety of sensory systems. The most detailed experiments have been performed in the visual system, where columns, or cortical slabs, have been described for different stimulus orientations and for each eye (see Chap. 20).

Towe[41] believes that the precise measurements necessary to test the columnar hypothesis have not been made and that the somatosensory submodalities and the "functional unit" that the columns represent have not been adequately defined. He argues that the following conditions, which are central to the original formulation of the columnar hypothesis, cannot be met— namely, that somatosensory cortex is constructed of columns of cells with (1) only one modality of input, (2)

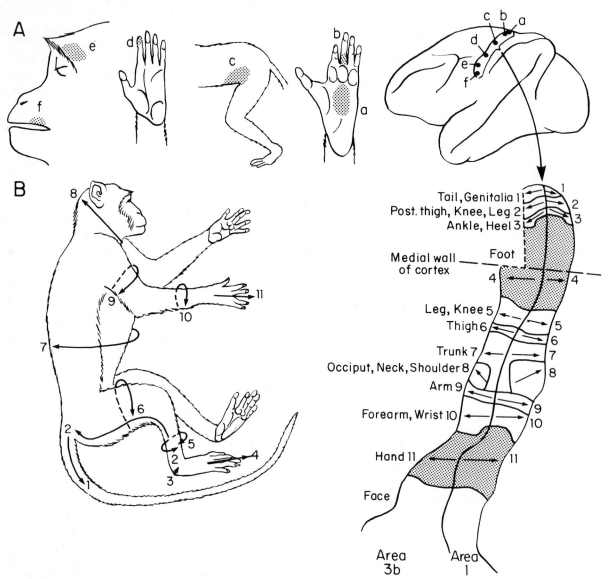

Figure 15–14 Topographic organization in SI of the monkey. A, Position in SI of 6 different neurons (a–f) whose receptive field locations are identified on the monkey. B, Multiple topographic representations of the body surface in Brodmann areas 3b and 1 as determined by extensive single neuron recordings. Arrows on the cortical map correspond to the progression on the body of the monkey of receptive field locations (1–11) indicated by similar arrows; i.e., the receptive fields encountered as one moves forward in region 11 of area 3b move from the wrist toward the finger tips. (*A*, after Jones, E.G.; Friedman, D.P.; Hendry, S.H.C. *J. Neurophysiol.* 48:545–568, 1982. *B,* after Nelson, R.J.; Sur, M; Felleman, D.J.; Kaas, J.H. *J. Comp. Neurol.* 192:611–643, 1980.)

Figure 15–15 Seven reconstructed electrode tracks (A–G) through different subareas of SI of the monkey in the plane of section shown by arrows in the inset. Tracks A,B,C, and F run perpendicular to the cortical surface, whereas D,E, and G run oblique to it. Single lines and shaded areas to the left of the tracks indicate the depths of cells and multineuron activity, respectively, responding to cutaneous stimuli; lines and shaded areas to the right of tracks locate cells and background activity, respectively, responding to deep stimuli. The receptive field locations of seven cells (1–7) encountered on track F are shown on the contralateral hand. Encircled numbers indicate Brodmann's areas. (After Powell, T.P.S.; Mountcastle, V.B. *Bull. Johns Hopkins Hosp.* 105:133–162, 1959.)

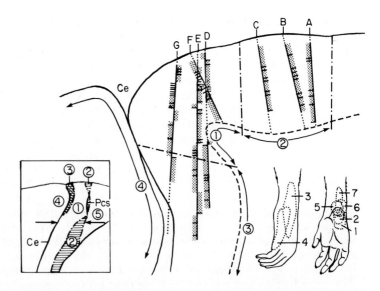

identical receptive fields, and (3) identical response latencies.

Thus, the model of columnar organization in somatosensory cortex may be a heuristic argument that, although it provides a qualitative description of much of the physiological data, is probably untrue in absolute detail.

Somatotopic Reorganization After Peripheral Lesions

An interesting series of recent experiments has revealed that the topographic organization of SI is not immutable.[20] If a restricted sector of somatosensory cortex (e.g., the hand region of areas 1 and 3) of the adult monkey is deprived of its normal inputs by transecting spinal nerves or dorsal roots or by amputating a finger, the affected cortex rapidly becomes largely or completely reactivated by inputs from adjoining and nearby skin locations. The same region of cortex has the capacity to reorganize in different ways. For example, transection of the median nerve causes the cortex that is normally devoted to the glabrous portion of the digits to respond to stimuli delivered to dorsal digit surfaces. If a single digit is removed, the same cortex becomes devoted to the remaining digits. Finally, if two digits are removed, part of the digital cortex may remain unresponsive. Many of these cortical reorganizations take place rapidly (within a half hour), as shown by reversible inactivation of a peripheral nerve with local anesthetics in the cat. On the other hand, the unresponsive cortex of monkeys with median nerve transections is still shrinking after several weeks. Therefore, both rapid and gradual changes occur in the so-

matosensory cortex of mammals deprived of normal afferent inputs. As discussed in Chapter 33, the short-term changes could result from the uncovering of dormant synapses, whereas the long-term changes may involve morphological alterations such as sprouting or changes in dendritic spines. Whether cortical reorganization is intrinsic or simply reflects subcortical changes is unclear, however, since peripheral damage produces organizational changes at every level of the somatosensory system, as we have shown in previous sections.

It has been suggested that the cortical topographic plasticity demonstrated in these deprivation studies is a mechanism that is available to the normal cortex. Deprivation, it is argued, can be considered to be an exaggerated form of disuse, and in situations of disuse, cortical areas involved with the unused digit would shrink, whereas those involved with the active digits might expand. For example, while threading a needle, the cortical representation of the glabrous surfaces of the thumb and index finger might expand at the expense of the representations of the other three fingers. Since neurons in expanded representations also have smaller receptive fields, the increased somatic acuity of the new representation could greatly facilitate the performance of the threading task.

Somatosensory Area SII

In contrast to the receptive fields of neurons in SI, a sizable number of those in SII are bilateral.[8, 47] Neurons with bilateral receptive fields related to proximal body parts are continuous across the midline, whereas those related to distal body parts on the limbs have disjunctive but bilaterally symmetrical receptive fields. However, many neurons representing the extremities have only contralateral receptive fields. Like SI, SII is topographically organized, but early studies indicated that both the ipsilateral and contralateral body surfaces are represented in overlaid register.[25] Adjacent body parts are represented on adjacent mediolateral strips, with the head strip most anterior and the foot strip most posterior (see Fig. 15–12).[8]

The degree of bilateral representation in SII remains uncertain. On one hand, the early reports[47] of a high percentage (90%) of neurons with bilateral fields have been attributed to the inclusion of neurons in the adjacent area 7b, all of which have bilateral receptive fields. On the other hand, the lower percentage (40%) reported by recent investigators[8] has been attributed to the lability of SII responses, especially the ipsilateral inputs, which apparently habituate rapidly with repeated stimulation.[25]

The modality properties of SII neurons are different from those of SI. Most (80–90%) SII neurons can be activated by light mechanical stimulation of the hairs or of the skin surface, and all adapt rapidly to maintained stimuli.[8, 47] Although the remaining neurons require more intense mechanical stimuli, they probably are not driven by deep mechanoreceptive inputs, since SII neurons apparently cannot be activated by either gentle joint rotation or mechanical stretch. Some SII neurons apparently prefer moving stimuli, although it is unclear whether this is a salient requirement of SII neurons. Finally, some SII neurons respond only when the monkey is actively palpating an object.

The second somatic area of the other well-studied mammal, the cat, seems to differ considerably from that of the monkey. First, feline SII neurons resemble those of SI in that the vast majority have only *contralateral* receptive fields; otherwise, they respond in similar ways (i.e., with rapidly adapting responses) to stimuli that effectively drive SII cells in the monkey. Second, SII of the cat may be composed of up to four separate subregions, which thus far have been only incompletely studied.

Somatic Inputs to Motor Cortex

In the unanesthetized primate, two thirds of the cells in primary motor cortex (Brodmann's area 4) respond to somatic stimulation. Most rostral cells can be driven by passive joint rotation, whereas those in more caudal zones respond to cutaneous stimulation, such as touching the skin or brushing the hairs.[35] As in SI, penetrations perpendicular to the cortical surface of area 4 encounter cells that respond to inputs of the same modality (i.e., cutaneous or deep) and receive inputs from the same joint or overlapping body surfaces—that is, they have similar receptive field locations. In one experiment, the somatic responses of individual area 4 neurons were recorded, and then weak electrical currents were passed through the recording microelectrode to elicit a movement. The elicited movement often was directed toward the somatic receptive field of the recorded neuron.[33] For example, electrical stimulation at the site of a neuron that had a receptive field on the thumb pad produced a flexion of the thumb. These data suggest that the somatic input could allow motor cortex to participate in a grasp reflex in which an object touching the palm elicits discharges in cells that project to motor neurons that innervate the muscles that close the hand (see Chap. 28).

Possible Functional Role of Somatosensory Cortex

The existence of orderly topographic maps on SI, SII, and possibly also motor cortex is an interesting observation but doesn't provide any insights into the role of these cortical regions in somesthesis. Such insights can be gained by requiring primates, both human and non-human, to perform haptic (active somatosensory) discriminations in the absence of all or part of somatosensory cortex and by recording neuronal activity in monkeys subjected to imaginative somatosensory stimuli.

Effects of Lesions in SI

In the monkey, damage to different subregions of SI produces different somesthetic deficits. After the appropriate lesions, monkeys were trained in a discrimination task that required them to pull one of two cylindrical levers according to whether the levers were hard or soft, small or large, rough or smooth, horizontal or vertical, convex- or concave-shaped, or square- or diamond-shaped.[31] On each trial, the monkey was allowed to palpate and compare the two levers freely. Initially, all animals preferred to use the hand ipsilateral to the lesion but could be trained to use the contralateral hand after a few days. All animals with selective lesions of area 3 required more time than normal animals to learn *every* task, and they failed to learn some tasks even after 1000 trials. After lesions of area 2, all monkeys also were severely impaired, some permanently, in learning the square-diamond and concave-convex discrimination but not in learning any other task. In contrast, lesions of area 1 produced modest but significant deficits only in learning the smooth-rough and hard-soft discriminations, but no deficits in learning the others. These data suggest that the animals with an area 1 lesion were impaired on those tasks characterized as involving discrimination of "texture," whereas those with area 2 lesions showed extreme impairment on tasks involving a discrimination of "angles." SI lesions produce deficits not only in spatial but also in temporal somesthetic sensibilities, since monkeys with postcentral lesions have difficulty in detecting and discriminating between rapidly oscillating indentations produced by a small probe placed on the glabrous skin.[25]

An exhaustive study of 93 human patients with a variety of cortical lesions confirms the importance of the postcentral gyrus of humans in the somatosensory discrimination of size and form.[32] In these experiments, size discriminability was tested by having subjects hold spheres of different sizes, and form discriminability was tested by holding rectangular parallelopipeds or spherical ellipses of different dimensions. The patients were allowed to palpate the test objects for 10 seconds and were then asked to choose either the largest sphere or the most oblong parallelopiped or ellipse. Although normal subjects differentiated two spheres whose diameters differed by less than 1.1 mm, patients with unilateral postcentral lesions required differences of at least 12 mm when using the contralateral hand; discriminations using the ipsilateral hand were normal. Similarly, normal subjects could just differentiate a 28-mm diameter sphere from an ellipse whose axes were 27.5 by 30.0 mm, whereas patients with postcentral lesions required the ellipsoid axes to be ≤ 21 by ≥ 50 mm. Such a deficit in form perception is called *astereognosis*. These deficits in both size and form discrimination occurred only if the anterior part of the middle third of the postcentral gyrus was involved; unlike the monkey, humans with lesions involving only the posterior part of the gyrus had no deficits. Human patients and monkeys with postcentral damage also suffer impaired *kinesthesis*—that is, the ability to appreciate the passive position and movements of the limbs as well as the force exerted by the limbs. In humans, the sensations of pain and temperature are preserved after postcentral lesions.

Neuronal Responses in SI

To perform the haptic discriminations just described, the CNS probably makes use of several features of the object. These might include the distance between adjacent irregularities (two-point discrimination), the orientation of stimulus edges, and the roughness as judged by the velocity sensitivity of the palpating hand. None of these properties is present in the discharge of single afferent fibers (recall Chap. 14) and thus must be a consequence of central somatosensory processing.

Almost 60% of SI cells, especially those in the anterior part of SI,[15] prefer stationary, punctate stimuli. Because these discharge characteristics so closely resemble those of cells in the thalamus, it is unclear how they could provide a further stage in the process of form recognition. Perhaps these SI neurons serve as a switchboard that relays the locations and timing of touch stimuli on the body surface to still "higher" processing centers. A small number of SI neurons could be involved in two-point discrimination, however, since they exhibit a larger, stronger surround inhibition than do similar cells in the VPL or the dorsal column nuclei.

A few SI cells (< 10%) do have more exotic stimulus requirements that suggest they are sensitive to other stimulus features such as form or movement.[14] Form detectors respond to a long edge that is pressed on the skin in a particular orientation. For example, the orientation favored by the cell in Figure 15–16A is horizontal; no response is elicited by a vertical edge. The orientation selectivity of this cell strongly resembles a similar selectivity of many cells in primary visual cortex for bars of light (see Fig. 20–8, Chap. 20). Movement detectors require movement of a stimulus over the skin; some respond to movement in all directions whereas others, like the neuron in Figure 15–16B (left column of records), prefer movement in a particular direction (in this case, upward). Note that stationary punctate stimuli delivered at different locations within the receptive field elicit no responses (Fig. 15–16B, right column of records). The receptive fields of direction-sensitive neurons are fairly large, often covering the entire hand or arm, and can be found on either the glabrous or hairy skin of the hand, forearm, and even the trunk. The characteristics of directionally selective somatosensory neurons resemble those of visual neurons that lie outside primary visual cortex in a "higher" visual processing area (see Fig. 20–17, Chap. 20).

The incidence of feature-detecting neurons increases as one explores more posteriorly in SI. For

A

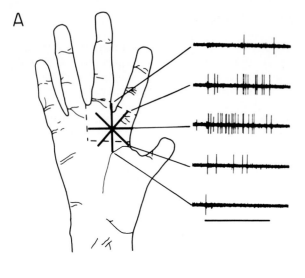

B

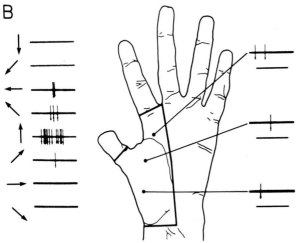

Figure 15–16 Examples of two neurons in SI with more exotic stimulus requirements. The cell in *A* is sensitive to the orientation of a vibrating edge, 0.7mm wide, placed on the first and second palmar whorls (receptive field extent shown by dashed lines). The best response was obtained when the edge was oriented perpendicular to the long axis of the hand. The cell in *B* is sensitive to distally moving stimuli delivered to the receptive field covering the thenar eminence, the first palmar whorl, and the base of the thumb. The best direction is for stimuli moving toward the finger tips (discharge patterns to the left). Stationary punctate stimuli delivered with the same probe at different loci (discharge patterns to the right) within the receptive field produced no response. (From Hyvärinen, J.; Poranen, A. *J. Physiol.* (Lond.) 283:539–556, 1978.)

example, area 2 contains a higher proportion of neurons responding to deep stimuli and of neurons with converging deep and cutaneous inputs than does area 3b. Also, area 2 has fewer cutaneous neurons, and their receptive fields generally are larger than those of area 3b—that is, they

cover several fingers, the whole palm, or more). Finally, some SI neurons outside area 3b have discontinuous receptive fields (e.g., on similar regions of adjacent fingers) that are not found in peripheral afferent fibers. Taken together, these observations suggest that there is an increasing complexity of all somatosensory processing in SI as one moves from area 3b to area 2, an interpretation that is generally consistent with the effects of selective SI lesions in the monkey.[31]

Other Somatosensory Areas

Somatosensory neurons that receive inputs from SI show a further diversity of responses. We have already seen that some neurons in SII have bilateral receptive fields, are motion-sensitive, and have responses that may require the animal to actively palpate an object. In the somatosensory area just dorsal to SII (area 7b; Fig. 15–12), only half of the neurons respond unequivocally to cutaneous stimuli, and most have bilateral receptive fields.[8] Many of the remaining neurons require activation from multiple modalities (e.g., cutaneous and proprioceptive or noxious, visual, or other nonsomatic stimuli); some of these neurons have labile responses that seem to be influenced by the animal's attention toward, or anticipation of, a noxious or novel somatic stimulus. Attempts to assess the role of SII and its neighboring structures by damaging them have not been very informative.

Neurons in area 5 also have complex stimulus requirements. Virtually all are insensitive to stationary point stimuli applied to the skin or deep tissues.[34] Instead, almost 80% require joint stimulation, either in isolation or with one or more other joints or with cutaneous stimuli. The few neurons with cutaneous receptive fields require moving stimuli like those required to drive neurons in area 2 of SI. Those neurons receiving combined inputs from cutaneous and deep receptors might convey information concerning the position of the limbs or the tension in their muscles as the skin is stimulated during the manipulation of objects. For example, the neuron in Figure 15–17 responded when either the right or left shoulder was adducted and exhibited a slightly enhanced late response when both were adducted together. When the skin of the palms and forearms was stimulated during the simultaneous adduction, the response was markedly enhanced although cutaneous stimulation alone produced no response. Such neurons might participate in *stereognosis*, an ability that is severely compromised by posterior

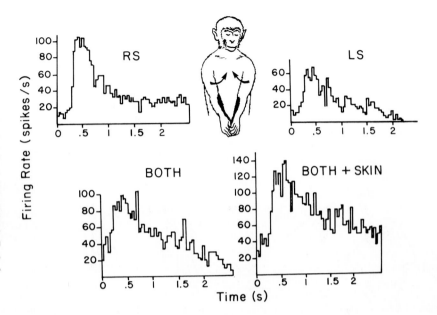

Figure 15–17 Response of a neuron in Brodmann's area 5 to shoulder rotation and its enhancement by skin contact. Adduction of either the right shoulder (RS) or the left shoulder (LS) alone elicits a brisk response. Adduction of both simultaneously (BOTH) causes an increase in the late (after about 700 ms) response. When the skin also is stimulated during bilateral shoulder rotations (BOTH + SKIN), both the early and the late responses are enhanced, although no response is elicited by skin stimulation alone. Adduction and/or skin stimulation was applied for the entire 2.5 seconds shown. This cell would be maximally active if the monkey held an object in front of him with both hands. (From Sakata, H.; Takaoka, Y.; Kawarasaki, A. *Brain Res.* 64:85–102, 1973.)

parietal lobe damage in both human and non-human primates (see Chap. 31).

Summary

The surface and depths of the body are represented in several topographic maps in the cerebral cortex. There are at least three separate maps within SI, each of which represents the entire contralateral body surface. As the recording electrode moves backward from the central sulcus, the receptive fields of encountered cells become larger and more neurons are driven by noncutaneous stimuli; also, more of the cutaneous neurons require stimuli that are moving or have a particular shape. SII, which also receives direct thalamic inputs, and its neighboring somatosensory regions contain even more neurons with inputs from ipsilateral body parts, and these neurons often have still larger receptive fields. Responses in SII seem less securely linked to the peripheral inputs, and a significant number of neurons require moving stimuli. Finally, neurons in area 5 prefer deep stimuli and often require convergent inputs from multiple modalities.

In general, then, there seems to be a gradual deterioration in somatic acuity and an increased emphasis on modality convergence as one moves backward from the central sulcus. Consistent with this distribution of somatosensory response properties in single neurons, damage to area SI causes clear deficits in all aspects of somesthesis, whereas damage to area 5 specifically affects stereognosis, while leaving other tactile capabilities intact.

TRIGEMINAL SYSTEM

Organization

The somatosensory representation of the body in the CNS is completed by inputs from the trigeminal nerve, which supplies the face, mouth, and head. The trigeminal nerve has three branches, each of which innervates different parts of the face, as shown in Figure 15–18. Although there is some overlap in the innervation of the three branches, it is much less than that of the spinal dermatomes. As with the spinal nerves, the somatosensory fibers of the trigeminal nerve are peripheral axons of pseudounipolar neurons located in a ganglion, in this case the semilunar ganglion. Most of the proximal axons of trigeminal somatic afferent neurons divide into an ascending and a descending tract, by which they reach two brain-stem nuclei that are rostral extensions of the dorsal horn of the spinal cord.

The most caudal nucleus, the spinal nucleus of the trigeminal nerve (Spinal V), has three subdivisions. The organization of the caudal subdivision resembles that of the dorsal horn, and it receives many nociceptive afferent fibers from the face; many cells in its outer (substantia gelatinosa) layers respond only to strong, nociceptive stimuli.[24] Surgical lesions of this caudal subdivision are

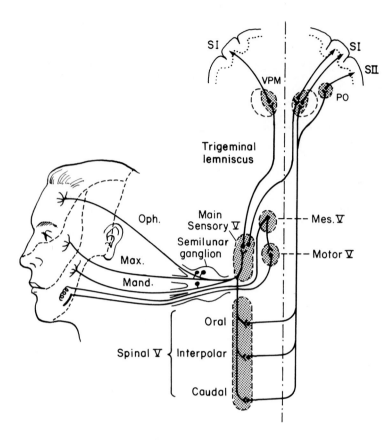

Figure 15–18 Anatomy of the trigeminal system. Information from three regions of the head is relayed via the three trigeminal branches—ophthalmic (Oph), maxillary (Max), and mandibular (Mand)—to two nuclei of the trigeminal nerve (V)—Main Sensory V and Spinal V, which have three subdivisions. These two nuclei provide both crossed and uncrossed projections to the ventral posteromedial nucleus (VPM) and the posterior nuclei of the thalamus (PO), which project in turn to cortical somatosensory areas SI and SII. Afferent fibers from spindles in the jaw muscles project to the mesencephalic nucleus of V (Mes V) and make monosynaptic connections with jaw motoneurons (in Motor V). (After Carpenter, M.B.; Sutin, J. *Human Neuroanatomy.* Copyright © 1983, The Williams & Wilkins Co., Baltimore.)

made to relieve chronic facial pain (i.e., trigeminal neuralgia). Much of the inner region of the caudal subdivision contains cells that respond to a variety of light contralateral mechanical stimuli. The more rostral oral and interpolar subdivisions receive mechanoreceptive and nociceptive inputs from the teeth and gums as well as from the mucous membranes of the mouth. The strong nociceptive projection from the tooth pulp, which is confined largely to the more caudal (interpolar) division, is bilateral. Because the spinal nucleus receives mostly smaller diameter fibers, it is considered to be the trigeminal homologue of the spinothalamic pathway.

The more rostral of the two brain-stem trigeminal somatosensory nuclei, the main sensory nucleus of the trigeminal nerve (Main Sensory V), receives primarily ipsilateral projections from low-threshold mechanoreceptors of the face and some mechanoreceptive afferent fibers from the teeth. Because these fibers have larger diameters, Main Sensory V is considered to be the trigeminal homologue of the dorsal column nuclei.

A separate trigeminal pathway delivers proprioceptive information from the jaw muscles and mechanore-

ceptive information from the gums to the CNS over the axons of monopolar neurons that reside in the mesencephalic nucleus of the trigeminal nerve (Mes. V). These primary afferent fibers project to the motor nucleus of the trigeminal nerve (Motor V), which in turn innervates the jaw muscles, thereby producing a monosynaptic proprioceptive reflex arc.

Information from the trigeminal nuclei reaches the thalamus via the trigeminal lemniscus, which runs with the medial lemniscus. These projections are primarily contralateral and are to subregions of the same nuclei that receive spinal somatic information—namely the medial portion of the ventral posterior nucleus (VPM) and the PO. The PO receives a large projection from the caudal subregion of Spinal V, whereas the VPM receives a heavy projection from the more rostral subregions of Spinal V and from Main Sensory V. Like the spinal somatic representation in the VPL, the trigeminal representation in the VPM is somatotopically organized so that the entire ventral posterior thalamus contains a complete somatotopic representation of the body surface (see Fig. 15–10).

Information from the VPM is relayed to the

ventrolateral portion of somatosensory cortex where neurons have large, primarily contralateral, facial receptive fields or bilateral oral receptive fields. Some of these cells display the inhibitory surrounds characteristic of other cortical somatosensory neurons. The cortical representation of the face is relatively enormous, owing to its high innervation density, and the representation of the mouth and tongue is further exaggerated relative to the rest of the face. Both SI and SII receive trigeminal inputs via the thalamus.

The trigeminal input to the rodent somatosensory cortex displays a unique organization.[50] Cells that respond to inputs from the mechanoreceptors at the base of the mystacial vibrissae, the whiskers, have a cortical distribution pattern that mimics the pattern of the vibrissae on the face. A section through layer IV of the rat somatosensory cortex (Fig. 15–19A), reveals a series of circles that are cross sections of cylindrical cellular aggregates known as *barrels*. Single cell recordings suggest that each barrel contains cells that respond to stimulation of a single whisker. This high degree of specificity between the peripheral receptor surface and both the anatomically and physiologically observed cortical organization makes this system ideal for the study of developmental plasticity in the CNS and the anatomical consequences of peripheral damage.

As with the hand cortex of primates, different peripheral manipulations of the vibrissae result in different types of cortical reorganization. For example, neonatal removal of a single row of whiskers (No. 2 in Fig. 15–19B) results in the conversion of the row of barrels corresponding to those whiskers into a single band of cells. Removal of the same number of whiskers in an arc across rows (No. 3 in Fig. 15–19B) results in only partial banding of cells. Examples of the effects of other whisker manipulations on cortical organization are shown in Nos. 4, 5, and 6 in Figure 15–19B. As with other neonatal manipulations in a variety of sensory systems, the major anatomical changes occur only if damage occurs during a brief *sensitive period*, which, for expression of the rodent whisker cortical representation, lasts less than 1 week.[20]

CONCLUSION

Our own experiences tell us that the somatosensory system is capable of an impressive variety of sensory feats. Not only can we accurately identify an unseen object by palpation but we can combine somesthesis with motor control to perform many very sophisticated movements such as playing a musical instrument. Perhaps it is the very versatility of the somatosensory system that has frus-

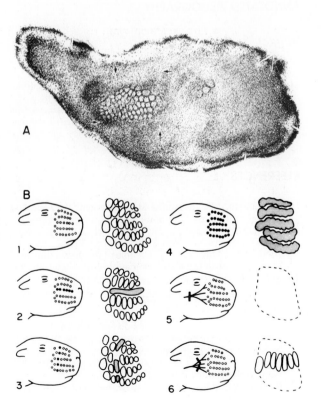

Figure 15–19 The barrel region of rodent somatosensory cortex. *A*, Horizontal section through layer IV of mouse somatosensory cortex displaying the circular patterns of cells (barrels) characteristic of the representation of the mystacial vibrissae. *B*, The effect of manipulations of the vibrissae of neonatal rats (shown to the left of each numbered experiment) on the organization of the barrel subfield (shown to the right): (1) normal, (2) removal of the middle row of vibrissae, (3) removal of an arc of vibrissae across rows, (4) removal of all rows of vibrissae, (5) total trigeminal nerve section, and (6) organization following section of selected trigeminal branches. (*A*, after Woolsey, T.A.; Van der Loos, H. *Brain Res.* 17:205–242, 1972. *B*, Modified, with permission, from Kaas, J.H.; Merzenich, M.M.; Killackery, H.P. *The Annual Review of Neuroscience*, vol. 6, © 1983 by Annual Reviews, Inc.)

trated attempts to understand how it performs its many functions. As is the case for all sensory systems, most is known about the behavior of primary somatosensory afferent fibers, which encode stimulus features rather simply. Most of what we know about the central mechanisms of somesthesis, however, involves anatomical organization (e.g., somatotopy, bands, and columns). Although this information may be clinically important, it is not clear how these structural features contribute to the processing of somatosensory information. For example, we do not yet know how each ascending somatosensory station contributes to any aspect of somesthesis; indeed, there seem to be only subtle differences between the response properties of many cells in SI and those of primary afferent fibers. The lesion studies tell us only if the CNS performs a somatosensory function without a structure; these studies provide no clue as to how that structure may accomplish the task. Thus, although we have a great deal of information about the somatosensory system, we are a long way from explaining even the most simple processes of somesthesis. As we shall see in Chapter 16, the mechanisms of pain sensation are even more elusive.

ANNOTATED BIBLIOGRAPHY

Iggo, A., ed. *Handbook of Sensory Physiology*, vol. 2. *Somatosensory System*. New York, Springer-Verlag, 1973.
  An excellent collection of older review articles on many aspects of somatosensory physiology.
Mountcastle, V.B. Central nervous mechanisms in mechanoreceptive sensibility. *Handbk. Physiol.*, 1:789–878, 1984.
  A comprehensive recent review of somatosensory mechanisms and functional anatomy by a major figure in somatosensory research; with over 650 references.

REFERENCES

1. Albe-Fessard, D.; Besson, J.M. Convergent thalamic and cortical projections—the non-specific system. In Iggo, A., ed. *Handbook of Sensory Physiology*, vol. 2. *Somatosensory System*, pp. 489–560. New York, Springer-Verlag, 1973.
2. Azulay, A.; Schwartz, A.S. The role of the dorsal funiculus of the primate in tactile discrimination. *Exp. Neurol.* 46:315–332, 1975.
3. Baker, M.A. Spontaneous and evoked activity of neurons in the somatosensory thalamus of the waking cat. *J. Physiol.* 217:359–379, 1971.
4. Berkeley, K.J. Multiple systems diverging from the dorsal column nuclei in the cat. In Rowe, M.; Willis, W.D., eds. *Neurology and Neurobiology*, vol. 14. *Development, Organization and Processing in Somatosensory Pathways*, pp. 191–202. New York, Alan R. Liss, 1985.
5. Brown, A.G. *Organization of the Spinal Cord*. New York, Springer-Verlag, 1981.
6. Bryan, R.N.; Coulter, J.D.; Willis, W.D. Cells of origin of the spinocervical tract in the monkey. *Exp. Neurol.* 42:574–586, 1974.
7. Burgess, P.R.; English, K.B.; Horch, K.W.; Stensaas, L.J. Patterning in the regeneration of type I cutaneous receptors. *J. Physiol.* (Lond.) 236:57–82, 1974.
8. Burton, H.; Robinson, C.J. Organization of the SII parietal cortex. Multiple somatic sensory representations within and near the second somatic sensory area of Cynomolgus monkeys. In Woolsey, C.N., ed. *Cortical Sensory Organization*, vol. 1. *Multiple Sensory Maps*, pp. 67–119, Clifton, New Jersey, Humana, 1981.
9. Carpenter, M.B.; Sutin, J. *Human Neuroanatomy*. Baltimore, Williams & Wilkins, 1983.
10. Craig, A.D.; Tapper, D.N. Lateral cervical nucleus in the cat: functional organization and characteristics. *J. Neurophysiol.* 41:1511–1534, 1978.
11. Darian-Smith, I. The sense of touch: performance and peripheral neural processes. *Handbk. Physiol.* Sec. 1, 3(2):739–788, 1984.
12. Fetz, E.E. Pyramidal tract effects on interneurons in the cat lumbar dorsal horn. *J. Neurophysiol.* 31:69–80, 1968.
13. Golovchinski, V.; Kruger, L.; Saporta, S.A.; Stein, B.E.; Young, D.W. Properties of velocity-mechanosensitive neurons of the cat ventrobasal thalamic nucleus with special reference to the concept of convergence. *Brain Res.* 209:355–374, 1981.
14. Hyvärinen, J.; Poranen, A. Movement-sensitive and direction and orientation-selective cutaneous receptive fields in the hand area of the post-central gyrus in monkeys. *J. Physiol.* (Lond.) 283:523–537, 1978.
15. Hyvärinen, J.; Poranen, A. Receptive field integration and submodality convergence in the hand area of the postcentral gyrus of the alert monkey. *J. Physiol.* (Lond.) 283:539–556, 1978.
16. Iggo, A. Cutaneous sensory mechanisms. In Barlow, H.B.; Mollon, J.D., eds. *The Senses*, pp. 369–408. Cambridge, Cambridge Univ. Press, 1982.
17. Jänig, W.; Spencer, W.A.; Younkin, S.G. Spatial and temporal features of afferent inhibition of thalamocortical relay cells. *J. Neurophysiol.* 42:1450–1460, 1979.
18. Jones, E.G. *The Thalamus*. New York, Plenum Press, 1985.
19. Jones, E.G.; Friedman, D.P.; Hendry, S.H.C. Thalamic basis of place- and modality-specific columns in monkey somatosensory cortex: a correlative anatomical and physiological study. *J. Neurophysiol.* 48:545–568, 1982.
20. Kaas, J.H.; Merzenich, M.M.; Killackery, H.P. The reorganization of somatosensory cortex following peripheral nerve damage in adult and developing mammals. *Ann. Rev. Neurosci.* 6:25–56, 1983.
21. Kruger, L.; Siminoff, R.; Witkovsky, P. Single neuron analysis of dorsal column nuclei and spinal nucleus of trigeminal in cat. *J. Neurophysiol.* 24:333–349, 1961.
22. Maunz, R.A.; Pitts, N.G.; Peterson, B.W. Cat spinoreticular neurons: location, responses and changes in responses during repetitive stimulation. *Brain Res.* 148:365–379, 1978.
23. Maynard, C.W.; Leonard, R.B.; Coulter, J.D.; Coggeshall, R.E. Central connections of ventral root afferents as demonstrated by the HRP method. *J. Comp. Neurol.* 172:601–608, 1977.
24. Mosso, J.A.; Kruger, L. Receptor categories represented in spinal trigeminal nucleus caudalis. *J. Neurophysiol.* 36:472–488, 1973.
25. Mountcastle, V.B. Central nervous mechanisms in mechanoreceptive sensibility *Handbk. Physiol.* 1:789–878, 1984.
26. Nelson, R.J.; Sur, M.; Felleman, D.J.; Kaas, J.H. Representations of the body surface in postcentral parietal cortex of *Macaca fasicularis*. *J. Comp. Neurol.* 192:611–643, 1980.

27. Poggio, G.F.; Mountcastle, V.B. A study of the functional contributions of the lemniscal and spinothalamic systems to somatic sensibility. *Bull. Johns Hopkins Hosp.* 106:266–316, 1960.

28. Poggio, G.F.; Mountcastle, V.B. The functional properties of ventrobasal thalamic neurons studied in unanaesthetized monkeys. *J. Neurophysiol.* 26:775–806, 1963.

29. Powell, T.P.S.; Mountcastle, V.B. Some aspects of the functional organization of the cortex of the postcentral gyrus of the monkey: a correlation of findings obtained in a single unit analysis with cytoarchitecture. *Bull. Johns Hopkins Hosp.* 105:133–162, 1959.

30. Price, D.D.; Hayes, R.L.; Ruda, M.; Dubner, R. Spatial and temporal transformations of input to spinothalamic tract neurons and their relation to somatic sensations. *J. Neurophysiol.* 41:933–947, 1978.

31. Randolph, M.; Semmes, J. Behavioral consequences of selective subtotal ablations in the postcentral gyrus of *Macaca mulatta. Brain Res.* 70:55–70, 1974.

31a. Rexed, B. A cytoarchitectonic atlas of the spinal cord in the cat. *J. Comp. Neurol.* 100:297–379, 1954.

32. Roland, P.E. Asterognosis: tactile recognition after localized hemispheric lesions in man. *Arch. Neurol.* 33:543–550, 1976.

33. Rosen, I.; Asanuma, H. Peripheral afferent inputs to the forelimb area of the monkey cortex: input-output relations. *Exp. Brain. Res.* 14:257–273, 1972.

34. Sakata, H.; Takaoka, Y.; Kawarasaki, A.; Shibutani, H. Somatosensory properties of neurons in the superior parietal cortex (area 5) of the rhesus monkey. *Brain Res.* 64:85–102, 1973.

35. Tanji, J.; Wise, S.P. Submodality distribution in sensorimotor cortex of the unanesthetised monkey. *J. Neurophysiol.* 45:467–481, 1981.

36. Tasker, R.R.; Organ, L.W.; Rowe, I.H.; Hawrylyshyn, P. Human spinothalamic tract—stimulation mapping in the spinal cord and brainstem. In Bonica, J.J.; Albe Fessard, D., eds. *Advances in Pain Research and Therapy I*, pp. 251–257. New York, Raven Press, 1976.

37. Torebjork, H.E.; Hallin, R.G. Perceptual changes accompanying controlled preferential blocking of A and C fiber responses in intact human skin. *Exp. Brain Res.* 16:321–332, 1973.

38. Torebjork, H.E.; Vallbo, A.B.; Ochoa, J.L. Intraneural microstimulation in man: its relation to specificity of tactile sensations. *Brain* 110:1509–1529, 1987.

39. Towe, A.L. Inhibition and occlusion in cortical neurons. In *Nervous Inhibitions* (proceedings of an international symposium), pp. 410–418, Oxford Pergamon Press, 1961.

40. Towe, A.L. Somatosensory cortex: descending influences on ascending systems. In Iggo, A., ed. *Handbook of Sensory Physiology*, vol. 2. *Somatosensory System*, pp. 701–718. New York, Springer-Verlag, 1973.

41. Towe, A.L. Notes on the hypothesis of columnar organization in somatosensory cerebral cortex. *Brain Behav. Evol.* 11:16–47, 1975.

42. Viereck, C.J. Interpretations of the sensory and motor consequences of dorsal column lesions. In Gordon, G., ed. *Active Touch*, pp. 139–159. Oxford, Pergamon Press, 1978.

43. Wall, P.D. Cord cells responding to touch, damage, and temperature of skin. *J. Neurophysiol.* 23:197–210, 1960.

44. Wall, P.D.; Noordenbos, W. Sensory functions which remain in man after complete transection of the dorsal columns. *Brain* 100:641–653, 1977.

45. Whitsel, B.L.; Petrucelli, L.M.; Sapiro, G. Modality representation in the lumbar and cervical fasciculus gracilis of the squirrel monkey. *Brain Res.* 15:67–78, 1969.

46. Whitsel, B.L.; Petrucelli, L.M.; Sapiro, G.; Ha, H. Fiber sorting in the fasciculus gracilis of the squirrel monkey. *Exp. Neurol.* 29:227–242, 1970.

47. Whitsel, B.L.; Petrucelli, L.M.; Werner, G. Symmetry and connectivity in the map of the body surface in somatosensory area II of primates. *J. Neurophysiol.* 32:170–183, 1969.

48. Willis, W.D. Nociceptive transmission in the primate spinal cord. In Rowe, M.; Willis, W.D., eds. *Neurology and Neurobiology*, vol. 14. *Development, Organization and Processing in Somatosensory Pathways*, pp. 333–346. New York, Alan R. Liss, 1986.

49. Willis, W.D.; Trevino, D.L.; Coulter, J.D.; Maunz, R.A. Responses of primate spinothalamic tract neurons to natural stimulation of the hindlimb. *J. Neurophysiol.* 37:358–372, 1974.

50. Woolsey, T.A.; Van der Loos, H. The structural organization of layer IV in the somatosensory region (SI) of mouse cerebral cortex. The description of a cortical field composed of discrete cytoarchitectonic units. *Brain Res.* 17:205–242, 1970.

Pain and Itch

INTRODUCTION

Pain is an almost universal experience. Not surprisingly, then, the mechanisms of the transduction of noxious stimuli and the pain sensation that is elicited have been the object of considerable attention not only by neurophysiologists but also by philosophers, theologians, physicians, and pain's unhappy victims. Pain has been associated

with injury, sin (both original and otherwise), and learning. Some even associate pain with morality. For example, Sherrington,[26] among others, felt that moral behavior in adults derived from internalization of childhood punishment administered for antisocial acts; however, persons that are congenitally unable to feel pain seem no less moral than those who have had the "advantage" of childhood discomfort and suffering.

Pain clearly plays an important role in our lives, as may be deduced from the many terms used to describe it: throbbing, burning, excruciating, piercing, lacerating, searing, splitting, stinging, aching, sore, sharp, dull, and so forth. More adjectives are devoted to pain than to any other sensory modality. Despite this rich vocabulary, pain is difficult to define or to describe to someone who has never experienced it, for pain is a subjective sensation. The International Association for the Study of Pain defines pain as "an unpleasant sensory and emotional experience associated with actual or potential tissue damage, or described in terms of such damage."[18]

Pain is evoked by a wide variety of stimuli that seem to have little in common, such as mechanical tissue distortion, high temperature, low pH, chemical mediators (i.e., neurally active substances secreted in response to injury) associated with inflammation, and hyperosmotic solutions. Sherrington[26] pointed out that all the stimuli capable of evoking pain are noxious because they are associated with actual or potential tissue injury. He postulated the existence of sensory receptors that sensed noxious agents and called them *nociceptors*. Following Sherrington's lead, we characterize the pathways that convey pain as nociceptive and describe the conscious sensation of pain as *nociception*.

MEASUREMENT

A painful experience is described in the same dimensions that are used to characterize other sensations, i.e., quality or submodality, location, intensity (including its time course), and affect (see Chap. 13). Before pain can be investigated in either the clinic or the laboratory, it must first be measured, a difficult task because of its subjective nature. Three approaches have been used to assess pain: (1) self-description, (2) physiological reaction, and (3) behavioral response.

The most common technique for pain assessment is self-description. In clinical practice the patient is asked, "Where does it hurt?"; "How much does it hurt?"; "Is it tender, burning?"; and so forth. In an attempt to control subjective variability, responses are standardized through the use of pain questionnaires that require the patient or subject to choose from a list of pain descriptors. An example is the McGill-Melzak Pain Questionnaire (Fig. 16–1), which elicits information on pain location, quality, history, and affect.[15] Pain intensity can be evaluated by asking the patient to select a descriptor from a list, such as "slight–moderate–severe–excruciating." To yield a more

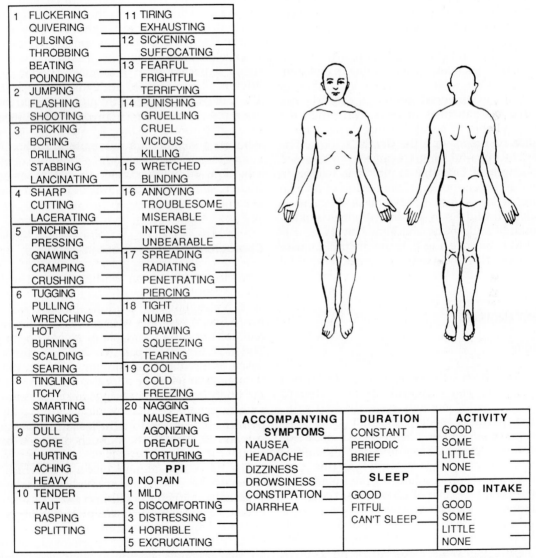

Figure 16–1 Part of the McGill-Melzak Pain Questionnaire. Items 1 through 10 and 18 and 19 elicit information on the character or submodality of the pain. Items 11 through 16 and 20 indicate affective (emotional) pain components. Item 17 concerns localization. PPI is the patient's pain intensity rating. Also included is information on accompanying symptoms and behaviors and a body map on which the patient draws the sites from which pain is felt to originate. (Redrawn with permission from the International Association for the Study of Pain, 909 NE 43rd Street, Seattle, Washington 98105.)

quantitative estimate, the subject must rank the intensity on a scale of 0 to 10, where 0 represents a complete absence of pain and 10 is the worst pain imaginable. There is a good correlation between the descriptor from the list and its assigned rank on the numerical scale.[16]

Pain also can be assessed by measuring the "involuntary" physiological sequelae that often accompany it. These largely autonomic responses (see Chap. 34) include emotional sweating, changes in cardiovascular function (heart rate, blood presure), and respiratory alterations. Unfortunately, such responses often are variable and habituate when the noxious stimulus is repeated; also, it is difficult to distinguish the response to pain from a response to other stresses.

Finally, pain can be assessed simply by observing a patient's behavior, such as crying in children. Facial expression is also a strong indicator; for example, an experienced dentist apparently can sense that his patient is in pain even when the patient is unable to speak because his mouth is otherwise occupied with the dentist's tools. The intensity of a painful stimulus can also be assessed by the length of time that a subject is willing to endure the discomfort (e.g., immersion of an arm in a tub of ice water). Behavioral assessment is particularly useful when self-description is limited, as in infants and young children. However, it is difficult to quantitate and to relate unambiguously to pain rather than to anxiety or some other mental state.

AFFERENT ENDINGS

Specificity

It has long been debated whether pain (1) is a specific sensory modality with its own afferent endings and central nervous system (CNS) pathways (the specificity theory), (2) results from overstimulation of afferent endings that, at more moderate stimulus levels, convey non-pain modalities (the overstimulation theory), or (3) is due to a particular pattern of simultaneous stimulation of afferent endings capable of encoding several sensory modalities (the pattern theory). Two lines of evidence suggest that ". . . the sensation of pain normally depends upon the activation of a discrete set of neural pathways made up of nociceptive afferent fibers, ascending tract cells, and neurons in the brain stem, thalamus and cerebral cortex."[33] First, the stimulation of certain unmyelinated and small myelinated fibers (C fibers and Aδ fibers)

cause humans to report the sensation of pain, while stimulation of the larger myelinated fibers leads to touch-like sensations. Also, infusion of the appropriate concentration of a local anesthetic in the vicinity of a nerve abolishes pain but preserves the sensation of touch because action potential conduction in small axons (unmyelinated C fibers and small unmyelinated Aδ fibers) is blocked before transmission in large axons (see Chap. 15).[10] On the other hand, application of mechanical pressure impairs transmission first in myelinated and then in unmyelinated fibers so that gentle pressure does not abolish pain. Thus, pain is associated with the activation of small-diameter afferent fibers. Second, high-frequency stimulation of large-diameter axons simply evokes a more intense (or higher-frequency) sensation of the modality expected from activation of that type of afferent fiber and does not evoke pain (see Chap. 15). For example, high-frequency stimulation of afferents innervating cutaneous tactile receptors gives rise to the feeling of sustained pressure, whereas a similar stimulus pattern applied to pacinian corpuscle axons leads to the sense of intense vibration. In neither case does a progressively higher stimulation rate cause the sensation to become one of pain.

Characteristics of Nociceptors

Nociceptive afferent endings are distributed widely throughout most body tissues (e.g., the skin, bones, muscles, tendons, blood vessels, kidney, viscera, etc). A few tissues lack pain endings and are therefore incapable of giving rise to pain. The most important pain-free region is the neural tissue of the brain; this situation permits the neurosurgeon to use only a local anesthetic to expose the brain (the brain coverings are richly endowed with nociceptors), thereby allowing the patient to perform sensory and motor tasks that permit the surgeon to evaluate how much damaged cortical tissue can be excised safely.

The particular afferent endings at which pain originates have been best studied in the skin. Most cutaneous nociceptors fall into two classes. Cutaneous Aδ mechanical nociceptors are innervated by small myelinated axons and discharge only when subjected to intense mechanical, but not to thermal or chemical, stimuli. Their receptive fields consist of three to 20 spots on the skin, each less than 1 mm in diameter.[22] As seen in Figure 16–2, they require a mechanical force whose strength is five to 1000 times greater than the threshold for neighboring mechanoreceptors.[8] Multimodal C fiber nociceptors, on the other hand, respond not only to intense mechanical stimuli at the same thresholds (Fig. 16–2) but also respond to noxious thermal and chemical stimuli. Their receptive fields consist of only one or a few spots. Their adequate thermal stimulus is high skin temperature of at least 45° C.[12] These endings are also activated by locally applied chemical agents such as histamine, acetylcholine, bradykinin, and K^+ and H^+.[11]

In addition to the two major nociceptors, a variety of other types of skin pain endings have been described. These include Aδ and C fiber nociceptors responding to very low temperatures,[13] Aδ fibers responding to noxious heat, and nonmultimodal C fiber endings responding only to strong mechanical force.[12]

Tissues other than skin contain a similar variety of Aδ and C fiber nociceptors. Skeletal muscle has endings that respond to strong mechanical force and to noxious chemicals and mediators (e.g., K^+, hyperosmolal saline, bradykinin, histamine, and serotonin). Some endings discharge when the muscle becomes hypoxic, especially when it is actively contracting, providing a physiological basis for the pain of claudication. Cardiac muscle has similar pain endings, including mechanical, chemical, and hypoxial nociceptors; the last presumably are responsible for the excruciating pain of angina pectoris associated with coronary ischemia. Respiratory nociceptors are located in the airways and in the alveolar region near pulmonary capillaries. Airway receptors respond to irritant gases, aerosols, and mechanical distention; alveolar endings can be excited by irritant gases, microemboli, congestion, and edema.[19] Finally, blood vessels in the lungs and elsewhere also contain pain endings. The nociceptors of the alimentary and genitourinary tracts are rather poorly characterized. Their adequate stimuli seem to be excessive mechanical distortion or high pressure and algesic chemicals. Chemical stimulation is particularly effective when the mucous coating protecting the luminal surface of the alimentary tract becomes too thin, as in a gastric or duodenal ulcer.

Sensitization

Compared with other afferent endings, nociceptors are characterized as high-threshold, since a relatively strong stimulus is required to cause them to discharge. However, their thresholds can be lowered by the application of certain agents that normally do not cause nociceptor excitation. This phenomenon is called *sensitization*, and the resulting pain, called *hyperalgesia*, is characterized by a reduced pain threshold or by spontaneous pain. For example, cutaneous Aδ mechanical nociceptors normally do not respond to heat; however, the repeated application of a 45° C probe eventually causes such endings to discharge.[7] Joint nociceptors also are silent in normal movement. However, the induction of joint inflammation promotes their discharge even during normal innocuous joint rotation, thus implicating sensitization as the likely basis of arthritic pain.[3]

Several endogenous substances, such as prostaglandins (especially PGE_2) and substance P, enhance the sensitivity of nociceptive endings to noxious stimuli without exciting these endings themselves.[35] Prostaglandins probably act directly on the nociceptive endings to reduce their thresholds. The analgesic properties of aspirin and similar salicylates seems to be due to their ability to inhibit prostaglandin synthesis. Substance P, on the other hand, may cause sensitization by increasing capillary permeability, thereby permitting blood-borne substances that exacerbate pain better access to nociceptive endings. Other agents reported to act as pain sensitizers are bradykinin and serotonin.

Fast and Slow Pain

We have seen that some nociceptive endings are innervated by Aδ axons whereas others are innervated by C fibers. The velocity with which

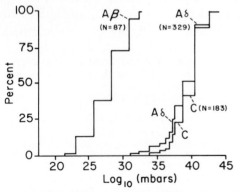

Figure 16–2 Response thresholds of single mechanoreceptors innervated by Aβ axons and nociceptors supplied by Aδ and C fibers in primate skin. The percentage of endings discharging is plotted against stimulus pressure. (After Georgopoulos, A.P. *J. Neurophysiol.* 39:71–83, 1976.)

Aδ afferent fibers conduct pain information into the CNS averages 20 meters per second, whereas the conduction velocity of C fibers is on the order of 1 meter per second. Thus, when a brief suprathreshold noxious stimulus is applied to a mixed population of nociceptors, two bursts of action potentials enter the CNS; because of the difference in their conduction velocities, the bursts reach their targets at different times proportional to the distance that must be travelled. If the distance is sufficiently long, such as when the stimulus is applied to the lower extremities or the distal forelimbs, the subject may report two distinct pains clearly separated in time. The initial sensation is called first or *fast* pain and the later response, second or *slow* pain (also see Chap. 15).

Summary

There are a variety of noxious stimuli and nociceptors and a variety of mechanisms through which they interact, as is summarized in Figure 16–3. Noxious stimuli are either physical (mechanical, thermal, electrical, photic) or chemical (e.g., mediators, pH, hypoxia), and may act in a variety of ways. A stimulus can either excite afferent endings directly or cause the release of a primary pain-related substance. The release of the primary agent could be local to affect nearby nociceptive endings either directly or indirectly by facilitating the action of other noxious agents; these secondary agents could, in turn, either affect the sensitivity of endings directly or increase local capillary permeability to allow other circulating agents better access to the endings. The release of a primary agent also could have a more distant effect if it diffused into the blood, was distributed by the circulation, and re-entered the tissue by transcapillary diffusion to stimulate endings at distant sites. A given stimulus can activate several of these mechanisms simultaneously. Finally, a stronger physical stimulus or a more concentrated chemical stimulus produces a larger response.

MODALITY

Quality of Pain

Pain is readily distinguishable from other sensory experiences and thus is appropriately considered as a separate sensory modality. Pain is not a unitary experience, however, since many types of pain can be differentiated. These different *submodalities* of pain are generally described in terms of various noxious stimuli—for example, burning, tearing, crushing, pricking—although each may arise from a variety of physical or chemical agents.

Basis of Submodalities

The several varieties of pain originate in different receptors and ascend into the CNS over different pathways. There are two major classes of nociceptive endings, which are innervated by either Aδ or C fibers. Pain originating in the Aδ fibers is characterized as "sharp" or "pricking," whereas pain originating in C fibers is usually described as "dull," "burning," or "aching." The association of submodality with afferent axons of specific diameters is based on several lines of evidence. First, low-intensity electrical stimulation sufficient to excite Aδ but not C fiber axons results in a sharp, tingling sensation that becomes painful with repetitive stimulation. Second, if pressure is applied to block myelinated but not unmyelinated axons (see Chap. 15), dull pain can still be felt but sharp pain is lost. In contrast, local anesthetics,

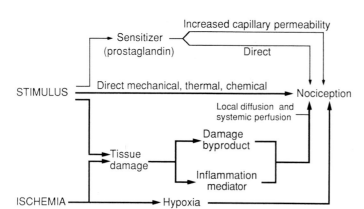

Figure 16–3 Mechanisms by which a noxious stimulus can excite nociceptors.

which affect unmyelinated axons before blocking myelinated axons, cause dull ache to be suppressed before sharp pain. These older observations that fast (Aδ) pain is sharp and slow (C fiber) pain is dull have been largely confirmed by more recent microneurography studies in humans.[27]

Within the "sharp" and "dull" submodalities, several further distinctions can be made (e.g., burning dull pain, aching dull pain, crushing dull pain), as the list of terms on the McGill-Melzak Pain Questionnaire indicates (see Fig. 16–1). The neurophysiological basis for these distinctions has not been well investigated, but two factors probably contribute. First, because nociception originates from a variety of afferent endings, a particular noxious stimulus excites a particular subpopulation of nociceptors. Second, noxious stimuli also excite nonnociceptive endings. Since pain endings have a higher threshold than other primary afferents, a cutaneous mechanical or thermal stimulus sufficient to excite nociceptors must first excite mechanoreceptors or thermoreceptors. This contribution to the pain motif by other sensory modalities provides an additional mechanism for discrimination within pain submodalities.

LOCALIZATION

Pain is localized by the same mechanisms as other sensory modalities, (i.e., a topographic organization of ascending pathways and projection back to the site of the primary afferent endings) (see Chap. 13). Because pain endings have a low innervation density, and pain pathways exhibit branching and convergence, pain is poorly localized when nociceptors are stimulted in isolation. For example, teeth contain primarily pain endings, and therefore patients with toothache often misidentify the offending tooth. Also, the pain resulting from the reflux of the contents of an acidic stomach into the lower esophagus (i.e., heartburn) is felt as a diffuse burning sensation, vaguely localized to the anterior chest. On the other hand, precise identification of the site of nociceptor stimulation is aided by the simultaneous activation of modalities with better localization. For example, a skin pin prick is well localized in part because of the simultaneous excitation of cutaneous tactile receptors at the stimulus site. Finally, sharp pain is better localized than dull pain because it better preserves its somatotopic organization in the CNS.

As with other sensory modalities, pain is incorrectly localized when its ascending pathway is stimulated unnaturally. Because of projection, pain caused by irritation of an axon is sensed as originating at the nociceptive endings innervated by that axon. For example, compression of a spinal dorsal root leads to pain projected to the corresponding cutaneous dermatome. Pain caused by pressure excitation of ascending spinal tracts due to syringomyelia, a spinal cord tumor, is felt as originating from the receptive field of the affected ascending axons. If a limb is removed, excitation of the central ends of severed axons is projected to a body part no longer physically associated with the sufferer.

Referred Pain

Pain displays an error of localization not exhibited by other sensory modalities: that is, *referred pain*. Referred pain is caused by excitation of nociceptors at one site, usually deep or visceral, that is sensed as originating at another site, usually superficial. For example, the pain of an acute myocardial infarction (heart attack) is often reported as a shooting pain in the left arm, and the pain of appendicitis may be felt as originating from the periumbilical abdominal surface. Although both referred and projected pain are errors of localization, referred pain results from the natural activation of a pain pathway, whereas projected pain is the result of unnatural and not necessarily noxious stimuli.

The mechanism of referred pain has been the subject of much speculation, most of which involves anatomical explanations. The superficial region to which pain is referred often is the dermatome that shares the same dorsal root as the visceral or deep afferents innvervating the structure from which the pain actually originates. For example, the pain associated with coronary hypoxia originates in cardiac afferents, which enter the spinal cord at T2–T4. The dermatomes corresponding to T2 (see Fig. 15–1) include the inner aspect of the left arm, the site to which this pain frequently is referred.

One explanation, first proposed by T.C. Ruch in an earlier edition of this textbook, postulates that many axons from visceral (or deep) nociceptors and axons from cutaneous nociceptors converge on the same second- or higher-order neurons (e.g., neuron B in Fig. 16–4). Therefore, when activity from the viscera reaches the cerebral cortex, the neurons responsible for conscious sensation cannot know whether the activity originated superficially or viscerally; thus, the cortical circui-

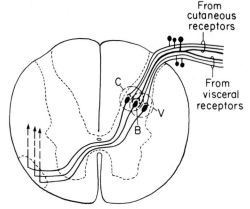

Figure 16–4 Anatomical bases of referred pain. C represents the second-order neuron pool excited by cutaneous nociceptors, V the pool excited by visceral nociceptors, and B a neuron excited or facilitated by afferents from both sources; one branching axon innervates both cutaneous and visceral nociceptive endings. Activity in either the C or V pool would result in correctly localized pain. Excitation or facilitation of neurons such as B would lead to referred pain as described in the text.

try arbitrarily assigns the site of origin to the skin, since past experience has taught the sufferer that the skin is the more common site of nociceptor stimulation. A second explanation suggests that visceral and superficial nociceptors use separate ascending pathways, but that the visceral pathway facilitates the superficial pathway. It is assumed that the cutaneous activity normally is too low to result in conscious sensation; however, when it is facilitated by input from deeper structures, the cutaneous barrage is transmitted with greater efficacy and exceeds the threshold for pain sensation. A third explanation assumes that many primary afferent nociceptive axons branch, with some branches innervating deep and visceral structures and others innervating skin (branching afferent in Fig. 16–4). When the deeper endings are stimulated, projection causes the skin to be identified as the source of the pain.[21] Also, excitation of the deeper branches would cause antidromic activation of cutaneous branches, possibly leading to the release of substances that could stimulate adjacent nociceptors. Although each or several of these explanations may apply, there is little firm physiological evidence about the actual mechanisms underlying referred pain. All three phenomena probably contribute to referred pain, with their relative importance depending on the particular circumstances.

In summary, pain may be localized correctly and precisely, particularly if it originates superfi-

cially and other modalities are activated simultaneously. On the other hand, if it originates in visceral or deep tissues or if nociceptors are activated in isolation, pain may be localizable only vaguely. Also, pain may be incorrectly referred to another region, or it may be referred and correctly localized simultaneously, thus appearing to arise from two regions simultaneously. Finally, activation of pain-conveying axons in the CNS may lead to erroneous localization by projection.

INTENSITY

The intensity of pain sensation depends on the same factors that determine the intensity of other sensory modalities: that is, the discharge frequency and recruitment of individual afferent receptors. In addition, pain probably more than any other modality is subject to peripheral and central modulation.

The relation between subjective pain intensity and primary afferent discharge has been studied by microneurography in human subjects. Figure 16–5 illustrates the effect of applying brief noxious thermal stimuli (39-51°C) to the skin of human volunteers.[28] Although individual nociceptive endings and individual subjects demonstrated considerable variability in their responses, the median neural and psychophysical responses (heavy lines in Fig. 16–5) both were linearly related to stimulus temperature once threshold had been exceeded; therefore, the subjective pain rating was proportional to the afferent nociceptive discharge. Recruitment is crucial as well, since the discharge of a *single* afferent ending does not result in pain or any other sensation. Spatial summation also seems to be required for the conscious appreciation of pain.

The expected correlation between pain sensation and stimulus intensity holds only in carefully controlled experiments. In fact, one of the characteristics of natural pain is lability. A stimulus that normally produces excruciating pain is barely felt under certain circumstances, as in the case of a wounded soldier who fights on oblivious of his wound or an athlete who continues to compete after receiving an injury that he notices only after the contest is over. In the absence of emotional and distracting stimuli, however, even a normally mild pain can cause great distraction (e.g., when one is attempting to sleep). Thus, central modulation clearly is involved in controlling the perception of pain.

NOCICEPTOR RESPONSES

PAIN RATINGS

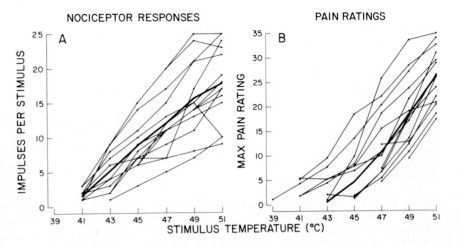

Figure 16–5 Responses to 5-s heat pulses applied to the skin of human subjects. *A,* The total number of action potentials in individual primary afferent nociceptors during the stimulus period as a function of stimulus temperature. *B,* The pain ratings by human subjects as a function of stimulus temperature using a 0–40 scale. Median values for the afferent fiber and psychophysical responses are denoted by the heavy lines. (After Torebjork, H.E.; LaMotte, R.H.; Robinson, C.J. *J. Neurophysiol.* 51: 325–339, 1984.)

Pain can also be modulated by peripheral mechanisms, since it has long been recognized that activation of innocuous sensory receptors (e.g., gently rubbing the skin near injured tissue) often ameliorates pain. This interaction between afferent inputs was formalized into the *gate control* theory of pain, which postulated that collaterals from large myelinated afferent axons associated with tactile sensibility produce presynaptic inhibition at spinal cord synapses formed by the small-diameter Aδ and C fibers associated with pain.[17] Thus, large-fiber afferent activity could control or "gate" the efficacy with which activity originating in nociceptors reaches the CNS.

Attempts to validate the gate control theory have been largely unsuccessful. For example, stimulation of large-diameter primary afferent fibers should produce presynaptic inhibition in Aδ nociceptive axons. When such experiments were performed, no presynaptic inhibition was found.[31] Whereas the details of the gate control theory are unlikely to prove correct, the theory has great historical importance because it directed the attention of neurophysiologists to the central mechanisms of pain modulation. We will see later in this chapter that electric stimulation of selected CNS sites inhibits synaptic transmission in nociceptive pathways, thereby producing analgesia; that small injections of morphine and similar substances at these sites also leads to analgesia; and that the CNS can synthesize neuromodulators (the endorphins) that have effects similar to those of morphine.

ADAPTATION

In comparison with the decrease in sensed intensity and the increase in sensory threshold fol-

lowing application of a maintained light touch or moderate temperature, pain sensation shows relatively little adaptation. For example, when a train of electrical pulses is applied to the human tooth pulp, a tissue whose afferent receptors are generally considered to be exclusively nociceptive, pain threshold rises by only 21% after 10 minutes of painful stimulation. The application of a noxious stimulus of sufficient intensity to injure tissue often leads to an increase in sensitivity called *hyperalgesia*, which is presumably due to release of substances that sensitize afferent endings when tissue is damaged.

CENTRAL PAIN PATHWAYS

Dorsal Horn

Nociceptive primary afferent axons enter the spinal cord along with other small-diameter afferents in the lateral division of the dorsal root. Upon entering the cord, the axons may join the tract of Lissauer and may ascend 1 to 3 segments before terminating in the dorsal horn. In the dorsal horn, nociceptive fibers terminate according to axon type (and thus pain submodality). There is disagreement concerning the exact site of termination: some investigators locate the termination of Aδ axons in lamina II and C fibers in lamina I, whereas others find the opposite.[33] Although nociceptor afferents terminate in laminae I and II, they may contact cells with somata in other laminae (e.g., cells of lamina V whose dendrites project to the more superficial laminae (see Fig. 15–4).

Several experiments indicate that the peptide *substance P* probably is one of the neurotransmitters released from the central terminations of primary nociceptor axons. First, electrical stimulation of peripheral C fibers causes release of substance P from a number of spinal synapses. This release is blocked by morphine administered in analgesic concentrations. Second, substance P is depleted from primary afferent terminals by peripheral nerve or dorsal root section.

Substance P is released not only from the central terminals of unmyelinated afferents, but from their peripheral terminals as well. As noted earlier, the peripheral administration of substance P causes nociceptor sensitization, probably because it increases capillary permeability, thereby facilitating the access of systemic algesic agents to nociceptive endings. Such sensitization leads to the interesting possibility that some nociceptors may be "self-recruiting," since activation of a single afferent neuron could lead to peripheral release of substance P, which in turn could produce a reduced excitation threshold for that neuron and its neighbors.

Substance P is also depleted by the application of capsaicin. Capsaicin, the active ingredient in chili peppers, has several actions on the nociceptive system.[6] When applied to peripheral nociceptors, it causes a burning pain that may last for several hours, presumably accounting for its appeal to devotees of spicy cooking. When injected systemically or applied to central terminations of afferent neurons, however, capsaicin leads to analgesia, probably because it depletes substance P. If administered systemically to a neonatal animal, capsaicin causes selective degeneration of unmyelinated and small myelinated afferent (but not efferent) neurons.

Substance P is not the only neurotransmitter that may play a role in synaptic transmission between primary and second-order nociceptive neurons. Other neuropeptides, which have been localized to small-fiber central terminals, include somatostatin, cholecystokinin, and vasoactive intestinal peptide.

Anterolateral System

Some of the spinal cord neurons contacted by nociceptor afferents send their axons to the opposite side of the spinal cord where they ascend in the contralateral anterolateral quadrant. Other nociceptor afferents contact one or more spinal interneurons, which in turn contact cells that ascend in the contralateral anterolateral quadrant. A relatively small number of ascending axons do not cross but ascend ipsilaterally. All of these ascending axons constitute the *anterolateral system* (Fig. 16–6).

Three types of spinal cord cells contribute to the anterolateral system. The first are specifically nociceptive cells that are activated by peripheral application of noxious mechanical, thermal, or chemical stimuli but are unaffected by touch, movement, and other innocuous stimulation. The second type respond to both noxious and innocuous stimuli applied to receptors innervated by Aβ, Aδ, and C fibers. Since the thresholds for the inputs to these cells range from very low to very high, they are often called *wide dynamic range* neurons; since they repond to a variety of stimulus energies, they are also known as *multimodal* neurons. The third type are excited only by innocuous stimuli, particularly mechanical (touch) and thermal. Thus the anterolateral system mediates not only pain (mainly contralateral), but other modalities as well.

The axons of the anterolateral system project to the thalamus via the spinothalamic tract, to the midbrain tectum via the spinotectal tract, and to the brain-stem reticular formation via the spinoreticular tract (see Chap. 15). The spinorecticular pathway relays excitation to the thalamus and other diencephalic structures and also to the region surrounding the cerebral aqueduct (Fig. 16–6). This region, the periaqueductal gray matter, is involved in pain modulation, as will be discussed later in this chapter.

Some spinothalamic tract neurons project exclusively to the thalamic intralaminar nuclei (IM, Fig. 16–6). Such cells generally respond only to noxious stimuli and have large receptive fields (sometimes encompassing the whole body or face). Since these neurons seem to be of ancient phylogenetic origin, they are said to constitute the *paleospinothalamic* tract. These neurons and those of the spinoreticular tract are ill-suited to localization and fine discrimination. They seem able to mediate only diffuse, burning pain and may subserve general pain responses such as arousal, affect, and autonomic reflexes. The remaining spinothalamic neurons project to specific thalamic projection nuclei (e.g., the ventrobasal complex [VBC], and the posterior thalamic nuclei [PO], Fig. 16–6). Of more recent origin, and thus called the *neospinothalamic* tract, these axons convey information on pain, thermal sensation, and crude touch. Many have small receptive fields. The neospinothalamic tract presumably mediates information required for pain localization and discrimination.

Thalamus and Cerebral Cortex

The neospinothalamic projection to the ventral posterolateral nucleus (VPL) is the best-studied

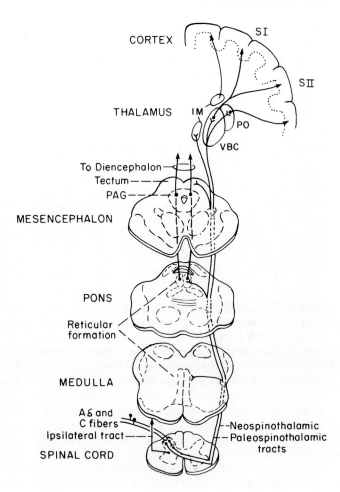

CORTEX SI
SII
THALAMUS IM
PO
VBC

To Diencephalon
Tectum
PAG

MESENCEPHALON

PONS

Reticular
formation

MEDULLA

Aδ and
C fibers
Ipsilateral tract
SPINAL CORD
Neospinothalamic
Paleospinothalamic
tracts

Figure 16–6 The anterolateral system, consisting of the neospinothalamic and paleospinothalamic tracts and their projections to the tectum, reticular formation, and periaqueductal gray matter (PAG), and to the thalamus, including the interlaminar nucleus (IM), the posterior nuclear group (PO), and the ventrobasal complex (VBC). The thalamus, in turn, projects to primary somatosensory cortical areas SI and SII.

ascending nociceptive pathway from the trunk and limbs. Axons conveying sensory information from the head via the trigeminal system terminate in the adjacent ventral posteromedial nucleus.[25] These two nuclei together form the ventrobasal complex, which projects, in turn, to primary cortical somatosensory areas SI and SII (Fig. 16–6). Only a small fraction of neurons in the ventrobasal complex are concerned with pain; most relay information regarding non-noxious modalities (Chap. 15). Surgical lesions of the ventrobasal complex to relieve intractable pain in humans causes a loss of sharp pain and cutaneous touch but leaves deep and aching pain sensation intact. Electrical stimulation of the ventrobasal complex evokes pain as well as other sensations.

The neospinothalamic tract also projects to the medial part of the posterior thalamic nuclei, which lie adjacent to VPL. These neurons project to the retroinsular cortex, a region adjacent to SII. Some of these cells respond to noxious stimuli but most have receptive fields that are large and sometimes

bilateral.[23] Thus this pathway could be involved in pain appreciation but not precise localization.

Paleospinothalamic cells terminate in the so-called nonspecific thalamic intralaminar nuclei (IM, Fig. 16–6), particularly the central lateral nucleus. Cells in IM also receive inputs (not shown) from the reticular formation, cerebellum, and basal ganglia, and are not somatotopically organized, and project diffusely to the cortex. The repeated application of the same noxious stimulus often elicits different neuronal responses. Based on all this evidence, IM probably is involved in general cortical arousal and, perhaps, the detection but not localization of novel stimuli.

Early views of the role of the cerebral cortex in pain sensation were based mainly on observations of the sequelae of traumatic head injury and on electrical stimulation of the human cortex during brain surgery. Such studies suggested that sensory cortex was reserved for the "discriminative" modalities, such as touch, while pain was appreciated at lower levels of the CNS without cortical involvement.[20] The current view, based on the last twenty years of neurophysiological and clinical investigation, is that the appreciation of pain probably is similar to the appreciation of other somatosensory modalities that use ascending spinothalamic and thalamocortical projections that are, at least in part, somatotopically organized. Like the somatosensory projections for discriminative and crude touch, the ascending pain projections are characterized by multiple pathways, the neospinothalamic and the paleospinothalamic systems. The major difference between pain and other somatic sensations is not the role of the somatosensory cortex but rather the additional ability of the CNS to control pain detection and perceived intensity through the release of specific chemicals, the neuromodulators.

ENDORPHINS AND PAIN MODULATION

Morphine, a derivative of opium, is a strong narcotic analgesic widely used to relieve severe pain. Morphine acts at synapses of the nociceptive pathways by binding to specific sites called *opiate receptors*. The binding of a morphine molecule to an opiate receptor decreases nociceptive synaptic excitability, thus reducing the sensation of pain. The analgesic action of morphine can be blocked by the narcotic antagonist *naloxone*, which attaches to opiate receptors to block morphine binding.

Several specific peptides, which are synthesized by and released from neurons, have properties similar to morphine, i.e., they bind to opiate receptors and produce an amelioration of pain (as well as other effects). Such substances are known as *opioid peptides* or *endorphins*, a contraction of "endogenous morphinelike substances" (see Chap. 61).

Endorphins depress synaptic excitability either pre- or postsynaptically, and thus behave like classical inhibitory neurotransmitters. However, the release of classical neurotransmitters is rapid and their action is brief (milliseconds to seconds), whereas the effect of endorphins may last minutes to hours since they bind tightly to their receptors and are more resistant to degradation. Although they are chemically similar to some neurohormones, endorphins exert their effect near their site of release rather than by diffusing into the circulation and affecting distant target organs. Therefore, endorphins are placed in a separate category, called *neuromodulators*, to emphasize their unique role in the modulation of synaptic excitability. As seen in Chapter 11, the same neuron may release both a neurotransmitter and a neuromodulator, and the two substances may have different effects.

Since their initial discovery in the 1970s, over 20 biologically active endogenous opioid neuropeptides have been identified ·and characterized, including β-endorphin, γ-endorphin, dynorphin, leu-enkephalin, and met-enkephalin. The action of endorphins is not limited to pain or to sensation or even to the nervous system, since they can influence the cardiovascular, respiratory, alimentary, and other organ systems. Indeed, opioid receptors have been found at many locations in the body, both in the CNS and elsewhere. Because the different opiate agonists and antagonists bind with different efficacies to the same and different receptors, it seems likely that several different opioid receptors exist; such receptors have been called μ, κ, σ, δ, and ε receptors.

The loci of action of algesic neuromodulators can be identified by injecting small amounts of morphine and similar substances at various CNS locations and noting their effect on responses to noxious stimuli. Morphine is particularly effective when applied to either the dorsal horn of the spinal cord or the periaqueductal-periventricular gray matter of the thalamus and mesencephalon.[34] Electrical stimulation of these same regions and of the raphe magnus nucleus in the medulla produces profound analgesia, sufficient to permit abdominal surgery. The analgesia may long outlast the stimulation, which can reduce pain sensitivity for up to 24 hours. The analgesic effect is specific for nociception; other sensory modalities, motor coordination, the general state of consciousness, and other CNS functions are unaffected. Also, the analgesia resulting from small morphine injections or electrical stimulation is reduced or eliminated by naloxone. Lesions of the nucleus raphe magnus or the dorsolateral columns, and administration of agents that deplete or block serotonin, also reduce the analgesia induced either by the injection of morphine into the periaqueductal gray or by electrical stimulation of the raphe nucleus.[32]

The current explanation of these data postulates that the periaqueductal-periventricular gray (PAG, Fig. 16–7) is the "integration center" of an endorphin-mediated analgesic system. The PAG region receives inputs from the hypothalamus, the limbic system (particularly the amygdala), and the insular cortex (see Chap. 32). Electrical stimulation of specific hypothalamic loci or the amygdala produces analgesia. In addition, the PAG receives inputs from the pontine and medullary reticular formations and other brain stem structures. Hence, the PAG is anatomically well situated to

Neurons of the PAG provide a major excitatory input to the nucleus raphe magnus of the medulla (rm, Fig. 16–7). These serotonergic raphe nuclei, in turn, send descending axons to the spinal cord, where they terminate in the outer laminae of the dorsal horn; their analgesic effect is mediated at least in part by endorphin-releasing segmental interneurons (Fig. 16–7). Such a pathway could account for the lack of analgesic effects produced by a variety of manipulations after raphe lesions and serotonin depletion.

SPECIFIC PAINS

Headache

Headache is a diffuse, dull pain arising from stimulation of cranial nociceptors. As determined by electrical stimulation in conscious patients undergoing neurosurgery, these nociceptors are located in the dural and intracerebral arteries, large veins and venous sinuses, the dura at the base of the brain, head and neck muscles, extracranial mucosa, and the skin of the scalp. Direct stimulation of the trigeminal, facial, glossopharyngeal, vagal, and second and third cervical nerves can also give rise to headache. In contrast, the parenchyma of the brain, the lining of the ventricles, the choroid plexus, and most of the dura and pia contain no nociceptors and are therefore insensitive to noxious agents or surgical manipulations.

Headache arising from deep nociceptors is often referred to the surface of the head. Nociceptors innervating structures above the tentorium, the fold of dura separating the cerebellum from the posterior portions of the cerebral hemispheres, are served by the trigeminal nerve and referred to the frontal, temporal, and parietal cranium. Nociceptors below the tentorium are innervated by the glossopharyngeal, vagal, and cervical nerves and are referred to the occipital region.

The most common causes of headache are (1) stimulation of extracranial, nonvascular nociceptors, (2) tension on the supporting structures of the brain, and (3) distension or irritation of cranial arteries. Headache from extracranial, nonvascular pain afferents can originate in the nasal mucosa, paranasal sinuses, orbit, extraocular muscles, and facial and neck muscles and, as elsewhere in the body, can be due to irritation, inflammation, muscle strain, and so forth. An excruciating headache is produced by excitation of nociceptors in the structures that support the brain. Normally the brain is almost completely supported by cerebro-

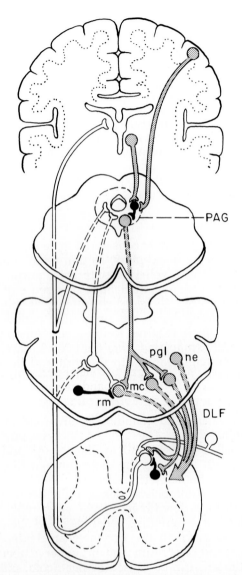

Figure 16–7 The descending pain-modulating system. An endorphin-mediated system originating in the periaqueductal-periventricular gray (PAG) projects to the nuclei raphe magnus (rm) and reticularis magnocellularis (mc), and thence via the dorsolateral funiculus (DLF) to the spinal cord. Other bulbospinal pathways potentially relevant to analgesia arise from nucleus paragigantocellularis (pgl) and noradrenergic (ne) cell groups. The solid cells denote endorphin-containing interneurons.

process neural signals related to emotion, stress, affect, general somatic sensation, and pain. At least some of the interneurons in the PAG region release endorphins that bind to postsynaptic opioid receptors; the action of such interneurons may help account for the profound analgesia resulting from periaqueductal stimulation or morphine injection.

spinal fluid (CSF); all but 40 grams of the total brain weight of about 1350 grams is buoyed up by the CSF. When CSF is lost—as, for example, after a lumbar puncture—more of the brain weight must be supported by suspensory structures, causing them to be stretched excessively. The resulting headache, which is exacerbated by sudden head movement and standing, is relieved by a recumbent posture and the gradual replacement of CSF over time. Pain of similar origin is caused by subdural hematomas and by tumors that displace the brain and stress its supporting tissue. Finally, stimulation of arterial wall nociceptors by excess distention has long been presumed to be the cause of three important types of headache: (1) the headache of systemic hypertension; (2) the cluster headache, a severe, repeating pain of limited duration (1 hour or less) that recurs every 1 to 3 days over 1 to 2 months; and (3) migraine headache, a pounding, unilateral, severe headache that often involves photophobia and nausea and sometimes is preceded by a sensory aura such as bright flashes.

Cluster and migraine headaches were once assumed to result from intracranial and/or extracranial vasodilation, sometimes preceded by vasoconstriction. Vasoconstriction, with the consequent reduced cerebral blood flow, was held responsible for the premonitory aura, whereas vasodilation was thought to cause the subsequent excitation of vascular nociceptors and the resulting pounding pain. Several lines of evidence seemed to support the vascular theory of migraine generation. Measurement of human cerebral blood flow with xenon tracer techniques shows marked hyperemia during migraine attacks in most (but not all) subjects. Also, migraine can be precipitated by alcohol and other nonspecific vasodilators. Finally, foods containing tyramine, such as aged cheese and pickled herring, also can precipitate a migraine. Tyramine is a sympathomimetic vasoactive amine that acts as a vasoconstrictor both through its direct excitation of vascular smooth muscle and by releasing stored norepinephrine. However, the vascular theory of migraine has been challenged in the past decade,[4] and it now seems that changes in vasomotor state occur in parallel with the migraine rather than produce it. Migraine is now thought to result from such conditions as spreading depression of cortical function, blood platelet disorders, aberrant immunological reactions, psychological disorders, and diet.

The putative role of diet in migraine headaches has been exploited by food faddists. "No clinical problem is more perplexing to the practicing physician than the relation of food and beverages to the production of migraine. A great deal of folklore surrounds the subject, and little scientific work has been done to clarify the issue. Furthermore, various misconceptions about diet and migraine are given impetus by articles in the popular press. Thus the physician may encounter patients who have constructed elaborate diets in the hope of escaping recurrent migraine attacks. Occasionally these diets assume ridiculous extremes; for example, one may find that a patient is living primarily on scallions, or bananas or onions."[4]

Toothache

Teeth are generally considered to be innervated exclusively by nociceptors, so that tooth stimulation results in a volley of pure pain, uncontaminated by innocuous modalities. The sense of mechanical force, elicited by probing, is due to mechanoreceptors in the supporting periodontal ligament fibers external to the tooth. Among tissues easily stimulated from the body surface, the tooth (perhaps along with the cornea of the eye) is unique in its single-minded dedication to nociception.

The teeth are innervated by axons mainly from the maxillary and mandibular divisions of the trigeminal nerve. The axons, which include unmyelinated C fibers and myelinated Aδ and Aβ fibers, gain access to the tooth through the root apex, then branch within the pulp, and terminate in the subodontoblastic plexus of the dental pulp (Fig. 16–8). A few branches may enter tubules in the dentine, where they penetrate up to one third of the distance to the dentine-enamel junction. Enamel and cementum contain no nerve fibers.

Toothache can be caused by heat, cold, inflammation, and mechanical probing of the dental pulp. In teeth with carious lesions that expose the dentine-enamel junction, pain also can be elicited by osmotic stimuli (e.g., biting into a candy bar) and by gentle drying (as with a dentist's air jet). High temperatures cause pain when the conducted heat raises the pulp temperature to noxious levels. Since the enamel protecting the pulp is a reasonably good insulator, heat sensitivity usually becomes a problem only if (1) dentine is exposed by recession of the gingiva, (2) a poorly insulated metallic restoration permits the ready conduction of heat to the pulp, or (3) pulp nociceptors are hyperexcitable due to inflammation. Cold causes pain probably through thermal contraction of the enamel, which leads to mechanical stresses trans-

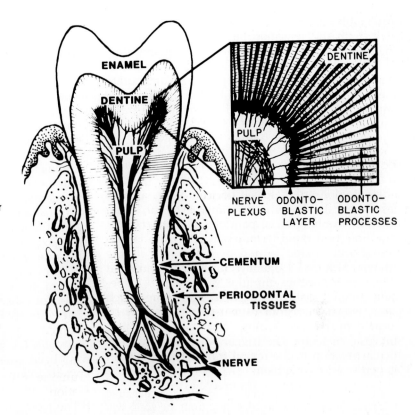

Figure 16–8 Innervation of the tooth. (After Sessle, B.J. *J. Dent. Res.* 66:962–981, 1987.)

mitted to the dentine and pulp, thereby exciting nociceptors.

Pulp inflammation leads to a particularly severe pain. This pain is likely due to both the accumulation of the mediators of the inflammatory process, many of which are potent activators of pain endings, and the unique microcirculation of tooth pulp. The blood vessels supplying the pulp enter the tooth with the nerve fibers through small openings in the root, then arborize to supply the whole pulp. Inflammation is associated with vasodilation of the arterioles, which in other tissue causes transcapillary fluid movement and swelling (edema). However, the rigid walls of the enamel and dentine do not allow the pulp to swell; instead, the arteriolar dilation causes an increase in pulp hydrostatic pressure to values near systemic arterial pressure, thereby producing mechanical distortion. Because the high intrapulpal pressure compresses the blood vessels draining the pulp, the circulation may become compromised, so that normal repair processes cannot correct the damage. In such cases the pain ceases only when the dentist removes the offending tissue (i.e., performs a pulpectomy or "root canal").

The reason for the great sensitivity of the dentine-enamel junction is unclear. The most popular current hypothesis suggests that fluid can move in the dentinal tubules and that such movement excites pulpal nerve endings.[1] This explanation could account for the pain associated with eating candy when a carious lesion has exposed the dentine-enamel junction. In this case, the high concentration of sugar in the mouth would cause increased osmotic pressure, which would draw fluid out of the dentinal tubules and activate pulp nociceptors. Similarly, the pain that accompanies drying would occur because the liquid-air interface causes capillary forces that move the fluid along the tubules.

Central Pain

Central pain originates in the CNS in the absence of any peripheral pathology. It may occur spontaneously or be triggered by innocuous or mildly noxious stimuli. The prototype of central pain is the sensory derangement that follows ischemic or hemorrhagic lesions of the posterolateral and posteromedial nuclei of the thalamus described above. Spontaneous pain may be absent, but when a noxious agent is applied, a severe, intolerable, explosive pain may be felt over a wide area, with the sensation bearing little relation to the intensity or location of the initiating stimulus. Surgical destruction of the offending thalamic nuclei using stereotaxic techniques may ameliorate the pain and produce a contralateral anesthesia temporarily, but both pain and somatic sensation usually return eventually.

Tic Douloureux

Tic douloureux ("spasm of pain"), or, as it is less poetically known, *trigeminal neuralgia*, is an intense, stabbing facial pain of rapid onset, brief duration, and rapid termination. The pain is unilateral and limited to regions innervated by the trigeminal nerve. Generally, an attack is initiated when an innocuous stimulus is applied to a specific, often highly restricted, trigger zone. Such sites, which vary from patient to patient, may be located on the cheek, nose, lip, chin, scalp, oral mucosa, or teeth. The trigger zone bears no particular relation to the region from which the pain is perceived to originate.

The pain of tic douloureux is an effective negative reinforcer. "Patients with triggering from the scalp will often refuse to brush their hair; shaving may be impossible for the man with a trigger zone in the upper lip or chin; triggering from the teeth or gingiva may preclude oral hygiene. Patients with pain triggered by chewing or swallowing may have insufficient oral intake adequately to maintain caloric requirements."[14]

Most patients with trigeminal neuralgia have a compression of the trigeminal root. It is hypothesized that the compression causes trigeminal degeneration or irritation, which eventually leads to bursts of epileptic-like discharges at central connections of the trigeminal system;[2] indeed, anticonvulsants are very effective in the treatment of this condition. For cases in which drugs are ineffective or have unacceptable side effects, a surgical release of the trigeminal compression or lesions of the trigeminal ganglion often produce effective pain relief.

Causalgia

Causalgia is a continuous burning pain that sometimes develops following a traumatic peripheral injury and may continue after healing is complete and no apparent peripheral damage remains. It was first reported as a sequel to gunshot wounds, but it can be induced by other peripheral traumas as well. The pain is greatly exacerbated by gently touching the affected area; in severe cases even contact by light clothing or a puff of air evokes intolerable pain. The characteristic that most distinguishes causalgia from the pains previously considered is the involvement of the autonomic nervous system (see Chapter 34). Sympathetic activation by excitement or other emotions greatly increases pain intensity. The sympathetic nervous system seems to be responsible for maintaining causalgia, since autonomic ganglion blocking agents, surgical sympathectomy, and peripheral antiadrenergics often relieve it. Other sympathetic activities in the affected region (e.g., sweating) are often also abnormal. Causalgia is only one of several *sympathetic dystrophies.*

There is considerable evidence that mammalian afferent nerve endings can be excited by the activation of adjacent sympathetic efferent endings, either directly by the release of norepinephrine or indirectly by the production of an intermediate neuroeffector substance. A sympathetic sensitivity has been demonstrated in developing axon sprouts following nerve injury[30] and may be characteristic of normal endings in uninjured tissue as well.[24]

In causalgia, it is unclear what changes in neuronal interaction are induced by the peripheral injury and why the abnormal interactions continue after the injury has healed. One possible explanation is illustrated in Figure 16–9, where the site of abnormal function is ascribed to the multimodal neurons of the dorsal horn, which are excited by afferents mediating both painful and innocuous modalities. It is proposed that immediately following the injury, multimodal or wide-dynamic-range (WDR) neurons become hypersensitive due to the intense bombardment from the C-nociceptors (Fig. 16–9A), and that this hypersensitivity remains after healing. Therefore, light touch, previously conveyed to conscious sensation only via other pathways, now ascends to consciousness over the anterolateral system and is, therefore, interpreted as pain (Fig. 16–9B). Furthermore, in the absence of cutaneous stimulation, the discharge of the mechanosensitive endings is maintained by sympathetic tone. Prior to the sensitization of the multimodal neuron, this sympathetically elicited discharge was subthreshold for sensation. After sensitization has lowered synaptic threshold, however, the low-level tonic discharge is sufficient to activate ascending nociceptive pathways (Fig. 16–9C) to produce continuous burning pain.

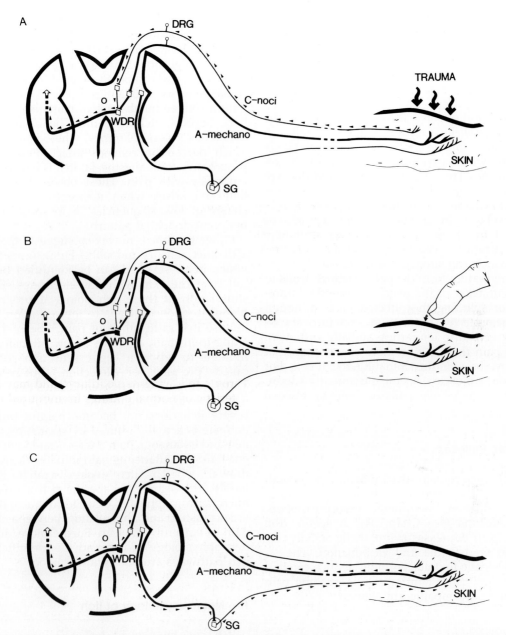

Figure 16–9 Schematic diagram of the development of causalgia following tissue injury. *A,* The immediate response to trauma is the discharge of cutaneous nociceptors (C-noci) leading to excitation of dorsal root ganglia (DRG) neurons and sensitization of dorsal horn multimodal units (here called wide-dynamic-range [WDR] neurons). *B,* The sensitized WDR neurons discharge in response to activation of low-threshold cutaneous mechanoreceptors (A-mechano). *C,* Tonic discharge in the sympathetic ganglia (SG) activating low-threshold cutaneous mechanoreceptors causes sympathetically maintained pain. (After Roberts, W.J. *Pain* 24:297–311, 1986.)

PAIN-INDUCED REFLEXES

In addition to its effect on conscious sensation, pain can initiate protective reflexes. These reflexes do not depend on sensation since they can be elicited in unconscious subjects; even in awake subjects, pain-induced reflexes sometimes are elicited at stimulus levels too low to be sensed. The nature of the reflex response depends on the pain submodality, and the site of origin, intensity, and duration of the stimulus. The efferent limb of the reflex includes both somatic and autonomic motor responses.

Somatic Reflexes

Brief application of a noxious stimulus to the body surface elicits a flexion-withdrawal reflex in which somatic muscle contraction causes the body part to be withdrawn from contact with the offending agent. This response is mediated through polysynaptic connections at the spinal level, as detailed in Chapter 24. Although the flexion-withdrawal reflex is involuntary, its efficacy can be controlled by descending influences; apprehension or learning can cause hyperreflexia while stoic determination can suppress withdrawal.

Pain originating from deeper structures leads to reflex contraction of the overlying muscle, immobilizing or splinting the affected part. Although nominally protective in function, such muscle contraction can lead to tenderness initially and eventually to pain that may be more severe than that caused by the stimulus initiating the reflex. For this reason, a muscle relaxant is frequently administered to help relieve pain of deep or visceral origin.

Autonomic Reflexes

A sharp, severe pain usually evokes a generalized sympathetic response: tachycardia, sweating, pupillary dilation, peripheral vasoconstriction, and piloerection (see Chap. 34). A severe dull crushing pain depresses sympathetic activity, resulting in bradycardia and vasodilation. The resulting systemic hypotension may reduce cerebral perfusion and cause fainting. Dull, crushing pain is frequently accompanied by nausea.

ITCH

Like other primitive, protopathic modalities, itch (or *pruritis*) is difficult to define. Generally, itch is any unpleasant sensation that evokes the desire to scratch. Itch can arise from skin or mucous membranes but not from deep or visceral structures. Itch, which is the single most common symptom in dermatology, may be due to a specific pathology or may occur without any clinically evident skin disease.[5] Itch is invariably disagreeable and can be sufficiently severe to cause mental depression and suicidal tendencies.

Itch is closely related to pain in that both are primitive sensations with negative affect. Several other similarities between itch and pain have led to the suggestion that itch is a submodality of pain. First, itch can be dissociated from all primary somatosensory modalities except pain. Patients with a congenital absence of pain also do not feel itch. Second, itching cannot be elicited in the cutaneous lesion of leprosy, where pain is lost but touch perception remains intact. Finally, under progressive local anesthesia, itch is blocked simultaneously with pain. These observations suggest that itch arises when nociceptive endings discharge at low frequencies or in some particular temporal or spatial pattern.

Other evidence, however, suggests that itch is a distinct sensory modality. First, locations from which itch is evoked can be identified by exploration of the skin with fine electrodes. Electrical stimulation of these itch points at progressively higher frequencies leads to the sensation of more intense itch rather than of pain.[29] Second, stimulating tooth pain afferents at low intensity causes humans to report a vague sensation sometimes called "pre-pain" or "detection," but never itch. Third, common experience indicates that pain and itch are easily distinguished, and that increasingly severe itch does not become painful unless the itch site is actually injured by scratching. Finally, itch and cutaneous pain are associated with different reflexes; cutaneous pain initiates flexion-withdrawal, whereas itch evokes scratch. Unfortunately, the search for specific itch axons, using microneurography and other techniques, has been unsuccessful, and the relation between itch and pain is currently unclear. Furthermore, the basic neurophysiology of itch and the identification of itch pathways are poorly understood.

The Neurophysiology of Itch

Itch can be evoked by either mechanical or chemical stimuli. Light mechanical pressure applied in the vicinity of a cutaneous itch ending leads to transient itching and reflex scratching. Sustained itching is due to chemical rather than mechanical stimuli.

The classical chemical mediator of itch is histamine. Superficial cutaneous injection of a small dose of histamine induces itch; injection of a larger dose leads to burning pain. The itch response is mediated by the H-1 receptor and perhaps other, as yet unidentified, receptors.[9] Blocking histamine receptors with antihistaminics often (but not always) prevents the development of itch. Histamine is not the sole chemical mediator, however, since itch can be induced in histamine-depleted skin by the application of spicules of the tropical legume cowhage (*Mucuna puriens*), the active ingredient in the itching powder once sold by trick and novelty shops (modern itching powder consists of fiber glass particles). The pruritic responses induced by cowhage have been ascribed to its protease activity. A number of other proteases—for example, trypsin, chymotrypsin, fibrolysin, ficin, papain, streptokinase–also cause itching.

Many other substances also induce itch, including kallikrein, bradykinin, vasoactive intestinal peptide, neurotensin, secretin, and substance P. In addition, prostaglandin E-1, serotonin, and Ca^{++}, among others, have been suggested as mediators or modulators of the itch response.[5] Little is known, however, about the actual mode of action of any of these substances.

Histological examination of skin beneath itch points indicates that itch receptors are superficial free nerve endings, located mainly at the dermal-epidermal border. These endings appear similar to cutaneous multimodal nociceptors. Pressure blocks applied to cutaneous nerves to inhibit first larger axons and then progressively smaller axons abolish itch and dull (slow) pain simultaneously, implying that itch is conveyed by unmyelinated C fibers.

Little is known about central pathways and the CNS mechanisms subserving itch. Presumably, itch and slow pain share common ascending tracts. Itch is abolished by spinal cord sections that abolish pain. Whether itch must reach cerebral cortex to evoke sensation or can be appreciated at a subcortical level is unclear. Because of its close association with pain, itch sensation might be expected to be blocked or ameliorated by CNS opiate receptor agonists. Actually, the opposite is the case: systemic administration of morphine causes itch while relieving pain.

Scratch

Scratch, a spinal reflex, is initiated by itch and reduces itch intensity. Why scratching relieves itch is not clear. One hypothesis is that stimulation of mechanoreceptors innervated by large axons inhibits protopathic sensibilities (i.e., those that produce suffering) mediated by small fibers, as postulated in the gate theory of pain. Another possibility is that the protopathic senses have mutually inhibitory ascending interactions, since

patients with severe pruritis often will scratch until they cause traumatic skin abrasions, preferring pain to itch. But, as with the studies of the mechanisms underlying pain sensation, the neurophysiological investigations of itch have barely scratched the surface.

MEANING OF PAIN

Pain is clearly useful in many situations. For example, superficial, sharp and brief pain elicits the flexion-withdrawal reflex, which minimizes tissue damage. Also, the negative affect associated with pain is an effective reinforcement for learned avoidance, thereby minimizing the probability of future encounters with the same noxious stimulus. Even deeper pains and hurts such as joint sprains, muscle stress, soft tissue wounds, etc., serve notice to their host that the affected part must be rested. It is no accident that most of the endogenous mediators of the inflammatory process associated with tissue injury either excite or sensitize nociceptive afferent endings.

But of what use is the chronic pain that seems to serve no purpose except to exacerbate misery—for example, migraine headache, arthritic pain, the pain of terminal cancer? Often, chronic pain is a disease in itself and may exist without a discernible organic basis. Insight into the role of chronic pain can be gained by examination of the lives of those born insensible or indifferent to pain. One well-studied case is that of a young Canadian woman who showed indifference to pain at a very early age. She had no other sensory or mental deficits, and in fact was highly intelligent. She was guided through the usual dangers of childhood, learning to avoid potentially injurious stimuli. In her late teens and early twenties, she began to demonstrate skeletal degeneration, particularly of joints and the spinal vertebrae, resulting in curvature of the spine and joint deformation. She became progressively more incapacitated through her mid-twenties, with degeneration and infection of the joints, and eventually died at age 28 of infection and bronchopneumonia. This case suggests that pain may contribute subliminally to motor responses such as posture. Assuming a particular upright posture for too long a time places stresses on the skeletal system which lead to bone distortion and degeneration. A supine posture assumed too long causes bed sores. Thus, the low-level discharge of nociceptors even at rates that are subthreshold for pain perception could lead to often unconscious motor actions that prevent potentially harmful stresses from becoming noxious. The loss of nociception removes this protection.

Chronic, intractable pain of organic origin represents this protective system gone awry. In chronic, intractable pain, the tissue damage or potential damage of which the individual is being warned is beyond his capacity to neutralize or interdict. In a sense, chronic physiological pain is an accident of nature, the price we must pay for the general benefits of nociception.

ANNOTATED BIBLIOGRAPHY

Perl, E.R. Pain and nociception. *Handbk. Physiol.* 3 (2):915–975, 1984.
A comprehensive review of all aspects of pain and nociception from the primary afferents to descending modulation, including a treatment of the endorphins (over 680 references).

REFERENCES

1. Brannstrom, M. Hydrodynamic theory of dentinal pain: sensation in preparations, caries, and the dentinal crack syndrome. *J. Endodont.* 12:453–457, 1986.
2. Calvin, W.H.; Loeser, J.D.; Howe, J.F. A neurophysiological theory for the pain mechanism of tic douloureux. *Pain* 3:147–154, 1977.
3. Coggeshall, R.E.; Hong, K.A.P.; Langford, L.A.; Schaible, H.G.; Schmidt, R.F. Discharge characteristics of fine medial articular afferents at rest and during passive movements of inflamed knee joints. *Brain Res.* 272:185–188, 1983.
4. Dalessio, D.J. Headache. In Wall, P.D.; Dubner, R., eds. *Textbook of Pain,* 277–292. Edinburgh, Churchill Livingston, 1984.
5. Denman, S.T. A review of pruritis. *J. Am. Acad. Dermatol.* 14:375–392, 1986.
6. Fitzgerald, M. Capsaicin and sensory neurons—a review. *Pain* 15:109–130, 1983.
7. Fitzgerald, M.; Lynn, B. The sensitization of high threshold mechanoreceptors with myelinated axons by repeated heating. *J. Physiol.* 265:549–563, 1977.
8. Georgopoulos, A.P. Functional properties of primary afferent units probably related to pain mechanisms in primate glabrous skin. *J. Neurophysiol.* 39:71–83, 1976.
9. Greaves, M.W.; Davies, M.G. Histamine receptors in human skin: indirect evidence. *Br. J. Dermatol.* 107:101–105, 1982.
10. Hallin, R.G.; Torebjork, H.E. Studies on cutaneous A and C fibre afferents, skin nerve blocks and perception. In Zotterman, Y., ed. *Sensory Functions of the Skin in Primates with Special Reference to Man,* 137–149. New York, Pergamon Press, 1976.
11. Kumazawa, T.; Perl, E.R. Primate cutaneous sensory units with unmyelinated (C) afferent fibers. *J. Neurophysiol.* 40:1325–1338, 1977.
12. LaMotte, R.H.; Campbell, J.N. Comparison of responses of warm and nociceptive C-fiber afferents in monkeys with human judgments of thermal pain. *J. Neurophysiol.* 41:509–528, 1978.
13. LaMotte, R.H.; Thalhammer, J.G. Response properties of high-threshold cutaneous cold receptors in the primate. *Brain Res.* 244:279–287, 1982.
14. Loeser, J.D. Tic douloureux and atypical facial pain. In Wall, P.D.; Dubner, R., eds. *Textbook of Pain,* 426–434. Edinburgh, Churchill Livingston, 1984.
15. Melzak, R. The McGill pain questionnaire: major properties and scoring methods. *Pain* 1:277–299, 1975.
16. Melzak, R. *Pain Measurement and Assessment.* New York, Raven Press, 1983.
17. Melzak, R.; Wall, P.D. Pain mechanisms: a new theory. *Science* 150:971–979, 1965.
18. Mersky, H., ed. Classification of chronic pain: descriptions of chronic pain syndromes and definitions of pain terms. *Pain,* Suppl. 3. New York, Elsevier, 1986.
19. Paintal, A.S. Vagal sensory receptors and their reflex effects. *Physiol. Rev.* 53:159–227, 1973.
20. Penfield, W.; Jasper, H. *Epilepsy and the Functional Anatomy of the Human Brain.* Boston, Little, Brown & Co., 1957.
21. Perl, E.R. Pain and nociception. *Handbk. Physiol.* 3 (2):915–975, 1984.
22. Perl, E.R. Characterization of nociceptors and their activation of neurons in the superficial dorsal horn: first steps for the sensation of pain. *Adv. Pain Res. Ther.* 6:23–51, 1984.
23. Poggio, G.F.; Mountcastle, V.B. A study of the functional contributions of the lemniscal and spinothalamic systems to somatic sensibility. *Bull. Johns Hopkins Hosp.* 106:266–318, 1960.
24. Roberts, W.J. A hypothesis on the physiological basis for causalgia and related pains. *Pain* 24:297–311, 1986.
25. Sessle, B. The neurophysiology of facial and dental pain: present knowledge, future directions. *J. Dent. Res.* 66:962–981, 1987.
26. Sherrington, C.S. *The Integrative Action of the Nervous System.* New Haven, Yale Univ. Press, 1906.
27. Torebjork, H.E. Afferent C units responding to mechanical, thermal and chemical stimuli in human non-glabrous skin. *Acta Physiol. Scand.* 92:374–390, 1974.
28. Torebjork, H.E.; LaMotte, R.H.; Robinson, C.J. Peripheral neural correlates of magnitude of cutaneous pain and hyperalgesia: simultaneous recordings in humans of sensory judgments of pain and evoked response in nociceptors with C-fibers. *J. Neurophysiol.* 51:325–339, 1984.
29. Tuckett, R.P. Itch evoked by electrical stimulation of the skin. *J. Invest. Dermatol.* 79:368–373, 1982.
30. Wall, P.D.; Gutnick, M. Ongoing activity in peripheral nerves: the physiology and pharmacology of impulses originating from a neuroma. *Exp. Neurol.* 43:580–592, 1974.
31. Whitehorn, D.; Burgess, P.R. Changes in polarization of central branches of myelinated mechanoreceptor and nociceptor fibers during noxious and innocuous stimulation of the skin. *J. Neurophysiol.* 36:226–237, 1973.
32. Willis, W.D. Control of nociceptive transmission in the spinal cord. In Autrum, H.; Ottoson, D.; Perl, E.R.; Schmidt, R.F., eds. *Progress in Sensory Physiology,* Vol. 3, pp. 1–159. New York, Springer-Verlag, 1982.
33. Willis, W.D. *The Pain System: The Neural Basis of Nociceptive Transmission in the Mammalian Nervous System.* Basel, Karger, 1985.
34. Yaksh, T.L.; Rudy, R.A. Narcotic analgesics: CNS sites and mechanisms of action as revealed by intracerebral injection techniques. *Pain* 4:299–359, 1978.
35. Yaksh, T.L.; Hammond, D.L. Peripheral and central substrates in the rostral transmission of nociceptive information. *Pain* 13:1–85, 1982.

Chapter 17

Robert A. Dobie
Edwin W Rubel

The Auditory System: Acoustics, Psychoacoustics, and the Periphery

INTRODUCTION

The sense of hearing, along with vision and olfaction, provides vertebrates with information about events occurring at a distance. In humans and in other animals with a poorly developed sense of smell, hearing is the only sense that constantly monitors the entire environment, receiving information from all directions. Even when an animal is asleep or directing its attention elsewhere, it can still respond to sounds (e.g., the sleeping mother is aware of her baby's slightest movement). The sense of hearing thus serves as a primitive alarm or warning system.

The vertebrate ear probably evolved from the lateral line organ (still seen in fishes and larval amphibians), which is specialized for detecting mechanical disturbances in a fluid environment. In the lateral line organ, highly specialized mechanoreceptors called hair cells transduce displacement of the fluid surrounding them to graded changes in membrane potential, and ultimately to excitation of afferent neurons. The actual subcellular elements responsible for the transduction are a group of tiny, actin-filled cilia called stereocilia, which protrude as a tuft from the top of the cell. Similar hair cells and subcellular elements are also responsible for transduction in the auditory end organs of the vertebrate inner ear and the vestibular end-organs (specialized for sensing head position and movement; see Chapter 27). In the auditory and vestibular organs, however, the hair cells detect the motion of fluids within the temporal bone; these fluids are in turn driven by sound waves and head movements, respectively. Together, these systems are often referred to as the acoustico-lateralis system.[16]

In non-mammalian vertebrates, the auditory end-organ has undergone major evolutionary changes related to the transition from an aquatic to a terrestrial environment.[29] The mammalian ear, by contrast, is surprisingly constant across species, displaying primarily quantitative rather than qualitative variations related to the specific behavioral and ecological needs of different species. For example, the mechanical characteristics of mammalian middle ear structures are specialized to enhance either high-frequency or low-frequency hearing,[48] while both the middle and inner ears of certain bats are constructed to provide exceptional sensitivity and resolution at the ultra-high frequency of the bats' sonar vocalizations.[3]

Although many mammals and birds use the auditory sense to receive communication signals from other members of their species, the ears of such animals do not appear to differ in structure or function from those of other, less vocal, species. However, there is usually a good match between the sound frequencies to which an animal is most sensitive and those frequencies that carry communication information by the animal's vocalizations.

The role of hearing in intraspecies communication reaches its zenith in humans. Loss of the ability to understand speech is by far the most serious consequence of acquired deafness. In congenital deafness, the ability to learn spoken language is severely diminished, so that the vast majority of congenitally deaf children never acquire intelligible speech. Despite the crucial role of the hearing sense in the acquisition and use of speech and language, the human ear, up to and including the auditory nerve, does not appear to differ in any important way from the general plan of the mammalian ear.

The structure and function of the adult mammalian ear is discussed in this chapter. Chapter 18 considers how the brain processes the afferent signals provided by the auditory nerve.

ANATOMICAL OVERVIEW

Figure 17–1 shows the location and structure of the mammalian ear. It is divided for convenience

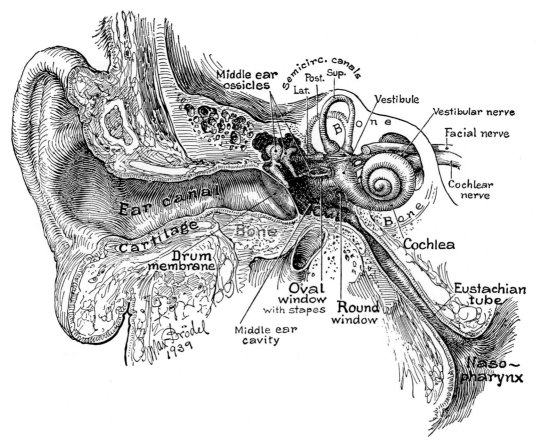

Figure 17–1 This schematic drawing combines a view of the coronal section through the ear canal and middle ear with a more diagrammatic representation of the eustachian tube, cochlea, and internal auditory canal. The middle ear muscles are not shown. (After Weaver; Lawrence. *Physiological Acoustics*, Princeton, Princeton University Press, 1954.)

of description into outer, middle, and inner parts. The outer ear includes the pinna (or auricle) and ear canal. The pinna is a skin-covered cartilaginous flap, often intricately shaped, and possesses muscles that permit it to be moved to a greater or lesser extent in different species. The ear canal, which is also lined by skin, is a roughly tubular structure with a cartilaginous outer portion and bony inner portion. The ear canal terminates at the tympanic membrane (or eardrum), the boundary between the outer ear and middle ear. The tympanic membrane is a trilaminar structure. Its outer lining is skin, its fibrous middle lamina lends rigidity to the drum membrane, and its inner layer is respiratory mucosa continuous with the lining of the middle ear.

The air-containing middle ear is connected to the nasopharynx by the eustachian tube, which provides aeration and is responsible for equilibration of pressure on the two sides of the eardrum. The middle ear is also continuous with additional air-filled, mucous membrane-lined spaces in the temporal bone (bullae or mastoid air cells, depending on the species). Although these spaces act as resonant cavities that affect very slightly the transmission of certain frequencies through the middle ear, their functional significance is otherwise obscure. A chain of three delicately suspended bones (ossicles) connects the tympanic membrane to the oval window. In order, they are named the malleus (hammer), incus (anvil), and stapes (stirrup); the stapes through its foot-plate is in direct contact with the fluid of the inner ear.

Sound entering the ear canal sets the tympanic membrane into vibration. This vibration is in turn mechanically coupled by the ossicular chain to the cochlear fluids. Two middle ear muscles, the tensor tympani and the stapedius, attach to the malleus and stapes, respectively; the stapedius in particular seems to play an active role in the regulation of sound transmission through the ossicular chain.

The inner ear contains the cochlea (also called the auditory labyrinth), the spiral end-organ of hearing. The inner ear also houses the vestibular labyrinth, a group of end-organs specialized for sensation of head position and movement (see Chapter 27). The cochlea can be thought of as a tunnel (about 1 inch long in humans) spiraling through the dense temporal bone. Figure 17–2 is a section along the axis of the cochlear spiral; the apex of the cochlea is shown at the top, and the base at the bottom. Although the axis of the spiral actually points anterolaterally rather than upward, the nomenclature for the cochlea has been based

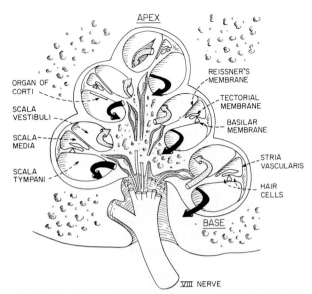

Figure 17–2 A section through the modiolus, or axis, of the cochlea shows its spiral orientation, with the three cochlear fluid compartments in each turn.

on diagrams like Figure 17–2; therefore, the terms "up" and "down" describe the dimension parallel to the cochlear axis. The terms "basal" and "apical" describe position along the cochlear spiral.

At every point along its two and a half turns, the cochlea has a roughly circular cross section; it is divided by two membranes into three fluid-filled compartments, each of which runs the full length of the cochlea. The uppermost compartment, called the scala vestibuli, contains perilymph, a fluid resembling cerebrospinal fluid. At the basal end of the cochlear spiral, the scala vestibuli opens into a space called the vestibule, whose fluid is set into motion by the movements of the stapes foot-plate. The lowermost of the three cochlear compartments, the scala tympani, also contains perilymph and is in fact continuous with the scala vestibuli through an opening at the apex of the cochlea called the helicotrema. At the basal end of the cochlea, the scala tympani terminates in a membranous partition (the round window), which is in contact with the air space of the middle ear. The middle compartment, the scala media (or cochlear duct), is separated from the other two by the basilar membrane below and by Reissner's membrane above. The scala media contains endolymph, a unique extracellular fluid with a high potassium concentration and a high positive electrical potential relative to perilymph.

Resting on the basilar membrane along the entire length of the cochlear duct is the organ of

Corti, which contains the auditory neuroepithelium. It is this organ that contains the receptor cells and supporting structures that transduce mechanical disturbances of the inner ear fluids into electrical signals in the auditory nerve. Viewed from above (i.e., from the scala vestibuli), the organ of Corti has three rows of outer hair cells and a single row of inner hair cells (about 15,000 total hair cells in the human cochlea). The afferent auditory nerve fibers run in the axis (modiolus) of the cochlea. Their cell bodies are located in the spiral ganglion within a bony ledge called the osseous spiral lamina. In the cat, and probably in other mammals as well, 90 to 95% of the 30,000 afferent neurons innervate inner hair cells. The auditory nerve also contains about 1000 efferent fibers, most of which innervate outer hair cells.

ACOUSTICS

Sound can be defined in a physical sense as a propagated change of density and pressure in an elastic medium. From some vibrating source (e.g., a tuning fork, a loudspeaker, or vocal cords) alternating waves of condensation and rarefaction spread out like ripples in a body of water. A sound source that oscillates back and forth like a tuning fork creates a periodic sound. If a tuning fork is struck, it oscillates at a single repetition rate, or frequency. The ensuing variations in sound pressure can be described by a sine (or cosine) function:

$$P = A \cdot \sin (2 \pi ft)$$

where P = instantaneous sound pressure, A = maximum sound pressure, f = the frequency of oscillation, and t = time. The frequency of such a sine wave or pure tone is the number of cycles per second, or Hertz (Hz). The period of a sine wave (the time between successive condensations or rarefactions) is simply the inverse of its frequency (i.e., seconds per cycle) and is measured in seconds, or more often milliseconds. For example, a 1000 cycle/second tone is said to have a frequency of 1000 Hz (1 kHz) and a period of 1 ms; the musical note middle C has a frequency of 256 Hz and a period of about 3.9 ms. The phase of a sine wave specifies its relationship in time to some temporal reference point or in some cases to other periodic sounds. If the frequency, phase, and intensity (see below) of a sine wave are given, one has completely and adequately described that sound; this is called a frequency domain description. Alternatively, one can completely describe a sound by specifying the instantaneous values of sound pressure as a function of time for as many cycles of the sound as desired; the latter is called a time domain representation.

Real sounds of biological interest are never pure tones. Some complex sounds (like sustained notes from a musical instrument, or speech sounds such as vowels) are periodic and can be described as the sums of many pure tones whose frequencies are integer multiples (called harmonics) of a single fundamental frequency. For example, an organ note with a fundamental frequency of 100 Hz will have harmonics (also called overtones) of 200, 300, 400 Hz, and so on. The relative strength of these overtones determines the tonal quality. Most real sounds, however, are not periodic. Although it is easiest to think of aperiodic sounds such as handclaps or consonants in the time domain (pressure as a function of time), these sounds can also be specified as sums of sine waves (i.e., in the frequency domain).

Intensity

As described earlier, the "amplitude" of a sound is usually specified by its pressure. It is important to realize, however, that sound waves consist of propagated disturbances not only of sound pressure but also of the molecules of the conducting medium. (The importance of this fact will be apparent in our subsequent discussion of middle ear function.) The intensity of sound is defined as the average rate of flow of sound energy across a given area perpendicular to the direction of propagation of the sound. It is therefore equal to the *product* of sound pressure and a quantity called volume velocity (the product of the average particle velocity and the cross-sectional area). By analogy to the flow of electricity, intensity is similar to electrical power and indeed is expressed in units of acoustic watts. Sound pressure is analogous to electrical potential or voltage, while volume velocity is analogous to electrical current.

Impedance

Acoustic impedance is defined as the ratio of sound pressure to volume velocity and represents the opposition to movement offered by an acoustic system (acoustic impedance is similar to electrical impedance, which is equal to the ratio of voltage to current). Sound is propagated in media that, like electrical conducting materials, have characteristic impedances. In an ideal medium with no boundaries, acoustic impedance is defined by density and elasticity alone. In air, for example, the molecules are far apart and compressible; relatively little sound pressure is required to cause high particle velocities. Because the ratio of pressure to volume velocity is low, air can thus be characterized acoustically as a low-impedance medium. Water, on the other hand, is a dense, high-impedance medium in which the ratio of pressure to volume velocity is high. The impedance of a conducting medium is not dependent on the frequency of sound being propagated; it is thus analogous to the resistive component of electrical impedance. "Bounded" media, like the cochlear fluids, have impedances that are higher than would be predicted by these parameters, because they are less compressible at their boundaries.

The above discussion refers to sound propagated in fluid (including gaseous) media; the situation is different for solid materials. Although sound waves of pressure and density can be propagated in solids, sound also can cause a solid object to vibrate or undergo deformation *as a whole*. For suitably suspended solid objects like the tympanic membrane and ossicles, these mechanical responses predominate. The opposition offered by such systems to applied sound is *mechanical impedance*: the ratio of sound pressure to the velocity of the system. Unlike acoustic impedance, the mass and stiffness of a system cause its mechanical impedance to vary with the frequency of sound.

If a system is massive, it has an increased opposition to movement (impedance), especially for high frequencies; low-frequency sounds are less impeded because their slower pressure changes allow more time to overcome inertia. Conversely, if a system is stiff, it preferentially impedes low-frequency sounds, because they require greater displacement for the same amount of sound energy, and the opposition offered by a stiff system (like a spring) is proportional to its displacement from a rest position. The relation of mechanical impedance to these factors is summarized in the following equation:

$$Z_f = \sqrt{R^2 + \left(2\pi fM - \frac{S}{2\pi f}\right)^2}$$

where Z_f = impedance for a given frequency (f), R = the resistive or frictional (non–frequency-dependent) component of impedance, M = mass, and S = stiffness.

Decibel

The loudest sound to which one can listen without discomfort is more than 1 trillion times (10^{12}) as intense as the softest sound that is audible. Because of this extremely wide operating (or *dynamic*) range, it is necessary to use a logarithmic measure for sound intensity. This logarithmic measure is called the decibel (dB), after Alexander Graham Bell, and is defined as

$$dB = 10 \log_{10}(I/I_0)$$

where I = the intensity of the sound of interest, and I_0 = the intensity of some reference sound. Since the ratio between sound intensities of painful and barely audible sounds is about 10^{12}, the human dynamic range can be restated as 120 dB [i.e., $10(\log_{10} 10^{12})$]. Since acoustic intensity is proportional to the square of sound pressure and since sound pressure is more conveniently measured than sound intensity or power, the more common formulation for the decibel is

$$dB = 20 \log_{10}(P/P_0)$$

When using a logarithmic relationship like this, it is essential to specify the value of P_0 and to understand that 0 dB does not mean the absence of sound. When dB = 0, P = P_0, and when P < P_0, dB will have a negative value. Unless otherwise stated, dB values usually refer to the sound pressure level (SPL) standard, for which P_0 = 20 micropascals, or 0.0002 dyne/cm². Other standards are also used. For example, dB HL (hearing level) measures sound pressure or intensity relative to the normal human threshold of hearing.

PSYCHOACOUSTICS

To appreciate the contributions and limitations that the peripheral components of the auditory system impose upon an animal's hearing, it is essential first to understand the acoustic processing capabilities of the entire organism. The study of the relationship of acoustic stimuli to behavior, specifically the ability to detect, discriminate, and identify sounds, is called psychoacoustics (a branch of psychophysics). A brief review of human psychoacoustics is presented so that the reader can better understand the relevance of the subsequent discussions of auditory physiology.

The levels of sound pressure required by a human to hear different frequencies (the minimum

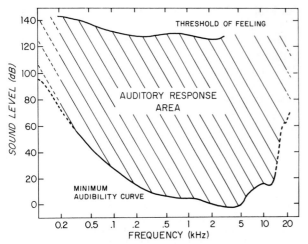

Figure 17–3 The lower curve shows the minimum audible sound–pressure level (re 20 μPa) for human subjects as a function of frequency. The upper curve shows the upper limits of dynamic range, the intensity at which sounds are felt or cause discomfort. (After Durrant, J. D.; Lovrinic, J. H. Introduction to psychoacoustics. In Durrant, J. D.; Lovrinic, J. H., eds. *Bases of Hearing Science*, 138–169. Baltimore, Williams & Wilkins, 1984.)

audibility curve) are shown in Figure 17–3. For any frequency, a range of intensities can be discriminated up to a level that is painful (threshold of feeling). As previously noted, the dynamic range for the best frequencies (2–4 kHz) is extremely large, about 120 dB. Within much of this range, intensity differences of one dB or less can be reliably detected, both for tones and for complex sounds. Frequency differences as small as 2 to 3 Hz can be discriminated for frequencies up to about 3 kHz. Not only can the auditory system discriminate very small changes in these parameters, it can do so across a whole range of stimulus frequencies and intensities. For example, small frequency and intensity changes can be detected for both very soft and very loud sounds. These capabilities should be kept in mind during our subsequent discussion of coding in the auditory nerve.

Identification of speech sounds is one of our most important abilities. Figure 17–4 shows the power contained at different frequencies for three vowels. The differences among the vowels, like the differences among the tones of different musical instruments playing the same note, is in the distribution and relative intensity of the different overtones, which are shaped by the positions of the tongue, palate, cheeks, and lips. The maxima in these curves of intensity as a function of frequency are called formants. Vowels are periodic sounds that can be indefinitely prolonged and have formants that do not change over the dura-

tion of the vowel. Consonants, on the other hand, are dynamic sounds that exhibit rapid changes in their formant structure over time. One can easily discriminate one vowel from another across a wide dynamic range from very soft to very loud sounds, and in the presence of white noise (random sound pressure fluctuations containing energy across a wide band of frequencies) nearly as intense as the vowels themselves. Vowels cannot be recognized simply by frequency *or* intensity but rather by *patterns* of intensity as a function of frequency. Although vowels are the simplest speech sounds

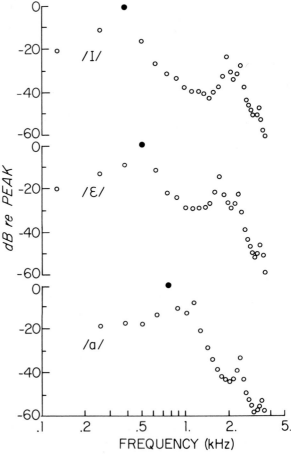

Figure 17–4 The steady-state spectra of three vowels /I/, /ɛ/, /ɑ/ as sounded in the words *beet*, *bet*, and *bat*, respectively are shown. Each has a fundamental frequency component of about 125 Hz (voice pitch), produced by the vibrations of the vocal cords, and a series of overtones whose frequencies are integer multiples of the fundamental. The positions of vocal tract structures, especially the tongue, vary to make different cavity resonances in the mouth and throat, enhancing some frequencies at the expense of others. The peaks in the spectra (called formants) are different for each vowel. (After Young, E. D.; Sachs, M. B. Representation of steady-state vowel in the temporal aspects of the discharge patterns of populations of auditory nerve fibers. *J. Acoust. Soc. Am.* 66:1381–1403, 1979.)

to discriminate, we will see that it is not simple to understand how they can be recognized on the basis of cochlear physiology alone.

SOUND TRANSMISSION TO THE INNER EAR

The outer ear and middle ear are essentially passive, linear mechanical systems, and the way in which they transmit sounds of different frequencies is predicted reasonably well by their physical properties (mass, stiffness, etc.). The combined properties of the outer and middle ear predict quite well the range of frequencies to which a given animal will be most sensitive.[34]

Outer Ear

The pinnas of some animals have considerable sound-collecting capabilities that are facilitated by strong muscular control. They can be moved to "focus" hearing in a particular direction, achieving much the same effect that a cupped hand behind the ear does for humans. The pinna is also a complex sound baffle that accentuates or attenuates certain frequencies, depending on the angle at which the sound waves are approaching the head. In man, the pinna is necessary for sound localization in the vertical plane, and for some sound localization in the horizontal plane when only one ear is functional.

The external ear canal acts essentially like a rigid tube closed at one end. It therefore resonates at a frequency whose wave length is four times the length of the ear canal (the dependence of the resonant frequency of a closed tube on its length is easily demonstrated by blowing over the lip of a bottle filled with different amounts of water). The resonant frequency of the adult human ear canal is about 3 kHz.

The combined effect of the pinna, the external ear canal, and the head on sound reaching the eardrum is shown in Figure 17–5. The graph shows the ratio of the sound pressure measured by a small microphone placed near the tympanic membrane to the sound pressure measured by a microphone placed in a free field in front of the speaker. A 15-dB improvement in pressure occurs between 2.5 and 4.0 kHz, largely due to the characteristics of the ear canal.

Middle Ear

As described earlier, air and water have very different specific acoustic impedances. Since

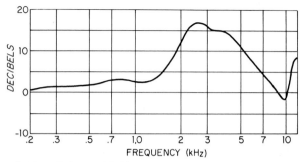

Figure 17–5 The difference between the sound levels produced at the tympanic membrane and those produced at the same point in space with the person absent is plotted as a function of the frequency of the constant sound source. The curve results mainly from ear canal and pinna resonance. (After Shaw, E. A. G. Transformation of sound pressure level from the free field to the eardrum in the horizontal plane. *J. Acoust. Soc. Am.* 56:1848–1861, 1974.)

sounds of interest to terrestial vertebrates travel in air while the hair cells are bathed in an aqueous medium, sound must traverse an air-water interface at the boundary of the middle ear and inner ear. Sound traveling in air has insufficient pressure to displace the densely packed water molecules. On the other hand, sound traveling from water to air has insufficient volume velocity to adequately displace the air molecules. The ratio of specific impedances of water to air is approximately 10,000. Because the transmission ratio across any acoustic interface is approximately 4 divided by the impedance ratio, only 0.04% of the sound energy will be transmitted across an air-water interface, in either direction. This impedance mismatch, which causes more than 99.9% of the acoustic energy to be reflected rather than transmitted, produces a transmission loss of approximately 34 decibels.

Fortunately, nature solved this problem by creating a middle ear, which acts as an impedance-matching transformer, converting the low-pressure/high-volume velocity excursions of sound in air to high-pressure/low-volume velocity excursions in the perilymphatic fluid of the inner ear. Impedance matching is accomplished in two ways (Fig. 17–6). First, the handle of the malleus is slightly longer than the long process of the incus, resulting in a lever ratio of approximately 1.3. Far more important, in humans the effective area of the tympanic membrane is about 17 times larger than the area of the stapes footplate. The equivalent increase in sound pressure from the eardrum to the footplate is the product of these two ratios (1.3 × 17), a 22-fold (or 28 dB) increase. The pressure increase is accompanied by an equivalent

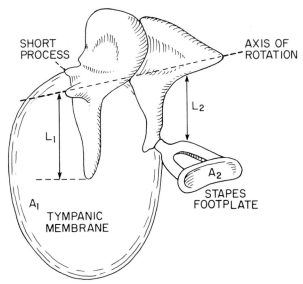

Figure 17–6 The impedance matching mechanisms of the middle ear. Both the difference in length between malleus handle (L₁) and incus long process (L₂) and the much larger ratio of areas of tympanic membrane (A₁) to stapes footplate (A₂) are shown. (After Abbas, P. J. Physiology of the auditory system. In Cumming, C. W., ed. Otolaryngology, Head and Neck Surgery, 2633–2677. St. Louis, C. V. Mosby, 1986.)

decrease in volume velocity, so that the total energy or power remains constant. (This must, of course, be true for any passive system according to the principle of conservation of energy.) Consequently, most of the energy that would have been lost or reflected back at the eardrum in fact crosses into the cochlear fluids.

This description is oversimplified but captures the function of the middle ear for those frequencies at which the mass and stiffness of the eardrum and ossicular chain are negligible. However, the picture is complicated by several factors. Above 2000 Hz, the tympanic membrane does not move as a unit and thus transmits energy less efficiently. In addition, the ossicular mass begins to impair transmission, and also a small amount of energy is dissipated by loose coupling between the individual ossicles. At low frequencies, the stiffness of the eardrum and ossicular chain can impair transmission. For example, unequal air pressure across the tympanic membrane, due to eustachian tube blockage and air absorption in the middle ear, can stiffen the tympanic membrane. Resonances of the middle ear cavity and of the mastoid and bulla cavities can also affect middle ear sound transmission.

Defects occurring anywhere from the outer ear to the stapes cause what are called conductive hearing losses, since they impair sound transmission into the inner ear. Examples of such problems

would be blockage of the ear canal with cerumen (ear wax), perforations in the tympanic membrane, ear infections that fill the middle ear with fluid, disruption of the joints between ossicles, and impairment of ossicular motion. Otosclerosis, for example, is a common hereditary hearing disorder in which the stapes footplate becomes immobilized by bone growth bridging the gap from the footplate to the surrounding bone of the otic capsule. All these forms of conductive hearing loss are potentially correctable by medical or surgical means. In contrast, hearing loss arising from disorders of the cochlea or auditory nerve (sensorineural hearing loss) is usually irreversible.

INNER EAR

Cochlear Duct

Figure 17–7 shows a schematic cross section of the cochlear duct. As described previously, the

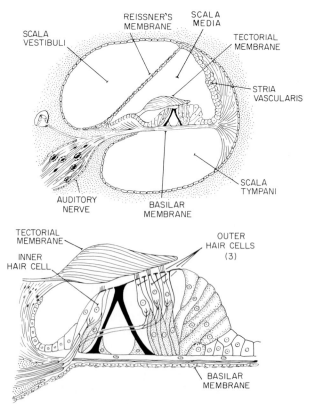

Figure 17–7 Cross section of cochlear duct. The boundaries of the scala media with the scala vestibuli and with the scala tympani are Reissner's membrane and the basilar membrane, respectively. (After Smith, C. A. The inner ear: its embryological development and microstructure. In Tower, D. B., ed. *The Nervous System*. Vol. 3, *Human Communication and its Disorders*, 1–18. New York, Raven Press, 1975.)

scala vestibuli and scala tympani contain perilymph, while the scala media contains endolymph. The high potassium concentration of the
endolymph is maintained by the stria vascularis,
which actively pumps K^+ ions into and Na^+ out
of the scala media. Reissner's membrane, which
separates the scala vestibuli and scala media, is
impermeable to charged ions (but not to water)
and protects the unique ionic and electrical composition of the endolymph. In contrast, the semi-
permeable basilar membrane permits perilymph
from the scala tympani to bathe the bodies of the
hair cells and supporting cells in the organ of
Corti. The tectorial membrane is also permeable,
so the stereocilia at the tops of the hair cells are
bathed in endolymph. The impermeable barrier
between endolymph and the scala tympani perilymph is at the cuticular plate, at the level of the
tops of the hair cells.[18] Since the hair cells have a
resting potential of approximately −70 mV and
the endolymph is maintained at +80 to +90 mV
by the high K^+ concentration, a very large net
electrochemical potential exists across the apical
surface of the hair cell.

The stereocilia of the three rows of outer hair
cells are attached to the tectorial membrane. Although gel-like, the tectorial membrane apparently
contains abundant collagen as its main protein
component, and thus probably possesses some
rigidity.[45] The single row of inner hair cells apparently is not directly attached to the tectorial membrane.[25]

The Traveling Wave

An impulsive sound (like a click) displaces the
stapes, creating a pressure gradient between the
scala vestibuli and the scala tympani (Fig. 17–8).
The result is a wave of displacement that begins
at the basal end of the basilar membrane and
travels towards its apex. This *traveling wave* displaces not only the basilar membrane but the
entire organ of Corti. These structures move together and the mechanical characteristics of each
are important in determining the response of the
whole; together, they are referred to as the *cochlear
partition*. The traveling wave begins at a high
velocity that decreases exponentially as it travels
toward the apex; in the human ear, it traverses
the cochlea in about 4 to 5 ms. It is not a sound
wave (the travel time for sound in the cochlea is
on the order of 20 μs) but is somewhat analogous
to the propagated disturbance that can be created
by whipping a rope attached at one end to a wall.
Since perilymph and endolymph, like all liquids,

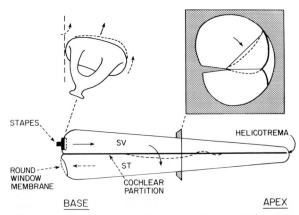

Figure 17–8 The generation of pressure gradients in the
cochlea. For ease of illustration, the cochlea is drawn "un-
coiled." Inward movements of the stapes footplate cause
compensatory outward movements of the round window
membrane. For static pressure changes and very low frequencies, the pressure is transmitted from the scala vestibuli
(SV) to the scala tympani (ST) via the helicotrema. For audible
frequencies, the cochlear partition is displaced in different
places for different frequencies as indicated by the dotted
line. (After von Békésy, G. *Experiments in Hearing.* Wever, E. G.,
trans. and ed. New York, McGraw-Hill Book Co., 1960.)

are incompressible and the cochlear scalae are
housed in rigid bone, any net displacement of the
cochlear partition toward the scala tympani (as
occurs during condensation) must be accompanied
by a compensatory outward movement of the
round window. Similarly, when the stapes moves
outward (in response to acoustic rarefaction), there
is a net displacement of the cochlear partition
toward the scala vestibuli, and the round window
membrane moves inward.

Frequency Tuning of the Basilar Membrane

For a continuous tone, the picture is more complex (Fig. 17–9). Two important points should be
noted from Figure 17–9. First, each moment-to-
moment change in sound pressure can be thought
of as a separate impulsive sound that sets up its
own traveling wave. Thus, a point very near the
base of the cochlea may be moving upward (toward the scala vestibuli) in response to the rarefaction phase of the sound, while at the same
instant a more apical point is moving downward
in response to a condensation phase that occurred
earlier. The portions of the cochlea responding to
the tone will all be vibrating at the frequency of
the tone, but not in phase (synchronized in time),
since the traveling wave is delayed by different
amounts at different points along the cochlear
partition. Figure 17–9 shows an example of the

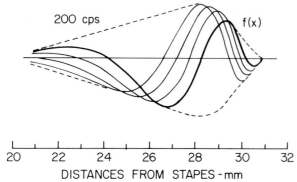

200 cps

f(x)

DISTANCES FROM STAPES - mm

Figure 17–9 Each solid curve indicates the displacement of the basilar membrane at a particular point in time in response to a 200 Hz tone. The darker curves occur later in time and show the progression of the traveling wave from base to apex of the cochlea. The dotted line indicates the envelope of displacement for this tone, i.e., the maximum displacement for each point along the basilar membrane. The actual excursions are many times smaller (relative to the length of the basilar membrane) than illustrated here. (After von Békésy, G. *Experiments in Hearing*. Wever, E. G., trans. and ed. New York, McGraw-Hill Book Co., 1960.)

displacement pattern of the basilar membrane in response to a pure tone at several successive moments in time (solid lines). Second, the amplitude of the vibrations is not the same at each point. The maximum excursions of each point along the basilar membrane can be described by an envelope (dotted line in Fig. 17–9) whose amplitude increases gradually to a maximum and then drops sharply. For different tones, the maxima occur at different positions. For low frequencies the envelope maxima are near the apex and for progressively higher frequencies the peak amplitude of the traveling wave shifts nearer the base of the cochlea.[47]

Complex sounds containing several different frequencies are represented in a similar manner along the cochlea. The high-frequency components produce peaks of vibration near the base, lower frequencies excite more apical regions, and so on. Thus, the cochlea acts like a spectrum analyzer, "dissecting out" the component frequencies of a complex sound to be separately displayed at different places along the basilar membrane.

Each successive point along the cochlear partition is most sensitive to a slightly different frequency (Fig. 17–10). This is explained primarily by changes in the structure (and impedance) of the basilar membrane along its length. The basilar membrane is narrow near the base and wide near the apex. More significantly, its stiffness decreases 100-fold from base to apex. These mechanical characteristics impart a degree of tuning to the cochlear partition, so that the basal portion of the

cochlea (with its hair cells and neural elements) responds best to high frequencies, and the apical part responds best to low frequencies. The principle of an orderly spatial array of responsive elements according to an increasing or decreasing order of frequencies is called the *place code*, or *place principle*; it is similar in all species of birds and mammals. Moreover, as will be discussed in Chapter 18, this orderly representation of frequency is preserved in the projection from the cochlea to the central nervous system and then at each successive level of the auditory pathways, where it is referred to as *tonotopic* organization. Along the cochlear partition, the distance from the apex to the position that responds maximally to a particular frequency is proportional to the logarithm of that frequency.

The place principle and the properties of the traveling wave were first established by the elegant experiments of Georg von Békésy (1960), for which he received a Nobel prize. His experiments were carried out on cadaver ears with relatively primitive instrumentation consisting of a microscope and stroboscopic illumination. In order to see the movements of the cochlear partition, he used sounds as loud as 130 to 140 dB SPL. Only recently have two new techniques emerged to measure the tiny excursions of the basilar membrane in response to near-threshold sounds. The Mossbauer technique measures Doppler shifts in radiation emitted by a gamma source placed on the basilar membrane; laser interferometry uses a mirror (10^{-8}g) placed on the membrane; the latter technique can measure movements as small as 0.01 nm. These new techniques have shown much sharper tuning than initially described by von Békésy.

A convenient way to describe the sharpness of tuning shown at any locus within the auditory system is to plot the minimum sound intensity

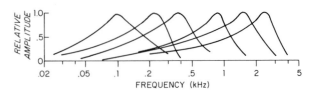

RELATIVE AMPLITUDE

FREQUENCY (kHz)

Figure 17–10 Each curve shows the response of a given point along the basilar membrane to tones of varying frequency. The curve farthest to the right is for a point near the midpoint of the cochlea and shows maximum response to about 2.5 kHz, gradually decreasing response to lower tones, and sharply reduced response to higher tones. The curves to the left are for progressively more apical locations. (From von Békésy, G. *Experiments in Hearing*. Wever, E. G., trans. and ed. New York, McGraw-Hill Book Co., 1960.)

required to obtain some criterion response at each frequency. Figure 17–11 shows the sound intensity required for a criterion displacement (0.35 nm) and velocity (0.04 mm/s) for a point on the basilar membrane near the cochlear base. These *tuning curves* from the live guinea pig (and similar curves from cats) show that the basilar membrane is exquisitely sharply tuned with high-frequency slopes exceeding 100 dB/octave.[19, 39] Active processes seem to be involved in the maintenance of this tuning, as its sharpness is considerably degraded by anoxia.

Two properties of the inner ear limit the range of audible frequencies. Very low frequencies (below about 20 Hz in man) are inaudible because the slow pressure changes of these sounds are transmitted from the scala vestibuli to the scala tympani through the helicotrema. This small opening between the two perilymphatic scalae serves primarily to protect the cochlear duct from excessive displacement during very slow or even static middle ear pressure changes, such as occur during changes in atmospheric pressure, nose-blowing, etc. Very high frequencies (above about 20 kHz in man) are inaudible because even the most basal portion of the basilar membrane is not stiff enough to respond to them. It is important to remember that, like the middle ear, mechanical properties of the inner ear are differentially specialized in different species. Some birds and mammals hear ultra-low frequencies (down to 0.1 Hz) while others, such as some bats, hear sounds in the ultrasonic range (e.g., up to 120 kHz).

It generally has been assumed that the place code is fixed throughout an organism's lifetime. However, recent experiments suggest that during development there is a systematic shift in the place code.[35] Early in development, the basal part of the cochlea is the first to mature, but it responds optimally to mid-range rather than high frequencies. As the inner ear matures, the base gradually encodes higher frequencies, and the best locations for middle or low frequencies shift toward the apex. These developmental findings suggest that the place code is not immutable even in the life of a single individual and suggests the possibility that other conditions such as aging or pathology may alter its organization.

Frequency Tuning by Hair Cells

Although the mechanical tuning properties of the cochlear partition in mammals appear sharp enough to account for the tuning described later for auditory nerve fibers and for behavioral responses, there is emerging evidence that the hair cells may also contribute to tuning. The stiffness gradient observed in cadaver ears explains only rather coarse tuning, while living hair cells are needed for the sharp tuning curves seen in Figure 17–11. In some reptiles, amphibians, and fish, the hair cells themselves may be the only tuned elements. In alligator lizards, for example, the basilar membrane moves as a unit; there is no traveling wave, and different points on the membrane do not respond maximally to different frequencies.[28] Instead, the height of the stereocilia bundles varies systematically with the position of individual hair cells along the basilar membrane,[27] and the preferred frequency of individual cells depends on the heights of the ciliary bundles (cells responding to low frequencies have long stereocilia, whereas cells tuned to higher frequencies have shorter stereocilia).[11] In addition, both bullfrogs and tur-

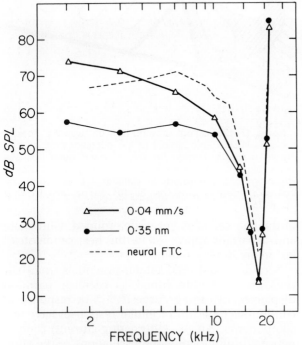

Figure 17–11 The solid lines show isoresponse curves for basilar membrane displacement (Δ) and velocity (●) for a point near the base of the cochlea. For each frequency, the sound pressure level needed to obtain a criterion response (0.35 nm and 0.04 mm/s, respectively) is plotted. The dashed line shows an isoresponse curve obtained from an auditory nerve fiber with a similar best frequency (about 18 kHz); the sound level needed for a small increase over resting spike rate is plotted. Neural frequency threshold curves are discussed later in this chapter (Auditory Nerve). All curves are from guinea pigs. (After Sellick, P. M.; Patuzzi, R.; Johnstone, B. M. Measurement of basilar membrane motion in the guinea pig using the Mossbauer technique. *J. Acoust. Soc. Am.* 72:131–141, 1982.)

tles have hair cells whose membranes exhibit a frequency-dependent electrical resonance, which appears to contribute to the acoustic tuning of the hair cells.[4, 22]

The hair cells of mammals and birds also demonstrate morphological features that are correlated with tuning. In chicks, the length, diameter, and number of stereocilia per cell vary systematically from basal to apical locations along the basilar membrane.[46] Stereociliary stiffness (and presumably responsiveness to high frequency tones) decreases from base to apex in guinea pigs.[44] The relationship between height of the stereociliary bundles and best frequency for a given cochlear region has also been confirmed in humans.[25] Thus, it seems likely that, in all vertebrate species, some tuning is contributed by the mechanical and/or electrical characteristics of individual hair cells.

Transduction and Synaptic Transmission

The traveling waves along the cochlear partition must somehow activate the hair cells. In the current theory, the basilar membrane and tectorial membrane can be considered as being hinged at different points on the medial wall of the cochlea; consequently, either upward or downward displacements of the cochlear partition create a shearing force that tends to bend the hair cell stereocilia bundle at the apical hair cell surface (Fig. 17–12). (Since the inner hair cells are not attached to the tectorial membrane, their stereocilia must be deflected by subtectorial fluid currents rather than directly by the tectorial membrane.) The stereocilia bundles appear to pivot in a relatively rigid fashion at their points of attachment to the apical surface of the hair cell. This motion opens and closes ionic channels that appear to be located near the tips of the stereocilia (see Fig. 17–13).[16] Recall that the stereocilia are exposed to the large positive endolymphatic potential, while the intracellular potential is of course negative. This large potential difference constitutes a "battery" (first described by Davis in 1958[5]), which drives potassium ions into the cell (unopposed by any concentration gradient; see Chapter 5) and depolarizes it. Depolarization of the hair cell membrane causes voltage-dependent calcium channels at the base of the hair cell to open. Calcium entry, in turn, presumably initiates fusion of synaptic vesicles with the synaptic specialization at the base of the hair cell. Neurotransmitter release will then effect spike initiation in the afferent neuron. The nature of the afferent transmitter that is released from the base

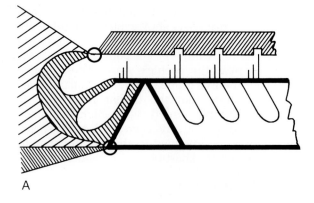

A

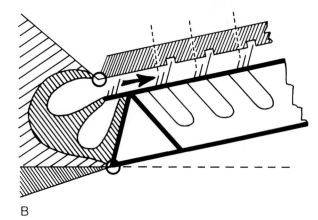

B

Figure 17–12 Diagram showing how an "upward" (toward scala vestibuli) displacement of the cochlear partition can create a shearing force tending to bend outer hair cell stereocilia in an excitatory direction. (After Ryan, A.; Dallas, P. Physiology of the inner ear. In Northern, J. L., ed. *Communicative Disorders: Hearing Loss*, 80–101. Boston, Little, Brown & Co., 1976.)

of the hair cell is as yet undetermined. Glutamate and aspartate appear to be the best candidates,[1] but many doubts remain.[8]

Sound-evoked intracellular potentials from hair cells are as highly tuned as cochlear partition displacement tuning curves.[9, 15, 36] In response to continuous tones, intracellular responses have two components. An AC (alternating current) component faithfully follows the oscillations of the stimulating tone while a DC (direct current) component contributes a net depolarization of the hair cell membrane. Presumably the DC component occurs because the stereocilia bundles display a non-linear behavior that produces more potential change when the stereocilia are deflected toward the tallest row than when they are deflected toward the shortest row.[17] A net depolarization, of course, enhances the probability that neurotransmitter will be released at the hair cell synapse.

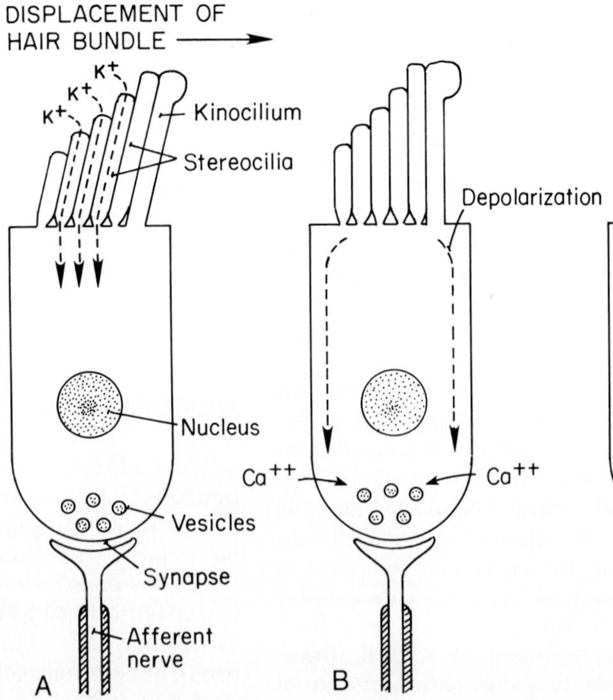

DISPLACEMENT OF
HAIR BUNDLE ⟶

Figure 17–13 Deflection of the hair bundle toward the tallest row of stereocilia opens poorly selective cationic channels near the sterocilia tips. *A,* Influx of potassium depolarizes the cell. *B,* Voltage-sensitive calcium channels open in turn, permitting *(C)* neurotransmitter release across the synapse to the afferent neuron. (After Hudspeth, A. J. The hair cells of the inner ear. *Sci. Am.* 248:54–64, 1983.)

The Role of Outer Hair Cells

Because they are present in all vertebrates and receive almost all of the afferent innervation of the cochlea, inner hair cells are considered to be the "true" (or at least the predominant) sensory receptors in the cochlea. Outer hair cells, on the other hand, receive scant afferent innervation. Although most terrestrial vertebrates have two types of hair cells, the difference in their afferent innervation is clearest in mammals.

It has been known for many years that exposures to excessively loud sounds or ototoxic drugs can produce widespread loss of outer hair cells with no loss of inner hair cells and afferent neurons. Animals so treated exhibit hearing losses of 40 to 50 dB (elevations of response thresholds) for certain frequencies.[37] Initially, this was thought to mean that very soft sounds were detected via stimulation of the outer hair cells. It now seems more likely that outer hair cells are needed to tune the cochlear partition, which in turn permits stimulation of *inner* hair cells by very soft sounds. In support of this idea, the sound intensity required to produce a criterion amount of basilar membrane movement increases with increasing damage to outer hair cells.[21]

Recent evidence suggests that the outer hair cells may be viewed as primarily effector or motor structures. Isolated guinea-pig outer hair cells contain contractile proteins (actin, myosin) and shorten in response to electrical depolarization, potassium, calcium and adenosine triphosphate, and acetylcholine.[2, 10, 13] Since the outer hair cells receive most of the efferent (cholinergic) innervation of the cochlea, it is possible that efferent activity regulates outer hair cell length, or tension, or the mechanical properties of their stereocilia, so as to alter the tuning properties of the cochlear partition. Indeed, electrical stimulation of the cochlear efferents appears to alter the micromechanical properties of the cochlea.[26]

Kim[20] has postulated an "active bidirectional transduction mechanism" linking the outer hair cells with the remainder of the cochlear partition. Since bending of stereociliary bundles induces graded changes in transmembrane potential (mechanical to electrical), and since electrical stimuli applied to outer hair cells induce motile responses (electrical to mechanical), it is possible that, especially for very soft tones near a cell's best frequency, an active resonance occurs. If this is true, sound-induced stereociliary deflection would open ionic channels, causing a relatively large depolarization (because of the large potential difference between endolymph and intracellular fluid). This depolarization would in turn cause the hair cell (or its hair bundle) to "push back" by shortening the contractile proteins, thus reinforcing the mechanical response of the cochlear partition.

Near-field (Gross) Cochlear Potentials

In some clinical and experimental situations, it is desirable to monitor the stability and health of the cochlea without entering it. Electrodes placed on the round window or in other locations relatively near the cochlea can detect a variety of sound-induced cochlear potentials (Fig. 17–14). The first of these, the cochlear microphonic (CM, Fig. 17–14), rather faithfully follows the wave form of the stimulating sound. Since it nearly disappears after lesions that destroy the outer hair cells but leave the inner hair cells intact, it is believed to come primarily from the outer hair cells and to represent a summation of the AC intracellular potentials discussed in the section on Transduction and Synaptic Transmission. A second stimulus-related potential is the summating potential (SP, Fig. 17–14), a DC potential that may be either positive or negative depending on the properties of the stimulus and the recording electrode. It is probably due to summation of the DC or "net" depolarizing hair cell potentials. Finally,

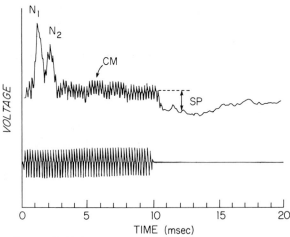

Figure 17–14 The lower trace shows the waveform of a 5-kHz tone–burst stimulus, while the upper trace displays an electrical potential response recorded by an electrode on the round window of a rat's cochlea. N_1 and N_2 are compound action potentials from synchronous auditory nerve activity; CM is the cochlear microphonic, or AC cochlear potential. The summating potential (SP), or DC cochlear potential, is clearly visible as the elevation of the entire response above the baseline level as seen after the tone burst ends. (After Møller, A. R. Anatomy and general physiology of the ear. In Møller, A. R. *Auditory Physiology*, 1–104. New York, Academic Press, 1983.)

compound action potentials (N_1 and N_2, Fig. 17–14) can be recorded that reflect synchronous activation of afferent neurons at the onset of a tone burst or other impulsive sound. Because the traveling wave moves much more quickly near the base of the cochlea, individual hair cells and their afferent neurons respond much more synchronously there than toward the cochlear apex. For this reason, gross potentials recorded outside the cochlea preferentially reflect the activity of synchronously responding hair cells and neurons in the base of the cochlea.

AUDITORY NERVE

The auditory nerve in man contains about 30,000 neurons[30]; all afferent fibers have their cell bodies in the spiral ganglion of the cochlea. About 95% have large cell bodies (type I neurons) whose dendrites pass radially (perpendicular to the cochlear axis) to form afferent synapses at the bases of inner hair cells. The remaining 5% (type II neurons) have small cell bodies and thin fibers that pass radially, then turn to form the outer spiral bundle, and eventually synapse on outer hair cells.[43] The great disparity in afferent innervation of the inner and outer hair cells has already been noted. Approximately 20 radial fibers form afferent synapses on each inner hair cell, while the sparse outer spiral fibers each branch to supply 10 to 60 outer hair cells (Fig. 17–15). The role of the few afferent fibers supplying outer hair cells is unknown because, to date, all auditory nerve fibers from which it has been possible to record sound-evoked spike activity and that were subsequently morphologically traced to their peripheral origins innervated inner hair cells.[32]

Frequency-Threshold Curves

Each afferent neuron can code information either by changes in the rate of spike discharge or by changes in the timing of spike discharges, or by both. Of course, information can also be coded by the spatial-temporal pattern of excitation of an array of neurons responding to a stimulus. We will first consider the discharge patterns of a single afferent neuron to sinusoidal tones of differing frequency and intensity.

A *frequency-threshold curve* plots the intensity required to produce a small increase in spike rate (over resting rate) at different frequencies. Typical frequency-threshold curves for auditory neurons are shown in Figure 17–16 (one is also shown in Fig. 17–11, along with basilar membrane tuning

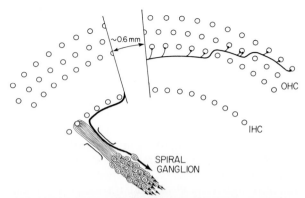

Figure 17–15 The afferent innervation of the cochlea is shown diagrammatically. The more numerous type I neurons converge on inner hair cells (IHC), whereas each type II neuron branches extensively to supply several outer hair cells (OHC) after running spirally along the organ of Corti. (From Spoendlin, H. The afferent innervation of the cochlea. In Naunton, R. F.; Fernandez, C. *Evoked Electrical Activity in the Auditory Nervous System*, 21–41. New York, Academic Press, 1978.)

curves). As might be expected from the discussion of the traveling wave, auditory afferents have frequency-threshold curves whose best frequencies correspond to the loci that they innervate along the basilar membrane. Like hair cell and basilar membrane tuning curves, they are sharply tuned to exclude frequencies above the fiber's best, or characteristic, frequency (CF). For low frequencies, the tuning is less sharp, and auditory nerve fibers respond to sounds well below the CF once stimulus intensity is raised 40 to 50 dB above CF threshold. Apparently, frequency-threshold curves of auditory nerves directly reflect the tuning present in the inner hair cells to which these afferent neurons connect.[9, 36]

Auditory nerve fibers have relatively similar response characteristics. Other than differences in CF, their frequency-threshold curves are more or less alike (although, as seen in Fig. 17–16, fibers with low CF tend to have somewhat broadly tuned curves). One important quantitative difference is in their unstimulated firing rate and response thresholds: most auditory nerve fibers have high unstimulated firing rates and respond to very soft sounds, while a few fibers have relatively low spontaneous rates and higher thresholds.

The frequency-threshold curves of auditory nerves are clearly sharp enough to permit the discrimination of very soft (near threshold) tones of only slightly different frequencies. Even for moderate to high intensity tones, which would activate a large fraction of auditory nerve fibers, the stimulating frequency could be deduced from the CF of the most apical fibers responding, because the frequency-threshold curves have such sharp high-frequency slopes. The general notion that the auditory system encodes stimulus frequency according to the spatial position of responding auditory neurons along the basilar membrane is called the *place* or *rate-place theory*, to indicate that the central nervous system analyzes firing rate (or changes in firing rate) according to the place of the fiber (along the cochlear partition) and its CF. As will be seen, this theory has some difficulties in accounting for our ability to discriminate differences in complex sounds containing many frequencies, such as vowels.

Suppression. A tone which by itself does not cause any measurable effect on the firing rate of a given auditory nerve fiber may nevertheless interfere with that fiber's response to tones to which it

Figure 17–16 Each of these frequency-threshold curves plots the sound intensity required to produce a minimal increase in spike rate (over spontaneous rate), as a function of stimulus frequency for a single auditory nerve fiber. Each neuron has a characteristic frequency to which it is most sensitive. Threshold is plotted in arbitrary units of dB attenuation from a standard (high-intensity) sound; thus − *100* represents a sound 100 dB less intense than the reference level. (After Kiange, N. Y. S.; Watanabe, T.; Thomas, E. C.; Clark, L. F. *Discharge Patterns of Single Fibers in the Cat's Auditory Nerve.* Research Monograph 35. Cambridge, Mass., MIT Press, 1965.)

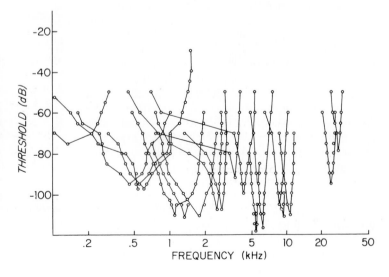

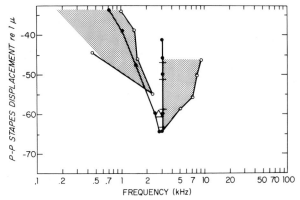

Figure 17–17 The line joining the solid circles shows a typical frequency-threshold curve. The open triangle represents an excitatory tone at the fiber's characteristic frequency, about 10 dB above threshold (plotted here as dB re stapes displacement, rather than in sound pressure level). Tones in the shaded area (enclosed by the open-circle line), although mostly outside the primary response area and thus unable to elicit any fiber response on their own, are able to *inhibit* the fiber's response when presented simultaneously with the excitatory tone. (After Nomoto, M.; Suga, N.; Katsuki, Y. Discharge pattern and inhibition of primary auditory nerve fibers in the monkey. *J. Neurophysiol.* 27:768–787, 1964.)

ordinarily does respond. This is most easily illustrated by considering a conventional frequency-threshold curve (solid circles, Fig. 17–17). The space above the curve indicates combinations of frequency and intensity to which the fiber will respond with an increase in discharge rate and can be considered to be that fiber's excitatory "response area." Tones in the cross-hatched area (mostly outside the excitatory response area), presented simultaneously with a normally excitatory tone, suppress the response to the latter tone. Such suppression tends to enhance the contrasts among the different frequencies present in a complex sound, since higher-intensity components would suppress the response to less intense components. Unlike the superficially similar phenomenon of lateral inhibition in the visual system, two-tone suppression cannot be explained by neural inhibition (no synaptic circuitry appropriate for the task exists in the cochlea), but rather by a nonlinearity measurable in the mechanical responses of the basilar membrane.[31] Obviously, this factor must be important in all cochlear responses to complex stimuli, although its effects are almost impossible to predict for stimuli containing more than a few components.

Intensity Coding

For steady-state tones, average firing rate increases as a function of stimulus intensity. Mam-

malian auditory nerve fibers show relatively steep increases in firing rate with sound intensity, and they reach their maximum firing rates (saturation) within about 40 dB of their thresholds. Figure 17–18 shows a series of such *rate-intensity functions* for a single auditory nerve fiber when it is stimulated with a tone at its CF (11.3 kHz) or at higher frequencies.

Above the saturation level, a fiber obviously cannot signal a change in stimulus intensity by a change in firing rate. Humans, however, are able to discriminate intensity differences from 0 to 120 dB, although it is clear that no single fiber can encode intensity over that range. An obvious solution would be to have fibers that operate at different stimulus levels; while each might only encode stimulus intensity over a 40-dB range, the ensemble of fibers would be able to encompass a much broader range (such a situation obtains for the rods and cones in the retina; see Fig. 19–18). Although the vast majority of auditory nerve fibers studied have very low thresholds (as in Fig. 17–18), a small population does have high thresholds and could encode stimulus intensity changes for loud sounds (they also tend to have less steep slopes to their rate-intensity functions). These high-threshold units have been overlooked in many investigations because they have very low spontaneous rates and therefore are somewhat harder to locate.[23]

The wide range of dynamic intensity coding

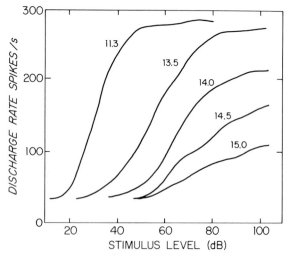

Figure 17–18 A single auditory nerve fiber's response to sounds of different frequencies. For its characteristic frequency (11.3 kHz), the spontaneous spike rate (about 40 spikes/s) is exceeded for tones above 10 dB SPL, but the neuron's dynamic range is only about 40 dB. (From Sachs, M. B.; Abbas, P. J. Rate versus level functions for auditory nerve fibers in cats: tone-burst stimuli. *J. Acoust. Soc. Am.* 56:1835–1847, 1974.)

could also be accomplished by spread of excitation recruiting additional afferents. Consider the fiber in Figure 17–18. Although its CF is 11.3 kHz, it also responds to 14.5 kHz if the sound is loud enough (i.e., >50 dB). A loud 14.5-kHz tone, therefore, excites not only fibers whose CFs are 14.5 kHz but also nearby fibers with similar CFs. If, like the fiber in Figure 17–18, the firing rates of the recruited neurons are not saturated, the firing rate could encode intensity over a considerable range (in this case, a range of >50 dB). This mechanism alone, however, cannot solve the dynamic range problem, because dynamic range and intensity discrimination are nearly as good for white noise, which contains all audible frequencies and stimulates the entire cochlea, as for pure tones!

It should be noted that most auditory nerve recordings have been performed in anesthetized animals whose efferent cochlear innervation and acoustic middle ear muscle reflex are depressed. We now show that these two systems may further improve the dynamic range and intensity-processing capabilities of the peripheral auditory apparatus.

The Efferent System. Efferent neurons that innervate the sensory cells of the inner ear have cell bodies in the superior olivary complex; each ear receives efferent fibers from both sides of the brainstem. These *olivocochlear* efferent fibers travel in the inferior vestibular branch of the 8th nerve and only join the auditory nerve just before it enters the cochlea. Most efferent fibers end on the cell bodies of outer hair cells; a smaller number form axodendritic synapses on the distal processes of afferent fibers innervating the inner hair cells. Efferent fibers respond to sound presented to either ear and display frequency tuning similar to that seen in afferent auditory neurons.[24]

Electrical stimulation of the crossed olivocochlear bundle causes about a 10-dB shift in rate-intensity functions of afferent nerve fibers.[49] In other words, when the efferent fibers are activated, the afferent fiber does not begin to change its firing rate until the stimulus tone is about 10 dB more intense than the fiber's normal threshold. Similarly, the intensity saturation level is 10 dB higher during efferent activation. As mentioned above, efferent activity may decrease sensitivity by changing the mechanical properties of the cochlea. If sound-evoked activity of the efferent system can cause similar changes (this has not been demonstrated), it would clearly provide a mechanism to shift the operating range of individual neurons. The central connections of the efferent system will be considered further in Chapter 18.

Acoustic Reflex. Loud sounds presented to either ear elicit a brisk contraction of both middle ear muscles (primarily the stapedius in man), which tenses the ossicular chain. This acoustic reflex reduces sound transmission through the middle ear and, like the efferent responses described above, indirectly reduces the sensitivity of individual auditory fibers. As might be expected from earlier discussions of middle ear mechanical factors, the reduction in sensitivity is greatest for low frequencies; in fact, sound transmission and afferent response for some high frequencies are even accentuated by the stiffening of the ossicular chain (Fig. 17–19).

The threshold sound intensity for eliciting the acoustic reflex is usually considered to be about 80 dB, but this reflex is quite sensitive to anesthesia. In awake cats, some modulation of stapedius tension, with concomitant effects on middle ear transmission, are present at much lower sound levels.[40] The neuroanatomical pathways mediating the stapedius reflex also will be discussed in Chapter 18.

Adaptation. Another phenomenon that may be related to setting the dynamic range for discrimination of many complex sounds is *adaptation*. When a tone is switched on, the neural discharge is high initially, and then drops to lower steady-state rates. This adaptation is not demonstrable in intracellular voltage records made from inner hair cells but is seen in the excitatory postsynaptic potentials in tetrodotoxin-blocked afferent fibers.[12] Thus, adaptation is a presynaptic phenomenon, probably attributable to the inability of the hair

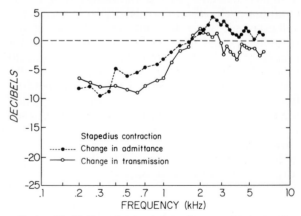

Figure 17–19 The effects of stapedius muscle contraction on the acoustical admittance (inverse of impedance) and on sound transmission through the middle ear are plotted as a function of frequency. Both admittance and transmission are decreased most for low frequencies. (After Møller, A. R. An experimental study of the middle ear and its transmission properties. *Acta Otolaryngol. (Stockholm)* 60:129–149, 1965.)

cell to release neurotransmitter fast enough to support sustained, high firing rates.

Rate-intensity functions such as those shown in Figure 17–18 usually reflect the steady-state, or adapted, firing rate. However, when the onset firing rate (or, more precisely, the probability of a spike within a few ms after tone onset) is plotted as a function of intensity, changes are seen over a much greater range than is seen for adapted fibers.[42] In other words, the dynamic range is greater if the initial rather than the adapted firing rate is considered. This observation has functional significance because most sounds of biological relevance, including speech, demand the encoding of rapid changes of amplitude rather than steady-state intensity levels.

Synchrony

Thus far, we have considered how frequency and intensity are encoded by changes in the absolute firing rates of afferent auditory neurons. However, a great deal of information is also encoded in the *timing* of discharges. For frequencies below about 5 kHz, auditory nerve fibers tend to synchronize their spike discharges with the period of a stimulus tone (phase-locking). As seen in Figure 17–20, individual neurons almost never discharge on every cycle of a pure tone, but rather every second or third cycle or so. When they *do* discharge, however, they fire only during that part of the cycle in which the basilar membrane is displaced upward (this direction of deflection leads to hair cell depolarization). If all action potentials occurred at exactly the same phase in the stimulus cycle, the intervals between successive spikes would always be integer multiples of the stimulus period. The interspike interval histograms for the fiber shown in Figure 17–21 indicate that indeed most intervals cluster around multiples of the stimulus period.[33] For each individual fiber responding to a low- to medium-frequency tone, the probability of firing is relatively fixed for

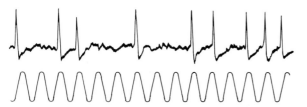

Figure 17–20 Spike discharges of a single auditory fiber (upper trace) in response to a 300-Hz tone (lower trace). (After Arthur, R. M.; Pfeiffer, R. R.; Suga, N. Properties of "two-tone inhibition" in primary auditory neurones. *J. Physiol. (Lond.)* 212:593–609, 1971.)

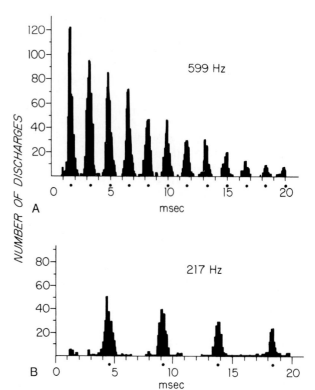

Figure 17–21 These interval histograms show the distribution of interspike intervals seen for different tones, in the responses of a single nerve fiber. *A*, For 599 Hz, the modal interval between spikes is about 1.67 ms, the period of the tone; the next-most-common interval is twice the period, and so on. Integer multiples of stimulus tone period are indicated by dots along the ordinate. *B*, The same pattern is seen for a 217-Hz tone. (From Rose, J. E.; Brugge, J. L.; Andeson, D. J.; Hind, J. E. Phase-locked response to low-frequency tones in single auditory nerve fiber of the squirrel monkey. *J. Neurophysiol.* 30:769–793, 1967.)

each cycle of the stimulating tone, resulting in a Poisson-type distribution of interspike intervals.

Little is actually known about the way in which the central nervous system makes use of these temporal cues for frequency discrimination. However, an ensemble of 8th-nerve fibers from the same region of the cochlea could converge on more central units to give an accurate measure of stimulus frequency up to about 1 kHz.[14] Brainstem units tuned to particular stimulus periods could in turn convert this temporal coding to a spike-rate representation, as has been demonstrated in the coding of interaural timing cues involved in binaural hearing (discussed in Chapter 18).

Temporal information in auditory nerve firing patterns is undoubtedly responsible for frequency discrimination in some patients with prosthetic devices. Many totally deaf persons can have a rudimentary form of hearing restored by a cochlear implant, which electrically stimulates the auditory nerve directly. With a single stimulating electrode, the same population of fibers is being activated, regardless of the stimulus frequency; therefore, any frequency discrimination must depend entirely on temporal cues in the discharge patterns of the auditory neurons. Patients with these implants can discriminate pulse trains of different frequencies up to about 1000 Hz by detecting differences in inter-pulse periods.[7] This provides an unambiguous demonstration of at least a limited use of temporal cues by the central nervous system.

Population Studies

Recent technical advances have made it possible to measure the responses of large numbers of auditory nerve cells in the same animal and to test the idea that all the information needed to discriminate sound is provided by the combination of which auditory nerve fibers are active (place coding) and the absolute firing rate of these fibers. While such recordings make it easy to see how the frequencies of simple signals are encoded, it is difficult to understand how broad-spectrum stimuli such as speech sounds are transmitted. Sachs and Young[38] have shown that the acoustic features that make different vowels distinguishable at normal conversational intensities (i.e., the

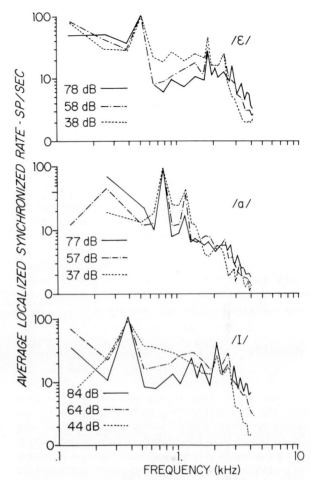

Figure 17–23 As in Figure 17–22, the ordinate arrays auditory nerve fibers according to characteristic frequencies, and the families of curves indicate responses of this population of neurons to vowel sounds at different intensities. (/ɛ/ is vowel sound, as in *bet*; /ɑ/, as in *bat*; /I/, as in *beet*.) However, the abscissa here is not spike rate, but average localized synchronized rate, a measure of the degree to which each neuron's response was phase-locked to the spectral components of the vowel stimulus. The formant peaks that characterize the stimuli are now apparent in the response profiles across a range of intensities. (After Young, E. D.; Sachs, M. R. Representation of steady-state vowels in the temporal aspects of the discharge patterns of populations of auditory nerve fibers. *J. Acoust. Soc. Am.* 66:1381–1403, 1979.)

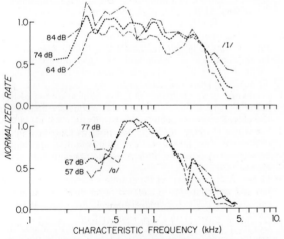

Figure 17–22 The upper family of curves depicts responses of a large number of auditory nerve fibers (each point along the frequency axis represents a fiber with that characteristic frequency) to the vowel /I/ (as in *beet*). For moderate conversational intensities (64–84 dB SPL), most fibers fire near their saturation rates (normalized rate = 1.0), and formant peaks (see Fig. 17–4) are indistinguishable. The lower family of curves shows the same phenomenon for the vowel /ɑ/ (as in *bat*). (After Sachs, M. B.; Young, E. D.; Miller, M. I. Speech encoding in the auditory nerve: implications for cochlear implants. *Ann. NY Acad. Sci.* 405:94–113, 1983.)

relative concentration of sound in particular frequency regions, or formants) are not distinguishable in the profile of firing rates across the population of auditory nerve fibers. In their experiment, recordings were made from a large population of auditory nerve fibers in cats, and the responses of each fiber to a variety of vowel sounds were determined. As shown in Figure 17–22, the firing rates of most fibers were saturated at ordinary conversational levels. However, when temporal coding was considered across the same array of auditory neurons, the vowel formants were easily distinguished in the response profile over a wide range of intensities (Fig. 17–23). Therefore, the auditory system may make use of temporal encoding of discharges to discriminate among complex stimuli such as vowels.

It should be remembered that the acoustic reflex and efferent regulation of cochlear function are abnormal in these anesthetized animals. In addition, most of the fibers were probably high–spontaneous rate/low-threshold fibers; high-threshold fibers, on the other hand, may permit encoding of vowel formants without requiring temporal coding. Actually, further studies have shown that for consonants, analyses of firing rates are about as good as synchrony (temporal) measures for detecting the acoustic features of a complex stimulus.[41] Synchrony measures, however, do appear to be more resistant to the effects of background noise and to reflect more accurately the psychophysical effects of background noise.[6]

SUMMARY

In the foregoing discussion we have attempted to relate physiology to perception at every level of the auditory periphery, from the pinna to the auditory nerve. Perceptual abilities, after all, can be no better than the quality of information transmitted at each stage of processing, and attempts to reconcile exquisite psychoacoustic performance with physiological measures have motivated and directed a great deal of physiological research in recent years. Chapter 18 relates both psychoacoustics and peripheral physiology to the anatomy and physiology of the central nervous system auditory pathways and considers the special problem of integrating information from the two ears.

ANNOTATED BIBLIOGRAPHY

Berlin, C.I., ed. *Hearing Science.* San Diego, College-Hill Press, 1986.
  An account of areas of rapid progress in the periphery, including inner and outer hair cell function, temporal coding, and cochlear fluid dynamics.
Pickles, J.O. *An Introduction to the Physiology of Hearing.* New York, Academic Press, 1982.
  This lucid and complete treatment of peripheral auditory physiology provides the interested reader with further details about issues discussed in this chapter.
Pickles, J.O. Recent advances in cochlear physiology. *Prog. Neurobiol.* 24:1–42, 1985.
  A well-written, somewhat technical review article summarizing recent advances in cochlear physiology.

von Békésy, G. Experiments in Hearing (research articles from 1928 to 1958). New York, McGraw-Hill Book Co., 1960.
  A classic, especially valuable for students interested in the history of auditory science.
Wever, E.; Lawrence, M. *Physiological Acoustics.* Princeton, Princeton University Press, 1954.
  A readable account of acoustics and outer and middle ear function, areas of study in which the ideas have not changed much since 1954.

REFERENCES

1. Bobbin, R.P. Glutamate and aspartate mimic the afferent transmitter in the cochlea. *Exp. Brain Res.* 34:385–393, 1979.
2. Brownell, W.E. Observations on a motile response in isolated outer hair cells. In Webster, W.R.; Aitkin, L.M., eds. *Mechanisms of Hearing.* Clayton, Victoria, Austl., Monash University Press, 1983.
3. Bruns, V. Structural adaptation in the cochlea of the horseshoe bat for the analysis of the long CF-FM echolocating signals, pp. 867–869. In Busnel, R.G.; Fish, J. F. eds. *Animal Sonar Systems.* New York, Plenum, 1980.
4. Crawford, A.C.; Fettiplace, R. The frequency selectivity of auditory nerve fibres and hair cells in the cochlea of the turtle. *J. Physiol.* 306:79–125, 1980.
5. Davis, H. A mechano-electrical theory of cochlear action. *Ann. Otol. Rhinol. Laryngol.* 67:789–801, 1958.
6. Delgutte, B.; Kiang, N.Y.-S. Speech coding in the auditory nerve. V. Vowels in background noise. *J. Acoust. Soc. Am.* 75:908–918, 1984.
7. Dobie, R.A.; Dillier, N. Some aspects of temporal coding for single-channel electrical stimulation of the cochlea. *Hearing Res.* 18:41–55, 1985.
8. Drescher, M.S.; Drescher, D.G.; Medina, S.E. Effect of sound stimulation at several levels on concentration of primary amines, including neurotransmitter candidates. *J. Neurochem.* 41:309–320, 1983.
9. Fettiplace, R.; Crawford, A.C. The coding of sound pressure and frequency in cochlear hair cells of the terrapin. *Proc. R. Soc. Lond. [Biol.]* 203:209–218, 1978.
10. Flock, A.; Flock, B.; Ulfendahl, M. Mechanisms of movement in outer hair cells and a possible structural basis. *Arch. Otorhinolaryngol.* 24:83–90, 1986.
11. Frishkopf, L.S.; DeRosier, D.J. Mechanical tuning of freestanding stereociliary bundles and frequency analysis in the alligator lizard cochlea. *Hearing Res.* 12:393–404, 1983.
12. Furakawa, T.; Hayashida, Y.; Masuura, S. Quantal analysis of the size of excitatory post-synaptic potentials at synapses between hair cells and afferent nerve fibers in goldfish. *J. Physiol.* 276:211–226, 1978.
13. Gitter, A.H.; Zenner, H.P.; Fromter, E. Membrane potential and ion channels in isolated outer hair cells of guinea pig cochlea. *ORL J. Otorhinolaryngol. Rel. Spec.* 48:68–75, 1986.
14. Godfrey, D.A.; Kiang, N.Y.-S.; Norris, B.E. Single unit activity in the posteroventral cochlear nucleus of the cat. *J. Comp. Neurol.* 162:247–268, 1975.
15. Goodman, D.A.; Smith, R.L.; Chamberlain, S.C. Intracellular and extracellular responses in the organ of Corti of the gerbil. *Hearing Res.* 7:161–179, 1982.
16. Hudspeth, A.J. Mechanoelectrical transduction by hair cells in the acousticolateralis sensory system. *Annu. Rev. Neurosci.* 6:187–215, 1983.
17. Hudspeth, A.J.; Corey, D.P. Sensitivity, polarity, and conductance change in the response of vertebrate hair cells

to controlled mechanical stimuli. *Proc. Natl. Acad. Sci. USA* 74:2407–2411, 1977.

18. Hunter-Duvar, I.; Landolt, I.; Cameron, R. X-ray microanalysis of fluid spaces in the frozen cochlea. *Arch. Otorhinolaryngol.* 230:245–249, 1981.
19. Khanna, S.M.; Leonard, D.G.B. Laser interferometric measurements of basilar membrane vibrations in cats. *Science* 215:305–306, 1982.
20. Kim, D.O. Functional roles of the inner- and outer-hair-cell sub-systems in the cochlea and brainstem. In Berlin, C.I., ed. *Hearing Science.* San Diego, College-Hill Press, 1986.
21. Leonard, D.G.B.; Khanna, S.M. Histological evaluation of damage in cat cochleas used for measurement of basilar membrane mechanics. *J. Acoust. Soc. Am.* 75:515–527, 1984.
22. Lewis, R.S.; Hudspeth, A.J. Voltage and ion dependent conductances in solitary vertebrate hair cells. *Nature* 304:538–541, 1983.
23. Liberman, M.C. Auditory-nerve response from cats raised in a low-noise chamber. *J. Acoust. Soc. Am.* 63:442–455, 1978.
24. Liberman, M.C.; Brown, M.C. Physiology and anatomy of single olivocochlear neurons in the cat. *Hearing Res.* 24:17–36, 1986.
25. Lim, D.J. Cochlear anatomy related to cochlear micromechanics. A review. *J. Acoust. Soc. Am.* 67:1686–1695, 1980.
26. Mountain, D.C. Changes in endolymphatic potential and crossed olivocochlear bundle stimulation alter cochlear mechanics. *Science* 210:71–72, 1980.
27. Mulroy, M.J. Cochlear anatomy of the alligator lizard. *Brain Behav. Evol.* 10:69–87, 1974.
28. Peake, W.T.; Ling, A.L. Basilar-membrane motion in the alligator lizard: its relation to tonotopic organization and frequency selectivity. *J. Acoust. Soc. Am.* 67:1736–1745, 1980.
29. Popper, A.N.; Fay, R.R. *Comparative Studies of Hearing in Vertebrates.* New York, Springer-Verlag, 1980.
30. Rasmussen, G.L. Studies of the eighth cranial nerve of man. *Laryngoscope* 50:67–83, 1940.
31. Rhode, W.S. Some observations on two-tone interaction measured with the Mossbauer effect. In Evans, E.F.; Wilson, J.P., eds. *Psychophysics and Physiology of Hearing,* 27–41. New York, Academic Press, 1977.
32. Robertson, D. Horseradish peroxidase injection of physiologically characterized afferent and efferent neurones in the guinea pig spiral ganglion. *Hearing Res.* 15:113–121, 1984.
33. Rose, J.E.; Brugge, J.L.; Anderson, D.J.; Hind, J.E. Phase-locked response to low-frequency tones in single auditory-nerve fibers of the squirrel monkey. *J. Neurophysiol.* 30:769–793, 1967.
34. Rosowski, J.J.; Carney, L.H.; Lynch, T.J., III; Peake, W.T. The effectiveness of external and middle ears in coupling power into the cochlea. In Allen, J.B.; Hall, J.L.; Hubbard,

A.; Neely, S.T.; Tubis, A. eds. *Peripheral Auditory Mechanisms, Lecture Notes in Mathematics, vol. 64* (Appendix Insert 3), New York, Springer-Verlag, 1986.
35. Rubel, E.W.; Lippe, W.R.; Ryals, B.M. Development of the place principle. *Ann. Otol. Rhinol. Laryngol.* 93:609–615, 1984.
36. Russell, I.J.; Sellick, P.M. Intracellular studies of hair cells in the mammalian cochlea. *J. Physiol.* 284:261–290, 1978.
37. Ryan, A.; Dallos, P. Effect of absence of cochlear outer hair cells on behavioral auditory thresholds. *Nature* 253:44–46, 1975.
38. Sachs, M.B.; Young, E.D. Encoding of steady-state vowels in the auditory nerve: representation in terms of discharge rate. *J. Acoust. Soc. Am.* 66:470–479, 1979.
39. Sellick, P.M.; Patuzzi, R.; Johnstone, B.M. Measurement of basilar membrane motion in the guinea pig using the Mossbauer technique. *J. Acoust. Soc. Am.* 72:131–141, 1982.
40. Simmons, F.B. Middle ear muscle activity at moderate sound levels. *Ann. Otol. Rhinol. Laryngol.* 68:1126–1143, 1959.
41. Sinex, D.G.; Geisler, C.D. Responses of auditory-nerve fibers to consonant-vowel syllables. *J. Acoust. Soc. Am.* 73:602–615, 1983.
42. Smith, R.L. Adaptation, saturation and physiological masking in single auditory-nerve fibers. *J. Acoust. Soc. Am.* 65:166–178, 1979.
43. Spoendlin, H. Neural connections of the outer hair cell system. *Acta Otolaryngol. (Stockh.)* 87:381–387, 1979.
44. Strelioff, D.; Flock, A. Mechanical properties of hair bundles of receptor cells in the guinea pig cochlea. *Soc. Neurosci.* Abstract 8, 1982.
45. Thalmann, I.; Thallinger, G.; Comegys, T.H.; Thalmann, R. Collagen—the predominant protein of the tectorial membrane. *ORL J. Otorhinolaryngol. Relat. Spec.* 48:107–115, 1986.
46. Tilney, L.G.; Saunders, J.C. Actin filaments, stereocilia, and hair cells of the bird cochlea: The length, number, width, and distribution of stereocilia of each hair cell is related to the position of the hair cell on the cochlea. *J. Cell Biol.* 86:244–259, 1982.
47. von Békésy, G. *Experiments in Hearing.* New York, McGraw-Hill Book Co., 1960.
48. Webster, D.M.; Webster, M. The specialized auditory system of kangaroo rats. In Neff, W.D., ed. *Contributions to Sensory Physiology,* vol. 8, pp. 161–196. New York, Academic Press, 1984.
49. Wiederhold, M.L.; Kiang, N.Y.-S. Effects of electric stimulation of the crossed olivocochlear bundle on single auditory-nerve fibers in the cat. *J. Acoust. Soc. Am.* 48:950–965, 1970.
50. Young, E.D.; Sachs, M.B. Representation of steady-state vowels in the temporal aspects of the discharge patterns of populations of auditory-nerve fibers. *J. Acoust. Soc. Am.* 66:1381, 1979.

Chapter 18

Edwin W Rubel
Robert A. Dobie

The Auditory System: Central Auditory Pathways

INTRODUCTION

In Chapter 17 we discussed the peripheral processing of auditory information, concluding with the discharge characteristics of auditory nerve fibers. In this chapter, we consider the processing of sound information by the central nervous system (CNS) and attempt to explain the transformations that take place between spike train activity in the 30,000 auditory nerve fibers and the perception of our rich acoustic environment.

To appreciate the difficulty of this processing task, we should reconsider the incredible versatility of our auditory capabilities. These begin with our primitive ability to detect the presence of a predator or prey in the immediate environment, independent of the illumination, direction of gaze, or wind conditions. Also we are able to use spectral and temporal variations in the acoustic energy to identify whether the intruder is close or distant, large or small, friend or foe, or a conspecific. To make such distinctions the acoustic information must be compared with stored auditory experiences, a task that requires more than the simple decoding of signals arriving on the eighth nerve. Localization of the spatial position of a sound also requires integration by CNS neurons, which must compare subtle differences in the energy arriving at the two ears. Such binaural processing, which characterizes most of our auditory pathways, is also useful for the perception of language, the highest level of human auditory processing. Here again, the raw afferent information that reaches the CNS is insufficient to account for language comprehension. Such a facility requires a stored auditory lexicon, which itself is largely independent of the specific nuances of speech sounds. A similar facility presumably underlies the understanding of complex communication signals in other species, such as whales and dolphins.

Our understanding of the CNS events that contribute to auditory detection, binaural localization, and language comprehension is rather primitive. There are, however, two organizing principles that characterize auditory, and in general terms all, sensory processing.

First, auditory information is processed by many parallel pathways. We noted in the preceding chapter that the discharge properties of the vast majority of eighth nerve fibers are quite similar, varying only in their best frequency. At the level of the cochlear nucleus where these axons terminate, this information is distributed to several subnuclei, which give rise to several ascending parallel pathways. Whereas each pathway appears to carry the full range of frequency information, each has a different set of cell types, which themselves have a variety of different inputs, targets, and response properties. From these subnuclei, and probably from subregions within each subnucleus, emanate pathways that undoubtedly transmit different kinds of information. For example, some cochlear nucleus cells drive neurons that receive inputs from both ears (binaural pathways), whereas others contribute to pathways that remain monaural as information ascends through successive levels of neural processing. There is, however, considerable "crosstalk" between these "parallel pathways," and it is unwise to consider them as completely separate.

Second, even along a particular pathway, auditory processing cannot be considered strictly serial or hierarchical. In the preceding chapter we saw that peripheral processing of sound is modified at the level of the middle ear by neural innervation of the tensor tympani and stapedius muscles and further modified at the inner ear by the olivocochlear (efferent) fibers innervating the hair cells. Similarly, at each level of the CNS, ascending information can be modified by descending information from more cephalad auditory regions. Unfortunately, how these centrifugal pathways modify afferent information is largely unknown.

In this chapter, we begin with an overview of the CNS pathways involved in hearing. Ascending pathways that originate at the eighth nerve and project toward the cerebral cortex (also called centripetal projections) will be considered first, followed by a discussion of descending (or centrifugal) projections. The general principle of tonotopic organization will then be considered, as will the anatomical and physiological properties of the major auditory regions. Equipped with this knowledge of the separate elements of the auditory system, we will discuss a few examples of auditory processing, (i.e., binaural interactions, prey localization in the mustache bat, and the middle ear and startle reflexes). Finally, we conclude with a brief section on development and plasticity in the central auditory pathways.

ORGANIZATION OF CENTRAL AUDITORY PATHWAYS

Most of our knowledge concerning the organization of nuclei and fiber tracts involved in the processing of acoustic information comes from detailed studies of the domestic cat.[14] However, the basic organization appears substantially unchanged in all mammals, and the brain-stem auditory pathways are similar in birds and reptiles as well. In some species, this phylogenetically conserved basic organization has been modified to accommodate their unique behavioral requirements. In this section, we first describe the basic organization, and then show how the brains of some animals provide unique opportunities to analyze particular aspects of acoustic processing.

Ascending Auditory Pathways

Auditory Nerve

The auditory nerve is composed of approximately 50,000 axons in the cat (approximately 30,000 in humans), of which 90 to 95% are myelinated and terminate distally on the bases of the inner hair cells. The remaining 5 to 10% are unmyelinated and transmit information from outer hair cells. These different innervation patterns suggest that inner and outer hair cells transmit different kinds of information. This assumption is supported by several facts. First, as noted in the previous chapter, the large myelinated axons are thought to innervate a single inner hair cell, whereas the unmyelinated axons contact a relatively large number of outer hair cells. Therefore, information from inner hair cells has a "private line" to the CNS, whereas information from outer hair cells shows considerable convergence. Second, information from outer hair cells is received by the brain about 1 to 2 ms later than that from inner hair cells. Finally, unless the central terminations of the unmyelinated axons ramify much more profusely and widely than those of myelinated axons, responses in CNS auditory neurons should be dominated by information from inner hair cells.

The myelinated and unmyelinated fibers gather together to form the auditory nerve that courses through the internal auditory meatus. The nerve enters the brain stem at an outgrowth of neural tissue called the acoustic tubercle. The acoustic tubercle contains the cell bodies and neuropil of the cochlear nucleus, where all auditory nerve fibers terminate.

Cochlear Nucleus

The auditory nerve terminates in three cytoarchitectonically distinguishable subnuclei, which were named for their relative positions in the human brain. The three subnuclei, called the anteroventral, posteroventral, and dorsal cochlear nuclei, seem to have gross functional differences as well. In different species the relative size of the subnuclei varies. For example, in humans the dorsal cochlear nucleus is so small that it is almost vestigial. The current view is that all auditory

nerve terminals are excitatory to the postsynaptic cells of the cochlear nucleus, but this is not certain. The three subnuclei of the cochlear nucleus give rise to three major fiber tracts, which constitute three brain-stem pathways with apparently distinct functions.

Brain-stem Pathways

Binaural Brain-stem Pathways. One pathway originates mainly from cell bodies in the anteroventral cochlear nucleus (AVCN) and dives down into the ventral part of the brain stem to form a large fiber bundle, the *trapezoid body*, which runs across the base of the brain (Fig. 18–1) to provide

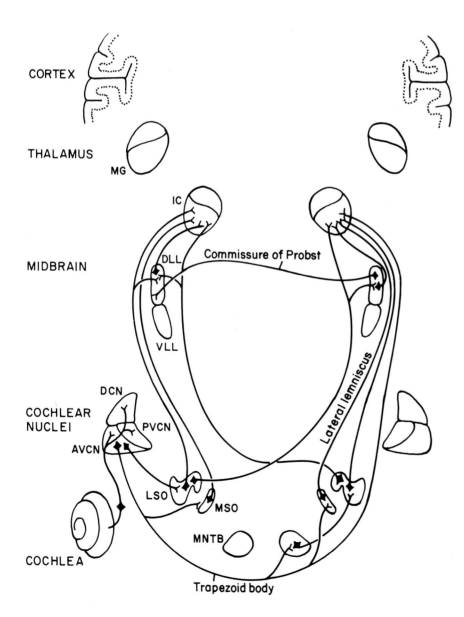

Figure 18–1 Binaural auditory pathways in the brain stem illustrated for the left cochlear nucleus. The connections from the other cochlear nucleus would form a mirror image. AVCN = anteroventral cochlear nucleus; PVCN = posteroventral cochlear nucleus; DCN = dorsal cochlear nucleus; LSO = lateral superior olive; MSO = medial superior olive; MNTB = medial nucleus of the trapezoid body; VLL = ventral nucleus of the lateral lemniscus; DLL = dorsal nucleus of the lateral lemniscus; IC = inferior colliculus; MG = medial geniculate nucleus. (Adapted by permission from Thompson, G. *Seminars in Hearing*, Vol. 4, pp. 81–95, Thieme Medical Publishers, Inc., New York, 1983.)

the major input to the *superior olivary complex*. The input from the anteroventral cochlear nucleus has two projection sites, which can be differentiated according to their preferred auditory frequencies; one, the *medial superior olivary nucleus* (MSO), receives input predominantly from the apical and middle turns of the cochlea (i.e., lower frequencies), whereas the other, the *lateral superior olivary nucleus* (LSO), receives inputs from the middle and basal turns (i.e., higher frequencies).

The superior olivary complex is the first site for binaural interactions. Single axons from the anteroventral cochlear nucleus are thought to innervate cells at corresponding positions in both the ipsilateral and the contralateral medial superior olivary nucleus. These neurons, therefore, receive inputs from both ears. The lateral superior olivary nucleus also receives binaural input. Neurons in the anteroventral cochlear nucleus not only make excitatory connections in the ipsilateral lateral superior olivary nucleus but also send an axon across the midline in the trapezoid body to a group of large cells near the midline called the *medial nucleus of the trapezoid body* (Fig. 18–1, MNTB). From this nucleus, axons project to the ipsilateral lateral superior olive. Thus, each lateral superior olive neuron receives a direct input from the ipsilateral ear via the cochlear nucleus and an indirect input from the contralateral ear through an intervening synapse in the medial nucleus of the trapezoid body.

The major targets of axons from the medial and lateral superior olivary nuclei are the dorsal nucleus of the lateral lemniscus (Fig. 18–1, DLL) and the inferior colliculus (IC) in the midbrain. The medial superior olivary nuclei project primarily ipsilaterally while the lateral superior olivary nuclei project to both the ipsilateral and contralateral inferior colliculi as well as to the lateral lemniscal nuclei. In addition, some axons from the anteroventral cochlear nucleus, which cross in the trapezoid body, continue up to the contralateral nuclei of the lateral lemniscus or all the way to the inferior colliculus (Fig. 18–1).

Intermediate Brain-stem Pathway. The second pathway, which is relatively poorly understood, originates primarily in the posteroventral cochlear nucleus (Fig. 18–2, PVCN) with some contribution from the anteroventral cochlear nucleus. Axons from these nuclei (solid lines) travel in the *intermediate acoustic stria* to innervate both the ipsilateral and contralateral periolivary nuclei, which surround the superior olivary complex. Additional axons ascend in the contralateral lateral lemniscus to innervate the lateral lemniscal nuclei and inferior colliculus.

Monaural Brain-stem Pathway. The third pathway originates in the dorsal cochlear nucleus (DCN), whose axons course across the midline in the *dorsal acoustic stria* on the dorsal aspect of the brain stem (see Fig. 18–2). They then join the contralateral lateral lemniscus and terminate in the inferior colliculus; some also terminate in the lateral lemniscal nuclei. Thus this pathway sends information from one ear directly to the contralateral inferior colliculus.

Auditory Forebrain Projections. As indicated in Figures 18–1 and 18–2, the three brain-stem pathways converge at the inferior colliculi, which communicate with each other via their commissure. From there, information ascends ipsilaterally in the *brachium of the inferior colliculus* to the main thalamic auditory area, the *medial geniculate body*. Axons from the medial geniculate body terminate in the ipsilateral temporal regions of the cerebral cortex. In lower mammals the auditory areas are primarily on the lateral surface of the cortex; in primates, they are buried in the lateral fissure on the superior aspect of the superior temporal gyrus. Cortical auditory regions in the two hemispheres are interconnected through the corpus callosum.

Descending Auditory Pathways

Sensory processing is often described as a hierarchical process with specific functions attributed to specific neural structures. This model is particularly inappropriate for describing auditory processing. The transmission of information from lower to higher structures is dramatically influenced by information flowing in the opposite direction. As was pointed out in Chapter 17, these descending influences are evident from the very beginning of auditory information processing, including the middle ear and inner ear.

Four major descending pathways are shown in Figure 18–3. At the brain-stem level, the periolivary nuclei project to both the ipsilateral and contralateral inner ear. This *olivocochlear pathway* is composed of two relatively distinct components.[9] One component originates in cell bodies lying lateral to the medial superior olivary nucleus and sends axons ipsilaterally to synapse on the distal dendrites of spiral ganglion cells just under the inner hair cells. The second originates in cell bodies situated in olivary and periolivary nuclei medial to the medial superior olive. It provides axons that synapse primarily on the base of contralateral outer hair cells. The functions of the olivocochlear projections were discussed in the previous chapter. A second descending system

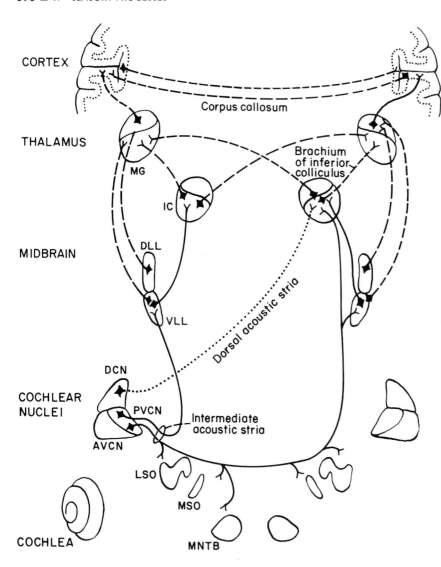

CORTEX

Corpus callosum

THALAMUS

MG

Brachium
of inferior
colliculus

IC

MIDBRAIN

DLL

Dorsal acoustic stria

VLL

DCN

COCHLEAR
NUCLEI

PVCN

Intermediate
acoustic stria

AVCN

LSO

MSO

COCHLEA

MNTB

Figure 18–2 Connections of the intermediate brain-stem pathway (solid lines) and monaural brain-stem pathway (dotted line). As in Figure 18–1, only projections from one side are shown. Forebrain auditory pathways are dashed (abbreviations as in Fig. 18–1). (Adapted with permission from Thompson, G. *Seminars in Hearing*, Vol. 4, pp. 81–95, Thieme Medical Publishers, Inc., New York, 1983.)

originates in the periolivary nuclei but ends in the cochlear nucleus. Cell groups situated lateral to the medial superior olive send axons primarily to the ipsilateral cochlear nucleus, whereas cells medial to the medial superior olive terminate mainly in the contralateral cochlear nucleus. A third descending pathway originates in the inferior colliculus and lateral lemniscal nuclei. The lateral lemniscal nuclei project to the superior olivary complex and the cochlear nucleus; the inferior colliculus projects to a variety of brain-stem auditory structures, such as the superior olivary complex, as well as some regions that are not considered part of the primary auditory pathways, such as certain pontine nuclei which transmit auditory information to the cerebellum and superior colliculus (see Chap. 20). Last, the auditory cortex

gives rise to at least two descending pathways. Axons descend to the divisions of the medial geniculate nucleus from which ascending cortical connections originate. Other cortical cells project to the ipsilateral inferior colliculus.

Summary

This discussion of the major connections of the central auditory pathways serves to illustrate two points. First, the ascending auditory pathways consist of multiple independent, parallel processing networks that extract, encode, and transmit information about different aspects of the acoustic energy reaching the two ears. Second, the descending pathways allow each structure up to and including the cerebral cortex to modulate the information in the ascending pathways.

TONOTOPIC ORGANIZATION

In Chapter 17 we noted that peripheral auditory processing is governed by the principle that micromechanical properties of the inner ear transform sound energy into a unique pattern of hair cell excitation. Hair cells in the apical cochlea respond best to low frequencies, basal hair cells prefer high frequencies, and cells in the middle of the cochlea prefer intermediate frequencies. Since this *place code* is a major factor in the discrimination of sounds of different spectral content, we might expect that the spatial encoding of frequency would be retained in the central auditory pathways.

Indeed, when Rose and colleagues[33] played pure tones to a cat, they discovered that each neuron of the cochlear nucleus was excited optimally by a single frequency (its *best frequency*), and that within each subdivision of the cochlear nucleus the best frequencies were organized in a regular order; neighboring cells preferred similar but slightly different frequencies. Figure 18–4 shows one of their recording electrode tracks through part of the cochlear nucleus and identifies the best frequencies of the neurons that were encountered. Within each subnucleus (e.g., Dc), best frequencies were encountered in a descending ordered

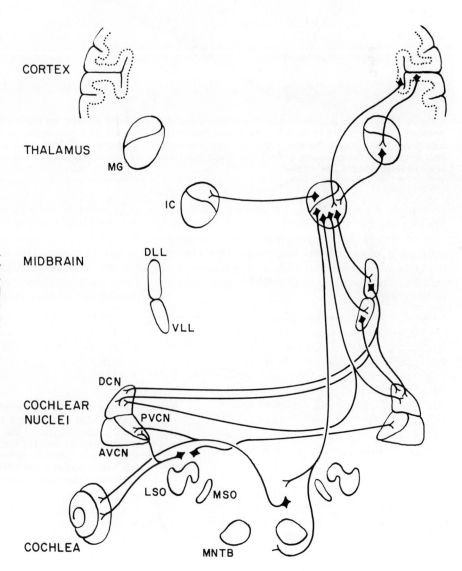

Figure 18–3 The major descending auditory pathways for one side of the brain (abbreviations as in Fig. 18–1). (Adapted with permission from Thompson, G. *Seminars in Hearing*, Vol. 4, pp. 81–95, Thieme Medical Publishers, Inc., New York, 1983.)

CORTEX

THALAMUS

MG

IC

DLL

MIDBRAIN

VLL

DCN

COCHLEAR
NUCLEI

PVCN

AVCN

LSO

MSO

COCHLEA

MNTB

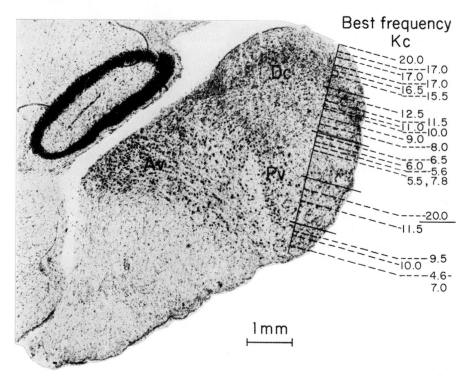

Best frequency
Kc

20.0
17.0
17.0
17.0
16.5
15.5
12.5
11.5
11.0
10.0
9.0
8.0
6.5
6.0
5.6
5.5 , 7.8

20.0
11.5

9.5
10.0
4.6-
7.0

1mm

Figure 18–4 Sagittal section through the cochlear nucleus of the cat showing an electrical penetration through the dorsal cochlear nucleus (Dc) and posteroventral cochlear nucleus (Pv). The best frequency (in kHz) of neurons encountered at successive points along the electrode tract are indicated to the right. Note that there is a systematic decrease in best frequencies until the electrode enters Pv, at which point a new decreasing sequence begins. (From Rose J.E. et al., *Bull. Johns Hopkins Hosp.* 104:211–251, 1959.)

sequence; as the electrode tip crossed the boundary into another subnuclear region (e.g., Pv), there was an abrupt change in frequency and the beginning of a new ordered sequence. This classical experiment showed that within each region of the cochlear nucleus there is an orderly representation of the cochlea.

In order to achieve a tonotopic organization in the cochlear nucleus each point on the cochlea must be represented as a two-dimensional sheet of neural tissue. This organization is illustrated in Figure 18–5. Several ganglion cells receive input from each position along the cochlea. This information can then be distributed to a laminar, sheet-

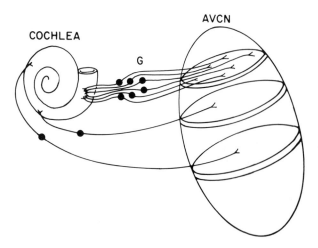

Figure 18–5 Schematic of the topographic projection of points on the cochlea to the anteroventral cochlear nucleus (AVCN). Each region of the cochlea is innervated by many spiral ganglion cells (G) whose central axons terminate as a sheet in the cochlear nucleus, forming an isofrequency lamina. These isofrequency lamina are stacked in order of their cochlear innervation to form a tonotopic organization.

like area at the level of the cochlear nucleus. These "isofrequency lamina" are then stacked to form the complete tonotopic representation of the cochlea.

The principle that the receptor surface is spatially "mapped" onto different areas of the brain is not unique to the auditory system. The same organizational principle has been seen in the somatosensory system (where it is called somatotopic organization; see Chap. 15) and will be seen in the visual system (where it is called retinotopic organization; see Chap. 20). In the auditory system the orderly anatomical arrangement of neurons according to their best frequencies is called *tonotopic organization*. Tonotopic organization has now been demonstrated in each of the auditory nuclei from the cochlear nucleus to the cerebral cortex and in a large variety of species, including both vertebrates and invertebrates. For vertebrates with elongated cochleas, such as reptiles, birds, and mammals, the following rules seem to govern the organization.

1. Connections between auditory nuclei are always in tonotopic register, regardless of whether the connections are ascending, descending, or at the same levels of the neuroaxis (i.e., cortico-cortical connections) and irrespective of their sign (excitatory or inhibitory).

2. When an auditory region is subdivided into subnuclei, each contains an entire tonotopic organization. For example, each subregion of the auditory cortex has a complete representation of the cochlea.

3. In binaurally activated areas, such as the medial superior olive, the tonotopic representation of each ear is in register so that each neuron is most sensitive to the same frequency played to either ear.

4. Although the frequency range varies from species to species, the basic tonotopic map is the same within a class of vertebrates and is virtually identical for individuals of the same species. Often, the similarity of maps between animals of the same species is so precise that a formula that predicts the best frequency of a neuron on the basis of its anatomical location can be accurate to within 200 to 300 Hz.[36]

Each frequency does not necessarily have an equal anatomical representation. Within a structure, the actual number of neurons devoted to a given frequency depends on the number of receptors responding to that frequency. When an animal is specialized to use a particular frequency band, both the cochlea and the central nervous system have a relatively larger amount of tissue devoted to that band, and the neural representation actually becomes magnified as one ascends the central pathways to the auditory cortex. For example, the mustache bat uses biosonar to navigate, and its auditory system is specialized to use 60-kHz echoes for this behavior. Its cochlea has a specialized region of high receptor density to encode this frequency. This region accounts for about 20% of the cochlea, but its central representation covers almost 50% of the auditory cortex. Although this is an extreme example, it is likely that other species that utilize specific frequencies for behavioral adaptations also show a neural "magnification factor" as one ascends the auditory pathways. A similar phenomenon characterizes other sensory systems. For example, the fovea has an increased representation in the central visual pathways (Chap. 20) as do the perioral regions and tips of the digits in the central somatosensory pathways (Chap. 15).

THE COCHLEAR NUCLEUS

As the cochlear nerve enters the brain, each axon divides into two main branches; one courses anterior to terminate in the *anteroventral* subdivision of the cochlear nucleus. The other branch runs posterior through the posteroventral subdivision where it usually synapses via en passant synapses and short collaterals. Most of these axons finally terminate in the *dorsal cochlear nucleus*. The tonotopic organization of each subdivision of the cochlear nucleus is established by the branching patterns of entering auditory nerve axons. Low-frequency fibers enter ventrally and branch immediately, whereas high-frequency fibers enter dorsally and branch later. The transmitter used by auditory nerve terminals has not been unequivocally identified; it is probably an excitatory amino acid, an aspartate-like compound.

While we describe the cochlear nucleus as being composed of three main subdivisions, several investigators have further subdivided each region on the basis of cell types or other anatomical features. In addition, there are marked variations between species in the relative sizes of each division.

To appreciate the contribution of the cochlear nucleus to auditory processing, we must briefly review the discharge properties of its inputs from the auditory nerve as revealed by several analysis techniques (Fig. 18–6). The discharge patterns of the vast majority of eighth nerve axons differ only in their best frequency. Most axons have a high rate of spontaneous activity (Fig. 18–6A) and sharp excitatory tuning curves (Fig. 18–6B). Those responding to frequencies below 5 to 6 kHz preserve the temporal characteristics of the stimulus by phase-locking (Fig. 18–6F). Rate-intensity func-

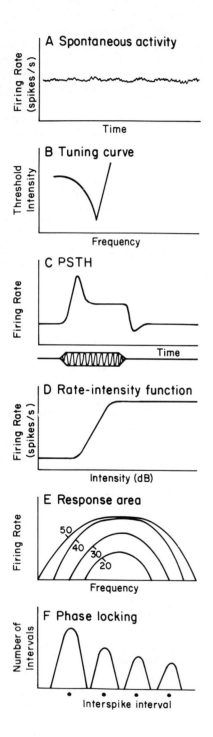

tions at a neuron's best frequency are usually monotonic and saturate at 30 to 40 dB above threshold (Fig. 18–6D). The temporal response characteristics of an auditory nerve axon to a tone burst are obtained by repeating a stimulus many times and grouping the action potentials into sequential time bins. The action potentials in each time bin are then added for all the stimuli to produce a post-stimulus-time-histogram (PSTH). The typical form of a PSTH for an auditory nerve fiber is shown in Figure 18–6C. Finally, one can determine the entire frequency "response area" (or frequency "receptive field," analogous to somatosensory and visual receptive fields; see Chaps. 15 and 20) of an auditory neuron by plotting its discharge rate as a function of frequency at a number of stimulus intensities. Such a plot, characteristic of all auditory primary afferents, is shown in Figure 18–6E.

In contrast to the similarities in the discharge properties of primary auditory afferents, cochlear nucleus neurons display a wide variety of response properties. For example, Figure 18–7 compares PSTHs in the auditory nerve with those found in the cochlear nucleus. Some cochlear nucleus responses are virtually identical to those of the nerve fibers (lowest cell), whereas others respond only to the onset of sound (second cell),

Figure 18–6 Typical analyses used to describe the response patterns of auditory neurons to simple acoustic stimuli. *A,* Spontaneous activity in the absence of acoustic stimuli. *B,* A tuning curve that plots the threshold for an increase in activity above the spontaneous rate as a function of the frequency of a pure tone acoustic stimulus. Each neuron has a "best frequency," indicated by the tip of the tuning curve; at higher or lower frequencies a louder stimulus is required to elicit a response. *C,* Post-stimulus time histogram (PSTH) shows the pattern of discharge rate as a function of time. A PSTH typically is constructed by repeating a short pure-tone stimulus (shown under graph) many times and averaging the number of discharges in successive time bins following stimulus onset. The typical auditory nerve response pattern (called primary-like) shows an initial high rate of activity immediately after stimulus onset, a decreased, steady rate which is maintained until the stimulus is turned off, and a brief period of inhibition at stimulus offset before the activity returns to spontaneous discharge rate. *D,* A rate-intensity function plots firing rate as a function of the intensity of a pure tone stimulus. Most auditory nerve fibers have monotonic rate-intensity functions as shown here. *E,* Average discharge rate is plotted as a function of stimulus frequency. The family of curves represent different stimulus intensities from 20 to 50 dB. *F,* Phase-locking of auditory nerve fibers can be seen by plotting the number of times that each interspike interval occurs against the interval between each pair of spikes. Note that discharges tend to be spaced at the period of the tone or at multiples thereof, which are indicated by dots on the abscissa.

Figure 18–7 Relationship of responses in the cochlear nucleus to morphologically identified cell types. The discharge pattern of an auditory nerve fiber (solid PSTH) to a moderate intensity, short (25–50 ms) tone burst (envelope below PSTH) at the fiber's best frequency is compared to various response patterns elicited by the same burst in the dorsal cochlear nucleus (DCN) and the cell morphologies to which the responses are believed to correspond. AVCN = anteroventral cochlear nuclei; PVCN = posteroventral cochlear nuclei. (From Kiang, N. Y.-S. In Tower, D., ed. *The Nervous System*, Vol. 3. *Human Communication and Its Disorders*, pp. 81–96. New York, Raven Press, 1975 with permission.)

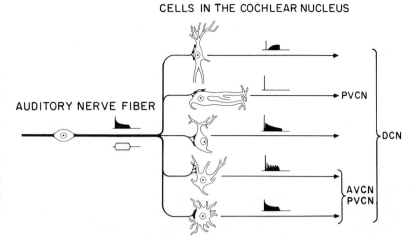

CELLS IN THE COCHLEAR NUCLEUS

AUDITORY NERVE FIBER

and still others show distinct temporal patterns of excitation and inhibition (first and third cells). These differences in the response properties of cochlear nucleus neurons are associated with particular subdivisions and with cell types with unique morphologies.[18]

Anteroventral Nucleus

The anteroventral subdivision of the cochlear nucleus contains two cell types that can be differentiated morphologically. One type is large, round or globular-shaped, with a few short, stubby dendrites that branch repeatedly. These are usually called "bushy cells" or spherical cells. The second type is large and stellate with several tapering dendrites of medium length.

The responses of bushy cells to tone bursts are strikingly similar to those described for auditory nerve fibers and, therefore, are called "primary-like" responses (see Fig. 18–7, lowest cell). A tone elicits a large initial response followed by a decline to a steady-state firing rate for the duration of the tone; when the tone is turned off, there is a short period of inhibition. These cells exhibit phase-locking, monotonic rate-intensity functions with a dynamic range of 20 to 40 dB, and tuning curves resembling those of auditory nerve fibers. The only consistent difference between the responses of auditory nerve fibers and bushy cells is that the latter often show "inhibitory sidebands"—that is, inhibition of spontaneous activity when tones are presented at frequencies and intensities adjacent to the excitatory tuning curve. Therefore, at this level, there is some integration of the events occurring in surrounding cells.

The striking similarity of neuronal responses in the anteroventral cochlear nucleus to those of the auditory nerve can be understood by considering the morphology of auditory nerve terminals. Auditory nerve fibers provide 2 to 4 large calyx-type endings onto the soma and primary dendrites of the bushy cells. These *end bulbs of Held* provide input from one or only a few hair cells. In vitro experiments using tissue slices containing the auditory nerve root and the cochlear nucleus indicate that bushy cells fire one action potential for each spike in the nerve up to very high rates of simulation;[10, 29] therefore, the synapse is extremely secure.

The second cell type of the anteroventral cochlear nucleus also has response characteristics similar to those of eighth nerve fibers, except that it discharges rhythmically after the initial transient portion of the PSTH (Fig. 18–7, 4th trace). This discharge pattern, which has been called a "chopper" response, is often associated with stellate cells.[30, 34]

It should not be concluded that *no* integration occurs in the anteroventral cochlear nucleus. Both primary-like and chopper cells receive several other synaptic connections that include GABAergic and noradrenergic inputs, which are presumed to be inhibitory; a cholinergic input; and probably other peptidergic inputs.[44] The sources of most of these other inputs have not been identified and their roles in information processing remain obscure.

Posteroventral Nucleus

The most common type of neuron in the posteroventral subdivision of the cochlear nucleus responds best to the *onset* of a stimulus. Thus, it faithfully phase-locks to low frequencies or rapid clicks (below 1 kHz) but responds only to the onset of a higher frequency tone. The neural network responsible for the dramatic cessation of activity in these neurons following the onset of a stimulus is not known, but must include only one or two interneurons since the response is inhibited very rapidly. These cells also often have broad tuning curves, suggesting that they receive inputs from cells located over considerable distances along the cochlea. Altogether, their response characteristics suggest that they can faithfully code rapid temporal changes in a complex acoustic stimulus or in the envelope of a complex stimulus. These response characteristics have been attributed to a cell whose soma and dendritic morphology resemble an octopus (Fig. 18–7, second cell); thus, it is called an *octopus cell*.

Other response types similar to those described in the anteroventral cochlear nucleus (i.e., chopper cells and primary-like cells) also exist in the posteroventral cochlear nucleus, but the cellular morphology associated with these responses is unknown.[30, 34]

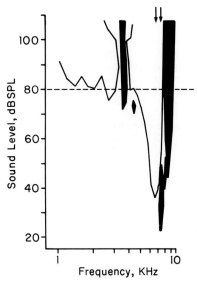

Figure 18–8 Complex tuning curve typical of one type of cell in the dorsal cochlear nucleus. Black areas show frequency-intensity combinations that are excitatory. Enclosed white areas show frequency-intensity combinations that inhibit spontaneous activity. The dotted horizontal line at 80 dB SPL is described in the text. (After Young, E.D. In Berlin, C.I., ed. *Hearing Science*, pp. 423–460, College Hill Press, 1984.)

Dorsal Nucleus

The dorsal cochlear nucleus is the most complex of the cochlear subnuclei. It has been subdivided into a number of regions with heterogeneous cell morphologies and, as noted above, receives descending inputs from a variety of other brain-stem regions including the inferior colliculus.

The response patterns of cells in the dorsal cochlear nucleus show a considerable variety, including the simple pattern characteristic of primary afferents and each of the other patterns shown in Figure 18–7. A consideration of the inhibitory responses further complicates the number of possible patterns.[49] For example, Figure 18–8 shows the inhibitory and excitatory regions for a neuron in the dorsal cochlear nucleus. At moderate sound pressure intensities (80 dB; dashed line), the cell is inhibited by low frequencies. Then, as the frequency is increased it is excited, then inhibited again, and then excited again as the frequency is raised further. At higher or lower intensities, the response pattern is different. Furthermore, responses of dorsal cochlear nucleus cells are dramatically altered by barbiturate anesthesia,[50] further complicating our understanding of information processing in this region.

UTILITY OF BINAURAL INFORMATION

In all the structures discussed so far, neural discharge has been influenced by either one ear or the other. Beginning with the superior olive, however, many neurons are influenced by stimuli delivered to both ears. Before examining such discharge patterns, it is helpful to consider the use we make of our binaural capabilities in order to better understand the binaural responses of neurons.

Binaural hearing allows us to perform two well-known auditory feats. First, we can localize sound in space. Binaural hearing is essential for the localization of a brief sound that varies in position around the horizontal meridian (measured in degrees azimuth). Some animals, such as the barn owl, also can use binaural cues to localize the position of a sound in three-dimensional space. Second, we can attend selectively to sound emanating from a particular location, such as the conversation of one person in a crowded, noisy room (the "cocktail party effect"). Both these feats are accomplished by using either differences in the timing or differences in the intensity of the sounds reaching the two ears.

Interaural time differences occur because sound

travels relatively slowly and there is a finite distance between the ears. The speed of sound in air is 340 m/s and the diameter of the human head is about 17.5 cm. Thus, a sound emanating from the 90-degree azimuth (along a line intersecting the ears) reaches the far ear approximately 660 μs after it reaches the near ear. For sound positions between 0 (the sagittal midline plane) and 90 degrees, the time differences will be proportionally less. This timing difference provides two related binaural signals. First, the onset of stimulation of the two ears will occur at slightly different times; this is important when sound undergoes an abrupt change in frequency or intensity. Second, a constant sound emanating from a single source will be phase-delayed to the far ear. For example, at the 90-degree azimuth, a pure tone of 500 Hz will be shifted in phase by about 120 degrees, and a tone of 740 Hz by about 180 degrees. Thus, for a steady, complex sound, different frequency components will be shifted in phase by different amounts even though the interaural time delay is constant. Interaural time delays are significant only at lower frequencies (less than 2 kHz).

Interaural sound pressure (or intensity) differences occur at higher sound frequencies because part of the sound energy is reflected when its wavelength is short compared to the size of the head; that is, the head is said to cast a "sound shadow." Therefore, when the sound source is displaced to one side of the head, the sound intensity at the far ear is less than that at the near ear. In other words, the spectral properties of complex sounds will differ at the two ears. In humans, interaural intensity differences can be as much as 20 dB at frequencies above 5 kHz. Below about 1 kHz, interaural intensity differences are quite small or nonexistent.

Thus, interaural time differences provide effective spatial information about low frequency sound components, whereas interaural intensity differences provide information about high frequency components. This dichotomy is called the *duplex theory* of sound localization.

The timing or intensity of sounds at the two ears depend not only on the location of the sound source but also on the position of the head and pinna. Thus, the activity of neurons associated with the neck and pinna musculature ultimately must influence the responses of neurons with binaural sensitivities.

SUPERIOR OLIVARY COMPLEX

The superior olivary complex consists of the medial and lateral superior olivary nuclei and the medial nucleus of the trapezoid body. The more diffusely organized periolivary nuclei are poorly understood and will not be considered in this chapter. Since the superior olivary nuclei are the first regions of the central auditory pathways to receive substantial input from both ears, they are thought to play an important role in sound localization and binaural processing. Both their anatomical organization and the response characteristics of their neurons are consistent with this suggestion.

Medial Superior Olive

The neurons of the medial superior olive lie in a thin sheet that is oriented dorsoventrally and rostrocaudally in the ventral brain stem. In cross section, this nucleus appears as a stack of loosely aligned cells (Fig. 18–9C). The major cell type has a bipolar dendritic tree; one dendrite extends toward the midline, and the other extends toward the lateral surface of the brain stem. Axons from the ipsilateral anteroventral cochlear nucleus innervate the lateral dendrites, whereas axons from the contralateral anteroventral cochlear nucleus cross the midline in the trapezoid body to terminate on the medial dendrites (Fig. 18–9A). In the mammal, it is not known whether a neuron in the anteroventral cochlear nucleus sends a collateral to each medial superior olive, but such a pattern does exist in the bird.[51]

Binaurally responsive cells are designated by two letters according to the sign of their responses (excitatory, E; inhibitory, I). The first letter is the sign of the response produced by stimulation of the contralateral ear, and the second is the sign produced by stimulation of the ipsilateral ear. Thus, an IE cell is inhibited by the contralateral ear and excited by the ipsilateral ear. Most cells in the medial superior olive are of the EE type. Their excitatory best frequency is the same for stimulation of either ear.

When tested with monaural stimuli, neurons in the medial olivary nucleus have primary-like responses similar to those in the anteroventral cochlear nucleus. Binaural stimulation, however, reveals that they also are sensitive to differences in the time at which sound arrives at the two ears. Figure 18–10 shows how the average discharge rate of a cell in the medial superior olive varies when a pure tone at the cell's best frequency is first presented to one ear and then is delayed by varying amounts before being delivered to the other ear. The highest discharge rate for this cell

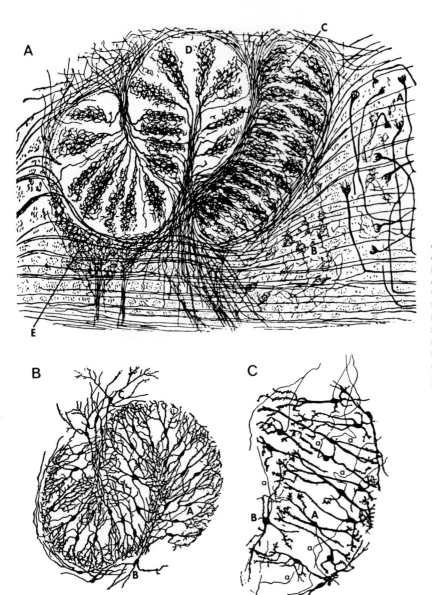

Figure 18–9 Ramon y Cajal's (1909) classical Golgi illustrations of the afferent axonal plexus *(A)*, and of cell types in lateral *(B)* and medial *(C)* superior olivary nuclei of young kittens. Abbreviations in *A*: A = medial nucleus of trapezoid body (note calyces of Held); B = periolivary cell region; C = medial superior olivary nucleus; D = lateral superior olivary nucleus; E = lateral nucleus of trapezoid body; F = fibers of trapezoid body. In *B* and *C*, capital letters identify different cell types and 'a' identifies axons.

resulted when the stimulus to one ear occurred about 0.5 ms or 1.5 ms later than the stimulus to the other. Delaying one stimulus was more effective than presenting both simultaneously or presenting the stimulus monaurally, even at an increased sound intensity. Finally, when the two ears were stimulated at the least favorable delay, the response was *less* than when the contralateral ear was stimulated alone. Different cells are responsive to different *characteristic delays* between stimuli presented to the two ears. Since an auditory stimulus reaches the two ears at different times whenever a sound comes from anywhere off the midline plane (i.e., off the 0-degree azimuth), neurons with different characteristic delays are thought to code the position of sound. When the temporal difference in binaural stimuli matches the characteristic delay of a neuron in the medial superior olive, a large response will occur; other neurons with different characteristic delays will be less active or inhibited.

Since we use interaural time differences to localize low frequencies, neurons in the medial superior olive seem uniquely suited to participate in low frequency sound localization. Generally, animals that have a low auditory frequency range

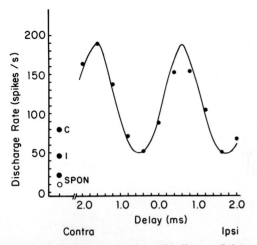

Figure 18–10 Discharge rate of a cell in the medial superior olive of the dog as a function of the interaural delay of a 444.5 Hz stimulus presented to both ears at 70 dB SPL. Points at the left indicate response to monaural contralateral stimulation (C), monaural ipsilateral stimulation (I), and no stimulation (SPON). Note that the unit's discharge rate is a periodic function of the interaural delay, with the maximum rate occurring when the stimulus to the ipsilateral ear is delayed approximately 600 µs or that to the contralateral ear is delayed about 1.5 ms. (After Goldberg, J. M.; Brown, P. B. *J. Neurophysiol.* 32:613–636, 1969.)

have a large medial superior olive, whereas those that hear predominantly high frequencies have a small one.

The avian nucleus laminaris, which is believed to be the homologue of the mammalian medial superior olive, has a unique anatomical organization that apparently causes interaural time differ-

ences to be represented topographically. As shown in Figure 18–11, the nucleus laminaris is a sheet of neurons, each of which has a distinct bipolar dendritic configuration; one dendritic tree extends dorsally and one ventrally. The dorsal dendrites receive input from the ipsilateral ear, and the ventral dendrites receive input from the contralateral ear. As seen in Figure 18–11, the ipsilateral input reaches the dorsal dendrites across the sheet of cells at the same time (all ipsilateral axons from the nucleus magnocellularis [NM] have the same length). In contrast, an axon from the contralateral NM innervates the ventral dendrites sequentially via short axon collaterals, such that the most lateral cell in the nucleus laminaris will be excited approximately 30 µs after the most medial cell. This anatomical arrangement of afferents could provide the substrate for differential stimulation of nucleus laminaris cells as a function of the position of a sound source in the contralateral hemifield. When a sound emanates from the midline, medial cells in the nucleus laminaris receive coincident dorsal and ventral activation. When a sound is moved to the 90-degree azimuth, the delay in the sound waves reaching the ear farthest from the sound source will be matched by the delay of action potentials from the ear closest to the sound source. Under these circumstances, coincident activation will occur only on the lateral neurons of the nucleus laminaris contralateral to the sound source. Thus, each sound location will produce a coincident arrival of action potentials to the dorsal and ventral processes at only one location in the contralateral nucleus laminaris.

Figure 18–11 Schematic diagram of primary auditory pathways in the avian brain stem. Auditory nerve fibers (VIIIn) terminate in the nucleus angularis (NA) and nucleus magnocellularis (NM). Axons from NM, the avian homologue of the mammalian anteroventral cochlear nucleus, terminate bilaterally in the nucleus laminaris (NL). Note that the terminal arborizations of these axons on the two sides of the brain differ. The ipsilateral axon is the same length across the mediolateral extent of NL while the contralateral axon forms a serial delay-line from medial to lateral.

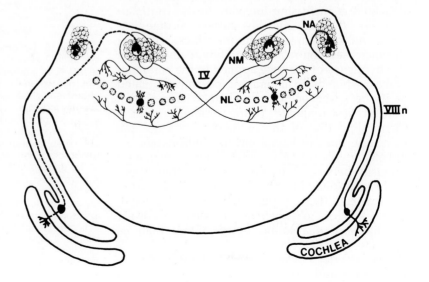

Lateral Superior Olive and Medial Nucleus of the Trapezoid Body

We will consider these two nuclei together because they appear to constitute one functional system in the mammalian brain.

The medial nucleus of the trapezoid body serves primarily as a high-fidelity relay nucleus between the contralateral anteroventral cochlear nucleus and the lateral superior olive. The response properties of its neurons seem virtually identical to those of anteroventral cochlear nucleus neurons; that is, they have V-shaped frequency tuning curves and primary-like or chopper type PSTHs (recall Fig. 18–7). This dominance of the cochlear nucleus input on responses in the medial nucleus of the trapezoid body is probably due to the fact that axons from the anteroventral cochlear nucleus are of extremely large diameter (i.e., rapidly conducting) and terminate in huge calyces that envelope the postsynaptic neuron (see Fig. 18–9A).

The postsynaptic neurons in the medial nucleus of the trapezoid body, in turn, send their axons to the ipsilateral lateral superior olive, where they make inhibitory synapses using glycine as the neurotransmitter.[44] In addition to this inhibitory input from the contralateral ear, most cells of the lateral superior olive also receive an excitatory input from the same frequency region of the ipsilateral anteroventral cochlear nucleus. Therefore, lateral superior olive neurons are of the IE type. Neurons in the lateral superior olive are extremely sensitive to the intensity differences of sounds presented to the two ears. Figure 18–12A shows that the response of an IE cell increases dramatically as ipsilateral sound intensity is increased (0 curve). When tones of the same frequency are delivered at increasing intensity to the contralateral ear, the rate-intensity curves become less steep and reach lower maximum values, reflecting the increased contralateral inhibition. By varying the frequency of the tone applied to the contralateral ear while the ipsilateral ear is being stimulated at its best frequency, investigators have shown that the best frequency of the inhibitory input is identical to that of the excitatory.[3, 38] Since the discharge rates of lateral olivary neurons are exquisitely sensitive to small differences in sound intensity between the two ears but rather insensitive to absolute intensities of sound delivered to the two ears (Fig. 18–12B), the discharge rates of lateral superior olivary neurons seem to encode the location of a sound source rather than its absolute intensity.

The superior olivary complex seems to be the level of the auditory system at which information

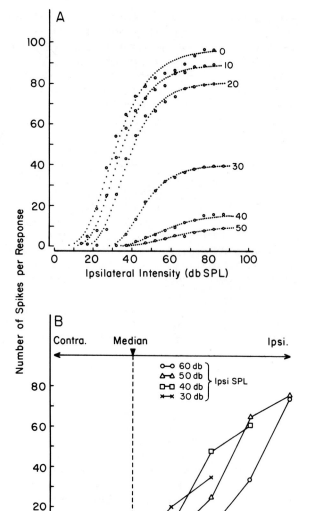

Figure 18–12 Family of binaural intensity functions for an IE cell in the lateral superior olivary nucleus of the cat showing sensitivity to interaural intensity differences. *A,* Each function was generated by presenting a contralateral tone at the characteristic frequency of the neuron (31.0 kHz) at the fixed intensity indicated and then varying the intensity of the simultaneously presented ipsilateral tone of the same frequency. Function marked 0 is for monaural ipsilateral stimulation. For any given level of stimulus to the ipsilateral (excitatory) ear (on abscissa), the response declines with increases in the level of the stimulus to the contralateral (inhibitory) ear. *B,* Data in *A* have been replotted to show variation in response as a function of interaural intensity difference (contralateral intensity relative to ipsilateral intensity) at different levels of the ipsilateral stimulus. The horizontal line at the top of the figure indicates the broad azimuthal ranges corresponding to the interaural intensity differences. (Data from Boudreau, J. C.; Tsuchitani, C. In Neff, W. D., ed. *Contributions to Sensory Physiology,* Vol. 4, pp. 143–213. New York, Academic Press, 1970.)

is distributed to the contralateral side of the brain. In this respect, the superior olivary complex functions somewhat like the optic chiasm.[7] Lesions of the superior olive entirely eliminate the ability to localize sounds, while lesions above this level (lateral lemniscus, inferior colliculus, medial geniculate or auditory cortex) disrupt sound localization only in the opposite hemifield.[16]

NUCLEI OF THE LATERAL LEMNISCUS

The dorsal and ventral nuclei of the lateral lemniscus are embedded in the ascending and descending axons of the lateral lemniscus. These nuclei, which are not well studied,[14] belong to two different ascending pathways. The major sources of afferent input to the dorsal nucleus are the nuclei belonging to the ventral binaural system, including the ipsilateral medial superior olive, both lateral superior olives, and the contralateral anteroventral cochlear nucleus. The response properties of dorsal nucleus cells reflect all of these inputs. In addition, they respond best to binaural stimuli. Ventral nucleus cells, on the other hand, receive their major afferent input from the contralateral ventral cochlear nucleus and respond accordingly. The dorsal and ventral nuclei project to many common nuclei, including the medial geniculate, the midbrain reticular formation, and the superior colliculus. One difference is that the dorsal nucleus projects to both inferior colliculi, whereas the ventral nucleus does not project to the contralateral inferior colliculus.

INFERIOR COLLICULUS

The inferior colliculus serves as the major integration and transmission relay for information ascending in the central auditory pathways. Because it is more accessible than other brain-stem auditory nuclei, it has received the most experimental attention. This attention, however, has revealed it to be an immensely complicated area, which has been divided into 10 or more subdivisions on the basis of morphological criteria.[26] Here, we distinguish just the three traditional subdivisions and consider only the physiological properties of cells in the largest subdivision, the *central nucleus*. These cells have sharp tuning curves, display increasing responses over a range of 20–40 dB above threshold, and show relatively poor phase locking. However, there is considerable variability. For example, some cells exhibit very broad tuning curves while others have unusually sharp tuning curves.

Most cells in the central nucleus of a variety of species are excited by contralateral stimuli. In the best studied animal, the cat, about 25% of the cells respond exclusively to stimuli delivered to the contralateral ear, and the other 75% are binaural. Of the latter, the majority with best frequencies above 3 kHz are excited by the contralateral ear and inhibited by the ipsilateral ear (EI); the majority with low best frequencies are excited by both ears (EE).[14] The relative proportions of the two cell types vary widely between species, according to the frequency range of their hearing. Thus, most rodents have a higher proportion of EI units, whereas the kangaroo rat, which has excellent low-frequency hearing, has a higher proportion of EE units.

Two kinds of evidence indicate that EE cells code the absolute temporal delay of stimuli reaching the two ears rather than the phase relationship between stimuli. First, the characteristic delay is independent of frequency.[48] Second, a characteristic delay can be demonstrated by adjusting the delay between noise bursts presented to the two ears; in this situation the characteristic delay reflects the delay between noise bursts rather than the relative phase shift, which varies as a function of frequency.[14] EE neurons respond best when stimulation of the ipsilateral ear lags that of the contralateral ear, which is the situation that occurs when a sound is in the contralateral hemifield. These neurons respond not only to the initial time disparity produced by the onset of sound but also to ongoing time disparities produced by a continuous sound located on one side of the head.

EI cells in the inferior colliculus behave like IE cells in the lateral superior olive (LSO). That is, almost all EI cells in the inferior colliculus are completely suppressed if the ipsilateral stimulus is of equal or greater intensity than the contralateral stimulus. Therefore, like the EE cells discussed above, EI cells respond when a sound source is in the contralateral hemifield, where it is louder to the contralateral ear.

To localize the rustling sound of its prey (e.g., mice) while it is perched or in midflight, the barn owl uses both interaural time and intensity differences.[20] The inferior colliculus of the barn owl contains neurons that respond selectively to stimuli emanating from specific spatial locations in the contralateral hemifield and are relatively unresponsive to signals from any other location (Fig. 18–13). These space-specific neurons respond only when both the interaural time and interaural in-

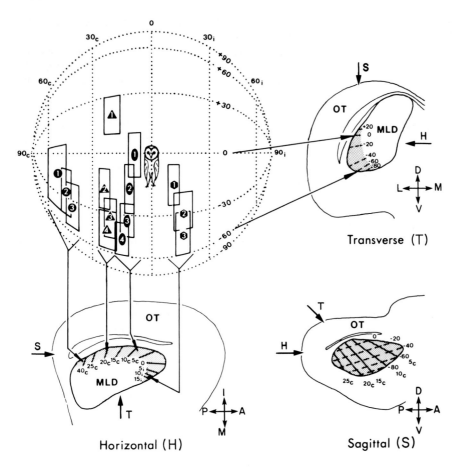

Figure 18–13 A neural map of auditory space in the inferior colliculus of the barn owl. The coordinates of auditory space are depicted as a globe surrounding the owl. Projected onto the globe are the best spatial locations for auditory stimuli (solid-line rectangles) for 14 units recorded in four separate penetrations. Neurons identified by the same symbol (diamond, hexagon, etc.) were recorded on the same penetration in an order indicated by the numbers. Locations of the four penetrations are indicated in the horizontal section by solid arrows. The transverse, horizontal, and sagittal sections through MLD (the avian inferior colliculus) illustrate the spatial map (stippled portion) topography. Isoazimuth contours, based on field centers, are shown as solid lines; isoelevation contours are represented by dashed lines. On each section, arrows identified as S, H, or T indicate planes of the other two sections. Solid, crossed arrows to the lower right of each section indicate directions: A = anterior; D = dorsal; L = lateral; M = medial; P = posterior; V = ventral; OT = optic tectum. (From Knudsen, E. I.; Konishi, M. *Science* 200:795–797, Copyright 1978 by the AAAS.)

Transverse (T)

Horizontal (H)

Sagittal (S)

tensity differences are appropriate for the particular location. Space-specific neurons are distributed so that the inferior colliculus has a topographical representation of all positions in the owl's auditory space. Successive cells encountered on an electrode penetration prefer progressively adjacent positions in auditory space (see Fig. 18–13). As a stimulus is moved from the midline plane to the 90-degree azimuth, neurons at progressively more posterior positions in the inferior colliculus are activated, and as the stimulus is elevated, progressively more dorsal neurons are excited.

The inferior colliculus is the main source of sensory input to the auditory forebrain, which is thought to be responsible for the conscious perception of sounds. The auditory forebrain is composed of thalamic regions receiving auditory input and the cortical areas with which they are reciprocally connected. This organization has been investigated most thoroughly in the cat.

In addition to serving as the major hub for ascending and descending auditory information, the inferior col-

liculus also delivers auditory information to structures outside of the main auditory pathways. Such structures include the deep layers of the superior colliculus, which also receive directional information from the visual system (see Chap. 20), and the pontine nuclei, which transmit auditory information to the cerebellum.

MEDIAL GENICULATE BODY

The medial geniculate body is the major thalamic relay for ascending auditory information (recall Fig. 18–2). It is usually subdivided into three main divisions (ventral, medial, and dorsal), each of which has a different cytoarchitecture and different auditory and nonauditory inputs.[46]

The *ventral division*, which is the most thoroughly studied division, has neuronal discharge characteristics that are similar to those of the auditory nuclei that have already been discussed.[5] Most neurons have sharp tuning curves, firing rates that increase monotonically with increasing stimulus intensity, and binaural inputs (either EE or EI). Most respond primarily to the onset of a pure tone, which suggests that they are more sensitive to dynamic than static acoustic information. Therefore, the ventral division faithfully transmits the frequency, intensity, and binaural characteristics of the acoustic stimulation.

A strict tonotopic organization is also preserved in the ventral division, which seems to result from the laminar organization of its cells and its dendrites, as shown in Figure 18–14. Morest[24] suggested that each lamina represented an isofrequency band whose elements exhibit little overlap with neighboring bands. Single neuron recording studies have confirmed this suggestion. The major ascending input to the ventral nucleus is from the central nucleus of the ipsilateral superior colliculus.

The *dorsal division* is a heterogeneous group of as many as 10 subnuclei with vague boundaries. The response patterns of its neurons are equally

Figure 18–14 A transverse section of the ventral division of the cat medial geniculate. The section was stained by the Golgi method to show cell bodies and dendrites. Note the laminar appearance with the dendrites oriented parallel to the lamina. The lamina form isofrequency sheets, and the tonotopic organization is perpendicular to the lamina. (From Morest, D. K. *J. Anat.* 98:611–630, 1964, with permission of Cambridge University Press.)

diverse. Although many can be driven by acoustic stimuli, their responses tend to be more irregular, their latencies more variable, and their tuning curves broader than those of neurons in the ventral division.[5] A consequence of the broad tuning is that the tonotopic organization is imprecise. Nevertheless, a complete frequency representation is repeated several times within this division. The activity of many cells are inhibited only by auditory stimulation. Also, the activity of some cells is influenced by other sensory modalities, particularly somesthetic. Their variable auditory responses and responses to other sensory modalities reflect the fact that inputs to these cells originate from the area surrounding the central nucleus of the inferior colliculus and from nonauditory areas of the midbrain tegmentum and cortex.

The auditory responses of neurons in the *medial division* are also diverse. Response latencies range from 7 to 70 ms, both broad and sharp tuning can be found, and all combinations of binaural interactions are observed. Many neurons have two distinguishing characteristics: (1) they are maximally excited by sounds of a particular intensity and suppressed by more intense sounds, and (2) they continue to respond for the duration of an auditory stimulus rather than only to its onset.[5] These characteristics suggest that neurons of the medial division signal the relative intensity and duration of an auditory stimulus rather than its spectral and temporal content. Finally, many medial division neurons are polysensory.

The medial division receives its auditory input from regions surrounding the superior olivary nucleus or the lateral lemniscal nuclei. It also receives a variety of other sensory and nonsensory inputs (for example, from the spinothalamic tract, vestibular nuclei, superior colliculus, and thalamic reticular nuclei).[46]

AUDITORY CORTEX

Woolsey[47] and his colleagues first mapped the auditory areas in the cat cerebral cortex by describing the cortical areas where electrical stimulation of the auditory nerve produced evoked potentials (a similar strategy is described for the visual system; see Chap. 20). Figure 18–15A shows Woolsey's summary, indicating five tonotopically organized cortical regions (AI, AII, SF, EP, and Ins) with other "auditory responsive" areas surrounding them. The tonotopic organization of each subdivision is indicated grossly by B, for the representation of the basal (high-frequency) cochlea, and A, for the apical (low-frequency) region. Intermediate frequencies are also represented. More recent studies based on single unit data and anatomical connections (Fig. 18–15 B, C) suggest four tonotopically organized cortical auditory areas (A, AI, P, and VP), again surrounded by a belt of "auditory responsive cortex."

The auditory cortex in primates, although not as well studied, has essentially similar features.[4]

The primary auditory cortex is buried in the depths of the Sylvian fissure on the superior aspect of the temporal gyrus. It has a tonotopic organization with low frequencies represented rostrally, and high frequencies caudally. The auditory cortex of humans can be distinguished by its densely packed, small cells in layer IV. Both clinical observations and modern imaging techniques indicate that high frequencies are represented deep in the Sylvian fissure, whereas lower frequencies are located more toward the top of the temporal gyrus.[4, 32]

When microelectrodes are driven through the cat auditory cortex perpendicular to its surface, each neuron encountered has a similar best frequency. Thus, like other sensory cortices (see Chaps. 15 and 20), there is a columnar organization of certain stimulus attributes, in this case frequency. If the microelectrode is driven perpendicular to the frequency columns and parallel to the cortical surface, the best frequencies either increase or decrease progressively until the boundary between one cortical area and another is reached. At the boundary, the orderly progression stops and may reverse. If the electrode moves along a frequency column but parallel to the cortical surface, similar best frequencies are again recorded, indicating that a band of cortex oriented perpendicular to the tonotopic progression receives input from the same place on the cochlea; i.e., it is an isofrequency band.

Connections of Auditory Cortex

The organization of the auditory areas of the cat cortex has been extensively studied by a combination of microelectrode recording and anatomical techniques.[12, 13, 22, 23, 46] Although cortical organization may be somewhat different in primates, the

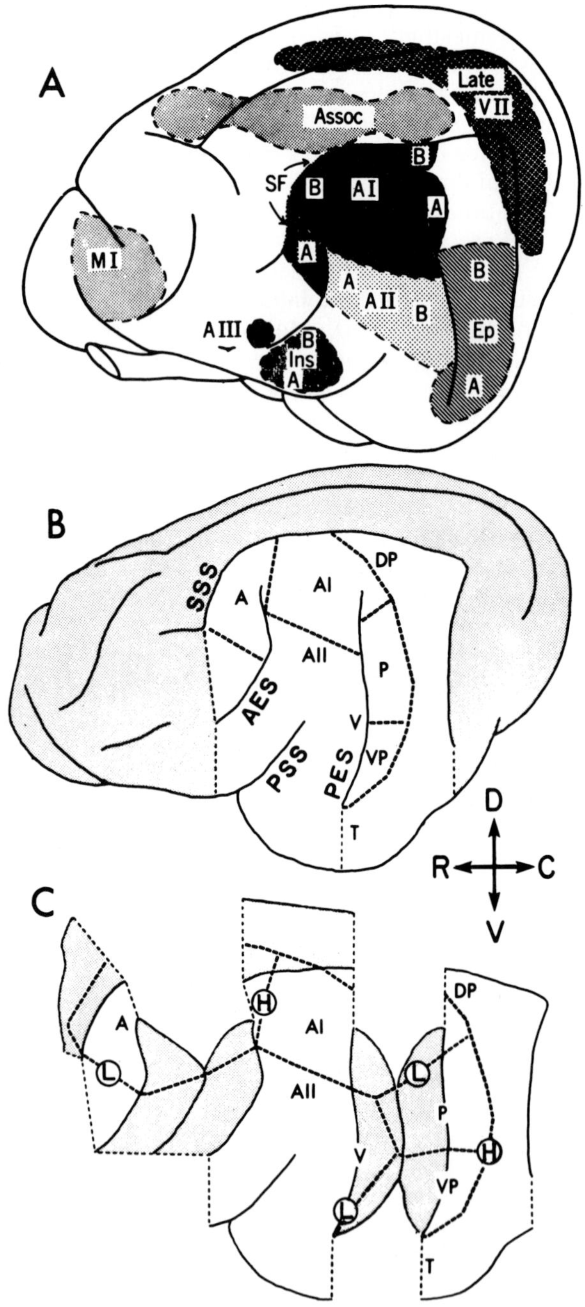

Figure 18–15 Auditory cortical fields in the left hemisphere of the cat. *A,* Summary diagram showing five tonotopically organized cortical regions: SF, AI, AII, EP, and Ins. The tonotopic organization of each area is indicated by A, representing apex (low frequency) of the cochlea, and B, representing base (high frequency) of the cochlea. Other regions that respond to sound but are not tonotopically organized are also shown. *B* and *C* show more recent parcellation of tonotopic cortical regions based on microelectrode recording and anatomical connections. Four tonotopic regions are defined: A, A1, VP, and P. In *C* the sulcus and the tonotopic organization of each area are shown, from high (H) to low (L) frequency. (From Imig, T. J.; Reale, R. A.; Brugge, J. F.; Morel, A.; Adrian, H. O. In Leporé, F.; Ptito, M.; and Jasper, H. H., eds. In *Two Hemispheres—One Brain: Functions of the Corpus Callosum,* pp. 103–115. New York, A. R. Liss, 1986.)

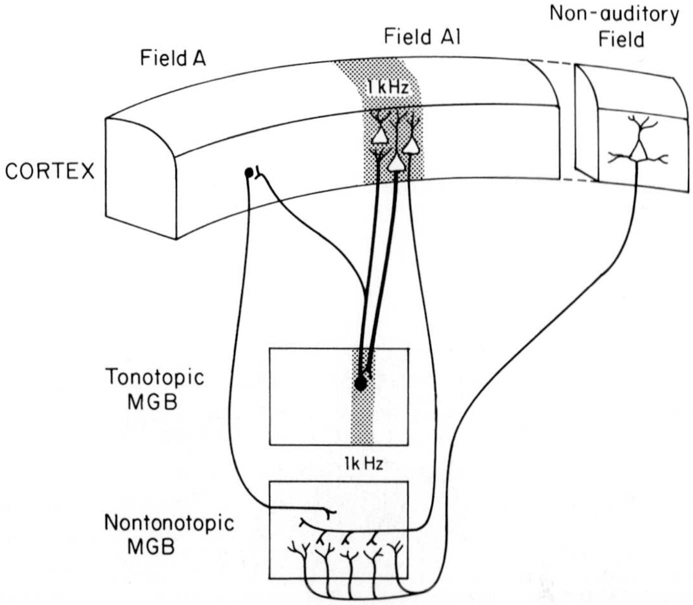

Figure 18–16 Schematic diagram showing patterns of connections between cortical auditory fields and both tonotopically and nontonotopically organized areas of the medial geniculate nucleus (MGB). Note the heavy reciprocal point-to-point connections between A1 and the same frequency region of the tonotopic MGB, the sparser projection from the tonotopic MGB to field A, and the diffuse projections of A, A1, and the non-auditory field to the nontonotopic MGB.

main principles are similar. In the cat, each of the four tonotopic cortical areas receives a major input from a subregion of the tonotopically organized ventral division of the medial geniculate body. This input presumably confers the tonotopic organization on the cortical areas. Each cortical area, in turn, sends a point-to-point projection back to the same thalamic region from which it received its frequency-specific innervation. These connections, therefore, form a frequency-specific geniculocorticogeniculate loop. In addition to its major cortical projection, each tonotopic region of the thalamus sends a minor projection to one or more of the other cortical auditory areas. Each cortical area also is reciprocally connected with one or more nontonotopic areas in the medial or dorsal division of the medial geniculate body. Figure 18–16 illustrates these complex feed-forward and feedback connections. For example, a 1-kHz region of the cortical field A1 makes a point-to-point reciprocal connection with a 1-kHz area in the ventral division of the medial geniculate body, which also has a minor projection to field A. The nontonotopic areas of the thalamus also project to the A1 cortical fields, but this projection is more diffuse, as is the reciprocal corticothalamic projection. Each cortical region also connects reciprocally with its homotopic area in the opposite hemisphere, with one or two other areas of the opposite hemisphere, and with all the other tonotopic areas in the same hemisphere. Like the thalamic connections, one or two of these connections are very strong and precisely tonotopic while others are more sparse and diffuse.

Recently, the functional organization of auditory cortex has been further elucidated by the discovery of "binaural bands" that extend parallel to the tonotopic gradient (perpendicular to each isofrequency band). If an electrode penetration is run perpendicular to an isofrequency band, the response of each neuron to contralateral ear stimulation may be enhanced by stimulation of the ipsilateral ear. If an electrode penetration is run adjacent to the first track, the responses of these cells to contralateral ear stimulation are suppressed by ipsilateral ear stimuli (Fig. 18–17). These *summation* and *suppression bands* are reminiscent of the ocular dominance columns in the visual cortex (Chap. 20). Concomitant anatomical studies show that cells in bands that receive dense projections from the opposite cortex display summation and that cells in interposed bands, which receive projections from the same hemisphere, show binaural suppression (see Fig. 18–17).

In summary, the anatomical and physiological organization of the cat auditory cortex suggests that multiple pathways emerge from separate areas of the thalamus to reciprocally innervate separate regions of their auditory cortex, and that each pathway processes different types of information. However, although each pathway has its own unique, dominant connections, each also communicates with the others.

Organization of Bat Auditory Cortex

The idea of separate cortical areas, each processing different functionally related signals, has

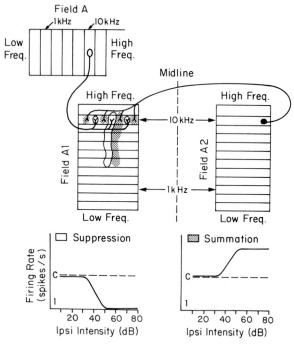

Figure 18–17 Schematic representation of cortico-cortical connections thought to be responsible for binaural summation (shaded) and binaural suppression (unshaded) bands in cerebral cortex. The 10-kHz band of field A1 on the left side receives alternating projections from the 10-kHz region of field A1 on the right side and the 10-kHz region of the ipsilateral field A. This pattern is repeated for each frequency (not shown here). The two lowest graphs show the responses of neurons in the shaded (right) and unshaded (left) regions. In each graph the response to contralateral stimulation alone is indicated by C; ipsilateral stimulation alone elicits no excitatory response. When the contralateral stimulus is paired with an ipsilateral stimulus, the response depends on the intensity of the ipsilateral stimulus. Increasing ipsilateral stimulus intensity causes a decreased response for "suppression" neurons in the unshaded regions and an increased response for "summation" neurons in the shaded regions.

been best revealed by making use of the unique prey-locating system of the mustache bat.[39, 40] During flight, the mustache bat emits a series of highly directional, ultrahigh-frequency, echo locating pulses. Each pulse is composed of harmonically related constant frequency portions followed by a short, frequency-modulated (FM) component. The second harmonic at 61 to 63 kHz is the most intense part of the vocalization. Obstacles or prey (flying insects) in the path of the bat cause a reflection (echo) of the biosonar pulse. This echo has several interesting properties, which the bat is able to detect. For example, the delay between the pulse and the echo is proportional to the distance of the object. The echo also contains

slightly different frequency components than the vocalizations because of a Doppler shift caused by the bat and target (e.g., insect) moving at different velocities. If the bat is closing in on his target, the echos will contain slightly higher frequencies than the pulse; if he is flying away from the target, the echos will be at slightly lower frequencies.

The mustache bat's auditory system has several specializations that allow it to optimize its tracking performance. Although its total frequency range is from around 7 to 120 kHz, about 30% of its primary auditory cortex is tuned to the narrow band of sounds between 61 and 63 kHz, the range that is crucial for the detection of the most intense part of its vocalizations. Second, another cortical area contains neurons whose responses depend on the precise frequency difference of the pulse and the echo (i.e., on the Doppler shift). Neurons sensitive to different Doppler shifts are topographically organized so as to represent systematic changes in the relative velocities of the bat and its target. Finally, neurons in a third area are differentially sensitive to the delay between the FM portions of the pulse and echo. These neurons respond poorly to the FM portions of either signal alone, but when the FM portion of the pulse is paired with the FM part of the echo *at a particular delay*, a vigorous response occurs. Thus, this area seems to contain neurons that respond to targets at specific distances.

Based on their work in the bat, Suga and his colleagues[39, 40] have proposed four rules for cortical processing: (1) cortical neurons are specialized for processing different parameters of a complex acoustic stimulus; (2) neurons specialized for processing different parameters are clustered in separate areas; (3) within the clusters, different neurons respond best to different values of the parameter and are arranged topographically according to systematic variations in the parameter; and (4) the components of an auditory signal that are most important to the organism have an expanded cortical representation. It is not yet known whether these rules provide general principles for auditory cortical organization in other mammals.

Lesion Studies of Auditory Cortex

Over the past 50 years, many investigators have attempted to determine the role of auditory cortex by removing it and examining the resulting deficits in behavior. This approach has allowed us to draw important conclusions about the functional importance of different regions. However, it must be remembered that all ablation studies are plagued by certain shortcomings. First, studies that destroy part of the brain test the capabilities of an animal *without* that brain area rather than the function of that area per se. Second, the deficit could be due to an absence of appropriate sensory integration but might equally well be due to deficits in sensory-motor integration, motivational factors, or purely motor capabilities. Whereas the latter deficit is relatively easy to demonstrate, the others are more difficult to dissociate from purely sensory dysfunction. Third, lesions of a given brain area produce effects beyond the tissue that was damaged. Other areas of the brain receiving direct, secondary, or tertiary connections from the lesioned area also will be affected. For example, regions supplying afferents to the damaged area will undergo some degree of *retrograde degeneration*, and the targets of the lesioned area may undergo *anterograde transneuronal atrophy* or cell loss. In addition, sprouting of terminals from other afferents to the damaged area may occur. Finally, it is difficult to remove specific auditory cortical regions without damaging neighboring ones. In view of these caveats, behavioral deficits produced by lesions serve only to implicate a region in the regulation of the affected behavior. Nevertheless, it is of interest to examine those behaviors that are disrupted by lesions of auditory cortex.

In general, after removal of the entire auditory cortex, animals can still learn simple frequency and intensity discriminations but are unable to identify the position of a particular stimulus within a more complex pattern (e.g., distinguish a tone burst pattern of high-low-high frequencies from one containing low-high-low frequencies). Cortical lesions also disrupt tasks that require temporal discriminations such as determining the number of tone bursts in a series or distinguishing acoustic stimuli of barely different durations.[27] Finally, cortical lesions affect behaviors based on stimuli delivered to the contralateral ear more than those to the ipsilateral ear. Although humans who have suffered unilateral cortical damage exhibit some improvement with time after the lesion, a moderate hearing loss in the opposite ear remains. In monkeys, bilateral lesions of the auditory cortex also produce a persistent 30- to 40-dB hearing loss of the middle frequencies.[11]

The anatomical and physiological data presented above suggest that the primary auditory cortex is involved in the localization of sound in the opposite hemifield. This suspicion has recently been confirmed by experiments in which cats were trained to localize the precise position of sound emanating from one of seven speakers positioned within ± 90 degrees of the midline plane.[16, 17] Unilateral lesions of the entire auditory cortex permanently disrupted the cat's ability to choose the correct speaker in the contralateral hemifield, although localization in the ipsilateral hemifield was unaffected. Furthermore, cats with removals of A1 cortex along isofrequency strips (for example, between 6 and 12 kHz) could not localize stimuli at those frequencies in the contralateral hemifield, whereas sound localization was normal at other frequencies. These experiments reveal that the ability to accurately localize sound requires an intact primary auditory cortex on the opposite side of the brain, and that this involvement is frequency specific. Sound location seems to be independently represented within each frequency representation of primary auditory cortex.

Other areas of the auditory cortex and surrounding auditory association areas have received relatively little attention. Physiological investigations reveal a variety of responses to acoustic as well as to other sensory stimuli, but these responses often are dramatically influenced by anesthesia and other "state" variables. Behavioral studies have also failed to yield a coherent picture of functional roles for these regions.

MOTOR CONSEQUENCES OF AUDITORY STIMULI: AUDITORY REFLEXES

Acoustic Middle Ear Reflex

As discussed in Chapter 17, contraction of the two muscles in the middle ear, the stapedius and the tensor tympani, limits movement of the tympanic membrane and the stapes, thereby damping the transmission of sound by the inner ear. This *middle ear reflex* protects the inner ear from intense sounds and attenuates loud sounds to improve discrimination. In addition, contraction of the middle ear muscles occurs prior to vocalizations in humans, probably to attenuate the low-frequency components of our own vocalizations so that our own speech is more intelligible. Contraction of the middle ear muscles is always bilateral and can be evoked by acoustic stimulation of either ear. Thus, in its bilateral action, the acoustic middle ear reflex resembles the pupillary light reflex (see Chap. 19). Finally, this reflex is useful clinically to separate peripheral from central dysfunction.

The neuronal pathways involved in the middle ear reflex involve a three- or four-neuron circuit from the cochlea to the motor neurons in the facial and trigeminal nuclei, which innervate the stapedius and tensor tympani muscles, respectively. The principal components of this pathway are (1) primary auditory nerve fibers, (2) neurons in the ventral cochlear nucleus, (3) neurons in or around both superior olivary nuclei, and (4) the motor nuclei of the facial (VII) and trigeminal (V) nerves. In addition to these direct pathways, there are longer latency pathways involving brain-stem auditory nuclei and the reticular formation and/or the red nucleus, which, in turn, project to the motor nuclei of V and VII. The relative roles of the direct and indirect pathways are unknown.

Acoustic Startle Reflex

Another important auditory reflex is the acoustic startle response, which usually causes us to react to a loud sound. Activation of the startle reflex produces a short latency motor response that travels down the spinal cord, beginning with muscle contraction in the neck and shoulders and culminating with contraction of muscles in the distal segments of the forelimbs and legs. In rats, the reflex time from the onset of the acoustic stimulus to electromyographic responses in the hindlimbs is only 7 to 8 ms. The descending input, which elicits the motor response, involves projections from the primary auditory neurons to the ventral cochlear nucleus and then to the contralateral lateral lemniscal nuclei.[6] Neurons in or near the lateral lemniscal nuclei project into the caudal portion of the pontine reticular formation, which sends axons down the reticulospinal tract to innervate spinal interneurons and occasionally motor neurons directly. Like the middle ear reflex, therefore, the startle reflex can involve a direct pathway of as few as four or five neurons. At each level, from the cochlear nucleus to the motor neurons, the direct pathway can be influenced by inputs from other neural networks. Indeed, the startle reflex can be modified by a variety of factors such as learning, state of the animal, or other "competing" sensory events. The startle reflex also produces a chain of autonomic responses whose neural substrate is largely unknown.

DEVELOPMENT AND PLASTICITY OF THE CENTRAL AUDITORY SYSTEM

Several properties of the auditory system make it well suited for understanding the factors that regulate both normal neuronal development and neural plasticity. First, the coding properties of auditory neurons can be precisely and quantitatively described on the basis of sound frequency, sound intensity, and binaural interactions. Second, the cell bodies that represent successive levels of processing (e.g., cochlear nucleus, superior olive, inferior colliculus) are spatially segregated, making them individually accessible. Third, there is an extensive literature on the normal development of both peripheral and central structures with which to compare the results of experimental manipulations.[31, 35, 45]

The age at which individuals begin responding to sound varies considerably from species to species. Some animals, such as humans, guinea pigs, and precocial birds, begin responding in utero or in ovo. For example, the human fetus responds to loud sounds by the beginning of the third trimester (25–26 weeks). Other species, such as most carnivores, rodents, and altricial birds, do not respond until after birth. Mice and rats, for example, first respond about 10 to 12 days after birth. Electrical stimulation of the auditory nerve can evoke cortical responses before sound-elicited responses are obtained. Therefore, the onset of auditory function is dependent on peripheral maturation, rather than the establishment of CNS connections.

Very soon after central connections are established, peripheral maturation can be documented by several observations. First, the cochlear microphonic response can be elicited by very loud, relatively low-frequency stimuli. Within 1 to 2 days an eighth nerve compound action potential can be recorded. As the middle and inner ear mature, response thresholds recorded from the auditory nerve or cerebral cortex decrease, and responses can be obtained to a broader range of frequencies. Tuning curves become progressively sharper over this same period, indicating that the ontogeny of frequency selectivity is due primarily to peripheral maturation.

Subtle changes in physiological properties also can be attributed to CNS maturation. For example, response latency decreases with CNS myelination. Also, as synapses mature, neurons are better able to follow rapidly repeating stimuli. In addition, immature tuning curves recorded from CNS neurons are, in part, due to the maturation of CNS connectivity.[38] In contrast, the topographic representation of frequency is established prior to the age at which auditory responses can first be detected. For example, IE neurons in the neonatal gerbil lateral superior olive have matching excitatory and inhibitory best frequencies.[38]

It is clear that normal input from the inner ear is required for normal maturation at all levels of the auditory system. For example, Levi-Montalcini[21] showed that removal of the embryonic anlagen of the inner ear (the otocyst) in the chick results in severe cell loss in the maturing cochlear nuclei. This loss is correlated with the age at which auditory nerve activity first elicits postsynaptic responses in the CNS.[15] Similarly, removal of the cochlea or silencing the electrical activity of the auditory nerve of neonatal chicks can cause a large number of CNS changes, including cell death, decreased protein synthesis, cell atrophy, alterations of enzyme activity, and dendritic atrophy.[1, 37] Figure 18–18 shows examples of the trophic influence of the inner ear on cochlear

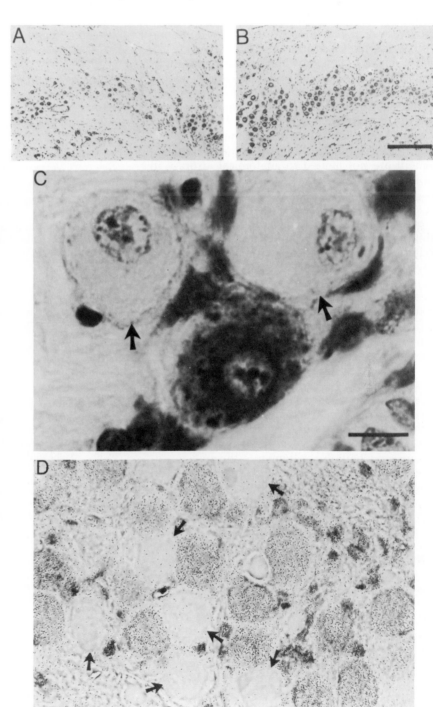

Figure 18–18 Photomicrographs showing rapid changes in the nucleus magnocellularis (NM) of the chick (avian homologue of the mammalian anteroventral cochlear nucleus; see Fig. 18–11) following destruction of the cochlea. A and B are low-magnification views of NM on the ipsilateral side (A) and contralateral side (B) of the brain just two days following the removal of the cochlea; approximately half of the ipsilateral neurons are gone or exhibit very pale staining. Scale bar = 0.2 mm. C, High-power view of the ipsilateral side showing that neurons in the process of transneuronal degeneration (arrows) have lost cytoplasmic staining for Nissl substance, indicating a severe depletion of cytoplasmic RNA. Scale bar = 10 mm. D, In this animal ³H-leucine was injected one half hour prior to death to study protein synthesis by NM neurons following elimination of eighth nerve activity three hours earlier. The cells without black silver grains (arrows) have already ceased protein synthesis and will degenerate. (A–C from Born, D. E.; Rubel, E. W. J. Comp. Neurol. 231:435–445, 1985. D Reprinted with permission from Born, D. E.; Rubel, E. W. J. Neurosci. 8:901–919, 1988 Pergamon Press plc.)

nucleus cells. The low-power photomicrographs (Fig. 18–18A,B) show considerable neuron loss in the cochlear nucleus (NM in Fig. 18–11) ipsilateral to inner ear removal in neonatal chicks. The cells that are destined to degenerate can be identified by loss of staining within 24 hours (Fig. 18–18C) and by reduced protein synthesis (cells without black grains over cytoplasm in Fig. 18–18D) within 3 to 6 hours after elimination of auditory nerve activity. Similar dramatic effects of peripheral manipulations on CNS auditory pathway development also have been shown in neonatal rats, mice, gerbils, and cats. In some cases, neonatal rearing with a conductive hearing loss of 30 to 40 dB produces atrophy of cells in the brain-stem auditory nuclei, presumably because of the reduction of afferent activity during development.[42]

Peripheral influences on auditory system development are not limited to anatomical atrophy in the brain-stem auditory nuclei. Changing the balance of inputs from the two ears can cause marked physiological alterations. For example, stimulation of the ipsilateral ear, which normally produces inhibition of inferior colliculus neurons, produces excitatory responses if the contralateral ear is removed soon after birth.[28]

These and other examples of peripheral influences on the structure and function of the auditory system are most clearly demonstrable in the immature organism. Similar manipulations on adult animals produce either less dramatic or no effects. There seems to be a restricted period during development, beginning at the time of synapse formation, when the establishment of normal mature central pathways depends critically on the amount or pattern of synaptic activity. An understanding of the biological basis of these critical or sensitive periods which also have been demonstrated for vision (see Chap. 20) will emerge through examining how the genome of the developing neuron is expressed and how expression can be modified by changes in the ionic fluxes that occur in the environment of developing neurons.

ANNOTATED BIBLIOGRAPHY

Irvine, D.R.F. *Progress in Sensory Physiology, Vol. 7: The Auditory Brainstem.* Berlin, Springer-Verlag, 1986.
  An up-to-date, comprehensive review of the structure, function, and connections of nuclei comprising the brain-stem auditory pathways.
Neff, W.D.; Diamond, I.T.; Casseday, J.H. Behavioral studies of auditory discrimination. In Keidel, W.D.; Neff, W.D., eds. *Auditory System*, pp. 307–400, vol. 5, pt. 3, *Handbook of Sensory Physiology.* New York, Springer-Verlag, 1975.
  An excellent review of the literature on the effects of cortical lesions on hearing.

Rubel, E.W. Ontogeny of structure and function in the vertebrate auditory system, pp. 135–237. In Jackson, M., ed., *Development of Sensory Systems.* vol. 9, *Handbook of Sensory Physiology.* New York, Springer-Verlag, 1978.
  A comprehensive review of the development of the ear and the auditory system.

REFERENCES

1. Born, D.E.; Rubel, E.W. Afferent influences on brain stem auditory nuclei of the chicken: neuron number and size following cochlea removal. *J. Comp. Neurol.* 231:435–445, 1985.
2. Born, D.E.; Rubel, E.W. Afferent influence on brain-stem auditory nuclei of the chicken: presynaptic action potentials regulate protein synthesis in nucleus magnocellularis neurons. *J. Neurosci.* 8:901–919, 1988.
3. Boudreau, J.C.; Tsuchitani, C. Cat superior olive S-segment cell discharge to tonal stimulation. In Neff, W.D., ed. *Contributions to Sensory Physiology,* Vol. 4, pp. 143–213. New York, Academic Press, 1970.
4. Brugge, J.F. Progress in neuroanatomy and neurophysiology of auditory cortex. In Tower, D., ed. *The Nervous System, Vol. 3. Human Communication and Its Disorders,* pp. 97–111. New York, Raven Press, 1975.
5. Calford, M.B. The parcellation of the medial geniculate body of the cat defined by the auditory response properties of single units. *J. Neurosci.* 3:2350–2364, 1983.
6. Davis, M.; Glendelman, D.S.; Tischler, M.D.; Glendelman, P.M. A primary acoustic startle circuit: lesion and stimulation studies. *J. Neurosci.* 2:791–805, 1982.
7. Glendenning K.K.; Hutson, K.A.; Nudo, R.J.; Masterton, R.B. Acoustic chiasm II: anatomical basis of binaurality in lateral superior olive of cat. *J. Comp. Neurol.* 232:261–285, 1985.
8. Goldberg, J.M.; Brown, P.B. Response of binaural neurons of dog superior olivary complex to dichotic tonal stimuli: some physiological mechanisms of sound localization. *J. Neurophysiol.* 32:613–636, 1969.
9. Guinan, J.J., Jr.; Warr, W.B.; Norris, B.E. Differential olivocochlear projections from lateral versus medial zones of the superior olivary complex. *J. Comp. Neurol.* 221:358–370, 1983.
10. Hackett, J.T.; Jackson, H.; Rubel, E.W. Synaptic excitation of the second and third order auditory neurons in the avian brain stem. *Neurosci.* 7:1455–1469, 1982.
11. Heffner, H.E.; Heffner, R.S. Hearing loss in Japanese Macaques following bilateral auditory cortex lesions. *J. Neurophysiol.* 55:256–271, 1986.
12. Imig, T.J.; Reale, R.A.; Brugge, J.F. The auditory cortex: patterns of corticocortical projections related to physiological maps in the cat. In Woolsey, C.N., ed. *Cortical Sensory Organization, Vol. 3, Multiple Auditory Areas,* pp. 1–41. Clifton, New Jersey, Humana Press, 1982.
13. Imig, T.J.; Reale, R.A.; Brugge, J.F.; Morel, A.; Adrian, H.O. Topography of cortico-cortical connections related to tonotopic and binaural maps of cat auditory cortex. In Leporé, F.; Ptito, M.; Jasper, H. H. eds. *Two Hemispheres—One Brain: Functions of the Corpus Callosum,* pp. 103–115. New York, A.R. Liss, 1986.
14. Irvine, D.R.F. *Progress in Sensory Physiology. Vol. 7: The Auditory Brain stem.* Berlin, Springer-Verlag, 1986.
15. Jackson, H.; Hackett, J.T.; Rubel, E.W. Organization and development of brain-stem auditory nuclei in the chick: ontogeny of postsynaptic responses. *J. Comp. Neurol.* 210:80–86, 1982.

16. Jenkins, W.M.; Masterton, R.B. Sound localization: effects of unilateral lesions in central auditory system. *J. Neurophysiol.* 47:987–1016, 1982.
17. Jenkins, W.M.; Merzenich, M.M. Role of cat primary auditory cortex in sound localization behaviour. *J. Neurophysiol.* 52:819–847, 1984.
18. Kiang, N.Y.-S. Stimulus representation in the discharge patterns of auditory neurons. In Tower, D., ed. *The Nervous System, Vol. 3. Human Communication and Its Disorders*, pp. 81–96. New York, Raven Press, 1975.
19. Knudsen, E.I.; Konishi, M. A neural map of auditory space in the owl. *Science* 200:795–797, 1978.
20. Konishi, M. Neuroethology of acoustic prey localization in the barn owl. In Huber, F.; Markl, H., eds. *Neuroethology and Behavioral Physiology*, pp. 303–317. Berlin, Springer-Verlag, 1983.
21. Levi-Montalcini, R. The development of the acousticovestibular centers in the chick embryo in the absence of the afferent root fibers and of descending fiber tracts. *J. Comp. Neurol.* 91:209–241, 1949.
22. Merzenich, M.M.; Andersen, R.A.; Middlebrooks, J.H. Functional and topographic organization of the auditory cortex. In Creutzfeld, O.; Scheich, H.; Schreiner, C., eds. *Hearing Mechanisms and Speech*, pp. 61–75, Berlin, Springer-Verlag, 1979.
23. Morel, A.; Imig, T.J. Thalamic projections to fields A, AI, P and VP in the cat auditory cortex. *J. Comp. Neurol.* 265:119–144, 1987.
24. Morest, D.K. The neuronal architecture of the medial geniculate body of the cat. *J. Anat.* 98:611–630, 1964.
25. Morest, D.K. The laminar structure of the medial geniculate body of the cat. *J. Anat.* 99:143–160, 1965.
26. Morest, D.K.; Oliver, D.L. The neuronal architecture of the inferior colliculus in the cat: defining the functional anatomy of the auditory midbrain. *J. Comp. Neurol.* 222:209–236, 1984.
27. Neff, W.D.; Diamond, I.T.; Casseday, J.H. Behavioral studies of auditory discrimination. In Keidel, W.D.; Neff, W.D., eds. *Auditory System*, pp. 307–400, vol. 5, pt. 3, *Handbook of Sensory Physiology*. New York, Springer-Verlag, 1975.
28. Nordeen, K.W.; Killackey, H.P.; Kitzes, L.M. Ascending projections to the inferior colliculus following unilateral cochlear ablation in the neonatal gerbil, *Meriones unguiculatus, J. Comp. Neurol.* 214:144–153, 1983.
29. Oertel, D. Synaptic responses and electrical properties of cells in brain slices of the mouse anteroventral cochlear nucleus *J. Neurosci.* 3:2043–2053, 1983.
30. Rhode, W.S.; Oertel, D.; Smith, P.H. Physiological response properties of cells labeled intracellularly with horseradish peroxidase in cat ventral cochlear nucleus. *J. Comp. Neurol.* 213:448–463, 1983.
31. Romand, R. *Development of Auditory and Vestibular Systems.* New York, Academic Press, 1983.
32. Romani, G.L.; Williamson, S.J.; Kaufman, L. Tonotopic organization of the human auditory cortex. *Science* 216:1339–1340, 1982.
33. Rose, J.E.; Galambos, R.; Hughes, J.R. Microelectrode studies of the cochlear nuclei of the cat. *Bull. Johns Hopkins Hosp.* 104:211–251, 1959.

34. Rouiller, E.M.; Ryugo, D.K. Intracellular marking of physiologically characterized cells in the ventral cochlear nucleus of the cat. *J. Comp. Neurol.* 225:167–186, 1984.
35. Rubel, E.W. Ontogeny of structure and function in the vertebrate auditory system. In Jackson, M., ed. *Development of Sensory Systems*, pp. 135–237, vol. 9, *Handbook of Sensory Physiology*. New York, Springer-Verlag, 1978.
36. Rubel, E.W.; Parks, T.N. Organization and development of brain stem auditory nuclei of the chicken: tonotopic organization of *N. magnocellularis* and *N. laminaris. J. Comp. Neurol.* 164:435–448, 1975.
37. Rubel, E.W.; Born, D.E.; Deitch, J.S.; Durham, D. Recent advances toward understanding auditory system development. In Berlin, C., ed. *Hearing Science*, pp. 109–157. San Diego, College Hill Press, 1984.
38. Sanes, D.H.; Rubel, E.W. The ontogeny of inhibition and excitation in the gerbil lateral superior olive. *J. Neurosci.* 8:682–700, 1988.
39. Suga, N. The extent to which biosonar information is represented in the bat auditory cortex. In Edelman, G. M.; Gall, W. E.; Cowan, W. M. *Dynamic Aspects of Neocortical Function*, pp. 315–373. John Wiley & Sons, New York, 1984.
40. Suga, N. Neural mechanisms of complex-sound processing for echolocation. *Trends in Neurosci.* 7:20–27, 1984.
41. Thompson, G. Structure and function of the central auditory system. *Semin. Hearing* 4:81–95, 1983.
42. Webster, D.B.; Webster, M. Effects of neonatal conductive hearing loss on brain stem auditory nuclei. *Ann. Otolaryngol.* 88:684–688, 1979.
43. Wenstrup, J.J.; Ross, L.S.; Pollak, G.D. Binaural response organization within a frequency-band representation of the inferior colliculus: implications for sound localization. *J. Neurosci.* 6:662–673, 1986.
44. Wenthold, R.J.; Martin, M.R. Neurotransmitters in the auditory nerve and central auditory system. In Berlin, C.I., ed. *Hearing Science*, pp. 341–369. San Diego, College Hill Press, 1984.
45. Willott, J.F. *The Auditory Psychobiology of the Mouse.* Springfield, Illinois, Charles C Thomas, 1984.
46. Winer, J.A. The medial geniculate body of the cat. *Adv. Anat. Embryol. Cell Biol.* 86:1–97, 1985.
47. Woolsey, C.N. Organization of cortical auditory system: a review and a synthesis. In Rasmussen, G.L.; Windle, W.F., eds. *Neural Mechanisms of the Auditory and Vestibular Systems*, pp. 165–180. Springfield, Illinois, Charles C Thomas, 1960.
48. Yin, T.C.; Kuwada, S. Neuronal mechanisms of binaural interaction. In Edelman, G.M.; Gall, W.E.; Cowan, W.M., eds. *Dynamic Aspects of Neocortical Function*, pp. 262–313. New York, John Wiley & Sons, 1984.
49. Young, E.D. Response characteristics of neurons of the cochlear nucleus. In Berlin, C. I., ed. *Hearing Science*, pp. 423–460. San Diego, College Hill Press, 1984.
50. Young, E.D.; Brownell, W.E. Responses to tones and noise of single cells in dorsal cochlear nucleus of unanesthetized cats. *J. Neurophysiol.* 39:282–300, 1976.
51. Young, S.; Rubel, E.W. Frequency specific projections of individual neurons in chick brain-stem auditory nuclei. *J. Neurosci.* 7:1373–1378, 1983.

The Visual System: Optics, Psychophysics, and the Retina

INTRODUCTION

Of all the senses, vision is clearly the one most important to the survival of most mammals. Although the other senses are heightened when an animal is suddenly deprived of sight, a sightless animal is at a greater disadvantage than one deprived of hearing, smell, or any of the other senses. In addition to its survival value, vision significantly enhances the quality of human life and our appreciation of the world around us. Although it is admittedly difficult to compare the enjoyment produced by a beautiful sunrise on Mount Rainier with that of either a good symphony or a wonderful French dinner, vision must certainly be ranked among the most pleasurable of senses.

Because of the obvious impact of vision on our daily existence, it is not surprising that philosophers, physicists, psychologists, and physiologists have long been fascinated by the neural mechanisms subserving vision. The visual system is particularly attractive for study because its adequate stimulus, light, can be easily controlled. Consequently, it is fair to say that of all the senses, vision has been the most intensively studied. Despite this intensive effort, however, we are still a long way from understanding how we see. Nevertheless, research on the mechanisms of vision has often led the way in sensory physiology, and many general concepts for sensory processing have been first or most clearly demonstrated in the visual system.

In the next two chapters, we deal with the properties of vision and the optical and neural mechanisms underlying visual processing. In this chapter we consider psychophysical experiments that document the amazing capabilities of the visual system and the role played by the eye's optics and the retina in the initial stages of image processing. In the next chapter, we follow the retinal signals into the central nervous system to examine what is known about the central processing of visual information.

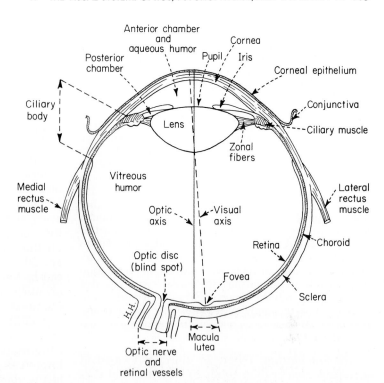

Figure 19–1 Horizontal section of the right human eyeball viewed from above.

OPTICS OF THE EYE

Focusing the Image of the World on the Retina. All eyes have an optical component whose major function is the formation of a visual image, and a neural component, the retina, that transduces the visual image into neural discharge patterns. A horizontal section of the human eye (Fig. 19–1) shows that light rays must pass through several optical elements before reaching the retina at the back of the eye. In order, they are the cornea, the aqueous humor, the lens, and the vitreous humor. The image of the visual world is brought to focus on the retina by the cornea and the lens, both of which act as convex lenses. The ability of these optical elements to bend (refract) and hence focus light rays depends on two factors. First, light is bent as it passes obliquely from a medium in

which it is traveling rapidly (e.g., air with a low index of refraction) into a denser medium in which it travels more slowly (such as the cornea and lens, which have higher indices of refraction, nearer to that of water). Second, if the interface between the media is curved, light rays incident on more convex surfaces will undergo greater bending. The index of refraction and radius of curvature for different surfaces of the relaxed human eye are summarized in Table 19–1. The refractive power of a spherical refracting surface like the cornea or lens can be calculated by the formula $(n' - n)/r$, where n and n' are the refractive indices of the object (first medium) and image (second medium) spaces, respectively, and r is the radius of curvature in meters. The unit in which refracting power is measured, the diopter (D), is the reciprocal of the focal distance of the lens in

Table 19–1 Optical Constants of a Typical Relaxed Normal Eye

Surface	Radius of Curvature (r) (mm)	Refractive Index		Distance from Anterior Surface of Cornea (mm)	Relaxed Refractive Power (D)
		Anterior (n)	*Posterior (n')*		
Anterior cornea	7.8	1.000 (air)	1.376	0	+ 48.2
Posterior cornea	6.8	1.376	1.336 (aqueous)	0.5	− 5.9
Anterior lens	10.0	1.336	1.416	3.6	+ 8.0
Posterior lens	− 6.0	1.416	1.336 (vitreous)	7.2	+ 13.3
Retina				24.0	

meters. It is apparent from Table 19–1 that the major refracting surface of the eye is the anterior surface of the cornea since it separates two media with widely disparate refractive indices. Any irregularities on this surface, therefore, will have major consequences for image formation.

The total refracting power of the eye is not equal to the simple sum of the powers of the four refractive surfaces of the cornea and lens (Table 19–1) for two reasons. First, the refractive index of the lens is not uniform but varies between 1.386 near its surfaces to 1.406 at its center. A lens with such a refractive index gradient is even more effective than one with the higher index. Indeed, it has been estimated that if a human lens were filled with a medium of uniform refractive index, it would require a refractive index of 1.416 to produce the same power as the actual lens. This equivalent value is used in Table 19–1. Second, the separation of the surfaces diminishes the cumulative refractive effect.[30] If the refractive powers of all four surfaces in Table 19–1 were simply added, the total power would be 63.6 D, whereas the actual total power of the relaxed human eye in fact is approximately 60 D.

Accommodation. When an object is located at distances greater than about 20 feet, the light rays that emanate from it and enter the eye are essentially parallel. These nearly parallel rays are adequately focused on the retina by the fixed refracting surfaces of the cornea and a relaxed lens. As objects approach closer than 20 feet, however, the emanating light rays entering through the pupil become increasingly divergent, and if the refracting surfaces were fixed, the light rays would be brought into focus behind the retina. Fortunately, the lens can vary its refraction because its thickness, unlike that of the cornea, can be adjusted by the ciliary muscle, a smooth muscle composed of circular and radial fibers. The radial fibers originate on the inside wall of the sclera and attach to the "zonal fibers," an elastic connective tissue by which the lens is suspended (Fig. 19–2). When the ciliary muscle (particularly its radial fibers) is relaxed, the elastic zonal fibers are under tension and pull on the equator of the lens, causing it to flatten and produce a reduced radius of curvature and reduced refraction. When the ciliary muscle contracts, tension is removed from the zonal fibers, allowing the lens to bulge, mostly at its anterior surface, and to increase its refraction. The ciliary muscle receives excitatory autonomic input from parasympathetic postganglionic neurons in the ciliary ganglion (see Chap. 34), which in turn are driven by preganglionic cells located in part within the Edinger-Westphal portion of the oculomotor nucleus. The elasticity of the zonal fibers

helps to restore the shape of the lens upon cessation of the parasympathetic drive, and the ciliary muscle can be further relaxed by activation of the cervical sympathetic nerves. The adjustment in lens power that the eye makes to see near objects clearly is called *accommodation*. Accommodation is a slow process, with the lens often requiring 0.5 s or more to change its shape as the eye shifts from distant to nearby viewing.

Most of the bulging of the lens occurs at its anterior surface. In infants, where it is most pliable, the lens is capable of about 12 to 14 D of adjustable refraction. A distance is reached, however, at which even the strongest contraction of the ciliary muscle will not bring the object to focus on the retina. In optically normal humans under age 30, this *near point* is about 10 cm, a value that can be verified empirically by attempting to focus on one's finger as it is brought gradually toward the nose. The near point begins to recede as early as childhood, but changes most noticeably in the mid-40s when the near point recedes beyond a comfortable reading distance. This condition, known as *presbyopia*, is usually attributed to a loss in the elasticity of the lens capsule.

Not all animals have accommodative mechanisms similar to those of humans. For example, some fish achieve adjustable refraction by sliding the lens toward the retina. Animals that dive must have a powerful, adjustable lens that can accommodate for the loss of corneal refraction under water. Humans, on the other hand, with their modest adjustable refraction, cannot see well under water. To improve that ability, divers wear a mask that traps air against the cornea, allowing it to refract light rays normally.

Refractive Errors. The position of the cornea and lens relative to each other and to the retina must be preserved with great accuracy. The fixed distance of the refractive surfaces from the retina is maintained because the inelastic scleral envelope is under a constant intraocular pressure of about 16 mm Hg. This pressure is controlled by the fluids that fill the eye. Although the volume of the vitreous humor remains essentially constant, the aqueous humor, which fills the anterior chamber (Fig. 19–1), is replaced every 2 to 3 hours. The aqueous humor is secreted by the epithelium of the ciliary body (Fig. 19–1) and flows from the posterior chamber to the anterior chamber through the pupil. From there, it is returned to the venous system via the canal of Schlemm after first passing through a trabecular network located in the angle between the cornea and the iris (Fig. 19–2). The resistance of the trabecular network regulates the

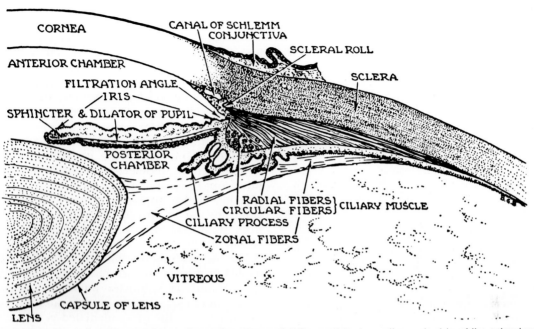

Figure 19–2 Detail of anterior segment of human eye. The radial fibers originate on the underside of the scleral roll. (After Fawcett, D.W. *Bloom and Fawcett: A Textbook of Histology,* 11th ed. Philadelphia, W. B. Saunders Co., 1986, by permission.)

flow out of the anterior chamber and thereby controls intraocular pressure. Problems with the secretion, flow, or absorption of the aqueous humor cause an increase in pressure, first in the aqueous and then in the vitreous humor. This condition, called *glaucoma,* restricts blood flow to the retina and ultimately results in blindness.

Small changes of only 0.3 mm in the length of the eye or 0.2 mm in the radius of corneal curvature produce a 1 D change in the eye's refracting power. Therefore, it is not surprising that about half of the human population has some problem with refraction. If the refractive media form the images of the visual world either in front of or behind the retina, the eye is said to be *ametropic* (a normal eye is *emmetropic*). If the images fall in front of the retina, it is most likely that the focal length is normal but the eye is too long. This condition, called *myopia,* usually develops at puberty and is frequently diagnosed when adolescents have difficulty seeing the blackboard. Myopia can be corrected by spectacles with concave lenses that cause the light rays to diverge slightly before striking the cornea so that they form an image on the retina (Fig. 19–3). If images of distant objects fall behind the retina, the eye is too short, or the refractive power of the eye is abnormally low (as occurs when blood sugar is low). This condition, called *hyperopia,* usually develops

shortly after birth and apparently is caused by genetic factors. Hyperopia occurs only infrequently and refractive errors of more than a fraction of a diopter are uncommon. Hyperopia can be corrected by wearing spectacles with convex lenses that provide additional refractive power to cause the light rays to converge more and thus to focus on the retina (Fig. 19–3).

In an ideal eye, the refractive surfaces of the lens and cornea would be spherical, with equal curvatures along all meridians. In many eyes, however, the corneal surface is not spherical, and there is a meridian of least curvature orthogonal to a meridian of greatest curvature. Consequently, parallel light rays passing through the meridian of greatest curvature will reach a focus before those passing through the meridian of lesser curvature. Ametropias of this type, called *astigmatism,* can be revealed when a subject views a chart of radial lines. At a given viewing distance, astigmates perceive lines of one orientation to be in focus (e.g., the horizontal line in Fig. 19–3), whereas lines at orthogonal orientations appear out of focus (e.g., vertical dashed line in Fig. 19–3). Astigmatism can be corrected by a complicated lens with different curvatures along different meridians. Astigmatic problems are especially amenable to correction by hard contact lenses, since as tears fill the space between the lens and the cornea, the

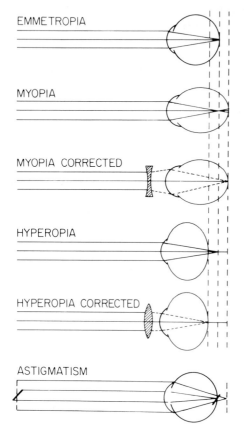

EMMETROPIA

MYOPIA

MYOPIA CORRECTED

HYPEROPIA

HYPEROPIA CORRECTED

ASTIGMATISM

Figure 19–3 Diagram of the normal (emmetropic) and three abnormal refractive states with optical corrections.

natural cornea is effectively eliminated so that the main refracting surface becomes the front surface of the contact lens, which need only be spherical.

Control of Light Intensity. The human visual system operates over a wide range of ambient light intensities ranging from the light levels emitted by a distant star to the more than 10 billion times brighter light intensities encountered in full daylight. A sensory system could solve this problem either by having many different receptors, each of which governs a narrow part of the entire intensity range, or by having just a few types of receptors whose operating characteristics can be adjusted for different intensities. The visual system implements a combined approach that uses optical means, a tradeoff between two major receptor types (rods and cones), and photochemical and neural mechanisms together to achieve the required large intensity range.

The amount of light permitted to impinge on the retina is controlled by the iris (Fig. 19–1), which, because of its abundant pigment, is impervious to light and thus forms an excellent dia-phragm. The human pupil constricts to a minimum diameter of 2.0 mm in the brightest light and dilates to about 8.0 mm in complete darkness. This roughly 16-fold change in area and, consequently, in light intensity (measured in luminance/square meter) is clearly inadequate to maintain a constant incident light over the entire 10^{10} range of light intensities. Therefore, the photochemical and neural processes discussed in later sections must provide most of the adjustment. Changes in pupil size are largely accomplished within seconds, however, and thus produce the most rapid adjustments. Hence, the iris provides a rather rapid, appreciable improvement in the ability to see in dim light and shields the retina from light too intense for its existing level of sensitivity.

The iris controls the diameter of the pupil with its dilator muscles, which have sympathetic innervation, and its sphincter muscles, which have parasympathetic innervation (see also Chap. 34). Constriction of the iris is mediated by the *pupillary light reflex*, whose efferent limb probably originates in the Edinger-Westphal nucleus, part of the third cranial nucleus. These preganglionic parasympathetic axons pass to the ciliary ganglion, where postganglionic fibers originate that innervate the sphincter of the iris. The afferent link in the reflex is thought to consist of optic nerve fibers that synapse in the pretectum (Chap. 20). Since pupillary constriction occurs in both eyes when light is presented to only one (the *consensual light reflex*), the pretectum probably sends bilateral projections to the Edinger-Westphal nucleus, which in turn provides both ipsilateral and contralateral preganglionic fibers. It is possible that, after the retina, the direct pathway of the pupillary reflex may involve as few as three additional neurons.

Optical Properties Affecting Visual Acuity

We are clearly capable of discriminating very fine details in a visual scene. This ability, known as *visual acuity*, can be measured in a variety of ways that are discussed later. Acuity is limited, in part, by the imaging system of the eye, which like most optical systems suffers from various aberrations. To provide as transparent an optical system as possible, neither the cornea nor the lens has any blood vessels. Both structures have rather modest metabolic requirements and receive most of their nutrients via the aqueous humor.

The transparency of the lens and the cornea depends on the condition of their constituent cells and their resident proteins. A *cataract,* or loss of transparency in the lens, is caused by the slow aggregation of proteins, which may be due to dietary deficiencies or deficits in carbohydrate metabolism. As may be surmised from Table 19–1, a person whose lens has been removed to alleviate a cataract problem requires a strong spectacle lens even for unaccommodated viewing.

Although its individual refracting elements may be of the highest optical quality, even a perfect optical system distorts the visual image because of the wave properties of light. When light travels past an opaque edge, some of the light bends around it to fall in the shadow. The consequence of this phenomenon, called *diffraction,* is that when a point source of light is viewed through a small aperture such as the pupil, the image consists of a central point of light surrounded by alternating dark and bright rings of decreasing intensity. The size of the rings depends inversely upon aperture size. Diffraction images produced by point sources that are close together, such as distant stars, will not appear distinct. In Figure 19–4, three "point" sources of light are viewed through a circular opening in front of a camera lens. Increasing the aperture to its full diameter (Fig. 19–4C) reveals that the rightmost diffraction pattern in Figure 19–4A was actually due to two closely spaced "points" of light.

The eye also suffers from chromatic and spherical aberrations. Chromatic aberration occurs because the index of refraction depends on the velocity of light in the medium and the velocity of

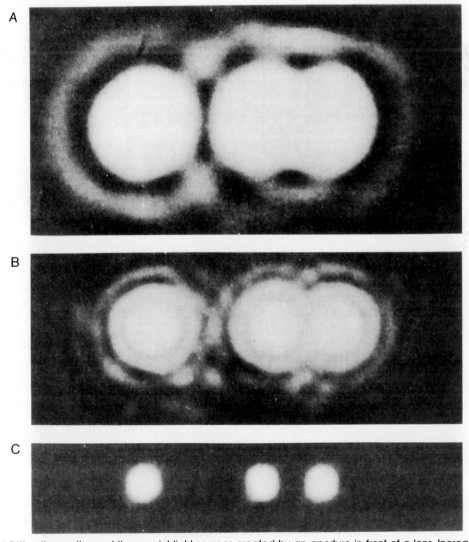

Figure 19–4 Diffraction patterns of three point light sources created by an aperture in front of a lens. Increasing the size of the aperture, as in A through C, decreases the amount of diffraction.

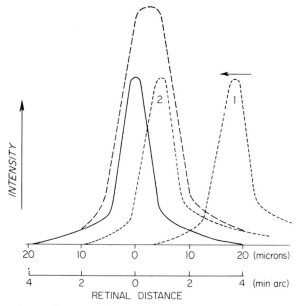

Figure 19–5 Spread of light intensity produced by a distant fine line over the human retina. As a second line (position one, small-dash intensity curve) is brought closer to the first (solid curve), a separation is reached (position two) in which the net intensity of the two lines is indistinguishable from that of a single brighter line (tallest curve). This light spread function sets the optical limit of resolution in the eye. Note that the real eye produces no diffraction rings.

light depends upon its wavelength. Thus different colors of light are brought into focus at different focal lengths. For example, the difference in refraction between the long wavelengths (red) and short wavelengths (blue) of light is about 1.5 D. However, chromatic aberration goes unnoticed, partly because the lens transmits very little light at wavelengths of less than 400 nm. Spherical aberration occurs because light rays entering a lens near its edge are refracted more than those entering near its center or axis. The human eye has relatively little spherical aberration because the cornea is cleverly constructed to be flatter at its periphery than at its center. Both chromatic and spherical aberrations become more serious as the pupil enlarges.

Since diffraction is minimized by a large pupil whereas chromatic and spherical aberrations are reduced by a small pupil, the best optical performance occurs for a compromise diameter between 2 and 3 mm, which is near the minimal pupil size. Even under these best possible conditions, the image of a point of light is spread over a considerable retinal surface. Figure 19–5 shows that the spread of light on the retina determined by recording the faint light reflected by a thin bright

line imaged on the retina[7] extends at least ±15 μm across the retina (solid curve in Fig. 19–5). If a second line is placed nearby (position 1, Fig. 19–5), the intensity distribution on the retina shows two humps, and the lines are perceived as distinct. However, when the line is moved closer (position 2), the summed light distribution (tallest, large-dash curve) is indistinguishable from that of a single brighter source. Based on these measurements, images separated by about 5 μm on the retina can be distinguished, provided of course that neural elements exist to detect the intensity humps.

Visual objects are usually measured not by the size (in micrometers) of their image on the retina but rather by the visual angle they subtend. For example, a thumbnail (1.5 cm across) viewed at arm's length (75 cm) will subtend a visual angle of 1° (i.e., $\tan^{-1}$ [1.5/75]), and by simple geometry, an angle of 1° on the retina as well; 1° on the human retina subtends about 300 μm. A visual degree may be subdivided into 60 minutes (1/60 of a degree) and 1 minute into 60 seconds. On the basis of such measures, the optical line spread was at least ±3 minutes of arc (Fig. 19–5, lower scale) although the bright line viewed by the subject subtended, in fact, only 0.2 minute of arc. As we shall see later, if acuity is assessed by the visibility of a pattern of fine stripes, the optical limit of resolution occurs when the width of the stripes and their separation is about 0.5 minute of arc.

FUNCTIONAL ANATOMY OF THE RETINA

Before we measure the capabilities of the human visual system, we must examine the anatomy of the retina, since its different parts deal with different aspects of the visual image. The anatomy of most vertebrate retinas is remarkably similar. Each consists of five types of neuron arranged in several distinct layers parallel to each other and to the surface of the retina, as diagrammed in Figure 19–6 for the monkey.[19] In all vertebrate retinas, light must travel through all of the retinal elements before striking the receptors, which are located at the back of the eye.

As described in Chapter 5, light energy is converted into electrical signals by two types of receptors. Each possesses an inner (IS) and outer (OS) segment joined by a nonmotile cilium (Fig. 19–6). (In the retina, the inner and outer designations are relative to the center of the eye.) The outer segments of most receptors are either cylindrical

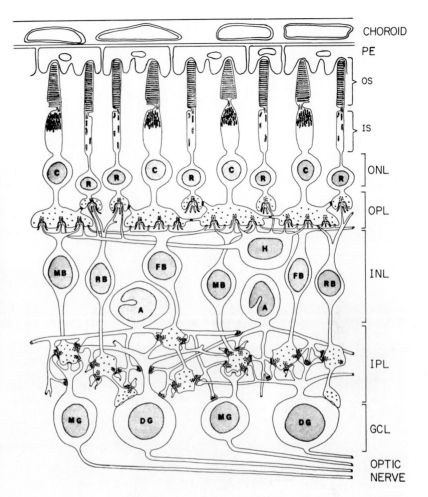

Figure 19–6 Schematic diagram of the macaque monkey retina. PE, pigment epithelium cells; OS and IS, outer and inner segments of receptors; R, rod; C, cone; ONL, outer nuclear layer; H, horizontal cell; OPL, outer plexiform layer; INL, inner nuclear layer; MB, midget bipolar; FB, flat bipolar; RB, rod bipolar; IPL, inner plexiform layer; A, amacrine cell; GCL, ganglion cell layer; MG, midget ganglion cell; DG, diffuse ganglion cell. (After Dowling, J. E.; Boycott, B. B. *Proc. R. Soc. Lond. [Biol.]* 166:80–111, 1966.)

or conical, as beautifully demonstrated in scanning electron micrographs (Fig. 19–7) from *Necturus*, a fresh water salamander whose retinal circuitry will be discussed extensively later. Because of their characteristic shapes, the receptors have come to be known as *rods* and *cones*. The morphological distinction is not always clear, however, as can be seen in the rodlike appearance of the outer segments of monkey foveal cones (Fig. 19–7, inset). The outer segments of rods and cones connect through inner segments to the receptor somata that make up the outer nuclear layer (Fig. 19–6, ONL).

Receptors make synaptic contact with two other retinal neurons. Connections to the *bipolar cells* (Fig. 19–6; MB, RB, or FB) direct the receptor signal radially, while connections with *horizontal cells* (H) direct it tangentially or laterally to neighboring bipolar cells, which can be quite distant. Synaptic interactions among these three elements occur in the outer plexiform layer (Fig. 19–6, OPL). Bipolar cells then synapse directly onto *ganglion*

cells (in the ganglion cell layer, GCL, in Figure 19–6) whose axons form the optic nerve that leaves the eye. As in the outer plexiform layer, bipolar cells can also contact other interneurons, the *amacrine cells* (Fig. 19–6A), which send their processes both to neighboring ganglion cells (Fig. 19–6, DG and MG) and to other amacrine cells. Connections between bipolar, amacrine, and ganglion cells all occur within the inner plexiform layer (Fig. 19–6, IPL). There is no anatomical interaction between the first (outer) and second (inner) synaptic processing stages (but see the interplexiform cell in Physiology of Retinal Neurons section). The three neurons that link the receptors and the retinal output cells all lie in an inner nuclear layer (INL) between the two plexiform layers with horizontal cell somata nearer the outer and amacrine cell somata positioned nearer the inner plexiform layer.

The retina appears to be inside out, since the receptors that transduce visual signals are located in the outer layer at the very back of the eye, with

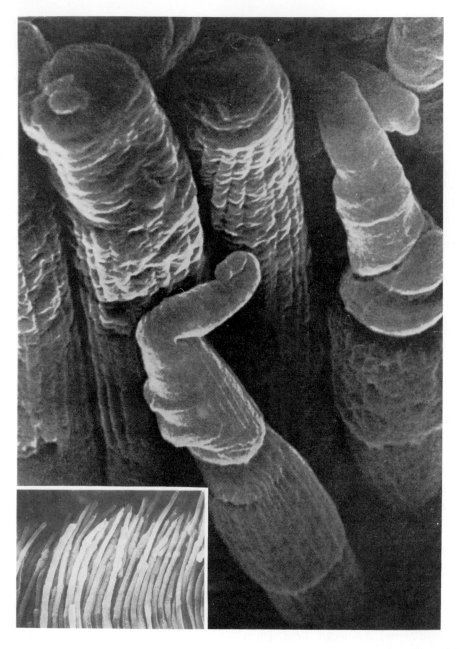

Figure 19–7 Scanning electron micrographs of rods and cones in *Necturus* (enlarged 5000 times) and foveal cones in the monkey (inset, ×2000). (After Werblin, F. S. *Sci. Am.* 228:71–79, 1973; Borwein, B., et al. *Am. J. Anat.* 159:125–146, 1980.)

their photosensitive outer segments aimed away from the incoming light. This apparent mistake by the celestial design committee actually makes good sense since, as mentioned in Chapter 5, the transducing part of the receptor must be continually sloughed off and renewed. So that this renewal process (and the blood vessels that must provide the metabolites to accomplish it) does not interfere with vision, the epithelial cells and their capillary supply are located at the very back of the eye in the pigment epithelium (PE) layer and the choroid

(Fig. 19–6), respectively. The inner retina, which receives its blood supply through vessels that enter at the optic disk, is nevertheless relatively transparent.

The choroidal epithelial cells are also pigmented to absorb the stray light that is not absorbed by the receptors. In nocturnal animals such as the cat, a reflecting layer called the tapetum lucidum is situated behind the photoreceptors. By reflecting the light that passes by the receptors, the tapetum gives them a second opportunity at absorption. The reflected light that is not absorbed on the second pass exits via the front of the eye and causes the eyes to glow when they are struck by car headlights.

A clue regarding the regional distribution of function in the retina can be obtained by examining the location, distribution, and connections of its two types of receptors. When an animal looks directly at an object, the image falls near the center of the retina. In many animals that require a high resolution of the visual scene for survival, the central retina has certain anatomical specializations. In the central area of primates, all of the intervening retinal elements are displaced laterally to allow the receptors direct access to the incoming light (Fig. 19–8I). This arrangement in humans creates a circular pit 300 to 700 μm in diameter (1 to 2.3° of visual angle) called the fovea. Only cones exist in the fovea; furthermore, these are the smallest of all retinal cones and are densely packed, with minimum separations of less than 0.3 μm (4 s of angle).[36] Anatomical studies by Polyak have shown that the great majority of foveal cones are each connected to a single bipolar cell (Fig. 19–6, MB) that in turn is connected to a single "midget" ganglion cell (Fig. 19–6, MG). Consequently, "through this chain, in and around the visual axis of the eye, an individual stimulus may be conveyed from each cone along its own private optic nerve fiber to the brain."[36] To accommodate these private lines, ganglion cells in the perifoveal region are densely packed and often are

stacked five or six deep (Fig. 19–8II), causing a thickening that encircles the fovea. Occasional rods (dark cells in ONL) begin to appear at the periphery of the fovea (the borders in Fig. 19–8I) and are quite numerous in the perifoveal thickening (Fig. 19–8II). In the human retina, rods are most densely packed about 3 mm off the fovea (i.e., some 10 to 15° from straight-ahead gaze). Therefore, aiming the eye directly at an interesting object allows it to be examined by the cones, whereas looking slightly to the side of an object causes its image to activate rods as well as cones.

A similar dense aggregation of small photoreceptors with rather direct connections to ganglion cells occurs in most mammals and birds. In the cat, this anatomical specialization is also roughly circular (called the area centralis), but in the rabbit it is a strip (called the visual streak) that is elongated along the horizontal meridian.

Over the whole human retina, there are 100×10^6 rods and 6×10^6 cones, but only 1×10^6 ganglion cells. Obviously, there must be considerable convergence of receptor and bipolar cells onto individual ganglion cells. The degree of convergence, however, varies from a minimum at the fovea to a maximum at the peripheral retina. Moreover, although rod and cone signals are kept separate at the outer plexiform layer, they can interact at the inner plexiform layer. Except for a small population of foveal ganglion cells that receive only cone inputs, virtually all other ganglion cells in the monkey and cat, especially those in

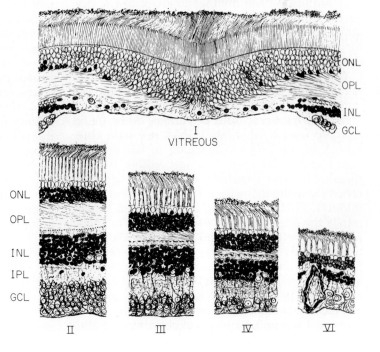

Figure 19–8 Regional variation of retinal structure in rhesus macaque. I, fovea; II, perifoveal region bordering fovea; III, more peripheral perifoveal region; IV, near peripheral retina; VI, far peripheral retina; other abbreviations as in Figure 19–6. (From Polyak, S. *The Vertebrate Visual System*. Chicago, University of Chicago Press, 1957.)

the peripheral retina, are driven by both rod and cone bipolar cells. Since the density of cones and ganglion cells decreases with retinal eccentricity whereas the density of rods remains relatively constant, not only does the convergence of receptors onto ganglion cells increase with eccentricity, but ganglion cells in the peripheral retina become increasingly dominated by rod inputs.

In summary, the anatomy of the retina allows us to make several deductions regarding its function. The peripheral retina summates visual stimuli over many rods and should therefore be sensitive in dim light but poor at resolving fine details. In contrast, the central retina, and the fovea in particular, summates inputs from only a few receptors and should be less sensitive in dim light but should be able to resolve details on the order of a single cone diameter (i.e., 14 s of arc in humans). Furthermore, since there are three cones with different spectral sensitivities, as we shall see later, the fovea should be particularly sensitive to color, whereas color vision should be less refined in the peripheral retina. These predictions of different functional roles for the central and peripheral retina are borne out in the following evaluations of the capabilities of the visual system.

CAPABILITIES OF THE VISUAL SYSTEM

Sensitivity to Light. As anyone who has gone from a sunlit street into a dark theater knows, the eyes require several minutes to adapt to the dim surroundings. Similarly, after the show it takes a while to adjust to the sunlight again. These adjustments of visual sensitivity are called *dark* and *light adaptation*, respectively.

The time course and magnitude of dark adaptation can be determined by first exposing (adapting) a subject to a large, highly intense light source and then requiring him, at various times thereafter, to adjust the intensity of a test light (presented in the extrafoveal region to fall on both cones and rods) until it is just visible (i.e., at threshold). The threshold to the test light decreases (sensitivity increases) gradually over many minutes after the adapting light has been turned off. If the adapting light was of a high intensity (e.g., 1600 millilamberts [mL] in Fig. 19–9), sensitivity increases along a curve that has two separate limbs.[24] An initial rapid decrease in threshold of about 100-fold (2 log units) reaches a plateau in about 8 to 10 minutes. A second, more gradual, decrease in threshold is somewhat greater (1000-fold, or 3 log units) and is largely complete after

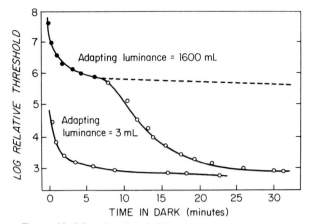

Figure 19–9 Relative thresholds as a function of time in the dark for a normal subject after high and low levels of light adaptation. Dashed curve shows approximate time course of adaptation when only cones are stimulated. Adapting light levels are expressed in millilamberts (mL). (Reproduced from Hecht, S., et al. *The Journal of General Physiology*, 1948, 31:459–472, by copyright permission of the Rockefeller University Press.)

a total of 30 minutes, although the threshold continues to fall very slowly for more than an hour. Since a relatively rapid pupillary dilation has already increased visual sensitivity by a factor of more than 10 before the first threshold measurement, the overall process of dark adaptation increases retinal sensitivity by at least a factor of 10^6.

Several different experiments suggest that this two-limbed curve is the combination of two separate dark adaptation processes. The initial rapid decrease can be attributed to the cone system. If the perifoveal test spot is colored, the subject perceives it as colored only during the initial rapid phase (solid dots, Fig. 19–9). If the test spot is shown directly on the fovea, where there are just cones, only the initial limb of the adaptation curve is obtained, and the latter part of the foveal curve follows the dashed time course of Figure 19–9. On the other hand, if adaptation is measured in a subject without cones, the initial limb of the curve is absent.

The second limb of the dark adaptation curve can be attributed to the rod system. If the adapting luminance is low (Fig. 19–9, 3-mL curve), the initial threshold is already low and drops along a single curve over a period of about 20 minutes. Under these conditions, a colored test light is always perceived as being colorless. Apparently, the low adapting luminance is insufficient to elevate the threshold to a level at which the cones can detect the test light, so the whole curve is determined by rods. In a rod monochromat, a rare individual who shows little evidence of cone function, the dark adaptation curve obtained after adaptation to a bright light shows no evidence of a cone limb. Dark adaptation has both biochemical and neural components, which are discussed later.

The detection of a visual stimulus depends not only on the level of light adaptation but also on the characteristics of the visual stimulus and its surroundings. For example, dim stimuli must either be presented for longer periods of time or be larger in size in order to be detected. This inverse relationship between intensity and either size or duration for threshold discriminations applies for central stimuli less than 5 minutes of arc in diameter and 30 ms in duration and for peripheral stimuli less than 30 minutes of arc and 200 ms duration.[1] Larger sizes and longer exposures have relatively little influence on threshold intensity.

The detectability and apparent brightness of a stimulus also depend upon the background against which it is displayed (e.g., the moon appears brighter by night than by day). If a test stimulus is presented against a background field of uniform light intensity (I), the increment of stimulus intensity (ΔI) necessary to distinguish it from the background is a constant fraction of I. This relationship, an example of Weber's law, which governs many sensory judgments, holds for a broad range of background intensities that span both rod and cone vision. The relationship does break down, however, at both high and low levels of light intensity, as well as for small or brief test flashes.

Spectral Sensitivity. The eye's sensitivity to a visual stimulus depends upon the wavelength of the light. If a subject must adjust a test stimulus to produce a constant brightness, less intensity is required at some wavelengths than at others. The plots of this relative sensitivity after dark adaptation (scotopic conditions) and light adaptation (photopic conditions) are shown in Figure 19–10. These spectral sensitivity curves have been adjusted (normalized) to the same peak sensitivity, but the peak sensitivity of the photopic curve is, in fact, three orders of magnitude less sensitive on an absolute scale. Figure 19–10 shows that the eye responds only to a limited band of wavelengths stretching from 400 nm to 700 nm, so that the *visible spectrum* ranges from ultraviolet to infrared radiation.[26]

Spectral sensitivity is determined by both optical and biochemical factors. At ultraviolet frequencies, short-wavelength radiation is absorbed mostly by the lens, which is slightly yellow in color, and to

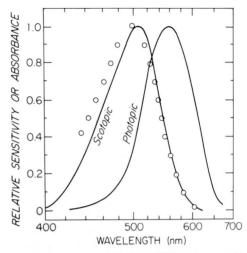

Figure 19–10 Human spectral sensitivity curves under scotopic and photopic conditions. Circles indicate the absorption spectrum of human rod pigment. (After Knowles, A. In Barlow, H. B.; Mollon, J. D., eds. *The Senses.* Cambridge, Cambridge University Press, 1982.)

a lesser extent by the cornea. The long-wavelength limit of the visual spectrum is determined by the absorption properties of the photopigments themselves. For the scotopic curve, only rods are involved, and the visual pigment in their outer segments is rhodopsin. One can obtain the spectral sensitivity curve for rhodopsin by macerating the retina, separating the rod outer segments by centrifugation, and shining dim light of different wavelengths through the suspension of rod particles. The resulting absorption spectrum of human rhodopsin (data points of Fig. 19–10) matches the scotopic psychophysical function quite well except at short wavelengths, where the filtering action of the lens operates. A virtually perfect fit is obtained in aphakic subjects whose lenses have been removed surgically because of cataracts.

Under bright conditions, the photopic spectral sensitivity curve can be attributed to cone pigments. The absorption spectra of cone pigments can be determined by microspectrophotometry in which a monochromatic beam of light of different wavelengths is passed through the outer segments of the receptors. Initial studies performed in the frog and carp showed that absorption spectra fell into three distinct groups. More recent studies on human cones[15] have also demonstrated three different absorption spectra with peak absorptions at approximately 420, 530, and 560 nm (Fig. 19–11). Although the peaks of the three pigments fall in the violet, yellow-green, and yellow portions of the spectrum, respectively, they are called the B

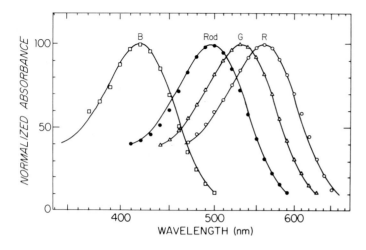

Figure 19–11 Mean absorbance spectra of human rods and cones. The B (open squares), G (open triangles), and R (open circles) cones have their peak absorbance at 419, 531, and 558 nm, respectively, whereas the rod spectrum peaks at 496 nm. (After Dartnall, H. J. A., et al. In Mollon, J. D.; Sharpe, L. T., eds. *Color Vision.* London, Academic Press, 1983.)

(blue), G (green), and R (red) pigments because of historical precedent. Once again, the absorption spectra in Figure 19–11 have been normalized; in fact, the B cone is less than one tenth as sensitive as the other two. Consequently, the photopic spectral sensitivity curve depends mostly on the combined responses of the G and R cones (compare Figs. 19–10 and 19–11). It is clear from Figure 19–11 that any particular wavelength of light will stimulate all three cone types to some extent. As expected, the spectral sensitivity of individual rods (Fig. 19–11) is similar to both the scotopic and rhodopsin absorbance curves of Figure 19–10.

The discovery of three cone pigments provided a physiological basis for the trichromatic theory of color vision. This theory—that there are three different cone pigments—was first articulated by Thomas Young in 1801 and later modified by Helmholtz. It explained the observation that normal individuals can match a sample hue of any wavelength or mixture of wavelengths by combining the correct intensities of three lights with primary hues (e.g., blue, green and red). For example, the sensation of yellow produced by a 570-nm light may be matched by a mixture with a high proportion of the red and green primaries. The sensation of white, on the other hand, may be produced by approximately equal intensities of all three primaries. Similar matches are possible for most sample hues. For the rest, however, a match can be obtained only when one of the primary colors is mixed with the sample itself.

Resolution of Stimuli in Time. The visual system can resolve stimuli that occur close together in time. To test this capability, a subject is required to adjust the frequency of a small (2-degree), repetitively flashing light until the light intensity appears to be continuous rather than repetitive.

The minimum frequency at which this subjective fusion occurs is called the *critical fusion frequency* (CFF). CFF is plotted in Figure 19–12 as a function of both the intensity (in log units) of the flickering stimulus and retinal eccentricity.[25] In the peripheral retina (20°), the CFF exhibits a modest increase at low stimulus intensities and a second increase at higher intensities, eventually reaching a maximum of only 20 Hz. In the fovea (0°, Fig. 19–12) on the other hand, the low-intensity limb of the curve is absent, whereas at high light levels, the CFF improves much more dramatically with increasing light intensity to reach a maximum of more than 50 Hz. Consequently, the low-intensity limb of the CFF curves obtained distant from the fovea can be attributed to the rods, whereas the high-intensity limb can be attributed to the cones. A high-intensity limb exists at both 5° and 20°

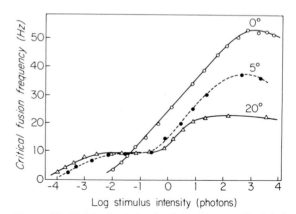

Figure 19–12 Effect of stimulus intensity on critical fusion frequency (measured in cycles/s or Hz) in the fovea (0°) and at 5° and 20° above fovea. (Reproduced from Hecht, S., et al. *The Journal of General Physiology*, 1933, 17:251–268, by copyright permission of the Rockefeller University Press.)

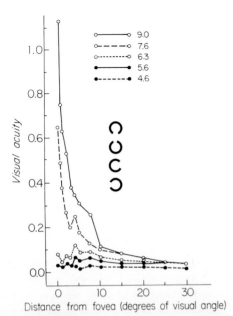

Figure 19–13 Visual acuity as a function of distance from the fovea at five different stimulus intensities ranging from 4.6 to 9.0 log micromicrolamberts. Visual acuity is expressed in arbitrary units on a relative scale. (After Mandelbaum, J.; Sloan, L. L. *Am. J. Ophthalmol.* 30:581–588, 1947.)

because some cones are present at those eccentricities. As shown later, the higher CFF for cones can be explained in part by the faster time course of their receptor potentials. The phenomenon of subjective fusion is employed in the making of movies, where the rate of occurrence of successive frames is kept above the CFF. In the early twentieth century, low frame rates produced the "flickers" of the Charlie Chaplin era.

Although Figure 19–12 indicates that the peripheral retina is poor at detecting rapidly flickering *small* stimuli, our own personal experience clearly indicates that *large* flickering stimuli are more easily detected in the periphery than in the fovea. For example, a flickering TV screen or fluorescent light is often easily detected out of the corner of one's eye.

Resolution of Stimuli in Space. The ability to discriminate very fine details is referred to as visual acuity and can be tested by a variety of means. One test requires a subject to identify the position of the gap in the letter *C* when it is presented in one of four different orientations. As seen in Figure 19–13, visual acuity depends on both the luminance of the *C* and its location relative to the fovea.[28] At the lowest luminance (Fig. 19–13, 4.6 curve), the *C* cannot be seen at all by the fovea; it is first seen when presented at an eccentricity of 4 degrees. The fovea is blind to low

levels of luminance because it lacks rods. As the *C* becomes brighter, however, foveal acuity improves dramatically (Fig. 19–13, 7.6 and 9.0 curves). In the foveal region, the smallest resolvable gap is on the order of 1 minute of arc and often 0.5-minute gaps can be detected. It seems likely that the excellent visual acuity in the pure cone fovea is attributable largely to the small size of foveal cones and the direct influence of single cones on single ganglion cells.

If, on the other hand, the *C* falls well off the fovea, increasing its intensity does not improve visual acuity appreciably. For example, since rods are most densely packed about 10 degrees to 15 degrees off the fovea, astronomers view dim stars by looking slightly to one side of them. Figure 19–13 shows, however, that the visual acuity at this distance from the fovea is very poor even if the object is quite bright (compare curves 4.6 and 9.0). The convergence of signals from many rods upon a single ganglion cell accounts for the poor acuity in the rod-dominated peripheral retina.

In an optometrist's office, visual acuity is usually tested by a Snellen chart, which has lines of letters designed on the assumption that the normal limit of resolution of a gap is 1 minute of arc. The gaps between individual components of the letters on a reference line are adjusted to subtend 1 minute of arc at the subject's viewing distance of 20 feet. The gaps between smaller letters are adjusted to subtend 1 minute of arc if the letters are viewed at lesser distances, say 15 feet, while gaps between larger letters would subtend 1 minute of arc if located at greater distances, say 100 feet. Acuity then is recorded as a ratio of the viewing distance (i.e., 20 feet) to the distance at which the just-discernible letters would subtend 1 minute of arc. Consequently, 20/20 represents normal vision; 20/100, rather poor vision; and 20/15, vision that is better than normal.

More recently, spatial resolution has been measured by determining a *contrast sensitivity function*.[46] In this test, a subject looks at an electronically generated grating, the spatial luminance profile of which is modulated sinusoidally (Fig. 19–14, inset). At each spatial frequency, the contrast of the grating is adjusted until it is barely visible. The reciprocal of this contrast threshold, the contrast sensitivity, indicates where the eye is most sensitive. As seen in Figure 19–14, the maximum sensitivity for humans lies at 5 to 6 cycles/degree (equivalent to a bar width of 5 to 6 minutes of arc), and the limit of grating resolution lies between 50 and 60 cycles/degree (36 to 30 s of arc). By comparison, the animal most used in visual

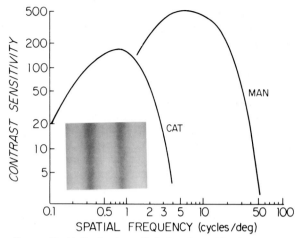

Figure 19-14 Contrast sensitivity function of the human and the cat. Contrast is defined as $(I_{max} - I_{min})/(I_{max} + I_{min})$ in which I_{max} and I_{min} are the maximum and minimum intensities of the stimulus grating (inset) and contrast sensitivity is the reciprocal of contrast threshold. (After Woodhouse, J. M.; Barlow, H. B. In Barlow, H. B; Mollon, J. D., eds. *The Senses.* Cambridge, Cambridge University Press, 1982.)

neurophysiology experiments, the cat, has a maximum sensitivity at about 0.7 cycle/degree and a visual acuity of about 5 cycles/degree (i.e., 6 minutes of arc).

Finally, some tests of acuity yield limits of resolution far below the 1 minute of arc obtained by all the aforementioned tests. For example, a thin wire can be seen against a bright sky when its diameter subtends only 0.5 s of arc. Also, two lines lying end to end can be adjusted to be colinear with a precision of about 5 s of arc. Such so-called *vernier acuity* appears remarkable since Polyak[36] has shown that the centers of human foveal cones are separated by at least 18 s of arc. The ability of the visual system to perform better than the rather coarse grain, or texture, of its transmitted retinal image is called *hyperacuity;*[45] the neural processes that allow vernier acuity and other hyperacuities to occur are not well understood.

The acuity discussed so far can be demonstrated by presenting the stimuli to one eye only. Frontal-eyed animals, however, actually view the visual world from two different positions; in humans, these positions are separated by about 6 cm. This arrangement facilitates seeing in depth (i.e., *stereopsis*) and helps provide the sense of the three dimensionality of objects. Our ability to discriminate differences in depth, *stereoacuity*, is also hyperacute, since under optimal conditions, disparities as small as 2 s of arc (differences in depth of

less than 4 mm at a 5-meter distance) can be detected.

Summary and Conclusions. Our visual capabilities depend on which part of the retina is viewing the visual scene. Indeed, it is for just that reason that we developed a sophisticated repertoire of eye movements to aim the eye. To make detailed discriminations in both space and time at high levels of illumination, we use the fovea, which contains only cones. On the other hand, in dim illumination there is insufficient light to activate the cones, so the visual system undergoes dark adaptation and mobilizes its rod photoreceptors. Not only are individual rods more sensitive to light, but also many rods in the peripheral retina converge on a single ganglion cell to further improve sensitivity, though at the expense of acuity.

Because the retina represents an evagination of the brain and contains several cell types with complicated interconnections, we can expect that retinal processing must play a large role in both limiting and facilitating our visual capabilities. Therefore, we now discuss the way in which the retina processes visual stimuli. We begin at the retinal output element, the ganglion cell, and then consider how each preceding element, in turn, contributes to ganglion cell behavior.

PHYSIOLOGY OF RETINAL NEURONS

The electrical activity of the retina has been studied by two radically different approaches. For cold-blooded animals, such as the fish and frog, the animal is decapitated, and the retina is removed and placed in a chamber with the appropriate nutrients. Such a preparation allows the visual test stimuli to be delivered directly to the photoreceptors and affords excellent stability for intracellular recordings; however, it lasts for only a few hours. In warm-blooded animals such as the cat, on the other hand, the animal is kept intact but is anesthetized and is often paralyzed by pharmacological agents to prevent eye movements. The visual stimulus is presented on a tangent screen that faces the animal, and the animal's own optics focus the image on the retina (see schematic illustration in Fig. 19-15). Because anesthesia relaxes both the lens and the iris musculature, an artificial pupil and the appropriate lenses must be placed in front of the eye to help focus the image. As shown in Figure 19-15, the tangent screen essentially replaces the animal's visual world and allows the experimenter to present different visual stimuli by means of a projector.

The images on the screen are cast on the retinal photoreceptors (exploded view in Fig. 19–15), which begin a sequence of neural events that ultimately leads to a change in electrical activity of the ganglion cells. Recordings from retinal neurons can be obtained by a microelectrode introduced through a small hole in the side of the eye. The intact eye preparation is particularly suitable for extracellular recording and may be maintained for several days.

Ganglion Cells. Although investigators had been studying retinal ganglion cells in a variety of invertebrate species (e.g., the eel and horseshoe crab) since the 1930s, Kuffler[27] was the first to show in a mammal that a very specific spatial pattern of visual images must fall on the retina to produce the best response in ganglion cells. This discovery, obtained in a cat preparation similar to that diagrammed in Figure 19–15, constituted a key breakthrough in visual neurophysiology and formed the foundation for virtually all subsequent work. Kuffler, as had others, found that even in darkness all ganglion cells had a "resting" repetitive discharge of action potentials of 20 to 50 spikes/s. When a small spot of light was turned on briefly at certain locations on the tangent screen (pluses in Fig. 19–15), one kind of ganglion cell exhibited a marked increase in discharge rate that was sustained while the spot was turned on. This discharge pattern is shown schematically in Figure 19–15 both as a train of idealized action potentials

(or spikes) and as an averaged instantaneous firing rate (i.e., the reciprocal of the interspike interval; see [+] trace, upper right). The shortest latency of the response to a bright spot is on the order of 20 to 30 ms. When the stimulus was turned off, the discharge rate dropped briefly below its prestimulus (or resting) rate before returning to it. In contrast, when the spot was projected on surrounding locations (minuses in Fig. 19–15), it caused a decrease in activity when turned on and a brief increase when turned off (see [−] trace, upper right). Spots presented at even more eccentric locations (zeroes in Fig. 19–15) caused no change in discharge at all (see [0] trace, upper right). That region of the tangent screen (or visual world) in which a visual stimulus can influence the response of a cell is called its *receptive field*. The receptive field of the ganglion cell in Figure 19–15 has a roughly circular central region that gives an *on*-response (increase in firing) to a light stimulus and a roughly donut- (or annulus-) shaped surrounding region that gives an opposite or *off*-response (decrease in firing) to a light stimulus. Kuffler also described a second kind of ganglion cell with a similar center-surround spatial configuration, but its center response to a light stimulus was a decrease in activity (like the off-response in Fig. 19–15), and its surround response was an increase in activity (like the on-response in Fig. 19–15). This latter cell, called an *off-center* ganglion cell, was encountered as often as the

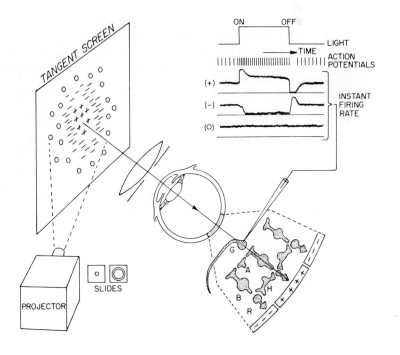

Figure 19–15 Schematic view of an experiment to plot the receptive field of a retinal ganglion cell. Small spots presented by a projector at different locations (+, −, and 0) in the visual world, i.e., the tangent screen, are focused by an external lens and pupil and the eye's own optics to fall on the outer segments of receptors (R in exploded view). A micropipette records the time course of discharge patterns (upper right) that result when the small spots are turned on and off at the different locations. H, B, A, and G are horizontal, bipolar, amacrine, and ganglion cells, respectively.

former *on-center* ganglion cell, and in 1953 it was believed that these two types accounted for most cat retinal ganglion cells.

The optics of the eye cause the visual world to fall in an orderly but inverted fashion on the retina. This "map" is called *topographic* because adjacent points in the visual world fall on adjacent points on the retina. Consequently, ganglion cells located on the peripheral retina respond to stimuli at the edges of the tangent screen (or the visual world), whereas ganglion cells on the central retina respond to stimuli presented near the visual axis. The size of retinal ganglion cell receptive fields varies with retinal eccentricity. The receptive field centers of ganglion cells in the area centralis (the feline equivalent of the fovea) are about 0.5 degree in diameter while those of peripheral ganglion cells increase gradually with eccentricity to values ranging from 2 to 8 degrees.[37] Because there are many ganglion cells (more than one quarter millíon in the cat and one million in the monkey), the receptive fields of several will overlap. A small spot of light, therefore, will drive several neighboring cells.

Since central and surround stimuli produce opposite changes in discharge rate (Fig. 19–15), a visual stimulus delivered to the surround region of a receptive field can be expected to reduce or antagonize the response to a stimulus delivered to the receptive field center. This center-surround antagonism is demonstrated in Figure 19–16A, which shows the neural response (average firing rate while the light is on) when the radius of a small spot centered on the receptive field is gradually increased. The ganglion cell shows an increasing discharge with expanding spot size (i.e., spatial summation) until the spot begins to encroach upon the surround, when the discharge reaches a maximum. As the spot becomes still larger, involving more of the surround, the discharge decreases to a low value. Indeed, in some ganglion cells, illumination of the entire receptive field produces no response at all (e.g., for large spot sizes, the discharge of the cell in Fig. 19–16A remains at the resting rate). Because of the center-surround antagonism, the preferred stimulus for an on-center retinal ganglion cell is a light spot surrounded by a dark annulus and for an off-center cell, vice versa. Therefore, ganglion cells have also been called contrast detectors, because they respond best to abrupt spatial gradients of light intensity.

The spatial organization of the antagonistic center and surround regions provides a clue regarding the roles of the other retinal neurons in producing

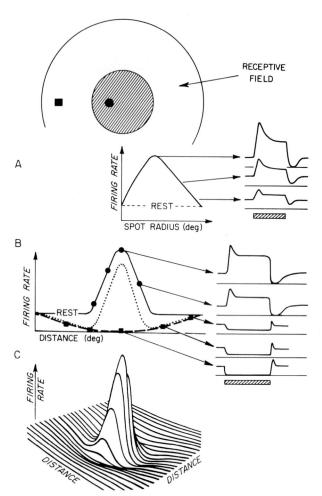

Figure 19–16 Demonstration of the center-surround organization of a ganglion cell receptive field. In *A*, the effect on firing rate of increasing the size of a spot centered on the receptive field; averaged instantaneous firing rates produced when different-sized spots are on (shaded bar) are shown to the right. In *B*, either the peripheral or central component of the receptive field has been disabled by adaptation with a colored light (see text), and the firing rate represents the discharge to a small light spot probe placed at different distances from the receptive field center. Instantaneous firing rates due to spot stimulation (shaded bar) of the central region (upper two curves) and surround (lower three curves) are shown to the right. The dotted curve is the algebraic (net) sum of the center and surround responses. In *C*, the net excitatory and inhibitory receptive field components are illustrated in two-dimensional space across the retina.

ganglion cell receptive fields. As seen in Figure 19–16A, it is difficult to stimulate one region and not the other, particularly at their apparent border. However, selective stimulation of either the center or surround is possible in animals with color vision. In the goldfish, for example, ganglion cells have receptive fields whose center and concentric surrounding regions each respond best to different wavelengths of light. For a ganglion cell with a center region sensitive to red light and a surround sensitive to green, the spatial extent of the red center in isolation can be obtained by first deactivating the surround mechanism by flooding the receptive field with an adapting green light to saturate its pigment.[42] The response of such an adapted cell to a small red spot positioned at different points in the receptive field is schematized in Figure 19–16B. In such a green-adapted cell, a spot on the periphery always produces an *on* response (upper averaged firing rate patterns to right of Fig. 19–16B) the magnitude of which increases along a bell-shaped curve to reach a maximum in the receptive field center (Fig. 19–16B, circles and solid curve). On the other hand, when the receptive field is flooded with a red light to deactivate its center, probing with a green spot always causes an *off* response (Fig. 19–16B, lower righthand firing rate patterns) the magnitude of which is also maximum at the center and has a bell-shaped spatial distribution that extends across the receptive field center (Fig. 19–16, squares and dashed curve). Thus, the excitatory center and inhibitory surround mechanisms do not occupy spatially separated regions but overlap over much of the receptive field. A similar overlapping spatial organization of center and surround has been demonstrated in the monkey[17] and in the cat by other techniques.[37]

The response of a light spot anywhere in the receptive field can now be generated simply by adding the *on* and *off* responses produced at that point. Such an addition, the dotted curve in Figure 19–16B, neatly produces the concentric center-surround organization of retinal ganglion cells. Because this spatial response profile is virtually identical when plotted in all radial directions from the receptive field center, the two-dimensional configuration of *on* and *off* regions across the retina resembles a sombrero (Fig. 19–16C).

In summary, many retinal ganglion cells in a wide variety of animals have an antagonistic center-surround organization. If only white light is available as a stimulus, the best probe to reveal the center mechanism is a small spot of light centered on the receptive field. On the other hand, a light annulus centered on the receptive field is the most effective stimulus for the surround mechanism. We will now consider how the other retinal neurons respond to these stimuli and contribute to the receptive fields of ganglion cells.

In contrast to ganglion cells, the visual responses of the other types of retinal neurons have been studied almost exclusively in cold-blooded animals, where investigators usually have concentrated on one or two cell types in each study. In 1970, however, Werblin and Dowling[44] described the response of *every* cell type in the retina of the fresh water salamander (*Necturus*) to selected visual stimuli. *Necturus* was chosen in part because the larger size of its retinal neurons allowed them to be penetrated easily with microelectrodes. Figure 19–17 shows the electrical responses obtained by intracellular recording from the five types of retinal cells in *Necturus*. Not only did Werblin and Dowling identify each cell by injecting it with a dye, but they also used electron microscopy to reveal its synaptic connections. Since the behavior of retinal neurons in *Necturus* is similar to that reported for individual cell types in other species, we use their responses as prototypes to describe the processing of information in the retina.

Before describing the responses of the other retinal elements, we briefly consider the responses of *Necturus* ganglion cells. In Figure 19–17, each cell is driven by three different visual stimuli selected to activate different parts of a ganglion cell's receptive field. The spot stimulus (Fig. 19–17, left column) is sufficiently small to drive primarily the receptive field center, whereas the annulus (Fig. 19–17, right column) with a 500-μm inner diameter preferentially drives its surround. The annulus of intermediate size (250 μm inner diameter) drives both the center and surround regions. Stimulation of its central region causes a depolarization in the lower ganglion cell (Fig. 19–17), which ultimately produces action potentials and a sustained discharge while the spot remains on. Stimulation of its surround causes a membrane hyperpolarization while the annulus is on and a burst of spikes when the light is turned off. The reduced firing for the stimulus of intermediate size was also noted by Kuffler and others in mammalian retinas and is due to a summation of the center and surround responses as indicated in Figure 19–16B. The spatial organization of the receptive field and the temporal firing pattern of this cell are identical to those already described for the cat ganglion cells of Figures 19–15 and 19–16.

In contrast with the sustained discharge of the lower ganglion cell, a second type of cell responds to central stimulation with only a transient depolarization that culminates in a burst of spikes (Fig. 19–17, top ganglion cell). Since this cell exhibits a transient burst when a surround stimulus is turned off and the 250-μm annulus elicits the sum of the central and surround response (i.e., a burst for both light on *and* off), it also has a spatial

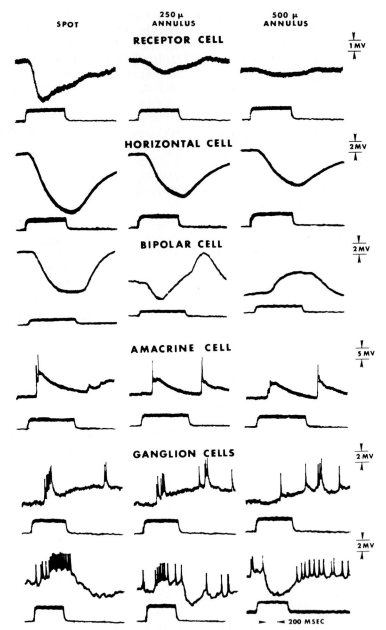

Figure 19–17 Intracellular responses (upper traces) in all five retinal neurons of *Necturus* to turning on (lower trace) either a small spot or an intermediate (250-μm inner diameter) or large (500-μm inner diameter) annulus centered on the receptive field. (From Werblin, F. S.; Dowling, J. E. *J. Neurophysiol.* 32:339–355, 1969.)

center-surround organization. Because of the difference in their *temporal* firing patterns, the lower cell has been called a *sustained ganglion cell* and the upper, a *transient ganglion cell*. Both sustained and transient cells may have either on- or off-center receptive fields so that four possible combinations of spatial and temporal firing properties are possible. Transient ganglion cells had also been noted by Kuffler in the cat.[27] We will return later to the functional significance of this dichotomy in temporal firing patterns, but first we consider how the other retinal neurons produce the receptive fields of ganglion cells.

Receptor Cells

Receptive Fields. A small spot of light falling on the receptive field of either a rod or a cone causes a membrane hyperpolarization. Stimuli delivered to the distant and even the near surround evoke little if any response, suggesting that the receptive field is quite small (Fig. 19–17). Furthermore, since the response to even the small annulus is hyperpolarization, there is no evidence of an antagonistic surround.

In other species, such as the turtle, however, the receptive fields of receptors are more complicated. First, the size of the receptive field exceeds the diameter of the receptor's outer segment, suggesting an interaction between receptors. Baylor and colleagues[2] demonstrated electrical coupling between turtle cones by penetrating neighboring cones and passing current between them. Coupling occurs between as many as five cones but only between those containing the same pigment. Rods in several species also are electrically coupled but over much greater distances than cones.[25a] In the toad and the turtle, under dim illumination a rod might receive up to 96% of its hyperpolarization from its neighbors.[8] Although electrical coupling in rods would contribute to their high sensitivity to low levels of light, the functional significance of electrical coupling in cones is unclear. Electrical coupling might also be present in primates since interreceptor contacts with membrane thickenings that resemble the gap junctions characteristic of some electrical synapses have been reported in the monkey.[19] Second, in turtle cones, the hyperpolarization caused by central stimulation is decreased by simultaneous stimulation of the surround. However, this surround effect is difficult to detect in other animals and is thought to be insufficient to account for the peripheral antagonism described later for bipolar cells (Kaneko, 1979).

Before continuing to the receptive field properties of the other retinal neurons, we now consider how receptors respond to changes in light intensity. These characteristics, which were not discussed when the photoreceptor was considered in Chapter 5, help us to account for some of the visual capabilities discussed earlier in this chapter.

Response to Light Intensity. The responses of rods and cones in *Necturus* to small light spots of increasing intensities are shown in Figure 19–18. If the light spot is presented to a dark-adapted retina, both rods and cones exhibit an increasing hyperpolarization with increasing spot intensity (Fig. 19–18, left column). For both receptor types, the amount of hyperpolarization increases along an S-shaped curve that spans about 3.5 log units (Fig. 19–18, DA curves, right column). There are two notable differences, however, in the rod and cone responses. First, rods begin responding to spots that are 10 to 30 times less intense than is required for a cone response. For example, the rod in Figure 19–18 exhibited a just noticeable

response to a stimulus equivalent to about 100 quanta (2.0 curve), whereas the cone threshold was more than 3000 quanta (3.5 curve). By sucking the outer segments of toad rods into a recording electrode to measure membrane current, Baylor and co-workers[3] demonstrated that rods are actually so sensitive that they respond to the absorption of a single photon of light. Second, for all spot intensities the rod response is slower than the cone response (Fig. 19–18, left column). The inability of the human retina to respond to dim stimuli flickering at high rates (recall Fig. 19–12) may in part be explained by the sluggish response of rods.

Only about 10% of the light incident on the eye is actually used by the receptors since roughly 60% is absorbed or scattered by the various ocular media and about 30% passes between or through the receptors without being absorbed.[37]

Control of Retinal Sensitivity. We saw earlier that the visual system can operate over an incredible range of ambient light intensities, ranging from complete darkness to full sunlight. In addition to the modest (16-fold) changes in sensitivity contributed by the pupil, adaptation also has chemical and neural components. Although the mechanisms of adaptation are not understood, a few qualitative observations are worth making.

Visual sensitivity clearly depends on the amount of photosensitive pigment available to capture the incident photons of light. When a person is exposed to bright sunlight, much of his rod photopigment (rhodopsin) is inactivated or "bleached." The improvement in sensitivity that occurs after entering a dark room, however, is not due only to the increased concentration of the resynthesized pigment available for light transduction, because visual threshold drops by more than 100-fold while only the final 10% of rhodopsin is being resynthesized.

A major factor in controlling retinal sensitivity appears to be a little-understood intracellular mechanism in the receptors themselves. Figure 19–18 (middle column) shows the response of a rod and a cone to a 2-s test spot that immediately followed an adapting background stimulus of 5.5 log units applied for 2 minutes.[34] After this adapting stimulus, the rod exhibited little change in membrane potential for spots of different intensities, so that a very shallow stimulus-response curve was obtained (Fig. 19–18, rightmost curve, right column). For the same background, however, the cone exhibited prominent changes in membrane potential for the range of spot intensi-

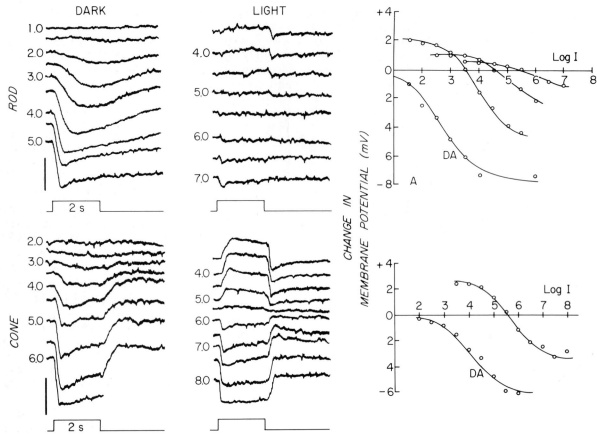

Figure 19–18 Intracellular rod and cone responses in *Necturus* to a full-field, 2-s test flash when the retina was dark adapted (DARK) or light adapted (LIGHT) to different backgrounds. The numbers preceding the responses in the two left columns approximately represent the log of the flux of incident absorbable quanta of light in the test flash (e.g., 3.0 means about 1000 quanta). Each successive curve reflects a change of 0.5 log unit. In the middle column, both the rod and cone were light adapted at a background intensity of about 5.5 log units. Vertical calibration bars represent 5 mV. The rod and cone operating curves in the right column are constructed by plotting the amplitude of individual responses such as those in the left two columns obtained at different background adaptation intensities, which are indicated by where the curves cross the x-axis. (After Normann, R. A.; Werblin, F. S. *J. Gen. Physiol.* 63:37–61, 1974.)

ties tested, and the stimulus-response curve was similar to that obtained during dark adaptation (Fig. 19–18, right column). Only at low adapting intensities (i.e., 3.5 log units) does the rod (Fig. 19–18, leftmost curve, right column) exhibit a response curve comparable with that obtained during dark adaptation. As the level of background illumination increases, therefore, rod responses to test intensities are progressively reduced, whereas the 3.5–log unit operating curve of cones retains its range and is merely shifted along the intensity axis, so that it always spans the intensity region around each new background intensity. Both the saturation of the rods and the shifting of the cone operating curves occur at background intensities below the levels that bleach substantial amounts of photopigment. As light

intensity increases, the cones effectively put on sunglasses, whereas the rods cannot and so become saturated. Indeed, rod monochromats (persons with no cones) must wear sunglasses to function out of doors. A similar shift in operating curves with background intensity can also be demonstrated for retinal ganglion cells, which exhibit a graded change in firing rate over an approximately twofold change in test spot intensity. Thus, these shifts in operating range contribute to the phenomena of light and dark adaptation over the middle of the 10^{10} operating range of human vision.

At the low end of the operating range, adaptation has an additional neural component since receptive fields of ganglion cells lose their surround inhibition when the retina is dark adapted. The mechanisms whereby surround inhibition is lost and those whereby individual neurons undergo shifts of operating range are not yet understood.

Bipolar Cells. The next element in the radial pathway linking receptors to ganglion cells is the bipolar cell, which comes in two varieties. In response to a central stimulus, one type of bipolar cell hyperpolarizes (like that illustrated in Fig. 19–17), whereas the other type depolarizes. Unlike the receptors, both hyperpolarizing and depolarizing bipolars have an antagonistic surround, but it is usually rather weak and, unlike that of ganglion cells, can be demonstrated only under special conditions. The bipolar cell in Figure 19–17, for example, was first hyperpolarized by a central stimulus in order to reveal the small antagonistic depolarizing response caused by the peripheral annulus. The anatomical extent of the dendritic field of a bipolar cell, as revealed by intracellular dye injections, is almost identical with the physiological extent of its receptive field center, suggesting that the center response is generated by a direct input from receptors overlying the dendritic tree. In contrast, the receptive field surround of the bipolar cell far exceeds its dendritic spread and, as described below, is thought to be mediated by horizontal cells. In fish, inputs from both rods and cones can synapse on the same bipolar cell. In the cat, however, bipolar cells are innervated exclusively by either cones or rods,[33] and signals from cones and rods can converge first at the ganglion cell.

Depolarizing and hyperpolarizing bipolar cells have different morphologies and synaptic processes, suggesting that they serve different types of ganglion cells. In both the carp and the cat, the dendrites of cone bipolar cells provide either flat or invaginating postsynaptic contacts with their overlying receptors. The "axons" of flat and invaginating bipolar cells each ramify in two different sublaminae of the inner plexiform layer. Each sublamina in turn receives the dendritic tree of either on- or off-center ganglion cells (Kaneko).[25a, 32] By penetrating and labeling bipolar cells in the carp, investigators have demonstrated that the axon of a depolarizing bipolar cell terminates in the sublamina that contains the dendritic trees of on-center ganglion cells, whereas the axons of hyperpolarizing bipolar cells appear to contact off-center ganglion cells. In the catfish, depolarizations produced by current injected into an on-center bipolar cell produced an increased discharge in nearby ganglion cells, whereas activation of a hyperpolarizing bipolar cell produced a depression of activity in other nearby ganglion cells.[31] These data suggest the existence of separate on- and off-channels to process visual information, one of several examples in which specific attributes of a visual stimulus are delivered to the central visual system over separate pathways.

More recent experiments have shown that the separation into on- and off-channels based on morphological and electrophysiological criteria is not absolute. In both the carp and the cat, invaginating bipolar cells may be of the depolarizing or hyperpolarizing variety, and, in the cat, some hyperpolarizing bipolar cells appear to contact on-center ganglion cells.[33]

Horizontal Cells. Like the receptor cells from which they receive their input, horizontal cells have a slightly depolarized membrane potential in the dark, and all hyperpolarize in response to white light. Horizontal cells can be divided into two groups according to their spectral response properties. Luminosity, or L-type, horizontal cells hyperpolarize for all wavelengths of light, whereas chromatic, or C-type, horizontal cells hyperpolarize for some wavelengths and depolarize for others.

Figure 19–17 shows that the receptive fields of horizontal cells, in contrast with those of receptors, are very large but, like those of receptors, display no evidence of an inhibitory surround. Many L-type horizontal cells in fish have receptive fields much in excess of their dendritic trees, a fact suggesting that, like receptors, they are electrically coupled. Indeed, electrical coupling has been demonstrated in fish and amphibia by showing that (i) current injected into one horizontal cell spreads to its neighbor with some decrement, (ii) a dye injected into one cell diffuses into its neighbors, and (iii) gap junctions are widely found between contiguous horizontal cells.[25a] Horizontal cells, then, process information from wide regions of the retina and seem to be ideal elements to produce the surround inhibition seen in bipolar cells.

A Model for Center-Surround Receptive Fields. On the basis of their light- and electron-microscopic studies, Dowling and Werblin[20] deduced a schematic wiring diagram for the *Necturus* retina (Figure 19–19). In the outer plexiform layer, receptors drive both bipolar and horizontal cells, while horizontal cells synapse laterally on adjacent bipolar cells. In the inner plexiform layer, bipolar cells synapse with both amacrine and ganglion cells, while amacrine cells drive ganglion cells and other amacrine cells and even feed back to bipolar cells. Finally, Dowling and Werblin[20] suggested that there were two types of ganglion cells; one

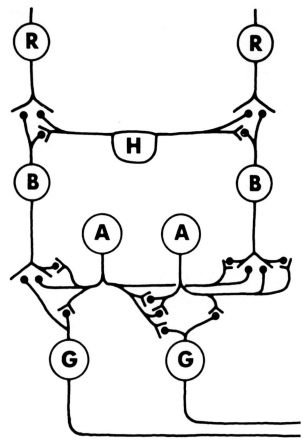

Figure 19–19 Schematic wiring diagram of the *Necturus* retina. (From Dowling, J. E.; Werblin, F. S. *J. Neurophysiol.* 32:315–338, 1969.)

was driven primarily by a bipolar cell input (left cell, Fig. 19–19), and a second was driven by an amacrine input. A similar diagram apparently can be used to describe all vertebrate retinas, the major difference being the relative numbers of the different types of synaptic contacts in the inner plexiform layer. In primate and cat retinas, the direct bipolar-to-ganglion-cell connections dominate, whereas in frog and pigeon retinas, there is much more amacrine cell involvement. Let us consider, then, how the left-hand pathway in Figure 19–19 may produce the center-surround organization of an on-center, sustained ganglion cell.

Figure 19–20 shows the changes of membrane potential in a receptor, horizontal, bipolar, and ganglion cell to a spot that is turned on (solid bar) in the receptive field center of the ganglion cell and an annulus that is turned on later (hatched bar) in the surround. The spot elicits a response only in the cone or cones that it covers. Since

horizontal cells sample large areas of the retina and require considerable spatial summation of stimuli, they are only weakly hyperpolarized by the small spot. The spot, however, causes a sizable depolarization in the underlying depolarizing bipolar cell that leads to depolarization and discharge in its target ganglion cell. When the annulus is turned on, there is no response in the central cone because of its small receptive field. However, the annulus falls on many other cones that synapse on the extensive dendritic tree of the horizontal cell. The net effect of these inputs is a large hyperpolarization in the horizontal cell that together with other horizontal cells is thought to cause the hyperpolarization in the bipolar cell. Indeed, in the carp hyperpolarization of horizontal cells of whatever type produces hyperpolarization in on-center bipolar cells and depolarization in off-center bipolar cells.[41] Finally, the hyperpolarization in the bipolar cell leads to membrane hyperpolarization of the ganglion cell to terminate its firing.

The center-surround receptive field, therefore, is first developed at the bipolar cell. It is remarkable that this elementary building block of visual processing is created without the benefit of an action potential. Apparently, the intraretinal distances are so short that it is efficient for cells to communicate by graded potentials. Only when visual information must be delivered from the retina over long distances into the central nervous system is it converted into trains of action potentials.

As indicated in Chapter 11, release of synaptic vesicles at all synapses thus far studied increases with depolarization in the presynaptic terminal. Therefore, equipped with pre- and postsynaptic responses to the spot and annulus, we can predict the nature of the transmitter at each of the synapses shown in Figure 19–20. Since hyperpolarization in the cone must reduce transmitter release and this causes depolarization in the bipolar cell, the postsynaptic effect of the cone-bipolar transmitter must be to cause hyperpolarization. In conventional terms, one would call this an inhibitory synapse, but it is preferable here to call it a sign-reversing synapse (I) because presynaptic hyperpolarization produces postsynaptic depolarization. In contrast, hyperpolarization in the cone causes hyperpolarization in the horizontal cell, suggesting that the cone-horizontal cell synapse is sign preserving (E). Similar logic can be used to deduce that the synapses from horizontal cell to bipolar cell and from bipolar cell to ganglion cell also must be sign preserving.

Unfortunately, rather little is known about specific transmitters in retinal neurons. Retinal transmitter substances have been identified by the use of neurotoxins and immunocytohistochemical techniques in which labeled antibodies to specific putative transmitters or their precursors can be detected in certain cells or retinal laminae.[6] Although such tests implicate just about every putative neurotransmitter in retinal function, few can be assigned definitely to any one retinal cell population or retinal circuit. Perhaps the best-characterized cells are the GABA-accumulating cone, horizontal, and amacrine cells of the goldfish. Curiously, one type of amacrine cell or another appears to contain every neurotransmitter candidate tested.

Amacrine Cells. Of all the retinal neurons, amacrine cells are the most difficult to study. Like the one shown in Figure 19–17, the vast majority of those in a variety of fish and amphibia exhibit a transient depolarization when stimuli are turned on and off anywhere in their receptive fields; frequently, the depolarization leads to spikes that are atypical and often abortive.[25a] The receptive fields of amacrine cells, like those of horizontal cells, are quite large, sum visual stimuli over large retinal areas, and generally show no evidence of an antagonistic surround. These so-called transient amacrine cells have no color sensitivity. A minority of amacrine cells (reported primarily in the carp) show sustained membrane potential changes that depend on the wavelength of the stimulus.

Because amacrine cells respond only as the light is turned on and off, Werblin[43] reasoned that they prefer a changing rather than a steady illumination and that they influence only transient ganglion cells. To test this proposal, he produced a windmill stimulator with light and dark vanes that could be spun to create a changing illumination in the receptive field surround of a ganglion cell. If the windmill was stationary, turning on a small central spot (solid bar) elicited depolarization in the underlying bipolar cell, BP_1 (Fig. 19–21). Because of

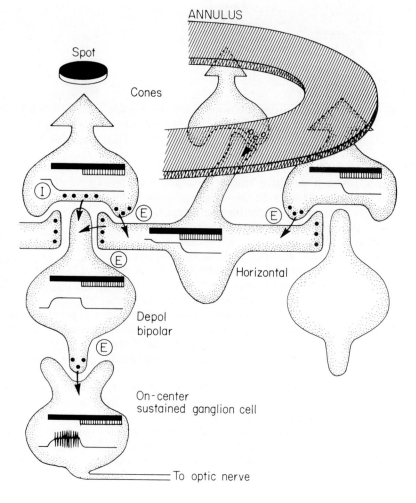

Figure 19–20 A hypothetical model that produces the center-surround organization of an on-center, sustained ganglion cell. The responses to the presence of a small spot of light (solid bars) and the subsequent appearance of an annulus (hatched bars) are shown for the ganglion cell and those retinal neurons that are stimulated and interact at the outer plexiform layer.

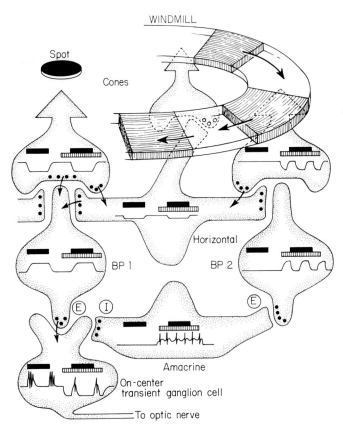

Figure 19–21 A hypothetical model that produces the center-surround organization of an on-center transient ganglion cell. The responses to the presence of a small spot of light (solid bars) and the spinning of a peripheral windmill (hatched bars) are shown for the ganglion cell and those retinal neurons that are stimulated and interact at both the inner and outer plexiform layers.

mechanisms like those demonstrated in Figure 19–20, the depolarization is less than would occur with no stationary peripheral (i.e., antagonistic) stimulus. The depolarization in the bipolar cell and subsequent repolarization when the spot turns off each produce transient depolarizations and action potentials in the target ganglion cell. The mechanism by which the *steady* depolarization in BP_1 produces the *transient* depolarizations in the ganglion cell is not known. When the windmill is spun (Fig. 19–21, hatched bar), there is a marked hyperpolarization in the ganglion cell and a significant reduction in its response to the spot (solid bar) stimulus. The hyperpolarization in the ganglion cell occurs without changes in the membrane potential of BP_1 and must therefore occur because of connections in the inner plexiform layer. Amacrine cells driven by the spinning windmill undergo transient depolarizations in response to each of the passing vanes (Fig. 19–21). It is likely, therefore, that the sustained hyperpolarization in the ganglion cell results from many such transient inputs arriving asynchronously at the ganglion cell from several amacrine cells. The amacrine cells would, in turn, be driven by peripheral bipolar cells such as BP_2, which would also depolarize

while each blade of the windmill was passing over its receptive field. The small light spot, of course, would have no effect on either the peripheral receptor or BP_2. Based on our previous logic, the synapse from the amacrine cell to the ganglion cell must be inhibitory.

Amacrine cells of birds and amphibia receive centrifugal optic nerve fibers originating in the midbrain. Although single-shock stimulation of the optic nerve of turtles produces EPSPs in amacrine cells and repetitive stimulation reduces their light response,[29] it is impossible from the present data to ascribe a functional role to the centrifugal system.

Interplexiform Cells. Recently, a sixth kind of retinal cell, believed to be present in all species, has been identified. Called *interplexiform cells* because they extend processes into both plexiform layers, these rare cells receive input exclusively from amacrine cells and synapse abundantly on horizontal cells or bipolar cell dendrites or both. Based on the effects of the intraretinal application of dopamine, the putative neurotransmitter of carp interplexiform cells, Dowling[18] has suggested that the interplexiform cells may act to diminish the surround inhibition mediated by horizontal cells.

(In some animals, dopamine also appears to disconnect electrically coupled horizontal cells.)

In conclusion, Figures 19–20 and 19–21 are presented to give the reader a feeling for how retinal neurons *could* be connected to produce ganglion cell receptive fields. It must however be stressed that in the cat, at least 23 different ganglion cell types and eight different cone bipolar cells have been identified on morphological grounds. If each of these different cell types eventually proves to have distinctive response properties, connections, and transmitters, as suggested by Sterling,[39] the actual retinal circuits may be much more complicated than those suggested in Figures 19–20 and 19–21.

PROJECTIONS FROM THE RETINA: X, Y, AND W PATHWAYS

Discharge Properties of X and Y Cells. By the early 1960s, the large majority of ganglion cells studied in a wide variety of animals had been demonstrated to have receptive fields with a concentric center-surround organization. Also, it was clear that the cells with center-surround receptive field organization could have either sustained or transient firing patterns. The functional significance of this dichotomy in temporal firing patterns was first pointed out in the cat by Enroth-Cugell and Robson.[21] To study the spatial interaction of a receptive field center and its surround, they subjected receptive fields to stimuli composed of gratings that had a periodic pattern (sine or square wave) of spatial luminance (like the pattern for testing visual acuity in Fig. 19–14). When the grating pattern was turned on and off while positioned at different locations on the receptive field, ganglion cells responded with one of two different response patterns. As shown in Figure 19–22, the upper ganglion cell exhibited a sustained increase in firing rate for some grating positions (like that of A) and a sustained decrease for others (like that of B) so that a grating position always could be found (such as in C) where turning the stimulus on and off produced little if any response. Such behavior suggested that these cells sum visual stimuli in a nearly linear fashion over their receptive fields. On the other hand, the lower ganglion cell exhibited a transient increase in rate when the grating was turned on and off at all positions in the receptive field, and no "null" position could be found; therefore, the firing pattern obtained when the grating pattern was in position C could not be explained by a linear summation of the responses due to gratings at A and B. Based on their linear and nonlinear summation properties, the two types were noncommittally classified as X and Y, respectively.[21]

Further tests on X and Y cells of the cat have revealed that they can also be differentiated by other, perhaps more interesting, functional criteria. First, the receptive field centers and surrounds of X cells are smaller than those of Y cells and are located more centrally on the retina.[13, 21] The smaller receptive field size of X cells can also be revealed by causing a periodic grating (like that in Fig. 19–22) to drift across the receptive field in order to determine how closely the bars may be spaced (i.e., their spatial frequency) and still elicit a distinct change in firing rate for each passing bar. By this test and by analogy with the data in Figure 19–14, X cells had a higher "visual acuity" than Y cells, because the best X cell responded to gratings with spatial frequencies of up to 4 cycles/degree (i.e., each bar was 1/8 degree, or about 8 minutes of arc wide), whereas the typical Y cell exhibited a modulation in firing only for spatial

Figure 19–22 Responses of X and Y ganglion cells (middle panels) to a stimulus grating of light and dark stripes turned on (upward deflection in lowest panel) and off at different locations within their receptive fields (upper panel).

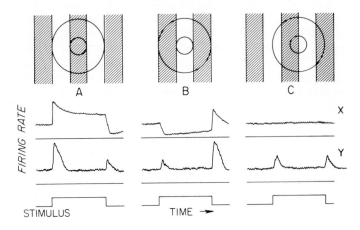

frequencies of less than about 1 cycle/degree. More recently, X cells that respond to almost 10 cycles/degree (bars subtend 3 minutes of arc) have been reported.[10] Like receptive field size, the visual acuity of both X and Y cells depends on distance from the area centralis. At the area centralis, the average acuity for an X cell is 7.5 cycles/degree compared with 3 cycles/degree for the infrequently encountered Y cell, while at an eccentricity of 20 degrees, average X and Y cell acuities drop to 2 and 0.8 cycles/degree respectively.[10]

Second, in response to discs moving across their receptive fields, X cells respond strongly to velocities of only 20°/s or less, whereas Y cells respond to targets moving at rates of up to 200°/s.[9] Although recent studies have shown that some X cells also can respond to high velocities, the mean preferred velocities of X cells are always significantly less than those of Y cells. Third, Y cells are excited by moving stimuli delivered well outside their conventionally delineated receptive fields, whereas X cells are not.[9] This *periphery effect* (more recently called the shift effect because it is effectively demonstrated by shifting a frame of light at the edges of the tangent screen) is one of the most unambiguous tests for distinguishing X from Y cells. Finally, although cats have poor color vision, the discovery of X and Y ganglion cells in the monkey[16] provided the opportunity to demonstrate that X cells have color sensitivity, whereas most Y cells are achromatic. (See Chapter 20 for further discussion.)

The characteristics of X and Y cells in response to these various stimuli are summarized in the first eight entries of Table 19–2. A comparison of these entries suggests that X and Y cells are concerned with different aspects of visual processing. X cells respond best to small, stationary stimuli falling on the area centralis. Consequently, they appear well suited to subserve visual tasks requiring high acuity. Y cells, on the other hand, prefer larger, moving stimuli falling on the peripheral retina; they seem better suited to be involved in the detection of visual motion.

Not only do X and Y ganglion cells appear to have different roles in visual processing but also the information that each conveys leaves the retina over separate pathways. Cleland and colleagues[9] reached this conclusion by simultaneously recording from ganglion cells and the cells to which they project in the dorsal lateral geniculate nucleus (LGN$_d$) of the thalamus. First, they demonstrated that LGN$_d$ neurons also can be identified as X- or Y-like by the same criteria used for ganglion cells. Then, after isolating a single LGN$_d$ cell, they located a ganglion cell with a similar receptive field location and demonstrated that its action potential reliably preceded one in the LGN$_d$ neuron at a relatively fixed, short latency. Of the few pairs of cells identified as connected by these criteria, the general picture emerged that a particular LGN$_d$ cell received an excitatory input from one or more neighboring ganglion cells of the same functional type. X ganglion cells drove X LGN$_d$ cells, Y ganglion cells drove Y LGN$_d$ cells and on- and off-center ganglion cells contacted on- and off-center LGN$_d$ cells, respectively. Furthermore, the conduction times measured during the simultaneous recordings fell into two nonoverlapping groups, with Y cells having conduction velocities on the order of 40 to 60 m/s compared with 10 to 20 m/s for X cells. In the monkey, Y ganglion cells also tend to have faster conduction velocities than X ganglion cells.[16] Consequently, X and Y information not only remain separate but Y information reaches its destination faster.

W Cells. In many of the early studies of cat retinal ganglion cells, there usually were a few cells that did not have a concentric center-surround receptive field configuration. In the most comprehensive study, Cleland and Levick[11] estimated that about 7.5% of their 960 sampled cells had unusual stimulus requirements. Examples of such rarely encountered cells include (i) a type with a continuously maintained activity that is suppressed by the presence of any form of contrast, (ii) a type that prefers a contrasting border confined to a small region of its receptive field, and (iii) a type that responds only when any object is moved in a particular direction across its receptive field (note that a concentric X or Y cell responds equally well to motion in any direction across its field).[40] Because all of these types have conduction velocities that are much slower than those of either X or Y cells, they have been named W cells. If conduction velocity is the only criterion used, however, some of the concentric center-surround units of the X type must be placed in the W category. Indeed, these reclassified cells do have discharge properties that are more "sluggish" than the briskly responding X and Y cells (see Rodieck[38] for more details). However, even if these sluggish X cells are moved to the W category, cells identified as W by electrophysiological criteria account for only 20% of all recorded cells. It has been argued that W cells are actually more plentiful but have been overlooked because of their exotic stimulus requirements and because their slow conduction velocities suggest that they must be very small.

Table 19–2 Properties of Cat Retinal Ganglion Cells

Property	X	Y	W
Response to steady spot	Sustained	Transient	Either
Latency to best visual stimulus	30 ms	20 ms	
Spatial summation over RF	Linear	Nonlinear	Not tested
Size of RF center*, [13, 21]	1.3° (0.6–2.9)†	3° (1–7)	Like Y cells
	0.4 (0.2–0.8)†	1.1 (0.5–1.9)	
RF location	More central	More peripheral	Uniform distribution
Acuity to moving grating (cycles/degree) (central→peripheral RFs)	7.5→2.0	3→0.8	Weak, if any, grating response
Velocity preference	Mostly low speeds (narrow tuning)	All speeds (broad tuning)	Mostly low, but depends upon type
Periphery effect	No	Yes	No
Color sensitivity (monkey)	Yes	No	1 subgroup
Conduction velocity (m/s)	10–20[12]	40–60	5–15
Anatomical cell type correlate	β	α	γ
Estimated %			
(a) Based upon anatomy[35] (central→peripheral)	55 (65→35)	3 (2→4)	42 (35→65)
(b) Based upon electrophysiology[23]	65→48	8→38	26→13
Destinations	Parvo LGN_d	Magno LGN_d	LGN_v
	Pretectum	Superior colliculus	Superior colliculus
		Pretectum	Pretectum

*No attempt was made to separate central and peripheral receptive fields.
†Mean or mode (and range).

In lower vertebrates such as the frog and rabbit, exotic ganglion cells account for a higher percentage of the total. Indeed, some ganglion cells in these species appear to have stimulus requirements (e.g., movement in a preferred direction or stimuli with preferred orientations) that (as we shall see in the next chapter) are first found at more central visual structures in higher mammals.

Morphological Differences in X, Y, and W Cells. Anatomical studies suggest that the three ganglion cell types distinguished by functional criteria also can be distinguished morphologically. Ganglion cells in Golgi-stained preparations of cat retinas can be classified into three main groups according to whether they have large perikarya and large dendritic trees (alpha cells), smaller perikarya and the smallest dendritic trees (beta cells), or the smallest perikarya but dendritic trees comparable to those of alpha cells (gamma cells).[5] In a heroic experiment, Cleland and co-workers[14] exhaustively searched small patches of cat retina with a microelectrode to locate all the Y cells, processed the retina histologically to locate all the alpha cells, and demonstrated a one-to-one correspondence between Y cells and alpha cells. In a more recent study, Y, X, and W ganglion cells, identified on the basis of conduction velocity, receptive field properties, and the level of maintained activity, were filled with horseradish peroxidase (HRP).[22] Although only 21 ganglion cells were adequately labeled, all Y cells had an alpha morphology, and all X cells had a beta morphology. The few W cells had gamma cell morphologies. If the correspondence between functional discharge properties and morphology holds, the anatomy suggests that depending on retinal location, X cells account for between 35% and 65% of all ganglion cells, Y cells only 2% to 4%, and W cells between 35% and 65%. If these percentages are accurate, neurophysiological studies have indeed underestimated the number of W cells.

Combined electrophysiological and anatomical studies led to the estimate that at every retinal location, the receptive field centers of from 7 to 20 X cells and from 3 to 6 Y cells overlap. For all the W cells (i.e., cells neither X nor Y) taken together, up to 60 have receptive field centers that overlap at a common retinal location.[35]

The morphological characteristics of X, Y, and W cells serve to further distinguish the three cell types (Table 19–2). Each type also projects to different destinations, a point that we pursue more fully in Chapter 20.

Summary. Some general conclusions can be drawn about the possible roles of X, Y, and W cells. X cells prefer small, stationary, colored (in the monkey) stimuli that fall on or near the fovea (area centralis). These cells have a high visual acuity and can therefore be expected to participate in the process that extracts detail from a visual scene. The information provided by X cells can be thought of as an abstract decomposition of the visual world into small regions of high contrast. Because the receptive field centers of up to 20 X cells are stimulated from a single point in visual space, the neural replica of the visual world may be represented several times in the discharges of different groups of X cells. For example, we have already seen that information from on- and off-center receptive fields is processed separately. Clearly, the limits of our visual acuity and our ultimate ability to abstract a perception from a visual stimulus depends on these cells. Y cells, on the other hand, seem suited for detecting larger, achromatic stimuli moving at a wide range of velocities across their receptive field centers or the more distant periphery; furthermore, they exhibit only a transient response to stationary stimuli. Because of their larger receptive fields and fewer numbers, Y cells would appear to provide only rather gross information about the spatial characteristics of movement in the visual world. Such information would be important not only for motion perception but also as a sensory stimulus to elicit an orienting movement toward a novel stimulus. Finally, because of their diverse stimulus requirements it is impossible to suggest a single unique role for W cells. As we shall see in the next chapter, however, several subclasses of the W cell with unique response properties may be part of pathways involved in the control of movement.

ANNOTATED BIBLIOGRAPHY

Barlow, H.B.; Mollon, J.D., eds. *The Senses*. Cambridge, Cambridge University Press, 1982.
A very readable treatment of the psychophysics of vision.
Kaneko, A. Physiology of the retina. *Annu. Rev. Neurosci.* 2:169–191, 1979.
A comprehensive review of the intracellular responses of all of the retinal neurons in a variety of species.
Lennie, P. Parallel visual pathways: a review. *Vision Res.* 20:561–594, 1980.
An excellent review of the properties of X, Y, and W retinal ganglion cells and their central projections.

REFERENCES

1. Barlow, H.B.; Mollon, J.D. Psychophysical measurements of visual performance, 114–132. In H.B. Barlow and J.D. Mollon, eds. *The Senses*. Cambridge, Cambridge University Press, 1982.
2. Baylor, D.A.; Fuortes, M.G.F.; O'Bryan, P.M. Receptive fields of cones in the retina of the turtle. *J. Physiol. (Lond.)* 214:265–294, 1971.
3. Baylor, D.A.; Lamb, T.D.; Yau, K.-W. Responses of retinal rods to single photons. *J. Physiol. (Lond.)* 288:613–634, 1979.
4. Borwein, B.; Borwein, D.; Medeiros, J.; McGowan, J.W. The ultrastructure of monkey foveal photoreceptors, with special reference to the structure, shape, size, and spacing of the foveal cones. *Am. J. Anat.* 159:125–146, 1980.
5. Boycott, B.B.; Wässle, H. The morphological types of ganglion cells of the domestic cat's retina. *J. Physiol. (Lond.)* 240:397–419, 1974.
6. Brecha, N. Retinal neurotransmitters: histochemical and biochemical studies, 85–129. In P.C. Emson, ed. *Chemical Neuroanatomy*. New York, Raven Press, 1983.
7. Campbell, F.W.; Gubisch, R.W. Optical quality of the human eye. *J. Physiol. (Lond.)* 186:558–578, 1966.
8. Cervetto, L.; Fuortes, M.G.F. Excitation and interaction in the retina. *Annu. Rev. Biophys. Bioeng.* 7:229–251, 1978.
9. Cleland, B.G.; Dubin, M.W.; Levick, W. E. Sustained and transient neurones in the cat's retina and lateral geniculate nucleus. *J. Physiol. (Lond.)* 217:473–496, 1971.
10. Cleland, B.G.; Harding, T.H.; Tulunay-Keesey, U. Visual resolution and receptive field size: examination of two kinds of cat retinal ganglion cell. *Science* 205:1015–1017, 1979.
11. Cleland, B. G.; Levick, W. R. Properties of rarely encountered types of ganglion cells in the cat's retina and an overall classification. *J. Physiol. (Lond.)* 240:457–492, 1974a.
12. Cleland, B.G.; Levick, W.R. Brisk and sluggish concentrically organized ganglion cells in the cat's retina. *J. Physiol. (Lond.)* 240:421–456, 1974b.
13. Cleland, B.G.; Levick, W.R.; Sanderson, K.J. Properties of sustained and transient ganglion cells in the cat retina. *J. Physiol. (Lond.)* 228:649–680, 1973.
14. Cleland, B.G.; Levick, W.R.; Wässle, H. Physiological identification of a morphological class of cat retinal ganglion cells. *J. Physiol. (Lond.)* 248:151–171, 1975.
15. Dartnall, H.J.A.; Bowmaker, J.K.; Mollon, J.D. Microspectrophotometry of human photoreceptors, 69–80. In J.D. Mollon and L.T. Sharpe, eds. *Color Vision*. London, Academic Press, 1983.
16. de Monasterio, F.M. Properties of concentrically organized X and Y ganglion cells of macaque retina. *J. Neurophysiol.* 41:1394–1417, 1978a.
17. de Monasterio, F.M. Center and surround mechanisms of opponent-color X and Y ganglion cells of retina of macaques. *J. Neurophysiol.* 41:1418–1434, 1978b.
18. Dowling, J.E. A new retinal neurone—the interplexiform cell. *TINS* 2:189–191, 1979.
19. Dowling, J.E.; Boycott, B.B. Organization of the primate retina: electron microscopy. *Proc. R. Soc. Lond. [Biol]* 166:80–111, 1966.
20. Dowling, J.E.; Werblin, F.S. Organization of retina of the mudpuppy, *Necturus musculosus*. I. Synaptic structure. *J. Neurophysiol.* 32:315–338, 1969.
21. Enroth-Cugell, C.; Robson, J.G. The contrast sensitivity of retinal ganglion cells of the cat. *J. Physiol. (Lond.)*, 187:517–552, 1966.
22. Fukuda, Y.; Hsiao, C.-F.; Watanabe, M.; Ito, H. Morphological correlates of physiologically identified Y-, X-, and W-cells in cat retina. *J. Neurophysiol.* 52:999–1013, 1984.
23. Fukuda, Y.; Stone, J. Retinal distribution and central projections of Y-, X-, and W-cells of the cat's retina. *J. Neurophysiol.* 37:749–772, 1974.
24. Hecht, S.; Schlaer, S.; Smith, E. L.; Haig, C.; Peskin, J. C. The visual functions of the completely color blind. *J. Gen. Physiol.* 31:459–472, 1948.
25. Hecht, S.; Schlaer, S.; Verrijp, C. D. Intermittent stimulation by light. II. The measurement of critical fusion frequency for the human eye. *J. Gen. Physiol.* 17:251–268, 1933.

25a. Kaneko, A. Physiology of the retina. *Annu. Rev. Neurosci.* 2:169–191, 1979.

26. Knowles, A. The biochemical aspects of vision, 82–101. In Barlow, H. B.; Mollon, J.D., eds. *The Senses.* Cambridge, Cambridge University Press, 1982.

27. Kuffler, S.W. Discharge patterns and functional organization of mammalian retina. *J. Neurophysiol.* 16:37–68, 1953.

28. Mandelbaum, J.; and Sloan, L.L. Peripheral visual acuity. *Am. J. Ophthalmology* 30:581–588, 1947.

29. Marchiafava, P.L.; Torre, V. The responses of amacrine cells to light and intracellularly applied currents. *J. Physiol. (Lond.)* 276:83–102, 1978.

30. Millodot, M. Image formation in the eye, 46–61. In H.B. Barlow; Mollon, J.D., eds. *The Senses.* Cambridge, Cambridge University Press, 1982.

31. Naka, K.I. Functional organization of catfish retina. *J. Neurophysiol.* 40:26–43, 1977.

32. Nelson, R.; Famiglietti, E.V., Jr.; Kolb, H. Intracellular staining reveals different levels of stratification for on- and off-center ganglion cells in cat retina. *J. Neurophysiol.* 41:472–483, 1978.

33. Nelson, R.; Kolb, H. Synaptic patterns and response properties of bipolar and ganglion cells in the cat retina. *Vision Res.* 23:1183–1195, 1983.

34. Normann, R.A.; Werblin, F.S. Control of retinal sensitivity. I. Light and dark adaptation of vertebrate rods and cones. *J. Gen. Physiol.* 63:37–61, 1974.

35. Peichl, L.; Wässle, H. Size, scatter and coverage of ganglion cell receptive field centres in the cat retina. *J. Physiol. (Lond.)* 291:117–141, 1979.

36. Polyak, S. *The Vertebrate Visual System.* Chicago, University of Chicago Press, 1957.

37. Rodieck, R.W. *The Vertebrate Retina: Principles of Structure and Function.* San Francisco, W. H. Freeman, 1973.

38. Rodieck, R.W. Visual pathways. *Annu. Rev. Neurosci.* 2:193–225, 1979.

39. Sterling, P. Microcircuitry of the cat retina. *Annu. Rev. Neurosci.* 6:149–185, 1983.

40. Stone, J.; Fukuda, Y. Properties of cat retinal ganglion cells: a comparison of W-cells with X- and Y-cells. *J. Neurophysiol.* 37:722–748, 1974.

41. Toyoda, J-I.; Kujiraoka, T. Analyses of bipolar cell responses elicited by polarization of horizontal cells. *J. Gen. Physiol.* 79:131–145, 1982.

42. Wagner, H. G.; MacNichol, E. E.; Wolbarsht, M.L. Functional basis for "on"-center and "off"-center receptive fields in the retina. *J. Opt. Soc. Am.* 53:66–70, 1963.

43. Werblin, F. S. Lateral interactions at inner plexiform layer of vertebrate retina: antagonistic responses to change. *Science* 175:1008–1010, 1972.

44. Werblin, F.S.; Dowling, J.E. Organization of retina of the mudpuppy, *Necturus musculosus.* II. Intracellular recording. *J. Neurophysiol.* 32:339–355, 1969.

45. Westheimer, G. The spatial sense of the eye. *Invest. Ophthalmol. Vis. Sci.* 18:892–912, 1979.

46. Woodhouse, J.M.; Barlow, H.B. Spatial and temporal resolution and analysis, 132–164. In Barlow H. B.; Mollon, J.D. eds. *The Senses.* Cambridge, Cambridge University Press, 1982.

The Visual System: Neural Processing Beyond the Retina

INTRODUCTION

In Chapter 19, we learned that the retina breaks down the visual world into many very small subregions called receptive fields. In primates, the visual world is represented by as many as one million receptive fields, some of which subtend only minutes of arc. Since we are unaware of this fragmented representation, the brain must be incredibly clever at converting this punctate neuronal picture into a coherent perception. The brain not only produces a snapshot of the visual scene, but also interprets the snapshot, attaching more significance to certain parts of it (e.g., a familiar face) than to others. Such manipulations of the afferent visual information are very complicated, and we have only just begun to understand how these and other aspects of visual perception might be realized. In this chapter, we explore how the information that leaves the retina is processed en route to, and within, the cerebral cortex, where perception is thought to occur. In addition, we examine the effect of visual signals on the generation of movements that orient the direction of gaze toward objects of interest and stabilize the visual world on the retina. In general we will limit our attention to data from cats and monkeys, since the neurophysiology of vision has been studied most extensively in cats and the visual system of monkeys is most similar to that of man.

OVERVIEW OF THE VISUAL PATHWAYS

Division of the Visual World

The visual world, or *visual field*, is everything that one can see without moving one's eyes. The basic organizing principle for the flow of retinal information entering the central nervous system is that the right half of the visual field projects to the left side of the brain and the left half of the visual field projects to the right side of the brain. Therefore, in animals with laterally placed eyes (e.g., rats), the axons in the optic nerve from each eye must cross the midline. This decussation occurs at the optic chiasm. In frontal-eyed animals such as primates, the division between the right and left half of the visual field is drawn at the vertical meridian that passes through the nose. Since the right visual field falls on the temporal retina of the left eye and the nasal retina of the right eye (Fig. 20–1B), the optic nerve (ON) fibers from the nasal retina of the right eye must cross to the left (i.e., contralateral) side of the brain at the chiasm (OX) but the fibers from the temporal retina of the left eye remain on the left (i.e., ipsilateral) side. In monkeys, about one half the optic nerve fibers from each eye cross the midline.

Three Visual Pathways

Once the axons from each eye have been appropriately directed at the chiasm, they head for several different structures. The targets of the different axons and the anatomical projections from those targets suggest that the visual system has three broad divisions.

In mammals, most of the retinal ganglion cells (80% in monkeys; 70% in cats) project to the *dorsal lateral geniculate nucleus* (LGN_d) of the thalamus (Fig. 20–1A). The LGN_d, in turn, sends a strong projection to the ipsilateral occipital lobe via the optic radiation. This primary visual cortex is known variously as V_1, area 17, (according to the anatomist Brodmann) or the *striate cortex*, so-called because the terminating LGN_d fibers are visible as a prominent stripe on a cross section of cortex. As mentioned in Chapter 19, the LGN_d in both cats and monkeys receives the axons of both X and Y retinal ganglion cells. Because the LGN_d appears to receive most of the X-cell axons, the *geniculo-striate pathway* is thought to play a prominent role in fine acuity and form vision.

The second strongest retinal projection is to the superficial layers of the *superior colliculus* in the midbrain (Fig. 20–1B). The superficial layers of the colliculus project to another thalamic nucleus, the pulvinar, which in turn projects to several cortical visual areas other than the striate cortex (i.e., to the *extrastriate cortex*). In both cats and monkeys, the superior colliculus receives both Y- and W-cell axons from the retina but little, if any, X-cell input.[21, 38] Consequently the *colliculo-pulvino-extrastriate pathway* is thought to be involved in visual-motion sensitivity and low-acuity form vision.

A third, much weaker, retinal projection terminates in the *pretectum* and *accessory optic nuclei* (Fig. 20–1C). The pretectum lies near the superior colliculus (also known as the optic tectum) and is composed of six separate nuclei that apparently subserve a variety of functions. In cats, the pretectum receives inputs from all three ganglion cell types,[34a] but in monkeys, only W-cell and Y-cell inputs have been identified thus far. The accessory optic nuclei comprise three cell clusters that lie in the rostral brainstem among the fascicles of the optic tract. The medial terminal nucleus (MTN) lies near the emergence of the oculomotor nerve between the substantia nigra and the medial lemniscus, whereas the lateral and dorsal terminal nuclei (LTN and DTN) occupy the lateral border of the brainstem medial and ventral to the LGN_d and lateral and ventral to the rostral pole of the superior colliculus, respectively. Anatomical and neurophysiological studies suggest that the accessory optic nuclei of cats receive W-cell and perhaps Y-cell inputs. The pretectum and accessory optic nuclei project only weakly, if at all, to the cortex. Instead, they send efferents to brainstem structures involved with movement (e.g., the inferior olive; see Chapter 29), including those involved with the pupillary light reflex. Therefore, the pretectum and accessory optic system constitute a *midbrain visuomotor pathway* that is involved in the generation of visually elicited movements. The deep layers of the superior colliculus will also be considered as part of the midbrain visuomotor pathway because they receive indirect inputs from a variety of sensory modalities and project to brainstem areas that control eye and head movements.

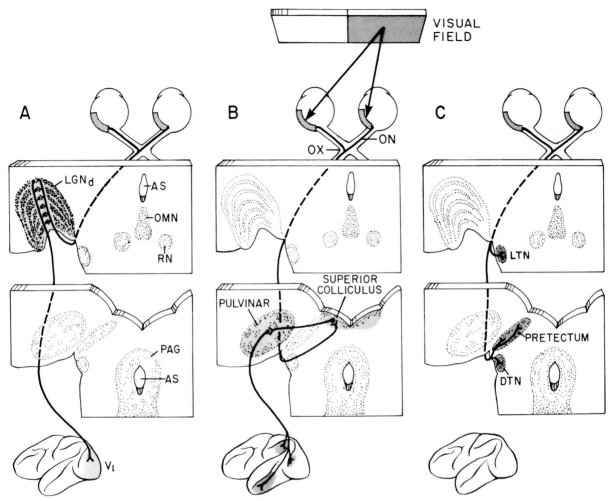

Figure 20–1 An overview of the anatomy of the visual system in which the geniculostriate (A), colliculo-pulvino-extrastriate (B), and midbrain visuomotor (C) pathways are presented separately. ON, optic nerve; OX, optic chiasm; LGN$_d$, Dorsal lateral geniculate nucleus; AS = aqueduct of sylvius; OMN = oculomotor nucleus; RN = red nucleus; PAG = periaqueductal gray; LTN = lateral terminal nucleus; DTN = dorsal terminal nucleus; V$_1$ = primary visual cortex.

Finally, a very small number of axons terminate immediately above the optic chiasm in a small nucleus in the medial hypothalamus called the *suprachiasmatic nucleus*. Little is known about the nature of the visual input in the suprachiasmatic pathways of cats and monkeys. In rodents, this pathway may be involved in regulating the metabolic, glandular, and sleep rhythms associated with a 24-hour day (see Chap. 66) since ablations of the suprachiasmatic nuclei abolish a variety of circadian rhythms that normally are synchronized by a diurnal cycle of light intensity.[34a]

In the following sections, we describe how visual information is processed through each of the three major pathways. After considering the two cortical pathways and the properties of the striate and extrastriate visual areas, we evaluate the various models for the cortical processing of visual information. Finally, we briefly examine the role of the midbrain visual structures in the generation of movement.

THE GENICULOSTRIATE PATHWAY

The Dorsal Lateral Geniculate Nucleus (LGN$_d$)

Anatomical Considerations. The termination of a retinal axon in the LGN$_d$ depends on its eye of origin. In both cats and monkeys, the LGN$_d$ is conspicuously laminated with alternating cellular and axonal layers. Through much of the monkey LGN$_d$, the cellular layers, when cut in coronal sections, resemble six nested horseshoes (numbered 1 to 6 from ventral to dorsal; Fig. 20–2, left side). The two most ventral LGN$_d$ layers (1 and 2) are called *magnocellular* because they have larger cells than the four dorsal layers (3–6), which are called *parvocellular*. The inputs from each eye are kept entirely separate; the nasal retina of the contralateral eye projects to layers 1, 4 and 6 (dashed line) and the temporal retina of the ipsilateral eye projects to layers 2, 3, and 5 (solid line). Although the lamination pattern is clear in monkeys, it is not as obvious in cats, in which the two dorsal laminae are well defined, but the number of ventral laminae are less certain.[21]

Because retinal fibers from each eye reach separate laminae, cells in the LGN$_d$ might be expected to respond to stimuli delivered to only one eye or the other. Early studies confirmed this expectation, but more recent work indicates that although LGN$_d$ cells are strongly activated through one eye, many cells in cats and a few in monkeys are also weakly inhibited by stimuli delivered through the other eye. The anatomical substrate and functional significance for this weak effect are unclear.

In cats, a subdivision of the LGN$_d$ lies just medial to its laminated portion. This medial interlaminar nucleus (Min) has at least two laminae, each receiving input from a different eye. A second nearby thalamic nucleus, the ventral LGN (LGN$_v$), which is called the pregeniculate nucleus in primates, also receives a weak retinal input that appears to be segregated for each eye.

Retinal axons to the LGN$_d$ are distributed not only according to their eye of origin but also according to the part of the retina from which they arise. By recording the visual responses of cells at

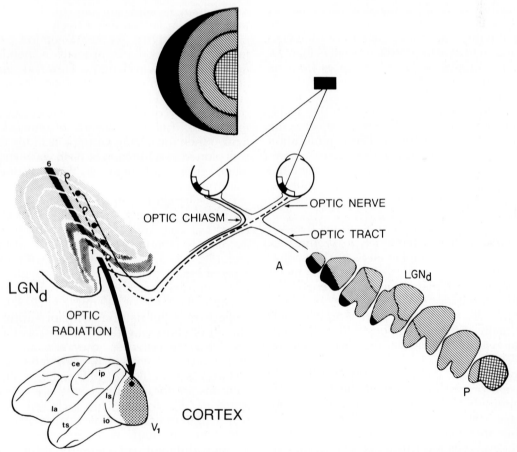

Figure 20–2 Anatomy of the geniculostriate pathways in monkeys. The left visual field, represented as concentric areas, projects to different parts of the right LGN$_d$ as indicated on the seven anterior (A) to posterior (P) coronal sections. A small rectangular stimulus in the right visual field projects to all six layers in the single section through the left LGN$_d$ *(black patches)*. Neurons from all six patches in turn project to a single point on the ipsilateral striate cortex, V$_1$. Abbreviations for cortical sulci: ls = lunate sulcus; io = inferior occipital; ip = intraparietal; la = lateral; ts = temporal sulcus; ce = central.

different LGN$_d$ locations, Malpeli and Baker[25] demonstrated that, in macaque monkeys, adjacent groups of cells respond to visual stimuli that are adjacent in the visual field and therefore fall on adjacent retinal loci. Therefore, there is an orderly *topographic map* of the visual field upon the LGN$_d$. Figure 20–2 shows, in seven coronal sections, the projection on the LGN$_d$ of the left visual field, which is divided into four concentric bands (Fig. 20–2, top) representing (1) the fovea (radius of 2.5 degrees), (2) the central 17 degrees excluding the fovea, (3) the rest of the binocular field, and (4) the monocular crescent (that area of the temporal visual field invisible to the contralateral eye because it lies in the shadow of the nose). The central retina projects to the posterior LGN$_d$, whereas the peripheral retina projects more anteriorly. The central 4 degrees of the visual field preempt 19% of the total LGN$_d$ volume, and the central 20 degrees 64% of its volume. Therefore, the volume of the LGN$_d$ devoted to the central retina (or central visual field) far exceeds that devoted to the peripheral retina. Indeed, in the part of the LGN$_d$ subserving retinal eccentricities greater than 20 degrees, the number of layers decreases from six to four. A measure of the relative representation of the fovea and periphery is given by the *magnification factor*, which is the ratio of the LGN$_d$ representation to the retinal representation at the same point of the visual field. The magnification factor is highest for the fovea and decreases precipitously within the first 10 degrees from the fovea, a change that, in part, reflects the progressive thinning of ganglion cells from the center to the periphery of the retina.

In summary, the retinal inputs from the two eyes are distributed to the LGN$_d$ to produce six separate, nested topographic maps of the contralateral central visual field. All six maps are in spatial register so that in a radial penetration traversing all the layers (black patches, Fig. 20–2, left), all the cells encountered have receptive fields with virtually identical locations within the contralateral visual field (e.g., at the solid rectangle, Fig. 20–2).

Receptive Field Properties. Apparently, very little visual processing occurs at the LGN$_d$. The vast majority of LGN$_d$ cells in both cats and monkeys have concentric center-surround receptive fields that are similar to those of retinal ganglion cells. In cats, cells located in the dorsal laminae behave like either X or Y ganglion cells, while those in the ventral laminae behave like W ganglion cells. However, LGN$_d$ and retinal ganglion cells show subtle differences in behavior. The receptive fields of LGN$_d$ cells are slightly larger than those of their retinal counterparts.[21] Also, X-like LGN$_d$ cells have a much stronger antagonistic surround, a feature that makes them less sensitive than ganglion cells to diffuse light stimuli falling on the entire visual field.[15] Finally, both X-like and Y-like LGN$_d$ cells are less sensitive to flickering stimuli than ganglion cells. Like their retinal counterparts, X-like LGN$_d$ cells have smaller somata and thinner axons than Y-like cells.

In monkeys, X and Y information is kept more segregated than in cats. Cells in the parvocellular layers respond to all visual stimuli like X ganglion cells (recall Table 19–2). Many cells in the magnocellular layers have most of the characteristics of Y ganglion cells, such as short conduction times, transient responses, and a lack of color sensitivity; some, however, exhibit linear summation as do X cells. Nevertheless, it is widely held that X and Y information is kept separate in the parvo- and magnocellular layers, respectively.

Color Sensitivity. As mentioned in Chapter 19, many ganglion cells in primates respond best to light of specific wavelengths. However, the effects of wavelength on neuronal responses were first studied at the LGN$_d$, in which DeValois and his colleagues[6] recorded the responses of LGN$_d$ cells to diffuse, full-field flashes of monochromatic light. When the average firing rate of the response was plotted as a function of flash wavelength, two major classes of neurons were distinguished: color-opponent cells and broad-band cells.

Color-Opponent Cells. Most LGN$_d$ cells exhibit an increase in firing rate to stimuli with wavelengths at one end of the visible spectrum and a decrease to stimuli with wavelengths at the other end (Fig. 20–3, A–D). Neurons with increased rates at long wavelengths can be divided statistically into two classes having peak rates to stimuli of either about 600 or 640 nm (Fig. 20–3, C and A, respectively). For neurons with increasing rates to stimuli of short wavelengths, the classification is less clear; nevertheless, they also are separated into two classes with peak firing rates at 440 and just over 500 nm (Fig. 20–3, D and B, respectively). Neurons in these four classes are called *color-opponent* cells because for each cell there is a wavelength of light that elicits no change in firing rate. Since the four classes have peak responses at four different wavelengths (near the blue, green, yellow, and red portions of the visual spectrum), these data support the second theory of color vision promulgated by Hering, who proposed the existence of four primary color mechanisms since yellow is as distinctive perceptually as red, green, and blue.

The cone types that feed information to color-opponent LGN_d cells can be deduced if one selectively bleaches different retinal pigments with intense adapting lights. When a cell like that in Figure 20–3B is subjected to an intense adapting red light, subsequent test flashes of different wavelengths evoke only increasing firing rates that peak at about 540 nm. When the same cell is pretreated with an intense green light, the test flashes elicit only decreasing firing rates, with a trough at 600 nm. The peak and trough correspond closely to the peaks in the absorption spectra of the simian green (G) and red (R) cones, respectively; therefore, it may be concluded that this cell

receives an excitatory input from G cones and an inhibitory input from R cones. DeValois and his colleagues[6] suggested that the receptive field of every color-opponent LGN_d cell receives a central input from one of the three cone types and a surround input from another. A cell with an excitatory or inhibitory response in the yellow portion of the spectrum would require a central or surround input from both R and G cones.

By using small stimulus spots of different wavelengths rather than diffuse flashes, Wiesel and Hubel[48] demonstrated that the opponent inputs from different cones are indeed spatially separate and localized to either the center or surround of the receptive field. The cell in Figure 20–3B, therefore, might be excited by a green light delivered to its receptive field center and inhibited by a red

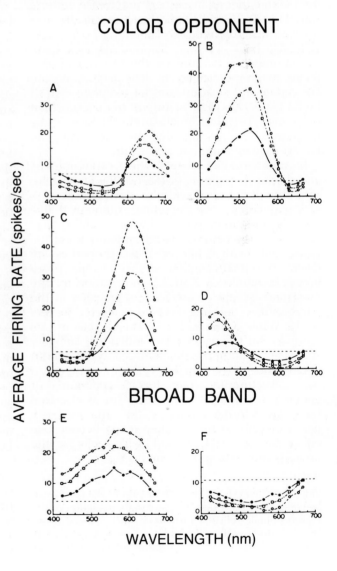

Figure 20–3 Color sensitivity of different cell types in monkey LGN_d. Average firing rate was determined in response to a full-field light flash of 12 different wavelengths equated for energy. The different curves (solid circle, open square, open circle) show the effect of increasing flash intensity in approximately 0.5-log unit steps. Dashed lines indicate prestimulus spontaneous firing rates. (From DeValois, R.L.; Abramov, I.; Jacobs, G. H. *J. Opt. Soc. Am.* 56:966–977, 1966.)

light on its surround. Such a cell with both color- *and* spatial-opponent processes is called *double-opponent*. Nearly all double-opponent LGN_d cells have one of the four possible R–G opponent combinations. Only a few cells respond to stimuli in the blue part of the spectrum. All double-opponent cells are found in the parvocellular layers, where they constitute the large majority of all cells. Since double-opponent cells have X-like characteristics, in particular the smallest receptive fields, it is clear that visual acuity (form vision) and color are linked at an early stage of visual processing. Indeed, in the carp, double-opponent receptive fields are found as early as the bipolar cell.[20]

Broad-Band Cells. The other general class of primate LGN_d cell responds to all wavelengths of light with either an increase or a decrease in firing rate (Fig. 20–3E and F) and hence is called a *non-opponent* or, more recently, a *broad-band cell.* These cells also have a center-surround spatial organization and they account for the majority of cells in the magnocellular layers. It is likely that both the center and surround of the receptive field of broad-band cells receive inputs from more than one cone type. Most broad-band cells have Y-like properties, suggesting that the processing of larger, moving stimuli is achromatic.

In summary, the receptive field configurations and discharge patterns of most of the LGN_d neurons thus far studied resemble closely those of X or Y ganglion cells of the retina. Although W-like cells also exist in both cat and monkey LGN_d, they have thus far received very little attention. In cats, LGN_d cells with X-like or Y-like properties are driven by retinal ganglion cells with similar properties. In monkeys, X and Y information for the most part is kept anatomically segregated in the parvocellular and magnocellular layers, respectively. While X-like and Y-like LGN_d cells in cats are not clearly separated into individual laminae, some segregation of the two cell types within laminae does exist. The LGN_d does not appear to be a structure in which further processing of ganglion cell information occurs. This is also true for color information, since the large majority (90%) of cells in the monkey retina behave like either the double-opponent X-like cells or the broad-band Y-like cells of the LGN_d.

It has been suggested, however, that the LGN_d provides a site where visual transmission to the cortex can be controlled. For example, electrical stimulation of the mesencephalic reticular formation (MRF) causes an increase in the on-going firing rate of most X-like and Y-like cells and reduces the surround inhibition of X-like

cells.[9] Since these effects mimic those produced by arousal, Fukuda and Stone[9] suggested that inhibition in the LGN_d may be controlled by the animal's state of alertness. Others, on the other hand, propose that the mesencephalon provides a signal related to an impending rapid eye movement (called a saccade). Such a signal (called an efference copy or corollary discharge) would act to reduce visual transmission during a saccade, thereby making us unaware of the visual world sweeping across the retina during rapid eye movements (like those made when one is reading this page). While this notion of "saccadic suppression" is attractive, it seems unlikely, since no change in the discharge rate of simian LGN_d cells occurs during spontaneous eye movements in the dark.

The Primary Visual Cortex (V₁)

The Topographic Map. The maps of the contralateral visual field in the LGN_d are relayed onto V_1 in a topographic fashion. As we saw in Figure 20–2, a single point (rectangle) in the right visual field projects to all six LGN_d layers, which in turn project onto a single locus in V_1. The topographic projection can be mapped by charting the cortical surface potentials evoked by exploratory punctate light flashes delivered to different locations in the contralateral visual field. These visual evoked potentials reflect the incoming afferent volleys and the resultant synaptic events that occur in the underlying cortical tissue (see Chapter 32 for explanation of the origin of cortical evoked potentials). As seen in Figure 20–4, the whole of the visible (lateral) surface of V_1 in macaque monkeys is devoted to only the central 7 to 9 degrees of vision.[42] The foveal part of the right visual field (center of right inset) is represented anteriorly and laterally. As the recording electrode moves posterior and dorsal along the cortical surface, the points in the visual field that produce evoked potentials move peripherally along the horizontal meridian (0 degrees in Fig. 20–4). The lower part of the vertical meridian (315°, 270°) is represented on the rostrodorsal cortical surface while the upper part of the vertical meridian (45°, 90°) is represented caudally (see Fig. 20–4). All the rest of the visual field is compressed to fit on the hidden medial surface of the occipital lobe.[4] In both the LGN_d and V_1, therefore, the space devoted to central vision far outweighs that devoted to peripheral vision. A similar representation, with the central visual field on the dorsolateral surface and the visual periphery on the medial surface of the striate cortex, obtains for all frontal-eyed animals.[45]

Organization of Inputs. Like other parts of the cerebral cortex, the primary visual cortex is divided

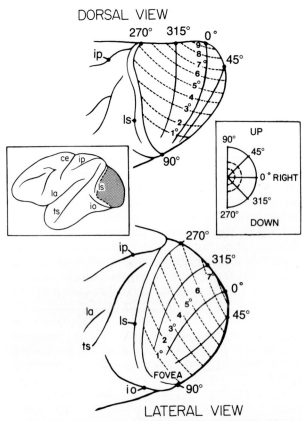

Figure 20–4 Topographic map showing the projection of the central portion of the monkey's right visual field onto the left occipital lobe (shaded region in left inset). Every point in the contralateral (right) visual field can be represented in polar coordinates (right inset) by an angle (solid lines: 0°, horizontal; 90–270°, vertical; etc.) and a radius (dotted lines), both expressed in degrees. The cortical representation of these points appears as a grid of intersecting angle (solid) and radius (dotted) lines on the dorsal and lateral surfaces of the occipital lobe. Abbreviations of sulci as in Fig. 20–2. (After Talbot, S. A.; Marshall, W. H. *Am. J. Ophthalmol.* 24:1255–1264, 1941; published with permission of The American Journal of Ophthalmology. Copyright by the Ophthalmic Publishing Company.)

into six cell layers that lie parallel to, and are numbered sequentially from, the cortical surface. Most efferent axons destined for other cortical areas have their cell bodies in the superficial layers (i.e., 2 and 3) while those axons that descend to the brainstem and thalamus have their cell bodies in the deeper layers (i.e., 5 and 6). Most thalamic afferents terminate in layer 4. Like the afferents to the LGN_d, LGN_d projections to V_1 are segregated according to cell type (i.e., X, Y, or W) and according to the eye through which they can be driven.

X-like and Y-like cells in the thalamus send their axons to different sublaminae within the six cor-

tical layers. The majority of parvocellular afferents (all X-like cells) terminate in the most superficial and deepest sublaminae of layer 4, whereas most magnocellular afferents (largely Y-like cells) are sandwiched in an intermediate sublamina. A smaller number of parvocellular and magnocellular afferents terminate in separate sublaminae of layer 6.[12] Although X-like and Y-like cells in the cat LGN_d are less neatly segregated, X-like and Y-like LGN_d afferents also terminate in different sublaminae of layer 4 (see Fig. 20–6 for examples). Information from both X-like and Y-like cells also reaches layer 6 in cats,[22] probably as collaterals from axons that also arborize in layer 4. Finally, V_1 in cats also receives W-like information through projections of the ventral LGN_d layers to cortical layers 1, 3, and 4.[22]

Not only do the afferents from X-like and Y-like cells terminate in different sublaminae, but the inputs from each eye are also kept separate. The segregation of the cortical inputs from the two eyes has been demonstrated most dramatically by injecting radioactive amino acids into one eye. After several days, some of the label is transported across the synapse at the LGN_d and all the way to the cortex, where one can see the radioactive terminals by painting histological sections with a photographic emulsion. In the section in Figure 20–5A, taken perpendicular to the cortical surface, the labeled radioactive terminals (seen as a high density of white exposed silver grains) lie in regular patches separated by unlabeled gaps along all of layer 4.[18] One can visualize the overall pattern of afferents across the cortex by reconstructing tangential sections through layer 4 and flattening them to a single plane, as seen in Figure 20–5B. Inputs from the injected eye appear as bands about 0.4-mm wide separated by unlabeled bands corresponding to the regions receiving inputs from the eye that was not injected.[18] This pattern of afferent input, which forms the anatomical substrate of the ocular dominance columns to be described later, can also be demonstrated, although less dramatically, in cats.[22]

Although ocular dominance bands in V_1 have been demonstrated in all Old World monkeys, they do not occur in all New World monkeys. Man, as befits his evolutionary status, does have ocular dominance bands. The reason why some animals have bands (or other means of segregating eye inputs) while others with similar visual capabilities (i.e., stereoscopic vision) have none remains a mystery.

Internal Circuitry. Cells in the visual cortex of both cats and macaque monkeys can be divided into three morphological classes. Two classes have

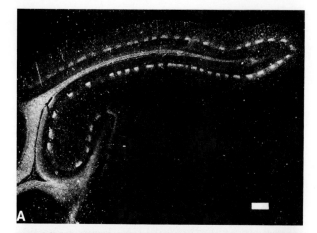

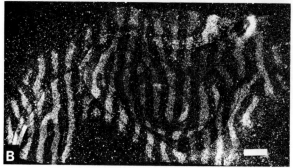

Figure 20–5 Anatomical demonstration of ocular dominance columns in monkey right striate cortex. Transverse *(A)* and assembled horizontal *(B)* sections through layer 4 show the distribution of labeled terminals following injection of radioactive amino acids into the right eye. White bars are 1 mm long. (After Hubel, D. H.; Wiesel, T. N. *Proc. R. Soc. Lond.* B 198:1–59, 1977.)

spiny dendrites but differ in having either pyramidal or stellate somata. Generally, these spiny cells have rather long axons that travel over considerable distances, usually across lamina.[9a, 23] Synapses formed by spiny neurons viewed in electron micrographs have round vesicles and prominent postsynaptic opacities, and thus are thought to be excitatory. Neurons in the third class have smooth (or sparsely spinous) dendrites, display a variety of shapes, and frequently have axons that terminate locally. Synapses formed by these cells show flattened vesicles and slight postsynaptic opacities, and are thought to be inhibitory. Indeed, some smooth stellate cells in V_1 of rats take up the putative inhibitory transmitter gamma-aminobutyric acid (GABA) (see Chapter 11) and exhibit immunoreactivity to the GABA synthetic enzyme glutamic acid decarboxylase (GAD). Examples of the laminar location of both spiny and smooth cells in cats as revealed by intracellular injection

of horseradish peroxidase (HRP) are shown in Figure 20–6.

Although the story on the circuitry of V_1 is far from being completely worked out, several general points applicable to the neurophysiological observations described later may be noted. (1) The neuron type most closely associated with the heaviest geniculocortical termination is the spiny stellate cell; there are no pyramidal cell bodies in the thalamic recipient zones of layer 4. (2) Spiny pyramidal and smooth stellate cells can be found in all other cortical layers, but spiny stellate cells are rare outside layer 4. (3) Cells in layer 4 project heavily to layers 2 and 3 and may project to the deeper layers (both 5 and 6) as well.[23] (4) Pyramidal neurons in layers 2 and 3 provide the bulk of the efferent connections to other visual cortical areas as well as a heavy projection, possibly via collaterals, to layer 5. (5) Large pyramidal cells in layer 5 project to the superior colliculus and pulvinar and other brainstem nuclei. In cats, at least some of these efferent cells and other layer-5 pyramidal cells synapse extensively in layer 6. (6) Pyramidal cells of layer 6 project to LGN_d, and at least some pyramidal cells send ascending axon collaterals to layer 4.[9a, 23]

These prominent and consistent connections tempt one to speculate about how neural information flows in V_1. Gilbert[9a] has suggested that excitatory visual information from the LGN_d enters largely through layer 4 and proceeds in a serial fashion to layers 2 and 3, layer 5, layer 6, and back to layer 4 (see Fig. 20–6, block diagram). While this processing scheme is consistent with much of our current knowledge about the anatomy, several caveats must be pointed out. First, layer 6 and layers 2 and 3 also receive thalamic inputs. Consequently, with the possible exception of those in layer 5, cells in all cortical layers could be directly influenced by afferents from the LGN_d. Second, Figure 20–6 (see also Lund et al.[23]) shows that although the somata of cells may be confined to a single layer, their dendrites frequently arborize in one or more adjacent layers. For example, the dendrites of cells in layers 4, 5, and 6 often extend into layer 1. Therefore many cells, particularly those in the deep layers, may sample information from axons that ramify in more superficial layers. Finally, our knowledge of the morphology of cells in V_1 is based on rather capricious staining techniques, such as Golgi impregnation and HRP labeling, which frequently do not reveal a cell's distal dendrites or all of its axonal processes. Consequently, we may now have only an incom-

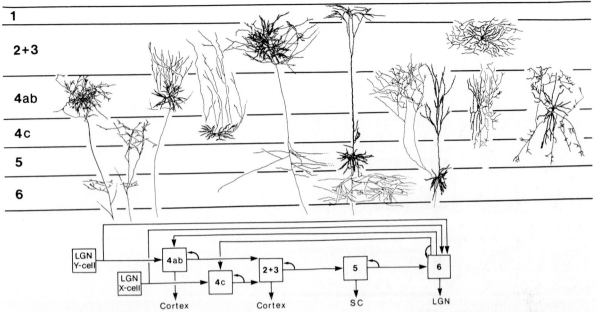

Figure 20–6 Morphology and block diagram of intracortical connections between neurons in cat striate cortex. The reconstructed cells shown in the drawing above are morphological examples of the cells indicated schematically in the corresponding boxes below. The three neurons to the right are smooth stellate cells; all others are spiny cells, either stellate or pyramidal (see text for further details). (After Gilbert, C. D. Reproduced, with permission, from the Annual Review of Neuroscience, Vol. 6. © 1983 by Annual Reviews, Inc.)

plete and sketchy picture of what may in fact be a more complicated circuitry. At the very least, however, the anatomy suggests that both serial and parallel processing must occur in V_1.

Activity of V_1 Neurons. Although the techniques for single-unit recording from the central nervous system were available in the early 1950s and it was clear that V_1 was the first cortical relay for visual information from the thalamus, early investigators were singularly unsuccessful in eliciting visual responses from cortical neurons. Two problems plagued early investigators. First, although stationary spot stimuli or even diffuse light flashes elicited brisk responses from retinal or LGN_d cells, such stimuli were largely ineffective in driving cortical neurons. Second, unlike most retinal ganglion and thalamic cells, cells in V_1 usually have little if any resting discharge, so that their presence is difficult to detect, especially without an adequate stimulus. It took a serendipitous finding by David Hubel and Torsten Wiesel in the late 1950s to establish that visual stimuli must be composed of moving edges if they are to drive (i.e., elicit responses from) cortical neurons. After several hours of unsuccessful attempts to drive a recalcitrant V_1 cell with a spot stimulus, they noticed that a brisk response was elicited when-

ever they changed the slide in the projector that provided the visual stimulus (recall Figure 19–15). Upon closer inspection, the slide was found to be cracked and it was the movement of the crack across the receptive field that was effective in driving the cell.

Receptive Field Organization and Cell Types. To find the receptive field of a cortical unit, the investigator sweeps a bar of light (or a dark bar on a light background), at various orientations, speeds, and directions, across a tangent screen (recall Figure 19–15) while advancing a recording electrode. Such a searching strategy is necessary because many cortical cells are silent if there is no appropriate visual stimulus. All V_1 neurons have receptive fields in the contralateral visual field. Once the approximate location of the receptive field has been determined, the bar is tilted to different orientations and moved orthogonal to its length across the receptive field. As seen for the cell recorded from monkey striate cortex (Fig. 20–7A), the response for each pass of the bar across the receptive field (indicated by dashed rectangle) is very sensitive to stimulus orientation, with the best response being obtained at one optimal orientation (Fig. 20–7A, fourth trace from top).[17] Most neurons respond best if the bar moves perpendic-

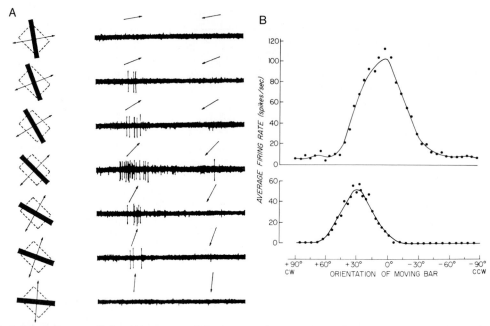

Figure 20–7 Orientation sensitivity of V₁ neurons. *A*, Responses of a complex cell in a monkey to various orientations of a black bar moving in different directions *(arrows)* over the receptive field *(dashed rectangle)*. *B*, Orientation tuning curves for a simple *(bottom curve)* and complex *(top curve)* cell in a cat. The plus (+) and minus (−) angles represent clockwise and counterclockwise bar rotations, respectively, from horizontal (0°). (*A*, After Hubel, D. H.; Wiesel, T. N. *J. Physiol (Lond.)* 195:215–243, 1968. *B*, After Henry G. H.; Dreher, B.; Bishop, P.O. *J. Neurophysiol.* 37:1394–1409, 1974.)

ular to its optimal orientation and, in addition, most prefer motion in a particular direction (e.g., right and up for the neuron of Figure 20–7A). The response component that depends on the direction of motion can be separated from the one that depends on orientation by use of a moving spot rather than a bar. However, relatively few V₁ neurons have direction but not orientation selectivity.

Figure 20–7B shows the so-called *orientation tuning curves* obtained by averaging the firing rates for a number of bar passes at each of several orientations for two typical neurons in cat striate cortex.[13] The lower cell responds over a narrower range of orientations and, therefore, is said to be more sharply tuned. The optimal orientation varies from cell to cell; for example, the upper cell in Figure 20–7B prefers a bar oriented at zero degrees whereas the lower one prefers a bar tilted by +30 degrees. When a large number of V₁ neurons are examined, all possible orientations are represented in approximately equal numbers. Therefore, at least at V₁, there appears to be no physiological correlate for the psychophysical observation that visual activity is better for objects oriented horizontally and vertically than for those oriented obliquely.

Once a neuron's optimal stimulus orientation has been ascertained, one can determine the on and off subregions of its receptive field by turning a thin bar of light lying parallel to the preferred orientation on and off at various locations within the receptive field. The responses to such stationary stimuli are always less than those to appropriately oriented moving stimuli. On the basis of such receptive field plots, Hubel and Wiesel[16] concluded that the great majority of cells in V₁ seem to fall into two groups. Like LGN_d and retinal ganglion cells, *simple cells* possess distinct on and off subregions: that is, illumination of part or all of the on subregion increases the cell's firing rate and illumination of the off subregion suppresses firing (in the few spontaneously active cells) and evokes a discharge when the bar is turned off. Simple cells differ in the detailed organization of their receptive fields. Figure 20–8A–D shows the receptive fields of several simple cells plotted with a thin bar of light; clearly the on (+) and off (−) subregions vary in size, configuration, and number. In contrast to simple cells, *complex cells* respond to a stimulus bar when it is turned either on *or* off anywhere within their receptive fields (Fig. 20–8, right). Complex cells, therefore, have overlapping on and off subregions, as shown schematically to the right of E in Figure 20–8. As for simple cells, the best stimulus for complex cells has an optimal orientation (note in Figure 20–8, right, the lack of response in the lower two spike recordings when the bar is horizontal or vertical). Although virtually all of the single-neuron data have been obtained from anesthetized animals, V_1 neurons in alert monkeys also prefer a moving elongated bar of a particular orientation and have simple or complex receptive fields.

The behavior of simple and complex cells can be further distinguished because each responds best to different attributes (also called trigger features) of the visual stimulus. First, simple cells require that a bar stimulus be precisely oriented, whereas complex cells are not so fussy. For example, the lower tuning curve in Figure 20–7B was obtained from a simple cell; the upper is from a complex cell. If the width of the orientation tuning curves at half-peak firing rate is used as an objective measure, the average width for a large population of simple cells in cats is 17 degrees compared with 28 degrees for complex cells.[13] In monkeys also, simple cells, on average, are more sharply tuned. Second, complex cells usually respond to stimuli moving at higher velocities than

do simple cells, but the dichotomy is not strict. Third, the receptive fields of simple cells generally are smaller than those of complex cells. For example, in cats the receptive fields of simple cells near the area centralis can have an overall width (perpendicular to the preferred axis) of less than 0.75 degree, with one of the subregions often being as small as 10 to 15 minutes of arc.[30] The receptive fields of complex cells, on the other hand, have an overall width of 1 to 3 degrees. In monkeys, simple cells subserving the fovea have subregions ranging from 10 minutes to 1 degree, with subregions in the central fovea of only 2 minutes.[27] As in the retina and LGN_d, the dimensions of the receptive fields tend to increase with distance from the area centralis. Finally, in both cats and monkeys many simple cells are silent in the absence of an appropriate visual stimulus, whereas most complex cells have a low (i.e., <10 spikes/s) resting discharge rate.

In both cats and monkeys, many simple but few complex cells are found in the parts of layers 4 and 6 receiving LGN_d afferents, whereas simple and complex cells are found in both more superficial and deeper layers.[9a] Also, the receptive fields of both simple and complex cells show differences

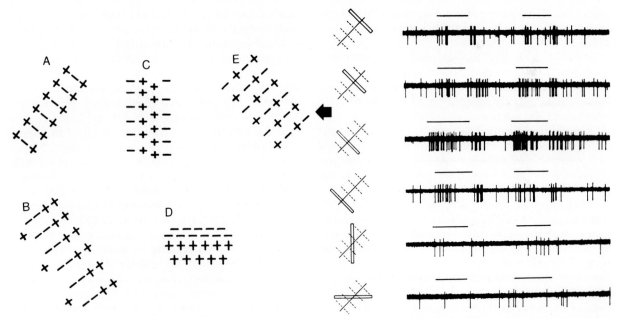

Figure 20–8 Receptive fields of different V_1 neurons. A–D are four examples of simple cells *(left)* that illustrate symmetric receptive fields with either on *(C)* or off *(A)* central subregions and asymmetric receptive fields *(B, D)*. The complex cell *(right)* responds when a light bar of appropriate orientation is turned on and off (horizontal line above action potentials indicates when bar is on) at all positions within the receptive field (column to the left of action potentials), producing the receptive field schematized in *E* with alternating on (+) and off (—) subregions. Note that the neighboring + and — subregions actually are coextensive. Like simple cells, the complex cell requires a bar with a specific orientation (see lower two recordings). (After Hubel, D. H.; Wiesel, T. N. *J. Physiol (Lond.)* 160:106–154, 1962.)

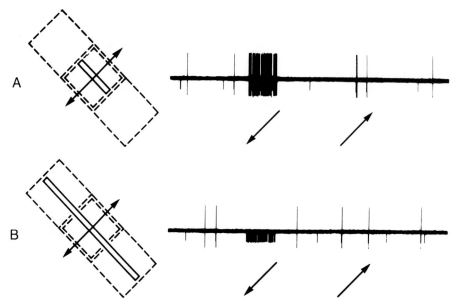

Figure 20–9 Effect of stimulus bar length on the simultaneously recorded responses of a hypercomplex (large action potentials) and complex (small action potentials) cell in monkey striate cortex. The receptive field of the complex cell *(single dashed lines)* is in register with that of the hypercomplex cell *(double dashed lines)* but is much smaller. Arrows indicate direction of movement. (After Michael, C. R. *J. Neurophysiol.* 42:726–744, 1979.)

according to which layer they are found in and hence where they project to. Although early studies suggested that simple cells were stellate and complex cells were pyramidal in shape, more recent studies indicate that this correlation is not perfect.

A third type of V_1 cell, with even more stringent stimulus requirements than either simple or complex cells, was discovered somewhat later. These *hypercomplex cells* also require moving bars, but bar orientation is even more critical (average half width tuning of only 11 degrees) than for simple cells. Furthermore, the length of the bar matters to hypercomplex cells but not to either simple or complex cells. This is shown dramatically in Figure 20–9, which shows a fortuitous, simultaneous recording from a *hypercomplex cell* (large action potentials) and a nearby *complex cell* (small downward action potentials).[28] Although a long bar (Fig. 20–9B) is effective at driving the complex cell whose receptive field is outlined by a single dashed line, only a bar less than a certain length (as in Fig. 20–9A) drives the hypercomplex cell. The receptive field of the hypercomplex cell, therefore, has clear inhibitory "end zones" that limit its borders to the double dashed square. Determination of the best stimulus for other hypercomplex cells often requires even more imagination—e.g., some require tongue- or L-shaped stimuli. Hypercomplex cells are seldom, if ever, found in sublayers receiving

LGN_d afferents and, at least in monkeys, are more prevalent in the superficial layers.

Binocularity. Another major difference between cells of the LGN_d and V_1 is that the activity of most cortical cells (more than 80% in cats and over 60% in monkeys) can be influenced strongly by appropriate stimuli falling on either eye. To determine the relative contributions from the two eyes, receptive fields are mapped separately for each eye with the other eye closed, and appropriate stimuli are delivered to each eye individually and to both together. By listening to the amplified action potentials played over a loudspeaker, Hubel and Wiesel[16] were able to divide striate cortex cells with receptive fields in or near the area centralis qualitatively into seven ocular dominance groups. The cells ranged from those dominated exclusively by the contralateral eye (group 1) to those dominated exclusively by the ipsilateral eye (group 7), with those driven equally by both eyes falling in group 4. The ocular dominance histogram of Figure 20–10 indicates that in cats the majority of both simple (open bars) and complex (hatched bars) cells are binocular to some extent, with slightly more cells favoring inputs from the contralateral eye. In monkeys, however, many more simple cells are monocular, particularly those in the layers receiving LGN_d afferents. Monkey complex and hypercomplex cells, like those in cats, are mostly binocular. *Since LGN_d cells are essentially*

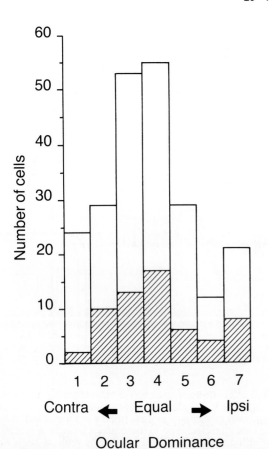

Figure 20–10 Ocular dominance histogram for simple *(open bars)* and complex *(hatched bars)* cells in cat V_1. Cells are driven either exclusively by one eye (categories 1 and 7) or by both with various degrees of effectiveness (categories 2 through 6). (After Hubel, D. H.; Wiesel, T. N. *J. Physiol (Lond.)* 160:106–154, 1962.)

Fig. 20–11) has two receptive field locations (b, b') that superimpose. On the other hand, other neurons, such as *A* and *C* in Figure 20–11, have the same receptive field location in the left eye (a, c) but their right eye receptive fields (a' and c') are displaced either to the right (cell *A*) or to the left (cell *C*). Similar retinal disparities can occur in the vertical and oblique directions as well. In cat striate cortex, the maximum disparities are less than 1.6 degrees,[8] and in monkeys less than 0.5 degree,[34] so it is not surprising that early investigators attributed them to experimental error. Since the maximum response for a binocular cortical cell occurs for stimuli falling simultaneously on each eye's receptive field, the best stimulus for cell *A* (Fig. 20–11) is a single target located behind the tangent screen (at a, a'), while cell *C* (Fig. 20–11) prefers a single target located in front of the tangent screen (at c, c'). A population of such

monocular, V_1 is the first structure in the geniculostriate pathway where the responses to stimuli falling on both eyes interact. For most V_1 cells, convergent input from the two eyes comes from the different layers of the ipsilateral LGN_d; however, for cells with receptive fields along the vertical meridian, inputs from the contralateral eye are provided, in part, by cortico-cortical connections through the corpus callosum.

The histogram in Figure 20–10 is based only on the excitatory inputs from the two eyes. Probably, even more cells would be "binocular" if inhibitory effects or subthreshold changes in the membrane potential were considered.

When receptive fields of binocular cells are plotted for each eye independently, they sometimes fall at exactly corresponding positions on the retina of each eye. After the eyes are appropriately focused on a tangent screen, such a cell (see B in

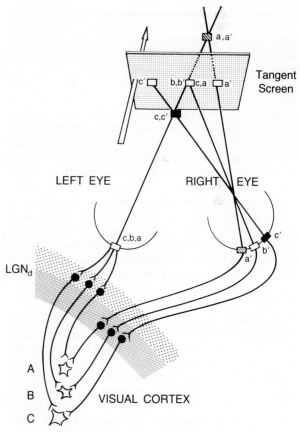

Figure 20–11 Diagram illustrating how three cortical neurons with different horizontal receptive field disparities would prefer stimuli located at different distances. Each neuron has the same receptive field location when plotted through the left eye (a, b, c on tangent screen) but different locations when plotted through the right (a', b', c'). See text for further discussion.

neurons *tuned* to a variety of convergent (like cell C) and divergent (like cell A) disparities could provide the first stages of the neural substrate for vision in depth, or *stereopsis*. Indeed, in monkeys, which have excellent stereopsis, the majority of cortical cells subserving central vision in area V_1 respond differentially to stimuli positioned at various depths in front of, behind, or at the plane where the animal is fixating.[34] In cats, disparity-sensitive cells can be either simple or complex and are found in every cortical layer.[8] Furthermore, in cats, the preferred stimulus orientation plotted for each eye independently also may differ by as much as 10 degrees, another feature that might help further in stereopsis or the fusion of images falling on both eyes.

In addition to those neurons tuned for a specific distance, other cells in both cats and monkeys exhibit an increased discharge for stimuli located anywhere behind the tangent screen and a decreased discharge for stimuli anywhere in front of it (called far neurons) or vice versa (called near neurons). These units could provide signals to elicit the vergence eye movements by which we view approaching or receding objects.

Cortical Columns. Cells in primary visual cortex are not arrayed haphazardly. Instead, electrode penetrations perpendicular to the cortical surface and parallel to the deep afferent fiber bundles sequentially encounter cells that prefer the same stimulus orientation. Such a penetration in V_1 of a cat is seen at the top of the electrode track illustrated in Figure 20–12.[16] The lines crossing the penetration indicate the preferred orientation of the cell (long line) or group of cells (short line) recorded there; all of the cells clearly prefer the same orientation. After the electrode traverses the white matter and the penetration runs oblique to the cortical surface and across the deep fiber bundles (shown dotted), the electrode encounters successive clusters of cells whose preferred orientations change continually (lower left in Fig. 20–12). If a tangential penetration is run parallel to layer 4 (as in inset, Fig. 20–12), clusters of cells are encountered whose preferred orientations change sequentially (i.e., either clockwise or counterclockwise) until all possible orientations have been traversed within about 1.0 mm of cortex. On the basis of these findings, Hubel and Wiesel[16] suggested that V_1 is organized in *orientation columns* that contain cells with the same preferred orientation. Each column might be a few cells wide (about 50 μm) but extends through all six cortical layers. Cells in adjacent columns have slightly different preferred orientations, as schematized in the inset in Figure 20–12.

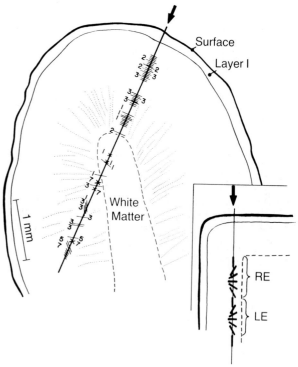

Figure 20–12 Reconstruction of an electrode penetration through cat V_1 with the preferred orientation of single cells *(long lines)* or a group of cells *(short lines)* indicated at their locations. Numbers refer to the ocular-dominance classification of the neuron or group of neurons. Afferent fibers from the LGN_d are indicated by an ×. Dotted lines indicate paths of deep fiber bundles. Inset shows schematic penetration along layer 4, which encounters first cells dominated by the right eye (RE) and then those dominated by the left (LE). (After Hubel, D. H.; Wiesel, T. N. *J. Physiol. (Lond.)* 160:106–154, 1962.)

Neighboring cells in striate cortex not only prefer stimuli with similar orientations but also are usually dominated by the same eye. In Figure 20–12, for example, all the cells on the initial perpendicular (top) part of the penetration are dominated by the contralateral eye (groups 1 to 3). In the oblique (lower left) part of the penetration, on the other hand, there are four shifts in ocular dominance that, in succession, include activity that is purely contralateral (group 1) in the white matter and deep cortical layers, activity dominated by the ipsilateral eye (group 7), activity dominated by the contralateral eye (ocular dominance groups 1 to 3) and, finally, activity dominated by the ipsilateral eye (groups 5 and 7). Penetrations made parallel to the cortical surface indicate that cells change their ocular allegiance (i.e., right (RE) to left (LE) eye in the inset in Fig. 20–12) roughly every 0.5 mm. Consequently, Hubel and Wiesel[16] suggested that V_1 is also organized in columns according to ocular dominance.

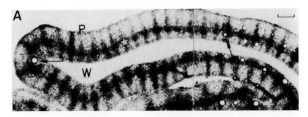

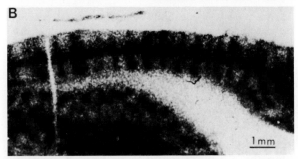

Figure 20–13 Anatomical demonstration of columns in a macaque monkey. A, Perpendicular section through striate cortex showing distribution of neurons active during monocular stimuli with all orientations. Columns extend from pia (P) to white matter (W). Arrows are not significant. B, Perpendicular section through area 17 showing distribution of 2-deoxyglucose–labeled neurons that respond when moving vertical stripes are presented to both eyes. (A, after Hendrickson, A. E., Wilson, J. R. *Brain Res.* 170:353–358, 1979. B, from Hubel, D. H.; Wiesel, T. N.; Stryker, M. P. *J. Comp. Neurol.* 177:361–379, 1978.)

The vertical organization of the ocular dominance and orientation columns has been confirmed recently by the 2-deoxyglucose method, which relies on the basic assumption that cells responding to visual stimuli are metabolically more active. Such active cells, therefore, take up an intravenously injected radioactive glucose analogue, but are unable to metabolize it for energy so that it remains trapped in the cells as a marker. The distribution of the trapped radioactive material thus provides a map of metabolic activity and hence of cell activity. Macaque monkeys subjected to either monocular or binocular stimuli with a single orientation show radioactive patterns of ocular dominance (Fig. 20–13A)[11] or orientation (Fig. 20–13B)[19] columns, respectively. The entire functional column through all cortical layers is revealed rather than just its parts in layers 4 and 6 (as shown in Fig. 20–5). Although Figures 20–5 and 20–13 suggest that the "columns" are arranged in neat bands, in many areas of V_1 they actually swirl through the cortex.[18] Furthermore, the exact configuration and relative relation between the two kinds of columns appears to be very complicated.

Hierarchical Processing. Consideration of the columnar organization and the receptive field properties of V_1 neurons led Hubel and Wiesel[16] to suggest that visual information in the cat striate cortex is processed in a hierarchical fashion. Because of their concentration in or near layer 4 and their relatively uncomplicated receptive field or-

ganizations, simple cells were posited to receive a direct input from the LGN_d.* In the model of Hubel and Wiesel, the receptive field of a simple cell with an on-center and symmetric off flanks (S_1 in Fig. 20–14) could be constructed on the basis of excitatory inputs from several on-center LGN_d cells (four are shown) whose receptive fields are displaced in the visual world along the preferred axis (orientation) of the simple cell. In some simple cells in monkeys, the off regions apparently are created by separate excitatory inputs from off-center LGN_d cells; the off regions of these simple cells are eliminated by the selective cooling of specific LGN_d layers that contain predominantly off-center cells.

Similarly, the receptive field of a complex cell located in either deeper or more superficial layers (CPLX in Fig. 20–14) might be generated by converging excitatory inputs from several simple cells (two are shown) with identical preferred stimulus orientations and only slightly displaced receptive fields. Finally, an appropriate combination of excitatory inputs from several complex cells could be configured to produce the receptive field of a hypercomplex cell.

---

*In the monkey, some layer 4 cells have nonoriented receptive fields resembling those of LGN_d neurons and could represent an additional intermediate stage of processing.

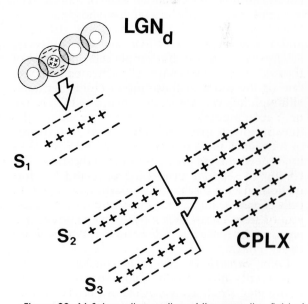

Figure 20–14 Schematic creation of the receptive field of a simple cell (S_1) from excitatory inputs of four LGN_d neurons and the receptive field of a complex cell (CPLX) from the receptive fields of two neighboring simple cells (S_2 and S_3). As in Figure 20–8, the adjacent + and − subregions of the schematic CPLX receptive field are actually coextensive.

Pharmacological experiments indicate that inhibitory mechanisms also are important in creating many of the features of cortical receptive fields. For example, when a potent GABA antagonist (bicuculline) is iontophoretically applied to either simple or complex cells to block their putative GABA-ergic inhibitory inputs, they lose their sharp tuning and respond equally to stimuli of all orientations.[39] Furthermore, inhibitory processes play an important role in establishing ocular dominance, since the application of bicuculline causes some exclusively monocular cells to become equally well driven by either eye.

The strict hierarchical model of visual processing has generated considerable controversy. Detractors of the model have extended it to its illogical extreme, where higher and higher order cells, presumably in other cortical areas, would be increasingly fussy about the precise attributes of the visual stimulus required to elicit a discharge. Eventually, some cells would be so specialized that they would respond only for one particular visual object, such as one's grandmother. Although we shall see later that a few such "grandmother" cells may indeed exist in the inferotemporal cortex, evidence generated after Hubel and Wiesel's elegant studies in the early 1960s (for which they received a Nobel prize in 1981) make it clear that a strictly serial hierarchical scheme is inadequate, even to explain the operation of V_1. For example, several authors have reported that complex as well as simple cells can be activated monosynaptically from the LGN_d. In addition, not only do X-like and Y-like inputs reach V_1 via separate sublayers, as already discussed, but simple cells in layer 4 of the cat can be monosynaptically activated by either one or the other of these inputs.[30] Furthermore, although it is very difficult to distinguish an X-like or Y-like input to a cortical cell by either the linearity of its spatial receptive fields, the degree of transiency in its response, or conduction velocity,[21] nevertheless there is some indication of a continued separate processing of X and Y information in the discharge properties of some V_1 cells. *Consequently, as suggested earlier by the anatomy, it seems almost certain that visual information is processed by parallel as well as serial pathways in the visual cortex.*

Color Sensitivity. There is considerable disagreement concerning the color sensitivity of cells in monkey striate cortex. A major problem is that most chromatic stimuli produce a simultaneous variation in both color *and* luminance. Nevertheless, it seems clear that many V_1 neurons (largely those in foveal and parafoveal regions have been studied) respond to color and not luminance. The majority of the color-sensitive neurons have nonoriented or simple receptive fields and many of these, like parvocellular LGN_d cells, exhibit both color and spatial opponency; red/green opponency is the most usual.[33] Some complex and hypercomplex cells also are color selective, although many fewer have been examined.

Some V_1 cells also exhibit a more complicated spatial and chromatic receptive field organization called *dual-opponent*.[27] A dual-opponent simple cell might respond to either a red light turned on or a green light turned off in one of its receptive field subregions, while turning the same stimuli on and off in the other subregion(s) might produce the opposite response. Because dual-opponent receptive fields have also been demonstrated in carp bipolar cells,[20] this receptive field organization may not reflect a cortical process.

Finally, although many V_1 neurons in macaque monkeys respond to chromatic stimuli, the majority of striate cortex cells respond to all wavelengths of light. The response of such *broad-band neurons* depends only on the intensity of the stimulating light and not on its wavelength, and they are therefore also called *luminosity neurons*. Broad-band neurons may have any type of spatial receptive field, but few are simple cells.

Summary. The receptive fields of V_1 neurons vary considerably according to the layer in which they are located. Cells in the layers receiving LGN_d input are mostly simple or nonoriented (in monkeys) and are frequently monocular. On the other hand, cells in the more superficial or deeper layers tend to be complex or hypercomplex and are invariably binocular. Although possible serial pathways exist between layer 4 and both the superficial and deep layers, anatomical and electrophysiological studies indicate that considerable parallel processing must also take place.

Although we know more about V_1 than any other primary sensory cortex, there are several reasons why a comprehensive model of processing in V_1 is premature. First, it is uncertain that all of the neuron types comprising the estimated 100×10^6 cells in monkey V_1 have been sampled. Second, most of those cells sampled, and hence many of the conclusions that are drawn, are from the neurons whose activity is easiest to record—e.g., the large cells in layer 5. Data from cells in layer 2 and the upper parts of layer 3 are much sparser. Third, virtually all the data are gathered from cells receiving input from the fovea and parafovea; we have very little idea of how the cortex handles information from the peripheral retina. Fourth, the trichotomy of simple, complex, and hypercomplex cells has already begun to break down as receptive field testing becomes more imaginative.

Although the trichotomy was a useful first description, its continued use requires a forced-choice procedure for cell classification that has resulted in considerable confusion. For example, a cell type has been found with the separate on and off receptive field subregions of a simple cell but with the stimulus-length–dependent response properties of a hypercomplex cell. Accurate cell identification is more than just a contest between those who favor division and those who favor consolidation, since cells with very specific stimulus requirements project to quite different structures.

THE COLLICULO-PULVINO-EXTRASTRIATE PATHWAY

Although most studies of the visual system have focused on the geniculostriate pathway, investigators have begun to examine pathways that reach the extrastriate (not V_1) cortex through the superior colliculus and pulvinar. This pathway transmits attributes of the visual scene that are quite different from those elaborated in the geniculostriate pathway.

The Superior Colliculus: Superficial Layers[37,49]

The superior colliculus is arranged in laminae identified from dorsal to ventral as superficial, intermediate, and deep. Few, if any, anatomical connections exist between the superficial layers and the intermediate and deep layers. Furthermore, the superfical and the deeper layers receive inputs from different sources and project to very different targets. The colliculus, therefore, can be considered as two separate structures. Its superficial layers are part of a visual pathway to the cortex and will be discussed now. The two deeper layers are involved in the midbrain visual system concerned with movement and will be discussed with the midbrain visuomotor pathways.

The retinal projection to the superior colliculus, like that to the LGN_d, is organized according to eye and retinal location. In both cats and monkeys, retinal ganglion cells project only to the superficial layers of both colliculi; the right visual field maps onto the left colliculus and vice versa. The map is topographic, with the central retina represented rostrally, the peripheral retina caudally, the upper visual field medially, and the lower visual field laterally. Once again, the fovea has a disproportionately large representation, with its central 10 degrees occupying more than 30% of the surface of the monkey colliculus. The superficial layers project to the pulvinar.

The superficial collicular cells of anesthetized cats and monkeys have similar stimulus preferences.[37] Most respond best to moving stimuli. In monkeys, high velocities (from 60° to 900°/s) are preferred, whereas in cats both high and low stimulus velocities are effective. Both species have neurons that are sensitive to stimulus direction; in cats, they comprise the majority (preferred direction away from the center of gaze), whereas in monkeys they are a distinct minority. Most cells also receive excitatory inputs from both eyes. Receptive fields appear to have a central activating region with a very weak suppressive surround and are larger than those of retinal ganglion cells at similar eccentricities. Most neurons also exhibit a transient discharge to small stationary light spots turned on and off. Finally, collicular cells in monkeys have *broad-band* chromatic properties.

The visual responses of superficial collicular cells, therefore, unlike those of the LGN_d, resemble only vaguely those of retinal ganglion cells. Their transient responses to stationary contrast and their movement and spectral sensitivities resemble those of Y ganglion cells; indeed, Y ganglion cells can be antidromically activated from the colliculus. On the other hand, their binocularity and direction selectivity probably derive from other inputs, possibly from striate cortex, which sends a strong projection to the superficial layers. Indeed, complex cells in cortical layer 5 can be antidromically activated from the colliculus and, as a population, they are strongly binocular and direction selective,[32] properties that are abolished in cats by a lesion of the ipsilateral visual cortex. In monkeys, on the other hand, cooling V_1 produces little effect on the behavior of superficial collicular neurons.[49]

Studies on alert-behaving monkeys indicate that cells in the superficial collicular layers have receptive field properties similar to those of anesthetized animals. In addition, some cells exhibit a greater neuronal discharge if the visual stimulus falling on their receptive fields is the target for a subsequent eye movement.[49] Such an enhanced visual response for a stimulus with behavioral "significance" occurs at neither the LGN_d nor V_1.

The Pulvinar

In both cats and monkeys, the pulvinar receives input from the superficial layers of the colliculus, strong descending input from V_1, and direct but modest input from the retina. The pulvinar has three subdivisions (lateral, medial, and inferior), each of which contains a complete topographic representation of the contralateral visual field. In cats, the lateral posterior nucleus (LP) is often treated as a separate nucleus although it is considered homologous to a subdivision of the primate pulvinar. The pulvinar probably projects almost exclusively to the cortex, where it provides inputs to nearly every extrastriate visual area.

The visual stimuli that are effective at eliciting discharges from cells in the pulvinar seem to reflect the influences of its various inputs. In monkeys, most pulvinar cells, like those of the superior colliculus, prefer moving stimuli delivered to both eyes; however, unlike those of the colliculus, many pulvinar cells are direction selective.[2] In both cats and monkeys, many pulvinar cells also exhibit orientation selectivity similar to that of cortical cells. In cats, some cells in the retinal-recipient zone have concentric receptive fields with Y-like or W-like properties. Finally,

pulvinar cells usually exhibit a variable visual response that often habituates to repeated stimuli.

The effects of pulvinar lesions upon vision are a subject of controversy. Reported deficits range from none to significant impairment of pattern discrimination and to abnormal eye movement scanning of a complex visual array. Because the prominent corticotectal pathway passes through the pulvinar, the effects of electrolytic lesions of the pulvinar must be interpreted with caution.

Summary

The LGN_d and pulvinar appear to relay different attributes of a visual stimulus to different cortical regions. The pathways through these two thalamic nuclei are summarized schematically in Figure 20–15. In both cats and monkeys, the LGN_d relays both X and Y information but the X information goes almost exclusively to striate cortex (V_1). In contrast, the pulvinar pathway relays Y and probably W information, predominantly to extrastriate

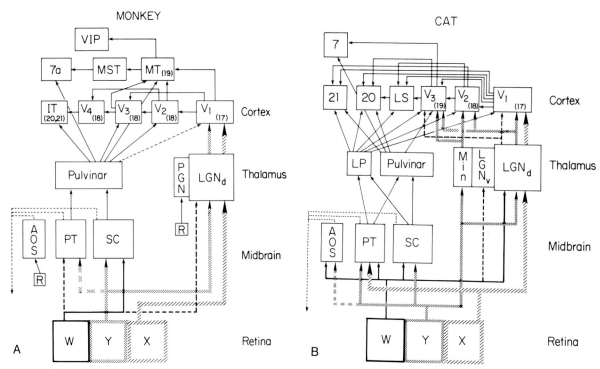

Figure 20–15 Block diagrams of the most prominent visual pathways in monkeys *(A)* and cats *(B)* starting with the three basic types of retinal ganglion cells. The connections are based on anatomical or electrophysiological studies or both. Pathways carrying X, Y, or W information are indicated by lines of different textures. For monkeys, R indicates a retinal input of unknown cell type. Dashed lines indicate weak or suspected pathways. AOS, accessory optic system; IT, inferotemporal cortex; LP, lateral posterior nucleus of pulvinar; LS, lateral suprasylvian cortex; LGN$_v$, ventral lateral geniculate nucleus; LGN$_d$, dorsal lateral geniculate nucleus; Min, medial interlaminar nucleus; MST, middle superior temporal cortex; MT, middle temporal cortex; PGN, pregeniculate nucleus; PT, pretectum; R, unspecified retinal source; SC, superior colliculus; VIP, ventral intraparietal cortex.

cortical areas that are discussed in the next section. Furthermore, in cats some of these extrastriate areas receive an additional Y-like input from the LGN_d and its associated nuclei; the number and amount of overlap of ascending inputs to extrastriate cortex is richer in cats than in any of the other mammals studied, including monkeys (Fig. 20–15). Finally, although W ganglion cells may represent up to 40% of all efferent retinal neurons, their destinations in the CNS are only poorly explored; they may be more important in visual processing than is indicated in Figure 20–15.

Although not shown in Figure 20–15, nearly every visual relay nucleus in the midbrain and thalamus receives a reciprocal descending input from the cortical area to which it projects. Only two of these descending inputs have been studied in any detail. We have already mentioned that a corticofugal pathway from layer 5 cells in V_1 has a considerable influence on neurons in the cat superior colliculus. In addition, striate cortex provides 35% or more of all synapses onto LGN_d neurons in both cats and monkeys. In both species, corticothalamic cells are confined to layer 6, and in cats, most are simple cells with sharp orientation tuning and direction selectivity; half have binocular receptive fields. It is surprising, therefore, that cells in the LGN_d, unlike those in the pulvinar or colliculus, reflect none of the receptive field properties characteristic of these simple cells, except for a very subtle binocular input. Indeed, cooling V_1 produces little, if any, change in the response of LGN_d neurons in either anesthetized cats or monkeys. Perhaps recording in the behaving monkey trained to perform specific visual discriminations would reveal a role for this prominent corticothalamic projection.

EXTRASTRIATE CORTICAL VISUAL AREAS

Although many of the attributes of a visual stimulus are coded in the discharge of single cells in V_1, the vast capabilities of visual sensation are thought to depend on "higher" cortical processing. Several extrastriate areas are classified as visual because they (1) receive projections from known cortical and subcortical visual areas, (2) have maps that represent part or all of the visual field, and (3) contain cells that respond to visual stimuli. Based on these criteria, up to half the entire cortex in monkeys (striate cortex alone occupies 17 to 20%) and 30% of the entire cortex in cats is somehow involved in vision. As many as 12 separate areas in monkeys and 14 in cats have

been identified. Although many of these areas are visible on the lateral surface of the cat brain, many in the monkey are buried in deep sulci.

Locations and Connections of Cortical Visual Areas

The locations of most of the extrastriate visual areas are shown in Figure 20–16,[43, 45] and their major connections are schematized in Figure 20–15. Unfortunately, no consistent set of designations for cortical visual areas exists. Some are numbered according to their anatomical proximity to V_1 (e.g., V_2, V_3, and V_4); others are named according to their cortical sulci or gyri (e.g., inferotemporal [IT], middle temporal [MT], middle superior temporal [MST], and lateral suprasylvian [LS]); and some bear Brodmann's numerical anatomical designations. This last scheme is particularly unfortunate because Brodmann's area 18 encompasses only V_2 in monkeys but comprises V_2, V_3, and V_4 in cats. The Brodmann designation, therefore, will be avoided wherever possible, although, for reference, it is provided in parentheses in Figure 20–15.

Figure 20–15 suggests that the anatomical connections of extrastriate visual areas are organized according to a loose hierarchy, with the flow of information originating in V_1. In both cats and monkeys, pathways pass visual information forward to successively adjacent anatomical areas; in both, there are also multiple parallel pathways. V_3, for example, receives inputs not only from the adjacent V_2 but from V_1 as well. Also, areas quite distant from V_1 (e.g., MT and areas 20 and 21 in cats) receive a direct V_1 input. Although not shown in Figure 20–15, reciprocal connections probably exist between all extrastriate visual areas.

In addition to their cortical inputs, most, if not all, extrastriate visual areas receive one or more thalamic inputs. All extrastriate visual areas in cats and many in monkeys receive projections from the pulvinar, which, in cats, includes the lateral posterior nucleus. Some extrastriate areas in cats, but few if any in monkeys, receive direct inputs from the LGN_d. Indeed, cats exhibit a much greater overlap of ascending inputs than any other mammal studied thus far.

Visual Maps

In contrast to V_1, which has a complete representation of the visual field, extrastriate areas have

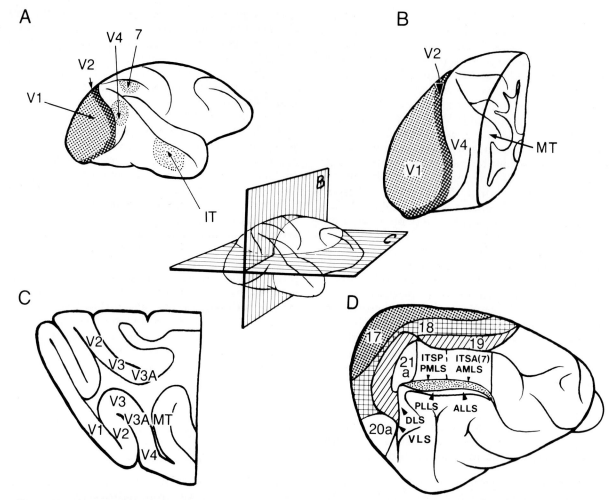

Figure 20–16 Location of cortical visual areas in monkeys *(A–C)* and cats *(D)*. The central inset illustrates how the sections *B* and *C* were taken through *A* to reveal cortical areas hidden in the depths of sulci. In cats, the lateral suprasylvian region consists of three pairs of subregions (AMLS, ALLS; PMLS, PLLS; VLS, DLS) buried in the banks of the suprasylvian sulcus and two subregions (ITSA and ITSP) on the middle suprasylvian gyrus. (After Van Essen, D. C.; reproduced, with permission, from The *Annual Review of Neuroscience*, Vol. 2. © 1979 by Annual Reviews Inc.; and Tusa, R. J.; Palmer, L. A.; Rosenquist, A. C. In Woolsey, C. N., ed. *Cortical Sensory Organization*. Clifton, N.J., Humana Press, 1981.

only partial representations.[43] In cats, for example, V_2 receives only the central 50 degrees of the binocular portion of the visual field. V_3 and parts of LS have a representation that emphasizes the horizontal meridian. Other parts of LS have extensive representations of the lower visual field, while areas 20a and 20b have mostly the upper. Also, the magnification factor in some extrastriate areas is quite different from that in V_1, so that the central and peripheral retina have nearly equal representation. Arguing that each area contains just sufficient field representation for the function it performs, Tusa and colleagues[43] conclude that areas 20a, AMLS, and ALLS are wired to emphasize the peripheral visual field while the remaining areas

are specialized for central vision. Although less well studied, several extrastriate visual areas in monkeys also have representations of only part of the contralateral visual field.[45a]

Behavior of Single Neurons

V_2. In both cats and monkeys, the receptive fields of V_2 cells are similar to those of V_1. In particular, simple, complex, and hypercomplex cells can be found and most are binocular. In both species, however, the receptive fields of V_2 cells are larger; for example, in cats, the maximum visual acuity to a drifting grating is 8 cycles/degree in V_1 but seldom above 1.5 cycles/degree in V_2.[29] Also, many V_2 cells in both cats (30%)[8] and monkeys (>80%)[34] are sensitive to retinal disparity. In cats, V_2 neurons predominantly respond to either distant or nearby stimuli (i.e., are far or near cells), whereas in monkeys, V_2 neurons prefer stimuli placed at a variety of specific distances from the eye (i.e., they are tuned for depth). It is impossible to conclude at this time, however, that V_2 might be more involved in stereopsis than V_1. Finally, color-opponent and some double-opponent cells, mostly of the red-green variety, are found in monkey V_2, but few have the oriented receptive fields characteristic of some color cells in V_1.

Cooling part of V_1 while recording from a topographically corresponding region of V_2 produces markedly different effects in cats and monkeys. In monkeys, cooling V_1 abolishes all visual responses in V_2. In cats, the same procedure reduces only the vigor of the responses, but the receptive field properties, including orientation and direction selectivity, remain intact. These data may be explained in part by Figure 20–15, which shows that in monkeys V_2 derives input solely from V_1, whereas in cats it also receives a direct thalamic input; apparently the thalamic input is sufficient to maintain the cell's visual response in the absence of V_1. Indeed, the larger receptive fields and sensitivity to higher stimulus velocities of cat V_2 cells probably reflects the documented inputs from Y-like thalamic cells in LGN_d and Min (Fig. 20–15).

V_3. In cats, V_3 cells prefer stimuli that are generally similar to those preferred by V_1 cells; however, V_3 cells have larger receptive fields and broader orientation tuning curves and prefer somewhat lower stimulus velocities. Most cells are activated equally by both eyes; some are direction selective. Nearly all V_3 cells apparently are either complex or hypercomplex. Many V_3 cells can be activated from the LGN_d at monosynaptic latencies that are characteristic of a W-like pathway. When V_1 and V_2 are cooled, the visual responses of V_3 cells are scarcely altered; therefore, the thalamic input to V_3, like that to V_2, appears to be more important than the cortical input.

The properties of monkey V_3 cells are a subject of controversy. Most are binocular, have larger receptive fields than V_1 or V_2 cells, are not direction selective, and are not color opponent. Some, perhaps most, are orientation selective.

In summary, the cells in V_2 and V_3 behave very much like those in V_1. From V_1 to V_3, however, the cells generally have progressively larger receptive fields and receive a more nearly equal input

from each eye. Unlike some of the cortical areas that we will now discuss, neither V_1, V_2, nor V_3 seems preferentially sensitive to any particular attribute of a visual stimulus.

Middle Temporal Cortex (MT). Cells in both MT and its presumed feline homologue (AMLS and PMLS of the Clare-Bishop area in the LS cortex (Figs. 20–15 and 20–16)) seem specialized to respond to motion.[26] Nearly all these neurons prefer moving rather than stationary stimuli, respond over a broad range of velocities (from 5°/s to 100°/s), and are direction selective. Figure 20–17 shows the direction selectivity of a typical MT neuron, which clearly prefers stimulus movement to the left and down. Furthermore, most MT cells are binocular and some respond to motion toward and away from the animal. Also, many respond to relative movement of the background and the test stimulus. Many neurons are orientation selective, although more broadly tuned than those in V_1. The receptive fields of most MT neurons are quite large and occasionally cross the vertical meridian. Finally, most show little or no sensitivity to stimulus shape or color. Consequently, in MT (lately also called V_5) and its feline homologue there is an enhanced sensitivity to stimulus motion and a reduced sensitivity to stimulus form.

MT projects to an adjoining area, the MST sulcus, which also contains direction selective neurons, but they have even larger receptive fields.[45a] MST, in turn, projects to area 7a of the parietal lobe (Fig. 20–15).

Area 7a. Much of our knowledge about the parietal lobe derives from recent studies on alert monkeys. The first cells described in area 7a discharged only when the animal looked at, or reached for, interesting objects such as bits of food. In subsequent experiments on monkeys trained to make a variety of eye and hand movements, many of these cells were found to increase their discharge for movements to, or in pursuit of, desirable targets (i.e., those associated with a reward), but not when the same movements were made spontaneously.[24] Closer examination revealed that this response associated with tracking could be attributed to a hidden visual response caused by the slight movement of the image of the target on the retina. Parietal neurons have large receptive fields that usually cover the contralateral visual field and may include the fovea; they are not fastidious about stimulus shape, orientation, color, or direction of movement. Like some cells already described in the superior colliculus, some area 7a cells respond most vigorously if the visual stimulus is to be the target of an ensuing

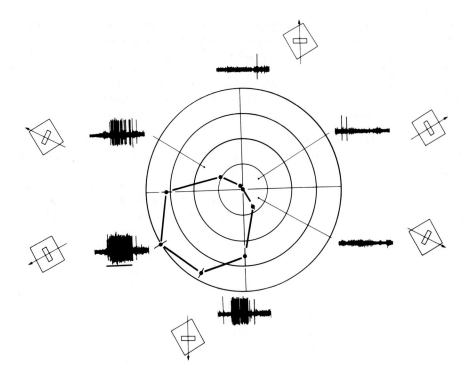

Figure 20–17 Direction selectivity of a neuron in middle temporal cortex (MT) demonstrated by moving a bar in various directions across its receptive field (3° × 3° square outline). The polar plot represents the average firing rate during five stimulus presentations in each of 12 directions of motion. Each circle represents a change of about 31 spikes/s. (After Maunsell, J. H. R.; Van Essen, D. C. *J Neurophysiol.* 49:1127–1147, 1983.)

saccade. These data suggest a role for the parietal lobe in visual attention, but it is not clear whether area 7a mechanisms actually shift visual attention (a motor role) or simply allow attention to change the efficacy of a visual message (a sensory role). Either interpretation is consistent with the principal deficit that follows posterior parietal lobe lesions in primates, i.e., profound inattention to all sensory (including tactile) stimuli delivered to the side contralateral to the lesion (see Chapter 31 for further discussion).

At least some area 7a neurons have visual responses similar to those in MT. For example, in a monkey fixating a small light spot, the area 7a cell illustrated in Figure 20–18[35] responded best to a bar moving to the left and up (B_2) and not at all to a bar moving to the right and down (B_1). In the anesthetized cat, similar direction-selective cells are found in the middle suprasylvian gyrus, the apparent feline homologue of area 7a. Some of the same area 7a cells that exhibit direction selectivity also fire vigorously when the monkey tracks a small target spot moving in a particular direction. For example, the cell in Figure 20–18 also discharges during eye movements to the right and down (A_1) but not in the opposite (A_2) direction. The response during tracking is not simply visual, for when the spot is turned off for brief periods during which the monkey continues to track, the discharge suffers hardly any decrement (Fig. 20–

$18A_3$). If the monkey is required to track a target spot moving across a stationary bar as in Figure $20–18C_1$, the same cell exhibits its most vigorous response. These data suggest that cells in area 7a respond differently when tracking occurs in the dark than when it takes place across a stationary background as in real life. Such firing patterns may be instrumental in generating a tracking eye movement. Indeed, both area 7a and MT project to pontine nuclei implicated in the control of eye movement. Alternatively, the discharge of cells such as that shown in Figure 20–18 may help distinguish actual motion of the environment from apparent motion due to one's own movement.

V_4. V_4 first attracted attention when it was reported to have more color-sensitive cells than other cortical areas have.[50] More recent studies, however, indicate that color-sensitive cells occur in areas V_1, V_2, and V_4 in about equal numbers and have qualitatively similar spectral sensitivities, so that color sensitivity alone is not a feature that distinguishes these three cortical areas.[36] Therefore, the original suggestion that V_4 has a unique role in the processing of spectral information should be viewed with caution.

Although most V_4 cells lack direction selectivity, about 50% are orientation selective like those of areas V_1 and V_2, and most are binocular. The receptive fields of V_4 neurons are much larger than those of V_2.

Inferotemporal Cortex (IT). Most neurons in the inferotemporal cortex respond to so many different stimulus attributes that it is often difficult to specify their adequate stimulus.[10] Most prefer movement, some are direction selective, and some are sensitive to stimulus shape, color, or texture or to combinations of all three features. Generally, the receptive fields of IT neurons are the largest yet discussed; they frequently include most of the contralateral visual field, part of the ipsilateral visual field, and invariably the fovea.

In contrast, a small subpopulation (10% or less) of IT neurons have extremely precise stimulus requirements and appear to respond only to specific objects such as a face.[5] Removal of any component or feature of the face, such as the eyes, nose, or anterior or posterior portion of the head, always reduces the response. The discovery of such cells underscores the necessity for imaginative testing, especially for neurons in the "higher" visual areas.

Effects of Cortical Lesions on Behavior

Although knowledge of the connections of an area and the behavior of its neurons helps to establish its putative role, the effect of an area's

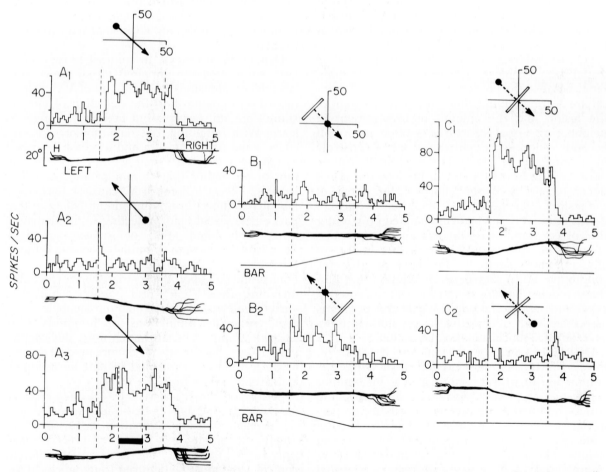

Figure 20–18 Averaged firing rates (displayed as time histograms) of a neuron in parietal cortex (area 7a) during smooth-pursuit eye movements and visual background motion. In the left column, the monkey tracks a small spot (black dot and solid arrow) of light in the dark in the neuron's preferred (A₁) and non-preferred (A₂) directions. When, during movement in the preferred direction, the target is briefly turned off (A₃, during horizontal solid bar between 2 and 3 seconds), smooth pursuit continues, as does the neuronal response, indicating that the firing is due to eye movement per se. In the middle column, movement of a bar of light when the animal is fixating causes a response for left and up movement (B₂), indicating that the same neuron also has visual sensitivity. In the right column, the animal tracks a spot across a stationary bar. Total visual field is 50 by 50 degrees; actual eye movements are indicated below histograms, and time is in seconds. (After Sakata, H.; Shibutani, H.; Kawano, K. *J. Neurophysiol.* 49:1364–1380, 1983.)

removal on the ability to perform visual tasks provides another important clue.

Most behavioral visual testing requires a subject to discriminate between two visual stimuli. In the simplest apparatus, a monkey is placed in a testing cage with two hollowed-out food wells that are covered by blocks painted with distinctive patterns or colors. The animal must learn to displace a particular block to obtain a hidden raisin. In a more sophisticated apparatus, different visual stimuli are projected onto two lit panels and the monkey must touch the "correct" one (cats use their noses) to receive a reward. Acuity can be tested by requiring the animal to distinguish between horizontal and vertical gratings with different spatial frequencies. Form, pattern, or shape discriminability can be tested by requiring subjects to distinguish between pairs of stimuli such as | vs. —, ○ vs. +, and △ vs. ▽.

For several reasons the effects of lesions on visual behavior must be interpreted with caution. First, because the borders of adjacent visual cortical areas are often indistinct, the precise extent of the lesion and the amount of involvement of neighboring visual structures is often in doubt. Second, since visual structures are extensively interconnected, a lesion in one usually causes degeneration in others. Third, because the visual deficit usually decreases with time, the timing of the testing after the lesion is important. Fourth, a deficit frequently can be demonstrated only under the correct stimulus conditions; for example, discrimination tasks performed near threshold usually are more affected than those at photopic conditions. Fifth, recovery from a lesion that produces a deficit usually is variable and depends on experience and retraining. Despite the problems associated with interpreting lesion studies, however, several conclusions can be drawn.

Deficits in Cats. Complete bilateral lesions of V_1 in cats produce virtually no deficits in visual discrimination or in the performance of visuomotor tasks such as placing a paw on an approaching surface or avoiding objects when walking. In particular, such cats do not lose the ability to discriminate between two small geometric forms.[7] Even lesions that include not only V_1 but most of V_2 produce only a 30 per cent decrement in visual acuity as revealed in grating, vernier, and angle acuity tasks.[3] Moreover, cats with combined lesions of areas V_1 and V_2 are able to learn form discriminations postoperatively as long as the discriminanda are bright. However, they have greater difficulty with pattern discriminations if the discriminanda are near visual threshold (i.e., dim).

Taken together, these observations indicate that the combined action of areas V_1 and V_2 is required for the processing of the finest details in spatial contours.

In contrast, after lesions that remove areas V_3, 20, and 21 and the LS cortex while sparing V_1 and V_2, cats seem to have great difficulty learning simple, large planimetric pattern discriminations. Since V_1 and V_2 in the cat are dominated by their inputs from the LGN_d (see Fig. 20–15) and the other cortical visual areas appear dominated by their pulvinar input, it has been suggested that the colliculo-pulvino-extrastriate pathway provides the first stage in simple, coarse form perception while the geniculostriate pathway is involved in fine form (acuity) perception. We have already seen that the responses of single neurons and data from cooling studies are consistent with this suggestion.

Deficits in Monkeys. In contrast to cats, monkeys suffer severely impaired visual acuity after large bilateral lesions involving all of V_1.[46] Brightness discrimination and the detection of moving stimuli are still possible but take much longer to learn. With retraining, destriate monkeys also recover some pattern vision and are best described as having a reduced visual acuity (of perhaps 10 min of arc rather than 1 min of arc; see Weiskrantz).[46] Focal lesions of V_1 cause a temporary inability to detect brief, dim stimuli falling in that part of the visual field projecting to the damaged locus. This blind spot, or *scotoma*, gradually shrinks with time and eventually becomes difficult to detect at all. Lesions of V_1 and extensive portions of the neighboring prestriate cortex (probably including areas V_2, V_3, and possibly V_4) abolish virtually all visual discrimination capabilities except crude discriminations of luminous flux. Clearly, lesions of V_1 are much more devastating in monkeys than in cats. The more modest deficits in the cat can be explained in part by its direct extrastriate inputs from the LGN_d.

Lesions of extrastriate visual areas often produce subtle behavioral deficits that can be revealed only by sophisticated discrimination tasks. Unfortunately, no single task or set of tasks is used universally; therefore, the function of different cortical areas must be deduced from deficits in the performance of tasks that differ significantly from one laboratory to another. In a few studies, however, monkeys have been required to perform the same tasks both before and after a variety of extrastriate lesions. In these cases, it is possible to evaluate the relative importance of several cortical areas for certain discriminations.

To distinguish between the functions of the inferior temporal and "prestriate" areas, investigators trained monkeys to select the correct stimulus of a single pair of stimuli (a single discrimination problem) and, further, to select the correct stimulus in each of eight pairs of stimuli presented randomly (a multiple discrimination problem). Trained monkeys with lesions of the prestriate area between the superior temporal and lunate sulci (an area that apparently includes parts of V_3, V_4, and MT; see Figs. 20–2 and 20–16) have difficulty in remembering how to do the single discrimination problem (a deficit in retention), and untrained monkeys have difficulty learning the single discrimination postoperatively (a deficit in acquisition). Monkeys with lesions of the inferotemporal cortex also have problems with single discrimination retention and acquisition, but they are less severe. On the other hand, such monkeys are much more handicapped than monkeys with prestriate lesions in the postoperative acquisition of a multiple discrimination. Inferotemporal lesions also impair the learning of a variety of other visual discrimination tasks while leaving visual sensory function intact. These data[10, 44] suggest that both areas are involved in visual discrimination learning but that the prestriate area is concerned with the sensory aspects of the problem, whereas the inferotemporal cortex is involved with the associative (putting together) or cognitive aspects.

To distinguish between the functions of the inferior temporal lobe and the posterior parietal lobe (including area 7a), investigators trained monkeys on a pair discrimination task and a "landmark" task. For the latter, monkeys were rewarded for choosing the covered food well located closer to a striped cylinder (the "landmark"), which was randomly positioned 5 cm from one food well and 20 cm from the other. Monkeys with inferior temporal lesions had difficulty with postoperative acquisition of the paired pattern discrimination but relatively little difficulty in postoperative retention of the landmark task. In contrast, monkeys with posterior parietal lesions had severe difficulty in retaining the landmark task but little problem with the postoperative acquisition of the pattern discrimination.[44]

These data suggest that the posterior parietal cortex is concerned with learning and remembering the spatial relations among objects. Earlier we discussed another deficit in visual spatial processing produced by posterior parietal lesions. That neglect of the contralateral "extrapersonal space,"[24] however, appears to be a deficit in visual attention rather than visual memory processes.

(See Chap. 31 for further discussion of parietal neglect.)

Possible Roles for Multiple Cortical Visual Areas

It would be gratifying to conclude this section with a discussion of how each visual area contributes to perception or movement. Unfortunately, the extrastriate areas are only poorly studied and the data are insufficient to reach firm conclusions. Nevertheless, some recent suggestions are worth considering.[1]

Neurons in V_1 seem to discharge for all the possible attributes of a visual stimulus. These include size, color, orientation, depth in space, and velocity and direction of motion. Furthermore, this information is probably available for every point in the visual world because V_1 contains a complete topographic map. Consequently, a complete perception of the visual world could probably be obtained if some neural executive were able to decode the composite neural image produced by all the cells in V_1. Because the subregions of receptive fields near the fovea are very small and many V_1 cells have stringent stimulus requirements (e.g., sharp orientation tuning), the percept of images falling near the fovea would be especially acute. Indeed, consistent with this suggestion, deficits in form vision and visual acuity are produced by lesions of V_1 in the primate.

If all the information required for visual perception is already encoded in V_1, what, then, is the role of the extrastriate visual areas? An early suggestion that they participate in perception by providing the substrate for the serial processing of information originating in V_1 seems unlikely for several reasons. Although a small percentage of extrastriate neurons in the inferotemporal cortex do respond to highly specific stimuli (e.g., faces or hands) as if their response represented a convergence or elaboration of the responses of many V_1 cells, the vast majority of cells in the inferotemporal cortex and other extrastriate areas do not have such response characteristics. Instead, neurons in extrastriate areas generally have *less* specific stimulus requirements than neurons in V_1. In particular, the sizes of the receptive fields of extrastriate neurons increase with their distance from V_1, so that cells in the parietal and temporal cortices have receptive fields that include most of the contralateral visual field. Also, extrastriate visual areas are generally characterized by incomplete or complicated topographic maps that often represent only parts of the visual field. Consequently, a strictly serial model of visual processing would leave portions of the visual world unrepresented. Finally, lesions of individual extrastriate areas have little effect on form vision and acuity.

Although the characteristics of extrastriate areas seem inappropriate for a detailed reconstruction of the visual scene, they do seem appropriate to effect a global reconstruction. Many visual percepts require bringing together information over considerable regions of the visual field, and these percepts frequently seem to require knowledge of only one attribute of the visual stimulus. For example, an object can be distinguished from its surroundings based on color alone (e.g., finding a lost golf ball) or its shape can be inferred from the extrapolation of its edges behind another object that obscures it. Zeki[50] was the first to suggest that the solving of such problems was the province of separate extrastriate areas, each specialized to extract information about some stimulus feature, such as color or orientation, in the examples just described. Some extrastriate areas do seem specialized for a specific stimulus feature (e.g., MT for visual motion), while others do not. Although V_4 may not be *the* extrastriate color area, a cortical color region apparently does exist, since lesions of the occipital lobe in humans produce a complete absence of all color sensation (achromatopsia). Extrastriate areas may not just operate in isolation, but they may provide feedback to enhance certain parts of the visual representation in V_1.

Visual information from the striate cortex seems to flow into extrastriate areas over two separate pathways, each involving different visual areas. A set of ventral pathways extends into the inferior temporal cortex, whereas a set of dorsal pathways extends into the parietal cortex. Van Essen and Maunsell[45a] concluded that the dorsal pathway involves MT, MST and 7a (see Fig. 20–15), whereas the ventral pathway includes V_4 and IT. On the basis of the lesion studies discussed above, Ungerleider and Mishkin[44] posited that the inferior temporal pathway participates mainly in the acts of noticing and remembering an object's qualities (recognizing *what* an object is), while the parietal pathway seems concerned with the perception of spatial relations among objects (identifying *where* an object is). Others believe that the existing data suggest that the inferior temporal pathway is involved in color and form analysis, whereas the parietal pathway is involved in the analysis of motion.[45a] Whatever their specific roles might finally prove to be, it is clear that the various extrastriate visual areas make a fundamental contribution to visual perception.

THE MIDBRAIN VISUOMOTOR PATHWAYS

The Superior Colliculus: Intermediate and Deep Layers

As mentioned previously, the intermediate and deep layers of the superior colliculus are anatomically and functionally distinct from the superficial layer. Whereas the superficial layer has been implicated in visual perception, the intermediate and deeper layers apparently participate in the control of movement.

Inputs. The intermediate and deep layers of the colliculus receive a variety of sensory information, including visual, auditory, somatosensory, vestibular, and proprioceptive signals. These signals have been evaluated mostly in anesthetized animals. In general, collicular neurons prefer a moving stimulus. The preferred movement is direction specific for cats but omnidirectional for monkeys.[37] In both cats and monkeys, the receptive fields are usually large, often involving sizable portions of the contralateral visual field. The visual responses are highly dependent on the level of anesthesia and tend to habituate upon repeated stimulation. Lesions of the visual cortex severely impair the visual responses of cells in the deep collicular layers, suggesting that their input is relayed through the cortex. Many cells in the intermediate and the deep layers respond to auditory and somatosensory stimuli. Some intermediate layer cells even respond to *both* auditory and visual stimuli, but only if each is positioned at nearly the same point in space (i.e., the receptive fields overlap).

Outputs. When their behavior is examined in an alert animal, many of the visually sensitive intermediate layer cells also respond to rapid shifts in the direction of gaze, called saccades.[49] Figure 20–19A shows the behavior of such a neuron.[41] About 80 ms after a target (T) jumps onto its visual receptive field, this unit discharges a brief burst of spikes (single arrow). Prior to the oblique saccade (note that it has both a horizontal [H] and vertical [V] eye movement component) that places the fovea on the target, a second, more vigorous burst occurs (double arrows). The first burst is of visual origin, for it persists when a target (T) jump of short duration elicits no saccade (Fig. 20–19B). In the manner of Y-like cells, a second burst occurs when the target jumps off the receptive field.

Cells in the deepest layers of the colliculus (Fig. 20–19C) display no early burst of visual origin but only a burst associated with the saccade. Each cell discharges only for saccades of a particular size and direction. For the cell of Figure 20–19D, the total number of spikes in the burst is plotted as a function of saccade size (radius) and direction (angle). This cell discharges most vigorously for 2-degree saccades directed toward about 330 degrees. Cells in the rostral colliculus prefer smaller saccades, whereas those caudal prefer larger saccades. This topographic map of preferred movement amplitude and direction in the deep layer corresponds to the sensory maps in the intermediate layers.

Many cells in the intermediate layers and some in the deep layers project (via the tectobulbar, tectopontine, and tectospinal tracts) to structures concerned with movement of the eyes, head, and pinnae. Accordingly, it is postulated that the superior colliculus is involved in the conversion of stimuli falling in an animal's visual, auditory, and somatosensory space into a motor command that orients the animal toward the stimulus. Indeed, in cats electrical stimulation of the colliculus produces eye, head, and body movements to the contralateral side, causing an animal to turn in contraversive circles, whereas unilateral collicular lesions produce ipsiversive circling and difficulty in following and reacting or orienting to stimuli on the contralateral side.[37] In monkeys, collicular stimulation reliably produces contralateral saccades and less reliably produces head movements. The motor abnormalities produced by collicular lesions often are rather subtle. For example, monkeys with collicular lesions exhibit delayed saccades that often fall slightly short of the target, an oculomotor deficit that is revealed only by careful behavioral testing.[49] In general, visual and motor deficits produced by collicular lesions become less severe in animals higher on the phylogenetic scale.

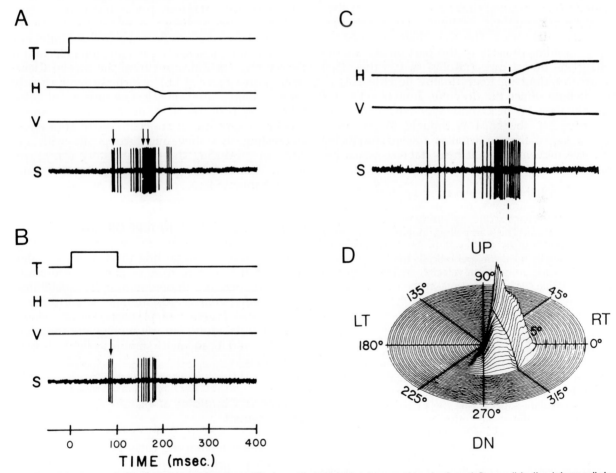

Figure 20–19 Discharge patterns of superior colliculus cells in a behaving monkey. In A and B, a cell in the intermediate layer discharges a burst of spikes (single arrow) after a target movement (T) and, in A, a second, more intense burst (double arrow) before an eye movement that places the eye on target. The eye movement is oblique, with both horizontal (H) and vertical (V) components. In C and D, a cell in the deep layers discharges the greatest number of spikes (ordinate in D) for small saccades with both rightward (upward deflection of H trace) and downward (downward deflection of V trace) components. (After Sparks, D. L. Brain Res. 156:1–16, 1978.)

Pretectum

As was already mentioned, the pretectum is a complicated structure composed of at least six subnuclei. At present, we know something about the functional role of only two of them. In cats, one of the subnuclei, the *nucleus of the optic tract* (NOT), receives direct visual inputs from the contralateral retina but also has a prominent input from the ipsilateral visual cortex (areas V_1, V_2). NOT neurons respond best to stimuli that occupy the entire visual field (i.e., full-field stimuli) and move in a temporal-to-nasal direction across the contralateral retina or in a nasal-to-temporal direction across the ipsilateral retina. NOT neurons respond to a wide range of stimulus velocities, from <0.1°/s to over 100°/s.[14] After disconnection of the ipsilateral cortex, NOT cells are driven only by slowly moving stimuli delivered to the contralateral eye, indicating that they acquire their ipsilateral input and their sensitivity to high velocities via the visual cortex. The NOT, like the accessory optic system, projects to the part of the inferior olive (the dorsal cap of Kooy) that is the major source of climbing fibers to the flocculus of the cerebellum. Since the flocculus is involved in the coordination of head and eye movements (see Chapter 27), the NOT is thought to provide a visual signal that contributes toward producing the eye movements that help stabilize visual images on the retina.

Although full-field movement occurs naturally only when one's head moves, one can simulate the motion of the entire visual world by sitting within a rotating patterned drum. This compelling motion of the entire environment causes the eyes to be "involuntarily" pulled along by the drum; periodically these slow movements, moving at the drum velocity, are interrupted by fast "resetting" eye movements in the opposite direction, producing a saw-toothed pattern of movement known as *optokinetic nystagmus* (OKN).

In cats, lesions of the pretectum or of its cortical pathway cause severe deficits in OKN. If the pretectum is destroyed bilaterally, OKN is completely abolished. On the other hand, bilateral lesions of the visual cortex abolish only the slow phase of OKN that results from monocular stimuli moving in a nasal-to-temporal direction. These data suggest that the pretectum, probably the NOT, and the visual cortex are essential elements in producing the optokinetic response. Indeed, in alert cats, NOT neurons respond vigorously during optokinetic nystagmus, and electrical stimulation through the recording electrode elicits a clear OKN pattern of eye movements.

Another pretectal subnucleus, the *nucleus olivarius*, appears to mediate the pupillary light reflex.

In both cats and monkeys, it receives a direct retinal input and projects either directly or through the nucleus of the posterior commissure to the visceral cell column of the oculomotor nucleus, which provides the preganglionic parasympathetic innervation for the pupil (see Chapter 34). The discharge of some cells in the pretectum increases with light intensity, and these cells probably constitute part of the afferent limb of the pupillary light reflex.

Accessory Optic System[40]

Most neurons in the accessory optic nuclei prefer large, textured visual patterns moving in specific directions across their large receptive fields. In general, the optimal speeds for exciting these neurons are very low, ranging from 0.5°/s to 15°/s, although some neurons prefer higher velocities. In the cat, the discharge of cells in the accessory optic nuclei probably reflects a combined influence from movement-sensitive W ganglion cells and a descending cortical input from the medial Clare-Bishop area (part of LS), which we have already seen contains direction-selective movement cells. Like the cells of the NOT, neurons in the various accessory optic nuclei project to the inferior olive, suggesting that they might cooperate with the NOT in producing the compensatory eye movements that stabilize visual images on the retina (see Chapter 27).

VISUAL DEVELOPMENT: NATURE OR NURTURE?

A basic question regarding any sensory modality concerns the relative roles of genetic factors and sensory experience in determining its capabilities. The visual system affords a unique opportunity to study development and plasticity because many aspects of its adult neurophysiology are well documented and its adequate stimulus—light—is easy to manipulate.

Behavioral Correlates of Visual Development

Behavioral testing indicates that many aspects of visual perception in infant humans, monkeys, and cats are rather poor.[28a] For example, visual acuity assessed by determining whether a child turns toward an "interesting" grating pattern of different spatial frequencies or a homogeneous surface of the same mean luminance is about 1 cycle/degree (recall Chapter 19) for the 1-month-old human infant. Visual acuity increases at about 1 cycle/degree for each month of age and approaches adult levels of 50 to 60 cycles/degree only after 5 years. In monkeys, a similar testing procedure reveals a neonatal acuity that is comparable to that of 3-week-old humans and develops about three times faster. In kittens, visual acuity, measured by training the animals to jump to a table on which a grating rather than a homogeneous pattern appears, improves gradually from 0.75 cycles/degree at 30 days to nearly adult values (5–7 cycles/degree) near the end of the fourth month. Stereopsis also is rather primitive at birth but develops rapidly within a period of 3 to 4 weeks when cats are about 5 weeks old and humans are about 5 months old. The rather gradual development of acuity and late development of stereopsis suggest that the neural processes underlying these capabilities evolve with time.

Neural Correlates of Visual Development

As might be expected from these behavioral data, the discharge properties of infant visual neurons differ from those of the adult. As soon as they can be tested, the receptive fields of both retinal ganglion cells and LGN_d cells are rather large. In both kittens and monkeys, the spatial resolution of LGN_d cells serving the central retina is low at 3 weeks of age and slowly attains adult values over a time course comparable to that for the behavioral development of visual acuity. For example, the mean resolution of monkey LGN_d cells improves from about 3 cycles/degree to about 20 cycles/degree in the first 200 days of life. The development of visual acuity, therefore, seems to be limited by the maturation of subcortical visual structures.[28a]

In V_1, a sizable number of cells in kittens under 3 weeks of age are unresponsive or lack the stimulus specificity (e.g., orientation tuning, spatial resolution) of adult cells. A few, however, do have adult orientation and direction selectivity. Many of the cells are binocular, but the locations of the receptive fields plotted for each eye can be very disparate. With normal visual experience, the vast majority of cortical neurons attain adult-like properties by about 6 weeks of age. The greatest increase in the number of synapses per cortical neuron occurs between the second and fifth weeks.

Effects of Visual Deprivation

The development of adult visual sensitivity that occurs between the third and sixth weeks of age may be a genetically predetermined postnatal process or, alternatively, may be shaped by the animal's visual experience. To assess experiential effects on visual development, investigators have raised animals under a variety of unusual visual conditions.

Monocular Deprivation. Many of the major concepts in visual development have resulted from studies of the behavioral, physiological, and anatomical consequences resulting from the restriction of vision to one eye, i.e., *monocular deprivation*.

Kittens with the lids of one eye sutured closed at birth (a reduction in light intensity of 4 to 5 log units) appear to be completely blind in the deprived eye when the lids are opened 3 months later. The cells in those LGN_d laminae receiving a projection from the deprived eye are markedly shrunken but the receptive field properties and visual responses of all LGN_d cells appear to be surprisingly normal, although there is some controversy on this point. In V_1, however, virtually every cell can be driven only through the normal eye, although most other receptive field properties appear normal. In contrast, monocular deprivation for 3 months in an adult animal produces an ocular dominance histogram that is essentially normal.[47]

To determine the *sensitive period* during which cortical cells are vulnerable to monocular deprivation, Olson and Freeman[31] subjected kittens to 10 to 12 days of contralateral eye closure at ages spaced regularly through the first 4 months of life. As can be seen in Figure 20–20, cells in V_1 are heavily dominated by the normal (ipsilateral) eye if monocular deprivation occurs between the 28th and 57th days of life (4–8 weeks). However, a clear, but markedly reduced, preference for the normal eye can be seen until the 16th week (109 days). During the fifth and sixth weeks, brief monocular closures of only 4 to 8 hours can produce a clear disruption of binocular inputs to cortical neurons.

The anatomical substrate for the shift in ocular dominance can be illustrated by injecting the normal eye with a transsynaptic orthograde label. In normal adult cats and monkeys, we have seen that geniculocortical terminations from each eye are segregated into bands of equal widths in layer 4 (see Fig. 20–5). After monocular deprivation, the bands are still discernible, but the width of the bands devoted to the deprived eye is markedly reduced while the width of the bands representing the good eye is expanded.

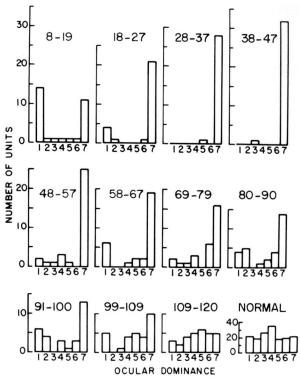

Figure 20–20 Demonstration of the sensitive period. Ocular dominance of V_1 cells in kittens subjected to monocular deprivation for about 10 days at the different ages indicated on histograms. Cells driven only by the deprived eye fall in group 1. (After Olson, C. R.; Freeman, R. D. *Exp. Brain Res.* 39:17–21, 1980.)

Binocular Competition. In view of the dramatic effects of monocular deprivation on V_1 neurons, one might expect the effects of *binocular deprivation* to be devastating. Surprisingly, although binocular deprivation produces many unresponsive cells, a significant proportion of V_1 cells respond normally. In particular, most responsive cells are binocular. To explain the less severe consequences of binocular deprivation, Wiesel and Hubel[47] suggested that afferent inputs from the two eyes compete for the allegiance of cortical cells. When one eye has a competitive advantage, as in monocular deprivation, its afferents from the LGN_d preempt their target cortical neurons; the apparent expansion of layer 4 afferents from the undeprived eye into the band formerly subserving the deprived eye is consistent with this idea. When eyes are at the same competitive disadvantage as in binocular deprivation, cortical cells remain binocular. The competition hypothesis has been tested by suturing the lids of one eye closed and making focal lesions in the retina of the open eye. Cortical

cells subserving the region of the retinal lesion (or artificial scotoma) often can be driven by the eye that had been closed whereas nearby cells with a normal input from the open eye are driven only by it.

A normal cortex develops only if visual stimuli are presented simultaneously to each eye. Alternating monocular deprivation from one eye to the other every 24 hours during the sensitive period causes an almost complete loss of binocularity in V_1, with roughly equal numbers of cells being dominated by each eye. Even differences of only several hours in the durations of the alternating exposures have been reported to produce a bias for the most exposed eye. Within the sensitive period, cortical ocular dominance can be shifted back and forth between eyes by alternating the monocular deprivations every week; furthermore, the few remaining binocular cells can have widely different (by 10 to 40 degrees) preferred orientations when plotted in each eye, suggesting that they have been caught in the process of changing their ocular allegiance. The neural process involved in establishing cortical ocular dominance, therefore, is both sensitive and rapid.

During the sensitive period, not only ocular dominance but the sensitivity to other stimulus attributes is malleable. For example, kittens raised wearing goggles that have horizontal or vertical stripes painted on the lenses develop V_1 cells whose preferred stimulus orientations correspond to the direction of the stripes. Since such kittens have only a few nonoriented or unresponsive cells, it appears that cells that may have been inclined toward other orientation preferences have been manipulated to prefer the exposed orientations. On the other hand, animals raised in low-frequency stroboscopic illumination to eliminate actual (but not perceptual) movement have low proportions of cortical cells with direction or orientation preferences.[28a] Finally, there are anecdotal reports that cortical cells can be reared to prefer stimuli that normal adult cells do not respond to at all (e.g., spots of light). However, these latter studies, which suggest that visual experience has an inductive role rather than a "tuning" role, are not compelling.

All in all, the experiments on visual deprivation suggest that during the sensitive period the connections for binocular vision are being established. Any condition that prevents the two eyes from normal interactions is devastating. One naturally occurring example is the convergent deviation (esotropia) that occurs in one eye of human infants. This early *strabismus*, if uncorrected, usually produces some deficits in visual acuity in the deviated eye and a loss of many binocular functions. Strabismus has been simulated in animals by using prisms to misalign the two retinal images so that fusion of the images from each eye does not occur. In cats and monkeys raised with such prisms, most cells in V_1 are monocular, with roughly equal numbers having allegiance to one eye or the other.

ANNOTATED BIBLIOGRAPHY

Gilbert, C.D. Microcircuitry of the visual cortex. *Annu. Rev. Neurosci.* 6:217–247, 1983.
A detailed account of the physiology and anatomy of primary visual cortex, V_1.

Mitchell, D.E.; Timney, B. Postnatal development of function in the mammalian visual system. Handbk. Physiol. 3(1):507–555, 1984.
An extremely well-written critical review of the development of vision in mammals; an excellent starting point for students interested in visual plasticity.

Rodieck, R.W. Visual pathways. *Annu. Rev. Neurosci.* 2:193–225, 1979.
A comprehensive review of the destinations of the different retinal ganglion cells.

Schiller, P.H. The central visual system. *Vision Res.* 26:1351–1387, 1986.
A delightful review of research on the central visual system, complete with pictures of many of the investigators and over 300 references.

Van Essen, D.C.; Maunsell, J.H.R. Hierarchical organization and functional streams in the visual cortex. *Trends Neurosci.* 6:370–375, 1983.
A description of the anatomy and some of the physiology of the cortical areas involved with vision, including a discussion of how the extrastriate areas could be organized to process different visual stimulus attributes. Over 60 references give the interested student a good start on the pertinent literature.

REFERENCES

1. Barlow, H.B. Why have multiple cortical areas? *Vision Res.* 26:81–90, 1986.
2. Bender, D.B. Receptive-field properties of neurons in the macaque inferior pulvinar. *J. Neurophysiol.* 48:1–17, 1982.
3. Berkley, M.A.; Sprague, J.M. Striate cortex and visual acuity functions in the cat. *J. Comp. Neurol.* 187:679–702, 1979.
4. Daniel, P.M.; Whitteridge, D. The representation of the visual field on the cerebral cortex in monkeys. *J. Physiol. (Lond.)* 159:203–221, 1961.
5. Desimone, R.; Albright, T.D.; Gross, C.D.; Bruce, C. Stimulus-selective properties of inferior temporal neurons in the macaque. *J. Neurosci.* 4:2051–2062, 1984.
6. De Valois, R.L.; Abramov, I.; Jacobs, G.H. Analysis of response patterns of LGN cells. *J. Opt. Soc. Am.* 56:966–977, 1966.
7. Doty, R.W. Survival of pattern vision after removal of striate cortex in the adult cat. *J. Comp. Neurol.* 143:341–370, 1971.
8. Ferster, D. A comparison of binocular depth mechanisms in areas 17 and 18 of the cat visual cortex. *J. Physiol. (Lond.)* 311:623–655, 1981.
9. Fukuda, Y.; Stone, J. Evidence of differential inhibitory influences on X- and Y-type relay cells in the cat's lateral geniculate nucleus. *Brain Res.* 113:188–196, 1976.
9a. Gilbert, C.D. Microcircuitry of the visual cortex. *Annu. Rev. Neurosci.* 6:217–247, 1983.
10. Gross, C.G. Inferotemporal cortex and vision. *Progr. Physiol. Psychol.* 5:77–123, 1973.
11. Hendrickson, A.E.; Wilson, J.R. A difference in ^{14}C deoxyglucose autoradiographic patterns in striate cortex between *Macaca* and *Saimiri* monkeys following monocular stimulation. *Brain Res.* 170:353–358, 1979.
12. Hendrickson, A.E.; Wilson, J.R.; Ogren, M.P. The neuroanatomical organization of pathways between the dorsal lateral geniculate nucleus and visual cortex in old world and new world primates. *J. Comp. Neurol.* 182:123–136, 1978.
13. Henry, G.H.; Dreher, B.; Bishop, P.O. Orientation specificity of cells in cat striate cortex. *J. Neurophysiol.* 37:1394–1409, 1974.
14. Hoffmann, K.-P. Neuronal responses related to optokinetic nystagmus in the cat's nucleus of the optic tract. In Fuchs, A.F.; Becker, W., eds. *Progress in Oculomotor Research.* New York, Elsevier/North-Holland, pp. 443–454, 1981.
15. Hubel, D.H.; Wiesel, T.N. Integrative action in the cat's lateral geniculate body. *J. Physiol. (Lond.)* 155:385–398, 1961.
16. Hubel, D.H.; Wiesel, T.N. Receptive fields, binocular interaction and functional architecture in the cat's visual cortex. *J. Physiol. (Lond.)* 160:106–154, 1962.
17. Hubel, D.H.; Wiesel, T.N. Receptive fields and functional architecture of monkey striate cortex. *J. Physiol. (Lond.)* 195:215–243, 1968.
18. Hubel, D.H.; Wiesel, T.N. Functional architecture of macaque monkey visual cortex. *Proc. Roy. Soc. Lond. (Biol.)* 198:1–59, 1977.
19. Hubel, D.H.; Wiesel, T.N.; Stryker, M.P. Anatomical demonstration of orientation columns in macaque monkey. *J. Comp. Neurol.* 177:361–379, 1978.
20. Kaneko, A.; Tachibana, M. Double color-opponent receptive fields of carp bipolar cells. *Vision Res.* 23:381–388, 1983.
21. Lennie, P. Parallel visual pathways: a review. *Vision Res.* 20:561–594, 1980.
22. LeVay, S.; Gilbert, C.D. Laminar patterns of geniculocortical projection in the cat. *Brain Res.* 113:1–19, 1976.
23. Lund, J.S.; Henry, G.H.; MacQueen, C.L.; Harvey, A.R. Anatomical organization of the primary visual cortex (Area 17) of the cat. A comparison with Area 17 of the macaque monkey. *J. Comp. Neurol.* 184:599–618, 1979.
24. Lynch, J.C. The functional organization of posterior parietal association cortex. *Behav. Brain Sci.* 3:485–534, 1980.
25. Malpeli, J.G.; Baker, F.H. The representation of the visual field in the lateral geniculate nucleus of the *Macaca mulatta. J. Comp. Neurol.* 161:569–594, 1975.
26. Maunsell, J.H.R.; Van Essen, D.C. Functional properties of neurons in middle temporal visual area of the macaque monkey. I. Selectivity for stimulus direction, speed, and orientation. *J. Neurophysiol.* 49:1127–1147, 1983.
27. Michael, C.R. Color vision mechanisms in monkey striate cortex: simple cells with dual opponent-color receptive fields. *J. Neurophysiol.* 41:1233–1249, 1978.
28. Michael, C.R. Color-sensitive hypercomplex cells in monkey striate cortex. *J. Neurophysiol.* 42:726–744, 1979.
28a. Mitchell, D.E.; Timney, B. Postnatal development of function in the mammalian visual system. *Hdbk. Physiol.* 3(1):507–555, 1984.
29. Movshon, J.A.; Thompson, I.D.; Tolhurst, D.J. Spatial and temporal contrast sensitivity of neurons in areas 17 and 18 of the cat's visual cortex. *J. Physiol. (Lond.)* 283:101–120, 1978.

30. Mustari, M.J.; Bullier, J.; Henry, G.H. Comparison of response properties of three types of monosynaptic S-cell in cat striate cortex. *J. Neurophysiol.* 47:429–454, 1982.

31. Olson, C.R.; Freeman, R.D. Profile of the sensitive period for monocular deprivation in kittens. *Exp. Brain Res.* 39:17–21, 1980.

32. Palmer, L.A.; Rosenquist, A.C. Visual receptive fields of single striate cortical units projecting to the superior colliculus in the cat. *Brain Res.* 67:27–42, 1974.

33. Poggio, G.F.; Baker, F.H.; Mansfield, R.J.W.; Sillito, A.; Grigg, P. Spatial and chromatic properties of neurons subserving foveal and parafoveal vision in rhesus monkey. *Brain Res.* 100:25–59, 1975.

34. Poggio, G.F.; Fischer, B. Binocular interaction and depth sensitivity in striate and prestriate cortex of behaving rhesus monkey. *J. Neurophysiol.* 40:1392–1405, 1977.

34a. Rodieck, R.W. Visual pathways. *Annu. Rev. Neurosci.* 2:193–225, 1979.

35. Sakata, H.; Shibutani, H.; Kawano, K. Functional properties of visual tracking neurons in posterior parietal association cortex of the monkey. *J. Neurophysiol.* 49:1364–1380, 1983.

36. Schein, S.J.; Marrocco, R.T.; de Monasterio, F.M. Is there a high concentration of color-selective cells in area V_4 of monkey visual cortex? *J. Neurophysiol.* 47:193–213, 1982.

37. Schiller, P.H. The superior colliculus and visual function. *Handbk. Physiol.* 3(1):457–505, 1984.

38. Schiller, P.H.; Malpeli, J.G. Properties and tectal projections of monkey retinal ganglion cells. *J. Neurophysiol.* 40:428–445, 1977.

39. Sillito, A.M.; Kemp, J.A.; Milson, J.A.; Berardi, N. A reevaluation of the mechanisms underlying simple cell orientation selectivity. *Brain Res.* 194:517–520, 1980.

40. Simpson, J.I. The accessory optic system. *Annu. Rev. Neurosci.* 7:13–41, 1984.

41. Sparks, D.L. Functional properties of neurons in the monkey superior colliculus: coupling of neuronal activity and saccade onset. *Brain Res.* 156:1–16, 1978.

42. Talbot, S.A.; Marshall, W.H. Physiological studies on neural mechanisms of visual localization and discrimination. *Am. J. Ophthal.* 24:1255–1264, 1941.

43. Tusa, R.J.; Palmer, L.A.; Rosenquist, A.C. Multiple cortical visual areas. In Woolsey, C.N., ed. *Cortical Sensory Organization.* Clifton, N.J., Humana Press, pp. 1–13, 1981.

44. Ungerleider, L.G.; Mishkin, M. Two cortical visual systems. In Ingle, D.J.; Goodale, M.A.; Mansfield, R.J.W., eds. *Analysis of Visual Behavior.* London, MIT Press, pp. 549–586, 1982.

45. Van Essen, D.C. Visual areas of the mammalian cerebral cortex. *Annu. Rev. Neurosci.* 2:227–263, 1979.

45a. Van Essen, D.C.; Maunsell, J.H.R. Hierarchical organization and functional streams in the visual cortex. *Trends Neurosci.* 6:370–375, 1983.

46. Weiskrantz, L. Behavioral analysis of the monkey's visual nervous system. *Proc. Roy. Soc. Lond. (Biol.)* 182:427–455, 1972.

47. Wiesel, T.N.; Hubel, D.H. Comparison of the effects of unilateral and bilateral eye closure on cortical unit responses in kittens. *J. Neurophysiol.* 28:1029–1040, 1965.

48. Wiesel, T.N.; Hubel, D.H. Spatial and chromatic interactions in the lateral geniculate body of the rhesus monkey. *J. Neurophysiol.* 29:1115–1156, 1966.

49. Wurtz, R.H.; Albano, J.E. Visual-motor function of the primate superior colliculus. *Annu. Rev. Neurosci.* 3:189–226, 1980.

50. Zeki, S.M. Uniformity and diversity of structure and function in rhesus monkey prestriate visual cortex. *J. Physiol. (Lond.)* 277:273–290, 1978.

Gustation and Olfaction

THE CONCEPT OF FLAVOR

The flavor of foods provides a rich and complex sensory experience that combines information from many modalities. Flavor arises from the sense of taste *(gustation)* and the sense of smell *(olfaction)*, with additional contributions from temperature and touch. These sensory inputs combine to create a percept that not only produces pleasure but also is crucial to biological survival. For example, the sensations that comprise the flavor of a substance reveal information about its consistency and chemical composition. These sensations affect salivary and pancreatic secretions and gut motility. Also, the flavor of foods determines their acceptance or rejection, a decision that is based on such factors as innate preference, appetite, and associative learning.

Human neonates and weanling rats show an *innate preference* for sweet substances. These substances, such as carbohydrates, are often high in calories and are, therefore, desirable foods. Neonates also show an innate aversion to bitter substances such as alkaloids, many of which are highly toxic.

Both rats and humans display *appetite*, or hunger, for substances that they require. For example, adrenalectomized rats or humans that either are salt-deprived or have adrenal cortex insufficiency will seek out and preferentially consume large quantities of sodium salt. Salt concentrations that are ordinarily aversive become palatable to these individuals, and their threshold for recognizing or detecting sodium chloride by taste is decreased. Furthermore, these individuals select a diet based on the flavor of the deficient substance. For example, a sodium-deficient rat will initially show a preference for lithium salts because they have the flavor of sodium salts, even though they produce nausea and do nothing to correct the deficiency. Appetites for phosphate, calcium, and thiamine have also been observed in a variety of diet-deficient animals.

Finally, the preference and appetite for certain foods is, in large part, a learned behavior, and this *associative learning* is often uniquely related to flavor. If an individual experiences abdominal distress after a meal, he will not soon eat the same meal. This association between food and discomfort is *conditioned flavor aversion learning*. An example of this type of learning is exhibited by cancer chemotherapy patients, who associate the nausea resulting from their treatments with each preceding meal. Eventually, these patients exhaust their supply of palatable foods and become anorexic. In rats, this aversion is specific to flavor;

when they are given either flavored water or water in combination with light or sound and are then poisoned, they develop a strong aversion only to flavored water, and not to "brightly lit" or "noisy" water. Furthermore, rats can form an association between the flavor of a food and subsequent abdominal distress even if a considerable delay occurs between feeding and illness. This long delay between the conditioned stimulus (flavor) and the unconditioned stimulus (illness) makes flavor aversion learning unique among forms of classical conditioning (see Chap. 33).

GUSTATION

Of the several modalities contributing to flavor, taste is the most influential in determining what we eat for several reasons. First, the peripheral receptors for taste lie directly in the path of the ingested food. Second, the taste of a substance is an indication of its dietary value (sweet, salt, bitter), but the smell of a substance is not (e.g., the horrendous odor of some pungent but nutritious cheeses), nor is its texture or temperature. Finally, taste and food are inextricably related. One can learn to enjoy the flavor of good-tasting food in a foul-smelling environment, but a bad taste means bad flavor regardless of the associated olfactory stimuli. Furthermore, background odors will not produce aversion learning when paired with later illness. Clearly, taste is the one sensation most uniquely associated with ingestion.

Qualities of Taste[3]

Before studying the mechanisms that underlie the sensations of taste, physiologists must understand what those sensations are, in order to design probes (stimuli) that adequately test the range of taste capabilities. Put simply, one cannot explain how the taste system does its job without first knowing what that job is. This type of knowledge is derived from psychophysical experiments.

It is traditionally taught that there are four distinct taste qualities: sweet, salt, sour, and bitter. Originally, Aristotle identified seven taste qualities that included astringent, pungent, and harsh along with the traditional four. Albrecht Haller, the master physiologist of the eighteenth century, identified 11, including rough, urinous and putrid.[13] The four taste qualities that survive today resulted from a process of elimination, in which many putative tastes were shown to be flavors resulting from the convergence of information

from several modalities. For example, an alkaline taste is thought to result from both olfactory and gustatory stimulation.

It is, of course, possible that other taste qualities exist but have not been identified. In an attempt to identify the qualities of taste objectively, Schiffman and Erickson[38] asked subjects to scale the similarity of the taste of pairs of substances. Each substance was then located as a point on a three-dimensional graph, with the distance between points indicating the similarity of the substances tasted (nearby substances had similar tastes). The result of this *multidimensional scaling* experiment is illustrated in Figure 21–1A. Substances that tasted

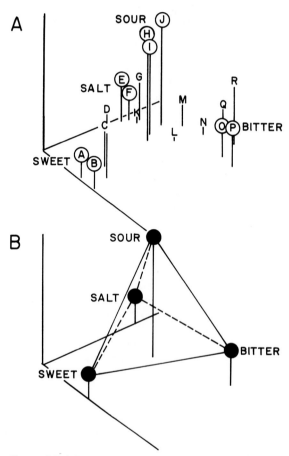

Figure 21–1 Two models of human gustatory quality. *A*, Multidimensional scaling of 19 taste stimuli. Distances between compounds indicate the similarity of their tastes. Circled substances have pure sweet, sour, salt, or bitter tastes. Taste stimuli are A, glucose; B, sucrose; C, saccharine; D, PbAc; E, LiCl; F, NaCl; G, NaAc; H, HCl; I, HNO_3; J, HAc; K, KCl; L, $CaCl_2$; M, K_2SO_4; N, $MgCl_2$; O, Q_2SO_4; P, QHCl; Q, Na_2CO_3; R, NaOH. *B*, Henning's taste model. Primary tastes define the corners of a tetrahedron with all tastes lying on its surface. (Modified, with permission from Schiffman, SS.; Erickson, R.P. *Physiol. Behav.* 7:617–633, Copyright 1971, Pergamon Press.)

either sweet, salty, sour or bitter (indicated by the circled letters) clearly tend to cluster in four spatially separate groups. Thus, one can see objective evidence of the existence of four distinct and separate taste qualities. Other lines of evidence also suggest that these are *primary*, or *basic*, taste qualities. For example, adaptation to a substance of one basic taste (e.g., sweet) will reduce the taste of all substances with that taste quality but will not affect the taste of substances with other basic taste qualities (e.g., sour). Thus, a variety of substances have one distinct and basic taste quality that is independent of the other three basic qualities.

Other substances shown in Figure 21–1A do not have one of the basic tastes; rather, they have tastes that fall roughly within the space bounded by the sweet, salt, sour, and bitter groups. Their taste, therefore, can be formed of mixtures of these taste qualities. KCl, for example, has a taste that is both salty and bitter. In Figure 21–1A, KCl (substance K) lies roughly between compounds with salty and bitter tastes. Adaptation to NaCl, a substance with a pure salt taste, causes KCl to taste purely bitter; that is NaCl *cross adapts* the salt taste. Thus, the taste of KCl is a demonstrable summation of salt taste and bitter taste (a mixture of primary tastes).

It is interesting to contrast Schiffman and Erickson's objective results[38] (Fig. 21–1A) with a simple geometric model of taste primaries (Fig. 21–1B) formulated by Henning.[15] Henning assumed that there are four taste primaries, analogous to color primaries, and that all tastes are constructed of a mixture of two or three of these primary tastes. These assumptions led to a representation of taste as a tetrahedron, with all possible taste sensations distributed about its surface. The similarity between the results of the objective scaling experiments and the four-taste model is striking.

Some substances, such as NaOH (substance Q in Fig. 21–1A) and $NaCO_3$ (substance R), clearly fall outside the space bounded by the four taste qualities; thus they have tastes that cannot be described in terms of those qualities. It is not clear whether these outlying data result from limitations in the experiments or from inadequacies in the four-taste model.

As the above discussion suggests, the four tastes of salt, sweet, bitter, and sour are independent and distinct taste qualities that, in various combinations, produce many of the sensations of taste. Therefore, physiologists have selected prototypical stimuli that taste salty (NaCl), sweet (sucrose or glucose), sour (HCl), or bitter (urea and quinine) to probe the mechanisms of gustation. It is impor-

tant to remember that these four qualities may not be either necessary or *sufficient* to describe all of what we taste.

The animals in which most of the neuronal data have been obtained may perceive different qualities. Even humans raised in different cultures may distinguish different qualities. For example, persons in some primitive societies identify only two taste categories: pleasant and unpleasant.

Taste Receptors[11, 27]

Anatomy

The peripheral gustatory receptors are located within oval structures called *taste buds*. Taste buds are distributed diffusely over the palate, the sides of the posterior oral cavity, the pharynx, and the larynx and are clustered in the fungiform, circumvallate (vallate), and foliate papillae of the tongue (Fig. 21–2A, B, and C, respectively).[6] On the human tongue, each type of papilla has a unique appearance and distribution. The circumvallate and the folliate papillae are large and contain hundreds of taste buds. They are located on the posterior third of the tongue, which is innervated by the lingual-tonsillar branch of the glossopharyngeal nerve (cranial nerve IX). The fungiform papillae, on the other hand, are small and contain fewer than six taste buds. The spots on the tongue in Figure 21–2D are fungiform papillae. The majority of these papillae are located at the margins of the anterior two thirds of the tongue, which is innervated by the chorda tympani nerve (cranial nerve VII). Pharyngeal and laryngeal taste buds are innervated primarily by special visceral afferents of the superior laryngeal branch of the vagus nerve (cranial nerve X), whereas the palate is innervated by the greater superficial petrosal branch of the facial nerve. The center of the tongue has only filliform papillae, which lack taste buds, and is therefore insensitive to taste stimuli, which are called *tastants*.

Psychophysical experiments suggest that individual human taste papillae respond to stimuli of only one basic taste quality if the stimuli are presented at low concentrations. Taste papillae and diffuse taste buds in different regions of the oral cavity seem to respond to different basic taste qualities. Sweets and salts have the lowest taste recognition thresholds on the margins of the anterior tongue, where the fungiform papillae reside. Sour stimuli seem to have the lowest thresholds in the region of the foliate papillae on the sides of

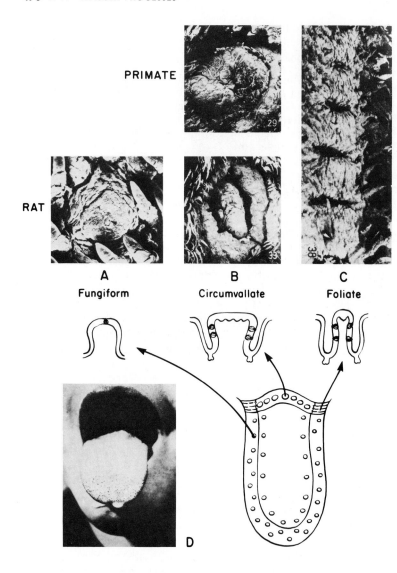

PRIMATE

RAT

A	B	C
Fungiform	Circumvallate	Foliate

D

Figure 21-2 Taste papillae. Surface of the tongue (lower right) shows the locations of the taste papillae displayed in the scanning electron micrographs above. *A*, Rat fungiform papilla surrounded by filiform papillae. *B*, Circumvallate papilla of rat (below) and primate (above). *C*, Rat foliate papillae. D. Tongue (of 16-year-old boy) stained for fungiform papillae. (*A* and *B* after Graziadei, P. P. C. In Pfaffmann, C., ed. *Olfaction and Taste*, Vol. 3, pp. 315–330. New York, Rockefeller Univ. Press, 1969. *C* after Bradley, R. M. In Beidler, L. M., ed. *Chemical Senses: Taste*, pp. 1–30. *Handbk. Perception*, Vol. 4, pt. 2. New York, Springer, Verlag, 1971. *D* after Beidler, L. M. In Carterette, E. C.; Freidman, M. P., eds. *Tasting and Smelling*, pp. 21–49. *Handbk. Perception*, Vol. 6A. New York, Academic Press, 1978. Line drawings from Murray, R. G. In Friedman, I., ed. The Ultrastructure of Sensory Organs, 1–81. New York, Elsevier, 1973.)

the posterior tongue. Bitter thresholds are lowest on the soft palate and the anterior tongue.

At higher stimulus concentrations, individual papillae are broadly responsive to several basic tastes. The sensitivity to moderate and strong concentrations of tastants is distributed differently. For example, application of strong concentrations of bitter tastants to the circumvallate papillae at the back of the tongue produces more intense taste sensations than similar applications to the anterior tongue. Consequently, the responses of different papillae and regions of the tongue depend on tastant concentration.

Figure 21–3A shows the morphology of a typical taste papilla, a fugiform papilla in the rat. The gourd-shaped aggregate of cells within it is a taste bud. A typical rat fungiform taste bud is 50 to 80

μm deep and 40 to 60 μm wide. It consists of modified epithelial cells that lie beneath a central taste pore and taste pit. Some cells lie only near the base of the taste bud and are therefore called basal cells; other, elongated cells extend the full length of the bud and terminate in the taste pore. The entire bud is innervated from below by a plexus of small (<1.0 μm) unmyelinated fibers woven in complex fashion throughout the taste bud. The unmyelinated fibers are the distal processes of primary afferents.

The taste bud contains four cell types (all called gemmel, or bud, cells) that are distinguishable on morphological grounds. Identifiable only in electron micrographs, these four types have been studied most extensively in lower mammals, but they apparently exist in primates as well. The *basal*

A

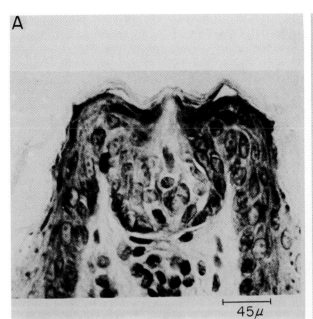

45μ

Figure 21–3 *A*, Light micrograph of a rat fungiform taste papilla and bud. *B*, Ultrastructure of a rabbit foliate taste bud. (1) Type I supporting cell; (2) type II cell; (3) type III receptor cell; (4) type IV basal cell; (S) Schwann cell apposed to nerve fiber. Note synaptic specialization at base of type III cell. (*A* after Beidler, L. M. In Carterette, E. C.; Freidman, M. P., eds. *Tasting and Smelling*, pp. 21–49. *Handbk. Perception*, Vol 6A. New York, Academic Press, 1978. *B* after Murray, R. G. In Friedmann, I., ed. *The Ultrastructure of Sensory Organs*, pp. 3–81. New York, Elsevier, 1973.)

B

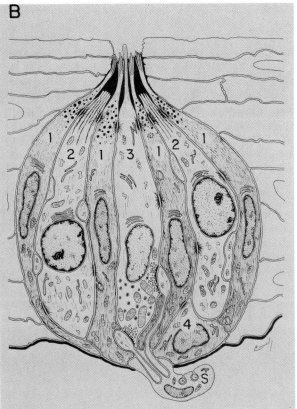

cell, type IV, is shown in Figure 21–3B, (#4). In the rabbit, there are one to four such basal cells among the 30 to 80 cells within a single foliate taste bud. Basal cells are thought to be undifferentiated stem cells from which the other gemmel cell types are derived. They and the other gemmel cells are distinguishable from cells outside the taste bud (perigemmel cells) by their relative lack of cytofilaments.

The remaining gemmel cells display tight junctions that effectively isolate the taste pit from all but the apical regions of the cells, where the receptive elements of the taste bud are believed to reside. Type I cells (#1 in Fig. 21–3B), which are the most numerous type (55%–75% of the taste bud cells), have tapered apical ends with numerous microvilli. Investigators viewing them by light microscopy originally thought these tapered ends to be hair-like taste receptors; however, although type I cells make intimate contact with nerve fibers, they lack vesicles and synaptic specializations. The ends do contain a dense granular material similar to that in the taste pore; therefore, the cells may be glial elements that maintain the

material in the taste pore. The remaining gemmel cells, type II and type III (#2 and #3, respectively, in Fig. 21–3B), both lack the apical, dense granules of type I cells. Like type I cells, type II cells contact nerve cells but have no synaptic specializations. In addition, type II cells, which represent 15 to 30% of the gemmel cells, are mitotic and may be capable of further differentiation. Their function is not clear. Type III cells, in contrast, have both presynaptic and postsynaptic specializations characteristic of synapses, suggesting that they are the taste receptors. Type III cells represent only 5% to 15% of the cells in the taste bud.

The entire structure of the taste bud is labile and depends on both the migration and differentiation of perigemmel cells to replace the degenerating gemmel cells. Beidler and Smallman demonstrated this process by injecting tritiated thymidine intraperitoneally in rats.[5] The thymidine was incorporated into premitotic perigemmel cells, and was used as a tracer to follow the process histologically. By sacrificing the animals at various times following the injection, it was demonstrated that the labelled thymidine migrated into the taste bud as the perigemmel cells migrated and differentiated into gemmel cells. If this process was interrupted by the systemic injection of colchicine, which blocks mitosis, the taste bud degenerated.

The integrity of the taste bud apparently is maintained by a trophic influence from the gustatory nerves. If the glossopharyngeal nerve is cut, the taste buds that it innervates degenerate following degeneration of the nerve endings. If the nerve regenerates, the taste buds regenerate following the return of the nerve endings. The basal cells return first, followed by the remaining gemmel cell types. This pattern is consistent with the hypothesis that the basal cells are derived from undifferentiated perigemmel epithelial cells, and become stem cells for types I, II, and III. Thus, a normal taste bud contains both morphologically distinct cell types and cells in various stages of development.

The constant cycle of taste bud growth, death, and renewal poses a significant problem for the nervous system. Each time a new receptor cell assumes its place, new synaptic contacts must be formed with taste fibers in such a way that the information available to the central gustatory neurons still defines taste quality unambiguously. There must, therefore, either be a match between the response of a receptor cell and the fiber that innervates it or a constantly evolving, seemingly chaotic taste code. One possible explanation is that taste fibers impose a limited sensitivity on their receptor cells. Unfortunately, when taste nerves are cross-transplanted from one region of the tongue to another, the nerves change their overall sensitivities, but the regions of the tongue do not. Thus, in this case at least, the sensitivities of the nerve fibers appear to be dictated by those of the receptors.

Responses of Taste Cells[35]

The response of gemmel cells to tastants has been determined in several mammals, including the rat, hamster, and mouse. In a typical experiment (Fig. 21–4A), a microelectrode is advanced through the taste pore until a negative, sustained potential of 20 to 80 mV is recorded. Each prototypical taste stimulus (e.g., 0.5 M sucrose, 0.5 M NaCl, 0.01 M HCl, and 0.02 M Q-HCl [quinine]) is then applied to the tongue and subsequently rinsed away with distilled water or a salt solution that mimics saliva (e.g., 41.4 mM NaCl). The penetrated cell is considered to be a *taste cell* if the application of the stimulus produces a change in membrane potential that is eliminated by rinsing

the tongue. It is not yet possible, however, to determine the type of the penetrated cell, because the injection of a dye apparently complicates the electron microscopy necessary to distinguish putative receptor (type III) cells.

Many cells in the rat taste bud respond to sapid stimuli. As shown for three typical cells in Figure 21–4B, the responses are slow and graded and may be depolarizing, hyperpolarizing, or depolarizing with a preceding hyperpolarization.[36] Furthermore, most cells show these changes in response to more than one of the prototypical taste stimuli, as depicted in Figure 21–4C, which shows the size of the responses of 38 taste cells. The majority of taste cells (90%) respond with depolarization to two or more of the prototypical stimuli (32%, 26%, and 32% are depolarized by four, three, and two stimuli, respectively), while 10% are depolarized by only one. These results suggest that individual receptor cells transduce the chemical stimuli associated with several basic taste qualities. The psychophysical adaptation experiments suggest that the basic taste qualities are independent; therefore, each receptor cell may possess several independent receptor sites.

The response of an individual taste cell depends on the solution with which the tongue is rinsed (or adapted to) between tastant applications. The average resting membrane potential of taste cells is approximately -40 to -43 mV when the cells are adapted to rinses of 41.4 mM NaCl. Adaptation with a distilled water rinse lowers the average resting potential. The response of a taste cell depends on its resting potential. Depolarizing responses tend to increase with water adaptation, and hyperpolarizing responses tend to decrease or become depolarizing.

The response thresholds for depolarization of individual rat taste cells by the four basic tastants also depend on the state of adaptation. Thresholds for taste cells adapted with 41.4 mM NaCl are 0.01 to 0.05 M for sucrose, 0.05 to 0.1 M for NaCl, 0.0005 to 0.001 M for HCl, and 0.001 to 0.005 M for Q-HCl. After water adaptation, the thresholds are considerably lower (i.e., 0.001 M for NaCl, 3×10^{-5} M for HCl, 1×10^{-5} M for Q-HCl). These latter values are lower than, or equal to, the absolute psychophysical thresholds obtained by whole-mouth stimulation in humans using the sip and spit method (e.g., 10^{-2} M NaCl, 10^{-3} M HCl, and 10^{-5} M Q-HCl). Therefore, taste thresholds are roughly equivalent to the thresholds of the taste receptors themselves.

Taste cells also convey information about stimulus concentration. As stimulus concentration in-

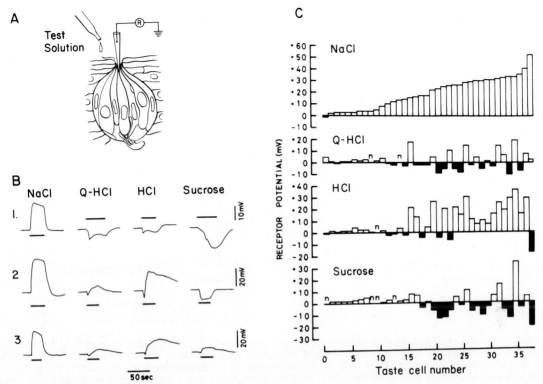

Figure 21–4 *A*, Sketch depicting a typical taste cell recording experiment. *B*, Intracellular potentials in three rat taste cells adapted to 41.4 mM NaCl and stimulated with four prototypical taste stimuli (0.5 M NaCl, 0.02 M Q-HCl, 0.01 M HCl, and 0.5 M sucrose). Stimulus duration indicated by horizontal bars. *C*, Changes in receptor potential from resting for 38 rat taste cells adapted and stimulated as in *B*. Open histograms indicate depolarizing responses or depolarizations following brief hyperpolarizations; closed histograms indicate hyperpolarizing responses alone. The letter "n" indicates no response. *B and C* after Sato, T.; Beidler, L. M. *Comp. Biochem. Physiol.* 75A:131–137, 1983.)

creases, the amplitude of the taste cell response increases.

Finally, the changes in taste cell membrane potential that accompany the application of taste stimuli are associated with changes in membrane resistance.[37] For example, in rat taste cells, depolarizing responses to NaCl (Fig. 21–5A and open circles in Fig. 21–5B) are accompanied by decreases in input resistance (Fig. 21–5A and open triangles in Fig. 21–5B), as measured by brief hyperpolarizing current pulses. In contrast, both depolarizing and hyperpolarizing responses to sucrose, HCl, and Q-HCl are associated with increases in input resistance as seen in the depolarizing response to sucrose in Figure 21–5C and D. In all cases, changes in the amount of depolarization parallel either increases or decreases in input resistance. Although these changes in input resistance suggest changes in ionic conductances, the mechanism of the transduction process is currently unknown (see Chap. 5).

Experimental Difficulties. At this point, we should consider the difficulties associated with testing the gustatory system, since these difficulties influence our interpretation of the results of all gustation experiments.

First, it is extremely difficult to deliver a pure chemical stimulus (without associated thermal or mechanical components) at a known concentration to a taste cell. Second, it is not clear what conditions simulate the absence of a taste stimulus. Rinses of both water and NaCl solutions are used. While distilled water surely contains no ionic taste stimuli, the mouth is normally bathed in NaCl solution (saliva). Therefore, both rinses represent true "resting" conditions for the receptor. However, as noted above, each has different effects on the resting potential and the response of taste cells. Third, stimulus concentration determines the presence (threshold) and amplitude of a response. In different experiments, the selected stimulus concentration may be that which produces either (1) a robust but submaximal response in taste cells, (2) a half-maximal response in whole taste nerve recordings, or (3) a moderate taste

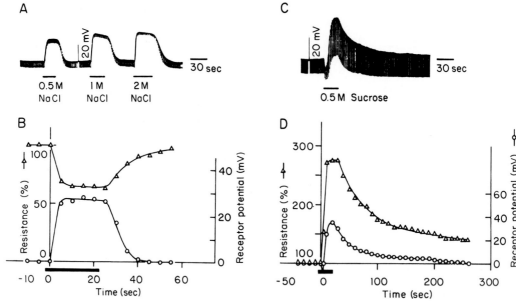

Figure 21–5 Responses and input resistance changes of rat taste cells to different tastants. *A* and *C*, Depolarizing responses of a cell adapted to 41.4 mM NaCl and stimulated with 0.5 M, 1 M, and 2 M NaCl *(A)* and 0.5 M sucrose *(C)*. Thickened traces reflect voltage transients produced by constant hypopolarizing current pulses applied through the electrode. Stimulus duration is indicated by horizontal bars. *B* and *D*, Time courses of membrane potential (open circles) and input resistance (triangles) changes produced by application of 0.5 M NaCl *(B)* and 0.5 M sucrose *(D)* during the period marked by the horizontal bar. (Modified, with permission, from Sato, T.; Beidler, L. M. *Comp. Biochem. Physiol.* 73A:1–10; Copyright 1982, Pergamon Press.)

sensation in humans. These three criteria produce similar test concentrations for each of the four basic stimuli. The test concentration of one basic stimulus, however, usually is very different from that of another. Finally, our knowledge of taste cell responses is restricted to the cells of lower mammals (e.g., the rat). It is only at the next level of processing, the afferent fibers innervating the taste buds, that we can begin to discuss the neurobiology of primate gustation. As we will see, there are significant species differences in the anatomy and physiology of the taste system.

Taste Pathways

Figure 21–6 shows the pathways and structures that are thought to subserve taste in primates (including humans) and rats and illustrates the neuroanatomy that applies in most mammalian species. Information from the taste receptors is carried into the central nervous system by fibers of the chorda tympani, the glossopharyngeal, and the vagus nerves (cranial nerves VII, IX, and X, respectively). These fibers are the peripheral axons of pseudounipolar, first-order neurons, which are located in the geniculate, petrosal, and nodose ganglia, respectively. The central axons travel to the medulla and pons, where they form the ipsi-

lateral solitary tract and eventually synapse on second-order gustatory neurons in the caudal brain stem, specifically those of the nucleus of the solitary tract.

In the primate,* most axons of the gustatory neurons in the nucleus of the solitary tract ascend with the ipsilateral medial lemniscus, as the central tegmental bundle, to synapse on third-order gustatory neurons in the small-celled portion of the ventral posteromedial nucleus of the thalamus (VPMpc). Most gustatory neurons of VPMpc, in turn, send axons via the ipsilateral limb of the internal capsule to fourth-order gustatory neurons of the frontal operculum and nearby parainsular cortex (primary gustatory cortex). The primary gustatory cortex then projects to fifth-order gustatory cells in the orbitofrontal cortex.

In the rat and many other subprimate mammals, the third-order gustatory neurons are located in the parabrachial region of the pons (parabrachial nucleus). Second-order gustatory neurons of the rodent nucleus of the solitary tract make a direct,

---

*There are considerable differences between Old World and New World monkeys in terms of the anatomy and physiology of the gustatory system. Because the brains of Old World monkeys appear to be anatomically closer to those of humans, they alone will be considered.

RAT   PRIMATE

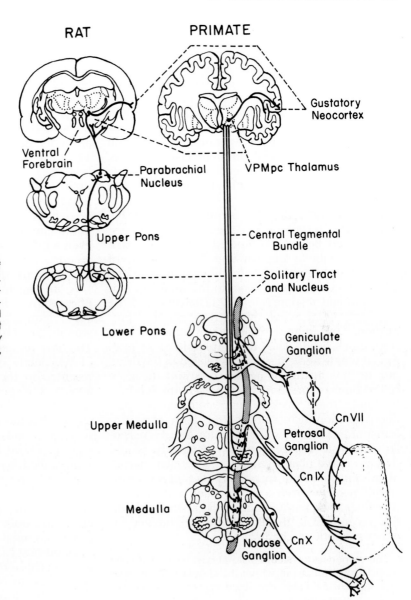

Gustatory
Neocortex

Ventral
Forebrain

Parabrachial
Nucleus

VPMpc Thalamus

Upper Pons

Central Tegmental
Bundle

Solitary Tract
and Nucleus

Lower Pons

Geniculate
Ganglion

Upper Medulla

Cn VII

Petrosal
Ganglion

Cn IX

Medulla

Nodose   Cn X
Ganglion

Figure 21–6 Comparative anatomy of taste pathways in higher primates and rats. The pathways from the tongue to the nucleus of the solitary tract are roughly comparable in both species. The neocortical taste region in orbitofrontal cortex is not shown. (After Pansky, B.; Allen, D. J. *Review of Neuroscience*, pp. 358–363. New York, Macmillan, 1980.)

obligatory, mostly ipsilateral projection to this region. The rat parabrachial nucleus sends fibers bilaterally to parts of VPMpc, which contains fourth-order gustatory neurons. Parabrachial axons also project heavily to ventral forebrain regions, including the lateral hypothalamus and amygdala (see Chap. 32). The fourth-order gustatory region of the rat VPMpc projects to a gustatory region in the rostral half of agranular insular cortex.

Thus, after the receptors, there are five levels of gustatory processing, represented by somewhat different structures in rats and primates. We now discuss the anatomy and physiology of these structures at each level.

Primary Taste Afferents

Each receptor structure is innervated by many taste fibers, which coalesce to form the three taste nerves. We will focus on the chorda tympani nerve which, because it innervates the front of the tongue, is easily stimulated, making it the most extensively studied of the three taste nerves.

Individual chorda tympani fibers branch several times before entering a papilla. On average, these branches innervate about five fungiform papillae, which are distributed over several millimeters on the tongue. This distribution has been demonstrated by stimulating individual papillae electrically and recording action potentials in individual chorda tympani fibers. Upon entering a fungiform papilla, each chorda tympani fiber branches and divides again within each taste bud. Thus, a single taste fiber probably receives inputs from several receptor cells in several different taste papillae. The response of a single fiber therefore represents the convergence of a great deal of sensory information.

To understand the transfer of information between taste receptors and their afferent nerves, Kimura and Beidler[18] compared the responses of rat taste cells to depolarizing NaCl stimuli of varying concentrations with the response of the whole chorda tympani nerve to the same stimuli. The summed responses of the taste cells and the response of the whole nerve were roughly proportional throughout the range of stimulus concentrations used. Since the amplitude of an integrated whole nerve response is related to the firing rate of its constituent nerve fibers and the taste cell response is presumed to represent the receptor response, these workers suggested that receptor depolarization increased the probability of neurotransmitter release by the receptor cell. Hyperpolarizing taste cell responses presumably would depress nerve fiber discharge by a similar mechanism. Since it is not technically feasible to record simultaneously from a single fiber and to drive that fiber with natural stimuli via a single identified receptor cell, there is no direct evidence for either synaptic mechanism.

The responses of whole taste nerves have also been recorded in humans in whom the chorda tympani was exposed during inner ear surgery. Threshold tastant concentrations for human chorda tympani responses are similar to psychophysical recognition thresholds in the same subjects. Also, the total neural activity of the human chorda tympani is well correlated with the subjectively reported stimulus intensity. Some human chorda tympani responses differ significantly from those of other species. For example, the chorda tympani fibers of many species (e.g., cats) respond briskly to distilled water stimuli following adaptation to saliva; these responses may be important for fluid homeostasis. Humans appear to lack these chorda tympani *water fiber* responses. These data suggest that one must be cautious in extrapolating to humans data obtained on the gustatory systems of other species.

Recordings from single afferent fibers, possible only in animal experiments, provide more details about the nature of the taste information that reaches the central nervous system. Figure 21–7A displays the responses of two single chorda tympani fibers from a monkey to a variety of stimuli. Several features of the discharge of taste afferents are illustrated. First, taste afferents have low (average 2 spikes/s), if any, spontaneous discharge rates in the absence of stimuli. Second, as with afferents subserving other modalities (see Chap. 14), a sensory stimulus often evokes an initial brisk response that then adapts to a lower rate upon continued presentation of the stimulus. This firing pattern is seen most easily in the robust responses to sucrose in the upper trace and to quinine in the lower trace of Figure 21–7A. Third, even a brisk response may reach only 45 spikes/s, whereas a weak response may be only 1 spike/s above the unstimulated rate.

A striking feature of the discharge of the two fibers shown Figure 21–7A is that they respond very vigorously to one of the prototypical taste stimuli, but only weakly, if at all, to the others. The fiber represented in the upper trace of Figure 21–7A clearly prefers sucrose, whereas that represented in the lower trace favors quinine. The averaged responses of a large number of such fibers to the four prototypical tastants (Fig. 21–7B) shows that there are four classes of fibers, each of which responds best to one of the four prototypical stimuli. Those fibers preferring sucrose and quinine (called *sucrose best* and *quinine best*) exhibit very little response to other taste stimuli and therefore are said to be narrowly tuned. Perhaps coincidentally, the sweet and bitter tastes of sucrose and quinine are the qualities that humans can discriminate most readily. Those fibers that respond best to NaCl, however, also respond well to HCl. Those responding best to HCl are likewise responsive to NaCl. These *salt best* and *sour best* fibers, therefore, are considered to be broadly tuned. Overall, chorda tympani fibers in the monkey tend to respond to one (34%) or two (35%) of the prototypical taste stimuli. Only a small percentage (6%) respond to all four.[33]

Rat chorda tympani fibers also have preferred tastants (Fig. 21–7B). Although the overall breadth of tuning is similar in the two species, the tuning to individual tastants is quite different. For example, the tuning of sucrose best and quinine best fibers is considerably broader in the rat than in the monkey, while that of salt best fibers is somewhat more narrow. Also, the relative number of fibers that are tuned to a given stimulus differs in the two species. In the monkey, 42% of chorda tympani fibers are salt best, followed by sweet

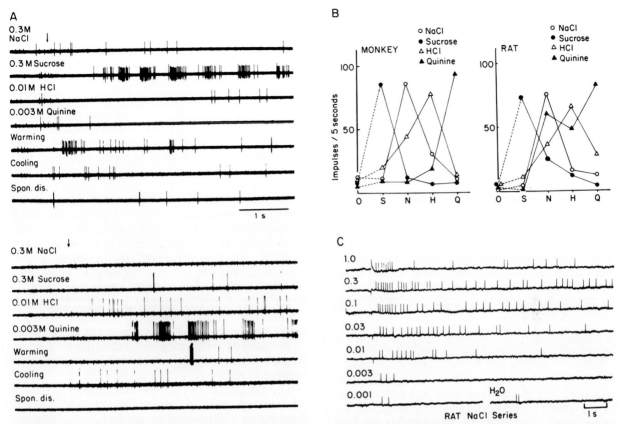

Figure 21–7 *A,* Discharge of two water-adapted cynomolgus monkey chorda tympani fibers to application (at arrow) of the steady stimuli listed; warming and cooling are 20°C temperature changes. Bottom traces indicate unstimulated discharges. *B,* Mean response profiles of cynomolgus monkey and rat chorda tympani fibers during application of taste stimuli as in *A.* The stimuli are 0.3 M sucrose (S), 0.3 M NaCl (N), 0.01 M HCl (H), and 0.003 M Q-HCl (Q) for the monkey; and 0.5 M sucrose (S), 0.1 M NaCl (N), 0.01 M HCl (H), and 0.02 M Q-HCl (Q) for the rat. (O) represents spontaneous discharge. Open circles = salt-best fibers; closed circles = sucrose-best fibers; open triangles = acid-best fibers; closed triangles = quinine-best fibers. *C,* Discharge of a single rat chorda tympani fiber in response to NaCl stimuli at different molar concentrations and in response to the water-adapting solution. (*A* modified from Sato, M.; Ogawa, H.; Yamashita, S. *The Journal of General Physiology,* 1975, Vol. 5, pp 23–26, by copyright permission of the Rockefeller University Press. *B* after Sato, M. In Denton, D. A.; Coghlan, J. P., eds. *Olfaction and Taste,* Vol. 5, pp. 23–26. New York, Academic Press, 1975. *C* modified from Pfaffmann, J., *The Journal of Neurophysiology,* Vol. 18, pp. 429–440, Copyright 1955, Pergamon Press.)

best (24%), quinine best (17%) and HCl best (15%). In the rat, while 48% of the chorda tympani fibers are salt best, 43% are HCl best, and only 7% are sucrose best.[9] Quinine best fibers are extremely rare in the rat chorda tympani. Thus, the same taste nerves in the two species code information somewhat differently. In particular, the chorda tympani of the monkey has more of the narrowly tuned sweet and bitter fiber types than does that of the rat.

The response of chorda tympani fibers, like those of gemmel taste cells, is related to the concentration of the taste stimulus.[31] For example, as the NaCl concentration is increased, the fiber of Figure 21–7C fires at increasingly higher rates. A similar increase in firing rate with concentration

occurs in other primary taste afferents. The increase in the whole nerve response to increasing stimulus concentration, therefore, reflects in part the increased spike frequency of individual fibers.

Since the firing rates of taste fibers code both taste quality and stimulus concentration and few fibers respond to only one quality for all concentrations, it is clear that the tuning of a fiber ultimately depends on stimulus concentration. For example, low concentrations of one tastant could produce the same discharge rate as higher concentrations of another, less effective tastant. A fiber may be unresponsive to all but one tastant at low stimulus concentrations but somewhat responsive to all taste qualities at higher concentrations. Therefore, the information carried by a single fiber is ambiguous and must be interpreted in the context of the activity of many such fibers. This ambiguity exists not only for chemical stimuli; many taste fibers also respond to thermal stimuli (Fig. 21–7A). Thus, these fibers provide the central nervous system with a wealth of information, which apparently can be extracted only if the discharge of many fibers is considered together. This decoding problem is considered in more detail in a later section.

The observations that we have made about the responses of chorda tympani fibers apply generally to the responses of fibers in other taste nerves, with some important qualifications. For example, the overall preference of the other taste afferents for the four prototypical taste stimuli often is very different from that of the chorda tympani. This is not surprising, since the oral cavity is both differentially sensitive to the various taste stimuli and regionally innervated. Also, other taste nerve fibers carry information not found in the chorda tympani. For example, some fibers of the superior laryngeal nerve of the rat respond to low-threshold mechanical stimuli.

Brain Stem Taste Nuclei[28]

Primary afferents communicating taste information terminate in the rostral part of the nucleus of the solitary tract. Within this nucleus, fibers of the chorda tympani project most rostrally, those of the vagus nerve most caudally, and those of the glossopharyngeal nerve to intermediate regions. Nontaste fibers from the oral cavity project to adjacent regions.

Both the effects of lesions and recordings from single neurons show that the nucleus of the solitary tract is involved in gustatory processing. In rats, for example, lesions rostral in the nucleus increase the threshold concentrations at which the animals will reject quinine. In both rats and alert monkeys, the gustatory responses recorded from neurons in the nucleus of the solitary tract are similar in most respects to those of the chorda tympani.[39] In particular, neurons in the nucleus of the solitary tract have similar thresholds for tastants and respond with increasing frequency to increasing stimulus concentration. However, they generally are more broadly tuned than chorda tympani fibers; for example, no neurons in the nucleus of the solitary tract of the monkey respond to only one prototypical stimulus, and 84% respond to either three or four. The broad tuning of neurons in the nucleus of the solitary tract may result from a convergence of more sharply tuned taste nerve fiber inputs.

Neurons in the nucleus of the solitary tract in the monkey display a chemotopic organization. Those lying rostrally are more likely to be sensitive to NaCl and glucose, whereas, those lying caudally are more likely to be sensitive to HCl. This distribution of neuron types is consistent with the pattern of afferent innervation in which rostral projections originate from the chorda tympani and the anterior tongue and caudal projections from the glossopharyngeal nerve and the posterior tongue. If the oral cavity of the monkey has the same differential sensitivity to tastants as that of humans, then the chemotopic and topographic representations are in register.

The nucleus of the solitary tract of the rat is not a simple relay for gustatory information, since the gustatory response of its neurons can be modulated by the physiological needs of the animal. For example, increases in blood glucose depress neuronal responses to sweet stimuli by an average of 43%. This effect is selective, since responses to bitter stimuli are decreased by only 3% under similar conditions.[10] In contrast, satiety has no effect on the response of neurons in the nucleus of the solitary tract in the monkey, suggesting that its neurons are *not* modulated by similar interoceptive cues.

The nucleus of the solitary tract relays gustatory information in two general directions. First, it projects to VPMpc, either directly in the primate or, in the rat, via an obligatory, intermediate synapse in the parabrachial nucleus. The thalamus, in turn, projects to the cortex, where the gustatory information may ultimately produce our perception of taste. Second, the nucleus of the solitary tract also projects to subcortical structures that may be involved in the control of feeding behavior and digestive processes. These structures are mainly in the caudal brain stem, including the medullary reticular formation and the surrounding motor and sensory nuclei of the medulla; in the rat the parabrachial nucleus also receives inputs. The medulla contains regions associated with ingestive behaviors such as swallowing (cranial motor nuclei V, VII, and X), salivation (inferior salivary nucleus), and vomiting (nucleus ambiguus and area postrema), as well as with a variety of digestive, cardiovascular, and respiratory functions. Gustatory responses have been recorded in the area postrema, salivary nucleus and efferent preganglionic autonomic fibers involved in salivary secretion.

Parabrachial Region[45]. In most nonprimate species, the parabrachial region of the pons receives a projection from the taste-responsive region of the nucleus of the solitary tract. The parabrachial nucleus is another locus of autonomic function, as it contains cell populations that are involved in arousal, cardiovascular regulation, and respiration. In most species, including primates, visceral afferents also project to this region.

The parabrachial nucleus of the rat contains taste-responsive neurons that project to the thalamus, hypothalamus, and amygdala; it also projects diffusely to the reticular formation, brain stem motor nuclei, and a variety of ventral forebrain structures such as the substantia innominata and the bed nucleus of the stria terminalis and, weakly, to the nucleus of the solitary tract and insular cortex. Thus, it potentially can influence a broad and functionally diverse subset of structures in the central nervous system.

Neurons in the parabrachial nucleus respond like those of the nucleus of the solitary tract but are even more broadly responsive to the prototypical taste stimuli. In the rabbit, they discharge both for gustatory stimulation and mouth and tongue movements and may be involved in the integration of sensory and motor information during feeding. Taste-sensitive neurons have not yet been recorded in the primate parabrachial nucleus.

Ventral Forebrain

The ventral forebrain contains the hypothalamus and a variety of limbic and other structures, which are involved in the control of appetitive behaviors. In both the rat and primate, it receives projections from the parabrachial nucleus and from cortical structures that are known to receive gustatory information. Both the hypothalamus and the amygdala of the limbic system have been implicated in feeding behavior and the processing of gustatory stimuli.

Damage or stimulation of the hypothalamus or amygdala in rats and primates (including humans) produces changes in salivation, gut secretion and motility, food and water intake and alters preferences for certain foods. The nature of these changes, however, depends on the site of the damage or stimulation (see Chap. 32). For example, lesions of the ventromedial hypothalamus or ventrobasal amygdala produce overeating (hyperphagia), whereas lateral hypothalamic and dorsomedial amygdalar lesions eliminate feeding and drinking (aphagia and adipsia, respectively). These lesions seem to affect the palatability of foods; for example, rats with ventromedial hypo-

thalamic lesions are pathologically hyperphagic when offered sweet foods but will not eat bitter foods. Also, cells in the amygdala and hypothalamus respond to taste stimuli, and some respond robustly when the animal merely sights food. Most of these cells also respond to a variety of other stimulus modalities. Like those in the rodent nucleus of the solitary tract, a few cells in the ventral forebrain of the primate alter their response to tastants according to the satiety of the animal.

Gustatory Thalamus

Gustatory information reaches the cortex via the small-celled portion of VPMpc, which in the primate, receives its input from the nucleus of the solitary tract, and in the rat and other lisencephalic species, from the parabrachial nucleus.

VPMpc is loosely organized according to the original source of its inputs from the three taste nerves. Neurons activated by orthodromic stimulation of the chorda tympani are located primarily in the rostro-medial VPMpc in the rat. Neurons responsive to glossopharyngeal stimulation lie mostly posterolateral to these. As in the nucleus of the solitary tract, indirect projections of the cranial nerve V, the lingual nerve, which carries non-taste information from the oral cavity, border the gustatory region.

In monkeys, bilateral lesions of medial VPM temporarily raise the threshold concentrations at which these animals reject quinine. Bilateral lesions of VPMpc in rats also raise quinine aversion thresholds and permanently eliminate preferences for saline, sucrose, and saccharine stimulation of VPMpc in rats produces behaviors that mimic the rejection responses that follow application of aversive tastants to the mouth. Finally, thalamic lesions in humans occasionally produce a loss of taste perception, or *ageusia*.

In both the rat and primate, the responses of VPMpc neurons to tastants seem unlike those of the peripheral taste nerves or the neurons of the nucleus of the solitary tract: (1) some VPMpc neurons are narrowly tuned (10% in the rat and 30% in the monkey respond to only one prototypical taste stimulus); (2) many thalamic taste neurons discharge at high spontaneous (i.e., unstimulated rates; and (3) some neurons show *decreases* in activity in response to gustatory stimulation.

Primary Gustatory Cortex

In primates, the primary gustatory cortex lies in or near the frontal operculum, anterior insula, and primary somatosensory cortex (see Fig. 21–6). This gustatory cortex was located by (1) recording potentials evoked by stimulation of the primary gustatory afferents, (2) placing in VPM an anatomical label that travels orthogradely, and (3) placing in the cortex an anatomical label (HRP) that travels retrogradely. In the rat, the primary gustatory cortex is the rostral, agranular insular cortex.

Lesions of the opercular cortex produce ageusia in humans and temporary increases in quinine rejection thresholds in monkeys. Lesions of the primary gustatory cortex in rats can produce a variety of effects, including impaired preference behavior in naive animals and a reduction in the normal innate aversion to new foods (neophobia).

Electrical stimulation in the opercular cortex of primates produces changes in gut motility and abdominal sensation. In addition, if stimulation is made contingent on the performance of a certain task, monkeys will readily learn and perform that task to receive stimulation as a reward. This self-stimulation behavior implies that the gustatory cortex is involved with motivation and reward.

The cortical representation of taste is sparse. Relatively few (less than 5%) of the neurons tested in the frontal opercular and parainsular cortex of the monkey respond to taste stimuli. Like those in the nucleus of the solitary tract, cortical gustatory neurons have low spontaneous activity, respond only with increases in activity, have thresholds similar to human psychophysical thresholds, and code increasing concentration with increased activity. Unlike those in the nucleus of the solitary tract, taste neurons in the frontal opercular cortex are more narrowly tuned to the prototypical taste stimuli.[40] While no neurons in the nucleus of the solitary tract respond to only one prototypical taste stimulus, 26% of the cortical taste neurons do so; 28% respond to only two stimuli, and 20% to three. Insular cortical neurons are even more narrow.y tuned to the prototypical taste stimuli than are those in the frontal operculum. Cells in the insular cortex receive both direct thalamic and frontal opercular input.

Cells of the primary gustatory cortex project back to VPMpc thalamus as well as to a variety of limbic structures. Also, in the monkey, these cells project to a second cortical gustatory region, the lateral orbitofrontal cortex.

Secondary Gustatory Cortex

The lateral orbitofrontal cortex in primates receives input from the gustatory region of the insular and frontal opercular cortex. It does not receive projections from VPMpc and therefore can be considered a secondary gustatory region. Like opercular and insular cortex, this region is not exclusively gustatory; it receives inputs from a number of other modalities (e.g., olfaction). Neurons in the orbitofrontal cortex are the most narrowly tuned of all central gustatory cells yet studied but seem to respond to only two of the prototypical taste stimuli: NaCl and sugar. Also they seem to be sensitive to the satiety of the animal, since in animals fed an excess of glucose, orbitofrontal neurons gradually become unresponsive to that stimulus.

The orbitofrontal cortex may be part of a system that associates sensory input with internal state. This association is fundamental to the regulation and motivation of behavior. The satiety-related cells of the orbitofrontal cortex and the ventral forebrain, with which the orbitofrontal cortex is anatomically and functionally connected (see Chaps. 31 and 32), may contribute to the regulation of feeding behavior by modulating the hedonic value of gustatory stimuli.

The lateral orbitofrontal cortex might be viewed as the highest identified level in a hierarchical system that extracts a limited number of distinct taste qualities from broadly responsive peripheral receptors. As gustatory information travels from the periphery to this cortical region, the response properties of single cells increasingly suggest the emergence of taste quality. The average tuning broadens as inputs from many taste fibers converge on cells of the nucleus of the solitary tract in both the rat and primate, but thalamic and cortical cells show a progressive narrowing of taste responsiveness that leads ultimately to the extremely narrow tuning of orbitofrontal neurons. The mechanisms that could produce the narrow tuning of cortical taste units from the broad tuning of receptor cells have long fascinated gustatory investigators.

The Coding of Taste[42]

It seems clear that taste quality is not coded in the discharge of distinctively separate classes of neurons (via labeled lines) at most levels of the gustatory system. To begin with, putative receptor cells respond to several taste qualities. How then do the four independent taste qualities that we all experience in our daily lives emerge? Rather than being encoded in the discharge of neurons responding to one and only one quality, information about a taste quality can be extracted from a population of neurons that respond somewhat to several stimuli but that respond best to one.

How might such an *across-neuron* analysis work? Figure 21–8A shows the response of a single neural element to a range of stimulus qualities and intensities. A particular response magnitude (e.g., R_1) is associated with a wide variety of qualities and intensities. To eliminate this ambiguity, one must compare the responses of several elements. Figure 21–8B shows the response of three neural elements (a, b, and c) to three stimulus qualities (I, II, and III) at three different stimulus intensities (strong, moderate, and weak). The tuning curves of these neural elements at strong stimulus intensities are depicted in Figure 21–8C. Theoretically, an individual would perceive quality I at strong stimulus intensities if elements a, b, and c had the excitation pattern seen in I in Figure 21–

8D. Similarly, quality II would be perceived if the excitation pattern in II, Figure 21–8D were present. Even if the stimuli are somewhat weaker (Fig. 21–8E), the same pattern of relative activity across the three elements still defines the stimulus quality; only the amplitude of the individual responses is decreased. However, when the stimulus intensity is extremely weak (Fig. 21–8F), the pattern is incomplete and the central nervous system cannot perceive the qualities unambiguously; quality II is actually undefined.

Thus, for other than extremely weak tastants, the across-fiber pattern defines stimulus quality (sweet, sour, bitter, salt) regardless of stimulus intensity (concentration). Across-element decoding probably is used to detect specific colors in the visual system, where the three cone receptors are broadly tuned to the wavelengths of light (see Fig. 19–11, Chap. 19).

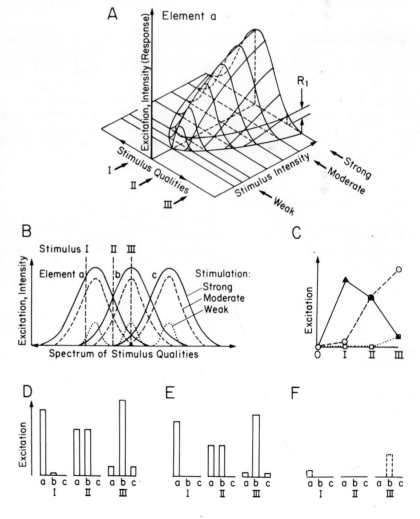

Figure 21–8 Across-element decoding. *A,* Response profile of a single element (corresponding to element a in *B–F*) to a range of stimulus intensities (including weak, moderate, and strong) and qualities (including I, II, and III). The R_1 response can be produced by a variety of intensities and qualities. *B,* Representative responses of elements a, b, and c to weak, moderate, and strong intensities over a range of stimulus qualities (including I, II, and III). *C,* Response profiles (tuning) of elements a, b, and c to stimulus qualities I, II, and III at strong stimulus intensity. Note the similarity to the data in Figure 21–7B. *D–F,* Histograms displaying the response of elements a, b, and c to stimulus qualities I, II, and III at strong intensity *(D)*, moderate intensity *(E)*, and weak intensity *(F)*. (After Somjen, G. *Sensory Coding in the Mammalian Nervous System*, pp. 83–98. New York, Meredith, 1972.)

OLFACTION

The sense of smell, called *olfaction,* serves a variety of functions. First, it is crucial for the perception of many flavors. Without olfaction, the flavors of coffee, chocolate, molasses, cranberry juice, and many other foods are indistinguishable.[26] Second, it participates in the communications between many animals (see Chap. 5). In humans, no innate chemical attractant (i.e., a human pheromone) has been identified; however, it is easy and natural for us to associate odors with certain experiences, which is why perfumes and deodorants facilitate many social interactions. Third, our ability to perceive unpleasant odors is crucial for avoiding tainted foods or detecting airborne contaminants in an increasingly polluted world. Although our lives are enriched by the perception of odors, only 20% of the almost 400,000 chemically different odorants (odor-producing stimuli) that have been tested are pleasing.

Qualities of Smell

As with taste, we must have some understanding of *what* we smell in order to study how the olfactory system produces the sensation. Odors are much more difficult to classify than tastes; nevertheless, the various classification schemes generated over the past 200 years (see Table 21–1) are rather similar. As he did for taste, Henning[14] suggested that there are smell primaries (he identified six), which are situated at the corners of an olfactory prism, and that all possible smells are distributed about the surface of the prism. Amoore and colleagues[1] thought that odorants with similar qualities have something in common. By surveying the literature and performing psychophysical experiments on humans, Amoore and his colleagues developed a list of seven primary odor qualities (see Table 21–1), five associated with odorants that have common molecular structures and two with odorants that have similar charge distributions. Thus, from Amoore's primaries came a stereochemical theory of odor, in which odorants of a given molecular size and shape, or a given electronic configuration (nucleophilic vs. electrophilic), bind to receptors with complementary sizes and shapes, or charges, producing a pattern of activity that indicates the appropriate primary odor quality.

While Henning's and Amoore's models of odor quality provide simple and appealing descriptions of the psychological dimensions of odor, there is little objective data to support the notion of primary odor qualities. The qualities of odors, unlike those of tastes are very difficult to describe, except with respect to their pleasantness or unpleasantness[13] (see Table 21–1); therefore, multidimensional scaling experiments like those described for taste are difficult to perform. Cross-adaptation experiments also produce confusing and often contradictory results because cross adaptation can occur between both similar and dissimilar odors and may not be reciprocal. Finally,

Table 21–1 Classifications of Odor Quality

Linnaeus (1752)	Haller (1763)	Zwaardemaker (1885)	Henning (1915)	Amoore (1962)	Examples from Zwaardemaker Classification
Aromatic	—	Aromatic	Fruity	—	Camphor, anise, citron
—	—	Ethereal	—	Ethereal	Beeswax, ether
Fragrant	—	Fragrant	—	—	Flowers, vanilla, balsam
—	Pleasant	—	Flowery	Floral	—
—	—	—	—	Pepperminty	—
—	—	—	—	Camphoraceous	—
—	—	—	Spicy	—	—
—	—	—	Resinous	—	—
Ambrosial	—	Ambrosial	—	Musky	Amber, musk
—	Intermediate	Empyreumatic	Burnt	—	Roasted coffee, tobacco smoke
Alliaceous	—	Alliaceous	—	—	Onion, iodine
Hircine	—	Hircine	—	—	Goaty odors, cheese, sweat
—	—	—	—	Pungent	—
Foul	—	Repulsive	Putrid	Putrid	Narcotics, bugs, coriander flour
Nauseating	Unpleasant	Foetid	—	—	Carrion, feces

*After Harper, R.; Smith, E. C. B.; Land, D. B. *Odor Description and Odor Classification: A Multidisciplinary Examination.* New York, Elsevier, 1968.

it was hoped that *specific anosmias* (lack of sensitivity to a specific odorant) were the result of missing receptor types for specific odor qualities, so that the missing olfactory sensations identified the odor primaries. Unfortunately, anosmias occur to so many odorants that this approach predicts an unwieldy number (perhaps 30) of primary qualities.

Thus, there is no consensus about the existence or identification of basic odor qualities and, therefore, no prototypical stimuli for olfactory research. Consequently, physiologists usually apply a variety of odor stimuli that span the spectra described in Table 21–1.

Sensitivity

Our sense of smell, commonly thought to be dull compared with that of other species, is surprisingly acute for some odorants. For example, human subjects can detect ethyl mercaptan (odor of skunk) at concentrations of 10^{-10}g/liter (10^{12} molecules/liter) of air, a sensitivity that exceeds that of a smoke detector. For this reason, ethyl mercaptan is used to odorize natural gas. At higher odorant concentrations, what we "smell" also involves other senses. For example, the piquant smell of ammonia is due largely to irritation of the free trigeminal nerve endings in our nasal passages. Anosmic individuals, who lack the olfactory sense, can still detect many common odorants, and people who lack trigeminal innervation are less sensitive to moderately strong odors. Thus, our experience of odor sensation is both sensitive, exceeding the limits of the best man-made devices, and complex, relying on multiple sensory signals.

The judgment of odor intensity by humans is reputed to be rather poor; yet for some odors, our ability to differentiate between two odorant intensities rivals that of a gas chromatograph. The perceived intensity of some odorants changes much more quickly with their concentration than does that of others. For example, a 180-fold increase in the concentration of propanol produces the same perceived intensity change as a 10^8-fold increase in the concentration of amyl butyrate. Therefore, variations in intensity can be differentiated much more easily for some odors than for others.

Our perception of odor depends on experiential factors such as expectations, health history, and age. For example, some normal observers reported the presence (a threshold judgment) of hydrogen sulfide (rotten-egg smell) when it was present at very low concentrations, whereas others required concentrations that were 45-fold higher. Expectation even creates olfactory hallucinations. For example, opening a "smelly" jar actually filled with distilled water or suggesting that light patterns can induce odors over television is sufficient to produce olfactory experiences in normal subjects.

Many odor perceptions seem to be stored securely as long-term memories. We have all encountered odors that immediately evoke associations with past experiences. Visual recognition memory is vastly superior to odor recognition over a period of days to weeks; however, after a year, visual recognition falls to near chance. In contrast, odor recognition is still as good as it was initially. As we shall see in Chapter 32, olfactory information reaches structures in the limbic system that are thought to be involved with memory.

Smell Receptors

Anatomy

In man, the transduction of olfactory information occurs in the olfactory cleft on the roof of the nasal cavity. The nasal cavity is divided into two passages, or nares, by a septum. As shown in Figure 21–9, each nasal passage consists of an anterior naris, several conchae (or folds) that are defined by the turbinate bones, and a posterior naris. The olfactory cleft is formed by the superior nasal concha; thus, much of the air inhaled during normal breathing (between 90 and 95%) bypasses the olfactory cleft as it travels to the trachea. Sniffing creates air currents, which aid in carrying odorant molecules upward into the secluded olfactory cleft.

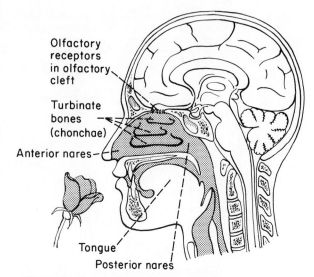

Figure 21–9 A sagittal section revealing the human right naris. (After Amoore, J. E; Johnson, J.W.; Rubin, M. *Sci. Am.* 210:42–49, 1984.)

The entire nasal cavity is coated with epithelium that secretes a viscous mucus. At the olfactory cleft, this nasal mucosa gives way to the olfactory epithelium, which contains the olfactory receptors.[23] As the mucus flows over the olfactory epithelium it is mixed with secretions from Bowman's glands and from sustentacular cells in the olfactory epithelium itself (see below). The layer of mucus, which is 20 to 50 μm thick in primates, flows at a rate of 10 mm/min in man, and the entire mucous sheet is replaced every 10 minutes. The rapid flow of mucus serves to sweep olfactants and airborne contaminants away from the olfactory epithelium.

The cellular basis of olfactory transduction has been introduced in Chapter 5. The olfactory epithelium consists of three cell types: receptor cells,

sustentacular cells, and basal cells. *Receptor cells*, which are long, slender bipolar neurons, lie among sustentacular cells and above basal cells, which lie at the base of the epithelium (Fig. 21–10). The olfactory epithelium overlies a deep vascular and glandular layer, the lamina propria. From its basal end, each receptor sends a single unmyelinated axon into the lamina propria, where axons from all the receptors aggregate to form the olfactory nerve. At its apical end, each receptor projects a rod-like dendrite to the surface of the epithelium, where it ends in a swelling called an *olfactory knob* (Fig. 21–10). From each knob, 1 to 20 cilia protrude into the mucous sheet, where they form a dense meshwork with cilia from neighboring receptors. Like sensory nerve endings, each olfactory knob contains many mitochondria, suggesting that the

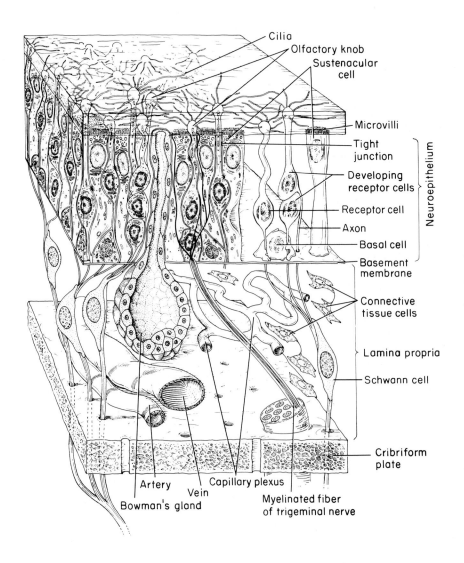

Figure 21–10 Anatomy of the olfactory epithelium and underlying lamina propria. (After Greep, R.D.; Weiss, L. *Histology*, 3rd ed. McGraw-Hill Book Co., 1973, p. 666.)

knob and cilia are the sites of olfactory transduction.

The olfactory epithelium covers an area of about 10 cm^2 in man; it covers over 170 cm^2 in the German shepherd, a dog known for its olfactory sensitivity. Furthermore, the average density of human olfactory receptor cells is 10^4/cm^2, whereas in the German shepherd it can be as high as 1.3 × 10^6/cm^2. Thus, olfactory acuity seems to be related to both the number and the density of receptor cells.

Sustentacular cells, or support cells, resemble either tanycytes (type I cells), which may have a transport and communication function in the central nervous system (see Chap. 64), or protoplasmic astrocytes, another type of glia (see Chap. 35). The apical surfaces of sustentacular cells contain over 1000 microvilli that increase their secretory surface. The basal processes of sustentacular cells often extend into the lamina propria. There, they are closely apposed both to glands and to capillaries, where they presumably obtain the ingredients for their mucous contribution, which includes mucopolysaccharides. In addition, sustentacular cells seem to serve as insulators, because their apical surfaces separate the receptor cells. Also, they form tight junctions with receptors that may limit the movement of extracellular ions. Finally, these cells can phagocytize degenerating receptor cells and may serve as a structural matrix. Thus, sustentacular cells may perform a variety of functions not unlike those of glia (Chap. 35).

The *basal cells*, which surround the receptor axons, are the stem cells for mature receptor neurons. Like taste receptors, olfactory receptors are constantly growing, degenerating, and being replaced. This turnover, which occurs over a period of 4 to 8 weeks, is not uniform across the olfactory epithelium at any one time but rather occurs in certain zones.[12] Each growth zone is composed of columns of receptor cells surrounding a sustentacular cell with basal cells below. The sustentacular cells, therefore, may guide the immature receptor cells to their final location. The continual recycling may protect the olfactory system from the damaging influences of the environmental irritants to which the receptors are constantly exposed. Many substances, such as zinc sulfate, cause degeneration of the olfactory epithelium.

Even in extraordinary circumstances, the damaged peripheral olfactory apparatus is capable of both functional and structural recovery. For example, after olfactory nerve transection, the regenerated receptor cells respond to olfactory stimuli and transmit olfactory information to second-order neurons. If their central target is removed, the receptor neurons form ectopic connection with other structures. If the termination site is replaced with transplanted neural tissue (e.g., from the cerebellum), the axons of receptor neurons form their characteristic and highly unusual synapses, called glomeruli (see below), with cells of the foreign tissue (e.g., cerebellar Purkinje cells).

Responses of Olfactory Receptors

Olfactants are both absorbed by and adsorbed to the olfactory mucosa as they flow over it. It is presumed that the sorbed olfactants diffuse to the receptor neuron, interact in some way with the chemoreceptive portions of its membrane, and then are either (1) desorbed back into the gas, (2) cleared by mucocilliary transport, (3) removed by the cardiovascular system, or (4) removed by pinocytosis or internalization by sustentacular or receptor cells, respectively, for enzymatic degradation.[19]

As with tastants, it is difficult to control and monitor olfactant delivery to receptors. Olfactants usually are presented by means of an *olfactometer*, a device that delivers a metered flow of a known concentration of olfactant across the olfactory mucosa. The olfactant usually is diluted in a gas and monitored with a chromatograph (Fig. 21–11A). Changes in the dilution of the olfactant in the gas or changes in the rate of flow alter the effective concentration of the olfactant at the receptor.

Several types of peripheral olfactory responses have been used to reveal information about the transduction process and the encoding of receptor discharge. These include the peripheral responses obtained by electro-olfactographic, intracellular, and extracellular recording (Fig. 21–11A).

The *electro-olfactogram* (EOG) is recorded through a large electrode in contact with the surface of the olfactory epithelium. When stimuli are applied to the olfactory mucosa, a slow negative transient response occurs (bottom traces, Fig. 21–11B). This EOG response is thought to reflect the extracellular current from summated generator potentials of receptor cells because (1) it can be recorded only from nasal epithelium containing such cells; (2) it is eliminated when the receptors undergo degeneration following olfactory nerve section; and (3) its time course and amplitude are well correlated with intracellularly recorded olfactory generator potentials. The EOG is used most often in nonmammalian vertebrates, where it has revealed some interesting properties of the transduction process.

A   B

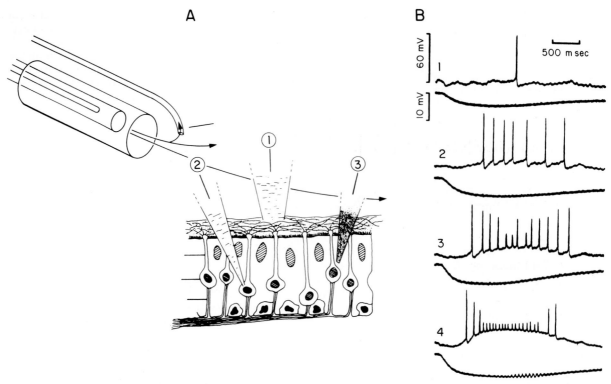

Figure 21–11 *A,* Sketch of a typical peripheral olfactory recording experiment. *B,* Response of an olfactory receptor cell (upper traces) and electro-olfactogram (EOG) (lower traces) in a salamander during stimulation with isoamylacetate; average resting potential was −62mV. Stimuli were presented in 0.2 to 0.3-s puffs at increasing concentrations from trace 1 to trace 4. (*A* modified from Lancet, D. *B* after Trotier, D.; MacLeod, P. *Brain Res.* 268:225–237, 1983.)

Intracellular recording (top traces, Fig. 21–11B)[46] from olfactory epithelial cells is extremely difficult because of their minute size and awkward location. Consequently, recordings have been obtained only in the salamander and lamprey. In these species, olfactory receptor cells have resting membrane potentials of −33 to −65 mV. In response to most olfactory stimuli, they depolarize with concurrent decreases in membrane resistance. Mature receptor cells discharge action potentials superimposed on the slow depolarization. Both the amplitude of membrane depolarization and the spike frequency increase with olfactant concentration.

EOG recordings suggest that depolarization of the receptor membrane probably is mediated by Na^+ and K^+ currents that require the presence of Ca^{2+}. For example, the elimination of extracellular Na^+ or its replacement with monovalent cations (e.g., Li^+, Rb^+, Cs^+, NH_4^+, TEA^+) reverses the sign of the EOG response to an odorant, presumably reflecting the remaining K^+ current.

Receptors undergo slow adaptation in depolarization when an olfactant is applied over an ex-

tended period. Higher olfactant concentrations produce slower decays, suggesting that the intrinsic transduction mechanism, or the clearance of the olfactant, is rate limited. The slow adaptation of receptor depolarization is counter-intuitive since psychophysical studies reveal that the human olfactory system adapts rapidly to odors.

When olfactants are applied to the olfactory mucosa, sustentacular cells respond with a long depolarization, a hyperpolarization, or a hyperpolarization followed by depolarization. These responses are secretory potentials, passively conducted potentials from receptor neurons, or a reflection of K^+ buffering.

Extracellular recording reveals that receptors typically discharge at low, irregular rates when unstimulated and produce a burst of spikes in response to olfactory stimuli.[32] Like gustatory receptors, olfactory receptors respond to a variety of olfactants of roughly equal concentrations (Fig. 21–12). However, their response patterns vary from one olfactant to another. Some olfactants, such as d-citronellol (DCI), produce short-latency responses, whereas others, such as pinacolone (PIN), produce responses at longer latencies. Some

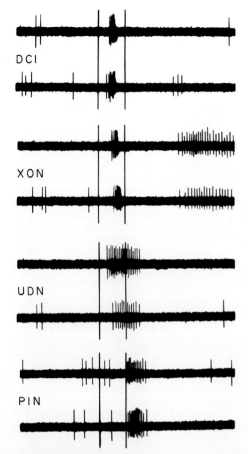

Figure 21–12 Extracellular responses of a frog olfactory receptor cell to successive, 2-s pulses of d-citronellol (DCI), cyclohexanone (XON), cycloundecanone (UDN), and pinacolone (PIN). All olfactants were presented at 20 to 30% saturation. (After Revial, M.F.; Sicard, G.; Duchamp, A.; Holley, A. *Chem. Senses* 7:175–190, 1982.)

(DCI) produce single, short, high-frequency bursts; others, such as cycloundecanone (UDN), produce longer, low-frequency discharges; others, such as cyclohexanone (XON), produce both an early and a late burst. Thus, the temporal pattern of the response as well as its frequency may be important in coding the properties of olfactory stimuli.

EOG recordings over the entire surface of the olfactory mucosa of salamanders show that different regions of the epithelium are sensitive to different olfactants.[21] For example, limonene, which has a citrus smell, produces a maximal EOG response more anteriorly in the nasal cavity than does propanol, a musky odorant. Thus, like the tongue, the nose seems to have for different olfactants a spatial map that is presumed to reflect the distribution of specific receptor cells. Mozell[25] suggested that as various odorants migrate across the mucosa,

each odorant is preferentially absorbed into the mucous layer at different spatial locations. Thus, each odorant may have its highest concentration at the location of its receptors. In this scheme, the nose functions as a chromatograph, with the affinity of the odorant for either the fixed phase (the mucous and receptors) or the moving phase (the inhaled gas) determining its distribution along the olfactory epithelium. For example, the anatomical distribution of butanol receptors is in register with the distribution of butanol along the mucosa. The match between odorant distribution and olfactory receptor location is reminiscent of the match between the distribution of sound frequencies in the cochlea and the distribution of the frequency tuning of hair cells (see Chap 17).

Smell Pathways

Figure 21–13 depicts the anatomy of three olfactory pathways in primates. (This description is an oversimplification; "it would perhaps be quicker to list those parts of the brain to which olfactory projections do *not* have access."[17]) For all three pathways, receptors in the olfactory epithelium send their unmyelinated axons (collectively, the first cranial nerve) through the cribriform plate of the ethmoid bone of the skull into the ipsilateral olfactory bulb. The olfactory bulb then projects either to the prepyriform cortex and the amygdala or to the septum and perhaps the nucleus accumbens via the lateral and medial olfactory tracts. Third-order olfactory neurons in the prepyriform/amygdala locations give rise to at least two different pathways. The first has an intermediate synapse in the substantia innominata before terminating in the lateral-posterior portion of orbitofrontal cortex (thin line, Fig. 21–13A). This cortex also receives a direct amygdalar olfactory input and an indirect input via a cortico-cortical relay in the pro-rhinal cortex. The second has a relay in the large-celled (magnocellular) portion of the mediodorsal nucleus of the thalamus (MDmc) before ending on fifth-order olfactory neurons in an adjacent portion of the orbitofrontal cortex, the central-posterior region (thick line, Fig. 21–13A). Finally, the third pathway through the septum-olfactory tubercle terminates in the lateral hypothalamus (dashed line, Fig. 21–13A).

In contrast to the Old World primates, most New World primates and nonprimate species have an additional set of two ascending pathways. The main output from the olfactory bulb involves the prepyriform cortex, MDmc, and the orbitofrontal cortex and therefore is similar to the principal olfactory pathway in humans (thick line, Fig. 21–

A

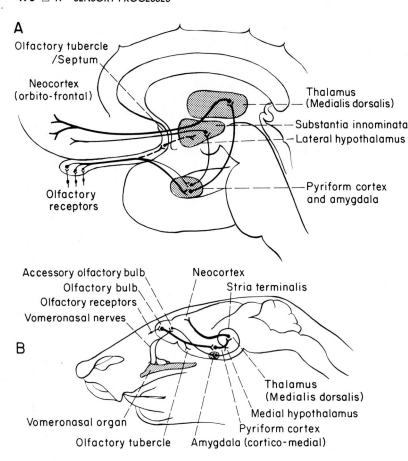

Olfactory tubercle /Septum

Neocortex (orbito-frontal)

Olfactory receptors

Thalamus (Medialis dorsalis)

Substantia innominata

Lateral hypothalamus

Pyriform cortex and amygdala

B

Accessory olfactory bulb

Olfactory bulb

Olfactory receptors

Vomeronasal nerves

Neocortex

Stria terminalis

Vomeronasal organ

Olfactory tubercle

Thalamus (Medialis dorsalis)

Medial hypothalamus

Pyriform cortex

Amygdala (cortico-medial)

Figure 21–13 Olfactory pathways in the Old World monkey *(A)* and in nonprimate mammals *(B)*. In *A*, three olfactory pathways are represented by different kinds of lines. In *B*, the main and accessory olfactory pathways are represented by lines of different thickness. (After Keverne, E. B. In Barlow, H. B.; Mollon, J. D., eds. *The Senses*, pp. 409–427. Cambridge, Cambridge Univ. Press, 1982.)

13B). As in higher primates, there is also a connection to the olfactory tubercle/septum region and the lateral hypothalamus. An additional pathway, shown in Figure 21–13B (thin line), originates from a second chemosensitive region of the nasal cavity called the *vomeronasal organ*, which is absent or vestigial in man and higher primates. The vomeronasal region projects via the vomeronasal nerves to the *accessory olfactory bulb*, which lies just posterior to the olfactory bulb. Second-order neurons in the accessory olfactory bulb project to third-order neurons in the corticomedial amygdala, which, in turn, projects via the stria terminalis to fourth-order neurons in the medial hypothalamus. This group of related structures is known collectively as the *accessory olfactory system*.

Olfactory Bulb

The two olfactory bulbs lie above the left and right nares and, in humans, below the frontal lobe. Each bulb receives axons from the ipsilateral olfactory nerve and, in higher primates, provides the second-order neurons of all central olfactory pathways. An olfactory bulb has five types of neurons with intricate connections and is, therefore, theoretically capable of sophisticated sensory processing (Fig. 21–14).

The olfactory bulb seems to consist of a direct through pathway with several interneurons that allow for lateral interactions. Olfactory receptor axons synapse on *mitral* and *tufted cells* in unique anatomical specializations called glomeruli (m and t, Fig. 21–14). At least 25,000 afferent axons converge on the dendrites of perhaps 25 mitral and 70 tufted cells. Since virtually all the mitral cells and many of the large tufted cells send their axons out of the olfactory bulb, they can be considered to be part of a direct through path. This direct path can be influenced by at least three stages of lateral interaction. First, *periglomerular cells*, which communicate with mitral and tufted cells by axonal and dendrodendritic synapses and which are possibly driven directly by receptors, allow synaptic interactions between adjacent glomeruli (pg, Fig. 21–14), and between receptor and mitral or tufted cell processes within glomeruli. Second, *granule*

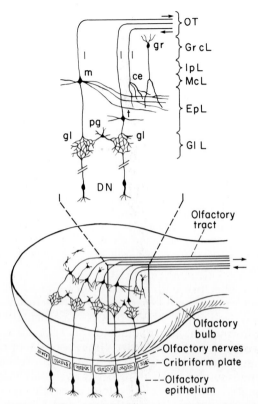

Figure 21–14 Gross organization of an olfactory bulb with a portion of its internal laminar structure shown expanded. Abbreviations: GlL = glomerular layer ; EpL = external plexiform layer; McL = mitral cell layer; IpL = internal plexiform layer; GrcL = granule cell layer; OT = olfactory tract; ON = olfactory nerve; pg = periglomerular cell; t = tufted cell; m = mitral cell; gr = granule cell; gl = glomerulus; ce = centrifugal efferent. (*Bottom* after Keverne, E. B. In Barlow, H. B.; Mollon, J. D., eds. *The Senses*, pp. 409–427. Cambridge, Cambridge Univ. Press, 1982. *Top* after Brodal, A. *Neurological Anatomy in Relation to Clinical Medicine*, p. 645. New York, Oxford Univ. Press, 1981.)

cells make dendrodendritic synapses with mitral and tufted cells (gr, Fig. 21–14). Both the periglomerular cells and the granule cells are thought to be inhibitory interneurons. Finally, the olfactory bulb contains many *short axon cells* and intrinsic tufted cells; the function of these interneurons is unknown.

Information flow in the olfactory bulb is influenced by centrifugal signals from other parts of the central nervous system. Structures such as the midbrain raphe and the locus ceruleus provide monoaminergic projections to the olfactory bulb, and the lateral preoptic area and the nucleus of the diagonal band of Brocca supply cholinergic inputs. These structures exert inhibitory influences on mitral and tufted cells, either directly or via granule cells. This inhibition might be related to

arousal, the internal state of the organism, or learning. For example, those mitral cells that respond selectively to food odors discharge more vigorously in hungry than in satiated rats. Also, ewes fail to identify their own lambs by odor if efferent projections to the olfactory bulb are disrupted pharmacologically.

Cells in the anterior olfactory nucleus receive ipsilateral projections from mitral cells and project to homotypic (similarly located) granule cells in the contralateral bulb. These cells, which also receive input from many central structures (e.g., pyriform cortex), thus may mediate a crossed inhibition between homotypical mitral cells. Crossed interactions may contribute to olfactory source localization, an important capability for predator avoidance and prey localization in animals. A bilateral interaction is required because the side of a unilateral stimulus cannot be perceived. Even humans can perceive the side of olfactory stimulation during birhinal stimulation if there is at least a 10% difference in concentration between the two nostrils or a 1-ms difference in arrival time.

The integrity of the olfactory bulb is crucial for olfaction. In primates, lesions of the olfactory bulb produce a frank anosmia, as do lesions of the olfactory nerve (common in human closed head injury). Rats in which all central olfactory structures *except* the olfactory bulb and pyriform cortex are destroyed are still capable of odor discrimination. In rats, remarkably little (perhaps 10%) of the olfactory bulb or its output, or both is required for simple olfactory detection and discrimination tasks.

If individual odorants are applied to the olfactory epithelium, certain regions of the bulb consistently seem more active than others.[43] For example, if a radioactive glucose analogue, 2-deoxyglucose (2DG), is given to animals before they are exposed to a certain olfactant, the radioactive label is incorporated into those olfactory bulb neurons that are most responsive to the olfactant (and therefore need the most glucose). The differential labeling cannot be explained as a topographic projection from olfactory receptors onto the olfactory bulb, since thousands of receptors, many of which are widely separated on the olfactory epithelium, converge on single mitral cells. Furthermore, the size of the focus activated by a particular olfactant depends on its concentration; higher odorant concentrations cause the foci to expand and involve more of the bulb. The relative amount of activity in the activated loci can be influenced by early learning.[20] For example, when neonatal rats are given a peppermint odor while being stroked to simulate maternal contact, they learn to prefer that odor; in animals that have undergone such learning there is both growth and an increased 2DG uptake in the olfactory bulb focus for that odor.

Responses of mitral or tufted cells to olfactory stimuli have been recorded in several species including the salamander and the hamster. Like receptors, most mitral cells respond to many odorants and, therefore, are also broadly tuned. When a brief pulse of odorant is delivered to the nasal epithelium, mitral or tufted cells show an initial increase or decrease or no change in tonic firing rate. In many cells, the initial responses are followed by responses of the opposite sign; that is, initially increasing responses are succeeded by a decrease or suppression of activity, whereas initially decreasing responses are followed by increased activity. Increases in olfactant concentration generally seem to change the character of the response. For example, an initially excitatory response may be more and more suppressed as concentration increases, or an initial inhibitory response may become excitatory. These changes with concentration are probably related to intrinsic inhibition, since receptor axons make only excitatory connections within the glomeruli, and intracellular recordings suggest little correlation between the changes in responses of cells in the bulb and those seen in the receptors themselves.

Meredith[22] has suggested that lateral inhibition and an expanding focus of activation underlie the complex changes in response described above. He proposed that cells that are initially excited and then become suppressed with increasing olfactant concentrations may be low-threshold cells that lie within the small foci revealed by 2DG at low concentrations. With increasing concentration, the 2DG foci expand to involve more peripheral, higher-threshold units that suppress the low-threshold units by lateral inhibition. Similarly, cells that are initially suppressed and then excited at higher concentrations could be high-threshold cells in the periphery of larger foci generated by higher concentrations. At low concentrations, these neurons are suppressed by low-threshold cells but are gradually more excited as concentration increases. This hypothetical scheme suggests that olfactory coding may be organized spatially; for each olfactant, low-threshold foci of activity are surrounded by concentrically oganized annuli of higher-threshold neurons, with low- and high-threshold units providing mutual inhibition.

We saw earlier that neonatal rats given a peppermint odor in association with stroking developed an increased 2DG uptake in foci for that odor. Mitral or tufted output cells in the region of increased 2DG uptake display more suppression and less excitation than do cells in control animals. The enhanced suppression is specific for peppermint and occurs only for cells in the peppermint foci. These electrophysiological changes may be the result of enlarged glomeruli, which occur only within the foci of increased 2DG uptake. Thus, learning is associated with an increase in metabolic activity, an expansion of cellular morphology, and a suppressed response only within the focus of the preferred odorant. The suppression is probably mediated via the granule cells, which themselves show increased metabolic activity when labeled with glycogen phosphorylase. Therefore, lateral inhibitory processes and morphological changes at restricted foci within the olfactory bulb may underlie a biologically significant form of early olfactory learning (i.e., the ability to identify an associated maternal odor, which facilitates sucking in neonatal rats).

The olfactory bulb and accessory olfactory bulb give rise to several separate central olfactory pathways, each of which may process a particular kind of information. To understand this specialization, investigators have recorded the responses of cells in individual pathways to determine whether changes in olfactory tuning occur as signals ascend into the central nervous system; alternatively they have interrupted the pathways and looked for behavioral deficits.

The Accessory Olfactory System

The accessory olfactory pathway provides olfactory input to the limbic system of many subprimate mammals and is thought to subserve the olfactory regulation of neuroendocrine and sexual function as well as a variety of other regulatory and social behaviors. Lesions of any part of the accessory olfactory system disrupt reproductive and maternal behavior or eliminate the pheromonal regulation of the timing of puberty, estrous cyclicity, marking behavior, and aggression. For example, exposure of an inseminated female mouse to olfactory stimuli from male mice of a different strain blocks pregnancy; however, if the vomeronasal organ is destroyed, the pregnancy is unaffected. These data suggest that the accessory olfactory system is important for the percepton of pheromones; its contribution to other olfactory perceptions has not been explored.

Major Olfactory Projections

Takagi[44] and his colleagues have attempted to distinguish the processing that occurs in three major olfactory pathways of primates (see Fig. 21–13) by recording the responses of neurons in the various olfactory nuclei to the same array of eight olfactants. Six of their stimuli represented five of the "qualities" proposed by Amoore (see Table 21–1); the remaining two were additional examples of one of these qualities, so that three of the stimuli evidently were difficult to discriminate. Figure 21–15 shows the number of odorants to which cells in each olfactory structure responded. Each histogram, therefore, represents the degree of olfactory tuning at that level

As mentioned above, the mitral cells of the olfactory bulb respond to many olfactants and, therefore, are broadly tuned. In the monkey, a majority of olfactory bulb cells responded to at least half of the olfactants (OB, Fig. 21–15). In sharp contrast, over 50% of the cells in the lateral hypothalamic area (LHA) responded to only one olfactant, even though three of the olfactants had

somewhat similar, camphoraceous odors. Thus, it was theorized that the hypothalamic pathway promotes a discrimination of odors, thereby facilitating the role of the lateral hypothalamus in regulatory behaviors (see Chap. 32).

Neurons in the lateral posterior orbitofrontal cortex (LPOF, Figure 21–15), seem to be very narrowly tuned, with 50% of the cells responding to only one olfactant and only about 77% of the cells responding to half of them. This narrow tuning, like that of cells in the lateral hypothalamus, occurs even through the tuning of posterior orbitofrontal afferent "relay" cells—in this case, those in the prepyriform (PPF) cortex and medial amygdala, (MA)—is almost as broad as that in the olfactory bulb (34% of the cells respond to at least half of the odors). Cells in this portion of the posterior orbitofrontal cortex also are far superior in discriminating between the similar camphoraceous odors; therefore Takagi and colleagues ar-

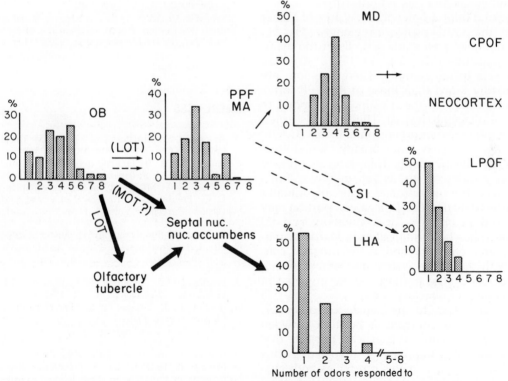

Figure 21–15 Tuning of neurons in three Old World primate olfactory pathways. For each structure, histograms represent the percentages of cells that responded to 1 or up to 8 of the test stimuli. Olfactory stimuli were selected from the following Amoore's "qualities": *camphoraceous,* dl-camphor, cineol, and borneol; *etherial,* 1,2-dichloroethane, isoamyl acetate; *burnt,* methyl cyclopentenolone; *fruity,* gamma-undecalactone; *pungent,* isovaleric acid. Pathways shown include *thick line,* pathway to lateral hypothalamic area (LHA) with relays including the septal nucleus (SN), nucleus accumbens (NA), and the olfactory tubercle (not shown); *thin line,* pathway to central posterior orbitofrontal cortex with relays in prepyriform cortex (PPF) and medial amygdala (MA); *dashed line,* pathway to lateral posterior orbitofrontal cortex (LPOF) with relays in PPF and MA, and also in substantia innominata (SI). (After Takagi, S.F. *Japan. J. Physiol.* 34:561–573, 1984.)

gued that the lateral posterior orbitofrontal region is involved in the conscious perceptual discrimination of different olfactants.

The effects of specific cortical lesions are consistent with this suggestion. Using odor cues alone, monkeys with lesions of only the lateral posterior orbitofrontal cortex have difficulty in discriminating camphor-smelling, bitter-tasting pieces of bread from other pieces with different odors and less repulsive tastes. Lesions of other orbitofrontal regions do not produce similar impairments.

Indeed, in a variety of species, the orbitofrontal cortex and its associated pathways have been implicated in olfactory discrimination. In dogs, ablation of the orbitofrontal cortex prevents differential classical conditioning to a variety of odors (i.e., the conditioning of several different responses to several different odorants). Such dogs, however, still can be taught to respond to a single odorant cue. Also, aphasic humans with frontotemporal lesions show deficits in olfactory discrimination, as do patients with lesions encroaching on the amygdala and prepyriform cortex.

The second orbitofrontal olfactory area of higher primates (the central posterior region; see Fig. 21–13) has not been well studied. Cells in the mediodorsal thalamus (MD, Fig. 21–15), a structure that projects to the central posterior region, are more broadly tuned than those of the PPF cortex and the MA (60% of the cells in MD respond to half of the olfactants tested). The few cells tested in the central posterior orbitofrontal cortex (CPOF, Fig. 21–15), are even more broadly tuned than those of the olfactory bulb. Therefore, this region may have a different, but as yet unknown, function. In fact, the orbitofrontal cortex of primates may subserve dual roles: (1) its lateral portion may be involved in olfactory discrimination; and (2) since it also has been implicated in learning, motivation, and emotion, and is reciprocally connected with the limbic system and other association areas, its central portion may be involved in the integration of olfactory cues.

Finally we return to the perception of flavor, which requires an intimate association between gustation and olfaction. It is at least possible that the orbitofrontal cortex may be involved in this perception, since in primates it receives both gustatory and olfactory input. In rats, the neocortical gustatory region also receives olfactory input and may have cells that respond to both olfactory and gustatory stimuli.

A CONCLUDING REMARK

The olfactory and gustatory systems are organized with multiple parallel pathways that subserve both the conscious perception of chemical sensation and the regulation of a variety of homeostatic behaviors with inherent emotional and motivational components. The close association of both modalities with the limbic system and ventral forebrain suggests that they are involved in unconscious functions requiring chemical sensation. Our chemical senses are not, as some have argued, denuded and degenerate; they are rich and alive but often influence functions of which we are not conscious.

ANNOTATED BIBLIOGRAPHY

Finger, T. E.; Silver, W. L., eds. *Neurobiology of Taste and Smell.* New York, Wiley Interscience, 1987.
  An excellent recent review of the chemical senses.
McBurney, D. H. Taste and olfaction: sensory discrimination. *Handbk. Physiol.* 3:1067–1086, 1984.
  A comprehensive review of the psychophysics of taste and olfaction.
Norgren, R. Central neural mechanisms of taste, *Handbk. Physiol Vol. III; The Nervous system.* 3:1087–1128, 1984.
  This review covers the anatomy and physiology of taste beginning at the solitary tract.

REFERENCES

1. Amoore, J. E.: Johnston, J. W.: Rubin, M. The stereochemical theory of odor. *Sci. Am.* 210:42–49, 1984.
2. Andres, K. Y. Der Feinbau der Regio Olfactoria von Makrosmatikern. *Z. Zellforsch.* 69:140–154, 1966.
3. Bartoshuk, L. M. History of taste research. In Carterette, E. C.; Freidman, M. P., eds. *Tasting and Smelling,* 3–18. *Handbk. Perception,* Vol 6A. New York, Academic Press, 1978.
4. Beidler L. M. Biophysics and chemistry of taste. In Carterette, E. C.; Freidman, M. P., eds. *Tasting and Smelling,* 21–49. *Handbk. Perception,* Vol 6A. New York, Academic Press, 1978.
5. Beidler, L. M.; Smallman, R. L. Renewal of cells within taste buds. *J. Cell Biol* 27:263–272, 1965.
6. Bradley, R. M. Tongue topography. In Beidler, L. M., ed. *Chemical Senses: Taste,* 1–30. *Handbk. Sensory Physiol.,* Vol. 4, pt. 2. New York, Springer-Verlag, 1971.
7. Brodal, A. *Neurological Anatomy in Relation to Clinical Medicine.* New York, Oxford Univ. Press, 1981.
8. Doving, K. B.; Pinching, A. J. Selective degeneration of neurons in the olfactory bulb following prolonged odor exposure. *Brain Res.* 52:115–129, 1973.
9. Frank, M. E.; Contreras, R. J.; Hettinger, T. P. Nerve fibers sensitive to ionic taste stimuli in chorda tympani of the rat. *J. Neurophysiol.* 50:941–960, 1983.

10. Giza, B. K.; Scott, T. R. Blood glucose selectively affects taste evoked activity in the rat nucleus tractus solitarius. *Physiol. Behav.* 31:643–650, 1983.
11. Graziadei, P. P. C. The ultrastructure of vertebrate taste buds. Pfaffmann, C., ed. *Olfaction and Taste, Vol. 3*, 315–330, New York, Rockefeller Univ. Press, 1969.
12. Graziadei, P. P. C.; Monti-Graziadei, G. A. Neurogenesis and neuron regeneration in the olfactory system of mammals. I. Morphological aspects of differentiation and structural organization of the olfactory sensory neurons. *J. Neurocytol.* 8:1–8, 1979.
13. Haller, A. von. *First Lines of Physiology*, 258–265. Edinburgh, Charles Eliot, 1786.
14. Henning, H. *Die Geruch.* Leipzig, Barth, 1924.
15. Henning, H. Die Qualitätenreihe des Geschmacks. *A. Psychol.* 74:203–219, 1916.
15a. Kessel, R.G., Kardon, R.H. *Tissues and Organs: A Text Atlas of Scanning Electron Microscopy*, 129. San Francisco, W.H. Freeman Co., 1979.
16. Keverne, E. B. Chemical senses: smell. In Barlow, H. B., Mollon, J. D., eds. The Senses, 409–427. Cambridge, Cambridge Univ. Press, 1982.
17. Keverne, E. B. Reply from E. B. Keverne. *Trends Neurosci.* 2:315–316, 1979.
18. Kimura, K.; Beidler, L. M. Microelectrode study of taste receptors of rat and hamster. *J. Comp. Cell Physiol.* 58:131–139, 1961.
19. Lancet, D. Vertebrate olfactory region. *Annu. Rev. Neurosci.* 9:329–355, 1986.
20. Leon, M. Plasticity of olfactory output circuits related to early olfactory learning. *Trends Neurosci.* 10:434–438, 1987.
21. Mackay-Sim, A.; Shaman, P.; Moulton, D. G. Topographic coding of olfactory quality: odorant specific patterns of epithelial responsivity in the salamander. *J. Neurophysiol.* 48:584–596, 1982.
22. Meredith, M. Patterned response to odor in mammalian olfactory bulb: the influence of intensity. *J. Neurophysiol.* 56:572–597, 1986.
23. Moulton, D. G.; Beidler, L. M. Structure and function in the peripheral olfactory system. *Physiol. Rev.* 47:1–52, 1967.
24. Moran D. T.; Rowley, J. C.; Jasek, B. W. Electron microscopy of human olfactory epithelium reveals a new cell type: The microvillar cell. *Brain Res.* 253:39–46, 1982.
25. Mozell, M. M. Evidence for a chromatographic model of olfaction. *J. Gen. Physiol.* 56:46–63, 1970.
26. Mozell, M. M.; Smith, B. P.; Smith, P. E.; Sullivan, R. J.; Swindler, P. Nasal chemoreception and flavor identification. *Arch. Otolaryngol.* 90:367–373, 1969.
27. Murray, R. G. The ultrastructure of taste buds. In Friedmann, I., ed. *The Ultrastructure of Sensory Organs*, 3–81. New York, Elsevier, 1973.
28. Norgren, R. Central neural mechanisms of taste. *Handbk. Physiol.*, 3:1087–1128, 1984.
29. Pansky, B.; Allen, D. J. *Review of Neuroscience*, 358–363. New York, Macmillan, 1980.
30. Patton, H. D. Taste and Olfaction. In Ruch, T.; Patton, H. D. eds. *Physiology and Biophysics*, 20th ed., Vol. 2: *The Brain and Neural Function*, 325–338. Philadephila, W. B. Saunders 1979.
31. Pfaffmann, C. J. Gustatory nerve impulses in rat, cat, and rabbit. *J. Neurophysiol.* 18:429–440, 1955.
32. Revial, M. F.; Sicard, G.; Duchamp, A.; Holley, A. New studies on odor discrimination in the frog's olfactory receptor cells. I. Experimental results. *Chem. Senses* 7:175–190, 1982.
33. Sato, M. Response characteristics of taste nerve fibers in macaque monkeys: comparison with those in rats and hamsters. In Denton, D. A.; Coghlan, J. P., eds. *Olfaction and Taste, Vol. 5*, 23–26. New York, Academic Press, 1975.
34. Sato, M.; Ogawa, H.; Yamashita, S. Response properties of macaque monkey chorda tympani fibers. *J. Gen. Physiol* 66:781–810, 1975.
35. Sato, T. Receptor potential in rat taste cells. In Antrum, H.; Ottoson, D., eds. *Progress in Sensory Physiology 6*, 1–37. New York, Springer-Verlag, 1986.
36. Sato, T.; Beidler, L. M. The response characteristics of rat taste cells to four basic taste stimuli. *Comp. Biochem. Physiol.* 73A:1–10, 1982.
37. Sato, T.; Beidler, L. M. Dependence of gustatory neural response on depolarizing and hyperpolarizing receptor potentials of taste cells in the rat. *Comp. Biochem. Physiol.* 75A:131–137, 1983.
38. Schiffman, S. S.; Erickson, R. P. A psychophysical model for gustatory quality. *Physiol. Behav.* 7:617–633, 1971.
39. Scott, T. R.; Yaxley, S.; Sienkiewicz, Z. J.; Rolls, E. T. Gustatory responses in the nucleus tractus solitarius of the alert cynomolgous monkey. *J. Neurophysiol.* 55:182–200, 1986.
40. Scott, T. R.; Yaxley, S.; Sienkiewicz, Z. J.; Rolls, E.T. Gustatory responses in the frontal opercular cortex of the alert cynomolgous monkey. *J. Neurophysiol.* 56:876–890, 1986.
41. Shepherd, G. M. *Neurobiology*, 203–266, New York, Oxford Univ. Press, 1983.
42. Somjen, G. *Sensory Coding in the Mammalian Nervous System*, 83–98, 154–155, 301–304. New York, Meredith, 1972.
43. Stewart, W. B.; Kauer, J. S.; Shepherd, G. M. Functional organization of rat olfactory bulb analyzed by the 2-deoxy-glucose method. *J. Comp. Neurol.* 185:715–734, 1979.
44. Takagi, S. F. The olfactory nervous system of the old world monkey. *Japan. J. Physiol.* 34:561–573, 1984.
45. Travers, J. B.; Travers, S. P.; Norgren, R. Gustatory neural processing in the hindbrain. *Annu. Rev. Neurosci.* 10:595–632, 1987.
46. Trotier, D.; MacLeod, P. Intracellular recordings from salamander olfactory receptor cells. *Brain Res.* 268:225–237, 1983.

Control of Movement

ALBERT F. FUCHS
Section Editor

Chapter 22

Albert F. Fuchs
Marjorie E. Anderson
Marc D. Binder
Eberhard E. Fetz

The Neural Control of Movement

The Spectrum of Movements
What Is Controlled During Movement?
How Are Movements Controlled?
Neural Structures Involved in Movement

Movement is crucial to the survival of all organisms. Even unicellular organisms such as *Escherichia coli* can move toward those chemicals critical for their metabolism (see Chap. 5). More complicated creatures such as coelenterates move only slowly but require motile tissues to attract and capture passing prey. Birds are not only capable of the incredible motor feat of flying but also rely on posturing and movement to attract a mate. Mammals are endowed with probably the greatest repertoire of motor behaviors. These range from the motility of the gut and the beating of the heart to the fine finger movements of an accomplished pianist. All these movements perform quite different functions, so it is not surprising that they are accomplished in different ways.

THE SPECTRUM OF MOVEMENTS

We are all aware that some movements seem to occur automatically, whereas others require effort and concentration. The simplest movement is the rapid response to a noxious sensory stimulus.

Such movements occur after a very short latency (<50 ms) and have a stereotyped trajectory. Examples include withdrawing your hand from a hot stove or lifting your foot from a tack. Because they are stereotyped and rapid (and therefore probably are controlled by relatively few neurons; Chap 24), these motor acts have been called *reflexes*. Reflexes usually are not under voluntary control, so that when the stimulus specific for the reflex occurs *unexpectedly*, the reflex movement is elicited automatically and cannot be modified. Indeed, many reflexes can be elicited in unconscious individuals or those in whom the cerebral cortex has been destroyed. Under special conditions, however, many reflexes can be modified and even prevented from occurring. For example, blinking, sneezing, and even withdrawing from a noxious stimulus (certainly if one is a fakir) can be suppressed with intense voluntary effort. However, some reflexes, such as the constriction of the pupils to a sudden bright light, are difficult if not impossible to influence by voluntary effort.

At the other end of the spectrum are movements that are completely under voluntary control. For volitional movements, such as painting, one can choose the direction, extent, trajectory, and timing of the movement. Furthermore, voluntary movements can be modified while they are occurring. In contrast with reflexes, voluntary movements are rarely triggered by any obvious sensory stimulus and are typically executed over longer periods of time. Furthermore, they may be affected by

503

such esoteric factors as attention and motivation and therefore may be under the influence of many neuronal structures scattered throughout the central nervous system. Most voluntary movements must be learned and require practice to become perfect. Once learned, a complicated motor behavior is apparently stored as a package called a *motor program* that can be called up whenever it is required. For example, one does not consciously control each part of the sequence of movements required to sign one's name, and an accomplished typist no longer thinks about the details of his finger placements.

A class of movements that lie intermediate to reflexes and completely voluntary movements are those involved in rhythmic behaviors such as breathing, chewing, and walking. These movements can be initiated and terminated voluntarily but, once initiated, they proceed automatically and are characterized by a repeated sequence or pattern of movements. As we shall see later (Chap. 26), rhythmic movements are controlled by neural networks in the spinal cord or brain stem that feature reciprocal connections between pools of neurons generating opposite behaviors (e.g., inspiration and expiration). These neural networks, called *pattern generators,* may be set in motion (or "triggered") by a sensory or a command signal. Once underway, most rhythmic behaviors can also be influenced by sensory stimuli. For example, the pattern of chewing is modified to compensate for the different textures of food.

Movements can therefore be considered to range from essentially automatic (the reflex) through semiautomatic to completely volitional. The nervous system controls these different movements in different ways according to their purpose. For example, the leg movements used in locomotion may be optimized for speed, whereas the delicate hand and finger movements of a watchmaker may require an exquisite control of position. We now consider the movement parameters that *could* be controlled so that we can better appreciate how motor systems are designed and can better interpret the various signals that are carried in the discharge of single neurons involved with movement.

WHAT IS CONTROLLED DURING MOVEMENT?[7]

Movement is produced by the contraction and relaxation of muscles. These muscular changes create forces that act on the load imposed by a body part (e.g., a limb) and any external object that must be moved. The parameters of the resulting movement (i.e., displacement [or position], velocity, and acceleration) are then determined according to the laws of mechanics.

The nervous system must apply neural signals that are appropriate to control the desired parameter. In some situations, the controlled parameter may be force (e.g., the delicate isometric contraction required to hold a fragile glass). In others, it may be position (e.g., the finger placements of a typist on a computer keyboard), or velocity, or acceleration. Alternatively, the controlled variable may involve a pair of movement parameters. For example, stiffness, which is the ratio of force to displacement, has been proposed as the controlled parameter in some limb movements.[2]

In addition to the control of these obvious movement parameters, other control strategies have been proposed. For example, control system theory, especially as applied to the field of robotics, has suggested that some movements may minimize the energy consumed, the time taken, or the variation of the movement about a given trajectory.[1]

These examples illustrate that different movements can be accomplished by controlling a wide variety of parameters. Furthermore, different control strategies can be exerted on the same muscle, according to how it is used. For example, the control of the jaw muscles during speaking is quite different from that during chewing, especially when one encounters foods with different textures (e.g., a raw carrot vs. mashed potatoes). Similarly, the control of the hand when performing a powerful grip is different from that required when writing.

Another example is the control of the joint angle between two parts of a limb. Muscles act reciprocally to produce movement in opposite directions around a joint. For example, flexion of the elbow (Fig. 22–1) is produced by contraction of the biceps and relaxation of the triceps, whereas extension of the elbow requires just the opposite pattern of muscle action. An entirely different pattern of neural signals, however, is required to co-contract both sets of muscles to stabilize the joint against movement in either direction. Joint angle may be established by controlling yet another movement parameter, muscle length. Since muscles are elastic, they exhibit the properties of springs. To produce different forearm positions, therefore, the neural signals could set the length of the opposing muscles, thereby producing, according to Hooke's law, the forces required to establish the correct joint angle.

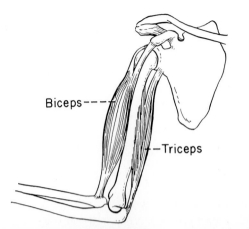

Figure 22–1 Schematic representation of the pair of antagonistic muscles (biceps and triceps) that bend the elbow.

HOW ARE MOVEMENTS CONTROLLED?[3, 5]

The control of movement is accomplished by signals generated in the central nervous system. Figure 22–2 shows the most simple scheme for the neural control of the angle of a limb joint. Movement is produced by muscular contractions con-

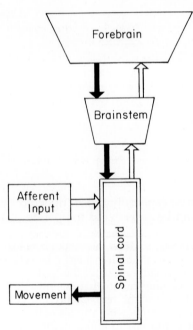

Figure 22–2 A simple schematic diagram illustrating the participants in the neural control of movement. Motor signals follow the solid arrows and sensory signals follow the open arrows. (After Wetzel, M.C.; Stuart, D.G. In *Mechanics and Energetics of Animal Locomotion*, 115–152. London, Chapman & Hall, 1977.)

trolled by a motor signal generated in the spinal cord. This spinal motor signal results from the integration of afferent signals from the limb (and other parts of the body) with motor command signals that descend from supraspinal brain structures such as the brain stem and forebrain. The descending supraspinal command signal is based, in part, on the same afferent information that influences the spinal cord directly.

In order to execute an accurate voluntary limb movement, the central nervous system takes advantage of afferent signals that reach it at many different levels. At the first level, a movement can be influenced by local cues generated in the limb that is moved. These local cues (e.g., tactile signals or signals about the state of the limb's muscles) converge with descending motor command signals onto common neuronal elements in the spinal cord to create the spinal motor signal. This interaction is schematized by the local feedback circuit shown in Figure 22–3A. At a second level, the same afferent information is delivered to the brain (Fig. 22–3A, central feedback), where it can influence the descending signal. Finally, the movement of the limb also is monitored by watching it. This third level can use afferent information from vision as well as a variety of other sensory modalities that are not shared with spinal interneurons.

Although all three levels usually act together, they can be thought of as parts of a hierarchy involved in the control of movement. Local feedback, the lowest level of the hierarchy, can have the most immediate influence upon the movement, but the raw afferent signals may not be sufficiently integrated with other afferent signals to provide anything more than a rather gross movement. Central feedback, the second level of the hierarchy, allows not only the convergence of various afferent signals but also interactions with voluntary command signals. Because this processing takes time, however, the influence of central feedback on the descending command signal is delayed. Finally, the feedback provided by other afferent modalities takes longer still, since it requires the most processing. These other afferent signals, however, may provide the most accurate control. For example, local and central feedback might allow one eventually to thread a needle or hit a nail with a hammer, but both tasks are greatly facilitated if the eyes are open.

The use of feedback provides a means by which an actual movement (or afferent signals describing the movement) can be compared with a command signal for the desired movement. The diagram of Figure 22–3A has three feedback loops. A system

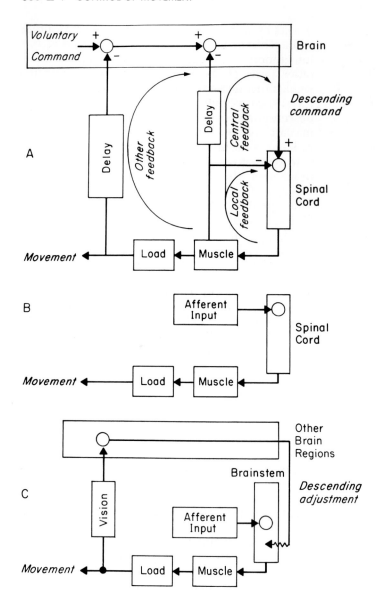

Figure 22–3 Schematic diagrams illustrating different strategies for the control of movement. *A*, A feedback system with both a local pathway and two longer pathways with delays. *B*, An open loop system with no feedback control. *C*, An open loop system in the brain stem whose elements are maintained in calibration by parametric control from other brain regions (see text for further explanation).

that uses such a control strategy is called a "closed loop" system.

Feedback generally improves the accuracy of a movement but delays, and possibly slows, its execution. Indeed, the brain often initiates and controls movement on the basis of sensory information that is distinctly out of date. In situations in which speed is paramount but the trajectory and accuracy of a movement are not, the nervous system eschews the feedback comparison of the actual with the desired movement. Instead, an afferent input always elicits a movement that can't be controlled. Such an "open-loop" system is depicted in Figure 22–3B. Some reflex-evoked movements, especially those triggered by noxious stimuli, can be considered to operate open loop.

Some reflexes, such as the vestibulo-ocular reflex (or VOR), however, are both fast *and* relatively accurate. If the head is turned suddenly, the VOR produces compensatory eye movements that minimize the slip of the visual world across the retina. Since the VOR produces these eye movements within 10 ms of the onset of head rotation, it is too fast to afford negative feedback and thus operates open loop. Nevertheless, the accuracy of the VOR must be maintained even if its neuronal elements are affected by disease and aging. An accurate VOR is maintained by a strategy called *parametric control*. Since an inappropriately adjusted VOR causes the visual world to slip across the retina during head rotation, the visual system monitors the amount of slip and adjusts the brain stem elements (or parameters) to eliminate it (Fig. 22–3C). Parametric control allows the VOR to be protected against gradual changes in its individual elements without the use of feedback control, thereby allowing the VOR to retain its rapid response (Chap. 27 gives further details).

For rapid voluntary movements, the delays inherent in feedback control can be overcome by computing the descending command signal in advance with as much accuracy as possible. Such *preprogramming* is a particularly effective strategy if the body part to be moved usually encounters a constant load, as in the rapid movements of the eyes and head. Preprogramming may also be effective when large limb movements are executed against predictable loads. Figure 22–4 shows 20° flexions of the wrist (upper record) against a spring load in a patient with a sensory neuropathy that has destroyed all of his large sensory fibers (i.e., those that transmit information about touch and

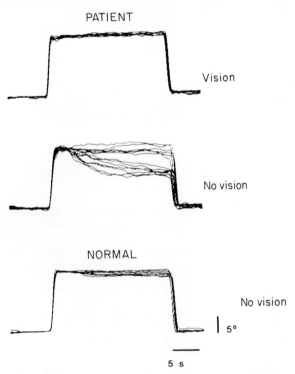

PATIENT

Vision

No vision

NORMAL

No vision

| 5°

5 s

Figure 22–4 Flexion and extensions of the wrist against an elastic load by a patient with a large-fiber sensory neuropathy and by a normal subject. These movements have a dynamic phase that flexes the wrist angle through 20° and a "hold" phase that keeps the wrist in its flexed position. The hold phase of movement made without vision is signficantly impaired in the patient but only modestly affected in the normal subject. (After Sanes, J.N., et al. *Human Neurobiol.* 4:101–114, 1985.)

the state of muscles, such as length. Even when blindfolded to remove the last remaining afferent signal about the movement, the patient made remarkably accurate initial movements to the desired position (Fig. 22–4, middle trace). When he tried to *hold* his wrist in its new position, however, he was generally unable to do so. In this patient, therefore, the motor signals for the dynamic phase of the movement appear to be preprogrammed, whereas the holding (or static) phase depends upon afferent feedback.

In addition to participating in movement, descending motor signals may also impinge upon sensory relay nuclei (e.g., in the brain stem) where they can affect the transmission of afferent information. Such a *corollary discharge* might impinge on neurons that relay afferent information, thereby providing a signal that helps them to distinguish sensory signals induced by external stimuli per se from those induced as a consequence of the movement. Corollary discharges may also contribute to the perception of muscular force or effort and may thus participate in sensations of heaviness.[4] Finally, corollary motor command signals can be used to construct an internal neural replica (or *efference copy*) of an impending movement. Such a neural copy of the movement would provide an important control signal in those motor systems in which the somatosensory feedback from muscles is inadequate (e.g., the system that controls the position of the eye).

NEURAL STRUCTURES INVOLVED IN MOVEMENT

Neural structures can be considered to deal with movement if they are involved in producing one of the various motor signals or processing the sensory signals required for accurate control. Four different experimental approaches have been used to implicate a brain structure in the control of movement. First, natural or experimentally produced lesions of a neural structure involved in movement are expected to produce deficits in some aspect of movement. Such deficits might range from a complete paralysis to an inability to suppress wild, involuntary movements (ballismus). Lesions of some structures may only affect the ability to make subtle modifications in a normal movement. Second, electrical stimulation of some putative motor structures can be expected either to produce an overt movement or to affect a movement that is in progress. Of course, the more the central nervous system is depressed by anesthesia or the further the structure is removed from the muscle involved, the less likely is its stimulation to produce an overt movement. Third, by recording from neurons in awake, moving animals, structures can be identified whose neuronal firing patterns are related to one or more of the movement parameters discussed above. As may be recalled from the discussion of corollary discharge, such neurons may or may not actually participate in motor control. Finally, anatomical techniques can be used to identify structures connected to those areas that have been identified by lesion, stimulation, and recording studies as being involved with movement. One or more of these types of evidence implicate several major subdivisions of the nervous system in the generation of movement; these regions and their interconnections are indicated schematically in Figure 22–5.

The most crucial part of the motor system is the *muscle* and its efferent neural element, the *motoneuron,* without which force could not be generated. Lesions of motoneurons produce a complete paralysis of movement, and suprathreshold electrical stimulation of the normal, intact motor nerve or its muscle invariably evokes muscle contraction. Sensory signals from the muscle and its body part

(i.e., the periphery) enter the *spinal cord* over sensory afferents, which may influence motoneurons directly or via only an additional interneuron or two. Through such "simple" circuits, restricted to the periphery and the spinal cord, stimuli that evoke reflexes can produce appropriate responses in motoneurons. The isolated spinal cord can mediate many reflexes, such as withdrawal from a noxious stimulus or the knee jerk, elicited by striking the patellar tendon below the knee. These *segmental* circuits, which provide rapid responses with minimal delays, are important for protective reflexes that do not require elaborate processing of afferent data. Furthermore, the spinal cord alone also is capable of generating the neural patterns required for many coordinated limb movements, such as those that stabilize three of a quadruped's limbs when the fourth is withdrawn or those that produce the crude sequential movements of the four limbs during locomotion (Chap. 26).

Motoneurons can be influenced not only by local signals originating in peripheral structures but also

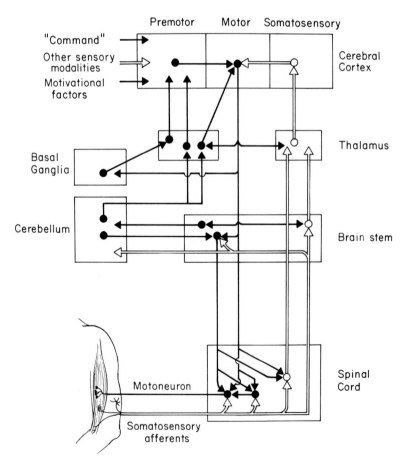

Figure 22–5 Major central nervous system structures and pathways involved in the neural control of movement. Open arrows trace sensory signals; solid arrows represent either motor or combined sensory and motor pathways.

by signals descending from the *brain stem* (Fig. 22–5). The large majority of these descending influences are mediated by the same spinal interneurons that are involved in spinal segmental reflexes (Chap. 26). Indeed, the descending signals must be tailored to make appropriate interactions with the current state of the reflex circuitry. Descending signals from the brain stem are issued from a variety of nuclei, including many that lie in a core of gray matter called the *reticular formation,* which runs the length of the brain stem from the mesencephalon to the medulla. Many of the brain stem nuclei in turn receive movement-related signals from the *sensorimotor cortex.* Even in the absence of inputs from these higher centers, however, the brain stem adds to the capabilities of the isolated spinal cord. For example, a decerebrate animal whose neuraxis has been transected at the level of the midbrain is able to right itself and make some postural adjustments. The sensory signals for these compensatory behaviors originate in the vestibular apparatus (not shown) and act through circuitry confined to the medullary brain stem. The compensatory eye movements generated by vestibular stimuli also appear to be normal. In addition, rhythmic motor behaviors such as respiration and mastication, which have their origins in the medullary and pontine reticular formations, respectively, are also intact.

The *sensorimotor cortex* plays a crucial role in generating those voluntary movements that are fashioned by both sensory input and motivational set (Chap. 28). Sensory input reaches the cortex via pathways involving the brain stem and *thalamus.* The relay in the brain stem also allows the afferent signals to influence descending motor signals (Fig. 22–5). Although most of the descending cortical signals affect the spinal cord indirectly through relays in the brain stem, some travel all the way to the spinal cord (the corticospinal system), where a few even influence motoneurons directly.

Finally, there are two movement-related structures that, because of their connections, seem strategically situated to shape motor behavior. Both the *basal ganglia* and *cerebellum* send prominent projections to the thalamus and thence to premotor and motor areas of the cerebral cortex. Both also receive major inputs from these and other cortical areas. Whereas the basal ganglia receive little, if any, direct sensory input to shape their activity, the cerebellum receives strong, short-latency sensory inputs from virtually every somatosensory receptor, as well as from the visual, vestibular, and auditory modalities. Although neither the basal ganglia (Chap. 30) nor the cerebellum (Chap. 29) is necessary for the production of *any* movement, they are both essential for postural adjustments and for ensuring that coordinated movements are smooth and accurate. The cerebellum may also be involved in the acquisition of new motor behaviors (Chap. 27).

The following chapters consider, in detail, the role of each of these structures in the generation of movement. As seen from Figure 22–5, however, all of the areas involved with movement have extensive reciprocal connections, making it impossible to attribute specific motor functions to individual structures. Nevertheless, the combined evidence from electrophysiological, anatomical, and clinical studies has led to some broad principles of motor organization that are discussed in this section on movement.

REFERENCES

1. Hogan, N. An organizing principle for a class of voluntary movements. *J. Neurosci.* 4:2745–2754, 1984.
2. Houk, J.C. Regulation of stiffness by skeletomotor reflexes. *Annu. Rev. Physiol.* 41:99–114, 1979.
3. Houk, J.C.; Rymer, W.Z. Neural control of muscle length and tension. *Handbk. Physiol.* Sec. 1, 2:257–323, 1981.
4. McCloskey, D.I. Corollary discharges: motor commands and perception. *Handbk. Physiol.* Sec. 1, 2:1415–1447, 1981.
5. Rack, P.M.H. Limitations of somatosensory feedback in control of posture and movement. *Handbk. Physiol.* Sec. 1, 2:229–256, 1981.
6. Sanes, J.N.; Mauritz, K.-M.; Dalakas, M.C.; Evarts, E.V. Motor control in humans with large-fiber sensory neuropathy. *Human Neurobiol.* 4:101–114, 1985.
7. Stein, R.B. What muscle variable(s) does the nervous system control in limb movements. *Behav. Brain Sci.* 5:535–577, 1982.
8. Wetzel, M.C.; Stuart, D.G. Activation and co-ordination of vertebrate locomotion, 115–152. In Alexander, R.M.; Goldspin, K.G., eds. *Mechanics and Energetics of Animal Locomotion.* London, Chapman and Hall, 1977.

Properties of Motor Units

In the preceding discussions of skeletal muscle physiology (Chap. 9), individual muscles were considered to be homogeneous aggregations of muscle fibers that produce force when they are electrically activated. We have seen that the rate of activation, the load, and the initial muscle position or length affect a muscle's capacity to generate forces. However, we have not as yet considered in detail how the nervous system controls muscles, nor have we considered the specializations of muscles that permit them to meet the varied demands of usage that are imposed upon them. In this chapter and those that follow, it will become clear that muscles are not just simple "motors" but rather are extremely refined and specialized machines that are controlled through the continuous flow of neural signals both to and from the central nervous system.

MECHANICAL DEMANDS IMPOSED ON MUSCLES

The primary function of skeletal muscle is to contract and thereby either produce movement about a joint ("shortening contractions"), "brake" and "dampen" movement about a joint ("lengthening contractions"), or brace and stabilize a joint

without producing movement ("isometric contractions"). In all cases, the muscles involved must be able to perform these functions at different speeds, at different levels of force, with a high degree of precision, and often for extended periods of time. Muscles acting around the human knee, for example, must be capable of stabilizing that joint to maintain upright posture in standing as well as flexing and extending the leg for walking, running, and jumping. These distinct tasks require the involved muscles to operate at very different speeds and to produce a wide range of forces.

It appears that the process of natural selection failed to develop a single type of skeletal muscle or muscle fiber endowed with all of the contractile characteristics required for the different demands just described. Instead, different types of muscle fibers have evolved that are specialized with respect to their mechanical properties. While muscles are still often classified as "fast" and "slow" or "red" and "white," based upon their having a preponderance of one "type" of muscle fiber (Chap. 9), with but few exceptions individual muscles are actually composed of several different fiber types. This design strategy permits a single muscle to participate effectively in a wide range of motor tasks.

MUSCLE FIBER TYPES

The various types of muscle fibers can be differentiated on the basis of their biochemical, mechanical, and morphological properties. Histochemical techniques have been particularly useful in revealing biochemical differences in muscle fiber proteins and their metabolic activities. Figure 23–1 shows the "histochemical profiles" of groups of contiguous muscle fibers from the lateral gastrocnemius and soleus muscles of the cat. The use of serial reconstructions allows one to trace individual fibers through many sections based upon their shapes and those of their neighbors. The identifi-

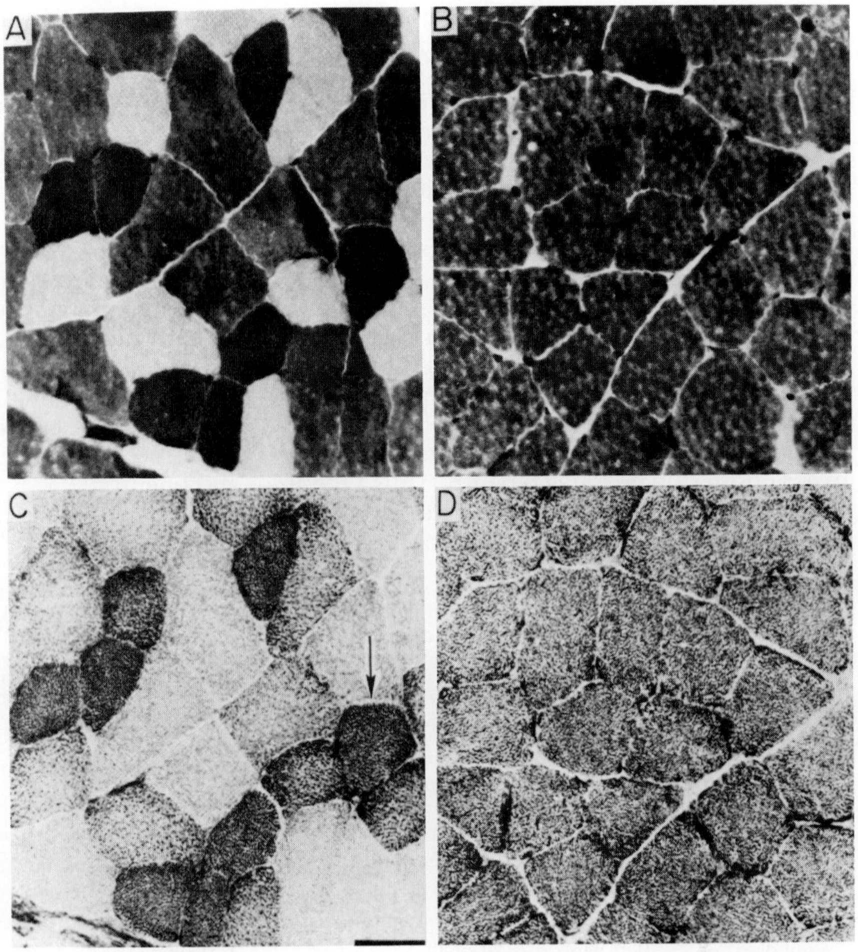

Figure 23–1 Histochemical profiles of muscle fibers in the heterogeneous lateral gastrocnemius (A,C) and homogeneous soleus (B,D) muscles of the cat hindlimb. A and B show sections stained for myofibrillar ATPase activity after incubation in acidic buffer (pH 4.65). C and D are sections stained for the oxidative enzyme NADH-D (cf. Table 24–1). Calibration bar indicates 100μ. (From Burke, R. E. Handbk. Physiol. Sec. 1, 2(1):345–422, 1981.)

cations are further aided by following the capillaries that appear as dark dots around the individual muscle fibers. In the four sections, the intensity of the staining within each fiber reflects the level of activity of a particular enzyme system or the concentration of a particular metabolite. The fibers of the cat soleus muscle, an unusual limb muscle composed of a single fiber type, are uniformly stained for both myosin ATPase activity (Fig. 23–1B) and the oxidative enzyme NADH dehydrogenase (Fig. 23–1D). The fibers of the lateral gastrocnemius, in contrast, exhibit a variability in staining for both enzymes that is more typical of mammalian muscles (Fig. 23–1,A and C).

Although these and other histochemical methods are notoriously capricious and a number of fibers with intermediate histochemical profiles have been described, most investigators agree that mammalian skeletal muscles are composed of three basic muscle fiber types: type I, type IIA, and type IIB.[2, 5] In addition to the distinguishing histochemical features, the three fiber types also exhibit morphological differences, although these are not as consistent across mammalian species.

The differences between type I fibers and either of the major groups of type II fibers are the most striking. The myofibrillar ATPase activity of type I fibers is much lower than that of any of the type II fibers in the same muscle, and the myosin present in type I fibers is immunologically distinct from that in type II fibers. The type I fibers are nearly completely dependent on aerobic metabolism as a source of ATP, whereas the type II fibers can operate anaerobically. Morphologically, type I fibers generally exhibit wider Z lines, more mitochondria, and less extensive sarcoplasmic reticula and t-tubular systems than do type II fibers.

The distinctions between the two groups of type II fibers are less pronounced, and some investigators have preferred to consider type II fibers as a single population that displays a wide range of properties. However, immunocytochemical analyses indicate that the distribution of myosin light chains differs significantly between types IIA and IIB.[4] This finding coupled with other histochemical and biochemical data summarized in Table 23–1 suggests that IIA and IIB fibers should probably be considered distinct types. The type IIA fibers have several features in common with both type I and type IIB fibers. Their similarities with type I fibers include the capacity to generate ATP aerobically, a rich capillary supply, and an abundance of myoglobin. Type IIA fibers are similar to the type IIB fibers with respect to their rich store of glycogen, high levels of myosin ATPase activity, and endowment of glycolytic enzymes. Further, as discussed later, several of the physiological properties of type IIA fibers are intermediate between those of type I and type IIB. Thus type IIA fibers represent perhaps the best compromise in muscle fiber design.

Table 23–1 Profiles of Predominant Muscle Fiber Types

Fiber Types	I	IIA	IIB
Histochemical features*			
Myofibril ATPase (pH 9.4)	Low	High	High
NADH dehydrogenase	High	Medium-high	Low
Succinic dehydrogenase	High	Medium-high	Low
Men.-α-GPD	Low	High	High
Glycogen	Low	High	High
Phosphorylase	Low	High	High
Neutral fat	High	Medium	Low
Biochemical features†			
AM ATPase, μmol/min/mg (pH 9.4)	0.04	0.16	0.27
Lactate dehydrogenase, μmol/min/g	105	220	450
Succinic dehydrogenase, μmol/min/g	2.0	2.5	0.7
Hexokinase, nmol/min/g	980	620	300
Myoglobin, mg/g	1.4	1.4	0.3
Morphology*			
Diameter	Small	Medium	Large
Capillary supply	Rich	Rich	Sparse
Z lines	Wide	Narrow	Narrow
Mitochondria	Rich	Moderate	Sparse
SR and t-tubular system	Least	More	Most

After Burke, R.E. *Handbk. Physiol.* Sec. 1, 2(1):345–422, 1981.
*Cat.
†Guinea pig.
Am = actomyosin; Men.-α-GPD = menadione-linked α-glycerophosphate dehydrogenase; SR = sarcoplasmic reticulum.

The major distinguishing histochemical, biochemical, and morphological features of the three fiber types are summarized in Table 23–1. The tripartite classification scheme presented here is valid for most skeletal muscles in a wide variety of mammalian species. However, each fiber type within a muscle exhibits a considerable range of properties. Moreover, the distinguishing properties of a fiber type vary somewhat in different animal species. It has been proposed that these variations may reflect both the unique functions of each muscle and the trophic influences exerted on muscle fibers by the motoneurons that innervate them. In this light, it is interesting to note that the closest "matching" of muscle fiber properties occurs among those innervated by the same motoneuron.

THE MOTOR UNIT

A motoneuron and the group of muscle fibers innervated by the branches of its axon are referred to as a *motor unit*.[9] In the skeletal muscles of adult mammals, each muscle fiber has but a single motor end-plate (Chap. 6) and is thus under the exclusive control of one motoneuron. Because the neuromuscular junction is normally an obligatory synapse, an action potential in a motoneuron reliably activates all of the fibers it innervates nearly synchronously. Consequently, the contraction of a motor unit constitutes the minimal and quantal element of motor output.

Although it is true that the motor unit functions as an indivisible quantal element, both the sizes and forces of these quanta vary widely both within individual muscles and across different muscles. As a general rule, the number of muscle fibers innervated by a single motoneuron, termed the innervation ratio, tends to be lower (i.e., fewer fibers/motoneuron) in muscles involved in skilled movements requiring fine control (e.g., hand muscles) than in muscles involved in more forceful and less precise movements (e.g., thigh muscles). For example, the human tibialis anterior muscle, an ankle flexor, is composed of approximately 270,000 muscle fibers that are innervated by some 450 motoneurons. Thus the average number of muscle fibers per motoneuron in this muscle is 600. In contrast, motor units of the intrinsic hand muscles are composed of about 100 fibers. Some of the smallest motor units are found in the extraocular muscles that control eye movements, which may have as few as ten muscle fibers per motoneuron, and in the muscles that control the actions of the ossicles of the inner ear, where each motoneuron may innervate only a single muscle fiber.

Neural and Mechanical Properties of Motor Units. Individual motor units differ not only in their innervation ratios, but also in a variety of neural and mechanical properties. Motor unit properties are studied in anesthetized animals either by stimulating single motoneurons with an intracellular microelectrode or by stimulating individual motor axons from subdivided ventral root filaments. In human subjects, motor units can be studied by stimulating motor axons within a muscle or by recording the electromyogram produced by individual motor units during voluntary contractions. In all cases, one measures the force produced by the motor unit's contraction, using some type of force transducer attached either to the parent muscle's tendon or to the appendage affected by the muscle's action. What emerges from such studies is that the different motor units in a single muscle display a remarkable range in the amount of force that they produce, the speed with which they contract, the conduction velocities of their motor axons, and their capacities for maintaining force output during sustained activity.

The mechanical properties of three representative motor units from a hindlimb muscle of a cat are presented in Figure 23–2. The motor unit labeled 1 produced the least amount of twitch and tetanic tension, contracted more slowly than the other two units and its motor axon had the lowest conduction velocity. At the other extreme, the unit labeled 3 produced nearly five times more tension than unit 1, contracted much more rapidly, and its motor axon had the highest conduction velocity of the three units. The values for the same properties of unit 2 lie between those of 1 and 3. These examples illustrate not only that force, contraction time, and axonal conduction velocity have a wide range in values for motor units of a single muscle but also that these parameters covary in a systematic manner. This systematic variation of motor unit neural and mechanical properties is illustrated in Figure 23–3, which examines a representative sample of motor units from the cat gastrocnemius muscles. In general, the units producing the smallest tensions contract relatively slowly and are innervated by relatively slowly conducting motor axons, whereas those units producing the largest tensions have the fastest contraction speed, and their motor axons have relatively higher conduction velocities. It is also clear that motor units that produce small forces generally greatly outnumber those that produce large forces.

Over the last 25 years, the mechanical properties of the motor units in many different muscles in several different species, including human, have been analyzed.[2] In nearly all of these studies, relationships between the various mechanical properties of motor units similar to those described above have been reported. Based largely on these relationships, motor units have been divided into distinct types. The typing of motor units is analogous to that described for muscle fibers, except that the criteria are derived from mechanical rather than chemical analyses.

The standard battery of physiological tests used to categorize motor units include those described (i.e., twitch tension, twitch contraction time, and axonal conduction velocity) and several others in which the mechanical responses of the units to repetitive stimuli are examined. The most discriminative properties for "unit typing" are tetanic tension, fatigue resistance, and the profile of the force records that units produce when stimulated at rates below their tetanic fusion frequency.[3] Measurements of these latter properties are included in the three representative motor unit profiles shown in Figure 23–4.

Of the 300 motor units composing the cat medial gastrocnemius muscle, approximately 25% have neural and mechanical properties similar to those of the motor unit shown on the left in Figure 23–4. Such motor units produce very small twitch tensions (average = 4 mN) and exhibit relatively

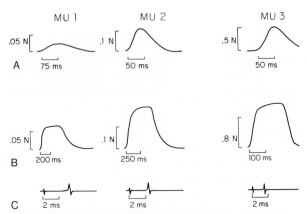

Figure 23–2 Properties of three different motor units from a single cat hindlimb muscle. *A,* Single electrical stimulus delivered to each of the isolated motor axons produces motor unit twitch contractions. *B,* The same three motor units were stimulated at their tetanic fusion frequencies. *C,* Electromyograms recorded in the muscle following a single electrical stimulus delivered to the individual motor axons. Notice that the motor unit at the left, which contracted more slowly than the other two units, also produced the least tension *(A and B)* and had the slowest axonal conduction velocity *(C).*

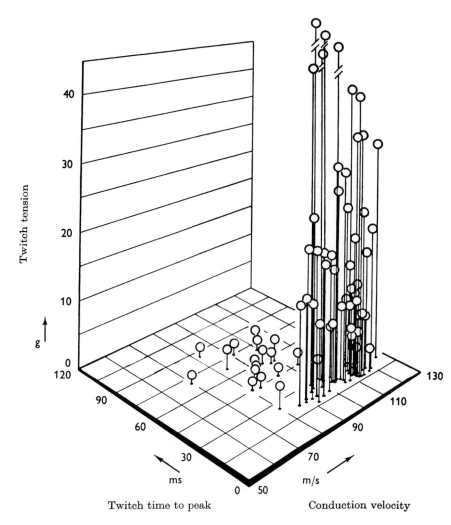

Figure 23–3 Properties of a representative sample of cat gastrocnemius motor units. Notice the wide range in the values for each of these properties. (After Burke, R. E. *J. Physiol. [Lond.]* 193:141–160, 1967.)

slow-twitch contraction times (> 50 ms). Hence, these units are called type S, or slow twitch. Since their tetanic tensions average about 40 mN, the total population of type S units accounts for only about 5% of the maximum tetanic force produced by the medial gastrocnemius muscle. However, when type S units are subjected to repetitive tetanic stimulation they can maintain their peak force output for long periods without any appreciable decrement.

Motor units with very different mechanical properties account for approximately 45% of the medial gastrocnemius population. As illustrated by the representative motor unit in the right hand column of Figure 23–4, units of this type generally contract more rapidly (average contraction time = 24 ms) than type S units and are quite powerful (average twitch tension = 80 mN; average tetanic tension = 640 mN). However, these units fatigue rapidly in that after only 2 minutes of intermittent

tetanic stimulation their force outputs are reduced by more than 75%. Consequently, these units are called type FF, for fast twitch, fatigable. Another characteristic of type FF units is the "sagging" appearance of their force output records during stimulation at rates below their tetanic fusion frequency. This "sag" in force output is not seen in the type S units described above (cf. Fig. 23–4B). Type FF motor units account for about 75% of the muscle's maximum force.

The third major class of motor units, comprising approximately 20% of the population in medial gastrocnemius, is called type FR (for fast twitch, fatigue resistant). These units, as shown by the example in the middle column of Figure 23–4, are relatively rapidly contracting (average contraction time = 25 ms), produce moderate forces (average twitch tension = 3 mN; average tetanic tension = 150 mN), and show relatively little fatigue, maintaining more than 75% of their

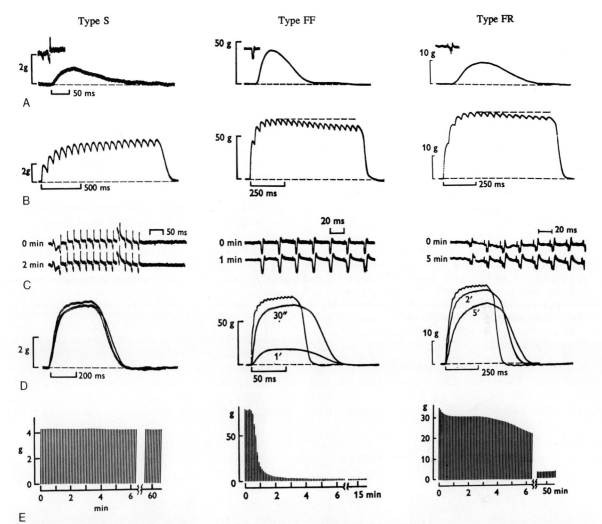

Figure 23–4 Mechanical and electrical responses of the three major types of motor units in the cat lateral gastrocnemius muscle: type S, type FF, and type FR. For each motor unit, *A* shows the twitch response with the electromyogram of the unit reproduced above the mechanical response and *B* shows the unfused tetanus. *C, D,* and *E* document the motor units' fatigue resistance. The motor units received short duration (330 ms) tetanic stimulation (40 stimuli/s) at a rate of 1/s for up to 60 min. *C* displays the electromyograms recorded during the tetani to show that neuromuscular transmission was not impaired by the stimulation regimen. *D* shows the tetanic force records selected at different times during this prolonged fatigue test. *E* plots the magnitude of the tetanic force as a function of time during the fatigue test (After Burke, R.E., et al. *J. Physiol. [Lond.]* 234:723–748, 1973.)

original force outputs in the standard fatigue test. The profile of unfused tetanus in type FR units features the same type of "sag" observed in type FF units.

The properties of these three major classes of motor units, which account for about 95% of those in the cat medial gastrocnemius muscle, are summarized in Table 23–2. The remaining 5% of the motor units in the medial gastrocnemius muscle display a range of mechanical properties that are intermediate between those of FF and those of FR motor units. They are relatively rapidly contracting

(average contraction time = 25 ms), display a "sag" in their profile of unfused tetanus, and are primarily distinguished by their fatigue resistance, which lies between that of FF and FR units (force is maintained at > 25% but < 75% of maximum initial force). This population of units is called FI (for fast twitch, intermediate fatigability).

Although the practice of classifying motor units into types has a number of advantages in comparative analyses, it should be done with caution. It is important to note that the values for each of the mechanical properties are distributed as a continuum within a group of

Table 23–2 Characteristics of Predominant Motor Unit Types in Cat Medial Gastrocnemius Muscle

Motor Unit Type	S	FR	FF
Muscle fiber type	I	IIA	IIB
Muscle fiber size (cross-sectional area, μm^2)	1980	2370	5290
Tetanic tension (mN)	10	150	640
Twitch contraction time (ms)	50	25	24
Profile of unfused tetanus	nonsagging	sagging	sagging
Fatigue index*	1	>0.75	<0.25
Innervation ratio	550	500	700
Specific tension (N/cm²)†	6	22	24

After Burke, R.E. *Handbk. Physiol.* Sec. 1, 2(1):345–422, 1981.
*Force after 2 min stimulation/initial force.
†Tetanic tension normalized to muscle fiber cross-sectional area.

motor units and that there is considerable "overlap" in the mechanical properties of the different unit types. Thus one cannot make motor unit type distinctions for a population of motor units based upon any single parameter alone; rather, several properties must be considered concurrently. Moreover, there are unique qualities associated with virtually every muscle's population of motor units that might be obscured by the practice of unit typing.

Histochemical Profiles of Motor Units. We saw earlier (Fig. 23–1) that like motor units, muscle fibers can also be classified into three different types. The parallelism in the two classification schemes prompts the obvious question of whether different motor unit and muscle fiber types are related. The range in myosin ATPase activities among different muscle fibers certainly presages the differences in twitch contraction times observed between type S and type F motor units. Similarly, comparisons of the oxidative and glycolytic enzyme reactivities of different fibers within a muscle clearly suggest that some (type I) appear well adapted for sustained activity, while others (type IIB) would be expected to fatigue rapidly when challenged with repetitive activation.

In order to examine the correlations between the properties of a motor unit and those of its constituent muscle fibers, one must first perform a complete analysis of the mechanical properties of a motor unit and then subject it to a histochemical analysis. The major difficulty in such an experiment lies in identifying the motor unit's fibers within the muscle. This identification can be accomplished by subjecting the motor unit to repetitive stimulation for up to an hour to deplete its muscle fibers of their glycogen store. The "glycogen-depleted" fibers can then be identified from muscle cross sections stained for the presence of glycogen. Figure 23–5 shows serial sections from the cat lateral gastrocnemius muscle in which two

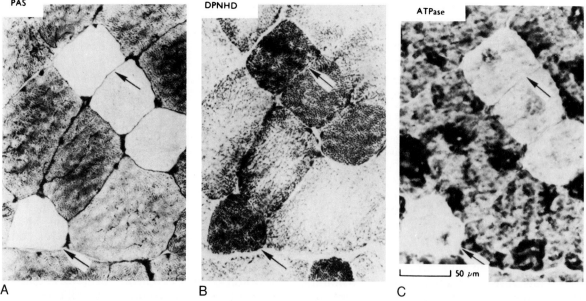

PAS   DPNHD   ATPase

A   B   C

Figure 23–5 Serial sections from a cat lateral gastrocnemius muscle in which a single unit was stimulated tetanically for a prolonged period. In A, the section has been stained to reveal the presence of glycogen. Notice the two "white" fibers that are devoid of any reaction product; these are two constituents of the motor unit that was stimulated. In B and C, the sections have been stained to reveal the levels of oxidative enzyme activity and myosin ATPase activity, respectively. The two fibers that belong to the stimulated motor unit (A) can be identified in these sections on the basis of their shapes and those of the neighboring fibers (After Burke, RE., et al. *J. Physiol [Lond.]* 234:723–748, 1973.)

fibers from a single motor unit, devoid of any glycogen reaction product, can be easily identified (Fig. 23–5A). The same glycogen-depleted fibers can be recognized in the subsequent sections that were stained for oxidative enzymes (Fig. 23–5B) and myosin ATPase activity (Fig. 23–5C). Not only does this type of experiment confirm that each motor unit is composed of a population of homogeneous muscle fibers, but it reveals several important correlations between the mechanical and biochemical properties of muscle fibers and yields new information on muscle unit anatomy.

Type S motor units are composed exclusively of type I muscle fibers. In medial gastrocnemius, as in most other cat muscles, type I fibers have small diameters, a characteristic that accounts in part (see later) for the small forces they produce. Type I fibers also have relatively low myofibrillar ATPase activity, a feature that is correlated with the relatively slow contraction speeds of these motor units. Finally, the high levels of oxidative enzyme activity and rich capillary supply probably account for their remarkable fatigue resistance.

A similar "match" is found between the mechanical characteristics of type FF motor units and the histochemical profiles of their constituent muscle fibers. FF motor units are composed exclusively of type IIB fibers. In cat hindlimb muscles, IIB fibers have large diameters, high levels of myofibrillar ATPase and glycolytic enzyme activity, but low levels of oxidative enzyme activity and a sparse supply of capillaries. These histochemical features are presumably related to the high force outputs, fast contraction speed and low fatigue resistance of this type of unit, respectively.

The third major type of motor unit, the FR unit, is composed of type IIA muscle fibers. These fibers have intermediate diameters and a rich capillary supply. Their high level of myofibrillar ATPase activity is consistent with their relatively fast contraction times. However, they have far greater levels of oxidative enzyme activity than do type IIB fibers, which presumably accounts for their greater fatigue resistance.

Histochemical-Mechanical Correlations in Muscle Fibers. It must be emphasized that the associations between certain mechanical properties of motor units and the histochemical properties of their muscle fibers are correlative but not necessarily causal. For example, twitch contraction time is determined by many factors (e.g, fiber length, E-C coupling kinetics, rate of cross-bridge formation, and the like) in addition to the speed of myofibrillar shortening. Thus one would not expect that the four- to fivefold range of contraction times in the motor unit population of a typical heterogeneous muscle could be ascribed to any single factor. Nonetheless, there is strong evidence that (i) the speed of sarcomere shortening is strongly correlated with the rate of ATP hydrolysis by myosin (cf. Chap. 8), and (ii) the qualitative assessment of myosin ATPase activity revealed by histochemistry is well correlated with results derived from quantitative biochemical analysis (cf. Table 23–1). Thus in general, myofibrillar ATPase histochemistry is a useful indicant of motor unit contraction speed.

Also implicit in the descriptions of the histochemical profiles of motor unit types is the suggestion that oxidative enzyme activity and fatigue resistance are correlated. Again, motor unit fatigue is a complex phenomenon involving several distinct processes, including alterations in the efficacies of neuromuscular transmission, muscle fiber action potential propagation, E-C coupling, and cross-bridge cycling. It is likely that all of these factors contribute to the differences in the fatigue resistance of motor units, but the relative importance of each is difficult to assess. It is quite clear, however, that within an individual muscle, the fatigue resistance of a motor unit is well correlated with the oxidative enzyme machinery of its constituent fibers.

Motor Unit Anatomy and Specific Tension. The "glycogen-depletion" technique described earlier, which permits a histochemical analysis of the fibers constituting a single motor unit, also can be used to reveal the distribution of the fibers of a single motor unit within the muscle. As shown in Figure 23–5, the fibers of a motor unit are only rarely contiguous, but rather are generally extensively intermingled with the fibers of many other motor units. The volumes or "territories" of a muscle containing the fibers of a single unit can be quite extensive, ranging from 15% to 30% of the entire muscle volume. However, within a motor unit territory, the density of fibers belonging to that unit is quite low, typically less than 5 per 100 fibers.

One can also count the number of glycogen-depleted fibers to obtain the innervation ratio and measure the mean diameter of the fibers to estimate the specific tension (force per unit area) of the motor unit's fibers. These and other less direct analyses performed in several different muscles of the cat hindlimb lead to the surprising conclusion that the force produced by a motor unit and its innervation ratio are not strongly correlated (cf. Table 23–2). Nor does it appear that the larger, faster-conducting motor axons innervate proportionately more muscle fibers than the smaller, slower-conducting ones. For example, the average innervation ratios of types S and FR motor units are nearly the same, whereas their average tetanic tensions differ by a factor of 15. Thus, it seems that the wide range in motor unit tetanic forces can be attributed mainly to systematic differences in muscle fiber diameter and in their specific tensions.

As shown in Table 23–2, the specific tensions of type I fibers are three to four times lower than those of type II fibers in the same heterogeneous (i.e., containing different fiber types; "mixed") muscle. This is a puzzling finding because the density of myofibrils does not appear to be lower in type I fibers, nor do the type I fibers in the homogenous soleus muscle show such low values. Moreover, in vitro analyses of single, type-identified muscle fibers from the cat reveal no differences in the specific tensions of different fiber types.[10] Thus, one can only speculate that perhaps some systematic difference in the mechanical attachments of type I fibers in mixed muscles reduces the efficacy of their force generation, or that perhaps type I fibers in mixed muscles are not fully activated even during tetanic stimulation.[10]

Correlative Motoneuron Properties. In addition to the correlations between the mechanical and histochemical properties of motor units described in the previous sections, there is a substantial body of evidence that many of the intrinsic properties of motoneurons also vary systematically with the properties of the muscle fibers they innervate.[18] Although the resting potentials and action potential thresholds do not appear to vary systematically among motoneurons innervating a single muscle,[13] most other conventionally measured properties (e.g., input resistance, rheobase, membrane time constant, and the duration of the spike afterhyperpolarization potential) do show significant differences between motoneurons innervating type F (FF, FR, and FI groups) motor units and those innervating type S motor units. Moreover, in many instances the differences between the motoneurons innervating type FF and type FR motor units are also significant (Table 23–3). These differences in the basic electrical properties of motoneurons influence the sequence of motor unit recruitment and the patterns of motor unit utilization, topics that will be deferred to a later chapter on the functional organization of motoneuron pools (Chap. 25).

Recently, histochemical procedures analogous to those employed to characterize muscle fiber types have been applied to motoneurons.[15] Comparisons were made of the relative oxidative enzyme activities (nicotinamide-adenine dinucleotide-diaphorase, NADH-D) of motoneurons innervating three different muscles in the rat, selected on the basis of the differences in their fiber-type composition. The mean level of NADH-D activity in motoneurons innervating the rat soleus muscle, which is composed predominantly (85%) of type I fibers, was significantly greater than that in motoneurons innervating the tibialis anterior muscle, which is composed of type II fibers with a preponderance of the type IIA group. Finally, the lowest mean level of NADH-D activity was found in motoneurons innervating the tensor fasciae latae muscle, which is almost exclusively (95%) composed of type IIB fibers. These data suggest that the relative levels of motoneuronal oxidative metabolic activity match those of the muscle fibers they innervate.

The morphological features of motoneurons also differ with respect to the type of muscle fibers they innervate. The average soma diameters, axon diameters, number of axon collaterals, extent of dendritic arborization, and total neuron surface areas in general are greater for motoneurons innervating type F motor units than for those innervating type S motor units (Fig. 23–6). The differences, however, are often small, and the values for the different groups overlap considerably as shown for the histogram of motoneuron soma diameters in Fig. 23–6.

Motor Unit Composition and Muscle Function. We have seen that the problem of using one muscle to accomplish a variety of tasks that may

Table 23–3 Properties of Motoneurons Innervating Different Motor Unit Types

Motor Unit Type	S	FR	FI*	FF
Motoneuron size (membrane area, μm^2)†	25×10^4	32×10^4		37×10^4
Motor axon conduction velocity (m/s)	86	100	104	99
Input resistance (MΩ)	1.6	0.9	0.7	0.6
Threshold current (rheobase, nA)	5	12	17	21
Membrane time constant (ms)	10.4	8.0	5.3	5.9
Afterhyperpolarization duration (ms)	161	78	63	65

After Zengel, J.E., et al. *J. Neurophysiol.* 53:1323–1344, 1985.
*Fast twitch with intermediate fatigability.
†Data from Burke, R.E., et al., *J. Comp. Neurol.* 209:17–28, 1982.

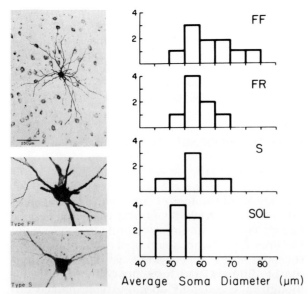

4
2
FF

4
2
FR

4
2
S

4
2
SOL

40   50   60   70   80

Average Soma Diameter (μm)

Figure 23–6 The morphology of motoneurons revealed by injecting horseradish peroxidase into their somata after identification of their motor unit type. Left, Montage photomicrographs of medial gastrocnemius motoneurons innervating different types of motor units. Right, Histograms of soma diameter for type-identified motoneurons of the cat gastrocnemius and soleus (SOL) muscles. (After Burke, R.E. *Handbk. Physiol, Sec. 1, 2(1):345–422, 1981.*)

differ in terms of speed, force, and duration has been solved in part by providing that muscle with a variety of motor units with different mechanical and biochemical properties. Type S motor units, which develop small forces and contract relatively slowly, are nonfatigable and thus are ideally suited for sustained muscular activity. Activating type FR motor units allows a muscle to produce larger forces more rapidly and yet still be quite resistant to fatigue. Finally, type FF units can be reserved for the most rapid and powerful actions that need be sustained for only brief periods. As was mentioned earlier in this chapter, the representation of the three basic motor unit types within the different skeletal muscles of the body varies widely. A few muscles are composed nearly exclusively of a single motor unit type, whereas some are composed of nearly equal proportions of all three unit types. As one would expect, the motor unit composition of a muscle generally reflects its usage. Those muscles involved primarily in sustained activity are composed predominantly of type S and type FR units, which are resistant to fatigue. Muscles used in powerful actions of short duration tend to have a greater proportion of type FF motor units.

The correlations between motor unit composition and muscle usage are well illustrated by comparing two of the cat's ankle extensor muscles, soleus and medial gastrocnemius. The soleus muscle is composed almost exclusively of type S motor units. Electromyographic and muscle force recordings from intact, freely moving animals show that the soleus muscle is often fully activated when the cat is just standing.[16] Thus when the animal is walking and jumping, there is little modulation of the force output of soleus. In contrast, only 25% of the motor units in the medial gastrocnemius are type S; correspondingly, this muscle shows only minimal activation while the cat is standing. However, the medial gastrocnemius becomes more active during locomotion, especially as an animal proceeds from walking to galloping. Finally, the medial gastrocnemius muscle becomes fully activated only during jumping, an activity that engages all of the FF motor units.

ADAPTATIONS OF MUSCLE FIBERS TO THE DEMANDS OF USAGE

In the interest of narrative expedience, motor unit and muscle fiber properties have been described as if they were static and immutable. In reality, muscle is perhaps the most adaptive of the body's tissues, responding to the changing demands placed upon it throughout the entire life cycle. The types of changes that occur in muscle fibers include alterations in morphology, metabolic activity, and under extreme conditions, even in the composition of myosin. Two distinct experimental strategies have been employed to study muscle fiber mutability. In the first, alterations in usage have been produced within the normal physiological range by physical conditioning such as endurance training or weight lifting. In the second, nonphysiological alterations in usage have been induced by denervation or chronic electrical stimulation. As is often the case in biological research, the inferences drawn from the experimental results using these two approaches have been quite different.

The predominant result of endurance training (such as long distance running) in both man and experimental animals is a dramatic increase in the oxidative metabolic capacity and fatigue resistance of all muscle fibers in the participating muscle groups. There is also a modest increase in fiber diameter, which is generally greater in the type I than in the type II fibers. There is little effect on the anaerobic metabolic machinery of the fibers, although some investigators report increased glycogen content. Most importantly, there is little reliable evidence for any conversion of type II to type I fibers with endurance exercise.[14]

Interval training with high force levels, such as in weight lifting, produces little change in the histochemical profiles of the involved fibers but results in marked fiber hypertrophy. Again, there is no evidence of significant conversions of one type of fiber to another.

If the level of muscle activity is dramatically reduced, as when a limb is immobilized, a marked atrophy results, particularly in type I and IIA muscle fibers. A significant loss of strength accompanies the atrophy both in experimental animals and human subjects, but again no significant interconversions of fiber type are induced.

The effects of nonphysiological alterations in muscle usage on muscle fiber properties are far more pronounced.[2] Chronic, low-frequency (10 stimuli/second) electrical stimulation of muscles has been used to mimic the normal activation pattern that occurs in muscles composed mainly of type I muscle fibers (e.g., cat soleus). The application of such stimuli to muscles composed primarily of type II fibers over a period of weeks to several months produces increased activity levels in oxidative enzymes, an accompanying increased fatigue resistance, and decreased fiber diameters. The most significance change, however, is the appearance of myosin with light-chain components characteristic of type I fibers, with an accompanying prolongation of contraction times. Taken together, these results constitute evidence that type II fibers have been converted to type I. The application of the same stimulation regimen to muscles composed primarily of type I fibers

produces little if any effect. It has also been reported that stimulating muscle nerves with short, high-frequency bursts to mimic the activation of type II fibers can produce increased glycolytic metabolism in muscle fibers, but the effects are less pronounced than those resulting from chronic low-frequency stimulation.

DEVELOPMENT OF MUSCLE FIBER TYPES

In addition to the strong influence that the levels and patterns of muscle usage exert on the morphological, physiological, and histochemical characteristics of muscle fibers, it is clear that genetic factors also play an important role in determining the fiber-type composition of skeletal muscles. In several species, type I and type II muscle fibers can be differentiated histochemically at birth, although they generally do not exhibit marked differences in physiological properties for several weeks postnatally. It has also been shown that the fiber compositions of muscles in monozygotic human twins are strikingly similar, much more so than in nonrelated subjects or even in dizygotic twins.[7]

The time course of histochemical differentiation of muscle fibers parallels that of the loss of polyneuronal innervation and the maturation of motoneurons. These data have led to the hypothesis that motoneurons "specify" muscle fiber types. Strong support for this notion comes from experiments in which the normal innervation of muscle is disrupted and later reestablished with either the original or a foreign nerve. Within several weeks

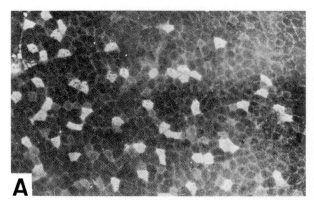 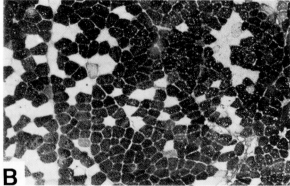

Figure 23–7 Comparison of the distribution of muscle fibers of individual motor units in normal and reinnervated rat tibialis anterior muscle. The muscle sections have been stained for the presence of glycogen following prolonged tetanic stimulation of a motor axon. *A,* Normal pattern of muscle fibers of a single motor unit. *B,* Pattern of muscle fiber distribution of a single motor unit following reinnervation of the muscle after ventral root transection. Notice the grouping of the motor unit's muscle fibers in the reinnervated muscle. (From Kugelberg, E., et al. *J. Neurol. Neurosurg. Psychiatry* 33:319–329, 1970.)

after innervation is restored, there is a profound change in the distribution of fiber types within the muscle. The normal pattern of extensive intermingling of fiber types is replaced by one featuring "type grouping," in which there are small aggregations of contiguous fibers of the same type. Studies employing the glycogen-depletion technique show that single motoneurons often innervate these contiguous fibers, suggesting that in the reinnervation process, motor axon branches remain close to the parent axon and innervate nearby muscle fibers (Fig. 23–7). Thus some of these contiguous muscle fibers appear to have been transformed to a different type by the reinnervating motoneuron. Similar conclusions have been reached from experiments in which a muscle composed predominantly of type I fibers is reinnervated by the nerve supplying a muscle composed predominantly of type II fibers and vice versa.[8]

ANNOTATED BIBLIOGRAPHY

Burke, R.E. Motor units: anatomy, physiology, and functional organization. *Handbook of Physiology. The Nervous System.* Sec. 1, 2(1):345–422, 1981.
Dr. Burke's scholarly review of motor units is the most comprehensive available. The reference section is particularly valuable for students interested in pursuing research topics in this field.
Henneman, E. Skeletal muscle: the servant of the nervous system. In Mountcastle, V. B., ed. *Medical Physiology,* 14th ed., vol. 1, St. Louis, C.V. Mosby, 1980.
This chapter by Professor Henneman provides a more extensive presentation of the material covered here as well as his insights on the functional organization and design of skeletal muscle.

REFERENCES

1. Burke, R.E. Motor unit types of cat triceps surae muscles. *J. Physiol. (Lond.)* 193:141–160, 1967.
2. Burke, R.E. Motor units: anatomy, physiology, and functional organization. *Handbk. Physiol.* Sec. 1, 2(1):345–422, 1981.
3. Burke, R.E.; Levin, D.N.; Tsairis, P.; Zajac, F.E. Physiological types and histochemical profiles in motor units of the cat gastrocnemius. *J. Physiol. (Lond.)* 234:723–748, 1973.
4. Gauthier, G.F.; Lowey, S. Distribution of myosin isoenzymes among skeletal muscle fiber types. *J. Cell Biol.* 81:10–25, 1979.
5. Henneman, E. Skeletal muscle: the servant of the nervous system. In Mountcastle, V. B., ed. *Medical Physiology*, 14th ed., vol. 1, 674–702. St. Louis, C.V. Mosby, 1980.
6. Henneman, E.; Olson, C.B. Relations between structure and function in the design of skeletal muscles. *J. Neurophysiol.* 28:581–598, 1965.
7. Komi, P.V.; Viitasalo, J.H.T.; Havu, M.; Thorstensson, A.; Sjoden, B.; Karlsson, J. Skeletal muscle fibers and muscle enzyme activities in monozygous and dizygous twins of both sexes. *Acta Physiol. Scand.* 100:385–392, 1977.
8. Kugelberg, E.; Edstrom, L.; Abbruzzese, M. Mapping of motor units in experimentally reinnervated rat muscle. *J. Neurol. Neurosurg. Psychiatry* 33:319–329, 1970.
9. Liddel, E.G.T.; Sherrington, C.S. Recruitment and some other factors of reflex inhibition. *Proc. R. Soc. London B* 97:488–518, 1925.
10. Lucas, S.M.; Ruff, R.L.; Binder, M.D. Specific tension measurements in single soleus and medial gastrocnemius muscle fibers of the cat. *Exp. Neurol.* 95:142–154, 1987.
11. McDonagh, J.C.; Binder, M.D.; Reinking, R.M.; Stuart, D.G. Tetrapartite classification of the muscle units of cat tibialis posterior. *J. Neurophysiol.* 44:696–712, 1980.
12. McPhedran, A.M.; Wuerker, R.B.; Henneman, E. Properties of motor units in a homogeneous red muscle (soleus) of the cat. *J. Neurophysiol.* 28:71–84, 1965.
13. Pinter, M.J.; Curtis, R.L.; Hosko, M.J. Voltage threshold and excitability among various sized cat hindlimb motoneurons. *J. Neurophysiol.* 50:644–657, 1983.
14. Saltin, B.; Gollnick, P.D. Skeletal muscle adaptability: significance for metabolism and performance. *Handbk. Physiol.* Sec. 10, 555–631, 1983.
15. Sickles, D.W.; Oblak, T.G. Metabolic variation among α-motoneurons innervating different muscle-fiber types. I. Oxidative enzyme activity. *J. Neurophysiol.* 51:529–537, 1984.
16. Walmsley, B.; Hodgson, J.A.; Burke, R.E. Forces produced by medial gastrocnemius and soleus muscle during locomotion in freely moving cats. *J. Neurophysiol.* 41:1203–1216, 1978.
17. Wuerker, R.B.; McPhedran, A.M.; Henneman, E. Properties of motor units in a heterogeneous pale muscle (m. gastrocnemius) of the cat. *J. Neurophysiol.* 28:85–99, 1965.
18. Zengel, J.E.; Reid, S.A.; Sypert, G.W.; Munson, J.B. Membrane electrical properties and prediction of motor-unit type of medial gastrocnemius motoneurons in the cat. *J. Neurophysiol.* 53:1323–1344, 1985.

Peripheral Motor Control: Spinal Reflex Actions of Muscle, Joint, and Cutaneous Receptors

Now that we have examined the properties of motor units (Chap. 23), which constitute the quantal elements of motor output, we can turn our attention to how the central nervous system controls their actions. This is by no means a simple problem, as evidenced by the remarkable complement of "neural machinery" dedicated to controlling and monitoring the discharge of motoneurons. Of particular importance for the moment-to-moment control of muscle action are the neural signals emanating from mechanoreceptors located in the skin, the joints, and the muscles themselves. While our sensations of touch, position, and movement are dependent on these same signals, we are relatively unaware of their simultaneous use by the spinal and supraspinal neurons involved in motor function. In general, afferent signals from the periphery effectively modify the "instructions" that are sent to skeletal muscles so that intended movements can be executed with precision (see Chap. 22). In this chapter, we will focus on how the afferent fibers innervating peripheral receptors influence the behavior of spinal motoneurons.

BASIC ANATOMY

The schematic diagram illustrated in Figure 24–1 depicts the basic neural linkages between the peripheral structures of a vertebrate limb and the spinal cord. The cell bodies of all of the afferent neurons are located in the dorsal root ganglia, external to the spinal cord and, with but few possible exceptions,[8] all of the afferent fibers enter the spinal cord via the dorsal root. The cell bodies of the motoneurons that innervate the muscles are located in the ventral gray matter of the cord and their axons exit through the ventral roots. The dorsal and ventral roots coalesce a few centimeters outside the cord to form the spinal nerves. The spinal nerves then give rise to peripheral nerves

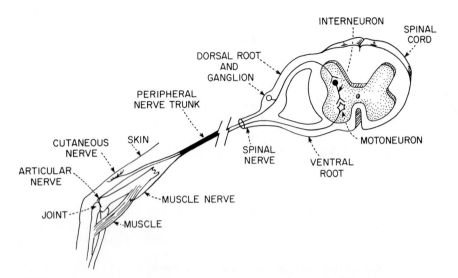

Figure 24–1 Components of the mammalian segmental motor system.

that innervate skeletal muscles, skin, joints, and other structures.

In the lumbar region of the cat's spinal cord, which has been the standard model for analysis of segmental motor systems, a dorsal root entering a single segment of the spinal cord may be composed of as many as 15,000 afferent fibers. Only half as many efferent fibers are generally found in the corresponding ventral root. Between the afferent fibers entering the cord and the motoneurons whose axons exit the cord, there are the cell bodies of approximately 300,000 to 400,000 spinal interneurons. The interneurons receive synaptic input from peripheral afferent fibers, descending neurons, other interneurons of both the same and other spinal segments, and even axon collaterals of the motoneurons themselves. Interneurons are

not only the principal integrative elements within the spinal cord, but (as will be discussed in Chap. 26) they form the interconnected networks that generate the basic motor patterns used in locomotion and other forms of coordinated muscle action.

CYTOARCHITECTURE OF SPINAL GRAY MATTER

On the right side of Figure 24–2, the gray matter of a mammalian spinal cord has been labeled using the lamination scheme devised by Rexed.[33] On the left side of the figure, a number of the distinct neuron groups are identified. The first six laminae of the cord are referred to as the posterior, or

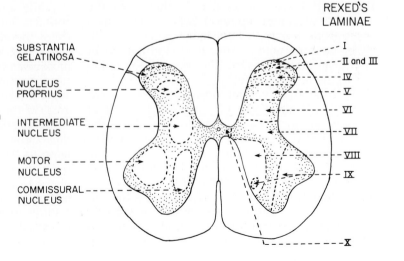

Figure 24–2 Transverse section of the human lumbar spinal cord.

dorsal, horn. Many of the neurons in these laminae receive direct synaptic input from peripheral afferent fibers and transmit these neural signals both to higher centers of the nervous system and to other neurons within the spinal cord. Lamina I is composed of a thin layer of neurons that receive primarily nociceptive input. The axons of some of these lamina I cells cross the spinal cord and give rise to the spinothalamic tract. Other cells in lamina I are propriospinal, projecting to other segments of the spinal cord. Laminae II and III are referred to collectively as the substantia gelatinosa and contain small neurons that are very densely packed. These cells are thought to be involved in controlling the transmission of nociceptive input (Chaps. 15, 16). Lamina IV is composed of several different groups of neurons. One group of large neurons, called the nucleus proprius, receives input from peripheral receptors subserving a variety of sensory modalities, and their axons ascend to the lateral cervical nucleus, to the dorsal (or posterior) column nuclei, and to the thalamus. Another group of neurons in lamina IV together with cells in lamina VI forms the portion of the spinothalamic tract that conveys non-nociceptive input to the brain. Finally, laminae V and VI house interneurons that project to the spinal motor nuclei and receive direct synaptic input from muscle, joint and cutaneous afferents, lamina IV interneurons, and several spinal-destined systems (Chap. 26) involved in motor control.

Most of the ascending projections to "motor areas" of the brain arise from neurons in laminae VI to IX. A group of large cells in lamina VI of spinal segments T1 to L3, called Clarke's column, generates the ipsilateral dorsal spinocerebellar tract, which transmits input from muscle and cutaneous afferents to the cerebellum (Chap. 29). This input is complemented by that from the contralateral ventral spinocerebellar tract, which arises from neurons at the edge of lamina VII. Other neurons within laminae IV through VI of the dorsal horn project to the reticular formation and to the inferior olive.

The anterior, or ventral, horn of the spinal cord consists of laminae VII, VIII, and IX. The size of the ventral horn varies considerably along the length of the spinal cord, with two prominent enlargements. These enlargements, one extending from C5 to C7 and the other from L5 to L7, are primarily due to the expansion of the ventral gray matter to accommodate the motoneurons that innervate the forelimb and hindlimb musculature, respectively. Motoneurons are found in two clusters; a medial group, called lamina IXm, consisting of motoneurons that innervate the axial (i.e., trunk and neck) muscles, and a lateral group, called lamina IX, that innervates the limb muscles.

Laminae VII and VIII are composed primarily of interneurons involved in motor control, as many of their axons project to motor nuclei in lamina IX. Lamina VII is most prominent in spinal segments featuring a well-developed lamina IX, whereas lamina VIII is most extensive in those segments containing an expanded lamina IXm, suggesting that lamina VII and VIII interneurons are involved in the control of limb and axial musculature, respectively. Finally, lamina X contains interneurons that project to the opposite side of the spinal cord, including some that convey input from peripheral afferent fibers.

GENERAL FEATURES OF AFFERENT INPUT

As the dorsal roots enter the spinal cord, they appear to segregate into a medial and a lateral portion. The fibers in the medial portion are generally larger in diameter than those in the lateral division, and their actual entry into the spinal cord is medial to the dorsal horn of the spinal gray matter. The smaller fibers of the lateral portion of each dorsal root enter the cord at the tip of the dorsal horn. Upon spinal entry, each fiber bifurcates to produce an ascending branch and a descending branch that run parallel to the axis of the cord. As shown in Figure 24–3, collaterals of these two branches have extensive arborizations and contact neurons within the central gray matter of the cord. It is thought that all dorsal root afferent fibers project both to higher centers in the nervous system and to spinal interneurons involved in the control of motoneuronal discharge.

While some afferent fibers make direct synaptic contacts with spinal motoneurons, the vast majority of the afferent fibers originating from muscle, skin, and joint receptors exert their actions on motoneurons indirectly, through one or more intercalated interneurons. Moreover, many of these same interneurons receive input from the various areas of the brain that comprise the descending or spinal-destined motor systems (Chap. 26). Despite the extensive convergence from many parts of the nervous system onto spinal interneurons and motoneurons, the inputs from peripheral afferent fibers alone can be sufficiently potent, in certain instances, to evoke discharge from motoneurons leading to muscle activation. This type of motoneuronal or muscle activation is referred to as a *spinal reflex*. Spinal reflexes persist in the ab-

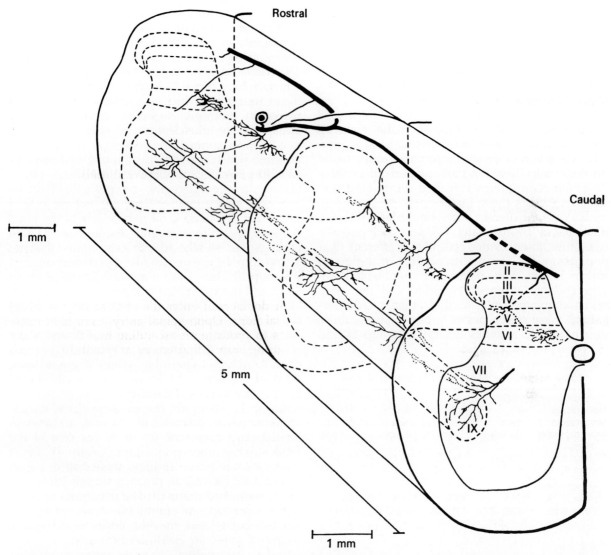

Figure 24-3 Morphology of a Ia afferent fiber of the cat. The dark lines show the fiber within the dorsal root and the ascending and descending branches of the fiber in the dorsal columns. Five regularly spaced collaterals are shown coursing down through the dorsal horn en route to the motor nuclei in lamina IX (indicated by enclosed cylinder). (From Brown, A.G.; Fyffe, R.E.W. *J. Physiol. [Lond.]* 274:111–127, 1978.)

sence of supraspinal structures and are highly stereotyped and reproducible. Moreover, since the underlying neural circuits involve relatively few neurons, spinal reflexes have provided an extremely useful experimental tool for examining the functional organization of neural networks. The neural substrates and physiological actions of several types of spinal reflexes will be presented later in this chapter.

AFFERENT FIBERS IN PERIPHERAL NERVES[4]

Muscle Afferents

More than two thirds of the fibers that make up a typical muscle nerve are afferents. Muscle afferents are generally divided into four groups based on their axon diameters, as summarized in Table 24–1. Fibers in groups I, II, and III are all myelinated, whereas those in group IV are unmyelinated. Group I afferent fibers are the largest (mean diameter of approximately 15 μm) and typically constitute about 25 per cent of the afferent fiber population. On average, two thirds of the group I fibers innervate the primary or annulospiral sensory endings of muscle spindles and the other third innervates Golgi tendon organs. To differentiate these two types of group I fibers, those that innervate spindles are called group Ia and

Table 24–1 Afferent Fibers in Peripheral Nerves

General Characteristics				
Fiber group	I	II	III	IV
Myelination	yes	yes	yes	no
Axon diameter (mean)	15 μm	8 μm	4 μm	<1 μm
Conduction velocity	90 m/s	48 m/s	24 m/s	1 m/s
Representation				
Muscle nerves	25%	15%	10%	50%
Receptors	ASE; GTO	FSE; RR; FNE PFR; PC	FNE	FNE
Cutaneous nerves	0	25%	25%	50%
Receptors		MD; DSR; BE; PC	DHR; FNE	FNE
Articular nerves	7%	23%	20%	50%
Receptors	GTO	RR; PJR; PC	FNE	FNE

ASE = annulospiral ending; GTO = Golgi tendon organ; FSE = flower spray ending; PFR = paciniform receptor; RR = Ruffini receptor; PC = pacinian corpuscle; FNE = free nerve ending; MD = Merkel's disk; DSR = dermal stretch receptor; BE = basket ending; DHR = down hair receptor; PJR = paciniform joint receptor.

those that innervate tendon organs, group Ib. Both spindles and tendon organs are slowly adapting mechanoreceptors whose afferent fibers discharge in response to changes in muscle length and force, respectively. The details of their anatomy and physiology will be presented later in this chapter.

Group II fibers are somewhat smaller (mean diameter of 8 μm) and account for about 15 per cent of the afferent fibers in a typical muscle nerve. A majority of group II fibers innervate the secondary (or flower-spray) endings of muscle spindles. The remainder innervate (1) Ruffini receptors, which behave much like tendon organs; (2) paciniform receptors, which are rapidly adapting and respond to both muscle stretch and light pressure applied to the muscle surface, aponeuroses, and neighboring connective tissue; (3) pacinian corpuscles, which are rapidly adapting and respond to light pressure and vibration (see Chap. 14); or (4) free nerve endings (i.e., receptors that have no accessory structures; see Chap. 5) that are located in muscle, its fascia, and particularly at the junctions between muscle and aponeuroses. These endings are slowly adapting receptors that respond to pressure and to muscle contraction.

Group III fibers have a mean diameter of about 4 μm and represent about 10 per cent of all muscle afferent fibers. Two thirds of the group III fibers innervate free nerve endings with properties similar to those innervated by group II fibers. Many of the group III afferents respond to noxious stimuli such as hypertonic saline. The remaining third of the group III fibers innervate the blood vessels that supply the muscle.

Group IV fibers are the smallest of the muscle afferents (mean diameter of < 1 μm), are unmyelinated, and constitute up to 50 per cent of the total afferent fiber population. Group IV fibers innervate free nerve endings about half of which respond exclusively to noxious stimuli (chemical, mechanical, and thermal). The remainder respond to the same types of stimuli but at intensities that are innocuous and may be encountered in the course of muscular exertion or exercise.

Articular Afferents

Joint capsules and ligaments between bones, periosteum, and interosseous membranes are innervated by articular nerves. They are composed exclusively of afferent fibers, which innervate many of the same types of receptors found in skeletal muscle. Again, about half of all the afferent fibers in articular nerves are unmyelinated, group IV fibers, many of which respond to noxious stimuli. Of the myelinated fibers, approximately 15% are group I, 45% group II, and 40% group III.

In most articular nerves, all the group I fibers are Ib afferents innervating tendon organs located in ligaments and interosseous membranes, but rarely in the joint capsules. Some of the group II fibers innervate the numerous Ruffini receptors located in joint capsules. These receptors respond to movement of the joints and also to forces applied to the joint capsule. Other group II afferents innervate paciniform joint receptors. Very few pacinian corpuscles are found in joint capsules but they account for most of the group II innervation in interosseous membranes and periosteum. The group III afferent fibers innervate free nerve endings, which are thought to be sensitive to pressure and noxious stimuli, and nerve endings on blood vessels.

Cutaneous Afferents

The composition of cutaneous nerves, along with a discussion of the transducing properties of the receptors they innervate, was presented in Chapter 14. As is the case with muscle and articular nerves, about half of all the fibers in cutaneous nerves are unmyelinated group IV afferents involved primarily in nociception. Cutaneous nerves generally have no group I afferents; half of their myelinated fibers are group II and half are group III. The group II fibers innervate the slowly adapting Merkel's disks in the epidermis, the slowly adapting dermal stretch receptors, a few pacinian corpuscles, and the "basket endings" of guard hair follicles. Nearly all of the group III fibers in cutaneous nerves innervate the rapidly adapting down hair receptors; the others innervate either blood vessels or slowly adapting nociceptors in the skin.

SEGMENTAL PROJECTIONS OF PERIPHERAL AFFERENT FIBERS

Studying Neural Circuits

The segmental projections of peripheral afferent fibers have been deduced using a variety of anatomical and physiological techniques. By injecting various diffusible marking dyes into the axons of afferent fibers, one can trace the course of their complex arborizations throughout the spinal cord

(see Fig. 24–3). One can then make inferences about functional connectivity based on the proximity of the axon terminals of the fiber to known groups of interneurons within the cord. Alternatively, one can construct a map of afferent projections within the cord by recording electrical field potentials at various sites consequent to stimulation of different peripheral nerves and even single afferent fibers. More precise "physiological mapping" can be performed by recording changes in the membrane potentials or spike discharge patterns of single interneurons and motoneurons produced by activating individual or groups of afferent fibers. Although corroboration of results using several of these techniques is desirable, such data are lacking for all but a few types of peripheral afferent fibers.

There are two major difficulties in generating "physiological maps" of neural connectivity. The first problem is ensuring that only a single type of afferent fiber (e.g., group Ia, group Ib, group II spindle, etc.) is producing the responses that are being measured. While a number of different techniques have been devised to obviate this problem,[20] in many cases it has not been possible to study the actions of identified afferent fibers in isolation. The second problem is to detect the small changes in either the membrane potential or the firing probability of the postsynaptic cell produced by the input. This problem has been more successfully circumvented by averaging the responses of the postsynaptic cell to repeated presentations of the same stimulus. These technical problems, of course, are not unique to studies on spinal circuitry; rather, they plague investigations at every level of the nervous system.

Group Ia Afferents

Of all the peripheral afferent fibers, the spinal projections of group Ia afferents have been studied most thoroughly. As depicted in Figure 24–3, the ascending and descending axon branches of a Ia fiber give rise to, at approximately 1-mm intervals, a series of collaterals that in turn arborize extensively in laminae V–VI, lamina VII, and lamina IX.[5] Intracellular recordings have shown that Ia afferent fibers generate excitatory postsynaptic potentials (EPSPs) in neurons located in all of these different laminae. The direct synaptic connections of Ia afferent fibers with motoneurons in lamina IX have received a great deal of attention and have provided the model synapse for studying chemical transmission in vertebrate central nervous systems (see Chap. 11).

Individual Ia afferents innervating muscle spindles produce EPSPs in from 80% to 100% of the motoneurons that innervate that same (homonymous) muscle. They also produce EPSPs in about 60% of the motoneurons that innervate muscles acting at the same joint with a similar action (i.e., synergist muscles). The EPSPs produced by a single afferent fiber in an homonymous motoneuron have an average magnitude of 100 μV, while those in the synergist motoneurons average 70 μV[17] (Fig. 24–4).

Ia afferent fibers also make contacts with motoneurons that innervate muscles that do not act in strict mechanical synergy.[11] Figure 24–5 summarizes the patterns of monosynaptic Ia connections with motoneurons that innervate the proximal muscles (hip and knee) of the cat hindlimb. This pattern of widespread connectivity indicates that muscles controlling the actions of different joints in the limb may be coordinated to some extent by input from their peripheral afferents.

The projections of Ia afferents to lamina VII produce EPSPs in a group of interneurons called Ia-inhibitory interneurons. As the name implies, these inhibitory interneurons produce inhibitory postsynaptic potentials (IPSPs) in the cells they contact. The principal projections of their axons are to motoneurons in lamina IX. As is the case for the projections of Ia afferents to motoneurons, the projections of Ia afferents to Ia-inhibitory interneurons are related to their muscle of origin. The schematic diagram in Figure 24–6 shows that activation of Ia afferent fibers from muscles that flex the knee (gracilis, posterior biceps, and semitendinosus) produces IPSPs (mediated by Ia-inhibitory interneurons) in motoneurons innervating knee extensor muscles (vasto-crureus and rectus).

This type of connectivity pattern is quite common in spinal circuits and is referred to as reciprocal innervation. Afferent fibers that produce EPSPs in one population of motoneurons generally produce IPSPs in motoneurons innervating muscles with antagonist actions.

The third major projection locus for Ia afferent fibers is the so-called intermediate nucleus of laminae V and VI. In this area, Ia afferents generate EPSPs in some interneurons that are also contacted by other groups of peripheral afferents from different sources (see below), as well as by cells from virtually every spinal-destined motor system (see Chap. 26). Some of these lamina V–VI interneurons produce IPSPs both in motoneurons and in cells of the dorsal spinocerebellar tract (DSCT); others produce EPSPs in motoneurons.[1, 28]

Group Ib Afferents

The spinal projections of group Ib afferents that innervate Golgi tendon organs differ from those of group Ia afferents principally by the absence of terminals in the motor nuclei of lamina IX (Fig. 24–7). This anatomical finding is consistent with the physiological finding that the actions of Ib afferents on motoneurons are always mediated through one or more interneurons. Some of these interneurons in laminae V–VI are the same cells that the group Ia afferents contact; others are the cells of origin of the spinocerebellar tracts. Ib afferents also arborize in lamina VII.[6]

It has been difficult to study the physiological actions of Ib fibers on motoneurons in isolation. Low-intensity electrical stimuli applied to muscle nerves can be used to selectively activate group I fibers; however, both group Ia and group Ib fibers are activated concurrently. Moreover, there are no

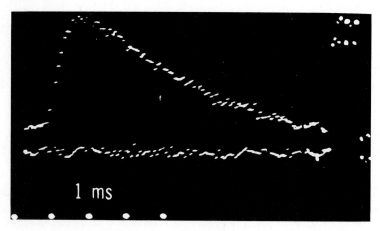

1 ms

Figure 24–4 Monosynaptic excitatory postsynaptic potential (EPSP) produced by a single Ia afferent fiber in a homonymous motoneuron. The records were obtained by averaging the membrane potential fluctuations following the occurrence of several hundred successive action potentials in the Ia afferent fiber (i.e., spike-triggered averages). The lower trace shows a spike-triggered average from the same afferent fiber after the electrode was moved just outside of the motoneuron's membrane. A 200 μV calibration pulse was added to the end of each sweep. (After Mendell, L.M. and Henneman, E. *J. Neurophysiol.* 34:171–187, 1971.)

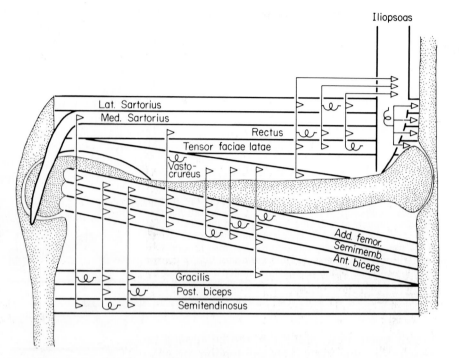

Figure 24–5 Distribution of heteronymous monosynaptic EPSPs produced by Ia afferent fibers in proximal muscles of the cat hindlimb. The Ia afferent fibers are represented by the coils; the motoneurons in which they produce monosynaptic EPSPs are indicated by triangles. (After Eccles, R.M. and Lundberg, A. *J. Physiol. [Lond.]* 144:271–298, 1958.)

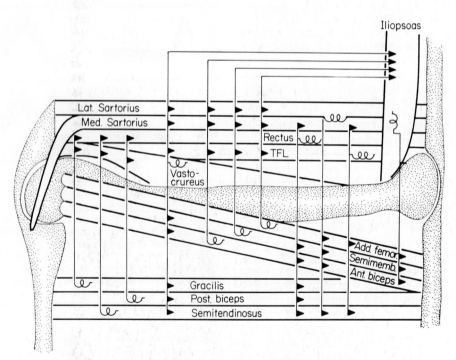

Figure 24–6 Distribution of disynaptic inhibitory postsynaptic potentials (IPSPs) evoked by Ia afferent fibers in proximal muscles of the cat hindlimb. The Ia afferent fibers are represented by the coils: the motoneurons in which they evoke disynaptic IPSPs are indicated by triangles. (After Eccles, R.M.; Lundberg, A. *J. Physiol [Lond.]* 144:271–298, 1958.)

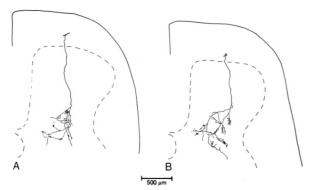

Figure 24–7 Reconstructions of the morphology of two adjacent collaterals (A and B) (see Fig. 24–3) from the descending branch of a Ib afferent fiber from the cat medial gastrocnemius muscle. The collaterals enter the gray matter at the top of the dorsal horn and run ventrally to lamina V before branching in lamina VI and the dorsal part of lamina VII. The perimeter of the spinal gray matter is indicated by the dashed line. (From Brown, A.G. Fyffe, R.E.W. *J. Physiol. [Lond.]* 296:215–228, 1979.)

known mechanical stimuli that can be applied to muscles to selectively activate tendon organs. Thus, until recently the actions of Ib afferents on motoneurons were inferred by observing changes in the "shapes" of synaptic potentials recorded in motoneurons as the strength of an electrical stimulus applied to a muscle nerve was gradually increased.[12]

The threshold of a nerve fiber to an externally applied electrical stimulus is directly related to its diameter; the largest fibers are activated at the lowest stimulus strengths (see Chap. 3). By convention, the magnitude of a brief (0.05 to 0.1 ms) electrical shock required to activate the lowest threshold fibers in a given nerve is referred to as the threshold stimulus, T. Thus, the largest group I fibers are activated at T, about 80% of all the group I fibers are activated when the stimulus intensity is increased by 50% (to 1.5T), and generally all of the group I fibers are activated at 2T. In some muscle nerves, a majority of the group Ia fibers tend to be slightly larger that most of the Ib fibers, and therefore lower stimulus intensities (1.0–1.3 T) activate more Ia than Ib fibers.[12, 20, 26]

As shown in Figure 24–8, the amplitude of a monosynaptic EPSP recorded in a motoneuron increases as the strength of the stimulus applied to the homonymous muscle nerve is increased from 1.4 to 1.8T, indicating that a greater number of group Ia fibers are activated at the higher strengths. Notice, however, the progressive change in the time course or "shape" of the EPSP; its falling phase becomes more rapid at the higher stimulus strengths. This change was the result of

the intrusion of an IPSP with a longer segmental latency (i.e., time from entry of the afferent fiber action potentials into the spinal cord to the onset of the synaptic potential). To demonstrate the presence of the IPSP, hyperpolarizing current was passed through the intracellular microelectrode to deflect the motoneuron's membrane potential to a level below the reversal potential for IPSPs (see Chap. 11). Under these conditions, a second positive peak appears in the synaptic potential less than 1 ms after the first, reflecting the "reversed" group Ib IPSP. The increased segmental latency of

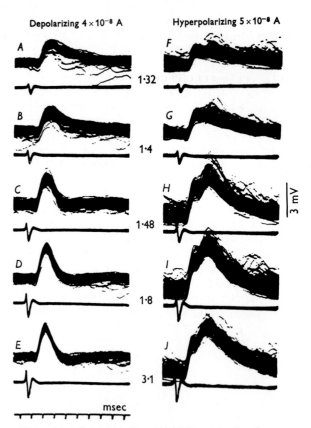

Figure 24–8 Demonstration of inhibition of cat motoneurons evoked by activating Ib afferent fibers. Intracellular recordings were made from a gastrocnemius motoneuron while the gastrocnemius muscle nerve was stimulated (ventral roots were cut to prevent antidromic invasion of the motoneuron). The traces on the left show that as the strength of the stimulus delivered to the muscle nerve was increased relative to group I threshold, the falling phase of the monosynaptic EPSP evoked by the Ia afferent fibers became more rapid. This suggests that an IPSP was evoked at the higher stimulus strengths. The traces on the right confirm this suggestion by showing that when the motoneuron was hyperpolarized beyond the reversal potential for the IPSP, a second wave of depolarization, the "reversed" IPSP, appears superimposed on the falling phase of the EPSP. (From Eccles, J.C.; Eccles, R.M.; Lundberg, A. *J. Physiol. [Lond.]* 138:227–252, 1957.)

this second part of the synaptic potential indicates that at least one interneuron is interposed between the incoming afferent fibers and the motoneuron.

More recently, two new techniques have been developed to record synaptic potentials in motoneurons produced by selective activation of group Ib afferents. With the first technique, one records the discharge of a single Ib afferent fiber from an intact dorsal root filament with one electrode while maintaining a simultaneous intracellular recording from a motoneuron with another electrode. One then determines the *average* change in the motoneuron's membrane potential for several milliseconds following each of many thousand consecutive Ib action potentials.[37] This general technique, called *spike-triggered averaging,* had been developed earlier to study the monosynaptic connections of Ia afferents to motoneurons (see Fig. 24-4).[29]

The second technique entails altering the excitabilities of group Ia fibers within a muscle nerve so that the Ib fibers can be selectively activated by low-intensity electrical shocks. Subjecting a muscle to high-frequency, low-amplitude vibration for periods of up to an hour causes each Ia fiber to generate spikes at the vibration frequency (200 to 300 Hz), while the Ib afferents remain generally quiescent. As a result of this high level of activity, the electrical excitability of the Ia afferent axons is decreased so that during their recovery period, a low intensity electrical shock applied to the muscle nerve selectively activates group Ib fibers whose electrical excitabilities are unaltered by the vibration.[20] Figure 24-9 shows a monosynaptic EPSP produced in a motoneuron by stimulating the

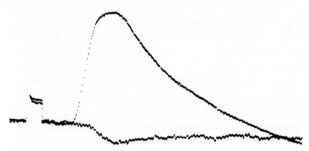

Figure 24-9 Monosynaptic Ia EPSP and disynaptic Ib IPSP recorded in a cat medial gastrocnemius motoneuron. The Ib IPSP was produced by stimulating the medial gastrocnemius muscle nerve at 1.3T after the muscle had been subjected to high-frequency vibration (200 Hz) for twenty minutes. The monosynaptic EPSP was produced by stimulating the muscle nerve at 2T. The 1-ms calibration pulse preceding the EPSP represents 1 mV; that preceding the IPSP represents 100 μV. The EPSP recorded was obtained by averaging the motoneuron's response to 32 successive stimuli; the IPSP record was the average response to 256 successive stimuli. (From Powers, R.K.; Binder, M.D. unpublished data.)

homonymous muscle nerve at 2T and a much smaller disynaptic IPSP produced in the same motoneuron by stimulating the muscle nerve at 1.3T after the muscle had been vibrated at 200 Hz for nearly half an hour.

In general, results obtained from experiments using these new techniques have corroborated those obtained using graded electrical shocks of muscle nerves: Ib afferent fibers from extensor muscles usually produce IPSPs in extensor motoneurons located throughout the limb, with segmental latencies indicating the presence of one or two interneurons in the pathway. The same Ib afferents generally produce EPSPs in flexor motoneurons, again with di- and trisynaptic latencies.[12, 37] The synaptic potentials produced by Ib afferents in flexor muscles are generally less pronounced, but follow the principle of reciprocal innervation in that they generate IPSPs in flexor motoneurons and EPSPs in extensor motoneurons. It is important to note, however, that regardless of the technique used to activate Ib fibers, a considerable number of exceptions to these general patterns have been observed.[28]

At first thought, it appears that the actions of Ia and Ib afferent fibers on spinal motoneurons are quite different. The Ia afferent projections seem to be highly "focused," producing monosynaptic excitation of motoneurons innervating synergist muscles and disynaptic inhibition of motoneurons innervating their antagonists around a single joint. In contrast, the Ib afferent fibers mediate all of their actions on motoneurons through at least one interneuron, their effects are often spread throughout the entire limb, and their connectivity patterns are not strictly reciprocal. However, by combining intracellular recordings from interneurons with the "vibration technique" described above, it has been shown that Ia and Ib afferents converge onto many of the same cells in laminae V-VI, and that virtually all of the actions of Ib fibers are facilitated by concurrent activation of Ia afferents.[1, 28] Thus, it is possible that Ia and Ib fibers work predominantly in concert at the segmental level and that the prominent, focused Ia "private pathway" (monosynaptic EPSPs and disynaptic IPSPs in antagonists) has, in addition, some other, as yet unknown, functions.

Group II Muscle Afferents

As described earlier, group II afferents in muscle nerves innervate several types of muscle receptors. However, only the actions of group II afferents innervating muscle spindles have been studied in isolation. The spinal projections of a group II afferent fiber from the triceps surae muscles (medial gastrocnemius, lateral gastrocnemius, and soleus) are shown in Figure 24–10. There are clusters of terminations in laminae IV–VI, as well as in the motor nuclei of lamina IX.[16] The presence of terminals in lamina IX suggests that this fiber innervates a muscle spindle, as group II spindle afferents make direct, excitatory connections with homonymous and synergist motoneurons, similar to those made by group I spindle afferents. Using the spike-triggered averaging technique, it has been shown that individual group II spindle afferents generate monosynaptic EPSPs in approximately 50% of their homonymous motoneurons and in fewer than 20% of their synergists.[36] The mean amplitude of these group II EPSPs is only 24 μV, much smaller than the comparable average produced by the Ia afferents (100 μV). A few examples of EPSPs and IPSPs produced in motoneurons by individual spindle group II afferents with di- and trisynaptic segmental latencies have been reported,[36] but these have not been systematically studied as yet.

The synaptic actions of group II muscle afferents have also been studied using graded electrical shocks applied to muscle nerves. As the strength of an electrical shock is increased from 1.5 to 5T, group II fibers are progressively recruited so that the changes in the synaptic potentials recorded simultaneously in a motoneuron can be attributed to the actions of group II fibers. These group II–generated synaptic potentials are often difficult to interpret because they are usually superimposed on the synaptic potentials produced by the lower threshold group I afferents. A further complication lies in the finding that the pattern of synaptic potentials evoked in motoneurons by electrical stimulation of group II fibers varies dramatically with the type of experimental preparation in which they are investigated.

Despite the uncertainties cited earlier, it is generally agreed that the predominant actions of group II muscle afferents are mediated through interneuronal networks that exert reciprocal actions on flexor and extensor motoneurons. However, the magnitude and polarity of the actions vary in different types of experimental preparations. In the intact anesthetized cat, stimulation of group II fibers in any hindlimb muscle nerve produces disynaptic EPSPs in motoneurons innervating flexor muscles and trisynaptic IPSPs in motoneurons innervating extensor muscles. In the decerebrate cat (neuraxis transected at the level of the inferior colliculus), the effects of activating group II afferent fibers are often absent, particularly in flexor motoneurons. However, if an additional lesion is made in the pons, the group II action on flexor motoneurons is restored and often even "reverses" to produce IPSPs in the place of EPSPs. In cats with complete transections of the spinal cord (i.e., spinal cats), group II afferents not only produce IPSPs in flexor motoneurons but they can also produce disynaptic EPSPs in extensor motoneurons.[1, 28]

These varied findings suggest the existence of parallel "alternative pathways" from group II afferents to motoneurons. Both flexor and extensor motoneurons may receive synaptic inputs from separate groups of inhibitory and excitatory interneurons activated by group II afferents. It is also possible that group II fibers innervating different types of muscle receptors may utilize different groups of interneurons to exert distinct actions on motoneurons. The fact that specific lesions alter the predominance of one pathway over another suggests that transmission of synaptic input from group II afferents to motoneurons may be under the control of supraspinal structures in the intact animal (see Chap. 26).

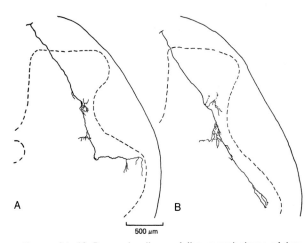

500 μm

Figure 24–10 Reconstructions of the morphology of two adjacent collaterals (*A* and *B*) (see Fig. 24–7) of a group II afferent fiber from the cat lateral gastrocnemius–soleus muscle nerve. The perimeter of the spinal gray matter is indicated by the dashed line. (From Fyffe, R.E.W. *J. Physiol. [Lond.]* 296:39–40P, 1979.)

Group III Muscle Afferents

The spinal projections of group III muscle afferents have not yet been studied in isolation. The only information available on their actions on motoneurons is based on studies using graded electrical shocks that activate group I, group II, and group III fibers concurrently. Increasing the strength of an electrical shock applied to a muscle nerve to activate group III fibers enhances the synaptic potentials produced in motoneurons by group II afferents as shown in Figure 24–11. The similarity of the synaptic actions of group III and group II muscle afferents suggests that they converge on common interneurons.[14] As is the case for group II afferents, the group III fibers may have multiple spinal pathways to both flexor and extensor motoneurons. However, our understanding of the role of group III afferents in the segmental control of motoneurons is severely limited by the absence of data on individual identified (with respect to receptor type) afferent fibers.

Group IV Muscle Afferents

Although approximately half of the afferents in many muscle nerves are unmyelinated group IV fibers, little is known of their spinal projections and physiological actions on motoneurons. Based almost entirely on results derived from stimulating group IV afferents in either cutaneous nerves or entire dorsal roots, it is generally assumed that they produce polysynaptic excitation of ipsilateral flexor motoneurons and inhibition of ipsilateral extensor motoneurons. Differences in the results obtained from decerebrate versus spinal cat preparations again suggest the existence of alternative

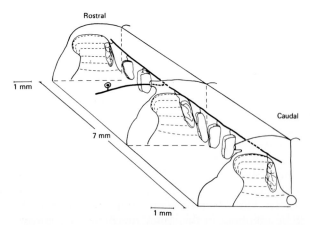

Figure 24–12 Morphology of a group II afferent fiber from a cat cutaneous nerve. The heavy lines show the fiber within the dorsal root and the ascending and descending branches of the fiber in the dorsal columns. Eight regularly spaced collaterals are shown coursing down through the medial border of the dorsal horn. (From Brown, A.G.; Fyffe, R.E.W.; Rose, P.K.; Snow, P.J. *J. Physiol. [Lond.]* 316:469–480, 1981.)

pathways from group IV afferents to motoneurons.

Cutaneous and Joint Afferents

Electrical stimulation of cutaneous and articular nerves in the cat hindlimb at strengths sufficient to activate group II and group III fibers generally evokes the same pattern of synaptic potentials observed with comparable stimulation of muscle nerves; that is, trisynaptic EPSPs in ipsilateral flexor motoneurons and trisynaptic IPSPs in ipsilateral extensor motoneurons. This multisynaptic response pattern is consistent with the anatomy of the representative cutaneous group II fiber shown in Figure 24–12. Its terminal arborizations are concentrated in laminae III–VII and, unlike the putative group II spindle afferent shown earlier (see Fig. 24–10), this group II cutaneous fiber had no direct terminations in the motor nuclei.

As described for high threshold muscle afferents, there appear to be alternative pathways from both cutaneous and joint afferents to motoneurons that are revealed by using different lesions to alter the experimental preparations. Based on these striking similarities, it has been proposed that group II and III muscle afferents, cutaneous afferents, and joint afferents share common reflex pathways to motoneurons. Since their dominant segmental action is excitation of ipsilateral flexor motoneurons, this diverse aggregation of afferent

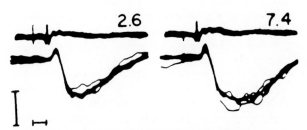

Figure 24–11 Convergence of the synaptic actions of group II and group III afferent fibers in a motoneuron innervating an extensor muscle (plantaris) of the cat hindlimb. The lower traces are intracellular recordings from the motoneuron; the upper traces, simultaneous extracellular recordings made from the dorsal roots near their point of entry into the spinal cord. In the traces on the left, the flexor digitorum longus muscle nerve was stimulated at 2.6 T to activate group II afferent fibers. The predominant effect is inhibition with a segmental latency, indicating the presence of one or more interneurons interposed between the afferent fibers and the motoneuron. Increasing the stimulus strength to 7.4 T activates group III fibers and, as shown on the right, the synaptic potential is augmented without any significant change in latency. Vertical calibration bar = 2 mV; Horizontal calibration bar = 2 mS. (After Eccles, R.M.; Lundberg, A. *Arch Ital. Biol.* 97:199–221, 1959.)

fibers has been referred to as the *flexor reflex afferents* (FRAs).[14]

After the administration of the monoamine precursors dopa (3,4-dihydroxyphenylalanine) and 5-HT (5-hydroxytryptophan) to mimic the actions of suprapinal neurons in spinal cats (see Chap. 26), the short-latency FRA effects described above are suppressed. Instead, one finds that stimulating the FRAs evokes long-lasting EPSPs (400–1000 ms) in ipsilateral flexor motoneurons after a segmental delay of 100 to 200 ms. Reciprocal actions of comparable duration and latency are observed in ipsilateral extensor motoneurons.[1, 28] These long-latency FRA effects on motoneurons are thought to be involved in the spinal mechanisms underlying locomotion and will be discussed in more detail in Chapter 26.

Not all the actions of cutaneous and joint afferents can be subsumed under the FRA rubric. For example, low-intensity electrical stimulation of nerves innervating the skin of the hindlimb often evokes EPSPs in the same motoneurons in which stimulation of groups II and III muscle afferents evokes IPSPs. Moreover, in the forelimb there are disynaptic, rather than trisynaptic, pathways from cutaneous afferents to motoneurons. Similarly, the actions of low threshold joint afferents depart from those of the FRA in that they converge on interneurons that mediate the oligosynaptic (di- and trisynaptic) input to motoneurons from Ia and Ib muscle afferents.[1, 28] Thus, it is likely that cutaneous and joint afferents are involved in actions other than their participation in the generalized "flexor response." Separation of the potentially diverse actions of the different components of the FRAs will require experiments using natural activation of receptors rather than nonselective electrical stimuli.

PRIMARY AFFERENT DEPOLARIZATION

Thus far, we have focused on the synaptic contacts that afferent fibers make with motoneurons and with interneurons that project to motoneurons. However, afferent fibers also project to other groups of dorsal horn interneurons that make synaptic contacts with the axon terminals of the afferent fibers themselves. The physiological action of these interneurons is to depolarize the terminals of the incoming afferent fibers. This phenomenon is referred to as primary afferent depolarization (PAD), and it is thought to underlie presynaptic inhibition (see Chap. 11).

PAD can be measured directly by means of intraaxonal recordings with microelectrodes, or indirectly by measuring the change in excitability of the terminals of afferent fibers to a local electrical stimulus. The indirect technique, called excitability testing, is depicted in Figure 24–13. First, from a peripheral nerve, one records the antidromic afferent volley consequent to delivering an electrical stimulus near the afferents' terminations within the gray matter of the spinal cord (S_1). Subsequently, one precedes ("conditions") a stimulus of the same intensity at S_1 with a stimulus delivered to another peripheral nerve (S_2). If the resultant antidromic volley is larger than that produced by the nonconditioned S_1 stimulus, then one infers that the conditioning stimulus (S_2) increased the

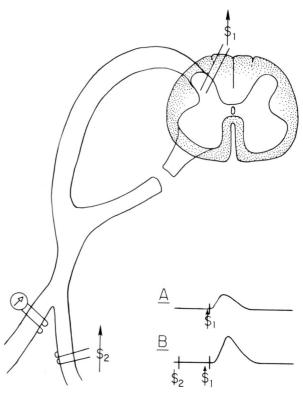

Figure 24–13 Technique for making indirect assessments of primary afferent depolarization in the spinal cord. *A,* Compound action potential recorded from a peripheral nerve consequent to a stimulus delivered within the gray matter of the spinal cord (S_1). *B,* When a stimulus of the same intensity at S_1 is preceded by a stimulus delivered to another peripheral nerve (S_2), the resulting antidromic volley is larger than that produced by a nonconditioned stimulus. This result indicates that stimulating S_2 increased the excitability of the afferent fiber terminals within the spinal cord, presumably by depolarizing them, such that a greater number of fibers are activated by S_1.

excitability of the afferent fiber terminals within the spinal cord so that the S_1 stimulus activated a greater number of afferent fibers. Presumably, the increase in afferent fiber excitability is attributable to depolarization of the terminals.

The interneurons that mediate PAD of cutaneous fibers are concentrated in laminae III and IV, while those producing PAD in group I muscle afferents are concentrated in laminae V and VI. There appear to be at least two interneurons in the PAD pathways, as the shortest PAD latencies are consistent with trisynaptic linkages. Both the first-order interneurons (those that receive contacts from peripheral afferents) and the last-order interneurons (those that make the axo-axonic synapses with the primary afferent terminals) generally reside in the same lamina.[28]

Figure 24–14 summarizes the known PAD circuits for peripheral afferent fibers in the cat spinal cord.[35] The relative magnitude of the PAD effects produced by the different groups of afferent fibers is indicated by the width of their "projection arrows." These widespread effects, particularly those to Ib and cutaneous afferents, are difficult to interpret. It has been suggested that PAD may be used to control the "gain" of segmental reflex pathways and perhaps even to "select" input from a given afferent fiber source during the execution of specific motor tasks. However, there is as yet no experimental evidence to support these speculations.

AFFERENT FIBER CONVERGENCE ON INTERNEURONS[1, 28]

Perhaps the most prominent feature of the segmental projections of the various classes of peripheral afferent fibers is their extensive convergence on common interneurons. Although this general finding has been evident from the earliest studies of polysynaptic pathways,[14] the predominance of "shared pathways" has not been incorporated into most heuristic models of the segmental motor system until the last few years. Instead, motoneurons have been viewed as the primary site of integration of the various afferent inputs.

For example, it has long been held that muscle spindles provide their homonymous motoneurons with an excitatory input whose magnitude is correlated with muscle length, while Golgi tendon organs provide them with an inhibitory input that is correlated with muscle force. It is now clear, however, that much of the afferent input from these two groups of receptors impinges on groups of common interneurons, some of which me-

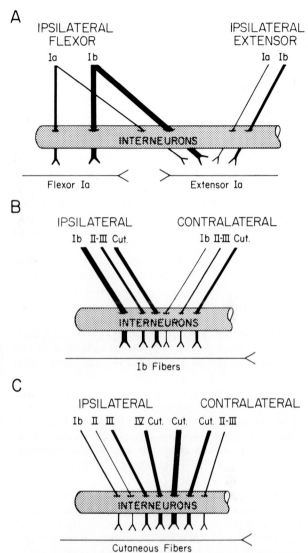

Figure 24–14 Summary of depolarization of Ia afferent fibers *(A)*, Ib afferent fibers *(B)*, and cutaneous afferent fibers *(C)* evoked by stimulating different groups of peripheral afferent fibers. The width of the arrows indicates the relative magnitude of the depolarizing action. Ia, Ib, II and III = the respective muscle afferent groups; CUT = cutaneous afferent fibers; IV cut. = unmyelinated cutaneous afferent fibers. (After Schmidt, R.F. *Ergeb. Physiol. Biol. Exp. Pharmacol.* 63:20–101, 1971.

diate excitatory and others inhibitory inputs to motoneurons. Furthermore, some of these same interneurons also receive input from spindle and tendon organ afferent fibers innervating many other muscles in the limb, as well as from joint and cutaneous afferents.[1, 28]

Perhaps it was the nature of the evidence that delayed the general recognition that spinal interneurons are the primary integrative sites in the spinal cord. Until recently there have been only a few intracellular recordings made from spinal interneurons,[1, 28] and these have generally been of limited utility because the projections of the recorded interneurons often were undetermined. In only a few cases have simultaneous recordings been made from an identified interneuron and a motoneuron to which it projected.[21] Instead, the convergence of different groups of afferent fibers on interneurons has been inferred by measuring the facilitation that one input exerts on the synaptic potentials produced in a motoneuron (or interneuron) by a second input. As shown schematically in Figure 24–15, this technique, called spatial facilitation, entails comparing the synaptic potentials produced by two sources of input activated concurrently with those evoked by either input system alone. In Figure 24–15A, the stimulation strengths applied to each of two peripheral nerves are adjusted to be subliminal for evoking a synaptic potential in the motoneuron when either of the peripheral nerves is stimulated alone. Stimulating the two peripheral nerves together, however, does evoke a synaptic potential in the motoneuron. From such results, one infers that the subliminal excitatory actions of the two inputs summate in common interneurons causing some of them to discharge and, in turn, to produce the EPSP in the motoneuron. In Figure 24–15B, each source of input evokes a clear EPSP in the motoneuron, but when they are stimulated together, the resulting EPSP is greater than the algebraic sum of the EPSPs they produce individually. This observation again suggests that there is a population of interneurons that are excited subliminally by both inputs, and that the concurrent action of the two inputs brings these common interneurons to spike threshold, causing them to contribute to the synaptic potential recorded in the motoneuron.

The spatial facilitation paradigm is also used to study the convergence of excitatory and inhibitory inputs on common interneurons, albeit with more caution. As illustrated in Figure 24–15C, input I evokes an EPSP in the motoneuron, whereas input II produces no detectable response. Stimulating the two inputs together evokes an EPSP in the motoneuron, but its amplitude is less than that of the EPSP evoked by input I alone. One may infer that these two inputs converge on common interneurons, but only after determining that input II does not produce either a conductance change in the motoneuron or presynaptic inhibition of the afferent fibers comprising input I.

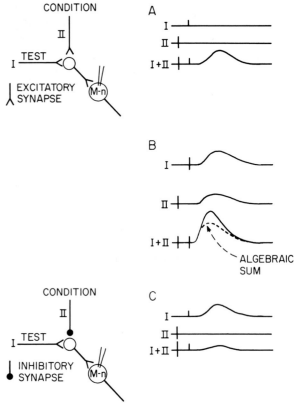

Figure 24–15 Technique of spatial facilitation used to demonstrate interneuronal convergence of segmental reflex pathways. In *A*, stimulating either source I (test) or II (condition) alone evokes no detectable synaptic potential in a motoneuron (M-n). However, when the two sources of afferent fibers are stimulated together, a clear EPSP is evoked. In *B*, both sources I and II generate EPSPs in the motoneuron, but when they are stimulated concurrently, the resultant EPSP is of greater magnitude than would be expected from a simple summation of the EPSPs they evoked individually. Both of these examples are indicative of an excitatory convergence onto a common pool of interneurons. In *C*, source I evokes an EPSP in the motoneuron. When I and II are stimulated together, the resulting EPSP is smaller than that evoked by source I alone. One can interpret this result as indicating that source II inhibits interneurons in the pathway from source I to the motoneuron, providing that source II does not produce a conductance change in the motoneuron or produce primary afferent depolarization (PAD) in source I. (After Baldissera, F., Hultborn, H., Illert, M. *Handbk Physiol.* Sec. 1, 2:509–595, 1981.)

SPINAL REFLEXES

In the previous sections we have emphasized that much of the input from peripheral receptors converges on common spinal interneurons, which in turn produce synaptic potentials in other interneurons and motoneurons. As will be described in Chapter 26, many of these same interneurons also receive inputs from neurons in the brain stem and cortex that are involved in the control of movement. In light of this extensive convergence, it is not surprising that the synaptic potentials evoked in interneurons and motoneurons by individual presynaptic neurons are generally quite small, often only 100 μV. Thus, a considerable degree of temporal and spatial summation of these inputs is required to activate a motoneuron (see Chap. 11). Nonetheless, in some instances, physiological activation of peripheral afferent fibers alone can result in sufficient synaptic input to generate action potentials in motoneurons. We refer to the short-latency, stereotyped activation of motoneurons by neural elements restricted to the periphery and spinal cord as spinal reflexes (see Chap. 22).

The characterization of a spinal reflex entails determining its adequate stimulus, observing which muscles generate the response, and measuring the time course or latency from the stimulus to the response. Having ascertained these characteristics, one can then proceed to identify the type or types of sensory receptors involved and the connections their afferent fibers make with the different groups of spinal neurons (i.e., the segmental reflex pathway). In the following sections, the different types of spinal reflexes and their segmental circuitry will be described. We will also relate these reflexes to the properties of the different peripheral receptors that are thought to be involved in their elaboration.

TENDON JERK REFLEX

The least complex of the spinal reflexes is the tendon jerk, which is illustrated schematically in Figure 24–16. Applying a brisk tap to the patellar tendon of a human subject just below the knee produces a small-amplitude, rapid stretch of the quadriceps muscles. The brief muscle stretch imparts an adequate stimulus to the annulospiral or primary sensory endings of muscle spindles within the quadriceps muscles, and thus one or more action potentials are evoked from nearly all of the group Ia fibers. As described earlier, activation of these group Ia fibers elicits EPSPs in each of their homonymous motoneurons. Through temporal and spatial summation of these Ia EPSPs, some of the homonymous motoneurons are depolarized to threshold, causing them to generate action potentials. The action potentials are then propagated to the muscle, resulting in a twitch

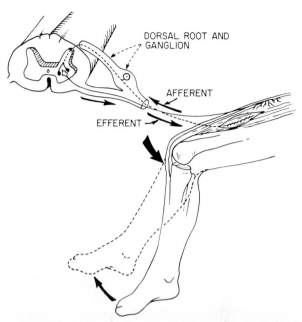

Figure 24–16 Schematic representation of the tendon jerk reflex evoked in the human quadriceps muscles.

contraction of the quadriceps that extends the knee. The tendon jerk reflex is highly focused in that the response is restricted to the muscles that are stretched.

By the 1920s it had been clearly demonstrated that muscle stretch is the adequate stimulus for the tendon jerk reflex and that the responsible receptors reside within the muscle itself (muscle proprioceptors).[22] Even after the nerves innervating the overlying skin, joints, and other muscles are severed in an experimental animal preparation, the reflex response to a brief muscle stretch persists. Cutting the ipsilateral dorsal roots, however, abolishes the tendon jerk, indicating that the tendon jerk is a true reflex and not due to mechanical activation of the muscle itself.

In the 1940s electrical recordings of neural discharge in the dorsal and ventral roots were used to analyze the circuitry underlying the tendon jerk reflex. As shown schematically in Fig. 24–17, the tendon jerk stimulus was simulated by imparting a very brief stretch to the medial gastrocnemius muscle of the cat with an electromechanical muscle-stretching device.[23] The muscle stretch activates many afferent fibers nearly synchronously so that an electrode on the dorsal roots records a large compound action potential (called an afferent volley) that reflects the summation of many individual action potentials. The reflex discharge of

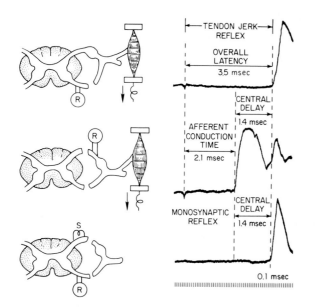

Figure 24–17 Experiment demonstrating that the tendon jerk reflex is mediated by a monosynaptic segmental pathway. The upper trace shows the reflex discharge recorded from a ventral root in response to a brief, tendon-jerklike stretch applied to the cat gastrocnemius muscle. The middle trace shows the afferent volley recorded from a dorsal root following the muscle stretch. The lower trace shows monosynaptic discharge recorded from the ventral root in response to electrical stimulation applied to the dorsal root. The latency of this monosynaptic reflex (1.4 ms) is equivalent to the difference between the overall latency of the tendon jerk reflex (3.5 ms) and the afferent conduction time (2.1 ms). (After Lloyd, D.P.C. *J. Neurophysiol.* 6:317–326, 1943.)

motoneurons can be recorded from the ventral roots that contain their axons. As shown in the top trace in Figure 24–17, the reflex discharge of motoneurons was recorded from the ventral root 3.5 ms following the muscle stretch. The middle trace (Fig. 24–17) indicates that the afferent volley arrived at the dorsal root 2.1 ms after the stretch, leaving only 1.4 ms for conduction of the afferent volley into the spinal cord, synaptic transmission, motoneuron activation, and conduction of the motoneuronal action potentials to the recording electrode on the ventral root (i.e., 3.5 ms overall latency − 2.1 ms afferent conduction time = 1.4 ms). The bottom trace (Fig. 24–17) shows that when the dorsal root was stimulated electrically, reflex discharge appeared in the ventral root after a delay of 1.4 ms. Thus, the reflex evoked by the muscle stretch had the same delay through the spinal cord as that of the shortest latency reflex response produced by electrical stimulation of dorsal root fibers. Since it had been previously established that the shortest latency pathway through

the spinal cord from dorsal root afferents to motoneurons is monosynaptic,[32] one can conclude that the tendon jerk reflex must be mediated by a monosynaptic spinal circuit. Moreover, the afferent conduction time from the muscle to the dorsal root (Fig. 24–17) indicates that the fibers responsible for the tendon jerk reflex must be group I.

As discussed earlier in this chapter, the group I afferent fibers in muscle nerves innervate either the annulospiral endings of the muscle spindles (group Ia) or the Golgi tendon organs (group Ib). It was long held that the muscle spindle afferents were responsible for the tendon jerk, based largely on the early demonstration that muscle spindles have a greater sensitivity to muscle stretch than do tendon organs.[25] This hypothesis was confirmed when it was shown that a tendon jerk–like stimulus (actually a brief stretch of < 60 μm applied to the tendon of the cat soleus muscle) activates nearly all the muscle's Ia afferent fibers, but none of the Ib or group II afferent fibers. Further, when the stretches were applied to the tendon with the dorsal roots intact, monosynaptic EPSPs equivalent to those evoked by electrical stimulation of the muscle nerve at group I strength were recorded in homonymous motoneurons. From these data, we can confidently infer that the tendon jerk stimulus activates Ia afferent fibers, which in turn produce monosynaptic reflex discharge of homonymous motoneurons, resulting in a twitch-like contraction of the muscle that they innervate.

The same population of group Ia fibers that mediates the tendon jerk reflex also makes monosynaptic connections with motoneurons that innervate synergist muscles. As described earlier, however, the afferent fibers make fewer contacts with synergist motoneurons, so that the EPSPs generated in them are smaller than those generated in the homonymous motoneurons. The resulting depolarization in the synergist motoneurons normally fails to bring them to threshold, and thus no reflex contraction occurs in the synergist muscles. On the other hand, applying stretches to a muscle and its synergist simultaneously enhances the reflex responses in both.

The Ia afferent fibers activated by a tendon jerk stimulus also impinge on Ia-inhibitory interneurons that, as described earlier in the chapter, inhibit motoneurons innervating the antagonist muscles acting at the same joint. Thus the tendon jerk reflex of a muscle is generally diminished when a concurrent stretch stimulus is delivered to its antagonist.

One can produce a reflex response analogous to the tendon jerk by substituting an electrical shock to a muscle nerve for the natural stretch stimulus. The response to an electrical shock is called the Hoffman, or H-reflex, and is widely used in neurological diagnosis. Group Ia muscle spindle afferent fibers can be stimulated electrically in human subjects when the peripheral nerve in which they reside lies near the skin surface. Applying a single shock through electrodes placed over the popliteal fossa, for example, produces a brisk reflex contraction of the soleus muscle that closely resembles that elicited by a tendon tap stimulus. The Ia fibers can be activated selectively because they are generally slightly larger (c. 10%) than the alpha motor axons in the same nerve and thus have a lower threshold to an externally applied electrical stimulus. As one would expect, the electrical activation of Ia fibers also leads to a transient inhibition of the antagonist muscles.

STRETCH REFLEX

If a more prolonged stretch is applied to a muscle, a reflex contraction of the stretched muscle ensues (i.e., a stretch reflex) that is of longer duration than the tendon jerk. Figure 24–18 compares the response of the soleus muscle in a decerebrate cat to small stretches under two different conditions. In the top trace (labeled REFLEX), the segmental circuitry was intact and it is clear that the muscle generated a considerable increase in force in response to the stretch. To determine how much of this force increase can be attributed to the stretch reflex, the same stretch stimulus was applied to the muscle after the dorsal roots had been cut to interrupt the segmental reflex pathways (labeled MECHANICAL). To produce the equivalent level of pre–stretch muscle activation in the de-afferented muscle, the ventral roots were stimulated tonically to activate the motor axons. Notice that the MECHANICAL response of the muscle initially is identical to that of the REFLEX, but then abruptly decreases (i.e., "yields"). The difference between these two responses (REFLEX ACTION) provides an estimate of the contribution of the reflex to the increase in muscle force accompanying the change in muscle length.[18, 22, 30] Clearly the stretch reflex dramatically decreases the severity of the yield and markedly increases the stiffness (force/unit length change) of the muscle. In effect, the stretch reflex acts to make the muscle behave more like a mechanical spring[18] (see Chap. 22).

Prolonged muscle stretches activate not only group Ia fibers but also other muscle afferents that influence the excitability of the motoneurons. For

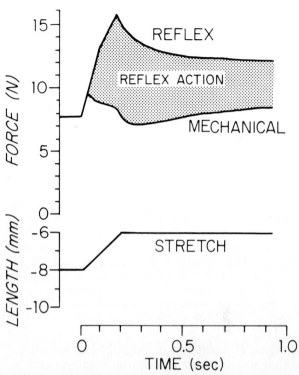

Figure 24–18 Demonstration of the stretch reflex in the cat soleus muscle. The upper traces show the change in muscle force consequent to stretching the muscle by 2 mm; the lower trace shows the length change. The trace labeled REFLEX was obtained with the spinal reflex circuits intact, whereas in the trace labeled MECHANICAL the ipsilateral dorsal roots had been cut. The shaded area indicates the difference in muscle force that can be attributed to reflex action. (After Houk, J.C., et al. In Taylor, A.; P. Chazka, A., eds. *Muscle Receptors and Movement.* Oxford Univ. Press, 1981.)

example, group II afferents that innervate the flower spray, or secondary, endings of the muscle spindles contribute to the stretch reflex through both monosynaptic and polysynaptic pathways. Moreover, in addition to the effects of their monosynaptic connections, group Ia afferents may influence homonymous motoneuron excitability through polysynaptic pathways. Group Ib afferents that innervate Golgi tendon organs are also activated by stretch stimuli but they tend to inhibit the stretch reflex (see Fig. 24–8) as will be discussed later. The details of these different stretch reflex circuits, as well as their relative efficacies, are unresolved issues and remain the subjects of considerable research and controversy.

The stretch reflex has an initial phasic component that occurs while the muscle length is changing and a smaller, sustained tonic component associated with the final "resting" length of the muscle (Fig. 24–18). These two components of the stretch reflex seem to reflect the receptor properties of the muscle spindles, which in turn influence the discharge patterns of the group Ia and group II afferent fibers that innervate them. In the following section the properties of muscle spindles will be discussed in relation to their role in stretch reflex mechanisms.

MUSCLE SPINDLES[3, 26, 27]

The mammalian muscle spindle is a highly developed sense organ that is found in abundance in nearly every skeletal muscle. Its name is derived from the spindle shape of its accessory structure, which consists of a parallel array of from 2 to 12 specialized muscle fibers whose mid-sections are enclosed by a collagenous fluid-filled capsule (Fig. 24–19). The spindle's muscle fibers, which are called intrafusal fibers because they lie within the spindle or fusiform capsule, are oriented parallel to the larger, extrafusal skeletal muscle fibers. Most of the intrafusal fibers are innervated by a separate group of small motoneurons called gamma, or fusimotor, neurons, whose somata are found intermingled among the homonymous alpha motoneurons that innervate the skeletal muscle (extrafusal) fibers.

The emergence of a separate motor system for muscle spindles appears to have been a relatively recent evolutionary advance. Muscle spindles are found throughout the vertebrate subphylum, but only mammals have gamma motoneurons. In the other vertebrate classes, the intrafusal fibers are innervated by collaterals of the alpha motoneuron axons. One still finds evidence of this shared innervation in many mammals, including humans. Motoneurons that innervate both skeletal and intrafusal muscle fibers are called beta motoneurons.

There are two major classes of intrafusal fibers. The longer intrafusal fibers (8–10 mm) are called nuclear bag, because their nuclei are clustered within their expanded equatorial regions. The shorter (4–5 mm), more slender intrafusal fibers are called nuclear chain, because their nuclei are arranged in a single file. A typical spindle has two nuclear bag fibers (called bag_1 and bag_2; see later) and four nuclear chain fibers, but the numbers of bag and especially chain fibers vary considerably from spindle to spindle.

The two types of sensory receptors have different locations and structural relationships within the spindle. The primary (or annulospiral) sensory ending is found at the equator of each spindle and coils around each of the intrafusal fibers within the spindle capsule. The primary ending is supplied by a group Ia afferent axon. One or more secondary, or flower spray, endings are generally found near the equator of the spindle. These receptors contact predominantly the intrafusal chain fibers and are supplied by group II afferent fibers. Both the primary and secondary endings are mechanoreceptors (see Chap. 5) that are sensitive to changes in the lengths of the intrafusal fibers they contact. Because the intrafusal fibers lie in parallel with, and are anchored to, the surrounding skeletal muscle fibers, they are stretched when their "host muscle" is stretched. Muscle stretch leads to a depolarizing receptor potential in the sensory endings, which in turn generates action potentials at the trigger zone of the afferent fiber. The firing rate of the afferent fiber increases with the amount of stretch imposed on the muscle (Fig. 24–20B). In contrast, muscle contraction results in shortening of the intrafusal fibers, which reduces the stretch applied to the receptor endings. Reducing the stretch on the receptor endings leads to a decrease in the receptor potential and a corresponding decrease in afferent fiber discharge (Fig. 24–20A).

The primary and secondary endings of the muscle spindle have similar static length sensitivities; that is, for a given change in muscle length they each provide a proportional change in the discharge rate of their afferent fibers. However, while the muscle length change is in progress, the primary ending displays a much greater dynamic response, as shown in Figure 24–21.

It is thought that the more pronounced dynamic sensitivity (i.e., the response to the rate of muscle length change) of the primary ending is a consequence of its more extensive association with the nuclear bag fibers within the spindle. The nuclear bag fibers seem to possess considerable mechanical friction and thus do not readily comply to imposed length changes. In general, mechanical systems with friction or viscosity in addition to elasticity (for steady, tonic response) are affected by the rate at which the stimulus is applied as well as by its magnitude. Thus the mechanical properties of the accessory structure (the nuclear bag fiber) "shape" the receptor's behavior.[27] Similar stimulus shaping by accessory structures occurs in many mechanoreceptors—for example, pacinian corpuscles (Chap. 5), vestibular hair cells (Chap. 27), and auditory hair cells (Chap. 17).

In addition to changes in muscle length, the primary and secondary endings of the muscle spindle are responsive to contractions of the intrafusal fibers. The equatorial regions of the intrafusal fibers (where the sensory endings are concen-

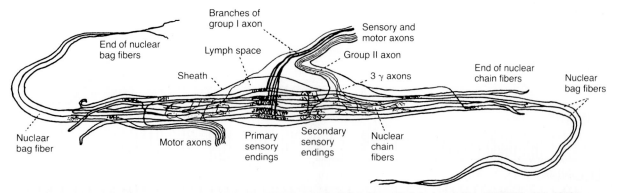

Figure **24–19** A dissected cat muscle spindle. (After Boyd, I.A. *Philos. Trans. R. Soc. Lond. [Ser. B]* 245:81–136, 1962.)

trated) are nearly devoid of myofibrils. The gamma motoneurons innervate the polar regions of the intrafusal fibers, which are well supplied with contractile elements (Fig. 24–19). Thus, when the polar regions of an intrafusal fiber contract, the equatorial region of the fiber is stretched, imparting an adequate stimulus to the sensory endings. It has been estimated that the maximal contraction of the intrafusal fibers of a muscle spindle provides a stretch stimulus to the sensory endings equivalent to that produced by extending the host muscle to its maximal physiological length.

Contraction of the intrafusal fibers can alter both the rate of afferent discharge at a given muscle length and the dynamic response of the sensory endings to changes in muscle length. These dif-

ferent effects are largely produced by two classes of gamma motoneurons.[10] Activation of one group, the dynamic gamma motoneurons, increases the sensitivity of the primary ending to changes in muscle length (Fig. 24–22). The dynamic gamma motoneurons appear to innervate the nuclear bag_1 fibers exclusively; and since the secondary endings do not normally contact the bag_1 fibers, dynamic gamma motoneurons influence only the behavior of the primary endings. The second group, the static gamma motoneurons, innervates both the nuclear bag_2 intrafusal fibers and the nuclear chain fibers. Thus, activation of static gamma motoneurons increases the rate of afferent discharge of both primary and secondary sensory endings at all muscle lengths. In some

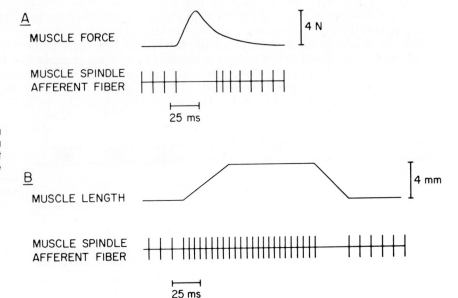

Figure **24–20** Responses of a muscle spindle afferent fiber to a twitch contraction of the parent muscle *(A)* and to changes in the muscle length *(B)*.

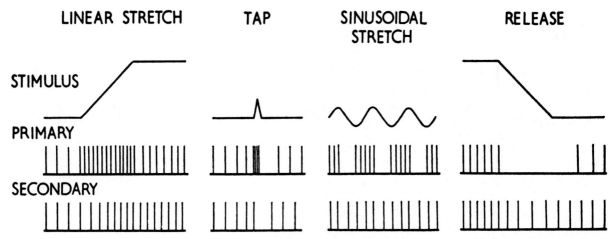

Figure 24–21 Comparison of the responses of primary (annulospiral) and secondary (flower-spray) muscle spindle endings to different types of muscle stretch stimuli. (From Matthews, P.B.C. *Physiol. Rev.* 44:219–288, 1964.)

cases, static gamma motoneurons actually seem to reduce the dynamic sensitivity of the sensory endings (Fig. 24–22). (Beta motoneurons can also be classified as static or dynamic based upon their actions on the spindles they innervate.)

FUNCTIONAL ROLE OF FUSIMOTOR SYSTEM[26, 27]

Since the discovery of the fusimotor system some 40 years ago, there has been a great deal of speculation and experimental work concerning its functional role in motor control. One simple hypothesis suggests that both the dynamic and static gamma motoneurons are co-activated with alpha motoneurons during the execution of movements so that the discharge of the muscle spindle can be maintained when the muscle shortens. If the muscle spindle serves to provide the central nervous system with muscle length information, a great deal of that information would be lost if the

spindle afferents were to "fall silent" during muscle contraction.

Recordings from muscle spindle afferents during twitch contractions of their muscle[19] supported the idea that the gamma system might serve to maintain afferent discharge in the face of muscle shortening. As shown in Figure 24–23A, the spindle afferent stopped firing during the rising phase of the twitch contraction indicating that the intrafusal fibers had shortened sufficiently to bring the sensory ending's receptor potential below its threshold. However, when a gamma motoneuron innervating intrafusal fibers of the same muscle spindle was stimulated concurrently, the spindle afferent's discharge did not fall silent during the active shortening phase of the contraction; in fact (Fig. 24–23B), its rate actually increased. Indeed, it has been demonstrated in man and in freely moving animals that muscle spindles do continue to discharge throughout a wide range of movements, suggesting that alpha and gamma motoneurons are, in fact, co-activated by the central

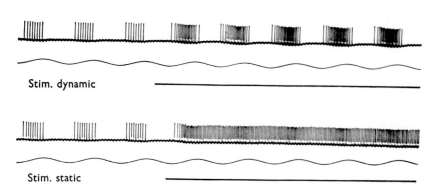

Figure 24–22 The effects of dynamic and static fusimotor stimulation on the response of a primary muscle spindle ending to sinusoidal changes in muscle length (1 mm peak-to-peak). The horizontal lines mark the periods of fusimotor stimulation at 100 stimuli per second. (From Crowe, A.; Matthews, P.B.C. *J. Physiol. [Lond.]* 174:132–151, 1964)

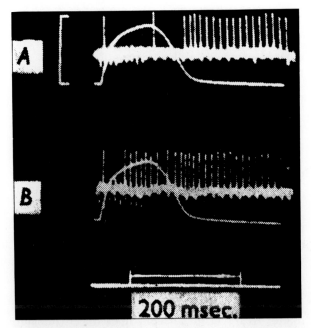

Figure 24–23 The effect of fusimotor stimulation on the response of a muscle spindle afferent to muscle contraction. In *A*, the afferent fiber showed a distinct pause in its discharge during the muscle contraction. However, as shown in *B*, the afferent fiber maintained its discharge during the contraction when concurrent fusimotor stimulation at 100 Hz was added. Vertical calibration bar = 100 g. (After Hunt, C.C.; Kuffler, S.W. *J. Physiol [Lond.]* 113:283–297, 1951.)

nervous system. Alpha and gamma motoneurons do not, however, appear to be rigidly co-activated during the execution of all movements, but rather exhibit some degree of independence.

In the freely moving conscious cat, the spindles in active muscles receive a tonic level of gamma input. However, the relative balance between static and dynamic gamma activation varies considerably in the execution of different types of movement.[31] For example, dynamic gamma motoneurons are often only minimally active during familiar, stereotyped movements such as those associated with normal walking but become much more active during more novel and unpredictable movements. The static gamma motoneurons, on the other hand, are quite active during the course of stereotyped movements, and their discharge is modulated in parallel with the level of muscle force.

SEGMENTAL INPUTS TO GAMMA MOTONEURONS[1]

It has proven quite difficult to obtain intracellular recordings from gamma motoneurons, owing

perhaps to their relatively small somata. Thus, much of our limited information on the reflex actions of peripheral afferent fibers on gamma motoneurons has come from analysis of their extracellular spike trains recorded from ventral root filaments or peripheral muscle nerves. Unlike alpha or beta motoneurons, gamma motoneurons do not receive monosynaptic input from any of the peripheral afferent fibers. Group I afferents produce both weak excitatory and weak inhibitory actions on gamma motoneurons mediated through one or more interneurons. More powerful reflex actions are exerted by group II afferents from muscle nerves, which generally excite flexor gamma motoneurons while evoking a mixture of excitatory and inhibitory effects on extensor gamma motoneurons. Electrical stimulation of flexor reflex afferents (FRAs) influences gamma motoneuron discharge through both short- and long-latency polysynaptic pathways. Generally, the short-latency pathways excite flexor gamma motoneurons and inhibit extensors, but as was the case for FRA input to alpha motoneurons, reversed effects are also common. In some preparations, the short-latency FRA pathways appear to exert different actions on dynamic and static gamma motoneurons. The long-latency FRA pathways influence both static and dynamic gamma motoneurons with the same pattern observed in alpha motoneurons (i.e., ipsilateral FRAs excite flexor gamma motoneurons, whereas contralateral FRAs excite extensor gamma motoneurons).

FLEXION REFLEXES

A gentle probing of the skin elicits a weak contraction of the underlying flexor muscles. Application of a noxious stimulus to the same area results in a more widespread and powerful activation of ipsilateral flexor muscles, which generally causes the limb to withdraw from the stimulus. Both of these responses are flexion reflexes. Flexion reflexes are particularly striking after spinal cord lesions in both man and experimental animals.

The flexion reflex is thought to be mediated by the FRAs which, as described earlier in this chapter, include groups II, III, and IV (unmyelinated) fibers found in muscle, joint, and cutaneous nerves.[14] The FRAs innervate mechanoreceptors, chemoreceptors, temperature receptors, and nociceptors. All of these FRAs transmit their effects to motoneurons through one or more spinal interneurons. In general, the FRAs promote excitation of ipsilateral flexor motoneurons and inhibition of ipsilateral extensor motoneurons as described earlier in this chapter. Some of the interneurons that are excited by the FRAs send axon collaterals to the contralateral dorsal horn via the anterior commissure. These collaterals excite other interneurons which generally excite contralateral extensor motoneurons and inhibit contralateral flexor motoneurons. This pattern of synaptic actions often results in extension of the contralateral limb which is referred to as the crossed-extension reflex. The crossed-extension reflex pattern presumably evolved to facilitate the contralateral limb's assuming additional weight-bearing in the event of a limb withdrawal from a harmful stimulus.

Although the flexion reflex displays considerable divergence in its actions, it is not without some semblance of topographic organization. The distribution of reflex action with respect to different flexor muscles in a limb is related to the location of the stimulus. This "mapping" is often referred to as local sign. For example, applying a noxious stimulus to the foot evokes strong activation of flexor muscles acting at the ankle, some activation of those at the knee, and little involvement of flexor muscles at the hip. In contrast, the same type of stimulus applied to the posterior compartment of the leg evokes strong knee and hip flexion, but little activation of flexor muscles controlling the ankle.

It is instructive to compare the properties of the flexion reflex to those of the far more simple tendon jerk reflex. The flexion reflex has a far longer latency than the tendon jerk, owing both to the slower conduction of the FRA fibers (as compared to the Ia fibers) and its polysynaptic spinal pathways. Whereas the tendon jerk reflex can be smoothly "graded" according to the intensity of the stimulus, the flexion reflex displays a relation between stimulus and response that is more complex; that is, a weak stimulus may evoke a barely detectable response, while one slightly stronger may evoke a powerful limb withdrawal. In addition, unlike the transient tendon jerk, the flexion reflex often persists beyond the duration of the stimulus, a phenomenon referred to as afterdischarge. This persistence in motor outflow is thought to be mediated by the sustained discharge of groups of interneurons involved in the reflex pathway. Finally, the degree of "focus" of these reflexes is quite different; the tendon jerk is restricted to a single muscle, whereas the flexion reflex may enjoin muscles throughout an entire limb.

LONG SPINAL REFLEXES

In quadripedal vertebrates, the flexion and crossed extension reflexes are not limited to one pair of limbs, but are transmitted up or down the spinal cord to influence the other pair of limbs. For example, flexion of a forelimb in a spinalized animal is often accompanied by extension of the ipsilateral hindlimb. Similarly, if the stimulus is powerful enough to evoke crossed-extension of the contralateral forelimb, one generally observes concomitant flexion of the contralateral hindlimb. This reflex pattern resembles normal locomotor action and may utilize the same spinal circuitry that underlies the coordinated activations of the limb musculature during walking and running (see Chap. 26).

The reflex coordination between pairs of limbs (called long spinal reflexes) involves the activation of propriospinal neurons located in the central gray matter of the cord. Their axons leave the gray matter to travel in adjacent fiber tracts (funiculi) for variable distances up or down the cord before re-entering the gray matter to synapse on last-order interneurons and motoneurons. Propriospinal neurons receive convergent synaptic input from peripheral afferents and axons from virtually every supraspinal structure involved in movement.[1]

CLASP-KNIFE REFLEX

The clasp-knife reflex (also called the "inverse myotatic reflex") is characterized by a sudden release of tension in muscles that are stretched beyond a critical point or threshold, such that the limb collapses on itself like the blade of a jackknife or clasp-knife. It is demonstrable only following lesions of the central nervous system and is one of the hallmarks of spasticity in humans. It is presently thought that the clasp-knife reflex is mediated by group II and group III afferent fibers innervating free nerve endings located in muscle fascia and aponeuroses, but further studies are needed to confirm this hypothesis.[34]

GOLGI TENDON ORGANS

Until quite recently, the clasp-knife reflex was attributed to the actions of group Ib afferent fibers from Golgi tendon organs. One of the prime reasons for linking the clasp-knife reflex to Golgi tendon organ afferents was the nature of their synaptic input to alpha motoneurons. As described earlier in this chapter, group Ib afferents generally produce di- and trisynaptic IPSPs in homonymous and synergist motoneurons and polysynaptic EPSPs in antagonist motoneurons. This pattern of connectivity leads to autogenetic inhibition; that is, the afferent fibers inhibit "their own" motoneurons. A second rationale for associating the clasp-knife reflex with the Golgi tendon organ was that the reflex typically occurs only after considerable force is exerted against a limb, and the earliest studies of the tendon organ also reported a rather high threshold to a muscle stretch (particularly in comparison with that of the muscle spindle receptors). However, more recent studies have considerably weakened these arguments, showing that both the discharge of Ib afferent fibers and the synaptic potentials they generate in motoneurons are finely graded and do not display the sharp "threshold" that is characteristic of the clasp-knife reflex. Moreover, the reflex can be mimicked by gently manipulating the surface of the muscle and tendon, stimuli that do not significantly activate Ib afferent fibers.[34]

While we can no longer ascribe a distinct reflex action to the Golgi tendon organs, it is clear that they provide the central nervous system with an accurate account of the amount of force produced by the muscle in which they reside. As depicted in Figure 24–24, tendon organs consist of collagenous fascicles enclosed in a capsule (500-1500 μm) with an intertwined sensory ending. They are usually found at the junctions between muscle fibers and the aponeurotic sheaths that extend throughout muscles, from the tendon of origin to the tendon of insertion. Belying their name, a minority of these receptors are actually found in the tendons themselves. An average of ten skeletal muscle fibers insert into the tendon organ capsule and these fibers alone impart the adequate stimulus to the receptor.

The tendon organ apparently operates like a mechanical strain gauge. Active and passive forces produced by any of the muscle fibers that attach to the receptor capsule are transmitted to the mechanosensitive terminals of the Ib axon, leading to depolarizing receptor potentials. The receptor is so sensitive that a twitch contraction of any one of the individual muscle fibers attached to the capsule will produce action potentials in the Ib afferent fiber.[15]

In most cases, each of the muscle fibers that attaches to a given tendon organ is part of a different motor unit (see Chap. 23). This suggests

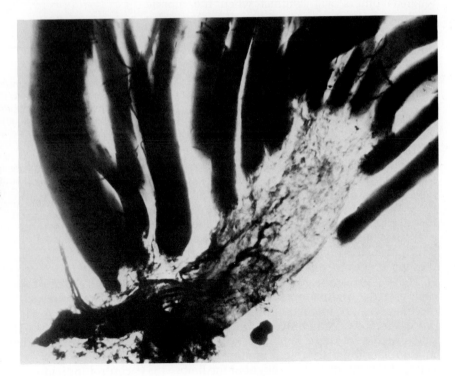

Figure 24–24 Photomicrograph of a cat Golgi tendon organ and the muscle fibers that insert on its capsule. (Courtesy of Dr. F.J.R. Richmond.)

that the coupling of muscle fibers to tendon organs does not require any special rearrangements of muscle fibers within the muscle; instead, the mix of fibers attached to the receptor simply reflects the local fiber composition of the muscle. (Recall from the discussion of motor unit anatomy in Chapter 23 that the fibers of a single motor unit generally account for fewer than 5 of 100 adjacent muscle fibers.) Because only the fibers that attach to the receptor are capable of exciting it, each tendon organ responds to the activity of only a small fraction of the motor units within a muscle. As shown in Figure 24–25, a tendon organ that responds vigorously to the contraction of single motor unit generating a force of less than 0.2 N may remain indifferent to the simultaneous contractions of many other motor units in the muscle even though together they generate a hundredfold more force![2]

Despite the limited sample of motor units that activate an individual tendon organ, the rate of discharge of Ib afferent fibers is nevertheless well correlated with the total level of force produced by their host muscle in most instances.[9] Presumably, this correlation occurs because the muscle fibers that attach to a tendon organ belong to motor units that are activated at different levels of force; some of the attached muscle fibers will contract when the muscle produces small forces, whereas others will contract when the muscle produces larger forces (see Chap. 25). Of course, the correlation between the force produced by the muscle and the rate of Ib afferent discharge is improved if one considers (or "averages") the signals emanating from several Ib afferents.

REFLEX FUNCTION

The preceding sections have presented the spinal reflexes from an analytical perspective rather than a functional one. Neurophysiologists have studied spinal reflexes in isolation in the hope of understanding the operation of simple neural circuits. These simple neural circuits, however, have turned out to be far more complex than anticipated. It is quite clear, for example, that very few peripheral or descending inputs (see Chap. 26) have "private access" to the final common pathway, the motoneuron. Most of the inputs to motoneurons are integrated through convergence onto common interneurons. Moreover, the outputs of many of these same interneurons diverge to modulate the discharge of motoneurons innervating different muscles throughout the limb. This

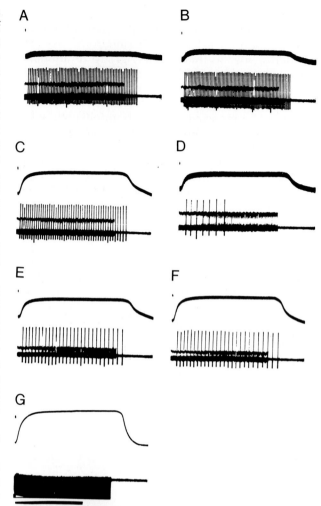

Figure 24–25. Responses of a single Ib afferent fiber from the cat soleus muscle to the forces (upper traces) produced by stimulating different groups of motor axons innervating the muscle. A–F show the afferent fiber's response to the forces produced by stimulating six different groups of motor axons. Stimulating other motor axons (G) produced far more force in the muscle than did those shown in A–F, but no discharge from this Ib afferent fiber. A calibration pulse precedes each tension record and indicates 0.5 N in A, B, and D., 1.0 N in C, E, and F; and 16.0 N in G. The ventral roots were stimulated at 100 Hz for 1.5 s. The small amplitude spike is the stimulus artifact. (From Binder, M.D. *Exp. Brain Res.* 43:186–192, 1981.)

convergence–divergence property of spinal interneurons suggests that spinal reflex circuits may be involved principally in coordinating the actions of the various muscles of the limb during movement (see Chap. 26). Thus, it is difficult to ascribe specific independent functions to the different spinal reflexes or to envision how their actions control individual muscles.

It is certainly true that segmental reflex pathways can produce dramatic changes in motor output, as when the body encounters noxious stimuli or physical obstructions during locomotion (flexion reflex). However, it is probably more common for the segmental pathways to produce subtle alterations in motor outflow to compensate for changes in the substrate against which the limbs exert forces, or for changes in the performance of the musculature (e.g., fatigue). We might better view the various segmental circuits as lending support and reinforcement to the patterns of motor output required to produce controlled movements described in Chapter 26.

ANNOTATED BIBLIOGRAPHY

Baldissera, F.; Hultborn, H.; Ilert, M. Integration in spinal neuronal systems. *Handbook of Physiology.* Sec. 1, 2 (1): 509–595, 1981.
Review of the extensive literature on the segmental actions of peripheral afferent fibers on spinal neurons, with emphasis on interneuronal networks.

Houk, J.C.; Rymer, W.Z. Neural control of muscle length and tension. *Handbook of Physiology.* Sec. 1, 2 (1):257–323, 1981.
The segmental motor system is considered from a control theory perspective. The authors make inferences on the functional organization of the spinal cord by examining the behavior of muscle in response to mechanical disturbances.

Matthews, P.B.C. *Mammalian Muscle Receptors and Their Central Actions.* London, Arnold, 1972.
This monograph is an exceptional work presented in a most engaging style. The author provides both historical perspective and comprehensive coverage of the subject matter.

REFERENCES

1. Baldissera, F.; Hultborn, H.; Ilert, M. Integration in spinal neuronal systems. *Handbk. Physiol.* Sec. 1, 2(1):509–595, 1981.
2. Binder, M.D. Further evidence that the Golgi tendon organ monitors the activity of a discrete set of motor units within a muscle. *Exp. Brain Res.* 43:186–192, 1981.
3. Boyd, I.A. The structural innervation of the nuclear bag fibres and the nuclear chain muscle fibre system in mammalian muscle spindles. *Philos. Trans. R. Soc. Lond. (Ser.B)* 245:81–136, 1962.
4. Boyd, I.A.; Davy, M.R. *Composition of Peripheral Nerves.* Edinburgh, Livingston, 1968.
5. Brown, A.G.; Fyffe, R.E.W. The morphology of group Ia afferent fibre collaterals in the spinal cord of the cat. *J. Physiol. (Lond.)* 274:111–127, 1978.
6. Brown, A.G.; Fyffe, R.E.W. The morphology of group Ib afferent fibre collaterals in the spinal cord of the cat. *J. Physiol. (Lond.)* 296:215–228, 1979.
7. Brown, A.G.; Fyffe, R.E.W.; Rose, P.K.; Snow, P.J. Spinal cord collaterals of type II slowly adapting units in the cat. *J. Physiol. (Lond.)* 316:469–480, 1981.
8. Coggeshall, R.E. Law of separation of function of the spinal roots. *Physiol. Rev.* 60:716–755, 1980.
9. Crago, P.E.; Houk, J.C.; Rymer, W.Z. Sampling of total muscle force by tendon organs. *J. Neurophysiol.* 47:1069–1083, 1982.
10. Crowe, A.; Matthews, P.B.C. Further studies of static and dynamic fusimotor fibres. *J. Physiol. (Lond.)* 174:132–151, 1964.
11. Eccles, J.C.; Eccles, R.M.; Lundberg, A. The convergence of monosynaptic excitatory afferents onto many species of alpha motoneurones. *J. Physiol. (Lond.)* 137:22–50, 1957.
12. Eccles, J.C.; Eccles, R.M.; Lundberg, A. Synaptic actions of motoneurones caused by impulses in Golgi tendon organs. *J. Physiol. (Lond.)* 138:227–252, 1957.
13. Eccles, R.M.; Lundberg, A. Supraspinal control of interneurones mediating spinal reflexes. *J. Physiol. (Lond.)* 147:565–584, 1959.
14. Eccles, R.M.; Lundberg, A. Synaptic actions in motoneurones by afferents which may evoke the flexion reflex. *Arch. Ital. Biol.* 97:199–221, 1959.
15. Fukami, Y. Responses of isolated Golgi tendon organs of the cat to muscle contraction and electrical stimulation. *J. Physiol. (Lond.)* 318:429–443, 1981.
16. Fyffe, R.E.W. The morphology of Group II muscle afferent fibre collaterals. *J. Physiol. (Lond.)* 296:39P–40P, 1979.
17. Henneman, E.; Mendell, L.M. Functional organization of motoneuron pool and its inputs. *Handbk. Physiol.* Sec. 1, 2 (1):423–507, 1981.
18. Houk, J.C.; Rymer, W.Z. Neural control of muscle length and tension. *Handbk. Physiol.* Sec. 1, 2 (1):257–323, 1981.
19. Hunt, C.C.; Kuffler, S.W. Further study of efferent small-nerve fibres to mammalian muscle spindles. Multiple spindle innervation and activity during contraction. *J. Physiol. (Lond.)* 113:283–297, 1951.
20. Jack, J.J.B. Some methods for selective activation of muscle afferent fibres. In *Studies in Neurophysiology. Essays in Honour of Professor A.K. McIntyre.* Cambridge; Cambridge Univ. Press, 1978.
21. Jankowska, E.; Roberts, W.J. Synaptic actions of single interneurones mediating reciprocal Ia inhibition of motoneurones. *J. Physiol. (Lond.)* 222:623–642, 1972.
22. Liddell, E.G.T.; Sherrington, C.S. Reflexes in response to stretch (myotatic reflexes). *Proc. R. Soc. Lond. (Ser. B)* 96:212–242, 1924.
23. Lloyd, D.P.C. Conduction and synaptic transmission of the reflex response to stretch in spinal cats. *J. Neurophysiol.* 6:317–326, 1943.
24. Lunberg, A.; Winsbury, G. Selective activation of large afferents from muscle spindles and Golgi tendon organs. *Acta Physiol. Scand.* 49:155–164, 1960.
25. Matthews, B.H.C. Nerve endings in mammalian muscle. *J. Physiol. (Lond.)* 78:1–53, 1933.
26. Matthews, P.B.C. *Mammalian Muscle Receptors and Their Central Actions.* London, Arnold, 1972.
27. Matthews, P.B.C. Muscle spindles: their messages and their fusimotor supply. *Handbk. Physiol.* Sec. 1, 2(1):257–323, 1981.
28. McCrea, D.A. Spinal cord circuitry and motor reflexes. *Exer. Sport Sci.* 14:105–141, 1986.
29. Mendell, L.M.; Henneman, E. Terminals of single Ia fibers: location, density, and distribution within a pool of 300 homonymous motoneurons. *J. Neurophysiol.* 34:171–187, 1971.
30. Nichols, T.R.; Houk, J.C. Improvement in linearity and regulation of stiffness that results from actions of stretch reflex. *J. Neurophysiol.* 39:119–142, 1976.

31. Prochaszka, A.; Hulliger, M.; Zangger, P.; Appenteng, K. "Fusimotor set": new evidence for α-independent control of γ-motoneurones during movement in the awake cat. *Brain Res.* 339:136–140, 1985.

32. Renshaw, B. Activity in the simplest spinal reflex pathways. *J. Neurophysiol.* 3:373–387, 1940.

33. Rexed, B. A cytoarchitectonic atlas of the spinal cord of the cat. *J. Comp. Neurol.* 100:297–379, 1954.

34. Rymer, W.Z.; Houk, J.C.; Crago, P.E. Mechanisms of clasp-knife reflex studied in an animal model. *Exp. Brain Res.* 37:93–113, 1979.

35. Schmidt, R.F. Presynaptic inhibition in the vertebrate central nervous system. *Ergeb. Physiol. Biol. Exp. Pharmacol.* 63:20–101, 1971.

36. Stauffer, E.K.; Watt, D.G.D.; Taylor, A.; Reinking, R.M.; Stuart, D.G. Analysis of muscle receptor connections by spike-triggered averaging. 2. Spindle group II afferents. *J. Neurophysiol.* 39:1393–1402, 1976.

37. Watt, D.G.D.; Stauffer, E.K.; Taylor, A.; Reinking, R.M.; Stuart, D.G. Analysis of muscle receptor connections by spike-triggered averaging. 1. Spindle primary and tendon organ afferents. *J. Neurophysiol.* 39:1375–1392, 1976.

Functional Organization of the Motoneuron Pool

Anatomy of Motoneuron Pools
Motor Unit Recruitment
The Size Principle
Recruitment Order and Motor Unit Force
Neural Mechanisms Underlying Orderly Recruitment
Altering Recruitment Order
Control of Muscle Force by Rate Modulation

The motoneuron has been called the "final common pathway" from the central nervous system to one of its principal effector organs, skeletal muscle.[27] What is meant by this term is that every part of the nervous system involved in the control of movement must act either directly or indirectly on motoneurons to produce muscle activation. Since individual muscles are generally innervated by a pool of more than 100 motoneurons, and most movements involve the coordinated action of many muscles, it would appear that the central nervous system faces an insurmountable accounting problem in orchestrating the smoothly graded, precisely controlled sequences of muscular activation that characterize the movements of all vertebrate animals. It turns out, however, that the nervous system is relieved of some of the details of muscle activation through the functional organization of the motoneuron pools themselves. As we shall see in this chapter, despite the enormous complexity of the synaptic circuitry involving peripheral, descending, and interneuronal inputs to motoneurons, motor output generally follows a simple, stereotyped pattern that effectively meets all of the varied demands of muscle usage.

ANATOMY OF MOTONEURON POOLS

The motoneuron pools (also called motor nuclei) of individual skeletal muscles generally form longitudinal columns of cells within the ventral lateral portion of the gray matter of the spinal cord. Motor nuclei were "mapped" initially using cell identification by chromatolysis following sections of individual muscle nerves.[33] More recently, horseradish peroxidase (HRP) has been injected into single muscles, where it is absorbed by the motor axons and transported retrogradely to the motoneuron cell bodies (somata). By staining serial sections of the spinal cord for the presence of HRP reaction product, the motoneurons innervating the injected muscle can be readily identified.

Figure 25–1 is taken from a study in which the left medial gastrocnemius (MG) muscle and the right soleus (SOL) muscle of a cat were injected with HRP.[7] The motoneuron pools of these synergist muscles extend over two spinal segments (the 7th lumbar and 1st sacral). By comparing the left and right sides of the spinal cord, one can infer that the MG and SOL motoneuron pools have a considerable degree of spatial overlap. Notice also that the density of labeled cells within any one of the histological sections is quite low. This is due to the fact that MG and soleus motoneurons are extensively intermingled with motoneurons that innervate several other hindlimb muscles. Surprisingly, this overlap of motor nuclei is not restricted to those innervating functionally related muscles. Despite the overlap, there remains a rough somatotopic relationship between the location of a motor nucleus within the spinal cord and the location of the muscle it innervates in the body. Moreover, even within a motoneuron pool, such as MG, there is some correspondence between the location of a motoneuron within the pool and the territory occupied by its motor unit within the muscle[3] (see Chap. 23).

The distributions of cell soma sizes for motoneurons innervating the medial gastrocnemius and soleus muscles are shown in Figure 25–2. Both alpha and gamma motoneurons (see Chap. 24) are

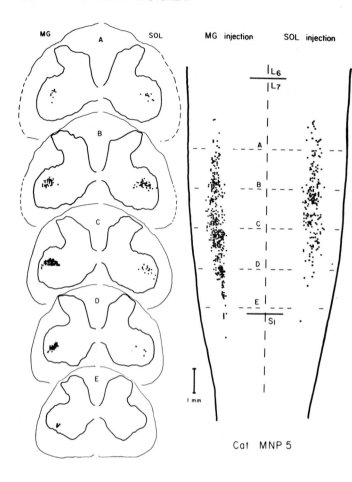

Figure 25–1 Anatomy of the medial gastrocnemius (MG) and soleus (SOL) motor nuclei in the cat spinal cord. *Right*: dorsal view of the cord showing the positions of MG *(left)* and SOL *(right)* cell bodies; the mid-line of the cord is indicated by the vertical dashed line with the boundaries between the L6, L7 and S1 spinal segments indicated by horizontal solid lines. *Left*: Cross sections of the cord at the levels *(A–E)* indicated by the horizontal dashed lines on the dorsal view figure *(right side)*. Dots represent MG *(left)* and SOL *(right)* motoneuron cell bodies located within 300 μm of the cross section. (From Burke, R.E.; Strick, P.L.; Kanda, K.; Kim, C.C.; Walmsley, B. *J. Neurophysiol.* 40:667–680, 1977.)

labeled with HRP and they are completely intermingled throughout the length of the motor pools. In both nuclei, motoneurons with soma diameters <40 μm are considered to be gamma motoneurons. Notice that SOL and MG have approximately the same number of alpha motoneurons with somata of 50 μm or less, but that there are many more MG cells that are larger than 50 μm (see Chap. 23).

MOTOR UNIT RECRUITMENT

By recording the electromyographic (EMG) potentials of individual motor units during the reflex activation of hindlimb extensor muscles in the cat (see Chap. 24), early studies revealed that increases in muscle force are produced by the abrupt recruitment of additional motor units.[10] Similarly, decreases in muscle force appeared to be the result of the abrupt derecruitment of motor units. It was also noted that motor units in the soleus are more easily recruited by reflex activation than units in the medial and lateral gastrocnemius muscles. Based on these observations, it was proposed that motor units display a range of functional thresholds that are correlated with the properties of their muscle units: Motor units in the slow-contracting soleus muscle have lower recruitment thresholds than units in the fast-contracting gastrocnemii. It was postulated that the mechanism underlying this orderly recruitment process was a systematic scaling of excitatory and inhibitory synaptic inputs to the relevant motoneuron pools.[10]

Recruitment and derecruitment of motoneurons in response to a reflex input are demonstrated in Figure 25–3. In this experiment, the action potentials produced by individual motor axons were recorded from subdivided, cut ventral roots.[21] A variable level of excitatory input to the motoneurons innervating the cat ankle extensors was provided through the stretch reflex (Chap. 24) by

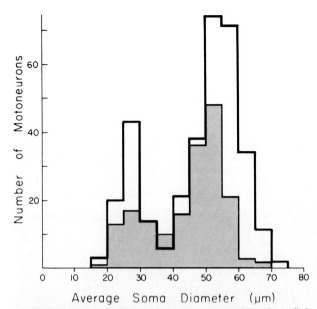

Figure 25–2 Comparison of the soma diameters of medial gastrocnemius (MG; white area) and soleus (SOL; shaded area) motoneurons labeled by horseradish peroxidase (HRP). Cells < 40 μm are assumed to be gamma motoneurons. Note that MG and SOL have roughly the same number of alpha motoneurons with cell bodies of 50 μm or less, but there are many more MG cells that are larger. (From Burke, R.E.; Strick, P.L.; Kanda, K.; Kim, C.C.; Walmsley, B. *J. Neurophysiol.* 40:667–680, 1977.)

increasing or decreasing the length of the triceps surae muscles (soleus and medial and lateral gastrocnemius). As shown in Figure 25–3, individual motor axons could be distinguished on the basis of the amplitude of their action potentials (recorded extracellularly) and the regularity of their interspike intervals. In the example shown in Figure 25–3, action potentials from the motor axons of five distinct units are distinguishable. The five units are numbered in the figure to indicate their recruitment order. Notice that the unit recruited last (unit 5) was the first to be derecruited as the strength of the stimulus was decreased and that, in fact, all five units are derecruited in precisely the reverse order of their recruitment. Repetition of this sequence of muscle stretch and release not only resulted in precisely the same recruitment-derecruitment order, but the amount of muscle stretch at which each unit was recruited ("recruitment threshold") was also remarkably stable and reproducible.

This experiment also demonstrates that the amplitudes of extracellularly recorded motor axon action potentials of units with low recruitment thresholds are generally smaller than those of units with higher recruitment thresholds.[20] Re-

turning to Figure 25–3, one can see that the unit recruited first (unit 1) had the smallest action potential, that the unit recruited second had a slightly larger action potential, and that the unit recruited last (unit 5) had the largest action potential. In similar experiments, the relationship between recruitment order and motor axon potential amplitude was examined for hundreds of pairs of motoneurons. In over 80% of the cases, the unit with the lower recruitment threshold had the smaller action potential.[21]

Further experiments of this type demonstrated that the phenomenon of orderly motor unit recruitment is not limited to segmental stretch reflexes or to extensor motoneurons. Using trains of electrical stimuli applied to areas in the brain stem, cerebellum, basal ganglia, and motor cortex of the cat, it was found that for over 80% of the pairs of motor units tested, the unit with the smaller motor axon potential was recruited before that with the larger potential, regardless of the source of excitation or of the type of muscle that the pair of motoneurons innervated.[35] Even the concurrent activation of an inhibitory input to the motoneuron pools did not appear to disrupt the normal pattern of recruitment and derecruitment.

THE SIZE PRINCIPLE

The findings described earlier were interpreted as evidence that motoneurons within a pool are recruited in order of increasing size so that cells with small somata and soma-dendritic surface areas are invariably activated before those with larger somata, regardless of the source of excitatory input to the spinal cord.[21] This hypothesis, referred to as the *size principle*, rested on two critical inferences: (1) that the amplitude of a unitary action potential recorded extracellularly from a ventral root filament is proportional to the diameter of the motor axon that produces it; and (2) that the diameter of a motor axon is directly related to the size of its cell body. The first of these inferences was the more secure and was based on earlier work demonstrating that in peripheral nerves the amplitude of an extracellularly recorded action potential is proportional to the diameter of the nerve fiber that generates it (see Chap. 3). The second inference was based largely on qualitative histological analyses which indicated that throughout the nervous system the diameter of an axon is related to the size of its cell body.[20, 21] More recently, the correlation between soma size and axon diameter in cat motoneurons

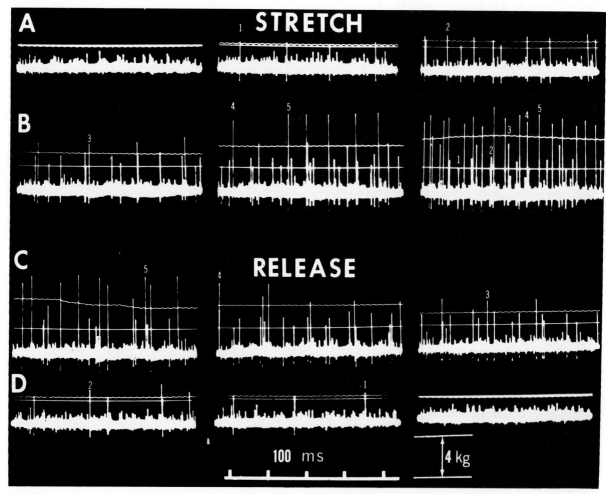

Figure 25–3 Responses of five motoneurons recorded from a ventral root filament to stretch and release of the triceps surae muscles. The distance between the two solid lines above the neural recordings indicates the level of force in the triceps. The numerals above the unit action potentials indicate the recruitment order of the five motoneurons. (From Henneman, E.; Somjen, G.; Carpenter, D.O. *J. Neurophysiol.* 28:560–580, 1965.)

has been confirmed using more precise measurements and statistical analysis.[9]

What are the consequences of a relatively invariant recruitment sequence based on motoneuron size? To address this important question, one must consider how the mechanical properties of muscle units vary with the size of the motor axons that innervate them (see Chap. 23). Recall that, in general, motor axons with relatively slow conduction velocities are associated with type S motor units, which produce small forces, contract relatively slowly, and are extremely resistant to fatigue. At the other end of the spectrum, motor axons with faster conduction velocities innervate type FF units, which produce large forces, contract rapidly, and show little fatigue resistance. Thus, as the level of excitatory drive to a motoneuron pool increases, motor units will be recruited in order of increasing force output, increasing contraction speed, and decreasing fatigue resistance. This scheme ensures a smooth gradation of forces as each new recruited unit adds its force increment to an increasing total level of force. It also ensures that the motor units that are used most frequently and for sustained efforts are appropriately those with the highest fatigue resistance.[22, 23]

RECRUITMENT ORDER AND MOTOR UNIT FORCE

An implicit prediction of the size principle is that the same recruitment order observed in anesthetized preparations would be manifest in freely moving animals and even during voluntary actions in man. To test this prediction, the recruitment order of motor units of the human first dorsal interosseus muscle was studied during voluntary activation.[28] Subjects were asked to gradually increase the force they exerted to adduct their index finger against a force transducer, while EMG potentials of single motor units were simultaneously recorded with an electrode inserted into the muscle (Fig. 25–4A). Since different motor units can be recognized on the basis of the amplitude and waveform of their EMG potentials, the recruitment order and recruitment force threshold of several units could be determined. Once a motor unit was recruited, the subjects were asked to maintain its discharge at a steady rate by observing the EMG potentials on an oscilloscope and listening to an audio monitor. By averaging the force transients following several hundred motor unit action potentials (i.e., spike-triggered averaging; see Chap. 24), an estimate of the twitch

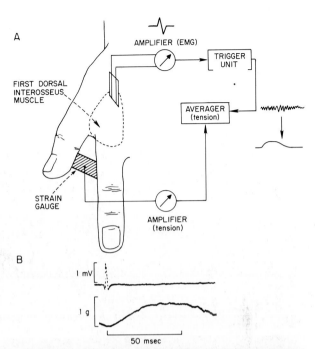

Figure 25–4 The use of spike-triggered averaging to study the recruitment order, twitch force, and contraction time of motor units in the human first dorsal interosseus muscle. *A*, Schematic diagram shows that motor unit action potentials are recorded with electrodes inserted into the muscle and displayed on an oscilloscope. The subject adducts the index finger against a strain gauge while maintaining tonic discharge from one or more motor units. The action potentials of a single motor unit are used to trigger an averager whose input signal is the amplified force record from the strain gauge. As shown in *B*, when the small force transients following several hundred consecutive action potentials from a motor unit are averaged, an estimate of the average twitch contraction of that unit emerges. The upper trace shows the average EMG waveform of the same motor unit. (From Stein, R.B.; French, A.S.; Mannard, A.; Yemm, R. *Brain Res.* 40:187–192, 1972; schematic after McComas, A. *Neuromuscular Function and Disorders,* 65. London, Butterworths, 1977.)

tension produced by the motor unit as well as its twitch contraction time was obtained[36] (Fig. 25–4B).

Data typical of this type of experiment are presented in Figure 25–5. As predicted by the size principle, motor units producing small twitch tensions generally had lower force recruitment thresholds than those producing larger tensions (Fig. 25–5B). A comparison of the two graphs in Figure 25–5 indicates that many of the motor units with the lowest force thresholds had slow contraction speeds as well.[37] In contrast, the highest threshold units all had fast contraction times and produced large twitch tensions. One can infer from these findings that type S and FR motor units

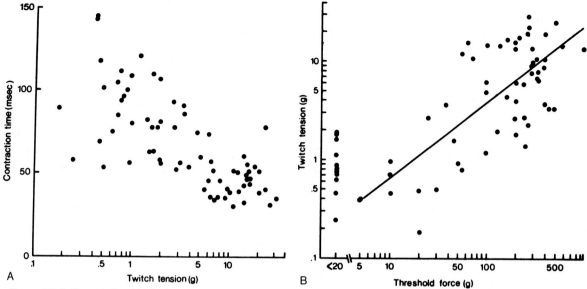

Figure 25–5 The relationship between the mechanical properties of motor units in the first dorsal interosseus muscle of humans and their recruitment threshold. *A,* There is a negative correlation between motor unit force and contraction time, indicating that the most forceful units are also the most rapidly contracting. *B,* There is a positive correlation between motor unit force and recruitment threshold (i.e., units producing small forces tend to be recruited before those producing large forces). Motor units with recruitment thresholds <20 g are stacked in a column. Comparison of *A* and *B* reveals that the units with the lowest recruitment thresholds tend to produce small forces and have a wide range in contraction speeds, whereas those units producing the largest forces are all rapidly contracting. (From Stephens, J.A.; Usherwood, T.P. *Brain Res.* 125:91–97, 1977.)

are recruited before the higher-threshold type FF units (cf. Chap. 23).

It has also been demonstrated in the cat that motor units are recruited strictly in order of increasing force.[39] The recruitment order of pairs of motoneurons innervating the same muscle was determined by recording their motor axon potentials from intact ventral root filaments. In addition, the isolated ventral root filaments were stimulated to measure the conduction velocity of the motor axons and the mechanical properties of the muscle units that they innervated. For each pair of motor units, the unit with the lower reflex threshold produced the smaller amount of tension. Further, when each unit was classified by "type" according to the standard criteria outlined in Chapter 23, recruitment invariably followed in the order type S–type FR–type FI–type FF. Thus, the motor units were recruited in order of decreasing fatigue resistance as well as in order of increasing force and contraction speed.

Not all the results from the experiment described earlier[39] are consistent with a stringent interpretation of the size principle. Although motor units were invariably recruited in order of increasing force, when their reflex thresholds were evaluated with respect to the conduction velocities of their motor axons, the lower threshold unit did not always have the lower conduction velocity as the size principle would dictate.[1] For pairs with at least one type S unit, the unit with the slower conduction velocity had the lower recruitment threshold in all but one of 22 cases. However, when both units in a pair were type F (i.e., FR, FI, or FF), recruitment was essentially random with respect to conduction velocity.[39] Thus, if the conduction velocity of a motor axon is a reliable indicant of motoneuron size, then motor unit recruitment cannot be strictly ordered with respect to motoneuron size at the upper end of the recruitment threshold range. This "problem" with the size principle had been anticipated in several motor unit studies in which it was found that motor axon conduction velocity and tetanic tension did not systematically covary within the type F motor unit population.[1, 12, 38]

NEURAL MECHANISMS UNDERLYING ORDERLY RECRUITMENT

The various experimental findings outlined earlier indicate that in both reflex and voluntary muscle activation, motor units are recruited in order of increasing force. Such an orderly recruitment sequence requires that the activation thresholds of the motoneurons within a pool are systematically different and are, in turn, highly correlated to the forces produced by their muscle units.

The activation or recruitment threshold of a motoneuron is determined by both its intrinsic properties and the magnitude of its various synaptic inputs. The key intrinsic properties are resting membrane potential, voltage threshold for spike initiation, and input resistance (R_N). While the resting potentials for cat soleus and medial gastrocnemius motoneurons display a rather wide range in values, as shown in Figure 25–6A, they apparently do not vary consistently with respect to the characteristics of the muscle units they innervate (i.e., type S or type F). Similarly, there appears to be no systematic variance in voltage thresholds for the same population of motoneurons (Fig. 25–6B) that could account for their recruitment order.[31] In contrast, the results of several studies indicate that the input resistances of motoneurons within a pool display at least a sixfold range in values, which do covary with a number of motor unit properties.[3, 14, 25, 26, 40]

In the initial formulation of the size principle, it was postulated that the lower activation threshold of the smaller of two motoneurons could be attributed to its higher input resistance (R_N), which in the steady state would lead to a greater depolarization for the same amount of synaptic current (voltage = current × resistance) and, thus, a greater susceptibility to discharge.[21] However, the input resistance of a cell depends upon three factors: its size, expressed in terms of the surface area of its soma and dendritic tree (A_N); its electrotonic architecture, expressed as electrotonic length (L_N); and its specific membrane resistance (r_m) (see Table 25–1 and Chap. 12). R_N will be strictly size-dependent only if L_N and r_m have the same values for all motoneurons within a pool. Estimates of the equivalent electrotonic length of cat motoneurons yield a mean value of about 1.5λ, which does not appear to vary systematically with other motoneuron properties.[5] However, measurements of R_N and estimates of motoneuron size (e.g., Fig. 25–7) suggest that r_m may vary substantially within a pool of motoneurons.[8, 26]

If input resistance were the critical factor determining motoneuron recruitment order, there would be a strong correlation between R_N and motor unit force within a motoneuron pool. However, the correlation between R_N and motor unit force does not appear to be very strong, accounting generally for only about 50% of the covariance of these two parameters.[14] Similarly, other intrinsic motoneuron properties such as membrane time constant, rheobase current, and the duration of the afterhyperpolarization (AHP) are all correlated

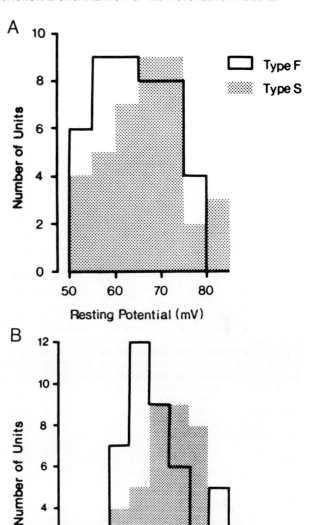

Figure 25–6 Distribution of resting potentials *(A)* and voltage thresholds *(B)* for cat triceps surae motoneurons. Shaded area denotes type S motor units, which have low recruitment thresholds; solid lines denote type F motor units. The distributions of values for both parameters are statistically equivalent for type S and type F units. (From Pinter, M.J.; Curtis, R.L.; Hosko, M.J. *J. Neurophysiol.* 50:644–657, 1983.)

to some extent with motor unit force, but in each case there is considerable "scatter" in the relationships.[14, 40]

Motoneuron rheobase, defined as the minimum amount of injected current required to generate an action potential, deserves special comment be-

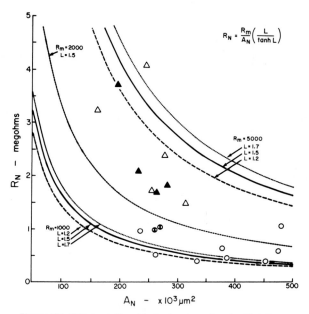

$$R_N = \frac{R_m}{A_N}\left(\frac{L}{\tanh L}\right)$$

Figure 25–7 The relationship between estimated total membrane surface area (A_N) and measured input resistance (R_N) for 19 cat ankle extensor motoneurons. The A_N estimates were based on anatomical reconstruction of the motoneurons following horseradish peroxidase (HRP) injections. The superimposed curves were generated using the equation $R_N = r_m / A_N(L_N/\tanh L_N)$ with 3 values of L_N (1.2, 1.5, and 1.7) and r_m (1000, 2000, and 5000 Ω-cm²). These results suggest that R_N is not strictly dependent on cell size and that r_m may vary over a wide range in different motoneurons. $\triangle$ = type S; $\blacktriangle$ = soleus; $\bigcirc$ = type FF; $\otimes$ = type FR.[8] (From Burke, R.E.; Dum, R.P.; Fleshman, J.W.; Glenn, L.L.; Lev-Tov, A.; O'Donovan, M.J.; Pinter, M.J. *J. Comp. Neurol.* 209:17–28, 1982.)

(EPSPs) were measured in a population of medial gastrocnemius motoneurons.[2] The input resistance of each motoneuron and the type of motor unit (F or S) that it innervated were also determined. Although the amplitudes of both the homonymous and the heteronymous synaptic potentials extended over more than a fivefold range, it is clear that they varied systematically with motoneuron input resistance (Fig. 25–9). Thus, it is difficult to determine from these data whether the variations in EPSP amplitude represented true differences in the magnitude of the Ia synaptic input within the motoneuron pool or can be accounted for simply on the basis of differences in the motoneurons' intrinsic properties.

Under steady-state conditions, the dependence of the amplitudes of synaptic potentials on the intrinsic properties of the cell in which they are recorded emerges from a straightforward application of Ohm's law: the amplitude of a synaptic potential recorded at the soma of a motoneuron is the product of the synaptic current that reaches the soma and the total resistance of the motoneuron (R_N). In the case of time-varying signals like an EPSP, there is considerable attenuation of current

cause it is a more direct index of excitability than input resistance. Rheobase covaries with R_N but, in addition, is influenced by membrane accommodation and properties of the spike-generating mechanism (voltage threshold), both of which may vary from motoneuron to motoneuron (see Chap. 12). As shown in Figure 25–8, motoneurons with the same input resistance may have rheobase values that extend over a threefold range. Nonetheless, the correlation of motoneuron rheobase with muscle unit tetanic tension is generally no stronger than that of R_N with unit tension.[14]

The failure of the experiments just described above to identify a single, intrinsic property of motoneurons that might be considered a "causal" agent in the establishment of orderly recruitment suggests that the magnitudes of the various excitatory and inhibitory synaptic inputs to motoneurons must also play an important role. To test this hypothesis, the amplitudes of homonymous and heteronymous (lateral gastrocnemius–soleus) Ia monosynaptic excitatory postsynaptic potentials

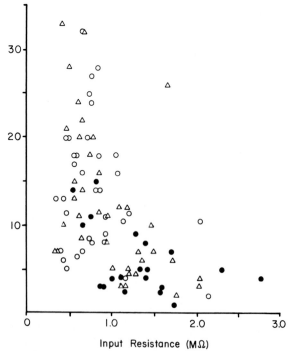

Figure 25–8 Relationship between input resistance and rheobase for cat MG motoneurons. Rheobase is defined as the minimum amplitude of a 50 ms current pulse injected into the motoneuron required to evoke an action potential. (From Powers, R.K.; Binder, M.D. *J. Neurophysiol.* 53:497–517, 1985.)

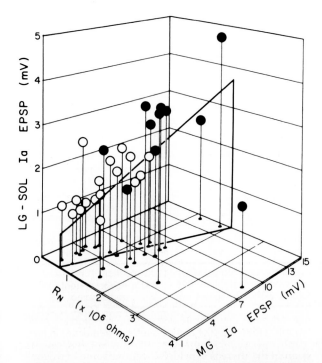

Figure 25–9 Relationships between the input resistance (R_N) of cat medial gastrocnemius motoneurons and the amplitudes of the composite homonymous (MG) and heteronymous (LG-SOL; lateral gastrocnemius–soleus) Ia excitatory postsynaptic potentials (EPSPs). Open circles denote type F motor units; filled circles denote type S units. The trapezoid drawn through the points represents the best fit of the data derived by the least-squares method. The correlation between the EPSP amplitudes ($r = 0.64$) is stronger than that between either of the EPSPs and R_N (0.44 for MG; 0.40 for LG-SOL). (From Burke, R.E. *J. Physiol. [Lond.]* 196:55–58, 1968.)

through the membrane capacitance (see Chap. 12) and, thus, input impedance (Z_N) is the more appropriate measure of the cell's intrinsic properties.

Statistical analysis can be used to help eliminate the confounding effect of input resistance on comparisons of synaptic potential amplitudes. Thus, although all three variables (R_N, MG Ia EPSP, and LG-SOL Ia EPSP) were correlated (Fig. 25–9), the strength of the correlation between the amplitudes of the two EPSPs is much stronger than that between either of the EPSP amplitudes and R_N. Moreover, when the motoneurons were divided into type F and type S subgroups, there was little correlation between R_N and EPSP amplitude in either subgroup, although the amplitudes of the two EPSPs remain correlated in both cases. The results of this analysis indicate that R_N itself cannot account for all of the variation in amplitude of these EPSPs, and therefore there must be other

factors involving the presynaptic neurons, the synaptic transmission process, and the transfer of synaptic current to the cells' soma that determine the amplitudes of Ia synaptic potentials within a pool of motoneurons.[2]

Table 25–1 summarizes the different factors that influence the amplitude of a synaptic potential recorded in the soma of a motoneuron. The amount of synaptic current (I_s) generated in a motoneuron depends on the number of synaptic boutons, their probability of releasing transmitter, the magnitude of the conductance change they produce on the postsynaptic membrane, and the synaptic driving force (see Chap. 11). The fraction of the synaptic current that reaches the soma of the motoneuron (I_N) depends both on the locations of the different boutons on the surface of the motoneuron and on the motoneuron's electrotonic characteristics (see Chap. 12). Finally, the synaptic current that reaches the soma (I_N) acts on the resistive load of the motoneuron (R_N) which, as discussed previously, depends on membrane resistivity, cell surface area, and the equivalent length of the motoneuron's soma and dendrites. Any or all of these factors may vary within a pool of motoneurons.

Regardless of which of the various factors that influence EPSP amplitude are at work (Table 25–1), the consequences are the same, namely that more depolarizing current generated by Ia afferent synapses reaches the somata and initial segments of some cells than of others. Using the voltage

Table 25–1 Factors That Determine the Amplitude of Synaptic Potentials Recorded in the Soma of a Motoneuron*

Synaptic Current (I_s):
$I_s = (N \times p) \times G (V_m - V_e)$
Number of boutons on motoneuron surface (N)
Probability of transmitter release at bouton (p)
Conductance change per bouton (G)
Synaptic driving potential ($V_m - V_e$)

Synaptic Current That Reaches Motoneuron Soma (I_N):
$I_N = I_s \times e^{-x/\lambda}$
Location of boutons (x)
Space constant (λ)

Motoneuron Input Resistance (R_N):
$R_N = r_m/A_N (L_N/tanh L_N)$
Motoneuron surface area (A_n)
Specific membrane resistance (r_m)
Equivalent electrotonic length of soma and dendrites (L_N)

*Under steady-state conditions, the amplitude of a synaptic potential at the soma of a motoneuron is derived from Ohm's law: $V_{EPSP} = I_N \times R_N$. Thus, all the factors listed above influence the EPSP amplitude by contributing to I_N or R_N or both. V_m = resting membrane potential; V_e = equilibrium potential for the synapse.

clamp technique (see Chap. 3), it has recently been demonstrated that the amplitudes of both single-fiber[13] and composite Ia EPSPs[19] vary systematically with the amount of synaptic current that reaches the soma. Moreover, in the case of composite EPSPs, the amount of synaptic current that reaches the soma also varies systematically with motoneuron input resistance.[19]

In summary, if one assumes that the monosynaptic reflex pathway is a valid model for studying motor unit recruitment, then the establishment of a hierarchy of recruitment thresholds within a pool of motoneurons apparently depends on both the intrinsic properties of the motoneurons and the synaptic efficacy of their inputs. Thus, the small, high-resistance motoneurons that produce small forces have lower recruitment thresholds than their larger counterparts because of their greater intrinsic excitability and because more synaptic current reaches their initial segments. A systematic variation of any of the factors listed in Table 25–1 could lead to a gradient of excitatory current across a motoneuron pool. The most likely candidates appear to be either a higher probability of transmitter release for synaptic boutons (see Chap. 11) on the low-threshold cells or perhaps a larger average conductance change per bouton on the low-threshold cells. At present, the available data are insufficient to differentiate between these and other possibilities.

ALTERING RECRUITMENT ORDER

It has been estimated that a single motoneuron may have 10,000 or more synaptic contacts on its soma and dendrites, representing the inputs from over 1000 different neurons. This remarkable diversity of influences on the motoneuron pool makes it all the more amazing that motor unit recruitment proceeds in the same orderly fashion regardless of whether the units are activated by reflex inputs or voluntary actions. One way to achieve a stereotyped recruitment order that is independent of the source of activation would be to have each synaptic system distribute its input to a motoneuron pool in precisely the same way. Thus, each source of synaptic input would exert the same qualitative effects (i.e., excitation or inhibition or both) on all the motoneurons, and the magnitudes of all these effects would exhibit a parallel scaling.[21, 35] Despite the intuitive appeal of this simple hypothesis, however, the available data fail to support it.

Table 25–2 Comparison of Synaptic Potentials in Cat Extensor Motoneurons with Different Recruitment Thresholds

Input Source	Low-threshold Motoneurons	High-threshold Motoneurons
Ia EPSPs[2]	larger	smaller
Reciprocal Ia IPSPs[6]	larger	smaller
Renshaw cell IPSPs[15]	larger	smaller
FRA IPSPs[4]	larger	smaller
Spindle group II EPSPs[30]	equal	equal
LVST neurons EPSPs[6]	equal	equal
CS neurons[11]	predominant IPSPs	predominant EPSPs
RS neurons[4, 11]	predominant IPSPs	predominant EPSPs
LT Cutaneous[4, 32]	predominant IPSPs	predominant EPSPs
Oligo. Group I[32]	predominant IPSPs	predominant EPSPs

EPSPs = excitatory postsynaptic potentials; IPSPs = inhibitory postsynaptic potentials; FRA = flexor reflex afferents; LVST = lateral vestibular spinal tract; CS = corticospinal; RS = rubrospinal; LT = low-threshold afferent fibers; Oligo. Group I = heteronymous group I afferents that do not produce monosynaptic EPSPs. Superior numbers refer to references in text.

Table 25–2 summarizes the results of the various studies in which the quantitative and qualitative distribution of synaptic potentials from identified input systems have been examined within a pool of motoneurons. In addition to the monosynaptic Ia system discussed earlier, several of the synaptic input systems show a simple quantitative scaling of synaptic efficacy across the motoneurons that is consistent with their presumed normal recruitment order (i.e., small before large; type S before type F). These systems include reciprocal Ia inhibition,[6] recurrent inhibition from Renshaw interneurons,[15] and polysynaptic inhibition from flexor reflex afferents (FRA).[4] However, several other systems display quite different patterns of input. For example, the amplitudes of monosynaptic EPSPs from group II muscle spindle afferents[30] and from the lateral vestibular nucleus[6] (see Chap. 26) bear no systematic relationships to the properties of the motoneurons they are recorded in or to the properties of the motor units innervated by the motoneurons. Thus, these two synaptic input systems appear essentially "neutral" with respect to recruitment order.

More problematic with respect to recruitment order are the inputs from the several excitatory oligosynaptic systems that have been investigated. Figure 25–10 shows synaptic potentials recorded in three different cat MG motoneurons studied during the same experiment.[32] The middle row of

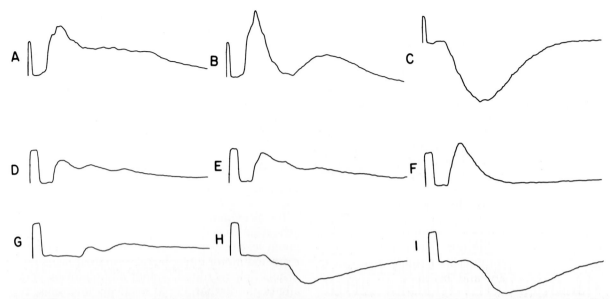

Figure 25–10 Comparison of the synaptic potentials generated by three different afferent input systems in cat medial gastrocnemius motoneurons. Each column shows the responses recorded from a single motoneuron and these three cells were studied in the same experiment. Top row (A–C) shows the synaptic potentials produced by 5T stimulation of the sural nerve (cutaneous afferents). Middle row (D–F) shows the heteronymous (LG-SOL; lateral gastrocnemius–soleus) composite Ia EPSPs. Bottom row (G–I) shows the synaptic potentials produced by 1.5T (group I) stimulation of the medial tibial nerve. A 5-mv, 1-ms calibration pulse precedes each record. (From Powers, R. K.; Binder, M.D. *J. Neurophysiol.* 53:497–517, 1985.)

tracings (D–F) illustrates the monosynaptic Ia EPSPs produced in these three motoneurons by stimulating the heteronymous, lateral gastrocnemius-soleus muscle nerves. The largest EPSP appeared in the motoneuron that had the highest input resistance (F); the smallest EPSP appeared in the cell that had the lowest input resistance (D). The top row of tracings (A–C) depicts the oligosynaptic potentials produced in the same three cells by stimulating the cutaneous sural nerve. Not only are there differences in the amplitudes of these synaptic potentials, but there are clear qualitative differences as well. One motoneuron (C) received predominantly inhibition from this input, the second a mixture of excitatory and inhibitory potentials (B), and the third motoneuron predominantly excitation from the same input source (A). The bottom row of records (G–I) displays the synaptic potentials produced in these cells by stimulating group I fibers (both Ia and Ib) in the nerves of several extensor muscles concurrently. These group I afferents do not make monosynaptic connections with MG motoneurons but instead contribute to the oligosynaptic interneuronal network discussed in Chapter 24. Notice that the motoneuron that received predominantly inhibition from the cutaneous afferents (C) was

strongly inhibited when these group I afferents were activated (I), whereas the motoneuron that received predominantly excitation from the cutaneous input (A) was also excited by activation of these group I afferents (G).

Despite the problems inherent in evaluating complex synaptic potentials with variable latencies, these findings suggest that the organization of some input systems appears to oppose the maintenance of the "usual" recruitment sequence. For example, activation of either of the oligosynaptic inputs described above leads to inhibition of low-threshold MG motoneurons and concurrent excitation of the high-threshold cells. Moreover, equivalent findings have also been reported from experiments in which the corticospinal and rubrospinal input systems have been activated[4, 11] (Table 25–2). This, of course, is not surprising since these descending inputs to motoneurons converge on interneurons involved in segmental reflex pathways (see Chaps. 24 and 26). It is also likely that other descending systems (see Chap. 26), whose synaptic efficacies within individual motoneuron pools have not yet been analyzed, may yield similar results.

The capacity of input systems that exert qualitatively different synaptic effects within a moto-

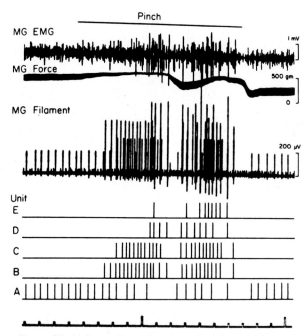

Figure 25–11 Recordings of motor unit activity and muscle force in the cat medial gastrocnemius (MG) muscle. Motor units were activated by muscle stretch and vibration with cutaneous input added by pinching the skin over the lateral portion of the ipsilateral ankle with a forceps (indicated above the top trace). The discharge of 5 MG motor axons was recorded from a small branch of the muscle nerve. The isolated activity of each unit is represented by the middle set of traces. Unit A was the only one activated by the muscle vibration alone (i.e., unit A had the lowest "normal" recruitment threshold), but its rate of discharge was reduced rather than increased when the cutaneous input was added. Moreover, the higher threshold units (B–E), which were not activated by muscle vibration, display two bursts of discharge in response to the cutaneous input. Small pulses on bottom trace mark 100 ms intervals. (From Kanda, K.; Burke, R.E.; Walmsley, B. *Exp. Brain Res.* 29:57–74, 1977.)

neuron pool to disrupt the usual recruitment order has been demonstrated both in cats and in human subjects. In Figure 25–11, motor units in the cat MG muscle were recruited using the tonic vibration reflex. Activating cutaneous afferents by pinching the ipsilateral ankle generally inhibited and derecruited low-threshold units, whereas higher-threshold units were concurrently recruited.[24] Similar results have been obtained from experiments in which recruitment thresholds of motor units in the human first dorsal interosseus muscle were determined during voluntary movements. Providing innocuous electrical shocks to the index finger generally increased the recruitment thresholds of low-threshold units and simultaneously decreased those of high-threshold units.[16]

There have been other reports of variability in the usual recruitment pattern of motoneuron pools, but several of these have been difficult to evaluate or reproduce. At this point, it is probably most prudent to conclude that major alterations in recruitment order can occur, but that under most conditions the synaptic input systems responsible for the usual pattern of recruitment appear to prevail.

CONTROL OF MUSCULAR FORCE BY RATE MODULATION

It was evident from the earliest studies of muscle activation that the amount of force produced at a given length depends not only on the number of recruited motor units, but also on their firing rates. When first recruited, spinal motoneurons generally begin discharging at relatively low rates, typically 6 to 12 impulses/s. At these minimum rates, the average force generated by the motor units is only 10% to 25% of their maximum tetanic tensions. Therefore, changes in the discharge rates of motoneurons (rate modulation) have a major role in determining muscle force production. In general, recruitment and rate modulation operate in parallel, as each depends on the level of excitatory synaptic drive to the motoneuron pool. However, the relative importance of these two mechanisms may vary, depending on the level of force produced by a muscle and the properties of its motor units.[18]

Figure 25–12 shows the steady firing rates of individual motor units recorded from the human extensor digitorum communis muscle as a function of the level of voluntary force produced.[29] Notice that all of the units, regardless of their recruitment thresholds, begin firing at approximately 8 impulses/s and that their firing rates then increase over a two- to threefold range. Another revealing feature of the data presented in Figure 25–12 is that the slopes of the force-firing rate relationships of the units with higher-force thresholds are greater than those of the units with lower-force thresholds. This finding suggests that in this muscle, rate modulation becomes the more important mechanism for controlling force output as motor output increases.

Rate modulation, like recruitment threshold, depends on both the intrinsic characteristics of the motoneuron and the efficacy and distribution of its synaptic inputs. As described in Chapter 12, motoneurons have a characteristic f-I (firing frequency-depolarizing current) relation, with a typ-

Figure 25–12 Rate modulation of motor unit discharge observed in the human extensor digitorum communis muscle during voluntary contractions. With very few exceptions, the subjects were unable to maintain steady firing of the units at less than 8 impulses/s. The firing rates of most units increased over a threefold range as the level of force generated by the muscle was increased. Notice that the force-firing rate slope is steeper for the higher threshold units. (After Monster, A.W.; Chan, H. *J. Neurophysiol.* 40:1432–1443, 1977.)

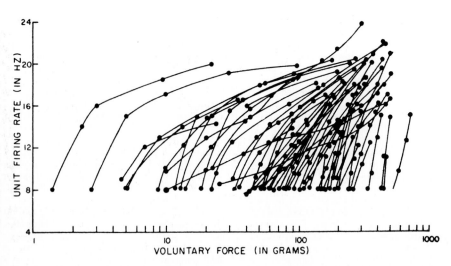

ical primary range slope of 1 to 3 impulses/s/nA of current. Since the amounts of synaptic current that reach the soma of a motoneuron from different sources appear to combine linearly,[17, 34] the discharge rate of each motoneuron is determined simply by the slope of its f-I curve and the sum of the synaptic currents that reach its soma and initial segment.

In conclusion, one can view the motoneuron pool as a functional entity charged with transducing neural commands into muscular force. We have seen that as the level of excitatory synaptic input to a motoneuron pool is increased, motoneurons are recruited in order of their activation thresholds. The discharge rates of the recruited motoneurons vary linearly according to their f-I relations, and the forces that their motor units produce will increase and decrease in parallel with the changes in firing rates. In effect, to control the amount of force generated by skeletal muscles, the central nervous system need only specify the level of excitation to the motoneuron pools; the motoneuron pools do the rest.

ANNOTATED BIBLIOGRAPHY

Burke, R.E. Motor units: anatomy, physiology, and functional organization. *Handbook of Physiology* Sec. 1, 2(1):345–422, 1981.
This comprehensive review of the literature provides a somewhat different perspective on neural mechanisms underlying orderly motor unit recruitment.
Henneman, E.; Mendell, L.M. Functional organization of motoneuron pool and its inputs. *Handbook of Physiology* Sec. 1, 2(1):423–507, 1981.
A review of the experimental literature pertaining to the size principle and its ramifications for the control of motor output.

Stuart, D.G.; Enoka, R.M. Motoneurons, motor units and the size principle. In Rosenburg, R.N., ed. *Clinical Neurosciences. Neurobiology.* New York, Churchill Livingstone, pp. 417–517, 1983.
This review emphasizes the seminal contributions that Professor Henneman's laboratory has made in this area of research and provides commentary on the status of several unresolved issues.

REFERENCES

1. Binder, M.D.; Bawa, P.; Ruenzel, P.; Henneman, E. Does orderly recruitment of motoneurons depend on the existence of different types of motor units? *Neurosci. Letters* 36:55–58, 1982.
2. Burke, R.E. Group Ia synaptic input to fast and slow twitch motor units of cat triceps surae. *J. Physiol. (Lond.)* 196:605–630, 1968.
3. Burke, R.E. Motor units: anatomy, physiology, and functional organization. *Handbk. Physiol.* Sec. 1, 2(1):345–422, 1981.
4. Burke, R.E.; Jankowska, E.; ten Bruggencate, G. A comparison of peripheral and rubrospinal synaptic input to slow and fast twitch motor units of triceps surae. *J. Physiol. (Lond.)* 207:709–732, 1970.
5. Burke, R.E.; ten Bruggencate, G. Electrotonic characteristics of alpha motoneurons of varying size. *J. Physiol. (Lond.)* 234:749–765, 1973.
6. Burke, R.E.; Rymer, W.Z.; Walsh, J.V. Relative strength of synaptic input from short latency pathways to motor units of defined type in cat medial gastrocnemius. *J. Neurophysiol.* 39:447–458, 1976.
7. Burke, R.E.; Strick, P.L.; Kanda, K.; Kim, C.C.; Walmsley, B. Anatomy of medial gastrocnemius and soleus motor nuclei in cat spinal cord. *J. Neurophysiol.* 40:667–680, 1977.
8. Burke, R.E.; Dum, R.P.; Fleshman, J.W.; Glenn, L.L.; Lev-Tov, A.; O'Donovan, M.J.; Pinter, M.J. An HRP study of the relation between cell size and motor unit type in cat ankle extensor motoneurons. *J. Comp. Neurol.* 209:17–28, 1982.
9. Cullheim, S. Relations between cell body size, axon diameter, and axon conduction velocity of cat sciatic α-motoneurons stained with horseradish peroxidase. *Neurosci. Letters* 8:17–20, 1978.

10. Denny-Brown, D. On the nature of postural reflexes. *Proc. R. Soc. Lond.* B104:252–301, 1929.

11. Endo, K.; Araki, T.; Kawai, Y. Contra- and ipsilateral cortical and rubral effects on fast and slow spinal motoneurones of the cat. *Brain Res.* 88:91–98, 1975.

12. Enoka, R.M.; Stuart, D.G. Henneman's "size principle": current issues. *TINS* 7:226–228, 1984.

13. Finkel, A.S.; Redman, S.J. The synaptic current evoked in cat spinal motoneurons by impulses in single group Ia axons. *J. Physiol. (Lond.)* 342:615–632, 1983.

14. Fleshman, J.W.; Munson, J.B.; Sypert, G.W.; Friedman, W.A. Rheobase, input resistance and motor-unit type in medial gastrocnemius motoneurons in the cat. *J. Neurophysiol.* 46:1326–1338, 1981.

15. Friedman, W.A.; Sypert, G.W.; Munson, J.B.; Fleshman, J.W. Recurrent inhibition in type-identified motoneurons. *J. Neurophysiol.* 46:1349–1359, 1981.

16. Garnett, R.; Stephens, J.A. Changes in the recruitment thresholds of motor units produced by cutaneous stimulation in man. *J. Physiol. (Lond.)* 311:463–473, 1981.

17. Granit, R.; Kernell, D.; Lamarre, Y. Algebraical summation in synaptic activation of motoneurons firing within the "primary range" to injected currents. *J. Physiol. (Lond.)* 187:379–399, 1966.

18. Harrison, P.J. The relationship between the distribution of motor unit mechanical properties and the forces due to recruitment and rate coding for the generation of muscle force. *Brain Res.* 264:311–315, 1983.

19. Heckman, C.J.; Binder, M.D. Analysis of effective synaptic currents generated by homonymous Ia afferent fibers in motoneurons of the cat. *J. Neurophysiol.* 60:1946–1966, 1988.

20. Henneman, E. Relation between the size of neurons and their susceptibility to discharge. *Science* 126:1345–1347, 1957.

21. Henneman, E.; Somjen, G.; Carpenter, D.O. Functional significance of cell size in spinal motoneurons. *J. Neurophysiol.* 28:560–580, 1965.

22. Henneman, E.; Olson, C.B. Relations between structure and function in the design of skeletal muscles. *J. Neurophysiol.* 28:581–598, 1965.

23. Henneman, E.; Mendell, L.M. Functional organization of motoneuron pool and its inputs. *Handbk. Physiol. Sec. 1,* 2(1):423–507, 1981.

24. Kanda, K.; Burke, R.E.; Walmsley, B. Differential control of fast and slow twitch motor units in the decerebrate cat. *Exp. Brain Res.* 29:57–74, 1977.

25. Kernell, D. Input resistance, electrical excitability and size of ventral horn cells in cat spinal cord. *Science* 152:1637–1640, 1966.

26. Kernell, D.; Zwaagstra, B. Input conductance, axonal conduction velocity and cell size among hindlimb motoneurons of the cat. *Brain Res.* 204:311–326, 1981.

27. Liddell, E.G.T.; Sherrington, C.S. Recruitment and some other factors of reflex inhibition. *Proc. R. Soc. Lond.* B97:488–518, 1925.

28. Milner-Brown, H.S.; Stein, R.B.; Yemm, R. The orderly recruitment of human motor units during voluntary isometric contractions. *J. Physiol. (Lond.)* 230:359–370, 1973.

29. Monster, A.W.; Chan, H. Isometric force production by motor units of extensor digitorum communis muscle in man. *J. Neurophysiol.* 40:1432–1443, 1977.

30. Munson, J.B.; Sypert, G.W.; Zengel, J.E.; Lofton, S.A.; Fleshman, J.W. Monosynaptic projections of individual spindle group II afferents to type-identified medial gastrocnemius motoneurons. *J. Neurophysiol.* 48:1164–1173, 1982.

31. Pinter, M.J.; Curtis, R.L.; Hosko, M.J. Voltage threshold and excitability among variously sized cat hindlimb motoneurons. *J. Neurophysiol.* 50:644–657, 1983.

32. Powers, R.K.; Binder, M.D. Distribution of oligosynaptic group I input to the cat medial gastrocnemius motoneuron pool. *J. Neurophysiol.* 53:497–517, 1985.

33. Romanes, G.J. The motor cell columns of the lumbo-sacral spinal cord of the cat. *J. Comp. Neurol.* 94:313–363, 1951.

34. Schwindt, P.C.; Calvin, W.H. Equivalence of synaptic and injected current in determining the membrane potential trajectory during motoneuron rhythmic firing. *Brain Res.* 59:389–394, 1973.

35. Somjen, G.; Carpenter, D.O.; Henneman, E. Responses of motoneurons of different sizes to graded stimulation of supraspinal centers of the brain. *J. Neurophysiol.* 28:958–965, 1965.

36. Stein, R.B.; French, A.S.; Mannard, A.; Yemm, R. New methods for analyzing motor functions in man and animals. *Brain Res.* 40:187–192, 1972.

37. Stephens, J.A.; Usherwood, T.P. The mechanical properties of human motor units with special reference to their fatigability and recruitment threshold. *Brain Res.* 125:91–97, 1977.

38. Stuart, D.G.; Enoka, R.M. Motoneurons, motor units and the size principle. In Rosenburg, R.N., ed. *Clinical Neurosciences. Neurobiology.* New York, Churchill Livingstone, 417–517, 1983.

39. Zajac, F.E.; Faden, J.S. Relationship among recruitment order, axonal conduction velocity, and muscle-unit properties of type-identified motor units in cat plantaris muscle. *J. Neurophysiol.* 53:1303–1322, 1985.

40. Zengel, J.E.; Reid, S.A.; Sypert, G.W.; Munson, J.B. Membrane electrical properties and prediction of motor-unit type of medial gastrocnemius motoneurons in the cat. *J. Neurophysiol.* 53:1323–1344, 1985.

Spinal and Supraspinal Control of Movement and Posture

In the preceding chapters (Chapters 23–25), the properties of motor units, the mechanisms and pathways by which receptors innervating muscles, joints, and skin influence motor output, and the functional organization of motoneuron pools have been presented. In this and subsequent chapters we will learn how different structures and areas within the central nervous system above the level of the spinal cord control motor output. This control is accomplished either by accessing neurons within the spinal cord directly, through the so-called spinal-destined systems, or indirectly, by modulating the discharge of those cells that compose the spinal-destined systems. Before examining these supraspinal motor systems, however, it is important to consider the intrinsic, functional organization of the spinal cord motor apparatus. Thus, we will briefly outline the basic patterns of muscle activation that produce locomotion and the contributions that the properties and interconnections of spinal neurons make to the generation and control of these patterns.

SPINAL PATTERN GENERATORS AND LOCOMOTION

Mammalian Locomotion

For some 200 years, physiologists interested in motor control have sought to understand how the neural signals that activate the different muscles of a limb during locomotion and other motor tasks are generated. For mammals, locomotion entails moving the proximal segment of each limb forward and backward cyclically, while the distal joints move through more complex trajectories, as illustrated in Figure 26–1. There is an initial sequential activation of the flexor and extensor muscles as each limb is lifted off the ground and thrust forward (*swing phase*). A second flexion-extension sequence follows during the *support phase* of the cycle, which begins when the limb recontacts the ground and flexes or yields under the weight of the body. The contralateral limb's cycle at the same girdle (pelvic or shoulder) displays a phase

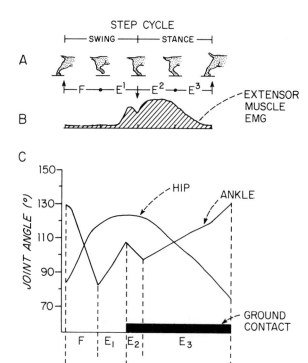

STEP CYCLE

A

B

EXTENSOR MUSCLE EMG

C

Figure 26–1 Analysis of the step cycle in a cat hindlimb. *A,* The swing phase of the cycle is initiated by flexion at the hip, knee, and ankle joints as the foot is lifted off the ground. The swing concludes with the first activation of the extensor muscles (E¹). The stance (support) phase of the cycle begins when the foot contacts the ground following the swing and features progressive activation of extensor muscles acting at the hip, knee, and ankle. Notice that during the second phase of extension (E²), the ankle joint angle decreases (see *C*) as the knee and ankle yield under the weight of the body. Extension of all three joints (joint angles increase) continues in the third and last part of the stance (E³), with extension of the hip persisting even until the onset of the swing phase. *B,* Typical EMG pattern (integrated) from hip, knee or ankle extensor muscles during the step cycle. *C,* Joint angle measurements at the hip and the ankle. (After Gosglow, G.E. et al. *J. Morphol.* 141:1–42, 1973; Grillner, S. *Handbk. Physiol.* Sec. 1, 2:1179–1236, 1981.)

shift of 180° (i.e., if we consider the step cycle to begin with a left limb touching the ground, the corresponding right limb will not do so until the left limb has completed half of its complete cycle). As the animal increases the speed of its locomotion, the duration of the entire step cycle decreases, but the decrease in the support phase is much more pronounced than that of the swing phase. At high speeds, however, the animal abruptly shifts into a gallop, a nonalternating gait in which the two hindlimbs touch the ground at approximately the same time.

Locomotion is accompanied by a characteristic

pattern of electromyographic (EMG) activity. In the cat, the extensor ("antigravity") muscles of the hip, knee, ankle, and foot generally display a burst of activity just prior to ground contact and sustained activation almost to the end of the support phase. The primary flexor muscles acting across the hip, knee, and ankle show a reciprocal pattern of activation. That is, they are active throughout the swing phase but are turned off during the stance or support phase of the step cycle. The knee flexors differ in that they often show a second burst of activity during the support phase, as do the toe flexor muscles, which may display sustained activity throughout the stance. Similar locomotive activity patterns have been reported in the hindlimbs of several other species, including man.

Spinal Mechanisms

It was thought at one time that a cascade or chain of reflexes was responsible for the rhythmic patterns of muscle activation observed during locomotion. In this scheme, a sensory input would trigger the reflex activation of a muscle to initiate a movement. The movement would, in turn, activate peripheral mechanoreceptors in muscles, skin, and joints, resulting in the reflex activation and inhibition of other muscles in the limb. This would be mediated through the spinal circuits outlined in Chapter 24. Thus, the alternation of flexor and extensor muscle activity required for locomotion and other rhythmic movements could be ascribed to the reciprocal connections of reflex afferents with spinal interneurons and motoneurons.

Although this "reflex chain" hypothesis for the generation of rhythmic movement is a simple and attractive one, it has been thoroughly refuted experimentally. In a variety of experimental animals, locomotive and other purposive movements can be generated in the absence of peripheral feedback and reflex action.[13] Furthermore, the spinal cord alone, isolated from supraspinal centers by spinal transection and from peripheral input by dorsal root section, can still produce rhythmic, alternating activation of ipsilateral antagonist muscle groups.

The lack of support for the reflex chain hypothesis has led to the widely held notion that networks of spinal interneurons, called central pattern generators (CPGs), produce the requisite signals to "drive" and coordinate the outputs of the various motoneuron pools. These putative

CPGs are thought to have several definitive properties. There must be separate but interconnected CPGs for muscles acting across the different joints in each limb, and, presumably, flexor and extensor CPGs should have mutually inhibitory synaptic connections. Each CPG must be able to perform without patterned input from the brain or periphery and to produce patterns of discharge during locomotion comparable to those of the flexor and extensor motor pools to which they project. Further, one would expect the burst periods of flexor and extensor CPGs to manifest the phase relationships and asymmetry observed in their respective motor outputs during the step cycle. In addition, the discharge of the CPGs should be subject to modification from peripheral afferent and descending synaptic inputs. Finally, and perhaps most significantly, it must be demonstrated that the putative CPGs are essential for the generation of rhythmic alternation of flexor and extensor motor output.

Spinal Interneurons

Although it is now generally agreed that interneurons within the spinal cord generate the rhythmic synaptic drive to motoneuron pools characteristic of locomotion, the identity of these interneurons remains unknown. Very few groups of spinal interneurons with known terminations and activity patterns related to locomotion have been studied as yet. Moreover, it is not clear whether any of the "known" interneurons are, in fact, components of the CPGs. Nonetheless, by studying the properties and interconnections of the identified groups of interneurons, it is possible to make a number of inferences about the probable operation of the CPGs.

Renshaw Cells. Among the most thoroughly studied types of spinal interneurons are the Renshaw cells that mediate recurrent inhibition of spinal motoneurons. Renshaw cells are located in the ventral horn of the spinal gray matter, medial to the motor nuclei. Their actions can be analyzed by stimulating alpha motoneurons while recording from perspective target spinal neurons. Renshaw cells project to and produce inhibitory postsynaptic potentials (IPSPs) in motoneurons (both alpha and gamma), Ia inhibitory interneurons, other Renshaw cells, and ventral spinocerebellar tract (VSCT) neurons. The most potent synaptic effects exerted by Renshaw cells on motoneurons are found in the homonymous and synergist pools. Renshaw cells receive their primary excitatory in-

put from the axon collaterals of homonymous and synergist alpha motoneurons. However, they also receive both excitatory and inhibitory input both from segmental reflex afferent pathways and from several supraspinal systems, including the cerebral cortex, red nucleus, and reticular formation.[3]

Ia Inhibitory Interneurons. Other well-studied spinal interneurons are Ia inhibitory interneurons that mediate Ia reciprocal inhibition (see Chap. 24). These interneurons are found in lamina VII, dorsomedial to the motor nuclei. They are excited primarily by the Ia afferent fibers that innervate muscle spindles in homonymous and synergist muscles. In addition, however, Ia inhibitory interneurons receive segmental excitation from flexor reflex afferents (FRAs) and several systems of supraspinal origin (see later in this chapter). Inhibitory inputs to these Ia inhibitory interneurons are derived from homonymous and synergist Renshaw cells and the Ia inhibitory interneurons that project to the antagonist motor pools (see Fig. 26–2). In turn, Ia inhibitory interneurons produce IPSPs in antagonist motoneurons and antagonist Ia inhibitory interneurons. These actions are assessed either directly, by recording simultaneously from the interneurons and their targets, or indirectly, by observing the effects exerted by Ia afferent fibers at disynaptic latencies on different prospective targets.

Interneuronal Networks. The network of reciprocal connections between Renshaw cells and Ia inhibitory interneurons associated with the motoneuron pools innervating a pair of antagonist muscles is shown schematically in Figure 26–2. Although this neural network is thought to be involved in shaping motor output rather than in generating it, it is instructive to consider the chain of neural events that ensues within this network consequent to either a "descending command" or a segmental reflex input from Ia afferent fibers. For example, a descending command of sufficient strength to evoke action potentials in extensor motoneurons initiates contraction of the extensor muscle and movement of the limb. Concurrently, the extensor Renshaw cells are activated by collaterals of the extensor motor axons to provide recurrent inhibition and a gradual reduction of the extensor motoneuron discharge. Activation of the extensor Renshaw cells also leads to inhibition of both the extensor Ia inhibitory interneurons and the flexor Renshaw cells, resulting in an increased excitatory drive to the flexor motoneurons through disinhibition. Thus, the action of Renshaw cells limits the extensor motoneuron discharge and promotes activation of antagonist motoneurons. The

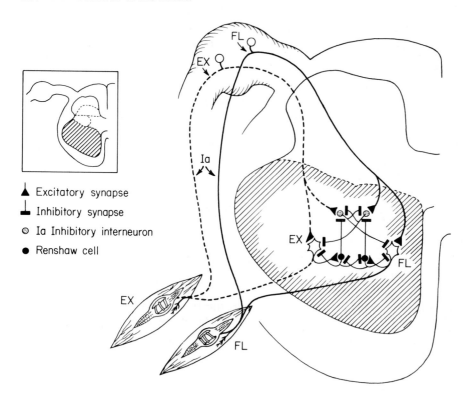

▲ Excitatory synapse

⊥ Inhibitory synapse

◎ Ia Inhibitory interneuron

● Renshaw cell

Figure 26–2 Schematic representation of the synaptic connections between Ia afferent fibers, Ia inhibitory interneurons, Renshaw cells, and flexor and extensor motoneuron pools.

activation of flexor motoneurons is further facilitated by the effects of the movement produced by the extensor muscle contraction. The antagonist flexor muscle will be stretched, resulting in increased discharge from its population of Ia afferent fibers (see Chap. 24). The increase in flexor Ia input not only increases the excitatory drive to the flexor motoneurons but also leads to increased activation of the flexor Ia inhibitory interneurons, further limiting the excitability of extensor motoneurons. Eventually the balance of excitatory and inhibitory synaptic "drives" will shift sufficiently to activate the flexor motoneurons and terminate the extensor motoneuron discharge. Subsequently, the same sequence of actions outlined above will be reiterated with the flexor and extensor roles reversed.

Although it has been demonstrated that, in fact, both Renshaw cells and Ia inhibitory interneurons are rhythmically active during locomotion, they do not appear to form essential components of the spinal CPGs. Systemic administration of the glycinergic antagonist strychnine effectively abolishes Ia and Renshaw cell IPSPs but does not eliminate the basic rhythmic output of motoneuron pools.[28] Nonetheless, it is clear that these reciprocal connections support and shape the rhythmic alternation of flexor and extensor motoneuron discharge required for locomotion, and it is easy to imagine how other similarly organized interneuronal networks might form the actual spinal CPGs.

FRA Interneurons. Although not as well characterized as the Renshaw cells and Ia inhibitory interneurons, interneurons in lamina VII of the lumbar spinal cord that mediate long-latency reflexes from the flexor reflex afferents (FRAs; see Chapter 24) have some of the requisite properties of CPGs, including sustained bursts of discharge, mutual inhibition, and reciprocal actions on antagonist motoneuron pools. After the administration of the monoamine precursors dopa (3,4-dihydroxyphenylalanine) and 5-HTP (5-hydroxytryptophan) to spinal cats, stimulating the FRAs evokes sustained discharge (400–1000 ms) in ipsilateral flexor and contralateral extensor motoneurons after a delay of 100 to 200 ms. These discharges are accompanied by inhibitory postsynaptic potentials (IPSPs) of comparable duration in the respective ipsilateral and contralateral antagonist motoneurons (Fig. 26–3). The existence of mutual inhibition between the interneuronal pathways transmitting these long-latency FRA effects to

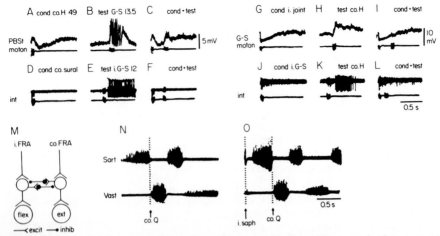

Figure 26–3 Analysis of the spinal network mediating the reciprocal activation of flexor and extensor muscles following flexor reflex afferent (FRA) stimulation. Upper traces in A–C are intracellular recordings from two flexor motoneurons (posterior biceps or semitendinosus muscles); those in G–I from two extensor motoneurons (gastrocnemius or soleus muscles). The upper traces in D–F and J–L are extracellular recordings from interneurons. The lower traces in A–L are extracellular recordings made from the dorsal root entry zone. Peripheral nerve stimuli were of sufficient strength to activate high-threshold muscle and cutaneous afferent fibers. N–O are simultaneous recordings made from flexor (sartorius) and extensor (vastus) muscle nerves. All of the records were obtained from spinalized cats following the administration of dopa. In flexor motoneurons (A–C), contralateral FRA stimulations evoked a long latency IPSP and ipsilateral FRA stimulation evoked a long latency EPSP (B) which was inhibited by preceding the ipsilateral FRA input with contralateral FRA input (C). Analogous mutual inhibition between the long-latency FRA pathways to extensor motoneurons is illustrated in G–I. Panels D–F and J–L display recordings from interneurons located dorsal to the motor nuclei that might mediate the patterns of PSPs observed in the corresponding flexor (A–C) and extensor (G–I) motoneurons. Panel M depicts a simple circuit diagram that might account for the results described in A–L. Each interneuron in the schematic is meant to represent a serial chain of neurons (i.e., there are several interneurons intercalated between the FRA and the motoneurons, as well as between the contralateral and ipsilateral FRA interneurons). Panels N and O display alternating bursts of flexor (sartorius–medial sartorius) and extensor (vastus–medial vastus) motoneuronal discharge elicited by short trains of stimuli to activate FRA. In N, stimulation of FRA in the contralateral quadriceps nerve (co.Q) during a period of spontaneous discharge in the flexor motor pool terminated the flexor discharge and initiated alternation between extensors and flexors. In O, stimulation of the ipsilateral FRA in the saphenous nerve (i.saph) followed by additional stimulation of the contralateral FRA (co.Q) generated alternating bursts of flexor and extensor motoneuronal activity. (After Baldissera, F.; Hultborn, H.; Illert, M. *Handbk. Physiol.* Sec. 1, 2:509–595, 1981.)

flexor and extensor motoneurons can be inferred from both reflex testing and interneuronal recordings. Some interneurons excited by the ipsilateral FRAs exhibit a sustained discharge that can be depressed by stimulating the contralateral FRAs. These interneurons are likely candidates for mediators of the long-latency reflex pathway from FRAs to ipsilateral flexor motoneurons. Others, which are excited by contralateral FRA stimulation and inhibited by ipsilateral FRA stimulation, may mediate FRA excitation to extensor motorneurons (Fig. 26–3).

Finally, this network of interneurons involved in the long-latency FRA pathway to motoneurons appears capable of mediating alternate activation of flexor and extensor muscles, a hallmark of CPGs. As shown in Figure 26–3, short trains of conditioning stimuli delivered to FRAs evoke a sustained series of alternating bursts of activity in flexor and extensor motor pools.

Although the experimental findings outlined above do not establish that the interneurons intercalated in the long-latency FRA pathway are the actual CPGs for locomotion, these interneurons appear to be the best candidates for "CPG components" described to date. It remains to be demonstrated that activity in these interneurons is essential for the generation of basic locomotive rhythms.

SUPRASPINAL SYSTEMS INFLUENCING MOVEMENT

The motor behavior of the limbs and trunk produced by most vertebrates with intact nervous systems is far more extensive than the patterns, such as locomotion, that can still be elicited when the neuraxis is severed between the spinal cord and the medulla. Furthermore, neurologically normal, alert individuals can initiate movement "voluntarily": that is, without an identifiable sensory stimulus. Most motor behavior in vertebrates, then, is initiated, or at least regulated, by neural signals in axons originating from cells in supraspinal structures. These spinal-destined or "descending" systems also modify the spinal reflex responses described in Chapter 24. Following transection of the brain stem or the spinal cord, some segmental reflexes will be enhanced (hyperreflexic) and others will be depressed (hyporeflexic).

The descending systems are made up of axons from neurons in several areas of the brain stem and cerebral cortex. Descending axons that reach the spinal cord may synapse on primary afferent fibers, on spinal interneurons, or directly on alpha or gamma motoneurons. Although some descending fibers act primarily to modulate activity in ascending neurons of the sensory systems, we focus, in this chapter, on actions that influence motoneurons and movement.

Anatomical Characteristics of Spinal-Destined Systems

Axons destined for the spinal cord originate from neurons located in the cerebral cortex, midbrain, pons, medulla, and even the cerebellum (Fig. 26–4). The course and termination within the spinal cord varies for axons from different structures, and this provides an anatomical basis for some of the functional differences between them.

Corticospinal Systems. Corticospinal fibers originate from neurons in layer V of the precentral and postcentral gyri of the cerebral cortex, including Brodmann's areas 4, 6, 3, 1, 2, and 5.[22] These include areas functionally designated as primary motor cortex (area 4), premotor cortex (lateral area 6), supplementary motor cortex (medial area 6), primary somatosensory cortex (areas 3, 1, and 2), and posterior parietal cortex (area 5), all of which are discussed in more detail in Chapter 28.

Corticospinal fibers travel through the midbrain in the ipsilateral cerebral peduncle, which they share with other corticofugal axons that do not reach the spinal cord. After passing through the pontine nuclei, they are grouped in the medullary pyramids, situated on the ventral surface of the medulla. As a consequence, the cortically originating fibers in the medulla have been named the *pyramidal tract*. It is important to note, however, that the pyramidal tract in the medulla contains not only corticospinal fibers destined for the spinal cord or for equivalent cranial nerve nuclei in the

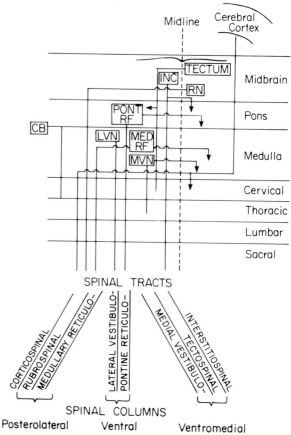

Figure 26–4 Summary diagram of the descending pathways from the brain stem and cerebral cortex to one side of the spinal cord. The predominant pathways (crossed or uncrossed) are indicated by the lines; smaller bilateral pathways are indicated by arrows. INC = interstitial nucleus of Cajal; RN = red nucleus; PONT RF = pontine reticular formation; MED RF = medullary reticular formation; LVN = lateral vestibular (Deiter's) nucleus; MVN = medial vestibular nucleus; CB = deep cerebellar nuclei. Lines extend to the most caudal level of the spinal projection.

brain stem but also fibers that will terminate on medullary neurons, such as those in the reticular formation, the dorsal column nuclei, and the inferior olivary nucleus. Thus, only some of the axons in the primate pyramidal tract are true corticospinal fibers.

Not only is the term *pyramidal tract* misused as a synonym for corticospinal tract; its existence has led to the complementary term *extrapyramidal system*, which is used in an even more confusing manner. The extrapyramidal system usually is equated with the basal ganglia. Although this group of nuclei is, indeed, "outside of" the pyramidal tract, the same could be said of most of the central nervous system! The term *extrapyramidal* was originally used to differentiate "extrapyramidal

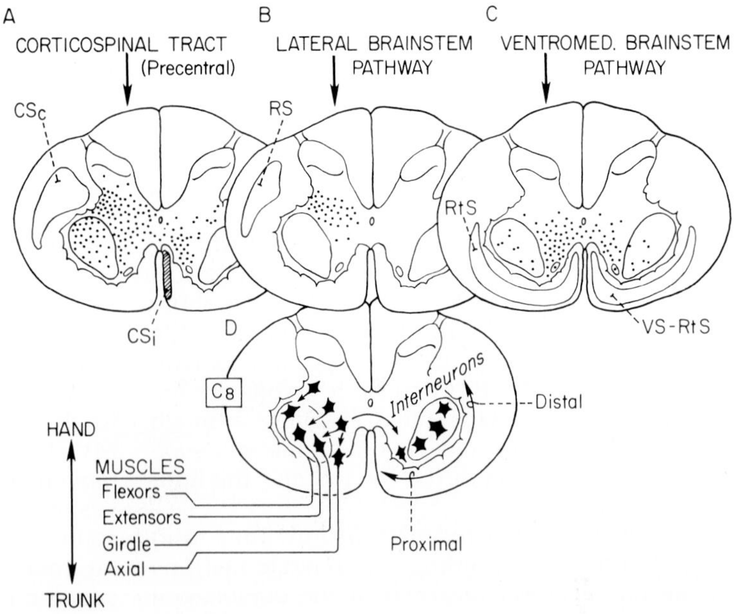

Figure 26–5 Spinal course and site of termination of descending fiber systems. *A,* Precentral corticospinal; *B,* rubrospinal; *C,* vestibulospinal and reticulospinal; *D,* spinal locations of motoneurons and interneurons to proximal and distal muscles of the limb. (After Lawrence, D.G.; Kuypers, H.G.J.M. *Brain* 91:15–36, 1968 with the permission of Oxford University Press.)

motor symptoms," which usually occur when the basal ganglia are damaged, from the "pyramidal syndrome," a combination of paresis and spasticity that was thought to result from the destruction of corticospinal fibers. It now is known, however, that much of the basal ganglia's influence on motor control is, in fact, exerted via the cerebral cortex and corticospinal fibers. Furthermore, we also know that exclusive destruction of corticospinal axons, and even the entire medullary pyramids, does not produce the spasticity usually associated with the pyramidal syndrome. We would advocate, then, that the terms *extrapyramidal system* and *pyramidal syndrome* be deleted from use and replaced by terms that more accurately describe the symptoms and syndromes to which they refer.

In carnivores and primates, most corticospinal axons cross the midline at the spinomedullary junction and descend through the spinal cord in the dorsolateral columns of the spinal white matter (CS_c, Fig. 26–5A). From this location they enter the spinal gray matter through its lateral aspect and terminate most densely in Rexed's laminae V–VIII of the dorsal and ventral horns. In primates, some reach lamina IX, the location of alpha motoneurons. In rodents, carnivores, and primates, the corticospinal tract descends the entire length of the spinal cord. In all species, at least a small component descends ipsilaterally in the ventral column (CS_i, Fig. 26–5A).

Some corticospinal axons give off multiple branches at both cervical and lumbar levels of the cord. These branches can be mapped by antidrom-

ically activating them with low-intensity microstimuli applied through microelectrodes that are moved through the spinal gray matter. In both the cat and the monkey, approximately 30% of corticospinal neurons with axonal branches projecting into the gray matter of the cervical enlargement also have been shown to send an axonal branch at least as far as the upper thoracic cord.[35]

Rubrospinal System. The rubrospinal system is made up of the axons of cells in the caudal (primarily magnocellular) part of the red nucleus in the midbrain. The axons cross the midline within the midbrain and occupy a ventrolateral position during their descent through the brain stem. In the spinal cord they are situated near the corticospinal fibers in the dorsolateral columns (Fig. 26–5B). The tract descends through the entire length of the cord in this position, with axons entering the gray matter from its lateral border and terminating most densely in the lateral portion of the intermediate zone (Rexed's laminae V–VII). In primates, some rubrospinal fibers synapse directly on motoneurons in the cervical and lumbosacral enlargements. As is the case for corticospinal fibers, some rubrospinal axons have branching collaterals in both the cervical and the lumbar gray matter.

Reticulospinal Systems. The reticular formation, from its rostral extent in the mesencephalon to its most caudal portion in the caudal medulla, is a source of several groups of descending axons.

Major groups of reticulospinal fibers originate from reticular neurons in the pontine tegmentum (n. pontis oralis and caudalis). These axons descend primarily in the ipsilateral ventral columns of the cord (Fig. 26–5C). Reticulospinal axons from neurons in the medullary reticular formation (n. reticularis gigantocellularis and ventralis) descend predominantly ipsilaterally in the ventral and ventrolateral columns. Reticulospinal axons tend to be large with correspondingly high conduction velocities, and at least some of them descend the entire length of the spinal cord. They terminate in the ventral horn, especially in medial portions of laminae VII and VIII. In contrast to axons in the corticospinal and rubrospinal systems, a much larger fraction of reticulospinal neurons sends axonal branches into both the cervical and the lumbar gray matter.[25]

The axons of other reticulospinal neurons from medullary levels, including serotonergic neurons in the raphe nuclei, descend in the dorsolateral columns and terminate more dorsally in the dorsal horn. These dorsally terminating systems have a more direct influence on transmission in sensory systems than they do on motoneurons (see Chapter 16).

Lateral Vestibulospinal System. Axons from neurons in the lateral vestibular nucleus (Deiter's nucleus) also descend through the entire length of the ventral spinal columns ipsilateral to their origin (Fig. 26–5C). As do axons of the reticulospinal system, they also terminate predominantly in medial portions of laminae VII and VIII of the ventral horn. In the cat, at least 50% of the lateral vestibulospinal fibers give off branches into both the cervical gray matter and the lumbar cord.[1]

Motoneurons and interneurons associated with proximal limb and axial musculature are located in the medial portions of the ventral horn and the intermediate gray matter (Fig. 26–5D). The fact that both lateral vestibulospinal and ventral reticulospinal fibers terminate predominantly in these medial regions means that these systems are in a position to exert strong influences on proximal and axial muscles.

Tectospinal and Medial Vestibulospinal Systems. Two major descending systems send the majority of their axons only to upper levels of the spinal cord. Tectospinal fibers, from the superior colliculus, cross the midline in the midbrain and descend into the ventral columns, but only through the upper cervical cord (see Fig. 26–4). Motoneurons innervating neck muscles are located in these segments of the cord, and the tectospinal system has a strong input to neck motoneurons.

The superior colliculus also provides a major input to neurons in the pontine reticular formation that influence eye movements, and the superior colliculus may play an important role in coordinating head and eye movements.

The medial vestibulospinal tract (MVST), which is derived primarily from neurons in the medial and descending vestibular nuclei, also projects mainly to cervical and thoracic segments of the spinal cord that control muscles of the neck and the upper back, respectively. The projection is bilateral, coursing through the medial longitudinal fasciculus in the medulla and the ventral columns in the cord. A smaller number of axons extends to thoracic and lumbar cord levels. Since most MVST neurons receive information signaling head movement from the semicircular canals of the vestibular labyrinth, this system has anatomical connections that are appropriate to use vestibular information in the control of head and trunk position (see also Chap. 27).

Spinal Sites of Action

Once descending axons have entered the spinal gray matter, there are several sites at which they can affect motor behavior. Examples of those sites that have been demonstrated in mammals are shown schematically in Figure 26–6.

Monosynaptic Actions on Alpha Motoneurons. Synaptic action directly on the alpha motoneuron (1 in Fig. 26–6) is the most obvious way in which descending fibers could affect motor function. Such a direct action would have the advantages of security and speed, since it would not require the time or convergence that would be necessary to reliably bring an interneuron to threshold. It also could be a selective way to activate particular motoneurons, without the risk that an intervening interneuron might branch to innervate a variety of motoneuron pools.

Direct, monosynaptic actions on alpha motoneurons have been demonstrated in primates for axons of corticospinal, rubrospinal, reticulospinal, and medial and lateral vestibulospinal systems. These monosynaptic connections are in the minority, however, and they have limited distributions. Corticospinal and rubrospinal axons, for example, contact alpha motoneurons directly only in primates, and when monosynaptic connections are present, they are primarily on motoneurons innervating distal muscles of the hand and foot.[7] These monosynaptic connections from corticospinal and rubrospinal axons are all excitatory.

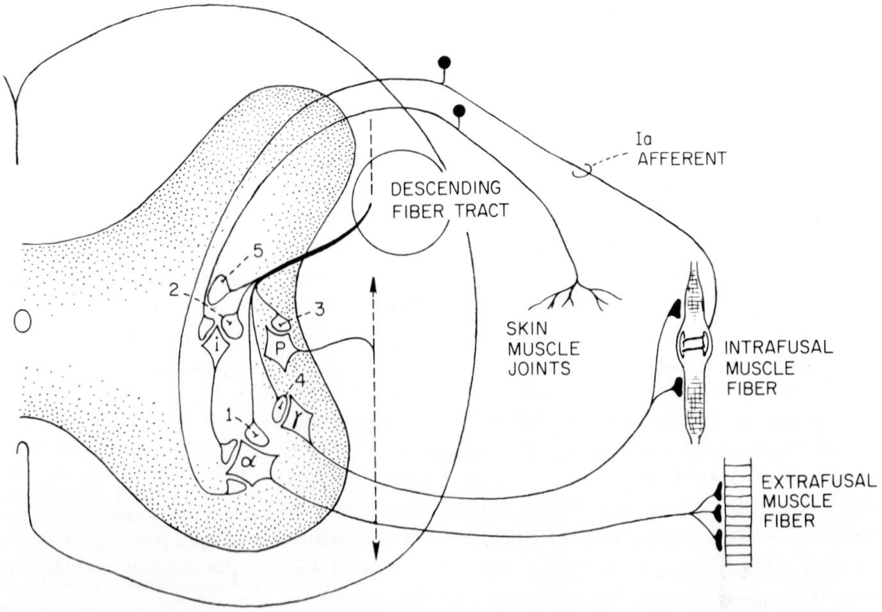

Figure 26–6 Schematic depiction of potential spinal sites for action of descending fibers. (1) Monosynaptically on alpha motoneurons (α); (2) on segmental interneurons (i); (3) on propriospinal neurons (p); (4) on gamma motoneurons (γ); (5) on afferent fiber axon terminals.

Lateral vestibulospinal and ventral reticulospinal axons also make monosynaptic connections with motoneurons. Moreover, they are present in the cat, as well as in the monkey. Again, however, the direct connections are not ubiquitous to all motoneuron pools. In the cat, for example, stimulation of the lateral vestibular nucleus produces EPSPs monosynaptically in neck and hindlimb extensor motoneurons, but direct synaptic actions were not found in extensor motoneurons innervating the forelimb or in any flexor motoneurons.[38] Moreover, within the hindlimb extensor motoneuron pools, monosynaptically produced EPSPs are prominent in motoneurons innervating muscles acting at the knee or the ankle, but not in those innervating toe muscles.[16, 38] On the other hand, the monosynaptic excitatory actions of ventral reticulospinal neurons tend to excite flexor motoneurons innervating forelimbs and hindlimbs and extensor motoneurons innervating hindlimb hip and toe muscles.[14, 16, 38] Thus, the reticulospinal and lateral vestibulospinal systems seem to complement each other, although the functional meaning of this organization is not clear.

The synaptic action of some vestibulospinal and reticulospinal axons is inhibitory, instead of excitatory, at the motoneuron. Some axons of the medial vestibulospinal system, which extends only as far caudally as thoracic cord levels, exert a direct inhibitory action on motoneurons innervating muscles of the neck and the back. IPSPs also are produced in some neck motoneurons by stim-

ulation in the caudal medullary reticular formation.

Even in those instances in which descending fibers do exert direct actions on motoneurons, the strength of that action is relatively small. As described in Chapter 24, the amplitude of the postsynaptic potential (PSP) produced by a single presynaptic axon can be determined by the method of spike-triggered averaging. This has been used to determine that the average amplitude of EPSPs produced in neck motoneurons by vestibulospinal axons is less than 40 μV; IPSPs produced monosynaptically from the medial vestibular nucleus generally have maximum amplitudes of less than 80 μV.[31] By comparison, the mean amplitude of EPSPs produced by single Ia afferents in homonymous motoneurons of the cat was found to be 143 μV in soleus motoneurons and 95 μV in medial gastrocnemius motoneurons.[32] The maximum aggregate EPSP produced in cat lumbosacral motoneurons by maximal stimulation of the lateral vestibulospinal tract usually is less than 2 mV in amplitude, whereas the maximum EPSP produced by stimulation of all homonymous Ia fibers often exceeds 5 mV.[9]

The amplitude of EPSPs from individual corticospinal fibers has been determined by spike-triggered averaging in only a few motoneurons in the monkey, in which corticospinal axons make monosynaptic connections with motoneurons. However, the amplitude of minimal EPSPs produced by weak cortical stimulation has been examined in motoneurons innervating distal limb muscles. On the average, the minimal EPSP produced by weak cortical stimulation was smaller than that produced by weak stimulation of Ia afferents in the homonymous muscle nerve.[27] Furthermore, maximum amplitude EPSPs evoked in forelimb motoneurons by cortical stimulation were smaller than maximal EPSPs produced by summed Ia afferents from homonymous and heteronymous muscles, although the contribution of homonymous Ia's, alone, may be smaller than that from corticospinal fibers.[7]

Although the amplitude of single EPSPs produced in motoneurons by corticospinal fibers may be small, corticomotoneuronal EPSPs show temporal facilitation with repetitive activation.[26] This facilitation, which is more prominent than that of EPSPs produced by repetitive activation of Ia afferents, is maximal at short interstimulus intervals (5 ms) and decays over a time course of about 10 ms. This means that the influence of direct corticomotoneuronal connections is enhanced when corticospinal neurons discharge a burst of spikes, which often is the case during natural motor activity (see Chap. 28).

In summary, some axons from supraspinal neurons do terminate directly on alpha motoneurons, but these terminations are sparse, are restricted to particular motoneuron pools, and usually do not produce PSPs large enough to bring the motoneuron to threshold for action potentials.

Actions on Interneurons and Spinal Reflex Arcs. As shown diagrammatically by synapses 2 and 3 in Figure 26–6, most descending axons terminate on spinal interneurons that, in turn, affect motoneurons either directly or through other spinal interneurons. Many of these are *segmental interneurons* (see i in Fig. 26–6), which also are excited by peripheral afferent fibers and act to produce a patterned reflex response in motoneurons in the same or nearby segments (see Chap. 24). Others are *propriospinal neurons* (see p in Fig. 26–6) that interconnect segments at greater distances. Some propriospinal neurons with long axons, in fact, could coordinate interlimb activity.

Lundberg and Jankowska and their associates have provided much of our knowledge of descending actions on interneurons.[10, 17, 18] By using a conditioning-testing technique similar to that described in Chapter 24, they demonstrated the convergence of descending and segmental inputs onto common interneurons and the effect of those interneurons on various motoneuron pools. In the example shown in Figure 26–7A, a test stimulus was given at the time of the arrow to the sural

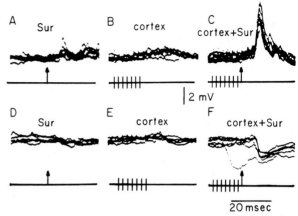

Figure 26–7 Evidence for convergence of peripheral and descending fibers on segmental interneurons. Intracellular records from a posterior biceps-semitendinosus motoneuron in *A–C* and from a gastrocnemius-soleus motoneuron in *D–F*. The sural (Sur) nerve was stimulated in *A* and *D*. The sensorimotor cerebral cortex was stimulated with a train of seven shocks in *B* and *E*. Cortical stimulation preceded sural stimulation in *C* and *F* and resulted in PSPs that were considerably larger than the sum of the PSPs produced by sural nerve and cortical stimulation alone. Vertical lines below records indicate the time of cortical stimuli; arrows indicate the time of sural nerve stimulation. (After Lundberg, A; Voorhoeve, P. *Acta Physiol. Scand.* 56:201–219, 1962.)

nerve (a cutaneous nerve). The stimulus intensity, however, was only sufficient to produce a submaximal EPSP in a motoneuron that could be activated antidromically (not illustrated) from the nerve to the posterior biceps and semitendinosis (PBST) muscles in the cat hindlimb. In Figure 26–7D, a similar stimulus to the sural nerve produced no clear PSP in a motoneuron activated antidromically from the nerve to gastrocnemius and soleus (GS) muscles. Shown in *B* and *E* are the small responses in the same motoneurons when trains of seven stimuli were applied to the contralateral sensorimotor cortex (see lines). In *C* and *F*, the shock to the sural nerve was preceded by the train of stimuli to the cortex (the conditioning stimulus), and the combined stimuli produced a large EPSP in the PBST motoneuron and a large IPSP in the GS motoneuron. The spatial summation provided by combined excitation of the cortical and segmental input brought two populations of interneurons to threshold. One exerted an excitatory action on PBST motoneurons, and the other exerted an inhibitory action on GS motoneurons. Facilitation of these interneurons also is enhanced by the temporal summation of corticospinal input provided when a train of stimuli is applied to the cerebral cortex.

The pattern of interaction of descending and segmental afferents can be determined if the test stimulus is applied selectively to different classes of peripheral afferent fibers, as described in Chapter 24. In fact, several descending systems often converge on interneurons shared with reflex arcs. The best studied of these interneurons is the Ia inhibitory interneuron, which was described earlier in this chapter. Figure 26–8 schematically summarizes known convergent inputs to Ia inhibitory interneurons. Ia inhibitory neurons to flexor motoneurons, for example, are excited monosynaptically by spindle afferents in ipsilateral extensor muscles and by axons of the ipsilateral lateral vestibulospinal tract and indirectly (through other excitatory interneurons) by axons of the ipsilaterally descending corticospinal and rubrospinal tracts and the contralaterally descending vestibulospinal tract. An axonal branch of the Ia inhibitory interneuron excited by Ia afferents from *flexor*

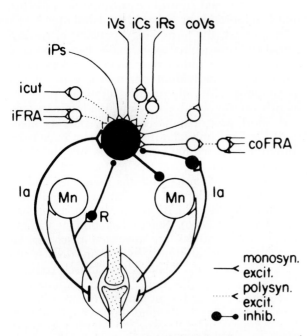

Figure 26–8 A circuit diagram depicting the convergent inputs to inhibitory interneurons excited by Ia afferents. Ia = primary spindle afferents; Mn = alpha motoneuron; R = Renshaw cells; iFRA = flexor reflex afferents ipsilateral to the Ia interneuron; icut = ipsilateral low threshold cutaneous afferents; iPs = ipsilateral propriospinal afferents; iVs = ipsilateral lateral vestibulospinal tract; iCs = ipsilateral (in the cord) corticospinal tract; iRs = ipsilateral rubrospinal tract; coVs = contralateral lateral vestibulospinal tract; coFRA = flexor reflex afferents from the contralateral limb. Excitatory neurons are depicted with open circles, inhibitory neurons are depicted with filled circles. (After Baldissera, F.; Hultborn, H.; Illert, M. *Handbk. Physiol.* Sec. 1, 2:509–595, 1981.)

muscles also *inhibits* the Ia inhibitory interneuron to flexor motoneurons, resulting in a disinhibition (facilitation) of the flexor motoneuron. Since Ia inhibitory interneurons are also contacted by several other spinal interneurons, including Renshaw cells, interneurons activated by cutaneous afferents, the flexor reflex afferents, and propriospinal neurons, the action of descending fibers on Ia inhibitory interneurons ensures that descending signals are integrated prior to the motoneuron with inputs from several segmental and intersegmental sources.

Some spinal interneurons are inhibited by descending fibers. In particular, a group of reticulospinal neurons in caudal and ventral portions of the medullary reticular formation sends axons through dorsal portions of the lateral spinal columns, and these inhibit spinal interneurons monosynaptically. Stimulation of these caudal reticulospinal neurons depresses polysynaptically produced reflexes elicited by excitation of Ib and FRA fibers.[10] This observation helps to explain why these reflexes are more depressed in decerebrate animals than in animals with high spinal transections. In part, this depression may be mediated by monoaminergic fibers, since administration of L-DOPA or 5-HTP also depresses the short-latency reflex effects produced by stimulation of cutaneous and Ib afferents.[2]

Descending systems not only affect segmental interneurons, they also synapse on propriospinal neurons that interconnect spinal segments. Short propriospinal neurons with cell bodies in C3–C4 have axons that project to the cervical enlargement. Long propriospinal neurons, which may have cell bodies as high as C3, send axons as far caudally as the lumbosacral cord in the cat.

Kuypers has emphasized, based on anatomical studies, that the short propriospinal neurons are located in the lateral intermediate spinal gray matter, the area of heaviest corticospinal and rubrospinal termination. Physiological experiments[18] have shown that an excitatory input exists to short propriospinal neurons from both the corticospinal and rubrospinal systems. Figure 26–9 demonstrates the convergence of descending and spinal inputs to a neuron at the C3–C4 level of the spinal cord that was classified as a propriospinal neuron by virtue of its antidromic excitation from the lateral funiculus at C5 (Fig. 26–9A). The intracellular records show EPSPs evoked by stimulation of the contralateral pyramidal tract (Fig. 26–9B), the contralateral red nucleus (Fig. 26–9C), and the deep radial nerve (Fig. 26–9D–G), a forelimb nerve whose axons enter the cord in the cervical enlarge-

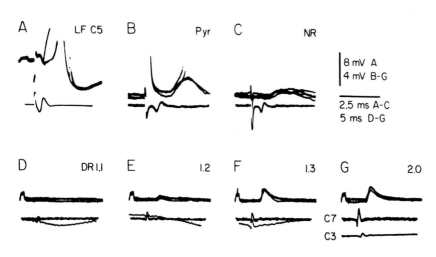

Figure 26–9 Intracellular records from a propriospinal neuron in the C3 cord. *A*, action potential (truncated) produced antidromically by stimulation of the lateral funiculus of the cord at C5. *B*, EPSP produced by stimulation of the contralateral medullary pyramid (Pyr). In two of the four superimposed trials, action potentials (truncated) were generated. *C*, small EPSP produced by stimulation of the contralateral red nucleus, NR. *D–G*, EPSPs produced by stimulation of the deep radial nerve at a stimulus intensity 1.1 *(D)* to 2.0 *(G)* times threshold for a potential recorded from the C7 dorsal root entry zone. (After Baldissera, F.; Hultborn, H.; Illert, M. *Handbk. Physiol.* Sec. 1, 2:509–595, 1981; and Baldissera, F.; Illert, M.; Tanaka, R. *Exp. Brain Res.* 31:131–141, 1978.)

ment. In addition, short propriospinal neurons may receive convergent input from the tectospinal system and reticulospinal axons from the medulla (not illustrated). As is the case for segmental interneurons, some propriospinal neurons are inhibited, instead of excited, by reticulospinal fibers.

Long propriospinal neurons, which are especially effective in activating motoneurons innervating proximal muscles around the trunk, shoulder, and hip,[36] are situated in the medial portion of the intermediate spinal gray matter. This is the primary locus of termination of vestibulospinal and ventrally descending reticulospinal axons (see Fig. 26–5). Control of these long propriospinal neurons, which may send axonal branches to motoneurons both in nearby segments and at a distance, would be an efficient mechanism for coordinating postural adjustment, which is controlled largely by proximal muscles.

The branching of axons from supraspinal structures is another mechanism that couples the descending control of muscles innervated by different spinal segments. As described previously, antidromic activation of descending neurons by low-intensity microstimulation within the spinal gray matter has been used to trace the terminal arborizations of these axons. Using this technique, it has been shown that at least 50% of the lateral vestibulospinal fibers with axonal branches into the cervical enlargement also have an axonal branch that extends as far as the lumbosacral cord. Many reticulospinal axons traveling in the ventral columns also have branches that terminate at both cervical and lumbosacral levels. Thus, both their action on long propriospinal neurons and their terminal axonal branches at many distant segmen-

tal levels make the lateral vestibulospinal and medial reticulospinal systems appropriate to influence the coupled control of proximal musculature that is important for postural adjustment.

In contrast, the termination patterns of corticospinal and rubrospinal axons make them especially appropriate to control fractionated, independent movement of individual extremities, especially at distal joints. Not only do they terminate predominantly on motoneurons, segmental interneurons, and short propriospinal neurons, but, as mentioned earlier, only about 30% of them have been shown to have axonal branches reaching both the cervical gray matter and the lumbar cord.

Actions on Gamma Motoneurons. Descending activity also could influence alpha motoneurons by changing the activity of gamma motoneurons (see 4 in Fig. 26–6). Since activation of gamma motoneurons causes an increased discharge in muscle spindle afferents, which in turn excite alpha motoneurons, this produces muscle contraction via the "spindle loop." In fact, gamma motoneurons, like alpha motoneurons, are affected both monosynaptically and polysynaptically by descending axons. For example, stimulation of the motor cortex has been shown to produce EPSPs in gamma motoneurons innervating baboon forearm muscles. Some of the EPSPs occurred soon enough after the corticospinal volley reached the cord to have been produced monosynaptically.[8] Other PSPs, including some IPSPs, had segmental delays that were long enough to allow an interneuron to be interposed. The lateral vestibulospinal system also affects gamma motoneurons both monosynaptically and through interneurons,[15] and stimulation of the rubrospinal and reticulospinal systems changes the firing of gamma motoneurons as well. In general, it appears that any cortical or brain-stem system that affects alpha motoneurons also affects gamma motoneurons and that the sign of the action (excitatory or inhibitory) is the same in the two motoneuron types. It is not known, however, whether the axons of individual cortical or brain stem neurons branch to influence both alpha and gamma motoneurons.

During voluntary isometric contraction of a muscle, coactivation of both alpha and gamma motoneurons is common. This *alpha-gamma coactivation* is demonstrated by an increase in both the electromyographic activity of the muscle—as a result of excitation of alpha motoneurons—and the discharge rate of the same muscle's spindle afferents, which could occur in the shortening muscle only if its gamma motoneurons were excited. In 1953, Merton[21] proposed that, since descending fibers could produce extrafusal muscle contraction by activation of gamma motoneurons and the spindle

loop, perhaps this "follow-up length servo mechanism" was a primary means of muscle activation. In man-made control systems, this servo-driven design is used to provide automatic load adjustment. In the case of the spindle loop, any change in load encountered by the moving limb would change the length of the contracting muscle, adding the length-induced changes in spindle discharge to the spindle activity produced by gamma motoneurons. The resulting change in excitatory drive to alpha motoneurons would, then, produce a muscle contraction adjusted for the load.

If extrafusal muscle contraction were to be initiated by gamma-driven muscle spindle discharge, then activity in the spindle afferents would have to precede electromyographic activity in the extrafusal muscle. In fact, this seldom is the case, as can be seen when simultaneous recordings are made of spindle afferent discharge from the peripheral nerve and EMG activity from the muscle. As is shown in Figure 26–10, the EMG activity usually begins before the increase in spindle affer-

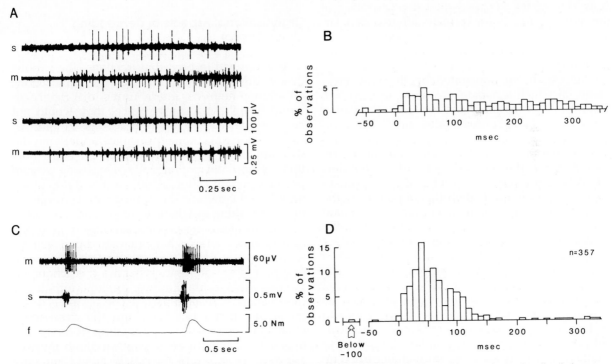

Figure 26–10 Relative timing of spindle afferent and extrafusal muscle activity during isometric muscle contraction in humans. *A,* Spindle afferent discharge (s) and electromyographic (EMG) activity (m) of the finger flexor muscle from which the spindle afferent originated. Records were collected with needle electrodes during two sustained isometric voluntary finger flexions. *B,* The fraction of sustained contraction trials in which spindle afferent discharge began at different times relative to EMG activity in the muscle from which the spindle afferent originated. The onset of EMG activity is defined as 0 time on the abscissa. The data are pooled from many spindles, and in 96% of the trials, EMG activity preceded the discharge of spindle afferents. *C,* Spindle afferent discharge and EMG activity during two brief isometric supinations of the hand. The isometric force is shown in the lower record (f). *D,* The time of spindle afferent discharge relative to the onset of EMG activity (0 ms) during brief isometric supination of the hand. Again, spindle afferent activity followed EMG activity in most trials. (From Vallbo, A.B. *J. Physiol.* 218:405–431, 1971.)

ent discharge, and thus could not be initiated by it. Gamma coactivation in such a situation could provide servo-adjusted, but not servo-initiated, control of extrafusal muscle contraction.

Although the discharge of spindle afferents increases during an isometric contraction or a contraction producing movement at slow to moderate speeds, discharge may *decrease* during rapid isotonic contraction.[30] This does not mean that alpha-gamma coactivation does not exist under these conditions; it simply means that the extrafusal shortening velocity is no longer matched or exceeded by gamma-induced intrafusal muscle contraction. Under these conditions, muscle length, rather than gamma control, dominates the spindle afferent firing rate.

Presynaptic Inhibition Produced by Spinal-Destined Systems. Stimulation of descending fibers exerts actions on primary afferent terminals, as well as on neuronal cell bodies. This action also is usually accomplished through spinal interneurons, however (see 5 in Fig. 26–6). The most frequent effect on axonal terminals—the presynaptic side of the synapse—is a depolarization of the terminal membrane (primary afferent depolarization [PAD]) that results in *presynaptic inhibition*. (See Chap. 11 for the mechanisms of presynaptic inhibition.)

PAD can be demonstrated by the Wall technique, in which a microelectrode is inserted into the dorsal gray matter and is used as a stimulating electrode to excite primary afferents.[37] The compound action potential produced by this test stimulus is conducted antidromically and can be recorded from the dorsal roots. If a stimulus intensity submaximal for exciting all of the primary afferent fibers is used, then the amplitude of the compound action potential elicited by constant intensity stimulation will vary as a function of the excitability of the primary afferent fibers. PAD produced by descending fibers, for example, will increase the excitability of the primary afferents. As a consequence, the compound action potential recorded from the dorsal roots and produced by local stimulation within the dorsal gray matter will be larger. If intra-axonal recording also is used, peripheral afferent fibers can be stimulated to identify the type of sensory information carried by the afferent in which PAD is produced by stimulation of supraspinal systems.

Stimulation of the sensorimotor cortex or red nucleus evokes PAD in terminals of Ib and cutaneous afferents.[6] Furthermore, PAD can be produced by stimulation of other peripheral afferents (see Chap. 24), and this PAD also is markedly facilitated by stimulation of corticospinal or rubrospinal systems.[17] Therefore, both the descending axons and the peripheral afferent fibers must share the same interneurons to produce PAD.

Stimulation of the vestibular nuclei and the medial longitudinal fasciculus (which contains some of the descending reticulospinal axons) evokes large PADs in Ia fibers, as well as in Ib and cutaneous afferents. In contrast, stimulation of the cerebral cortex or the red nucleus produces only minor PADs in Ia afferents. Instead, stimulation of corticospinal or rubrospinal systems markedly *depresses* the presynaptic inhibition of Ia afferents produced polysynaptically by activation of the Ia axons themselves.

The functional role of PAD and presynaptic inhibition is not known. Since movement causes activity in afferent fibers (e.g., Ia, Ib, joint, cutaneous), one possibility is that presynaptic control from supraspinal neurons helps to control the magnitude (or gain) of this segmental feedback influence on motoneurons.

Organizational Aspects of Descending Motor Control

As described in preceding sections, the various descending systems do not exert the same actions on all motoneurons, interneurons, or primary afferent fibers. As a consequence, they would not be expected to exert uniform actions on all muscles. The organizational principles underlying descending fiber actions are of interest both functionally and developmentally. Does one system produce certain actions, and another different actions? And how are the appropriate connections made, resulting in whatever organization exists?

Most studies of descending actions on motoneurons or spinal reflexes have examined their actions with respect to one of two organizational schemes. The first is somatotopic: What are the actions on motoneurons to hindlimb vs. forelimb muscles or to proximal vs. distal muscles? The second is based on action at a joint: What are the descending effects on motoneurons innervating flexor vs. extensor muscles? (In classifying hindlimb muscles as extensors or flexors, the classification commonly used is that developed by Sherrington,[33] who palpated muscles during stimulation of the cutaneous branch of the peroneal nerve in the decerebrate cat and classified those that contracted as flexors and those that relaxed as extensors.) A third organizational scheme, based on motor unit type, was added when it was discovered that individual muscles are composed of motor units with very different functional properties (Chap. 23).

All spinal-destined systems show a somatotopic organization at their origins, as well as in the course of their descent to the spinal cord. In the red nucleus, neurons with axons destined for the cervical cord are situated ventromedial to those innervating lumbar segments. In the lateral vestibular nucleus, neurons that can be excited antidromically from the ventral spinal columns at cervical, but not lumbar, levels are concentrated dorsal to those activated antidromically from both the lumbar and the cervical cord. The identification of neurons that give axonal branches to multiple separated segments, however, means that the somatotopic separation certainly is not rigid.

An organization based on action around a joint is clear only for some systems. As described earlier, the lateral vestibulospinal system has a strong bias toward excitation of extensor motoneurons and reciprocal inhibition of flexor motoneurons.[38] Although early reports emphasized the excitatory action of rubrospinal fibers on flexor motoneurons, this certainly is not exclusive or even as strongly biased as the lateral vestibulospinal action. And reticulospinal effects are even more mixed with respect to flexors and extensors. The strong facilitatory action of the lateral vestibulospinal system on extensors undoubtedly is related to the important roles of the vestibular apparatus and extensor (antigravity) muscles in maintaining a stable upright posture.

As described in Chapter 23, motor units within a muscle may be differentiated on the basis of the contraction speed, fatigability, maximum force, and histochemical profile of their muscle fibers. There are several types of fast-twitch motor units, with varying fatigue and histochemical characteristics (FF, FR, and FI). Slow-twitch (S) motor units contract slowly, produce less maximum force, and are resistant to fatigue.

Some descending systems have different effects on the various motor unit types. For example, PSPs produced by stimulation of the red nucleus have larger excitatory components (EPSPs) in type F motoneurons and larger inhibitory components (IPSPs) in type S motoneurons.[4] There is no correlation, however, between motor unit type and the amplitude of the EPSP produced monosynaptically by descending axons in the ventral quadrant of the cord, which includes axons from the lateral vestibular nucleus.[5] The PSP pattern produced by red nucleus stimulation is similar to that produced by stimulation of the cutaneous sural nerve, a

finding consistent with the documented convergence of both inputs onto common short-axoned propriospinal neurons.

The actions of other descending systems have not been compared in motoneurons individually classified as to motor unit type, but the action of the corticospinal system has been compared in motoneuron pools innervating muscles containing different dominant fiber types ("fast-twitch" and "slow-twitch" muscles). In general, corticospinal excitation is greater in motoneurons innervating monkey medial gastrocnemius (a predominantly fast-twitch muscle) than in those innervating soleus (a predominantly slow-twitch muscle), and a similar bias is found in the action of the corticospinal system on the rapidly contracting vs. slowly contracting heads of cat triceps brachii muscles.[29]

Many muscles, in fact, may be subdivided grossly into compartments that have quite different patterns of activity during common motor actions such as walking. Whether or not the actions of different descending systems on motoneurons innervating a particular muscle are organized with respect to complex functional schemes remains to be determined. Do particular corticospinal fibers, for example, excite motoneurons to both flexor and extensor muscles around a joint? And are these corticospinal cells activated during motor actions requiring stabilization of that joint by cocontraction of flexor *and* extensor muscles? Other corticospinal neurons might be selectively active during movements requiring reciprocal excitation of flexor and inhibition of extensor muscles, and these presumably would have quite different patterns of action on different motoneuron pools.

Functional Studies of Spinal-Destined Systems

As described earlier (see Fig. 26–5), descending fibers coursing through the dorsolateral columns of the cord (corticospinal and rubrospinal) terminate more densely in different parts of the spinal gray matter than do the fibers located more medially in the ventral and ventrolateral cord (vestibulospinal and reticulospinal). Because of this anatomical diversity, Kuypers hypothesized that the functional roles of the lateral and medial systems also would differ. He and his colleagues tested this hypothesis by observing the motor functions left intact after relatively selective lesions of the different systems in cats and monkeys.[19, 20]

The capabilities of the ventrally descending systems were assessed after corticospinal and rubrospinal fibers were transected. Corticospinal (and corticobulbar) fibers were interrupted without damage to other descending systems by careful bilateral transection of the pyramidal tracts in the medulla. In monkeys with this lesion, balance, ambulation, grasping the cage, climbing, guidance and stabilization of the arm during reaching, and use of the hand independently of the proximal arm all recovered within four to six weeks.[19] In contrast, the persistent motor problems that reflected functions for which corticospinal fibers were essential were (1) slowness and fatigability of all movements, (2) absence of individual finger movements, and (3) difficulty in releasing the grip on food. When the rubrospinal tract was interrupted in the medulla (without pyramidotomy), slowness and weakness of the hand also were observed, but these deficits lasted less than a week. If, however, pyramidotomy and rubrospinal transection were combined, the impairment of distal movements was exaggerated, and food was obtained by a circumduction at the shoulder followed by a raking movement of the hand.[19]

Functional changes produced by lesions that destroyed the medial pathways but spared corticospinal and rubrospinal fibers were much different. Selective destruction of the medial pathways is difficult, but lesions were made in the dorsomedial medullary tegmentum to interrupt pontine and rostral medullary reticulospinal, tectospinal, medial vestibulospinal, and lateral vestibulospinal fibers. (In some cases, the underlying medial lemniscus also was damaged.) Following these lesions, animals retained their ability to make rapid, discrete hand and finger movements to retrieve food morsels from small wells, consistent with the major termination of intact corticospinal and rubrospinal fibers in cord regions innervating distal musculature. In contrast, animals showed a flexor-biased posture of the trunk and limbs, coarse proximal ataxia during reaching movements, and unsteadiness in gait.

Persistent motor deficits following selective lesions such as those described can show only the motor functions for which the interrupted systems are *necessary*. They cannot be used to infer the entire range of motor activity in which a system would have participated, had it been intact. With this in mind, however, the data of Lawrence and Kuypers,[19, 20] together with those of other investigators who have made lesions of the pyramidal tract, point to the major importance of the corticospinal and rubrospinal tracts in controlling rapid, fine movements of the distal extremities. The medial reticulospinal, vestibulospinal, and tectospinal systems seem most important for the control of movement and stabilization at proximal joints.

Neuronal Activity During Movement

If descending fibers are necessary for different types of motor function, then what specific information is provided by these fibers? This has been investigated by recording from neurons with spinal-destined axons during movements made by awake animals that were usually highly trained to perform very specific motor tasks. From such data it is possible to learn something about the conditions under which a set of neurons is activated, whether their activation precedes (and therefore might cause) a particular muscle activity, and whether there is a strong correlation between the neuron's discharge and particular kinematic parameters of the movement, such as position, velocity, and acceleration.

The activity of corticospinal neurons has been examined extensively and is described in detail in Chapter 28. Of the remaining descending systems, only the activity of neurons in the red nucleus has been recorded during highly controlled movements.

Rubrospinal neurons exhibit a burst of activity in association with movement (Fig. 26–11). If the motor task is chosen carefully to maximize the neuron's change in activity, the cell's burst of discharge precedes the onset of the movement by up to 120 ms (about 60 ms in advance of EMG activity) and lasts for a period of time that is proportional to the duration of the movement. During the movement, the cell's burst frequency is proportional to movement velocity over a considerable range of velocities, and the total number of spikes in the burst (the integral of the discharge rate over the duration of the burst) is proportional to the amplitude of the movement.[12] Furthermore, most red nucleus neurons show their greatest discharge during movements of distal portions of the limbs, in agreement with the predictions of the earlier anatomical and lesion studies.[11]

The only motor act during which reticulospinal and vestibulospinal neurons have been studied is locomotion. Furthermore, the animals were not intact, trained animals but were decerebrate cats in which locomotion was induced by stimulation of the mesencephalic locomotor region (MLR), an area in the mesencephalic tegmentum. About two thirds of the lateral vestibulospinal neurons studied during MLR-induced locomotion showed ac-

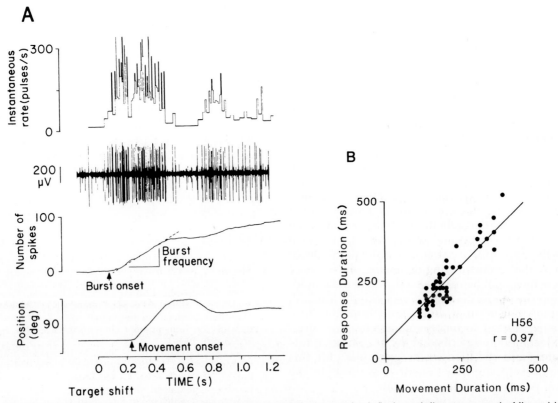

Figure 26–11 Activity of a magnocellular red nucleus neuron during a monkey's flexion rotation movement of the wrist and fingers. A, The upper plot shows the instantaneous discharge rate of the large action potentials in the second trace. The number of spikes is integrated over time in the third record, the slope of which is the mean frequency of discharge during the burst. Angle of the manipulandum grasped by the monkey's hand is shown in the bottom record. The movement started 240 ms after the tracking target shifted to a new position (target shift). The neuronal burst began 160 ms prior to the beginning of the movement. B, Relationship between burst duration and movement duration. (After Gibson, A.R.; Houk, J.C.; Kohlerman, N.J. *J. Physiol.* 358:527–549, 551–570, 1985.)

tivity that varied systematically during the step cycle.[23] For almost 80% of these, the peak discharge rate occurred slightly prior to the activity of extensor muscles in the ipsilateral hindlimb. Since stimulation of the lateral vestibular nucleus produces EPSPs primarily in extensor motoneurons, this pattern of vestibulospinal activity would be appropriate to facilitate extensor activity. In fact, stimulation in the lateral vestibular nucleus during the step cycle does enhance extensor muscle activity, but only if applied during the support phase—the time when the foot is in contact with the ground and the extensor muscles normally are active (see Fig. 26–1).

Although a phasic pattern of activity occurs in vestibulospinal neurons, their phasic activity is not *necessary* to generate the patterned muscle activity that occurs during locomotion. As described earlier, phasic locomotor activity occurs in

motoneurons deprived of all supraspinal input by spinal transection. Furthermore, if the cerebellum has been removed in decerebrate animals exhibiting MLR-induced locomotion, modulation of vestibulospinal activity is abolished. The discharge of vestibulospinal neurons that can be produced by passive movement of the limbs also disappears after removal of the cerebellum.[23] Since the cerebellum provides a major synaptic input to vestibulospinal neurons, it is likely that sensory input elicited by limb movement produces, via the cerebellum, the phasic pattern of activity in Deiter's neurons. Furthermore, this activity is appropriate to facilitate extensor activity during the support phase of the step cycle.

Rubrospinal neurons also show a phasic modulation of activity during the step cycle,[24] but in contrast to vestibulospinal neurons, about 75% of the rubrospinal cells studied had peak activity in the late support or the swing phase of locomotion, when flexor muscles are most active. Since electrical stimulation of the red nucleus has been reported to produce excitation of some proximal flexor motoneurons, this pattern of rubrospinal activity is appropriate to enhance flexor activity during the swing phase. As was the case for vestibulospinal neurons, however, the modulated rubrospinal activity was abolished by removal of the cerebellum, although locomotion persisted. The rubrospinal activity that was correlated with locomotion was not affected when movement at distal joints was disturbed, but it was changed if hip movement was changed (delayed or accelerated) manually. Thus, rubrospinal activity during locomotion in decerebrate animals also may be modulated by spinocerebellar information, and it is of the appropriate pattern to facilitate the flexor activity that occurs during the swing phase.

Reticulospinal neurons also show periodic modulation during locomotion. Furthermore, forelimb stepping can occur without input from the lumbosacral "spinal locomotor generator" for hindlimb locomotion described at the beginning of this chapter, and the "locomotor generator" for this forelimb stepping may consist of neurons in the lateral pontine reticular formation.[34]

SUMMARY

Neurons in many areas of the brain send axons to the spinal cord and influence the output of alpha motoneurons. Most of this information, however, is first integrated with that from peripheral somatosensory receptors by convergence onto spinal interneurons. Furthermore, supraspinal systems also influence the activity of gamma motoneurons and the presynaptic terminals of primary afferent fibers. The various supraspinal systems differ, however, in the anatomical organization of their spinal connectivity and in their effects on motor function. Lesion studies have demonstrated that corticospinal and rubrospinal systems are especially important for the independent control of distal motor function, whereas vestibulospinal and reticulospinal systems are more critical for balance and proximal motor control. We are only beginning to understand, however, the precise signals carried by spinal-destined axons from the brain and the way that these interact with segmentally derived information to control posture and movement.

ANNOTATED BIBLIOGRAPHY

Baldissera, F.; Hultborn, H.; Illert, M. Integration in spinal neuronal systems. *Handbook of Physiology* Sec. 1, 2:509–595, 1981.
A detailed review of the integration of segmental and descending information at the spinal cord.
Grillner, S. Control of locomotion in bipeds, tetrapods, and fish. *Handbook of Physiology* Sec. 1, 2:1179–1236, 1981.
A review of spinal and supraspinal mechanisms involved in the control of locomotion.
Kuypers, H.G.J.M. Anatomy of the descending pathways. *Handbook of Physiology*, Sec. 1, 2:597–666, 1981.
An anatomical review that includes comparative aspects of different spinal-destined systems.

REFERENCES

1. Abzug, C.; Maeda, M.; Peterson, B.W.; Wilson, V.J. Cervical branching of lumbar vestibulospinal axons. *J. Physiol. (Lond.)* 243:499–522, 1974.
2. Anden, N.E.; Jukes, M.G.M.; Lundberg, A.; Vyklicky, L. The effect of dopa on the spinal cord. I. Influence on transmission from primary afferents. *Acta Physiol. Scand.* 67:373–386, 1966.
3. Baldissera, F.; Hultborn, H.; Illert, M. Integration in spinal neuronal systems. *Handbk. Physiol.* Sect. 1, 2:509–595, 1981.
4. Burke, R.E.; Jankowska, E.; ten Bruggencate, G. A comparison of peripheral and rubrospinal synaptic input to slow and fast twitch motor units of triceps surae. *J. Physiol. (Lond.)* 207:709–732, 1970.
5. Burke, R.E.; Rymer, W.Z.; Walsh, J.V., Jr. Relative strength of synaptic input from short-latency pathways to motor units of defined type in cat medial gastrocnemius. *J. Neurophysiol.* 39:447–458, 1976.
6. Carpenter, D.; Lundberg, A.; Norrsell, U. Primary afferent depolarization evoked from the sensorimotor cortex. *Acta Physiol. Scand.* 59:126–142, 1963.
7. Clough, J.F.M.; Kernell, D.; Phillips, C.G. The distribution of monosynaptic excitation from the pyramidal tract and from primary spindle afferents to motoneurons of the baboon's hand and forearm. *J. Physiol. (Lond.)* 198:145–166, 1968.
8. Clough, J.F.M.; Phillips, C.G.; Sheridan, J.D. The short-latency projection from the baboon's motor cortex to fusimotor neurones of the forearm and hand. *J. Physiol. (Lond.)* 216:257–279, 1971.
9. Eccles, J.C.; Eccles, R.M.; Lundberg, A. Synaptic actions on motoneurones in relation to the two components of the group I muscle afferent volley. *J. Physiol. (Lond.)* 136:521–546, 1957.
10. Engberg, I.; Lundberg, A.; Ryall, R.W. Reticulospinal inhibition of transmission in reflex pathways. *J. Physiol. (Lond.)* 194:201–223, 1968.
11. Gibson, A.R.; Houk, J.C.; Kohlerman, N.J. Magnocellular red nucleus activity during different types of limb movement in the macaque monkey. *J. Physiol. (Lond.)* 358:527–549, 1985.
12. Gibson, A.R.; Houk, J.C.; Kohlerman, N.J. Relation between red nucleus discharge and movement parameters in trained macaque monkeys. *J. Physiol. (Lond.)* 358:551–570, 1985.

13. Grillner, S. Control of locomotion in bipeds, tetrapods, and fish. *Handbk. Physiol.* Sect. 1, 2:1179–1236, 1981.

14. Grillner, S.; Lund, S. The origin of a descending pathway with monosynaptic action on flexor motoneurons. *Acta Physiol. Scand.* 74:274–284, 1968.

15. Grillner, S.; Hongo, T.; Lund, S. Descending monosynaptic and reflex control of gamma-motoneurones. *Acta Physiol. Scand.* 75:592–613, 1969.

16. Grillner, S.; Hongo, T.; Lund, S. The vestibulospinal tract. Effects on alpha-motoneurons in the lumbosacral cord in cat. *Exp. Brain Res.* 10:94–120, 1970.

17. Hongo, T.; Jankowska, E.; Lundberg, A. The rubrospinal tract. III. Effects on primary afferent terminals. *Exp. Brain Res.* 15:39–53, 1972.

18. Illert, M.; Lundberg, A.; Tanaka, R. Integration in descending motor pathways controlling the forelimb in the cat. 3. Convergence on propriospinal neurones transmitting disynaptic excitation from the corticospinal tract and other descending tracts. *Exp. Brain Res.* 29:323–346, 1977.

19. Lawrence, D.G.; Kuypers, H.G.J.M. The functional organization of the motor system in the monkey. I. The effects of bilateral pyramidal lesions. *Brain* 91:1–14, 1968.

20. Lawrence, D.G.; Kuypers, H.G.J.M. The functional organization of the motor system in the monkey. II. The effects of lesions of the descending brain-stem pathways. *Brain* 91:15–36, 1968.

21. Merton, P.A. Speculations on the servo-control of movement. In Wolstenholme, G.E.W., ed. *The Spinal Cord*, 247–255. London, Churchill Livingstone, 1953.

22. Murray, E.A.; Coulter, J.D. Organization of corticospinal neurons in the monkey. *J. Comp. Neurol.* 195:339–365, 1981.

23. Orlovsky, G.N. Activity of vestibulospinal neurons during locomotion. *Brain Res.* 46:85–98, 1972.

24. Orlovsky, G.N. Activity of rubrospinal neurons during locomotion. *Brain Res.* 46:99–112, 1972.

25. Peterson, B.W.; Maunz, R.A.; Pitts, N.G.; Mackel, R.G. Patterns of projection and branching of reticulospinal neurons. *Exp. Brain Res.* 23:333–351, 1975.

26. Porter, R. Early facilitation at corticomotoneuronal synapses. *J. Physiol. (Lond.)* 207:733–745, 1970.

27. Porter, R.; Hore, J. Time course of minimal corticomotoneuronal excitatory postsynaptic potentials in lumbar motoneurons of the monkey *J. Neurophysiol.* 32:443–451, 1969.

28. Pratt, C.; Jordan, L. Ia inhibitory interneurons and Renshaw cells as contributors to the spinal mechanisms of fictive locomotion. *J. Neurophysiol.* 57:56–71, 1987.

29. Preston, J.B.; Whitlock, D.G. A comparison of motor cortex effects on slow and fast muscle innervations in the monkey. *Exp. Neurol.* 7:327–341, 1963.

30. Prohazka, A.; Stephens, J.; Wand, P. Muscle spindle discharge in normal and obstructed movements. *J. Physiol. (Lond.)* 287:57–66, 1979.

31. Rapoport, S.; Susswein, A.; Uchino, Y.; Wilson, V.J. Synaptic actions of individual vestibular neurones on cat neck motoneurones. *J. Physiol. (Lond.)* 272:367–382, 1977.

32. Scott, J.G.; Mendell, L.M. Individual EPSPs produced by single triceps surae Ia afferent fibers in homonymous and heteronymous motoneurons. *J. Neurophysiol.* 39:679–692, 1976.

33. Sherrington, C.S. Flexion-reflex of the limb, crossed extension reflex, and reflex stepping and standing. *J. Physiol. (Lond.)* 40:28–121, 1910.

34. Shimamura, M.; Kogure, I.; Fuwa, T. The role of the paralemniscal pontine reticular formation in forelimb stepping of thalamic cats. *Neurosci. Res.* 1:393–340, 1984.

35. Shinoda, Y.; Zarzecki, P.; Asanuma, H. Spinal branching of pyramidal tract neurons in the monkey. *Exp. Brain Res.* 34:59–72, 1979.

36. Vasilenko, D.A. Propriospinal pathways in the ventral funicles of the cat spinal cord: their effects on lumbosacral motoneurones. *Brain Res.* 93:502–506, 1975.

37. Wall, P.D. Excitability changes in afferent fibre terminations and their relation to slow potentials. *J. Physiol. (Lond.)* 142:1–21, 1958.

38. Wilson, V.J.; Yoshida, M. Comparison of effects of stimulation of Deiters' nucleus and medial longitudinal fasciculus on neck, forelimb, and hindlimb motoneurons. *J. Neurophysiol.* 32:743–758, 1969.

The Vestibular System

INTRODUCTION

The vestibular sense is different from the senses discussed in previous chapters. For the five "special" senses—vision, audition, taste, smell, and touch—an adequate stimulus frequently is at-tended to and produces a conscious sensation. In contrast, we are seldom aware of movements of the head or the static orientation of the head in space, two situations that provide the adequate stimulus for the vestibular receptors. For example, the bobbing up and down of the head during locomotion usually goes unnoticed. Indeed, it is only when the vestibular system receives an unusual stimulus, such as when we somersault down a hill or ride deep swells on the sea, that a vestibular sensation is generated. Patients who have gradually undergone a complete loss of vestibular function may even be unaware of their deficit, suggesting that there are essential differences between the vestibular and other senses.

Although vestibular sensation is less prominent than that of the special senses, signals from the vestibular apparatus are essential for postural adjustments and the generation of movement. Vestibular signals help stabilize the head, which, since it houses all of the special sense organs, must be kept oriented toward objects of interest. In addition, if the head moves, vestibular signals can rotate the eye within the head to change the direction of gaze and aim the retina. Finally, vestibular signals also cause the appropriate synergistic action of leg and trunk muscles to provide a stable body platform from which the eye and head movements can be initiated. In most situations, movements of the body, head, and eyes are carefully coordinated, as can be appreciated by observing the acrobatic act of a bird landing on a telephone wire. Furthermore, these compensatory movements occur unconsciously; for example, if one stumbles while walking, the head is righted without any apparent voluntary effort.

Vestibular information can reach the muscles responsible for these various movements over short pathways that may consist of as few as three neurons. Furthermore, the muscle responses to a specific vestibular stimulus are stereotyped. Therefore, the vestibulocollic (neck), vestibulo-ocular, and vestibulospinal connections may be considered reflex pathways similar to those in the spinal cord (Chap. 24). Unlike the spinal reflexes, however, the vestibular reflexes have been examined in alert animals so that the neural signals carried by these pathways are well known. This understanding of the functional neuroanatomy of vestibular reflexes has made it possible to examine how "descending" inputs, like those from the cerebellum, can affect reflex behavior.

In this chapter, we first consider the peripheral vestibular apparatus, which is common to all vestibular reflexes. Then we evaluate how vestibular signals contribute to each of these reflexes individually. Finally, we show how the cerebellum can influence the behavior of one such reflex, the vestibulo-ocular reflex.

THE VESTIBULAR PERIPHERY

The Sensory Receptor. Much as sound vibrations are ultimately transduced into electrical activity by *hair cells* in the cochlea (Chap. 17), so also head movements are transduced by hair cells in the vestibular apparatus. Two types of vestibular hair cells can be distinguished by their shapes and their afferent and efferent innervations (Fig. 27–1A). They are similar, however, in having many small *cilia* that protrude from their apical surfaces (Fig. 27–1B). All vestibular hair cells have a single true cilium called the *kinocilium*, and a regular array of up to 50 smaller *stereocilia*. As illustrated in Figure 27–1, this bundle of "hairs" is morphologically polarized in two ways. First, the kinocilium is situated on one side of the hair bundle; second, the stereocilia become progressively shorter as one moves away from the kinocilium. As shown in Chapter 5, deflection of the hair bundle toward the kinocilium causes depolarization of the hair cell while deflection away from the kinocilium causes hyperpolarization. Depolarization of the hair cell causes release of an excitatory transmitter that produces an increased discharge rate in the afferent nerve fiber that innervates it. The preferred direction for a stimulus applied to a hair cell, therefore, is from the shortest stereocilium to the kinocilium.

Like many other mechanoreceptors (e.g., the pacinian corpuscle or the hair cells of the ear), vestibular hair cells are driven indirectly through an accessory structure (Chap. 5). In general, the hairs are inserted into an overlying gelatinous substrate. In the lateral line organ of fishes, the gelatinous substrate resembles a tongue that protrudes directly into the water from the side of the

head and body and is bent as the fish swims. In the other vestibular organs of fish and all those of mammals, the hair cells and their accessory structure are enclosed in a cartilaginous or bony cavity in the skull which is filled with an internally generated fluid. Consequently, head movement affects hair cells indirectly. The size, shape, and location of the cavities determine how the endolymph moves during head movement and, therefore, how the endolymph affects the motion of the gelatinous substrate and ultimately the cilia. To sense the different aspects of head movement, the vestibular apparatus has developed a number of different substrates and different cavities that confer a distinctive responsiveness upon their resident hair cells.

Location of the Receptors. Figure 27–2 shows the location of the various cavities that house the hair cells in the human temporal bone. Three of the cavities are called the *semicircular canals*, although they in fact are complete ring-shaped tubes that communicate at their most medial aspect via a fourth chamber called the *utricle*. Before entering the utricle, one side of each canal expands to twice its diameter to form an *ampulla* that houses the sensory epithelium (the crista ampullaris) containing the hair cells (also see Fig. 27–3). The diameter of the ring ranges from 4 to 7 mm, and the unexpanded diameter of the cross section of the canal is about 0.5 mm. The three canals lie in roughly orthogonal planes. In man, the lateral canal is tipped backward by about 25° from the horizontal plane, the anterior canal (also called the superior canal) is oriented anterolaterally at about 41° from the sagittal plane, and the posterior canal is oriented posterolaterally at an angle of about 56° with the sagittal plane (see Fig. 27–5A). As seen in Figure 27–2, the *utricle* and *saccule* (together called the *otolith organs*) are larger chambers on the ventromedial aspect of the vestibular apparatus. With the head in its normal, upright position, the sensory epithelia (the maculae) of the utricle and the saccule lie approximately in the horizontal and sagittal planes, respectively. The entire peripheral apparatus (canals, utricle, and saccule) is called the *vestibular labyrinth*.

The Adequate Stimulus for the Canals. The way in which the hair cells of the semicircular canals are stimulated by head movement can be discerned by peeling away the wall of an ampulla as illustrated in Figure 27–3. Across the bottom of the ampulla is a saddle-shaped structure containing a single layer of hair cells. Astride the sensory epithelium is a gelatinous tongue called the *cupula* into which the cilia of the hair cells protrude.

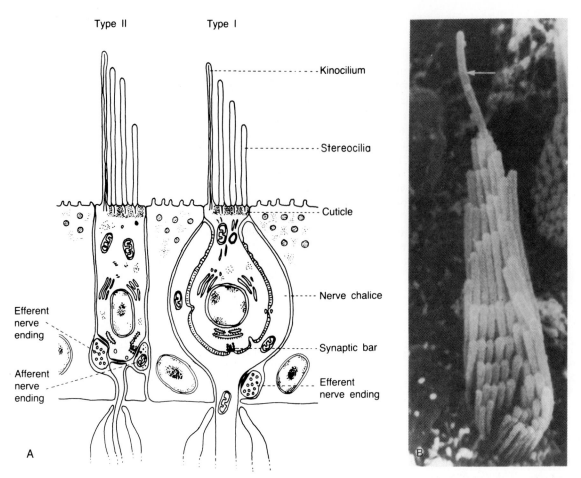

Figure 27–1 Vestibular hair cells. *(A)* Structure and innervation of type I and type II hair cells. (After Wersäl, J. *Handbook of Sensory Physiology,* Vol. 6, pt. 1. New York, Springer-Verlag, 1974.) *(B)* Scanning electron micrograph of the apical surface of a hair cell in the guinea pig utricle after removal of the otolithic membrane. (After Lindeman, H.H. *Adv. Otorhinolaryngol.* 20:405–433, 1973.) Kinocilium indicated by arrow, which is 0.8 μm long.

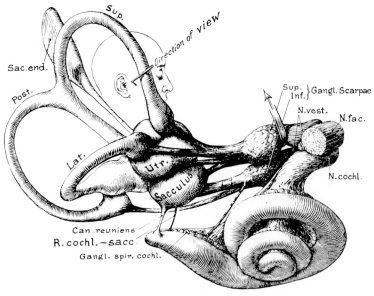

Figure 27–2 Location and innervation of the vestibular end-organs in the human temporal bone. (After Hardy, M. *Anat. Rec.* 59:403–418, 1934). The vestibular nerve (N. vest.) is composed of axons of bipolar cell somata lying in the superior (Sup.) and inferior (Inf.) vestibular (Scarpa's) ganglia. Distal processes of bipolar cells divide into branches to innervate the three canals and the two otolith organs. Endolymph produced in the sacculus endolymphaticus (Sac. end.) reaches the cochlea through the canal reuniens via the otolith cavities. The facial (N. fac.) and cochlear divisions of the 8th nerve (N. cochl.) are also shown.

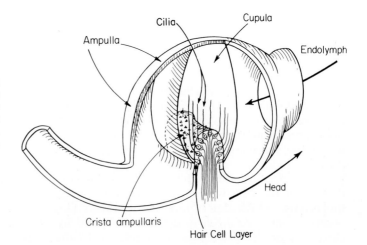

Figure 27–3 Schematic diagram of the ampulla and its contents. The cilia of hair cells (only one cilium per receptor is shown) insert into the gelatinous cupula, which is deflected owing to the endolymph motion resulting from the indicated head movement. (After Lindemann, H.H. *Ergeb. Anat. Entwickl.* 42:1–113, 1969.)

During a counterclockwise rotation of the head (Fig. 27–3), the endolymph is left behind and applies a clockwise force to the cupula. The cupula, which is attached to the walls of the ampulla all the way around, billows like a sail and applies a shear to the cilia to activate the hair cells (Chap. 5). Since during a head rotation the force to deflect the cupula is provided by the inertia of the endolymph, the adequate stimulus for activating the hair cells is *head acceleration.* As the head accelerates to a constant velocity, the cupula initially undergoes an increasing deflection. After constant velocity is achieved, however, the friction of the canal walls gradually causes the endolymph to be dragged along at the velocity of the head, so that over a period of about 15 s the cupula returns to its neutral position.

That head acceleration is necessary to stimulate the canals can be confirmed by rotating a blindfolded subject on a turntable. While the subject is being accelerated from rest to some constant velocity, he senses that he is rotating. Once constant velocity has been achieved and while it is maintained, his perception of motion gradually disappears over about 20 s, a period that reflects both central processing as well as the return of the cupula to its neutral position. The perception of rotation returns only when the turntable is decelerated to rest, driving the cupula in the opposite direction.

The Adequate Stimulus for the Otolith Organs. As in the semicircular canals, the cilia of the hair cells in both the utricle and saccule insert into a gelatinous cap as shown schematically in Figure 27–4. In the utricle, the sensory epithelium lies on the floor of the chamber with the cilia pointed

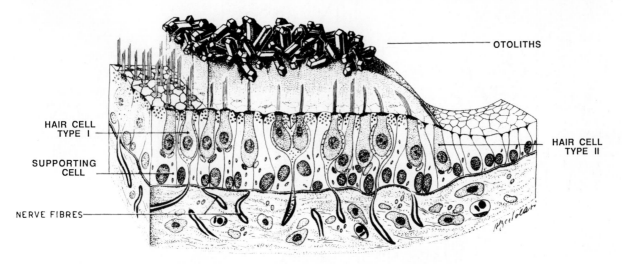

Figure 27–4 Schematic drawing of the macula. The cilia of types I and II hair cells insert into a substrate impregnated with calcium carbonate particles. (Modified with permission from Iurato, S. *Submicroscopic Structure of the Inner Ear.* Copyright 1967, Pergamon Books, Ltd.)

mostly vertically. In the saccule, the sensory epithelium is attached to the medial wall, with the cilia pointed horizontally. Unlike the cupula, the gelatinous caps that crown the utricular and saccular epithelia do not extend to the roofs of their respective chambers. Furthermore, they are impregnated with small (1 to 5 μm) calcium carbonate particles (called *otoliths*) that have a higher density than the surrounding endolymph. Consequently, when the head assumes a static *pitch* position (turning about the interaural axis) or *roll* position (turning about a naso-occipital axis), a steady deflection of the cilia will result. Therefore, unlike the canals, which detect movement, the otolith organs seem specialized to detect maintained inclinations or the tilt of the head relative to the force of gravity. Since the force of gravity produces a constant, unidirectional acceleration, it may also be said that *the otoliths sense linear acceleration whereas the canals sense angular acceleration.*

Direction Selectivity. As we discussed earlier (Fig. 27–1), the location and length of the cilia of a hair cell reveal its preferred direction of stimulation. In Figure 27–5, arrows pointing from the smallest stereocilium toward the kinocilium of a representative hair cell at that location show the direction of head movements that best stimulate the canals (*A*) and the otolith organs (*B*). All the hair cells of an individual canal are "aimed" in the same direction, either toward the utricle for the horizontal canals or away from the utricle for the anterior and posterior canals. The pair of anterior and posterior canals on opposite sides of head and the two horizontal canals are roughly coplanar with opposite preferred directions. Therefore, any complex angular acceleration of the head can be broken down into three component directions because the canals themselves, rather than individual hair cells, have different orientations. On the other hand, the hair cells in the utricle and saccule have a wide variety of preferred directions. In the utricle, the hair cells face the striola, a curved dividing ridge running through the middle of the spoon-shaped macula (shown dotted in Fig. 27–5B), whereas in the saccule, the hair cells face away from its "s"-shaped striola (dotted line, Fig. 27–5B). Therefore, a complex tilt of the head activates each otolithic hair cell quite differently because each has a slightly different orientation. In short, in the otolith organs directionality is conferred by hair-cell orientation while in the ampulae it is conferred by canal orientation.

ACTIVITY IN THE VESTIBULAR NERVE

The vestibular receptors are innervated by bipolar sensory neurons with cell bodies lying in the vestibular ganglion (Fig. 27–2). The ganglion (also called Scarpa's ganglion) is divided into two

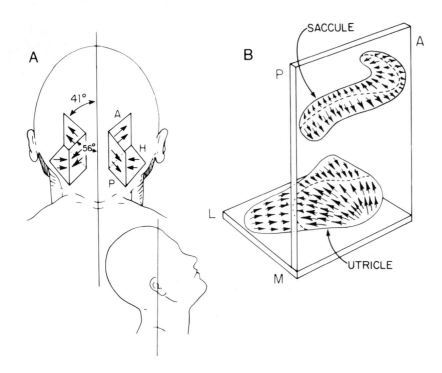

Figure 27–5 Preferred stimulus directions for hair cells in the semicircular canals, *(A)*, and the utricle and saccule *(B)*. On the sensory epithelium, the arrows point from the smallest stereocilium toward the kinocilium of a representative cell at that location. In *A*, the head is tilted slightly backward for easier viewing of canal planes.

parts; a superior division innervates the horizontal and anterior canals and most of the utricle, and an inferior division innervates the posterior canal and the saccule. The proximal processes of the bipolar cells make up the vestibular portion of the eighth cranial nerve. Extracellular recordings from axons innervating the hair cells have revealed the nature of the signals that travel in the primary vestibular afferents.

Canal Afferents. With the head at rest, primary canal afferent nerve fibers are characterized by a high steady discharge rate that averages 90 spikes/s in the monkey[14] and 36 spikes/s in the cat.[5] This high resting rate has several functional consequences. First, it allows stimuli in both the preferred and opposite directions to be coded, because they can cause an increase and decrease in discharge, respectively, about the resting firing rate. Second, a high resting rate theoretically allows a very low sensory threshold since the afferent fibers do not have to be brought to their firing threshold in order to report a sensory stimulus. In fact, no evidence of a threshold for angular acceleration can be detected in canal afferent discharges in either the monkey or cat, in agreement with psychophysical experiments that demonstrate human perceptual thresholds to angular acceleration to be as low as $0.1°/s^2$. If you started to turn your swivel chair at a constant acceleration of $0.1°/s^2$, you would be moving so slowly that it would require almost 90 s to complete a single revolution. Finally, the high resting rates cause the target structures receiving primary afferent input (mostly the vestibular nuclei of the medulla, see below) to be under a continual neural barrage. If half of the 20,000 fibers in the simian vestibular nerve serve the canals, canal afferents alone generate almost a million spikes per second with the head stationary. We will see later that the removal of this barrage produces serious behavioral consequences.

Earlier we saw that the adequate stimulus for the canals is angular acceleration. Figure 27–6A shows the discharge pattern of a representative canal afferent recorded in an anesthetized monkey that was rotated on a turntable from rest to a constant velocity. The discharge rate increased while the animal was accelerating and then returned slowly to the resting rate once the constant angular velocity was attained. These data help explain why a human on a similar turntable gradually loses the sensation of motion during constant-velocity rotation. For small-magnitude stimuli, accelerations (+A, Fig. 27–6A) and decelerations (−A, Fig. 27–6A) elicit roughly equal

but opposite changes in firing rate. As is suggested by Figure 27–6A, in which a doubling of the applied stimulus from 12.4 to $24.8°/s^2$ approximately doubled the response, firing rate increases almost linearly with acceleration.

To determine the time course of the response, a constant, angular acceleration is applied until the firing rate stops changing (Fig. 27–6B). Under these conditions, the firing rate (FR) increases along an exponential time course (Fig. 27–6B, dotted line) that can be described by the relation

$$FR = C(1 - e^{-t/T_1}) \qquad (1)$$

where T_1, the time constant, averages 6 s for the squirrel monkey[14] and 3.8 s for the cat[5] and where C is a constant of proportionality. For most natural head movements toward novel stimuli, the head usually accelerates for less than a few seconds. For such brief accelerations, the exponential rise in firing rate can be approximated by a linear increase in firing rate (dashed line in Fig. 27–6B). Since during a step change in head acceleration, the head experiences a linear increase in velocity (Fig. 27–6B), the time course of the firing rate of canal afferents is identical to the velocity of the head. Therefore, for brief accelerations, *the firing rate of canal afferents encodes head velocity.* Canal afferents from the three canals have quantitatively similar discharge patterns but of course respond most vigorously to accelerations in the plane of their own canal.

Since the adequate stimulus for the canals is head acceleration, the peripheral vestibular apparatus somehow performs the equivalent of a mathematical integration in time to produce the firing patterns of afferent fibers. How this integration is accomplished can be understood by developing a physical model, usually referred to as the *torsion pendulum model*, based on the hydromechanical properties of the endolymph-cupula system. As was discussed earlier, an angular acceleration applied to the head temporarily leaves the endolymph behind because of its inertia. Since the canals are moving relative to the stationary endolymph, the cupula deflects, bending the cilia of the hair cells. This bending will be opposed by the elastic restoring force of the cupula and the frictional forces that resist fluid motion in the small lumen of the canal. The displacement, θ, of the endolymph (and hence of the cupula) can be related to the angular acceleration of the head $\left(\dfrac{d^2h}{dt^2}\right)$ by the differential equation

$$M\frac{d^2\theta}{dt^2} + R\frac{d\theta}{dt} + K\theta = H\frac{d^2h}{dt^2}$$

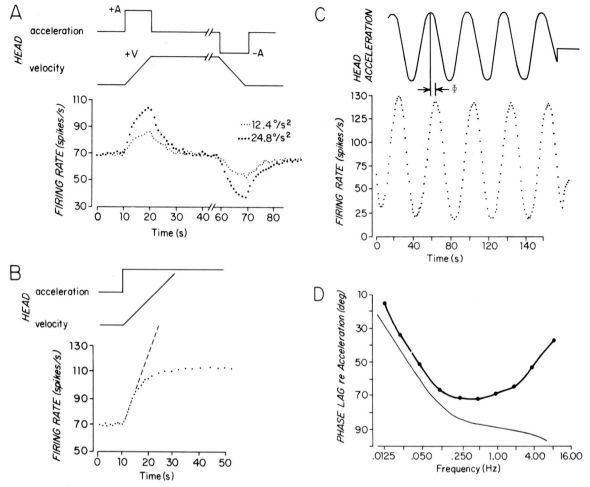

Figure 27–6 Response of canal afferents to angular acceleration of the head. Instantaneous firing rate (dotted curves) of an 8th nerve fiber during a brief pulse of two different angular accelerations *(A)* and a step of angular acceleration *(B)*. For t < 5 s, the firing rate in *B* (dotted curve) can be approximated by a ramp increase in firing rate (dashed line). *(C)* Instantaneous firing rate of an 8th nerve fiber during sinusoidal angular accelerations at 0.025 Hz. Phase lag Φ of unit activity relative to peak angular acceleration as a function of frequency is plotted in D (●——●) for many 8th nerve afferents. (After Fernandez, C.; Goldberg, J. *J. Neurophysiol.* 34:661–675, 1971; Goldberg J.; Fernandez, C. *J. Neurophysiol.* 34:635–660, 1971). Thin curve in *D* represents the phase lag expected from a simple torsion pendulum with T_1 = 6 s, T_2 = 3 ms (see discussion in text of hydromechanical properties of the canals).

where M represents the moment of inertia of the endolymph-cupula complex, R the viscous damping of the lumen walls, and K the cupular elasticity. Since the system is heavily damped and (R/M) >> (K/M), the solution for this equation is made up of the sum of two exponential functions, which, for a step change (H) of head acceleration, reduces to

$$\theta(t) = H(1 - A_1 e^{-t/T_1} - A_2 e^{-t/T_2})$$

where T_2 = (M/R) and T_1 = (R/K). T_2, which can be deduced from hydrodynamic principles, has a value of about 5 ms, whereas T_1 is about 5 s. Since $T_1 >> T_2$, the solution of θ(t) is dominated by T_1, reducing that solution to a single exponential similar to Equation 1.

Consequently, the firing of canal afferents reflects the motion of the cupula, which is determined by its hydromechanical situation in the semicircular canal. For slight deviations from the torsion pendulum model, the reader is referred to Fernandez and Goldberg.[11]

Otolith Afferents. As might be expected from the preferred directions of the hair cells that they innervate (Fig. 27–5B), afferent fibers from the utricle and saccule respond both when the animal undergoes pitch and when it experiences roll. Figure 27–7A shows how the average discharge rate varies with tilt angle for a saccular afferent. Clockwise roll and forward pitch from the animal's perspective are represented as positive angles. With the animal in its normal upright position (i.e., 0°), this particular saccular afferent discharged at its lowest rate. Its rate increased for both pitch and roll tilts until it reached a maximum with the animal upside down (i.e., 180°). Recalling that the long axis of saccule hair cells is roughly horizontal, one can see from the inset in Figure 27–7B that the hair cell innervated by this afferent must be positioned anterolaterally in order to be sensitive to both pitch and roll. The orientation of this hair cell during various roll tilts is shown on Figure 27–7A. The curves fitted to the individual data points of Figure 27–7A show that the firing rate of this afferent increased as the cosine of the angle in both pitch and roll; therefore, the firing rate signaled the orientation of the force of gravity relative to the preferred direction of the hair cell.

If this "effective" force of gravity is considered to be positive when cilia are deflected toward the kinocilium (e.g., −180° and +180°, Fig. 27–7A) and negative for opposite deflections (all angles from −90° to +90°), then Figure 27–7B shows that the firing rate is a linear function of the effective force of gravity. The response of a utricular afferent to pitch and roll is shown in Figure 27–7C and D. Although this unit exhibits an unusual, somewhat nonlinear relation between discharge rate and the effective force of gravity (Fig. 27–7D), it still displays a roughly sinusoidal relation between firing rate and tilt in both pitch and roll (Fig. 27–7C). The reader is left to verify that the hair cell innverated by this afferent has the position indicated in the inset (Fig. 27–7D). Neither these units nor any others exhibit any apparent discontinuity as the curves of Figure 27–7B and D pass

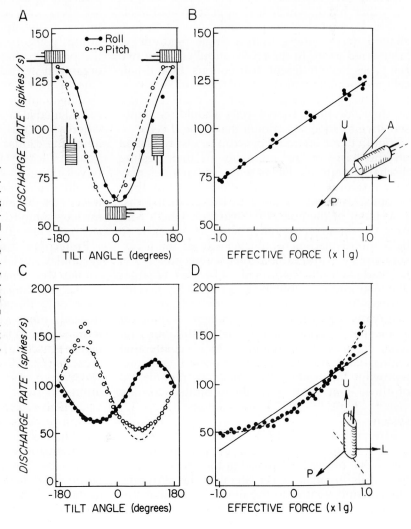

Figure 27–7 Responses of macular afferent fibers to different head positions. *A,* The discharge rate of a saccular afferent fiber due to tilting head sideways (roll) or forward-backward (pitch). Positive angles are clockwise roll or forward pitch. Curves are best-fitting trigonometric functions. *B,* Plot of relative response of fiber shown in *A* vs. calculated shearing force produced by action of gravity on saccular otoliths. *C* and *D,* Similar plots for response of a utricular afferent fiber. Insets in *B* and *D* show the position of hair cells innervated by the afferents. (After Fernandez, C.; Goldberg, J.M.; Abend, W.K. *J. Neurophysiol.* 35:978–997, 1972.)

through zero effective force, suggesting that otolith neurons do not possess a sensory threshold. Indeed, humans can detect linear accelerations as small as 5×10^{-4}g. An elevator moving at this acceleration would require almost 40 s to travel between floors. In summary, otolith afferents report static tilts or linear accelerations such as those produced by the force of gravity. However, both otolith organs also respond to dynamic changes in linear acceleration such as those experienced in a moving elevator or an accelerating bus.

Two Afferent Pathways? The sensory epithelia serving the otolith organs and the canals are remarkably similar. Each has type I and type II hair cells that are distinguishable by their morphologies and innervation patterns. As seen in Figure 27–1A, type II hair cells, which are present in all species, are cylindrical and are innervated by a plexus of nerve endings derived from small and medium-sized afferent axons. Type I hair cells, on the other hand, which first appear in mammals and birds, are goblet-shaped and receive a chalice-shaped nerve ending from a single medium- to large-sized axon. In the crista ampullaris, the two cell types have different distributions; type II hair cells are located on the flanks of the crista, whereas type I hair cells are located on its crest. Because of their location, type I cells should be more sensitive to adequate stimuli than type II cells.

Eighth nerve afferent fibers can be distinguished on the basis of the regularity of their resting discharge patterns. Fibers with regular rates may be considered to be tonic afferents since their response is closely related to the predicted displacement of the accessory structure. Fibers with irregular rates, on the other hand, are sensitive to both the displacement and velocity of the accessory structure and exhibit conspicuous adaptation to steady stimuli. The irregular fibers are of larger diameter and probably innervate type I hair cells, whereas regular fibers are smaller and appear to innervate type II hair cells. Although it is unclear at present whether their central destinations remain separate, it is tempting to speculate that regular and irregular fibers represent separate pathways for vestibular information, reminiscent of those that have been demonstrated for the visual (Chaps. 19 and 20) and auditory (Chap. 17) systems.

Efferent Innervation. The vestibular nerve contains a modest number of efferent fibers (about 200 to 300 of the total 10,000 in the cat) that arise from cells in small nuclei lateral to the abducens nucleus and ventral to the medial vestibular nucleus. Efferent fibers terminate directly on type II hair cells but end only on the afferent terminals of type I hair cells (Fig. 27–1A). They are thus strategically positioned to influence transmission of vestibular information close to its source. The function of the efferent fibers, unfortunately, is not known.

CENTRAL PROCESSING OF VESTIBULAR INFORMATION

Terminations of the Vestibular Nerve. The vestibular portion of the eighth nerve projects almost exclusively to the ipsilateral vestibular nuclei, which lie on the floor of the fourth ventricle at the pontomedullary junction (Fig. 27–8). Traditionally, the vestibular nuclei have been divided into four principal nuclei (superior, medial, lateral, and descending) and several smaller (so called "minor") nuclei with alphabetical designations (e.g., x-group, y-group, and so forth).[7] The projections of the primary afferents to the vestibular nuclei are organized according to the receptor they innervate. In the monkey, fibers from all three canals terminate in overlapping areas in the superior, rostral medial, and ventral lateral vestibular nuclei, whereas the utricle and saccule project most heavily to the caudal medial and the descending vestibular nuclei.[48] In the cat, the y-group also receives a direct saccular input. As a general but not absolute rule, the semicircular canals project most heavily to the more rostral parts of the vestibular nucleus complex, whereas the otolith organs project more heavily to the caudal parts.

Projections from the Vestibular Nuclei. The vestibular nuclei, like the dorsal column nuclei, contain the somata of secondary sensory neurons; unlike the dorsal column nuclei, however, they have only meager projections to the thalamus. Instead, most axons from the vestibular nuclei terminate in (i) motor nuclei innervating the extra-ocular muscles (Fig. 27–8, Ab,Tro, OMN) or neck muscles (Fig. 27–8, cervical cord), (ii) areas of the spinal cord and brain stem thought to provide inputs to these motoneuron pools, and (iii) the vestibulocerebellum (Fig. 27–8, Floc). In the following sections, we first consider the behavior of vestibular circuits involved with the eye muscles and then those concerned with the head, trunk, and limbs. Finally, we address the role of the vestibulocerebellum in modifying signals traveling in the vestibulo-ocular circuits.

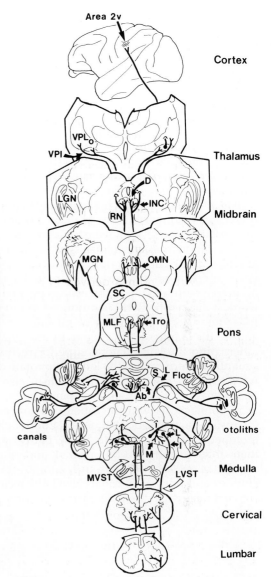

Figure 27–8 Simplified diagram of the major vestibular pathways. S, L, M and D (superior, lateral, medial, and descending vestibular nuclei); Ab, Tro, and OMN (abducens, trochlear, and oculomotor extraocular muscle motor nuclei); MVST and LVST (medial and lateral vestibulospinal tracts); LGN and MGN (lateral and medial geniculate nuclei); MLF (medial longitudinal fasciculus); Floc (flocculus); RN (red nucleus); INC (interstitial nucleus of Cajal); D (nucleus Darkshewitsch); SC (superior colliculus); VPI and VPL$_o$ (thalamic nuclei ventralis posterior inferior and ventralis posterior lateral, pars oralis). (Courtesy of T. Langer.)

VESTIBULO-OCULAR CIRCUITS

Pathways from the Canals to the Extraocular Muscles. Fibers from the vestibular nuclei terminate in all three oculomotor nuclei (Fig. 27–8, Ab, Tro, and OMN) and in brain stem nuclei related to eye movements such as the nucleus of Darkshewitsch and the interstitial nucleus of Cajal (Fig. 27–8, D and INC, respectively). Projections to the oculomotor nuclei originate largely from the rostral vestibular nuclei,[32] so that the vestibulo-ocular connections are concerned mostly with adjustments to head movements detected by the canals.

The vestibulo-ocular pathways are precisely organized, with specific canals being connected through synaptic relays in the vestibular nuclei to specific extraocular muscles.

To appreciate this organization we must first understand the organization of the extraocular muscles and their innervation. Each eye is moved by six extraocular muscles, each of which rotates the eye in a particular direction. Each of the six extraocular muscles is innervated by separate motoneuron pools. The lateral and medial rectus muscles, which attach at the horizontal equator of the eyeball, act as an antagonistic pair. The lateral rectus causes temporal rotation; the medial rectus, nasal horizontal rotation. Their motoneurons are located in the 6th and 3rd cranial nerve nuclei, respectively. The superior rectus elevates and the inferior rectus depresses the eye; their motoneurons are located in separate pools within the 3rd cranial nucleus. Finally, the superior and inferior oblique muscles act to rotate the eye about its visual axis (the former producing intorsion, the latter extorsion); their motoneurons lie in the 4th and 3rd cranial nerve nuclei, respectively.

The pathways connecting the canals to the extraocular muscles can be traced in the anesthetized cat by recording the postsynaptic potentials that stimulation of individual canal nerves produces in ocular motoneurons. Such experiments show that each canal is connected only to those motoneurons that produce an eye movement opposite to the head rotation that stimulates that canal. For example, if the right posterior canal nerve is stimulated to create an excitatory input similar to that occurring naturally when the head is pitched backward, a disynaptic EPSP is produced in motoneurons innervating the contralateral inferior rectus and a disynaptic IPSP occurs in motoneurons innervating the contralateral superior rectus; this synaptic pattern causes the left eye to rotate downward. Since a roll rotation of the head also stimulates the posterior canals, they have similar reciprocal connections with motoneurons innervating the ipsilateral superior oblique and inferior oblique motoneurons, respectively. These pathways and their interneurons in the medial and superior vestibular nuclei are shown in Figure 27–9A. The anterior and horizontal canals also have reciprocal connections with their functionally appropriate

muscles; all connections are disynaptic, with interneurons in the superior, medial, and ventral lateral vestibular nuclei. The only exception to this pattern is the inhibitory input to medial rectus motoneurons, which is polysynaptic. Thus each canal gives rise to two excitatory and two inhibitory pathways. Eight of the twelve pathways travel in the medial longitudinal fasciculus (MLF), which is a prominent fiber bundle that runs on the dorsomedial surface of the brain stem (Fig. 27–8); the others travel in the brachium conjunctivum or the ascending tract of Deiters, which lies near the MLF.

Many of the neurophysiological and nearly all of the behavioral experiments on vestibulo-ocular connections have been done during horizontal (yaw) rotation, which activates the circuits shown in Figure 27–9B. The pattern of vestibular input onto abducens motoneurons is crossed excitation and uncrossed inhibition. Medial rectus motoneurons, on the other hand, receive a direct uncrossed excitation via the ascending tract of Deiters and a double-crossed excitation via the MLF from recently discovered interneurons (called internuclear neurons) within the abducens nucleus itself. The functional role of the yaw rotation circuits can be demonstrated by hypothetically subjecting the system depicted in Figure 27–9B to a leftward head rotation (from the cat's perspective). This stimulus causes an increase in the discharge of afferents innervating the left horizontal canal and a decrease in afferent activity from the right canal. The increase in left canal activity is communicated via a crossed excitatory connection to the contralateral

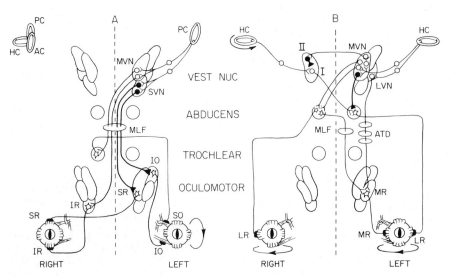

Figure 27–9 The shortest pathways from the posterior *(A)* and horizontal *(B)* semicircular canals to extraocular motoneurons in the cat. Filled neurons with filled triangular terminals are inhibitory. Semicircular canals: HC, horizontal; PC, posterior; AC, anterior. Oculomotor nucleus subdivisions: IO, inferior oblique; SR, superior rectus; IR, inferior rectus; MR, medial rectus. Extraocular muscles: SO, superior oblique; IO, inferior oblique; SR, superior rectus; IR, inferior rectus. Fiber tracts: MLF, medial longitudinal fasciculus; ATD, ascending tract of Deiters. (After Fuchs, A.F. In Towe, A.L.; Luschei, E.S., eds. *Handbook of Behavioral Neurobiology*, 303–366, 1981.)

(right) abducens nucleus. Increased activity in this motor nucleus causes the right lateral rectus muscle to contract, producing a rightward rotation of the right eye. Contralaterally projecting internuclear neurons that originate in the right abducens nucleus are also excited. Together with uncrossed excitatory neurons travelling in the ascending tract of Deiters, they produce contraction of the left medial rectus muscle and rightward rotation of the left eye. To facilitate this rightward rotation, the activity in the left lateral rectus is reduced by two mechanisms. First, the increased activity of the left canal causes an active inhibition via inhibitory interneurons in the left vestibular nuclei; second, the decreased activity in the right horizontal canal produces a disfacilitation via the crossed excitatory connection onto the same left abducens motoneuron. Taken together, these pathways to the two horizontal rectus muscles serve to rotate both eyes in a direction opposite the head rotation so that the direction of gaze remains stable.

Circuits like those in Figure 27–9 clearly meet the criteria for reflexes described for the spinal cord (Chap. 24) since as few as three neurons (a three-neuron arc) link the receptors to the muscle, and both electrical and adequate stimulation of vestibular afferents elicits a stereotyped motor response. Therefore, these brain stem pathways have been called *vestibulo-ocular reflexes* (VORs). All of the vestibulo-ocular pathways operate like the horizontal VOR (the reader should confirm this for Fig. 27–9A) so that, acting together, they can produce compensatory eye movements for any direction of head rotation.

Commissural Pathways Linking Opposite Canals. The successful operation of each VOR depends upon reciprocal inputs from complementary canals on opposite sides of the head. For example, each abducens motoneuron in Figure 27–9B receives both a crossed excitatory and an uncrossed inhibitory input from different interneurons in opposite vestibular nuclei. Furthermore, there is good evidence that these interneurons themselves receive reciprocal inputs from the two horizontal canals, because electrical stimulation of the ipsilateral vestibular nerve produces monosynaptic excitation of neurons in the vestibular nucleus, whereas stimulation of the contralateral vestibular nerve produces a polysynaptic inhibition.[45] The crossed inhibition is mediated by commissural connections (indicated schematically in Fig. 27–9B) from the corresponding contralateral vestibular nucleus. As might be expected from the connections in Figure 27–9B, transection of the commissural pathway causes an immediate increase in resting discharge rate and a decrease in sensitivity of VOR interneurons.[30] Thus, the vestibular system uses the first available central synapse to compare the signals

from the two vestibular end-organs. It is this differential signal that drives the motoneurons. Indeed, as discussed below, many of the neurological problems associated with vestibular disorders can be traced to an imbalance of inputs from the opposite canals.

The Normal VOR in Behaving Monkeys. A clue as to the neural processing that must go on in the circuits of Figure 27–9 can be obtained by measuring the VORs during active head rotation in an alert animal. The operation of the horizontal VOR under natural conditions was first studied by Bizzi and his colleagues[4] who trained monkeys to discriminate as to which of two targets had appeared in its peripheral visual field. To perform these discriminations, an animal usually moves both its head and eyes toward the peripheral target. If its head is held immobile, the eyes reach the target with a very rapid movement, called a *saccade*. A saccade lasts only 60 ms if the target is at a distance of 30° as in Figure 27–10A. If the head is free to move, as in Figure 27–10B, the same eccentric target usually first elicits an eye movement (E) that proceeds along a saccadic trajectory; then after 20 to 40 ms, the head (H) begins to move in the same direction. The head moves at a much slower speed and by the end of the 60-ms saccade has moved only about 5°. However, the usual 30° saccade has been foreshortened by about

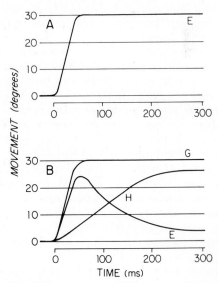

Figure 27–10 Head-eye coordination in the monkey. *(A)* The head is held, allowing only a horizontal eye movement (E) to occur. *(B),* The head is free and after the initial rapid eye movement, the head movement in space (H) produces a compensatory eye movement in the head (E), which keeps the eye movement in space (G) on target. (After Bizzi, E. Handbk. Physiol. 2:1321–1336, 1981.)

5°, so that the sum of the head movement in space plus the eye movement in the head (together called the gaze, G) places the eye on target. Although the eye is now on target, the head continues to turn, and the eyes must rotate in the opposite direction to compensate for the head movement, so that the monkey's gaze remains on target. As may be seen from the measurement of gaze, which shows almost no variation, the compensation is remarkably good. The final head position, which is reached after about 250 ms, is aimed only slightly short of the target so that the eye eventually must return almost to its initial straight-ahead position.

Several manipulations of this experimental paradigm can be performed to demonstrate that the compensatory eye movements following the saccade are largely of vestibular origin. When one eliminates the possibility of visual feedback by extinguishing the target light before any movement begins, the compensatory eye movements are qualitatively unchanged. The production of eye movements owing to the possible stimulation of neck afferents during head rotation is also inconsequential, since rotating the animal's body while keeping its head fixed produces hardly any compensatory eye movements. On the other hand, rotating the whole monkey in complete darkness while its head is held fixed to its body produces almost perfect compensatory eye movements that maintain the eye stable in space. Finally, plugging the semicircular canals to prevent their normal operation eliminates the compensatory eye movements. Therefore, the compensatory eye movements in Figure 27–10B do indeed result from stimulation of the canals and thus represent the normal operation of the horizontal VOR. The vertical VOR also operates in many normal situations. For example, it is the vertical VOR that helps produce the compensatory eye movements that allow a commuter to read the ads and graffiti on the walls of swaying subway cars.

The efficacy of the VOR can be measured by the ratio of the size of the compensatory eye movement to the size of the head rotation or, more usually, the ratio of eye to head velocity. Even in complete darkness, this ratio, called the gain of the VOR, is about −1 for the monkey. Therefore, when the monkey turns his head in a well-lighted environment, the VOR will stabilize the visual world on the retina so that visual images do not slip across the retina and produce blurring.

For almost all other animals, the gain of the VOR is less than −1. In humans, the gain is estimated at between −0.6 and −0.8, whereas in the cat, it lies between −0.8 and −0.9. If either of these animals turns its head in the light, however, the gain of the compensatory eye movements improves to −1. It is thought that the residual movement of images across the retina, which remains because of the inadequate VOR, elicits supplementary pursuit eye movements, which, coupled with the VOR, produce perfect compensation.

How are the compensatory eye movements of the VOR produced? To answer this question requires attention to the quantitative details of the reflex—a systems analysis approach that we can illustrate in the following paragraphs. If, during head rotation, the vestibular labyrinth produced a signal related to the dynamic head position, such a signal could simply be forwarded with a change in sign to the appropriate extraocular muscles along the precisely organized VOR pathways described above. We have already seen, however, that eighth nerve activity originating in the canals is proportional to head velocity. To obtain a signal proportional to head position, therefore, the neural machinery between the eighth nerve and the extraocular muscles must perform the equivalent of a mathematical integration upon the head velocity signal. Because the hydromechanics of the cupula and endolymph already have produced one mathematical integration from angular acceleration to velocity, *the entire VOR can be thought of functionally as performing a double integration in time.*

If H is the angle through which the head rotates, the head moves with a velocity of $d/dt(H) = dH/dt$, and an acceleration of $d/dt(dH/dt) = d^2H/dt^2$. The VOR produces an eye movement nearly equal to −H. The adequate stimulus for the canals, d^2H/dt^2, is integrated by the cupula-endolymph to produce a head velocity signal (i.e., $\int d^2H/dt^2 \, dt = dH/dt$) that must be integrated again to produce head position (i.e., $\int dH/dt \, dt = H$). A negative H is obtained by crossing the connections from the canals to the extraocular muscles.

Firing Patterns in VOR Pathways. Our early knowledge about neural processing in the horizontal VOR was derived from extracellular unit recordings from the decerebrate but otherwise unanesthetized cat.[46] In this preparation, head rotations are applied by placing the cat on a turntable so that its whole body, particularly its head, can be rotated in the horizontal plane about a vertical axis (yaw rotations). As shown in Figure 27–11A, a sinusoidal whole-body oscillation applied by the turntable results in eye movements (E) that are also sinusoidal and roughly opposite head position (H) as would be expected from the data in the alert monkey (Fig. 27–10). To produce

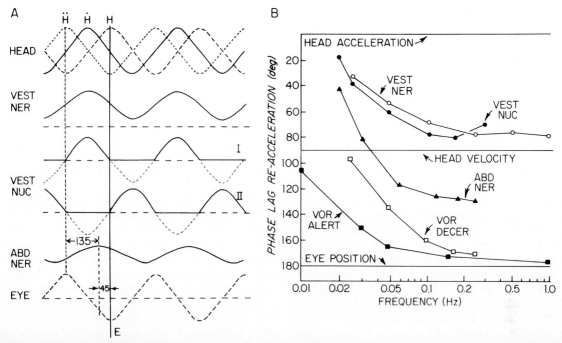

Figure 27–11 Signals in the horizontal VOR pathways of the decerebrate cat during sinusoidal yaw rotation. *A*, This schematic diagram indicates how the adequate stimulus, head acceleration (Ḧ) is processed through the vestibular nerve (VEST NER), vestibular nuclei (VEST NUC) and abducens nerve (ABD NER) to produce a compensatory eye movement (E). The stimulus frequency is about 0.2 Hz. Horizontal dashed lines refer either to zero firing rate or straight-ahead eye position. *B*, The phase lags of the same signals relative to acceleration are considered for a wide range of sinusoidal frequencies.

an eye position signal (or, equivalently, a negative head position signal), we must start with the adequate stimulus, head acceleration. As you recall from calculus, the integral of a sine function is the negative cosine function. Consequently, the integration of a sinusoidal head acceleration signal (Ḧ) by the cupula-endolymph produces the head velocity signal (Ḣ) shown in the HEAD panel. A second integration of Ḣ then produces the appropriate negative sine function for head position (H). As seen in the HEAD panel, each successive integration can be thought of as introducing a delay (or phase lag) of one fourth of a cycle, or 90°. The delay of neural information is usually measured as a phase lag (in degrees) relative to head acceleration, so that a firing rate that is modulated sinusoidally with a phase lag of 90° is proportional to head velocity. Neural signals with lags intermediate to the three signals in the HEAD panel have components of each, e.g., a signal with a phase lag of 70° is related mostly to head velocity but is slightly related to head acceleration as well. Neural signals with intermediate phase lags result from an imperfect integration.

If the VOR can be considered to delay the acceleration signal to produce the compensatory

eye movement, where and how does this delay occur? This question has been addressed in the decerebrate cat by measuring the firing patterns of primary afferents, vestibular nucleus neurons, and ocular motoneurons that participate in the horizontal VOR during sinusoidal yaw acceleration. Typical firing rates for one frequency (about 0.2 Hz) are shown schematically in Figure 27–11A and the phase lags relative to head acceleration for all frequencies are shown in Figure 27–11B. During sinusoidal head accelerations, canal afferents of the cat exhibit a complete sinusoidal modulation in firing rate (Fig. 27–11A, VEST NER), which between 0.2 and 1.0 Hz has a phase shift of about 80° (Fig. 27–11B, VEST NER). Since this frequency range is representative of the head movements of the alert cat, this analysis again shows that the firing of canal afferents closely resembles that expected of a head velocity signal, i.e., nearly a 90° phase lag. At lower frequencies, however, the phase lag decreases, indicating that the peripheral integrator is less effective. A similar analysis for canal afferents in the monkey (Fig. 27–6C,D) shows that the peripheral integrator is also less effective at higher frequencies (greater than 1 Hz).

In contrast to canal afferents, cells in the vestibular nuclei have low resting rates and therefore respond to rotation over only part of the stimulus cycle. Two types of vestibular nucleus neuron can be differentiated according to their firing patterns. Type I cells exhibit an increase in discharge for head rotations that produce an increase in discharge of their ipsilateral vestibular nerve (compare VEST NUC I with VEST NER, Fig. 27–11A); during head rotations that increase discharge in the contralateral vestibular nerve (decrease in the ipsilateral), the firing of type I cells is driven to zero (the response appears half-wave rectified). In contrast, type II cells exhibit precisely the opposite behavior since they increase their activity during head rotations that excite the contralateral eighth nerve and are silent during ipsilateral eighth nerve excitation. Type II cells could therefore serve as the inhibitory interneuron in the commissural pathway discussed earlier and diagrammed in Figure 27–9B. The discharge patterns of both type I and type II cells are in phase with head velocity (Fig. 27–11A). Furthermore, type I units have phase shifts that are indistinguishable from those of canal afferents over the entire frequency range between 0.1 and 1.0 Hz (compare VEST NER with VEST NUC, Fig. 27–11B). Therefore, at least in the decerebrate cat, very little, if any, additional phase lag is introduced at the vestibular nuclei.

Perhaps the phase lag is introduced entirely at the motoneuron itself. An abducens motoneuron must have a discharge pattern that resembles the variation in horizontal eye position (E, Fig. 27–11A). However, the motoneuron signal produces the eye movement through a muscle which, since it has both viscosity and elasticity, delays the motoneuron drive. To compensate for these delays, the motoneuron firing must precede (or lead) the eye movement by about 45° during 0.2-Hz oscillations and therefore lags head acceleration by about 135° (ABD NER, Fig, 27–11A). Even though the compensatory eye movements in the decerebrate cat (Fig. 27–11B, VOR DECER) exhibit progressively less phase lag as the frequency decreases, the motoneurons always lead eye position by 30° to 45°.

The alert cat also suffers a decrement in phase lag at low frequencies (Fig. 11B, VOR ALERT). Furthermore, both cats and primates experience a decrease in low-frequency VOR gain. This deterioration in VOR performance poses no problem, however, because visually elicited eye movements assist the VOR in stabilizing the retinal image motion resulting from slow head rotations.

A comparison of the phase lags of their neural activities makes it clear that the direct unmodified input from the vestibular nucleus alone cannot account for abducens nerve activity. Figure 27–11B shows that at 0.2 Hz, activity in the abducens nerve is delayed by about 45° relative to activity in the vestibular nucleus. Therefore, the three-neuron arc linking the canals to the extraocular muscles through the vestibular nucleus is functionally inadequate to produce accurate compensatory eye movements. If this pathway acted alone, eye movements would lag head movements by 135° rather than about 180°.

The phase lag of abducens nerve activity lies intermediate to a head velocity and eye (or head) position signal. Robinson[40] therefore suggested that abducens nerve activity could be generated if both a velocity signal (from the vestibular nuclei) and a position signal (from an unspecified location) converged upon motoneurons. In addition to their direct connections, signals from the vestibular nuclei can reach the abducens nucleus via more circuitous routes involving the nearby reticular formation or other brain stem nuclei (e.g., the nucleus prepositus hypoglossi in the medulla). The position signal, then, might be generated in these more circuitous pathways that would perform a neural integration of the velocity signal. While there is only scant neurophysiological evidence that this *VOR integrator* actually exists, it seems clear that the VOR must perform this function somehow to produce accurate compensatory eye movements.

A Model Integrator. A VOR integrator can, in fact, be generated theoretically with a rather modest number of constraints. Since neurons in the reticular formation are well endowed with axon collaterals that apparently end locally, an integrator could be constructed by interconnecting a net of neurons with recurrent collaterals. Figure 27–12 shows a very primitive net of only six neurons that was constructed simply by flipping a coin to determine whether or not they are connected. In the connectivity matrix of Figure 27–12A, a *1* in a box indicates the neuron in the column is connected with the neuron in the row. For example, neuron No. 2 is connected to Nos. 1 and 4 as well as to itself. Let us assume that a brief input volley, perhaps from the vestibular nuclei, causes each neuron to discharge a spike at OT (Fig. 27–12B), but in order for a neuron to discharge again, it must have at least two active synaptic inputs. If T represents the conduction time in every collateral plus the time required to raise the target neurons to their thresholds, then every cell except No. 5, which receives only one collateral, will fire a second

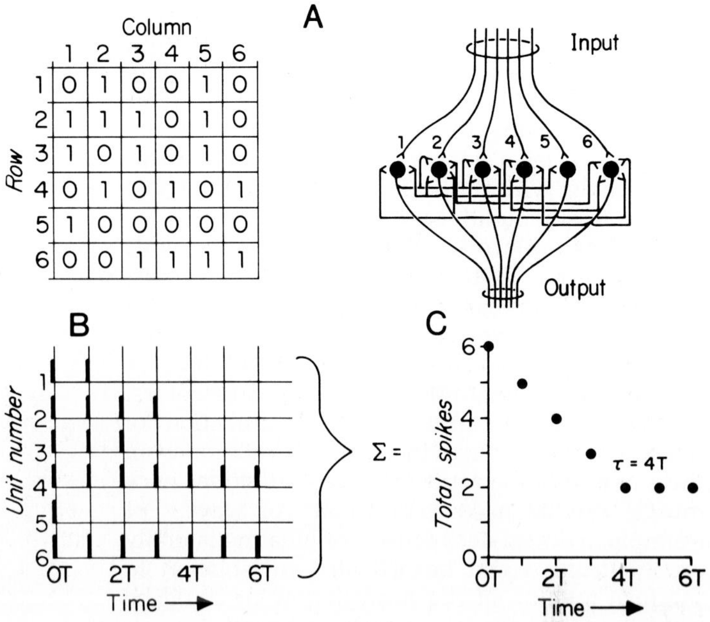

Figure 27–12 A model neural integrator. In *A*, a circuit of six neurons with recurrent collaterals (right) is connected according to the matrix generated by a coin flip (left). *B* indicates the action potentials generated by individual neurons in the circuit, while *C* shows the total number of spikes produced by all six cells as a function of time (see text for details).

spike at t = T. The failure of No. 5 to fire a second spike prevents No. 1 from firing a third spike since one of its two collateral inputs is from No. 5. As can be seen in Figure 27–12B, this gradual attrition of neurons continues until only Nos. 4 and 6 continue firing because of their reciprocal connections. The time course of the output of the entire net, taken as the total number of spikes at any time, is shown in Figure 27–12C. If the net performed a perfect integration, it would continue to discharge 6 spikes at each time interval forever, since the perfect integral of the brief (pulselike) input volley is a step. Actually, this net represents an imperfect or "leaky" integrator, because its output decreases (decays) with time to a low value after only 3T, where T is only 1 or 2 ms. In fact, a suitable neural integrator to account for the phase shifts in Figure 27–11B would also not be perfect but would have a "memory," or time constant, of 1 to 2 s. It is easy to imagine how such a time constant could be obtained by a similar but more complicated net with more neurons, variable feedback times (i.e., an interneuron or two in the collateral pathways), different spike threshold requirements, and the like.

We have seen that an appropriate model for the decerebrate cat is one in which head velocity information from a VOR interneuron in the vestibular nucleus and eye position information, generated perhaps in the reticular formation, converge at the motoneuron. A more complicated situation obtains in the alert, behaving animal, where VOR interneurons in both the cat[31] and the monkey[44] discharge not only with head velocity but also with eye position when the head is fixed. There-

fore, in the normal animal, at least part of the integrated signal is already present in the discharge of the VOR interneuron, presumably because the integrator also feeds back into the vestibular nucleus (for more details see Fuchs).[13]

VESTIBULOSPINAL CIRCUITS

Vestibular signals not only ascend in the brainstem to produce compensatory eye movements but also descend to the spinal cord to influence head movements and aid in the maintenance of posture and balance. As discussed in Chapter 26, activity generated in the vestibular labyrinth can influence spinal motoneurons by two major descending pathways originating in the vestibular nuclei. The medial vestibulospinal tract (MVST) originates mostly in the medial and descending vestibular nuclei, with a modest contribution from the lateral vestibular nucleus. It provides both crossed and uncrossed fibers to the cervical and upper thoracic cord, where it connects both mono- and disynaptically with motoneurons innervating neck and axial musculature. The lateral vestibulospinal tract (LVST) originates almost exclusively in the lateral vestibular nucleus. It contains only ipsilateral fibers that terminate at all levels of the cord and provide mostly polysynaptic inputs to limb motoneurons. Although the dichotomy is not absolute, the MVST is largely concerned with motoneurons supplying neck and axial muscles, and the LVST is involved mainly with limb motoneurons. The two descending pathways have different functional roles because neck muscle motoneurons receive signals mainly from the semicircular canals, whereas limb motoneurons probably receive mostly otolith inputs.

The Vestibulocollic Reflex. Like the circuits underlying the VOR, those subserving the transmission of vestibular information to the neck muscles have been revealed by recording motoneuron PSPs induced by electrical stimulation of 8th nerve branches supplying different vestibular receptors. Unlike extraocular muscles, the contraction of which always produces the same eye rotation from the same initial eye position, an individual neck muscle can be used in a variety of ways. For example, a dorsal neck muscle originating laterally on the trunk could be activated in isolation to produce rotation of the head in yaw or could be co-contracted with its contralateral mate to produce dorsiflexion of the head. Therefore it is not surprising that many different canals provide inputs to a single neck motoneuron.

In the cat, most dorsal neck motoneurons receive disynaptic canal inputs, often from all six canals.[52] A motoneuron innervating the biventer cervicis or complexus muscle receives crossed disynaptic excitation from the contralateral horizontal canal and uncrossed disynaptic inhibition from the ipsilateral horizontal canal via axons that descend in the MLF (Fig. 27–13A). Therefore, these pathways are remarkably similar to those that project to the lateral rectus and underlie the horizontal VOR. The same motoneuron also receives disynaptic excitation from both anterior canals and disynaptic inhibition from both posterior canals (Fig. 27–13B). These short, three-neuron pathways are used to produce the *vestibulocollic reflexes* (VCRs).

If a counterclockwise head rotation (from the animal's perspective) is applied to the cat of Figure 27–13A, right neck motoneurons will be both excited from the contralateral canal and disinhibited from the ipsilateral canal through separate interneurons in the vestibular nuclei. The resulting activation of the neck motoneuron produces a muscle shortening that turns the head clockwise to counteract the head rotation and stabilize the head in space. On the other hand, if the head is pitched forward, thus activating the anterior canals while deactivating the posterior canals (Fig. 27–13B), the neck motoneuron will be excited from both anterior canals and disinhibited by both pos-

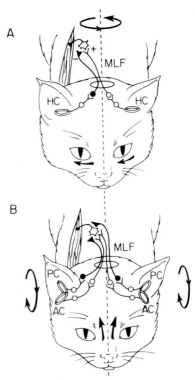

Figure 27–13 The three neuron pathways from the semicircular canals to dorsal neck motoneurons in the cat. The interneurons, which lie in the vestibular nuclei, have axons that travel in the descending medial longitudinal fasciculus (MLF) and lateral vestibulospinal tracts. HC, horizontal canal; PC, posterior canal; AC, anterior canal. Paired arrows indicate the eye movements that result from yaw *(A)* and pitch forward *(B)* head rotation.

terior canals. This pattern of synaptic inputs would cause the corresponding neck muscles on both sides of the head to shorten, producing a backward pitch of the head to return it to its upright position. Indeed, if electrical stimulation is delivered to individual canal nerves in the alert cat, the predicted compensatory head rotations do result.[49] Although almost nothing is known about vestibulocollic circuitry in the monkey, electrical stimulation studies[49] suggest that the functional connections must be quite similar to those of the cat.

Compared with the VOR, relatively little is known about neural processing in the VCR. When a decerebrate cat is placed on a sinusoidally oscillating turntable, the left dorsal neck muscles contract in response to angular acceleration to the right and relax with rotation to the left. The phase lag of neck motoneurons as assessed by the modulation of neck muscle EMG activity varies with frequency, but is relatively constant at 150° relative to head acceleration between 0.05 and 0.2 Hz (Fig.

27–14). Note that this operating range is lower than that of abducens motoneurons (Fig. 27–11B), as might be expected from the greater inertial load imposed by the head. Although the direct experiment has not yet been done, strong circumstantial evidence suggests that the interneurons of the VCR, like those of the VOR, also discharge with head velocity in the decerebrate cat.[53] As seen in Figure 27–14, there is a substantial phase lag between activity in the vestibular nucleus and neck motoneurons that cannot be accounted for by the direct VCR connections. Therefore, like the VOR, the VCR needs an additional pathway to perform a mathematical integration of vestibular nucleus activity.[10]

As in the VOR, vestibular information can reach the spinal cord by indirect routes through the reticular formation. Reticulospinal information flows in either a lateral tract originating in the medulla or a medial tract originating in the pons (see Chap. 26). In the cat, stimulation of the vestibular nerve produces polysynaptic potentials in medullary, reticulospinal neurons,[36] and stimulation in this medullary reticular formation produces mono- and polysynaptic potentials in neck motoneurons.[38] Therefore, it has been suggested that reticular circuits may also be involved in the

mathematical integration of vestibulocollic information.

In both the VOR and VCR, the most direct pathways from the canals to the motoneurons travel over the MLF. Although transection of the ascending MLF severely impairs both the horizontal and vertical VOR in the alert monkey, destruction of the descending MLF paradoxically has little effect on the VCR in the decerebrate cat.[53] The VCR, therefore, apparently relies much more on indirect pathways that use the reticulospinal tracts.

The Cervicocollic Reflex. Since turning the head also activates muscle and joint receptors in the neck, the vestibulocollic circuits must interact with segmental reflex circuitry. If the head of our cat in Figure 27–13 is subjected to the same counterclockwise rotation with its body held immobile, the right dorsal neck muscle will lengthen and elicit a reflexive clockwise rotation of the head. This reflex, the *cervicocollic reflex* (CCR), is basically a neck-stretch reflex that assists the VCR in stabilizing the head.[37] Because neck motoneurons receive both a monosynaptic input from low-threshold (presumably spindle) afferents and a monosynaptic input from the vestibular nuclei, one site of interaction of the VCR and CCR is at the motoneuron itself. Other possible sites of interaction include the vestibulospinal interneurons themselves, some of which are activated by neck rotation, and segmental interneurons participating in longer loop pathways.

The Vestibulomembral Reflexes. Nearly all of the vestibulospinal control of limb musculature is exerted over the lateral vestibulospinal tract. Stimulation of Deiters's nucleus causes polysynaptic, and occasionally monosynaptic, excitation of ipsilateral extensor motoneurons serving the forelimb and hindlimb. Stimulation also produces polysynaptic inhibition of ipsilateral flexor motoneurons (see also Chapter 26).[53] Consequently, if a cat is tilted with its right side down, its right vestibular nerve will be activated, thereby causing extension of the right limbs. A simultaneous decrease in activity in the left vestibular nerve will produce relaxation of the left limbs. This *tonic labyrinthine reflex* arrests the animal's fall and preserves postural stability.

Although the LVST does transmit some canal information to spinal motoneurons (especially to those supplying the neck muscles, Fig. 27–13), the current view is that Deiters's nucleus is concerned mostly with the processing of tilt information. Like eighth nerve afferents (see Fig. 27–7), Deiters's neurons in the decerebrate cat respond to various

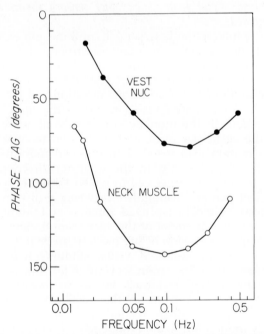

Figure 27–14 Comparison of the phase lag relative to head acceleration of neurons in the vestibular nucleus (VEST NUC) and neck motor units (NECK MUSCLE) as a function of the frequency of sinusoidal yaw rotation. (After Ezure, K.; Sasaki, S. *J. Neurophysiol.* 41:445–458, 1978.)

combinations of pitch and roll tilts.[43] These responses result from otolith input since they persist when all six semicircular canals are rendered inoperative by plugging them with bone chips. Since Deiters's neurons receive mono- and polysynaptic inputs from the utricular and saccular nerves,[19] otolith information can influence limb motoneurons through as few as three or four synapses. These reflex pathways can be activated in the decerebrate cat by applying sinusoidal linear accelerations along the animal's horizontal and vertical axes. Once again a comparison of the phase lags of otolith afferent activity with those of the EMGs of forelimb extensor muscles indicates that otolith drive relayed directly to spinal motoneurons via the LVST cannot be the only signal influencing motor output.[1] Some additional circuits, probably involving the reticulospinal pathways, must also be involved.

Neck Reflexes. When the head is tilted, not only are afferents from the otolith organs stimulated to produce vestibulomembral responses, but afferents from receptors in the upper three joints of the cervical vertebrae have a pronounced effect on limb musculature. As described above, tilting the whole animal to the right (no neck movement relative to the head) produces vestibulomembral reflexes that extend the right limbs and flex the left. If the head is held fixed and the body is turned to the right at the neck, the left limbs extend and the right limbs flex, producing a *tonic neck reflex*. Consequently, during natural head movements to the right, when both labyrinth and neck receptors are activated, the *tonic neck* and *vestibulomembral* reflexes tend to cancel each other, permitting the animal to make head movements from a stable body base. The site of interaction between neck and otolith signals is unknown, but such an interaction could be effected by utilizing descending propriospinal pathways or by the convergence of the two signals upon neurons of the lateral vestibular nucleus, many of which respond to both whole body tilt and neck rotation.[6]

Although we have focused on the static role of the LVST in posture, it may also have a role in dynamic postural adjustments since the firing rates of LVST neurons are modulated during locomotion (Chap. 26). The role of these rhythmic modulations, however, is unclear.

Vestibular Interactions with Other Sensory Modalities. It is well known that the vestibular system is aided in its role of postural and head stabilization by other sensory modalities. If a standing person is facing forward in a bus that suddenly accelerates, he tilts backward, thereby stimulating tactile afferents in the feet, ankle joint receptors,

proprioceptive afferents in ankle muscles, otolith afferents in the labyrinth, and motion sensors in the visual system. Signals from these various sensors generate compensatory reflexes that tilt the subject forward. In recent behavioral experiments to investigate the contribution of extravestibular signals to postural stability, subjects were tested on a movable platform that dissociates the usual normal relation between vestibular and other sensory cues (e.g., a backward tilt of the body can artificially be linked with a backward rather than normal forward tilt of the visual environment). Nashner[35] has demonstrated that shifts in postural equilibrium caused by platform movement elicit EMG responses in leg muscles with latencies as short as 100 ms and that altering the motion of the visual surroundings affects the earliest components of the EMG response. However, extravestibular reflexes that tend to destabilize stance are rapidly attenuated after a few trials. For example, such a destabilizing reflex occurs when toe-up rotation of the platform about the ankle joint of an erect subject stretches the gastrocnemius muscle, eliciting a long-latency stretch reflex (a functional stretch reflex, see Chap. 28) that throws the subject further backward. *Thus when other sensory inputs are in conflict with the vestibular input, the vestibular signal seems to act as the reference postural signal with which other sensory inputs are compared.*

The interaction between labyrinthine and extravestibular signals can occur in structures as peripheral as the vestibular nuclei. Neurons in the vestibular nuclei of the cat respond to linear acceleration of the head in one direction and movement of the entire visual world in the opposite direction. In the monkey, apparently all neurons in the vestibular nucleus that respond to angular acceleration also respond to full-field movement of the visual world in the opposite direction.[50] Therefore, the target cell to which these neurons project cannot distinguish between a discharge caused by clockwise head rotation and counterclockwise movement of the visual world. Since in humans the illusion of self-motion can be induced by a moving full-field visual stimulus (e.g., a person sitting in a train feels he is moving when a train on the next track begins to move), the discharge of neurons in the vestibular nucleus may provide signals related not only to actual body motion but to apparent body motion as well. In addition, activity originating in a broad spectrum of cutaneous, joint, and muscle receptors reaches the vestibular nuclei by brain stem pathways and also via both the fastigial nucleus and the anterior lobe of the cerebellum. For example, electrical stimulation of peripheral nerves reveals that somatosensory and labyrinthine signals converge upon the same vestibulospinal cells.[42]

CORTICAL PROJECTIONS OF VESTIBULAR INFORMATION

The previous sections indicate that afferent vestibular signals are transmitted rather directly to brain stem and spinal motoneurons to effect compensatory eye, head, and body movements. In contrast with this strong projection to motor centers, and unlike other sensory modalities, vestibular information has only a very weak thalamocortical projection. Even modern anatomical studies reveal only sparse connections from the vestibular nuclei to the thalamus;[23] also electrical stimulation of the eighth nerve drives only the occasional thalamic neuron.[24] Units in the ventroposterior nucleus (both VPI and VPL_o, see Fig. 27–8) discharge in relation to head velocity and hence may be driven by vestibular nuclear units with similar sensitivity.[8] Like many cells in the vestibular nucleus, some thalamic cells with vestibular sensitivity can be driven by full-field visual or somatosensory stimuli. Electrical stimulation of the vestibular nerve of the monkey also produces evoked potentials in cortical area 2V (at the lower end of the intraparietal sulcus; Fig. 27–8) and area 3a. Like the thalamic relay cells, a small population of neurons in area 2V also responds to head rotation and full-field visual stimulus motion in opposite directions.[8] In the cat, neurons in the anterior suprasylvian sulcus, the area homologous to simian area 2V, respond not only to yaw (or pitch or both) rotation but also to displacement of one or several limbs, to visual stimuli, and to dynamic or static deflection of the head.[3] Perhaps the modest parietal vestibular area is part of the circuitry involved in comparing discordant motion information from the different afferents to produce various sensations of motion. On the other hand, the vestibular projection to area 3a, which is a staging area for somatic inputs to area 4 (see Chap. 28), is most likely involved with the descending control of movement.

MAINTENANCE AND MODIFICATION OF THE VOR

The efficacy of many reflexes is controlled by comparing their response (output) with the stimulus that produces it (input) and maintaining the difference between input and output as small as possible. For the VOR, the objective is to minimize the difference between head and eye velocity so that the visual world remains stable on the retina. "Closed loop" control in which the output is fed back to be compared with the input, however, is impossible for the VOR, which responds as early as 10 ms after a head rotation, a time too short to allow any feedback to occur (e.g., it takes a retinal ganglion cell alone at least 20 to 30 ms to respond to a light stimulus). Therefore, the VOR is said to be an "open loop" system, the efficacy of which depends on an accurate setting of the gains of its individual elements (e.g., the interneuron in the vestibular nucleus). Although these gains could be laid down at birth, it seems more reasonable to suppose that they are continually adjustable so that the VOR can adapt to changing conditions such as an increase in the inertia of the head during growth, a change of the visual input with the wearing of spectacles, and the death of neurons due to age and injury. Such neuronal adjustments (called parametric feedback) would be deemed appropriate if they reduced the slip of the visual world across the retina. An assessment of the effectiveness of the adjustment requires many head rotations; therefore, the maintenance of an accurate VOR gain is an ongoing, gradual process.

In addition to maintaining an appropriate gain over the long term, the VOR must also be capable of rapid changes in gain. We would be at a great disadvantage if the VOR were always "hard wired" to produce a compensatory eye movement every time we moved our heads. For example, although the VOR is useful while a passenger in a bouncing bus looks out the window at a distant scene, it must be actively suppressed (i.e., the gain must be driven to zero) when he shifts his gaze to something moving with him such as a newspaper lying in his lap. Since it is possible to shift rapidly from outside viewing to reading and back again, the VOR also is capable of rapid gain modifications that are independent of, and superimposed on, the long-term processes that maintain VOR gain.

Role of the Cerebellum. The archicerebellum, in particular the flocculus, would seem to have the requisite afferent inputs and the necessary projections to the brain stem to effect both the maintenance and rapid modification of VOR gain.

Signals related to head rotation, motion of the visual world, and eye movement all project to the flocculus. First, the flocculus receives a strong input from the ipsilateral vestibular nuclei in all mammals, and in most mammals, it receives a direct input from the ipsilateral vestibular nerve. Second, based largely on work in the rabbit, the flocculus receives visual information over both climbing fiber and mossy fiber pathways. The best visual stimulus to elicit the climbing fiber response is a contrast-rich visual pattern (e.g., a painting by Jackson Pollock) that is moving at slow velocities of less than 1°/s.[47] The climbing fiber responses are produced by a direct pathway (only three neurons) that includes a retinal projection to the nucleus of the optic tract (see Chap. 20) and a subsequent connection to the flocculus subdivision of the inferior olive (see also Fig. 27–17). Visual information also reaches the flocculus over mossy fiber pathways that can drive simple spike discharges of Purkinje cells in the monkey. Third, mossy fibers in the flocculus of the alert monkey discharge with eye movement.[28]

Unlike the situation in other parts of the cerebellum, Purkinje cells in the flocculus project directly to the brain stem rather than to the deep cerebellar nuclei. In the monkey, floccular Purkinje cells project almost exclusively to the ipsilateral vestibular nuclei. In both the cat[2] and rabbit,[18] this projection is thought to inhibit interneurons of the VOR, because PSPs evoked by 8th nerve stimulation in various ocular motoneurons are depressed at short latencies by flocculus-conditioning stimuli. Taken together, this evidence suggests that the flocculus is strategically placed to compare vestibular, eye-movement, and visual information and to produce an output that modifies transmission through VOR pathways.

Rapid Modification of the VOR. To test the possibility that the flocculus is involved in rapid modification of the VOR, Lisberger and Fuchs[29] recorded from Purkinje cells in the flocculus in monkeys trained to suppress the VOR while undergoing horizontal sinusoidal head rotations. In the dark, when head oscillations produce the usual compensatory eye movements, the typical floccular Purkinje cell shown in Figure 27–15 (identified by its occasional climbing fiber response at the dots) shows no modulation in firing rate (COMPENSATION condition). However, within fractions of a second after the animal has eliminated the compensatory eye movements by fixating a target moving with him (like a person reading a newspaper on a bus), the simple spike activity of the same Purkinje cell becomes clearly modulated (SUPPRESSION condition). Figure 27–15 also depicts the direct pathways from the horizontal canal to the abducens nucleus. The insets show the firing patterns of each neuron in the pathway for one cycle of head rotation during the compensation (dashed lines) and suppression (heavy lines) conditions. During suppression of the VOR, the modulation of firing rate in Purkinje cells (F) resembles that recorded in the ipsilateral eighth nerve (A). Therefore, during suppression of the VOR (the solid lines in the insets), the averaged firing rate of the Purkinje cell (F) and the vestibular nerve activity (A) impinging on the VOR interneuron (I) are in phase but are of opposite sign since the synapse of the Purkinje cell from the flocculus is inhibitory. Consequently, the interneuronal discharge (A-F) reaching the abducens motoneurons is essentially unmodulated and the eye (E) remains stationary. However, during the VOR in the dark (dashed lines in the insets), the flocculus inhibition (F) is constant and does not check the modulation of firing produced in the VOR interneuron by the vestibular nerve input. The modulated interneuron activity (A-F) then drives the motoneurons and helps produce the compensatory eye movements of the VOR (E, dashed line).

Clearly, the flocculus must receive another input that tells it when to act on the VOR. Indeed, the same Purkinje cell that increases its discharge when the head rotates to the right also exhibits an almost equal increase in discharge if the head is immobile and the eyes rotate to the right in pursuit of a moving target. During the VOR or while a rotating monkey fixates a target that is stationary in space, rotation of the head to the right produces eye rotation to the left. Consequently, the Purkinje cell receives roughly equal but opposite influences owing to the rightward head but leftward eye velocity, so its net discharge exhibits no change at all. Lesions of the flocculus severely impair not only the ability of a monkey to suppress its VOR but also its ability to follow a smoothly moving target with its eyes. Therefore, the eye movement–related discharge of Purkinje cells appears to be an important component in the rapid modification of the VOR.

Maintenance of an Appropriate VOR. To study the mechanisms involved in the maintenance of the VOR, it is necessary to create a situation that simulates an inappropriate setting of the VOR gain. Gonshor and Melvill Jones[16] succeeded in forcing gradual changes in VOR gain by requiring human subjects to wear spectacles containing reversing prisms for periods of up to 27 days. Under these conditions, a head movement to the left produced an apparent movement of the visual world to the left, rather than the usual movement to the right. Consequently, the normal VOR was completely inappropriate to stabilize the visual world on the retina. However, the gain and phase of the VOR gradually adapted to the altered visual scene. Within 3 days (Fig. 27–16), the gain dropped to a low value of only 0.2. Over the next

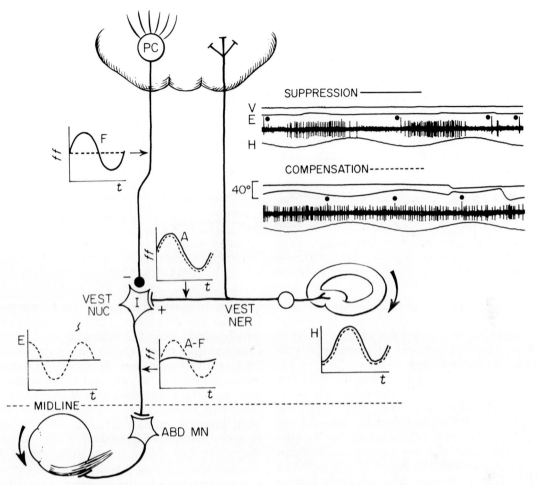

Figure 27–15 Schematic diagram of the horizontal VOR pathway and its inhibitory input via the flocculus. The single-unit recordings show the behavior of a flocculus Purkinje cell (PC) while a monkey is undergoing sinusoidal head rotations (at 0.9 Hz) in the dark *(compensation)* and while it is suppressing the VOR by fixating a target rotating with it *(suppression)*. V and E represent vertical and horizontal eye positions, respectively, and H represents horizontal head position. The dots indicate complex spikes. In the circuit diagram, the head rotation, H(t), causes eye movements, E(t), appropriate to either the suppression (solid line) or compensation (dashed line) condition. A single cycle of averaged neural activity during suppression (solid line) and compensation (dashed line) is traced through the primary afferents (A) and the flocculus (F) to show a hypothetical discharge (A-F) in the vestibular nucleus interneuron (I).

two weeks, the gain remained low (between 0.3 and 0.4), but the phase gradually changed so that the eye, rather than moving 180° out of phase with head position, eventually led head position by an average of 60° (Fig. 27–16, 14-day record). On occasional cycles, a head movement even elicited a small eye movement in the same direction, i.e., the phase lag was zero, so that the VOR had undergone a reversal of sign. These gain and phase changes were appropriate since the adapted eye movements nearly stabilized the altered visual scene on the retina. After the prisms were removed (dashed lines, Fig. 27–16), the phase lag was restored to normal within hours, while the

gain recovered over a time course similar to its original adaptation.

Experiments on animals confirm that such changes in VOR gain are both adaptive and enduring. Monkeys fitted with telescopic glasses, which produce apparent movement of the visual world that is either less ("minifying" lenses) or greater (magnifying lenses) than normal during head rotation, display decreases or increases, respectively, of the VOR gain. As in humans, these changes in gain require about two or three days. If, after the glasses are removed, the head is immobilized to prevent the occurrence of the visual slip that signals an inappropriate VOR gain,

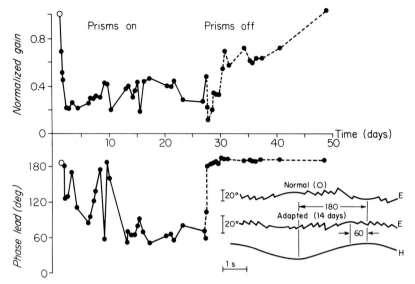

Figure 27–16. Effects of long-term reversal of the visual image on the gain and phase of the VOR in humans. Inset shows that the normal VOR is exactly out of phase with head position, i.e., the eye movement, E, leads the head movement, H, by 180 deg. After a subject wears reversing prisms for 14 days, the VOR gain is significantly reduced (note change in E calibration) and E leads H by only 60 deg. The graph shows the time course of gain and phase consequent to donning and subsequently removing the prisms. The gains have been normalized to the preadapted value (i.e., the normal VOR gain was taken as 1). (After Melvill Jones, G. *Philos. Trans. R. Soc. Lond.* [Biol.] 178:319–334, 1977.)

the recovery to a normal gain is slowed considerably, and the adaptations are retained for several days. Therefore, the long-term change in VOR gain is an excellent example of neural plasticity (see Chap. 33 for others).

Ito[20] was the first to deduce that the flocculus could be involved in VOR plasticity. He envisioned the flocculus as a "side path" from the labyrinth to the VOR interneurons whereby an adjustment of Purkinje cell sensitivity to the vestibular input could alter the gain of the VOR. In his model, diagrammed in Figure 27–17, the VOR gain ($\dot{E}/\dot{H}$) is equal to G-H, where G is the gain of the direct path and H is the gain of the floccular "side path." The adequacy of the gain adjustment is assessed by the residual slip of visual images over the retina as monitored by the visual climbing fiber input, which reports slow, full-field stimulus motions. Drawing upon Marr's theory of the role of cerebellar cortex (Chap. 29), Ito further suggested that a persistent error signal arriving over the climbing fibers could gradually modify the

efficacy of vestibular mossy fiber synapses on floccular Purkinje cells. This "learning process," which would alter the side path gain, H, would continue until there was no longer any visual slip. To test Ito's suggestion, Robinson[41] removed the entire cerebellum of cats and demonstrated that they no longer exhibited gain changes when outfitted with reversing prisms. Furthermore, the low VOR gain of animals adapted prior to surgery returned to normal immediately after removal of the cerebellum. More recent experiments on the monkey indicate that it is the flocculus per se that is necessary for adaptive changes of the VOR.[29]

Although the deficits produced by cerebellar removal strongly implicate the flocculus in VOR plasticity, investigators disagree on the location of the modifiable synapses. Ito[21] argued that the plastic changes in the rabbit occur at the mossy fiber–Purkinje cell synapse. On the other hand, Miles,[34] working with monkeys, could find no evidence of altered discharge patterns in the Purkinje cells of animals adapted to "minifying" or

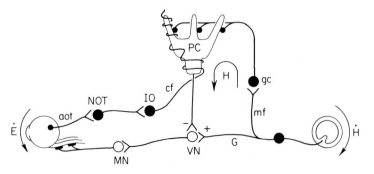

Figure 27–17. Circuitry involved in long-term modification (plasticity) of the VOR by the flocculus. This model suggests that visual information originating in the accessory optic system (accessory optic tract, aot) and relaying in the nucleus of the optic tract (NOT) and the inferior olive (IO) provides a climbing fiber (cf) input, which "educates" the synapses of vestibular granule cells (gc) on Purkinje cells (PC) to alter transmission (H) through the flocculus and thereby modify the overall VOR gain (G-H). (From Ito, M. *Brain Res.* 40:81–84, 1972.)

magnifying lenses. Miles and Lisberger[34] therefore concluded that the flocculus is not the locus of the modifiable synapses but instead provides the control signal that modifies a synapse in the brain stem VOR pathways. For more details on this interesting controversy, the reader is referred to the two articles[21, 34] cited previously. Whatever the outcome of this controversy, studies of the neural substrate of rapid and gradual modifications of the VOR have demonstrated that the cerebellum is involved not only in rapid control of motor responses but also in motor learning, or plasticity.

Compensation After Labyrinthectomy. Another example of plasticity in the vestibular system is the recovery of function that follows destruction of one labyrinth. As discussed earlier, a balanced input from both vestibular labyrinths is necessary for normal functioning of the VOR and indeed of all vestibular reflexes. Unilateral destruction of one labyrinth eliminates the resting discharge in the deafferented eighth nerve and hence produces a tonic imbalance within the central pathways. The result is an exaggeration of the spontaneous ocular and postural reflexes that would normally be elicited by stimulation of the intact side. For example, an abortive VOR is elicited in which the eyes are driven slowly toward the side of the lesion because of the unchecked input from the intact side. As the eyes are driven toward their extreme deviations, a fast eye movement drives (resets) them in the opposite direction whereupon the inexorable smooth eye movement resumes. This sawtooth pattern of eye movements is called *nystagmus*.

A common clinical tool to test the integrity of VOR pathways involves eliciting vestibular nystagmus by injecting warm or cold water into one ear canal. The injection of these liquids causes *caloric nystagmus* by creating temperature gradients in the endolymph of the canals on one side, thereby artificially producing a steady deflection of their cupulae and an imbalance of vestibular inputs.

Nystagmus and some other abnormalities, such as decreased extensor tone in the ipsilateral extremities, disappear during the first several days after the lesion. In cats with lesions of the labyrinth, the disappearance or amelioration of symptoms, called *compensation*, is associated with a recovery of some resting activity in the depressed type I neurons in the vestibular nucleus on the injured side. In addition, the efficacy of the inhibitory commissural pathways is increased,[39] so that type I neurons on the side with the lesion can undergo changes in firing rate during horizontal angular acceleration because of increases and decreases of inhibitory influences from the intact contralateral canal (see Fig. 27–9B). The neural mechanisms underlying the renewed activity in the vestibular nucleus are unknown, but its occurrence provides a basis for the compensation of postural and oculomotor asymmetries.[15]

The simultaneous destruction of both labyrinths produces severe postural and oculomotor problems that are also compensated in time. The mechanisms responsible for the recovery of head-eye coordination after bilateral eighth nerve lesions have been studied in monkeys.[9] These mechanisms include an increase in gain of the normally ineffective neck-ocular reflexes and the development of centrally programmed compensatory eye movements.

SUMMARY

Vestibular hair cells constantly provide the central nervous system with information about the orientation and movement of the head in space. We are usually unaware of vestibular stimuli, just as we are unaware of the kinesthetic and position sense mediated by muscle proprioceptors and joint afferents. Rather than producing overt perceptions, the major role of vestibular information is to serve as the sensory input for motor reflexes that stabilize the eyes, head, and body in space. To stabilize the eyes, head acceleration is first integrated by the canal end-organ and then integrated again by neural circuits to produce the requisite commands for ocular motoneurons. To stabilize the head and body, the vestibular system probably uses similar but less well understood neuronal mechanisms. All of the compensatory vestibular motor reflexes also appear to be facilitated by other sensory inputs. Thus, the major role of the vestibular system is to integrate information about the attitude and motion of the head with visual, proprioceptive, and somatosensory signals to maintain posture, balance, and ocular stability. The vestibular system appears to be aided in this function by the vestibulocerebellum, which not only facilitates the rapid modification of the various reflexes but also maintains their long-term calibration in the face of neurological, optical, and other insults.

ANNOTATED BIBLIOGRAPHY

Wilson, V.J.; Melvill Jones, G. *Mammalian Vestibular Physiology.* New York, Plenum Press, 1979.
   A detailed account of the vestibular system, including its behavior, neurophysiology, and psychophysics.

REFERENCES

1. Anderson, J.H.; Soechting, J.F.; Terzuolo, C.A. Dynamic relations between natural vestibular inputs and activity of forelimb extensor muscles in the decerebrate cat. I. Motor output during sinusoidal linear accelerations. *Brain Res.* 120:1–15, 1977.
2. Baker, R.; Precht, W.; Llinas, R. Cerebellar modulatory action on the vestibulo-trochlear pathway in the cat. *Exp.Brain Res.* 15:364–385, 1972.
3. Becker, W.; Deeke, L.; Mergner, T. Neuronal responses to natural vestibular and neck stimulation in the anterior suprasylvian gyrus of the cat. *Brain Res.* 165:139–143, 1979.
4. Bizzi, E. Eye-head coordination. In *Handbk. Physiol.* 2:1321–1336, 1981.
5. Blanks, R.; Estes, M.; Markham, C. Physiologic characteristics of vestibular first-order canal neurons in the cat. II. Response to constant angular acceleration. *J. Neurophysiol.* 38:1240–1268, 1975.
6. Boyle, R.; Pompeiano, O. Convergence and interaction of neck and macular vestibular inputs on vestibulospinal neurons. *J. Neurophysiol.* 45:852–868, 1981.
7. Brodal, A. Anatomy of the vestibular nuclei and their connections. In Kornhuber, H.H., ed., *Handbook of Sensory Physiology*, vol. 6, p. 239–352. New York, Springer-Verlag, 1974.
8. Büttner, U; Lang, W. The vestibulocortical pathway: neurophysiological and anatomical studies in the monkey. In Granit, R.; Pompeiano, O., eds., *Reflex Control of Posture and Movement, Progress in Brain Res.* 50:581–588, 1979.
9. Dichgans, J.; Bizzi, E.; Morasso, P.; Tagliasco, V. Mechanisms underlying recovery of eye-head coordination following bilateral labyrinthectomy in monkeys. *Exp. Brain Res.* 18:548–562, 1973.
10. Ezure, K.; Sasaki, S. Frequency-response analysis of vestibular-induced neck reflex in cat. I. Characteristics of neural transmission from horizontal semicircular canal to neck motoneurons. *J. Neurophysiol.* 41:445–458, 1978.
11. Fernandez, C.; Goldberg, J. Physiology of peripheral neurons innervating semicircular canals of the squirrel monkey. II. Response to sinusoidal stimulation and dynamics of peripheral vestibular system. *J. Neurophysiol.* 34:661–675, 1971.
12. Fernandez, C.; Goldberg, J.M.; Abend, W.K. Response to static tilts of peripheral neurons innervating otolith organs of the squirrel monkey. *J. Neurophysiol.* 35:978–997, 1972.
13. Fuchs, A.F. Eye-head coordination, 303–366. In Towe, A.L.; Luschei, E.S., eds. *Handbook of Behavioral Neurobiology*, New York, Plenum Press. 1981.
14. Goldberg, J.; Fernandez, C. Physiology of peripheral neurons innervating semicircular canals of the squirrel monkey. I. Resting discharge and response to constant angular accelerations. *J. Neurophysiol.* 34:635–660, 1971.
15. Goldberg, J.M.; Fernandez, C. The vestibular system. In *Handbk. Physiol.* 3:977–1022, 1984.
16. Gonshor, A.; Melvill Jones, G. Extreme vestibulo-ocular adaptation induced by prolonged optical reversal of vision. *J. Physiol. (Lond.)* 256:381–414, 1976.
17. Hardy, M. Observations on the innervation of the macula sacculi in man. *Anat. Rec.* 59:403–418, 1934.
18. Highstein, S.M. Synaptic linkage in the vestibulo-ocular and cerebello-vestibular pathways to the VIth nucleus in the rabbit. *Exp. Brain Res.* 17:301–314, 1973.
19. Hwang, J.C.; Or, T.H.; Cheung, Y.M. Response of central vestibular neurons to utricular stimulations in cats. *Exp. Brain Res.* 40:346–348, 1980.
20. Ito, M. Neural design of the cerebellar motor control system. *Brain Res.* 40:81–84, 1972.
21. Ito, M. Cerebellar control of the vestibulo-ocular reflex—around the flocculus hypothesis. *Annu. Rev. Neurosci.* 5:275–296, 1982.
22. Iurato, S. *Submicroscopic Structure of the Inner Ear.* Oxford, Pergamon Press, 1967.
23. Lang, W.; Büttner-Ennever, J.A.; Büttner, U. Vestibular projections to the monkey thalamus: an autoradiographic study. *Brain Res.* 177:3–17, 1979.
24. Liedgren, S.R.C.; Milne, A.C.; Rubin, A.M.; Schwartz, D.W.F.; Tomlinson, R.D. Representation of vestibular afferents in somatosensory thalamic nuclei of the squirrel monkey (*Saimiri sciureus*). *J. Neurophysiol.* 39:601–611, (1976).
25. Lindeman, H.H. Studies on the morphology of the sensory regions of the vestibular apparatus. *Ergeb. Anat. Entwickl.* 42:1–113, 1969.
26. Lindeman, H.H. Anatomy of the otolith organs. *Adv. Otorhinolaryngol.* 20:405–433, 1973.
27. Lisberger, S.G.; Fuchs, A.F. Response of flocculus Purkinje cells to adequate vestibular stimulation in the alert monkey: fixation vs. compensatory eye movements. *Brain Res.* 69:347–353, 1974.
28. Lisberger, S.G.; Fuchs, A.F. Role of primate flocculus during rapid behavioral modification of vestibuloocular reflex. II. Mossy fiber firing patterns during horizontal head rotation and eye movement. *J. Neurophysiol.* 41:764–777, 1978.
29. Lisberger, S.G.; Miles, F.A.; Zee, D.S. Signals used to compute errors in monkey vestibuloocular reflex: possible role of flocculus. *J. Neurophysiol.* 52:1140–1153, 1984.
30. Markham, C.H.; Yagi, T.; Curthoys, I.S. The contribution of the contralateral labyrinth to second order vestibular neuronal activity in the cat. *Brain Res.* 138:99–109, 1977.
31. McCrea, R.A.; Yoshida, K.; Berthoz, A.; Baker, R. Eye movement related activity and morphology of second order vestibular neurons terminating in the cat abducens nucleus. *Exp. Brain Res.* 40:468–473, 1980.
32. McMasters, R.; Weiss, A.; Carpenter, M. Vestibular projections to the nuclei of the extraocular muscles. Degeneration resulting from discrete partial lesions of the vestibular nuclei in the monkey. *Am. J. Anat.* 118:163–194, 1966.
33. Melvill Jones, G. Plasticity in the adult vestibulo-ocular reflex arc. *Philos. Trans. R. Soc. Lond.* [Biol.] 178:319–334, 1977.
34. Miles, F.A.; Lisberger, S.G. Plasticity in the vestibulo-ocular reflex: a new hypothesis. *Annu. Rev. Neurosci.* 4:273–299, 1981.
35. Nashner, L.M. Adaptation of human movement to altered environments. *TINS* 5:358–361, 1982.
36. Peterson, B.W.; Filion, M.; Felpel, L.P.; Abzug, C. Responses of medial reticular neurons to stimulation of the vestibular nerve. *Exp. Brain Res.* 22:335–350, 1975.
37. Peterson, B.W.; Goldberg, J.; Bilotto, G.; Fuller, J.H. Cervicocollic reflex: its dynamic properties and interaction with vestibular reflexes. *J. Neurophysiol.* 54:90–109, 1985.
38. Peterson, B.W.; Pitts, N.G.; Fukushima, K.; Mackel, R. Reticulospinal excitation and inhibition of neck motoneurons. *Exp. Brain Res.* 32:471–489, 1978.
39. Precht, W.; Shimazu, H.; Markham, C.H. A mechanism of central compensation of vestibular function following hemilabyrinthectomy. *J. Neurophysiol.* 29:996–1010, 1966.

40. Robinson, D.A. Models of oculomotor neural organization. In Bach-y-Rita, P.; Collins, C., eds., *The Control of Eye Movement*, New York, Academic Press, 1971.
41. Robinson, D.A. Adaptive gain control of vestibuloocular reflex by the cerebellum. *J. Neurophysiol.* 39:954–969, 1976.
42. Rubin, A.M.; Liedgren, S.R.C.; Ödkvist, L.M.; Milne, A.C.; Fredrickson, J.M. Labyrinthine and somatosensory convergence upon vestibulospinal neurons. *Acta Otolaryngol.* 86:251–259, 1978.
43. Schor, R.H.; Miller, A.D.; Tomko, D.L. Responses to head tilt in cat central vestibular neurons. I. Direction of maximum sensitivity. *J. Neurophysiol.* 51:136–146, 1984.
44. Scudder, C.A.; Fuchs, A.F. Projections from the vestibular to abducens nuclei as revealed by spike-triggered averaging. *Soc. Neurosci. Abstr.* 7:777, 1981.
45. Shimazu, H.; Precht, W. Inhibition of central vestibular neurons from the contralateral labyrinth and its mediating pathway. *J. Neurophysiol.* 29:467–492, 1966.
46. Shinoda, Y.; Yoshida, K. Dynamic characteristics of responses to horizontal head angular acceleration in vestibuloocular pathway in the cat. *J. Neurophysiol.* 37:653–673, 1974.
47. Simpson, J.I.; Alley, K.E. Visual climbing fiber input to rabbit vestibulo-cerebellum: a source of direction-specific information. *Brain Res.* 82:302–308, 1974.
48. Stein, B.; Carpenter, M. Central projection of portions of the vestibular ganglia innervating specific parts of the labyrinth in the rhesus monkey. *Am. J. Anat.* 120:281–318, 1967.
49. Suzuki, J-I.; Cohen, B. Head, eye, body and limb movements from semicircular canal nerves. *Exp. Neurol.* 10:393–405, 1964.
50. Waespe, W.; Henn, V. Neuronal activity in the vestibular nuclei of the alert monkey during vestibular and optokinetic stimulation. *Exp. Brain Res.* 27:523–538, 1977.
51. Wersäll, J.; Bagger-Sjöbäck, D. Morphology of the vestibular sense organ, 123–170. In Kornhuber, H.H., ed. *Handbook of Sensory Physiology*, vol. 6. New York, Springer-Verlag, 1974.
52. Wilson, V.J.; Maeda, M. Connections between semicircular canals and neck motoneurons in the cat. *J. Neurophysiol.* 37:346–357, 1974.
53. Wilson, V.J.; Peterson, B.W. Vestibulospinal and reticulospinal systems. *Handbk. Physiol.* Sec. 1, 2:667–702, 1981.

Motor Functions of Cerebral Cortex

INTRODUCTION

Primary and Secondary Motor Cortex Areas

Anyone who has seen the victim of a cerebral stroke can appreciate the importance of motor cortex in performing normal movements. Typically, such patients are incapable of using one or more contralateral limbs and frequently cannot speak fluently. The loss of control is most notable in the muscles of the hands and face. Furthermore, many patients are unable to regain normal control of the paralyzed muscles, indicating that other regions cannot compensate to perform these motor functions.

Clinical and experimental evidence indicates that the cerebral cortex plays a crucial role in both the programming and the execution of normal voluntary movements. The relevant cortical areas can be broadly divided into two groups: *primary motor cortex*, which has relatively direct anatomical and functional relation to muscles and is important for the normal *execution* of movements, and *secondary* motor cortical areas that are synaptically more remote from the periphery and are more involved in *programming* movements under particular circumstances. The locations of these cortical regions in monkeys and humans are shown in Figure 28–1. The largest of these areas, the *primary motor cortex* (MI) lies on the precentral gyrus, in Brodmann's area 4. Electrical stimulation of the primary motor cortex produces specific and repeatable movements, with the lowest stimulus thresholds of any cortical area. MI contains a somatotopic map of the representation of body muscles. Electrical stimulation of secondary areas generates movements more rarely; these movements are more complex and variable and require stronger stimulus currents. These secondary motor areas include the *premotor cortex* (PMC), which lies anterior to MI, in the lateral portion of Brodmann's area 6, and the *supplementary motor cortex*, or supplementary motor area (SMA), which lies in the medial portion of area 6, largely in the medial bank of the sagittal sulcus. Both of these regions are also somatotopically organized and are interconnected with each other and MI. Still further rostral, in Brodmann's area 8, are the *frontal eye fields*, which are concerned with movements of

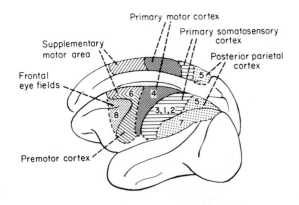

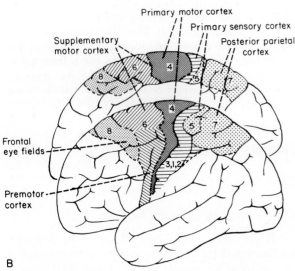

Figure 28–1 Major motor areas of cerebral cortex in *(A)* macaque monkey and *(B)* man. Numbers refer to Brodmann's designation of cortical regions with different cytoarchitectonic features. (From Brodmann, K. Vergleichende Lokalisationslehre der Grosshirnrinde. Leipzig, Barth, 1909.)

the eyes. Finally, the *posterior parietal cortex,* comprising Brodmann's areas 5 and 7, appears to be involved in programming directed movements of the limbs and eyes to targets in space.

In addition to electrical stimulation, other lines of experimental evidence support a functional distinction between primary and secondary motor cortical areas. *Lesions* in primary motor cortex of primates tend to produce paresis or paralysis, obvious deficits in generating muscle activity. Lesions in secondary motor regions produce more subtle deficits, called apraxias, which are the inability to perform certain types of movements under particular conditions. Finally, the recording of *neuronal activity* during movements has revealed that many primary motor cortex cells discharge

strongly in relation to muscle activity, whereas neurons in secondary motor areas tend to be active under particular conditions that give rise to movements. This chapter reviews the evidence that primary motor cortex is predominantly involved in the execution of movements whereas secondary motor areas are more concerned with motor programming. The major emphasis is on primary motor cortex, which has been studied most extensively and is best understood.

The Pyramidal and Extrapyramidal Systems

The cerebral cortex may influence movements through two major output pathways: the pyramidal and the extrapyramidal systems. While these two systems normally work together to control movement and posture, their functions are distinguished by neurologists who classify motor deficits as "pyramidal" or "extrapyramidal." The *pyramidal system* consists of cortical neurons whose axons descend through the medullary pyramids; the vast majority continue to the spinal cord via the corticospinal tracts, but some terminate in the brain stem reticular formation. The *extrapyramidal system* comprises the remaining motor systems, including the descending tracts from brain stem to spinal cord, the basal ganglia and the cerebellum. Inclusion of many different centers in the extrapyramidal system limits the usefulness of this classification, as discussed in Chapter 26.

Organization of Pyramidal Tract Neurons

The corticospinal tract exists only in mammals and increases in size along the phylogenetic scale.[24, 36a] In humans, each pyramid contains about one million fibers. About 60% of the pyramidal tract arises from cells in precentral cortex (approximately 30% from Brodmann's area 4 and 30% from area 6); the remaining 40% of the pyramidal tract arises from postcentral cortex. Figure 28–2 illustrates the course of precentral and postcentral pyramidal tract fibers in the macaque. The precentral pyramidal tract neurons originate in both primary and secondary motor areas and terminate preferentially in the more ventral laminae of the cord, which contain motoneurons and interneurons involved in movement. The number and proportion of corticospinal fibers terminating among motoneurons becomes progressively greater from prosimians, through monkeys and apes, to humans. Postcentral pyramidal tract cells tend to terminate more dorsally in the spinal cord,

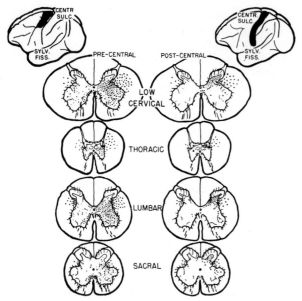

Figure 28–2 Course and termination of pyramidal tract fibers from precentral and postcentral cortex of the monkey. Heavy dots and line segments show the course of fibers descending in lateral and ventral corticospinal tracts. Light dots show sites of termination of corticospinal fibers. (From Kuypers, H. G. J. *Brain* 83:161–184, 1960.)

The cortical origin of the pyramidal tract has been documented most clearly by retrograde transport of horseradish peroxidase (HRP) from the spinal cord or the pyramids. In the monkey, corticospinal cells labeled with HRP are extensively distributed over several cortical fields, including Brodmann's areas 8, 6, and 4 precentrally, and areas 3, 1, 2, 5, and 7 postcentrally (Fig. 28–3A). Sagittal sections of the cerebral cortex reveal that the somata of pyramidal tract neurons all reside in layer 5 of the cortex (Fig. 28–3B). The labeled corticospinal cells appear to be grouped in clusters, which on surface reconstructions tend to form mediolateral bands.[18]

Most of the pyramidal tract fibers are small— 90% are less than 4 μm in diameter—and about half are unmyelinated. Thus the majority of pyramidal tract fibers are slowly conducting. The large Betz cells in motor cortex contribute about 2% of all pyramidal tract fibers. In the primate, single pyramidal tract cells typically distribute terminals to many motoneuron pools.[31]

PRIMARY MOTOR CORTEX (MI)

Somatotopic Organization

Effects of Stimulation on Muscles

among interneurons relaying peripheral input from afferent fibers to motoneurons and to higher centers. This postcentral component of the pyramidal tract may be more involved in regulating sensory function by presynaptic and postsynaptic modulation of transmission of afferent impulses.

The cortical motor areas have been delimited and mapped by observing the movements evoked by electrically stimulating the cortex. The primary motor cortex (MI), which has the lowest threshold sites, is organized somatotopically much like the

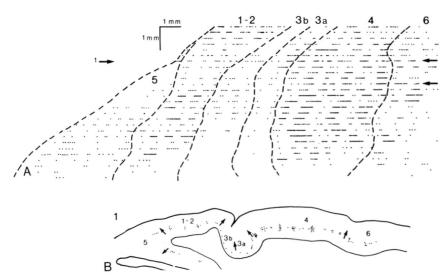

Figure 28–3 Cortical location of pyramidal tract neurons labeled by retrograde transport of horseradish peroxidase (HRP) from contralateral thoracic spinal cord of the cynomolgus monkey. *A,* Location of labeled cells projected on flattened cortical surface. *B,* Representative sagittal section through cortex at arrow in A. Numbers denote Brodmann's areas. (From Jones, E. G.; Wise, S. P. *J. Comp. Neurol.* 175:391–438, 1977.)

primary somatosensory cortex. Thus stimulating adjacent cortical sites evokes movements of adjacent body parts. The representation of the somatic musculature in MI of the monkey is summarized in Figure 28–4A. The motor map in the precentral gyrus provides the most extensive and detailed

cortical representation of limb muscles. The anatomical distortion of the simunculus figure (Fig. 28–4A) indicates that a proportionately larger region of cortex is devoted to those body parts capable of finer motor control, such as the tongue, digits, and toes. The greater cortical area devoted

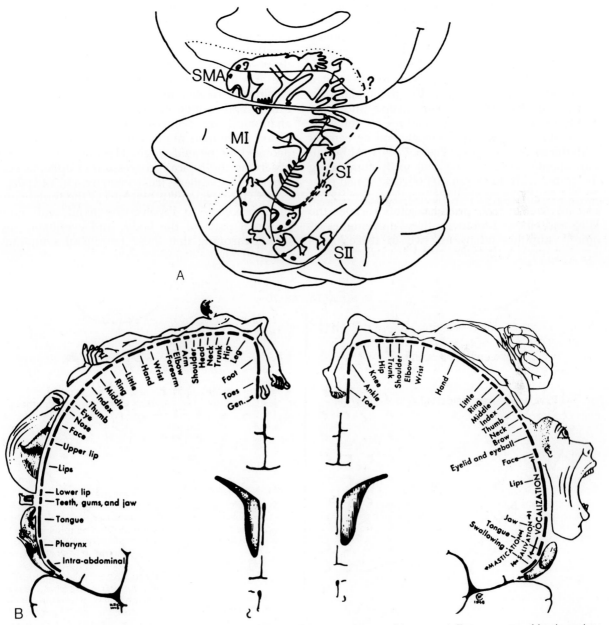

Figure 28–4 Somatotopic organization of motor cortical areas in monkeys and humans. *A,* The summary of body regions activated by cortical surface stimulation in an anesthetized macaque monkey. MI = primary motor cortex; SMA = supplementary motor cortex; SI = primary somatosensory cortex; SII = secondary sensory cortex. *B,* Relative size and location of primary motor cortex areas devoted to different body parts in humans. Sensory input is shown at left; motor output representation is shown at right. (*A,* After Woolsey, C. N. In Harlow, H. F.; Woolsey, C. N. *Biological and Biochemical Bases of Behavior,* Madison, Wis., Univ. of Wisconsin Press, 1958; *B,* reprinted with permission of Macmillan Publishing Company from Penfield, W.; Rasmussen, A. T. *Cerebral Cortex of Man. A Clinical Study of Localization of Function.* Copyright © 1950 by Macmillan Publishing Company, renewed 1978 by Theodore Rasmussen.)

to these regions contains a larger number of inter-neurons concerned with controlling this output and also provides more efferent cells that project to the brain stem and spinal cord.

Many of the fine movements evoked by motor cortex stimulation in the monkey are mediated by the pyramidal tract. This is illustrated by experiments in which the pyramidal tract was sectioned. Figure 28–5 summarizes the results of an experiment in which Woolsey and colleagues[38] sectioned the pyramidal tract unilaterally in monkeys and then mapped the movements evoked by stimulating the cortex on both sides. Movements elicited from the normal side, with the pyramidal tract intact (Fig. 28–5, right), were diverse and commonly involved distal finger and toe movements. The cortical distribution is in general agreement with the summary map of Figure 28–4A. Although movements could also be evoked by stimulating the cortex with the pyramidal tract sectioned (Fig. 28–5, left), these movements were more restricted and involved mainly proximal joints such as the knee and elbow. Moreover, these movements required stimulus intensities two to three times

higher than those of corresponding points on the side with the intact pyramidal tract. This experiment demonstrates that the pyramidal tract is important in mediating outputs from motor cortex to motoneurons of distal muscles.

A comparable somatotopic motor map for human cortex was obtained by Penfield and colleagues,[25] who stimulated the surface electrically during neurosurgical procedures to identify the functions of cortical sites. As in the monkey, the most medial portions of human motor cortex are devoted to the leg and the more lateral regions to the arm and face, as indicated in Figure 28–4B. Again, the relative area of cortex devoted to the thumb and tongue greatly exceeds their relative anatomical size. Some of this topographic organization of human motor cortex had already been deduced by the neurologist Hughlings Jackson by observing the typical progression of epileptic seizures. The so-called ''Jacksonian march'' of epileptic activity begins with twitches of the thumb and then proceeds to involve the hand and more proximal parts of the body, and eventually the face. Reasoning that these peripheral symptoms

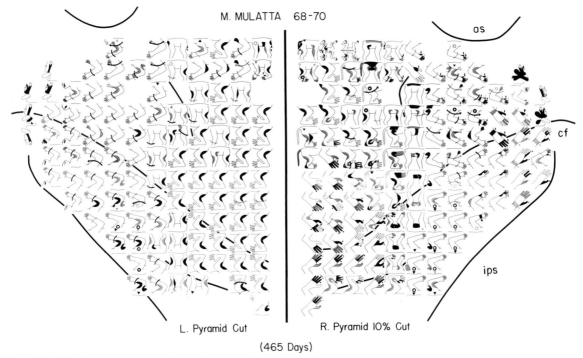

Figure 28–5 Figurine map illustrating movements evoked by electrical stimulation of pre- and postcentral cortex of macaque monkey with a unilateral pyramidal tract section. Figurines are located at the cortical points that evoked movements of the joints indicated. Pyramidal tract from left hemisphere was sectioned and from right hemisphere was largely intact. c.f. = central fissure; i.p.s. = intraparietal sulcus; a.s. = arcuate sulcus. (After Woolsey, C. N.; Gorska, T.; Wetzel, A.; Erickson, T. C.; Earls, F. J.; Allman, J. M. *Brain Res.* 40:119–123, 1972.)

reflect the spread of epileptic activity through adjacent cortical areas, Jackson deduced much of the relative cortical representation of the musculature shown in Figure 28–4B.[25b]

Conscious human patients whose precentral gyrus is electrically stimulated often report that the evoked movements are involuntary. Precentral stimulation evokes simple, stereotyped motor responses, rather than any conscious impulse to move, indicating that the primary motor cortex is relatively close to the output of the motor system. At many sites of human motor cortex electrical stimulation produces relaxation of musculature rather than muscle contraction.

In addition to evoking involuntary movements and arresting voluntary movements, electrical stimulation of motor cortex may also evoke somatic sensations. About a third of the precentral cortex sites stimulated by Penfield and Boldrey[25] evoked experiences of tingling or numbness— similar to the sensations produced by stimulating postcentral cortex. These sensations could be evoked from precentral sites even after ablation of adjacent postcentral regions that normally process somatic sensation (see Chap. 15), suggesting that precentral cortex may also be involved in somatic sensation.

Because cortical stimulation evokes movements that involve many muscles, it has been argued that movements rather than muscles are represented in motor cortex. In this view, the primary motor cortex sends command impulses that signal brain-stem and segmental circuitry to generate coordinated activity of many motoneuron pools. The alternative view is that motor cortex cells ultimately control individual muscles, and that their activity is coordinated by inputs from higher centers. To test these alternatives, Chang, Ruch, and Ward[5] dissected ankle muscles in anesthetized monkeys and measured the tensions produced by repetitive stimulation of different cortical sites. Each muscle could be made to contract from stimulation of a wide region of cortex, but the sites of lowest threshold were more localized. Stimulation of some low-threshold sites evoked contraction of specific muscles in isolation, whereas stimulation of other sites caused several muscles to contract together, suggesting that MI can affect both single and multiple muscles.

Effects of Stimulation on Motoneurons

The effects of corticospinal projections have been analyzed in more detail by recording intracellularly the excitatory postsynaptic potentials (EPSPs) evoked in motoneurons by cortical stimulation.[26, 27] In primates, some corticospinal cells make monosynaptic connections with motoneurons. Figure 28–6A illustrates corticomotoneuronal EPSPs recorded in motoneurons innervating arm muscles of the baboon.[26] As the stimulus intensity increases and recruits more corticomotoneuronal cells, the amplitude of the EPSPs also increases (Fig. 28–6B). Above a certain stimulus intensity the size of the EPSP increases no further, suggesting that the whole "colony" of corticomotoneuronal cells projecting to that motoneuron has been recruited. To estimate the cortical extent of the motoneuron's colony, Phillips and Porter[26] measured the current spread of the cortical stimuli by measuring the threshold intensity required to evoke a response in single pyramidal tract cells as a function of distance from the lowest threshold point. The threshold typically increased as the square of the distance, as shown in Figure 28–6C. Using this relation, the extent of the cortical colonies was estimated to range from 2 to 10 mm^2 for different motoneurons. EPSPs in distal motoneurons reached their maximum at lower stimulus intensities than those in proximal motoneurons, suggesting that the cortical colonies of distal motoneurons were more localized in the cortex. Moreover, motoneurons of distal muscles also received larger maximal EPSPs than those of proximal muscles. Thus, motoneurons of distal muscles received a greater net synaptic input from cortex, and this originated from a smaller cortical area.

The spatial distribution of cortical colonies was determined by mapping the cortical points from which minimal corticomotoneuronal EPSPs could be evoked in hindlimb motoneurons of the monkey.[16] The cortical distributions of these colonies were irregular in shape, typically between 3 and 7 mm^2, and the colonies projecting to motoneurons of different muscles overlapped extensively. Different motoneurons of the same pool sometimes received monosynaptic EPSPs from different cortical areas, suggesting that some corticomotoneuronal cells may project to specific motoneurons within the pool. Cortical stimulation also evoked disynaptic inhibitory postsynaptic potentials (IPSPs) in motoneurons; these inhibitory effects are mediated in part by the Ia inhibitory interneurons, which receive monosynaptic input from the cortex.

The relative magnitudes of the monosynaptic inputs to motoneurons descending from motor cortex and those arriving from Ia muscle spindle afferents suggest that the cortical and reflex inputs are somewhat different for each muscle. The size

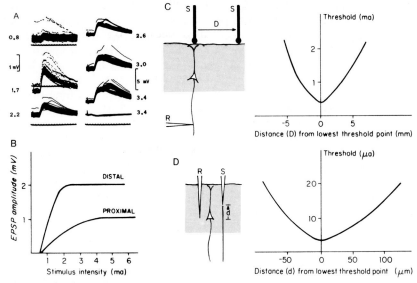

Figure 28–6 Effects of cortical stimulation. *A*, corticomotoneuronal excitatory postsynaptic potentials (CM-EPSPs) evoked in a motoneuron by increasing intensity of cortical surface stimulation. *B*, Size of CM-EPSP as function of stimulus intensity for motoneuron of distal and proximal muscle. *C*, Spread of cortical surface stimulus as measured by the threshold intensity required to activate pyramidal tract cell; typical curve of threshold in milliamperes, as a function of distance (D) of the stimulating electrode (S) from the best point (for 0.2 ms anodal pulse). *D*, Spread of intracortical microstimulation as measured by threshold intensity required to activate cortical cell; typical curve of threshold in microamperes, as a function of distance (d) of the stimulating electrode (S) from the best point (for 0.2 ms cathodal pulse). R represents recording electrode. (*A, B,* and *C* from Phillips, C. G.; Porter, R. *Prog. Brain Res.* 12:222–242, 1964 [*B* and *C* modified]. *D* after Stoney, S. D.; Thompson, W. D.; Asanuma, H. *J. Neurophysiol.* 31:659–669, 1968.)

of the maximal EPSP from these two sources is represented in Figure 28–7A by the width of the arrows converging onto motoneurons innervating hand and finger muscles of the baboon. The extensor digitorum communis (EDC) motoneurons receive the largest corticomotoneuronal EPSPs, which exceed the magnitude of the Ia-EPSPs from EDC muscle afferents. Other motoneurons, such as those of the hypothenar muscles (Uh), receive greater net synaptic input from muscle spindles than from cortical inputs.

Not only does electrical stimulation of MI produce monosynaptic EPSPs in primate motoneurons, but repetitive cortical stimulation, which mimics the bursts of activity recorded in corticospinal cells, enhances the effectiveness of EPSPs beyond simple summation. Figure 28–7B shows that repetitive stimulation (200/s) of the Ia muscle afferent produced algebraic summation of successive Ia-EPSPs, whereas repetitive stimulation of motor cortex produced an increase in the size of successive corticomotoneuronal EPSPs. This increase was not due simply to recruitment of additional corticospinal cells, since the amplitudes of the descending volleys remained constant. Neither could the increase be due to recruitment of excitatory spinal interneurons, since the latency and shape of the EPSPs did not change, but only their

amplitude. Such facilitation of corticomotoneuronal EPSPs enhances the effectiveness of high-frequency activity, which typically occurs at the onset of movement (see Fig. 28–14).

Intracortical Microstimulation

The experiments described above employed cortical surface stimulation, which can spread over considerable distances, limiting the spatial resolution of the effective sites. To evoke discrete responses from more circumscribed intracortical sites, Asanuma and Rosen[1] delivered intracortical microstimulation through microelectrodes. This technique has provided three-dimensional maps of low-threshold output sites in the depth of the cortex. To quantify the effective spread of stimulus currents, Asanuma and colleagues[31a] measured the stimulus intensities required to activate a pyramidal tract neuron as a function of the distance of the stimulating electrode. As the electrode was moved from the lowest threshold point, threshold intensities increased as the square of the distance (see Fig. 28–6D). The effective radius of a 10-µa current pulse was found to be 80 to 90 µm.

Using intracortical microstimulation to evoke activity in a given muscle, Asanuma and Rosen[1] found that the low-threshold points for individual muscles were located in a cylindrical volume of cortex perpendicular to the cortical surface (see Fig. 28–12). The diameters of these columnar zones ranged from 0.5 to several millimeters. These observations suggested a columnar organization of output cells to specific muscles, analogous to the columnar arrangement in sensory cortex of cells with inputs from particular receptors. However, in addition to activating cells directly, repetitive microstimulation is very effective in evoking responses transsynaptically, via fibers that are activated directly. This could explain why the low-threshold points for evoking muscle responses with repetitive stimulation have the same distribution as the afferent and efferent fibers. These fibers could activate the corticospinal output cells, which are located in cortical layer V (see Fig. 28–3B).

Cortical Connections

Cortical Neurons and Their Targets

The operations of the motor cortex can best be understood in the context of its input and output connections. On the basis of cell morphology and fiber distributions, anatomists have distinguished five layers within motor cortex. Unlike other cortical areas, primary motor cortex lacks a prominent layer IV containing granule cells, so the precentral gyrus is also called "agranular cortex." Drastically

simplified, cortical cells can be classified into two basic types: pyramidal and nonpyramidal. *Pyramidal* cells, so called because of their pyramidal somata, have a long apical dendrite directed toward the cortical surface and short basal dendrites issuing from the base of the soma. Their extensive dendritic tree suggests that they integrate synaptic input from several layers. Axons of the pyramidal cells leave the cortex for other cortical and subcortical targets, making them the main cortical output cell; in addition, axon collaterals of pyramidal cells usually provide local intrinsic connections. *Nonpyramidal* cells, on the other hand, have small dendritic trees that sample input from a more restricted region; because their axons remain within the cortex and arborize locally, they are interneurons. Nonpyramidal cells include stellate, basket, and granule cells, some of which are inhibitory.

As illustrated in Figure 28–8, pyramidal cells in each cortical layer send axons to specific targets. The pyramidal cells in layers II and III project to other cortical regions. The more superficial cells project to ipsilateral cortical areas, notably to secondary motor areas (SMA and PMC) and to postcentral sensory cortex; the deeper cells in layer III send axons through the corpus callosum to the motor cortex of the opposite hemisphere. These two superficial cortical layers in turn receive their input from other cortical regions. Thus, layers II and III are largely concerned with intercortical

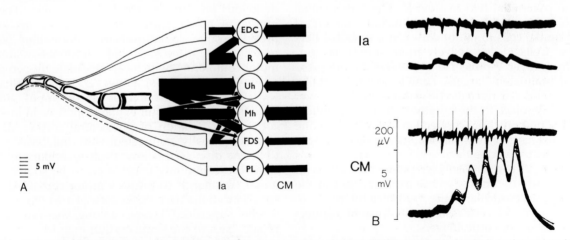

Figure 28–7 Comparison of monosynaptic excitatory postsynaptic potentials (EPSPs) evoked in forelimb motoneurons from cerebral cortex (CM) and from Ia afferent fibers. *A*, Total synaptic input to different types of motoneurons. Thickness of arrows represents the relative size of maximal EPSP. EDC = extensor digitorum communis; R = remaining dorsiflexors of wrist; Uh = intrinsic hand muscles innervated by ulnar nerve; Mh = intrinsic hand muscles innervated by median nerve; FDS = flexor digitorum sublimis; PL = palmaris longus. *B*, High-frequency stimulation evokes facilitation of successive corticomotoneuronal (CM) EPSPs *(bottom)* but not Ia EPSPs *(top)*. Upper records show that arriving volleys, recorded on cord dorsum, are constant. (*A* from Clough, J. F. M.; Kernell, D.; Phillips, C. G. *J. Physiol. [Lond.]* 198:145–166, 1969. *B*, from Phillips, C. G.; Porter, R. *Prog. Brain Res.* 12:222–242, 1964.)

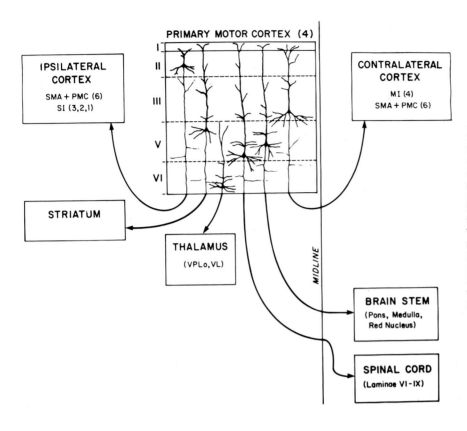

PRIMARY MOTOR CORTEX (4)

IPSILATERAL
CORTEX

SMA + PMC (6)
SI (3,2,1)

CONTRALATERAL
CORTEX

MI (4)
SMA + PMC (6)

STRIATUM

THALAMUS
(VPLo, VL)

MIDLINE

BRAIN STEM
(Pons, Medulla,
Red Nucleus)

SPINAL CORD
(Laminae VI-IX)

Figure 28–8 Targets of projections from pyramidal cells in different layers of primary motor cortex. Cells projecting to targets across the midline are shown at right. Numbers refer to Brodmann's architectonic designations. VPLo = nucleus ventralis posterior lateralis oralis; VL = nucleus ventralis lateralis; SMA = secondary motor area; PMC = premotor motor cortex. (From Jones, E. G. In Jones, E. G.; Peters, A., eds. *Cerebral Cortex*, vol. 5, 113–184. New York, Plenum Press, 1986.)

communication between somatotopically related areas.

Most of the outputs to subcortical targets arise from pyramidal cells in layer V. The corticospinal cells lie deepest in layer V and include the largest pyramidal cells in motor cortex, the so-called Betz cells. Cells in successively more superficial portions of layer V project to successively more proximal brain-stem targets, namely to the medulla, pons, and red nucleus; the most superficial layer V cells project to the striatum. A few pyramidal cells send branching connections to more than one of these targets, but the majority project to only one site.

Layer VI contains smaller pyramidal cells with long apical dendrites, whose axons project to the thalamus, particularly the ventrolateral nucleus. Many layer VI cells also send recurrent connections to upper cortical layers.

Sources of Inputs to Primary Motor Cortex

The major inputs to primary motor cortex (MI) arise from other cortical areas and from thalamic nuclei. Interconnections between *ipsilateral* cortical regions are diagrammed in Figure 28–9. These

cortical areas are all reciprocally connected in a somatotopically organized manner. The most massive connections are those that interconnect homologous pre- and postcentral cortex sites. These inputs to motor cortex provide cutaneous and proprioceptive somatosensory information (and motor cortex, in turn, provides information about motor commands, which can modulate the activity of sensory cortex cells). However, postcentral cortex is not the only source of sensory input, since somatic stimuli can still evoke responses in motor cortex after ablation of postcentral cortex. MI is also reciprocally connected with the SMA and PMC. As discussed below, these secondary motor areas are commonly considered to be sources of motor commands to primary motor cortex. Afferent connections from other cortical areas converge in the superficial layers, among the cells that project back to the same cortical regions.

Contralateral motor cortical regions are connected through the corpus callosum. The corpus callosum preferentially interconnects those regions related to limb girdle and axial musculature (face and trunk regions), suggesting that it helps to integrate motor control of these midline structures. The hand and foot areas of motor cortex (like those of

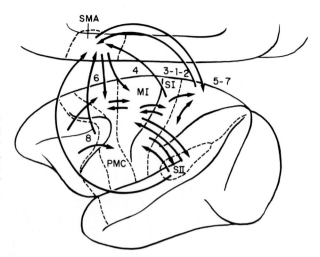

Figure 28–9 Major interconnections between cortical areas involved in motor control. Numbers indicate Brodmann's areas and arrows indicate direction of projection. Regions are reciprocally interconnected in a somatotopic manner. SMA = supplementary motor area. (After Wiesendanger, M. In Towe, A. L.; Luschei, E. S., eds, *Handbook of Behavioral Neurophysiology*, vol. 5, pp 401–491. New York, Plenum Press, 1981.)

sensory cortex) are not interconnected with the contralateral side. The supplementary and premotor cortex areas are also interconnected with their contralateral counterparts. Surprisingly, sectioning the corpus callosum does not noticeably disrupt motor coordination of proximal or distal limb muscles.

The main *thalamic* input to primary motor cortex comes from the ventrobasal nucleus (nucleus ventralis lateralis caudalis [VLc] and nucleus ventralis posterolateralis oralis [VPLo], using the nomenclature of Olszewski[23]), which relays information about peripheral events arriving from the spinothalamic tract as well as central commands arriving from the cerebellar nuclei. Recent evidence reveals that the primary and secondary motor cortical areas each receive their thalamic inputs from separate nuclei, which in turn transmit inputs from different sources. As schematized in Figure 28–10, the VPLo and VLc nuclei relay impulses from the rostral dentate nucleus and from spinothalamic cells to primary motor cortex; nucleus ventralis lateralis oralis (VLo) transmits input from the basal ganglia (globus pallidus and substantia nigra) to the supplementary motor cortex, and nucleus X relays information from the caudal dentate nucleus to the lateral premotor cortex. The existence of separate subcortical input pathways to each of the three motor areas implies that the basal ganglia and caudal dentate nucleus could influence pri-

mary motor cortex only indirectly, via relays through their secondary motor fields.

In addition to these "specific" motor relay nuclei, the thalamus also provides input to motor cortex from "nonspecific" nuclei such as the intralaminar and reticular nuclei; these are thought to regulate the general excitability of cortical neurons (Chap. 15).

Somatic Sensory Input to Motor Cortex

Most cells in motor cortex receive somatic input, which can be readily demonstrated by their responses to peripheral stimulation. In the unanesthetized primate, two thirds of the MI cells respond clearly and consistently to adequate somatic stimulation. Most cells can be driven by passive joint rotation, usually flexion or extension of one or more joints. Other motor cortex cells respond to cutaneous stimulation, such as touching skin or brushing hairs; their receptive fields are similar in size to those of postcentral cortex cells. A few precentral cells receive both deep and cutaneous input. About a third of the precentral neurons show no clear somatosensory responses, although a few respond to visual or auditory stimuli.

Cells encountered in a vertical electrode penetration through the motor cortex tend to respond to input of the same modality (deep or cutaneous) and to receive input from the same joint or overlapping receptive fields. This tendency for motor

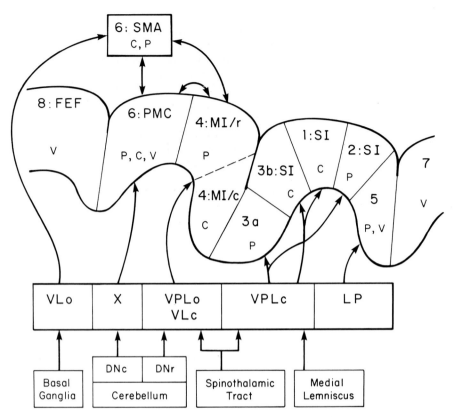

Figure 28–10 Schematic sagittal section through pre- and postcentral cortex, showing Brodmann's cytoarchitectonic areas and their thalamic inputs. The predominant sensory modality represented in each area is symbolized by C (cutaneous), P (proprioceptive), or V (visual). Note the separate thalamocortical pathways to the three motor cortical areas. Primary motor cortex (MI) is connected reciprocally with ventralis posterolateralis oralis (VPLo) and ventralis lateralis caudalis (VLc); supplementary motor area (SMA), with ventralis lateralis oralis (VLo); and premotor cortex (PMC), with area X. Input from cerebellum is relayed via the caudal and rostral components of the dentate nucleus (DNc and DNr, respectively). (After Jones, E. G. In Jones, E. G.; Peters, A., eds. *Cerebral Cortex*, vol. 5, 113–114. New York, Plenum Press, 1986. Schell, G. R.; Strick, P. L. *J. Neurosci.* 4:539–560, 1984.)

cortex cells in a vertical "column" to have similar input resembles the columnar organization of sensory cortex cells and is readily understood as a consequence of the radial distribution of the afferent axons.

The input from peripheral receptors is distributed in motor cortex in a somatotopic map, which is in register with the motor output map shown in Figure 28–4. This map summarizes the source of somatic input to most neurons at each cortical site; however, cells with inputs from other regions and modalities are considerably intermingled in motor cortex—more so than in sensory cortex. This heterogeneity of inputs has allowed investigators to emphasize different features of the distribution of cells with sensory inputs in motor cortex. For example, in awake macaques, Murphy and colleagues[22] found that neurons driven by stimulation of different forearm regions are distributed in a set of roughly concentric rings (Fig. 28–11). In the central region most cells respond to input from the fingers; surrounding rings contain cells responsive to wrist, elbow, and shoulder. These groupings are by no means exclusive, since neurons receiving input from adjacent joints were substantially intermingled.

In monkeys, the inputs from cutaneous and deep receptors project to two separate regions of area 4. Neurons in a rostral zone of precentral cortex respond predominantly to passive joint movements, while cells in a more caudal zone respond primarily to cutaneous input.[32, 33] This segregation of proprioceptive (p) cells rostrally and cutaneous (c) cells caudally is illustrated in the schematic sagittal section through pre- and postcentral cortex in Figure 28–10. This figure also indicates the predominant somatic modality represented in each Brodmann's area. Primary motor and sensory cortical areas are composed of alternating zones with predominantly proprioceptive and cutaneous input, and are bounded rostrally and caudally by regions that receive visual input (v).

Input-Output Relations

Woolsey's finding[37] that the somatotopic map of peripheral input to motor cortex was in broad register with the map of output effects evoked by cortical stimulation suggested a close functional relationship between input and output, which has been confirmed on the level of single neurons.

Using extracellular microelectrodes Rosen and Asanuma[29] compared the sensory responses of individual cortical neurons in the precentral hand area of the squirrel monkey with the movements evoked by microstimulation at the same sites. Figure 28–12 summarizes their results for a series of penetrations in the thumb area. Microstimulation evoked different thumb movements at different sites; the low-threshold sites for evoking flexion were distributed in an output zone oriented perpendicular to the cortical surface. Similar zones could be demonstrated for extension, adduction, and abduction. The sensory responses of neurons recorded in each zone were closely related to the output; for example, cutaneous receptive fields were located on the part of the thumb lying in the

direction of the movement evoked by the stimulation. Similarly, cells driven by passive thumb extension were found in the zone whose stimulation evoked extension. Similar close relations between the sensory responses of cells and the movement evoked by microstimulation were also observed for the concentric regions illustrated in Figure 28–11.

These input-output relations have a simple functional interpretation. The cutaneous input from receptors that would be directly activated by movement could function to assist in grasp reflexes and placing reflexes by increasing activity of the relevant output cells; indeed, the motor cortex plays an important role in such reflexes. Similarly, the proprioceptive input from stretch

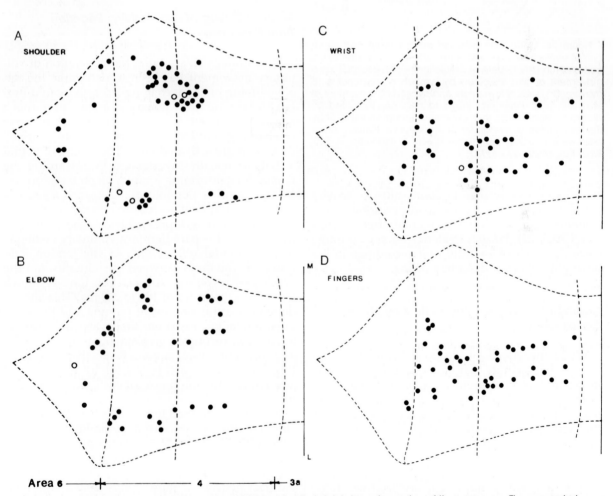

Figure 28–11 Distribution of sensory input from different forelimb joints in motor cortex of the macaque. The precentral gyrus was unfolded onto a flat plane, and boundaries between Brodmann's areas represented by vertical dashed lines. M = medial; L = lateral. Many points also represent output effects on the same joints evoked by microstimulation. (From Murphy, J. T.; Kwan, H. C.; MacKay, W. A.; Wong, Y. C. *J. Neurophysiol.* 41:1120–1131, 1978.)

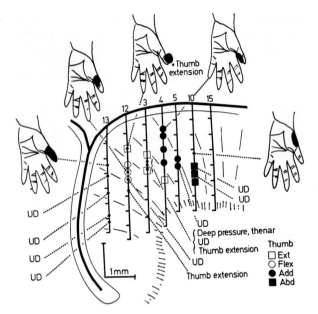

Figure 28–12 Relation between sensory input and motor output evoked by intracortical microstimulation in precentral cortex of monkey. Repetitive stimulation at 5 μa evoked thumb movement at sites indicated by symbols corresponding to movements listed at lower right and evoked no movement at sites with bars. Peripheral inputs to cells are indicated by diagrams representing their cutaneous receptive fields, or by the description of adequate proprioceptive stimulus; some cells were undriven (UD). (From Rosen, I.; Asanuma, H. *Exp. Brain Res.* 14:257–273, 1972.)

receptors to cells that facilitate muscle activity could function to overcome load perturbations in a manner similar to the segmental stretch reflex (see Chap. 24). Further evidence for such a transcortical stretch reflex from unit recordings in behaving monkeys is described below.

Coding of Movement Parameters by Motor Cortex Cells

While the effects of electrical stimulation and lesions demonstrate that motor cortex plays a role in controlling movements, these techniques cannot reveal which parameters of movement it controls. The movement variables that are coded in the activity of precentral cortex cells can be investigated only by observing neuronal activity in awake, behaving animals. In such "chronic unit recording" studies, the activity of single neurons is recorded with a movable microelectrode. Relevant behavioral responses are trained through operant conditioning techniques, which gradually "shape" the animal's behavior by selectively re-

warding those responses that are closest to the desired final behavior. Thus, chronic recording studies can document the activity of cells under particular behavioral conditions designed to test hypotheses about the functions of the recorded neurons. For example, a repeatable movement in response to sensory signals is ideal for investigating the relative timing of cell activity involved in generating a simple voluntary response. On the other hand, to determine whether motor cortex cells are preferentially involved in coding limb position or force one can train the animal to move different loads through the same displacement—a task designed to dissociate the force and displacement. Experiments can also be designed to demonstrate the neural changes involved in preparing to make a particular movement.

Relative Timing of Cell Activity: Reaction Time Responses

One strategy in studying the generation of voluntary movement has been to determine the relative timing of changes in neural activity in different motor structures when the animal performs a simple movement. Those structures with neurons discharging earlier could presumably "drive" those that are activated later. The timing of neural activity is usually investigated by rewarding the animal for performing a rapid, stereotyped movement as soon as possible after detecting a sensory cue. An example of such a *reaction-time* task is the rapid release of a depressed lever when a light is turned on. This paradigm is a voluntary analogue of a segmental reflex, such as the tendon jerk; however, the delay between the stimulus and the behavioral response—the reaction time—is much longer, on the order of 100 to 200 ms. This latency depends on the amount of training and the stimulus modality. Reaction times are increasingly longer for responses evoked by proprioceptive, auditory, and visual stimuli, respectively. Some of this difference is accounted for by the different afferent conduction times for each sensory modality.

In monkeys trained to release a bar after a visual stimulus, many precentral pyramidal tract cells change their activity 10 to 100 ms prior to electromyographic (EMG) activity in the agonist wrist muscles. The changes in activity of these pyramidal tract neurons is more tightly linked with initiation of muscle activity than with occurrence of the light, indicating that they are more related to the movement than the sensory cue.[10]

The routing of neural impulses between a reaction time stimulus and the voluntary response remains largely unresolved. Experiments using the reaction time task have not revealed a simple sequential activation of successive centers. The problems are illustrated by experiments designed to determine whether cerebellar cells might precede activation of motor cortex neurons. Figure 28–13 shows that the onset times of neural discharge in the cerebellar nuclei and motor cortex during the same motor responses are distributed over several hundred milliseconds and exhibit considerable overlap. Similar overlapping distributions characterize the onset of activity of cells in other motor regions. Thus, serial activation of different motor centers cannot be established by measuring neural onset times, because the cells within each region—including motoneuron pools—are recruited over times much longer than the conduction times between regions. Moreover, the duration of most movements greatly exceeds

the conduction times, allowing any recurrent loops between regions to be traversed repeatedly during a single movement. Consequently, the appealing notion that initiation of movement involves sequential activation of cells in hierarchically related centers has proved difficult to confirm experimentally. Indeed, the distributions illustrated in Figure 28–13 suggest that cells in different regions are recruited more or less in parallel rather than in a strictly serial manner.

Relation to Active Force

The activity of motor cortex cells during movements could potentially code a variety of movement parameters. To determine whether activity of pyramidal tract neurons is more strongly related to limb position or to the force required to move the limb, Evarts[10] trained monkeys to move a handle through the same displacement while lifting different loads. The discharge of most motor cortex cells was more closely related to the force exerted or to changes in force than to displacement of the wrist. Similarly, when the monkey was required to hold the handle steady against different externally applied forces, pyramidal tract neurons again discharged in proportion to the isometric force exerted.

The fact that activity of motor cortex neurons covaries with force is still not sufficient evidence to prove that they are causally involved in generating force. Such a causal linkage can be demonstrated by showing that some of the pyramidal tract neurons also have excitatory effects on agonist muscles.[12] Such effects have been demonstrated by the technique of spike-triggered averaging, in which the action potentials of a cortical neuron are used to trigger a computer that averages the muscle (EMG) activity occurring in the time interval after the spike. Some pyramidal tract neurons produce a postspike increase in average activity of co-activated limb muscles, as shown in Figure 28–14C. The magnitude and latency of this "postspike facilitation" suggest that this neuron is a corticomotoneuronal (CM) cell and that the extensor carpi ulnaris (ECU) is one of its target muscles. Typically, the discharge of CM cells facilitates several co-activated muscles, indicating that they exert a divergent influence on a "muscle field."

The discharge of a CM cell facilitates the activity of its target muscles in proportion to its firing rate. Figure 28–14 illustrates the activity of a typical CM cell during wrist movements against an elastic load (which requires a force proportional to dis-

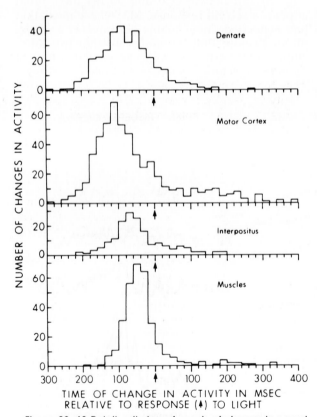

Figure 28–13 Relative timing of onsets of change in neural discharge in motor cortex, cerebellum, and forelimb muscles during active wrist movements in a monkey. Although the distributions show differences in timing, they overlap extensively. (From Thach, W. T. *J. Neurophysiol.* 41:654–674, 1978.)

placement). The average of many extension responses (Fig. 28–14*B*) indicates that this CM cell fires with a phasic burst of activity at movement onset, when extra force is required to overcome inertial and elastic loads. The initial phasic discharge of CM cells begins about 80 ms before activity of their target muscles. This lead time is considerably longer than the latency of the postspike facilitation, which peaks about 10 ms after the spike; therefore, much of this early discharge helps bring the motoneurons to firing threshold. The initial burst is followed by a tonic discharge during the period that the monkey exerts a constant force. The tonic firing rate of CM cells increases with the level of maintained force.[6] Besides facilitating their target muscles, the discharge of some CM cells also exerts inhibitory effects on antagonists of the target muscles.

While motor cortex output cells discharge in proportion to active force, this relationship is not invariant in all behavioral conditions. Pyramidal tract neurons fire more strongly in relation to finely controlled wrist and finger movements than to rapid ballistic movements.[13] Similarly, CM cells that facilitate finger muscles are more active during a precision grip task than during a power grip task, although in the latter condition their target

muscles are even more active.[21] These observations are consistent with the behavioral effects of lesions, which suggest that motor cortex cells are particularly important in controlling fine finger movements; these neuronal recordings indicate further that pyramidal tract neurons are much less involved in rapid forceful movements.

Load Compensation Responses

The accurate control of voluntary movement involves a continual balance between centrally originating command signals and sensory feedback from the periphery. When a movement is suddenly resisted by an unexpected increase in the load, several neural mechanisms are activated to increase the motor output and overcome the perturbation. The stretch of an active muscle by a sudden load increase produces a series of muscle responses. The initial electromyographic response, called M1, is mediated by the segmental stretch reflex and has a latency of 25 to 30 ms in man (Fig. 28–15). Between 50 and 90 ms there often appears a second response, M2, whose mediation has generated considerable controversy and experimentation. Phillips suggested that this long-latency EMG response could be mediated by CM

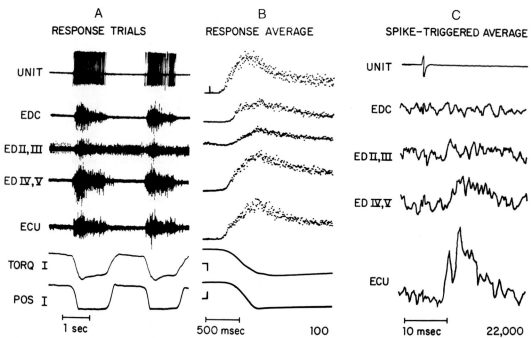

Figure 28–14 Response of motor cortex output cell during alternating flexion and extension of wrist. *A*, From top, activity of cortical unit and coactivated extensor muscles, wrist torque, and position. *B*, Averages of activity and movement parameters for 100 extension responses. *C*, Spike-triggered averages of rectified electromyographic (EMG) activity show postspike facilitation of extensor carpi ulnaris (ECU). (From Fetz, E. E.; Cheney, P. D. *J. Neurophysiol.* 44:751–774, 1980.)

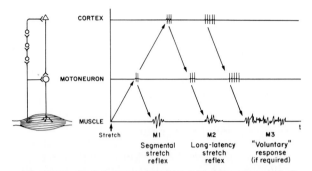

Figure 28–15 Schematic diagram of pathways mediating segmental and transcortical contributions to electromyographic (EMG) responses evoked by muscle stretch.

tional burst when a transient load increase opposed flexion (B); moreover, the cell's activity dropped following a load decrease, which assisted the flexion (C).[8] Such responses in CM cells would produce changes in muscle activity that would help to overcome the load change. While motor cortex output clearly contributes to the "functional stretch reflex," other pathways and mechanisms may also be involved. Finally, subjects prepared to resist the perturbation often show a still later response (M3), which is associated with a voluntary muscle contraction (Fig. 28–15).

Preparatory Set

Prior to execution of a voluntary movement, cortical cell activity may also change in preparation to make the response. For example, a driver waiting at a red light is prepared (or set) to step on the gas pedal when the light changes. Experiments designed to investigate the neural mechanisms involved in such a "motor set" have typically involved monkeys trained to respond after two successive stimuli: the first is an instructional stimulus which indicates the type of movement to be

cells that form part of a transcortical stretch reflex, analogous to the segmental reflex.[25b] Indeed, in monkeys generating active limb movements, unexpected load increases that stretch the agonist muscle evoke excitatory responses in task-related pyramidal tract neurons. For example, the pyramidal tract neuron in Figure 28–16 fired with active elbow flexion (A) and responded with an addi-

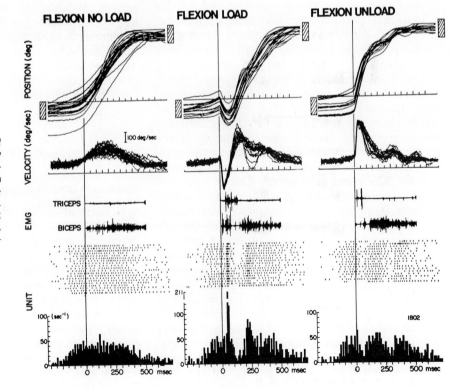

Figure 28–16 Responses of motor cortex cell and elbow muscles in a monkey during normal elbow flexion *(A)* and with brief increases *(B)* and decreases *(C)* of the flexion load. (From Conrad, B.; Matsunami, K.; Meyer-Loman, J.; Wiesendanger, M.; Brooks, V. B. *Brain Res.* 71:507–514, 1974.)

made to a subsequent trigger stimulus. For example, Evarts and Tanji[11] used a red light to instruct the monkey to pull a handle when the handle was subsequently perturbed, and a green light to instruct the monkey to push the handle. During the long delay between the visual instruction stimulus and the proprioceptive trigger stimulus caused by the handle perturbation, many motor cortex cells changed their firing rates, suggesting that they were involved in preparing to make the correct movement. Moreover, the response of many pyramidal tract neurons to the perturbation of the handle depended on which movement the monkey was prepared to make, suggesting that the peripheral input to these cells was also influenced by the set. The changes in the proprioceptive responses to the perturbation were usually those that would assist in the subsequent movement.

Coding of Movement Parameters by Population Responses

While most neuronal recording experiments to date have concerned the responses of single neurons in relation to movement parameters, the generation of a movement clearly involves large populations of cells. Thus the coding of response parameters must involve the coordinated activity of neuronal populations. Recording simultaneously from a group of motor cortex neurons in monkeys making wrist movements, Humphrey and colleagues[15] found that an appropriately weighted average of their firing rates could match the time course of force, and the population average matched force better than the discharge of any single neuron. Moreover, by using different sets of weighting factors, the discharges of the same cells could be made to match also the position trajectory, or the rates of change of position or force. The weighting factors derived from one set of movements would predict the time course of these parameters in subsequent movements. The match between the predicted and the observed movement parameters improved with the number of task-related motor cortex cells included in the weighted average. Figure 28–17C illustrates the firing rates of five motor cortex cells during one cycle of wrist movement. Figure 28–17D shows that the fit between the weighted unit activity and the time course of the force improves as more neurons are included in the weighted average (*bottom to top*). This happens because each additional cell is added with a weight that would further decrease the remaining error. More re-markably, with different weighting factors, the activity of the same population could be used to match a variety of different movement parameters.

The discharge patterns of neuronal populations have also been used to match the direction of limb movement in space by a kind of vector addition. Georgopoulos and colleagues[14] trained monkeys to move a handle from the center of a circle to any of 12 equidistant points on its circumference. They found that many motor cortex neurons fired with movements in several directions, but that neurons usually showed the greatest average activity with movement in one "preferred" direction. They assigned a vector to each neuron that pointed in its preferred direction. Assuming that for every direction of movement, the neuron made a vector contribution (in its preferred direction) that was proportional to its mean firing rate during that movement, they found that the sum of the vectors of the neuronal population pointed in the direction of the limb movements. Again, the direction predicted by the vector sum of the population discharge became more accurate as more cells were included.

These studies indicate that close correlates of various movement parameters may be derived from the activity of neuronal populations. Moreover, the appropriate function of the population activity matches the mechanical parameters with increasing accuracy as the number of cells included is increased. Although such functions of population discharge can provide good descriptive matches with movement variables, this does not provide a causal explanation of how the nervous system controls these parameters.

Effects of Motor Cortex Lesions

Spasticity and Paralysis

Lesions of motor cortex in humans may result from cortical strokes, missile wounds, or surgical excisions required to treat epilepsy or to remove tumors. Damage to primary motor cortex initially produces flaccid paralysis of the body parts represented in the affected region. Within one to two weeks the ability to move the proximal joints is often regained. At the same time spasticity appears, becoming most severe and permanent in the distal muscles; spastic muscles are hyperactive and show exaggerated stretch reflexes. The most severe and long-lasting deficits occur in the extensors of the wrist and fingers.

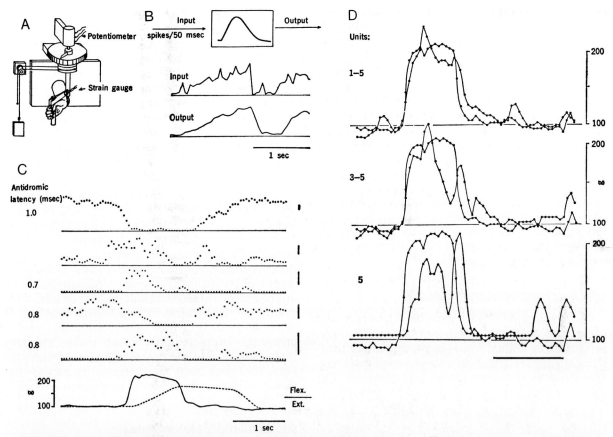

Figure 28–17 Population coding of movement parameters. *A*, Monkey's task involved alternating wrist movements against a constant load. *B*, Firing rate of cortical neurons was smoothed by a filter as shown. *C*, Smoothed firing rates of five simultaneously recorded motor cortex cells during a single trial (bars = 10 spiles/sec); the antidromic latencies of the four pyramidal tract neurons are given at left. Bottom traces show force (solid line) and displacement (dashed line). *D*, Matches between the force trace and the weighted averages of different numbers of cells: *Bottom*, single cell (No. 5); *middle*, three cells (Nos. 3–5); *top*, all five cells (Nos. 1–5). (From Humphrey, D. R.; Schmidt, E. M.; Thompson, W. D. *Science* 170:758–762. Copyright 1970 by the AAAs.)

Damage to primary motor cortex also impairs the ability to move joints independently. For example, after a lesion in the foot area, the ankle typically cannot be extended without concomitant flexion of the knee. Also, after damage to the precentral hand area, the fingers can be flexed only as a group. These muscle "synergies" permanently replace the normal independent control of joints.

In contrast with clinical cases, which often involve additional complications, experimental ablation or cooling studies in animals provide more controlled conditions for testing cortical function.[9] The effects of cortical ablation are less severe in carnivores than in primates. After total decortication, a cat or dog can regain the ability to stand and walk, although it permanently loses tactile placing reflexes (the ability to correctly place the

paw on a surface following tactile contact) and hopping reactions (the ability to reposition the paw when moving relative to the surface). In carnivores, subcortical centers are sufficient to generate righting responses and locomotion, but the cortex is necessary to perform learned responses appropriate to the external environment.

Primates are more severely and permanently affected by cortical damage. Ablation of the precentral hand area in the chimpanzee initially produces complete paralysis of the hand. After a month, crude grasping responses return, but the movements remain slow and inept and the fingers show persistent spastic flexion. Removal of the entire motor cortex produces maximal paralysis and spasticity, with hypertonia in flexors of the arm and extensors of the leg. In higher primates (apes and humans), lesions of motor cortex also produce a positive Babinski reaction. While tactile stimulation of the foot pad normally evokes plantar (downward) flexion of the toes, after lesions of precentral cortex or the pyramidal tract the same stimulus evokes upward movement of the foot and toes. Babinski's sign has become a classic clinical test for damage to the pyramidal system.

Recovery from cortical ablation depends upon several factors. In primates the consequences of cortical removal can be considerably less severe if the cortex is ablated in stages which are separated by months or years. Following such gradual decortication, some monkeys recover the ability to make rudimentary movements. Also, recovery from cortical injury is considerably greater when the damage occurs early in life. An infant monkey deprived of its precentral cortex compensates almost completely for the loss.

Effects of Pyramidal Tract Lesions

The corticospinal tract may be selectively and completely sectioned at the medullary pyramids, producing an animal whose remaining motor functions are subserved by the extrapyramidal system. The behavioral consequences of pyramidal tract section were extensively described by Tower:[2]

The most conspicuous result of unilateral pyramidal lesion in the monkey is diminished general usage and loss of initiative in the opposite extremities. . . . When both sides are free to act, initiative of almost every sort is delegated to the normal side, but if the normal side is restrained, the affected side can, with sufficiently strong excitation, be brought to act.

The affected side shows a paresis—a loss of fine control of movement, particularly in the distal muscles. The skillful use of the hand in precise movements is entirely lost. Although the limb can still be used in postural activity and in reaching and grasping, the movements are clumsy and require considerable attention and effort.

Unilateral pyramidal section also produces hypotonia in muscles of the opposite side. Contralateral muscles exhibit abnormally low resistance to passive stretch and slowed tendon reflexes. Reflexes evoked by cutaneous stimulation also have higher thresholds. Reactions to pinprick and abdominal reflexes either are abolished or are harder to elicit and slower. Contact placing is abolished and proprioceptive placing is more difficult to elicit. The ability to reach for and grasp objects remains but is much clumsier. Also, the ability to manipulate objects and release a grasp is impaired.

Pyramidal lesions in the chimpanzee produce similar but more severe deficits. Discrete control of the digits is profoundly impaired, and the grasp reflex becomes so hyperactive that it is impossible to release objects. The chimpanzee, like humans, shows the classic Babinski sign after pyramidal section.

MOTOR FUNCTIONS OF OTHER CORTICAL AREAS

A goal-directed movement, such as reaching for an object, requires that the activities of many muscles be coordinated in a manner appropriate to environmental conditions. While primary motor cortex is involved mainly in *executing* movements, other cortical areas are more involved in the prior stages of *programming* the limb movements to be appropriate for the particular context or goal. Programming the patterns of muscle activity is largely accomplished by three interconnected cortical areas: the supplementary motor cortex, the premotor cortex, and the posterior parietal cortex (see Figs. 28–1 and 28–9). Each region appears to specialize in different aspects of movement control.[36]

Supplementary Motor Cortex

The supplementary motor area (SMA) lies anterior and medial to primary motor cortex, largely in the medial surface of the hemisphere. Movements evoked by electrical stimulation of the SMA require higher intensities and longer stimulus trains than those evoked from precentral cortex. Like MI, the SMA is somatotopically organized, with the head anterior and hindlimb posterior. The evoked movements are often more complex and prolonged than the simple muscle contractions obtained from precentral cortex. For example, evoked responses may involve orientation of the limb or body, or coordinated movements such as opening the hand. These movements may outlast the duration of the stimulus and sometimes are elicited on both sides of the body. Neurosurgeons stimulating the SMA in awake patients have evoked vocalization with associated facial movements, coordinated movements of the limbs, and also inhibition of voluntary movements.

Excision of the SMA in human patients has resulted in transient speech deficits, or aphasias, which typically disappear after several weeks. The loss of the SMA also results in a persistent slowness in generating repetitive movements. Lesions of the SMA in monkeys has produced interesting apraxias of reaching and bimanual coordination. Monkeys with unilateral ablation of the SMA cannot coordinate their hands in a bimanual task. Faced with a horizontal plastic plate containing holes stuffed with raisins, a normal monkey quickly presses the raisins out with one hand and catches them with the other hand cupped below. After a unilateral lesion of the SMA the two hands cannot be coordinated independently, but instead move together in a similar manner, as if the intact SMA now controlled both hands. Sectioning the corpus callosum abolishes the bimanual deficit, suggesting that each SMA normally communicates with both hemispheres.[3]

Combined unilateral lesions of the SMA and premotor cortex abolish the ability of a monkey to reach around a transparent barrier to obtain a visible slice of apple with its contralateral hand.[19] Instead, the monkey persists in reaching for the apple directly and repeatedly hits the plastic plate. The same monkey has no problems performing the task using its other hand, indicating that its comprehension and motivation are intact. This apraxia lasts for at least two years. In contrast, a monkey with a lesion of primary motor cortex can reach around such a barrier, albeit clumsily.

Single unit recordings in the SMA of conscious monkeys also suggest that SMA cells may play a role in coordinating movements of the two limbs. Many cells fire in a similar way for comparable movements of the ipsilateral and contralateral arm. Such bilateral responses are found for cells related to distal as well as proximal joint movements. Some SMA cells respond to somatic stimulation; these are often driven by manipulation of multiple joints or bilateral somatic stimuli.

That the SMA is involved in programming sequences of movements has been revealed by measuring cerebral blood flow in conscious human subjects. Increases of cortical neural activity is correlated with localized increases in blood flow, which can be detected by monitoring the circulation of radioactive xenon with an array of radiation detectors around the scalp.[28] Subjects performing the simple motor task of squeezing a spring between the thumb and forefinger showed clear increases in blood flow in both the primary motor and sensory cortex contralateral to the active hand. When they performed a more complex sequence of finger movements, touching the thumb successively with each of the other fingers in turn, their blood flow increased over the SMA bilaterally as well as over sensorimotor cortex. Furthermore,

when the subjects simply *thought* about the sequential movements without performing them, regional blood flow again increased over the SMA, but not over sensorimotor cortex, suggesting that the SMA is involved in programming complex sequential movements.

Premotor Cortex

The premotor cortex (PMC) lies anterior to primary motor cortex on the lateral surface of the hemisphere in area 6 (see Fig. 28–1). PMC is comparable in size to MI in monkeys, but in humans it is almost six times larger than MI. Unlike MI and the SMA, PMC makes only a minor contribution to the corticospinal tract; instead, its descending output is directed largely to the medullary reticular formation. Electrical stimulation of PMC is much less likely to evoke movements than stimulation of MI, or even of the SMA. The elicited movements often involve the proximal musculature and require much higher stimulus intensities than those evoked from MI or the SMA.

In humans, lesions involving PMC produce weakness in shoulder and hip muscles and difficulty in abduction and elevation of the contralateral arm. Limb movements are also slower, as evidenced by delayed muscle activation. PMC lesions may also produce an inability to move the two arms simultaneously in a coordinated fashion (called movement-kinetic apraxia).

Like some cells in MI, many neurons in PMC discharge when a monkey is preparing to make a particular movement. In the experiment illustrated in Figure 28–18,[35] a monkey was trained to depress the one key of a set that was illuminated (the instructional stimulus), but only after another small light (the trigger stimulus) had been turned on. Most of the task-related cells in PMC, like those in MI, responded during the execution of the movement; others responded to the instructional cues (see also Fig. 31–13). However, certain PMC cells exhibited a sustained increase in discharge throughout the delay between the instructional cue and the trigger stimulus. This set-related discharge was often directionally specific, occurring only when the monkey was preparing to move in a particular direction. This discharge was not simply related to a subliminal excitation of agonist motoneurons, since it typically stopped with the onset of the movement. PMC contains a higher proportion of cells related to motor set than does MI. Similar "set-related" cells also have been observed in the SMA and prefrontal cortex.

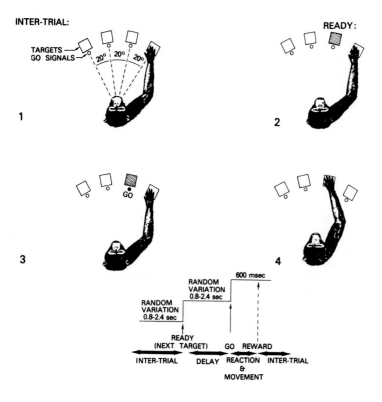

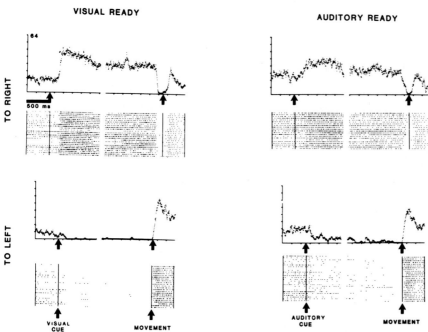

Figure 28–18 Set-related activity in a premotor cortex cell. *Top*: Behavioral paradigm. The monkey rested its hand on one of four keys during the intertrial period (1), observed the ready light appear at another key (2), then at a random time saw a go signal (3) and moved to the illuminated key (4). *Bottom*: Set-related discharge in this premotor cell occurred whenever the monkey was prepared to move to the right, whether the cue was visual (left) or auditory (right). This cell was also inhibited after instructions to move to the left *(bottom)*. (From Weinrich, M.; Wise, S. P. *J. Neurosci.* 2:1329–1345, 1982.)

Frontal Eye Fields

The frontal eye fields lie anterior to the PMC in Brodmann's area 8 (see Fig. 28–1). Efferent projections from the frontal eye fields travel via the internal capsule to pontine regions controlling eye movements and to the superior colliculus, which also is involved in the generation of saccadic eye movements (see also Fig. 20–19).

Electrical stimulation of the frontal eye fields in primates and carnivores typically causes conjugate movement of both eyes to the opposite side; stimulation at some sites may cause the eyes to move obliquely upward or downward, with the horizontal component away from the stimulated side. Stimulation of a particular site in the frontal eye fields evokes saccades with a characteristic direction and amplitude, which are largely independent of the stimulus parameters and the initial position of the eyes in the orbit. Thresholds for evoking saccades are raised if the subject is fixating or tracking a visual target. Stimulation of the frontal eye fields may also evoke head movement to the opposite side and can sometimes produce autonomic responses such as dilatation of both pupils and lacrimation.[2]

Unilateral ablation of area 8 produces a sustained deviation of both eyes toward the side of the lesion and impairs voluntary eye movements to the opposite side. Unilateral lesions of area 8 may also produce deviation of the head to the side of the lesion. Bilateral ablation of the frontal eye fields can impair the ability to gaze laterally in either direction. Although the eyes are capable of following a moving target, they always drift back to the central position. Bilateral ablation can also destroy the ability to attend visually to objects in the peripheral visual field. These deficits are usually temporary, and eye movements often recover within several days after the lesion.

The activity of neurons in the frontal eye fields is related to eye or head movements. Many cells fire before or after voluntary saccades in a given direction.[4] Cells which discharge before voluntary saccades in a particular direction have been recorded at the sites with lowest thresholds for evoking those saccades. While many cells exhibit a burst of firing immediately before the saccade, certain "visuomovement" cells exhibit sustained activity between presentation of a visual target and the saccade to the target, reminiscent of the behavior of certain PMC neurons. Other frontal eye field neurons discharge with gaze in a certain direction, as well as pursuit in this direction. A third type of cell in Brodmann's area 8 is related exclusively to head movements.

Taken together, this evidence suggests that the frontal eye fields are involved in generating voluntary gaze and saccades. However, their role is not completely analogous to that of area 4 for limb movements, since lesions of the frontal eye fields cause only temporary impairment of eye movements. The frontal eye field cells are probably involved in mediating visually evoked saccades, but they apparently exert their influence in conjunction with or through the superior colliculus (see also Chap. 20).

Posterior Parietal Cortex

In addition to the anterior premotor regions described above, the posterior parietal cortex (areas 5 and 7) also appears to be involved in programming limb movements. Humans with lesions of posterior parietal cortex show classic signs of an apraxia—an inability to make directed limb movements in a particular context, in the absence of motor weakness. Their symptoms can be described as a sensory and motor neglect of the opposite hemifield of extrapersonal space. They cannot reach accurately for objects and appear to neglect information from the contralateral hemifield. Such neglect is particularly pronounced with parietal lesions of the nondominant hemisphere (see Chap. 31).

Mountcastle and colleagues[20] have shown that the activity of certain neurons in areas 5 and 7 of monkeys are specifically related to active reaching movements. Some neurons discharged only when the monkey reached for a desirable object, such as food, in its immediate extrapersonal space, but did not fire during similar limb movements that were not directed toward acquiring an object of interest. A related class of cells fired preferentially during active manipulation of objects of interest. Area 7 contains comparable oculomotor cells that fire specifically when the monkey moves its eyes toward an object of interest, but not during spontaneous eye movements. These workers suggested that these posterior parietal neurons may generate a motor command to acquire objects of interest, either manually or visually, in extrapersonal space. (An alternative interpretation is described in the discussion of the parietal lobe in Chap. 31).

DISTRIBUTED CORTICAL MOTOR FUNCTION

This chapter has adopted the view that primary motor cortex is preferentially involved in the execution of movements, whereas secondary motor areas are more involved in programming. While this dichotomy provides a useful framework, the view that these functions are exclusively localized in separate cortical regions must be recognized as overly simplistic. Some of the evidence discussed suggests that generation of movement involves continual interactions between different cortical regions. First, the neural connections between primary and secondary cortical areas are reciprocal, not serial, and each area has its own subcortical input and output connections. Second, although the various motor regions have neurons with response properties consistent with their proposed function, each area actually contains a mixture of cell types, all of which are present to some extent in the other areas. Cortical neurons with similar response properties might be interconnected and act together. Such a distributed representation could explain the recovery of some functions that are temporarily lost after lesions. Finally, the relative timing of cell responses in different regions during a reaction time movement largely overlap, suggesting that all areas are activated more or less simultaneously. Thus, attributing the functions of motor programming and execution to different cortical regions is more a matter of relative degree than an absolute dichotomy. In a similar way, cortical and subcortical centers must also interact continuously and cooperatively in producing movements.

BIBLIOGRAPHY

Recommended Reviews

Evarts, E.V. Role of motor cortex in voluntary movement. *Handbook of Physiology.* 2:1083–1121, 1981.
Review of motor cortex cell activity in relation to movements and transcortical reflexes.

Humphrey, D.R. Cortical control of reaching. In Talbott, R.E.; Humphrey, D.R., eds. *Posture and Movement*, 51–112. New York, Raven Press, 1979.
The roles of secondary motor cortex areas in programming limb movements.

Phillips, C.G.; Porter, R. Corticospinal neurones. Their role in movement. Monographs of the Physiological Society, London, Academic Press, 1977.
A comprehensive review of the role of motor cortex and pyramidal tract in control of movement.

REFERENCES

1. Asanuma, H.; Rosen, I. Topographical organization of cortical efferent zones projecting to different forelimb muscles in the monkey. *Exp. Brain Res.* 14:243–256, 1972.

2. Bucy, P.C., ed. *The Precentral Motor Cortex.* Urbana, Ill., Univ. of Illinois Press, 1949.

3. Brinkman, C. Supplementary motor area of the monkey's cerebral cortex: Short- and long-term deficits after unilateral ablation and the effects of subsequent callosal section. *J. Neurosci.* 4:918–929, 1984.

3a. Brodmann, K. Vergleichende Lokalisationslehre der Grosshirnrinde. Leipzig, J.A. Barth, 1909.

4. Bruce, C.J.; Goldberg, M.E. Physiology of the frontal eye fields. *Trends Neurosci.* 7:436–441, 1984.

5. Chang, H.T.; Ruch, T.C.; Ward, A.A., Jr. Topographic representation of muscles in motor cortex in monkeys. *J. Neurophysiol.* 10:39–56, 1947.

6. Cheney, P.D.; Fetz, E.E. Functional classes of primate corticomotoneuronal cells and their relation to active force. *J. Neurophysiol.* 44:775–791, 1980.

7. Clough, J.F.M.; Kernell, D.; Phillips, C.G. The distribution of monosynaptic excitation from the pyramidal tract and from primary spindle afferents to motoneurones of the baboon's hand and forearm. *J. Physiol. (Lond.)* 198:145–166, 1968.

8. Conrad, B.; Matsunami, K.; Meyer-Loman, J.; Wiesendanger, M.; Brooks, V.B. Cortical load compensation during voluntary elbow movements. *Brain Res.* 71:507–514, 1974.

9. Denny-Brown, D. *The Cerebral Control of Movement.* Springfield, Ill., Charles C Thomas, 1966.

10. Evarts, E.V. Relation of pyramidal tract activity to force exerted during voluntary movement. *J. Neurophysiol.* 31:14–27, 1968.

11. Evarts, E.V.; Tanji, J. Reflex and intended responses in motor cortex pyramidal tract neurons of monkey. *J. Neurophysiol.* 39:1069–1080, 1976.

12. Fetz, E.E.; Cheney, P.D. Postspike facilitation of forelimb muscle activity by primate corticomotoneuronal cells. *J. Neurophysiol.* 44:751–772, 1980.

13. Fromm, C.; Evarts, E.V. Relation of motor cortex neurons to precisely controlled and ballistic movements. *Neurosci. Lett.* 5:259–266, 1977.

14. Georgopoulos, A.P.; Kalaska, J.F.; Crutcher, M.D.; Caminiti, R.; Massey, J.T. The representation of movement direction in the motor cortex: single cell and population studies. In Edelman, G.; Gall, W.E.; Cowan, W.M., eds. *Dynamic Aspects of Neocortical Function*, 501–529. New York, John Wiley & Sons, 1984.

15. Humphrey, D.R.; Schmidt, E.M.; Thompson, W.D. Predicting measures of motor performance from multiple spike trains. *Science* 170:758–762, 1970.

16. Jankowska, E.; Padel, Y.; Tanaka, R. Projections of pyramidal tract cells to α-motoneurones innervating hind-limb muscles in the monkey. *J. Physiol.* 249:637–667, 1975.

17. Jones, E.G. Connectivity of the primate sensory-motor cortex. In Jones, E.G.; Peters, A., eds. *Cerebral Cortex*, vol. 5, 113–184. New York, Plenum Press, 1986.

18. Jones, E.G.; Wise, S.P. Size, laminar and columnar distribution of efferent cells in the sensorimotor cortex of monkeys. *J. Comp. Neurol.* 175:391–438, 1977.

19. Moll, L.; Kuypers, H.G.J.M. Premotor cortical ablations in monkeys: contralateral changes in visually guided reaching behavior. *Science* 198:317–319, 1977.

20. Mountcastle, V.B.; Lynch, J.C.; Georgopoulos, A.; Sakata, H.; Acuna, C. Posterior parietal association cortex of the monkey: command functions for operations within extrapersonal space. *J. Neurophysiol.* 38:871–908, 1975.

21. Muir, R.B.; Lemon, R.N. Corticospinal neurons with a special role in precision grip. *Brain Res.* 261:312–316, 1983.

22. Murphy, J.T.; Kwan, H.C.; MacKay, W.A.; Wong, Y.C. Spatial organization of precentral cortex in awake primates.

III. Input-output coupling. *J. Neurophysiol.* 41:1120–1131, 1978.

23. Olszewski, J. *The Thalamus of the Macaca Mulatta: An Atlas for Use with the Stereotaxic Instrument.* Basel, Karger, 1952.

24. Patton, H.D.; Amassian, V.E. The pyramidal tract: its excitation and functions. *Handbk. Physiol.* 2:837–861, 1960.

25. Penfield, W.; Rasmussen, A.T. *Cerebral Cortex of Man. A Clinical Study of Localization of Function.* New York, Macmillan, 1950.

25a. Phillips, C.G. Motor apparatus of the baboon's hand. *Proc. R. Soc. Lond. [Biol.]* 173:141–174, 1969.

25b. Phillips, C.G.; Porter, R. Cortical spinal neurones. Their role in movement. Monographs of the Physiological Society. London, Academic Press, 1977.

26. Phillips, C.G.; Porter, R. The pyramidal projection to motoneurones of some muscle groups of the baboon's forelimb. *Prog. Brain Res.* 12:222–242, 1964.

27. Porter, R. The corticomotoneuronal component of the pyramidal tract: corticomotoneuronal connections and functions in primates. *Brain Res. Rev.* 10:1–26, 1985.

28. Roland, P.E.; Larsen, B.; Lassen, N.A.; Skinhoj, E. Supplementary motor area and other cortical areas in organization of voluntary movements in man. *J. Neurophysiol.* 24:91–100, 1980.

29. Rosen, I.; Asanuma, H. Peripheral afferent inputs to the forelimb area of the monkey cortex: input-output relations. *Exp. Brain Res.* 14:257–273, 1972.

30. Schell, G.R.; Strick, P.L. The origin of thalamic inputs to the arcuate premotor and supplementary motor areas. *J. Neurosci.* 4:539–560, 1984.

31. Shinoda, Y.; Yokota, J-I.; and Futami, T. Divergent projection of individual corticospinal axons to motoneurons of multiple muscles in the monkey. *Neurosci. Lett.* 23:7–12, 1981.

31a. Stoney, S.D.; Thompson, W.D.; Asanuma, H. Excitation of pyramidal tract cells by intracortical microstimulation: effective extent of stimulating current. *J. Neurophysiol.* 31:659–669, 1968.

32. Strick, P.L.; Preston, J.B. Two representations of the hand in area 4 of a primate. *J. Neurophysiol.* 48:139–159, 1982.

33. Tanji, J.; Wise, S.P. Submodality distribution in sensorimotor cortex of the unanesthetized monkey. *J. Neurophysiol.* 45:467–481, 1981.

34. Thach, W.T. Correlation of neural discharge with pattern and force of muscular activity, joint position and direction of intended next movement in motor cortex and cerebellum. *J. Neurophysiol.* 41:654–676, 1978.

35. Weinrich, M.; Wise, S.P. The premotor cortex of the monkey. *J. Neurosci.* 2:1329–1345, 1982.

36. Wiesendanger, M. Organization of secondary motor areas of the cerebral cortex. *Handbk. Physiol.* 2:1121–1147, 1981.

36a. Wiesendanger, M. The pyramidal tract: its structure and function. In Towe, A.L.; Luschei, E.S., eds. *Handbook of Behavioral Neurophysiology*, vol 5, Motor Coordination, 401–491. New York, Plenum Press, 1981.

37. Woolsey, C.N. Organization of somatic sensory and motor areas of the cerebral cortex. In Harlow, H.F.; Woolsey, C.N., eds. *Biological and Biochemical Bases of Behavior.* Madison, Wis., Univ. of Wisconsin Press, 1958.

38. Woolsey, C.N.; Gorska, T.; Wetzel, A.; Erickson, J.C.; Earls, F.J.; Allman, J.M. Complete unilateral section of the pyramidal tract at the medullary level in *Macaca mulatta. Brain Res.* 40:119–123, 1972.

The Cerebellum

INTRODUCTION

It is clear to the clinician who sees patients with cerebellar disease that the cerebellum exerts a profound influence on the control of posture and movement. Damage to the cerebellum does not produce paralysis or an inability to move. Rather, cerebellar lesions produce an inability to maintain postural stability, to make movements of appropriate speed and amplitude such that the desired target position is reached at the proper time, and to modify motor output appropriately.

Early clinical observations prompted lesion studies that were designed to dissect out the effects of damage to different portions of the cerebellum. These were followed by anatomical and electrophysiological studies of the circuitry of the cerebellum and most recently by combined behavioral and electrophysiological studies of the activity of cerebellar neurons when awake animals make movements under particular behavioral conditions. The anatomical and acute electrophysical studies have given us a detailed picture of the intrinsic "wiring diagram" of the cerebellum, as well as information about the sources and destinations of information entering or leaving the cerebellum. Some special electrophysiological characteristics of cerebellar neurons also have been identified. We are only beginning, however, to identify the information that is coded by the discharge of cerebellar neurons in awake, behaving animals and the behavioral conditions under which this information is changed.

GROSS ANATOMY

The cerebellum is situated in the posterior fossa, caudal to the tentorium, which separates it from the occipital lobe of the cerebral cortex. It is attached to the pons and medulla by three cerebellar peduncles: superior (brachium conjunctivum), middle (brachium pontis), and inferior (restiform body). It is made up of a deeply folded cerebellar cortex wrapped around the connecting white matter and the deep cerebellar nuclei, six groups of cells embedded in the white matter, three on each side of the midline (Fig. 29–1). The pairs of nuclei are the medial (fastigial), intermediate (interpositus), and lateral (dentate). In humans, the intermediate nucleus is divided into the emboliform and globose nuclei.

As illustrated schematically in Figure 29–2, deep transverse fissures in the cerebellar cortex divide it into the *anterior lobe* (anterior to the primary fissure), the *posterior lobe* (between the primary fissure and the posterolateral fissure), and the *flocculonodular lobe* (posterior to the posterolateral fissure). More shallow transverse fissures divide the cortex into lobules and, finally, into folia. Longitudinal indentations, which are especially prominent in the human cerebellum, separate the

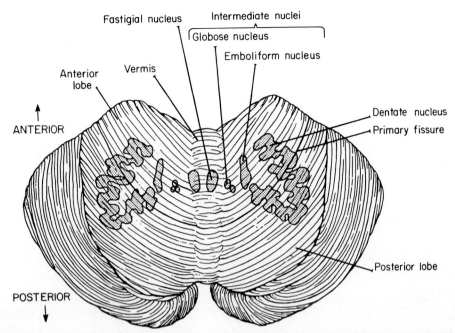

Figure 29–1 A see-through superior view of the human cerebellum. The superficial cerebellar cortex is divided transversely into lobes by major fissures such as the primary fissure. Numerous shallow transverse indentations, shown by the thin lines, divide it into folia. The cerebellar nuclei (fastigial nucleus; intermediate nucleus, divided into the globose and emboliform nuclei; and the dentate nucleus) are situated below the cerebellar cortex, among the fibers that enter and leave the cerebellum. (After Werner, J.K. *Neuroscience: A Clinical Perspective.* Philadelphia, W. B. Saunders, 1980.)

cerebellar hemispheres on each side from a central ridge, the *vermis* (Fig. 29–1). In the vermis and paravermal regions adjacent to it, the lobules are numbered from I to X, beginning anteriorly.

MICROANATOMY

Although the afferent and efferent connections vary for different parts of the cerebellum, the basic neuronal cell types and the laminar organization are the same in each lobe and lobule. Knowledge of the types of cerebellar neurons in each cortical lamina has made it possible to use the electrical signals recorded at different depths to deduce the synaptic relationships and "circuitry" of the cerebellum.[12]

The cerebellar cortex contains five major neuronal cell types that are organized into three layers (Fig. 29–3). The granular layer is the thickest and deepest, lying adjacent to the white matter. Its name is derived from the billions of *granule cells* within this layer, which also contains a smaller number of *Golgi cells*. The outermost layer, the molecular layer, is made up largely of the parallel fibers (bifurcating axons of granule cells) and the dendrites through which they pass. The molecular layer also contains the somata of two neuronal cell types, the *basket* and *stellate cells*. Separating the granular and molecular layers is the *Purkinje cell* layer, which is only one cell deep. The dendrites of Purkinje cells extend toward the cortical surface and divide into an elaborate dendritic tree within the molecular layer. Although this dendritic tree branches profusely, it does so primarily in one plane, creating a dendritic sheet oriented perpendicular to the long axis of the folium and thus to the course of the parallel fibers passing through it.

The Purkinje cell is the only neuron whose axon leaves the cerebellar cortex. Its axon terminates in one of the deep cerebellar nuclei or in the vestibular nuclei of the brain stem (Fig. 29–4).[40] The projection of Purkinje cell axons to nuclear cells tends to be in a longitudinal pattern, with axons of vermal Purkinje cells going primarily to the fastigial nucleus, those of the intermediate cortical zone to the interpositus, and those from the lateral zone of the cerebellar hemispheres to the dentate nucleus. Nuclear neurons also are contacted by afferent fibers entering the cerebellum.

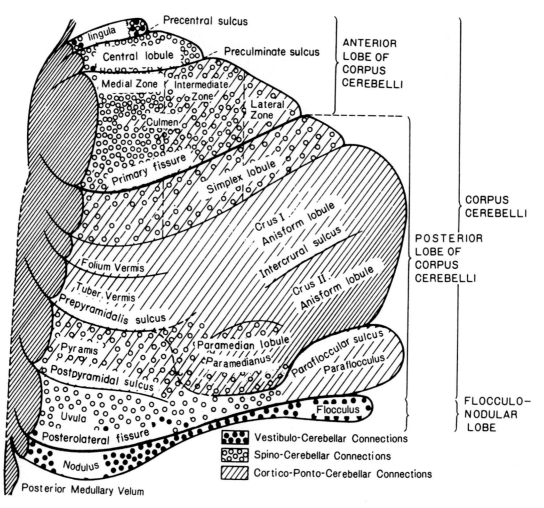

Figure 29–2 Unfolded right half of the cerebellum. The three major lobes, divided by the primary and posterolateral fissures, are labeled at the right. More shallow sulci divide the cortex transversely into lobules. Longitudinal dashed lines depict the three longitudinal zones: medial (vermis), intermediate (paravermal), and lateral. Major areas of termination of afferents from the spinal cord, vestibular apparatus, and pontine nuclei (carrying information from the cerebral cortex) are shown by the symbols. (After Dow, R. S. *Biol. Rev,* 17:179–220, 1942.)

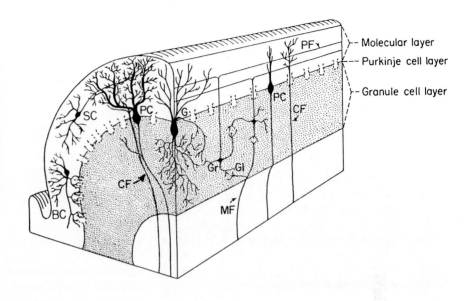

Figure 29–3 A portion of a single cerebellar folium. BC, basket cell; CF, climbing fiber; G, Golgi cell; Gl, glomerulus; Gr, granule cell; MF, mossy fiber; PC, Purkinje cell; PF, Parallel fiber; SC, stellate cell. Parallel fibers course through the dendrites of Golgi, Purkinje, basket, and stellate cells in their course along the transverse length of a folium. Climbing fibers branch in the anterior-posterior plane of the cerebellar cortex to entwine the dendrites of a few Purkinje cells.

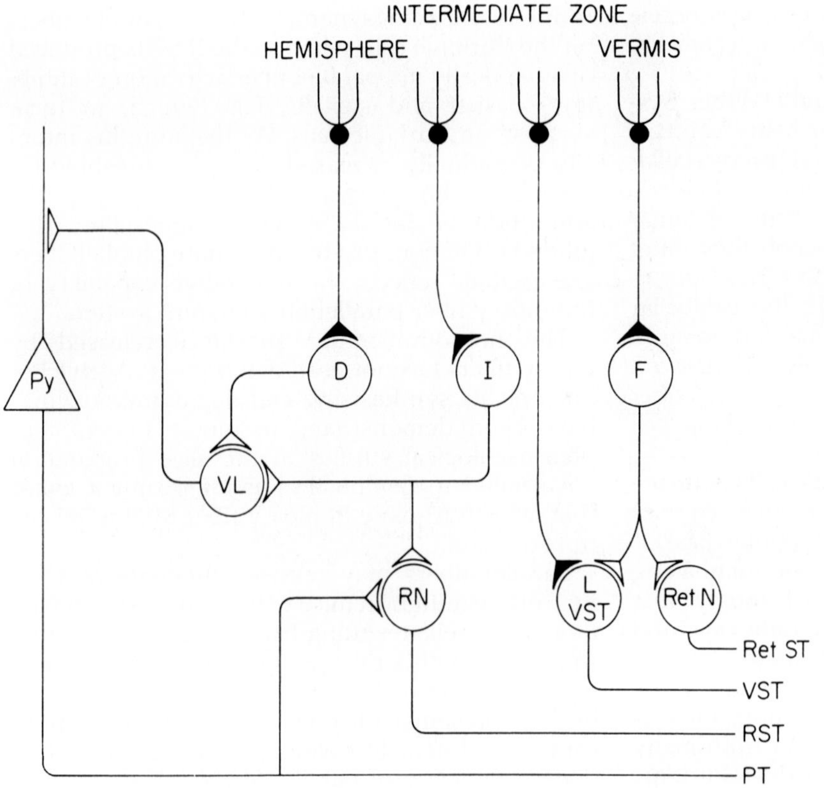

Figure 29–4 Simplified circuit diagram of relationship between cerebellar Purkinje cell output, deep cerebellar nuclei, thalamus, cerebral cortex, and brain stem suprasegmental motor pathways. D, lateral cerebellar nucleus; I, intermediate cerebellar nucleus; F, medial cerebellar nucleus; Py, pyramidal tract neuron in cerebral cortex; VL, ventral lateral nucleus of the thalamus; RN, red nucleus; LVST, lateral vestibular nucleus; Ret N, reticular nuclei; Ret ST, reticulospinal tract; VST, vestibulospinal tract; RST, rubrospinal tract; PT, pyramidal tract.

It is the cells of the deep cerebellar nuclei whose axons leave the cerebellum. Few of these axons reach the spinal cord or cranial nerve motor nuclei directly. Instead, the cerebellum must influence motoneurons primarily through direct or indirect connections to corticospinal, reticulospinal, vestibulospinal, or other systems destined to reach motoneurons (Fig. 29–4). The axons of some projection neurons, especially those from the dentate and interpositus nuclei, ascend to the thalamus, where they synapse on neurons projecting to motor and premotor regions of cerebral cortex. The axons of cerebellar nuclear neurons also form a major input to rubrospinal neurons in the red nucleus. Finally, the axons of many cerebellar nuclear cells, especially those from the interpositus and fastigial nuclei, project to reticulospinal and vestibulospinal neurons in the pons and medulla.

CEREBELLAR AFFERENT SYSTEMS

Information reaches the cerebellum over four distinct afferent systems: the *mossy fiber system*, the *climbing fiber system*, and two *monoaminergic systems*. Since the action of these afferent systems differs, especially at the Purkinje cell, it is necessary to consider the role of each in cerebellar function.

Mossy Fiber–Granule Cell–Parallel Fiber System. Mossy fibers originate from many sources, including the spinal cord, several precerebellar nuclei of the reticular formation, the pontine nuclei, and primary afferent fibers from the vestibular apparatus. Mossy fibers branch when they enter the cerebellum and send axon collaterals both to the deep cerebellar nuclei and to the cerebellar cortex. In the cerebellar cortex, they terminate in the granule cell layer, where they synapse on granule and Golgi cells. It is the parallel fiber axons of granule cells that reach the Purkinje cell dendrites.

The parallel fiber axons of granule cells bifurcate in the molecular layer. Each branch courses through the dendritic trees of Purkinje, basket, and stellate cells, extending for about 3 mm along the long axis of a folium. Since each mossy fiber projects to about 600 different granule cells and each parallel fiber synapses with about 300 Purkinje cells, there is considerable divergence in the mossy fiber afferent system. In addition, there is extensive convergence of information from many mossy fiber afferents on an individual Purkinje cell. Each granule cell receives input from about four mossy fibers, and in the cat, each Purkinje cell is contacted by about 80,000 of the 400,000 parallel fibers that pass through its dendritic tree.[38, 39] This would mean that one Purkinje cell would receive information from over 300,000 mossy fibers.

Mossy fibers excite granule cells, and parallel fibers excite Purkinje cells. The action of parallel fibers can be studied by stimulating a group of them through electrodes positioned on the surface of the folium and recording the response of Purkinje cells situated less than 3 mm down its length (Fig. 29–5A). As shown in Figure 29–5B, local surface stimulation (LOC) produces an initial depolarization (EPSP) followed by a hyperpolarization (IPSP) of the Purkinje cell. The origin of these potentials can be determined by comparing the response produced by stimulating the beam of parallel fibers that directly contacts the dendrites of the Purkinje cell under study with the potentials generated by stimulating a set of parallel fibers situated farther around the curvature of the same folium. The latter set does not pass through the dendritic tree of the Purkinje cell being studied, but it does contact dendrites of basket and stellate cells whose axons reach the Purkinje cell. Using these techniques, it can be shown that the EPSP is due to the monosynaptic action of parallel fibers on the Purkinje cell, whereas the IPSP is produced disynaptically via parallel fiber activation of inhibitory basket and stellate cells, which, in turn, synapse on Purkinje cells. As the stimulus intensity is gradually increased from just threshold for the EPSP (40) to twice that intensity (80), the amplitude of the EPSP grows gradually. This graded excitation, produced as more parallel fibers are excited, reflects the integrative capability of the mossy fiber-parallel fiber afferent system.

The neurotransmitter substance released by mossy fibers has not been determined. Acetylcholine and its synthesizing and degrading enzymes have been demonstrated in mossy fibers,[24] but pharmacological studies in the slice preparation have failed to demonstrate evidence for a major role of acetylcholine in mossy fiber to granule cell transmission.

Parallel fibers may release glutamate as their neurotransmitter. In mice with decreased numbers of granule cells resulting from either mutations or irradiation with x-rays, glutamate levels are much lower than normal, as is the uptake of glutamate by synaptosomal fractions.[43] Iontophoretic application of glutamate excites Purkinje cells, just as does stimulation of parallel fibers.[25] The action of iontophoretically applied glutamate occurs postsynaptically, because it persists after synaptic

Figure 29-5 Experimental analysis of the actions of climbing fibers and parallel fibers on Purkinje cells. *A,* Experimental set-up, with an intracellular recording microelectrode in a Purkinje cell (P) and stimulating electrodes locally on the surface of the same folium (LOC) and in the inferior olivary nucleus (IO) of the contralateral medulla. The LOC electrode is used to excite parallel fibers (PF), which are the axons of granule cells (GC). Climbing fibers (CF) are stimulated at their origin in the inferior olive. MF, mossy fiber; ML, molecular layer; PL, Purkinje cell layer; GL, granule cell layer. *B,* Postsynaptic potentials recorded intracellularly from the Purkinje cell when parallel fibers are stimulated through the LOC electrode with increasing stimulus intensities (40 to 80). Superimposed traces are from successive trials at the same stimulus intensity. *C,* Postsynaptic potentials recorded intracellularly from a Purkinje cell when the inferior olive was stimulated. Superimposed records are from successive trials, in which the stimulus intensity was varied. (*A* and *C* from Eccles, J. C.; Llinas, R.; Sasaki, K. *J. Physiol.* 182:268–296, 1966; *B* from Eccles, J. C.; Llinas, R.; Sasaki, K. *Exp. Brain Res.* 1:161–183, 1966.)

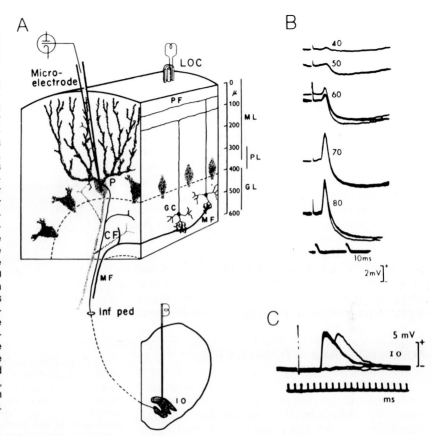

transmission has been blocked by adding magnesium and lowering calcium.[17] Finally, Ca-dependent release of glutamate has been demonstrated in synaptosomal preparations from the molecular layer and in cerebellar slice preparations.[27]

Climbing Fiber System. The second major type of afferent fiber reaching all portions of the cerebellum is the *climbing fiber*. Climbing fibers are the axons of cells in the contralateral inferior olivary nucleus of the medulla. The olivary axon gives a branch to the cerebellar nuclei and has branches to one or more lobules of the cerebellar cortex. Each of the cortical branches ends as a climbing fiber, which branches extensively and makes many synaptic contacts (at least 300 on frog Purkinje cells) on the major branches of the dendritic tree of each of the Purkinje cells that it contacts (Fig. 29-5A). A single Purkinje cell, however, receives terminals from only one climbing fiber axon.

The action of climbing fibers on Purkinje cells can be determined by stimulating the contralateral inferior olive. Fig. 29-5C shows the synaptic potentials in a Purkinje cell that had been damaged by electrode impalement and no longer could

generate action potentials. The large depolarization is the EPSP produced by the climbing fiber input. In contrast with the EPSP produced by parallel fibers, the amplitude of the climbing fiber–induced EPSP is not graded as the stimulus intensity is changed. Instead, the EPSP occurs in an all-or-none manner in the superimposed records from successive trials, as would be expected if it is produced when the single climbing fiber contacting this Purkinje cell either is or is not activated. Multiple peaks occur in the EPSP when there are multiple action potentials in the climbing fiber. The very large number of synaptic contacts made by a single climbing fiber on a Purkinje cell produces an EPSP of much greater magnitude than that caused by the action of a single parallel fiber (compare B and C, Fig. 29-5).

Climbing fibers may release aspartate as their neurotransmitter. Aspartate is transported retrogradely from the cerebellar cortex to the inferior olive by climbing fiber axons, and aspartate levels are reduced in the cerebellar cortex if the inferior olive is destroyed by a disease in humans, inherited olivopontocerebellar atrophy,[41] or by treatment with 3-acetylpyridine, a neurotoxin that acts by inducing the cellular synthesis of abnormal nucleotides.[34]

Monoaminergic Afferent Systems. In addition to the classic climbing fiber and mossy fiber afferent systems described previously, some afferent fibers entering the cerebellum exhibit the fluorescence that characterizes the presence of norepinephrine or serotonin.[18] The adrenergic afferents originate in the locus coeruleus of the pons; the serotonergic fibers come from various brain stem raphe nuclei. Both types of monoaminergic fibers have terminations in both the molecular and granule cell layers of the cerebellar cortex. Iontophoretic administration of norepinephrine reduces the spontaneous discharge rate of Purkinje cells. The hyperpolarization of the Purkinje cell, however, is accompanied by a *decrease*, instead of an increase, in membrane conductance, implicating a mechanism different from that underlying the classic IPSP.[45] (See Chapter 11 on synaptic transmission.)

ACTION POTENTIALS IN PURKINJE CELLS

Purkinje cells, unlike the vast majority of neurons, generate two types of responses to synaptic input. As shown in the intracellular record of Figure 29–6A, one is a simple, single-peaked action potential, the *simple spike*, and the other is a multipeaked action potential, the *complex spike*. By stimulating sources of mossy fibers or the inferior olive, the source of climbing fibers, it has been shown that simple spikes are produced by activation of the mossy fiber–granule cell–parallel fiber pathway, whereas complex spikes are produced by activation of climbing fibers and hence are sometimes called *climbing fiber responses* (CFRs).

The complex and simple spikes result from voltage-gated Ca and Na channels that are distributed along different parts of the cell membrane. As described in Chapter 12, in vitro experiments using a cerebellar slice preparation have demonstrated that action potentials generated in the soma–initial segment region are Na dependent and are blocked by tetrodotoxin. Those generated in the dendrites, however, are Ca dependent and resistant to tetrodotoxin. Because climbing fibers produce a large, synchronous depolarization of the dendrites, they activate the dendritic Ca channels and initiate action potentials there that produce the complex spike. Parallel fibers, however, each produce a small synaptic current that must sum at the initial segment to produce an action potential. As a consequence, parallel fiber input

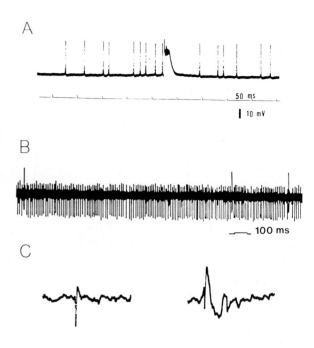

Figure 29–6 Records from cerebellar Purkinje cells. *A,* Intracellular records from an anesthetized cat showing short-duration simple spikes and a single long-duration complex response. Complex responses are evoked by climbing fiber connections; simple spikes, by input arriving over the mossy fiber–granule cell pathway. *B,* Extracellular records from an awake monkey. Simple spikes are primarily negative (downward) and occur at high frequency. Complex spikes have a marked positivity (upward) and occur at a low frequency. *C,* High speed records of a simple spike (left) and a complex spike, or climbing fiber response (right). (*A,* from Martinez, F.; Crill, W. E.; Kennedy, T. T. *J. Neurophysiol.* 34:348–356, 1971. *B* and *C* from Thach, W. T. *J. Neurophysiol.* 31:785–797, 1968.)

leads to Na-based simple spikes whose frequency is graded as a function of the summed synaptic current from many parallel fiber synapses. Regardless of the ionic mechanism by which action potentials are generated in the somadendritic membrane, however, they are conducted along the Purkinje cell axon by conventional Na-dependent spikes.

Why should Purkinje cells have two types of responses to synaptic input? Although this question cannot be answered at present, one hypothesis is that the entry of Ca during complex spikes initiated in the dendrites leads to a long-term effect on the future behavior of Purkinje cells. Calcium, for example, influences both protein phosphorylation and the conductance of certain membrane channels. It is possible, then, that the climbing fiber–induced complex responses could result in long-term changes in Purkinje cell responses to parallel fiber excitation and resulting long-term changes in motor performance.

INTERNEURONS OF THE CEREBELLAR CORTEX

In addition to the direct input from parallel and climbing fibers, Purkinje cell activity also is influenced by the three types of inhibitory interneurons in the cerebellar cortex: *Golgi cells, basket cells,* and *stellate cells* (Fig 29–3). Golgi cells, which are excited by mossy, climbing, and parallel fibers, exert an inhibitory action on granule cells and modulate transmission in the mossy fiber–granule cell pathway. Basket and stellate cells, which are situated in the molecular layer, have dendritic trees perpendicular to the axis of the folium and the course of parallel fibers. Both basket and stellate cells are excited by parallel fibers, and the axons of both cell types inhibit Purkinje cells. Basket cell inhibitory synapses are primarily on Purkinje cell somata and initial segments, whereas stellate cell synapses are on the dendrites. Basket and stellate cell axons run perpendicular to the long axis of the folium, and basket cell axons form axonal basketlike arborizations around approximately six Purkinje cells. Thus activation of a beam of parallel fibers would initially produce excitation of the Purkinje cells contacted by that beam and subsequently produce inhibition, via stellate and basket cells, of those same Purkinje cells *and* their neighbors to each side (recall Figure 29–5B). This type of lateral inhibition could serve to sharpen the spatial distribution of Purkinje cells excited by activity in a beam of parallel fibers.

γ-Aminobutyric acid (GABA) is a likely candidate for the neurotransmitter used by cerebellar inhibitory interneurons,[9] although taurine also has been implicated as a possible transmitter at the stellate cell synapse on Purkinje cells.[36]

PURKINJE CELL–NUCLEAR CELL INTERACTIONS

Purkinje cells inhibit their target neurons, in both the cerebellar and the vestibular nuclei. This important discovery was made by Ito and his colleagues, who stimulated the cerebellar cortex while recording intracellularly from neurons in the lateral vestibular[21] or the deep cerebellar nuclei.[22] Stimulation of the cerebellar cortex produced IPSPs in nuclear cells at short, monosynaptic latencies, and spontaneously occurring action potentials were interrupted during the hyperpolarization.

The pharmacological properties of the Purkinje cell inhibitory synapse on neurons of the lateral vestibular nucleus (LVN) have been studied extensively. Purkinje cell inhibition of LVN neurons is mimicked by the iontophoretic application of GABA, and both Purkinje cell and GABA-induced hyperpolarizations have similar reversal potentials and are blocked by picrotoxin. Immunoreactive techniques have shown that GABA is present in the terminals of Purkinje cell axons, and the GABA concentration in the dorsal portion of the lateral vestibular nucleus, the portion contacted directly by Purkinje axons, is reduced markedly when the cerebellar cortex is ablated chronically.

Some Purkinje cells show immunoreactivity for various other potential neurotransmitters, especially peptides.[6, 8] Although the diversity of immunoreactivity profiles in different Purkinje cells suggests a potential diversity of function, these substances all seem to have inhibitory actions on cells tested pharmacologically in the lateral vestibular nucleus.[7]

THE NATURE OF AFFERENT INFORMATION

The cerebellum has access to information from each sensory modality. In addition, it receives input from areas of the central nervous system, such as the cerebral primary motor cortex, that could provide information about signals produced centrally. Different portions of the cerebellum, however, receive their primary input from different sources. It is this differential distribution of afferent information together with a differential distribution of output from various parts of the cerebellum that leads to the variation in symptoms that can be seen following damage to selective portions of the cerebellum.

Somatosensory Information. Stimulation of the skin, stretch of muscles, or movement of joints produces changes in the discharge of Purkinje cells located primarily in paravermal portions of the anterior and posterior lobes ipsilateral to the site of stimulation. In some Purkinje cells, the stimulus may produce changes in both the simple and complex spike activity. In others, the discharge rate of only simple or complex spikes may be affected.

Information from the hindlimbs and lower trunk reaches the cerebellum directly over mossy fibers from the dorsal and ventral spinocerebellar tracts. The cuneo- and rostral spinocerebellar tracts serve a similar role for information from the upper portion of the body. Axons in the dorsal spinocerebellar (DSCT) and cuneocerebellar tracts are activated either by proprioceptive afferents (group Ia, Ib, and II muscle afferents or joint receptors) or by exteroceptive (cutaneous) afferents. They have relatively discrete receptive fields (e.g., a few interrelated muscles or a small patch of skin). DSCT axons originate from the cells of Clarke's column in the thoracic and lumbar cord. Since primary afferent fibers make very large synaptic contacts on cells in Clarke's column, where they produce unitary EPSPs as large as 5 mV,[26] the firing of DSCT neurons is dominated by peripheral sensory input.

In contrast, axons in the ventral spinocerebellar tract (VSCT) and its forelimb equivalent, the rostral spinocerebellar tract, have larger receptive fields and receive convergent input from both proprioceptive and exteroceptive afferents. Furthermore, they also receive input from descending systems such as the vestibulospinal or corticospinal tracts. A major input to VSCT cells comes from spinal interneurons that receive convergent segmental and descending inputs and also project to spinal motoneurons. It has been suggested, therefore, that the ventral and rostral spinocerebellar systems signal the general state of spinal interneuronal pools.[31]

In addition to the direct spinocerebellar paths to the cerebellum, somatosensory information carried by mossy fibers also reaches the cerebellum by indirect routes, primarily through the reticular formation. The paramedian and lateral reticular nuclei in the medulla carry information from exteroceptive and proprioceptive afferents. In addition, however, the same reticular neurons also receive input from the cerebral cortex. As a consequence, somatosensory information reaching the cerebellum through mossy fiber axons from the reticular nuclei not only arrives more indirectly than does that from the spinocerebellar tracts; it also has been integrated with information from other sources.

Somatosensory information also reaches the anterior and posterior lobes of the cerebellum through climbing fibers. In particular, cells in the dorsal and medial accessory olivary nuclei are activated by somatic stimulation. Olivary neurons can be quite sensitive to externally imposed stimuli; complex spikes in Purkinje cells of the anterior lobe of cats decerebrated or anesthetized with sodium pentobarbital may be produced by joint displacements as small as 50 μm.[44] However, olivary neurons in alert animals are much less sensitive to somatic stimulation produced as a result of the animal's own movements than they are to passively imposed stimuli.[14] It will be of interest to investigate further whether, during active movement, climbing fibers provide a signal that denotes a discrepancy between the expected and the actual somatic stimulus.

Within the anterior and rostral posterior lobes, there is a general somatotopic organization of information from different portions of the body. Stimulation of hindlimb afferents affects rostral portions of the anterior lobe, whereas information from forelimb afferents is distributed more caudally, and from facial afferents, still more caudally. Within this general organization, however, there is considerable irregularity in the mapping of contiguous portions of the body onto the cerebellar cortex.

Auditory, Visual, and Vestibular Information. Auditory or visual stimuli produce changes in the discharge of Purkinje cells in vermal portions of the cerebellar cortex, especially in lobules VI and VII of the posterior lobe. At least some of the auditory input comes over mossy fibers from the cochlear nuclei; primary cerebral auditory cortex also may contribute via the pontine nuclei. Purkinje cell responses are not particularly sensitive to the frequency of the auditory stimulus; rather, they reflect information about stimulus location or movement.[1, 3]

Visual information reaches lobules VI and VII largely through mossy fibers from the pontine nuclei. Pontine cells receive visual input both from visual areas of the cerebral cortex and from superficial layers of the superior colliculus. In the cat, pontine visual neurons have large receptive fields, often greater than 20°, and respond best to moving stimuli.[16] Climbing fibers to lobules VI and VII also mediate visual information, which may reach the inferior olive via the superior colliculus or the pretectum.[51]

Although some vestibular information also reaches the cerebellar vermis, the major destination of vestibular information is the so-called vestibulocerebellum, which consists of the flocculus, nodulus, uvula, and portions of the paraflocculus and lingula. The vestibulocerebellum receives mossy fiber input directly from primary vestibular afferent fibers, as well as indirectly from neurons in the vestibular nuclei of the brain stem. These afferents are excited by movement of the head. Those from the semicircular canals relay signals proportional to angular head velocity, whereas those from the otolith organs primarily signal head position (details in Chapter 27). In addition, climbing fiber responses can be evoked in Purkinje cells by vestibular stimulation.[42]

A major climbing fiber input to neurons in the flocculus is visual. This originates from olivary neurons in the dorsal cap of Kooy, whose discharge is affected by visual stimuli via neurons in the pretectal nuclei. The most effective visual stimuli are complex visual patterns moving slowly across large portions of the visual field.[4] As described in Chapter 27, this visually derived information may be especially important in adjusting eye movement during the vestibulo-ocular reflex.

Information from the Cerebral Cortex. In higher vertebrates, especially humans, the major source of afferents to the cerebellum is the pontine nuclei and the major inputs to the pontine nuclei come from the ipsilateral cerebral cortex. Although pontocerebellar fibers distribute to most portions of the cerebellar cortex, they are the primary source of mossy fiber inputs to the lateral portions, the cerebellar hemispheres. The most extensive corticopontine projections are from motor, premotor, and somatosensory portions of the cerebral cortex (areas 4, 6, 3, 1, and 2) and from posterior parietal areas 5 and 7.

Anatomical studies have shown that the motor cortex also is the primary source of direct cortical input to the inferior olive. Stimulation of other portions of the cerebral cortex (especially parietal regions) also produces climbing fiber responses in the cerebellar cortex, but these are produced by cortical output to brainstem systems that, in turn, project to the olive.

Since cortical output reaches the cerebellum via both mossy fiber and climbing fiber systems, stimulation of the cerebral cortex evokes both simple and complex spikes in Purkinje cells. The complex spike, however, usually occurs considerably later than the increase in simple spike discharge. This is a common finding, as well, when both simple and complex spikes are elicited by stimulating primary afferent systems.

MOTOR DISORDERS ASSOCIATED WITH CEREBELLAR LESIONS

Deficits in Normal Movement. Before attempting to understand how the neuronal circuitry and information described above might contribute to motor control, it is useful to examine more carefully the characteristics of the motor disturbances seen in humans or animals with cerebellar lesions. Because the afferent and efferent connections vary for different portions of the cerebellum, damage

to selective portions of the cerebellum produces different signs of motor dysfunction. For example, damage to lateral portions of the hemispheres, which receive a major input from the cerebral cortex, produces deficits in the ability to move the arm voluntarily from one position to another. On the other hand, damage to the flocculonodular lobe, which receives major inputs from the vestibular system, produces deficits in the control of stable eye and trunk position. The basic deficit is the same, however. There is poor temporal and spatial control of motoneuron output.

Gordon Holmes[19] provided the classic description of motor dysfunction in humans with lesions of the cerebellum. He concluded that patients with cerebellar disease showed (i) delays in the initiation and termination of movement (Fig. 29–7), (ii) errors in the direction and smoothness of movement (Fig. 29–8), (iii) an inability to continue repetitive movements (Fig. 29–9), (iv) a defect in the synergy of movement at different joints, and (v) instability of posture.

Postural instability may be manifest as the inability to actively maintain a steady position of the trunk, the limbs, or even the eyes. If a subject with cerebellar disease is asked to stand with his eyes closed and his arms raised, his body will sway and his arms will oscillate. As he attempts to walk, he may display a staggering, irregular pattern of gait. An inability to maintain a stable eye position results in nystagmus, a repeated alternating slow conjugate eye movement in one direction followed by a rapid movement of the eyes in the opposite direction (Chap. 27).

Dow[11] showed that isolated lesions of the flocculus caused monkeys to lose their ability to stand or walk without swaying and falling. They could use their limbs to feed, however, without apparent tremor. That the flocculonodular lobe is especially important in maintaining postural stability is also supported by the findings in young children with

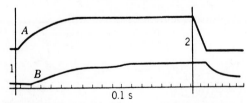

Figure 29–7 Myograms of voluntary contraction by a patient with a unilateral cerebellar lesion. A, Normal hand; B, affected hand. Vertical lines 1 and 2 are synchronization marks for the two records. Note slower start, weaker contraction, and delayed relaxation in the lower record. (From Ruch, T. In Stevens, S. S., ed. *Handbook of Experimental Psychology.* New York, John Wiley & Sons, 1951.)

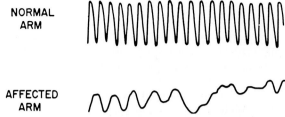

Figure 29–8 Tracings of rapidly alternating supination and pronation of arms in a patient with a unilateral cerebellar lesion. By comparison with the normal arm, movements made with the affected arm initally were slow and quickly deteriorated further. (From Holmes, G. *Lancet* 203:59–65, 1922.)

medulloblastoma, a type of tumor that arises from the roof of the fourth ventricle. The flocculus and nodulus, which are wrapped under the cerebellum and positioned just above the fourth ventricle, are compressed first as the tumor expands. In early stages of the disease, these children show few signs of motor disturbance when they are tested lying in bed, but when they sit or stand, they are unable to maintain balance and walk. They also have a prominent nystagmus.

Damage to the cerebellum also results in disorders of voluntary limb movement. The delays in the initiation of movement described in humans by Holmes also can be demonstrated in monkeys in which the dentate nucleus has been lesioned, either transiently by cooling or permanently by electrolytic coagulation. If the animal has been trained to move its arm rapidly in response to a sensory cue such as a light, the reaction time (time from turning the light on to movement initiation) is prolonged when dentate function is disrupted. Electromyographic recording of the muscle activity shows that the delay in movement initiation is due to a delay in both the reduction of background activity in the antagonists and in the initiation of activity in the agonists.

Following cerebellar damage, the termination of a movement also is abnormal. When a limb is

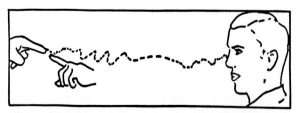

Figure 29–9 Record of tremulous movement obtained by having the patient move his finger from his nose to the finger of the examiner at the left. (From Ruch, T. In Stevens, S. S., ed. *Handbook of Experimental Psychology*. New York, John Wiley, and Sons 1951.)

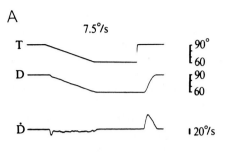

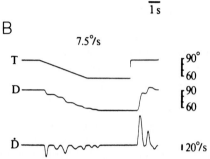

Figure 29–10 Elbow movements made in a visuomotor tracking task. In *A* and *B*, subjects are asked to flex and then extend the elbow so that a visually displayed signal of elbow angle (D) matches a visually displayed target trajectory (T). The velocity of the elbow movement (Ḋ) is shown by the bottom record in each section. *A*, Normal subject. *B*, Subject with cerebellar lesion. Note the relatively smooth trajectory of the movement made by the normal subject after acquiring the target velocity, compared to the step-wise movements of irregular velocity made by the patient with cerebellar damage. (From Beppu, H., et al. Visuomotor tracking in cerebellar ataxia. *Brain* 107:787–809, 1984.)

extended toward a target, agonist activity is prolonged, and the onset of antagonist activity used in stopping the limb is delayed. This results in an overshoot and an oscillation in limb position until the target position is acquired. Furthermore, the movement to a target may not have the smooth trajectory characteristic of movement made by a normal individual. Instead, it may occur in a series of steps, resulting in a tremulous approach to the target (Fig. 29–9).

Discontinuous movements of the limbs or the eyes also occur when a subject tracks a moving target. In Figure 29–10, the performance of a person with cerebellar disease is compared to that of a normal subject on a task in which the subject is asked to match a signal indicating elbow angle to the position of a visual target.[5] The normal subject makes a single rapid movement to acquire the target and then maintains target velocity with little problem. The patient, on the other hand, makes an initial rapid movement to acquire the target, but then is unable to maintain target velocity and must resort to many small rapid movements to keep up. Similar irregularities in eye position can be seen when a subject with cerebellar damage is asked to visually track a moving target.

Deficits in Motor Plasticity. The tests described above all reveal deficits in the ability to *produce* a normal movement, force, or steady position. Damage to the cerebellum also has been shown to interfere with a subject's ability to *modify* movement for a new situation. This disturbance in *motor plasticity*, or motor learning, has been demonstrated in both humans and experimental animals with lesions of the cerebellum.

The abilities of normal subjects and those with cerebellar damage to modify the stretch reflex in the lower extremity have been compared.[35] If normal subjects lean forward suddenly when standing on a flat surface, the gastrocnemius-soleus muscles (ankle plantar flexors) are stretched. The reflexly produced EMG activity (the functional stretch reflex, FSR) that occurs in the stretched muscles at a latency of about 100 ms is appropriate to restore the center of gravity to a position over the supporting base. The FSR is also produced if, instead, the muscles are stretched by rapidly rotating the supporting surface in the toe upward direction, such as might occur if you were standing on a boat deck that suddenly tilted. In this case, however, the resultant contraction of the gastrocnemius-soleus muscles accentuates the posterior rotation of the body and tends to make the person fall backward. If normal subjects repeatedly are subjected to upward toe rotation, the amplitude of the FSR decreases on successive trials, an example of reflex plasticity. Subjects with cerebellar dysfunction, on the other hand, continue to make the destabilizing FSR response even after many trials.

Damage to the cerebellum also reduces a subject's ability to modify the vestibulo-ocular reflex (VOR). This reflex, which was described in detail in Chapter 27, usually acts to produce an eye rotation of opposite direction but equal amplitude as a head movement. Such a counterrotation of the eyes keeps the visual world stable on the retina. If subjects wear optical lenses that distort the amplitude or direction of movement of the visual world when the head moves, the VOR amplitude or direction will not be appropriate to stabilize the visual target on the retina. If cerebellar function is intact and the head is moved repeatedly under these distorting conditions, the VOR gradually changes its gain so that the visual world once again is nearly stable on the retina. If a cat whose cerebellum has first been removed is required to wear similar optical devices, it does not undergo a compensatory modification of its VOR gain. These experiments, which again implicate the cerebellum in motor plasticity, are described in detail in Chapter 27.

Damage to the cerebellum not only prevents an appropriate modification of reflexes, it also impairs the adjustment of movements that usually would be considered "voluntary." Again, an example of this comes from the oculomotor system. Normally, the two eyes are directed to stationary visual targets by rapid conjugate movements called *saccades*. To test the role of the cerebellum in the adjustment of saccadic amplitude, saccadic dysmetria (saccades of inappropriate amplitude) has been produced by cutting the tendons of the horizontal extraocular muscles of one eye in monkeys with or without cerebellar lesions.[37] The tendons regrew partially but reattached at new positions on the globe. Since conjugate saccadic eye movements are accomplished by sending the same motoneuron signals to the muscles of each eye, the incomplete and less effective muscle attachments caused the weakened eye to move less than the normal eye. When the weakened eye was patched, the nervous system received visual input only from the normal eye. The resulting neural output to the normal eye caused it to reach the target, but the same neural drive to the weakened eye caused it to fall short. If, however, the patch was switched to the normal eye and visual input was received through the weakened eye, the visual input indicated that the saccadic command was insufficient to bring the eye to the target. In monkeys with the cerebellum intact, after the normal eye was patched for several days the saccadic amplitude increased until the weakened eye landed on target. Now, however, the saccades of increased amplitude carried the occluded normal eye past the target. The increased neural signal to the muscles of both eyes reflected the neural changes necessary to "repair" the dysmetria. If the cerebellar vermis and fastigial nucleus were destroyed, however, the dysmetria produced by tenectomy was not "repaired" when visual input was restricted to the impaired eye.

Deficits in Associative Learning. Cerebellar lesions not only impair a subject's ability to adapt the functional sensorimotor response to a stimulus; they also disrupt associative learning. This has been studied using the classically conditioned nictitating membrane response of the rabbit as a model. A puff of air to the cornea or a slight shock to the periorbital area (the unconditioned stimulus,

UCS) normally elicits contraction of the retractor bulbi muscles, posterior displacement of the eye, and passive movement of the nictitating membrane across the eye (the unconditioned response, UR). If the UCS is preceded repeatedly by a stimulus that normally is behaviorally neutral, such as a light or a tone (the conditioned stimulus, CS), the animal produces the nictitating membrane response (the conditioned response, CR) to the CS alone. Lincoln and his colleagues[28] were the first to report that large lesions of the cerebellum diminished an animal's ability to develop the classically conditioned response. More recently, Yeo and associates[52, 53] reported that the conditioned nictitating response did not develop following lesions of the anterior interpositus nucleus or the hemispheral portions of lobule VI.

THE DISCHARGE OF CEREBELLAR NEURONS IN ALERT ANIMALS

The challenge to understanding how the cerebellum controls movement lies in determining how different signals, arriving over mossy fiber and climbing fiber afferents and carrying information about different peripheral or central events, interact at the cerebellum to determine the activity of Purkinje cells and cells in the cerebellar nuclei. This problem has been addressed by recording from cerebellar neurons in awake animals during a variety of behavioral conditions.

Activity of the Deep Cerebellar Nuclei. Thach[48] was the first to record from cerebellar nuclear neurons in awake monkeys. He found that nuclear cells discharge at mean rates of over 20 to 30 spikes/s as the animal sits quietly, without overt movement. Since Purkinje cell synaptic input from the cerebellar cortex is inhibitory, either the tonic discharge of cerebellar nuclear cells must be maintained by an excitatory drive from other sources, such as the collaterals of mossy and climbing fibers that enter the nuclei,[20] or the nuclear cells must have intrinsic membrane mechanisms that cause an overriding repetitive discharge.

Nuclear cells change their activity in association with movements or isometric contraction. A and B in Figure 29–11 depict the activity of two cells in the interpositus nucleus before and after flexion (left) or extension (right) of the wrist. In this figure, each dot represents an action potential and each row of dots, a single behavioral trial aligned, in time, on the initial change of force on the handle that the monkey moved (dashed vertical line). Both nuclear cells exhibit a change in discharge

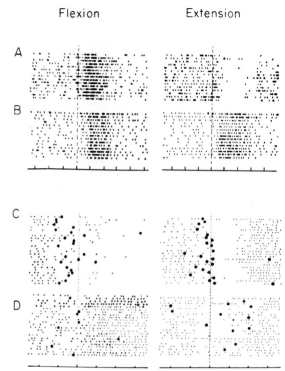

Figure 29–11 Discharge of cerebellar nuclear and Purkinje cells during visually triggered wrist flexion (left) and extension (right). A and B, Data from two neurons in the nucleus interpositus. C and D, Data from two Purkinje cells in the intermediate portion of the cerebellar cortex. Dashed lines represent the time of initial force change on the handle when the monkey begins the movement. Each dot represents an action potential and each horizontal row of dots, a behavioral trial. In C and D, large dots represent complex spikes; small dots, simple spikes. Time marks: 50 ms. (A and B from Thach, W. T. J. Neurophysiol. 33:527–536, 1970. C and D from Thach, W. T. J. Neurophysiol. 33:537–547, 1970.)

during the movement. To examine the role that changes in cerebellar activity might have in controlling movement, Thach examined the time at which these changes in discharge occurred.[49] If the movement was triggered by a light, the discharge of most dentate neurons began to change before the initial change in EMG activity of wrist muscles and, thus, potentially could be involved in initiating the movement. In contrast, most cells in the interpositus nucleus changed their discharge too late to be involved in movement initiation. Instead, the activity of these cells was timed more appropriately to code feedback information about the motor command or the consequences of the movement. Indeed, Strick[46] reported that if the forearm was displaced *passively* by movement of the handle that a monkey was grasping, the discharge of cells in the interpositus nucleus changed earlier, on the average, than did that of cells in the dentate nucleus.

To examine further the variables that might be coded in the discharge of cerebellar neurons, Thach used a paradigm in which the wrist position, the required muscle activity pattern, and the direction of the subsequent movement could be dissociated.[50] Monkeys were trained to grasp a handle and move the hand sequentially between three wrist positions in response to a visual cue. During some cycles of movement, a load that opposed flexion and assisted extension was applied to the handle, and during other cycles, a load was applied in the opposite direction. Thus, to acquire and maintain a given wrist position, the animal would have to contract flexor muscles during some cycles, but adjust the contraction of extensor muscles in others. During the hold period between each movement, the activity of interpositus neurons was most highly correlated with the load, and thus with the pattern and amount of muscle activity required to hold the limb position. The firing rate showed little correlation, however, with the absolute position of the wrist or the direction of movement that would be required to achieve the next position. The activity of many dentate neurons, on the other hand, was more highly correlated with the direction of the next required displacement, and less with muscle activity.

Strick[46] hypothesized that if dentate neurons were especially involved in the preparation for or initiation of movement, then an instructional cue should influence the discharge of dentate neurons more than that of neurons in the interpositus nucleus. He used lights to signal to a monkey in advance whether a handle should be pushed or pulled when it was perturbed. The perturbation produced a change in arm position and a short latency change in the discharge of many dentate and interpositus neurons. For interpositus neurons, the magnitude of the response to perturbation was independent of the prior instruction for active *push* or *pull*. For two thirds of the dentate cells, however, the neuron's response to the perturbation was influenced markedly by the prior instruction, although the tonic firing rate prior to the perturbation was unchanged.

Thus, the discharge of dentate neurons appears to be influenced by "preparatory set" and to be tightly coupled with preparation for and initiation of a movement. In contrast, the activity of neurons in the interpositus carries information that may reflect the motor command issued by the central nervous system or the somatosensory consequences of the movement. Feedback information

through interpositus could still influence the metrics of the movement or the rate at which it was completed, however, even if it did not influence the initiation of the movement.

Purkinje Cell Activity. As described earlier, Purkinje cells generate simple spikes, due to input from the mossy fiber–granule cell system, and complex spikes, due to climbing fiber input. Simple spikes occur at high frequencies. Figure 29–6B shows the discharge of a Purkinje cell in the paravermal posterior lobe of an awake monkey sitting quietly. Simple spikes, which are primarily negative (downward) in these extracellular records, occur at an average rate of about 70 spikes per second. The simple spike activity, then, occurs at a frequency high enough to code kinetic and kinematic variables, such as rate of change of force or position. Climbing fiber discharge, on the other hand, occurs too infrequently to code any rapidly changing parameters of position or force. In Figure 29–6B, the complex spikes are the ones with the prominent positivity (because their dendritic origin is at a distance relative to the probable perisomatic location of the extracellular recording electrode). In the resting animal, their mean firing rate seldom exceeds 1 to 2 spikes/s, and even when the animal moves or an adequate stimulus is applied, climbing fibers discharge at a maximum rate of only several spikes per second.

The simple spike activity of Purkinje cells does change in association with movement. In Figure 29–11, C and D show the activity of two Purkinje cells in the anterior lobe during flexion and extension movements of the wrist. In these illustrations, simple and complex spikes are denoted by small and large dots, respectively. Both Purkinje cells show changes in simple spike activity that begin before or during the movement. For some Purkinje cells in the anterior lobe, the simple spike activity varies as a function of movement velocity during a visuomotor wrist flexion–extension tracking task.[32] In the posterior vermis, which plays a role in the control of visually guided eye movements, the simple spike discharge of many cells changes as a function of maximum eye velocity.[47] For some Purkinje cells, the simple spike discharge rate also varies as a function of steady limb or eye position. Thus the simple spike activity does code both static and dynamic aspects of position or force.

Complex spike activity also may change in association with movement, although clear changes in complex spike activity often have not been detected during self-paced active movements that are accompanied by robust changes in simple spike activity.[48] The large dots in Figure 29–11 C and D depict the occurrence of complex spikes in two Purkinje cells. For cell C, and perhaps for cell D, the occurrence of a complex spike is temporally correlated with the wrist movement.

Several hypotheses have been proposed regarding the role of complex spikes. Since complex spikes usually are followed by a brief pause in simple spike activity, one possibility is that they reset the excitability of the Purkinje cells. Another hypothesis is that the complex spike signals the presence of an error in performance. A related hypothesis is that the climbing fiber response or some cellular response that it induces may act in some way to modify the Purkinje cells' response to future inputs. This, then, could provide a cellular mechanism for motor plasticity (learning).

Gilbert and Thach[15] examined the complex and simple spike activity of Purkinje cells when monkeys were required to adapt their motor behavior. The animals were trained to flex or extend their wrists to return a lever to a fixed position when it was displaced by an applied load. When the same load was applied repeatedly and the monkey made accurate movements to the target zone, complex spike activity occurred at a low rate and at random times. When a novel load was applied, however, the complex spike activity increased and became temporally linked to the displacement. This change in complex spike activity persisted until the animal had adjusted its motor response to make a relatively accurate movement in response to the new load, and then the complex spike activity returned to a low-frequency pattern that was unrelated to the task. The movement-related pattern of simple spike activity also changed during the trials in which the motor response was being adjusted, but when the motor adjustment was complete, the simple spike activity remained at its new level.

Gilbert and Thach were attempting to test a hypothesis first proposed by Marr[33] and then modified by Albus.[2] Marr and Albus proposed that the response of Purkinje cells to parallel fiber inputs could be modified if for some period of time, climbing fiber input was synchronized with parallel fiber activation of the same Purkinje cell. The mechanisms by which changes in synaptic effectiveness might occur were not specified, but they could include a change in the density or sensitivity of postsynaptic receptors at the parallel fiber synapses or in their effect on postsynaptic membrane channels. These might be induced, for example, by calcium entry into Purkinje cells during the dendritic spikes produced by climbing fibers.

The data described by Gilbert and Thach showed that there was a concurrent change in complex and simple spike activity during motor adjustment and that only the change in simple spike activity persisted after accurate motor performance was achieved. They could not determine, however, whether the climbing fiber activity *induced* the change in simple spike activity, as proposed by Marr and by Albus. Other investigators have attempted to test the parallel fiber synapse modification hypothesis more directly by pairing stimulation of climbing fibers with stimulation of parallel fibers or a source of mossy fiber input (the vestibular nerve). The Purkinje cell's response to parallel fiber input, alone, was tested before and after paired CF–parallel fiber stimulation. Changes in the response to parallel fiber input were considered support for the climbing fiber–induced plasticity model. Unfortunately, the data from such experiments have been conflicting. Some investigators found no support for the hypothesis when it was tested in isolated, perfused cat brain.[30] Others[23] reported a depression in the parallel fiber–evoked Purkinje cell response in rabbit flocculus studied in vivo. Furthermore, this depression appeared selective for the particular parallel fiber source that had been paired with climbing fiber activation.[13]

If climbing fiber activity were necessary only to induce plasticity but not to maintain it, then once the plastic changes have occurred, elimination of the climbing fiber input should not affect them. Llinas and his co-workers[29] and Demer and Robinson[10] have tested this hypothesis by making permanent or reversible lesions of the inferior olive in animals in which modification of postural responses or the vestibulo-ocular reflex had been induced previously. In each case, elimination of normal olivary function produced an immediate reduction in the adaptive motor changes. This means that, although climbing fiber discharge could play a role in the induction of the modified motor response, it appears to be important in the maintenance of the modified response, as well.

Data from cerebellar lesion studies support the hypothesis that the cerebellum plays some role in motor learning, and some data support the hypothesis that climbing fiber input is important in inducing or maintaining adaptive responses when a motor error is present. The cellular mechanisms underlying such adaptive motor behavior, however, remain to be determined. It is not really certain whether cellular changes occur *within* the cerebellum, or whether cerebellar output induces such changes elsewhere.

SUMMARY

The cerebellum is not necessary for the production of movement. Damage to the cerebellum, however, disrupts the production of accurate, smooth, coordinated movement and stable posture. Cerebellar lesions also impair motor plasticity. The deep cerebellar nuclei, which connect the cerebellum with other portions of the nervous system, show changes in activity that could influence both the initial and the ongoing phases of movement. Neurons of the cerebellar nuclei are excited by collaterals of both mossy and climbing afferent fibers and inhibited by the axons of Purkinje cells of the cerebellar cortex. Purkinje cells exhibit two types of action potentials. The tonic high-frequency, simple spike discharge produced by the mossy fiber afferents is modulated in association with changes in motor activity and in many cells is correlated with position or velocity. Complex spikes are produced by climbing fibers from cells of the inferior olive and occur at low rates. Their activity may be linked to an error in motor performance, and it has been hypothesized that they may induce cellular changes in Purkinje cells that induce or support motor learning.

ANNOTATED BIBLIOGRAPHY

Bloedel, J.R., Courville, J. Cerebellar afferent systems. Handbk. Physiol. Sec. 1, 2(2):735–830, 1981.
An extensive review of anatomical and physiological data regarding the sources and cerebellar terminations of the various cerebellar afferent systems, the types of information carried by each system, and their effects on neurons in the cerebellar cortex and deep nuclei.

Brooks, V.B.; Thach, W.T. Cerebellar control of posture and movement. Handbk. Physiol. Sec., 2(2):877–946, 1981.
A review of literature regarding the effects of cerebellar lesions in humans and experimental animals and the activity of cerebellar neurons studied during movements.

Ito, M. *The Cerebellum and Neural Control*. New York, Raven Press, 1984.
The most recent comprehensive review of anatomical, electrophysiological, and behavioral data regarding the cerebellum and its role in the control of movement.

REFERENCES

1. Aitken, L.M.; Boyd, J. Responses of single units in cerebellar vermis of the cat to monaural and binaural stimuli. *J. Neurophysiol.* 38:418–429, 1975.
2. Albus, J.S. A theory of cerebellar function. *Math. Biosci.* 10:25–61, 1971.
3. Altman, J.; Bechterev, N.; Radionova, E.; Shmigidina, G.; Syka, J. Electrical responses of the auditory area of the cerebellar cortex to acoustic stimulation. *Exp. Brain Res.* 26:285–298, 1976.
4. Barmack, N.H.; Hess, D.T. Multiple-unit activity evoked in dorsal cap of inferior olive of the rabbit by visual stimulation. *J. Neurophysiol.* 43:151–164, 1980.
5. Beppu, H.; Suda, M.; Tanaka, R. Analysis of cerebellar motor disorders by visually guided elbow tracking movement. *Brain* 107:787–809, 1984.
6. Chan-Palay, V.; Nilaver, G.; Palay, S.L.; Beinfeld, M.C.; Zimmerman, E.A., Wu, J.Y.; O'Donohue, T.L. Chemical heterogeneity in cerebellar Purkinje cells: existence and coexistence of glutamic acid decarboxylase-like and motilin-like immunoreactivities. *Proc. Natl. Acad. Sci. USA* 78:7787–7791, 1981.
7. Chan-Palay, V.; Ito, M.; Tongroach, P.; Sakurai, M.; Palay, S. Inhibitory effects of motilin, somatostatin, (leu)enkephalin, (met)enkephalin, and taurine on neurons of the lateral vestibular nucleus: interactions with gamma-aminobutyric acid. *Proc. Natl. Acad. Sci. USA* 79:3355–3359, 1982a.
8. Chan-Palay, V.; Palay, S.L.; Wu J.Y. Sagittal cerebellar microbands of taurine neurons: Immunocytochemical demonstration by using antibodies against the taurine-synthesizing enzyme cystein sulfinic acid decarboxylase. *Proc. Natl. Acad. Sci. USA* 79:4221–4225, 1982b.
9. Curtis, D. R.; Felix, D. The effect of bicuculline upon synaptic inhibition in the cerebral and cerebellar cortices of the cat. *Brain Res.* 34:301–321, 1971.
10. Demer, J.L.; Robinson, D.A. Effects of reversible lesions and stimulation of olivocerebellar system on vestibuloocular reflex plasticity. *J. Neurophysiol.* 47:1084–1107, 1982.
11. Dow, R.S. Efferent connections of the flocculonodular lobe in *Macaca mulatta. J. Comp. Neurol.* 68:297–305, 1938.
12. Eccles, J.C.; Ito, M.; Szentagothai, J. *The Cerebellum as a Neuronal Machine*. New York, Springer-Verlag, 1967.
13. Ekerot, C.-F.; Kano, M. Longterm depression of parallel fibre synapses following stimulation of climbing fibres. *Brain Res.* 342:357–360, 1985.
14. Gellman, R.; Houk, J.C.; Gibson, A.R. Somatosensory properties of the inferior olive of the cat. *J. Comp. Neurol.* 215:228–243, 1983.
15. Gilbert, P.F.C.; Thach, W.T. Purkinje cell activity during motor learning. *Brain Res.* 128:309–328, 1977.
16. Glickstein, M.; Stein, J.; King, R.A. Visual input to the pontine nuclei. *Science* 178:1110–1111, 1972.
17. Hackett, J.T.; Hou, S.-M.; Cochran, S.L. Glutamate and synaptic depolarization of Purkinje cells evoked by parallel fibers and by climbing fibers. *Brain Res.* 170:377–380, 1979.
18. Hokfelt, T.; Fuxe, K. Cerebellar monoamine nerve terminals, a new type of afferent fibers to the cortex cerebelli. *Exp. Brain Res.* 9:63–72, 1969.
19. Holmes, G. The cerebellum of man. *Brain* 62:1–30, 1939.
20. Ito, M.; Kawai, N.; Udo, M.; Mano, N. Axon reflex activation of Deiters' neurones from the cerebellar cortex through collaterals of the cerebellar afferents. *Exp. Brain Res.* 8:249–268, 1969.

21. Ito, M.; Yoshida, M. The origin of cerebellar-induced inhibition of Deiters' neurons. I. Monosynaptic initiation of the inhibitory postsynaptic potentials. *Exp. Brain Res.* 2:330–349, 1966.

22. Ito, M.; Yoshida, M.; Obata, K.; Kawai, N.; Udo, M. Inhibitory control of intracerebellar nuclei by the Purkinje cell axons. *Exp. Brain Res.* 10:64–80, 1970.

23. Ito, M.; Sakurai, M.; Tongroach, P. Climbing fibre responsiveness and glutamate sensitivity of cerebellar Purkinje cells. *J. Physiol.* 324:113–134, 1982.

24. Kan, K.-S.K.; Chao, L.-P; Forno, L.S. Immunohistochemical localization of choline acetyltransferase in the human cerebellum. *Brain Res.* 193:165–171, 1980.

25. Krnjevic, K.; Phillis, J.W. Iontophoretic studies of neurones in the mammalian cerebral cortex. *J. Physiol.* 165:274–304, 1963.

26. Kuno, M.; Miyahara, J.T. Factors responsible for multiple discharge of neurons in Clarke's column. *J. Neurophysiol.* 31:624–638, 1968.

27. Levi, G.; Gordon, R.D.; Gallo, V.; Wilkin, G.P.; Balazs, R. Putative acidic amino acid transmitters in the cerebellum. I. Depolarization-induced release. *Brain Res.* 239:425–445, 1982.

28. Lincoln, J.S.; McCormick, D.A.; Thompson, R.F. Ipsilateral cerebellar lesions prevent learning of the classically conditioned nictitating membrane/eyelid response. *Brain Res.* 242:190–193, 1982.

29. Llinas, R.; Walton, K.; Hillman, D.E.; Sotelo, C. Inferior olive: its role in motor learning. *Science* 190:1230–1231, 1975.

30. Llinas, R.; Yarom, Y.; Sugimori, M. Isolated mammalian brain in vitro: new technique for analysis of electrical activity of neuronal circuit function. *Fed. Proc.* 40:2240–2245, 1981.

31. Lundberg, A. Function of the ventral spinocerebellar tract—a new hypothesis. *Exp. Brain Res.* 12:317–330, 1971.

32. Mano, N.; Yamamoto, K. Simple-spike activity of cerebellar Purkinje cells related to visually guided wrist tracking movement in the monkey. *J. Neurophysiol.* 43:713–728, 1980.

33. Marr, D. A theory of cerebellar cortex. *J. Physiol.* 202:437–470, 1969.

34. Nadi, N.S.; Kanter, D.; McBride, W.J.; Aprison, M.H. Effects of 3-acetylpyridine on several putative neurotransmitter amino-acids in the cerebellum and medulla of the rat. *J. Neurochem.* 28:661–662, 1977.

35. Nashner, L.M.; Grimm, R.J. Analysis of multiloop dyscontrols in standing cerebellar patients. *Prog. Clin. Neurophysiol.* 4:300–319, 1978.

36. Okamoto, K.; Kimura, H.; Sakai Y. Evidence for taurine as an inhibitory neurotransmitter in cerebellar stellate interneurons: selective antagonism by TAG (6-aminomethyl-3-methyl-4H,1,2,4-benzothiadiazine-1,1-dioxide). *Brain Res.* 265:163–168, 1983.

37. Optican, L.M.; Robinson, D.A. Cerebellar-dependent adaptive control of primate saccadic system. *J. Neurophysiol.* 44:1058–1076, 1980.

38. Palkovits, M.; Magyar, P.; Szentagothai, J. Quantitative histological analysis of the cerebellar cortex in the cat. III. Structural organization of the molecular layer. *Brain Res.* 34:1–18, 1971.

39. Palkovits, M.; Magyar, P.; Szentagothai, J. Quantitative histological analysis of the cerebellar cortex in the cat. IV. Mossy fiber–Purkinje cell numerical transfer. *Brain Res.* 45:15–29, 1972.

40. Palkovits, M.; Mezey, E.; Hamori, J.; Szentagothai, J. Quantitative histological analysis of the cerebellar nuclei in the cat. I. Numerical data on cells and on synapses. *Exp. Brain Res.* 28:189–209, 1977.

41. Perry, T.L.; Currir, R.D.; Hansen, S.; McLean, J. Aspartate-taurine imbalance in dominantly inherited olivocerebellar atrophy. *Neurology* 27:257–261, 1977.

42. Precht, W.; Simpson, J.; Llinas, R. Responses of Purkinje cells in rabbit nodulus and uvula to natural vestibular and visual stimuli. *Pflugers Arch.* 367:1–6, 1976.

43. Rohde, B.H.; Rea, M.A.; Simon, J.R., McBride, W.J. Effects of x-irradiation–induced loss of cerebellar granule cells on the synaptosomal levels and the high affinity uptake of amino acids. *J. Neurochem.* 32:1431–1435, 1979.

44. Rushmer, D.S.; Roberts, W.J.; Augter, G.K. Climbing fiber responses of cerebellar Purkinje cells to passive movement of the cat forepaw. *Brain Res.* 106:1–20, 1976.

45. Siggins, G. R.; Oliver, A. P.; Bloom, F. E. Cyclic adenosine monophosphate and norepinephrine: effects on transmembrane properties of cerebellar Purkinje cells. *Science* 171:192–194, 1970.

46. Strick, P.L. The influence of motor preparation on the response of cerebellar neurons to limb displacements. *J. Neurosci.* 3:2007–2020, 1983.

47. Suzuki, D.A.; Noda, H.; Kase, M. Visual and pursuit eye movement–related activity in posterior vermis of the monkey cerebellum. *J. Neurophysiol.* 46:1120–1139, 1981.

48. Thach, W.T., Discharge of Purkinje and cerebellar nuclear neurons during rapidly alternating arm movements in the monkey. *J. Neurophysiol.* 31:785–797, 1968.

49. Thach, W.T. Discharge of cerebellar neurons related to two maintained postures and two prompt movements. I. Nuclear cell output. *J. Neurophysiol.* 33:527–536, 1970.

50. Thach, W.T. Correlation of neural discharge with pattern and force of muscular activity, joint position, and direction of intended next movement in motor cortex and cerebellum. *J. Neurophysiol.* 41:654–676, 1978.

51. Weber, J.T.; Partlow, G.D.; Harting, J.K. The projection of the superior colliculus upon the inferior olivary complex: an autoradiographic and horseradish peroxidase study. *Brain Res.* 144:369–377, 1978.

52. Yeo, C.H.; Hardiman, M.J.; Glickstein, M. Classical conditioning of the nictitating membrane response of the rabbit. I. Lesions of the cerebellar nuclei. *Exp. Brain Res.* 60:87–98, 1985.

53. Yeo, C.H.; Hardiman, M.J.; Glickstein, M. Classical conditioning of the nictitating membrane response of the rabbit. III. Connections of cerebellar lobule HVI. *Exp. Brain Res.* 60:114–126, 1985.

The Basal Ganglia

INTRODUCTION

The basal ganglia, like the cerebellum, exert profound influences on movement. Unlike the cerebellum, however, the basal ganglia receive no direct input from primary sensory systems; information used by the basal ganglia is derived primarily from the cerebral cortex. The motor deficits that follow damage to the basal ganglia vary, depending on the particular types of cells that are affected. The symptomatology may be an inability to suppress involuntary motor activity, or it may be an inability to generate rapid voluntary movements. In humans, the naturally occurring diseases that cause damage to the basal ganglia generally destroy neurons with particular biochemical properties. To understand how the basal ganglia produce their contributions to motor control, we eventually will have to make sense of both the biochemical properties and the neural signals of the complex interconnection of cells that make up the basal ganglia.

ANATOMICAL COMPOSITION

The term *basal ganglia* commonly is used to include several cell groups located at the base of the rostral brain stem and telencephalon (see Figure 30–1). The *caudate* (Cd) nucleus and *putamen* (Put) are cytologically similar and are separated only by the fibers of the internal capsule. They have common phylogenetic origins and are referred to as the (neo)striatum. The *globus pallidus* (GP) lies ventromedial to the putamen and has two segments, the external (lateral; GPe) and the internal (medial; GPi). (In carnivores, the homologue of the internal globus pallidus is called the entopeduncular nucleus.) The *subthalamic* (ST) nucleus is situated along the medial border of the internal capsule in the diencephalon, ventral to the thalamus. The *substantia nigra* (SN) lies caudal to the subthalamic nucleus in the mesencephalon, where it overlies the cerebral peduncle. The substantia nigra is divided into a pars reticulata (SNr), which is cytologically similar to the globus pallidus, and a pars compacta (SNc), which contains cells that are heavily pigmented in primates and give the substantia nigra its name.

In addition to the system just described, there is a second system with similar cytological and histochemical characteristics.[20, 23, 38] The striatal component of this system includes the nucleus accumbens and part of the olfactory tubercule (the ventral striatum), and the pallidal component (ventral pallidum) is situated ventral to the main, or dorsal, pallidum described above. The primary connections of this portion of the basal ganglia are with the limbic system, and they will not be dealt with further in this chapter.

CIRCUITRY

Input to the Basal Ganglia

In contrast to the cerebellum, the basal ganglia receive no sensory input directly from the periphery, the spinal cord, or primary sensory nuclei in the brain stem. Instead, most comes from the cerebral cortex. Some additional input originates from cells in the centromedian-parafascicular nuclear complex of the thalamus and the dorsal raphe and pedunculopontine regions of the brain stem.

A

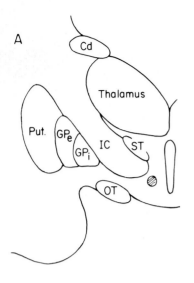

B

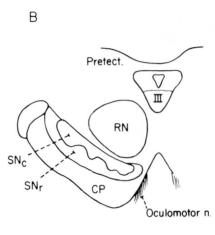

Figure 30–1 Components of the basal ganglia. *A*, Transverse section through the left side of the diencephalon showing the caudate nucleus (Cd), putamen (Put), external globus pallidus (GP$_e$), internal globus pallidus (GP$_i$), and subthalamic nucleus (ST). Other diencephalic structures shown include the thalamus, the internal capsule (IC), and the optic tract (OT). *B*, Transverse section through the left midbrain showing the location of the pars compacta (SN$_c$) and pars reticulata (SN$_r$) portions of the substantia nigra. Other mesencephalic structures shown include the pretectal region (Pretect.), the oculomotor nucleus (III), the red nucleus (RN), the cerebral peduncle (CP), and the oculomotor nerve.

Most afferent information enters the basal ganglia via the striatum (Fig. 30–2). In general, afferents from the entire cerebral cortex reach the striatum in a topographically organized manner, although no region of the striatum receives input from only a single cortical region. The putamen is the recipient of the heavy input from the sensorimotor cortex, and the caudate nucleus is the major destination of afferents from the remainder of the frontal and parietal lobes, the mesial aspects of the cortex, and the temporal and occipital lobes. The projections are ipsilateral, except for those from Brodmann's cortical areas 4, 6, and 8, which are bilateral.

Individual striatal neurons, including the ubiquitous medium spiny projection neurons whose axons leave the striatum, receive convergent input from both the cerebral cortex and the intralaminar thalamic nuclei. Stimulation of either the cortex or the thalamus tends to produce an initial excitatory postsynaptic potential (EPSP) followed by an inhibitory postsynaptic potential (IPSP) in striatal cells.[35] The excitatory action, at least from the cerebral cortex, probably is due to the presynaptic release of glutamate. The succeeding IPSP is at least partially due to the action of inhibitory striatal neurons. These could include axon collaterals of inhibitory striatal projection neurons as well as interneurons whose axons do not leave the striatum. Figure 30–3B shows the extensive local axon collateral system of a medium spiny striatal projection neuron that had been injected intracellularly with horseradish peroxidase. The axonal branches within the striatum stay primarily within the territory of the cell's own dendritic array, which is shown in Figure 30–3A.

Although the striatum is the major destination of afferent fibers from outside of the basal ganglia, some fibers from the cerebral cortex terminate in the subthalamic nucleus and some from the pedunculopontine (PPN) and dorsal raphe regions of the brain stem terminate in the globus pallidus or substantia nigra.

Efferent Connections of the Basal Ganglia

The globus pallidus and the substantia nigra are the output nuclei of the basal ganglia (see Fig. 30–4).

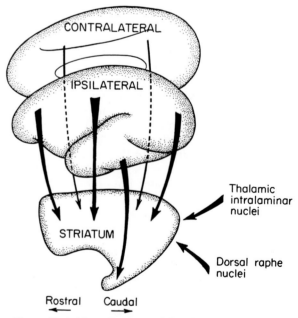

Figure 30–2 Afferent systems of the striatum. Demonstrated afferents originate from the ipsilateral and contralateral cerebral cortex and from ipsilateral thalamic intralaminar nuclei and dorsal raphe nuclei of the midbrain.

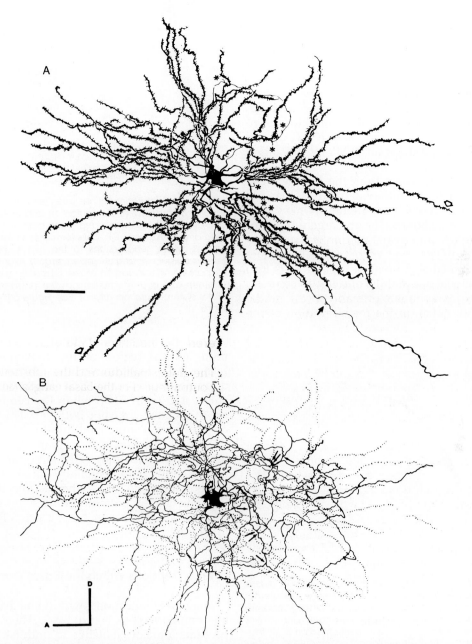

Figure 30–3 Dendritic and axonal arborizations of a medium spiny striatal neuron. *A*, Somatodendritic morphology, with heavy lines depicting the spine-studded dendrites *(open arrow)* and thin lines depicting the axons *(solid arrows)*. Asterisks indicate the points of origin of axon collaterals. *B*, The axon collateral plexus *(solid lines)* of the same cell superimposed on the dendritic arborization *(dotted lines)*. Arrows point to the origin of the main axonal collaterals. Calibration = 40 μ. Anterior and dorsal orientations are indicated in *B*. (From Preston, R.J.; Bishop, G.A.; Kitai, S.T. *Brain Res.* 183:253–263, 1980.)

In particular, neurons of the internal pallidal segment (GPi) and the pars reticulata portion of the substantia nigra (SNr) send axons to the thalamus and the superior colliculus and back to the PPN region.

The synaptic action of both pallidal and nigral output axons is inhibitory. This has been demonstrated both at the thalamus[47, 48] and at the superior colliculus.[8] In some cases, the same cell sends axonal branches to the thalamus, the superior colliculus, and even the PPN region.[2, 22] The putative neurotransmitter at these inhibitory synapses is gamma-aminobutyric acid (GABA), and its action is to produce an IPSP in the target neurons.

In the thalamus, axons from the basal ganglia usually terminate on different neurons than do those from the cerebellum. In the monkey, the destinations of pallidal axons are largely the ventral anterior (VA) and rostral portion of the ventral lateral (VLo) nuclei, and nigral axons terminate largely in the medial portions of VA and VL (VLm) and in the medial dorsal (MD) nucleus. In contrast, cerebellar axons terminate primarily in the caudal portion of VL (VLc), in the rostral portion of the

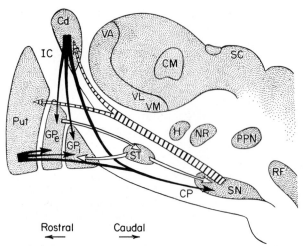

Figure 30–5 Interconnections between nuclei of the basal ganglia. Both striatal nuclei (putamen and caudate nucleus) send axons to the substantia nigra and both pallidal segments *(solid arrows)*. Axons of GP$_e$ cells terminate in the subthalamic nucleus, which sends axons to the two sources of basal ganglia output, GP$_i$ and SN *(open arrows)*. Axons from cells in the pars compacta region of the substantia nigra go to the putamen and caudate nucleus as the dopaminergic nigro-striatal system *(hatched arrows)*. See Figure 30–4 for abbreviations.

ventral posterolateral (VPLo) nucleus and in a more medial region of the thalamus (area X). Intracellular recordings show that most thalamic neurons have *either* an IPSP when GPi or SNr is stimulated *or* an EPSP when the cerebellar nuclei are stimulated, but not both.[47, 48] This separation of information must continue to the cerebral cortex, since injection of retrograde tracers such as horseradish peroxidase (HRP) into the medial premotor cortex (supplementary motor area [SMA]) labels neurons in VLo, whereas injection of HRP into the lateral premotor or the primary motor cortices labels neurons in area X or VPLo, but *not* in VLo.[43]

Interconnections Within the Basal Ganglia

The interconnections within the basal ganglia are summarized in Figure 30–5. Their complexity presents a major challenge to understanding how information that enters this system through the striatum or the subthalamic nucleus is processed before output leaves via neurons in GPi or SNr. Not only are the interconnections complex in an anatomical sense but they also must be pharmacologically complex, since almost every putative neurotransmitter is present in the basal ganglia.[19] In fact, many putative neurotransmitters are pres-

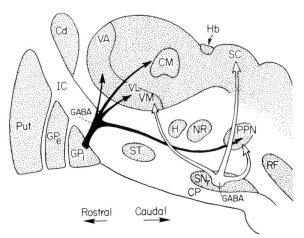

Figure 30–4 Efferent systems of the basal ganglia. Axons leaving the basal ganglia originate from cells in the internal globus pallidus (GP$_i$) and the pars reticulata of the substantia nigra (SN$_r$). Pallidal axons *(solid arrows)* terminate in the ventral anterior (VA), ventrolateral (VL), and centromedian (CM) thalamic nuclei and in the pedunculopontine region (PPN) of the mesencephalic-pontine tegmentum. Nigral axons *(open arrows)* terminate more medially in the thalamic ventromedial nucleus (VM) and in the superior colliculus (SC) and PPN. Both pallidal and nigral axons release gamma-aminobutyric acid (GABA). Other structures: cd = caudate nucleus; CP = cerebral peduncle; Put = putamen; GP$_e$ = external globus pallidus; IC = internal capsule; ST = subthalamic nucleus; H = Forel's field; NR = red nucleus; Hb = habenula; RF, ponto-medullary reticular formation.

ent in higher concentrations in the basal ganglia than in any other part of the nervous system!

The Neostriatum and Its Output. The neostriatum (caudate nucleus and putamen) is a densely packed group of neurons. Most of these have dendrites with numerous spines that provide the major sites of synaptic contact. Although the neostriatum appears relatively uniform when processed with many classical staining methods, it appears patchy when processed histochemically to demonstrate particular neurotransmitter- or neuromodulator-related compounds. The patch (striosome) vs. matrix appearance of the caudate nucleus first was emphasized when it was processed to show antiacetylcholinesterase (AChE), the enzyme that acts in the hydrolysis of acetylcholine.[17] Further investigation has shown that the AChE-poor striosome patches are rich in opiate receptors and enkephalin-like reactive elements.[16, 18, 41] Cells in the striosome patches also label retrogradely with HRP injected into the dopamine-containing pars compacta region of the substantia nigra (see below).[16] Furthermore, afferents from limbic portions of the cerebral cortex terminate heavily in the striosome patches. In contrast, the AChE-rich matrix surrounding the patches contains fibers and cells that react for somatostatin, has a high concentration of a particular type of dopamine receptor (D2 receptor), and is the primary site of termination of fibers from the sensorimotor cortex. Furthermore, HRP injected into the pars reticulata portion of SN retrogradely labels neurons in the matrix, but not in the striosome patches.

What does this patchy histochemical nature of the neostriatum mean in terms of the way information is processed? The answer to this remains to be determined, but since histochemical "patchiness" is a characteristic of many other portions of the nervous system as well, it is a question of more generalized importance. If, as is suggested by preliminary data,[40] the dendrites of the medium spiny striatal projection neurons do not cross from patch to matrix or vice versa, then the patches and the matrix may define discrete "information channels" through the striatum. One possibility is that the patches are especially involved with limbic information that interacts with the pars compacta portion of the substantia nigra, whereas the matrix portion is more involved with information from nonlimbic cortex and the pallidal and SNr regions that connect with neocortical-destined portions of the thalamus (see below). On the other hand, the ACh- and somatostatin-containing interneurons of the striatum may provide communication between patch and matrix units and permit a more extensive integration of what otherwise would appear to be separate channels of information processing.

The many medium spiny neurons of the striatum are its output cells. At least three putative neurotransmitters—GABA, enkephalin, and substance P—are found in these projection neurons, and there is some evidence that individual neurons contain both GABA and one of the two neuropeptides.[39] As shown by the solid arrows in Figure 30–5, striatal efferent axons terminate in both parts of the globus pallidus and in the substantia nigra. They exert an inhibitory synaptic action that is blocked by GABA blockers such as picrotoxin.[42] Some or all of the efferent neurons also may release an excitatory neurotransmitter, however, because if the action of GABA is first blocked by the administration of picrotoxin, stimulation of the striatum excites nigral neurons.[32] Although it is possible that this excitatory action is produced by the neuropeptides, this has not been clearly demonstrated.

The Subthalamic Nucleus. The axons of subthalamic neurons also stay within the basal ganglia. As shown by the open arrows in Figure 30–5, they terminate in GPi or SNr. Subthalamic cells receive input from the cerebral cortex and are also the major targets of GPe axons. Since input to GPe comes from the striatum, the subthalamic nucleus is a point at which processed information from the striatum can interact with cortical input prior to action on basal ganglia output neurons in SNr and GPi.

The Dopaminergic Nigrostriatal System. One of the systems of the brain studied most intensively in recent years has been the dopaminergic nigrostriatal system. Dopamine is present in neurons of the pars compacta portion of the substantia nigra, and it is these neurons, together with some other monoamine-containing neurons of the central nervous system, that degenerate in Parkinson's disease. Interest in this system became especially intense when it was discovered that the administration of L-dopa, a precursor of dopamine, could partially alleviate the motor symptomatology of parkinsonism.

As shown by the hatched arrows in Figure 30–5, dopamine-containing axons of SNc neurons terminate primarily in the caudate nucleus and putamen. (Axons of dopamine-containing neurons located ventromedial to SNc terminate in the nucleus accumbens.) The synaptic action of dopaminergic nigrostriatal neurons is still a matter of some controversy. Intracellular recordings have shown that stimulation of the substantia nigra produces a late depolarization of striatal medium spiny neurons. This occurs even after decortication and subsequent degeneration of corticostriatal axons that might otherwise have been activated by stimulus spread to the nearby cerebral peduncle or medial lemniscus.[49] It also has been reported that dopamine applied iontophoretically produces a depolarization, as measured intracellularly in striatal neurons. When action potentials of striatal neurons have been recorded extracellularly, however, iontophoretic application of dopamine in the striatum usually has been reported to *reduce* the rate of action potentials. The discrepancy can be at least partially explained by the observation that iontophoretically applied dopamine reduces the amplitude of EPSPs produced by stimulation of corticostriatal fibers,[5] and this could result in a decrease in striatal discharge rate. It is not known whether dopamine is acting to reduce the release of glutamate from the presynaptic axons or whether it acts to block the synaptic action of glutamate on the postsynaptic membrane. Dopaminergic nigrostriatal fibers also may reduce the activity of striatal output neurons by activating inhibitory striatal interneurons. Thus, although nigrostriatal activation or the application of dopamine may produce a direct depolarization of striatal neurons, the ultimate action of dopamine on the discharge of striatal neurons usually is a *reduction* in the rate of action potentials.

THE BASAL GANGLIA AND MOTOR CONTROL

The abnormalities of movement that result from damage to the basal ganglia vary, depending on the neurons that are destroyed. Damage to the striatum or subthalamic nucleus results in involuntary movements that are completely suppressed only when the individual is sleeping. Striatal damage produces *athetosis* or *chorea*. In athetosis, involuntary movement is superimposed on hypertonic background motor unit activity, a combination that produces slow, writhing, ceaseless movements. The abnormal movements usually involve the hands and less frequently the lips, the tongue, or occasionally the neck and foot. Involuntary movements also occur in chorea, but in this condition they are superimposed on a hypotonic background, a combination that results in rapid, jerky alternating movements in irregular and unpredictable sequences.

In Huntington's disease, in which chorea is the characteristic motor symptom, it is specifically the medium spiny output neurons of the striatum that degenerate. The striatal interneurons, characterized histochemically by reactivity for AChE or somatostatin, and the dopaminergic nigrostriatal fibers remain intact.[36] Huntington's chorea, then, apparently results from the removal of striatal actions on GP and SN. As discussed earlier, at least some of this synaptic action is inhibitory and is mediated by GABA.

When the subthalamic nucleus is destroyed, involuntary motor activity occurs in proximal limb muscles and produces *ballismus*, a violent flinging of the limbs. If only one subthalamic nucleus is destroyed, the involuntary movement is confined to the contralateral side and is known as *hemiballismus*.

Isolated damage to basal ganglia output neurons in GPi or SNr is uncommon in humans but it has been produced in animals. Animals with GPi lesions do not show obvious deficits in their general motor behavior, but when measured carefully, their movements are slower than normal.[29, 46] Figure 30–6A, for example, shows the averaged rectified electromyographic activity and the times of start switch release and target contact for an animal performing an arm-reaching movement in a reaction-time task. After destruction of the globus pallidus with the neurotoxic agent kainic acid (dashed lines) the muscle activity changes more slowly than in prelesion trials (solid lines), and the movement time, from start switch release to target contact, is prolonged. The reaction time is not prolonged, however, and neither the sequence of muscle activity nor the movement strategy used by the individual animal is changed (Figure 30–6B).

In humans, the dopaminergic nigrostriatal system is the portion of the basal ganglia that most often undergoes pathological changes. These changes result in Parkinson's disease, in which the cardinal motor symptoms are bradykinesia, rigidity, and tremor. Similar motor signs are produced in monkeys or humans if the nigrostriatal system is destroyed by neurotoxic agents such as 6-OH dopamine or N-methyl-4-phenyl–1,2,3,6-tetrahydropyridine (MPTP).[37] MPTP is an extremely toxic compound that mistakenly has been obtained by some street drug users.

Bradykinesia, or slow movement, is a hallmark of parkinsonism. Voluntary movements of both the limbs and the eyes may be slowed. When subjects are tested in a reaction time paradigm, the most consistent finding is that the movement time, from the initiation to the completion of the movement, is prolonged. The reaction time often is lengthened as well. Furthermore, individuals

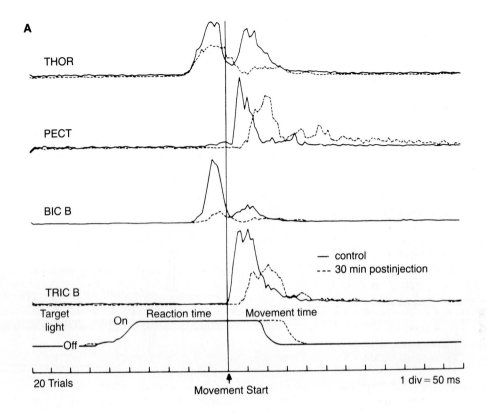

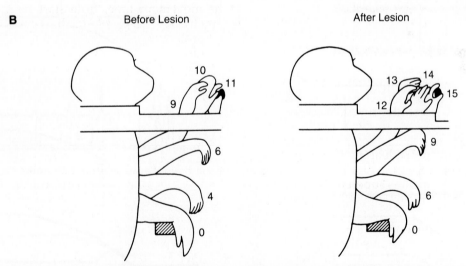

Figure 30–6 Changes in electromyographic activity and movement induced by a lesion of the globus pallidus. *A,* Rectified electromyographic (EMG) activity of the thoracic paraspinal (THOR), pectoralis (PECT), biceps brachii (BIC B), and triceps (TRIC B) muscles and the state of the target light before *(solid lines)* and after *(dashed lines)* injection of an excitotoxin (kainic acid) into the globus pallidus contralateral to the moving arm. The muscle activity is depressed and the movement time (from movement start to the time at which the target light is contacted and returns to "off") is prolonged, but the reaction time is not changed following the lesion. All records are averages of 20 trials, aligned on the start of the movement. *B,* Tracings of head and limb positions of the same animal before (left) and after (right) the lesion. The numbers refer to the film frame traced *after* release of the start switch. Frame interval = 10 ms. (From Horak, F.B.; Anderson, M.E. *J. Neurophysiol.* 52:290–304, 1984.)

with bradykinesia do not vary the velocity of movements as much as normal people do when movement amplitude is changed. As shown in Figure 30–7A, if a normal subject makes rapid movements of increasing amplitude, the velocity of the movement also increases. As a consequence, the larger movements have only slightly longer durations than do the small ones. To achieve the higher velocity of movement, the magnitude of the EMG activity in the muscles acting reciprocally around the joint is adjusted (or scaled). In contrast, movements of increasing amplitude made by a patient with parkinsonism occur at the same initial velocity, and the duration is prolonged until the target is achieved (Fig. 30–7B).[21]

Rigidity is a resistance of the muscle to stretch. The rigidity that is characteristic of Parkinson's disease is due to the unusual tonic activity of alpha motoneurons when the muscle should be at rest and the pronounced increase in muscle activ-

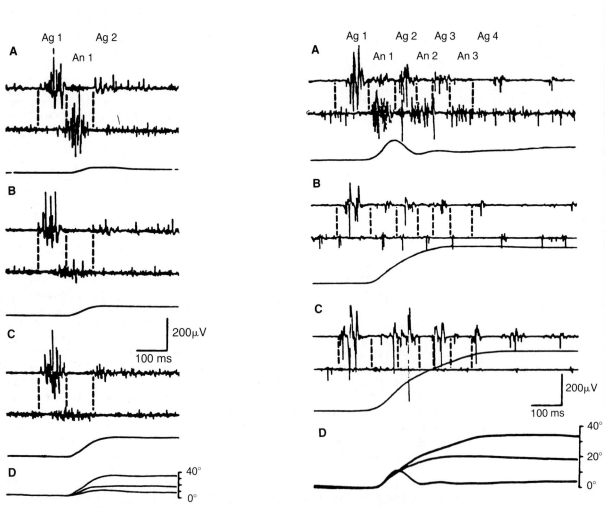

Figure 30–7 Attempted ballistic elbow movements of normal individuals and those with Parkinson's disease. *A,* Elbow flexion movements of 10 degrees (A), 20 degrees (B), and 40 degrees (C). For each movement, EMG records are shown from the biceps brachii *(upper trace,* Ag 1 and Ag 2) and triceps brachii *(lower trace,* An 1) muscles, together with the record of elbow position. In D, the position traces for the three movements are superimposed to illustrate the increase in initial velocity with movements of increasing amplitude. *B,* Elbow flexion movements made by a 68-year-old man with Parkinson's disease. The records are as in *A,* but the agonist (Ag, biceps) and antagonist (An, triceps) muscles must have multiple bursts of activity and the initial velocity of the movement is quite constant, irrespective of the amplitude of the movement (D). (From Hallett, M.; Khoshbin, S., *Brain* 103:301–314, 1980.)

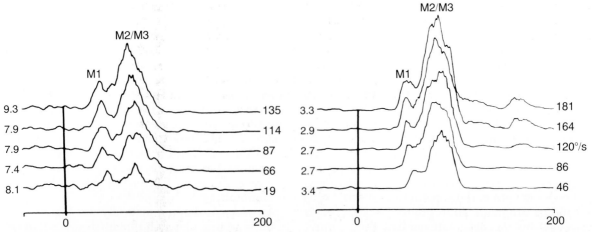

Figure 30–8 Electromyographic activity of wrist flexor muscles when the muscle is stretched. *Left*: Normal subject, rectified, averaged EMG recorded during extension displacements of the wrist at increasing mean velocity (numbers at the right of each record). M1 is the initial segmental response and M2/M3 constitutes the long latency stretch response. *Right*: Subject with Parkinson's disease. Note the enhancement of the long latency response. Numbers on the left show the background EMG level as a percentage of that recorded during maximum voluntary contractions. (Courtesy of William Bedingham; From Lee, R.G.; Murphy, J.T.; Tatton, W.G. *Long-latency myotatic reflexes in man: Mechanisms, functional significance, and changes in patients with Parkinson's disease or hemiplegia*. In Desmedt, J. E., Ed. *Adv. Neurol.* 39:489–508, 1983, with permission of Raven Press, New York.)

ity when the muscle is stretched. The increased response to stretch is the result of polysynaptic, not monosynaptic, pathways. Figure 30–8 shows rectified and averaged EMG responses from wrist flexor muscles that were stretched at different velocities. The initial response, labeled M1 in Figure 30–8, is similar for normal subjects (left) and for subjects with parkinsonism (right). The later response (M2/M3), however, is markedly enhanced in subjects with parkinsonism. This enhancement of the "long latency stretch reflex," with little or no increase in the short latency response, differentiates subjects with parkinsonism from those with spasticity, another type of hyperreflexia that is common after a cerebral infarct (stroke). In spasticity, it is the early M1 response, which probably represents the monosynaptic stretch reflex, that characteristically is enhanced (see Chap. 25 for discussion of M1, M2, M3).

The tonic motor unit activity in subjects with parkinsonism may be phasic instead of continuous. The phasic activity produces a tremor, usually at a frequency of 4 to 6 beats per second. When it is superimposed on rigidity, it may cause an irregular "cogwheel" character to the resistance encountered when the muscle is stretched. The tremor usually disappears during an intentional movement.

In addition to problems with control of the background activity of muscles and the rate at which their activity can be changed, subjects with

parkinsonism also may have disturbances of more complex aspects of motor control. Several studies have reported that subjects with parkinsonism have more difficulty than normal subjects using prior information to control a movement. For example, normal subjects shorten their reaction times in a choice reaction time task if they are given information in advance as to which of the two motor behaviors will be required when the "go" signal appears. If they are tracking a visual target that moves continuously to predictable positions and at predictable velocities (for example, a repetitive sinusoidal displacement), normal subjects also can move in phase with the target, instead of lagging target position by a visual reaction time delay. In each case, it is assumed that the information from a prior cue or earlier cycles of the regular target displacement can be used to mobilize centrally generated motor programs. In contrast, moving to an unpredictable target would require "on line" use of sensory information to generate or activate the required motor output with a visual reaction time delay.

Although individuals with parkinsonism also can reduce the delay in their motor performance when they are given prior information, the reduction is much less dramatic than the changes shown by normal individuals. This is true for saccadic eye movements made to predictable vs. random targets[7] and for limb movements made in a choice reaction time task[6] or in a continuous tracking task in which the target moves in regular vs. predictable patterns.[14] It is unclear from these data whether the inability of the subjects with parkinsonism to produce dramatic reductions in motor delay with predictive information is simply due to the slowness of their movements, whether it is an indicator that they cannot generate or select appropriate central motor programs, or whether it is due to an especially strong reliance on visual guidance. The latter has been demonstrated in monkeys whose arm tracking was studied when basal ganglia function was reversibly interrupted by cooling the globus pallidus.[31] The hypothesis that subjects with parkinsonism cannot select central motor programs appropriately is supported, however, by the observations that they cannot perform two different movements simultaneously[4] or choose the appropriate postural adjustment patterns from the two that normally occur in the legs when subjects are subjected to postural instabilities.[30]

In summary, both naturally occurring and experimentally produced lesions provide evidence that the striatum and globus pallidus are important in suppressing unwanted movements and enabling desired movements to occur at maximum velocities. Destruction of the dopaminergic nigrostriatal system, which modulates striatal and, indirectly, pallidal activity, results in changes in movement speed, motor unit activity at rest, and perhaps the use of prior information to control "motor set" or to choose or modify central motor programs.

NEURONAL ACTIVITY IN THE BASAL GANGLIA

Resting Activity in the Absence of Movement

Neurons in both nuclei of the striatum (caudate nucleus and putamen) have very little activity in the awake animal that is not moving. Action potentials occur intermittently at mean rates of less than 2 spikes/s in most striatal neurons.[1, 10, 34] In contrast, cells in the globus pallidus or the pars reticulata portion of the substantia nigra discharge at mean rates of greater than 50 spikes/s. Since cells in GPi and SNr provide the basal ganglia's inhibitory output to the rest of the nervous system, target neurons in the thalamus, superior colliculus, and PPN must be subjected to tonic inhibition in the awake, resting animal.

Changes in Activity During Movement

Since most information enters the basal ganglia through the striatum and leaves through GPi or SNr, it is informative to examine the activity of neurons in these nuclei during various behaviors. Hikosaka and his colleagues[24, 25, 26, 28] have studied cells in the caudate nucleus, SNr, and the superior colliculus in animals trained to make saccadic eye movements under different conditions. Several investigators have studied the activity of neurons in the putamen and the globus pallidus or SNr during limb and orofacial movements.[3, 9, 10, 11, 12, 15, 34, 44]

As shown diagrammatically in Figure 30–9, there is a burst of action potentials in a set of cells in the caudate nucleus when the animal makes a saccadic eye movement to a target that appears in the contralateral visual field.[28] SNr neurons that send axonal branches to the superior colliculus, however, have a pause in their discharge during the same kind of visually triggered saccadic eye movement.[26] The pause in SNr discharge would be the expected result of a burst of activity in the inhibitory striatonigral axons. Since the SNr axons also are inhibitory, a pause in their activity results

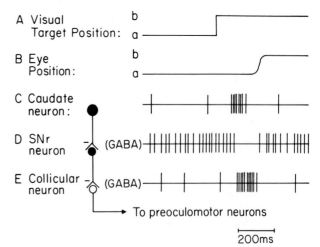

Figure 30–9 Changes in activity of the basal ganglia and their target structures during a saccadic eye movement. *A,* Position of the horizontally displaced visual target. (*a* is the leftward and *b* the rightward position for the target in *A* and for the eyes in *B.*) *B,* Horizontal eye position. *C,* Action potentials from a neuron in the caudate nucleus. *D,* Action potentials from a neuron of the substantia nigra, pars reticulata (SNr) that is inhibited by GABA released from the caudate neuron. *E,* Action potentials from a neuron in the superior colliculus that is inhibited by GABA released from the SNr neuron. (After Hikosaka, O.; Wurtz, R.H. *J. Neurophysiol.* 49:1285–1301, 1983; and Hikosaka, O.; Sakamoto, M., *Exp. Brain Res.* 63:659–662, 1986.)

in a burst in the activity of neurons in the superior colliculus. Because activity in collicular neurons can initiate the neural activity that produces saccadic eye movements, a rapid saccadic eye movement ensues.[26] Consistent with the interpretation that the pause in SNr activity leads to the saccade is the observation that local injection into SNr of muscimol, a GABA agonist that inhibits SNr neurons, results in spontaneous saccades toward the contralateral visual field.[27]

Similarly, sets of neurons in the putamen have a burst of action potentials prior to movements of the arm, leg, or mouth.[9] There is at least a rough somatotopic organization in the putamen, and cells active during leg movements are situated most dorsally, with those active during arm and facial movements in successively ventral positions.

In the globus pallidus, many neurons change their activity during limb movements.[3, 10, 12, 15] This change is not always a decrease in activity, however. Often it is an increase, which would result in *increased inhibition* of target neurons, largely in the thalamus. This difference between the movement-related activity of nigral and pallidal neurons is not well understood, but it may be related to the fact that the changes in muscle activity occurring before and during limb movement are much more complex than those during eye movement. For example, limb segments have a much larger mass than does the eyeball, and accurate movement of the limb often requires both a burst of activity in the agonist muscles and subsequent "braking" activity in antagonistic muscles. Furthermore, adjustment in the activity of muscles at other joints usually is necessary during limb movements to provide postural stabilization or adjustment. The eye, on the other hand, has a small mass and a constant center of rotation, and its movement is terminated largely by the elastic forces of tissues in the orbit, without the necessity for active contraction of the antagonists.

The relation between changes in GPi discharge and the kinematic characteristics of the movement has been examined during flexion and extension movements of the forearm to specific target elbow angles. For some neurons, the magnitude of the change in discharge is correlated with the direction, amplitude, or velocity of the movement.[15] Thus the cells' change in discharge may play a role in determining these kinematic parameters.

The discharge of cells in the basal ganglia also may be related, however, to the conditions under which the movement is made. For example, nigral neurons whose discharge reliably decreases before or during eye movements to a visual target show no change in activity when the animal makes eye movements to the same position without target illumination.[25] Furthermore, many cells show a greater decrease in discharge if the eye movement is made to a "remembered" target position, that had been illuminated earlier, than they do when the movement is to a visual target that is illuminated as a trigger for the movement and stays lit until the movement is completed.[24] Neurons in the putamen also show this kind of "set-related" effect on their movement-related changes in discharge.

Thus, basal ganglia neurons in the striatum, both segments of the pallidum, and SNr exhibit phasic changes in discharge during movements. These changes may be related to the direction, amplitude, and velocity of the movement, and they may be especially pronounced when the movement is made under particular behavioral conditions.

The Activity of Dopaminergic Nigrostriatal Neurons

Since it is the dopaminergic nigrostriatal neurons that degenerate in Parkinson's disease, and since profound changes in motor function are the cardinal signs of the disease, one might predict that dopaminergic nigrostriatal neurons would exhibit profound changes in discharge before or during movement. Instead, the firing of SNc neurons often does not change at all during movement. When there is a change, the discharge may increase or decrease for long periods of time and clearly is not coding kinematic parameters of the movement.[11, 44] Although, for some dopaminergic neurons, a brief burst of action potentials may follow the appearance of a behaviorally important visual or auditory cue,[45] the dopamine released by these cells may be more important for a general neurohormonal action than for the transmission of discrete neuron-to-neuron information.[13] This interpretation is consistent with the fact that systemic administration of L-dopa, a dopamine precursor that crosses the blood-brain barrier and is converted to dopamine, is relatively efficient in controlling the symptomatology of Parkinson's disease. Furthermore, fetal dopaminergic tissue that is implanted in the ventricles of animals can reverse many of the abnormalities exhibited by animals in which the endogenous nigrostriatal system has been destroyed.

SUMMARY

Although the basal ganglia are a pharmacologist's delight and clearly play important roles in motor function, we are far from understanding any of those roles. We do know that the basal ganglia present a unique situation in which successive inhibitory connections provide a double-negative mode of operation from the striatum to basal ganglia target neurons in the thalamus and superior colliculus. Many unanswered questions remain regarding both cellular mechanisms and the information content of the cellular activity. For example, we do not know what makes the cells in GPi and SNr fire at such a rapid rate, since their primary synaptic inputs (from GABA-ergic striatal axons) are inhibitory. And we really do not know the nature of the critical information carried by neurons of the basal ganglia. It seems unlikely that this information is related only to motor function, because the basal ganglia have connections—both afferent and, indirectly, efferent—with large portions of the cerebral cortex.

The basal ganglia and the cerebellum are two subcortical structures that are associated with the control of movement. Although earlier investigators stressed a model in which outputs from the basal ganglia and cerebellum converged at the thalamus to derive a major input to the precentral motor cortex,[33] it now is clear that information from the basal ganglia or the cerebellum has different cortical destinations and that each uses different routes to exert a different influence on motor function.

ANNOTATED BIBLIOGRAPHY

McKenzie, J.S.; Kemm, R.E.; Wilcock, L., eds. The Basal Ganglia: Structure and Function. *Advances in Behav. Biol.* 27, 1984.
  A collection of recent papers on the anatomy, physiology, and pharmacology of the basal ganglia.
Parent, A. *Comparative Neurobiology of the Basal Ganglia.* New York, Wiley & Sons, 1986.
  An excellent up-to-date review of comparative anatomical and histochemical literature, with some functional data.

REFERENCES

1. Anderson, M.E. Discharge patterns of basal ganglia neurons during active maintenance of postural stability and adjustment to chair tilt. *Brain Res.* 143:325–338, 1977.
2. Anderson, M.E.; Yoshida, M. Axonal branching patterns and location of nigrothalamic and nigrocollicular neurons in the cat. *J. Neurophysiol.* 43:883–895, 1980.
3. Anderson, M.E.; Horak, F.B. Influence of the globus pallidus on arm movements in monkeys. III. Timing of movement-related information. *J. Neurophysiol.* 54:433–448, 1985.
4. Benecke, R.; Rothwell, J.C.; Dick, J.P.R.; Day, B.L.; Marsden, C.D. Performance of simultaneous movements in patients with Parkinson's disease. *Brain* 109:739–757, 1986.
5. Bernardi, G.; Calabresi, P.; Mercuri, N.; Stanzione, P. Dopamine decreases the amplitude of excitatory postsynaptic potentials in rat striatal neurones. *Advances in Behav. Biol.* 27:161–173, 1984.
6. Bloxham, C.A.; Mindel, T.A.; Frith, C.D. Initiation and execution of predictable and unpredictable movements in Parkinson's disease. *Brain* 107:371–384, 1984.
7. Bronstein, A.M.; Kennard, C. Predictive ocular motor control in Parkinson's disease. *Brain* 108:925–940, 1985.
8. Chevalier, G.; Deniau, J.M.; Thierry, A.M.; Feger, J. The nigro-tectal pathway. An electrophysiological reinvestigation in the rat. *Brain Res.* 213:253–263, 1981.
9. Crutcher, M.D.; DeLong, M.R. Single cell studies of the primate putamen. I. Functional organization. *Exp. Brain Res.* 53:233–243, 1984.
10. DeLong, M.R. Activity of pallidal neurons during movement. *J. Neurophysiol.* 34:414–427, 1971.
11. DeLong, M.R.; Crutcher, M.D.; Georgopoulos, A.P. Relations between movement and single cell discharge in the substantia nigra of the behaving monkey. *J. Neurosci.* 3:1599–1606, 1983.
12. DeLong, M.R.; Crutcher, M.D.; Georgopoulos, A.P. Primate globus pallidus and subthalamic nucleus: Functional organization. *J. Neurophysiol.* 53:530–543, 1985.
13. Fabre, M.; Rolls, E.T.; Ashton, J.P.; Williams, G. Activity of neurons in the ventral tegmental region of the behaving monkey. *Behav. Brain Res.* 9:213–235, 1983.
14. Flowers, K. Some frequency response characteristics of parkinsonism on pursuit tracking. *Brain* 101:19–34, 1978.
15. Georgopoulos, A.P.; DeLong, M.R.; Crutcher, M.D. Relations between parameters of step-tracking movements and single cell discharge in the globus pallidus and subthalamic nucleus of the behaving monkey. *J. Neurosci.* 3:1586–1598, 1983.
16. Gerfen, C.R. The neostriatal mosaic: Compartmental organization of projections from the striatum to the substantia nigra in the rat. *J. Comp. Neurol.* 236:454–476, 1985.
17. Graybiel, A.M.; Ragsdale, C.W., Jr. Histochemically distinct compartments in the striatum of human, monkey and cat demonstrated by acetylthiocholinesterase staining. *Proc. Natl. Acad. Sci. USA* 75:5723–5726, 1978.
18. Graybiel, A.M.; Ragsdale, C.W., Jr.; Yoneoka, E.S.; Elde, R.H. An immunohistochemical study of enkephalins and other neuropeptides in the striatum of the cat with evidence that the opiate peptides are arranged to form mosaic patterns in register with the striosomal compartments visible by acetylcholinesterase staining. *Neurosci.* 6:377–397, 1981.
19. Graybiel, A.; Ragsdale, C. Biochemical anatomy of the striatum. In Emson, P.C., ed. *Chemical Anatomy.* New York, Raven Press, pp. 427–504, 1983.
20. Haber, S.N.; Groenewegen, H.J.; Grove, E.A.; Nauta, W.J.H. Efferent connections of the ventral pallidum: Evidence of a dual striatopallidofugal pathway. *J. Comp. Neurol.* 235:322–335, 1985.
21. Hallett, M.; Khoshbin, S. A physiological mechanism of bradykinesia. *Brain* 103:301–314, 1980.
22. Harnois, C.; Filion, M. Pallidofugal projections to thalamus and midbrain: A quantitative antidromic activation study in monkeys and cats. *Exp. Brain Res.* 47:277–285, 1982.

23. Heimer, L.; Switzer, R.D.; Van Hoesen, G.W. Ventral striatum and ventral pallidum. Components of the motor system? *Trends Neurosci.* 5:83–87, 1982.

24. Hikosaka, O.; Wurtz, R.H. Visual and oculomotor functions of monkey substantia nigra pars reticulata. I. Relation of visual and auditory responses to saccades. *J. Neurophysiol.* 49:1268–1284, 1983a.

25. Hikosaka, O.; Wurtz, R.H. Visual and oculomotor functions of monkey substantia nigra pars reticulata. II. Visual responses related to fixation of gaze. *J. Neurophysiol.* 49:1254–1267, 1983b.

26. Hikosaka, O.; Wurtz, R.H. Visual and oculomotor functions of monkey substantia nigra pars reticulata. IV. Relation of substantia nigra to superior colliculus. *J. Neurophysiol.* 49:1285–1301, 1983c.

27. Hikosaka, O.; Wurtz, R.H. Modification of saccadic eye movements by GABA-related substances. II. Effects of muscimol in the monkey substantia nigra pars reticulata. *J. Neurophysiol.* 53:292–308, 1985.

28. Hikosaka, O.; Sakamoto, M. Cell activity in monkey caudate nucleus preceding saccadic eye movements. *Exp. Brain Res.* 63:659–662, 1986.

29. Horak, F.B.; Anderson, M.E. Influence of globus pallidus on arm movements in monkeys. I. Effects of kainic acid–induced lesions. *J. Neurophysiol.* 52:290–304, 1984.

30. Horak, F.B.; Nashner, L.; Nutt, J. Postural instability in Parkinson patients: Motor coordination and sensorimotor integration. *Society for Neurosci. Absts.* 10:634, 1984.

31. Hore, J.; Vilis, T. Arm movement performance during reversible basal ganglia lesions in monkeys. *Exp. Brain Res.* 39:217–228, 1980.

32. Kanazawa, I.; Yoshida, M. Electrophysiological evidence for the existence of excitatory fibres in the caudatonigral pathway in the cat. *Neurosci. Lett.* 20:301–306, 1980.

33. Kemp, J.; Powell, T.P. The connexions of the striatum and globus pallidus: Synthesis and speculation. *Phil. Trans. Roy. Soc. Lond.* B262:441–457, 1971.

34. Kimura, M. The role of primate putamen neurons in the association of sensory stimuli with movement. *Neurosci. Res.* 3:436–443, 1986.

35. Kocsis, J.D.; Sugimori, M.; Kitai, S.T. Convergence of excitatory synaptic inputs to caudate spiny neurons. *Brain Res.* 124:403–413, 1977.

36. Kowall, N.W.; Ferrante, R.J.; Martin, J.B. Patterns of cell loss in Huntington's disease. *Trends Neurosci.* 10:24–28, 1987.

37. Langston, J.W.; Ballar, P.; Tetrud, J.W.; Irwin, I. Chronic parkinsonism in human due to a product of meperidine-analog synthesis. *Science* 219:979–980, 1983.

38. Nauta, W.J.H.; Smith, G.P.; Faull, R.L.M.; Domesick, V.B. Efferent connections and nigral afferents of the nucleus accumbens septi in the rat. *Neuroscience* 3:385–401, 1978.

39. Oertel, W.H.; Mugniani, E. Immunocytochemical studies of GABAergic neurons in rat basal ganglia and their relationship to other neuronal systems. *Neurosci. Lett.* 47:233–238, 1984.

40. Penny, G.R.; Wilson, C.J.; Kitai, S.T. The influences of neostriatal patch and matrix compartments on the dendritic geometry of spiny projection neurons in the rat as revealed by intracellular labeling with HRP combined with immunocytochemistry. *Soc. Neurosci. Abst.* 10:514, 1984.

41. Pert, C.B.; Kuhar, M.J.; Snyder, S.H. Opiate receptor: Autoradiographic localization in rat brain. *Proc. Natl. Acad. Sci. USA* 73:3729–3733, 1976.

42. Precht, W.; Yoshida, M. Blockage of caudate-evoked inhibition of neurons in the substantia nigra by picrotoxin. *Brain Res.* 32:229–233, 1973.

43. Schell, G.; Strick, P. Origin of thalamic input to the arcuate premotor and supplementary motor areas. *J. Neurosci.* 4:539–560, 1984.

44. Schultz, W.; Ruffieux, A.; Aebischer, P. The activity of pars compacta neurons of the monkey substantia nigra in relation to motor activation. *Exp. Brain Res.* 51:377–387, 1983.

45. Schultz, W. Responses of midbrain dopamine neurons to behavioral trigger stimuli in the monkey. *J. Neurophysiol.* 56:1439–1461, 1986.

46. Trouche, E.; Beaubaton, D.; Amato, G.; Legallet, E.; Zenatti, A. The role of the internal pallidal segment on the execution of a goal directed movement. *Brain Res.* 175:362–365, 1979.

47. Ueki, A. The mode of nigro-thalamic transmission investigated with intracellular recording in the cat. *Exp. Brain Res.* 49:116–124, 1983.

48. Uno, M.; Ozawa, N.; Yoshida, M. The mode of pallido-thalamic transmission investigated with intracellular recording from cat thalamus. *Brain Res.* 33:493–507, 1978.

49. Wilson, C.J.; Chang, H.T.; Kitai, S.T. Origins of postsynaptic potentials evoked in identified rat neostriatal neurons by stimulation in substantia nigra. *Exp. Brain Res.* 45:157–167, 1982.

Higher Functions

Association Cortex

INTRODUCTION

The part of the brain that shows the greatest increase in size with phylogeny is the cerebrum, which consists mostly of a thin layer of cells, the cerebral cortex, and their associated fibers. The cerebral cortex is completely absent in fish and is only rudimentary or poorly developed in amphibians, reptiles, and birds. Mammals, on the other hand, have a well-developed cerebral cortex. In humans, it contains over two and a half billion neurons[46] and covers an area of about one square meter. Most of this expansive structure has neither a purely sensory nor a purely motor function. Rather, it is known as association cortex because originally it was thought to associate sensory information from the primary sensory areas and to relay that integrated information to the motor areas. Lower mammals, such as rats, have little or no association cortex, and cats have substantially less than do primates. Thus, a vastly increased association cortex clearly sets humans apart from lower mammals. It has been argued, therefore, that much of the sophistication and flexibility that distinguishes human behavior from that of other animals arises from this unique structure. Our intellect, personality, and capacity for speech all derive in part from the function of association cortex.

In our discussion of sensory and motor systems, we saw that large parts of the cortex are involved primarily in sensation or movement. Cortical areas originally were designated as *motor* if electrical stimulation of these regions in anesthetized subjects produced movement. Motor areas, which are considered in detail in Chapter 28, include the primary motor region of the precentral gyrus (Brodmann's area 4), the supplementary motor and premotor regions (area 6), and the frontal eye field (area 8). Lesions in area 4 produce paresis and hypertonia, especially in distal muscles, while lesions in areas 6 and 8 produce more subtle apraxias (impairments in movement). Cortical areas were originally designated as *sensory* if electrical stimulation of these regions in alert subjects produced sensory experiences and if, in anesthetized subjects, short latency potentials could be evoked in these regions by peripheral sensory stimulation. Cortical sensory areas include the primary sensory regions, which receive direct inputs from the thalamic nuclei that receive input from the primary ascending pathways; they in-

clude area 17 (vision), areas 1, 2, and 3 (somatic sensation), areas 41 and 42 (audition), and area 43 (taste). The primary olfactory region includes part of area 28 but is composed mostly of prepyriform and periamygdaloid cortex. The location of the primary vestibular sensory region in humans is unknown, but in other primates it lies in part at the posterior border of the somatosensory region. Lesions made in primary sensory areas cause sensory *agnosias* (a loss of the ability to interpret sensory stimuli). The locations of the cortical motor and primary sensory areas in humans and monkey are indicated in Figure 31–1.

In contrast to these motor and sensory regions, *association areas* are "silent" regions that, under nembutal anesthesia, reveal no sensory-evoked potentials and yield no movements when electrically stimulated. In animals that are awake or under anesthesia other than nembutal, however,

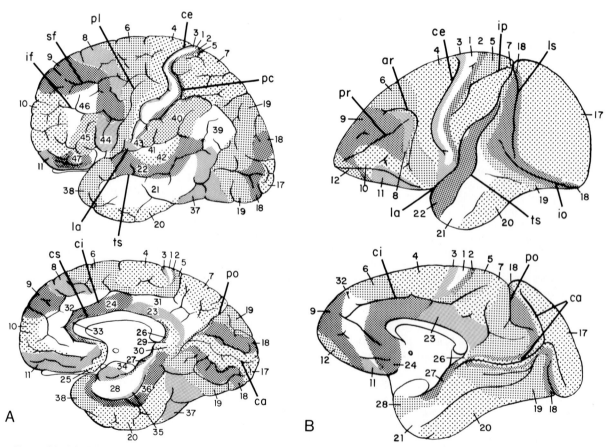

Figure 31–1 Architectonic maps, according to Brodmann, of the cerebral cortex of humans *(A)* and monkey (Cercopithecus) *(B)*. ar = Arcuate sulcus; ca = calcarine sulcus; ce = central sulcus; ci = cingulate sulcus; cs = callosal sulcus; if = inferior frontal sulcus; io = inferior occipital sulcus; ip = intraparietal sulcus; la = lateral sulcus (sylvian fissure); ls = lunate sulcus; pc = postcentral sulcus; pl = precentral sulcus; po = parieto-occipital sulcus; pr = principal sulcus; ts = superior temporal sulcus.

much of association cortex yields electrical responses to a variety of sensory stimuli.

It is convenient to divide association areas into four regions, on the basis of the cortical lobes in which they are found. The *prefrontal association cortex* consists of the rostral tip of the cerebral hemisphere, which lies anterior to premotor cortex. The *parietal association cortex* lies in the parietal lobe between the somatosensory and the visual areas of the parietal and the occipital lobes, respectively. The *temporal association cortex* occupies the portion of the cerebral hemisphere that is lateral to the sylvian fissure and primary auditory cortex, and anterior to visual cortex. A fourth cortical association area occupies the deep medial region of the limbic lobe; it is considered separately in Chapter 32.

The functional roles of the various association areas have been deduced largely from studies of behavioral deficits that follow traumatic or surgically induced cortical lesions in humans. Such studies are often anecdotal and qualitative and lead to varying and sometimes contradictory conclusions. Some of the problems associated with interpreting human studies have been obviated by measuring the performance of monkeys trained to perform complex behavioral tasks before and after controlled surgical lesions. In such preparations the investigator also can record the neuronal firing patterns associated with a given behavior. However, nonhuman primates require extensive training to acquire many of the behaviors that are sensitive to lesions of association cortex and, of course, no amount of training will make them as behaviorally sophisticated as humans. For example, many human association areas are involved in the comprehension and generation of language, functions that clearly cannot be studied in monkeys. Also, homologous regions in different species may subserve very different functions, and lesions of association cortices in a monkey may produce behavioral deficits that are very difficult to relate to those seen in human patients with similar lesions. Indeed, as can be seen from Figure 31-1, monkeys have fewer cytoarchitectonically defined cortical subdivisions.

Nevertheless, some consistent conclusions can be drawn about the involvement of the various association cortices in different behaviors. We begin by considering the parietal and temporal association areas. Both of these association cortices have been discussed previously with regard to auditory and visual processing (Chaps. 18 and 20). We then consider the prefrontal association areas, without which independent, directed, and purposeful behavior becomes difficult. Finally, we describe the lateralization of cerebral function and examine the suggestion that each hemisphere is uniquely involved in controlling certain behaviors.

PARIETAL ASSOCIATION CORTEX

Afferent and Efferent Connections

The parietal lobe lies posterior to the central sulcus and rostrodorsal to the sylvian fissure (lateral sulcus). It is divided into an anterior region that is composed of primary somatosensory cortex (Brodmann's areas 1, 2, and 3) and a posterior portion (areas 5 and 7 above, 39 and 40 below) that is considered to be association cortex. In the monkey, parietal association cortex is divided by the intraparietal sulcus into a superior lobule, classified cytoarchitectonically as Brodmann's area 5, and an inferior lobule, classified as Brodmann's area 7 (Fig. 31–1B).

The sources of their major inputs suggest that areas 5 and 7 are largely concerned with different sensory modalities. The strongest cortical input to area 5 originates from the adjacent primary somatosensory cortex (areas 1, 2, and 3), which in primates receives both somatosensory and vestibular information. The thalamic input to area 5 arises mostly from the nucleus lateralis posterior (LP). Area 7, on the other hand, receives its cortical input from prestriate visual areas (e.g., MST in Fig. 20–15) and its major thalamic input from the pulvinar (Chap. 20). The inferior parietal association area also receives long association fibers from the superior temporal gyrus and other association areas such as prefrontal cortex (areas 8, 45, and 46) and cingulate cortex (areas 23 and 24).

In contrast to the segregation of their afferent inputs, both posterior parietal cortices apparently have widespread, often common, projection sites.[50] These include common, but not always overlapping, destinations in the cortex (premotor, prefrontal, cingulate, and insular), the basal ganglia (putamen and caudate nuclei) (see Chap. 30), and the subthalamus. There are some differences, however. Many of the cortical efferents of area 5 terminate in premotor and motor cortex (area 4), areas involved with the generation of somatic movement (Chap. 28). Although areas 5 and 7 both project to temporal cortex, area 7 seems to project to subregions that are ultimately destined for limbic structures (Chap. 32), whereas area 5 does not. In addition, area 7, but not area 5, projects to the parahippocampal gyrus and receives subcortical inputs from locus ceruleus and the raphe nuclei. These data suggest that area 5 may be more involved with the processing of somatosensory information to produce movement, whereas area 7 may primarily process visual information in order to influence not only movement but also arousal, attention, and emotion. The extensive anatomical connections of the posterior parietal lobe with other association areas suggest, however, that the participation of several association cortices may be required for a given function.

Effects of Lesions: Neglect

Monkeys and humans who have suffered lesions of the posterior parietal lobe commonly neglect parts of the body that are contralateral to the lesion and neglect events occurring in the contralateral portion of the external world (also called extrapersonal space). Humans with posterior parietal damage (particularly on the right side) may groom and dress only the ipsilateral side of their bodies and, in extreme cases, may deny that the contralateral arm and leg belong to them, insisting that someone else's limb is in bed with them.[7] Patients with posterior parietal lesions may fail to eat from the contralateral side of their plates, and those with lesions of the right side may read "cowboy" as "boy." Also, when asked to draw a daisy or a clock face, such patients draw petals only on the ipsilateral half of the daisy (Fig. 31–2A) and place all the hours on the ipsilateral side of a clock. Finally, when asked to bisect a line, they do so asymmetrically (Fig. 31–2B). Although neglect seems less profound in nonhuman primates, monkeys with posterior parietal lesions also tend to ignore events occurring in their contralateral extrapersonal space.

Contralateral neglect also occurs in the case of mental images. In one study,[1] two patients with lesions involving the right parietal lobe were asked to describe from memory the main square in Milan, looking either from the front of a cathedral across the square or from the opposite side of the square toward the cathedral. Regardless of which direction the patients imagined themselves to be facing, the structures to their left were always omitted. Thus, the patients remembered the features of both sides of the square, but could generate an internal image of only the ipsilateral features.

Neglect may also result from lesions of areas other than the posterior parietal cortex.[20] These include

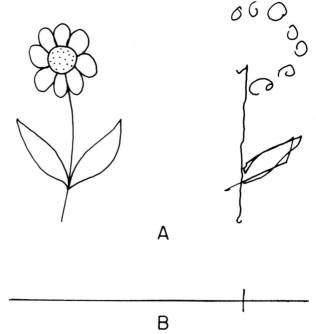

A

B

Figure 31–2 Examples of hemispatial neglect in a patient with a right parietal lobe lesion. In *A*, the patient was asked to copy the flower on the left, and in *B*, to bisect the line. (After Heilman, K.M.; Watson, R.T.; Valenstein, E. In Heilman, K.M.; Valenstein, E., eds. *Clinical Neuropsychology*. Oxford, Oxford Univ. Press, 1985.)

cortical areas such as the temporo-parietal-occipital junction and prefrontal cortex, limbic areas such as the cingulate gyrus, and subcortical areas such as the thalamus and mesencephalic reticular formation. However, because all these areas are in anatomical contact with the posterior parietal cortex, they are all perhaps part of a common functional system.

Neglect encompasses a variety of deficits, which may be motor, sensory, cognitive, or attentional in nature. Clear evidence for several of these deficits has been presented in monkeys or humans with posterior parietal lesions.

Motor Deficits

Among the most striking disorders of posterior parietal lesions in monkeys and humans is the inability to reach accurately for objects. In monkeys trained to reach out with the contralateral arm to touch small, illuminated buttons on a perimeter, posterior parietal lesions cause decreases both in the number of correct responses and in the accuracy of those responses.[27] Monkeys whose posterior parietal cortex is inactivated by cooling also exhibit an increase in the time between the lighting of a target and its extinction by the monkey's response (i.e., the reaction time). In such an experiment (Fig. 31–3), a monkey is trained to initiate a trial by pressing an illuminated center button, whereupon its light goes out and one of six peripheral buttons spaced at 10-degree intervals becomes illuminated.[66] The monkey must then press the illuminated peripheral button within 1.5 s for a reward. As seen in Figure 31–3,

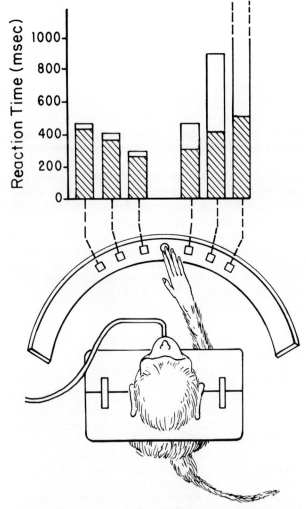

Figure 31–3 A reaching task illustrating motor deficits caused by reversible parietal lobe damage. The monkey is trained first to press the center button and then rapidly to touch whichever peripheral button is subsequently turned on. In normal monkeys, increasing the button eccentricity causes modest, symmetrical increases in reaction times to either side (hatched histograms). Cooling the left parietal lobe (open histograms) increases the reaction times to buttons in the right but not the left hemifield. (Modified with permission from Stein, J. In Gordon, G., ed. *Active Touch. The Mechanisms of Recognition of Objects by Manipulation: A Multidisciplinary Approach*, 79–90. Copyright 1978, Pergamon Books, Ltd.)

cooling area 7 on the left side caused a marked increase in reaction times when the monkey reached its right hand into the right visual field. This deficit was not simply a visual one, however, because an animal trained to respond to an illuminated peripheral button by pushing a centrally located lever rather than reaching for the button showed no prolongation of reaction time when area 7 was cooled.

Cooling of area 7 has no effect on the performance of a tactile task in which a monkey is required to discriminate between a rough and a smooth handle on two reward boxes. On the other hand, selective cooling of area 5 causes these tactile manipulations to be more clumsy. The monkey cannot make the appropriate discrimination with its contralateral hand (it even has difficulty finding the handles) and cannot open the boxes even when it can see them. These data again suggest that area 5 may be more involved with tactile reaching, and area 7 with "visual reaching."

Another clever experiment[70] demonstrated that the lack of movement (*hypokinesia*) in an extremity contralateral to a posterior parietal lobe lesion is not necessarily due to a sensory deficit. In this study, monkeys were trained to respond to tactile stimulation of one leg by moving the opposite arm. After being subjected to left parietal lesions, monkeys with right-sided neglect responded with the left arm to stimuli delivered to the right leg but did not respond with the right arm to stimuli delivered to the left leg. These observations indicate that right-sided sensory stimuli were perceived but that there was a lack of intent to move into the right extrapersonal space.

Posterior parietal lesions are also accompanied by ataxia (failure of muscle coordination), which can be quite severe. For example, monkeys often are unable to pick up a piece of banana even under visual control and after the hand is placed in contact with it.[30]

Another clear motor deficit can be demonstrated during eye movement tracking. If a small spot on which a trained monkey is fixating jumps suddenly to the side, the eyes jump (make a saccade) onto the target after a delay (reaction time) of about 200 ms. Lesions of the posterior parietal lobe cause the saccadic reaction time to increase by about 50 ms without affecting its accuracy. Similarly, if a fixated spot moves smoothly to the side at a slow, constant velocity, a normal animal makes a small catch-up saccade after 200 ms, and the eyes then move smoothly with the target. After posterior parietal lesions, the amount of time required for the eye to catch the target and move smoothly with it is greater than 800 ms. Thus, the time required for the initiation of a variety of eye movements is increased. In some human patients with parietal lobe lesions, the eyes never move smoothly in pursuit of a smoothly moving target, because they never achieve target velocity. This effect is true for eye movements in response both to single targets (i.e., smooth pursuit) and to full visual field motion (i.e., optokinetic nystagmus) (see Chap. 20).

Sensory Deficits

The effects of posterior parietal lesions on sensation per se are usually difficult to separate from the associated motor deficits; consequently, reports from different laboratories are not always consistent. In general, the most striking disorders follow removal of the entire posterior parietal lobule (both areas 5 and 7) and possibly some of the adjacent extrastriate cortex (area 19) as well. Monkeys with such posterior parietal lesions cannot learn to discriminate between two grades of sandpaper of similar grit, although they can easily learn to discriminate between ones of widely disparate roughness.[75] Unilateral removal of the parietal lobe also impairs the detection of flutter vibrations, since it elevates the detection threshold and abolishes the capacity to discriminate between different vibratory frequencies.[28] Finally, complete removal of the posterior parietal lobule impairs both the learning and retention of a task in which a single pair of test objects or multiple pairs of test objects must be discriminated by palpation.[56a] It should be stressed, however, that the deficits produced by posterior parietal lesions on these somatosensory tasks are much less severe than those produced by lesions of primary somatosensory cortex (areas 1, 2, and 3) (see Chap. 15).[61]

Human patients with posterior parietal lobe damage often display inability to recognize objects by palpation alone. This form of *astereognosis* is not, however, simply a problem with somatosensory discrimination. Patients with posterior parietal lesions can make quite sophisticated discriminations in the size, shape, and texture of tactually presented objects[59] but are unable to match one tactually presented object with an identical object presented in any array of similar objects. Thus, these can be seen as deficits of ideation, in which the mental image of the object is disrupted, and not simply as tactual imperceptions.

Lesions of the posterior parietal cortex have very little effect on visual discrimination tasks unless the tasks have a strong spatial component (see next section).

Deficits in Spatial Perception

It has been argued that the deficits in movement and sensation produced by posterior parietal lesions are secondary to a deficit in spatial perception. In this scenario, the brain receives accurate sensory information but is unable to place it into the correct spatial coordinate system. This inability could account for both inaccurate reaching and incorrect interpretation of tactile cues.

Monkeys and humans with posterior parietal lesions exhibit many behaviors that can be attributed to impaired spatial perception. First, they have problems in finding routes. When monkeys with such lesions are released in the middle of a room, they take longer than normal animals to return to their cages. Humans with posterior parietal lesions cannot find their way home from the doctor's office or to and from the bathroom in a hospital ward. Patients with parietal lobe damage also have trouble reading and constructing maps. Although they can recognize individual items in a room, they cannot relate them to each other by drawing a floor plan.

Second, human and nonhuman primates with parietal lobe damage perform poorly on maze tasks. For example, monkeys with lesions have difficulty relearning a task in which they must guide a stylus from the center of a maze to an exit for a reward. They also make many more errors (e.g., they enter incorrect alleys) than do normal monkeys.[35] In an imaginative variation on this task, monkeys were given ring-shaped sweets that had been threaded on wires with right-angle bends of various complexity. They had 45 s to move the sweet along the wire until it was free and could be eaten. Monkeys with posterior parietal lesions took almost three times longer than normal monkeys to accomplish this task.[51]

Third, some patients report that stimuli presented on the side contralateral to a parietal lobe lesion are perceived as occurring on the ipsilateral side. This illusion, called *allesthesia*, has been explained as a disorder of the body schema for the contralateral body half. Events that occur in contralateral space are transferred to the intact, normal ipsilateral-half schema because the contralateral schema is deficient.

Fourth, some patients have difficulty dealing with three-dimensional space, so that they lose the ability to draw even crudely in perspective.

For example, they may draw a bicycle with the frame and handle bars all in the same plane,[49] suggesting that they fail to appreciate the sizes and interrelationships of the various parts of an object.

Finally, monkeys with parietal lobe lesions consistently have problems with a task in which a neutral stimulus (called a landmark) such as a metal cylinder is used to indicate which of two food cups the monkey must select for a reward. In this task, the monkey must select the cup nearest the landmark on the first series of trials; on the second, it must select the cup farthest from the landmark (this is called a reversal). Criterion is 28 correct choices in 30 trials for the initial discrimination and each of 7 reversals. The data in Figure 31–4A were obtained for a series of trials following the seventh reversal. Although normal monkeys (N) made a relatively high number of errors on the first discrimination and reversal, these declined steadily to the values in Figure 31–4A as they learned the task. In contrast, monkeys with posterior parietal lesions continued to make frequent errors throughout the task, showing little improvement with subsequent reversals; their performance on the seventh reversal was considerably impaired (P in Fig. 31–4A). This deficit is specific to parietal lobe lesions, since animals with prefrontal (F) and inferotemporal (T) lesions behaved normally.

The landmark deficit is not due to a disorder in visual discrimination, because monkeys with posterior parietal lesions can perform an object reversal task in which they are required to reverse responses between food wells covered by a small cube and a cylinder (Fig. 31–4B). Nor is the deficit due to difficulties in distinguishing intrinsic, or egocentric, spatial coordinates (such as right and left), because monkeys with parietal lesions perform well on place reversal tasks (Fig. 31–4C), which require that they reverse responses to the right or the left food well. In contrast, object reversal and place reversal tasks are sensitive to the effects of inferotemporal and frontal lesions, respectively. The deficits in landmark discrimination, which characterize parietal lobe lesions, suggest an inability to use external referents for spatial orientation. It is this *allocentric*, or extrapersonal, spatial orientation that the parietal association cortex is thought to subserve.

Attentional Deficits

Apparent motor and sensory problems may also be attributable to deficits in attention. Although it is difficult to develop an experimental situation that allows attention to be measured quantitatively, the two paradigms described below illustrate aspects of attention that are affected by posterior parietal lesions.

Many of the contralateral sensory deficits produced by posterior parietal lesions (see earlier discussion) disappear with time or are not easily revealed. Patients with such lesions can then perceive stimuli presented to either side of the body; however, when stimuli are presented to both sides of the body simultaneously, they display hemi-inattention, ignoring the stimulus contralateral to the injured parietal lobe. This hemi-inattention is called *extinction*. It can be demonstrated for all sensory modalities (although only weakly for audition) and is one of the most reliable diagnostic tests for posterior parietal lesions in human and nonhuman primates.

In an ingenious paradigm, Posner and co-workers[54] demonstrated that patients with posterior parietal lesions are deficient in pure shifts of attention that occur with no overt movement. Basically, they measured *covert* shifts of attention through a reaction time task in which the patients had to detect a visual target falling on one of two peripherally placed display panels located on

Figure 31–4 Effects of lesions of the frontal (F), temporal (T) and parietal (P) lobes on the learning of a landmark (A), object (B), and place (C) discrimination reversal task. Histograms indicate the average number of errors made in learning the seventh discrimination reversal (criterion was 28 correct in 30 consecutive responses). Histogram labeled N represents normal performance. (After Pohl, W. *J. Comp. Physiol. Psychol.* 82:227–239, 1973. Copyright 1973 by the American Psychological Association. Adapted by permission of the author.)

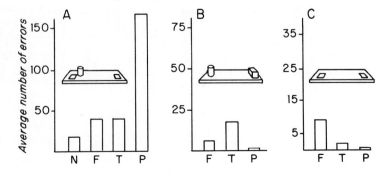

either side of a central panel on which they were supposed to fixate. At various intervals before the target appeared, one of the small peripheral panels was brightened briefly, thereby providing a cue. The target appeared on the "cued" side 80% of the time and on the "uncued" side the rest of the time. When the target appeared on the cued side (Fig. 31–5), reaction times were roughly equivalent regardless of whether the target was presented ipsilateral (CI) or contralateral (CC) to a parietal lobe lesion; however, when the target appeared on the uncued side, reaction times were much greater if the target was contralateral (UC) to a parietal lesion than if it was ipsilateral (UI) to it. This defect is somewhat reminiscent of extinction and suggests that parietal lobe patients predominantly have difficulty in disengaging attention from their ipsilateral (good) side to shift attention to their contralateral (bad) side. They seem to be able to attend to the cue, or to the target, regardless of location, and to be able to use the cue to anticipate the target. They have significant difficulty only if, while attending the ipsilateral cue, they must shift attention to a contralateral target. The effect of brain injury on disengagement of attention seems unique to the parietal lobe; it does not occur in patients with frontal or temporal lobe lesions.[54]

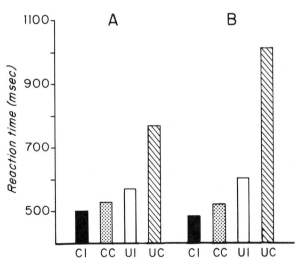

Figure 31–5 Reaction times of patients with right *(A)* and left *(B)* parietal lobe damage to a cued reaction time task that creates a covert shift of attention. Half a second before they must detect a visual target falling on one of two display panels, patients are given a cue on the same or opposite side of the target panel (cued and uncued conditions, respectively), either ipsilateral or contralateral to the side of the lesion. The four conditions are cued ipsilateral (CI), cued contralateral (CC), uncued ipsilateral (UI), and uncued contralateral (UC). (After Posner, M.I.; Walker, J.A.; Friedrich, F.J.; Rafal, R.D. *J. Neurosci.* 4:1863–1874, 1984.)

Neuronal Correlates of Parietal Lobe Function

Records of the discharge of single neurons in alert monkeys also provide information about the function of parietal cortex.

Neurons in areas 5 and 7 respond to sensory stimulation of the body surface. Those in area 5 prefer somatosensory stimuli and over two thirds can be driven by joint rotation and about one fifth by deep or cutaneous touch. They are driven less reliably than neurons in areas 1, 2, and 3, and they have much larger receptive fields. They often prefer moving stimuli. Those in area 7, on the other hand, prefer visual stimuli.[39] Posterior parietal neurons have large receptive fields that usually cover the contralateral visual field and may include the fovea. They are not fastidious about stimulus shape, orientation, color, or direction of movement.[30]

Some neurons in both areas 5 and 7 seem to be involved with movement per se, increasing their discharge when the animal reaches for a reward or a target associated with a reward. Mountcastle and colleagues[39] demonstrated these "projection" neurons by training a monkey to reach out on cue and touch a lighted button that rotated at arm's length along a circular perimeter. Such neurons are not activated by sensory stimuli and do not discharge when the animal makes aggressive or aversive movements with the same trajectory. Also, the firing pattern of projection neurons is independent of the specific arm trajectory. Figure 31–6A shows that the averaged firing rates associated with movements to three different contralateral locations are similar, suggesting that the firing rate does not contain detailed motor instructions but is related to the whole act of projecting the arm to contralateral extrapersonal space. When the monkey was trained to direct not its arm but its eyes, by making a saccade to a target in its contralateral visual field, other neurons in its posterior parietal cortex emitted a burst of spikes before and during the movement (Fig. 31–6B).[30a] These saccade-related neurons bear a strong resemblance to arm projection neurons in that they generally discharge for targeted saccades of all sizes directed into the contralateral field but not for contralateral spontaneous saccades, and their discharge is unrelated to visual stimuli.[30a]

The firing patterns of many cells in posterior parietal cortex appear to have both visual and motor components. One cell type increases its discharge while a monkey fixates a food object with its eyes or fixates a spot of light whose

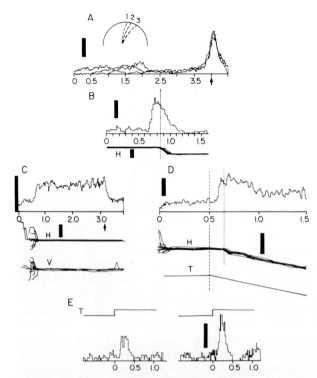

Figure 31-6 Representative discharge patterns of neurons in parietal cortex during motor and sensory events in contralateral extrapersonal space. *A,* Average discharge (instantaneous firing rate) of an arm projection neuron while a monkey reaches for targets at three different locations (inset); activity aligned at instant that target is touched (arrow). *B–D,* Discharge of three different eye movement–related neurons during horizontal (H) saccadic eye movements in *B;* during fixation—i.e., constant horizontal (H) and vertical (V) eye position—in *C;* and during horizontal smooth pursuit (H) of a slowly moving target (T) in *D.* Downward movements correspond to either the left or down direction; calibration bars represent 20 degrees in *B* and *C* and 10 degrees in *D.* In *C,* arrow indicates delivery of reward. *E,* Response of a parietal neuron to appearance (upper trace) of a target spot (T) in its receptive field while the monkey looked at another fixation spot (left) and when it was required to detect the dimming of the target spot (right). All time marks are in seconds; bars associated with all histograms indicate 50 spikes/s. (*A* and *B* after Mountcastle, V. B.; Lynch, J. C.; Georgopoulos, A.; Sakata, H.; Acuna, C. *J. Neurophysiol.* 38:871–908, 1975. *C* and *D* after Lynch, J. C.; Mountcastle, V. B.; Talbot, W. H.; Yin, T. C. T. *J. Neurophysiol.* 40:362–389,1977. *E* after Bushnell, M. C.; Goldberg, M. E.; Robinson, D. L. *J. Neurophysiol.* 46:755–772, 1981.)

dimming it must detect for a reward (Fig. 31-6C). Once the reward is obtained (arrow, Fig. 31-6C), the activity of such cells returns to resting values although the eyes remain fixed on target. These fixation cells also discharge when the fixated spot moves slowly in the visual field. During such smooth pursuit, a second type of cell also increases its activity (Fig. 31-6D). Many of these pursuit

cells also discharge when visual stimuli move across a stationary target that the monkey is fixating (see Fig. 20–18, Chap. 20). On the basis of all these observations, Mountcastle and coworkers[39] proposed that the saccade and tracking cells of area 7 and the projection cells of areas 5 and 7 participate in the initiation and execution of acts into extrapersonal space.

Another group of workers[30a] recorded from similar saccade and tracking neurons but found that every cell that discharged in association with eye movement also responded to an irrelevant visual stimulus delivered in the absence of movement. They also demonstrated that the visual response of many cells was enhanced if the visual stimulus had a special significance, such as being the target for a saccade or a reaching movement. Even if the animal did not make an overt response to the target, but simply signaled the dimming of a peripheral spot while maintaining fixation, the visual response was enhanced. Figure 31–6E compares the response to a small spot falling on the cell's receptive field with the enhanced response to the same spot whose dimming the monkey had to detect. From these data it was concluded that the cells of posterior parietal cortex are instrumental not in initiating movement but in controlling attention. Both this hypothesis and Mountcastle's alternative are consistent with the interpretations of the behavioral data discussed earlier; i.e., that the parietal lobe is involved in either physical or cognitive orientation by directing either overt movement or attention.

Another physiological correlate of parietal lobe function in humans is relative regional cerebral blood flow, which has been measured in human subjects during a variety of motor tasks in both intrapersonal and extrapersonal space.[57, 58]

A radioactive isotope injected into the internal carotid artery is distributed to those brain areas that require increased blood flow. Areas with increased blood flow presumably are those that have increased metabolic needs as a result of the elevated activity of their neurons. Increased local blood flow results in increases in local radiation which are detected by 254 separate sensors, all focused on different cortical regions.[57]

In one motor task, subjects were required to touch their thumbs sequentially with each of their fingers and to repeat this sequence many times. This task was intrapersonal because no external frame of reference was required to accomplish it. An extrapersonal motor task required subjects to move their fingers to different boxes in a maze in response to verbal commands. Figure 31–7 displays

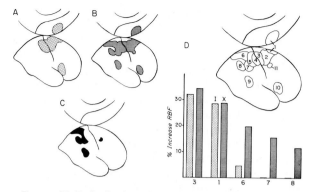

Figure 31–7 Cortical regional blood flow during finger movements in intra- and extrapersonal space. Increases significant at p < 0.05 are shown in *A* for sequential touching of finger to thumb (intrapersonal), and in *B* for finger exploration of a maze (extrapersonal). In *C, B* has been subtracted from *A,* leaving regions activated only during the extrapersonal task. In *D,* histograms show percentage of increase in regional blood flow during the extrapersonal (X, heavy shading) and intrapersonal (I, light shading) tasks in cortical regions identified in inset: 1 = supplementary motor; 2 = premotor; 3 = hand motor; 4 = hand sensorimotor; 5 = hand sensory; 6 = superior parietal; 7 & 8 = inferior parietal; 9 = auditory; 10 = inferior frontal; and 11 = frontal eye field. (After Roland, P.E.; Larsen, B.; Lassen, N.A.; Skinhøj, E. *J. Neurophysiol.* 43:118–136, 1980; and Roland, P.E.; Skinhøj, E.; Lassen, N.A.; Larsen, B. *J. Neurophysiol.* 43:137–150, 1980.)

the areas of the right cerebral cortex that showed a significant (p <0.05) increase in regional blood flow for the sequential finger movements in intrapersonal space (Fig. 31–7A) and the areas active during maze finger movements in extrapersonal space (Fig. 31–7B). Figure 31–7C is a composite that eliminates the areas common to both movement tasks, indicating that the task in extrapersonal space has recruited the posterior parietal lobe. A comparison of the percentage increase in regional blood flow during exploration of extrapersonal (X) and interpersonal (I) space reveals that motor cortex (region 3) and supplementary motor cortex (region 1) are recruited during both tasks, whereas the various divisions of posterior parietal cortex (regions 6, 7, and 8) are recruited only during the extrapersonal task (Fig. 31–7D).

TEMPORAL ASSOCIATION CORTEX

Afferent and Efferent Connections

The temporal lobe can be divided into three functionally separate areas. The superior temporal gyrus (Brodmann's areas 22, 41, and 42) is concerned primarily with audition. Areas 41 and 42 (see Fig. 31–1A) are considered to be primary auditory cortex, as they receive direct thalamic inputs from the medial geniculate nucleus (see Chap. 18). These areas project in turn to the neighboring area 22, which, as we shall see later in this chapter, is crucial for understanding speech. Area 22 also receives inputs from prestriate visual areas, allowing it access to visually perceived words and symbols. The major projection of area 22 is to areas in the frontal lobe; it also projects to the parietal lobe and cingulate gyrus. The inferior temporal gyrus (Brodmann's areas 20, 21, and 37), on the other hand, is concerned primarily with vision. It receives cortical inputs from prestriate areas (Chap. 20) and subcortical visual input from the pulvinar; it also receives input from the parietal lobe. Inferotemporal cortex projects back to these areas and also makes reciprocal connections with the frontal lobe. Other inferotemporal projections are to limbic structures (amygdala and entorhinal cortex) and the basal ganglia (caudate and putamen). Finally, the mediobasal portion of the temporal lobe (primarily anterior) contains the amygdala, hippocampus, and other structures considered to be part of the limbic system (Chap. 32), as well as prepyriform cortex and other cortical areas associated with olfaction (Chap. 21).

The close proximity and strong connections of the lateral temporal association cortex with limbic structures make it difficult to damage one without damaging the other. Consequently, most lesions in humans, especially the surgical cortical removals to control intractable temporal lobe epilepsy, usually involve the amygdala and hippocampus in varying degrees. Over the years, studies of animals and human patients have revealed deficits that can be attributed rather specifically to lesions of limbic structures (Chap. 32). In this chapter, therefore, we consider only those temporal lobe deficits that seem to be related primarily to damage of the lateral surface of the temporal lobe.

Auditory Discrimination Learning

Although the temporal lobe receives direct inputs from the auditory thalamus, lesions of the temporal lobe produce deficits in only the most complex auditory tasks. For example, patients with unilateral temporal lobe resections can easily discriminate between two tones of similar (500 Hz ± 5 Hz) pitch, two tones of similar duration (differing by as little as 50 ms), and two similar rhythmic patterns.[36] However, patients whose right temporal lobes have been removed have difficulty distinguishing between complex sounds that differ only in their harmonic structure or between two tonal patterns (of three to five notes) that differ in only one note (Fig. 31–8A). Destruction of the left temporal lobe occasions only modest, if any, impairment in these musical discriminations.

Experiments in monkeys suggest that complex auditory discriminations are sensitive to lesions that involve the superior temporal gyrus but spare the primary auditory cortex.[67] In these experiments, the animal was trained to perform a simple discrimination between a bell and a buzzer and also a more complicated discrimination (Fig. 31–8B) in which it responded if two successive trains of clicks had the same frequency (response condition, R) and refrained from responding if the frequencies were different (no response condition, NR). After bilateral lesions of the superior temporal lobe, all monkeys readily performed the simple discrimination. When performing the more complex discrimination, however, they invariably responded even if the frequencies were different (Fig. 31–8B). Moreover, they were unable to relearn the discrimination even if the frequencies were made very disparate. The animals could make simple frequency discriminations and perform delayed discrimination tasks; hence, the auditory deficit was not a simple sensory or memory loss. The deficit was restricted to audition, for the animals with superior temporal (ST) lesions readily distinguished similar trains of visual flash stimuli (Fig. 31–8).

The persistence of simple binaural auditory discrimination skills after unilateral temporal lobe lesions should not be too surprising, since each temporal lobe receives information from both ears. Perhaps the clearest auditory deficits consequent to temporal lobe lesions are revealed when auditory stimuli are presented simultaneously to both ears (a dichotic listening task), a condition that appears to create a competitive situation. Musical (or nonverbal) ability can be tested by brief (4-s) presentations of a pair of four different melodies played by the same instrument, one melody to each ear; thereafter, all four are presented binaurally and the subject must identify which two of the four he has heard.[62] Verbal ability, on the other hand, can be tested by presenting three pairs of digits (e.g., 2, 3; 3, 5; 4, 7) briefly, one digit to each ear, and then asking the subject to report which digits were heard. Among patients with unilateral temporal lobe epilepsy, which presumably affects neuronal processing in that part of cortex, those with right temporal lobe epilepsy identify significantly fewer melodies than those with left. In contrast, those with left temporal lobe epilepsy identify fewer digits than those with right temporal lobe epilepsy. These data suggest that to some extent, the discrimination of verbal and nonverbal auditory information depends on different sides of the brain; this cortical "laterality" of some cognitive functions is considered in more detail later in this chapter.

Visual Discrimination Learning

Human and nonhuman primates with temporal lobe damage have difficulty performing visual as well as auditory tasks. Figure 31–8B shows the impairment that occurs in the visual equivalent of the auditory experiment just described. Monkeys with inferotemporal lesions had difficulty distinguishing temporal patterns of light, but not of sound, and could not refrain from responding if the frequencies of two successive trains of light flashes differed.

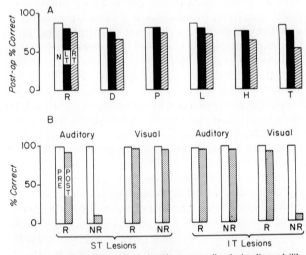

Figure 31–8 Auditory and visual sensory discrimination ability following temporal lobe lesions. *A,* Ability of human patients with right (hatched) and left (solid) temporal lobectomy to distinguish between two tones of similar pitch (P), duration (D), or loudness (L); two similar rhythmic patterns (R); two complex sounds of differing harmonic structure (H); and two tonal patterns differing in only one note (T). Patients with right temporal lobectomy always made more errors than those with left, with the most obvious differences in the tonal (p < .001) and harmonic (p < .01) tasks. N (open histogram) is normal behavior. *B,* Ability of monkeys with superior temporal (ST) and inferotemporal (IT) lesions to discriminate auditory (clicks) or visual (flash) stimuli occurring at different frequencies. Monkeys were trained to respond when the discriminanda were the same (R histograms) and not to respond when they were different (NR histograms). Pre = preoperative; Post = postoperative. (*A* after Milner, B. In Mountcastle, V.B., ed. *Interhemispheric Relations* and *Cerebral Dominance*, 177–195. Baltimore, Johns Hopkins Press, 1962. *B* after Stepien, L.S.; Cordeau, J.P.; Rasmussen, T. *Brain* 83:470–489, 1960.)

In another demonstration of deficits in visual performance, Holmes and Gross[21] presented pairs of symbols at random on two translucent panels and trained monkeys always to touch one particular symbol (or discriminandum) for a reward. As shown in Figure 31–9, normal monkeys (N) learned a variety of visual discriminations, with the number of errors during the learning process reflecting the severity of the task. Monkeys with bilateral lesions of the inferotemporal lobe (IT) had much more difficulty (that is, they made many more errors and consequently took longer) learning to discriminate between two completely different objects or patterns. In contrast, lesions of the lateral striate visual areas (S) had little, if any, effect on these tasks. The inferotemporal deficit has been found with a great variety of discriminanda, including those that differ in a single parameter such as hue, brightness, or size, and those differing in several parameters such as two-dimensional patterns and three-dimensional objects.[17] The deficit is not a total inability to learn all discriminations, since normal monkeys and those with lesions behaved comparably when the task involved discriminating between two identical objects that differed in orientation by 180 degrees (Fig. 31–9C). If, however, identical objects differed in orientation by only 30 degrees, discrimination learning after inferotemporal lesions was again impaired (Fig. 31–9D).

Monkeys with inferotemporal lesions not only have difficulty learning a new visual discrimination, but also discriminations learned prior to the lesion are not remembered. Although few animals can perform a previously learned discrimination immediately after their operations, most are able to relearn the task: i.e., they are still able to perform the task, but the memory for the original task is lost. To quantify this memory loss, experimenters count the number of errors during the relearning process. If the monkeys make fewer errors than they did while learning the task initially, they probably have retained some of their visual habit. If, however, they make the same or a greater number of errors during relearning, they probably have little or no visual habit remaining. As summarized in Figure 31–10, Chow and Survis[6] found that normal animals (N) could still discriminate between a diamond and a disk with perfect retention (i.e., no errors) when tested more than 7 weeks after learning the task. In contrast, animals with inferotemporal lesions (Trained—IT) made more errors while relearning the task than they did while learning it initially, and thus had poor, if any, retention. Animals that were "overtrained" (given additional trials after they had initially learned the task) showed considerably improved retention of the task following the inferotemporal lesion (Overtrained—IT).

Thus, it has been demonstrated that inferotemporal lesions in monkeys affect both the retention of a previously learned visual discrimination and the learning of a new one. The magnitude of the deficits, however, depends on the amount of training, the severity of the task, and the choice of reinforcement; a combination of negative (electrical shock) and positive (food) reinforcements often improves the performance of monkeys with temporal lobe lesions.

Deficits owing to inferior temporal lobe lesions are purely visual.[17] The learning and retention of olfactory, tactile, auditory (see Fig. 31–8B) and gustatory discriminations are unaffected. Also, the deficit is not purely sensory, since measures such as minimum separable acuity (Chap. 19), critical fusion frequency, and detection of a brief flash are

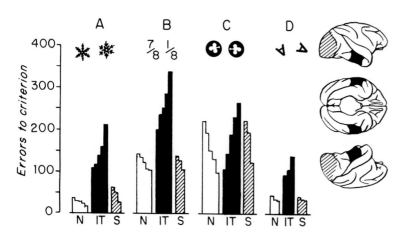

Figure 31–9 Object discrimination learning in five monkeys with bilateral lesions of inferotemporal cortex (IT), three with bilateral lesions of lateral striate cortex (S), and five normal animals (N). The discriminanda used in four different tasks, A–D, are shown above the histograms. Typical IT (black areas) and S (hatched areas) lesions are shown on the brain at the right. Criterion was 45 correct in 50 consecutive responses. (After Holmes, E.G.; Gross, C.G. J. Neurosci. 4:3063–3068, 1984.)

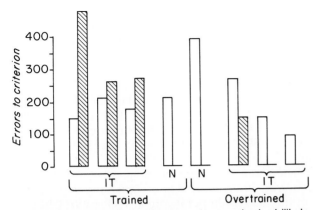

Figure 31–10 Effect of overtraining on a monkey's ability to retain an object discrimination task after inferotemporal (IT) lesions. Two normal (N) and three monkeys with IT lesions were trained to discriminate between a diamond and a disc; criterion was 27 correct in 30 consecutive responses ("Trained" histograms). "Overtrained" monkeys were given 500 additional trials. Open histograms = preoperative; hatched histograms = postoperative. (After Chow, K.I.; Survis, J. *Arch. Neurol. Psychiatr.* 79:640–646, 1958.)

normal. Animals with inferotemporal lobe lesions often learn such tasks slowly but eventually achieve normal performance. Finally, the deficit is permanent, since after years of postoperative training experience, animals with inferotemporal lesions still acquire new discriminations more slowly than normal animals with the same experience.

Similar deficits in visual discrimination occur in human subjects who through disease or surgery have suffered more extensive and less well-defined temporal lobe lesions. Purely perceptual visual changes are so slight that the patient is rarely aware of them. Also, as in the operated monkey, only rather difficult visual discrimination tests reveal a deficit. For example, humans with damage of the right temporal lobe cannot make judgments that require perception of incompletely represented figures. They have difficulty categorizing, by sex and age, drawings of human faces that have been transformed so that shadows are rendered in black and highlights in white. Also, they have difficulty choosing which of four fragmented concentric circles is different (Fig. 31–11A) or guessing the identity of a target item (Fig. 31–11B) from cumulatively presented, fragmented visual clues.[34]

Although right temporal lobe resection produces only a mild deficit in visual perception, it produces a severe defect in visual memory. To reveal this deficit, Milner[38] showed patients 12 photographs of unfamiliar faces that they had to identify 90 minutes later from a larger array of 25 photographs. Figure 31–12 shows that patients with right temporal lobe damage (R) identified a significantly lower percentage of faces than did normal subjects (N). Furthermore, they performed more poorly than patients with either left temporal lobe lesions (L) or frontal lobectomies (F). When the 12 sample photographs and the larger array of 25 photographs were presented with "No delay," the performance of patients with right and left temporal lobe damage was not significantly different. The modest impairment in immediate facial recognition displayed by patients with right temporal lobe lesions can be attributed to their inability to distinguish similar intricate visual patterns (see Fig. 31–11A).

The disorder of visual memory is not specific to faces but reflects a more general inability to distinguish and remember patterns with common structural attributes. To demonstrate this, a patient is first shown 20 cards, each of which contains a line drawing of either a geometric or nonsense figure. After this initial presentation, seven more pre-

Figure 31–11 Examples of tasks requiring the perception of incompletely represented figures. Patients with right temporal lobe lesions have difficulty identifying the non-matching figure in *A* and guessing the target item in *B* from the fragmented clues. (Modified with permission from Miller, L. *Neuropsychologia* 25: 359–369. Copyright 1985 Pergamon Press plc and from Meier, M.J.; French, L.A. *Neuropsychologia* 3:261–272. Copyright 1965 Pergamon Press plc.)

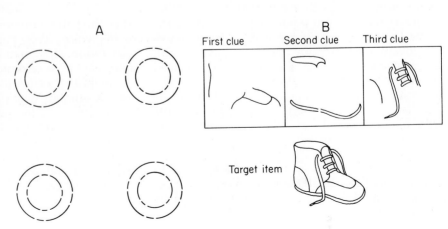

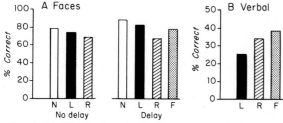

Figure 31–12 Visual and verbal recall in patients with left and right temporal lobe damage. In *A*, subjects were asked to recall 12 faces from an array of 25 either immediately (No delay) or 90 minutes later (Delay). Right temporal lobectomy patients (R) performed significantly worse than normal subjects (N, p < .001), left temporal patients (L, p < .001), and frontal lobectomy patients (F, p < .01) on Delay but not on No delay tasks. In *B*, patients were required to perform two verbal memory tasks after 90-minute delays. Patients with left temporal lobe epilepsy (L) scored significantly lower than either patients with right temporal lobe epilepsy (R, p < .001) or patients with frontal lobe damage (F). (*A* after Milner, B. Neuropsychologia 6:191–209, 1968. *B* After Milner, B.: In Mountcastle, V.B., ed., *Interhemispheric Relations and Cerebral Dominance*, 177–195. Baltimore, Johns Hopkins Press, 1962.

sentations of 20 cards each are carried out and the patient is required to identify those he has seen in the initial presentation (four nonsense and four geometric drawings of the original set recur in each trial). Patients with right temporal lobe damage correctly identify fewer recurring figures than do patients with left temporal damage. The deficit is especially dramatic for the recall of the nonsense figures: right temporal patients identify one-third fewer drawings.[25]

In contrast to patients with right temporal lobe damage who have difficulty remembering similar intricate visual patterns, patients with left temporal lobe damage have difficulty recalling verbal material. Although the deficit is most pronounced after a lobectomy, even patients with left temporal lobe epilepsy recall significantly fewer items from a short prose passage than do patients with similar lesions in the right temporal lobe or in other parts of the brain; they also are able to recall fewer previously learned word pairs. Figure 31–12B, which displays a composite measure of delayed verbal recall (number of story items plus number of correct word associations after a delay of 90 minutes), shows that the performance of patients with left temporal lobe epilepsy is significantly worse than that of patients with comparable frontal lobe damage or right temporal lobe epilepsy.[36]

In summary, unilateral temporal lobectomy in humans produces highly specific deficits that depend on the side of removal. Left temporal lobe damage, anterior to the auditory and speech-re-

lated regions, affects verbal memory, while right temporal lobectomy affects nonverbal memory. Examples of the latter include analysis of changes in unfamiliar melody patterns and discrimination of unfamiliar faces and of nonsense figures. Such tasks tend to employ highly idiosyncratic stimuli that are too rich in detail to be described easily with words. Other cognitive functions that are more strongly dependent on one side of the brain than on the other are discussed later in this chapter.

There seems to be general agreement that tasks requiring short-term memory are more dependent on the lateral temporal cortex. Well-controlled lateral lesions in monkeys affect their abilities to perform auditory and visual discriminations that require the short-term retention of two auditory (see Fig. 31–8) or visual (see Fig. 31–9) discriminations. Also, humans with excisions of the lateral aspect of the left temporal lobe have difficulty recalling, after a very brief delay, the essential elements from two orally presented segments of prose. These deficits are not exacerbated as the excision encroaches more upon medial temporal structures, indicating that verbal short-term memory may also depend on lateral temporal cortex.[45] On the other hand, deficits in long-term memory following temporal lobe removals can be attributed primarily to *combined* damage of the hippocampus and amygdala[40] (see also Chap. 32).

Prosopagnosia

An infrequent but strikingly bizarre deficit caused by lesions involving the temporal lobe is an inability to recognize the faces of friends or even one's own face in a mirror. This deficit, called *prosopagnosia*,[9] is not a perceptual one, since the patient can usually discriminate unfamiliar faces (i.e., he can match different photographs). Also, the patient can identify a friend by the sound of his voice. Prosopagnosia is not confined to recognition of human faces alone; these patients may also fail to recognize their own cars or different makes of cars, clothes of the same type, or specific animals within a group. The latter condition, in which a farmer cannot recognize specific members of his herd or an experienced bird watcher cannot recognize different kinds of birds, is sometimes called *zooagnosia*. The common characteristic of stimuli that reveals prosopagnosia is that they possess a visual ambiguity (i.e., a stimulus group is composed of different members with similar characteristics). Therefore, prosopagnosia is considered to be a disorder of visually triggered contextual memory.

Although early experiments suggested that prosopagnosia resulted from damage to the right hemisphere, recent studies, especially those using modern imaging techniques, suggest that most patients with prosopagnosia have bilateral lesions. These lesions usually involve the occipitotemporal region, which appears to be concerned with form vision (see Chap. 20); in this regard, it is interesting to note that many patients with prosopagnosia also have cerebral color blindness, or *achromatopsia*, a finding that implies that form and color may be processed in common structures. On the other hand, patients with bilateral lesions of the occipitoparietal region (part of the suggested visual motion pathway; see Chap. 20) are not similarly afflicted. The part of the temporal lobe involved specifically with the recall of unfamiliar faces, as described earlier, does not appear to be involved in the recognition of familiar faces, since patients with lesions in those areas easily recognize the faces of family and friends.

PREFRONTAL ASSOCIATION CORTEX

Afferent and Efferent Connections

The frontal lobe is considered to be all of the cortex rostral to the central sulcus and medial to the sylvian fissure. Traditionally, it is divided on cytoarchitectonic grounds into three regions: an agranular region containing primary motor cortex (area 4) and premotor cortex (area 6), a transitional "dysgranular" region containing area 44 and the frontal eye field (area 8), and a granular region known as prefrontal association cortex (areas 9–14, and 45–47) (see Fig. 31–1). Granular prefrontal association cortex can be further subdivided into an orbital region on the medial inferior surface above the bony orbits (areas 11–14 and 47) and a dorsolateral region (areas 9, 10, 45, and 46). Some authors include the dysgranular cortex as part of the prefrontal cortex, and so shall we.

The dorsolateral and orbital prefrontal areas appear to have somewhat different afferent and efferent connections. The dorsolateral surface receives inputs from a variety of sensory modalities via long intracortical pathways. These include visual inputs from prestriate cortex, inferior parietal cortex (there is a superior parietal projection as well), and the inferior temporal gyrus. Also, there are auditory inputs from the superior temporal gyrus. In addition, there is a thalamic input from the parvocellular portion of nucleus medialis dorsalis, but the nature of the information flowing in this pathway is not known. The orbital surface also receives an input from the middle and inferior gyri of the temporal lobe and a projection from the magnocellular portion of nucleus medialis dorsalis. Part of the thalamic pathway relays olfactory signals from the prepyriform cortex (see Chap. 21), and part relays information from such limbic structures as the amygdala, septum, and mesencephalic tegmentum.

Efferents from the prefrontal association area connect, in turn, with many of the structures that provide prefrontal afferents. In particular, the prefrontal region provides the strongest neocortical connections to the basal forebrain structures involved in emotional behavior or visceral control. Prefrontal target structures include the cingulate and hippocampal gyri, the inferior temporal lobe, and, indirectly, the amygdala. Furthermore, the prefrontal region is apparently the only cortical area with direct connections to the hypothalamus. Also, there are direct connections to the caudate nucleus, putamen, nucleus medialis dorsalis, and, in general, to those cortical areas that provide afferents. All of these connections arise from both the dorsolateral and orbital surfaces.

Experiments on Monkeys

Delayed Response Tasks: Effects of Lesions

Although responses to the immediate environment constitute much of an animal's behavior, many responses, although evoked by the current situation, owe their direction to sensory information gained previously. This capacity can be assessed experimentally by the delayed response test, which may take a variety of forms. In the simple delayed response test, a monkey is allowed to see a piece of food being placed in one of two covered wells, after which an opaque screen, which obscures the wells, is lowered in front of the animal for a certain amount of time. After this delay, the screen is raised and the animal is allowed to obtain the reward in the baited well. With training, a normal monkey makes successful choices at least 90% of the time after delays of up to 45 s. After bilateral prefrontal ablations, even delays as short as 2 s make a successful response a matter of chance.[23] Most animals with lesions are unable to relearn the task, even with extensive retraining for over a year. Deficits in delay tasks have also been observed after lesions of structures directly connected to the prefrontal cortex, such as nucleus medialis dorsalis, the caudate nucleus, and the hippocampus. They do not occur following extirpation of either the parietal association area or the lateral temporal lobe. Failure to perform this *delayed response* task successfuly cannot be attributed to global learning deficits, because monkeys with prefrontal lesions retain the ability to perform visual discriminations on the basis of size, form, or brightness and can learn new visual discriminations.[22]

Monkeys with prefrontal lesions are also deficient in performing *delayed alternation* tasks in which the animal must remember which well was baited and then must choose the *other* to obtain a reward.

The deficits in the various delayed response tasks are often interpreted as a failure in short-term memory. While delayed response clearly involves memory, several observations suggest that other factors are involved. Although animals with prefrontal lesions can perform the tasks when there is *no* delay, a momentary distraction produced by incidental noises reduces correct behavior to chance. Also, if monkeys are maintained in darkness during the delay period, they perform at preoperative levels.[31] These data suggest that prefrontal cortex does not store short-term memories but may shield those memories from interference (i.e., retroactive inhibition, as is provided by environmental stimuli). Finally, motivation is an important additional factor in these tasks, since specialized retraining techniques and the administration of drugs that stimulate appetite also improve performance on delayed response tasks.

Prefrontal lesions also affect tasks that require shifts of behavioral set.[11] In a *go/no go* task, for example, an animal is trained to respond when one stimulus is presented and to refrain from responding to another stimulus (recall the behavioral situation of Fig. 31–8B). Monkeys with prefrontal lesions are unable to refrain from responding to the *no go* stimulus; that is, they are locked into a single behavioral strategy of responding. Also, following frontal lobe lesions, monkeys trained to transfer their response from one stimulus to another on cue tend to persist in their choice of the formerly correct but now inappropriate stimulus. This failure to overcome a response tendency is called *perseveration*. Perseveration can be seen in delayed response tasks and in tasks requiring sequential processing of information, and it is a central feature of frontal lobe dysfunction in both human and nonhuman primates.

Delayed Response Tasks: Neuronal Correlates

Many neurons in prefrontal cortex have firing patterns that seem to be related to the events of the delayed response task. In an experiment by Fuster[11] a monkey faced two response buttons, one of which was illuminated briefly to provide a cue. After a delay, the monkey had to press the illuminated (delayed response) or darkened (delayed alternation) button to obtain a reward. Most of the neurons studied (Fig. 31–13) showed a change in activity associated with the occurrence of the cue. The response seemed less related to the specific attributes of the cue, which could be

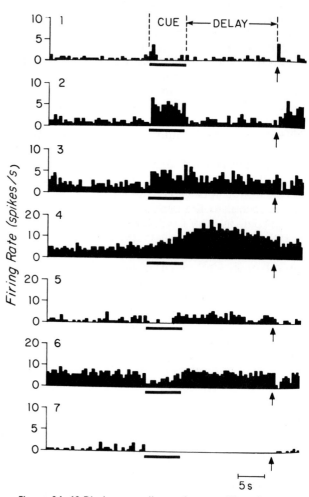

Figure 31–13 Discharge patterns of seven different neurons in monkey prefrontal cortex during a delayed response task in which a 7-s cue (horizontal bar) was followed by an 18-s delay (end of delay marked by arrow). Each histogram represents the average of five trials. (After Fuster, J.M. *J. Neurophysiol.* 36:61–78, 1973.)

auditory or visual, than to the animal's intent to respond behaviorally to the cue. Some of these neurons also exhibited an increase (neurons 3 and 5) or decrease (neuron 7) in activity during the delay between the cue and the response. A smaller number did not respond immediately after the cue but exhibited a sustained change in firing during the delay period between the cue and the response (neuron 4). Niki and Watanabe[43] found that for many units the response during the delay period depended on the attributes of the cue rather than on the impending movement. For example, the discharge of these units changed if the cue was green or red or located on the right or on the left. These experiments suggest that some prefrontal neurons may be involved in short-term memory for abstract information the behavioral relevance of which has been established through training (e.g., the characteristics of a cue that indicate a rewarded response).

An electrical potential that seems to reflect the short-term memory phenomenon can be seen in human frontal lobe electroencephalographic recording (Chap. 32) during certain behavioral tasks. For example, if a subject is presented with a click followed at a regular interval by a light that must be extinguished rapidly for a reward, a slow, negative potential develops in the interval between the auditory and visual stimuli as the subject learns the association between the response and the reward. If, sometime later, the extinction of the light is no longer rewarded, the negative potential subsides, indicating that it depends heavily on the contingency between the response and the reward. This potential is called the *contingent negative variation*[73] and is thought to result from the summated activity of many prefrontal neurons. Its timing, gradual acquisition, and decay suggest that it may represent a short-term memory trace for task-related stimuli. Thus, the contingent negative variation shows a striking similarity to the cue-related neuronal activity seen in monkeys.

Disorders of Emotion and Affect

As might be expected from the strong prefrontal connections to limbic structures, lesions of prefrontal cortex have profound effects on behaviors related to emotion, such as general activity level, conditioned circulatory responses, and experimental neuroses.

In monkeys, ablation of the entire prefrontal lobule or certain of its subareas induces a state of hyperactivity characterized by incessant, stereo-typed walking or pacing, much like that of a caged tiger. The amount of hyperactivity depends on the environment; it ceases in darkness and is proportional to the level of illumination and ambient noise. Bilateral lesions of area 13 (the posterior part of the orbital cortex) produce nearly maximal hyperactivity, although hyperactivity is also produced by lesions of area 9. The hyperactivity develops only slowly after the lesion and is preceded by a period of hypoactivity (e.g., apparent apathy, drooping of the head, sluggishness of movement, and blankness of expression) that can last up to two or three weeks after surgery. Once hyperactivity begins, it persists indefinitely.

Conditioned circulatory responses can be elicited in a normal monkey by giving it a warning tone 60 s before it receives a strong electrical shock. When the tone and the shock are repeatedly paired, the monkey's arterial blood pressure gradually increases and remains elevated during the 60-s interval preceding the shock. This is an example of classical conditioning, which is discussed in Chapter 33. After prefrontal lobectomy, the anticipatory increase in blood pressure disappears.[60]

Chimpanzees can be made to exhibit behaviors that strongly resemble neurotic behavior in humans by requiring them to perform difficult discrimination problems near threshold. Jacobsen[22] described a chimpanzee that, whenever it made an error, flew into violent temper tantrums, rolled on the floor, and defecated and urinated. Eventually the animal had to be forced into the experimental apparatus. After bilateral lobectomy, however, its behavior changed profoundly. It then entered the experimental room willingly and worked with alacrity. Mistakes and failures to obtain food caused no emotional manifestation, although many more errors were made than before the operation. This anecdotal observation provided much of the rationale for the "psychosurgical" prefrontal lobotomies introduced in 1935 by Egas Moniz, a Portuguese neurologist. These procedures helped make agitated mental patients and hyperactive children more tractable but, as we shall see later, produced other, undesirable sequelae.

In summary, lesions of prefrontal cortex in monkeys consistently affect their abilities to perform a variety of delayed response tasks. The discharge patterns of many neurons in the prefrontal area suggest that they are involved in some aspect of delayed response behavior. Also, prefrontal lesions seem to affect the emotional response to stress.

Experiments on Humans

The behavior of patients with frontal lobe lesions has been observed closely since the mid-nineteenth century. On the basis of accumulated data, clinicians now agree that such patients consistently display an inability to (1) initiate behaviors, (2) switch from one behavioral strategy to another, (3) use mistakes in ongoing behavior to alter performance, (4) suppress incorrect responses, (5) string together a number of simple behaviors to produce a more complicated one, and (6) deal with distractions.[68] Some of these deficits are easily revealed in the patient's behavior in normal situations, whereas others can be demonstrated only by having the patient do complicated problem-solving tasks.

Effects of Lesions: Emotional Deficits

A lack of initiative and an inability to monitor personal performance are revealed in the dramatic change in personality of patients with frontal lobe lesions. Immediately after the lesion, patients become apathetic or somnolent and develop a masklike facial expression; some also display mutism and catatonic attitudes or akinesia (i.e., an absence of movement). After several days, variable changes in behavior occur. These include tactlessness, lability, changes in social habits, distractibility or impaired concentration, disturbances in judgment, emotional shallowness, and a tendency toward irresponsible or inappropriate behavior. A classic character change is *Witzelsucht*, a tendency toward frivolous and sometimes stupid and tedious joking, often at the expense of others. Such personality changes may so alter a patient that he seems a different and sometimes unacceptable person to his relatives and friends.

The study of frontal deficits in man is usually taken to begin with the case of Phineas P. Gage, an efficient and capable foreman who, in 1848, was injured when a tamping iron was blown through the frontal region of his brain. His physician described the brain-injured Gage as "fitful, irreverent, indulging at times in the grossest profanity (which was not previously his custom), manifesting but little deference to his fellows, impatient of restraint or advice when it conflicts with his desires, at times pertinaciously obstinate yet capricious and vacillating, devising many plans for future operations which are no sooner arranged than they are abandoned in turn for others appearing more feasible . . . his mind was radically changed so that his friends and acquaintances said he was no longer Gage."[18]

Effects of Lesions: Problem Solving

The frontal lobes appear to be less concerned with the generation of cognition than with its regulation and integration. As late as 1957, an exhaustive study of patients with frontal lobe lesions revealed no deficits on the "intelligence" tests than available.[74] Since then, investigators have developed several tests that are relatively specific for frontal lobe deficits. The tasks involved are so complicated, however, that failure to perform them does not specifically indicate one of the deficits enumerated earlier, but rather can be attributed to any or several of them.

The test that most reliably distinguishes a patient with a prefrontal lesion from normal subjects or patients with other brain lesions is the *Wisconsin Card Sorting Test* (Fig. 31–14).[37] In this task the patient receives a pack of response cards that he must match to one of four cue cards that differ in the color, shape, and number of their geometric designs. The examiner secretly selects a sorting strategy and the patient must guess the strategy and sort the cards accordingly. For example, suppose that the examiner has decided that the correct response is a shape match but the subject guesses a number match and places the first response card in front of cue card #2. The examiner indicates that the response is "wrong," so the subject places the second response card in front of cue card #4 and the examiner indicates a correct response. After the subject has made 10 consecutive correct responses, the matching principle is changed and the subject must modify his strategy accordingly. Patients with unilateral epileptic foci in the lateral prefrontal cortex (LF in Fig. 31–14) make more errors than do either normal subjects or patients with foci in other association cortices. After the lateral prefrontal foci are removed, the patients make even more mistakes. Most of the mistakes made by prefrontal epileptics both before and after the surgery are *perseverative errors*; that is, the subject doggedly continues to make his former response although the matching principle has been changed. Such responses indicate inabilities to shift problem-solving strategies and to correct ongoing mistakes.

Certain maze tests also reveal frontal lobe deficits. The first test of any kind to reveal dysfunction resulting from frontal lobe damage, and the one still most frequently used, is the *Porteus Maze Test*. It consists of a series of printed mazes that require advanced planning of the route to be followed, anticipation of possible blocked passages, and completion of any planned actions to permit the

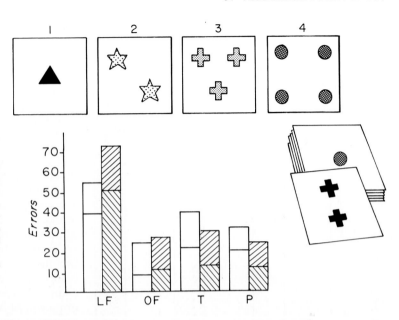

Figure 31–14 Mean number of errors made in the Wisconsin card sorting test, which requires subjects to place each of 128 cards in a pack in front of the appropriate cue card (top row) according to instructions for matching number, color (indicated by different shadings) or form. Subjects included patients with epileptic foci in the lateral frontal (LF), orbitofrontal (OF), temporal (T), and parietal (P) cortices. Histograms show performance before (open histograms) and after (hatched histograms) removal of the foci. Lower portions of histograms indicate perseverative errors (i.e., responses that would have been correct if the immediately preceding sorting criterion still applied). (After Milner, B. *Arch. Neurol.* 9:90–100, Copyright 1963, American Medical Association.)

subject to draw a line out of the maze without retracing, crossing any borders, or lifting the pencil. There is no time limit for completion of the maze and subjects receive scores according to the difficulty of the completed maze. Lobotomized patients invariably earn lower test scores and make more qualitative errors (crossing lines or lifting the pencil) than do normal subjects, although they understand and can recite the rules.[53] Such patients repeat identical errors, often on successive trials in the maze and, unlike normal subjects, show no improvement with repeated attempts. Failure in this maze test reveals an inability to learn by one's mistakes, to organize events in temporal succession (i.e., to plan ahead), and to suppress incorrect responses. On the other hand, the inability of patients with parietal lobe damage to perform maze tasks can be attributed to deficits in spatial perception.

Patients with frontal lobe damage also have problems with perceptual constancy; these problems can be demonstrated with the use of ambiguous figures such as the double Neckar cube (Fig. 31–15). If a normal subject views this drawing for 1 minute, he experiences about 15 reversals of perspective (N in Fig. 31–15). In the same period, patients with bilateral lesions of frontal cortex (BF) experience more reversals (about 20) than do normal subjects or patients with other types of bilateral (BNF) brain damage. Patients with unilateral lesions, both frontal and non-frontal, undergo fewer reversals than do normal subjects over this period.[69] Thus, it appears that patients with bilateral frontal lobe lesions display an unusual lability

in their perceptions, switching from one mental image of the ambiguous figure to the other. This lability may contribute to their inability to develop effective problem-solving strategies and to maintain attention in the face of distractions.

One of the clearest deficits produced by frontal lobe lesions occurs in a test involving visual search. In such a task, a subject is asked to match an object that appears in the center of his visual field with one of 48 objects scattered randomly throughout his visual field. The time required for patients with a unilateral prefrontal gunshot wound (LF and FR in Fig. 31–16A) to find a match in the field contralateral to the lesion is prolonged by three- to sixfold. As shown in Figure 31–16A, bilateral prefrontal damage (BF) produces a more than ninefold increase of search times in both half fields, while lesions to the temporal (T), parietal (P), and occipital association (O) cortices produce relatively modest deficits.[69] Patients with frontal lobe lesions also may ignore or be inattentive to stimuli presented contralateral to the lesion, a phenomenon reminiscent of the hemi-inattention that characterizes parietal lobe damage. In addition, patients with frontal lobe lesions make abnormal scanning eye movements when observing a visual scene. As can be seen in Figure 31–16B, much of the directedness of the scanning eye movements is lost in these patients, who make seemingly aimless eye movements.

Hemi-inattention can be produced in monkeys by focal lesions of the frontal eye fields (Brodmann's area 8), which are considered to be a "premotor" area for

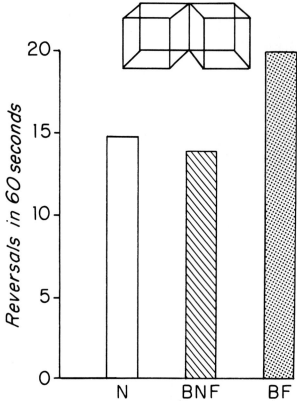

Figure 31-15 Average number of apparent reversals of Necker cubes (inset) over a 60-s period of continuous inspection by normal subjects (N) and patients with either bilateral frontal (BF) or bilateral nonfrontal (BNF) lesions. (After Teuber, H.L. In Warren, J.M.; Akert, K., eds. *The Frontal Granular Cortex and Behavior*. New York, McGraw-Hill Book Co., 1964.)

eye movement. Consequently, deficits in visual search in human patients after frontal lobe lesions may reflect damage to the frontal eye fields. Apparently, however, some patients with frontal lobe lesions who have no area 8 involvement also experience problems with visual search.

Some efforts have been made to attribute certain deficits to specific subregions of the frontal lobes. In human patients, lesions of the dorsolateral convexity, including area 9, appear to impair performance on tests of integrated cognitive function (e.g., the Wisconsin Card Sorting Test and Porteus Maze Test) more than do orbital lesions. Similarly, in monkeys lesions involving the banks and depths of the principal sulcus (see Fig. 33–1B), including area 9, are those that produce deficits in delayed alternation. Orbitofrontal lesions in monkeys and in humans seem more likely to alter emotional state. Also, electrical stimulation of this region produces autonomic responses in monkeys and powerful emotional reactions in humans.

These observations have led some to suggest that orbitofrontal cortex should be considered part of the limbic system, with which it is reciprocally connected.

Comparison of Deficits in Monkeys and Humans

If allowance is made for the richer and more diversified nature of human behavior, the similarities between the deficits produced by frontal lobe lesions in monkeys and humans are striking. Both monkeys and human patients with frontal lobe lesions exhibit an increased tendency toward perseveration, have difficulty with tasks that require the linking together of a temporal sequence of behaviors, and display significant emotional changes. Some individual tests, however, reveal impairment in monkeys with lesions (e.g., delayed response task) but pose no problem for humans with frontal lobe damage. This difference is not surprising because humans probably solve many of these problems with strategies that are not available to monkeys; thus, similar tasks pose very different problems for nonhuman and human primates. It is also possible that structurally homologous neural tissue serves somewhat different functions in the two species.

ASYMMETRICAL ORGANIZATION OF CERTAIN COGNITIVE FUNCTIONS

In ordinary life, normal people show very little evidence of a functional asymmetry in their brains. Perhaps the only obvious example is that most people are able to do things much better with one hand than with the other. The first indication of a possible asymmetry in brain function came with Broca's observation in 1851 that speech disorders are associated primarily with lesions of the left hemisphere. Since then, a variety of behaviors have been shown to be more affected by the absence of one hemisphere than the other.

These cortical asymmetries have been demonstrated by two essentially different approaches. In the first, one of the hemispheres is incapacitated and the behavior of the remaining hemisphere is tested. This partial or complete inactivation can occur following natural lesions (e.g., strokes) or surgical lesions for the relief of epilepsy. Temporary unilateral inactivation of one hemisphere can also be accomplished by injecting the anesthetic sodium amytal into the internal carotid artery on

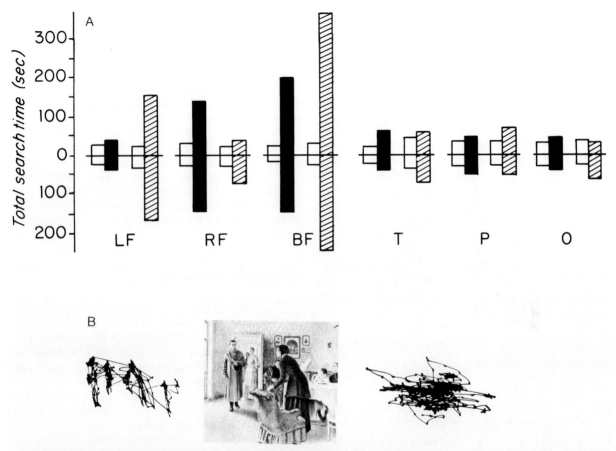

Figure 31–16 Deficits in visual searching produced by frontal lobe lesions. *A* The total search time required to find a test object in an array of objects by a normal subject (open histograms) and patients with gunshot wounds of left (LF), right (RF), or bilateral (BF) frontal lobes, temporal (T), parietal (P), or occipital (O) lobes. Histograms indicate search times for test objects in each quadrant of the visual field—upper and lower left (black histograms) and upper and lower right (hatched histograms). *B* Two-dimensional eye movements of a normal subject (left) and a patient with a frontal lobe lesion (right) during a 3-min observation of the picture. Both subjects were asked to identify the clothes that the people were wearing. (*A* after Teuber, H.L. In Warren, J.M.; Akert, K., eds. *The Frontal Granular Cortex and Behavior.* New York, McGraw-Hill Book Co., 1964. *B* after Pribrains, K.H.; Luria, A.R. *Psychophysiology of the Frontal Lobes.* New York, Academic Press, 1973.)

one side. Such injections produce a contralateral hemiparesis, hemianesthesia, and hemianopsia for 6 to 10 minutes while the ipsilateral hemisphere remains conscious, although partly depressed, and in control of the contralateral body.[71]

Cortical asymmetries can also be revealed in patients whose cerebral commissures (the corpus callosum) have been cut to relieve intractable epilepsy. Remarkably, this procedure, which interrupts the largest tract in the brain, produces few observable postoperative abnormalities, and those it does produce (e.g., lack of concentration) do not distinguish commissurotomy patients from those with other cortical insults. Within a week, even those deficits disappear, so that these individuals appear normal on a routine neurological

examination. However, careful psychophysical testing of interhemispheric integration reveals that, in many situations, each hemisphere operates as an independent entity and each is involved principally in somewhat different functions.

Our discussion of sensory and motor systems in previous chapters indicated that each cerebral hemisphere is responsible primarily for the opposite side of the body. For example, sensory stimuli applied to the left side of the body first influence neurons in the right cerebral hemisphere. This effect is particularly obvious for vision, where a sagittal plane through the nose effectively bisects the visual world into left and right hemifields that project to the right and left occipital lobes, respectively (Chap. 20). A similar lateralization of function occurs for movement, since the muscles on one side of the body are controlled by signals from the contralateral hemisphere (Chap. 28). Therefore, if the right arm is to manipulate stimuli presented in the left hemifield, sensory information from the right cortex must cross the midline in the cerebral commissures to influence the left motor cortex. Thus, to determine whether the right hemisphere of a commissurotomized patient participates in a particular function, one would present a sensory stimulus to his left visual field and ask him to respond by using or signaling with his left hand.

It is difficult to present stimuli selectively to one visual field and not to the other. It can be accomplished by having the subject fix his gaze on a designated point on a viewing screen and flashing stimuli individually to the left and right halves of his visual field, using tachistoscopic exposure times of one tenth of a second or less. The brief exposure times ensure that stimuli intended for one hemifield will not be projected to the other through scanning movements of the eyes.

Hemispheric Similarities

Using the testing strategy outlined earlier, one can demonstrate that each individual hemisphere of a commissurotomized patient is capable of awareness, judgment, and voluntary motor control in a variety of situations. For example, pictures of scenes or events presented to either hemifield are perceived to have the appropriate personal, social, or political meanings.[65] Also, a familiar object presented to either visual hemifield can usually be identified by the ipsilateral hand when it is presented in a hidden array of other common objects. This result indicates that visual and haptic experiences can be integrated by each hemisphere. Not only is the subject able to make sensory judgments by handling objects with either hand, but he is also able to manipulate them in a manner appropriate to their use or purpose (e.g., demonstrate the use of a tool). Furthermore, each hand is capable of mimicking hand postures that have been flashed into the ipsilateral visual field, indicating that each hemisphere is capable of performing some tasks that require complex sensory-motor integration.

The independence of each hemisphere can be elegantly demonstrated by a double visual reaction time task. In this task, a commissurotomized patient fixates a central spot and responds as quickly as possible to pairs of visual stimuli presented simultaneously in both peripheral fields. For ex-

ample, if the left hand is required to touch the red button of a leftward red/green pair and, at the same time, the right hand is required to touch the black button of a rightward black/white pair, normal subjects take 40% longer to press both buttons than they take to press one button of a single pair with one hand alone (i.e., the double discrimination takes longer). Commissurotomy patients, however, perform the double discrimination task as rapidly as normal subjects perform a single discrimination task. This observation suggests that the commissurotomy creates two functionally complete and independent hemispheres, and that normal subjects experience inter-hemispheric interference when required to perform a dissociated bilateral task.[12]

Hemispheric Differences

The same testing strategy reveals that although a visual stimulus can be recognized by the right hemisphere, it is usually named by the left hemisphere. The composite stimulus used to reveal this asymmetry (Fig. 31–17) is formed of the right half of one complete visual image and the left half of another (i.e., a chimera) and is presented so that each half-image reaches the contralateral hemisphere.[29] In 98% of the trials in which this stimulus was presented, commissurotomized patients reported the perception of a single unitary image (e.g., a face). The image that was reported depended on the method by which the subject was allowed to make the report. If the subject was asked to point to the stimulus, 90% of the time he identified the face that corresponded to the left half-face (L) that projected to his right hemisphere (Fig. 31–17, right histograms). On the other hand, if required to name the stimulus, he showed a greater tendency to name the face that corresponded to the right half-face (R) that projected to his left hemisphere (Fig. 31–17, left histograms).

In similar experiments, if two letters, numbers, or pictures of objects are flashed simultaneously—one to the right and one to the left visual field—the subject generally admits to seeing only the right-field (i.e., left hemisphere) stimulus (Fig. 31–18A). However, if he is asked to retrieve the pictured object from behind a screen using his left hand, he will consistently select the object that matches the left-field stimulus.[64] When asked to confirm verbally which item was selected by the left hand, he will incorrectly name the right-field stimulus. Taken together, these observations suggest that if a task requires only visual recognition,

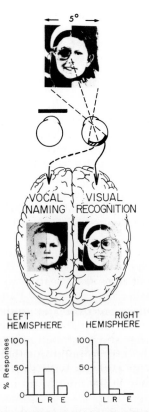

Figure 31–17 A chimeric stimulus presented tachistoscopically through one eye of a commissurotomized patient as he fixates a central point produces separate but complete perceptions in the two hemispheres. Histograms indicate the percentage of times that a subject points at (right set of histograms) or names (left set of histograms) the left (L) or right (R) half stimulus, or neither (E). (After Levy J.; Trevarthen, C.; Sperry, R.W. *Brain* 95:61–78, 1972.)

Figure 31–18 Responses of a commissurotomized patient to the tachistoscopic presentation of stimuli to the left and right visual fields, which reach the right and left hemispheres, respectively. *A,* When stimuli are presented simultaneously to the right and left hemifields, the subject can name the right-field stimulus but retrieves the left-field stimulus. *B,* The subject can read and understand the names of objects presented in the left hemifield and can retrieve the appropriate items with his left hand but cannot name them. (After Sperry R.W.; Gazzaniga, M.S.; Bogen, J.E. In Vinken, P.J.; Bruya, G.W., eds. *Handbook of Clinical Neurology,* vol. 4, 273–290. Elsevier Science Publishers (Biomedical Division), 1969.)

the right hemisphere makes a decision based on stimulus form; if a verbal response is required, the left hemisphere dictates the response.

Left Hemisphere Dominance

The left hemisphere not only is dominant for naming but also is crucial for all aspects of language. Its importance is demonstrated most dramatically by the injection of the anesthetic sodium amytal. An injection that reaches the left hemisphere produces a transient total loss of speech, but language comprehension is largely intact.[71] On the other hand, an injection that reaches the right hemisphere rarely produces a deficit in the expression or comprehension of language, although speech may be slurred.

The hemispheric dominance for language is somewhat dependent on whether a person is right- or left-handed. Using the sodium amytal technique, Rasmussen and Milner[56] showed that 96% of 140 right-handed patients had speech representation in the left hemisphere. Of 122 left-handed subjects, 70% had speech in the left hemisphere, 15% had speech in the right hemisphere, and 15% apparently had speech representation in both hemispheres. Some ambidextrous patients have bilateral cortical speech representation.

Although the isolated right hemisphere has limited speech capabilities, it can comprehend many spoken or written words. This comprehension is measured by flashing a written word briefly (tachistoscopically) in the left visual field and requiring the subject to use his left hand to retrieve

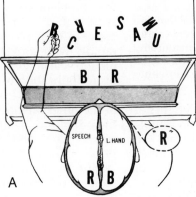
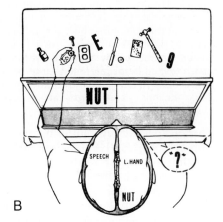

the object described by the word (Fig. 31–18B). The right hemisphere is best at understanding short names of familiar objects, but it can also understand all grammatical classes of words that occur with moderate frequency in normal language.[77] It is also capable of more sophisticated language comprehension. For example, a patient can use his left hand to retrieve items based on abstract descriptions or definitions of their use, such as a cup in response to "holds water" or a magnifying glass for "makes things bigger."[42] The right hemisphere has a vocabulary estimated to be that of a 10- to 16-year old person[77] and is less powerful at semantic association than the left hemisphere. It may, however, have considerable latent capabilities that are normally suppressed by the left hemisphere. For example, children who acquire left hemispheric lesions early in life recover much of their capacity for speech, a recovery that presumably is mediated by the intact right hemisphere. Even adults show considerable language recovery after left hemispherectomy.[63] Therefore, in the intact organism, normal language function may involve considerable interplay between the cortices. Nevertheless, it seems clear that the left hemisphere is dominant for speech and has a superior comprehension of language.

The left hemisphere is also superior for most functions that are associated with language skills. For example, persons who have *aphasia* (difficulty in speaking) also suffer from *alexia* (inability to read aloud or silently). Some aphasics can express their ideas in writing; most, however, have *agraphia* (impaired writing), although the hand may be well coordinated in other motor tasks.

Some rare patients even have an *alexia without agraphia* pure blindness for words which they nevertheless can write). This effect appears to result from a partial disconnection of the right and left hemispheres (often a lesion of the splenium of the corpus callosum) combined with a right hemianopsia, such that words that are visually perceived by only the right hemisphere cannot reach the left hemisphere to be interpreted semantically. Since both hemispheres are intact, spoken words still have meaning, and words can be expressed in either writing or speech.

Curiously, patients suffering surgical left hemispherectomy can barely speak but have no trouble singing a song or reciting a nursery rhyme. On the other hand, after the right hemisphere of intact subjects is anesthetized by intracarotid injection of sodium amytal, these persons sing in a monotone, and although changes in pitch are well timed, they are generally inaccurate. Although the songs are "sung" without melody, every word is enunciated

correctly.[16] Patients with lesions of the right hemisphere exhibit a similar *aprosody* (lack of intonation) during normal speech.

Cortical Subareas for Different Aspects of Language

Language disturbances owing to cortical lesions may occur in many subtle varieties.[69a] Classically, the aphasias are divided into two broad categories involving deficits in either language expression or comprehension. In *expressive aphasia* (also called nonfluent, or motor, aphasia), speech is slow and labored. Only short phrases, clichés (i.e., automatisms), or expletives are uttered. Characteristically, small grammatical words and the endings of nouns and verbs are omitted, so speech has a telegraphic style. When asked to describe a trip, such a patient may say "Seattle go." In extreme cases, the patient is speechless. Patients with expressive aphasia recover the understanding of language quickly and are often aware of their speech disorder and frustrated or depressed by it. Broca was the first to point out that postmortem examination of patients with expressive aphasia reveals damage in the inferior frontal gyrus of the left hemisphere (Fig. 31–19). Broca's area lies immediately anterior to the speech area of motor cortex that controls the muscles of the jaw, tongue, palate, larynx, and face.

In *receptive aphasia* (also called fluent, or sensory, aphasia), the comprehension of language is severely impaired. Patients with normal auditory perception for non-speech sounds may interpret the speech of others as an incomprehensible noise. Some patients can understand single words and slowly spoken, familiar, and redundantly conveyed ideas but often misunderstand less familiar words and connected discourse. Associated with this problem in comprehension is an abnormal speech that is itself incomprehensible. Although patients with receptive aphasia may speak very rapidly with correct rhythm, grammar, and articulation, close attention reveals that the speech is profoundly abnormal. For example, such a patient may say "Before I was in the one here, I was over in the other one. My sister had the department in the other one."[14] Such patients also fail to use the correct word and replace it with circumlocutory phrases (e.g., "what you use to cut with" for "knife") and empty words (e.g., "thing"). Finally, patients with receptive aphasia may substitute one word or phrase for another (e.g., "hammer" for "paper"). Such substitutions are called *paraphasias*. Receptive aphasia was first described in 1874 by

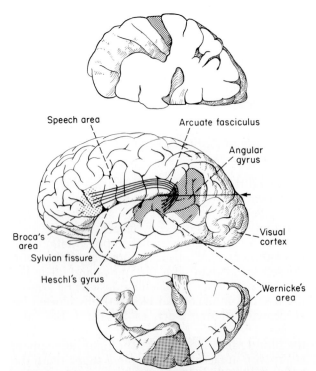

Figure 31–19 Temporal and frontal language areas of the left human hemisphere and their connecting fibers. On the complete brain, Broca's area is indicated by sparse stipple, and the temporal language area, which includes Wernicke's area and the angular gyrus, is indicated by dense stipple. Drawings above and below that of the complete brain represent horizontal sections through the left (below) and right (above) temporal language area, revealing the larger left planum temporale (stippled area) that occurs in 65% of normal human brains. (Modified from Geschwind, N. Language and the brain. 226:76–83, Copyright © 1972 by Scientific American, Inc. All rights reserved.)

Wernicke, who demonstrated that his two patients had diffuse vascular lesions involving the superior temporal gyrus on the left side (see Fig. 31–19). Wernicke's area is adjacent to the primary auditory cortex (Heschl's gyrus) and adjacent to the angular gyrus, which is thought to be an area that associates visual, tactile, and auditory information.

Lesions of the angular gyrus, or the neighboring supramarginal gyrus, produce alexia with agraphia. These patients can neither read nor write and have difficulty converting visually, tactually, or orally presented symbols into language. For example, they cannot spell, nor can they recognize words that are presented one letter at a time.

A current notion is that language comprehension is accomplished in the left temporal cortex. Sensory information about the spoken or written word reaches Wernicke's area from the adjacent auditory cortex or nearby visual cortex, respectively. In the case of visual information, there is presumably an intermediate stage of elaboration and interpretation in the angular gyrus. Information about the comprehended language is delivered to Broca's area, where an oral response is organized and shipped to the adjacent motor areas for speech.

In many normal human brains, the left hemispheric dominance for language has a visible anatomical correlate. An extension of Wernicke's area on the upper surface of the temporal lobe (Fig. 31–19), the planum temporale (stippled area on horizontal sections of Fig. 31–19), has been shown to be larger on the left side in 65% of the brains studied, equal in 24%, and smaller in 11%. The left planum is also one third longer, on average, than the right.[15] In some patients the right planum is completely absent.[72] A similar study of the inferior frontal gyrus revealed little, if any, right-left asymmetry.[72] Similar asymmetries are present in chimpanzees, which are capable of manipulating symbols in a manner approaching a rudimentary language.

Connections between the temporal and frontal language areas are made via the arcuate fasciculus, a fiber bundle that runs posteriorly in temporal cortex, turning near area 39 (the angular gyrus) to run anteriorly through parietal and frontal cortex (see Fig. 31–19). If the arcuate fasciculus is destroyed, the patient can comprehend words but cannot repeat them (*conduction aphasia*). This effect can be demonstated by asking questions that require simple yes-no responses. While these patients can respond to the questions, they cannot repeat them. Some researchers believe that this problem is related to short-term verbal memory deficits (see discussion on temporal association cortex earlier in this chapter). Patients with conduction aphasia also make many speech-sound substitutions and occasionally display *apraxia* (inability to understand and execute learned motor commands).

Apraxia often accompanies both sensory and conduction aphasia. In its most common expression, *ideomotor apraxia*, afflicted patients can perform many sophisticated movements automatically, but not if they are commanded to do so (volitional movement is impaired). These problems are associated primarily with lesions to the left hemisphere but are usually expressed bilaterally. Geschwind has suggested that this type of apraxia can result from an inability to process and transfer verbal commands to the appropriate motor centers. Thus, ideomotor apraxia is very similar to sensory or conduction aphasia, except that the destination for this information is premotor cortex rather than Broca's area.

The simple anatomical dichotomy that places expression in the frontal lobe and comprehension in the temporal lobe is probably oversimplified, according to the results of electrical stimulation.[44] In order to avoid damage to important language areas during cortical surgery, the cortex is electrically stimulated in alert patients while they are naming objects on a screen, reading and completing sentences, or recalling objects presented earlier. Stimulation at one site may interrupt only one of these language functions, whereas stimulation at an adjacent site (within 5 mm) may affect another or have no effect at all. In multilingual patients, the site for naming in one language can be separate from that for naming in the other. Thus, not only do the different language functions appear to be discretely and differentially organized, but they all are intermixed in the frontal and temporal lobes, suggesting that neither lobe processes only a single aspect of language.

The perception of species-specific vocalization by Japanese macaques may also be mediated by the temporal cortex, with the left hemisphere playing a dominant role. These monkeys were trained to discriminate two subtypes of their coo vocalizations by breaking mouth contact with a feeding spout to avoid a shock whenever one of the subtypes was presented. Unilateral ablation of the left but not right superior temporal gyrus, including auditory cortex, resulted in an initial impairment in the discrimination. Bilateral temporal lesions initially completely abolished the ability of the animals to discriminate these coos and impaired that discrimination for up to 9 months.[19]

Left Hemisphere Superiority in Other Functions

Most patients with no left hemisphere, and most patients with aphasia, have a reduced capacity for mathematical computation *(acalculia)*. Commissurotomy patients required to do calculations with the right hemisphere may succeed in matching numbers with their left hands or adding 1 to numbers below 10; however, they fail when required to add or subtract two or more numbers or to perform simple multiplication and division.[64]

Superiority of Right Hemisphere in Spatial Constructive Skills

Several different tests indicate that the isolated right hemisphere is superior in spatial constructive skills. For example, soon after surgery, commissurotomy patients can copy objects such as Greek crosses or two-dimensional representations of three-dimensional objects more accurately with the left hand than with the right hand.[2] The left hand is also better at matching designs by arranging a set of multicolored blocks (Kohs blocks), or at drawing figures such as Neckar cubes, and shows a tendency to take over such tasks from the blundering right hand.[3] The right hemisphere is superior at tactually exploring a shape (with the left hand) and remembering the spatial relationships of its parts. This capability is revealed by requiring a subject to feel a complicated three-dimensional wooden object with his hand and then to match it with several two-dimensional unfolded visible objects with similar forms (Fig. 31–20).[69a] Most of these differences are transient, however, disappearing with time following the original insult.

The right hemisphere of commissurotomy patients also is better at distinguishing a whole object from one of its parts. To demonstrate this capability, a subject first feels (Fig. 31–21A) or sees (Fig. 31–21B) the arc of a circle and then must select

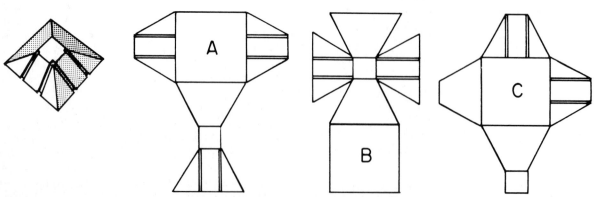

Figure 31–20 Example of a spatial constructive task in which a subject palpates a three-dimensional object (at left) and then matches it to one of the three unfolded visible patterns. (The correct answer is *C*.) (After Levy, J. In Kinsbourne, M.; Smith, W.L., eds. *Hemispheric Disconnection and Cerebral Function*, p. 168, 1974. Courtesy of Charles C Thomas, Publisher, Springfield, Illinois.)

Figure 31–21 Perception of part-whole relations in commissurotomized patients. Subject must match whole circles to arcs by combining visual and tactile information (*A* and *B*) or tactile information alone *(C)*. Histograms No. 1 illustrate the average percentage of correct responses in five subjects using the right (hatched) and left (solid) hands. Histograms Nos. 2 and 3 show the results of control tasks in which the subject matched a complete test circle to three other complete circles (No. 2) or a test arc to other arcs (No. 3). Dashed line at 33⅓% indicates the level of a chance performance. (After Nebes, R.D. *Cortex* 7:333–349, 1971.)

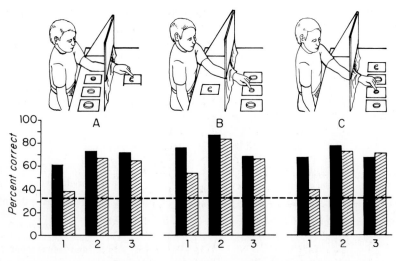

from among three whole circles of different sizes the one from which the arc was taken.[41] The match may be made on the basis of sight and touch (Fig. 31–21A, B) or touch alone (Fig. 31–21C). The left hand (solid histograms) does significantly better than the right on all three variations of the task (histograms 1). If, rather than performing a part-whole match, the subject must match a whole test circle to three other complete circles (histograms 2), or match a test arc to three other arcs (histograms 3), he performs this simple matching task equally well with either hand. In a similar test, a subject is required to find which one of several fragments of various patterns will complete a pattern that has one piece missing. For example, he touches designs represented by raised lines on a photogravure plate to select the pattern that would complete a visually presented sample pattern. Again, in this test commissurotomized subjects achieve a higher percentage of correct scores with their left hands, although the right hand always operates at above chance.[76]

In many of these spatial constructive tasks, each hand explores the test object in a way that suggests that the two hemispheres solve the same problems by employing different strategies. The right hand tends to work more slowly, with the subject often giving verbal labels to aspects of the pattern such as the number of elements, the length of lines, or the size of angles. In contrast, the left hand is 15% faster and, when performing tasks with that hand, the subject remains both confident and silent.[76]

Since sign language uses visual constructive skills, it might be reasoned that signing requires the right hemisphere.[8] However, errors in signing can be produced by sodium amytal injections into the left carotid (in this

case, it was impossible to inject into the right). Interestingly, speech and signing recovered at different rates, so that test objects were correctly identified verbally before they were correctly signed. Thus, the left hemisphere specialization for language may be related more to semantic content (meaning and symbolism) than to the production of a specific motor act.

Summary

When one hemisphere is functionally disconnected from the other by means of anesthesia or commissurotomy, one hemisphere seems more effective than the other for certain functions or behaviors. These cerebral dominances are summarized in Figure 31–22. We have seen that some of the hemispheric differences are more striking than others. For example, whereas it is clear that the right hemisphere has little, if any, capacity for speech, it has considerable capacity for language comprehension, although the left hemisphere is still superior. Also, although the right hemisphere has better visual constructive skills than does the left (recall Fig. 31–21), the left hemisphere always operates at better than chance, but its visual constructive capabilities seem to be the result of a somewhat different perceptual or cognitive strategy.

In the intact brain, many of the most direct and profound interactions between neurons are interhemispheric. When the two hemispheres are isolated, information may be processed in a way that is unique to this unusual situation. Figure 31–22 is, then, a representation of some of the potentialities of the injured human brain and does not seek to represent limitations of the individual

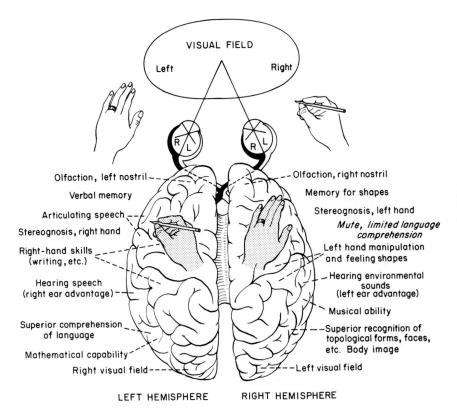

Olfaction, left nostril —
Verbal memory
Articulating speech —
Stereognosis, right hand
Right-hand skills
(writing, etc.) —
Hearing speech
(right ear advantage) —
Superior comprehension
of language
Mathematical capability
Right visual field — — — —

— — Olfaction, right nostril
Memory for shapes
Stereognosis, left hand
*Mute, limited language
comprehension*
— Left hand manipulation
and feeling shapes
— Hearing environmental
sounds
(left ear advantage)
— Musical ability
— — Superior recognition of
topological forms, faces,
etc. Body image
— — Left visual field

LEFT HEMISPHERE   RIGHT HEMISPHERE

Figure 31–22 Schematic summary of lateralized functions revealed by psychological tests of commissurotomized and other patients.

hemispheres of a normally functioning, integrated cerebral cortex.

Experiments on patients with cortical lesions have produced some amazing results. People with one hemisphere disconnected from the other can lead remarkably normal lives. Each hemisphere can operate independently, with two apparently completely separate domains of conscious awareness. Restricted lesions can affect one behavior without affecting closely related behaviors (e.g., alexia without agraphia). As we have tried to indicate in this chapter, association cortex is one of the most fascinating, but poorly understood, structures in the central nervous system.

ANNOTATED BIBLIOGRAPHY

Trends Neurosci. 7:403–427, 1984.
   A good review of the frontal lobes, from single-unit activity in animals to metabolic measurements of frontal lobe function in man.
Trevarthen, C. Hemispheric specialization. *Handbook of Physiology*, Sec. 1, 3(2):1129–1190, 1984.
   A comprehensive review, with almost 700 references, of the literature on hemispheric dominance. Although a handbook article, it is nevertheless quite readable.
Lynch, J. C. The functional organization of posterior parietal association cortex. *Behav. Brain Sci.* 3:485–534, 1980.

A spirited, readable discussion of the function of parietal cortex by several authorities in the field. Includes a comprehensive bibliography and a detailed discussion of single-unit behavior in trained monkeys.

REFERENCES

1. Bisiach, E.; Luzzatti, C. Unilateral neglect of representational space. *Cortex* 14:129–133, 1978.
2. Bogen, J. E. The other side of the brain: dysgraphia and dyscopia following cerebral commissurotomy. *Bull. Los Angeles Neurol. Soc.* 34:73–105, 1969.
3. Bogen, J. E.; Gazzaniga, M. S. Cerebral commissurotomy in man. Minor hemisphere dominance for certain visuospatial functions. *J. Neurosurg.* 23:394–399, 1965.
4. Brodmann, K. *Vergleichende Lokalisationslehre der Grosshirnrinde in ihren Prinzipen dargestellt auf Grund des Zellenbaues.* Leipzig, Barth, 1909.
5. Bushnell, M. C.; Goldberg, M. E.; Robinson, D. L. Behavioral enhancement of visual responses in monkey cerebral cortex. I. Modulation in posterior parietal cortex related to selective visual attention. *J. Neurophysiol.* 46:755–772, 1981.
6. Chow, K.I.; Survis, J. Retention of overlearned visual habit after temporal cortical ablation in monkey. *Arch. Neurol. Psychiat.* 79:640–646, 1958.
7. Critchley, M. *The Parietal Lobes.* New York: Hafner, 1966.
8. Damasio, A.; Bellugi, U.; Damasio, H.; Poizner, H.; Van Guilder, J. Sign language aphasias during left-hemisphere amytal injection. *Nature* 322:363–365, 1986.

9. Damasio, A. R. Prosopagnosia. *Trends Neurosci.* 8:132–135, 1985.

10. Fuster, J. M. Unit activity in prefrontal cortex during delayed-response performance: neuronal correlates of transient memory. *J. Neurophysiol.* 36:61–78, 1973.

11. Fuster, J. M. The prefrontal cortex and temporal integration. In Peters, A.; Jones, E. G., eds. *Cerebral Cortex*, vol. 4: *Association and Auditory Cortices.* New York, Plenum Press, 1985.

12. Gazzaniga, M. S.; Sperry, R. W. Simultaneous double discrimination response following brain bisection. *Psychonomic Sci.* 4:261–262, 1966.

13. Gazzaniga, M. S.; Bogen, J. E.; Sperry, R. W. Dyspraxia following division of the cerebral commissures. *Arch. Neurol.* 16:606–612, 1967.

14. Geschwind, N. Language and the brain. *Sci. Amer.* 226:76–83, 1972.

15. Geschwind, N.; Levitsky, W. Left/right asymmetries in temporal speech region. *Science* 161:186–187, 1968.

16. Gordon, H. W.; Bogen, J. E. Hemispheric lateralization of singing after intracarotid sodium amylbarbitone. *J. Neurol. Neurosurg. Psychiat.* 37:727–738, 1974.

17. Gross, C. G. Inferotemporal cortex and vision. *Prog. Physiol. Psychol.* 5:77–123, 1973.

18. Harlow, J. M. Passage of an iron rod through the head. *Boston Med. Surg. J.* 39:389–393, 1848.

19. Heffner, H. E.; Heffner, R. S. Temporal lobe lesions and perception of species-specific vocalizations by macaques. *Science* 226:75–76, 1984.

20. Heilman, K. M.; Watson, R. T.; Valenstein, E. Neglect and related disorders. In Heilman, K. M.; Valenstein, E., eds. *Clinical Neuropsychology,* 243–293. Oxford, Oxford Univ. Press, 1985.

21. Holmes, E. J.; Gross, C. G. Effects of inferior temporal lesions on discrimination of stimuli differing in orientation. *J. Neurosci.* 4:3063–3068, 1984.

22. Jacobsen, C. F. Functions of frontal association area in primates. *Arch. Neurol. Psychiat.* 33:558–569, 1935.

23. Jacobsen, C. F. Studies of cerebral functions in primates. I. The functions of the frontal association areas in monkeys. *Comp. Psychol. Monogr.* 13(3):1–60, 1936.

24. Jones, E. G.; Powell, T. P. S. An anatomical study of converging sensory pathways within the cerebral cortex of the monkey. *Brain* 93:793–820, 1970.

25. Kimura, D. Right temporal-lobe damage. *Arch. Neurol.* 8:48–55, 1963.

26. Kimura, D. Left-right differences in the perception of melodies. *Q. J. Exp. Psychol.* 16:355–358, 1964.

27. LaMotte, R. H.; Acuna, C. Defects in accuracy of reaching after removal of posterior parietal cortex in monkeys. *Brain Res.* 139:309–326, 1978.

28. LaMotte, R. H.; Mountcastle, V. B. Disorders in somesthesis following lesions of parietal lobe. *J. Neurophysiol.* 42:400–419, 1979.

29. Levy, J.; Trevarthen, C.; Sperry, R. W. Perception of bilateral chimeric figures following hemispheric deconnexion. *Brain* 95:61–78, 1972.

30. Lynch, J. C. The functional organization of posterior parietal association cortex. *Behav. Brain Sci.* 3:485–534, 1980.

30a. Lynch, J. C.; Mountcastle, V. B.; Talbot, W. H.; Yin, T. C. T. Parietal lobe mechanisms for directed visual attention. *J. Neurophysiol.* 40:362–389, 1977.

31. Malmo, R. B. Interference factors in delayed response in monkeys after removal of frontal lobes. *J. Neurophysiol.* 5:295–308, 1942.

32. Meier, M. J.; French, L. A. Lateralized deficits in complex visual discrimination and bilateral transfer of reminiscence following unilateral temporal lobectomy. *Neuropsychologia* 3:261–272, 1965.

33. Mesulam, M.; Van Hoesen, G. W.; Pandya, D.; Geschwind, N. Limbic and sensory connections of the inferior parietal lobule (Area PG) in the rhesus monkey: a study with a new method for horseradish peroxidase histochemistry. *Brain Res.* 136:393–414, 1977.

34. Miller, L. Cognitive risk-taking after frontal or temporal lobectomy. I. The synthesis of fragmented visual information. *Neuropsychologia* 23:359–369, 1985.

35. Milner, A. D.; Ockleford, E. M.; Dewar, W. Visuo-spatial performance following posterior parietal and lateral frontal lesions in stumptail macaques. *Cortex* 13:350–360, 1977.

36. Milner, B. Laterality effects in audition. In Mountcastle, V. B., ed. *Interhemispheric Relations and Cerebral Dominance,* 1962. Baltimore, Johns Hopkins Press, 177–195.

37. Milner, B. Effects of different brain lesions on card sorting. *Arch. Neurol.* 9:90–100, 1963.

38. Milner, B. Visual recognition and recall after right temporal-lobe excision in man. *Neuropsychologia* 6:191–209, 1968.

39. Mountcastle, V. B.; Lynch, J. C.; Georgopoulos, A.; Sakata, H.; Acuna, C. Posterior parietal association cortex of the monkey: command functions for operations within extrapersonal space. *J. Neurophysiol.* 38:871–908, 1975.

40. Murray, E. A.; Mishkin, M. Severe tactual as well as visual memory deficits follow combined removal of the amygdala and hippocampus in monkeys. *J. Neurosci.* 4:2565–2580, 1984.

41. Nebes, R. D. Superiority of the minor hemisphere in commissurotomized man for the perception of part-whole relations. *Cortex* 7:333–349, 1971.

42. Nebes, R. D. Direct examination of cognitive function in the right and left hemispheres. In Kinsbourne, M., ed. *Asymmetrical Function of the Brain,* 99–137. Cambridge, Cambridge Univ. Press, 1978.

43. Niki, H.; Watanabe, M. Prefrontal unit activity and delayed response: relation to cue location versus direction of response. *Brain Res.* 105:79–88, 1976.

44. Ojemann, G. A. Brain organization for language from the perspective of electrical stimulation mapping. *Behav. Brain Sci.* 2:189–230, 1983.

45. Ojemann, G. A.; Dodrill, C. B. Verbal memory deficits after left temporal lobectomy for epilepsy. Mechanism and intraoperative prediction. *J. Neurosurg.* 62:101–107, 1985.

46. Pakkenberg, H. The number of nerve cells in the cerebral cortex of man. *J. Comp. Neurol.* 62:101–107, 1985.

47. Pandya, D. N.; Yeterian, E. H. Architecture and connections of cortical association areas. In Peters, A.; Jones, E. G., eds. *Cerebral Cortex,* vol. 4: *Association and Auditory Cortices,* 3–61. New York, Plenum Press, 1985.

48. Pandya, D. N.; Kuypers, H. G. J. M. Cortico-cortical connections in the rhesus monkey. *Brain Res.* 13:13–36, 1969.

49. Paterson, A.; Zangwill, O. L. Disorders of visual space perception associated with lesions of the right cerebral hemisphere. *Brain* 67:331–358, 1944.

50. Petras, J. M. Connections of the parietal lobe. *J. Psychiat. Res.* 8:189–201, 1971.

51. Petrides, M.; Iversen, S. D. Restricted posterior parietal lesions in the rhesus monkey and performance on visuo-spatial tasks. *Brain Res.* 161:63–71, 1979.

52. Pohl, W. Dissociation of spatial discrimination deficits following frontal and parietal lesions in monkeys. *J. Comp. Physiol. Psychol.* 87:227–239, 1973.

53. Porteus, S. D.; Kepner, R. DeM. Mental changes after bilateral prefrontal lobotomy. *Genetic Psychol. Monogr.* 29:3–115, 1944.

54. Posner, M. I.; Walker, J. A.; Friedrich, F. J.; Rafal, R. D. Effects of parietal injury on covert orienting of attention. *J. Neurosci.* 4:1863–1874, 1984.

55. Pribram, K. H.; Luria, A. R. *Psychophysiology of the Frontal Lobes.* New York, Academic Press, 1973.

56. Rasmussen, T.; Milner, B. Clinical and surgical studies of the cerebral speech areas in man. In Zulch, K. J.; Creutzfeld, O. D.; Galbraith, G. C., eds. *Otfrid Foerster Symposium on Cerebral Localization,* 238–257. Berlin, Springer-Verlag, 1975.

56a. Ridley, R. M.; Ettlinger, G. Tactile and visuospatial discrimination performance in the monkey: the effects of total and partial posterior parietal removals. *Neuropsychologia* 13:191–206, 1975.

57. Roland, P. E.; Larsen, B.; Lassen, N. A.; Skinhøj, E. Supplementary motor area and other cortical areas in organization of voluntary movements in man. *J. Neurophysiol.* 43:118–136, 1980.

58. Roland, P. E.; Skinhøj, E.; Lassen, M. A.; Larsen, B. Different cortical areas in man in organization of voluntary movements in extrapersonal space. *J. Neurophysiol.* 43:137–150, 1980.

59. Roland, P. E. Astereognosis. *Arch. Neurol.* 33:543–550, 1976.

60. Ruch, T. C. The association areas—the homotypical cortex. In Ruch, T. C.; Patton, H. D., eds. *Physiology and Biophysics I. The Brain and Neural Function,* pp. 589–625. Philadelphia, W. B. Saunders, 1979.

61. Semmes, J.; Turner, B. Effects of cortical lesions on somatosensory tasks. *J. Invest. Dermatol.* 69:181–189, 1977.

62. Shankweiler, D. Effects of temporal-lobe damage on perception of dichotically presented melodies. *J. Comp. Physiol. Psychol.* 62:115–119, 1966.

63. Smith, A. Speech and other functions after left (dominant) hemispherectomy. *J. Neurol. Neurosurg. Psychiat.* 29:467–471, 1966.

64. Sperry, R. W.; Gazzaniga, M. S.; Bogen, J. E. Interhemispheric relationships: the neocortical commissures; syndromes of hemisphere disconnection. In Vinken, P. J.; Bruyn, G. W., eds. *Handbook of Clinical Neurology,* vol. 4, pp. 273–290. Amsterdam, North-Holland, 1969.

65. Sperry, R. W.; Zaidel, E.; Zaidel, D. Self recognition and social awareness in the disconnected hemisphere. *Neuropsychologia* 17:153–166, 1979.

66. Stein, J. Effects of parietal lobe cooling on manipulative behavior in the conscious monkey. In Gordon, G., ed. *Active Touch. The Mechanisms of Recognition of Objects by Manipulation: A Multidisciplinary Approach,* pp. 79–90. Oxford, Pergamon Press, 1978.

67. Stepien, L. S.; Cordeau, J. P.; Rasmussen, T. The effect of temporal lobe and hippocampal lesions on auditory and visual recent memory in monkeys. *Brain* 83:470–489, 1960.

68. Stuss, D. T.; Benson, D. E. Neuropsychological studies of the frontal lobes. *Psychol. Bull.* 95:3–28, 1984.

69. Teuber, H. L. The riddle of frontal lobe function in man. In Warren, J. M.; Akert, K., eds. *The Frontal Granular Cortex and Behavior,* pp. 410–444. New York, McGraw-Hill, 1964.

69a. Trevarthen, C. Hemispheric specialization. *Handbk. Physiol.* Sec. 1, 3(2):1129–1190, 1984.

70. Valenstein, E.; Heilman, K. M.; Watson, R. T.; van den Abell, T. Nonsensory neglect from parietotemporal lesions in monkeys. *Neurology* 32:1198–1201, 1982.

71. Wada, J.; Rasmussen, T. Intracarotid injection of sodium amytal for the lateralization of cerebral speech dominance. *J. Neurosurg.* 17:266–282, 1960.

72. Wada, J. A.; Clarke, R.; Hamm, A. Cerebral hemispheric asymmetry in humans. *Arch. Neurol.* 32:239–246, 1975.

73. Walter, W. G.; Cooper, R.; Aldridge, V. J.; McCallum, W. C.; Winter, A. L. Contingent negative variation: an electric sign of sensorimotor association and expectancy in the human brain. *Nature* 203:380–384, 1964.

74. Weinstein, S.; Teuber, H. L. Effects of penetrating brain injury on intelligence test scores. *Science* 125:1036–1037, 1957.

75. Wilson, M.; Stamm, J. S.; Pribram, K. H. Deficits in roughness discrimination after posterior parietal lesions in monkeys. *J. Comp. Physiol. Psychol.* 53:535–539, 1960.

76. Zaidel, D.; Sperry, R. W. Performance on the Raven's colored progressive matrices test by subjects with cerebral commissurotomy. *Cortex* 9:34–39, 1973.

77. Zaidel, E. Linguistic competence and related functions in the right cerebral hemisphere of man following commissurotomy and hemispherectomy. Ph.D. Thesis. Pasadena, California Institute of Technology, 1973.

Chapter 32

Jo Ann E. Franck
Philip A. Schwartzkroin
James O. Phillips
Albert F. Fuchs

The Limbic System

INTRODUCTION

In previous chapters, we considered the neurophysiology of behaviors and sensations that can be described in objective terms. A large part of our existence, however, involves subjective experiences such as the emotions of rage, fear, and pleasure. In Chapter 31 we saw that some of these experiences require the participation of a variety of neocortical association areas. Many, however, also involve a specific group of cortical and subcortical neural structures that traditionally have been referred to collectively as the *limbic system*.

Three major components of the limbic system—the amygdala, the septum, and the hippocampus—are the primary focus of this chapter. We will examine how they integrate a broad range of afferent information and use that information to influence a diverse set of neural structures involved in autonomic, endocrine, and sensory functions. We will consider the hypothalamus, the common destination for many limbic outputs, in detail. We will study behaviors, such as emotion, by dissecting them into measurable components such as hormonal secretions. As we shall see, the neurophysiological bases for emotion and other regulatory behaviors are poorly understood.

Anatomy of the Limbic System

The limbic system consists of the limbic lobe of the cerebral cortex and the subcortical nuclei with which it is associated. The limbic lobe was originally defined on purely anatomical grounds by Broca,[7] who called it the "great limbic lobe" because its constituent regions have a common origin along the border (or, in Latin, *limbus*) of the foramen of Monro. The limbic lobe includes the cingulate, subcallosal, and parahippocampal gyri (including the subiculum) as well as the hippocampal formation, which includes the dentate gyrus and the hippocampus. The remainder of the limbic system is made up of the amygdala, septum, parts of the basal ganglia, and much of the diencephalon (i.e., hypothalamus, epithalamus, and anterior thalamus). In the phylogenetically or embryologically primitive brain, the hippocampus, septum, amygdala, and hypothalamus sit adjacent to one another, and the pathways between them are direct and easily identified. As the forebrain develops, the individual components of the limbic system and the fiber tracts that connect them become twisted and stretched in a way that obscures their earlier simple configuration.

Viewed from its medial aspect, the mature limbic system resembles the letter C (Fig. 32–1).[8] The amygdala is located at one end of the "C," deep in the temporal lobe. A major output of the amygdala, the stria terminalis, sweeps up along the floor of the lateral ventricle and arches foward and down to terminate predominantly in the bed nucleus of the stria terminalis (not shown), and also in both the preoptic region and the hypothalamus (Fig. 32–2),[8] just ventral and caudal to the septal nuclei. A second amygdalar output is a ventral amygdalofugal pathway that projects to the preoptic region, the hypothalamus, the dorsomedial thalamus, and the septal region, where it termi-nates in the nucleus accumbens septi, adjacent to the caudate nucleus of the basal ganglia; the nucleus accumbens projects in turn to the putamen of the basal ganglia (see Chap. 30).

The hippocampus, whose temporal pole lies just posterior and lateral to the amygdala, arises adjacent to the parahippocampal gyrus in the temporal lobe and loops above the ventricular system to end behind the septal nuclei. The fornix, which is the principal fiber pathway associated with the hippocampus, gathers fibers all along the temporal-to-septal extent of the hippocampus and, like the stria terminalis, dives into the diencephalon. Fibers from the hippocampus travel in the portion of the fornix that passes rostral to the anterior commissure (precommissural fornix) and distribute primarily to the septal region (see Figs. 32–1 and 32–2). This part of the fornix, in turn, carries fibers that originate in the septal region and project back to the hippocampus and dentate

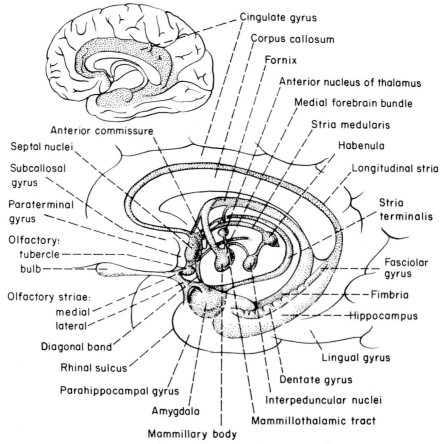

Figure 32–1 Semischematic drawing of the anatomical relations of the parts of the limbic system and many of its associated structures, as seen in a medial view of the right hemisphere (inset). (After Carpenter, M.B.; Sutin, J. *Human Neuroanatomy*, 618, 640. Baltimore, Williams & Wilkins Co., © 1983.)

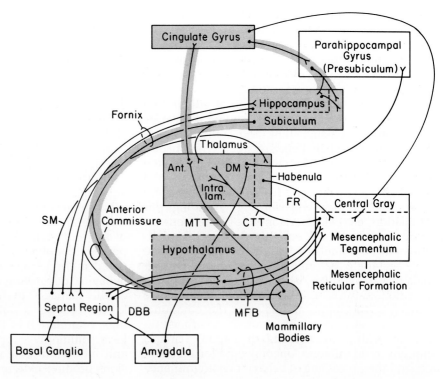

Figure 32–2 Schematic block diagram of the major limbic system structures and their interconnections. Shaded boxes and pathways identify Papez's circuit to explain emotional experience. Abbreviations: CTT = central tegmental tract; DBB = diagonal band of Broca; FR = fasciculus retroflexus; MFB = medial forebrain bundle; MTT = mammillothalamic tract; SM = stria medularis. (After Carpenter, M.B.; Sutin, J. *Human Neuroanatomy*, 631. Baltimore, Williams & Wilkins Co., © 1983.)

gyrus. Other fibers of the fornix originate in the subiculum, which lies adjacent to the hippocampus. Fibers from the subiculum travel in both the precommissural fornix and the part of the fornix that passes posterior to the anterior commissure (postcommissural fornix). These terminate primarily in the mammillary bodies (hypothalamic structures), but also in the anterior thalamus and mesencephalic reticular formation (see Figs. 32–1 and 32–2).

Most of the structures that receive hippocampal efferents are also connected with each other. The mesencephalic reticular formation is reciprocally connected with several diencephalic structures, such as the hypothalamus, and projects to the thalamus via the central tegmental tract (CTT, Fig. 32–2). The hypothalamus also is reciprocally connected to the septal region. The septal region projects back to the mesencephalon either directly, via the medial forebrain bundle (MFB), or indirectly, via the stria medullaris (SM) to the habenula and then via the fasciculus retroflexus (FR). Finally, the mammillary bodies project to the anterior thalamus via the mammillothalamic tract (MTT).

Many of the subcortical structures related to the limbic system also project back to the cortex. The anterior thalamus projects to the cingulate gyrus (see Fig. 32–2). The cingulate gyrus, in turn, projects to a variety of cortical and subcortical struc-

tures, including part of the septum (the nucleus accumbens), the mesencephalic reticular formation, and the parahippocampal gyrus. The parahippocampal gyrus also houses the presubiculum, which lies just inferior to the subiculum. The presubiculum receives fibers not only from the cingulate gyrus but also from the dorsomedial thalamus; it projects to the hippocampus and subiculum (see Fig. 32–2).

Introduction to Limbic Functions

Early anatomists suggested a variety of functions for limbic structures. Broca[7] thought that the limbic lobe played a role in olfaction because it received projections from the olfactory bulb. Papez,[38] on the other hand, proposed that parts of the limbic system, including the limbic lobe, were involved with the feelings of emotion and emotional behavior. At that time there were two competing theories concerning emotion. One suggested that the feelings of emotion resulted from the sensory feedback from behaviors associated with emotion; for example, a tight stomach and clenched fist produced sensations that resulted in feelings of anger. The second suggested that emotional behavior and emotional feelings were produced independently and without the need of sensory feedback. In this scheme, both the hypothalamus and thalamus received cortically processed sensations. The hypothalamus used this information to produce a behavioral response, whereas the thalamus created a "feeling" that was projected back to the cortex.

Papez[38] constructed an anatomical circuit that united these ascending and descending influences to produce a self-sustaining and complete emotional experience. In Papez's circuit (the shaded elements in Fig. 32–2), activity originating in the cortex was relayed via the hippocampus to the hypothalamus, which was responsible for the elaboration of emotional behavior (e.g., sweating, pupil dilation, etc.) via its autonomic connections (to be discussed later). A subregion of the hypothalamus, the mammillary bodies, then relayed the same activity to the cingulate gyrus via the anterior thalamus to produce the feelings associated with the emotion. Intracortical connections returned the activity to the hippocampus, and the emotional experience was sustained through reverberating activity within the circuit. In Papez's circuit, thought could influence ongoing emotional behavior, and the processing of behavior could influence thoughts and feelings without requiring peripheral sensory feedback. Indeed, at that time, reverberating circuits were thought to be the substrate of many cortical functions (e.g., memory).

Papez's speculations were soon supported by lesion experiments in monkeys. Klüver and Bucy[22] showed that large temporal lobe lesions, involving many of the elements of Papez's limbic circuit, produced profound emotional deficits characterized by the complete elimination of aggressive behavior. Humans with similar lesions seemed to alternate between placidity and transient fear. The Klüver-Bucy syndrome also included a constellation of other deficits such as visual neglect (the significance of visual stimuli is lost), hyperorality (objects are taken into the mouth and chewed or "smoked" regardless of appropriateness or palatability), and hypersexuality (indiscriminate sexual contacts are attempted in a manner that is not so much goal-directed as it is automatic).

Yakolev and later MacLean[25] expanded the concept of the limbic system to include behaviors and structures related to, but not originally encompassed by, Papez's conception of the limbic circuit. They proposed that the limbic system included *all* those regions of the central nervous system that were functionally interposed between brain regions that interpret the external world (e.g., neocortex) and those that deal with internal regulation (e.g., hypothalamus and brain stem). This limbic system, which included the elements of Figure 32–2 as well as neocortical association areas (not shown), would have important roles, not only in emotional responses but also in learning and memory, body homeostasis, pain perception, and cardiac and respiratory function.

Indeed, a variety of experiments on humans and animals support MacLean's expanded view. In humans suffering from idiopathic epilepsy, bilateral removal of portions of the temporal lobes, including the hippocampus (and probably the amygdala as well), results in severe memory deficits, suggesting that the hippocampus is part of the neural substrate for memory. Bilateral lesions of the amygdala in many mammals produce hyper- or hypophagia (over- or undereating), adipsia (no drinking), rage or placidity, or hypersexual behavior, depending on the size and exact location of the damage. In humans, bilateral amygdalar lesions produce a state of reduced emotional feelings. Stimulation of the amygdala in awake monkeys commonly suppresses all ongoing behavior and induces a state of quiet vigilance. Septal lesions produce extreme aggression in animals, and stimulation of the septal nuclei acts as a reinforcer for a wide range of behaviors; indeed, rats will voluntarily work to deliver electrical stimuli to their septal nuclei and completely ignore food and drink. Taken together, these data suggest that limbic structures are involved in driving many behaviors. Behavioral changes similar to most of those described can also be obtained by stimulating or damaging the target areas of the limbic system (e.g., the hypothalamus).

In summary, the concept that certain structures constitute a single limbic system originated because the structures are linked anatomically and were implicated in psychophysiological theories of emotion. Eventually, all anatomically related structures that were believed to process sensory events and imbue them with affect, or to produce autonomic responses, were included. Today, the limbic system is thought to participate not only in emotion but also in physiological and behavioral homeostasis. Some argue that this diversity of functions does not fit the notion of a single, unified system and, therefore, the component parts of the limbic system are best considered as individual structures concerned with specific behaviors. Since the various behaviors are interrelated, however, the construct of a limbic system is a convenient shorthand notation when describing the relations between functions or between the structures that subserve them. In this chapter, therefore, we will continue the historical precedent of referring to each structure as part of the limbic system.

HYPOTHALAMUS

Limbic regulation of behavior and homeostasis is effected, in part, through the hypothalamus. Hypothalamic nuclei directly control the neuroendocrine system through connections to the pituitary and also to brain stem regions controlling cardiovascular, respiratory, and thermoregulatory function. Therefore, the hypothalamus has often been referred to as the "head ganglion" of the autonomic nervous system. As seen in the inset in Figure 32–3, the hypothalamus lies at the base of the brain, immediately dorsal to the pituitary gland.[19] On the basis of their morphology and apparent functional roles, four hypothalamic

zones have been identified: the lateral, periventricular, posterior, and medial zones. The nuclei included in each of these zones and the general functions they subserve are summarized in Table 32–1.

Lateral Hypothalamus

The lateral zone of the hypothalamus includes the lateral hypothalamic area and the lateral preoptic area. These regions are ill-defined, primarily because many of the fiber bundles that travel between the brain stem and forebrain (collectively called the medial forebrain bundle; see Fig. 32–2) pass through the lateral hypothalamus and obscure nuclear boundaries.

In 1949, Hess[15] reported that stimulation of the lateral hypothalamus in the cat produces rage behavior. The rage reaction was accompanied by

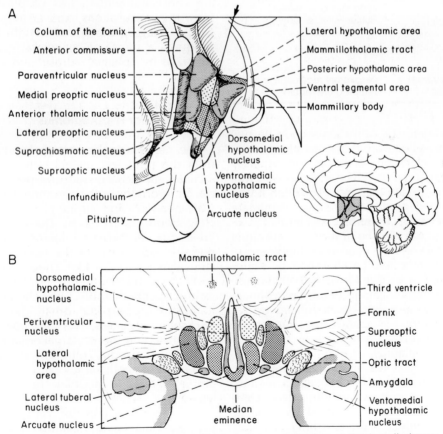

Figure 32–3 Medial view of the brain (inset) to show the location and structure of the hypothalamus, its relation to the pituitary and thalamus (A), and the location of its main nuclear groups (B). Arrow in A indicates plane of section shown in B. (Modified and reprinted by permission of the publisher from Kandel, E.R.; Schwartz, J.H. *Principles of Neural Science*, 615, 616. Copyright 1983 by Elsevier Science Publishing Co., Inc.)

Table 32–1 The Four Hypothalamic Zones and Their Functions

Region	Functions
Lateral Zone	
Lateral preoptic area	Motivation (self-stimulation)
Lateral hypothalamic nuclei	Emotion (rage)
	Food and water intake regulation
Periventricular Zone	
Periventricular nuclei and arcuate nuclei	Anterior and posterior pituitary regulation
Suprachiasmatic nuclei	Diurnal rhythm
Parts of:	
Premammillary nucleus	
Ventromedial nuclei	
Paraventricular nuclei	
Anterior nuclei	
Supraoptic nuclei	
Medial Zone	
Medial preoptic nuclei	Temperature regulation
Dorsomedial nuclei	Sexual differentiation
Parts of:	Control of sexual behavior
Anterior nuclei	Food intake regulation
Ventromedial nuclei	Nutrient biochemistry maintenance
Posterior Zone	
Posterior nuclei	Cardiovascular regulation
Supramammillary nuclei	Respiratory regulation
Mammillary bodies	Temperature regulation

the general sympathetic activation seen in normal angry animals—piloerection, rise in blood pressure, cardiac acceleration, sweating, and pupillary dilation. The aggressive behavior, however, was not directed toward another animal. Rather, it was often self-directed or directed toward inanimate objects. The behavior, therefore, was termed *sham rage*. Sham rage can also be induced by ablations of the prefrontal cortex, a neocortical limbic structure; this sham rage can be abolished by knife-cuts immediately posterior to the hypothalamus, which interrupt hypothalamic–brain stem connections. Sham rage also can be modulated by removal or stimulation of the amygdala, suggesting that sham rage requires a pathway from the amygdala to the hypothalamus.

In another study, Olds and Milner[34] reported that electrical stimulation of hypothalamic regions in the rat produces quite a different effect. They discovered that rats can be trained to perform tasks when the only reward is a brief electrical stimulation of certain brain regions. Indeed, animals learned to stimulate their own brains with such frightening determination that they often neglected food and water. One of these "pleasure

centers" was the lateral hypothalamus, and the best sites generally corresponded to the distribution of fibers traveling from the brain stem in the medial forebrain bundle (including its terminal fields in the septum, amygdala, and hippocampus). Humans will also self-stimulate homologous areas and report feelings of extreme well-being, regardless of the "reality" of their environment. These discoveries led to the concept of specific neural substrates for pleasure in the brain and revolutionized theories of motivation and learning. Thirty years after this discovery, however, the neural mechanism of these "pleasure centers" is still elusive. The paradoxical observations that both rage and pleasure can be elicted by stimulation of similar lateral hypothalamic sites may be explained by the heterogeneity of this region and the fact that numerous long-fiber bundles pass through it.

Lesions of the lateral hypothalamus produce deficits in regulatory behaviors and sensations that are consistent with a loss of pleasure, derived from those sensations, and an impaired motivation. Damaging the lateral hypothalamus in rats produces a transient behavioral syndrome characterized by aphagia, adipsia and global sensory neglect.[13]

Periventricular Hypothalamus

The periventricular hypothalamus, which controls the pituitary, includes the periventricular, arcuate, suprachiasmatic, and premammillary nuclei and portions of the ventromedial, paraventricular, anterior, and supraoptic hypothalamic nuclei. Some neurons in these nuclei secrete "releasing" or "release-inhibiting" factors into the portal blood vessels of the median eminence, as shown in Figure 32–4A.[8] These factors then promote (or inhibit) anterior pituitary secretion of trophic agents that affect peripheral target organs to ensure appropriate organ development and internal homeostasis. The subsequent secretion of hormones from specific target organs (e.g., thyroid and adrenal glands), is, in turn, detected by receptors in the hypothalamus (and other parts of the limbic system) to modulate further hormone release.

The specific actions of certain nuclei in the periventricular hypothalamus have been determined. For example, the suprachiasmatic nucleus appears to be responsible for maintaining an internal biological clock that establishes diurnal rhythms. Diurnal rhythms are altered by varia-

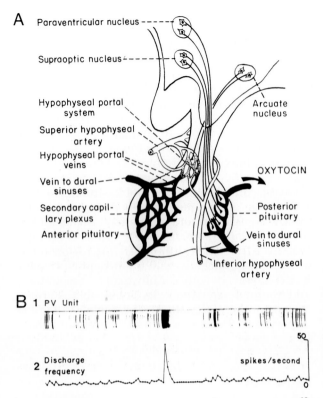

A
Paraventricular nucleus
Supraoptic nucleus
Hypophyseal portal system
Superior hypophyseal artery
Hypophyseal portal veins
Vein to dural sinuses
Secondary capillary plexus
Anterior pituitary
Arcuate nucleus
OXYTOCIN
Posterior pituitary
Vein to dural sinuses
Inferior hypophyseal artery

B
1 PV Unit
2 Discharge frequency — 50 spikes/second — 0
3 Intramammary pressure — 10 mm Hg — 0
10 seconds

Figure 32–4 *A,* Diagram of the anterior and posterior pituitary. Release or release-inhibiting factors are synthesized in hypothalamic nuclei such as the arcuate and are released into the hypophyseal portal system. These factors promote or inhibit secretion of hormones from the anterior lobe of the pituitary. In contrast, hormones that are secreted by the posterior pituitary are released directly from axons of hypothalamic neurons into the systemic circulation; in particular, oxytocin produced in the paraventricular nuclei is released from the posterior pituitary. *B,* Increased firing of a paraventricular (PV) neuron (1 and 2) is related to an increase in intramammary pressure (3) and lactation. (*A* after Carpenter, M.B.; Sutin, J. *Human Neuroanatomy,* 567. Baltimore, Williams & Wilkins Co., © 1983; and *B* after Cross, B.A. In Ganong, W.; Martini, L., eds. *Frontiers in Neuroendocrinology,* 13–171, Oxford University Press, 1973.)

tions both in light (there is a direct input to the suprachiasmatic nucleus from the retina; see Chap. 20) and in neuroendocrine production. Lesions of the suprachiasmatic nuclei of rats abolish their diurnal activity cycle. On the other hand, the large cells in the supraoptic and paraventricular (distinct from the periventricular) nuclei of the hypothalamus synthesize vasopressin and oxytocin. These hormones are transported down the axons of the cells to the posterior pituitary (Fig. 32–4A) and released directly into the bloodstream

to control water balance, uterine contractions, and lactation. For example, large bursts of activity by a neuron (1 and 2 in Fig. 32–4B)[10] in the paraventricular hypothalamus of a nursing rodent result in oxytocin secretion, which leads to increases in intramammary pressure (3 in Fig. 32–4B) to produce ejection of milk.[10]

If the periventricular hypothalamus and pituitary gland are isolated from the rest of the brain, basal endocrine function remains surprisingly normal. Such an animal, however, is unable to produce an appropriate endocrine response to *changes* in its environment. This additional level of endocrine control, which requires reaction to external stimuli, seems to be mediated by other, probably limbic, structures.

The periventricular hypothalamus has relatively few direct connections with telencephalic or brain stem regions. Therefore, the regulation of activity of nuclei in this region must be effected through intrinsic connections with other hypothalamic nuclei, probably via the medial hypothalamic regions, which are interconnected extensively with limbic structures (i.e., the amygdala, septum, and hippocampus) and other hypothalamic nuclei.

Medial Hypothalamus

The medial hypothalamus includes the medial preoptic nuclei, the anterior hypothalamic nuclei, the ventromedial nuclei, and the dorsomedial nuclei. Neurons in the anterior hypothalamic nuclei play a role in temperature regulation. In rats, many such neurons alter their firing rates when core temperature (or local hypothalamic temperature) is reduced. Chemical or electrical activation of the anterior hypothalamus also alters respiration and thereby may control the retention or loss of heat by altering the respiratory rate.

The medial hypothalamic zone is involved in sexual differentiation and the regulation of sexual behavior. Certain anatomical features of the medial hypothalamic nuclei, such as dendritic branching patterns, synaptic organization, and neuron number, differentiate male and female rats. Exposure of the medial hypothalamus to gonadal hormones during development seems crucial for these anatomical sexual differences since, if left to grow without exposure to hormones, the hypothalamus develops the anatomical features characteristic of females regardless of the animal's genetic sex.

The ventromedial hypothalamus controls food intake, and neurons in these nuclei can detect nutrient levels in the blood. A milestone in the investigation of hypothalamic function was the demonstration that lesions in the ventromedial hypothalamic region produce hyperemotionality and dramatic hyperphagia in rats.[16] Taken together with the aphagia produced by more lateral damage, these data spawned theories that the ventromedial and lateral hypothalamic nuclei were "satiety" and "hunger" centers, respectively, and that the interaction between these centers was responsible for the "set point" controlling body weight. However, Powley[39] demonstrated that in the rodent the ventromedial obesity syndrome was not so much a result of destroying a satiety center as it was a disruption of the neural control of nutrient biochemistry, since ventromedial lesions also produced hyperinsulinemia, increased gastric secretions, and gluconeogenesis. This work underscores the importance of examining the response of the whole organism to hypothalamic manipulations.

Posterior Hypothalamus

The posterior hypothalamus includes the posterior hypothalamic nuclei, the supramammillary nuclei, and the mammillary bodies. These regions project to brain stem nuclei involved with cardiovascular, respiratory, and temperature regulation, and receive, in turn, input from brain stem areas as well as from the septum, hippocampus, and anterior thalamus.

Both hypo- and hypertension can be produced by electrically or chemically stimulating the posterior hypothalamus. Furthermore, some neurons in these regions change their discharge rates in response to blood pressure and blood gas levels and thus can monitor cardiac output. The posterior hypothalamus also appears to determine the set point for normal body temperature. For example, bacterial pyrogens (e.g., *Escherichia coli*) placed in the anterior and posterior regions, which regulate temperature, produce a constellation of behaviors (e.g., shivering), that raise core temperature (produce fever). Interaction of the set-point mechanism of the posterior hypothalamus with the anterior hypothalamic "thermostat" controlling heat loss effectively regulates temperature in both health and disease.

Summary

The hypothalamus consists of four functional areas. The lateral zone interacts with diffuse brain

systems by virtue of its position in the midst of substantial fiber tracts traveling between caudal brain stem regions and rostral forebrain and limbic sites. It has been implicated in the control, at least in part, of emotional tone, pleasure, and appetite. The periventricular zone receives input primarily from the medial hypothalamus and projects to the pituitary. It is involved in the release of many hormones that allow an animal to cope with a variety of changes in its environment. The medial hypothalamus forms an interface between the lateral and periventricular regions and receives the heaviest connections from limbic structures. Various parts of the medial hypothalamus are involved in temperature regulation, sexual differentiation, and the maintenance of blood nutrient biochemistry. Finally, the posterior region interacts most significantly with brain stem autonomic nuclei involved with cardiovascular, respiratory, and thermal regulation. Its access to the hypothalamus, therefore, affords the limbic system the opportunity to influence a wide variety of behaviors.

AMYGDALA

Anatomy

On functional grounds, the amygdala can be divided into three subdivisions: corticomedial, basolateral, and central. Each of these subdivisions contains multiple subnuclei that may have different connections and roles. Figure 32–5[37] illustrates the complicated neural connections of the three subdivisions, which will now be considered individually.

Corticomedial Amygdala

The corticomedial amygdala (Fig. 32–5), which includes the cortical (CO) and medial (M) nuclei, is involved primarily in the processing of olfactory information, since the axons of the mitral cells in the olfactory bulb synapse in both nuclei (see Chap. 21). As might be expected, it is smaller in relation to other central nervous system structures in species that are less reliant on olfactory information.

The corticomedial amygdala has extensive reciprocal connections with hypothalamic nuclei (Fig. 32–5B), especially those involved in the control of pituitary function. Like the hypothalamic neurons in those nuclei, cells in the cortical and medial nuclei have receptors for gonadal and adrenal steroids. Thus, the detection of circulating andro-

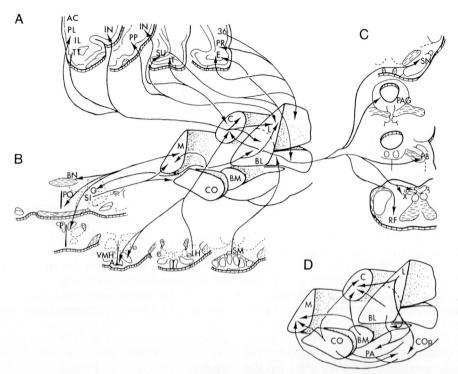

Figure 32–5 Afferent and efferent relations of the amygdala with cortex *(A)*, hypothalamus *(B)*, and brain stem *(C)*. Known intrinsic connections are shown in *D*. Abbreviations: A = arcuate nucleus; AC = anterior cingulate cortex; BL = basolateral amygdala; BM = basomedial amygdala; BN = bed nucleus stria terminalis; C = central amygdala; CO = cortical amygdala; COp = posterior cortical amygdala; E = entorhinal cortex; G = globus pallidus; IL = infra-limbic cortex; IN = insular area; L = lateral amygdala; LH = lateral hypothalamus; M = medial amygdala; P = paraventricular nucleus; PA = periamygdaloid cortex; PAG = periaqueductal grey; PB = parabrachial nucleus; PL = pre-limbic cortex; PP = pre-pyriform cortex; PR = perirhinal cortex; RF = reticular formation; S = nucleus of solitary tract; SI = substantia inominata; SM = supra-mammillary nucleus; SN = substantia nigra; SU = subiculum; TT = taena tecta; VMH = ventromedial hypothalamus; X = dorsal motor nucleus of the vagus nerve; 36 = area 36. (Modified to include new data and efferent connections from Ottersen, O.P. In Ben-Ari, Y., ed. *The Amygdaloid Complex*, 91–104. Elsevier Science Publishers B.V. (Biomedical Division) Amsterdam, 1981.)

gens, estrogens, and corticosteroids by amygdalar neurons may play a role in the feedback control of hormone secretion from the pituitary (by way of the hypothalamus) as well as in steroid-induced differentiation and behavior.

In most mammals there is a close functional relation between olfaction and the control of reproductive hormones and behavior. Pheromones (species-specific chemical messengers) exert a profound influence on sexual behavior via the olfactory system (see Chap. 5), and similar sex attractants have been isolated in primates.[27] Many mammalian species, such as the rat, have separate (accessory) olfactory systems to relay this specialized information to limbic structures (see Chap. 21).

Basolateral Amygdala

The basolateral subdivision (Fig. 32–5A), which consists primarily of the basolateral (BL) and lateral (L) nuclei, is involved mostly in the integration of converging input from the parietal and temporal association areas (see Chap. 31). There is considerable species variation in the pattern of cortical input to the amygdala, with higher species receiving progressively more neocortical input. With the exception of the olfactory cortex, no primary sensory areas of the cortex project to the amygdala. However, both the basolateral and the lateral nuclei receive indirect sensory inputs.[48] Indeed, somatosensory, auditory, gustatory, and visual stimuli all affect the discharge of cells in the lateral and basolateral nuclei of the rat, with some cells responding to multiple modalities. Although some regions are segregated roughly according to modality, there is a great deal of modality overlap in the lateral nucleus.

Central Amygdala

The central division contains only the central nucleus, which has a unique connectivity and neurochemistry that has been appreciated only with the advent of immunocytochemical techniques. Many neurons in the central nucleus contain peptidergic neurotransmitter candidates such as corticotropin-releasing factor, somatostatin, neurotensin, cholecystokinin, and enkephalin. Although their role in the amygdala is unclear, these peptides are known to have specific functions in other systems. For example, some are the releasing factors or release-inhibiting factors that are synthesized in hypothalamic neurosecretory cells and released into the portal blood to promote synthesis and release of pituitary hormones.

Like the more lateral regions of the amygdala, the central nucleus receives sensory information from temporal cortical association areas. The central nucleus also receives projections from the brain stem, primarily those from the parabrachial nuclei related to taste.[2, 31] As can be seen in Figure 32–5C, the central amygdala (C) projects to regions in the brain stem including the periaqueductal gray, the parabrachial nucleus, the nucleus of the solitary tract, the reticular formation, and the dorsal motor nucleus of the vagus.[40] These connections with autonomic centers provide a substrate for the well-documented control exerted by the central amygdala over heart rate, respiration, gastric function, and reaction to pain (described later).

Efferents from the central nucleus include axons from many of the peptidergic cell groups that contain neurotensin, corticotropin-releasing factor, or somatostatin. Enkephlain-containing cells appear to be intrinsic neurons. Fibers from the central nucleus are also found around all brain stem monoaminergic nuclei.[40]

The connections of its various subnuclei allow the amygdala to modulate a variety of complex physiological and behavioral processes. We have suggested that the corticomedial amygdala is involved with olfaction and hypothalamic function, that the basolateral division plays a role in the processing of complex sensory information, and that the central region is related to brain stem autonomic control. Although these emerge as the primary functions for each region, the central and basolateral divisions also interact with the hypothalamus, and the central nucleus also exchanges information with the cortex. Furthermore, the nuclei of the amygdala have extensive interconnections with each other (Fig. 32–5D), so that it seems unlikely that the primary functions of each nucleus are performed in isolation. However, selective activation or inactivation of certain amygdalar nuclei, as seen in the next section, indicates that they seem preferentially involved in different behaviors.

Physiology and Function

Investigations of amygdalar physiology have been hampered by several major factors. First, the cognitive and experiential factors that appear to be such an important part of amygdalar function are difficult to control. For example, completely opposite emotional responses (e.g., anxiety versus aggression) can be obtained from electrical stimulation of the same region of the human amygdala, depending on the state and concerns of the individual. Second, destroying (or electrically or chemically stimulating) a part of the amygdala invariably injures (or stimulates) passing fibers from other regions. Last, the study of the amygdala is complicated by the fact that it is involved in very complex behaviors that have many components. For example, the role of the amygdala in learning or emotional behavior must include its involvement in the control over cardiovascular responses that accompany some forms of learning and many emotions. Despite these caveats, the amygdala has definitely been implicated in the control of a variety of processes, some of which we will now discuss.

Homeostatic Control

Electrical stimulation of the central nucleus in the rat produces a constellation of physiological and behavioral changes. Ongoing behaviors cease, and the animal exhibits mouthing or chewing movements accompanied by cardiovascular and gastric changes characteristic of the physiological "state" produced by aversive stimuli or stressful situations. If stimulation is continual, gastric ulcers can result. Conversely, lesions of the central nucleus reduce the incidence of gastric ulcers and sympathetic arousal. These effects are probably mediated by projections of the central nucleus to the brain stem parabrachial nucleus and dorsal nucleus of the vagus nerve, regions that control gastric and cardiovascular function.

The central nucleus also is required for the conditioning of autonomic responses to previously neutral stimuli that have been associated with aversive situations. Figure 32–6 shows data obtained from a rabbit undergoing aversive conditioning. In this situation, the rabbit was presented a 5-s tone as a conditioning stimulus (CS) for several trials; the CS was then paired with an eyelid shock (an unconditioned stimulus, UCS) for several more trials.[20, 21] Heart rate decreased during the pairing of the CS and the UCS and remained

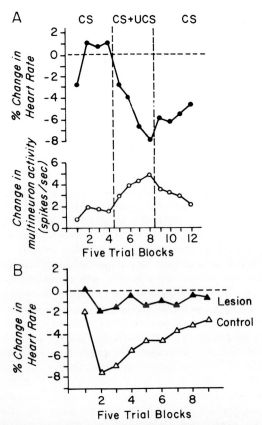

Figure 32–6 Role of the central nuclei of the amygdala in Pavlovian fear conditioning in the rabbit. *A,* Changes in heart rate and central nucleus multineuron activity that occur when the conditioned stimulus, in this case a tone, is presented both before and after (CS) and in combination with an eyelid shock (CS + UCS). During paired CS + UCS trials, heart rate decreases and neural activity increases. *B,* Lesions of the central nucleus prevent the changes that occur with aversive conditioning. (After Kapp, B.S.; Gallagher, M.; Frysinger, R.C.; Applegate, C.D. In Ben-Ari, Y., ed. *The Amygdaloid Complex,* 355–366. Elsevier Science Publishers B.V. (Biomedical Division) Amsterdam, 1981.)

steroids, and lesions confined to the region of highest adrenal steroid uptake increase the secretion of adrenocorticotropin (ACTH). In contrast to the apparently inhibitory effect exerted by the medial and cortical nuclei on adrenal steroid secretion, stimulation of lateral portions of the amygdala increases ACTH secretion. However, the effects of amygdalar stimulation on ACTH release depend, in part, on the existing state of the animal. For example, in conscious, freely moving cats, electrical stimulation of the lateral or corticomedial portions increases adrenal steroid output when prestimulation basal levels of ACTH are low but decreases steroid output when pre-existing levels are high.

Sensory Modulation of Behavior

The influence of the amygdala on autonomic and reproductive function seems to depend, in part, on the sensory information that reaches its lateral and central portions via cortical pathways. The inferior temporal, superior temporal, and inferior parietal cortices, which receive information from visual, auditory, and somatosensory cortices, all project to the lateral and central regions of the amygdala.[46, 48] Indeed, some single neurons in the amygdala respond to several sensory modalities.[2, 3] Mishkin and Aggleton[28] concluded that the cortical association areas that project to the amygdala are the final stop in the processing of the physical properties of sensory stimuli into a coherent perception, and that the links between these perceptions and nonsensory events associated with them, such as autonomic arousal, are made in the amygdala. Numerous experiments suggest that the sensory stimuli most effective in altering the pattern of neural discharge in the amygdala have an emotional or acquired discriminative quality. For example, neurons in the anterior amygdala of monkeys respond differently to rewarded and non-rewarded visual stimuli in a food-reinforced discrimination task.[36] Also, some neurons in the amygdala of awake monkeys apparently respond selectively to faces,[24] as do some neurons in the inferotemporal association cortex (see Chaps. 20 and 31). This complex visual stimulus presumably has special emotional significance.

Amygdalar stimulation in humans with intractable epilepsy can induce mental phenomena or auras that can have visual, auditory, visceral, somatosensory, and emotional qualities. The most common of these phenomena include hallucinations, feelings of déjà vu, and alimentary sensa-

depressed when the CS was presented alone (Fig. 32–6A). The changes in heart rate were accompanied by a parallel increase in multi-neuron activity recorded from sites in the central nucleus (Fig. 32–6A). Furthermore, the decrease in heart rate produced by conditioning (Control, Fig. 32–6B) was abolished after lesions of the central nucleus, suggesting a causal relation between altered cellular activity in the central amygdala and the conditioned autonomic response.

The amygdala also plays a role in maintaining homeostasis through its connections with hypothalamic neurons, which in turn project to brain stem regions or to the pituitary. Neurons in the corticomedial amygdala have receptors for adrenal

705 VI HIGHER FUNCTIONS

tions. These observations illustrate the variety of sensory experiences associated with the amygdala.

Motivation and Emotion

The amygdala plays a crucial role in allowing animals to relate to and learn from environmental stimuli. Numerous investigators have found that rats, cats, and monkeys with lesions of the lateral and/or central amygdala have difficulty learning to avoid noxious stimuli. These deficits may be attributable to a lack of "fear," resulting from the loss of central amygdalar control over autonomic arousal. The amygdala, however, also has a high concentration of endogenous opiates, and animals with amygdalar lesions may simply have a different perception of pain. Amygdalar lesions also produce a deficit in the formation of associations between stimuli (such as a bell) and subsequent desirable consequences (such as the acquisition of food).

Animals whose amygdalas have been destroyed exhibit a general lack of affect and no longer respond appropriately to their environments. For example, free-roaming monkeys with amygdalar lesions become social isolates. Because they are placid, they lose their position in the dominance hierarchy of the group; if they don't withdraw from the group on their own, they are often forced to leave.

A similar "flat" affect is found in patients with amygdalectomies, a treatment of last resort in clinical cases of severe aggressive behavior. Amygdalectomy does indeed eliminate aggressive behavior, but it also eliminates responses to situations that should produce positive emotional experiences. Consistent with the lesion studies, electrical stimulation of the human amygdala can produce aggressive feelings. The emotional experiences produced by such stimulation depend on the state of the patient, who may feel anger, fear, anxiety, or pleasure. Each feeling is accompanied by the constellation of physiological responses expected if that "state" were actually occurring.

A variety of mood-altering drugs influence the discharge of neurons in the amygdala. As previously mentioned, the lateral amygdala is rich in certain opioid receptors that may be important in the perception of pain. Injection of naloxone, an opiate antagonist, into the amygdala produces many of the autonomic responses that constitute the "morphine withdrawal syndrome" (e.g., heart rate changes, shivering). Anxiolytics such as diazepam (Valium) depress the activity of single neurons in the amygdala. Presumably these drugs act here, as they do elsewhere in the nervous system, by binding to GABA$_A$ receptors and potentiating the inhibitory actions of the neurotransmitter gamma-aminobutyric acid (GABA). Binding sites for these anxiolytics are found in high concentrations on neurons in the lateral and basolateral amygdala (Fig. 32–7), suggesting that these neurons are one location where Valium exerts its therapeutic action.

Learning

We have already noted that rats, monkeys, and humans with lesions of the amygdala seem to have particular difficulty in producing an appropriate response to appetitive or aversive stimuli. It is difficult, however, to determine whether the stimulus has simply lost its usual significance (a deficit in affect) or whether the association between the stimulus and the behavior is impaired (a deficit in learning). In the former case, a positive reward (such as food) may lose its hedonic value or a negative reward (such as pain) may not be perceived as offensive; the animal may simply "not care" about these consequences.

Several lines of evidence suggest that the amygdala is involved in true learning.[17] Large lesions of the amygdala in rats generally produce deficits in active avoidance tasks such as running away at

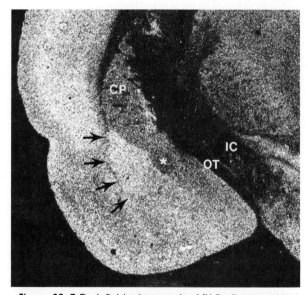

Figure 32–7 Dark-field micrograph of ³H-flunitrazpam binding to rat amygdala. "Receptors" for this Valium-like benzodiazepine appear as bright dots. Arrows mark the lateral border of the lateral and basolateral amygdala. There are more binding sites in the lateral and basolateral than in the central (asterisk) amygdala. Abbreviations: CP = caudate/putamen; OT = optic tract; IC = internal capsule.

the presentation of a tone that signals the occurrence of a shock; however, such lesions apparently do not compromise performance on passive avoidance tasks such as not entering a particular region of a maze during the presentation of a stimulus that has been paired with a shock. With considerable training, an amygdalectomized animal can learn to actively avoid aversive stimuli. Such animals may have problems only in initiating behaviors, thereby explaining why active avoidance tasks seem to be more compromised than passive avoidance tasks. However, rats and monkeys with amygdalar lesions also continue to perform previously learned behaviors that are no longer rewarded, whereas normal animals do not. Thus, animals with amygdalar lesions not only have difficulty initiating a response, they also can't learn to suppress it once it becomes inappropriate. Similar problems with perseveration result from lesions of orbital prefrontal association cortex (see Chap. 31), which has intimate connections with the amygdala and the overlying temporal pole.

Another form of learning, the conditioned emotional response, also is impaired in amygdalectomized animals. The conditioned emotional response can be demonstrated in a normal rat trained to press a bar for food. If a tone that previously has been paired with a shock is presented, a rat will stop bar pressing even though the tone (and the shock) are completely unrelated to bar pressing or food delivery. During the tone, the rat also exhibits physiologic signs of autonomic arousal (e.g., piloerection, increased heart rate; see Chap. 34), signs that constitute a conditioned emotional response. After a rat has sustained a lesion of the amygdala, it will continue bar pressing throughout the tone and will show no signs of autonomic arousal. Such deficits in aversive learning are most closely associated with basolateral lesions, but can be ameliorated by subsequent corticomedial amygdalar lesions.

While manipulation of the amygdala disrupts learning in a variety of tasks, the role of this region in learning retention—or memory—is problematic. Lesions of the amygdala alone often do not disrupt the performance of aversive learning tasks that were acquired before surgery. However, if a monkey learns either a visual or tactile discrimination task, its performance on these tasks is impaired after removal of *both* the amygdala and hippocampus.[29] In humans the global amnesia consequent to temporal lobe injury or surgical removal may depend as much on the loss of the amygdala as on hippocampal damage. The role of limbic structures in memory is considered more fully in a subsequent section on the hippocampus.

Summary

The amygdala (Fig. 32–8A) is involved with a variety of behaviors. The corticomedial amygdala interacts with phylogenetically "old" regions of the brain, such as the olfactory bulb, brain stem, and hypothalamus. The lateral amygdala interacts with "newer regions," such as the cortex, and the central amygdala interacts with both. The amygdala samples all physical aspects of the environment and appears to attach motivational and autonomic significance to the integrated sensory stimuli.

Simplistically, the lateral amygdala and parts of the central amygdala may compare stimuli as they are received and determine whether they are "meaningful," based on *immediate* consequences or past stimulus encounters. For example, a subordinate monkey in a natural environment gradually learns that the best response to a challenge by a dominant member of the troop is submission (perhaps accompanied by autonomic arousal and "fear"). Lesions of the lateral or central amygdala or both in the submissive animal, like those produced in the free-roaming monkeys described above, might therefore result in both a failure to recognize the significance of the dominant animal (acquired through past encounters) and an inabil-

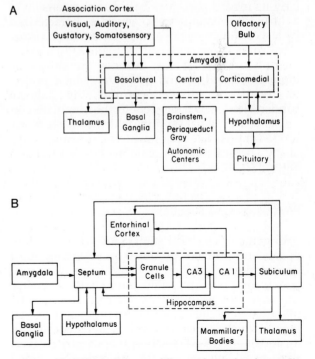

Figure 32–8 Block diagram of the most prominent connections of the amygdala *(A)* and the septum and hippocampus *(B)*.

ity to appreciate the consequences (aggression or pain) of his foolhardy behavior. Thus, the corticomedial and other parts of the central amygdala may serve as "effectors" by operating on instructions from the lateral amygdala to promote a complete and appropriate response.

SEPTUM

Anatomy

The septum forms the medial wall of the lateral ventricles immediately anterior to the hippocampus. Many septal efferents terminate in the hippocampus, which, in turn, provides a large reciprocal projection. This intimate septo-hippocampal relationship has been explored extensively in the rodent brain, where septo-hippocampal input mediates the prominent "theta" rhythm, a low-frequency (4–10 Hz) oscillation of the gross electrical potential (the electroencephalogram [EEG] described below) recorded from the hippocampus. Human hippocampi do not generate the theta (or rhythmic slow) activity so characteristic of many mammals, and, interestingly, no satisfactory septal homologue has been found in the human brain.

Recent work suggests that the septum, particularly its medial part, may be a dorsal extension of a diffuse forebrain system, which is characterized by large, cholinergic neurons that establish prominent connections with cortical regions.[26] This cholinergic forebrain system includes the substantia innominata and ventral pallidum, which project to frontal and parietal neocortex, the diagonal band region, which projects to the cingulate cortex, and the medial septum, which projects to the hippocampus. The major anatomical relationships of the septum and hippocampus are displayed in Figure 32–8B.

Physiology and Function

In general, electrical stimulation of the septum of cats and primates produces an inhibitory effect on autonomic function (a decrease in heart rate and respiration), probably through its extensive connections with hypothalamic regions.[18] On the other hand, septal lesions in rats produce a "septal rage," which is characterized by hyperemotionality and an increased reactivity (apparently an aversion) to all sensory stimuli. Animals with septal lesions are extremely aggressive and intractable to normal handling, but once the rage and hyperreactivity subside, they are actually more docile than normal animals.

Although the septum is intimately linked with the hippocampus, it does not appear to participate in all hippocampal functions. For example, the autonomic changes and rage characteristic of animals with septal lesions are not produced by hippocampal damage. On the other hand, lesions of either structure produce similar deficits in learning. Finally, the medial septum may play a role in hippocampal sensory processing and learning, as the pacemaker for rodent hippocampal theta activity.

HIPPOCAMPUS

Intrinsic Circuitry and Connections

In contrast to the amygdala, the hippocampus is a relatively simple neural tissue. It is composed of two principal cell types arranged in discrete layers, with a highly organized, laminar pattern of afferents and efferents. Although the hippocampus is a long structure, extending from the septum to the temporal lobe (the hippocampus of higher species gets progressively pushed into the temporal lobe), its intrinsic organization is remarkably similar throughout its length. This intrinsic organization is shown in the schematic drawing of the transverse section in Figure 32–9A.[45]

The principal output neurons of the hippocampus are the pyramidal cells. The larger pyramidal cells, which extend out from the dentate hilus through the lateral curvature of the hippocampus, are collected in the CA3 region, or the regio inferior; the smaller pyramidal cells, which extend from the regio inferior to the midline, are in the CA1 region, or the regio superior. The CA1 and CA3 pyramidal cells are arranged in a single layer that extends laterally from the midline and curves around to insert between the dorsal and ventral blades of the dentate gyrus (DG, Fig. 32–9A). The dentate gyrus contains the second major cell type of the hippocampus, the granule cells. The unipolar granule cells have dendrites that extend into the molecular layer of the dentate gyrus and axons that enter the CA3 region of the hippocampus. Dentate granule cells make only intrinsic connections.

The intrinsic organization of the hippocampus is rather simple (see Fig. 32–8B). The major pathway through the hippocampus consists of a three-neuron feed-forward circuit. Much of the input to

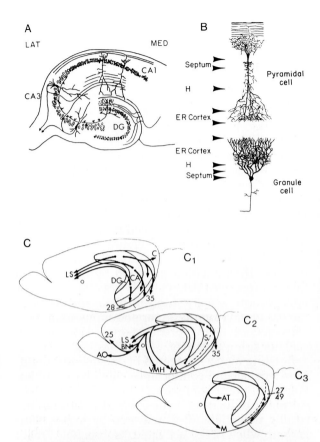

Figure 32-9 Intrinsic circuitry and input and output connections of the rat hippocampus. *A,* Principal hippocampal cells and their locations (pyramidal CA1 and CA3 cells and granule cells in the dentate gyrus [DG]). *B,* Sites at which afferents arising within (H) and outside (entorhinal cortex [ER] and septum) the hippocampus synapse on pyramidal and granule cell dendrites. *C,* Main efferent hippocampal targets; outputs of the subiculum (S), a major hippocampal efferent destination, are also shown. Other abbreviations: AO = anterior olfactory nucleus; AT = anterior thalamus; BN = bed nucleus of the stria terminalis; C = cingulate area; CA = pyramidal cell layer of hippocampus; LS = lateral septum; M = mammillary bodies; VMH = ventromedial hypothalamus; Brodmann's area 22 = presubiculum; Brodmann's area 28 = entorhinal area; Brodmann's area 35 = perirhinal area; Brodmann's area 49 = parasubiculum. (After Swanson, L.W.; Cowan, W.M. *Journal of Comparative Neurology* 172:49–84, 1977.)

the hippocampus impinges on the dentate granule cells, which project entirely to the ipsilateral CA3 region. The CA3 cells, in turn, project out of the hippocampus (principally to the septum) and also send a collateral axon, the so-called Schaffer collateral, into the CA1 region, where they terminate on the apical dendrites of CA1 pyramidal cells. The axons of CA1 pyramidal cells project primarily to an adjacent cortical region, the subiculum.

Afferent input to the granule cell dendrites is organized in a highly laminar pattern, as shown in Figure 32–9B.[9] Fibers from the entorhinal cortex (ER, Fig. 32–9B) synapse on the outer two thirds of the granule cell dendrites. Fibers from the hippocampus (H) terminate beneath the entorhinal fibers, closer to the granule cell bodies. Finally, septal afferents synapse almost exclusively on the most proximal part of the dendrites. Afferent input from these same sources, but not necessarily the same cells, also reaches the pyramidal cells directly, where it displays the same dendritic distribution (Fig. 32–9B).

The efferent targets of the hippocampus are much fewer in number than those of the amygdala. Much of the efferent output of the hippocampus is communicated to the rest of the brain through connections with the entorhinal and subicular cortices. The connections from the pyramidal cell layer of the hippocampus (CA) to the subiculum (S)[5] and entorhinal cortex (Brodmann's area 28) are shown in C_1, Figure 32–9C. In addition, the hippocampus projects to the lateral septum (LS). Most of the hippocampal projections to other areas are relayed through the subiculum (C_2, Fig. 32–9C) and entorhinal cortex (C_3, Fig. 32–9C).

Physiology and Function

The hippocampus does not exert extensive autonomic influence. Like the amygdala, the hippocampus apparently does process complex sensory information; unlike the amygdala, however, it is not primarily involved in imparting motivational significance to those stimuli. What use the hippocampus does make of the sensory information is unclear. In the last several decades, however, several lines of research have implicated it in learning and memory processes and in the generation of epilepsy. The unique, simple structure of the hippocampus has enabled scientists to trace these processes down to electrical activity in its individual cells or cell membranes, as we will see in the following section.

Cellular Physiology

The pyramidal cells of the hippocampus are among the most intensively studied central nervous system neurons. Their large size, homogeneous appearance, and orderly morphology have made them inviting targets for microelectrodes, both in vivo and in vitro. Each subregion of the hippocampus (CA1, CA3, and the dentate gyrus) contains a characteristic cell population with relatively unique features. The richness of their intrin-

A

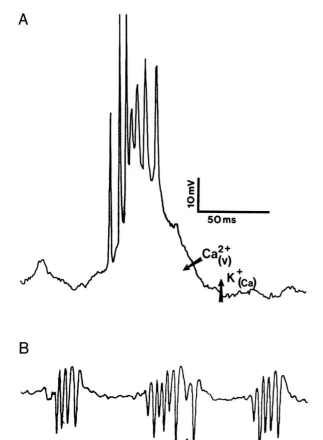

10 mV

50 ms

Ca²⁺(v)

K⁺(Ca)

B

10 mV

50 ms

Figure 32–10 Bursting in CA3 pyramidal cells. *A,* Intracellular recording of a spontaneously occurring burst of action potentials in a normal cell. Action potentials ride on a large depolarization produced by a voltage-sensitive calcium influx $G_{Ca^{2+}(v)}$. Repolarization is produced by a potassium efflux triggered by the incoming calcium, $G_{K(Ca^{2+})}$. *B,* Extracellular recording from several CA3 cells that have been induced to fire synchronously; three bursts are shown.

sic and synaptically modulated conductances (see Chap. 12) enables these cells to exhibit a wide variety of discharge patterns.

Intrinsic Properties. The pyramidal cells in the CA3 region discharge bursts of spikes spontaneously. The spikes are sodium- and/or calcium-dependent action potentials riding on large underlying depolarizations,[50] which are the result of an intrinsic, voltage-sensitive, slow calcium conductance (Fig. 32–10A)[47] (recall Chap. 12). The bursts are curtailed by a calcium-dependent outward potassium conductance, which is activated

by an increase in intracellular calcium. This potassium conductance is critical for repolarizing the cell, since the Ca^{2+} conductance is inactivated very slowly, if at all, and therefore tends to keep the cell depolarized. Under normal conditions, the intrinsic burst or pacemaker activity is not synchronized in different CA3 neurons. If one records the extracellular potentials of several CA3 cells (multineuron activity) under such conditions, the asynchronous action potentials produce no summated spike potentials. If CA3 cells are made to fire synchronously by electrical stimulation of hippocampal afferents, the action potentials from neighboring CA3 cells add to produce clear bursts of spike potentials, as revealed in the extracellular multineuron recording of Figure 32–10B.[47] Synchronous burst activity can also be produced by blocking the inhibition provided by hippocampal interneurons (see below). This great tendency for epileptiform activity (uncontrolled synchronous bursts) to develop in the hippocampus could be a consequence of the predilection of normal CA3 cells for firing in bursts.

Bursting may be triggered by the low-threshold, transient calcium conductance (see Chap. 12) that is found in hippocampal pyramidal neurons. This calcium conductance is inactivated at or near the resting potential and is activated by hyperpolarization; when it is activated, subsequent small depolarizations, which open the calcium channels, could lead to a burst discharge. Thus, in a paradoxical way, hyperpolarizing inhibition may promote bursting activity in these neurons.

Hippocampal pyramidal cells have a host of other ion conductances that are often activated near the threshold for action potentials. These conductances determine the inter-spike or inter-burst intervals, and thus control cell excitability. In addition to voltage-sensitive, depolarizing calcium conductances, there seems to be a noninactivating sodium conductance that adds to cell depolarization when a neuron is slightly depolarized as, for example, by an excitatory postsynaptic potential (EPSP). Activation of the potassium conductances described in Chapter 12 (e.g., the "delayed rectifier," the calcium-dependent potassium conductance, the A-current, and the M-current) hyperpolarizes pyramidal cells and acts to balance the depolarizations produced by both sodium and calcium conductances.

Synaptic Connections. Several features of the afferent synapses suggest that hippocampal inputs are very powerful. First, the feed-through circuit, from entorhinal cortex to dentate granule cells to CA3 to CA1, is entirely excitatory. Second, a great many of the afferent synapses in hippocampus are formed *en passant*, so that a single axon contacts

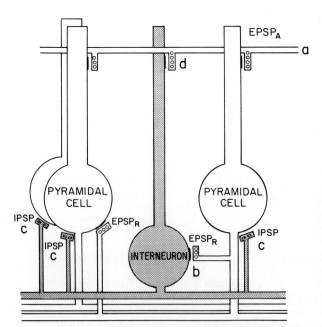

Figure 32–11 Simplified local circuit inhibitory and excitatory interactions among pyramidal cells of the hippocampus. Excitatory afferents (a) contact pyramidal and interneuron (d) cell dendrites to produce excitatory postsynaptic potentials (EPSP$_A$). Pyramidal cells excite interneurons (b) to produce recurrent excitatory postsynaptic potentials (EPSP$_R$) in inhibitory interneurons which, in turn, inhibit further pyramidal cell discharge (c). (After Schwartzkroin, P.A. In Delgado-Escueta, A.V.; Ward, A.A.; Woodbury, D.M.; Porter, R.J., eds. *Advances in Neurology, vol. 44. Basic Mechanisms of the Epilepsies: Molecular and Cellular Approaches*, 991–1010. New York, Raven Press, 1986.)

changes. First, a large increase in chloride conductance produces the initial component of the IPSP and significantly reduces cell input resistance to create a shunt for other currents (e.g., those from EPSPs). The inhibitory terminals and their receptors are strategically placed on the pyramidal cell soma, initial segment, and proximal shafts of primary dendrites, where the inhibitory "shunting" would be most effective (see Chap. 12). Second, a later opening of potassium channels contributes to a subsequent, longer-lasting component of the IPSP. Thus, recurrent inhibition makes excitation self-limiting. Many interneurons need not first be activated by pyramidal cells since they have extensive dendritic trees, which receive excitatory input directly from incoming afferent fibers (d in Fig. 32–11). This feed-foward inhibitory circuit is activated at lower stimulus thresholds, and appears to be far more powerful than feedback inhibition.

Recordings from isolated dendrites of pyramidal cells have shown that synaptic inhibition also directly controls the excitability of dendrites.[4] Such control is essential for hippocampal neurons, since the dendrites may be able to generate not only calcium-mediated but also sodium-mediated action potentials.

In summary, hippocampal inhibition can be either feedback or feed-forward, can be mediated either at somatic or dendritic sites, can involve multiple postsynaptic receptors, and can be mediated by a variety of ionic conductances.

Neurochemistry. Some insight into the role of a structure and its various afferent and efferent connections may be gleaned by considering its resident neurotransmitters. The relatively simple structure of the hippocampus makes it possible to examine many putative neurotransmitter substances in the in vitro "slice" preparation.

For the transverse hippocampal slice preparation, intact brains are removed from animals immediately after sacrifice. The hippocampus is isolated, and thin (e.g., 320 μm) sections are cut in the plane shown in Figure 32–9A and placed immediately in a chamber containing a warm bathing medium that is continually bubbled with O_2. For several hours in this preparation, stable intracellular recordings can be made under visual guidance, putative transmitters can be applied iontophoretically, and the ionic concentrations of the bathing medium can be modified.

The role of GABA has already been described. While the transmitters used by hippocampal pyramidal and granule cells are most likely excitatory amino acids such as glutamate or aspartate, their identity is still a matter of some controversy. However, a variety of putative neurotransmitter

many target cells (EPSP$_A$, Fig. 32–11).[43] Third, pyramidal cells are electrically compact, with an average electrotonic length of 1.0. Therefore, even distant synapses are remarkably effective at activating these neurons. Indeed, by progressively cutting the input to pyramidal cell dendrites, Andersen[1] estimated that activation of only 3% of the afferents can cause a pyramidal cell to fire.

Despite the apparent efficacy of the excitatory synaptic input and intrinsic depolarizing conductances, the spontaneous firing rate of hippocampal neurons is quite low because an excitatory synaptic event is rapidly curtailed by a long and powerful inhibitory postsynaptic potential (IPSP), which allows the pyramidal cell to discharge only one action potential.[11] The IPSP is the result of recurrent inhibition produced by the excitation of an adjacent inhibitory neuron by a discharging pyramidal cell (EPSP$_R$ at b, Figure 32–11). The inhibitory neuron synapses back onto pyramidal cells (IPSP at c, Fig. 32–11), where it releases GABA to produce at least two separate conductance

Table 32–2 Neuroactive Substances in the Hippocampus

Neurotransmitter	Receptor Type	Synapse	Ionic Effect	Pyramidal Cell Effect
Acetylcholine	Muscarinic	Septum–hippocampus	Blocks K$^+$ M current Decrease $G_{K^+(Ca^{2+})}$	Slow depolarization Increasing firing frequency
	Nicotinic	Septum–hippocampus	?	?
Norepinephrine	Alpha Beta Both adenylate cyclase linked	Locus ceruleus– hippocampus	Modulates $G_{K^+(Ca^{2+})}$	Slow hyperpolarization Slow depolarization
Dopamine	Adenylate cyclase linked	Substantia nigra and ventral tegmentum– hippocampus	Increases $G_{K^+(Ca^{2+})}$	Slow hyperpolarization
Serotonin	5HT$_1$ and 5HT$_2$—both adenylate cyclase linked	Midbrain raphe– hippocampus	?	Decreased excitation
Glutamate/Aspartate	Quisqualate	CA3–CA1; entorhinal cortex–CA1	Increases $G_{K^+GNa^+}$	Fast depolarization
	Kainate	Dentate–CA3	Increases $G_{Na^+GK^+}$	Fast depolarization
	NMDA	CA3–CA1; entorhinal cortex–CA1	Increases $G_{Na^+}G_{Ca^{2+}}$	Slow depolarization
GABA	GABA$_A$	Interneuron–pyramidal cell body	Increases G_{Cl^-}	Rapid inhibition
	GABA$_B$	Interneuron–pyramidal cell dendrites	Increases G_{K^+}	Late inhibition
		Interneuron–afferent axon terminals (presynaptic inhibition)	Decreases $G_{Ca^{2+}(V)}$	Reduces transmitter released at afferent terminal, decreasing afferent effect on pyramidal cell

Each putative neuron transmitter is identified according to the synapse at which it is found, its postsynaptic receptor, its ionic effect, and its effect on pyramidal cell membrane potential.

substances have been identified in neurons projecting to the hippocampus. These include acetylcholine, norepinephrine, dopamine, serotonin, and several peptides. Table 32–2 lists a variety of neuroactive substances found in the hippocampus, the receptors they are thought to interact with, their source, and both their ionic effect and the final effect on pyramidal cell firing. The roles of many of these transmitters are unknown, but it has been suggested that they might be neuromodulators with actions both slow in onset and long in duration (see Chap. 11).

The neuromodulator that has been best studied and whose role is most secure is the cholinergic input from the medial septum.[4] Here, acetylcholine primarily produces a direct excitation of pyramidal cells by blocking the potassium-mediated M-current.[14] This action, which is slow in onset, long-lasting, and mediated by muscarinic receptors, may be involved in maintaining neuronal discharge. When a pyramidal cell is steadily depolarized, it tends to fire rapidly at first, but its firing rate gradually decreases. If cholinergic agonists are applied, the neuron continues to fire at

a high rate, presumably because the potassium channels have been prevented from opening. The slow modulation of excitability by acetylcholine could be used to synchronize activity in a large group of neurons, thereby providing the pacemaker mechanism for septally induced hippocampal theta rhythm in rats.

Synchronous Discharge and Field Potentials

The Electroencephalogram. Electrodes placed on or in the skin of the head record a surface electroencephalogram (EEG), which consists of very small changes in electrical potential that result from the activity of cortical neurons. The EEG potential is conducted through (and greatly attenuated by) the meninges, skull, and scalp. Usually the EEG is recorded from many electrodes placed bilaterally over the four cerebral lobes. In the normal waking or sleeping state, the bulk of the surface EEG potential is thought to reflect synchronous, summated activity of cortical pyramidal cells that underlie the active recording electrode.

The resulting potential is generally of low frequency (up to a few tens of Hz) and amplitude (tens to hundreds of μV).

The dominant frequency and amplitude characteristics of the surface EEG vary systematically with states of arousal. Four different EEG patterns are commonly described: beta (13–20 Hz, lowest amplitude) is associated with intense mental activity; alpha (8–13 Hz, low amplitude) occurs with an awake but relaxed condition; theta (4–10 Hz, low amplitude) is seen briefly in stage 1 and rapid eye movement (REM) sleep; delta activity (0.5–4 Hz, high amplitude) accompanies deep sleep, stages 3 and 4. Figure 32–12[42] displays typical surface EEG patterns during various states of arousal.

The human surface EEG provides a valuable non-invasive clinical tool for the assessment of brain function. Computer-enhanced EEG imaging techniques allow clinicians to compare the distribution of different patterns of EEG activity across the cortical surface to aid in the diagnosis of diseases.

EEG potentials can also be recorded from deep cortical structures. A relatively large electrode is driven into the brain, and the recorded potentials are low-pass filtered to eliminate the recording of individual action potentials. In this way, one can

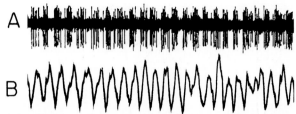

Figure 32–13 Simultaneous recording of the activity of single hippocampal pyramdal cells *(A)* and EEG theta activity *(B)*. (After Vanderwolf, C.H.; Leung, L.-W.S. In Seifert, W., ed. *Neurobiology of the Hippocampus,* 275–302. New York, Academic Press, 1983.)

record a hippocampal EEG that can be used to study both limbic function and the mechanisms of the EEG.

The limbic system of most non-primate mammals displays a state-dependent EEG characterized by either an irregular pattern or a largely sinusoidal, slow regular pattern, known as theta rhythm (Fig. 32–13B).[47] Like the surface theta rhythm, limbic theta rhythm has a frequency to 4 to 10 Hz, but it is larger in amplitude. It seems to result from synchronized bursts of limbic neuronal activity (Fig. 32–13A) that occur at times when the surface EEG displays high-frequency activity. Limbic theta rhythm is associated with a number of active, purposeful behaviors such as rearing, jumping, swimming, and following objects with head movements. Animals also seem to synchronize their sensory input with the limbic theta rhythm by engaging in synchronized exploratory behaviors such as sniffing and movement of the vibrissae (see later).

These hippocampal potentials provide a model for the study of the generation of all cortical potentials. The hippocampus is an excellent structure to study because a number of special factors produce unusually large potentials there. First, the cell bodies are lined up (layered) with the soma-dendritic axis of each cell oriented parallel to that of its neighbor. Second, activation of afferent fibers, which make en passant synapses, inevitably excites a large number of pyramidal cells. Third, the extracellular space around hippocampal cells is unusually small, particularly near the cell bodies. This sparse extracellular space results in a higher extracellular resistance, which produces a greater voltage for the same extracellular current.

To understand the mechanism of the hippocampal EEG, we must consider the general rules that govern the generation of all extracellular potentials. Recall that when an EPSP is produced on a dendrite, current flows from the extracellular

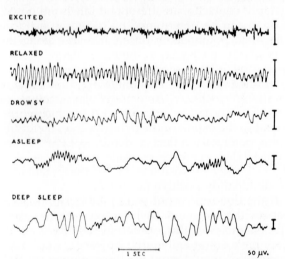

EXCITED

RELAXED

DROWSY

ASLEEP

DEEP SLEEP

1 SEC   50 μv.

Figure 32–12 Electroencephalographic records during excitement, relaxation, and varying degrees of sleep. In the fourth trace, the runs of 14/s rhythm superimposed on slow waves are called "sleep spindles." Note that excitement is characterized by a rapid frequency and small amplitude and that varying degrees of sleep are marked by increasing irregularity and by the appearance of "slow waves." (From Jasper. In Penfield and Erickson. *Epilepsy and Cerebral Localization,* 1941. Courtesy of Charles C Thomas, Publisher, Springfield, Illinois.)

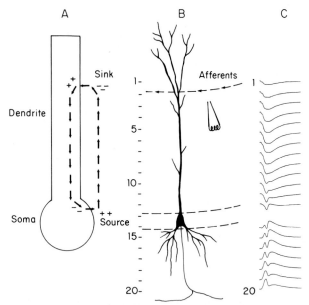

Figure 32–14 The generation of extracellular field potentials produced by an excitatory input to the dendrite of a neuron. *(A)* the current flow in a schematic neuron. If the neuron in *(B)* receives excitatory afferents to the indicated dendritic location, the field potentials in *(C)* will be recorded at the 20 locations along the neuron. (See text for further details.)

space into the cell, as shown in Figure 32–14A (the convention for the direction of current flow is to follow the positive charge). With movement of charge into the cell at the point of EPSP generation, a circuit is established that uses both intracellular and extracellular pathways. In effect, the cell becomes a dipole with a predominantly positive charge at one end (the site of EPSP generation) and a predominantly negative charge at the other (the soma). The flow of current across the resistance of the extracellular space produces a voltage change that can be recorded extracellularly and that reflects the cellular dipole. Since the extracellular resistance is normally rather small (especially compared with the resistance across a cell membrane), the voltage change associated with these extracellular currents is on the order of microvolts rather than the millivolt intracellular potentials.

Potentials like the hippocampal theta rhythm are thought to result from the synchronous activation of the afferents to a group of cells, so that all of the individual postsynaptic cellular potentials occur at nearly the same time. The resultant extracellular currents produce voltage changes that are summed at the recording electrode. In the case of the theta rhythm, activation of hippocampal cells may result from a synchronous septal input (as discussed earlier).

Some surface cortical potentials also are the result of synchronous activation of either intrinsic cortical or thalamocortical afferents to cortical pyramidal cells. Synchronous activity can be elicited by brief natural stimuli such as a flash of light or a click. The resulting *evoked potentials* can be used to plot the distribution of sensory inputs on the cortical surface, as was done for the visual inputs to the occipital cortex in Chapter 20, Figure 20–4. Alternatively, potentials can be evoked (in animals) by electrical stimulation of the cortical afferents. The resulting potentials have been used to infer the synaptic and functional relationships between elements in a small patch of cortex.

Field Potentials in the Hippocampus. The field potentials that result from the activation of a bundle of afferent fibers synapsing on the distal dendrites of CA1 pyramidal neurons are displayed in Figure 32–14. These excitatory afferents produce EPSPs in several pyramidal neurons. At the distal synapse, current (positive charge) enters the cell (as in A), leaving the extracellular region relatively negative. This region of current flow is called a *sink*. The current then flows through the dendrite and leaves the cell at the soma, where an extracellular positivity, called a *source*, develops. The spatial relationship of sources to sinks gives us some information about the nature and location of the synaptic input. If, for example, the site of synaptic contact of an afferent bundle is unknown, it often can be determined by recording extracellular field potentials at different points along the axis of the postsynaptic neuron. Figure 32–14C shows recordings made at 20 different positions (identified in Fig. 32–14B) following stimulation of an afferent pathway. An extracellular negativity (a sink) is recorded in positions 1–13, which correspond to the extent of the dendritic tree; therefore, these potentials reflect a dendritic EPSP. If the extracellular recording is made near the level of the cell body in positions 15–20, the potential is predominantly positive (a source).

If the dendritic input were inhibitory, the location of the sink and source would be reversed. The diagnosis of the location of inhibitory inputs does have some ambiguity, however, since a somatic IPSP would produce the same source-sink configuration as a dendritic EPSP. Therefore, field potentials have a limited utility, and an absolute determination of synaptic sign and location usually necessitates intracellular recordings.

The synaptic drive from an afferent barrage may produce action potentials as well as postsynaptic potentials. During action potentials, recordings at the cell body reveal a brief positivity (corresponding to current flowing out from the EPSP) followed by a large extracellular negativity (a sink) associated with synchronous spike generation and current influx. In the hippocampus, such an extracellular negativity has been termed the *population spike* and has been shown to be an envelope of action potentials discharged by neurons in the region immediately surrounding the extracellular recording electrode. The hippocampal population spike, which may exceed 10 mV, is unusual, however, since field potentials usually are produced much more effectively by slower synaptic events, which summate more easily and do not require a precise synchrony of current fluxes.

Role in Behavior

We have seen that the special anatomy of the hippocampus makes it an attractive model system to study how ionic conductances, neurotransmitters and current flow can alter activity in single cells or groups of neurons. However, we also have some ideas about the role of the hippocampus as a whole in a variety of behaviors.

Sensory Processing

Neurons in the CA3 field of the cat respond equally well to auditory, visual, tactile, and olfactory stimuli.[49] These neuronal responses habituate rapidly to repeated presentations of one stimulus modality without an appreciable alteration of the response to the others. It has been concluded, therefore, that CA3 neurons respond, not to the qualitative aspects of a stimulus (as amygdalar neurons do), but rather to its "novelty." In contrast, neurons in the CA1 field often respond to only one sensory modality and to the complex features of that stimulus; therefore, the discharge of CA1 cells is not determined solely by their CA3 input.

Some hippocampal CA1 pyramidal cells in freely moving rats respond to specific spatial cues.[23] Such neurons increase their firing rates only when the animal is in a particular spatial area, as exemplified in Figure 32–15.[32] Each spot in Figure 32–15A indicates the position of a rat's head whenever the recorded CA1 neuron discharged a spike. Clearly, this unit discharged reliably when the animal was in the upper left-hand quadrant of the diamond-shaped platform. Such units are said to have "place" fields by analogy with the receptive fields of the visual system. In Figure 32–15B, head position was plotted every 4 seconds whether the unit fired or not; the resulting pattern shows that

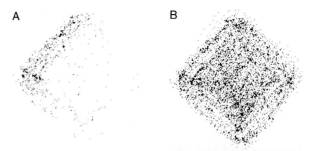

Figure 32–15 Response pattern of a CA1 "place cell" in the hippocampus. *A,* Locations of a light affixed to a rat's head whenever the CA1 cell discharges a spike. *B,* Position of the light every 4 s whether the neuron discharged or not. The illustration was obtained over a 25-minute period as the rat was gently pushed around a small diamond-shaped platform. (After O'Keefe, J. In Seifert, W., ed. *Neurobiology of the Hippocampus,* 375–403. New York, Academic Press, 1983.)

the selective firing of this place cell was not due to the animal's preference for a particular location on the platform. In different environments, individual cells can have single or multiple place fields whose configuration and sizes can be altered according to the behavioral situation.

Various sensory stimuli also induce theta activity in the immobile rat. As mentioned earlier, the theta rhythm is present during purposeful walking or running, or during any behavior that involves large axial movements such as turning or twisting[47]; therefore, it may also have a movement component.

Homeostatic Control

Unlike the amygdala, the hippocampus does not have extensive connections with brain stem nuclei or the hypothalamus and does not seem to exert major control over autonomic and homeostatic functions. The rather modest effects of stimulation or ablations of the hippocampus on heart rate, respiration, and so forth probably are routed through the subiculum or septum (see Fig. 32–8B).

The hippocampus probably is involved with the pituitary-adrenal system, because hippocampal neurons have receptors for adrenal steroids. Steroids released from the adrenals in response to stress may influence pyramidal cells, either to regulate feedback control over secretion or to mediate steroid-induced behaviors and/or other physiological responses. Both adrenal steroids and the pituitary factor controlling their secretion, ACTH, affect hippocampal cellular activity. The exact nature of hippocampal control of adrenal output may depend on the state of the animal. In stressful situations, hippocampal stimulation decreases adrenal steroid secretion, while the identical stimulation in a resting animal increases steroid levels. Thus, like the amygdala, the hippocampus appears to impose a situation-specific control on the more absolute regulation exerted by the hypothalamus.

Learning and Memory

An intriguing property of hippocampal synaptic physiology is the ease with which postsynaptic events can be potentiated with repeated stimulation. In 1973, Bliss and Lomo[6] reported that high-frequency stimulation of fibers from the entorhinal cortex induces a long-term enhancement of excitability, called *long-term potentiation*, in dentate granule cells. Since hippocampal removal in humans and animals produces profound effects on memory and on the ability to learn new tasks, it was hypothesized that the enhancement of synaptic efficacy in the hippocampus could be a substrate for learning and memory (see Chap. 33).

Although it is unclear whether long-term potentiation is a mechanism for memory, the involvement of the hippocampus in memory function is unquestioned. The most dramatic demonstration of memory impairment following hippocampal damage is the now-classic work on the patient H. M.[44] In an attempt to control intractable epilepsy, H. M. received a bilateral resection of the hippocampus, amygdala, and surrounding temporal lobe tissue. (In modern surgical treatment of epilepsy, the resection is always unilateral.) H. M. suffered a global amnesia for events that occurred within two to three years before the surgery and had permanent postoperative learning and memory deficits. In contrast, he was quite facile at recalling events, people, and places that were learned more than three years prior to surgery. On the basis of these observations, it was hypothesized that the hippocampus is important for the consolidation of memory but probably is not the site of long-term memory storage. These findings also suggest that consolidation is a continuous process, for which the hippocampus is crucial. As mentioned earlier, however, recent studies on monkeys suggest that not only the hippocampus but also the amygdala must be damaged to produce the full memory deficits associated with temporal lobe removal.[29]

A large volume of work has focused on defining the specific role of the hippocampus in learning and memory. There are many kinds of memory,

and hippocampal lesions do not disrupt all kinds equally. For example, O'Keefe and Nadel[33] proposed that the hippocampus is crucial for the formation of spatial memory, and that it creates a cognitive "map" of the environment. Hippocampal damage in rats and other animals, including humans, often produces an apparent inability to use remembered spatial cues to guide behaviors, while leaving performance on other memory tasks (e.g., delayed alternation) unimpaired. The existence of "place" neurons in the rat[32] (Fig. 32–15) was seen as further support that the hippocampus was involved in spatial processing. Whereas some place neurons respond to novel environments or locations, others respond primarily to locations that have been reinforced during learning trials, suggesting that the discharge of some reflects the learning or experience of the animal. While the behavior of place neurons is intriguing, cells of this type represent only a small fraction of the neurons in the hippocampus.

In contrast, similar experiments on rats led Olton and his colleagues[35] to suggest that the hippocampus processes memory that is contextual rather than purely spatial. Olton called this *working memory*, which he defined as memory for those aspects of a task that have meaning only in the context of that task. For example, remembering that you obtained a puff pastry from the second shelf of your refrigerator would be a working memory process if there were only one puff pastry on that shelf. However, that memory would not help you to get any other kind of food from your refrigerator. Olton contrasted this type of memory with *reference memory*, which he defined as memory for those aspects of a task that are not task specific but are common strategies for the performance of similar tasks. An example of reference memory would be remembering that you pull a handle to open the refrigerator door, a behavior that would help you to obtain a variety of foods, including the puff pastry.

Rats with hippocampal lesions can retain global strategies for problem solving, such as how to run around in a maze (reference memory), but apparently cannot retain information specific to individual trials, such as which arm of a maze they had just explored or whether that arm still contains a food pellet. Unlike normal rats, therefore, they tend to make consistent errors, such as returning to previously explored portions of mazes. Certain aspects of the deficits produced by hippocampal lesions in animals are also identifiable in humans with similar lesions. For example, humans with bitemporal lobectomy, such as the previously discussed H. M., learn and remember procedures for performing tasks, such as writing mirror images, but are unable to declare that they have ever practiced such skills before. This selective effect on *declarative* but not *procedural* memory is reminiscent of Olton's distinction between working and reference memory.

LIMBIC EPILEPTOGENESIS

"All you healthy people do not even suspect what happiness is, that happiness which we epileptics experience during the second before an attack. In his Koran, Mohammed assured us that he saw paradise and was inside. All clever fools are convinced that he is simply a liar and a fraud. Oh no!! He is not lying! He really was in paradise during an attack of epilepsy, from which he suffered as do I. I do not know whether this bliss lasts seconds, hours, or months, yet take my word, I would not exchange it for all the joys which life can give."—Fyodor Dostoyevski.

A unique property of limbic structures is the propensity of their neurons to generate epileptic discharges. While certainly not a "function" of the limbic system, this epileptogenicity is so marked that it has attracted investigators in search of the causes of seizure disorders. The phenomenon of epileptogenesis (experimental and clinical) deserves discussion because it is the predominant idiopathic abnormality of limbic system structures.

Limbic (or temporal lobe) epilepsies are often referred to as *psychomotor* or *partial-complex* seizures. Clinically, this form of epilepsy is often characterized by emotionally or experientially charged auras or feelings such as those described by Dostoyevski. If the epileptic activity in the temporal lobe is restricted to a focus, motor manifestations of the abnormality may be limited to relatively benign behavioral automatisms (such as chewing movements). If the epileptic activity spreads throughout the brain, however, the motor seizures that are commonly associated with the term *epilepsy* will be triggered. Limbic seizures are the most difficult of all epileptic disorders to treat effectively with drugs and therefore are the most frequent candidates for surgical interventions.

Hippocampal Epileptogenesis

Epilepsies in humans and animals are characterized by two pathological hallmarks: cell hyperexcitability and firing synchrony. Because of the relatively simple structure and marked epileptogenicity of the hippocampus, preparations involving the hippocampus are often used in attempts to define the neuronal properties that underlie hyperexcitability and synchrony in epileptic foci.

We have already seen that normal CA3 pyramidal cells intrinsically generate bursts of action potentials and function as "pacemakers," which drive bursting activity in CA1 neurons. In the normal hippocampus, synaptic inhibition and intrinsic hyperpolarizing potentials limit cell excitability so that bursts do not occur in response to synaptically induced excitation, and the frequency of spike (or intrinsic burst) firing is kept well below the threshold for epileptogenesis. Removal of synaptic inhibition by drugs that block IPSPs (such as penicillin or bicuculline), however, increases cell excitability and induces epileptiform activity both in vivo and in vitro. Increased excitability may also result from enhanced excitatory neurotransmission, intrinsic membrane instabilities, loss of potassium conductances, and increased cell firing, which results in increases in external potassium concentration.

Morphological Correlates of Epilepsy

In the brain of epileptic humans, most of the demonstrated abnormalities have involved cell loss or rearrangement. Dentate granule cells observed at biopsy or autopsy have truncated dendritic trees or dendrites that are deformed and denuded of spines. Such morphological changes could alter the cable properties of the neurons and lead to increased excitability. Another pattern of damage is the classic "Ammon's horn sclerosis" in which a temporal lobe that has been removed to ameliorate seizures has patches of missing pyramidal cells in either the CA1 or CA3 region. Following this cell death, considerable anatomical rearrangement may occur, including the growth of recurrent excitatory collaterals, which may synchronize cell firing. Ligands for endogenous glutamate receptors in the hippocampus are toxic and can produce a pattern of cell death that is remarkably like the Ammon's horn sclerosis seen in humans.[30] In animals with such lesions, the remaining hippocampal neurons, studied in vitro, appear hyperexcitable or "epileptic."[12] Epileptic activity may also be the result of a loss of GABAergic neurons in the epileptic foci since experimental blockade of GABAergic transmission induces seizure activity.[41]

SUMMARY

We have considered three major components of the limbic system: the amygdala, the septum, and the hippocampus. These regions differ from each other both morphologically and functionally, and in describing their actions we have tried to convey a sense of the diversity that characterizes individual limbic structures. Despite their diversity, these limbic structures are united by common functional threads. All appear to be involved in the comparison of external stimuli with internal states or maps. All show a high degree of plasticity and are involved in learning, motivation, and memory. All are *unusually* prone to pathology (e.g., epilepsy).

Although early theorists underestimated the complexity of the relations between limbic structures, their ideas about the collective functions of those structures have stood the test of time. As they originally suggested, limbic structures do lie at the interface between our perception of sensory events and our reactions to those events. They are involved in allowing us to register the significance of events and to use our experiences to adapt to our environment. Thus, the structures of the limbic system play an essential role in our attempts to impose order, meaning, and significance on the physical world.

ANNOTATED BIBLIOGRAPHY

Delgado-Escueta, A.; Ward, A.; Woodbury, D.; Porter, R., eds. *Advances in Neurology*, Vol. 44. *Basic Mechanisms of the Epilepsies*. New York, Raven Press, 1986.
A major volume on all aspects of the epilepsies.
Isaacson, R.; Pribram, K., eds. *The Hippocampus*, Vols. 3 and 4. New York, Plenum Press, 1986.
Review articles on all aspects of hippocampal function and physiology.
Jasper, H.; Van Gelder, N., eds. *Basic Mechanisms of Neuronal Hyperexcitability*. New York, Alan R. Liss, 1983.
Articles by the most noted researchers in epilepsy on the basic cellular mechanisms of pathologic neuronal firing.
Seifert, W., ed. *Neurobiology of the Hippocampus*. New York, Academic Press, 1983.
Also review articles on hippocampal physiology, with special concentration on mechanisms of learning and memory.

REFERENCES

1. Andersen, P. Organization of hippocampal neurons and their interconnections. In Isaacson, R. L., Pribram, K. H., eds. *The Hippocampus, Vol. I. Structure and Development*, 155–175. New York, Plenum Press, 1975.
2. Azuma, S.; Yamamoto, T.; Kawamura, Y. Studies on gustatory responses of amygdaloid neurons in rats. *Exp. Brain Res.* 56:12–22, 1984.
3. Ben-Ari, Y.; Le Gal La Salle, G.; Champagnat, J. C. Lateral amygdala unit activity: I. Relationship between spontaneous and evoked activity. *Electroenceph. Clin. Neurophysiol.* 37:449–461, 1974.
4. Benardo, L. S.; Masukawa, L. M.; Prince, D. A. Electrophysiology of isolated hippocampal pyramidal dendrites. *J. Neurosci.* 2:1614–1622, 1982.
5. Benardo, L. S.; Prince, D. A. Cholinergic excitation of mammalian hippocampal pyramidal cells. *Brain Res.* 249:315–331, 1982.
6. Bliss, T. V. P.; Lomo, T. Long-lasting potentiation of synaptic transmission in the dentate area of the anesthetized rabbit following stimulation of the perforant path. *J. Physiol.* 232:331–356, 1973.
7. Broca, P. Anatomie comparé des circonvolutions cérébrales le grand lobe limbique et la scissure limbique dans le série des mammifères. *Rev. Anthropol.* 1:385–498, 1878.
8. Carpenter, M. B.; Sutin, J. *Human Neuroanatomy*. Baltimore, Williams & Wilkins, 1983.
9. Cotman, C. W.; Nadler, J. V. Reactive synaptogenesis in the hippocampus. In Cotman, C. W., ed. *Neuronal Plasticity*, pp. 227–272. New York, Raven Press, 1978.
10. Cross, B. A. Unit responses in the hypothalamus. In Ganong, W., Martini, L., eds. *Frontiers in Neuroendocrinology*, 13–171. New York, Oxford Univ. Press, 1973.
11. Dingledine, R.; Langmoen, I. A. Conductance changes and inhibitory actions of hippocampal recurrent IPSPs. *Brain Res.* 185:277–287, 1980.
12. Franck, J. E.; Schwartzkroin, P. A. Do kainate-lesioned hippocampi become epileptogenic? *Brain Res.* 329:309–313, 1985.
13. Grossman, S. P.; Dacey, D.; Halaris, D. E.; Collier, T.; Routtenberg, A. Aphagia and adipsia after preferential destruction of nerve cell bodies in hypothalamus. *Science* 202:537–539, 1978.
14. Halliwell, J. V.; Adams, P. R. Voltage-clamp analysis of muscarinic excitation in hippocampal neurons. *Brain Res.* 250:71–92, 1982.
15. Hess, W. R. *Das Zwischenhirn: Syndrome, Lokalisationen, Funktionen*. Basel, Schwabe, 1949.
16. Hetherington, A. W.; Ranson, S. W. Experimental hypothalamic-hypophyseal obesity in the rat. *Proc. Soc. Exp. Biol. Med.* 41:465–466, 1939.
17. Isaacson, R. L. *The Limbic System*. New York, Plenum Press, 1982.
18. Kaada, B. R. Somatomotor, autonomic and electrocorticographic responses to electrical stimulation of rhinencephalic and other structures in primates, cat and dog. *Acta Physiol. Scand.* 24(Suppl. 83):1–285, 1951.
19. Kandel, E. R.; Schwartz, J. H. *Principles of Neural Science*. New York, Elsevier, 1985.
20. Kapp, B. S.; Frysinger, R. C.; Gallagher, M.; Haselton, J. Amygdala central nucleus lesions: effect on heart rate conditioning in the rabbit. *Physiol. Behav.* 23:1109–1117, 1979.
21. Kapp, B. S.; Gallagher, M.; Frysinger, R. C.; Applegate, C. D. The amygdala, emotion and cardiovascular conditioning. In Ben-Ari, Y., ed. *The Amygdaloid Complex*, 355–366. Amsterdam, Elsevier, 1981.
22. Klüver, H.; Bucy, P. C. Preliminary analysis of functions of the temporal lobes in monkeys. *Arch. Neurol. Psychiat.* 42:979–1000, 1939.
23. Kubie, J. L.; Ranck, J. B. Sensory-behavioral correlates in individual hippocampus neurons in three situations: space and context. In Seifert, W., ed. *Neurobiology of the Hippocampus*, 433–447. New York, Academic Press, 1983.

24. Leonard, C. M.; Rolls, E. T.; Wilson, F. A. W.; Baylis, G. C. Neurons in the amygdala of the monkey with responses selective for faces. *Behav. Brain Res.* 15:159–176, 1985.
25. MacLean, B. Some psychiatric implications of physiological studies on fronto-temporal portions of limbic system (visceral brain). *Electroenceph. Clin. Neurophysiol.* 4:407–418, 1952.
26. McKinney, M.; Coyle, J. T.; Hedreen, J. C. Topographic analysis of the innervation of the rat neocortex and hippocampus by the basal forebrain cholinergic system. *J. Comp. Neurol.* 217:103–121, 1983.
27. Michael, R. P.; Keverne, E. B.; Bonsall, R. W. Pheromones: isolation of male sex attractants from a female primate. *Science* 172:964–965, 1971.
28. Mishkin, M.; Aggleton, J. Multiple functional contributions of the amygdala in the monkey. In Ben-Ari, Y., ed. *The Amygdaloid Complex*, pp. 409–420. Amsterdam, Elsevier, 1981.
29. Murray, E. A.; Mishkin, M. Severe tactile as well as visual memory deficits follow combined removal of the amygdala and hippocampus in monkeys. *J. Neurosci.* 4:2565–2580, 1984.
30. Nadler, J. V.; Perry, B. W.; Gentry, C.; Cotman, C. W. Degeneration of hippocampal CA3 pyramidal cells induced by intraventricular kainic acid. *J. Comp. Neurol.* 192:333–359, 1980.
31. Norgren, R. Taste pathways to hypothalamus and amygdala. *J. Comp. Neurol.* 166:17–30, 1976.
32. O'Keefe, J. Spatial memory within and without the hippocampal system. In Seifert, W., ed. *Neurobiology of the Hippocampus*, pp. 375–403. New York, Academic Press, 1983.
33. O'Keefe, J.; Nadel, L. *The Hippocampus as a Cognitive Map.* Oxford, Oxford Univ. Press, 1978.
34. Olds, J.; Milner, P. Positive reinforcement produced by electrical stimulation of septal area and other regions of rat brain. *J. Comp. Physiol. Psychol.* 47:419–427, 1954.
35. Olton, D. S.; Becker, J. T.; Handlemann, G. E. Hippocampus, space, and memory. *Behav. Brain Sci.* 2:313–365, 1979.
36. Ono, T.; Fukuda, M.; Nishino, H.; Sasaki, K.; Muramoto, K.-I. Amygdaloid neuronal responses to complex visual stimuli in an operant feeding situation in the monkey. *Brain Res. Bull.* 11:515–518, 1983.
37. Ottersen, O. P. The afferent connections of the amygdala of the rat as studied with retrograde transport of HRP. In Ben-Ari, Y.Y., ed. *The Amygdaloid Complex*, 91–104. Amsterdam, Elsevier, 1981.
38. Papez, J. W. A proposed mechanism of emotion. *Arch. Neurol. Psychiat.* 38:725–743, 1937.
39. Powley, T. L. The ventromedial hypothalamic syndrome, satiety, and a cephalic phase hypothesis. *Pharmacol. Rev.* 84:89–126, 1977.
40. Price, J. L.; Amaral, D. G. An autoradiographic study of the projections of the central nucleus of the monkey amygdala. *J. Neurosci.* 1:1242–1259, 1981.
41. Ribak, C. E. Contemporary methods in neurocytology and their application to the study of epilepsy. In Delgado-Escueta, A. V.; Ward, A. A., Jr.; Woodbury, D. M.; Porter, R. J., eds. *Advances in Neurology, Vol. 44. Basic Mechanisms of the Epilepsies; Molecular and Cellular Approaches,* pp. 739–764. New York, Raven Press, 1986.
42. Ruch, T. C.; Patton, H. D. *Physiology and Biophysics I. The Brain and Neural Function.* Philadelphia, W. B. Saunders, 1979.
43. Schwartzkroin, P. A. Hippocampal slices in experimental and human epilepsy. In Delgado-Escueta, A. V.; Ward, A. A.; Woodbury, D. M.; Porter, R. J., eds. *Advances in Neurology, Vol. 44. Basic Mechanisms of the Epilepsies: Molecular and Cellular Approaches,* 991–1010. New York, Raven Press, 1986.
44. Scoville, W. B.; Milner, B. Loss of recent memory after bilateral hippocampal lesions. *J. Neurol. Neurosurg. Psychiat.* 20:11–21, 1957.
45. Swanson, L. W.; Cowan, W. M. An autoradiographic study of the organization of the efferent connections of the hippocampal formation in the rat. *J. Comp. Neurol.* 172:49–84, 1977.
46. Turner, B. H.; Mishkin, M.; Knapp, M. Organization of the amygdalo-petal projections from modality-specific cortical association areas in the monkey. *J. Comp. Neurol.* 191:515–544, 1980.
47. Vanderwolf, C. H.; Leung, L.-W. S. Hippocampal rhythmical slow activity: a brief history and the effects of entorhinal lesions and phencyclidine. In Seifert, W., ed. *Neurobiology of the Hippocampus*, 275–302. New York, Academic Press, 1983.
48. Van Hoesen, G. W. The differential distribution, diversity and sprouting of cortical projections to the amygdala in the rhesus monkey. In Ben-Ari, Y., ed. *The Amygdaloid Complex,* 77–90. Amsterdam, Elsevier, 1981.
49. Vinogradova, O. S. Functional organization of the limbic system in the process of registration of information: facts and hypotheses. In Isaacson, R. L.; Pribram, K. H., eds. *The Hippocampus,* Vol. II. New York, Plenum Press, 1975.
50. Wong, R. K. S.; Schwartzkroin, P. A. Pacemaker neurons in the mammalian brain: mechanisms and function. In Carpenter, D. O., ed. *Cellular pacemakers, Vol. I: Mechanisms of Pacemaker Generation,* 237–254. New York, John Wiley & Sons, 1982.

Physiological Bases of Learning, Memory, and Adaptation

INTRODUCTION

Not so many years ago, students were taught that the mammalian central nervous system (CNS) is fixed, or "hard-wired." According to this view, the CNS achieves its final form early in life and thereafter is a relatively inflexible system, incapable of showing more than minor adaptations to trauma or injury. The reason for this point of view is that most neurons, unlike many other cell types in the mammalian organism, do not divide after they are "born." Rather, once formed and differ-

entiated, neurons migrate to their assigned places in the CNS, establish appropriate contacts, and neither multiply nor change their functions during the rest of their lives. Any trauma leading to the injury or death of a neuron, therefore, has serious consequences, since the neuron cannot be replaced. Thus, one faces the rather frightening situation of being born with the full complement of brain cells, and then experiencing environmental insults which, over the course of one's life, gradually deplete that complement of neurons.

In contrast to this view, recent investigations have demonstrated that even the adult brain, while not able to manufacture new cells, is able to adapt to environmental pressures by changing the interactions between neurons. In retrospect, it seems obvious that the CNS must be capable of undergoing some such changes since it can learn new behaviors, adapt to changing environmental situations, acquire memories, and grow and mature with the organism. Thus, the mammalian CNS clearly is capable of *plasticity*.

Plasticity in the mammalian CNS seems rather limited, however, when compared with the varieties of plasticity demonstrated by sub-mammalian species. For example, lower vertebrates may regenerate new limbs with a complete nerve supply, recover completely from spinal cord transections, or re-establish nervous system connections with sensory receptors.[7, 24] Although mammals cannot recover from comparable injuries, they do exhibit a variety of other plasticities, particularly those involved in learning complex behaviors. While some plastic capabilities appear to be quite different in higher and lower species, the mechanisms that govern them may be similar. Since lower animals have a simpler CNS, they serve as attractive model systems that allow a closer examination of plastic changes than is possible in the more complex, and less accessible, mammalian brain. Therefore, many examples of plasticity discussed in this chapter are taken from studies of lower species.

No one would question that the phenomena of learning in mammals and regeneration in invertebrates and fish are clear examples of plasticity. In the most general sense, however, plasticity can be taken as any behavioral modification that results from an environmental change. By this definition, plasticity must include other, more subtle phenomena. For example, the visual pathway in the mammalian CNS is genetically programmed to develop with very specific connections and to respond in a stereotyped way to a given visual input. It is therefore said to be hard-wired. Yet, even neuronal responses in the visual pathway are modified according to "experience." For example, when an intense light stimulus is presented repeatedly in the same part of the visual field, the visual responses of some neurons gradually decrease with each successive stimulus. This decrement in response, called *habituation*, is a basic form of plasticity that also can be demonstrated in many other systems. Another subtle example of plasticity occurs in a pure motor response. Following repeated contractions produced either by voluntary effort or by stimulation of the motor nerve, the force generated by a muscle gradually decreases. This muscular *fatigue*, which is particularly noticeable in certain types of muscle fibers (see Chap. 23), can be attributed to the history of muscular activity. Examples such as habituation and fatigue indicate that most CNS neuronal systems exhibit some form of plasticity, and that few, if any, are strictly hard-wired.

The mechanisms underlying these and other forms of plasticity have been studied at several different levels. We will consider three of them. First, we will examine changes in cellular activity that may be correlated with behavioral changes and identify parts of the mammalian CNS required for such modifications. Second, we will explore the cellular and subcellular mechanisms thought to be involved in the plastic changes accompanying some forms of learning and memory. Third, we will examine the morphological changes that occur in some neurons not only as an organism recovers from injury, but also when normal function requires changes in neural behavior. Finally, we will briefly consider the striking changes that occur in the developing nervous system, since some of the mechanisms underlying development may also operate during plastic alterations in the adult CNS.

CNS CORRELATES OF BEHAVIORAL CHANGES

Experimental Problems

Two principal approaches have been used to identify the parts of the brain associated with plasticity. First, a particular cell population or brain structure can be destroyed or otherwise rendered nonfunctional in animals that have already learned or are learning a particular behavior. As discussed in earlier chapters, such disruptions can be produced by permanent lesions, such as excision or electrocoagulation, or by reversible lesions, such as cooling, injecting an inhibitory neurotransmitter, or treating the cells with tetrodotoxin to eliminate sodium-dependent action potentials. Unfortunately, disruption studies may provide ambiguous data, since most brain regions are imbedded within a complex meshwork of other neurons and fibers, which can also be affected by the lesion. Furthermore, as pointed out in earlier chapters, lesion studies not only assess the role of the damaged structure itself but also measure the capability of the remaining nervous system to compensate for its removal.

Second, the activity of single neurons can be recorded while the animal is learning a behavioral task. In previous chapters, we saw that neurons can be implicated in the generation of simple behaviors if their firing patterns are related to some aspect of the behavior. In simple reflexes, the sensory receptor and motoneuron show tight correlations of discharge rate with the sensory stimulus and motor response, respectively. A neuronal correlation with behavior was particularly well exemplified in the vestibulo-ocular reflex (Chap. 27). When one records from neural structures that are several synapses distant from either the sensory input or the motor output, however, the correlation becomes less strict and may be complicated by motivational and attentional factors. Since it is at such structures, which accomplish the complex integration from sensory to motor signals, that most plasticity must occur, it is difficult not only to identify the site of plasticity but also to be certain that one is measuring its neuronal correlate. One can appreciate this difficulty simply by comparing the gross brain area devoted to "pure" sensory or motor function with the much larger remaining area involved with integrating sensory and motor information. As outlined in Chapter 31, a large portion of the primate neocortex is composed of so-called "association" regions, which receive inputs from many cortical and subcortical sensory structures and provide outputs to numerous neural targets.

Even if the discharge of a neuron *is* correlated with a behavior (or its modification), the neuron might not be directly involved in that behavior. The discharge may simply be a "corollary" signal that *reflects* the behavior but is destined for another site to indicate that the behavior is occurring. For example, during the rapid eye movements used in reading, we experience no blurring of the visual perception of the printed page. It is thought that the neural signal that instructs the eye to move reaches not only the motoneurons controlling the eye muscles but also other brain regions involved with visual perception per se. This *efference copy* signal presumably inhibits the visual input that occurs during the eye movements (see the section on LGN_d in Chap. 20 for further discussion). Thus, although it may be possible to demonstrate a correlation between cellular activity and a change of behavior, it is far more difficult to establish that the cellular activity *causes* the behavioral change. Despite these caveats, neurophysiological experiments have convincingly demonstrated plasticity in a number of behaviors in a variety of species.

Plasticity in Simple Behaviors

Sherrington[39] was the first to describe a plastic change in mammals when he observed that certain limb flexion reflexes decreased in amplitude with repeated stimulus presentations. Later, Thompson and Spencer[44] showed that this habituation was caused by a decrease in the excitatory synaptic input onto motoneurons mediating the flexion. The modern era of investigation of the neural correlates of these and similar behaviors was initiated by investigators who focused on simple invertebrate systems in the hope that the principles gained in understanding plasticity in simple systems could be extrapolated to similar processes in the more complex mammalian brain.

Conditioning in Invertebrates. One of the most productive lines of research on plasticity has been carried out by Kandel and colleagues[19, 22, 25] on the nervous system of the sea slug, *Aplysia* (Fig. 33–1). In particular, they concentrated on the neural mechanisms involved in the reflex withdrawal of the gill and siphon when the siphon is touched. The neuronal circuitry involved in this reflex (Fig. 33–1A) includes a sensory neuron (SSN) that is activated by siphon stimulation, a motoneuron (MN) pool that innervates the gill and siphon, and connecting interneurons (IN). Stimulation of the siphon skin activates the sensory neurons, which

excite the motoneurons both monosynaptically and polysynaptically via the pool of interneurons.

This elementary reflex withdrawal can be altered (or *conditioned*) by manipulating the sensory stimuli to the siphon or to other parts of the animal. The simplest maneuver is to stimulate the siphon skin repetitively, which leads to a gradual decrease or *habituation* of the gill withdrawal reflex. At the cellular level, this habituation is correlated with a decrease in the amplitude of the excitatory post-synaptic potential (EPSP) produced in interneurons and motoneurons by stimulation of the sensory neurons (see inset, Fig. 33–1A). The decreased synaptic drive onto the gill motoneurons results in a reduced gill withdrawal. This alteration of synaptic efficacy and the reduced gill response may last for hours, days, or even weeks, depending on the pattern of stimulation, and is specific to the sensory cells that mediate the reflex.

Another, more complex, behavioral change, *sensitization*, can also be demonstrated in the gill withdrawal reflex. In *Aplysia*, sensitization is an *increase* in reflex output when a strong or noxious stimulus is presented to another part of the organism. For example, stimulation of the tail produces a facilitation of the gill withdrawal reflex; this facilitation, like habituation, may last for weeks. In Figure 33–1B, the EPSP in a gill motoneuron was first completely habituated (note similarity of *before* in B and *after* in A) but then restored after repeated noxious stimulation of the tail. As shown diagrammatically in Figure 33–1B, this facilitation occurs because the sensitizing stimulus to the tail activates a specific population of facilitating interneurons (FIN) that enhance transmitter release from the presynaptic terminals of the siphon sensory neurons (SSN) onto both the interneurons and motoneurons involved in the gill withdrawal reflex.

An even more complicated behavioral change, *classical conditioning*, can also be demonstrated in the gill withdrawal reflex. A conditioned reflex response is established by presenting a conditioned stimulus (CS), which by itself elicits little response, followed immediately by an unconditioned stimulus (UCS), which alone elicits a clear response. After repeated pairings of the CS and UCS, the CS delivered alone produces a clear response. The CS is then designated as CS^+. Classical conditioning occurs only if there is a predictable temporal relation between the CS and UCS. If the conditioned stimulus is repeatedly presented at random relative to the UCS (i.e., is unpaired), it will not produce a clear response

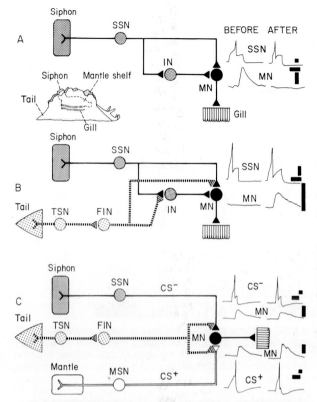

Figure 33–1 Schematic illustrations of the circuits in the sea slug *Aplysia* involved in the gill withdrawal reflex and its modification. *A*, The basic circuit, which exhibits *habituation*. *B*, The basic circuit plus a facilitating pathway from the tail exhibits *sensitization*. *C*, The basic circuit plus the additional circuits required to explain *classical conditioning*. Insets on the right show intracellular potentials in the siphon sensory neuron (SSN) and siphon motoneurons (MN) before and after the three conditioning behaviors. *C* also shows intracellular potentials in the mantle sensory neuron (MSN). The cells were tested 1 week and 1 hour after conditioning in *A* and *C*, respectively. Vertical calibration bars represent 5 mV, horizontal bars represent 50 ms. Other cells include: IN = interneuron; FIN = facilitating interneuron; TSN = tail sensory neuron. See text for a detailed discussion. (After Kandel, E.R. In Kandel, E.R.; Schwartz, J.A., eds. *Principles of Neuroscience*, 816–833. Copyright 1985 by Elsevier Science Publishing Co., Inc.)

when delivered alone; such a conditioned stimulus is designated as CS⁻. For example, if a conditioned stimulus to the mantle shelf is paired (i.e., as a CS⁺) with the UCS to the tail, subsequent stimulation of the mantle shelf alone produces an enlarged EPSP (Fig. 33–1C, lower inset). Similarly, pairing siphon stimulation with tail stimulation also enhances the EPSP produced by subsequent siphon stimulation (not shown). If, however, stimulation of the siphon is presented at random times (i.e., as a CS⁻) relative to the UCS, subsequent

stimulation of the siphon alone does not produce an enhanced EPSP (Fig. 33–1C), upper inset). This *differential conditioning* experiment demonstrates that classical conditioning occurs only when the CS predicts the presentation of the UCS. As with sensitization, this example of classical conditioning is thought to involve interneurons that facilitate the transmission of sensory information from the mantle to gill motoneurons.

Conditioning in Mammals. Comparable studies of classical conditioning in the mammalian CNS have been carried out in the alert cat. Woody and colleagues[50–52] investigated changes in properties of cortical neurons during the development of an eye-blink conditioned reflex. The UCS was a tap on the glabella (the most prominent point of the forehead on the midline between the eyes) and the CS⁺ was an auditory signal (a click). Within about 10 to 20 training sessions, each involving 150 trials of paired CS⁺ and UCS, the CS⁺ click presented alone elicited an eye blink in over 70 per cent of the click presentations. Another unpaired (CS⁻) auditory signal (a hiss) presented randomly did not elicit a blink.

The excitability of neurons in the auditory association cortex of the suprasylvian gyrus increased as conditioning developed; the threshold currents passed through extracellularly placed electrodes were less for neurons in conditioned than unconditioned animals. Furthermore, the thresholds of cells that responded only to clicks were lower than those responding to the hiss, suggesting that conditioning had preferentially made the click units more excitable.

Another readily studied reflex is the extension of the nictitating membrane (NM), a transparent third inner eyelid found in birds, cats, and rabbits but vestigial in man, in response to an air puff (UCS) delivered to the cornea.[6, 45, 46] Before conditioning, multineuron activity recorded in the deep (i.e., output) cerebellar nuclei of the rabbit exhibited only a minimal response to an auditory tone presented at random times (i.e., as a CS⁻) relative to the air puff (Fig. 33–2A, first record). When the same auditory stimulus (CS⁺) was paired with the UCS, the activity increased significantly after one day of conditioning and increased still further after the second day. As seen in Figure 33–2A (lower records), the increase was caused principally by the addition of activity in the interval between the CS⁺ and UCS, indicating that the CS⁺, which originally had no effect on neuronal firing, had established an influence on the neuron's behavior. Since changes in the neuronal discharge typically

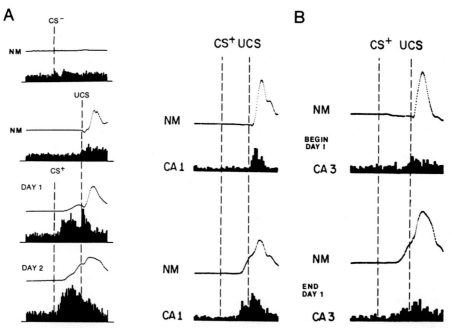

Figure 33–2 Classical conditioning of the nictitating membrane (NM) response in the rabbit and its neural correlates in the deep cerebellar nuclei *(A)* and hippocampus *(B)*. An upward deflection of NM indicates nictitating membrane closure. The conditioned stimulus (CS) was a tone and the unconditioned stimulus (UCS) was an air puff; the vertical dashed lines indicate their onset. *A*, Multineuron response histograms obtained from the medial dentate/lateral interpositus region during unpaired presentations of the tone (CS⁻) and UCS and after 1 and 2 days of paired tone (CS⁺) and UCS presentations. These histograms are averaged over the entire day of conditioning indicated. Each bar in the histogram is 9 ms in duration. *B*, Multineuron responses in the CA1 and CA3 cell layers at the beginning of conditioning are compared with the responses after 1 day of conditioning. Neuronal activity histograms and NM responses are averages of eight trials. Each bar on the histogram is 15 ms in duration. (After Berger, T.W.; Thompson, R.F. *Brain Res.* 145:323–346, 1978; and Thompson, R.F.; Clark, G.A.; Donegan, N.H.; Lavond, D.G.; Lincoln, J.S.; Madden, J. IV; Mamounas, L.A.; Mauk, M.D.; McCormick, D.A.; Thompson, J.K. In Lynch, G.; McGaugh, J.L.; Weinberger, N.M., eds. *Neurobiology of Learning and Memory.* New York, Guilford Press, 137–164, 1984.)

preceded changes in NM response by 30 to 40 ms and the conditioned behavioral response occurred only when the conditioned neuronal response was present, these data suggest an obligatory link between the change in cerebellar activity and the occurrence of the new, learned behavior. Indeed, when the cerebellar nuclei are destroyed, the rabbit *cannot* learn this simple conditioned behavior, indicating that the cerebellum *is necessary* for learning this task (see Chap. 29 for additional discussion of the cerebellum's role in plasticity).

Similar changes in neuronal activity occur in the pyramidal neurons of the hippocampus during nictitating membrane conditioning. For example, after only eight paired CS⁺-UCS presentations, pyramidal cell activity increased above spontaneous levels and continued to grow with subsequent conditioning. (See Chap. 32 for more details about the hippocampus.) As shown for the two cells in Figure 33–2B, the major growth occurs in the time interval between the CS⁺ and the UCS.

Also, the neuronal discharge precedes the behavioral (NM) response by many milliseconds. Unlike lesions of the cerebellum, destruction of the hippocampus does not impair the rabbit's ability to undergo this simple nictitating membrane conditioning. However, the hippocampus is essential in other simple conditioning paradigms that place greater demands on memory. In all the paradigms discussed thus far, the UCS has been turned on before the CS⁺ is turned off. If the UCS is turned on shortly *after* the CS⁺ is turned off, a conditioned response also develops, but this "trace conditioning" does not survive a bilateral ablation of the hippocampus.

Neural Structures Subserving Learning in Mammals

Even before these studies on the neural substrates of conditioned reflex responses, attempts had been made to localize the structures involved in learning in rats. Lashley[27] sought the neural representation for the memory of an event (the memory "engram") by testing rats on a variety of behavioral tasks after damaging parts of their neocortices. Small cortical lesions, regardless of their location, did not disrupt the rat's ability to remember a task that required choosing the correct arm of a maze based on a visual cue. Only large lesions affected the learned maze behavior. The effect of a lesion was proportional to its size, so Lashley concluded that the engram for different behaviors is *distributed* throughout the cortex.

In contrast, Olds and co-workers,[33] who recorded neuronal activity in many parts of the rat brain during many different behaviors, concluded that the engram for some learned behaviors can be *localized* to specific brain structures. The behaviors that they tested included spontaneous exploration, eating and drinking, and visual discrimination and memory tasks in a maze. As in the conditioning experiments in the rabbit, cells in the rat hippocampus often exhibited changes in firing while the animal was learning a new behavior. Neurons in other areas also exhibited similar changes.

Figure 33–3 illustrates a typical experiment; the recording site was in the thalamus. The experimental design required the rat to consume a food pellet (UCS) within a brief period or lose it. An auditory tone was presented either randomly (CS⁻) with respect to the food presentation or paired with the food presentation so that the tone (CS⁺) predicted the delivery of the food. When, as in Figure 33–3A, the tone was presented randomly, it elicited a modest increase in neural activity that gradually habituated (Fig. 33–3B). When the same tone was paired with the delivery of the food, it gradually elicited a more robust response (Fig. 33–3C) that did not habituate (not shown). Furthermore, the conditioned neural response was correlated with the animal's movement in anticipation of the food, indicating that it was learning the predictive relation between the tone CS⁺ and the reward.

To determine whether the observed changes in hippocampal activity represent the substrate of the learned response or are merely correlated with it, rats trained on a memory task were subjected to hippocampal lesions. Figure 33–4A shows a task designed to test spatial memory. First, a rat is exposed to visual cues that are related to each of the arms of a maze, including the arm that is baited (G). The cues are then removed and, after a variable delay, the rat is allowed to choose an

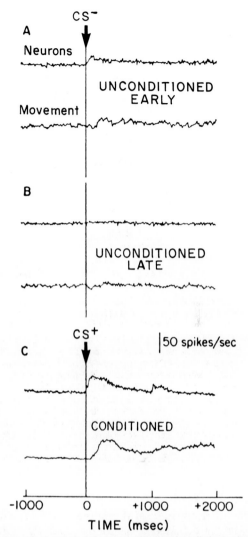

Figure 33–3 Neuronal and behavioral activity typical of conditioned responses in the rat. For three different conditions, a stimulus delivered at zero time produces changes in averaged neuronal activity *(upper traces)* and averaged movement activity *(lower traces)*. A randomly presented conditioned stimulus (CS⁻) initially evokes responses *(A)* that habituate after repeated testing *(B)*. In *C*, if the stimulus is conditioned (CS⁺), it produces larger responses. (Note that the second response at about 1 s occurs when the food pellet is delivered.) Recording electrode is in the posterior thalamus. Each record is the average of about 200 trials accumulated from the beginning of a procedure (i.e., habituation or conditioning) to its end. (After Olds, J.; Disterhoft, J.F.; Segal, M.; Kornblith, C.L.; Hirsh, R. *J. Neurophysiol.* 35:202–219. Copyright 1972 by The American Physiological Society.)

arm. Even if the cues are presented for only 30 s and unexpected detours are imposed, normal rats are able to make the correct choice after delays of up to 30 minutes. In contrast, rats with hippocampal lesions always take longer to learn the task, and some never do.

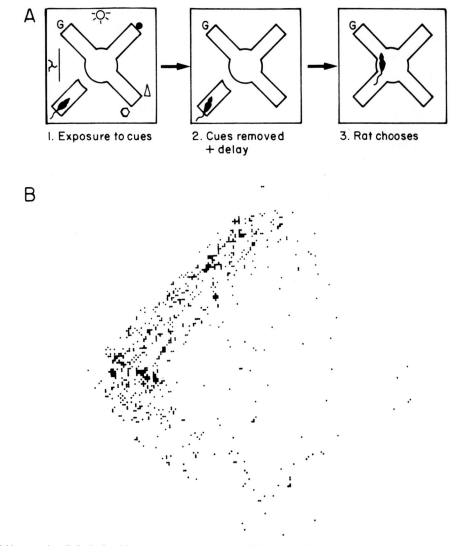

Figure 33–4 Neuronal activity in the hippocampus associated with a spatial memory task in the rat. *A,* In 1 the rat observes the cues at each arm of the maze that allow him to recognize the arm with the reward (G). The animal must retain this knowledge for a time after the cues are removed (2) and must ultimately choose the correct arm (3). *B,* The locations of a light affixed to an alert rat's head are monitored and shown as dots whenever a CA1 neuron of the hippocampus discharges a spike. The illustration was obtained over 25 minutes with position sampled at 4 times a second as the rat was gently pushed around a small platform. (After O'Keefe, J. In Seifert, W., ed. *Neurobiology of the Hippocampus.* New York, Academic Press, 375–403, 1983.)

The unique behavior of some single neurons in the hippocampus further implicates it in spatial memory processes.[32] For example, when a rat is gently pushed around a small platform, some CA1 neurons discharge only when the animal is at, or passes through, a certain area. In Figure 33–4B, we have repeated Figure 32–15, which illustrates such a "place neuron"; this cell discharges only when the rat is in the L-shaped region in the upper left quadrant of the platform.[32] Neurons

with such "place fields" could provide the rat with information about its location in space to help it solve a spatial task.

To test whether the hippocampus is specifically involved in spatial memory, Olton and colleagues[34] baited every arm of a multi-arm maze, and the rat was allowed a short time period, and sometimes a limited number of choices, in which to obtain all of the food. To optimize its search, therefore, the rat had to remember those arms from which it had already received food. Even after Olton removed all the cues outside the maze and rearranged the arms between choices, normal rats reliably found the remaining baited arms. After hippocampal lesions, rats could not perform this memory task, which, unlike the one described above, seems to require them to remember which arms have been explored previously rather than where they were. Perhaps the hippocampus does not contain a pure spatial map but rather a map that associates places with times (i.e., an event map). Whatever its exact role, the hippocampus *is* clearly necessary for certain memory tasks in the rat.

These experiments on rats have implicated both the hippocampus and the posterior thalamus (see Fig. 33–3) in memory processing. The same structures also participate in memory processing in humans, but each seems involved in a different kind of memory.[42] Patients subjected to bilateral removals of the hippocampus, amygdala, and temporal cortices for treatment of intractable epilepsy (such as the well-studied patient HM) can learn normally (e.g., memorize 120 colored slides) but forget what they have learned more quickly than do normal subjects.[38] On the other hand, patients with thalamic lesions (such as NA, who was pithed by a fencing foil, which destroyed his dorsomedial thalamic nucleus) learn slowly but retain what they have learned for normal periods of time. It has been suggested, therefore, that patients with extensive lesions of the limbic system are unable to convert short-term memories into long-term memories, whereas patients with lesions of the posterior thalamus have difficulty in acquiring short-term memories but have no difficulty in consolidating what is acquired into long-term memory. In addition to these *anterograde amnesias*, lesions of both the limbic system and the thalamus also produce a *retrograde amnesia* for events that occurred a few years before the lesion.

MEMBRANE AND SYNAPTIC CHANGES UNDERLYING PLASTICITY

Possible Synaptic Mechanisms

One of the earliest, and still influential, hypotheses of the neuronal basis of learning was proposed by Hebb in 1949.[23] He suggested that repetitive use of a synapse enhances the efficacy of that synapse *if* the discharge of the postsynaptic cell is correlated with a particular input signal. This means that conjunction of postsynaptic firing and the presynaptic input would be the requirement for strengthening specific connections. The "Hebb synapse" is a hypothetical construct that has only recently found concrete support in the studies of the mechanisms underlying long-term potentiation (see below).

At about the same time, Lloyd[29] suggested a different model for neuronal plasticity. He discovered that high-frequency (100–500 Hz) tetanic stimulation of Ia afferents (see Chap. 24) for several seconds was followed by an enduring increase in the size of ventral root responses to subsequent Ia volleys. The duration of this potentiation depends on the duration of the tetanus.[41] The effect of a 7-minute tetanus lasts for almost 15 minutes, whereas a tetanus of 20 minutes causes enhancement lasting up to 2 hours (Fig. 33–5). Post-tetanic potentiation (PTP) that lasts for hours might be a possible mechanism for short-term memory, but its dependence on prolonged bursts of synchronous high-frequency stimulation of afferents has no clear counterpart in learning paradigms. For this reason, PTP is not a likely mechanism underlying learning and memory. PTP seems to be an entirely presynaptic phenomenon. One hypothesis is that it is the result of an activity-dependent calcium accumulation in the presynaptic terminal, which leads to an increased transmitter release.

Possible Membrane Mechanisms

The ability of a cell to respond to a stimulus and the nature of that response are determined by the various voltage- and transmitter-dependent conductances in its membrane (see Chap. 12). Thus, the two possible synaptic mechanisms just described must somehow result from changes in post- and presynaptic membrane conductances, respectively. There are many possibilities. For example, conductance changes in the presynaptic terminal could alter the duration and/or amplitude of the action potential invading the terminal, thereby affecting the coupling between the action potential and the release of neurotransmitter. At the synapse itself, additional or new receptors could be formed or receptors could be redistributed (as in the denervation supersensitivity discussed below). Finally, as in many invertebrate systems, postsynaptic potassium conductances could decrease.

Another set of possible mechanisms is revealed by recent studies of the proteins that make up the membrane channels and the pathways by which

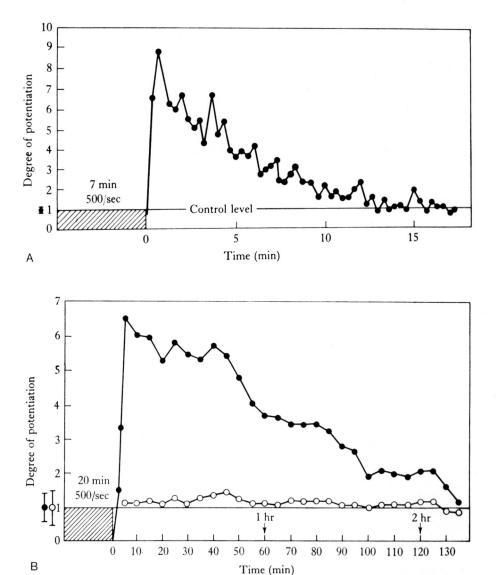

Figure 33–5 Prolonged post-tetanic potentiation (PTP) following 7 *(A)* and 20 *(B)* minutes of tetanus. The stimulation was delivered to the tibial nerve, and whole nerve potentials were recorded from the L7 ventral root. Degree of potentiation represents the amplitude (as a multiple of the control response amplitude) of the monosynaptic reflex ventral root potential to constant test stimuli. Open circles in lower panel are control data taken from the ventral root of the contralateral nontetanized side. (After Spencer, W.A.; April, R.S. In Horn, G.; Hinde, R.A., eds. *Short-Term Changes in Neural Activity and Behavior*, 433–474. Copyrighted and printed by the Cambridge University Press, 1970.)

the proteins are synthesized. Enduring changes in the protein states of membranes can be effected by altering the activities of protein kinases, some of which have been shown to be regulated by calcium, calmodulin, and/or cyclic nucleotide second-messenger systems.[19] The synthesis of new proteins, initiated by the transcription of previously "dormant" DNA sequences, may also change the nature of the neuronal membrane, thus altering both pre- and postsynaptic function. These alterations in the molecular structure of a cell may occur not only in response to modulator substances as mentioned above but also in response to changes in intra- and/or extracellular ion concentrations or in response to cell activity.

Plasticity not only can be effected at the level of *individual* neurons but also may be a consequence of changes in interactions among *groups* of neurons. For example, within a neuronal circuit there may be recruitment of cells into a discharge pattern, increased or decreased involvement of interneurons in response to a particular stimulus, or even a change in the participation of glial elements.

Mechanisms of Conditioning in *Aplysia*

The conditioning of gill withdrawal in *Aplysia* provides a model system for the study of mechanisms underlying neuronal plasticity. The simplest alteration, habituation, is the result of decreased transmitter release upon repeated activation of the presynaptic cell. The stimulation rate required for habituation is far lower than that required for PTP and is probably within the range of normal physiological function. Repetitive stimulation of the siphon afferents (the presynaptic neuron in Fig. 33–6A) produces a cumulative inactivation of presynaptic Ca^{2+} channels as well as transmitter depletion, leading to a decrease in transmitter released onto the gill motoneuron (postsynaptic cell in Fig. 33–6A). Habituation is independent of postsynaptic membrane changes and requires no interaction with other neurons.

In the more complex sensitization phenomenon, stimulation of the tail activates interneurons that release serotonin onto the presynaptic terminals of the siphon afferents. The serotonin combines with a receptor on the terminal membrane and activates adenylate cyclase with a consequent increase of the second messenger cyclic adenosine monophosphate (cAMP) (Fig. 33–6B).[22, 25] The cAMP in turn activates a protein kinase, which phosphorylates a membrane protein of the K^+ channel; the result is closure of the K^+ channel.

The closure of K^+ channels increases the duration of presynaptic action potentials, a change that allows more Ca^{2+} to enter and more transmitter to be released for each spike produced in the siphon afferent.

An even more complex series of cellular events occurs in the *classical (differential) conditioning* of gill withdrawal (Fig. 33–6C). In this behavior, the facilitatory action of serotonin released from interneurons by tail stimulation (as just described) is enhanced if the mantle sensory afferent is active just before serotonin is released from the tail interneurons. It is postulated that activity in the mantle afferent allows some calcium to enter the terminal; Ca^{2+} in turn, acts through calmodulin (not shown) to increase adenylate cyclase activity so that the serotonergic input due to tail stimulation produces more cAMP (Fig. 33–6C). In other words, Ca^{2+} entry during action potentials in the mantle sensory neuron "primes" adenylate cyclase to make the cellular events associated with tail stimulation more effective. The key to the differential conditioning, then, is the coordinated timing (or pairing) of stimuli that activate the mantle sensory neuron and the facilitating serotonergic interneuron from the tail. As predicted by Hebb,[23] coordinated activity in both the pre- and postsynaptic elements appears to strengthen (or facilitate) the synaptic effect.

Membrane and cellular changes involving Ca^{2+} and calmodulin-regulated adenylate cyclase activity have also been demonstrated for conditioned reflexes in other molluscs. For example, Alkon[3] has studied the mechanisms underlying the response of *Hermissenda*, another aquatic slug, to combined visual and gravitational stimuli. These molluscs move reflexively toward a spot of light in their environment (a phototaxic response). On the other hand, when they are subjected to a repeated vigorous rotation simulating oceanic turbulence, they become more and more immobile and attach to the walls of their container as if to avoid being dislodged. If a light stimulus (CS^+) is paired with rotation (UCS), the phototaxic response gradually decreases; eventually the animals tend to stay put even when the light stimulus is presented. This conditioned response results from an interaction between visual and gravitational pathways. The phototaxic response is subserved by a three-neuron circuit composed of an excitatory photoreceptor, an excitatory interneuron, and a motoneuron. Gravitational stimuli excite a mechanoreceptor that excites an inhibitory photoreceptor, which synapses on the excitatory photoreceptor. Rotation thus enhances the light-

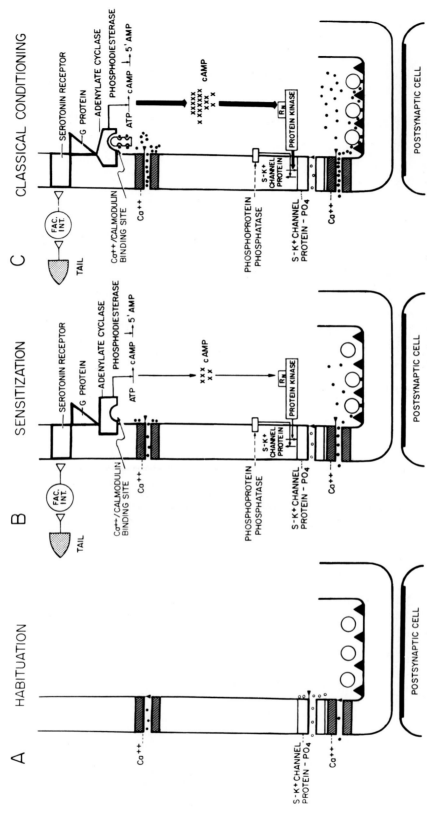

Figure 33–6 Cellular mechanisms of habituation, sensitization, and classical conditioning of the gill and siphon withdrawal reflex in *Aplysia*. The details of these membrane and molecular mechanisms are described in the text. (After Hawkins, R.D.; Kandel, E.R. In Lynch, G.; McGaugh, J.L.; Weinberger, N.M., eds. *Neurobiology of Learning and Memory*. New York, The Guilford Press, 385–404, 1984.)

induced depolarization in the inhibitory photoreceptor and thereby increases its inhibition of the light response in the excitatory photoreceptor and decreases the phototaxic response. The changes in membrane properties and current (in particular the reduction of K^+ currents) that conditioning produces in the inhibitory photoreceptor can be simulated by the intracellular injection of a Ca^{2+}-calmodulin–dependent protein kinase.[1] Therefore, Ca^{2+}-dependent protein phosphorylation could mediate the long-term modulation of specific K^+ channels as a step in the generation of this conditioned behavioral change.

Similar mechanisms may operate during conditioning in vertebrates. For example, in parts of the motor cortex known to be involved in eye-blink conditioning in cats, intracellular injection of a Ca^{2+}-dependent protein kinase followed by depolarization produces changes of membrane resistance in some cells similar to those seen in the *Hermissenda* inhibitory photoreceptor.[50] Also, CA1 neurons of a hippocampal slice taken from rabbits that have undergone classical conditioning of the nictitating membrane response discharge action potentials with a decreased afterhyperpolarization, a finding that suggests that conditioned cells have undergone a decrease in K^+ conductances.[15]

Mechanisms Underlying Long-Term Potentiation (LTP) in Mammalian Hippocampus

The most intensively studied example of plasticity in the mammalian CNS is long-term potentiation (LTP), a phenomenon first described in the dentate region of the hippocampus in anesthetized rabbits[10] and later in awake, unrestrained rabbits.[11] Like PTP, LTP is initiated by repetitive activity at a synapse. In the hippocampus, repetitive stimulation of an afferent pathway to the dentate granule cells or to CA3 or CA1 pyramidal cells (see Chap. 32 for details on hippocampal circuitry) leads gradually to an increased probability of postsynaptic discharge to stimulation of the conditioned pathway. Figure 33–7 shows potentiation of the response of a CA1 pyramidal cell.[4] In Figure 33–7A, test stimulation of the stratum radiatum fibers ending on the apical dendrites of CA1 cells elicits only EPSPs. After only five bursts of stimuli, each consisting of 16 shocks at 100 Hz, the same test stimulus evoked an action potential (Fig. 33–7B). This increase in synaptic efficacy occurs following repetitive stimulation within a "physiological" range (as low as 3 to 5 Hz, but up to 400

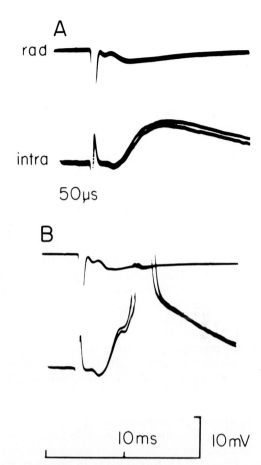

Figure 33–7 Long-term potentiation of transmission across the apical dendritic synapses of a CA1 cell in the hippocampus. *A,* Extracellular (rad) responses near the activated synapses and intracellular (intra) responses to stimulation of radiation fibers (100 μA for 50 μs). *B,* Similar responses to the same stimulus recorded about 10 min after five periods of tetani, each consisting of 16 shocks at 100 Hz. (From Anderson, P.; Avoli, M.; Hralby, O. *Exp. Brain Res.* (Suppl.) 9:315–324. Copyright, 1984 by Springer-Verlag.)

Hz), requires only a few stimuli, and lasts for long periods of time (hours, days, and even weeks in intact animals).

Synaptic Effects Underlying LTP. Both pre- and postsynaptic mechanisms may be involved in LTP. Several different experimental approaches indicate that LTP in rats is associated with an increase in the release of presynaptic excitatory neurotransmitters, such as glutamate. In one experiment, one barrel of a multicannula assembly is used to infuse radioactive (^{3}H) glutamine into the hippocampus, and a second barrel is used to collect the ^{3}H-glutamate that is subsequently synthesized and released. Following the induction of LTP, the release of glutamate is markedly increased,[16] a change that is similar to that seen during PTP but is much more enduring (days and weeks vs. hours).

Postsynaptic changes are also noted during other forms of hippocampal long-term potentiation. For example, passage of a transmembrane depolarizing current in a CA1 cell, when timed to occur in conjunction with a weak synaptic input, causes a gradual increase in the size of the EPSP produced by that synaptic input.[49] This potentiation is reminiscent of that produced at the Hebb synapse, where appropriately timed postsynaptic potentials are essential for the facilitation process.

The major critical events in LTP occur in the synapses activated by the repetitive stimulation. However, in a hippocampal slice preparation, LTP has also been observed as the result of interactions between two excitatory monosynaptic inputs. To demonstrate this, Barrionuevo and Brown[5] adjusted the strengths of the stimuli to two afferent pathways so that one generated a weak and the other a strong synaptic response in CA1 neurons. Although repetitive activation of either pathway alone had no long-term effect on the strength of the weaker input, stimulation of both together produced an enhancement of the synaptic efficacy of the weak input that persevered for hours.

Intracellular Mechanisms of LTP. LTP appears to be associated with an increase in free intracellular calcium, for intracellularly injected chelating agents block LTP.[30] Antagonists to the excitatory transmitter, N-methyl-D-aspartate (NMDA), also block LTP, probably by blocking Ca^{2+} entry[13] (see Chap. 12). The precise role of calcium in LTP is still unknown, but it may activate second-messenger systems, as it does in invertebrate conditioning. For example, high-frequency stimulation of the perforant pathway, which in the intact animal produces LTP, causes a 2.5-fold increase in cAMP levels in the dentate gyrus of rat hippocampal slices.[43]

Although there are clear and appealing similarities between the mechanisms that probably underlie LTP in mammals and those that underlie conditioning in invertebrates, LTP unfortunately cannot be correlated with alterations in behavior. In *Aplysia*, cellular plasticities and behavioral conditioning are tightly linked. In the mammalian brain, on the other hand, the synaptic alterations associated with hippocampal LTP have not been clearly associated with short-term memory or with any learned behavior.

MORPHOLOGICAL CORRELATES OF PLASTICITY

Changes in Normal Animals

Morphological changes of synapses have been reported in both invertebrates and vertebrates in different learning paradigms. In *Aplysia*, conditioned neurons display an increased number of dendritic swellings. Also following LTP produced either in rat hippocampal slices or intact animals, the numbers of synapses (onto inhibitory interneurons or pyramidal neurons) increase. Furthermore, the dendritic spines apparently change from a cupped to a rounded shape,[12] a configuration that may uncover additional receptor sites. Other studies of LTP indicate that the participating hippocampal neurons (those with dendrites in the termination zone of the repetitively stimulated perforant pathway) have wider spine stalks and heads,[18] alterations that would make them electrotonically shorter and, therefore, more effective at transmitting synaptic signals.

Morphological changes occur not only in association with conditioning and LTP but also in normal animals subjected to unusual surroundings. For example, rats exposed to socially complex and challenging environments (i.e., they are raised with litter mates and objects to explore in their cages) have thicker neocortices than animals reared in isolation; the increased thickness is due to an unusually elaborate dendritic branching pattern.[21]

Some of the morphological changes that accompany learning behaviors may be controlled by trophic circulating substances. In songbirds, for example, changes in circulating hormone levels alter dendritic (and synaptic) complexity, and perhaps even the number of cells in the nucleus hyperstriatum ventrale, the brain region that controls song "learning."[14] This finding in an intact, complex nervous system is consistent with the results of many studies in developing and cultured systems that show that nerve cells are responsive to trophic factors. For example, nerve growth factor (NGF), a specific and well-characterized trophic substance, is critical for the survival and normal development of sympathetic and sensory mammalian neurons in vitro.[28] For these neurons, NGF appears to affect cell viability and encourage the growth of neurites. Other critical cell features, such as the cell's neurotransmitter phenotype, also may be induced by trophic factors. For example, factors released by targets of sympathetic neurons can determine whether a particular neuron uses acetylcholine or norepinephrine as its neurotransmitter.[9] Cells grown in culture and exposed to a medium conditioned by the target organ can be switched between the two transmitter systems (see Chap. 34). Finally, the ability of injured neural systems to repair themselves may be facilitated by exposure to certain types of growth factors.[31]

Responses to Injury

Although this chapter focuses primarily on changes in cell properties that are correlated with behavioral changes such as learning, an examination of some plasticities associated with responses to injury is instructive. The means by which cells respond to trauma illustrate adaptive capabilities of the nervous system and perhaps represent exaggerations of the plastic processes involved with normal changes in behavior.

In contrast to the relatively subtle morphological changes associated with plasticity in intact animals, quite spectacular alterations in cell morphology result from lesions of the nervous system, especially in young animals. A dramatic example is the increase in the number of receptors on a target cell after removal of presynaptic input (deafferentation). The result is that the denervated cell becomes more sensitive to the remaining available neurotransmitter. This *denervation supersensitivity* has been clearly demonstrated at the neuromuscular junction, where increases in acetylcholine receptors are found across the entire surface of the muscle membrane after the axon of the innervating motoneuron has been cut.

More subtle changes in neuron morphology occur following other neural insults. In some systems, deafferentation leads to postsynaptic loss of dendritic spines and an aberrant growth of the dendritic tree, as if the cell were attempting to "find" new inputs.[36] In epileptic foci,[48] fine dendrites and spines are lost. In neurologically mutant mice, which have abnormal afferent inputs, the dendritic trees of cerebellar Purkinje cells are distorted.[40]

The most studied responses to trauma are *regeneration* and *sprouting*. In general, regeneration is the re-growth and establishment of contacts by a neuron following the destruction of its distal processes (usually the axon); other afferent inputs to the neuron are usually unaffected. Sprouting occurs when one afferent system is removed and another grows additional processes (or sprouts) that fill some of the synaptic sites vacated by the missing input.

Regeneration. After trauma, it would be useful functionally if the injured neurons could re-grow and regain their original synaptic contacts. For example, when the spinal cord is severed, many long descending and ascending tracks are interrupted, leaving the organism paralyzed (and anesthetic) in its lower extremities. Although many of the severed descending fibers are capable of growth, most fail to cross the gap in the cord (Fig. 33–8A). Some do cross the cut and proliferate on the distal side of the lesion, but they do not reach their appropriate termination sites.

Numerous procedures to help neuronal processes bridge the lesion gap have been developed. As shown schematically in Figure 33–8B, spinal cord fibers can use bridges made of peripheral nerves[2] or of synthetic materials. Since glia form a barrier at the gap, efforts have been made to prevent glial scarring. Although all these procedures have been successful in encouraging processes to grow, the regenerating axons never succeed in reaching their appropriate targets.

Sprouting. Not only does regeneration occur in the transected spinal cord, but sprouting occurs as well. When the spinal cord is transected, local afferent inputs below the level of the transection sprout to fill the vacant synaptic sites left by the degenerating long descending and ascending tracts (Fig. 33–8C). The sprouting fibers increase the size of their terminal arborization and make much more extensive synaptic contact on cells in the spinal cord.[20] The spasticity (increased muscle tone) often seen in animals with spinal cord transections may be due to this sprouting of segmental afferent systems.

The circumstances that encourage sprouting are difficult to specify. It appears, however, that the number of synaptic contacts made by an afferent system in one location can be altered by eliminating its other target sites. In the visual system, for example, the retina projects to several structures, including the lateral geniculate nucleus, the superior colliculus, and other midbrain nuclei (see Chap. 20). In the hamster, removing the lateral geniculate nucleus causes more terminal arbors to be formed in the superior colliculus and other

A   Aborted spinal cord regeneration

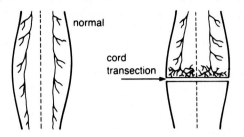

B   Peripheral nerve bridge

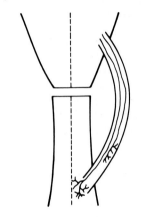

C   Sprouting in spinal cord

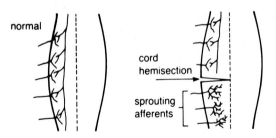

Figure 33–8 Schematic illustrations of the effects of spinal cord transections. *A,* Abortive elaboration of terminals. *B,* Regeneration facilitated by a bridge made of peripheral nerves. *C,* Sprouting in the denervated but not in the intact spinal cord.

visual relay stations.[37] Thus, synaptic sprouting can be directed in much the same way that the pruning of a rose bush encourages flowering in a remaining branch.

Sprouting is partly controlled by the postsynaptic cell. When a cell loses some of its synaptic inputs (as, for example, by a lesion of a particular afferent system), it may accept new sprouting synapses, but not necessarily from every source.

The rules that govern the formation of synapses during the initial development of synaptic connections in the immature brain continue to apply under conditions of sprouting. In particular, trophic factors released by the target cell apparently invite sprouting and regulate both the number and the specificity of synaptic inputs.

Sprouting and regeneration occur not only in the spinal cord, but in the brain as well. Tsukahara[47] has demonstrated sprouting in the cerebellar and cortical afferent systems that converge on the cells of the red nucleus of the mesencephalon. Inputs from the cerebellum normally terminate close to the cell body of red nucleus neurons and, when stimulated, produce an EPSP that rises rapidly (single-arrow recording, Fig. 33–9A), as predicted by the electrotonic equivalent cylinder model (see Chap. 12). In contrast, efferents from the cortex to red nucleus neurons terminate on more distal dendritic branches and, when stimulated, give rise to a slower EPSP (double-arrow recording, Fig. 33–9A). Immediately after the cerebellar inputs are transected, the red nucleus cell loses its fast-rising EPSPs. Several days after the lesion, however, stimulation of the afferents from the cortex produces a complicated postsynaptic potential that has not only a late, slow component but also an initial, fast-rising component (single-arrow recording, Fig. 33–9B) that resembles that produced previously by cerebellar stimulation. These data suggest that, following the cerebellar lesion, the cortical fibers sprout to preempt synaptic sites on the proximal dendrites and somata previously occupied by the cerebellar afferents (Fig. 33–9B). Analysis of the electrotonic properties of the altered red nucleus cells, and histological verification of the termination of cortical afferents on the somata of red nucleus cells confirmed this suggestion.

Sprouting may also occur during the changes that accompany *normal* learning in a behaving animal. During acquisition of a learned flexion of the hindlimb, the EPSP produced in red nucleus cells changes in a way that suggests a movement of the cortical input toward the soma. The location and number of synaptic contacts made by a particular afferent system may change, therefore, depending on the use of that system and its involvement in learning the behavior. Such changes may occur relatively rapidly and may provide an effective mechanism for alterations in synaptic efficacy.

Transplantation of Neural Tissue. A special kind of regeneration is currently capturing the imagination of many neurobiologists. Neural tissue from certain areas of an immature (i.e., fetal)

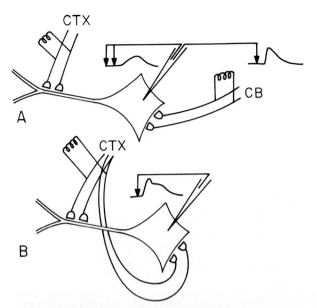

Figure 33–9 Sprouting in the red nucleus of the cat. *A,* Stimulation of the cortical input (CTX) to a red nucleus cell produces a slowly rising excitatory postsynaptic potential (EPSP), double arrows, whereas cerebellar stimulation (CB) produces a rapidly rising EPSP (single arrow). *B,* Removal of the cerebellar input eventually leads to a cortically evoked EPSP with both slow and fast components, suggesting a cortical occupation of the vacant cerebellar sites. (After Tsukahara, N. In Cotman, C.W., ed. *Synaptic Plasticity.* New York, Guilford Press, 201–229, 1984.)

mammalian CNS is transplanted into the brain of a lesioned host (either adult or immature) of the same species.[8] Not only do the transplanted cells survive and grow processes that establish contacts (often appropriate) with postsynaptic elements, but the damaged host animal also exhibits some behavioral recovery. In theory, then, such transplantation could be used to repair injured systems.

The effects of transplantation on the recovery of learning and memory are illustrated by the work of Björklund and Stenevi.[8] They transplanted cholinergic neurons (from the septal region of fetal rats) to an adult rat's brain in which the normal cholinergic septal (fimbrial) input to the hippocampus had been disrupted by the destruction of the medial septal nucleus. Prior to the transplant, adult rats with such septal lesions have difficulty performing forced alternation tasks in a T maze where an animal receives food only for selecting the arm that did not previously contain a reward. Cholinergic neurons from a fetal donor transplanted into the hippocampus of such impaired adults survive, grow processes, and establish synaptic contacts, as demonstrated by acetylcholinesterase histochemistry.[26] Furthermore, these ap-

parent synaptic contacts are indeed functional, since electrical stimulation of the transplanted cells produces long-lasting depolarizations of CA1 cells in the adjacent hippocampus. The effects of the stimulation are mediated by acetylcholine, because they are blocked by the muscarinic antagonist atropine and potentiated by physostigmine.[8] Finally, such transplantation leads to a partial recovery of the animal's maze performance; however, animals with implants required eight weeks to learn an alternation task that normal animals learn in two weeks, and their performance never returned to normal.[17] No recovery was seen after transplants of grafts from the locus ceruleus, a region that is rich in noradrenergic neurons.

The cholinergic grafts evidently produce recovery not by establishing normal septal connections but rather by acting as a cellular source of acetylcholine, whose presence seems critical for establishing many behaviors attributed to the hippocampus. The mere presence of neurotransmitters provided by such transplanted cellular factories also appears to ameliorate the behavioral deficits caused by other brain lesions. For example, aging rats placed in a water maze have difficulty using and remembering spatial cues to find an escape platform hidden below the surface of a pool of opaque water.[8] Similar deficits are produced by septal (i.e., cholinergic) lesions in young rats. The transplantation of septal cholinergic neurons into the aging animals alleviates their deficiencies in performing the water maze task[8] and in other tasks requiring memory. Similarly, grafting dopaminergic neurons into the striatum of the basal ganglia in experimental animals with parkinsonian symptoms (see Chap. 30 for a discussion of the role of the basal ganglia) produces a significant relief of those deficits.

DEVELOPMENT

The successes of neural transplantation provide dramatic evidence that the CNS is capable of considerable plasticity. Successful transplantation, however, requires that the donor cells come from an immature (usually fetal) source, although the host or recipient of the transplanted tissue may be of virtually any age. This special efficacy of immature tissue suggests that a consideration of the mechanisms underlying normal development may reveal some of the factors that operate during plastic changes in the adult. Although it is beyond the scope of this chapter to discuss the many changes occurring in the developing nervous system,[35] let us consider some of them briefly.

It seems clear that young individuals have a better capacity for acquiring behaviors than do older persons. The electrophysiological and anatomical phenomena that are probably related to learning also are more marked in the young animal. For example, LTP is greater in younger animals. Also, regeneration and sprouting are particularly robust in the immature nervous system. In addition, tissue from an immature CNS serves as a very effective bridge to guide regenerating fibers across lesions. Finally, an immature host is better able to support grafted tissue and to establish new and effective circuits than is an aged host.

The immature nervous system is plastic, at least in part, because its nerve cells have the potential for plasticity encoded in their genes and these genes are expressed (or "read out") during the early stages of development. In contrast, the mature nervous system lacks this capability because the expression of the genome has been fundamentally altered. The genetic information is present but is not utilized. With the tools of molecular neurobiology, it may one day be possible to intervene in the process of gene regulation in mature cells in such a way that they once again can express their potential for plasticity. Specific, more detailed examples of development in the auditory and visual systems can be found at the ends of Chapters 18 and 20.

ANNOTATED BIBLIOGRAPHY

Cotman, C.W., ed. *Synaptic Plasticity*. New York, Guilford Press, 1985.
A very readable volume that covers most of the examples of plasticity discussed in this chapter in more detail and with extensive references.
Lynch, G.; McGaugh, J.L.; Weinberger, N.M., eds. *Neurobiology of Learning and Memory*. New York, Guilford Press, 1984.
A comprehensive review of the phenomenology and neurophysiology of learning and memory in a variety of systems from Aplysia *to humans.*

REFERENCES

1. Acosta-Urquidi, J.; Alkon, D.L.; Neary, J.T. Ca^{++}-dependent protein kinase injection in a photoreceptor mimics biophysical effects of associative learning. *Science* 224:1254–1257, 1984.
2. Aguayo, A.J. Axonal regeneration from injured neurons in the adult mammalian central nervous system. In Cotman, C.W., ed. *Synaptic Plasticity*. New York, Guilford Press, 457–484, 1985.
3. Alkon, D.L. Associative training of *Hermissenda*. *J. Gen. Physiol.* 64:70–84, 1974.
4. Anderson, P.; Avoli, M.; Hralby, Ø. Evidence for both pre- and postsynaptic mechanisms during long-term potentiation in hippocampal slices. *Exp. Brain Res.* (Suppl.)9:315–324, 1984.
5. Barrionuevo, G.; Brown, T. Associative long-term potentiation in hippocampal slices. *Proc. Nat. Acad. Sci. USA* 80:7347–7351, 1983.
6. Berger, T.W.; Thompson, R.F. Neuronal plasticity in the limbic system during classical conditioning of the rabbit nictitating membrane response. I. The hippocampus. *Brain Res.* 145:323–346, 1978.
7. Bernstein, J.J.; Wells, M.R.; Bernstein, M.E. Spinal cord regeneration: synaptic renewal and neurochemistry. In Cotman, C.W., ed. *Neuronal Plasticity*. New York, Raven Press, 49–71, 1978.
8. Björklund, A.; Stenevi, U., eds. *Neural Grafting in the Mammalian CNS*. Amsterdam, Elsevier, 1985.
9. Black, I.B.; Adler, J.E.; Dreyfus, C.F.; Jonakait, G.M.; Katz, D.M.; LaGamma, E.F.; Markey, R.M. Neurotransmitter plasticity at the molecular level. *Science* 225:1266–1276, 1984.
10. Bliss, T.V.P.; Gardner-Medwin, A.R. Long-lasting potentiation of synaptic transmission in the dentate area of the unanesthetized rabbit following stimulation of the perforant path. *J. Physiol. (Lond.)* 232:357–374, 1973.
11. Bliss, T.V.P.; Lømo, T. Long-lasting potentiation of synaptic transmission in the dentate area of the anesthetized rabbit following stimulation of the perforant path. *J. Physiol. (Lond.)* 232:331–356, 1973.
12. Chang, F.-L.F.; Greenough, W.T. Transient and enduring morphological correlates of synaptic activity and efficacy change in the rat hippocampal slice. *Brain Res.* 309:35–46, 1984.
13. Collingridge, G.L.; Kehl, S.J.; McLennan, H. Excitatory amino acids in synaptic transmission in the Schaffer collateral-commissural pathway of the rat hippocampus. *J. Physiol. (Lond.)* 334:33–46, 1983.
14. DeVoogd, T.; Nottebohm, F. Gonadal hormones induce dendritic growth in the adult avian brain. *Science* 214:202–204, 1981.
15. Disterhoft, J.F.; Coulter, D.A.; Alkon, D.L. Conditioning-specific membrane changes of rabbit hippocampal neurons measured in vitro. *Proc. Nat. Acad. Sci. USA* 83:2733–2737, 1986.
16. Dolphin, A.C.; Errington, M.L.; Bliss, T.V.P. Long-term potentiation of the perforant path in vivo is associated with increased glutamate release. *Nature* 297:496–498, 1982.
17. Dunnett, S.B.; Low, W.C.; Iversen, S.D.; Stenevi, U.; Björklund, A. Septal transplants restore maze learning in rats with fornix-fimbria lesions. *Brain Res.* 251:335–348, 1982.
18. Fifkova, E.; Anderson, C.L. Stimulation-induced changes in dimensions of stalks of dendritic spines in the dentate molecular layer. *Exp. Neurol.* 74:621–627, 1981.
19. Goelet, P.; Castellucci, V.F.; Schacher, S.; Kandel, E.R. The long and the short of long-term memory—a molecular framework. *Nature* 322:419–422, 1986.
20. Goldberger, M.E.; Murray, M. Recovery of function and anatomical plasticity after damage to the adult and neonatal spinal cord. In Cotman, C.W., ed. *Synaptic Plasticity*. New York, Guilford Press, 77–110, 1985.
21. Greenough, W.T.; Volkmar, F.R.; Juraska, J.M. Effects of rearing complexity on dendritic branching in frontolateral and temporal cortex of the rat. *Exp. Neurol.* 41:371–378, 1973.

22. Hawkins, R.D.; Kandel, E.R. Steps toward a cell-biological alphabet for elementary forms of learning. In Lynch, G.; McGaugh, J.L.; Weinberger, N.M., eds. *Neurobiology of Learning and Memory*. New York, Guilford Press, 385–404, 1984.

23. Hebb, D.O. *The Organization of Behavior*. New York, John Wiley & Sons, 1949.

24. Jacobson, M.; Baker, R.E. Development of neuronal connections with skin grafts in frogs: behavioral and electrophysiological studies. *J. Comp. Neurol.* 137:121–142, 1969.

25. Kandel, E.R. Cellular mechanisms of learning and the biological basis of individuality. In Kandel, E.R.; Schwartz, J.A., eds. *Principles of Neuroscience*. New York, Elsevier, 816–833, 1985.

26. Kromer, L.F.; Björklund, A.; Stenevi, U. Innervation of embryonic hippocampal implants by regenerating axons of cholinergic septal neurons in the adult rat. *Brain Res.* 210:153–171, 1980.

27. Lashley, K. In search of the engram. *Proc. Soc. Exp. Biol. (N.Y.)* 4:454–482, 1950.

28. Levi-Montalcini, R.; Angeletti, P.U. Essential role of the nerve growth factor in the survival and maintenance of dissociated sensory and sympathetic embryonic nerve cells in vitro. *Devel. Biol.* 7:653–659, 1963.

29. Lloyd, D.P.C. Post-tetanic potentiation of response in monosynaptic reflex pathways of the spinal cord. *J. Gen. Physiol.* 33:147–170, 1949.

30. Lynch, G.; Larson, J.; Kelso, S.; Barrionuevo, G.; Schottler, F. Intracellular injections of EGTA block induction of hippocampal long-term potentiation. *Nature* 305:719–721, 1983.

31. Nieto-Sampedro, M.; Cotman, C.W. Growth factor induction and temporal order in central nervous system repair. In Cotman, C.W., ed. *Synaptic Plasticity*. New York, Guilford Press, 407–455, 1985.

32. O'Keefe, J. Spatial memory within and without the hippocampal system. In Seifert, W., ed. *Neurobiology of the Hippocampus*. New York, Academic Press, 375–403, 1983.

33. Olds, J.; Disterhoft, J.F.; Segal, M.; Kornblith, C.L.; Hirsh, R. Learning centers of rat brain mapped by measuring latencies of conditioned unit responses. *J. Neurophysiol.* 35:202–219, 1972.

34. Olton, D.S.; Becker, J.T.; Handelmann, G.E. Hippocampus, space and memory. *Behav. Brain Sci.* 2:313–365, 1979.

35. Purves, D.; Lichtman, J.W., eds. *Principles of Neural Development*. Sunderland, Mass., Sinauer Assoc., 1985.

36. Rutledge, L.T. Effects of cortical denervation and stimulation on axons, dendrites, and synapses. In Cotman, C.W., ed. *Neuronal Plasticity*. New York, Raven Press, 273–289, 1978.

37. Schneider, G.E. Mechanisms of functional recovery following lesions of the visual cortex or superior colliculus in neonatal and adult hamsters. *Brain Behav. Evol.* 3:295–323, 1970.

38. Scoville, W.B.; Milner, B. Loss of recent memory after bilateral hippocampal lesions. *J. Neurol. Neurosurg. Psychiatr.* 20:11–21, 1957.

39. Sherrington, C.S. *The Integrative Action of the Nervous System*, 2nd ed. New Haven, Yale Univ. Press, 1947.

40. Sotelo, C. Dendritic abnormalities of Purkinje cells in the cerebellum of neurologic mutant mice (Weaver and Staggerer). In Kreutzberg, G.W., ed. *Advances in Neurology*, Vol. 12. New York, Raven Press, 335–351, 1975.

41. Spencer, W.A.; April, R.S. Plastic properties of monosynaptic pathways in mammals. In Horn, G.; Hinde, R.A., eds. *Short-Term Changes in Neural Activity and Behavior*. Cambridge, Cambridge University Press, 433–474, 1970.

42. Squire, L.R. The neuropsychology of human memory. *Annu. Rev. Neurosci.* 5:241–273, 1982.

43. Stanton, P.K.; Sarvey, J.M. The effect of high-frequency electrical stimulation and norepinephrine on cyclic AMP levels in normal versus norepinephrine-depleted rat hippocampal slices. *Brain Res.* 358:343–348, 1985.

44. Thompson, R.F.; Spencer, W.A. Habituation: a model phenomenon for the study of neuronal substrates of behaviors. *Psychol. Rev.* 173:16–43, 1966.

45. Thompson, R.F.; Berger, T.W.; Berry, S.D.; Clark, G.A.; Kettner, R.E.; Lavond, D.G.; Mauk, M.D.; McCormick, D.A.; Solomon, P.R.; Weisz, D.J. Neuronal substrates of learning and memory: hippocampus and other structures. *Adv. Behav. Biol.* 26:115–129, 1982.

46. Thompson, R.F.; Clark, G.A.; Donegan, N.H.; Lavond, D.G.; Lincoln, J.S.; Madden, J., IV; Mamounas, L.A.; Mauk, M.D.; McCormick, D.A.; Thompson, J.K. Neuronal substrates of learning and memory: a "multiple-trace" view. In Lynch, G.; McGaugh, J.L.; Weinberger, N.M., eds. *Neurobiology of Learning and Memory*. New York, Guilford Press, 137–164, 1984.

47. Tsukahara, N. Synaptic plasticity in the red nucleus and its possible behavioral correlates. In Cotman, C.W., ed. *Synaptic Plasticity*. New York, Guilford Press, 201–229, 1985.

48. Westrum, L.E.; White, L.E.; Ward, A.A., Jr. Morphology of the experimental epileptic focus. *J. Neurosurg.* 21:1033–1046, 1964.

49. Wigstrom, H.; Gustafsson, B.; Huang, Y.-Y.; Abraham, W.C. Hippocampal long-term potentiation is induced by pairing single afferent volleys with intracellularly injected depolarizing current pulses. *Acta Physiol. Scand.* 126:317–318, 1986.

50. Woody, C.D. Studies of pavlovian eye-blink conditioning in awake cats. In Lynch, G.; McGaugh, J.L.; Weinberger, N.M., eds. *Neurobiology of Learning and Memory*. New York, Guilford Press, 181–196, 1984.

51. Woody, C.D.; Knispel, J.D.; Crow, T.J.; Black-Cleworth, P.A. Activity and excitability to electrical current of cortical auditory receptive neurons of awake cats as affected by stimulus association. *J. Neurophysiol.* 39:1045–1061, 1976.

52. Woody, C.D.; Alkon, D.L.; Hay, B. Depolarization-induced effects of Ca^{2+}-calmodulin dependent protein kinase injection, in vivo, in single neurons of cat motor cortex. *Brain Res.* 321:192–197, 1984.

Section VII

ALBERT F. FUCHS
Section Editor

Emotive Responses and Internal Milieu

Chapter 34

Harry D. Patton

The Autonomic Nervous System

DISTINCTIVE FEATURES OF THE AUTONOMIC NERVOUS SYSTEM

Earlier chapters focus attention on reflex systems in which the effector organ is skeletal muscle and the response is skeletal movement. Such reflex arcs are termed *somatic reflexes*. Smooth muscle, glands, and the conducting tissue of the heart also receive motor nerve supplies that when reflexly activated, alter the functional state of the innervated organ; such reflexes are termed *autonomic reflexes*.

Autonomic nerve discharge to smooth muscles and glands has an important role in visceral and glandular responses to environmental changes; for example, reflex alteration of arteriolar diameter, mediated over autonomic motor fibers supplying vascular smooth muscle, is at least partly responsible for the shifting of blood from one vascular bed to another in accordance with physiologic demand. Similarly, although not initiating the beat of the heart, reflex discharges over the autonomic nerves supplying the cardiac pacemaker modulate and regulate the rate of beating, so that the varying demands upon the pumping system are automatically met. Examples of the regulation of visceral and glandular structures by autonomic reflex arcs are encountered frequently in the study of physiology. This chapter concerns the general properties and organization of the autonomic nervous system.

The distinction between autonomic and somatic motor outflows is based on both anatomic and functional grounds. Before a detailed description of the autonomic system, a brief account of its unique and distinctive properties is appropriate. Anatomically, the autonomic outflow differs from the somatic outflow in the location of the motoneuron soma. In the somatic division, the cell bodies of the motoneurons are located exclusively within the central nervous system, in the anterior spinal horns or in the motor nuclei of cranial nerves in the brain stem. With one exception, the adrenal medulla, smooth muscle, and glands receive direct motor innervation from cell bodies situated in ganglia outside the central nervous system. Thus the typical autonomic reflex chain contains one synaptic junction between the outflow from the central nervous system and the effector organ.

The centrally located penultimate neuron whose axon feeds the ganglion is appropriately termed the *preganglionic neuron*; its axon is typically myelinated. The ultimate neuron, originating in the ganglion, is called, somewhat less appropriately, the *postganglionic neuron*; its axon is unmyelinated. Figure 34–1 contrasts diagrammatically somatic and autonomic reflex arcs. It should be noted that the afferent sides of these arcs are indistinguishable; indeed, one and the same afferent pathway may feed both autonomic and somatic outflows. The preganglionic neuron may be considered the homologue of the interneuron of somatic arcs. Just as the interneuron, by its numerous intersegmental connections with motoneurons, tends to cause diffuse efferent discharge over several spinal segments, so the preglanglionic neuron, by connection with several ganglia, may distribute the postganglionic efferent discharge.

A second fundamental difference between autonomic and somatic reflex arcs lies in the site at which inhibition occurs. In somatic arcs, inhibition is exerted by one neuron upon another, but never by a nerve cell upon the effector muscle cell. Relaxation of a skeletal muscle is accomplished by inhibition within the spinal cord of the motoneurons that excite it. This is *central inhibition*. In autonomic reflex arcs, presynaptic fibers may inhibit preganglionic neurons, but in addition, some autonomic postganglionic fibers inhibit the action of the effector organs that they innervate. A good example of such *peripheral* or *neuroeffector inhibition* is the action of impulses in the vagus nerve upon the heart—the excitability of the pacemaker cells

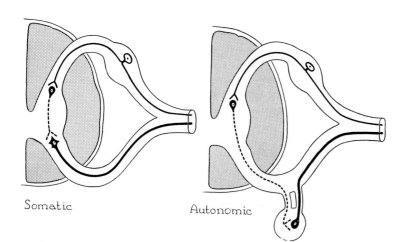

Somatic   Autonomic

Figure 34–1 Diagram illustrating a somatic reflex arc *(left)* and an autonomic reflex arc *(right)*.

is so diminished that the heart rate declines. Intense vagal stimulation may result in temporary cardiac standstill.

ORGANIZATION OF AUTONOMIC OUTFLOW

It is convenient to divide the autonomic outflow into two divisions distinguished by the location of their preganglionic cell bodies. The *sympathetic*, or *thoracolumbar, division* originates from preganglionic neurons in the thoracic and upper lumbar spinal segments, and its axons leave the spinal cord via the corresponding ventral roots. The *parasympathetic*, or *craniosacral, division* originates from preganglionic neurons in certain cranial nerve nuclei and in the second, third, and fourth sacral segments of the spinal cord. The cranial outflow leaves the brain with the appropriate cranial nerves, and the sacral parasympathetic outflow emerges over the S_2, S_3, and S_4 ventral roots.

Sympathetic, or Thoracolumbar, Outflow.[60] The cell bodies of the preganglionic neurons are found in the intermediolateral gray matter of spinal segments T_1 to L_2 or L_3; the axons of these neurons emerge in the corresponding ventral roots and enter the spinal nerve, where they part from the somatic motor fibers via the *white ramus communicans* to reach the paravertebral sympathetic ganglion chain.

This chain contains one ganglion for each segmental nerve, except in the cervical region, where individual ganglia become variably fused to form two or three ganglia—the superior, middle, and inferior cervical ganglia. The superior ganglion is the largest and gives rise to the postganglionic sympathetic supply to the head. The inferior ganglion is often fused with the first thoracic ganglion into a dumbbell-shaped structure called the stellate ganglion. The remaining thoracic as well as the lumbar and sacral ganglia are segmentally arranged.

On entering the ganglionic chain, a preganglionic fiber may pursue one of three courses:

(i) It may pass up or down the chain to establish synaptic connections with postganglionic neurons in ganglia belonging to more superior or inferior segments. In this way, ganglia of cervical, lower lumbar, and sacral segments lacking white rami receive input from the spinal cord. The fibers connecting the ganglia in the chain are largely composed of preganglionic fibers following this course. Each such fiber connects with many postganglionic cells situated in several ganglia; a single white ramus may connect with as many as eight

or nine segmental ganglia. Consequently, a discharge of discrete central origin is spread diffusely over several segments in a manner reminiscent of the multisegmental discharge of somatic arcs that contain one or more spinal interneurons. The axons of the postganglionic cells in the paravertebral chain pass through the *gray ramus communicans* to enter the corresponding segmental nerve, where they reach the autonomic effectors of the skin and subcutaneous structures (cutaneous and deep blood vessels, sweat glands, pilomotor smooth muscle). *Each spinal nerve receives a gray ramus from its corresponding ganglion.* The postganglionic neurons in the lower thoracic, the lumbar, and the sacral paravertebral ganglia are destined for peripheral (as opposed to visceral) effectors, which they reach through the gray rami and the segmental mixed nerve trunks. The upper thoracic ganglia contain not only neurons supplying thoracic peripheral effectors via their gray rami but also other neurons whose axons supply visceral structures of the chest (heart, bronchioles, vascular smooth muscle) via branches directly to these structures. The most important of these latter are the cardiac accelerator nerves, which derive from the stellate and upper four or five thoracic ganglia and run directly to the heart.

(ii) The preganglionic fibers may pass without interruption through the chain into the splanchnic nerves to reach the celiac or other ganglia lying in the prevertebral sympathetic plexus, which invests the abdominal aorta and its major branches down to the iliac arteries. The prevertebral ganglia are sometimes called the collateral ganglia. The postganglionic neurons of the prevertebral plexus supply fibers to the smooth muscle of the abdominal and pelvic viscera, to the glands of the gut, to the blood vessels of the abdominal viscera, and so forth.

(iii) Some preganglionic fibers of the splanchnic nerve directly innervate the secretory cells of the adrenal medulla. The adrenal medulla is the only sympathetic effector organ known to be directly innervated by preganglionic fibers.

Parasympathetic, or Craniosacral, Outflow. The cranial portion of the parasympathetic outflow originates from preganglionic neurons situated in brain stem nuclei of cranial nerves III, VII, IX, and X. The axons of these neurons travel with these nerves to supply postganglionic neurons in ganglia within or near the thoracic and abdominal viscera. The sacral parasympathetic outflow arises from preganglionic neurons, mostly in the third and fourth sacral spinal segments but sometimes also in the second and fifth. The axons of these neurons emerge with the corresponding ventral roots, but separate from the somatic efferent fibers to form the *nervi erigentes*, or pelvic nerves, which constitute the preganglionic parasympathetic supply to the genitalia and the autonomic effectors of the pelvic cavity. The parasympathetic ganglia containing the postganglionic neurons are usually situated in or near the organ innervated.

Because parasympathetic neurons are located in or close to the effector, their axons are short; most of the span between brain (or cord) and the effector is made up of long preganglionic axons. In the sympathetic nervous system, the reverse relation usually holds—i.e., there are relatively short preganglionic axons (e.g., from spinal cord to ganglion chain) and longer unmyelinated postganglionic axons (e.g., from ganglion chain to arterioles in the foot).

A further distinction is that the individual sympathetic ganglia are linked together by the interganglionic commissural fibers, whereas the parasympathetic ganglia are separated and unlinked. Consequently, parasympathetic effects are often more spatially discrete than are sympathetic effects.

INTERACTIONS BETWEEN SYMPATHETIC AND PARASYMPATHETIC INNERVATIONS

Some organs receive dual innervation, from the parasympathetic and the sympathetic nervous systems. In some of these organs, the two innervations exert antagonistic effects, and consequently, the function of the organ depends on the balance between the two competing regulators. The pacemaker of the heart is an example; it receives excitatory innervation from sympathetic accelerator fibers and inhibitory innervation from parasympathetic vagal fibers. The heart rate thus accelerates or slows in proportion to the dominance of the tonic discharge over the two regulating systems. A similar "Yin and Yang" control mechanism pertains to the smooth muscle of the gut, but here the functional roles of sympathetic and parasympathetic are reversed: the sympathetic fibers inhibit, and the parasympathetic fibers stimulate peristalsis. In the pupil, the two innervations are both excitatory, but supply muscles that are antagonistic; the parasympathetic innervation (cranial nerve III) excites the circular (constrictor) muscles and the sympathetic fibers (superior cervical sympathetic ganglion) excite the radial (dilator) muscles. Pupil size thus varies according to the balance between the two regulating systems.

Sometimes in dually innervated organs, the two nerve supplies are synergistic. Both sympathetic and parasympathetic impulses to the salivary glands stimulate salivary secretion, although the composition of the resultant secretion is somewhat different.

Finally, many autonomic effectors receive innervation from only one division of the autonomic nervous system. Examples are the smooth muscle of cutaneous vessels, the pilomotor muscles, and the sweat glands, which receive only excitatory sympathetic supply. The lacrimal glands receive a solitary excitatory parasympathetic supply (cranial nerve VII); sympathetic fibers to the lacrimal gland are vasoconstrictor only and do not directly affect secretion of tears.

From the foregoing, it should be clear that there is no simple way to codify autonomic innervation. This is doubly unfortunate because a clear and detailed knowledge of autonomic innervation is essential to modern medical practice, which bristles with a multitude of medicaments aimed at one or another autonomic regulatory system. The student is thus perforce faced with the prospect of memorizing the innervation of each organ and its functional significance.

Although rote memorization is the only sure way to master the intricacies of autonomic anatomy and physiology, a broad semiphilosophical concept first enunciated by Walter B. Cannon serves as a useful mnemonic device. Cannon contended that the actions of the parasympathetic nervous system tend to build up or to conserve bodily energy stores, whereas those of the sympathetic nervous system are bodily reactions appropriate to environmental circumstances that require emergency action and depletion of body stores: what Cannon called the "fight or flight" reaction.

As an illustration, imagine a student attending an eight o'clock lecture in neurophysiology. In these circumstances, his muscles are relaxed, he is free of stress (except for the difficulty of staying awake), and he is, in general, in a torpid state of postprandial comfort. In this idyllic nonstressful situation, his vegetative functions are dominated by the parasympathetic nervous system. The heart rate is slow and regular, the blood pressure is at a low-normal level, peristalsis and gastrointestinal secretion are busily extracting nutriments from his recently ingested breakfast. A major portion of the blood volume occupies the dilated visceral vascular bed, where it is available to carry absorbed nutriments to the liver, which in turn is busily engaged in storing of glycogen. Blood flow to the inactive muscles is curtailed in favor of visceral perfusion. All of these physiological conditions are promoted by parasympathetic action and by relative quiescence of the sympathetic nervous system.

Imagine next the bodily changes that might be deemed physiologically appropriate if some dire or life-threatening emergency, such as a fire alarm or an earthquake, supervenes. In this altered environmental situation, the physiological priorities shift dramatically toward behavioral patterns that are dominated by sympathetic discharge (here, more "flight" than "fight") and by relative abeyance of parasympathetic actions. For example, digestive activities, previously a conservative virtue, now become an extravagance preempting a lion's share of blood volume urgently needed elsewhere for vital muscular action. Prudent postponement of digestion is brought about both by decreased vagal discharge and by increased sympathetic discharge, which curtails both gastrointestinal mobility and secretion and constricts the visceral vascular bed. Visceral vasoconstriction, along with increase in the rate and force of contraction of the heart, elevates arterial pressure and shifts blood into the peripheral beds. Sympathetic impulses to the liver promote glycogenolysis, dumping glucose into the blood. Other sympathetic actions include pupillary dilation, sweating, and piloerection, all common accompaniments of emotional excitement but of less obvious biological utility in coping with the emergency than are the vascular changes.

The thoughtful student will detect in Cannon's "fight or flight" thesis and in his book *The Wisdom of the Body* an unsettling undercurrent of teleology. Teleology usually leads either to contradictions or to conclusions that are so arbitrary as to be of little predictive or explanatory value. Attempts to carry the "wisdom of the body" theme to extremes quickly lead to conflicts. For example, it is hard to find anything biologically praiseworthy in the concomitant occurrence of piloerection and sweating and of cutaneous vasoconstriction, which are conflicting thermoregulatory responses. Equally dubious is the adaptive value of emotive pupillary dilation. Cannon's doctrine is thus less useful as explanation than as a mnemonic device for learning the competitive roles and actions of sympathetic and parasympathetic functions. For the latter purpose, it serves admirably, and the contrasting functions of the two systems usually can readily be recalled by conjuring up the images of animals in states of conservative contentment and of dramatic emergency.

AUTONOMIC INNERVATION OF VARIOUS STRUCTURES

The autonomic innervation of some important visceral structures may now be summarized (see also Fig. 34–2). More detailed accounts appear in monographs.[40, 48, 49, 60, 78]

Lacrimal Glands
Parasympathetic. Preganglionic neurons originate in the superior salivatory nucleus; axons pass with the VIIth nerve, the nervus intermedius, and the greater superficial petrosal and vidian nerves to reach *postganglionic neurons* in the sphenopalatine ganglion. Their axons pass in the maxillary division of the Vth nerve to the lacrimal glands. *Function:* vasodilation and secretion.

Sympathetic. Preganglionic neurons originate in the intermediolateral cell column of the upper thoracic spinal segments; axons ascend the sympathetic chain to reach *postganglionic neurons* in the superior cervical ganglion. Their axons ascend in the carotid plexus, the deep petrosal and vidian nerves, and the maxillary division of the Vth nerve to the glands. *Function:* vasoconstriction.

Eye
Parasympathetic. Preganglionic neurons originate in the oculomotor nucleus; axons travel in the oculomotor nerve to reach *postganglionic neurons* in the ciliary ganglion. Their axons traverse the short ciliary nerve to reach the ciliary muscle and the constrictor (circular) muscle of the iris. *Function:* pupillary constriction (miosis), accommodation for near vision.*

Sympathetic. Preganglionic neurons originate in the upper thoracic segments; axons ascend the sympathetic chain to reach *postganglionic neurons* in the superior cervical ganglion. Their axons pass via the carotid plexus and the ophthalmic division of the Vth nerve to the dilator muscles of the iris, the smooth muscle of the levator palpebrae superioris, the ciliary muscle and the blood vessels of the retina, orbit, and conjunctiva. In lower mammals, e.g., the cat, fibers also supply the nictitating membrane. *Function:* pupillary dilation (mydriasis), vasoconstriction, contraction of the nictitating membrane, elevation of the lid and possible accommodation for far vision.*

---

*During accommodation to near vision, parasympathetic oculomotor nerve discharge causes contraction of the ciliary muscle. This action releases tension on the lens capsule, permitting the lens to assume a more nearly spherical shape and thus to shorten its focal distance. Accommodation to distant vision occurs when parasympathetic discharge ceases, so that the ciliary muscle relaxes and tension of the zonula fibers is restored and the lens is flattened. However, stimulation of the superior cervical sympathetic ganglion causes further flattening of the lens even when the parasympathetic nerves are cut. The change amounts to about 1.5 diopters. It is uncertain whether the effect is due to a change in size of the ciliary body solely as a result of vasoconstriction or whether sympathetic fibers have an inhibitory effect on ciliary muscle.[1, 51]

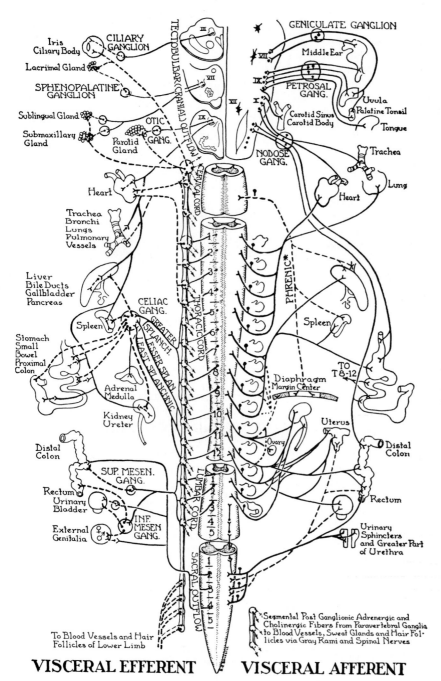

Figure 34–2 Afferent and efferent innervation of visceral structures. *Blue*, cholinergic neurons; *red*, adrenergic neurons; *solid lines*, efferent preganglionic fibers; *broken lines*, efferent postganglionic fibers. Afferent fibers from diaphragm are drawn in broken lines to emphasize that it is a somatic structure even though it lies in the visceral cavity.

Salivary Glands[18, 46, 66]

Parasympathetic. Preganglionic neurons of outflow to submaxillary and sublingual glands originate in the superior salivatory nucleus; those of the outflow to the parotid glands originate in the inferior salivatory nucleus. From the former, axons pass in the facial nerve and through the chorda tympani to the submaxillary and sublingual ganglia. Axons from the inferior salivatory nucleus pass via the tympani branch of the IXth nerve to the lesser superficial petrosal nerve and thence to the otic ganglion. Axons of the *postganglionic neurons* pass from the submaxillary and sublingual ganglia to their respective glands and from the otic ganglion via the auriculotemporal branch of the Vth nerve to the parotid gland. *Function:* vasodilation and secretion.

Sympathetic. Preganglionic neurons originate in upper thoracic segments; axons ascend the chain to reach *postganglionic neurons* in the superior cervical ganglion. Their axons run along the external carotid and external maxillary arteries. *Function:* vasoconstriction and secretion.

Heart[49, 60]

Parasympathetic. Preganglionic neurons originate in the dorsal motor nucleus of the vagus; axons pass through the vagal trunk to reach *postganglionic neurons* in ganglia found in the cardiac plexus and in the walls of the atria. Parasympathetic nerve endings reach the sinoatrial and atrioventricular nodes and the atrium. It is disputed whether they reach the coronary circulation or the ventricles. *Function:* cardiac deceleration and coronary dilatation.

Sympathetic. Preganglionic neurons originate in the intermediolateral column of the upper four or five thoracic spinal segments; axons pass with the corresponding ventral roots and white rami to the sympathetic chain to reach *postganglionic neurons* in the upper four or five thoracic ganglia and in the cervical ganglia. Axons from the cervical ganglia form the superior, middle, and inferior cardiac nerves that run to the cardiac plexus, where they are joined by varying numbers of thoracic cardiac nerves from the thoracic ganglia. Some postganglionic neurons arise in ganglia along the course of the cardiac nerves and in the cardiac plexus and receive their input from preganglionic fibers which run through the chain without synaptic interruption. Distribution of terminals extends to coronary vessels, the pacemaker, the conduction system, and both the atrial and the ventricular myocardium. *Function:* cardiac acceleration, increased contractility, and coronary constriction.

Lungs[60]

Parasympathetic and Sympathetic. Origins of innervation are similar to those for the heart except that the sympathetic preganglion portion originates in the T_2 to T_6 segments. Both parasympathetic and sympathetic fibers enter the pulmonary plexus, in which the parasympathetic ganglia are embedded. Postganglionic fibers of both supply terminals in bronchioles and blood vessels. *Function:* the parasympathetic impulses constrict and the sympathetic dilate the bronchioles. De-

spite demonstrable vascular endings, there is little evidence of significant vasomotor regulation of pulmonary vessels.

Abdominal Viscera, Glands, and Vessels[60]

Parasympathetic. Vagal *preganglionic* fibers traverse the prevertebral plexus without interruption to reach *postganglionic neurons* in the intrinsic plexus of the visceral organ. *Function:* stimulation of peristalsis and gastrointestinal secretion.

Sympathetic. Preganglionic neurons originate in the lower seven or eight thoracic segments and in the upper lumbar segments; axons run via the splanchnic nerves to reach the *postganglionic neurons* in the prevertebral ganglionic plexus. Postganglionic endings are supplied to the visceral blood vessels and the smooth muscles of the viscera. Innervation of the adrenal medulla is preganglionic. *Function:* vasoconstriction and inhibition of peristalsis; contraction of gastrointestinal sphincters; secretion in adrenal medulla. Sympathetic supply to the liver causes glycogenolysis.

Pelvic Viscera[60]

Parasympathetic. Preganglionic neurons originate in the second, third, and fourth sacral segments; axons form nervi erigentes, which reach ganglia in the walls of the organs innervated. *Postganglionic neurons* supply the uterus, tubes, testes, erectile tissue, sigmoid colon, rectum, and bladder. *Function:* contraction of bladder wall and lower colon; erection. The significance of uterine innervation is unknown.

Sympathetic. The origin and path of *preganglionic neurons* are the same as those for innervation of abdominal viscera. *Postganglionic axons* run in the hypogastric nerves to the blood vessels, lower colon and rectum, and seminal vesicles. *Function:* contraction of internal vesical sphincter; vasoconstriction; ejaculation of semen; inhibition of peristalsis in lower colon and rectum.

Peripheral Vessels and Cutaneous Effectors

Sympathetic. These structures receive only sympathetic innervation by *postganglionic* fibers with cell bodies in the ganglion chain. Their axons join the segmental nerves via the gray rami, to be distributed to the skin and deep vessels. *Function:* vasoconstriction in both cutaneous and deep vessels (existence of an additional system of vasodilator fibers to the vessels of muscle has been postulated from indirect evidence[2, 5, 17]); secretion in sweat glands; excitation of pilomotor muscles.

GENERAL PRINCIPLES OF AUTONOMIC REGULATION

Junctional Transmission in the Autonomic Nervous System. Figure 34–3 shows semidiagrammatically the results of two experiments that emphasize a fundamental contrast between neuromuscular transmission in the somatic and in the autonomic nervous systems. The record in A shows the nerve action potential, the muscle action potential and the muscle tension of a skeletal

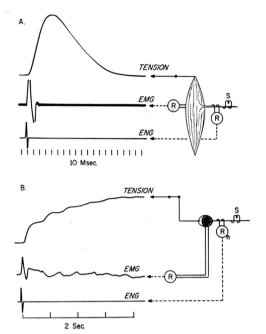

Figure 34–3 Neuromuscular transmission in somatic and autonomic effector systems. *A,* Simultaneously recorded electroneurogram, electromyogram, and mechanical response of skeletal nerve muscle preparation following single shock to motor nerve. *B,* Simultaneously recorded electroneurogram, electromyogram, and mechanical response of nictitating membrane following single shock to postganglionic motor nerve. Note difference in time scales. (Partly after Eccles and Magladery, *J. Physiol.* 90:31–99, 1937.)

muscle following a single shock to its motor nerve. Each nerve fiber in the trunk responds to the stimulus with a single brief action potential; the resultant compound potential, recorded close to the neuromuscular junction, is brief. Shortly thereafter the action potential of the muscle is recorded as a brief solitary event followed closely by the muscle twitch, which is recorded as a change in muscular tension. The arrangement of stimulating and recording electrodes to obtain Figure 34–3B is similar; but the muscle is the smooth muscle of a cat's nictitating membrane, which is supplied by a motor nerve composed of sympathetic axons (from the superior cervical ganglion). As in the somatic motor nerve, the nerve action potential is brief and solitary, but the electromyogram shows a series of somewhat asynchronous deflections that persist long after the nerve fibers have repolarized. The tension record shows, further, that each muscle action potential is associated with increments of contraction, so that the resulting tension curve is prolonged and bumpy, resembling an unfused tetanus of skeletal muscle.

Persistence of electrical and mechanical activity

long beyond the duration of the excitatory nerve impulses is typical of autonomic neuroeffectors and persuasively suggests that the nerve exerts its action on effectors by liberating a chemical transmitter agent and thus initiating changes of intracellular second messengers that remain and continue to act on the effector response after the nerve action has ceased. Observations of this sort led to the theory of humoral transmitters. It has been pointed out that somatic neuromuscular transmission is also accomplished by the liberation of a chemical transmitter (acetylcholine) that depolarizes the end-plate, but because the transmitter is destroyed rapidly, its action is brief. At autonomic junctions, the slow destruction and prolonged action of the transmitters make the chemical nature of transmission much more immediately obvious.

Cholinergic Fibers. Humoral transmission in an autonomic neuroeffector system was first clearly demonstrated by Otto Loewi.[45] Because cardiac inhibition resulting from vagal stimulation far outlasts the period of nerve stimulation, Loewi suspected a humoral transmitter. In the experiment illustrated in Figure 34–4, fluid perfusing a donor frog heart (*D*) was used to perfuse a second, recipient heart (*R*). When the vagus nerve supplying heart *D* was stimulated, cardiac arrest occurred; shortly thereafter heart *R* also stopped beating, an event implicating an inhibitory chemical agent liberated into the perfusion fluid at the vagal endings in heart *D* and then carried to heart *R*. Loewi noncommittally termed the vagal inhibitory transmitter *Vagusstoff*.

Identification of Loewi's Vagusstoff followed from Dale's studies[14] on the pharmacologic actions of choline and its esters. He noted that the acetyl ester of choline is *parasympathomimetic*, i.e., when

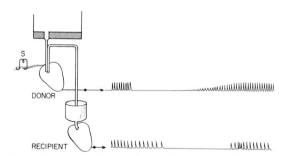

Figure 34–4 Loewi's experiment demonstrating the humoral mechanism of vagal inhibition of the heart. Stimulation of the vagus supplying the donor heart released a chemical inhibitory mediator which not only arrested the donor heart but, after diffusion into the perfusion fluid, also arrested the recipient heart. (After Bain. *Q. J. Med Exp. Physiol. Med. Sci.* 22:269–274, 1932.)

injected into the blood stream, acetylcholine acts upon autonomic effectors, including the heart, in a manner similar to or mimicking the action exerted on these effectors by their respective parasympathetic nerves. The drug atropine blocks the action of acetylcholine on smooth muscle and similarly blocks the action of parasympathetic nerves on their effectors. The drug eserine, on the other hand, potentiates the action of acetylcholine by inactivating the enzyme acetylcholinesterase, which splits acetylcholine into the relatively inert choline and acetic acid. Eserine also potentiates the effect of parasympathetic nerve stimulation. Such observations provided presumptive evidence that Loewi's Vagusstoff—as well as the transmitter at all other parasympathetic postganglionic endings—is either acetylcholine or a closely related substance.

Subsequent investigations have revealed that acetylcholine is also the transmitter agent liberated by autonomic preganglionic fibers, sympathetic as well as parasympathetic. The brief action of the transmitter in ganglia (and at the somatic neuromuscular junction) results from high concentrations of cholinesterase at these sites, so that the liberated transmitter is destroyed within the refractory period of the postjunctional cell.

Nerve fibers that liberate acetylcholine transmitter are called *cholinergic fibers*. In summary, these include somatic motor fibers, all autonomic preganglionic fibers, and all parasympathetic postganglionic fibers. In addition, the sympathetic postganglionic fibers supplying sweat glands are cholinergic.

Sympathetic Postganglionic Mediators. Sympathetic postganglionic endings liberate a humoral transmitter that is destroyed relatively slowly. The proof was first provided by Cannon and his associates[9] when they analyzed the mechanism of the cardiac acceleration that accompanies exercise and emotional excitement. Even in animals in which the heart was completely denervated by section of the vagi and the cardiac accelerator nerves, these investigators observed that struggling, excitement, or physical exercise induced a prompt (1 minute) increase in the heart rate of 80 to 100 beats per minute. Such acceleration results partly from the liberation into the circulation of epinephrine by the adrenal medulla. However, adrenal secretion does not entirely account for the response. Cannon found that after the adrenals were removed or denervated a moderate (25 to 30 beats per minute) but delayed (three minutes) increase in heart rate followed emotional excitement. Delayed emotional tachycardia persisted in

animals subjected to hypophysectomy and bilateral abdominal and cervical sympathectomy. However, complete removal of the abdominal and thoracic sympathetic chains abolished the response.

These experiments implicated an extra-adrenal humoral agent released into the blood stream during exercise or excitement and capable of exerting a sympathetic-like (acceleratory) influence on the heart. Derivation of this substance from sympathetic nerve endings was indicated by experiments in which sympathetic nerves were stimulated electrically. When injected into the blood stream, perfusates collected from organs during stimulation of their sympathetic nerves caused cardiac acceleration and increased blood pressure. Years earlier, Elliott had presciently observed that epinephrine, the secretion of the adrenal medulla, is a *sympathomimetic agent*; i.e., it mimics the action of sympathetic postganglionic stimulation. Although the actions of the sympathetic mediator and of epinephrine were very similar, there were some differences, and Cannon cautiously termed the mediator *sympathin*. It is now known that both the adrenal medulla and the sympathetic postganglionic endings secrete at least two catecholamines—epinephrine and norepinephrine. Although closely related structurally, these two substances do not invariably exert identical actions on effector organs; a full catalogue of their pharmacologic properties can be found in textbooks of pharmacology. Norepinephrine is the sympathetic postganglionic mediator in mammals and corresponds to Cannon's sympathin. Adrenal medullary secretion, on the other hand, is mixed. It is principally epinephrine in humans.

Nerve fibers secreting epinephrine or norepinephrine or both are called *adrenergic fibers*. Most sympathetic postganglionic fibers are adrenergic. A notable exception is the sympathetic postganglionic innervation of the sweat glands, which is cholinergic and readily blocked by atropine. Other sympathetic postganglionic cholinergic systems have been postulated but are not so well documented.

In passing, one may note that the discovery of the adrenergic nature of sympathetic postganglionic fibers renders less anomalous the absence of a peripheral synapse formed by the adrenal medullary chromaffin cells. Indeed, the adrenal medullary cells and the sympathetic postganglionic neurons are similar, since they secrete the same substance and derive from the same embryologic tissues.

As is also the case for many central neurons,

many postganglionic autonomic neurons release small peptides in addition to conventional neurotransmitters.[8a, 12] Such peptides (neuropeptide Y, vasoactive intestinal peptide, cholecystokinin, and many others) have definite actions on effector cells as well as on the neurons themselves. In addition, ATP and possibly adenosine are released from many autonomic fibers. Because they coexist and are released together with the conventional transmitters, these additional agonists have been termed cotransmitters.[8, 8a] At present, the functional importance of the cotransmitters for autonomic physiology is still being worked out. This subject is one that will see considerable exploration in the coming years.

Receptors for Autonomic Transmitters.[62, 63] As in the somatic nervous system, autonomic transmitters exert their characteristic actions by combining with specific receptor molecules in the postjunctional membrane. In nearly every case, activation of the receptor is coupled via internal second messenger systems to cytoplasmic chemical reactions that also give rise to permeability changes in the postjunctional membrane. The nature of the induced permeability change depends not only on the transmitter but also on the nature of the postjunctional receptor. It follows that one and the same transmitter may exert totally different effects on organs having different receptors. Thus the action of acetylcholine on skeletal muscle is depolarizing and excitatory because it combines with a nicotine receptor molecule in the end-plate membrane to form a molecular complex that increases end-plate permeability to Na^+ and K^+ (Chap. 6). The same transmitter, acetylcholine, released from vagal endings on the heart is inhibitory, as discussed below, for in heart muscle the

receptor–transmitter complex initiates an increased permeability to K^+, with resultant hyperpolarization and depressed excitability. Similarly, the nature of the receptor determines the response of a junction to drugs that are agonists or antagonists of acetylcholine. At the neuromuscular junction, the action of acetylcholine is blocked by curare, which competes with the transmitter for the receptor sites. The muscarinic acetylcholine receptors of the heart are relatively insensitive to curare but are readily blocked by the drug atropine. Judicious use of pharmacologic agents thus requires knowledge not only of transmitters but also of the specific receptors of different tissues (Table 34–1).

Nicotinic and muscarinic ACh receptors can be distinguished by their chemical structure, pharmacology, and actions. The responses of nicotinic receptors to acetylcholine are all excitatory and are characterized by evanescence, the response lasting only milliseconds. In this receptor, the ACh binding sites and the ion channel opened by ACh are all part of a single molecule (Chap. 6). The receptors can be activated by nicotine in low concentration but are blocked with higher concentrations, and they are also blocked by curare and related compounds. The ACh receptors of skeletal muscle and those in autonomic ganglia responsible for the fast (see later) excitatory response to preganglionic volleys are classed by the foregoing criteria as nicotinic receptors. However, the end-plate and ganglionic nicotinic receptors are not identical molecules, since they react somewhat differently to a number of pharmacological agents. For example, ganglionic transmission is readily blocked by tetraethylammonium and hexamethonium, drugs that have much less effect on neuromuscular

Table 34–1 Autonomic Receptor Types

ACh Receptors

Type	Agonists	Antagonists	Location
Muscarinic	Muscarine Pilocarpine Carbamylcholine	Atropine Scopolamine	All parasympathetic effector systems
Nicotinic	Nicotine Carbamylcholine Metacholine	Curare Tetraethyl-ammonium ions Hexamethonium Pentolinium	Skeletal muscle end-plate autonomic ganglion cells

Adrenergic Receptors

Type	Agonists	Antagonists	Location
Alpha	Norepinephrine Phenylephrine Methoxamine	Phenoxybenzamine Phentolamine	Most sympathetic effectors that are excited (except heart)
Beta	Isoproterenol Methoxyphenamine	Propranolol	Most sympathetic effectors that are inhibited and heart

transmission.[50] Conversely, cobra neurotoxin and α-bungarotoxin are powerful blockers of neuromuscular transmission but have little effect on ganglionic receptors.[44, 55]

The muscarinic receptor is characterized by a prolonged response to acetylcholine that may last for seconds (Figs. 34–3 and 34–4); the response may be accompanied by either excitation or inhibition, depending on the organ. Muscarinic receptors are insensitive to curare but are readily blocked by atropine. Examples of muscarinic cholinergic receptors are those in cardiac muscle, exocrine glands, and smooth muscle as well as many in the central nervous system.

Similarly, there exist at least two classes of adrenergic receptors that are distinguished by their distinctive responses to various pharmacological blocking agents[21] (Table 34–1). These receptors are named α and β receptors. Alpha receptors are found in vascular smooth muscle (vasoconstriction), pupillary radial (dilator) muscle, sphincter muscles of gut and bladder (constriction), and in the spleen (contraction). They are blocked by phenoxybenzamine, phentolamine, and the ergot alkaloids. Beta receptors are selectively blocked by propranolol. They are responsible for the adrenergic acceleration of heart rate and the increase of strength of cardiac contraction and for the inhibition of gastrointestinal motility during sympathetic discharge. Beta receptors are often divided into subcategories on the basis of varying sensitivity to blocking agents. Recent work has shown that the adrenergic receptor subtypes, defined pharmacologically, also correspond to profound differences in their effects on internal second messenger systems. Thus α receptors activate phospholipase C, whereas β receptors activate adenylate cyclase.

Some cell membranes are adorned with a surprisingly complex mixture of receptor molecules. Sympathetic ganglion cells, for example, have both nicotinic and muscarinic acetylcholine receptors as well as an α-adrenergic receptor.[42, 54, 59] Acetylcholine elicits both a "fast" EPSP (1.5 to 2.0 ms synaptic delay) and a "slow" EPSP that has a synaptic delay of 200 to 300 ms and a duration of up to 30 seconds.[15, 16, 42, 43] The fast response is blocked by curare; the slow response, which is accentuated by repetitive presynaptic activation, is blocked by atropine. The adrenergic receptor is activated by epinephrine, norepinephrine, or dopamine (a precursor of norepinephrine) and gives rise to a slow IPSP (synaptic delay, 30 to 100 ms); it is blocked by dibenamine, an α-adrenergic blocking agent.[42] The immediate input to adrenergic sites appears to be the ganglionic chromaffin cells,

also known as small, intensely fluorescent cells (SIF cells) because they fluoresce brightly when the ganglion is treated with formaldehyde.[36] These cells, which serve as ganglionic interneurons between preganglionic fibers and ganglion cells, have muscarinic cholinergic receptors, but their axons release catecholamines, mainly dopamine,[5, 42] which is responsible for the late IPSP. The functional significance of slow potentials is not entirely clear, but they probably play a role in modulation of ganglionic transmission during repetitive preganglionic input. It has been suggested[42, 43] that the SIF cells may be part of a wiring scheme for reciprocal innervation comparable to the inhibitory interneurons in spinal reflex pathways. The ganglion cell innervation, which is discussed further in a later section, is shown diagrammatically in Figure 34–11.

Synthesis and Storage of Autonomic Neurotransmitters.[4, 12, 23, 24, 31, 38] The major points of ACh synthesis are summarized diagrammatically in Figure 34–5. Acetylcholine is synthesized intracellularly from choline and acetyl CoA; the reaction is catalyzed by choline acetyltransferase (CAT):

$$Choline + acetyl\ CoA \xrightarrow{CAT} ACh + CoA$$

Choline acetyltransferase is synthesized in the cell soma and reaches the terminals, where acetylcholine synthesis occurs, by axoplasmic transport.[32] Choline is derived from extracellular fluid, much of it from acetylcholine liberated during activity and hydrolyzed in the cleft by acetylcholine esterase. The uptake of choline by endings is accomplished by a carrier-mediated high-affinity uptake system that is dependent on Na^+ and ATP. The uptake system normally recovers for recycling about 60% of the choline released by hydrolysis.[11] The uptake mechanism can be blocked by the drug hemicholinium; then the synthesis drops drastically, ACh stores rapidly become depleted, and ganglionic transmission fails. Choline is the rate-limiting factor in ACh synthesis.

Acetyl CoA is synthesized within the mitochondria almost entirely from pyruvate, where it then condenses with oxaloacetate to give citrate. Citrate leaves the mitochondrion by a permease and is used to resynthesize acetyl CoA in the cytoplasm. Cytoplasmic choline acetyltransferase can then synthesize ACh.[31] It is not completely understood how cytoplasmic ACh becomes packaged in the synaptic vesicles. The steps in the release of ACh at autonomic junctions by action potentials are believed to be the same as those described in

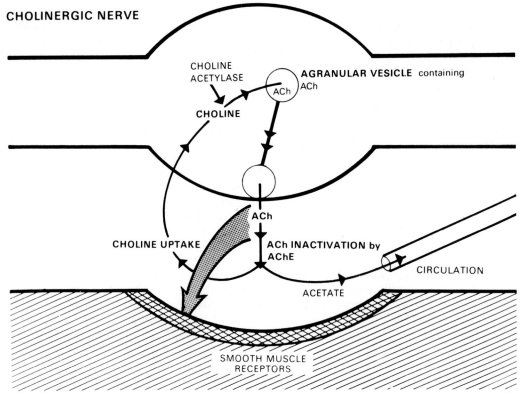

CHOLINERGIC NERVE

Figure 34–5 Diagram of the synthetic cycle in cholinergic nerve endings. ACh—acetylcholine; AChE—acetylcholinesterase. (From Burnstock. *Pharmacological Reviews*, Vol. 24:509, © 1972 by Am. Soc. for Pharm. & Exp. Ther.)

Chapter 6 for somatic motor endings, i.e., depolarization, Ca^{2+} entry, and exocytosis.

The transmitters in sympathetic postganglionic fibers are catecholamines, chiefly norepinephrine, and dopamine. Although the adrenal medulla carries the synthesis one step further to epinephrine, this substance is found in only small quantities in nerve endings. Sympathetic endings on smooth muscle effectors lack clearly defined end-plates of the kind found in skeletal muscle; rather, the transmitter is concentrated in a series of nodular varicosities that occur at 3 to 5 μm intervals along the terminal fiber as it courses in close approximation to the smooth muscle cells. The varicosities are packed with mitochondria and vesicles, many of which contain an electron-dense material. These membrane-bound dense core vesicles, which are readily distinguished from the clear ACh-containing vesicles of motor nerve fibers and cholinergic autonomic endings, are believed to be packages of catecholamines. Labeled norepinephrine injected intravenously accumulates within 30 minutes in dense core vesicles.[79] The localization of

catecholamines in the varicosities can also be visualized in freeze-dried tissues treated with formaldehyde, a procedure that triggers condensation of catecholamines to form fluorescent compounds. This technique has proved of great value in tracing adrenergic pathways both in the central and peripheral nervous system, in determining details of adrenergic innervation of receptors and as an assay of the regional distribution of catecholamines. Sympathetic postganglionic neurons and axons fluoresce only moderately, but the terminals—and especially the varicosities—glow brilliantly like tiny lights on a wire. The vesicles seem to be manufactured in the soma and transported to the varicosities by axoplasmic transport; when postganglionic nerves are compressed by a ligature, dense-core fluorescent vesicles accumulate proximal to compression as if they were being dammed up.[13] However, since the calculated rate of transport is only 5 to 10 mm a day, it seems unlikely that this transport mechanism plays a significant role in replenishing transmitter lost during neural action; it may supply the protein structure of the vesicle, whereas the synthesis of transmitter during nerve action occurs in the varicosities and is then stored in the vesicles.

The synthesis of norepinephrine is more complex than that of acetylcholine, several alternate pathways being available (see also Chaps. 59 and 65). The multistep enzyme-controlled synthetic process is a favored target of pharmacological manipulation and provides the adventuresome drug-oriented physician with multiple opportunities to exercise his talents.[12] The starting substance for norepinephrine synthesis (Fig. 34–6) is tyrosine, which is derived from dietary protein or from conversion of dietary phenylalanine in the liver. Adrenergic nerve endings actively take up tyrosine from body fluids. The first step in the synthetic process is hydroxylation of tyrosine to form dihydroxyphenylalanine, or DOPA; this reaction is catalyzed by the cytoplasmic enzyme tyrosine hydroxylase. Hydroxylation is the rate-limiting step in synthesis, and inhibitors of tyrosine hydroxy-

lase cause significant depletion of norepinephrine stores. Although the enzyme is found in cell bodies, it is most concentrated in the cytoplasm of the terminals, another reason for doubting that axoplasmic transport of transmitter from cell body to terminal is a major or essential factor in maintaining transmitter stores. The activity of tyrosine hydroxylase is inhibited by catecholamines, so that the rate of synthesis is regulated by the cytoplasmic level of norepinephrine. During neural activity that would otherwise deplete neuronal stores of transmitter, the enzyme activity accelerates, synthesis increases, and transmitter stores are maintained.[75]

The remaining steps in the synthesis are shown in Figure 34–6. All of the enzymes involved are found in the cytoplasm except for dopamine β-hydroxylase, which is mainly confined in the granular vesicles along with the end product, norepinephrine. The granules also contain ATP and a protein, chromagranin. The outer mem-

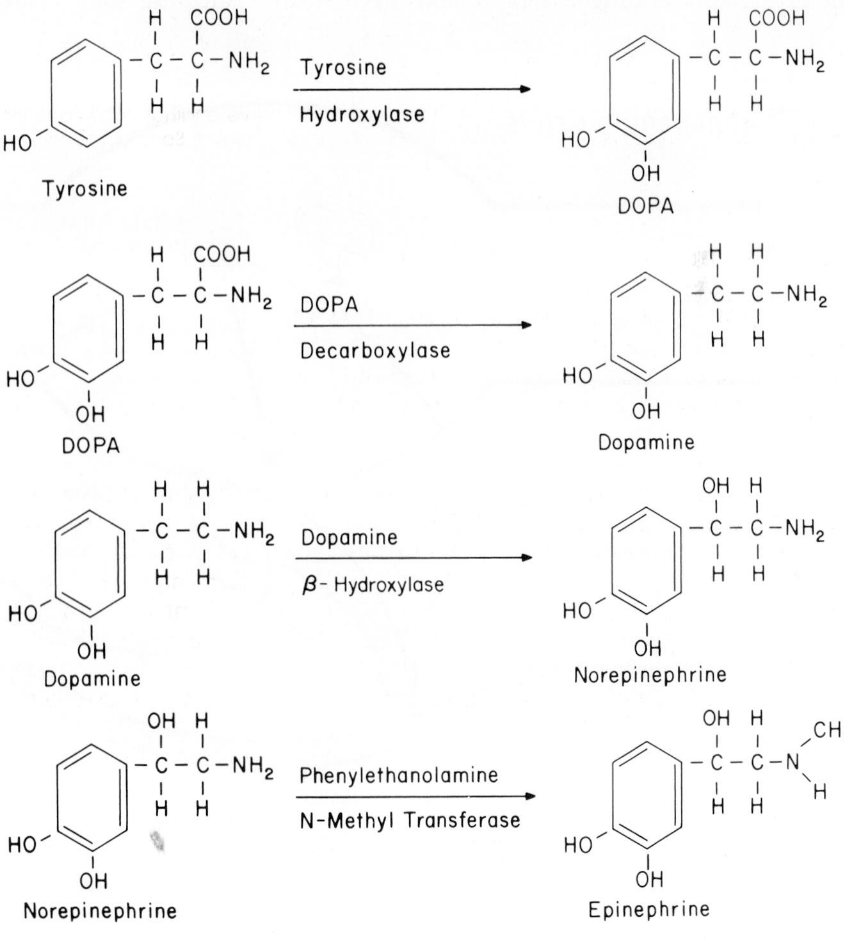

Figure 34–6 Steps in the synthesis of catecholamines.

branes of the mitochondria contain an enzyme, monoamine oxidase (MAO), that degrades dopamine and norepinephrine, so that their incorporation in the vesicle is important in preventing the destruction of transmitter by MAO. The drug reserpine blocks the uptake of norepinephrine by the vesicles, an action that leaves the transmitter exposed to the destructive enzyme; the result is transmitter depletion and blockade.

Release of transmitter from adrenergic nerve endings is presumably by a Ca^{2+}-mediated process of exocytosis similar to the transmitter release mechanism of other nerve fibers. Since the process of exocytosis presupposes a dumping of the vesicular contents into the junctional cleft, it is perhaps not surprising that neural action liberates not only norepinephrine but also ATP and dopamine β-hydroxylase in proportions similar to those found in intact vesicles.[25, 26, 76] This somewhat untidy and prodigal discharge not only of the product but also of some of the synthetic tools is a strong argument for exocytosis.

Released norepinephrine is largely recovered by the presynaptic endings. The reuptake mechanism is apparently coupled to the Na^+ gradient since it fails in either low Na^+ or when the Na-K pump is blocked by ouabain. Reuptake is also competitively blocked by cocaine, an action that explains its well-known potentiation of the consequences of sympathetic nerve discharge. The reuptake mechanism is probably the main mechanism for limiting the duration of action of released norepinephrine. Although some released transmitter is degraded by an extraneuronal enzyme, catechol-*O*-methyl transferase (COMT), the uptake mechanism recovers intact about 70% of the released transmitter.[41] The recovered molecules are resequestered in vesicles and can be reused. Injected exogenous norepinephrine that has been labeled appears within 30 minutes in microsomal fractions of sympathetic terminals containing granular vesicles.[27, 61] The main features of norepinephrine synthesis, storage, and release are illustrated diagrammatically in Figure 34–7.

Development of Transmitter Specificity in Sympathetic Neurons.[56, 58] In view of the intricacy of the transmitter synthetic machinery, one might suppose that its development is genetically pre-

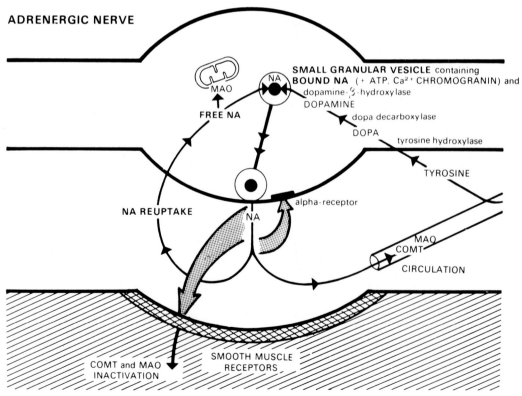

Figure 34–7 Diagram of the synthetic cycle in adrenergic nerve endings. COMT—catechol-O-methyl transferase; MAO—monamine oxidase; NA—norepinephrine or noradrenaline. (From Burnstock. *Pharmacological Reviews*, Vol. 24:509, © 1972 by Am. Soc. for Pharm. & Exp. Ther.)

determined and unalterable. In keeping with this prediction, sympathetic ganglion cells isolated from the superior cervical ganglion of the newborn rat and grown in pure tissue culture[58] begin within a few days to synthesize norepinephrine and dopamine from labeled tyrosine added to the medium. At the same time, the cultured cells display a rapidly developing capacity to accumulate labeled norepinephrine from the medium, a property that can be blocked with cocaine. The stored label appears in fractions of homogenized tissue containing granular or dark-core vesicles. Treatment of the cultures with reserpine does not alter uptake into the cells but as in vivo, prevents incorporation of norepinephrine into the vesicles; the storage life of incorporated catecholamines is markedly reduced, since the extravesicular material is available to mitochondrial monoamine oxidase and is destroyed. Finally, cells isolated in culture release stored catecholamines when they are depolarized (by elevating extracellular K^+). It thus appears that the ability to synthesize, to take up, and to store norepinephrine is an innate property of sympathetic ganglion cells that is expressed in the absence of any modulating influence from other tissue components.

Despite these seeming manifestations of metabolic predestination, appropriate manipulation of the culture conditions alters or even reverses the propensity of ganglion cells to synthesize catecholamines and converts them to effective factories for production of acetylcholine.[57] When non-neuronal cells, such as cardiac myocytes, are added to the culture, the developing neurons promptly (48 hours) display choline acetyltransferase activity and acetylcholine synthesis, while DOPA-decarboxylase activity and catecholamine production drop. The shift is accompanied by a decrease in the numbers of dense-core vesicles and appearance of clear vesicles resembling those found in cholinergic terminals.[57] In the mixed culture preparations, the neurons form functional cholinergic synapses with one another,[56] a phenomenon that is absent in the isolated cell preparations. Occurrence of synaptogenesis between ganglion cells having nicotinic cholinergic receptors is in keeping with earlier observations that cholinergic, but not adrenergic, fibers can innervate structures endowed with ACh receptors. Functional junctions between neurons and myocytes also occur. Cardiac myocytes, like cardiac muscle cells, have both adrenergic (excitatory) and cholinergic (inhibitory) receptors. Studies of junctional transmission in neuronal-myocyte cultures indicate that discharge of some neurons hyperpolarizes, while discharge of other neurons depolarizes, their target myocytes. The former are blocked by atropine, the latter by β-adrenergic blocking agents. In still other cells, stimulation produces an atropine-sensitive hyperpolarization followed by a propranolol-sensitive depolarization, suggesting that these neurons synthesize and release both acetylcholine and norepinephrine.[22]

The critical factor in inducing cholinergy is chemical, for a cell-free medium in which nonneural cells (fibroblasts, cardiac myocytes, skeletal muscle) have been grown is capable of initiating the shift. The chemical messenger is a protein that has not yet been identified. It thus appears that the developing cell is potentially capable of developing either cholinergic or adrenergic synthetic mechanisms, the final decision depending on environmental factors. The freedom of choice is itself variable; cultures of explants from rats older than 3 weeks are adrenergic[65] and resistant to alteration by addition of non-neural cells or their products. The susceptibility of neurons to cholinergic induction is also diminished by inducing them to discharge impulses either by electrical stimulation or by depolarizing them with elevated extracellular K^+. When the cell-free medium taken from heart or skeletal muscle cultures is added to "exercised" ganglionic neuronal cultures, the cells remain primarily adrenergic. The refractoriness of the activated neurons depends on extracellular Ca^{2+} during the stimulation period.[72]

The sweat glands are a unique effector system in which a reversal of transmitter identity during development has been observed in vivo.[40a] We have already noted that the sympathetic neurons innervating the sweat glands (sudomotor neurons) are perversely cholinergic and that sweat glands have muscarinic ACh receptors. Accordingly, sudomotor fibers in adult animals display strong AChE staining but do not fluoresce when treated with formaldehyde as do noradrenergic fibers. In rats, the sweat glands and their innervation develop postnatally, providing investigators a unique opportunity to follow the developmental history of the exceptional transmitter identity of the sudomotor innervation. In newborn (7 days) rats, the sympathetic fibers in the footpad (where the glands are due to develop) are clearly noradrenergic; they are filled with small granular vesicles that fluoresce brilliantly when treated with formaldehyde. AChE staining is absent. By 14 days, the glands reach a mature state, and coincidentally, AChE becomes pronounced while the number of small granular vesicles and the intensity of the catecholamine fluorescence are reduced. By

21 days, the glands and their innervating fibers are mature and granular vesicles and fluorescence are absent. It thus appears that the sympathetic postganglionic neurons destined to innervate the sweat glands are at the outset noradrenergic and only develop their nonconformist cholinergy as the glands develop and innervation is completed.

Action of Autonomic Transmitters on Their Effectors.[8, 33, 59] Some of the techniques that have proved so effective in unraveling the mechanisms of synaptic and junctional transmission in the somatic nervous system are less easily performed, and even less easily interpreted, in autonomic neuroeffector systems. The small size of smooth muscle cells makes them difficult (but not impossible) to explore with intracellular microelectrodes. Many smooth muscle and gland cells are electrically coupled one to another by gap junctions or other low-resistance pathways so that sizable portions of the organ behave as functional syncytia. Coupling is a frustrating impediment to the investigator who attempts to use voltage clamps or to determine reversal levels for junctional potentials. These difficulties have been most notably circumvented in cardiac muscle, from which intracellular records can be readily obtained.[35] The special problem of maintaining successful intracellular records from a spontaneously beating heart, the movements of which tend to dislodge the electrode or damage the impaled cell, was solved by the ingenious "dangle" electrode technique of Woodbury and Brady[80] who mounted the capillary pipette on a fine flexible wire. When the electrode is placed on the cardiac surface, the motion works it through the cell membrane, after which the electrode rides freely with the movement.

In the spontaneously beating heart, the beat originates in the sinoatrial node, a small nodule of specialized tissue in the wall of the right atrium. Intracellular recordings from this tissue show rhythmically recurring slow depolarizations (the "pacemaker" potential). When depolarization proceeds to threshold, an action potential is generated (Fig. 34–8A). No pacemaker potential is seen in recordings from atrial fibers (Fig. 34–8B). The pacemaker depolarization probably results from a gradual rise of Ca^{2+} permeability and possibly a decrease of K^+ permeability during diastole (see Chap. 37).[70]

Figure 34–8C shows intracellular recordings from pacemaker tissue before, during, and after vagal stimulation. During the stimulation period, the membrane became hyperpolarized, and the rhythmically recurring depolarizations were abolished. With cessation of vagal stimulation, the pacemaker cell slowly depolarized as the transmitter action decayed, until threshold was reached and an action potential was generated. During the first few beats, the rate of rise of the pacemaker potential was slow, so that the heart rate remained depressed; also, the duration of the action potential was curtailed. In subsequent beats, the recorded potentials gradually resumed the prestimulation configuration.

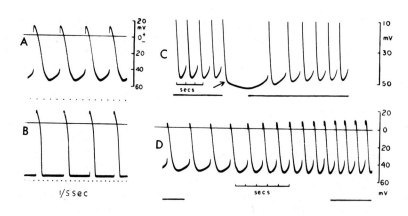

Figure 34–8 Intracellular recordings from frog atrial cells beating normally during vagal inhibition and during sympathetic acceleration. *A,* Record from pacemaker cell in normally beating heart; note pacemaker potential (slow depolarization preceding each beat). *B,* Record from atrial fiber in normally beating heart; note absence of pacemaker potential (flat baseline preceding spike). *C,* Records from pacemaker cell during vagal stimulation; gain is high so that peaks of action potentials and zero reference lines are not shown. During vagal stimulation, indicated by break in bottom line, cell becomes hyperpolarized, and pacemaker potentials and spikes are in abeyance. Note decreased slope of pacemaker potential in first two beats after recovery. *D,* Records from pacemaker cell during sympathetic stimulation, indicated by break in bottom line. Slope of pacemaker potential increases, and rate accelerates. (Reproduced from Hutter and Trautwein, *The Journal of General Physiology,* 1956, 39:715–733, by copyright permission of The Rockefeller University Press.)

The hyperpolarization during vagal stimulation reflects an increased permeability of the pacemaker membrane to K^+, so that the membrane potential is driven toward the K^+ equilibrium potential. In accord with this hypothesis, acetylcholine increases total membrane conductance[68] and increases K^+ fluxes[34] in atrial pacemaker cells. Also in accord with this interpretation is the shortening of the action potential during vagal inhibitory action (note the first spike after the period of arrest in Fig. 34–8C). This shortening is a manifestation of rapid repolarization that in pacemaker tissue, as in nerve and skeletal muscle cells, is the consequence of increased permeability of the membrane to K^+.

Figure 34–8D shows the effect of sympathetic stimulation on potentials recorded intracellularly from a pacemaker cell. The firing level remains constant, but the slope of the pacemaker potential increases. As a result, the threshold voltage is reached more rapidly, and the rate of firing increases accordingly. Simultaneously, the "overshoot" of the spike increases, so that the overall amplitude of the action potential is greater. These events are satisfactorily explained if it is assumed that the sympathetic transmitter increases the number of Ca channels opening during diastole.[69]

Intracellular recordings from autonomic effectors have been successfully achieved in a variety of other tissues;[8, 33, 47] some examples are shown in Figure 34–9. In general, inhibitory actions are accompanied by hyperpolarization and excitation by depolarization; an exception is the salivary gland, in which stimulation of the parasympathetic secretory fibers of the chorda tympani or the sympathetic secretory fibers produces hyperpolarization.[46] In some effectors, e.g., the vas deferens and the retractor penis muscle, small spontaneous depolarizations at rest have been recorded;[4, 8, 33] these are presumed to be mediated by spontaneous quantal release of transmitter comparable to the miniature end-plate potentials of skeletal muscle. No generalizations can be made about the ionic mechanism of junctional potentials; some appear to be the result of development of selective ionic channels, but others may be the consequence of metabolic changes that affect ionic pumps.[42, 47]

Junctional potentials of smooth muscle often have exceedingly long latent periods and long durations relative to their counterparts in skeletal muscle. Responses of smooth muscle to parasympathetic stimulation usually have delays of 100 ms or more and last up to a second.[4, 8] Responses to sympathetic volleys have a somewhat livelier

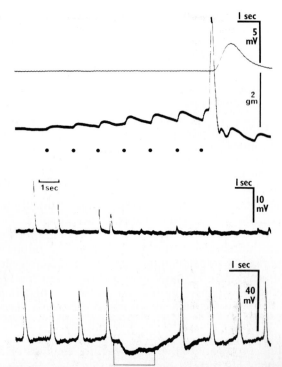

Figure 34–9 Junctional potentials in smooth muscle. *Above,* Intracellularly recorded responses of guinea pig vas deferens to stimulation of postganglionic sympathetic nerves (black dots). The upper trace shows the mechanical response. Note slow time scale. (After Burnstock. In *The Peripheral Nervous System,* J.I. Hubbard, ed., New York, Plenum Press, 1974). *Center,* Intracellular records of spontaneous activity of guinea pig vas deferens. The smaller potentials probably represent spontaneous release of packets of norepinephrine. (After Burnstock and Holman, *J. Physiol. [Lond.]* 160:446–460, 1962.) *Below,* Intracellular records from guinea-pig taenia coli muscle. Spontaneous action potentials were blocked, and the cell hyperpolarized during stimulation of sympathetic fibers at 60 pulses per second between arrows. (After Bennett et al., *J. Physiol. (Lond.),* 1966, 182, 527–540.)

tempo, but even these are slow compared to skeletal muscle or to the nicotinic responses of ganglion cells. The sluggishness of effector responses to autonomic nerve volleys is one reason to suppose that transmission in these systems may involve additional time-consuming steps over and above those accounting for the much shorter delays in nerve-skeletal muscle junctions. Considerations of this kind, coupled with studies of the slowly developing but enduring effects of circulating hormones on their target cells,[64] have led to the concept of internal second messengers in synaptic and neuroeffector transmission. The recent progress in this field makes necessary a brief summary of the concept of second messengers and the experimental data supporting it.

Second Messengers; the Cyclic Nucleotides in Junctional Transmission.[28, 29, 53] At many synapses and neuroeffector junctions, the receptor molecule in the postjunctional membrane becomes, when combined with transmitter, a channel or pore through which one or another ion can move; the altered ionic permeability affects the excitability of the postsynaptic cell, leading to either excitation or inhibition. Greengard[28] calls this the "receptor-ionophore" model, because the receptor itself becomes the ionophore when combined with transmitter. Examples are motoneuron postsynaptic potentials, skeletal muscle end-plate potentials, and the nicotinic cholinergic response of sympathetic ganglion cells described earlier in this chapter. The "receptor–second messenger" model proposes that the receptor is not an ionophore but rather a membrane-transmembrane protein that couples to intracellular GTP binding proteins. When the receptor is activated by the transmitter, it triggers a series of intracellular reactions to produce a "second messenger" that in turn can activate membrane enzymes that alter membrane proteins, often producing either permeability changes or a change in membrane pumps. Such an indirect mechanism understandably has a longer latency. Further, because the termination of the postsynaptic effect depends on degradation of the intra-cellular second messenger and reversal of the membrane changes, such effects may have long durations, often measured in seconds rather than in milliseconds.

Attention has focused on the cyclic nucleotide adenosine 3′,5′-monophosphate (cyclic AMP) as a second messenger in nerve and smooth muscle cells, because this substance accumulates intracellularly following repetitive prejunctional neural activity or when the postsynaptic elements are treated with exogenous transmitter. Formation of cyclic AMP from 5′ AMP and ATP is catalyzed by a membrane-bound enzyme, adenylate cyclase, which is activated by the transmitter; significantly, antidromic stimulation of the postsynaptic element does not stimulate cyclic AMP production. Cyclic AMP activates another enzyme, protein kinase, which phosphorylates membrane proteins; it is this action that presumably alters membrane permeability or ion pumps. Termination of the process depends on degradation of cyclic AMP by phosphodiesterase and reversal of the protein phosphorylation by a membrane-bound phosphatase. The entire sequence of events is shown in Figure 34–10.

In some cases, the second messenger is guanosine 3′,5′-monophosphate, and the regulated enzyme is guanosine cyclase rather than adenosine

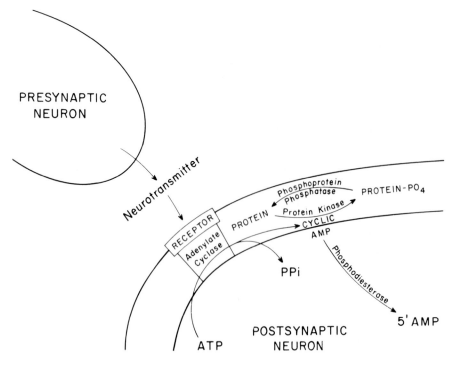

Figure 34–10 Diagram of second messenger hypothesis involving cyclic AMP. Phosphorylation of membrane proteins is postulated to change membrane permeability or to alter metabolic pumps. (From Greengard, P. *Adv. Cyclic Nucleotide Res.* 5:585–602, 1975, with permission from Raven Press, New York.)

cyclase. In yet other cases, the second messengers are the products of phosphatidyl inositide turnover by the enzyme phospholipase C. Cells may have more than one of these enzymes, which may be sensitive to different transmitters and thus result eventually in phosphorylation of different proteins with different effects on cell function. This complex enzymatic control of membrane function is well exemplified in sympathetic ganglion cells. As described earlier, ganglion cell membranes have three receptors: a curare-sensitive cholinergic nicotinic receptor responsible for the early EPSP, an atropine-sensitive muscarinic cholinergic receptor responsible for the late EPSP, and an α-adrenergic (probably dopaminergic) receptor responsible for the late IPSP.[42] The early EPSP rather clearly depends on an ionophore receptor mechanism, but Greengard and Kebabian[30] have presented evidence that the late potentials depend on cyclic nucleotide mechanisms. The late EPSP is mediated by a muscarinic cholinergic mechanism involving unknown internal second messengers and the late IPSP is mediated by a dopaminergic cyclic AMP mechanism.[3, 30, 37] These relationships are represented diagrammatically in Figure 34–11.

A role for cyclic nucleotides has been postulated in a variety of neural and neuromuscular events, including central synaptic transmission,[7] chronotropic regulation of cardiac muscle,[69, 71] transmitter release from presynaptic fibers,[74] neurotransmitter synthesis[52] and microtubular function.[67] The evidence supporting these manifold proposed functions is variable and not always complete.[3] In fact, the protean manifestations of cyclic nucleotides are a surprise to physiologists, for they bespeak a degree of nonspecificity that complicates establishing their direct role in any one specific function. Another problem is the difficulty of reconciling the time course of biochemical and pharmacological events with the precise and rigid temporal requirements for junctional transmission established by electrophysiological methods. It seems likely that second messengers may be involved less in the immediate mechanisms of synaptic transfer than in slow modulation of receptor-ionophore mechanisms in accordance with the past history of the junction.[73] The second messenger systems are also pivotal in the actions of endocrine hormones treated later (see Chap. 59ff).

DENERVATION SUPERSENSITIVITY[10, 19]

When an autonomic effector is denervated, it becomes increasingly sensitive to chemical agents. This sensitivity is most pronounced when the organ is directly denervated by section of its *postganglionic* nerves (Cannon's law of denervation). Such denervation supersensitivity was first described by Budge, who produced Horner's syndrome in rabbits. *Horner's syndrome*, which results

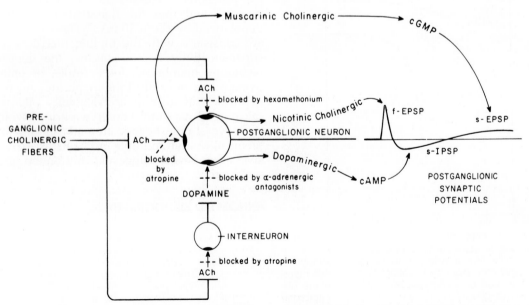

Figure 34–11 Diagrammatic representation of postulated events in sympathetic ganglion transmission. The dopaminergic interneurons are the SIF cells of the ganglion. (From Greengard and Kebabian, *Fed. Proc.* 33:1059–1061, 1974.)

from interruption of the sympathetic supply to the face, consists of pupillary constriction (miosis), drooping of the eyelid (ptosis), and flushing of the face owing to loss of vasoconstrictor tone.

Since the postganglionic sympathetic supply to the face originates in the superior cervical ganglion, which receives a preganglionic sympathetic input from the fibers ascending the cervical chain, Horner's syndrome may be experimentally produced either by dividing the cervical chain to interrupt the preganglionic fibers, or by a transection of the postganglionic fibers emerging from the ganglion. Budge found that, when a preganglionic section on one side and a postganglionic section on the other were performed, the resultant Horner's syndrome was initially symmetrical bilaterally. With the passage of time, however, the pupil on the side of the postganglionic denervation was larger than the one on the preganglionically denervated side, and the discrepancy was intensified when the animal was frightened or subjected to emotional excitement. Budge could not explain the phenomenon of the paradoxical pupil; but it is now known that denervation supersensitivity to circulating epinephrine (released into the blood stream during emotional excitement) accounts for the paradoxical pupil as well as for a number of similar phenomena in other denervated organs.

A quantitative study of denervation supersensitivity is illustrated in Figure 34–12. The response

of the nictitating membrane to a standard dose of epinephrine was measured on successive days following postganglionic denervation on the right side. Both nictitating membranes underwent a gradually increased sensitivity to epinephrine as evidenced by the amplified responses, but the sensitivity was much more prominent in the membrane postganglionically denervated. On the 14th day, the right superior cervical ganglion was removed, and the sensitivity of the related membrane increased, approaching that displayed by the left membrane. If denervation is caused by crushing of the nerves so that they may regenerate, supersensitivity occurs but wanes as the regeneration fibers reestablish connections with the muscle cells.

The mechanism of denervation supersensitivity is not entirely clear. It may be partly due to proliferation of receptor molecules in denervated cells comparable to the proliferation of acetylcholine receptors in denervated skeletal muscle (Chap. 6). In sympathetic ganglion cells of the heart, sensitivity to iontophoretically applied acetylcholine is confined to the subsynaptic membrane, but within a few days after denervation, the entire neuronal surface becomes highly sensitive.[39] Another factor believed important is the loss of uptake mechanisms accompanying degeneration of presynaptic fibers along with, in some instances, loss of enzymes that destroy transmitter, both being factors that might be expected to potentiate responses to circulating transmitter. It has been reported that the number of nexuses or gap-junctions increases in denervated smooth muscle[77] and that the denervated cells become depolarized by about 10 mV.[20] These observations are consistent with the finding that denervation supersensitivity is not specific; the denervated nictitating membrane, for example, becomes supersensitive not only to epinephrine and norepinephrine but also to acetylcholine, pilocarpine, calcium, potassium, norepinephrine tyramine, and several nonphenolic aromatic amines. It thus seems likely that multiple mechanisms are responsible as originally suggested by Cannon and Rosenblueth.[10]

ANNOTATED BIBLIOGRAPHY

Bennett, H.R. *Autonomic Neuromuscular Transmission.* New York, Cambridge University Press, 1972.
  Synaptic transmission in the autonomic nervous system.
Burnstock, G. The changing face of autonomic neurotransmission. *Acta Physiol. Scand.* 126:67–91, 1986.
  A stimulating introduction to the variety of cotransmitters and nonadrenergic, noncholinergic transmission in the autonomic nervous system.

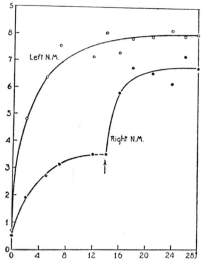

Figure 34–12 Contractile responses of denervated nictitating membrane to epinephrine. *Ordinates,* Amplitude of isotonic contraction (cm) in response to 10 μg of epinephrine. *Abscissa,* Days after initial denervation. At day 0, left membrane denervated postganglionically, right membrane denervated preganglionically. At day 14 *(arrow),* right membrane denervated postganglionically. (From Hampel, *Am. J. Physiol.* 111:611–621, 1935.)

Geffen, L.B.; Jarrot, B. Cellular aspects of catecholaminergic neurons. *Handbk. Physiol. 1(1):521–571, 1977.*
A thorough review of the biochemistry and biophysics of catecholamine synthesis and release from autonomic neurons.

Greengard, P. *Cyclic Nucleotides, Phosphorylated Proteins, and Neuronal Function.* New York, Raven Press, 1978.
The chemistry and physiology of second messengers.

Holman, M.E.; Hirst, G.D.S. Junctional transmission in smooth muscle and the autonomic nervous system. *Handbk. Physiol.* 1(1):417–461, 1977.
A review of autonomic synaptic transmission; an excellent bibliography.

Pick, J. *The Autonomic Nervous System.* Philadelphia, J.B. Lippincott, 1970.
Anatomy of autonomic nervous system in mammals and lower forms.

REFERENCES

1. Alpern, M. Accommodation. In Davson, H. *The Eye.* vol. 3. New York, Academic Press, 1962.
2. Barcroft, H.; Swan, H.S.C. *Sympathetic Control of Human Blood Vessels.* Baltimore, Williams & Wilkins, 1953.
3. Beam, K.G.; Greengard, P. Cyclic nucleotides, proteins and synaptic function. *Cold Spring Harbor Symp. Quant. Biol.* 40:157–168, 1976.
4. Bennett, H.R. *Autonomic Neuromuscular Transmission.* Cambridge, Cambridge University Press, 1972.
5. Björklund, A.; Cegrell, L.; Falck, B.; Ritzén, M.; Rosengren, E. Dopamine-containing cells in sympathetic ganglia. *Acta Physiol. Scand.* 78:334–338, 1970.
6. Blair, D.A.; Glover, W.E.; Greenfield, A.D.M.; Roddie, I.C. Excitation of cholinergic vasodilator nerves to human skeletal muscles during emotional stress. *J. Physiol. (Lond.)* 148:633–647, 1959.
7. Bloom, F.E. Cyclic nucleotides in central synaptic function. *Fed. Proc.* 38:2203–2207, 1979.
8. Burnstock, G. The co-transmitter hypothesis, with special reference to the storage and release of ATP with noradrenaline and acetylcholine. In Cuello, A.C., ed. *Co-transmission,* 151–163. Macmillan Press, London, 1982.
8a. Burnstock, G. The changing face of autonomic neurotransmission. *Acta Physiol. Scand.* 126:67–91, 1986.
9. Cannon, W.B.; Rosenblueth, A. *Autonomic Neuro-effector Systems.* New York, Macmillan, 1937.
10. Cannon, W.B.; Rosenblueth, A. *The Supersensitivity of Denervated Structures.* New York, Macmillan, 1949.
11. Collier, B.; Katz, H.S. Acetylcholine synthesis from recaptured choline by a sympathetic ganglion. *J. Physiol. (Lond.)* 238:639–655, 1974.
12. Cooper, J.A.; Bloom, F.E.; Roth, R.H. *The Biochemical Basis of Neuropharmacology,* 5th ed., New York, Oxford University Press, 1986.
13. Dahlström, A. Observations on the accumulation of noradrenaline in the proximal and distal parts of peripheral adrenergic nerves after compression. *J. Anat. (Lond.)* 99:677–689, 1965.
14. Dale, H.H. The action of certain esters and ethers of choline, and their relation to muscarine. *J. Pharmacol. Exper. Ther.* 6:147–190, 1914.
15. Eccles, R.M. Intracellular potentials recorded from a mammalian sympathetic ganglion. *J. Physiol. (Lond.)* 130:572–584, 1955.
16. Eccles, R.M.; Libet, B. Origin and blockade of the synaptic responses of curarized sympathetic ganglia. *J. Physiol. (Lond.)* 157:484–503, 1961.
17. Eliasson, S.; Lindgren, P.; Uvnäs, B. Representation in the hypothalamus and the motor cortex in the dog of the sympathetic vasodilator outflow to the skeletal muscles. *Acta Physiol. Scand.* 27:18–37, 1953.
18. Emmelin, N. Nervous control of salivary glands. *Handbk. Physiol.* Sec. 6, 2:595–632, 1967.
19. Fleming, W.W. Supersensitivity in smooth muscle. *Fed. Proc.* 34:1969–1970, 1975.
20. Fleming, W.W.; Urquilla, P.R.; Taylor, D.A.; Westfall, D.P. Electrophysiological correlations with postjunctional supersensitivity. *Fed. Proc.* 34:1981–1984, 1975.
21. Furchgott, R.F. Pharmacological characteristics of adrenergic receptors. *Fed. Proc.* 29:1352–1361, 1970.
22. Furshpan, E.J.; McLeish, P.R.; O'Lague, P.H.; Potter, D.D. Chemical transmission between rat sympathetic neurons and cardiac myocytes developing in tissue cultures: evidence for cholinergic, adrenergic and dual-function neurons. *Proc. Nat. Acad. Sci.* 73:4225–4229, 1976.
23. Geffen, L.B.; Jarrot, B. Cellular aspects of catecholaminergic neurons. In *Handbk. Physiol.* 1(1):521–571, 1977.
24. Geffen, L.B.; Livett, B.C. Synaptic vesicles in sympathetic neurons. *Physiol. Rev.* 51:98–157, 1971.
25. Geffen, L.B.; Livett, B.G.; Rush, R.A. Immuno-histochemical localization of protein components of catecholamine storage vesicles. *J. Physiol. (Lond.)* 204:593–605, 1969.
26. Gewirtz, G.P.; Kopin, I.J. Release of dopamine-beta-hydroxylase with norepinephrine during cat splenic nerve stimulation. *Nature* 227:406–407, 1970.
27. Gillis, C.N. The retention of exogenous norepinephrine by rabbit tissues. *Biochem. Pharmacol.* 13:1–12, 1964.
28. Greengard, P. *Cyclic Nucleotides, Phosphorylated Proteins, and Neuronal Function.* New York, Raven Press, 1978.
29. Greengard, P. Cyclic nucleotides, phosphorylated proteins, and the nervous system. *Fed. Proc.* 38:2208–2217, 1979.
30. Greengard, P.; Kebabian, J.W. Role of cyclic AMP in synaptic transmission in the mammalian nervous system. *Fed. Proc.* 33:1059–1067, 1974.
31. Hebb, C. Biosynthesis of acetylcholine in nervous tissue. *Physiol. Rev.* 52:918–957, 1972.
32. Hebb, C.O.; Waites, G.M.H. Choline acetylase in antero- and retro-grade degeneration of cholinergic nerve. *J. Physiol. (Lond.)* 132:667–671, 1956.
33. Holman, M.E.; Hirst, G.D.S. Junctional transmission in smooth muscle and the autonomic nervous system. *Handbk. Physiol.* 1(1):417–461, 1977.
34. Hutter, O.F. Ion movements during vagus inhibition of the heart. In *Nervous Inhibition.* Florey, E., ed. New York, Pergamon Press, 1961.
35. Hutter, O.F.; Trautwein, W. Vagal and sympathetic effects on the pacemaker fibers in the sinus venous of the heart. *J. Gen. Physiol.* 39:715–733, 1956.
36. Jacobowitz, D. Catecholamine fluorescence studies of adrenergic neurons and chromaffin cells in sympathetic ganglia. *Fed. Proc.* 29:1929–1944, 1970.
37. Kebabian, J.W.; Bloom, F.E.; Steiner, A.L.; Greengard, P. Neurotransmitters increase cyclic nucleotides in postganglionic neurons: immunocytochemical demonstration. *Science* 190:157–159, 1975.
38. Krnjević, K. Chemical nature of synaptic transmission in vertebrates. *Physiol. Rev.* 54:418–540, 1974.
39. Kuffler, S.W.; Dennis, M.J.; Harris, A.J. The development of chemosensitivity in extrasynaptic areas of the neuronal surface after denervation of parasympathetic ganglion cells in the heart of the frog. *Proc. R. Soc. (Biol.)* 177:555–563, 1971.
40. Kuntz, A. *The Autonomic Nervous System.* Philadelphia, Lea & Febiger, 1953.
40a. Landis, S.C. Development of cholinergic sympathetic

neurons: evidence for transmitter plasticity in vivo. *Fed. Proc.* 42:1633–1638, 1983.

41. Langer, S.Z. The metabolism of ³H noradrenaline released by electrical stimulation from the isolated nictitating membrane of the cat and from the vas deferens of the rat. *J. Physiol (Lond.)* 208:515–546, 1970.

42. Libet, B. Generation of slow inhibitory and excitatory postsynaptic potentials. *Fed. Proc.* 20:1945–1956, 1970.

43. Libet, B.; Tosaka, T. Slow inhibitory and excitatory postsynaptic responses in single cells of mammalian sympathetic ganglia. *J. Neurophysiol.* 32:43–50, 1969.

44. Lee, C.Y. Mode of action of cobra venom and its purified toxins. In Simpson, L.L., ed. *Neuropoisons, Their Pathophysiological Actions.* New York, Plenum Press, 1971.

45. Loewi, O. Über humorale Übertragbarkeit der Herznervenwirkung. *Arch. Ges. Physiol.* 189:239–242, 1921.

46. Lundberg, A. Electrophysiology of salivary glands. *Physiol. Rev.* 38:21–40, 1958.

47. Marshall, J.M. Modulation of smooth muscle activity by catecholamines. *Fed. Proc.* 36:2450–2455, 1977.

48. Mitchell, G.A.G. *Anatomy of the Autonomic Nervous System.* Edinburgh, A. & S. Livingstone, 1953.

49. Mitchell, G.A.G. *Cardiovascular Innervation.* Edinburgh, A. & S. Livingstone, 1956.

50. Moe, G.K.; Freyburger, W.A. Ganglionic blocking agents. *Pharmacol. Rev.* 2:61–95, 1959.

51. Morgan, M.W., Jr.; Olmsted, J.M.D.; Watrous, W.G. Sympathetic action in accommodation for far vision. *Am. J. Physiol.* 128:588–591, 1940.

52. Morgenroth, V.H., III; Hegstrand, L.R.; Roth, R.H.; Greengard, P. Evidence for involvement of protein kinase in the activation by adenosine 3',5'-monophosphate of brain tyrosine 3-monooxygenase. *J. Biol. Chem.* 250:1946–1948, 1975.

53. Nathanson, J. Cyclic nucleotides and nervous system function. *Physiol. Rev.* 57:157–256, 1977.

54. Nishi, S. Ganglionic transmission. In *The Peripheral Nervous System*, Hubbard, J.I., ed. New York, Plenum Press, 1974.

55. Obata, K. Transmitter sensitivities of some nerve and muscle cells in culture. *Brain Res.* 73:71–88, 1974.

56. O'Lague, P.H.; MacLeish, P.R.; Nurse, C.A.; Claude, P.; Furshpan, E.J.; Potter, D.D. Physiological and morphological studies on developing sympathetic neurons in dissociated cell culture. *Cold Spring Harbor Symp. Quant. Biol.* 40:399–407, 1976.

57. Patterson, P.H.; Chun, L.Y. The influence of non-neuronal cells on catecholamine and acetylcholine synthesis and accumulation in cultures of dissociated sympathetic neurons. *Proc. Nat. Acad. Sci.* 71:3607–3610, 1974.

58. Patterson, P.; Reichardt, L.F.; Chun, L.L.Y. Biochemical studies on the development of primary sympathetic neurons in cell culture. *Cold Spring Harbor Symp. Quant. Biol.* 40:389–397, 1976.

59. Phillis, J.W. *The Pharmacology of Synapses.* New York, Pergamon Press, 1970.

60. Pick, J. *The Autonomic Nervous System.* Philadelphia, J.B. Lippincott, 1970.

60a. Potter, D.D.; Landis, S.C.; Furshpan, E.J. Adrenergic-cholinergic dual function in cultured sympathetic neurons of the rat. In Elliott, K.; Lawrenson, G., eds. *Development of the Autonomic Nervous System*, Ciba Found. Symp. 83, 123–138. London, Pitman Books, 1981.

61. Potter, L.T.; Axelrod, J. Subcellular localization of catecholamines in tissues of the rat. *J. Pharmacol. Exper. Therap.* 142:291–298, 1963.

62. Rang, H.P. Acetylcholine receptors. *Quart. Rev. Biophys.* 7:283–399, 1974.

63. Rang, H.P.; Bülbring, E.; Cuthbert, A.W.; Potter, L.T., eds. *Drug Receptors.* London, Macmillan, 1972.

64. Robinson, G.H.; Butcher, R.W.; Sutherland, E.W. *Cyclic AMP.* New York, Academic Press, 1971.

65. Ross, D.; Johnson, M.; Bunge, R. Development of cholinergic characteristics in adrenergic neurons is age dependent. *Nature* 267:536–539, 1977.

66. Schneyer, L.H.; Young, J.A.; Schneyer, C.A. Salivary secretion of electrolytes. *Physiol. Rev.* 52:720–777, 1972.

67. Sloboda, R.D.; Rudolph, S.A.; Rosenbaum, J.L.; Greengard, P. Cyclic AMP-dependent endogenous phosphorylation of microtubule-associated protein. *Proc. Nat. Acad. Sci.* 72:177–181, 1975.

68. Trautwein, W.; Kuffler, S.W.; and Edwards, C. Changes in membrane characteristics of heart muscle during inhibition. *J. Gen. Physiol.* 40:135–145, 1956.

69. Tsien, R.W. Adrenaline-like effects of intracellular iontophoresis of cyclic AMP in cardiac Purkinje fibres. *Nature* 245:120–122, 1973.

70. Tsien, R.W.; Hess, P. Excitable tissues. The heart. In Andreoli, T.E.; Hoffman, J.F.; Fanestil, D.D.; Schultz, S.G., eds. *Physiology of Membrane Disorders*, 2nd ed., 469–490. New York, Plenum Medical Book Co. 1986.

71. Tsien, R.W.; Giles, W.; Greengard, P. Cyclic AMP mediates the effects of adrenaline on cardiac Purkinje fibres. *Nature* 240:181–183, 1972.

72. Walicke, P.A.; Campenot, R.B.; Patterson, P.H. Determination of transmitter function by neuronal activity. *Proc. Nat. Acad. Sci.* 74:5767–5771, 1977.

73. Weight, F.F. Modulation of synaptic excitability. *Fed. Proc.* 38:2078–2079, 1979.

74. Weiner, N. Multiple factors regulating the release of norephinephrine consequent to nerve stimulation. *Fed. Proc.* 38:2193–2202, 1979.

75. Weiner, N.; Cloutier, G.; Bjur, R.; Pfeffer, R.I. Modification of norepinephrine synthesis in intact tissue by drugs and by short-term adrenergic nerve stimulation. *Pharmacol. Rev.* 24:203–243, 1972.

76. Weinshilboum, R.M.; Thoa, N.B.; Johnson, D.G.; Kopin, I.J.; Axelrod, J. Proportional increase of norepinephrine and dopamine-β-hydroxylase from sympathetic nerves. *Science* 174:1349–1351, 1971.

The Milieu of the Central Nervous System

INTRODUCTION

Thus far we have considered how perception might occur, where cognitive functions are localized, and how movement might be generated. All of these behaviors have been explained as the action of neurons on neurons. In order for these interactions to proceed efficiently, however, neurons must be protected from damage, must be insulated from unwanted electrical signals, must reside in the appropriate ionic milieu, and must receive energy to support metabolism and neuronal signaling. Such support functions depend on several extraneuronal elements within and around the central nervous system (CNS). These include the cerebrospinal fluid (CSF), the extracellular space, and neuroglia, which are non-neuronal cells that constitute about 50% of the cellular volume of the CNS. In this chapter, we will consider how clever the celestial design committee was in providing an environment in which neuronal function can proceed efficiently.

THE HYDRAULIC SHOCK ABSORBER: THE CEREBROSPINAL FLUID

Three distinct layers of connective tissue, called the meninges, cover the brain. The thick dura mater is the outer layer adjacent to the bony encasement of the CNS. The pia mater is the inner delicate covering of the brain and superficial blood vessels. Sandwiched between the pia and dura mater is the arachnoid mater. The meninges usually are referred to simply as the dura, arachnoid, and pia. A potential space between arachnoid and dura is filled only during pathological conditions (e.g., subdural hematoma). The subarachnoid space, between the arachnoid and pia, contains the CSF, in which the brain and spinal cord are suspended (Fig. 35–1). The displaced CSF reduces the effective weight of the brain (1400–1500 g in air) to less than 50 g. This decreased brain weight minimizes brain damage during sudden accelerations (force = mass × acceleration) of the head.

The CSF also fills the ventricular system of the cerebral hemispheres and the brain stem (Fig. 35–2). Each cerebral hemisphere has a lateral ventricle connected to the midline third ventricle, which communicates with the fourth ventricle under the cerebellum by the narrow cerebral aqueduct of Sylvius. The ventricular system is lined with a single layer of columnar epithelial cells called the *ependyma*. In each ventricle, highly vascularized clusters of cells called the choroid plexus secrete CSF. The CSF flows from the lateral ventricles to the third ventricle and finally to the fourth ventricle, where three openings (the two lateral foramina of Luschka and the medial foramen of Magendie) allow the CSF to enter the subarachnoid space around the brain and spinal cord. It then flows along the brain stem and over the cerebral hemispheres to the arachnoid granulations located in

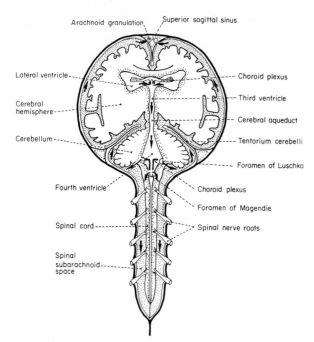

Figure 35–1 Diagrammatic representation of the nervous system showing relationship of brain and spinal cord to the cerebral ventricles and subarachnoid space. (After Millen, J.W.; Woollam, D.H.M. *The Anatomy of the Cerebrospinal Fluid.* London, Oxford University Press, 1962.)

the sagittal venous sinuses, where it is absorbed (see Fig. 35–1).[22] The adult human CNS contains about 140 ml of CSF, which is secreted at about 0.35 ml/min or 500 ml/day.[12, 15] The blood flow through the choroid plexus (3 ml/min/g) is voluminous, exceeding even that through the kidney. Two thirds of the CSF is formed by choroid plexuses in the brain ventricles. The ependyma cells lining the ventricular cavities and the pia on the brain surface separate the CSF from the extracellular fluid. Both of these epithelial surfaces allow free exchange of the CSF and its solutes with the extracellular fluid. Part of the CSF probably comes from the extracellular fluid that fills the spaces between CNS cells.

Secretion of Cerebrospinal Fluid

The components of CSF that originate in the blood must pass through the capillary endothelium, the surrounding extracellular matrix, and a single layer of choroid plexus epithelial cells.[12] The endothelium of the capillaries of the choroid plexuses is fenestrated, in contrast with the relatively tight connections between capillary endothelial cells in the rest of the brain (see later). In a choroid

plexus neither the capillary endothelium nor the loose extracellular matrix presents a barrier to fluid and ion movement.[42, 43] Thus we need only consider the epithelial cells of the choroid plexus to understand the formation of CSF.

Transport across all epithelia occurs at two sites. Water and solutes move through the epithelial cells themselves (transcellular transport) and the space between cells (paracellular transport) (see Chap. 2). The choroid epithelial cells secrete CSF by the transcellular transport of sodium, chloride, and bicarbonate ions, and water from the capillary or basal side of the epithelial cells to the ventricular or apical side (Fig. 35–3).[48] This net transport of water and solute requires an asymmetric location of membrane mechanisms for ion movement. As in all transport epithelia, the transport properties of the basolateral (i.e., capillary) surface of

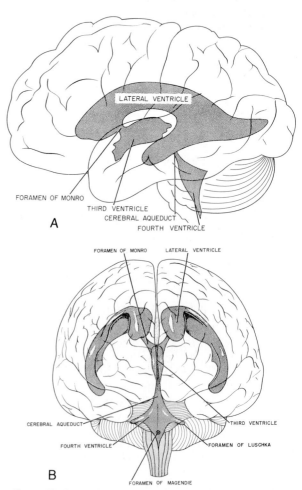

Figure 35–2 Location of the ventricular system in the brain shown in a lateral (A) and frontal (B) view. (From Curtis, B.A.; Jacobson, S.; Marcus, E.M. *An Introduction to the Neurosciences.* W.B. Saunders, 1972.)

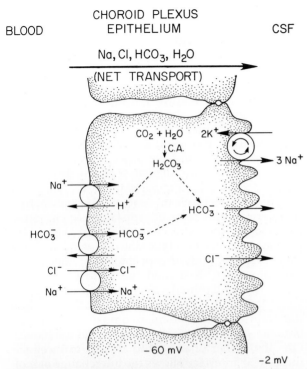

BLOOD

CHOROID PLEXUS
EPITHELIUM

CSF

Figure 35–3 Membrane transport mechanisms present in epithelial cells of the choroid plexus. C.A. = carbonic anhydrase.

choroidal cells differ from those of the apical brush border.

First, consider sodium movement. The apical brush border of choroidal epithelial cells contains classical Na^+,K^+ ATPase molecules, which exchange three Na^+ for two K^+ using ATP as an energy source (see Fig. 35–3). As expected, cardiac glycosides such as ouabain block sodium secretion when applied to the ventricular but not to the capillary surface of the choroid plexus. Since the application of ouabain to the ventricular surface also blocks all CSF formation, active sodium secretion is necessary for the formation of CSF. The sodium ions transported by the Na^+,K^+ ATPase enter the choroidal cell on its capillary surface by two mechanisms: a Na^+-H^+ exchange and a Na^+-Cl^- cotransport system[42] (see Fig. 35–3).

To preserve electroneutrality, chloride and bicarbonate must accompany the transport of sodium. Chloride-sensitive microelectrodes reveal that $[Cl^-]_i$ is about 24 mM, twice that expected if Cl^- were distributed passively.[42] The membrane potential of choroidal cells is about −60 mV. Under these conditions the concentration gradients tending to move chloride into the cell are less than the electrical forces driving it out. The

Na-Cl cotransport system, therefore, moves chloride ions uphill against an electrochemical gradient (secondary active transport). Similar mechanisms occur in the small intestine and the proximal tubule of the kidney (see Chaps. 2 and 55). This cotransport system is blocked by the diuretic furosemide.[40]

Bicarbonate is the other anion transported from the blood to the CSF. The basolateral surface of choroidal epithelial cells contains a bicarbonate-chloride exchange molecule. Because the intracellular chloride concentration is greater than expected from passive distribution, chloride will leave the cell through the basolateral membrane by flowing down its electrochemical gradient in exchange for bicarbonate. Another source of intracellular bicarbonate is the hydration of CO_2 to carbonic acid by the enzyme carbonic anhydrase (see Fig. 35–3). Acetazolamide, a carbonic anhydrase inhibitor, reduces the rate of CSF secretion by almost 50% by limiting the formation of hydrogen ions available for Na^+-H^+ exchange.[40, 43] Because the intracellular activity of bicarbonate and chloride is higher than expected from passive distribution across the membrane, both ions will diffuse down their electrochemical gradient through channels in the apical membrane to the CSF.

The epithelial cells are attached to each other by continuous tight junctions called *zonula occludens*.[32] These attachments are somewhat leaky and allow small molecules and water to pass between epithelial cells. This leaky paracellular pathway for transport through the intercellular space of the choroid epithelium is similar to that in the gallbladder and proximal tubule of the kidney (see Chap. 55).

The choroid plexus receives both a cholinergic and adrenergic innervation, suggesting a neural role in the regulation of CSF secretion.[14] Although the precise mechanisms are unknown, compounds that often act as second messengers affect CSF transport mechanisms. For example, cAMP appears to increase bicarbonate permeability in the apical border.[37]

Absorption of Cerebrospinal Fluid

The bulk flow of CSF carries it through the subarachnoid space around and over the brain to the arachnoid or pacchionian granulations (Fig. 35–4B). These are outpouchings of the arachnoid that protrude into the sagittal venous sinus, which lies in the dura between the two cerebral hemi-

A

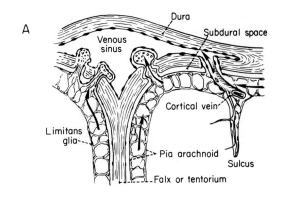

B

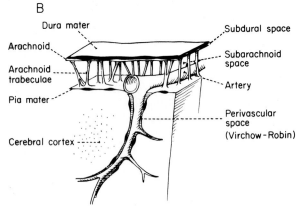

Figure 35–4 *A,* Diagram showing the relationship of the meninges to the arachnoid granulation in the venous sinuses. *B,* Diagram showing the relationship of brain arteries and arterioles to the meninges and subarachnoid space. (*A* after Carpenter, M.B.; Sutin, J. *Human Neuroanatomy,* 8th ed. © 1983, the Williams & Wilkins Co., Baltimore. *B* from Patton, H.D.; Sundsten, J.W.; Crill, W.E.; Swanson, P.D. *An Introduction to Basic Neurology,* Philadelphia, W.B. Saunders, 1976.)

spheres. Little CSF absorption occurs until the CSF pressure exceeds 70 mm H_2O.[15] This pressure threshold for absorption suggests the existence of some type of one-way cellular valve in the arachnoid granulations that allows CSF and even red cells to enter the venous blood in the sagittal sinus but prevents a reverse flow. Electron microscopic examinations, however, have revealed neither structural valves nor the pores necessary for passage of large molecules. All we know about arachnoid granulation cells is that they are attached to each other by tight junctions and contain vesicular profiles large enough to hold red blood cells. It is possible that the bulk flow transport across cells of the arachnoid granulation is by the sequential pinocytosis and exocytosis of large vacuoles.[15]

NUTRIENT SUPPLY TO THE BRAIN

Although pial and ependymal surfaces of the brain receive some nutrients from the CSF, deeper structures must rely on blood flow through capillaries. The feeder arteries run within the subarachnoid space that follows the arteriole supply for short distances (Virchow-Robin space) as the vessels enter the brain (Fig. 35–4A). The capillary endothelium, a continuous homogeneous basement membrane, and the processes of numerous glia (see below) are all that separate the plasma from the extracellular space of the CNS. Nevertheless these structures present an effective barrier that permits only selected substances to pass from the blood to the extracellular fluid.[5, 28] When Ehrlich injected trypan blue into a vein, the dye stained all the body except the brain; injection of dye into the CSF stained the brain blue. These experiments identified a *blood-brain barrier* that prevents the ready transfer of large molecules into the brain from the vascular compartment (trypan blue rapidly combines with albumin in the blood). On the other hand, the brain extracellular fluid—CSF interface allows the interchange of molecules as large as albumin. Although the blood-brain barrier to free molecular movement is present in capillaries[36] throughout most of the brain, some small CNS regions are stained after trypan blue injection. These structures include the area postrema of the fourth ventricle, the subfornical organ, the epiphysis, the neurohypophysis, the median eminence, and the pineal gland.[32]

The endothelium of brain capillaries serves as an epithelial transport surface that allows passage of nutrients and water and removes waste products but prevents movement of large molecules. The amount of water and solutes entering the interstitial fluid of the brain from capillaries is uncertain, but it is likely that some net transport occurs. This transport serves as a relatively small additional source of CSF, since the extracellular fluid exchanges freely with the CSF.

The major barrier to transport between the blood and extracellular fluid is the tight intercellular circumferential junctions (zonula occludens) joining the endothelial cells of the brain capillaries and this contrasts with the fenestrated choroid capillary endothelium. Substances such as lanthanum or horseradish peroxidase will not pass between the endothelial cells of brain capillaries.[36] Transport also cannot occur by pinocytosis, since brain capillary endothelial cells have only a few vesicle profiles which actually are tubular invaginations that do not traverse the cell.[6]

Although the endothelial cells of brain capillaries form a "tight"* epithelial surface, paracellular transport of water and small solute molecules can occur. Paracellular transport arises from sparse pores about 7 nm in diameter. An osmotic load or anoxia will increase the paracellular movement of water. This is one mechanism responsible for dangerous brain swelling.[15]

The abluminal surface of brain capillary endothelium (brain side) contains Na^+, K^+ ATPase molecules which, as expected, use one molecule of ATP to extrude three sodium ions for the influx of two potassium ions (Fig. 35–5). This ion pump removes sodium ions entering the cell through ion selective channels on the luminal surface. These sodium channels have a conductance of 5 to 10 pS and have a density of $50/\mu m^2$. They are blocked by amiloride but not by tetrodotoxin, which blocks the voltage-sensitive sodium channels in excitable cells. Electroneutrality is maintained primarily by the transport of chloride ions. They enter the luminal surface of the endothelial cell in exchange for bicarbonate ions and move through the cytoplasm into the interstitial space through chloride selective channels on the abluminal membrane. Carbon dioxide from the brain diffuses into the endothelial cytoplasm, and carbonic anhydrase accelerates the formation of carbonic acid. The resulting bicarbonate ions drive the bicarbonate-chloride exchange.

A comparison of Figures 35–3 and 35–5 shows the similarities and differences between the brain capillary endothelium and the epithelium of the choroid plexus. Both transport surfaces have Na^+, K^+ ATPase molecules asymmetrically concentrated in the abluminal cell membrane. The differences are the same that distinguish leaky from tight epithelial surfaces in the rest of the body. The choroid plexus is designed for large secretory rates. Its leaky paracellular transport brings along water and some ions to maintain isotonicity. The paracellular route in the brain capillaries provides significantly less net flow of water coupled to transcellular active ion transport. Sodium ions are

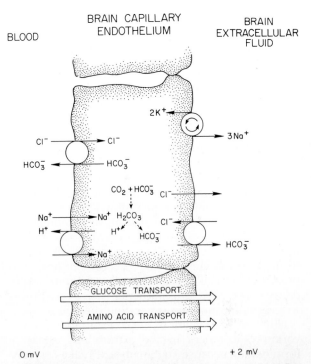

Figure 35–5 Membrane transport processes present in the endothelial cells of brain capillaries.

cotransported with chloride on the luminal surface of the choroid cells but they enter capillary endothelial cells through amiloride-sensitive ion channels.

Glucose, the primary source of energy for neurons and other cells, must pass through the blood-brain barrier throughout the brain. Indeed, plasma is the sole source of brain glucose, whose concentration in the extracellular fluid is about two thirds that of plasma (see Table 35–1). The transport of glucose has an exceedingly high capacity.[10] D-Glucose combines with a carrier protein on the luminal surface of the endothelial cell and is translocated across the cell membrane (see Fig. 35–5). The carrier site then dissociates from the glucose and recycles to transport other glucose molecules. Glucose transport across brain capillary endothelia is not coupled with sodium as across intestinal epithelia. Like all carrier-mediated transport mechanisms, the glucose transport exhibits saturation kinetics, stereospecificity, and competitive inhibition (see Chap. 2). As at the red cell membrane, glucose transport across capillary endothelia in the brain is not dependent on insulin.[10]

Amino acids are transported into the extracel-

---

*Tight and leaky epithelial membranes refer to the ratio of the junctional conductance to the transcellular conductance, not the absolute conductance. We do not know these values for brain capillaries, but their endothelium has all the structural and transport characteristics found in other tight epithelial surfaces.

lular fluid of the brain by similar facilitated transport mechanisms. They are needed for the synthesis of proteins and peptides and for the synthesis of neurotransmitters such as serotonin, histamine, and dopamine. Three different facilitated transport systems carry neutral, acidic, and basic amino acids.[10]

EXTRACELLULAR SPACE

Neuroscientists have debated the size of the extracellular fluid space in the brain for 30 years. Early electron micrographs showed it to be extremely small, and some anatomists doubted that it existed at all. However, recent physiological measurements using radioactive tracers, improved fixation techniques for microscopic studies, and precise electrical measurements of brain impedance show that the extracellular space constitutes 15 to 20% of the brain (excluding the CSF).

Direct measurement of the concentrations of the various substances in the extracellular space is formidable, because its small size makes it very difficult to sample the fluid. Ion sensitive electrodes can measure the extracellular concentrations of Na^+, K^+, Ca^{2+}, and Cl^-. Since extracellular fluid exchanges freely with CSF over the pia surface and at the ependyma lining of the ventricles, the entire spectrum of extracellular ion concentrations can be estimated by direct chemical assays of the CSF. These results are listed in Table 35–1. Note the much lower protein concentration in the CSF. Although the concentrations of ions are similar to those found in an ultrafiltrate of the blood, there are small differences. In the CSF, $[K^+]_o$ and $[Ca^{2+}]_o$ are lower, but $[Mg^{2+}]_o$ is higher than in blood serum. Also, glucose is somewhat lower in CSF than in blood.

It is crucial that the concentration of the extracellular fluid constituents in the brain be closely maintained. The electrical properties of neurons and synaptic mechanisms are affected by extracellular ion concentrations (see Chaps. 11 and 12). The resting potential of cells is generated primarily by a selective membrane permeability to potassium ions, and minor changes in $[K^+]_o$ could either hyperpolarize or depolarize neurons. Such fluctuations would markedly alter both the excitable properties of neurons and synaptic transmission. For example, partially depolarized neurons might be more permeable to calcium ions, thereby affecting their repetitive firing properties or transmitter release mechanisms. Fortunately, extracellular fluid in the brain is so well regulated that its K^+ concentration does not change even if plasma $[K^+]_o$ is doubled. Since brain $[K^+]_o$ is controlled by homeostatic mechanisms (see below), systemic hyperkalemia causes serious alterations in cardiac electrical activity long before any neurological dysfunction occurs.

The extracellular fluid is the highway for the movement of molecules that maintain CNS function. Nutrients and waste products pass through the extracellular space on their way to and from the vascular space. The extracellular fluid allows hormones to carry information to the nervous system from the rest of the body and neurotransmitters and neuromodulators to influence many different cells. Although the extracellular space itself does not present a significant barrier to the movement of these molecules, it does contain an extracellular glycoprotein matrix whose role is unknown. The effect of the size and tortuosity of the extracellular space and the role of the glia is discussed later when we consider the special problem of controlling $[K^+]_o$.

Table 35–1 Components of the Cerebrospinal Fluid (CSF) and Blood*

Component	CSF	Blood
Water Content (%)	99	93
Protein (mg/dl)	35	7000
Glucose (mg/dl)	60	90
Osmolarity (mOsm/ℓ)	295	295
Na^+ mM	138	138
K^+ mM	2.8	3.5
Ca^{2+} mM	2.1	4.8
Mg^{2+} mM	2.3	1.7
Cl^- mM	119	102
pH	7.33	7.41

*From Fishman, R. A. *Cerebrospinal Fluid in Diseases of the Nervous System*, Philadelphia, W. B. Saunders, 1980.

NEUROGLIA

Virchow recognized that the nervous system contained an interstitial substance, which he called neuroglia or *nerve glue*. We now know that the glue is primarily cellular and that the brain contains two general classes of cells: neurons and neuroglia. The supporting neuroglia (or glia) are either small (microglia) or large (macroglia).[20, 32] The rod-shaped microglia are probably of mesodermal origin. They are the phagocytic representatives of the reticuloendothelial system in the CNS. Here we discuss only the macroglia, which are divided into two cell types: oligodendroglia and astrocytes.

Oligodendroglia

Oligodendroglia lie between groups of myelinated axons in both the white and gray matter of the brain and also around neurons as satellite cells. The cytoplasm of oligodendroglia also surrounds small unmyelinated neurons. Finally, oligodendroglia form the myelin sheaths around axons in the CNS (Fig. 35–6), a role similar to that of Schwann cells in the peripheral nervous system. Whereas each Schwann cell forms the myelin for only a single peripheral axon, one oligodendroglial cell forms the myelin for up to 35 central axons.[32] As shown in Chapter 3 for peripheral axons, myelinated central axons are capable of saltatory conduction, which increases conduction velocity. This occurs because the high lipid content of the myelin decreases the amount of electrical charge necessary to depolarize the internodal region of the axon, thereby increasing the charge available for depolarizing the nodes of Ranvier farther down the axon.

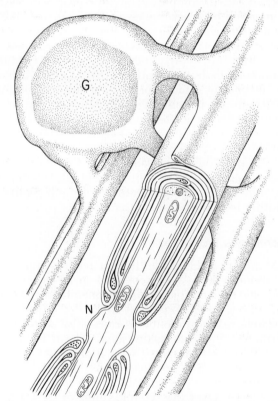

Figure 35–6 Relationship of oligodendroglia cell (G) to myelinated axons in the central nervous system. (N) Node of Ranvier. (After Bunge, M.B., Bunge, R.P.; Ris, H. *The Journal of Biophysical and Biochemical Cytology*, 1961, 10:67–94 by copyright permission of The Rockefeller University Press.)

Both oligodendroglia and Schwann cells seem to control the presence of ion channels in axons surrounded by the myelin. Recent patch clamp recordings have revealed voltage-dependent ion channels in Schwann cells.[9, 17] The function of these channels is unknown, but a provocative suggestion is that the myelin-forming cells furnish channels for the axon. In normal myelinated axons there are few sodium channels in the membrane between the nodes of Ranvier. In less than ten days following experimentally induced demyelination, sodium channels appear in the internodal region, and conduction becomes continuous.[3] Apparently, the myelin provides a signal to prevent the insertion of sodium channels into the internodal region. These poorly understood relationships between myelin and axon are excellent examples of how the environment surrounding a cell can control its physiological properties.

Astrocytes

The astroglia, or astrocytes, are star-shaped macroglia (Fig. 35–7) that can be identified in electron micrographs by the locations of their processes. Some processes are closely apposed to neuronal cell bodies and dendrites in regions between the thousands of synaptic terminals. Other astrocytic processes form endfeet that cover the basal lamina of brain capillaries and arterioles. The endfeet and cell bodies of astrocytes also form the glia limitans, which separates the neurons on the outer surface of the brain from the pia mater.[32]

The cytoplasm of astrocytes contains many fibrils. Astrocytes located in the white matter have a higher density of fibrils and are called *fibrous* astrocytes. Those with fewer fibrils are called *protoplasmic* astrocytes. A specific antibody against acidic glial fibrillary protein allows the precise identification of astrocytes by immunocytochemistry.[20]

All astrocytes are connected by gap junctions that are similar structurally to those intracellular pathways found in other parts of the body. As at other gap junctions (see Chap. 11), astrocytic gap junctions allow electrical current and relatively large molecules to pass from one astrocyte to another. Some gap junctions are also present between astrocytes and oligodendroglia.

Suggested Roles for Neuroglia

Over the last hundred years, a variety of roles have been suggested for neuroglia. The earliest suggestion was that neuroglia provide some sort

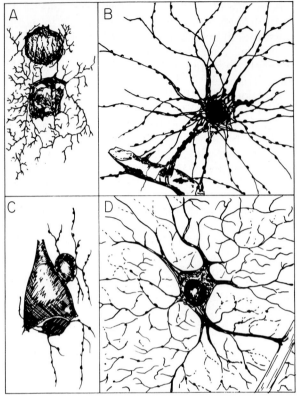

Figure 35–7 Types of glial cells present in the central nervous system A, microglia; B, fibrous astrocyte; C, oligodendroglia; D, protoplasmic astrocyte. (After Penfield, W. *Cytology and Cellular Pathology of the Nervous System*, New York, Hoeber, 1932.)

of structural support for the adult brain. A more recent and more imaginative suggestion implicated glia in the developing CNS, where a special type of glia called radial glia extend from the ventricular surface to the cortical surface.[35] Apparently, developing neurons that are formed on the ventricular side of the cerebral mantle migrate along the radial glia to their final destination in cerebral cortex. Radial glia are thought to provide directional guidance rather than to act as girders for the nervous system.

After the brain is damaged irreversibly, it is replaced primarily with a heavy growth of glia called astrocytic scar. This thick layer of glia probably restricts the regrowth of central axons and is assumed to have a deleterious effect on CNS nerve regeneration.[38] In experimental animals, CNS axonal growth increases if a millipore filter provides an avenue for nerve growth across the scar. (See Chap. 33 for a discussion of neuronal regrowth.)

The presence of astrocytic capillary endfeet and the close approximation of other astrocytic proc-

esses to neurons suggest that astrocytes may have a nutritive function.[20] Various nutrients supplied by the blood could be transported to neurons through the astrocyte cytoplasm. This mechanism is probably not necessary, however, since the capillary endfeet do not form a tight cellular sheath around capillaries, and there is physiological evidence that even large molecules can pass between the endfeet in the extracellular space.

Since the spaces between synapses are occupied by astrocytic processes, astrocytes may serve to isolate certain inputs and prevent untoward synaptic interactions. Synaptic terminals constitute a large amount of the neuropil in the CNS; each neuron receives thousands of synaptic boutons. Consequently, the small extracellular space could cause the concentration of neurotransmitters to build up, producing an unwanted, persistent effect, or desensitization of receptor molecules (see Chap. 6). If these effects do occur, the response properties of the neurons will be altered. Recent studies suggest that the precise local interaction of neurotransmitter effects and cellular responses plays a role in complex behavioral processes such as memory and learning (see Chap. 33).

Glia remove neurotransmitters from the extracellular fluid. Bowery and colleagues[4] have shown that glia of sympathetic ganglia concentrate gamma-aminobutyric acid (GABA) from the interstitial space by a factor of 200. The mechanism for GABA influx is sodium dependent, and uptake occurs at levels above 1 μM, well below the concentration required for a physiological effect on neurons. Once inside the glia, GABA can be metabolized through transamination and the Krebs cycle. Glial cells also concentrate glutamate, a putative excitatory neurotransmitter, and glycine, another inhibitory transmitter.

Membrane Properties

Mammalian glial cells have a resting potential of about −90 mV, which is 20 to 30 mV more negative than a typical neuronal membrane potential and is nearly identical to E_K.[33, 39] The large negative resting potential is caused by a selective permeability to potassium ions, as can be demonstrated by changing $[K^+]_o$. If the glial cell is permeable only to potassium ions, the membrane potential can be predicted by the Nernst equation (Fig. 35–8). Over the 100-fold range of $[K^+]_o$ from 0.3 mM to 30 mM, the measured membrane potential is predicted by the Nernst equation (Fig. 35–8B, solid line). Because the membrane potential follows the Nernst prediction over so wide a range

A

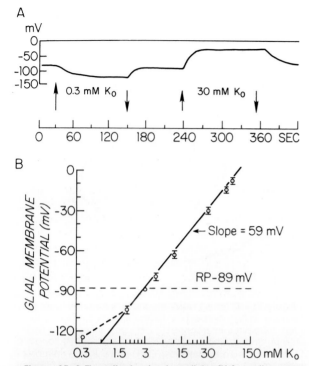

B

Figure 35–8 The effects of extracellular $[K^+]_o$ on the membrane potential of glia (astrocytes). *A* is a recording from a glial cell in the optic nerve during reduction of $[K^+]_o$ to 0.3 mM and during an increase in $[K^+]_o$ to 30 mM. ↑ = onset of perfusion; ↓ = return to normal $[K^+]_o$. *B* is a plot of glial membrane potential as function $[K^+]_o$. Solid line = the change predicted by the Nernst equation (After Kuffler, S.W.; Nicholls, J.G.; Martin, A.R. In *From Neuron to Neuron*, 2nd ed. Sunderland, Mass., Sinauer, 1984.)

of $[K^+]_o$, the permeability to potassium must be much greater than that to other ions.[18, 19, 21, 24–27] Indeed, changes in $[Na^+]_o$ or $[Cl^-]_o$ do not affect glial resting potentials.

The concentration of potassium channels in astrocytes is not uniform over the cell. The astrocytic endfeet that make contact with the capillaries and the pia surface have much higher potassium permeability than the rest of the cell.[23] Figure 35–9 shows the effects on the soma membrane potential of brief increases in $[K^+]_o$ at different parts of a cultured astrocyte. The magnitude of the depolarization is a function of the local potassium conductance (ion channel density). The variation in the density of potassium channels over the cell does not affect the resting potential but does have implications in the control of extracellular ion concentrations by glia (see below). This increased permeability to potassium at endfeet on arterioles may be a factor in the local control of blood flow. Cerebral arterioles show local dilatation in response to increased $[K^+]_o$.[30]

Recently, the use of patch electrodes on tissue-cultured astrocytes has revealed individual voltage-dependent sodium, potassium, and calcium channels.[2, 17] We do not know the role of these channels in the function of astrocytes. One suggestion is that channel proteins synthesized in glial cells are incorporated in turn into the excitable membranes of neurons. Tissue culture could be causing a dedifferentiation of glial cells, and some of the channels in cultured astrocytes might not even be present in vivo. Another possibility is that, under special circumstances, the ion channels of glial cells might be involved in the generation of electrical signals; however, thus far, active responses have never been recorded from glial cells. The influx of calcium ions into astrocytes might also serve as a second messenger for as yet unidentified glial functions.

As noted in Chapter 12, gap junctions between neurons of the mammalian CNS are rare. They do, however, occur commonly between macroglia. The interglial gap junctions allow fluorescent dyes to pass from cell to cell and provide a low-resistance pathway for intercellular ionic current. Astrocytes, therefore, are often referred to as an electrical syncytium. No gap junctions have been identified between neurons and glia.

Control of Extracellular Potassium Concentration

During the repolarization phase and the afterhyperpolarization of each neuronal action poten-

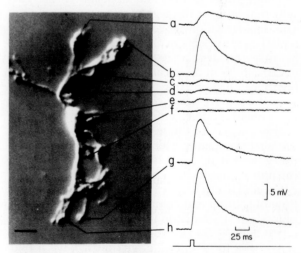

Figure 35–9 *A*, Photomicrograph of a recently dissociated astrocyte from the salamander. Scale bar = 20 μm. *B*, Transient depolarizations a–h are recorded from the soma following a brief increase in $[K^+]_o$ at the labeled sites. The bottom trace indicates the time course of the potassium ejection pressure. (From Newman, *Science* E.A. 233:453–454. Copyright 1986 by the AAAS.)

tial, a small amount of potassium leaves the neuron and enters the intercellular cleft. Even though the efflux of potassium from normally active neurons is small (a few picomoles of ions per cm^2 of membrane per spike), a local increase in $[K^+]_o$ can be detected following a few action potentials with ion-selective electrodes.[13] Rapid, repetitive activation of neurons raises the interstitial potassium concentration still further. This increase in interstitial potassium must be cleared to prevent depolarization of neurons and synaptic terminals; otherwise an inactivation of spike generating mechanisms and the uncontrolled release of neurotransmitters would result.

Three different mechanisms could remove local increases in potassium from the extracellular fluid.[16, 24, 41] They are diffusion, flow of potassium current through glia, and active transport back into cells. Calculations using reasonable models for the diffusion process reveal that diffusion alone is at least one hundred times too slow to remove the increase in $[K^+]_o$. Both transcellular current flow and active transport therefore provide the main mechanisms that prevent dangerous increases in $[K^+]_o$.

The transcellular glial current flow, which regulates extracellular potassium, is called *spatial buffering*[24] (Fig. 35–10). The gap junctions between glial cells allow the free intercellular movement of small ions like potassium. At -90 mV, the glial resting potential is close to E_K. Increased neuronal activity raises local $[K^+]_o$ (stippling in Fig. 35–10B) to about 10 to 12 mM and depolarizes the glial cells that lie near the region of increased activity. The depolarization moves toward the Nernst potential (about -56 mV) but does not reach it, because the depolarized glia are interconnected electrically to nondepolarized cells by the intercellular gap junctions (Fig. 35–10C). Since the depolarized cells are somewhat negative to the local E_K, potassium ions enter the cell and serve as charge carriers to the less depolarized regions of the glial syncytium. This current flow depolarizes the glia at a distance from the increased $[K^+]_o$, causing potassium to leave the distant cells. The extracellular electrical circuit is completed by the flow of sodium and chloride ions. In this manner, potassium is redistributed in the extracellular space. For the transcellular transport of potassium across glia to be effective, glia must have a high permeability to potassium. Recent experiments reveal that the astrocytic endfeet, next to capillaries and the pia surface, have a high density of potassium channels,[23] thereby allowing more potassium to leave the glia cell at the endfeet. Since voltage-dependent potassium channels exist in as-

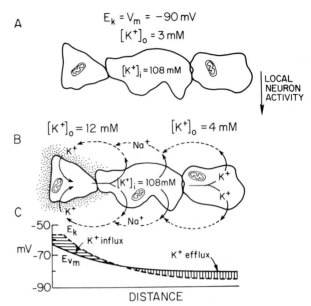

Figure 35–10 Transcellular mechanism for removal of local increases in $[K^+]_o$ caused by neuronal activity. Diagram shows three astrocytes interconnected by gap junctions in resting state (A) and (B) when activity in neurons raises $[K^+]_o$ locally to 12 mM (stippled area, left side). C is a plot of E_K and membrane potential (E_m) as a function of distance along the glial syncytium shown in B. The difference between E_m and E_K causes an influx of potassium in the stippled area and an efflux of potassium at a distance. Extracellular current is carried by sodium ions (shown here) and by chloride ions. (After Orkand, R.K. *Ann. N.Y. Acad. Sci.* 481:269–272, 1986.)

trocytes, they also could facilitate the spatial buffering of potassium. Active transport would cause the uptake of the redistributed potassium by neurons and glia.

During normal brain activity, the $[K^+]_o$ does not rise above 10 to 12 mM. The mechanisms responsible for controlling extracellular potassium are effective below this concentration. In the laboratory, prolonged stimulation that evokes excessive firing of neurons or the direct application of potassium to the cortex raises $[K^+]_o$ to values greater than 30 mM and a self-sustained efflux of potassium spreads across the cortex, depolarizing all neurons and probably releasing many synaptic transmitters and modulators. This is called *spreading depression*. Indirect evidence suggests that the occurrence of spreading depression in the visual cortex explains the visual aura preceding migraine headaches.

It has been suggested, but not proved, that the glia at the site of an old brain injury (called the *glia scar*) are not as efficient as normal glia in the spatial buffering of potassium ions.[34] For example, if the membranes of these altered glia exhibit an increase in membrane permeability to other ions, the accumulation of excess extracellular potassium will be associated with a smaller local glial depolarization and, therefore, a decreased removal of potassium via spatial buffering. Such a mechanism could explain the tendency for epileptic seizures, which often arise from the region of an astrocytic scar. It is known that epileptic discharges occur more easily when $[K^+]_o$ is increased.

REFERENCES

1. Abbott, N.J. Are glial cells excitable after all? *Trends Neurosci.* 8:141–142, 1985.
2. Bevan, S.; Chiu, S.Y.; Gray, P.T.A.; Ritchie, J.M. The presence of voltage-gated sodium potassium and chloride channels in rat cultured astrocytes. *Proc. R. Soc. Lond. (Biol.)* 255:299–313, 1985.
3. Bostock, H.; Sears, T.A. The internodal axon membrane: electrical excitability and continuous conduction in segmental demyelination. *J. Physiol. (Lond.)* 280:273–301, 1978.
4. Bowery, N.G.; Brown, D.A.; White, R.D.; Yamini, G. (^{3}H)-γ-Aminobutyric acid uptake into neuroglial cells of rat superior cervical sympathetic ganglia. *J. Physiol.* 293:51–74, 1979.
5. Bradbury, M.W.B. The blood-brain barrier. Transport across the cerebral endothelium. *Circ. Res.* 57:213–222, 1985.
6. Bundgaard, M. Pathways across the vertebrate blood-brain barrier: morphological viewpoints. *Ann. N.Y. Acad. Sci.* 481:7–18, 1986.
7. Bunge, M.B.; Bunge, R.P.; Ris, H. Ultrastructural study of remyelination in an experimental lesion in adult cat spinal cord. *J. Biophys. Biochem. Cytol.* 10:67–94, 1961.
8. Carpenter, M.B.; Sutin, J. *Human Neuroanatomy*, 8th ed. Baltimore, Williams & Wilkins, 1983.
9. Chiu, S.Y.; Shrager, P.; Ritchie, J.M. Neuronal-type Na$^+$ and K$^+$ channels in rabbit cultured Schwann cells. *Nature* 311:156–157, 1984.
10. Crone, C. The blood-brain barrier as a tight epithelium: Where is information lacking? *Ann. N.Y. Acad. Sci.* 481:174–185, 1986.
11. Curtis, B.A.; Jacobson, S.; Marcus, E.M. *An Introduction to the Neurosciences.* Philadelphia, W.B. Saunders, 1972.
12. Cutler, R.W.P.; Sperrell, R.B. Cerebrospinal fluid: a selective review. *Ann. Neurol.* 11:1–10, 1982.
13. Dietzel, I.; Heinemann, U. Dynamic variations of the brain cell microenvironment in relation to neuronal hyperactivity. *Ann. N.Y. Acad. Sci.* 481:72–84, 1986.
14. Edvinson, L.; Lindvall, M.; Owman, C.; West, K.A. Autonomic nervous control of cerebrospinal fluid production and intracranial pressure. In Wood, J.H., ed. *Neurobiology of Cerebrospinal Fluid*, vol. 2, 661–676. New York, Plenum Press, 1983.
15. Fishman, R.A. *Cerebrospinal Fluid in Diseases of Nervous System.* Philadelphia, W.B. Saunders, 1980.
16. Gardner-Medwin, A.R. A new framework for assessment of potassium-buffering mechanisms. *Ann. N.Y. Acad. Sci.* 481:287–302, 1986.
17. Gray, P.T.A.; Ritchie, J.M. Ion channels in Schwann and glial cells. *Trends Neurosci.* 12:411–413, 1985.
18. Kuffler, S.W. Neuroglial cells. Physiological properties and a potassium mediated effect of neuronal activity on the glial membrane potential. *Proc. R. Soc. Lond. (Biol.)* 168:1–21, 1967.
19. Kuffler, S.W.; Nicholls, J.G. The physiology of neuroglial cells. *Ergeb. Physiol.* 57:1–90, 1966.
20. Kuffler, S.W.; Nicholls, J.G.; Martin, A.R. A cellular approach to the function of the nervous system. In *From Neuron to Neuron*, 2nd ed., 323–360. Sunderland, Mass., Sinauer, 1984.
21. Kuffler, S.W.; Nicholls, J.G.; Orkand, R.K. Physiological properties of glial cells in the central nervous system of amphibia. *J. Neurophysiol.* 36:855–868, 1966.
22. Millen, J.W.; Woollam, D.H.M. *The Anatomy of the Cerebrospinal Fluid.* London, Oxford University Press, 1962.
23. Newman, E.A. High potassium conductance in astrocyte endfeet. *Science* 233:453–454, 1986.
24. Orkand, R.K. Introductory remarks: glial-interstitial fluid exchange. *Ann. N.Y. Acad. Sci.* 481:269–272, 1986.
25. Orkand, R.K. Signalling between neuronal and glial cells. In Sears, T.A., ed. *Neuronal-Glial Cell Interrelationships*, pp. 147–157. New York, Springer-Verlag, 1982.
26. Orkand, R.K. Glial cells. *Handbk. Physiol.* Sec. 1, 1:855–875, 1977.
27. Orkand, R.K.; Nicholls, J.G.; Kuffler, S.W. Effect of nerve impulses on the membrane potential of glial cells in the central nervous system of amphibia. *J. Neurophysiol.* 29:788–806, 1966.
28. Pardridge, W.M. Brain metabolism: a perspective from the blood brain barrier. *Physiol. Rev.* 63:1481–1535, 1983.
29. Patton, H.D.; Sundsten, J.W.; Crill, W.E.; Swanson, P.D. In *Introduction to Basic Neurology.* Philadelphia, W.B. Saunders, 1976.
30. Paulson, O.B.; Newman, E.A. Does the release of potassium from astrocyte endfeet regulate cerebral blood flow? *Science* 237:896–898, 1987.
31. Penfield, W. *Cytology and Cellular Pathology of the Nervous System.* New York, Hoeber, 1932.
32. Peters, A.; Palay, S.; Webster, H. de F. *The Fine Structure of the Nervous System: The Neurons and Supporting Cells.* Philadelphia, W.B. Saunders, 1976.
33. Picker, S.; Pieper, C.F.; Goldring, S. Glial membrane potentials and their relationship to $[K^+]_o$ in man and guinea pig. A comparative study of intracellularly marked normal, reactive, and neoplastic glia. *J. Neurosurg.* 55:347–363, 1981.
34. Pollen, D.A.; Trachtenburg, M.C. Neuroglia: gliosis and epilepsy. *Science* 167:1252–1253, 1970.
35. Rakic, P. Neuron-glia relationship during granule cell migration in developing cerebellar cortex. A Golgi and electromicroscopic study in *Macacus rhesus. J. Comp. Neurol.* 141:283–312, 1971.
36. Reese, T.S.; Karnousky, M.J. Fine structural localization of a blood-brain barrier to exogenous peroxidase. *J. Cell Biol.* 34:207–217, 1967.
37. Saito, Y.; Wright, E.M. Regulation of bicarbonate transport across the brush border membrane of the bull-frog choroid plexus. *J. Physiol.* 350:327–342, 1984.
38. Sears, T.A., ed. *Neuronal-Glial Cell Interrelationships.* New York, Springer-Verlag, 1982.
39. Takato, M.; Goldring, S. Intracellular marking with Lucifer Yellow CH and horseradish peroxidase of cells electrophysiologically characterized as glia in the cerebral cortex of the cat. *J. Comp. Neurol.* 18:173–188, 1979.
40. Vogh, B.P.; Langham, M.R., Jr. Effect of furosemide and bumetanide on cerebrospinal fluid formation. *Brain Res.* 221:171–183, 1981.
41. Walz, W.; Hertz, L. Functional interactions between neurons and astrocytes. II. Potassium homeostasis at the cellular level. *Prog. Neurobiol.* 20:133–183, 1983.
42. Wright, E.M. Transport processes in the formation of the cerebrospinal fluid. *Rev. Physiol. Biochem. Pharmacol.* 83:1–34, 1978.
43. Wright, M.; Saito, Y. The choroid plexus as a route from blood to brain. *Ann. N.Y. Acad. Sci.* 481:214–219, 1986.

Index

Page numbers in *italics* refer to illustrations; page numbers followed by the letter t refer to tables.

Neuromuscular junction, anatomy of, 132–133, *134*
  invertebrate, 153
  studies on, history of, 132–133
Neuromuscular transmission, 130–155
  acetylcholine cycle in, 135
  acetylcholinesterase in, 132, 135
  agonists in, 132, 140–142
  antagonists in, 132
  blocking agents for, 141–142
  chemical, 131–132, *131*. See also *Acetylcholine; Neurotransmitters.*
  end-plate potential in, 132
  in myasthenic disorders, 151–152
  postsynaptic response in. See *Postsynaptic response.*
  versus junctional transmission in autonomic system, 743–744, *744*
Neuronotrophic factor, 1162
Neurons. See also *Interneurons; Motoneurons.*
  A region of, *267, 268*
  action potential initiation in, 266–269, *267–268*
  aminergic, 1188–1189
  antidiuretic hormone–producing, 1174–1175, *1174–1175*
  B region of, 268–269, *268*
  cable properties of, 269–273, *271–273*
  communication among, 231–261
  conductance in, 93–94, *94,* 277–281
    calcium, 279, 281t
      high-threshold, 279–280, *280,* 281t
      low-threshold, persistent, 279, 281t
      transient, 280–281, *280,* 281t
    potassium, A type, 278–279, *279,* 281t
      calcium-sensitive, 277–278, *278,* 281t
    sodium, 281, 281t
  environment of, 233, *233,* 763–764, *763*
  in circadian rhythm pacemaker, 1244–1245, *1245*
  inverse, in spinal cord, 320
  morphology of, 231–232, *232,* 264–265, *265*
  neurotransmitters affecting, 281–283, *282*
  peptidergic, 1188–1189
  postganglionic. See *Postganglionic neurons.*
  preganglionic. See *Preganglionic neurons.*
  regeneration of, after injury, 731–732, *732*
  repetitive firing of, 273–277, *274–276*
  sprouting of, after injury, 731–732, *732*
  structure of, versus function, 265–266, *265*
  trigger zone of, 266
Neuropeptide(s). See also specific neuropeptide, e.g., *Enkephalins; Endorphins; Substance P.*

Neuropeptide(s) *(Continued)*
  as neuromodulators, 356–357, *357,* 1190
  as neurotransmitters, 234, 249, 1184–1201
    autonomic, 746
    coexistence of, 746, 1190
    effects of, 1190–1191, *1190,* 1191t
    evidence for, 1189–1190
    features of, 1185, *1185,* 1186t–1187t, 1187
    integrative function of, 1191–1192
  definition of, 1185
  diversity of, 1187–1188, *1188–1189*
  peptidergic versus aminergic neurons and, 1188–1189
  structure of, 1185, *1185,* 1186t–1187t, 1187
  synthesis of, 1187–1188, *1188–1189*
Neuropeptide Y, structure of, 1186t
Neurophysins, 1175
Neurosecretion, definition of, 1206
Neurotensin, in gastric acid secretion, 1439t, 1445
  in ileogastric reflex, 1431
  splanchnic circulation affected by, 917
  structure of, 1186t
Neurotoxins, as acetylcholine receptor blocking agents, 141–142
  voltage-gated ionic channels and, *83,* 84
Neurotransmitters. See also specific transmitter.
  action of, on smooth muscle, 219, 221–222
  autonomic, cotransmitters and, 746
    early studies on, 744–745, *744*
    heart contractility and, 793–794
    heart rate and, 793–794
    postganglionic mediation by, 745–746
    receptors for, *744,* 746–747, 746t, *755*
    second messengers and, 754–755, *754–755*
    specificity of, 750–752
    synthesis of, 747–750, *748–750*
    versus somatic, 744
  coexistence of different types of, 249, 746, 1190
  criteria for, 243, 245
  enzymatic destruction of, 245
  glucocorticoid effects on, 1514–1515
  in absorption of food, 1439, 1439t
  in amygdala, 702
  in cerebellar climbing fibers, 637–638
  in cerebellar mossy fibers, 636–637
  in digestion, 1439, 1439t
  in excitatory transmission, 245–246
  in feeding regulation, 1549
  in gastrointestinal tract, 1425, 1425t
  in gonadotropin-releasing hormone regulation, 1292–1293, 1298
  in gonadotropin secretion regulation, 1321
  in hippocampus, 709–710, 710t

Neurotransmitters *(Continued)*
  in inhibitory transmission, 246, *247,* 248
  in milk ejection reflex, 1415–1416, *1416–1417*
  in pain perception, 353–354
  in plasticity of nervous system, 727, *728,* 729–731
  in pubertal hormone secretion, 1278
  in retina, 434–435
  in salt regulation, 1054–1055, 1057
  in somatic system, versus autonomic system, 744
  in synaptic transmission, 248–249, 250–253, *250, 252*
  inactivation of, 245
  ionic channel gating by, *138–139,* 139–140
  neuropeptides as. See under *Neuropeptides.*
  of basal ganglia, 652–654
  peptidergic versus aminergic neurons and, 1188–1189
  release of, 144–152
    calcium ion in, 148–150, *149*
    from autonomic axons, 152
    from synaptic vesicles, 147–148
    in botulism, 151
    in tetanus, 151
    miniature end-plate potential in, 145–146, *146*
    presynaptic action potential in, 145, *145*
    quantal content facilitation/depression in, 150–151
    quantal hypothesis in, 145, 146–147
    repetitive stimulation of, 150–151
    sensory receptor regulation of, 103
    synaptic delay in, 150
    time course of action in, 150
  removal of, from brain extracellular fluid, 766
  retention of, by blood-brain barrier, 1217
  reuptake of, 245
  splanchnic circulation affected by, 917
  supersensitivity to. See *Denervation supersensitivity.*
  thyrotropin-releasing hormone control by, 1496
  transport of, sodium-dependent, *40,* 41
  voltage-dependent conductance effects of, 281–283, *282*
Neurotrophic factors, 1161–1163, *1161*
Neurotropic response, of growing axons, 1162
Newborn, sex hormone secretion in, 1273, *1275*
Nexus. See *Gap junction.*
n-gate, in potassium channel activity, 67–68
Nickel ion, as calcium channel blocker, 149